Principles of Fourier Analysis

SECOND EDITION

TEXTBOOKS in MATHEMATICS

Series Editors: Al Boggess and Ken Rosen

PUBLISHED TITLES

ABSTRACT ALGEBRA: A GENTLE INTRODUCTION
Gary L. Mullen and James A. Sellers

ABSTRACT ALGEBRA: AN INTERACTIVE APPROACH, SECOND EDITION
William Paulsen

ABSTRACT ALGEBRA: AN INQUIRY-BASED APPROACH
Jonathan K. Hodge, Steven Schlicker, and Ted Sundstrom

ADVANCED LINEAR ALGEBRA
Hugo Woerdeman

APPLIED ABSTRACT ALGEBRA WITH MAPLE™ AND MATLAB®, THIRD EDITION
Richard Klima, Neil Sigmon, and Ernest Stitzinger

APPLIED DIFFERENTIAL EQUATIONS: THE PRIMARY COURSE
Vladimir Dobrushkin

A BRIDGE TO HIGHER MATHEMATICS
Valentin Deaconu and Donald C. Pfaff

COMPUTATIONAL MATHEMATICS: MODELS, METHODS, AND ANALYSIS WITH MATLAB® AND MPI, SECOND EDITION
Robert E. White

DIFFERENTIAL EQUATIONS: THEORY, TECHNIQUE, AND PRACTICE, SECOND EDITION
Steven G. Krantz

DIFFERENTIAL EQUATIONS: THEORY, TECHNIQUE, AND PRACTICE WITH BOUNDARY VALUE PROBLEMS
Steven G. Krantz

DIFFERENTIAL EQUATIONS WITH APPLICATIONS AND HISTORICAL NOTES, THIRD EDITION
George F. Simmons

DIFFERENTIAL EQUATIONS WITH MATLAB®: EXPLORATION, APPLICATIONS, AND THEORY
Mark A. McKibben and Micah D. Webster

ELEMENTARY NUMBER THEORY
James S. Kraft and Lawrence C. Washington

PUBLISHED TITLES CONTINUED

SPORTS MATH: AN INTRODUCTORY COURSE IN THE MATHEMATICS OF SPORTS SCIENCE AND
SPORTS ANALYTICS
Roland B. Minton

TRANSFORMATIONAL PLANE GEOMETRY
Ronald N. Umble and Zhigang Han

TEXTBOOKS in MATHEMATICS

Principles of Fourier Analysis

SECOND EDITION

Kenneth B. Howell

The University of Alabama in Huntsville

USA

CRC Press
Taylor & Francis Group
Boca Raton London New York

CRC Press is an imprint of the
Taylor & Francis Group, an **informa** business

A CHAPMAN & HALL BOOK

CRC Press
Taylor & Francis Group
6000 Broken Sound Parkway NW, Suite 300
Boca Raton, FL 33487-2742

First issued in paperback 2022

© 2017 by Taylor & Francis Group, LLC
CRC Press is an imprint of Taylor & Francis Group, an Informa business

No claim to original U.S. Government works

Version Date: 20161108

ISBN 13: 978-1-03-247700-8 (pbk)
ISBN 13: 978-1-4987-3409-7 (hbk)
ISBN 13: 978-1-315-18149-3 (ebk)

DOI: 10.1201/9781315181493

Library of Congress Cataloging-in-Publication Data

Library of Congress Cataloging-in-Publication Data
Names: Howell, Kenneth B.
Title: Principles of Fourier analysis / Kenneth B. Howell.
Description: Second edition. | Boca Raton : CRC Press, 2017.
Identifiers: LCCN 2016031351 | ISBN 9781498734097
Subjects: LCSH: Fourier analysis–Textbooks. | Mathematical analysis–Textbooks.
Classification: LCC QA403.5 .H69 2017 | DDC 515/.2433--dc23
LC record available at https://lccn.loc.gov/2016031351

Visit the Taylor & Francis Web site at
http://www.taylorandfrancis.com

and the CRC Press Web site at
http://www.crcpress.com

Contents

Preface   xiii

Sample Courses   xv

I   Preliminaries   1

1   The Starting Point   3
- 1.1   Fourier's Bold Conjecture 3
- 1.2   Mathematical Preliminaries and the Following Chapters 5
- Additional Exercises 6

2   Basic Terminology, Notation and Conventions   7
- 2.1   Numbers 7
- 2.2   Functions, Formulas and Variables 8
- 2.3   Operators and Transforms 12

3   Basic Analysis I: Continuity and Smoothness   15
- 3.1   (Dis)Continuity 15
- 3.2   Differentiation 22
- 3.3   Basic Manipulations and Smoothness 25
- 3.4   Addenda 27
- Additional Exercises 34

4   Basic Analysis II: Integration and Infinite Series   37
- 4.1   Integration 37
- 4.2   Infinite Series (Summations) 41
- Additional Exercises 48

5   Symmetry and Periodicity   49
- 5.1   Even and Odd Functions 49
- 5.2   Periodic Functions 51
- 5.3   Sines and Cosines 53
- Additional Exercises 56

6   Elementary Complex Analysis   57
- 6.1   Complex Numbers 57
- 6.2   Complex-Valued Functions 59
- 6.3   The Complex Exponential 61
- 6.4   Functions of a Complex Variable 66
- Additional Exercises 71

7 Functions of Several Variables   **73**
7.1 Basic Extensions 73
7.2 Single Integrals of Functions with Two Variables 78
7.3 Double Integrals 82
7.4 Addenda: Proving Theorems 7.7 and 7.9 84
Additional Exercises 90

II   Fourier Series   93

8 Heuristic Derivation of the Fourier Series Formulas   **95**
8.1 The Frequencies 95
8.2 The Coefficients 96
8.3 Summary 99
Additional Exercises 100

9 The Trigonometric Fourier Series   **101**
9.1 Defining the Trigonometric Fourier Series 101
9.2 Computing the Fourier Coefficients 107
9.3 Partial Sums and Graphing 115
Additional Exercises 117

10 Fourier Series over Finite Intervals (Sine and Cosine Series)   **121**
10.1 The Basic Fourier Series 121
10.2 The Fourier Sine Series 123
10.3 The Fourier Cosine Series 125
10.4 Using These Series 126
Additional Exercises 127

11 Inner Products, Norms and Orthogonality   **129**
11.1 Inner Products 129
11.2 The Norm of a Function 131
11.3 Orthogonal Sets of Functions 132
11.4 Orthogonal Function Expansions 134
11.5 The Schwarz Inequality for Inner Products 135
11.6 Bessel's Inequality 137
Additional Exercises 141

12 The Complex Exponential Fourier Series   **145**
12.1 Derivation 145
12.2 Notation and Terminology 147
12.3 Computing the Coefficients 149
12.4 Partial Sums 150
Additional Exercises 151

13 Convergence and Fourier's Conjecture   **155**
13.1 Pointwise Convergence 155
13.2 Uniform and Nonuniform Approximations 161
13.3 Convergence in Norm 169
13.4 The Sine and Cosine Series 173
Additional Exercises 175

14 Convergence and Fourier's Conjecture: The Proofs   **179**
   14.1 Basic Theorem on Pointwise Convergence 179
   14.2 Convergence for a Particular Saw Function 186
   14.3 Convergence for Arbitrary Saw Functions 195
   14.4 A Divergent Fourier Series 196

15 Derivatives and Integrals of Fourier Series   **201**
   15.1 Differentiation of Fourier Series 201
   15.2 Differentiability and Convergence 206
   15.3 Integrating Periodic Functions and Fourier Series 210
   15.4 Sine and Cosine Series 214
   Additional Exercises 216

16 Applications   **219**
   16.1 The Heat Flow Problem 219
   16.2 The Vibrating String Problem 226
   16.3 Functions Defined by Infinite Series 234
   16.4 Verifying the Heat Flow Problem Solution 243
   Additional Exercises 247

III   Classical Fourier Transforms   **249**

17 Heuristic Derivation of the Classical Fourier Transform   **251**
   17.1 Riemann Sums over the Entire Real Line 251
   17.2 The Derivation 253
   17.3 Summary 255

18 Integrals on Infinite Intervals   **257**
   18.1 Absolutely Integrable Functions 257
   18.2 The Set of Absolutely Integrable Functions 261
   18.3 Many Useful Facts 261
   18.4 Functions with Two Variables 268
   Additional Exercises 276

19 The Fourier Integral Transforms   **279**
   19.1 Definitions, Notation and Terminology 279
   19.2 Near-Equivalence 281
   19.3 Linearity 283
   19.4 Invertibility 284
   19.5 Other Integral Formulas (A Warning) 286
   19.6 Some Properties of the Transformed Functions 287
   Additional Exercises 294

20 Classical Fourier Transforms and Classically Transformable Functions   **297**
   20.1 The First Extension 298
   20.2 The Set of Classically Transformable Functions 302
   20.3 The Complete Classical Fourier Transforms 304
   20.4 What Is and Is Not Classically Transformable? 308
   20.5 Finite Duration and Finite Bandwidth Functions 310
   20.6 More on Terminology, Notation and Conventions 313
   Additional Exercises 314

21 Some Elementary Identities: Translation, Scaling and Conjugation **319**
   21.1 Translation 319
   21.2 Scaling 327
   21.3 Practical Transform Computing 328
   21.4 Complex Conjugation and Related Symmetries 332
   Additional Exercises 335

22 Differentiation and Fourier Transforms **339**
   22.1 The Differentiation Identities 339
   22.2 Rigorous Derivation of the Differential Identities 346
   22.3 Higher Order Differential Identities 349
   22.4 Anti-Differentiation and Integral Identities 351
   Additional Exercises 356

23 Gaussians and Gaussian-Like Functions **359**
   23.1 Basic Gaussians 359
   23.2 General Gaussians 364
   23.3 Gaussian-Like Functions 368
   Additional Exercises 373

24 Convolution and Transforms of Products **375**
   24.1 Derivation of the Convolution Formula 375
   24.2 Basic Formulas and Properties of Convolution 377
   24.3 Algebraic Properties 379
   24.4 Computing Convolutions 382
   24.5 Existence, Smoothness and Derivatives of Convolutions 388
   24.6 Convolution and Fourier Analysis 392
   Additional Exercises 395

25 Correlation, Square-Integrable Functions and the Fundamental Identity **399**
   25.1 Correlation 399
   25.2 Square-Integrable/Finite Energy Functions 403
   25.3 The Fundamental Identity 412
   Additional Exercises 416

26 Generalizing the Classical Theory: A Naive Approach **419**
   26.1 Delta Functions 419
   26.2 Transforms of Periodic Functions 426
   26.3 Arrays of Delta Functions 429
   26.4 The Generalized Derivative 432
   Additional Exercises 444

27 Fourier Analysis in the Analysis of Systems **447**
   27.1 Linear, Shift-Invariant Systems 447
   27.2 Computing Outputs for LSI Systems 454
   Additional Exercises 461

28 Multi-Dimensional Fourier Transforms **463**
   28.1 Basic Definitions 463
   28.2 Computing Multi-Dimensional Transforms 466
   Additional Exercises 470

29 Identity Sequences   **471**
    29.1 An Elementary Identity Sequence 471
    29.2 General Identity Sequences 473
    29.3 Gaussian Identity Sequences 477
    29.4 Verifying Identity Sequences 481
    29.5 An Application (with Exercises) 485
    29.6 Laplace Transforms as Fourier Transforms 487
    Additional Exercises 489

30 Gaussians as Test Functions and Proofs of Important Theorems   **491**
    30.1 Testing for Equality with Gaussians 491
    30.2 The Fundamental Theorem on Invertibility 492
    30.3 The Fourier Differential Identities 495
    30.4 The Fundamental and Convolution Identities of Fourier Analysis 501

IV   Generalized Functions and Fourier Transforms   **509**

31 A Starting Point for the Generalized Theory   **511**
    31.1 Starting Points 511
    Additional Exercises 514

32 Gaussian Test Functions   **515**
    32.1 The Space of Gaussian Test Functions 515
    32.2 On Using the Space of Gaussian Test Functions 519
    32.3 Other Test Function Spaces and a Confession 521
    32.4 More on Gaussian Test Functions 522
    32.5 Norms and Operational Continuity 529
    Additional Exercises 535

33 Generalized Functions   **537**
    33.1 Functionals 537
    33.2 Generalized Functions 540
    33.3 Basic Algebra of Generalized Functions 547
    33.4 Generalized Functions Based on Other Test Function Spaces 553
    33.5 Some Consequences of Functional Continuity 553
    33.6 The Details of Functional Continuity 559
    Additional Exercises 564

34 Sequences and Series of Generalized Functions   **567**
    34.1 Sequences and Limits 567
    34.2 Infinite Series (Summations) 574
    34.3 A Little More on Delta Functions 577
    34.4 Arrays of Delta Functions 579
    Additional Exercises 583

35 Basic Transforms of Generalized Fourier Analysis   **587**
    35.1 Fourier Transforms 587
    35.2 Generalized Scaling of the Variable 592
    35.3 Generalized Translation/Shifting 597
    35.4 The Generalized Derivative 605
    35.5 Transforms of Limits and Series 613

35.6 Adjoint-Defined Transforms in General 614
35.7 Generalized Complex Conjugation 621
Additional Exercises 623

36 Generalized Products, Convolutions and Definite Integrals   **629**
36.1 Multiplication and Convolution 630
36.2 Definite Integrals of Generalized Functions 639
36.3 Appendix: On Defining Generalized Products and Convolutions 643
Additional Exercises 646

37 Periodic Functions and Regular Arrays   **649**
37.1 Periodic Generalized Functions 649
37.2 Fourier Series for Periodic Generalized Functions 655
37.3 On Proving Theorem 37.5 663
Additional Exercises 671

38 Pole Functions and General Solutions to Simple Equations   **673**
38.1 Basics on Solving Simple Algebraic Equations 674
38.2 Homogeneous Equations with Polynomial Factors 677
38.3 Nonhomogeneous Equations with Polynomial Factors 689
38.4 The Pole Functions 693
38.5 Pole Functions in Transforms, Products and Solutions 700
Additional Exercises 705

V The Discrete Theory   **707**

39 Periodic, Regular Arrays   **709**
39.1 The Index Period and Other Basic Notions 709
39.2 Fourier Series and Transforms of Periodic, Regular Arrays 711
Additional Exercises 720

40 Sampling, Discrete Fourier Transforms and FFTs   **721**
40.1 Some General Conventions and Terminology 721
40.2 Sampling and the Discrete Approximation 722
40.3 The Discrete Approximation and Its Transforms 725
40.4 The Discrete Fourier Transforms 737
40.5 Discrete Transform Identities 741
40.6 Fast Fourier Transforms 747
Additional Exercises 756

Tables, References and Answers   **761**
Table A.1: Fourier Transforms of Some Common Functions   **763**
Table A.2: Identities for the Fourier Transforms   **767**
References   **769**
Answers to Selected Exercises   **771**

Index   **783**

Preface

Let's be clear about what this book is and is not. It is not a book on the applications of Fourier analysis; it is a book on the *mathematics* of the Fourier analysis used in applications. The overriding goal is for the reader to gain a sufficiently deep understanding of the mathematics to confidently and *intelligently* employ "Fourier analysis" in the reader's own work, be it in optics, quantum mechanics, electronics, systems analysis, vibrational analysis, the analysis of partial differential equations or any of the many other areas in which "Fourier analysis" can be usefully applied. And when I say "*intelligently* employ", I mean more than simply memorizing and blindly using formulas from this book. "Intelligent employment" requires understanding what the formulas are really saying, when they can be used, how they can be used, and how they should *not* be used.

Got the idea?

In this rather large book, we will develop the mathematics of what I, the author, consider to be the four core theories of Fourier analysis — the classical theory for Fourier series, the classical theory for Fourier transforms, the generalized theory for Fourier transforms and the theory for discrete Fourier transforms — and we will see how the theories of each are related to the others (ultimately discovering that the classical and discrete theories are special cases of the generalized theory). Relatively little mathematical background is required on the part of the reader. A basic knowledge of calculus, differential equations and linear algebra should suffice. On the other hand, those who have had more advanced courses in real analysis, complex analysis or functional analysis really should be on the lookout for places where that theory could be useful in developing the material in this text. In particular, those acquainted with the Lebesgue integral and the analysis of analytic functions on the complex plane should be able to simplify some of the more involved proofs presented here and may even be able to extend some of the discussions. Indeed, the proof of one small lemma (lemma 35.8 on page 599) had to be left as an exercise for those familiar with Cauchy's integral formula from complex analysis.

Notice the appearance of the word "proof". While I will happily use nonrigorous arguments to motivate and enlighten, I also feel strongly that important claims in a text such as this must be supported by mathematically rigorous arguments that can be understood by the reader. There are too many other texts (especially in Fourier analysis) in which this is not done, and in which the authors make claims that are inaccurate or just plain false. Good proofs keep us honest. Where convenient and enlightening, I've tried to incorporate the proofs into the narrative. Where less convenient, the proof of a claim usually follows the statement of the claim in the traditional manner. Of course, a few carefully chosen proofs are left as exercises. And some of the proofs are — let's face it — long and hard. I won't apologize for including these; some important things just don't come easy.

On the other hand, not every reader needs to tackle every proof. Beginning students, especially, need to understand the gist of material without becoming bogged down in detailed discussions devised simply so that some fact can be verified under every possible condition. Accordingly, I've attempted to arrange the material so that the particulars of the longer, less enlightening and downright tedious proofs can be skipped easily and with relative safety. (True, many of the shorter and highly enlightening proofs can also be skipped by just going straight from the introductory word "*PROOF*" to the little black rectangle denoting the end of the proof — but slackers who do that endanger their souls.)

As much as possible, I've written this book to serve as a text and reference for everyone who uses or may use Fourier analysis. That is why such minimal mathematical prerequisites have been imposed. This text should serve both beginning students who have seen little or no Fourier analysis, as well the more advanced students who are somewhat acquainted with the subject but need a deeper understanding (see the *Sample Courses* described following this preface). Because of the general analysis developed here, this book could also be useful in a more general "applied analysis" course. Parts of it should even be of interest to professionals who are already experts in Fourier analysis because the generalized theory presented here (in part IV) extends the better-known theory normally presented.

By the way, you may notice that this is the second edition of this text. I've taken this opportunity to correct the many embarrassing little errors that were found in the first edition, and I've condensed, expanded and rearranged various bits and pieces from that edition. In addition, I've added a few things that I thought would be interesting or particularly relevant, including:

- A brief discussion (in exercises 11.11 and 11.12) of the Haar wavelets.

- A section (section 14.4) in which a continuous, periodic function with a divergent Fourier series is constructed (showing both the necessity of the assumption of piecewise smoothness in certain theorems and the fact that such counterexamples may be rather strange functions).

- A brief discussion of the classic sampling theorem for band-limited functions (in chapter 20).

- A short chapter (chapter 28) on multi-dimensional Fourier transforms.

- A section (section 29.6) relating the Fourier and Laplace transforms (with a derivation of the classical integral formula for the inverse Laplace transform).

Is this a complete guide to the mathematics of Fourier analysis? Of course not! Time and space considerations, along with the limited prerequisites, make that impossible. Instead, please view this tome as providing a starting point and "brief" overview of the mathematics of Fourier analysis. Interesting topics were left out. If I left out a topic of particular interest to you, I am sorry. I certainly left out topics of interest to me.

I also suspect that this text is not error free. Many errors from the first edition have been corrected, but more doubtlessly lurk in these pages. If you find an error not previously reported, please let me know.

Finally, let me thank the many people who helped make this second edition possible, especially the many students who suffered through earlier versions of this book and advised me on what to keep, change, correct or toss. For their aid, patience and insight, I am truly grateful.

Kenneth B. Howell
(howellkb@uah.edu)

Sample Courses

In the ideal world of the author, this book would be the main text for a two-semester sequence in Fourier analysis (possibly supplemented with some material on wavelets or applications of particular interest to the class). More realistically, it can serve as the text for a number of one-term courses on Fourier analysis. Here are brief descriptions of suggested material for two such single-semester courses: an "introductory course" and an "intermediate course" in Fourier analysis. The suggestions are based on courses regularly taught by the author. Naturally, individual instructors should make adjustments based on the needs, background, abilities and interests of their own students.

The introductory course is for undergraduates in engineering, science (especially physics and optics) and mathematics who have had little or no prior exposure to Fourier analysis, but know they will be needing it. For this course, consider covering the following:

Part I: Preliminaries
All of chapters 1 through 6. Cover this material quickly and leave the material in chapter 7 to be discussed as the need arises. If you want, delay covering chapter 6 on complex variables until just before starting chapter 12.

Part II: Fourier Series
All of chapter 8.
All of chapter 9.
All of chapter 10.
Sections 1 through 4 of chapter 11.
All of chapter 12.
Sections 1, 2 and 4 of chapter 13.
If time allows: sections 1 and 4 of chapter 15, along with sections 1 and 2 of chapter 16.

Part III: Classical Fourier Transforms
All of chapter 17.
Sections 1, 2 and 3 of chapter 18 (go through section 3 rather quickly).
All of chapter 19 (skip the proofs in the last section).
All of chapter 20.
Sections 1, 2 and 3 of chapter 21.
Sections 1 and 3 of chapter 22.
Sections 1 and 2 of chapter 23 (briefly discuss the transforms in section 3).
All of chapter 24.
Section 1 and, perhaps, section 2 of chapter 25 .
All of chapter 26.

LSI Systems, Multi-Dimensional Transforms or Discrete Transforms
All of chapter 27 or all of chapter 28 or all of chapters 39 and 40 (depending on time and the interests of the class and instructor).

The intermediate course is for upper-level and graduate students in engineering, science and mathematics. While they are not expected to have taken the introductory course, these students can be expected to have had some previous introduction to some elements of Fourier analysis. For this course, try covering the following:

Part I: Preliminaries
Chapters 1 through 6 (cover these quickly and leave the material in chapter 7 to be discussed as the need arises).

Part III: Classical Fourier Transforms
Sections 1, 2 and 3 of chapter 18 (return to section 4 as necessary later on).
All of chapter 19 (possibly skipping the proofs in the last section).
All of chapter 20.
Sections 1, 2 and 3 of chapter 21 (consider including section 4, also).
Sections 1, 2 and 3 of chapter 22.
Sections 1 and 2 of chapter 23 (briefly mention the transforms in section 3).
All of chapter 24.
All of chapter 25.
Sections 1 through 4 of chapter 29.
Sections 1 and 2 (and, perhaps, 3 and 4) of chapter 30.

Part IV: Generalized Functions and Fourier Transforms
All of chapter 31.
Sections 1 through 4 (and, possibly, 5) of chapter 32.
Sections 1 through 4 (and, possibly, 5) of chapter 33.
All of chapter 34.
Sections 1 through 5 of chapter 35.
Sections 1 and 2 of chapter 36.
Sections 1 and 2 of chapter 37.

Part V: The Discrete Theory
All of chapter 39.
All of chapter 40.

And what about those sections and chapters not listed above? Well, that additional material is for the reader to peruse outside of class as need or interest suggests.

Part I

Preliminaries

1

The Starting Point

You may already know that Fourier analysis is "stuff you do with the Fourier series and the Fourier transform". You may even realize that Fourier series and Fourier transforms are useful because they provide formulas for describing functions found in many applications — formulas that are often more easily manipulated and analyzed than other formulas and equations describing these functions.

On the other hand, you may not know anything about Fourier analysis. If so, then I'll just tell you this: It involves things called "Fourier series" and "Fourier transforms", and it is useful in many applications because these series and transforms provide convenient formulas for the functions arising in these applications.

Whether or not you have seen any "Fourier analysis", let us take a brief look at one of the historical starting points of the subject. It will help illustrate how these "Fourier formulas" might be helpful, and will provide us with a good starting point for our own studies.

1.1 Fourier's Bold Conjecture

In the early 1800s, Joseph Fourier (along with others) was attempting to mathematically describe the process of heat conduction in a uniform rod of finite length, subject to certain initial and boundary conditions. Fourier's approach required that the temperature $u(x)$ at position x in the rod at some fixed time be expressed as

$$u(x) \;=\; a_0 \;+\; a_1 \cos(cx) \;+\; b_1 \sin(cx) \;+\; a_2 \cos(2cx) \;+\; b_2 \sin(2cx)$$
$$+ \; a_3 \cos(3cx) \;+\; b_3 \sin(3cx) \;+\; \cdots$$
(1.1)

where c is π divided by the rod's length, and the a_k's and b_k's are constants to be determined after plugging this representation for u into the equations modeling heat flow. (Precisely how they are determined will be discussed in chapter 16, where you will also discover that I've simplified things here a bit. For one thing, the a_k's and b_k's are actually functions of time.)

Fourier's approach was successful, and that idea of representing a function in terms of sines and cosines eventually led to the development of a lot of incredibly useful mathematics.

What Fourier did with the function $u(x)$ was very similar to what we normally do with a three-dimensional vector $\mathbf{v}$. Basically, $\mathbf{v}$ is just some entity possessing "length" and "direction". Rarely, though, are vector computations done directly using a vector's length or direction. In practice such computations are normally done using the vector's *components* (v_1, v_2, v_3). For example, the length of $\mathbf{v}$ is usually computed using the component formula

$$\|\mathbf{v}\| = \sqrt{(v_1)^2 + (v_2)^2 + (v_3)^2} \quad .$$

These components are the coefficients in the unique representation of $\mathbf{v}$ as a linear combination of vectors from the standard basis $\{\mathbf{i}, \mathbf{j}, \mathbf{k}\}$,

$$\mathbf{v} = v_1\mathbf{i} + v_2\mathbf{j} + v_3\mathbf{k} \quad . \tag{1.2}$$

Because every vector can be so represented and because $\{\mathbf{i}, \mathbf{j}, \mathbf{k}\}$ is a particularly nice basis (because all of its elements are orthogonal to each other and have unit length), most vector manipulations can be reduced to fairly simple computations with the three separate components of each vector. Indeed, it's hard to imagine doing vector analysis without using these components.

?►Exercise 1.1:   *What is the geometric definition for the dot product of two vectors? What is the component formula for the dot product? Which of these two formulas do you normally use to compute $\mathbf{v} \cdot \mathbf{w}$?*

The similarities between formulas (1.1) and (1.2) are significant. In each, a fairly general entity — the vector $\mathbf{v}$ in formula (1.2) and the function $u(x)$ in formula (1.1) — is being expressed as a (possibly infinite) linear combination[1] of "basic entities". In formula (1.2), these basic entities are $\mathbf{i}$, $\mathbf{j}$ and $\mathbf{k}$, the standard basis vectors for three-dimensional space, while in formula (1.1) the basic entities are sines and cosines.[2] In a sense, formula (1.1) says that the function $u(x)$ can be expressed in "component form" $(a_0, a_1, b_1, a_2, b_2, a_3, b_3, \ldots)$, and suggests that some manipulations involving $u(x)$ can be reduced to simpler computations involving these components.

This gives us our starting point. We will start with a goal of developing a theory for manipulating and analyzing functions that is analogous to the theory we already use for manipulating and analyzing vectors in two- and three-dimensional space. For our "basis functions", we will use sines and cosines. This assumes, of course, that all functions of reasonable interest can be expressed as linear combinations of sines and cosines. This is a bold assumption. Moreover, at this point, we have no real reason to believe it is valid! So, perhaps, we should refer to it as:

Fourier's Bold Conjecture
Any "reasonable" function can be expressed as a (possibly infinite) linear combination of sines and cosines.

If Fourier's conjecture is valid, then we should be able to simplify many problems (such as, for example, the problem of mathematically predicting the temperature distribution along a given rod at a given time) by expressing the unknown functions as linear combinations of well-known sine and cosine functions. With luck, the coefficients in these linear combinations will be relatively easy to determine, say, by plugging the expressions into appropriate equations and solving some resulting algebraic equations.

Naturally, it is not all that simple. For one thing, I cannot honestly tell you that Fourier's conjecture is completely valid, at least not until we better determine what is meant by a function being "reasonable". But the conjecture turns out to be close enough to the truth to serve as the starting point for our studies, and determining the extent to which this conjecture is valid will be one of our major goals in this text. And, of course, whenever possible, we will want to find out how

---

[1] Recall: If $\{\phi_1, \phi_2, \phi_3, \ldots\}$ is any collection of things that can be multiplied by scalars and added together, then a *linear combination* of the ϕ_k's is any expression of the form

$$c_1\phi_1 + c_2\phi_2 + c_3\phi_3 + \cdots$$

where the c_k's are constants. Unless otherwise stated, a linear combination is always assumed to have a finite number of terms. When we add the adjective "possibly infinite", however, we are admitting the possibility that the expression has infinitely many terms.

[2] Since $\cos(0cx) = 1$ for all x, we can view the a_0 term in formula (1.1) as being $a_0 \cos(0cx)$.

to compute the "components" of any given (reasonable) function and how to use these components in the manipulations of interest to us (e.g., differentiation, finding solutions to various differential equations and evaluating functions).

Since sines and cosines are periodic functions, it is logical to first consider periodic functions. This will lead to the classical Fourier series (discussed in part II of this book). We will also see that the analysis developed for periodic functions can be applied to functions defined on finite intervals. In trying to stretch the analysis to nonperiodic functions on the real line, we will discover the classical Fourier transform (part III). Continuing along those lines will eventually lead to generalized functions and the generalized Fourier transform (part IV). Finally, we will consider the adaptations we must make so that we can deal with functions known only by sets of data taken by measurement. This will lead to the discrete theory of Fourier analysis (part V).

By the way, do not expect Fourier analysis to simply be "vector analysis with functions". Frankly, as the subject material evolves, the analogy between Fourier analysis and vector analysis will seem more and more tenuous to most readers.

1.2 Mathematical Preliminaries and the Following Chapters

The theory of Fourier analysis did not spring fully developed from the minds of the mathematically ignorant. Likewise, we cannot pretend to study Fourier analysis without having some understanding of the mathematics underlying the subject.

Presumably, you are already reasonably proficient with the basics of calculus (computing and manipulating derivatives, integrals and infinite series) as well as the basics of linear algebra, and you nod knowingly at statements like "the domain of a function f is the set of all x for which $f(x)$ is defined." Still, a little review would be wise if only to ensure that we are all using the same notation and terminology. More importantly, though, the development and intelligent use of Fourier analysis requires a better understanding and appreciation of certain basic mathematical concepts than many beginning students have yet had reason to cultivate. So, in the next few chapters (the rest of part I of this text), we will briefly review some of the mathematics we will need, emphasizing issues you might not have considered so deeply in your previous studies.

If you are impatient to begin the study of Fourier analysis, don't worry. It's not necessary to cover everything in part I before starting on part II or part III. After all, most of part I is supposed to be a review! You should have seen most of this material before (in some form), and you can always return to the appropriate sections of this review as the need arises. Just make sure you understand the material in the next chapter (primarily on notation and some conventions we will be following); carefully skim through the chapter after that, and then quickly skim through the rest of part I. Then plan on returning to the appropriate sections as the need arises.

Additional Exercises

1.2. *Show that the standard components of a vector are the dot products of the vector with the corresponding basis vectors. That is, show that, if*

$$\mathbf{v} = v_1\mathbf{i} + v_2\mathbf{j} + v_3\mathbf{k} \quad ,$$

then

$$v_1 = \mathbf{v} \cdot \mathbf{i} \quad , \qquad v_2 = \mathbf{v} \cdot \mathbf{j} \qquad and \qquad v_3 = \mathbf{v} \cdot \mathbf{k} \quad .$$

(Analogous formulas will be developed in part II of this text for computing the "components" of functions.)

2

Basic Terminology, Notation and Conventions

We begin our review of the mathematical preliminaries by discussing how we will describe some of the basic entities of Fourier analysis — numbers, functions and operators. Perhaps the most important part of this discussion is in determining just what will be meant by the phrase "f is a function" and by the notation "$f(x)$". Pay close attention to this discussion even if you think you know what a "function" is. It turns out that people in different disciplines have developed slightly different views as to the meaning of this word. That is one reason a text on Fourier analysis by a mathematician specializing in functional analysis will often look quite different from a corresponding text by an electrical engineer specializing in signals and systems. These differences cause few problems for those who understand the differences, but they can lead the unwary into making substantially more work for themselves and even, on occasion, to making foolish errors in computations. Moreover, if we do not all agree on exactly what a function is and what $f(x)$ denotes, then we will find it very difficult to develop clear, precise and brief notation for the manipulations we will be doing with these things. And if we cannot adequately describe these manipulations, then the rest of this text might as well be written using grunts and hand waves.

2.1 Numbers

The set of all real numbers, also called the real number line, will be denoted by either $(-\infty, \infty)$ or $\mathbb{R}$ depending on how the spirit moves us. If $-\infty \leq \alpha < \beta \leq \infty$, then (α, β) denotes the open interval between α and β (i.e., the set of all x where $\alpha < x < \beta$), and $[\alpha, \beta]$ denotes the closed interval (i.e., the set of all x where $\alpha \leq x \leq \beta$). Of course, for the closed interval $[\alpha, \beta]$, neither α nor β can be infinite. Furthermore, both α and β must be finite whenever (α, β) is identified as a *finite* or *bounded* interval.

For brevity, let us agree that whenever a phrase such as "the interval (α, β)" is encountered in this text, it may automatically be assumed that $-\infty \leq \alpha < \beta \leq \infty$.

The set of all complex numbers, denoted by $\mathbb{C}$, will also play an important role in our computations. A brief review of elementary complex analysis is given in chapter 6.

2.2   Functions, Formulas and Variables
Basics

Here are the standard definitions for function, domain and range commonly found in elementary introductions to mathematical analysis: A *function* f on some set $\mathcal{I}$ of real numbers is a mapping from that set into the set of real or complex numbers. That is, for each real number x in $\mathcal{I}$, f defines a corresponding value $f(x)$. The function is said to be *real* or *complex valued* according to whether all the $f(x)$'s are real or complex values. In this book you should assume that any function under discussion is complex valued unless it is otherwise explicitly stated or obviously implied by other assumptions. (As already noted, a brief review of complex numbers and complex-valued functions will be given in chapter 6. No real harm will be done if, until then, you visualize all functions as being real valued.) The *domain* of f is the set of all values of x for which $f(x)$ is defined, and the *range* of f is the set of values that $f(x)$ can assume.

Most of the time, we will be concerned with functions defined on some given interval of the real line. If no interval is explicitly stated or obviously implied by other conditions, then you may assume that the functions under consideration are defined on the entire real line.

Typically, a function f is described (or defined) by stating its domain and a formula for computing the value of $f(x)$ for all "relevant values of x". (For now, "all relevant values of x" should be taken as meaning "all x in the domain of the function", though we'll soon see that this is not always quite the case.) For our purposes, a formula for f is any set of instructions for determining the value of $f(x)$ for each relevant value of x. Sometimes the formula will be a simple expression involving well-known functions (e.g., $(3 + x)^2$ or $\sin(2\pi x)$). Other times the formula may be a collection of simple formulas with each valid over a different interval. For example, the *ramp function* is the function on $(-\infty, \infty)$ given by the formula

$$\text{ramp}(x) \;=\; \begin{cases} 0 & \text{if } x < 0 \\ x & \text{if } 0 \le x \end{cases} \quad .$$

We should also expect formulas involving integrals and infinite summations, such as

$$f(x) \;=\; \int_{t=0}^{x} 3t^2 \, dt \qquad \text{and} \qquad g(x) \;=\; \sum_{k=1}^{\infty} \frac{1}{(1+k)^k} \sin(k\pi x) \quad .$$

Obviously, we will not be able to evaluate some of these formulas for particular values of x using elementary techniques.

Although functions are often identified with formulas, you should realize that the two are not truly the same. For example, $2x$ and $x + x$ are two different formulas, but they certainly describe the same function. That is what we mean when we write $2x = x + x$.

Within the formulas for functions are *variables*, symbols used to show how given values are manipulated to evaluate the indicated function at those given values. It is important to recognize that there are different types of variables and that the context in which a given variable appears determines what it represents. Consider, for example, the expression

$$f(x) \;=\; \int_{t=0}^{x} 3t^2 \, dt \quad . \tag{2.1}$$

It contains two variables, x and t. The x can be considered a true variable. It represents values that can be "inputted" into the function or formula. In a particular application, x can be replaced by a specific number, say 4, giving us the value of $f(x)$ when $x = 4$,

$$f(4) \;=\; \int_{t=0}^{4} 3t^2 \, dt \;=\; t^3 \Big|_{0}^{4} \;=\; 64 \quad .$$

On the other hand, if we try to assign t the value 4 in equation (2.1), then we get

$$f(x) = \int_{4=0}^{x} 3 \cdot 4^2 \, d4 \quad ,$$

which makes no sense at all. This t cannot be assigned a value. It is being used to describe the function being integrated and has no meaning outside the integral. Such variables are called *internal* or *dummy* variables.[1]

Along the same lines, the precise meaning of "$f(x)$", where f is some function, also depends on the context in which it is used. We will have three slightly different meanings assigned to this notation.

First of all, $f(x)$ will denote the numerical value of f at x. In this sense, $f(x)$ is a number. (This is the standard definition found in many textbooks.)

We will also use $f(x)$ to represent any formula for computing the numerical value of $f(x)$ for every x in the domain of f. In other words, we won't quibble over whether $f(x) = x^2$ indicates a value or is a formula for computing the values.

Finally, let us agree that $f(x)$, as well as any formula defining f, can denote the function. So, instead of saying

> The derivative of f, where $f(x) = x^2$, is f', where $f'(x) = 2x$.

and

> Consider the function g given by the formula $g(x) = \sin(2\pi x)$.

we will often just say

> The derivative of x^2 is $2x$.

and

> Consider the function $\sin(2\pi x)$.

We are simply agreeing that, at times, we will not explicitly distinguish between "a function" and "a description of the function". This agreement does violate conventions stated in some math texts, but it does agree more with common usage in most disciplines (and many math texts), and it will greatly simplify the discussion in this text.

Exactly which of these three interpretations should be applied to an appearance of "$f(x)$" should be clear from the context.

Another way of denoting $f(x)$ is $f|_x$. This notation will be particularly convenient when we start dealing with operators and transforms.

Keep in mind that changing the symbol used as the variable in a function does not change the function.[2] For example, if $f(x) = x^2$ for $-\infty < x < \infty$, then defining $g(s)$ to be s^2 for $-\infty < s < \infty$ does not introduce a new function. f and g are the same function because, for every real value a, $f(a) = a^2 = g(a)$. On the other hand, replacing the variable in a function's formula with a nontrivial formula involving another variable definitely does give us a different function. For example, substituting $2s$ for the x in $f(x) = x^2$ results in a new function, $h(s) = f(2s) = 4s^2$. f and h are not the same function because, in general,

$$h(a) = 4a^2 \neq a^2 = f(a) \quad .$$

---

[1] The distinction between true and dummy variables is not always clear cut. Consider the expression $\frac{d}{dx}x^2\big|_{x=3}$.

[2] We are talking about *function definition* and not computations using formulas. Suddenly changing, without adequate warning, the symbol being used for a particular variable in a series of computations can easily render your results totally meaningless!

A Pragmatic Approach to Domains and Function Equality

Often we must deal with functions that are not well defined at a few isolated points in the intervals of interest. Sometimes this is because the formula defining the function has ambiguities. Other times this is because of inherent discontinuities in the function. In practice, though, we are only concerned with the behavior of a function over intervals, not at isolated points. Because of this, we can take a rather pragmatic point of view concerning these functions and adopt the following convention:

Convention (irrelevance of function values at isolated points)
Let f and g be two functions on an interval (a, b). If $f(x) = g(x)$ for all but a finite number of x's in (a, b), then f and g are viewed as the same function over that interval.[3,4]

To a great extent, this convention concerns how we use formulas to define functions. A few examples may help clarify the matter.

!▶ *Example 2.1:*   *A trivial example is given by*

$$f(x) = \frac{x^2 - 1}{x - 1} \quad ,$$

which is undefined for $x = 1$. In applications, however, most of us would feel justified in "simplifying" $f(x)$,

$$f(x) = \frac{x^2 - 1}{x - 1} = \frac{(x + 1)(x - 1)}{x - 1} = x + 1 \quad ,$$

and then ignoring the fact that the original formula for $f(x)$ was not defined for $x = 1$. In other words, "for all practical purposes" we would agree that

$$\frac{x^2 - 1}{x - 1} = x + 1 \quad .$$

!▶ *Example 2.2 (unit step functions):*   *Two unit step functions u and h are given by*

$$u(x) = \begin{cases} 0 & \text{if} \ x \le 0 \\ 1 & \text{if} \ 0 < x \end{cases} \quad \text{and} \quad h(x) = \begin{cases} 0 & \text{if} \ x < 0 \\ 1 & \text{if} \ 0 \le x \end{cases} \quad .$$

These two formulas differ only at the one point $x = 0$, where u equals 0 and h equals 1. According to the above convention, the solitary difference between u and h at the one point can be ignored, and we can view u and h as being the same function on the real line.

One reason we can ignore the values of a function at isolated points is that the basic manipulations of Fourier analysis are based on integration, and, for integrals, the value of a function at a single point (or a finite set of points) is truly irrelevant. For example, if v is either of the above-defined step functions, then

$$\int_{-1}^{2} v(x) \, dx = \int_{-1}^{0} 0 \, dx + \int_{0}^{2} 1 \, dx = 0 \big|_{-1}^{0} + x \big|_{0}^{2} = 2 \quad .$$

---

[3] Those who know about equivalence classes should realize that, with this convention, we are defining an equivalence relation ($f \sim g$ whenever $f(x) = g(x)$ for all but a finite number of x's in (a, b)) and then identifying functions with their corresponding equivalence classes.

[4] Those who know about Lebesgue integration can extend this convention to *If f and g are two functions on (a, b) that differ only on a set of measure zero, then f and g may be viewed as the same function over (a, b).*

The value of $v(0)$ is completely irrelevant to the computation of this integral. We would have gotten exactly the same result if $v(0) = 827$ as long as we still have $v(x) = 0$ when $x < 0$ and $v(x) = 1$ when $0 < x$.

This convention also corresponds to the way we normally use functions to describe events around us. Functions such as the step function are used to describe phenomena involving very rapid changes during very brief periods — so brief that it is impractical to accurately describe the phenomena during these brief periods. For example, when an incandescent light is turned on, it takes time for the filament to heat up enough to produce light. But when you walk into a room and turn on a light, the filament heats up so quickly that a function like the step function — which is zero for $t < 0$ and some fixed value for $0 < t$ — is usually adequate for describing the light output. In such cases, we don't really care about the exact light output at the exact instant we activate the lights. And if we do care (maybe we are studying the rate at which the lamp's filament heats up), then we should not try to describe the phenomenon using a simple step function.

While the value of a function at an individual point is irrelevant, the values of the function over intervals on either side of that point are quite relevant. We will see how this affects the way we deal with discontinuities in the next chapter.

Notice how this convention affects our notion of two functions being equal. By the convention, the statement that two functions f and g (given by formulas $f(x)$ and $g(x)$) are *equal* over an interval (α, β), which we will also write as

$$f = g \quad \text{(or as } f(x) = g(x)) \quad \text{over} \quad (\alpha, \beta) \quad,$$

means the following:

1. If (α, β) is a finite interval, then, numerically, $f(x) = g(x)$ for all except some finite number (possibly zero) of x's between α and β.

2. If (α, β) is an infinite interval, then, in the sense just described, $f = g$ on every finite subinterval of (α, β).

!►Example 2.3: *By our convention,*

$$\frac{x^2 - 1}{x - 1} = x + 1 \quad over \quad (-\infty, \infty) \quad,$$

even though the formula on the left-hand side is not well defined for $x = 1$.

This convention also modifies our concept of a function's domain. We can now accept a function f as being defined over an interval even if it (or the formula defining it) is not well defined at a few isolated points on that interval. More precisely, the statement that f is *defined on* (α, β) will mean that:

1. If (α, β) is a finite interval, then the value of $f(x)$ is defined for all except some finite number (possibly zero) of x's between α and β.

2. If (α, β) is an infinite interval, then f is defined, in the sense just described, on every finite subinterval of (α, β).

!►Example 2.4: *Recall the cotangent function,*

$$\cot(x) = \frac{\cos(x)}{\sin(x)} \quad.$$

This is defined for every real value of x except $x = 0, \pm\pi, \pm 2\pi, \pm 3\pi, \dots$ (where $\sin(x) = 0$). Since each finite subinterval of $(-\infty, \infty)$ can only contain a finite number of such points, we will say that $\cot(x)$ is defined on $(-\infty, \infty)$.

About Delta Functions

(This is for those of you who are acquainted with the (Dirac) delta function. If you don't know about the delta function, skip to the next section.)

You may wonder how the (Dirac) delta function δ — which is often visualized as being zero everywhere on the real line except at $x = 0$, where it is "infinite" — fits into our discussion. The answer is simple: It doesn't. Despite its name, the (Dirac) delta function δ is *not* a function, at least not in the sense being considered here.

The problem is that the important properties of the delta function cannot be derived simply from an expression of the form

$$\delta(x) = \begin{cases} 0 & \text{if } x \neq 0 \\ +\infty & \text{if } x = 0 \end{cases} .$$

(This will be verified rigorously at the start of part IV.) Invariably, some additional (and often mathematically questionable) property must be specified (such as " $\int_{-\infty}^{\infty} \delta(x)\,dx = 1$ "). Consequently, "the delta function" falls outside of the "classical" theory of functions we are now discussing.

Later (in part IV and, to a lesser extent, in chapter 26) we will develop the mathematics for dealing with the delta "function". It is an important part of Fourier analysis, and is well worth the wait. Until then, though, we will not have the mathematics to justify any use of the delta function.

2.3   Operators and Transforms
Basic Concepts

Any mathematical entity that changes one function into another function is called either an *operator* or a *transform*. (The two terms are equivalent, and which term is used for a particular entity is largely a matter of tradition.) For example, the differential operator D is defined by

$$D[f] = f' .$$

For some specific f's,

$$D[x^2] = 2x \quad \text{and} \quad D[\sin(2\pi x)] = 2\pi \cos(2\pi x) .$$

Note that here the symbol x is being used both as a dummy variable to describe the function being differentiated (inside the "[ ]") and as a true variable. Thus,

$$D[x^2]\big|_3 = 2x\big|_3 = 2 \cdot 3 = 6 ,$$

while

$$D[x^2]\big|_3 \neq D[3^2] = D[9] = 0 !$$

It is often more convenient to use different symbols for the two variables. The reader may recall the Laplace transform $\mathcal{L}$, which is defined by the formula

$$\mathcal{L}[f]\big|_s = \mathcal{L}[f(t)]\big|_s = \int_{t=0}^{\infty} f(t)\,e^{-st}\,dt \tag{2.2}$$

where f and s are, respectively, any suitable function and any suitable value. In particular, for $s > 2$,

$$\mathcal{L}[e^{2t}]\big|_s = \int_{t=0}^{\infty} e^{2t} e^{-st}\,dt = \int_{t=0}^{\infty} e^{-(s-2)t}\,dt = -\frac{e^{-(s-2)t}}{s-2}\bigg|_{t=0}^{\infty} = \frac{1}{s-2} .$$

Here it is particularly important to realize that the function being "plugged into" the Laplace transform is the function described by the formula e^{2t}. Any other formula describing this function could have been used. Also, the actual symbol used as the variable in the formula (as well as for the dummy variable in the integration) is totally irrelevant. Using x instead of t,

$$\mathcal{L}\left[e^{2x}\right]\big|_{s} = \int_{x=0}^{\infty} e^{2x} e^{-sx}\, dx = \int_{x=0}^{\infty} e^{-(s-2)x}\, dx = -\frac{e^{-(s-2)x}}{s-2}\bigg|_{x=0}^{\infty} = \frac{1}{s-2} \quad.$$

In principle, we could use the same symbol for both variables,

$$\mathcal{L}\left[e^{2x}\right]\big|_{x} = \frac{1}{x-2} \quad.$$

In practice, though, this would surely lead to confusion in computing Laplace transforms.

Later on, much of our work will involve extensive manipulations of various transforms of many functions. In doing these manipulations, say, for a transform $\mathcal{T}$, keep in mind that $\mathcal{T}[f(x)]$ is shorthand for

$$\mathcal{T}[f] \quad \text{where} \quad f \text{ is the function described by the formula } f(x) \quad.$$

Changing the symbol used for the variable in the formula (here, the x in $f(x)$) does not change the function described by that formula and so, does not change the transform of that function. Thus,

$$\mathcal{T}[f(x)] = \mathcal{T}[f(t)] \quad.$$

On the other hand, as noted earlier, replacing the symbol used in the formula with a nontrivial formula involving any other symbol does change the function and, thus, changes the transform of that function.

?▶ Exercise 2.1: *Consider the Laplace transform as defined above, and let $f(t) = e^{2t}$. Show that $\mathcal{L}[f(2x)] \neq \mathcal{L}[f(t)]$ by computing $\mathcal{L}[f(2x)]\big|_{s}$ and comparing it to $\mathcal{L}[f(t)]\big|_{s}$ (computed above).*

Like functions, operators and transforms have "formulas" and "domains". For a given operator, the domain is the set of all functions on which the operator can operate, and a formula is just an expression telling us how to compute the result of an operator operating on any function in its domain. Typically, as in formula (2.2), the operator's formula describes how to manipulate the formula for any "input function" — the $f(x)$ in (2.2) — to get the formula for the corresponding "output function" — the $\mathcal{L}[f]\big|_{s}$ in (2.2).

The specification of the domain of an operator should always be part of the definition of the operator. Unfortunately, violations of this rule are commonplace. If no domain for a particular operator $\mathcal{T}$ is given, then any function f for which $\mathcal{T}[f]$ "makes sense" can usually be assumed to be in the domain of $\mathcal{T}$. For example, although it was not stated, the domain for the differential operator D is the set of all functions on $(-\infty, \infty)$ for which the derivative is defined as a function on $(-\infty, \infty)$.

?▶ Exercise 2.2: *What would be a reasonable domain for the Laplace transform?*

Unfortunately, it may not always be clear when "$\mathcal{T}[f]$ makes sense", and we will see examples of how easy it is to make serious errors by assuming that some particular function is in an operator's domain when, in fact, it is not. Determining the appropriate domains for the operators in Fourier analysis will be an important issue.

Linear Transforms

Many operators of interest are linear. Recall that an operator $\mathcal{T}$ is *linear* if and only if the following holds:

> If f and g are in the domain of $\mathcal{T}$, and a and b are any two (possibly complex) constants, then the linear combination $af + bg$ is also in the domain of $\mathcal{T}$. Furthermore,
>
> $$\mathcal{T}[af + bg] = a\mathcal{T}[f] + b\mathcal{T}[g] \quad .$$

Of course, if f, g and h are three functions in the domain of a linear operator $\mathcal{T}$, and a, b and c are three constants, then, since $af + bg$ is in the domain of $\mathcal{T}$, so is the sum of $af + bg$ with ch, $af + bg + ch$. Furthermore,

$$
\begin{aligned}
\mathcal{T}[af + bg + ch] &= \mathcal{T}[(af + bg) + ch] \\
&= \mathcal{T}[af + bg] + c\mathcal{T}[h] = a\mathcal{T}[f] + b\mathcal{T}[g] + c\mathcal{T}[h] \quad .
\end{aligned}
$$

Continuing along these lines leads to the following completely equivalent definition of an operator $\mathcal{T}$ being linear:

> Whenever $\{f_1, f_2, \ldots, f_N\}$ is a finite set of functions in the domain of $\mathcal{T}$, and $\{c_1, c_2, \ldots, c_N\}$ is a finite set of (possibly complex) constants, then the linear combination $c_1 f_1 + c_2 f_2 + \cdots + c_N f_N$ is also in the domain of $\mathcal{T}$. Furthermore,
>
> $$\mathcal{T}[c_1 f_1 + c_2 f_2 + \cdots + c_N f_N] = c_1 \mathcal{T}[f_1] + c_2 \mathcal{T}[f_2] + \cdots + c_N \mathcal{T}[f_N] \quad .$$

!▶ **Example 2.5:**   *Consider the differential operator D with the set of all differentiable functions on $(-\infty, \infty)$ as its domain. From calculus we know that, if f and g are functions with derivatives on $(-\infty, \infty)$ and a and b are any two constants, then the linear combination $af + bg$ is differentiable on $(-\infty, \infty)$ and*

$$(af + bg)' = af' + bg' \quad .$$

In other words, if f and g are in the domain of D, and a and b are any two constants, then the linear combination $af + bg$ is in the domain of D and

$$D[af + bg] = aD[f] + bD[g] \quad .$$

Thus, D is a linear operator.

?▶ **Exercise 2.3:**   *Is the Laplace transform a linear operator (use the domain from exercise 2.2)?*

3

Basic Analysis I:
Continuity and Smoothness

One good way to generate errors and embarrass yourself is to use a formula or identity without properly verifying its validity under the circumstances at hand. This seems particularly easy to do in Fourier analysis, and it is not at all unusual to see differential identities and integral formulas from the theory of Fourier analysis being used with functions that are neither differentiable or integrable. (It's especially disturbing when such abuses occur in textbooks.) The results range from questionable to disastrously wrong.

We, of course, will try to avoid such mistakes. So we must be able to identify when the various results derived in this text are valid and when they are *not*. To simplify this process, functions are commonly classified according to pertinent properties that they may or may not satisfy. For example, a function f on some interval (α, β) is classified as being *bounded* over that interval if there is a finite value M such that

$$|f(x)| \leq M \qquad \text{whenever} \quad \alpha < x < \beta \quad .$$

If no such $M < \infty$ exists, then f is said to be *unbounded* (over the interval).

In the next few chapters, we will briefly review some of the basic elements of function analysis (i.e., "calculus") that will be especially important in later discussions. In this particular chapter, the emphasis is on how smoothly function values vary near each point where they are defined, and on how this smoothness affects some of the manipulations we might wish to do.

3.1   (Dis)Continuity

You surely remember that a function f is continuous at a point x_0 if $f(x_0)$ and $\lim_{x \to x_0} f(x)$ both exist[1] and

$$\lim_{x \to x_0} f(x) = f(x_0) \quad .$$

Let's now look at what can happen when a function is not continuous.

---

[1] Unless otherwise indicated, any statement that a certain limit exists should be understood to mean that the limit converges to some *finite* (possibly complex) number. We are thus excluding "limits converging to infinity", such as $\lim_{x \to 0} x^{-2}$.

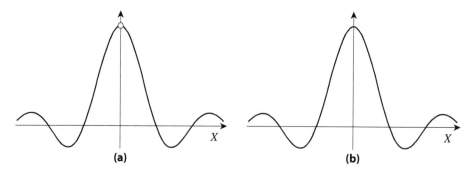

Figure 3.1: The sinc function (a) with the trivial discontinuity at $x = 0$ and (b) with the trivial discontinuity removed.

Discontinuities

Let f be some function on (α, β), and let x_0 be a point in the interval (α, β). If f is not continuous at x_0, then it must have one of three types of discontinuities at x_0 — trivial, jump or "bad" — as described below.

Trivial Discontinuities

The function f has a *trivial discontinuity* (also called a *removable discontinuity*) at x_0 if the limit of $f(x)$ does exist as x approaches x_0, but either this limit does not equal $f(x_0)$ or $f(x_0)$ does not even exist according to the definition given for the function. A classic example is the *sinc* (pronounced "sink") function on $(-\infty, \infty)$, which, typically, is given by the formula[2]

$$\operatorname{sinc}(x) \;=\; \frac{\sin(x)}{x} \quad .$$

While this formula is indeterminate at $x = 0$, we see that, using L'Hôpital's rule,

$$\lim_{x \to 0} \frac{\sin(x)}{x} \;=\; \lim_{x \to 0} \frac{\frac{d}{dx}\sin(x)}{\frac{d}{dx}x} \;=\; \lim_{x \to 0} \frac{\cos(x)}{1} \;=\; 1 \quad .$$

But recall our discussion in the previous chapter. As far as we are concerned, the value of a function at a single point is irrelevant, and (re)defining its formula at any single point (or any finite number of points on any finite interval) does not change that function. This means we can "remove" the discontinuity in the sinc function by appropriately (re)defining $\operatorname{sinc}(x)$ to be 1 when $x = 0$,

$$\operatorname{sinc}(x) \;=\; \begin{cases} \dfrac{\sin(x)}{x} & \text{if } \; x \neq 0 \\[2mm] 1 & \text{if } \; x = 0 \end{cases} \quad .$$

The graphs of the sinc function with the trivial discontinuity at $x = 0$ and with this discontinuity removed are sketched in figure 3.1.

Likewise, any other function f with a trivial discontinuity at some point x_0 can have that discontinuity removed by (re)defining $f(x_0)$ to be $\lim_{x \to x_0} f(x)$. Since redefining a function's formula at isolated points does not change the function as far as we are concerned, let us agree that, if any function is initially defined or otherwise described with a finite number of trivial discontinuities on any finite interval, then those trivial discontinuities are automatically assumed to be removed.

---

[2] Warning: Some texts define the sinc function by $\quad \operatorname{sinc}(x) \;=\; \dfrac{\sin(2\pi x)}{x} \quad .$

!▶ **Example 3.1:** *For all* $-\infty < x < \infty$, *let*

$$f(x) = \frac{\sin(2\pi x)}{\sin(\pi x)} \quad . \tag{3.1}$$

This function is clearly continuous at any x *other than* $x_0 = 0, \pm 1, \pm 2, \ldots$. *On the other hand, when* x *is an integer, both the numerator and denominator are zero. Using L'Hôpital's rule to evaluate the limits at these points, we find that*

$$\lim_{x \to x_0} \frac{\sin(2\pi x)}{\sin(\pi x)} = \frac{2\pi \cos(2\pi x_0)}{\pi \cos(\pi x_0)} = \begin{cases} +2 & \text{if} \quad x_0 \text{ is even} \\ -2 & \text{if} \quad x_0 \text{ is odd} \end{cases} \quad .$$

Thus, since trivial discontinuities are assumed to be removed, formula (3.1) is understood to mean

$$f(x) = \begin{cases} \dfrac{\sin(2\pi x)}{\sin(\pi x)} & \text{if} \quad x \text{ is not an integer} \\ +2 & \text{if} \quad x = 0, \pm 2, \pm 4, \ldots \\ -2 & \text{if} \quad x = \pm 1, \pm 3, \pm 5, \ldots \end{cases} \quad . \tag{3.2}$$

The above example illustrates the fact that, typically, trivial discontinuities arise because of limitations in the formula used to describe the function. "Removing the trivial discontinuities" then amounts to giving a more complete or precise formula for the function, and our agreement that "all trivial discontinuities are assumed removed" is simply an agreement that a more complete formula (such as formula (3.2)) will be assumed whenever we state a less precise formula (such as formula (3.1)).

?▶ **Exercise 3.1:** *Verify that, if* g *is given by*

$$g(x) = \frac{\sin(2\pi x)}{x} \quad ,$$

then $g(0) = 2\pi$. *(Remember, trivial discontinuities are assumed to be removed.)*

Jump Discontinuities

The function f has a *jump discontinuity* at x_0 if the left- and right-hand limits of the function at x_0,

$$\lim_{x \to x_0^-} f(x) \quad \text{and} \quad \lim_{x \to x_0^+} f(x) \quad ,$$

both exist but are not equal (see figure 3.2). The *jump* in f at x_0 is the difference

$$j_0 = \lim_{x \to x_0^+} f(x) - \lim_{x \to x_0^-} f(x) \quad .$$

Clearly, such a function cannot be made continuous by (re)defining the function at the jump discontinuity. We could, for reasons of aesthetics (again, see figure 3.2), (re)define the value of a function at a jump discontinuity to be the midpoint of the jump,

$$f(x_0) = \frac{1}{2}\left[\lim_{x \to x_0^+} f(x) + \lim_{x \to x_0^-} f(x) \right] \quad ,$$

but this will not appreciably simplify the mathematics of interest to us. Since this is the case and since we have already agreed that the value of a function at a single point is irrelevant, we will simply not worry about the value of a function at a jump. And if the value of a function is accidentally specified at a jump, we will feel free to ignore that specification.

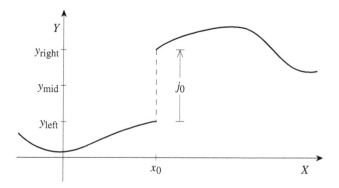

Figure 3.2: Generic jump discontinuity in f at x_0 with $y_{\text{left}} = \lim_{x \to x_0^-} f(x)$, $y_{\text{right}} = \lim_{x \to x_0^+} f(x)$ and $y_{\text{mid}} = $ "midpoint of the jump".

!▶ *Example 3.2 (the step function):*   *One of the simplest examples of a function with a jump discontinuity is the* unit step function

$$\text{step}(x) \;=\; \begin{cases} 0 & \text{if } x < 0 \\ 1 & \text{if } 0 < x \end{cases} \quad .$$

Note that $\text{step} = u = h$ *where u and h are the functions from example 2.2.*[3]

Bad Discontinuities

Any discontinuity that is neither trivial nor a jump will be considered a *bad discontinuity*. Some functions with bad discontinuities at $x = 0$ have been (very crudely) sketched in figure 3.3. The classical theory of Fourier analysis is not well suited for dealing with functions having such discontinuities. Because of this, little will be said about these functions until the generalized theory is discussed in part IV.

Classifying Functions Based on Continuity
Continuous Functions

A function f is *continuous* on an interval (α, β) if and only if it is continuous at each point in the interval. Remember that, if any finite subinterval of (α, β) contains a finite (but not infinite[4]) number of trivial discontinuities, then all trivial discontinuities are automatically assumed to have been removed.

!▶ *Example 3.3:*   *The function from example 3.1,*

$$f(x) \;=\; \frac{\sin(2\pi x)}{\sin(\pi x)} \quad,$$

is continuous on the real line.

---

[3] The unit step function is also known as the Heaviside step function and is commonly denoted by either u or h. That notation, however, would become confusing for us since we'll be using these symbols for so many other things.

[4] In this book, we will concern ourselves only with functions initially possessing at most a finite number of trivial discontinuities in any given finite interval. More advanced readers should be aware that functions with infinitely many trivial discontinuities (and no other discontinuities) can still be treated as continuous so long as the set of all discontinuities "has measure zero".

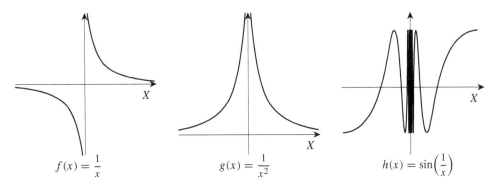

Figure 3.3: Three functions with bad discontinuities at $x = 0$.

Even though a function is continuous on a given interval, it might still be rather poorly behaved near an endpoint of the interval. For example, even though the function $^1\!/_x$ is continuous on the finite interval $(0, 1)$, it is not bounded. Instead, it "blows up" around $x = 0$. When we want to exclude such functions from discussion when (α, β) is a finite interval, we will impose the condition of "uniform continuity", as defined in the next paragraph.

Let (α, β) be a finite interval. A function f is *uniformly continuous on* (α, β) if, in addition to being continuous on (α, β), its one-sided limits at the endpoints,

$$\lim_{x \to \alpha^+} f(x) \qquad \text{and} \qquad \lim_{x \to \beta^-} f(x) \quad ,$$

both exist.

?▶ Exercise 3.2: *Why is* $(x - 1)^{-1}$ *not uniformly continuous on* $(0, 1)$?

Let us observe that, if f is continuous on any interval (α, β), finite or infinite, and if $\alpha < a < b < \beta$, then f is continuous over the finite subinterval (a, b). Moreover, since f is continuous at a and b, the one-sided limits

$$\lim_{x \to a^+} f(x) \qquad \text{and} \qquad \lim_{x \to b^-} f(x)$$

both exist. Thus, f is uniformly continuous over (a, b). This fact is significant enough to be recorded in a lemma for future reference.

Lemma 3.1
Let f be continuous on any interval (α, β), and let $\alpha < a < b < \beta$. Then f is uniformly continuous over the finite subinterval (a, b).

The next two lemmas describe two properties of uniformly continuous functions. The first should seem pretty obvious if you think about sketching a uniformly continuous function. The second provides an alternate definition of uniform continuity that will be useful for some of the more theoretical work we may be doing later.

Lemma 3.2
Any function that is uniformly continuous on a finite interval is also a bounded function on that interval.

Lemma 3.3 *(alternate definition of uniform continuity)*
A function f is uniformly continuous on a finite interval (α, β) if and only if, for each positive value ϵ, there is a corresponding positive value Δx_ϵ such that

$$|f(x) - f(\bar{x})| < \epsilon$$

whenever $\{x, \bar{x}\}$ is a pair of points in (α, β) with

$$|x - \bar{x}| < \Delta x_\epsilon \quad .$$

It might be noted that the alternate definition of uniform continuity indicated in lemma 3.3 can be used to define uniform continuity on infinite intervals as well as finite intervals.

The boundedness of uniformly continuous functions on finite intervals can probably be accepted as fairly obvious. The validity of lemma 3.3 may not be so obvious and should be proven before the lemma is used. However, it will be a while before we need this lemma, and, while the proof is terribly interesting (to some), it is also somewhat lengthy. So let us place this proof in an addendum to this chapter (see page 31) to be reviewed at a more appropriate time.

Discontinuous Functions

Fourier analysis would be of very limited value if it only dealt with continuous functions. Still, we won't be able to deal with every possible discontinuous function. We will have to restrict our attention to discontinuous functions we can reasonably handle. Typically, the minimal continuity requirement that we can conveniently get away with is "piecewise continuity" over the interval of interest. Occasionally, the requirements can be weakened so that we can deal with some functions that are merely "continuous over some partitioning of the interval".

Because it is the more important, we will describe "piecewise continuity" first.

Let f be a function defined on an interval (α, β). If (α, β) is a finite interval, then we will say f is *piecewise continuous* on (α, β) if and only if *all* of the following three statements hold:

1. f has at most a finite number (possibly zero) of discontinuities on (α, β).

2. All of the (nontrivial) discontinuities of f on (α, β) are jump discontinuities.

3. Both $\lim_{x \to \alpha^+} f(x)$ and $\lim_{x \to \beta^-} f(x)$ exist (as finite numbers).

If, on the other hand, (α, β) is an infinite interval, then f will be referred to as *piecewise continuous* on (α, β) if and only if it is piecewise continuous on each finite subinterval of (α, β).

It is important to realize that a piecewise continuous function is not simply "continuous over pieces of (α, β)". To see this, let (α, β) be a finite interval, and let $x_1, x_2, \ldots, x_N$ be the points in (α, β) — indexed so that $x_1 < x_2 < \cdots < x_N$ — at which a given piecewise continuous function f is discontinuous . These points partition (α, β) into a finite number of subintervals

$$(\alpha, x_1) \quad , \quad (x_1, x_2) \quad , \quad (x_2, x_3) \quad , \quad \ldots \quad , \quad (x_N, \beta) \quad ,$$

with f being continuous over each of these subintervals. But the second and third parts of the definition also ensure that

$$\lim_{x \to \alpha^+} f(x) \quad , \quad \lim_{x \to x_1^-} f(x) \quad , \quad \lim_{x \to x_1^+} f(x) \quad , \quad \lim_{x \to x_2^-} f(x) \quad , \quad \ldots \quad , \quad \lim_{x \to \beta^-} f(x)$$

all exist (and are finite). Thus, not only is f continuous on each of the above subintervals, it is uniformly continuous on each of the above subintervals.[5]

---

[5] It may be more descriptive to say that these functions are "piecewise *uniformly* continuous on finite intervals", but nobody does. By the way, some authors use the term "sectionally continuous" instead of "piecewise continuous".

?► Exercise 3.3:   *Show, by example, that there are functions continuous on a finite interval, say,* (0, 1), *that are not piecewise continuous on that interval.*

"Continuity over a partitioning" is simply piecewise continuity without the uniformity. More precisely, we'll say that a function is *continuous over a partitioning* of an interval (α, β) if and only if that function has at most a finite number of (nontrivial) discontinuities on each finite subinterval of (α, β).

By the way, the "partitioning of the interval" being referred to is the partitioning of (α, β) into subintervals,

$$\ldots \quad , \quad (x_1, x_2) \quad , \quad (x_2, x_3) \quad , \quad (x_3, x_4) \quad , \quad \ldots$$

by the points $\ldots, x_1, x_2, x_3, \ldots$ at which f is discontinuous (with the indexing chosen so that $\ldots < x_1 < x_2 < x_3 < \ldots$).

!► Example 3.4:   *The function* $f(x) = \frac{1}{x}$ *has only one discontinuity on the real line, at* $x = 0$. *Since* $\frac{1}{x} \to \pm\infty$ *as* $x \to \pm 0$, *the discontinuity is neither trivial nor a jump. Hence, this function is continuous over a partitioning of* $(-\infty, \infty)$. *In particular, it is continuous over partitioning consisting of the subintervals*

$$(-\infty, 0) \quad \text{and} \quad (0, \infty) \quad .$$

However, because the discontinuity at $x = 0$ *is neither trivial nor a jump, this function is not piecewise continuous on the real line.*

Equality of (Dis)continuous Functions

Considering our pragmatic approach to the equality of functions, it may be worthwhile to re-examine this concept when the two functions are piecewise continuous over an interval or even just continuous over a partitioning of that interval.

Lemma 3.4
Let f and g be two functions defined on an interval (α, β), and assume $f = g$ on this interval *(in the sense described in section 2.2). Then (after removal of all trivial discontinuities):*

1. *If* f *is continuous on* (α, β), *then so is* g. *Moreover,* $f(x) = g(x)$ *for every* x *in* (α, β).

2. *If* f *is piecewise continuous on* (α, β), *then so is* g. *Moreover,* $f(x) = g(x)$ *for every* x *in* (α, β) *at which* f *is continuous.*

3. *If* f *is continuous on a partitioning of* (α, β), *then* g *is continuous on the same partitioning. Moreover,* $f(x) = g(x)$ *for every* x *in* (α, β) *at which* f *is continuous.*

The proof of this lemma is straightforward and left as an exercise.

?► Exercise 3.4:   *Prove lemma 3.4.*

Endpoint Values

On occasion we will have a function f that is uniformly continuous on some finite open interval (α, β), and we will want to discuss something regarding "the value of $f(x)$ at one of the endpoints".

Strictly speaking, the values $f(\alpha)$ and $f(\beta)$ may not be well defined either because $f(x)$ was not originally defined for $x = \alpha$ or $x = \beta$, or because f is not continuous at one or both of these points. Still, if we are restricting our attention to the behavior of f just over the interval (α, β), then we really do not care about the values of the function outside that interval. So let us agree that, whenever we are restricting our attention to a function f over a finite interval (α, β) over which f is uniformly continuous, then, by $f(\alpha)$ and $f(\beta)$, we mean

$$f(\alpha) = \lim_{x \to \alpha^+} f(x) \quad \text{and} \quad f(\beta) = \lim_{x \to \beta^-} f(x) \quad .$$

3.2 Differentiation

In Fourier analysis, we often must deal with derivatives of functions that are not, strictly speaking, differentiable. To understand why this is not a contradiction, let us carefully review the terminology.

Differentiability

A function f is *differentiable at a point* x if and only if

$$\lim_{\Delta x \to 0} \frac{f(x + \Delta x) - f(x)}{\Delta x} \tag{3.3}$$

exists. If f is differentiable at every point in a given interval (α, β), then f is said to be *differentiable on the interval* (α, β) or, if we want to be very explicit, *differentiable everywhere on* (α, β).

Observe that, if a function is differentiable at a point or on some interval, then that function must also be continuous at that point or on that interval. On the other hand, there are many continuous functions that are not everywhere differentiable. It is also worth recalling the geometric significance of differentiability and the above limit, namely, that the statement "f is differentiable at x" is equivalent to the statement "the graph of f has a single well-defined tangent at x". Moreover, the limit in expression (3.3) gives the slope of this tangent line.

?►Exercise 3.5: *Verify that $|x|$ is continuous, but not differentiable, at $x = 0$.*

Derivatives

For each point x at which f is differentiable, the *derivative of f at x*, denoted by $f'(x)$, is the *number* given by the limit in expression (3.3),

$$f'(x) = \lim_{\Delta x \to 0} \frac{f(x + \Delta x) - f(x)}{\Delta x} \quad . \tag{3.4}$$

Suppose f is differentiable at all but a finite number (possibly zero) of points in each finite subinterval of (α, β). Then formula (3.4) also defines another function on (α, β), called, naturally, the *derivative* of f on (α, β) and commonly denoted by f' (or $^{df}/_{dx}$ or $^{df}/_{dt}$ or ...). Notice that the derivative of a function can exist on an interval even though the function is *not* differentiable everywhere on that interval. In fact, as our next example shows, it is possible for the derivative to be continuous (after removing the trivial discontinuities) even though the function, itself, has a nontrivial discontinuity.

!▶ Example 3.5: *The step function,*

$$\text{step}(x) = \begin{cases} 0 & \text{if } x < 0 \\ 1 & \text{if } 0 < x \end{cases} ,$$

is clearly differentiable everywhere on $(-\infty, \infty)$ *except at the point* $x = 0$ *where the step function has a nontrivial jump discontinuity. It should also be clear that*

$$\text{step}'(x) = \begin{cases} 0 & \text{if } x < 0 \\ 0 & \text{if } 0 < x \end{cases} .$$

The discontinuity at $x = 0$ *is a trivial one. Removing this discontinuity gives*

$$\text{step}'(x) = 0 \quad \text{for} \quad -\infty < x < \infty ,$$

which is a continuous function on the entire real line even though the step function is not differentiable on the real line.

?▶ Exercise 3.6: *What is the derivative of* $|x|$ *?*

Remember, "differentiability" (i.e., "differentiability at every point on a given interval") is a much stronger condition than "the derivative exists". Do not assume a function f is differentiable on an interval just because f' exists on the interval. This is important because we will be using and deriving a number of formulas involving derivatives of differentiable functions. In general, these formulas will not be valid for functions that are not differentiable everywhere. Using these formulas without checking that the functions involved are suitably differentiable can lead to serious (and embarrassing) errors.

!▶ Example 3.6: *You surely recall that*

$$\int_\alpha^\beta f'(x)\, dx = f(\beta) - f(\alpha)$$

whenever f *is differentiable on an interval containing* α *and* β *. If we ignore the requirement that* f *be differentiable, then, (mis)using this equation with* $f(x) = \text{step}(x)$ *and recalling that* $\text{step}'(x) = 0$ *we obtain*

$$0 = \int_{-2}^3 0\, dx = \int_{-2}^3 \text{step}'(x)\, dx = \text{step}(3) - \text{step}(-2) = 1 \quad !$$

Smoothness
Smooth Functions

To be *smooth* over an interval (α, β), a function f must satisfy two conditions:

1. f must be differentiable (and, hence, continuous) everywhere on (α, β), and

2. f' must also be a continuous function on (α, β).

!▶ Example 3.7: *The function* $|x|$ *is not smooth on any interval containing the origin since, as was seen in exercise 3.5,* $|x|$ *is not differentiable at* $x = 0$ *.*

!►Example 3.8:   *Even though the derivative of the step function is continuous on the real line (after removing the trivial discontinuity, see example 3.5), the step function, itself, is not smooth on any interval containing the origin because it has a jump discontinuity at $x = 0$.*

The graph of a smooth, real-valued function looks like a smoothly curving line. Typically, the graphs of nonsmooth functions contain nontrivial discontinuities (as with the step function at $x = 0$) or else have sharp corners (as with $|x|$ at $x = 0$).

From the definition it is clear that a smooth function is differentiable. And, if you were to test a random sampling of known differentiable functions, it may appear as if all differentiable functions are smooth. This, however, is not true. There are differentiable functions that are not smooth (see exercise 3.17 on page 35).

Uniform Smoothness

Let (α, β) be a finite interval. A function f is *uniformly smooth* on (α, β) if and only if

1. f is smooth on (α, β) , and

2. both f and f' are uniformly continuous on (α, β) .

(This also defines uniform smoothness for a function on an infinite interval, provided the definition of uniform continuity is the alternative definition given in lemma 3.3 — with the word "finite" replaced by "infinite".)

!►Example 3.9:   *Consider the function $f(x) = x^{1/2}$ over the interval $(0, 1)$. Both f and its derivative, $f'(x) = \frac{1}{2}x^{-1/2}$, are clearly continuous everywhere on $(0, 1)$. In fact, f is uniformly continuous on $(0, 1)$ (You verify this!). But*

$$\lim_{x \to 0^+} f'(x) = \lim_{x \to 0^+} \frac{1}{2}x^{-1/2} = \infty \quad .$$

So f' is not uniformly continuous on $(0, 1)$, and hence, f is not uniformly smooth on the interval $(0, 1)$.

?►Exercise 3.7:   *Show that $x^{1/2}$ is uniformly smooth on (α, β) whenever $0 < \alpha < \beta < \infty$.*

Piecewise Smoothness

A function is said to be *piecewise smooth* over a finite interval (α, β) if and only if (α, β) can be partitioned into a finite number of subintervals over which the function is uniformly smooth.

If (α, β) is an infinite interval, then a function is *piecewise smooth* over (α, β) if and only if it is piecewise smooth over every finite subinterval of (α, β) .

?►Exercise 3.8:   *Sketch the graphs of some functions that are piecewise smooth over the real line. Also, sketch the graphs of some functions that are not piecewise smooth over the real line.*

Higher Order Smoothness

A function f on some interval is said to be *twice differentiable* (on that interval) whenever it and its derivative, f', are both differentiable. Notice that, for f' to be differentiable, it must first be continuous. So a twice-differentiable function is automatically a smooth function.

To continue along these lines, let k be any positive integer. We will refer to a function f as being *k-times differentiable* on an interval (or k^{th} *differentiable* or k^{th} *order differentiable* on some interval) if and only if it and all of its derivatives up to order $k-1$ are differentiable on that interval. To extend the observation made in the previous paragraph, note that whenever f is k-times differentiable on some interval, then $f, f', f'', \ldots$ and $f^{(k-2)}$ must all be smooth functions on that interval.

Ultimately, a function and all of its derivatives may be differentiable on an interval, which then means that the function and all of its derivatives are smooth functions on that interval. Such functions are said to be either *infinitely differentiable* or, equivalently, *infinitely smooth* on the interval.

?▶ Exercise 3.9   **a:** *Verify that $f(x) = x^{4/3}$ is differentiable but not twice differentiable on the real line.*

    b: *Give an example of a function that is twice differentiable but not third-order differentiable on the real line.*

    c: *Give an example of an infinitely differentiable function on the real line.*

3.3   Basic Manipulations and Smoothness
Scaling, Shifting and Linear Combinations

Scaling, shifting and forming linear combinations are operations that arise naturally when using functions. Since we will be using these operations extensively, and since we have already recalled what a "linear combination" is (footnote 1 on page 4), let us briefly review what it means to scale and shift a function.

Scaling

Let γ be a number and f a function. "Scaling by γ" can refer to either of two operations. It can mean that the function is being multiplied by γ (i.e., $f(x)$ is replaced by $\gamma f(x)$). More commonly, in this text at least, *scaling* by γ means that the variable is being multiplied by γ (i.e., $f(x)$ is replaced by $f(\gamma x)$). In this case, γ is usually assumed to be a nonzero real number. Notice that the behavior of the scaled function, $f(\gamma x)$, around the point $x = a$ corresponds to the behavior of the original function, $f(x)$, around the point $x = \gamma a$. Conversely, the behavior of the original function, $f(x)$, around the point $x = a$ corresponds to the behavior of the scaled function, $f(\gamma x)$, around the point x where $\gamma x = a$ (that is, the point $x = {}^a/_\gamma$). It follows then, that the graph of $f(\gamma x)$ is a horizontally compressed version of the graph of $f(x)$ when $1 < \gamma$, and is a horizontally expanded version of the graph of $f(x)$ when $0 < \gamma < 1$.

?▶ Exercise 3.10:   *Let*

$$f(x) = \begin{cases} x(1-x) & if \ \ 0 < x < 1 \\ 0 & otherwise \end{cases}.$$

Sketch and compare the graphs of $f(x)$ and $f(\gamma x)$ for the following cases:

   a: $1 < \gamma$ *(say, $\gamma = 2$)*   **b:** $0 < \gamma < 1$ *(say, $\gamma = {}^1/_2$)*

   c: $\gamma = -1$

?► Exercise 3.11:   *In general, what happens to the graph of $f(\gamma x)$*

   a: *as $\gamma \to \infty$?*

   b: *as $\gamma \to 0^+$?*

   Assume f is defined on all of $\mathbb{R}$.

Shifting and Translation

A function $f(x)$ is said to be *shifted* or *translated* by a fixed real number γ when the variable x is replaced by $x - \gamma$. The graph of the resulting shifted function can be obtained from the graph of the original function by shifting the original graph horizontally by a distance of $|\gamma|$. If $\gamma > 0$, the shift is to the right. If $\gamma < 0$, the shift is to the left.

?► Exercise 3.12:   *Let*

$$f(x) = \begin{cases} x(1-x) & \text{if } 0 < x < 1 \\ 0 & \text{otherwise} \end{cases}.$$

Sketch the graphs of $f(x)$ and $f(x - \gamma)$ for the following cases:

   a: $0 < \gamma$ *(say, $\gamma = 2$).*

   b: $\gamma < 0$ *(say, $\gamma = -2$).*

In each case, be sure to compare the graph of $f(x - \gamma)$ with the graph of $f(x)$.

Smoothness under Basic Operations

Suppose we have a function $f(x)$ that satisfies any of the properties discussed thus far in this chapter (boundedness, continuity, piecewise continuity, differentiability, etc.) over an interval (α, β), and let γ be any nonzero real number.

    It is easy to see that the scaled function $f(\gamma x)$ also satisfies the same properties as $f(x)$, but over the interval (a, b) where

$$(a, b) = \begin{cases} \left(\dfrac{\alpha}{\gamma}, \dfrac{\beta}{\gamma}\right) & \text{if } 0 < \gamma \\[2ex] \left(\dfrac{\beta}{\gamma}, \dfrac{\alpha}{\gamma}\right) & \text{if } \gamma < 0 \end{cases}.$$

    It should also be clear that the translation of f by γ, $f(x - \gamma)$, satisfies the same properties as f, but over the interval $(\alpha + \gamma, \beta + \gamma)$.

    Finally, suppose we have a collection of functions, $\{f_1, f_2, \ldots\}$, and that, on the interval (α, β), all of these functions satisfy any one of the conditions discussed thus far (e.g., all are bounded or all are smooth on the interval). Then it should be clear that any *finite* linear combination of these f_k's also satisfies that property over the interval (α, β).

    On the other hand, if g is defined to be an infinite linear combination of the f_k's,

$$g(x) = c_1 f_1(x) + c_2 f_2(x) + c_3 f_3(x) + \cdots \quad ,$$

then there is no general assurance that g satisfies any of the properties satisfied by all the f_k's. Indeed, an infinite linear combination of the f_k's is actually an infinite series of functions that might not even converge to any sort of a function. This will be one of our concerns when we deal with such linear combinations.

3.4 Addenda

Some of the proofs in this text will involve technical issues that are best discussed only when the need arises. For want of a better place, we'll discuss some of those issues here. If you've not yet reached those proofs, you may just want to give the following material a quick glance so you'll know where to return when you do reach those proofs.

A Refresher on Limits

Presumably, you already have a good intuitive notion of what is meant by the equivalent statements

$$f(x) \to L \quad \text{as} \quad x \to x_0 \qquad \text{and} \qquad \lim_{x \to x_0} f(x) = L \quad ,$$

as well as such standard variations as

$$\lim_{x \to x_0^+} f(x) = L \quad , \qquad \lim_{x \to \infty} f(x) = L \quad \text{and} \qquad \lim_{x \to x_0} f(x) = \infty \quad .$$

For most of this text, your intuitive notion of these concepts should serve quite well, provided you also recall such basic limit theorems from elementary calculus as

If

$$\lim_{x \to x_0} f(x) \qquad \text{and} \qquad \lim_{x \to x_0} g(x)$$

both exist and are finite, then

$$\lim_{x \to x_0} \left(f(x) g(x) \right) = \left(\lim_{x \to x_0} f(x) \right) \left(\lim_{x \to x_0} g(x) \right) \quad .$$

You should also realize that, suitably rephrased, these theorems hold for complex-valued functions of complex variables, as well as for functions of two or more variables.

On occasion, however, we may need to employ a certain "limit test", which the reader may have forgotten. This is a fundamental test for showing both that the limit of $f(x)$ exists as x approaches a finite point x_0 and that

$$\lim_{x \to x_0} f(x) = 0 \quad . \tag{3.5a}$$

This last statement is, of course, equivalent to

$$\lim_{x \to x_0} |f(x)| = 0 \quad . \tag{3.5b}$$

Recall what expression (3.5b) really says. It says that, by setting the value of x "suitably close to x_0", we will force the value of $|f(x)|$ to be "correspondingly close to zero". This, then, is what our test needs to show, namely, that we can force $|f(x)|$ to be as close to zero as desired by simply choosing x to be "close enough" to x_0. (Remember, "close" means "within some small but non-zero distance".) Traditionally, ϵ denotes how close to zero we desire $f(x)$ to be, and Δx (or δ) denotes how close we need x to be to x_0 to ensure that $f(x)$ is within the desired distance, ϵ, of zero. In these terms, what we need to show can be stated as:

For every given $\epsilon > 0$, there is a corresponding $\Delta x > 0$ such that

$$|f(x)| < \epsilon \qquad \text{whenever} \qquad 0 < |x - x_0| < \Delta x \quad .$$

Thus, a basic test for showing both that the limit of $f(x)$ as $x \to x_0$ exists and that

$$\lim_{x \to x_0} f(x) = 0$$

is to explicitly show that, *for every choice of $\epsilon > 0$, there is a corresponding (positive) value for Δx such that*

$$|f(x)| < \epsilon \qquad whenever \qquad 0 < |x - x_0| < \Delta x \quad .$$

This "test" should look vaguely familiar. It is, in fact, the standard definition of expression (3.5b). Admittedly, it is rarely used to actually compute a limit. Still, on occasion, we will find it necessary to return to this basic test/definition.

While on the subject, let's recall the basic tests/definitions for a few other limits:

1. *If x_0 is a finite point and L is a finite value, then the statements*

    $$\lim_{x \to x_0} f(x) = L \qquad and \qquad \lim_{x \to x_0} |f(x) - L| = 0$$

    mean the same thing.

2. *If L is a finite value and f is a function on the real line, then*

    $$\lim_{x \to \infty} f(x) = L$$

    if and only if, for every $\epsilon > 0$, there is a corresponding finite real value X such that

    $$|f(x) - L| < \epsilon \qquad whenever \qquad X < x \quad .$$

3. *If x_0 is any finite point, then*

    $$\lim_{x \to x_0} |f(x)| = \infty$$

    if and only if, for every $M > 0$, there is a corresponding $\Delta x > 0$ such that

    $$|f(x)| > M \qquad whenever \quad 0 < |x - x_0| < \Delta x \quad .$$

Some Useful Inequalities

At various points in our work, we will need to determine "suitable upper bounds" for various numerical expressions. At some of these points, the inequalities discussed below will be invaluable.

Two basic inequalities will be identified. You are probably well acquainted with the first one, the triangle inequality, though you may not have given it a name before. You may not be as well acquainted with the second one, the Schwarz inequality. It is somewhat more subtle than the triangle inequality and will require a formal proof. Both, it should be mentioned, are fundamental inequalities in analysis and have applications and generalizations beyond the simple formulas discussed in this section.

The Triangle Inequality

Let A and B be any two real numbers. If you just consider how values of $|A|$, $|B|$, $|A + B|$ and $|A| + |B|$ depend on the signs of A and B, then you should realize that

$$|A + B| \leq |A| + |B| \quad . \tag{3.6}$$

This inequality is called the *triangle inequality*. The reason for its name is explained in chapter 6 (see page 58), where it is also shown that this inequality holds when A and B are complex numbers as well.

There are two other inequalities that we can immediately derive from the triangle inequality. The first is the obvious extension to the case where we are adding up some (finite) set of numbers $\{A_1, A_2, A_3, \ldots, A_N\}$. Successively applying the triangle inequality,

$$
\begin{aligned}
|A_1 + A_2 + A_3 + \cdots + A_N| &\leq |A_1| + |A_2 + A_3 + \cdots + A_N| \\
&\leq |A_1| + |A_2| + |A_3 + \cdots + A_N| \\
&\leq \cdots \quad ,
\end{aligned}
$$

we are, eventually, left with the inequality

$$
|A_1 + A_2 + A_3 + \cdots + A_N| \leq |A_1| + |A_2| + |A_3| + \cdots + |A_N| \quad ,
$$

which can also be called the triangle inequality.

The derivation of the other inequality requires a smidgen of cleverness. Let A and B be any two numbers (real or complex) and observe that, by the original triangle inequality and elementary algebra,

$$
|A| = |A - B + B| \leq |A - B| + |B| \quad .
$$

Subtract $|B|$ from both sides and you have

$$
|A| - |B| \leq |A - B| \quad .
$$

For future reference, we'll summarize our derivations in the following lemma and corollary.

Lemma 3.5 (triangle inequality)
Given any finite set of numbers (real or complex) $\{A_1, A_2, A_3, \ldots, A_N\}$, *then*

$$
|A_1 + A_2 + A_3 + \cdots + A_N| \leq |A_1| + |A_2| + |A_3| + \cdots + |A_N| \quad .
$$

Corollary 3.6
Let A *and* B *be any two real or complex numbers. Then*

$$
|A| - |B| \leq |A - B| \quad .
$$

These inequalities will often be used with functions that are either nondecreasing or nonincreasing. Observe that, if f is a nondecreasing function on the real line (i.e., $f(a) \leq f(b)$ whenever $a \leq b$), then the above inequalities immediately imply that

$$
f(|x + y|) \leq f(|x| + |y|) \quad \text{and} \quad f(|x| - |y|) \leq f(|x - y|) \quad .
$$

On the other hand, if g is a nonincreasing function (i.e., $f(a) \geq f(b)$ whenever $a \leq b$), then we have

$$
f(|x + y|) \geq f(|x| + |y|) \quad \text{and} \quad f(|x| - |y|) \geq f(|x - y|) \quad .
$$

?►Exercise 3.13: *Let* α *be a positive value, and let* x *and* b *be any two real numbers. Verify the following inequalities:*

a: $e^{\alpha|x-b|} \leq e^{\alpha|x|} e^{\alpha|b|}$ **b:** $e^{\alpha|x-b|} \geq e^{\alpha|x|} e^{-\alpha|b|}$

c: $e^{-\alpha|x-b|} \geq e^{-\alpha|x|} e^{-\alpha|b|}$ **d:** $e^{-\alpha|x-b|} \leq e^{-\alpha|x|} e^{\alpha|b|}$

(Note: We'll use these particular inequalities later.)

The Schwarz Inequality (for Finite Sums)

The Schwarz inequality is a generalization of the well-known fact that, if $\mathbf{a}$ and $\mathbf{b}$ are any two two- or three-dimensional vectors, then

$$|\mathbf{a} \cdot \mathbf{b}| \leq \|\mathbf{a}\| \|\mathbf{b}\| \quad .$$

In component form, with $\mathbf{a} = (a_1, a_2, a_3)$ and $\mathbf{b} = (b_1, b_2, b_3)$, this inequality is

$$\left| \sum_{k=1}^{3} a_k b_k \right| \leq \left(\sum_{k=1}^{3} |a_k|^2 \right)^{1/2} \left(\sum_{k=1}^{3} |b_k|^2 \right)^{1/2} \quad .$$

This inequality, suitably generalized, is the one generally referred to as the Schwarz inequality.[6]

Theorem 3.7 (Schwarz inequality for finite summations)
Let N be any integer, and let $\{a_1, a_2, a_3, \ldots, a_N\}$ and $\{b_1, b_2, b_3, \ldots, b_N\}$ be any two sets of N numbers (real or complex). Then,

$$\left| \sum_{k=1}^{N} a_k b_k \right| \leq \left(\sum_{k=1}^{N} |a_k|^2 \right)^{1/2} \left(\sum_{k=1}^{N} |b_k|^2 \right)^{1/2} \quad . \tag{3.7}$$

PROOF: Suppose we can show

$$\sum_{k=1}^{N} |a_k| \, |b_k| \leq \left(\sum_{k=1}^{N} |a_k|^2 \right)^{1/2} \left(\sum_{k=1}^{N} |b_k|^2 \right)^{1/2} \quad . \tag{3.8}$$

Then inequality (3.7) follows immediately by combining the above inequality with the triangle inequality,

$$\left| \sum_{k=1}^{N} a_k b_k \right| \leq \sum_{k=1}^{N} |a_k b_k| = \sum_{k=1}^{N} |a_k| \, |b_k| \quad .$$

So we only need to verify that inequality (3.8) holds.

Consider, first, the trivial case where either

$$\sum_{k=1}^{N} |a_k|^2 = 0 \quad \text{or} \quad \sum_{k=1}^{N} |b_k|^2 = 0 \quad .$$

In this case, all the a_k's or all the b_k's clearly must be 0, and the statement of inequality (3.7) reduces to the obviously true statement that $0 \leq 0$.

Now consider the case where

$$\sum_{k=1}^{N} |a_k|^2 > 0 \quad \text{and} \quad \sum_{k=1}^{N} |b_k|^2 > 0 \quad .$$

For convenience, let

$$A = \left(\sum_{k=1}^{N} |a_k|^2 \right)^{1/2} \quad \text{and} \quad B = \left(\sum_{k=1}^{N} |b_k|^2 \right)^{1/2} \quad .$$

---

[6] It's also referred to as Schwarz's inequality or the Cauchy–Schwarz inequality or even the Cauchy–Buniakowsky–Schwarz inequality — depending on the generalization and the mood of the author.

Using elementary algebra, we see that

$$0 \leq \sum_{k=1}^{N} (B\,|a_k| - A\,|b_k|)^2$$

$$= \sum_{k=1}^{N} \left[B^2\,|a_k|^2 - 2AB\,|a_k|\,|b_k| + A^2\,|b_k|^2 \right]$$

$$= B^2 \sum_{k=1}^{N} |a_k|^2 - 2AB \sum_{k=1}^{N} |a_k|\,|b_k| + A^2 \sum_{k=1}^{N} |b_k|^2$$

$$= B^2 A^2 - 2AB \sum_{k=1}^{N} |a_k|\,|b_k| + A^2 B^2$$

$$= 2AB \left[AB - \sum_{k=1}^{N} |a_k|\,|b_k| \right] \quad .$$

Thus, since $2AB$ is positive,

$$0 \leq AB - \sum_{k=1}^{N} |a_k|\,|b_k| \quad .$$

And so,

$$\sum_{k=1}^{N} |a_k|\,|b_k| \leq AB = \left(\sum_{k=1}^{N} |a_k|^2 \right)^{1/2} \left(\sum_{k=1}^{N} |b_k|^2 \right)^{1/2} \quad . \qquad \blacksquare$$

Uniform Continuity

Here we discuss the proof of lemma 3.3 on page 20 on uniform continuity. For convenience, the lemma will be broken into two smaller lemmas, lemmas 3.9 and 3.10 below. Also, to reduce the number of symbols, let's just consider proving the lemma assuming that $(\alpha, \beta) = (0, 1)$. (We can always extend the arguments to cases involving arbitrary intervals by the use of scaling and shifting.)

The proof of each part of lemma 3.3 employs something you should recall from calculus. For reference, I'll remind you of that something in the next lemma.

Lemma 3.8
Let α and β be two real numbers and assume $\{a_1, a_2, a_3, \ldots\}$ is a sequence of real numbers with $\alpha \leq a_n \leq \beta$ for each n. Assume, further, that $\{a_1, a_2, a_3, \ldots\}$ is either a nondecreasing sequence (i.e., $a_n \leq a_{n+1}$ for each n) or is a nonincreasing sequence (i.e., $a_n \geq a_{n+1}$ for each n). Then this sequence converges and

$$\alpha \leq \lim_{n \to \infty} a_n \leq \beta \quad .$$

Here is the first part of lemma 3.3:

Lemma 3.9
Let f be uniformly continuous on $(0, 1)$, and let ϵ be any fixed positive value. Then there is a

corresponding positive value Δx_ϵ *such that*

$$|f(x) - f(\bar{x})| < \epsilon$$

for each pair of points x *and* $\bar{x}$ *in* (α, β) *that satisfies*

$$|x - \bar{x}| < \Delta x_\epsilon \quad .$$

Since f is assumed to be uniformly continuous on $(0, 1)$, we can assume

$$f(0) = \lim_{x \to 0^+} f(x) \qquad \text{and} \qquad f(1) = \lim_{x \to 1^-} f(x) \quad .$$

If you recall what these limits mean (see the refresher on limits earlier in this addendum), you will realize that the above lemma is trivially true with " $(0, 1)$ " replaced by " $(0, \beta)$ " for some suitably small positive value β. What we will do is to construct a sequence of intervals $(0, b_0)$, $(0, b_1)$, $(0, b_2)$, ... such that the above lemma is obviously true when " $(0, 1)$ " is replaced by each " $(0, b_n)$ ", and then show that one of those $(0, b_n)$'s must be the entire interval $(0, 1)$.

PROOF (of lemma 3.9):   For each pair of integers n and k with $n > 0$ and $0 \le k \le 2^n$, let

$$\delta_n = \frac{1}{2^n} \qquad \text{and} \qquad x_{n,k} = k\delta_n = \frac{k}{2^n} \quad .$$

(Observe that $x_{n,0} = 0$ and $x_{n,2^n} = 1$.) Now, choose b_n to be the largest $x_{n,k}$ such that the following statement is true:

$$|f(x) - f(y)| < \epsilon$$

whenever

x and y are points in $[0, x_{n,k}]$ satisfying $|x - y| < \delta_n \quad .$

Since this statement holds trivially when $k = 0$, we are guaranteed that each b_n exists. Also, since the largest $x_{n,k}$ for each n is $x_{n,2^n} = 1$, we must have $b_n \le 1$ for each n. Observe, moreover, that if one of the b_n's, say, b_N, equals 1, then the claim of the lemma immediately follows (with $\Delta x_\epsilon = \delta_N$). So all we need to verify is that $b_n = 1$ for some n. We will do this by showing that it is impossible for $b_n < 1$ for all n. Our arguments will use the results from the following exercise.

?▶ Exercise 3.14:   *Let* N *be a fixed positive integer. Using the above definition for the* b_n's *:*

a: *Show that* $b_N \le b_{N+1}$. *(Suggestion: Let* K *be the integer such that* $b_N = x_{N,K}$. *Then verify that* $b_N = x_{N+1,2K}$ *and that* $x_{N+1,2K} \le b_{N+1}$.)

b: *Show that, as long as* $b_N < 1$, *there must be a pair of points* s_N *and* t_N *with*

$$b_N - \delta_N \le s_N < t_N \le b_N + \delta_N \tag{3.9}$$

such that

$$|f(s_N) - f(t_N)| \ge \epsilon \quad . \tag{3.10}$$

So now let's assume $b_n < 1$ for every positive integer n, and see why this assumption cannot be valid.

From the first part of the above exercise, we know that the b_n's form a nondecreasing sequence in $[0, 1]$. As noted in lemma 3.8, every such sequence converges to some value in $[0, 1]$. Let

$$b_\infty = \lim_{n \to \infty} b_n \quad .$$

Now let $s_1, s_2, \ldots$ and $t_1, t_2, \ldots$ be the points described in the second part of the above exercise. From inequalities (3.9) and the fact that $\delta_n = 2^{-n} \to 0$ and $n \to \infty$, we see that

$$\lim_{n \to \infty} s_n = \lim_{n \to \infty} b_n = b_\infty \quad \text{and} \quad \lim_{n \to \infty} t_n = \lim_{n \to \infty} b_n = b_\infty \quad .$$

Thus, by the continuity of f,

$$\lim_{n \to \infty} f(s_n) = f(b_\infty) \quad \text{and} \quad \lim_{n \to \infty} f(t_n) = f(b_\infty) \quad .$$

Combining this with inequality (3.10) gives us

$$\epsilon \leq \lim_{n \to \infty} |f(s_N) - f(t_N)| = |f(b_\infty) - f(b_\infty)| = 0 \quad ,$$

which is certainly impossible because ϵ is a positive value. Consequently, our assumption that $b_n < 1$ for every positive integer n cannot be valid. There must be a positive integer n for which $b_n = 1$. ∎

Here is the other part of lemma 3.3:

Lemma 3.10
Let f be a function on $(0, 1)$, and assume that, for each $\epsilon > 0$, there is a corresponding Δx_ϵ such that

$$|f(x) - f(\bar{x})| < \epsilon$$

whenever x and $\bar{x}$ is a pair of points in (α, β) with

$$|x - \bar{x}| < \Delta x_\epsilon \quad .$$

Then f is uniformly continuous on $(0, 1)$; that is, f is continuous on $(0, 1)$, and

$$\lim_{x \to 0^+} f(x) \quad \text{and} \quad \lim_{s \to 1^-} f(x)$$

both exist.

The continuity of f on $(0, 1)$ should be obvious if you recall the definitions of continuity and limits. Showing that the limits at $x = 0$ and $x = 1$ exist is a bit trickier. Here's a brief outline of the proof that $\lim_{x \to 0^+} f(x)$ exists, assuming f is real valued. For a complex-valued function, apply the following to the real and imaginary parts separately.

PROOF (outline only, that $\lim_{x \to 0^+} f(x)$ exists): First of all, using the assumptions of the lemma, it can be easily verified that we can choose a sequence of positive values $\delta_1, \delta_2, \delta_3, \ldots$ satisfying all of the following:

1. For each positive integer n,

$$|f(x) - f(\bar{x})| < \frac{1}{n}$$

whenever x and $\bar{x}$ is a pair of points in $(0, 1)$ such that

$$|x - \bar{x}| < \delta_n \quad .$$

2. $\delta_1 \geq \delta_2 \geq \ldots \geq \delta_n \geq \delta_{n+1} \geq \ldots .$

3. $\delta_n \to 0$ as $n \to \infty$.

Note that, in particular,

$$|f(x) - f(\delta_1)| < 1 \qquad \text{whenever} \quad 0 < x \leq \delta_1 \quad .$$

Thus, for each x in $(0, \delta_1)$,

$$L_1 < f(x) < U_1$$

where

$$U_1 = f(\delta_1) + 1 \qquad \text{and} \qquad L_1 = f(\delta_1) - 1 \quad .$$

For each successive positive integer n greater than 1, choose U_n to be the smaller of

$$f(\delta_n) + \frac{1}{n} \qquad \text{and} \qquad U_{n-1} \quad ,$$

and choose L_n to be the larger of

$$f(\delta_n) - \frac{1}{n} \qquad \text{and} \qquad L_{n-1} \quad .$$

It is easily verified that, for each $n > 2$ and x in $(0, \delta_n)$,

$$L_1 \leq L_{n-1} \leq L_n \leq f(x) \leq U_n \leq U_{n-1} \leq U_1 \quad .$$

This tells us that the U_n's form a bounded, nonincreasing sequence of real numbers, while the L_n's form a bounded, nondecreasing sequence of real numbers. Consequently, both series converge. Denote the limits of these two sequences by U_∞ and L_∞, respectively, and then observe that

$$0 \leq U_\infty - L_\infty = \lim_{n \to \infty} [U_n - L_n]$$

$$\leq \lim_{n \to \infty} \left[\left(f(\delta_n) + \frac{1}{n} \right) - \left(f(\delta_n) - \frac{1}{n} \right) \right] = \lim_{n \to \infty} \frac{2}{n} = 0 \quad .$$

So $U_\infty = L_\infty$. From this and the fact that $L_n \leq f(x) \leq U_n$ for each $n > 2$ and x in $(0, \delta_n)$, it immediately follows that

$$\lim_{x \to 0^+} f(x) = U_\infty \quad . \qquad \blacksquare$$

Showing that $\lim_{x \to 1^-} f(x)$ exists is just as easy.

Additional Exercises

3.15. *Verify the validity of each of the following statements:*

a. *If f is uniformly continuous on a finite interval (α, β), then f is piecewise continuous on (α, β).*

b. *If f is both continuous and piecewise continuous on a finite interval (α, β), then f is uniformly continuous on (α, β).*

c. *If f is piecewise continuous on (α, β), then f is continuous over a partitioning of (α, β).*

d. If f and g are both continuous on (α, β), and a and b are any two constants, then the linear combination $af + bg$ is continuous on (α, β).

e. If f and g are both uniformly continuous on a finite interval (α, β), and a and b are any two constants, then the linear combination $af + bg$ is uniformly continuous on (α, β).

f. If f and g are both piecewise continuous on (α, β), and a and b are any two constants, then the linear combination $af + bg$ is piecewise continuous on (α, β).

g. If f and g are both continuous over a partitioning of (α, β), and a and b are any two constants, then the linear combination $af + bg$ is continuous over a partitioning of (α, β).

3.16. *Several functions defined on $(-\infty, \infty)$ are given below. For each, sketch the graph over the real line and state whether the given function is bounded, continuous, piecewise continuous, or continuous over a partitioning of $\mathbb{R}$.*

a. $\sin(x)$ **b.** $\dfrac{1}{\sin(x)}$ **c.** $\operatorname{sinc} x$

d. $\operatorname{step}(x)$ **e.** e^x **f.** $\tan(x)$

g. the stair function, where $\operatorname{stair}(x) = $ the smallest integer greater than x

3.17. *Consider the function*

$$f(x) = x^2 \sin\left(\frac{1}{x}\right) \quad .$$

a. Verify that this function is continuous at $x = 0$ and that $f(0) = 0$.

b. Sketch the graph of this function over any interval containing $x = 0$.

c. Obviously, this function is differentiable at every $x \neq 0$. Show that it is also differentiable at $x = 0$ by computing

$$f'(0) = \lim_{\Delta x \to 0} \frac{f(0 + \Delta x) - f(0)}{\Delta x} \quad .$$

(Thus f is differentiable everywhere on $(-\infty, \infty)$.)

d. Compute $f'(x)$ assuming $x \neq 0$.

e. Show that f is not smooth on any interval containing $x = 0$ by showing that

$$\lim_{x \to 0} f'(x) \neq f'(0) \quad .$$

In fact, you should show that $\lim_{x \to 0} f'(x)$ does not even exist! (Suggestion: Try computing this limit using $x_n = (n2\pi\alpha)^{-1}$ with $n \to \infty$ and various different "clever" choices for α. You might even try to sketch the graph of $f'(t)$.)

4

Basic Analysis II:
Integration and Infinite Series

The importance of integration and infinite series to Fourier analysis cannot be overstated. Indeed, we'll find that the basic entities of classical Fourier analysis — Fourier transforms and Fourier series — are constructed using integrals and infinite series.

4.1 Integration
Well-Defined Integrals and Area

In this text, any reference to "an integral" will invariably be a reference to a definite integral of some function f over some interval (α, β),

$$\int_\alpha^\beta f(x)\,dx \quad . \tag{4.1}$$

A number of integration theories have been developed, and the precise definition of expression (4.1) and of "integrability" depends somewhat on the particular theory. For our purposes, any of the theories normally used in the basic calculus courses[1] will suffice.

Whichever theory is used, if f is a real-valued, piecewise continuous function on (α, β), then, geometrically, expression (4.1) represents the "net area" of the region between the X–axis and the graph of $f(x)$ with $\alpha < x < \beta$. That is,

$$\int_\alpha^\beta f(x)\,dx \ = \ \text{Area of region } \mathcal{R}_+ \ - \ \text{Area of region } \mathcal{R}_-$$

where $\mathcal{R}_+$ and $\mathcal{R}_-$ are the regions above and below the X–axis indicated in figure 4.1a. The corresponding total area, of course, is given by

$$\int_\alpha^\beta |f(x)|\,dx \ = \ \text{Area of region } \mathcal{R}_+ \ + \ \text{Area of region } \mathcal{R}_- \quad .$$

It should be clear that

$$\left| \int_\alpha^\beta f(x)\,dx \right| \ \le \ \int_\alpha^\beta |f(x)|\,dx \quad . \tag{4.2}$$

---

[1] This is usually a variant of the Riemann theory. The more advanced students acquainted with the Lebesgue theory, of course, should be thinking in terms of that theory.

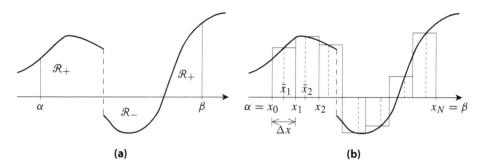

Figure 4.1: (a) Regions above and below the X–axis and (b) the Riemann sum approximation for $\int_\alpha^\beta f(x)\,dx$ when f is real valued.

(We will have much more to say about $\int_\alpha^\beta |f(x)|\,dx$ and inequality (4.2), especially when (α, β) is an infinite interval, in chapter 18.)

If, instead, f is a complex-valued, piecewise continuous function, then f can be written as $f = u + iv$ where u and v are real-valued functions (see chapter 6). We then have

$$\int_\alpha^\beta f(x)\,dx \ = \ \int_\alpha^\beta [u(x) + iv(x)]\,dx \ = \ \int_\alpha^\beta u(x)\,dx \ + \ i \int_\alpha^\beta v(x)\,dx \ ,$$

with the integrals of u and v representing the "net areas" between their graphs and the X–axis. We should note that inequality (4.2) is also true when f is complex valued. This will be verified in chapter 6 (see, in particular, the section on complex-valued functions starting on page 59).

Because of the central role that integration plays in Fourier analysis, it will be important to ensure that our integrals (equivalently, the "net areas" represented by the integrals) are well defined. This means that we must be able, in theory at least, to find the areas of the regions $\mathcal{R}_+$ and $\mathcal{R}_-$, and that both of these areas must be finite.[2]

Certainly, no matter which theory of integration is used, $\int_\alpha^\beta f(x)\,dx$ is well defined whenever (α, β) is a finite interval and f is piecewise continuous on (α, β). In this case, the areas of $\mathcal{R}_+$ and $\mathcal{R}_-$ are clearly finite and, for each positive integer N, the total net area can be approximated by a corresponding N^{th} *Riemann sum*,

$$R_N \ = \ \sum_{k=1}^{N} f(\bar{x}_k)\,\Delta x \tag{4.3}$$

where (see figure 4.1b)

$$\Delta x \ = \ \frac{\beta - \alpha}{N} \ ,$$

and, for $k = 0, 1, 2, \ldots, N$,

$$x_k \ = \ \alpha + k\,\Delta x$$

and $\bar{x}_k$ is some conveniently chosen point on the closed interval $[x_{k-1}, x_k]$ at which f is well defined (i.e., where f is continuous).

Geometrically, each term in the Riemann sum is the "signed" area of the kth rectangle in figure 4.1b, with the sign being positive when the rectangle is above the X–axis (i.e., when $f(\bar{x}_k) > 0$) and negative when the rectangle is below the X–axis (i.e., when $f(\bar{x}_k) < 0$). Clearly, as $N \to \infty$, the

---

[2] It is sometimes possible to use clever trickery to "cancel out infinities" and seemingly obtain a finite "net area" when both $\mathcal{R}_+$ and $\mathcal{R}_-$ have infinite areas. Avoid using these tricks until you really understand them. Besides, the generalized theory, which we'll develop in part IV, will provide more general and much safer ways of dealing with such situations.

R_N's will converge to a finite value, and this finite value is the net area represented by the integral,

$$\int_\alpha^\beta f(x)\,dx = \lim_{N\to\infty} R_N = \lim_{N\to\infty} \sum_{k=1}^{N} f(x_k)\,\Delta x \quad.$$

?▶ Exercise 4.1 (for the more ambitious):   *Prove that the R_N's defined by expression (4.3) converge to a finite number as $N \to \infty$. Remember: (α, β) is finite and f is piecewise continuous on (α, β). (Try to first prove this assuming f is uniformly continuous on (α, β).)*

For the classical theory of Fourier analysis (part II and part III of this text), we will usually limit our discussions to functions that are at least piecewise continuous over the intervals of integration. This will ensure that the integrals over finite intervals are well defined. An additional property, absolute integrability, will be introduced and used in part III to identify integrals on infinite intervals that are well defined.

Integral Formulas

We will be using a number of integral formulas in our work, most of which should be well known from basic calculus. For example, you surely recall that no one really calculates an integral via Riemann sums. Instead, we use the fact that, as long as f is uniformly smooth on a finite interval (α, β),

$$\int_\alpha^\beta f'(x)\,dx = \text{``}f(\beta) - f(\alpha)\text{''} \quad. \tag{4.4}$$

Notice the quotes around the right-hand side of this equation. As written, this formula assumes f is continuous at the endpoints of (α, β). Often, though, we will be dealing with functions that have jump discontinuities at the endpoints of the intervals over which we are integrating. In these cases, the correct formula is actually

$$\int_\alpha^\beta f'(x)\,dx = \lim_{x\to\beta^-} f(x) - \lim_{x\to\alpha^+} f(x) \quad. \tag{4.5}$$

For convenience, this will often be written as

$$\int_\alpha^\beta f'(x)\,dx = f(x)\big|_\alpha^\beta \quad, \tag{4.6}$$

where it is understood that

$$f(x)\big|_\alpha^\beta = \lim_{x\to\beta^-} f(x) - \lim_{x\to\alpha^+} f(x) \quad.$$

Because we will often be integrating functions that are not smooth, let us state and verify the following slight generalization of the above.

Theorem 4.1
Let f be continuous and piecewise smooth on the finite interval (α, β). Then

$$\int_\alpha^\beta f'(x)\,dx = f(x)\big|_\alpha^\beta \quad. \tag{4.7}$$

PROOF (partial):   First of all, if f' has no discontinuities, then f is uniformly smooth on (α, β) and, from elementary calculus, we know equation (4.7) holds.

If f' has only one discontinuity in (α, β), say, at $x = x_0$, then f is uniformly smooth on (α, x_0) and (x_0, β). Thus,

$$\int_\alpha^\beta f'(x)\,dx \; = \; \int_\alpha^{x_0} f'(x)\,dx \; + \; \int_{x_0}^\beta f'(x)\,dx$$

$$= \; \left[\lim_{x \to x_0^-} f(x) - \lim_{x \to \alpha^+} f(x) \right] + \left[\lim_{x \to \beta^-} f(x) - \lim_{x \to x_0^+} f(x) \right]$$

$$= \; \lim_{x \to \beta^-} f(x) - \lim_{x \to \alpha^+} f(x) + \lim_{x \to x_0^-} f(x) - \lim_{x \to x_0^+} f(x) \quad . \qquad (4.8)$$

But, because f is continuous everywhere on (α, β),

$$\lim_{x \to x_0^-} f(x) - \lim_{x \to x_0^+} f(x) \; = \; f(x_0) - f(x_0) \; = \; 0$$

and so, equation (4.8) reduces to equation (4.7).   ∎

Extending this to the cases where f' has more than one discontinuity is left as an exercise.

?▶ Exercise 4.2:   *Extend the above proof of theorem 4.1 to the following cases:*

a: f' *has exactly two discontinuities on* (α, β) .

b: f' *has any finite number of discontinuities on* (α, β) .

It's worth glancing back at example 3.6 on page 23 to see what foolishness can happen when formula (4.7) is used with a function that is *not* continuous (see, also, exercise 4.6).

As a corollary, we have the following slight generalization of the classic integration by parts formula. This formula will be important when we discuss differentiation in Fourier analysis.

Theorem 4.2 (integration by parts)
*Assume f and g are both continuous and piecewise smooth functions on a finite interval (α, β).
Then*

$$\int_\alpha^\beta f'(x)g(x)\,dx \; = \; f(x)g(x)\Big|_\alpha^\beta \; - \; \int_\alpha^\beta f(x)g'(x)\,dx \quad . \qquad (4.9)$$

PROOF: Clearly, the product fg will also be piecewise smooth and continuous on (α, β). By theorem 4.1 and the product rule,

$$f(x)g(x)\Big|_\alpha^\beta \; = \; \int_\alpha^\beta (f(x)g(x))'\,dx \; = \; \int_\alpha^\beta f'(x)g(x)\,dx \; + \; \int_\alpha^\beta f(x)g'(x)\,dx \quad ,$$

which, after cutting out the middle and rearranging things slightly, is equation (4.9).   ∎

Occasionally, we will need to approximate fairly general integrals. The following well-known (and easily proven) theorem can often be useful in such cases.

Theorem 4.3 (mean value theorem for integrals)
Let f be a uniformly continuous, real-valued function on the finite interval (α, β). Then there is an $\bar{x}$ with $\alpha \le \bar{x} \le \beta$ such that

$$f(\bar{x})\,[\beta - \alpha] \; = \; \int_\alpha^\beta f(x)\,dx \quad .$$

The value $f(\bar{x})$ in this theorem is commonly referred to as the mean (or average) value of the function f over the interval and is simply the height of the rectangle having the interval (α, β) as its base and having the same net area as is under the graph of f over the same interval (see figure 4.2).

Other well-known formulas from integral calculus will be recalled as the need arises.

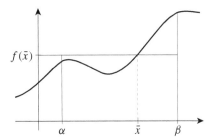

Figure 4.2: Illustration for the mean value theorem.

4.2 Infinite Series (Summations)

For mathematicians (and others indoctrinated by mathematicians — like you), an *infinite series* is simply any expression that looks like the summation of an infinite number of things. For example, you should recognize

$$\sum_{k=1}^{\infty} \frac{1}{k} = 1 + \frac{1}{2} + \frac{1}{3} + \frac{1}{4} + \frac{1}{5} + \frac{1}{6} + \cdots$$

(with the "$\cdots$" denoting "continue the obvious pattern") as the famous harmonic series.

In Fourier analysis, we must deal with infinite series of numbers, infinite series of functions and, ultimately, infinite series of generalized functions. Here, we will review some basic facts concerning infinite series of numbers. Later, as the need arises, we'll extend our discussions to include those other infinite series.

Basic Facts

Let $c_0, c_1, c_2, \ldots$ be any sequence of numbers, and consider the infinite series with these numbers as its terms,

$$\sum_{k=0}^{\infty} c_k = c_0 + c_1 + c_2 + \cdots \quad .$$

The index here, k, started at 0. In practice, it can start at any convenient integer M. For any integer N with $N \geq 0$ (or, more generally, with $N \geq M$), the Nth *partial sum* S_N is simply the value obtained by adding all the terms up to and including c_N,

$$S_N = \sum_{k=0}^{N} c_k = c_0 + c_1 + c_2 + \cdots + c_N \quad .$$

The *sum* (or *value*) of the infinite series, which is also denoted by $\sum_{k=0}^{\infty} c_k$, is the value we get by taking the limit of the Nth partial sums as $N \to \infty$,

$$\sum_{k=0}^{\infty} c_k = \lim_{N \to \infty} S_N = \lim_{N \to \infty} \sum_{k=0}^{N} c_k \quad .$$

This assumes, of course, that the limit exists. If this limit does exist (and is finite), then the series is said to be *convergent* (because the limit of partial sums converges). Otherwise, the series is said to be *divergent*. In the special cases where the limit is infinite (or negatively infinite), we often say that the series *diverges to infinity* (or to *negative infinity*).

A convergent series $\sum_{k=M}^{\infty} c_k$ can be further classified as being either "absolutely convergent" or "conditionally convergent". It is *absolutely convergent* if $\sum_{k=M}^{\infty} |c_k|$ converges, and it is *conditionally convergent* if $\sum_{k=M}^{\infty} c_k$ converges but $\sum_{k=M}^{\infty} |c_k|$ does not. Basically, if a series converges absolutely, then its terms are decreasing quickly enough to ensure the convergence of the series. On the other hand, a conditionally convergent series converges because its terms tend to cancel themselves out. Unfortunately, the pattern of cancellations for such a series depends on the arrangement of the terms, and it can be shown that the sum of any conditionally convergent series can be changed by an appropriate rearrangement of its terms. By contrast, the sum of an absolutely convergent series is not affected by any rearrangement of its terms. For this reason (and other reasons we'll discuss later), it is usually preferable to work with absolutely convergent series whenever we are fortunate enough to have the choice.

Let's note a few facts regarding an arbitrary infinite series of numbers $\sum_{k=M}^{\infty} c_k$ that are so obvious that we will feel free to use them later without comment:

1. If we do *not* have $c_k \to 0$ as $k \to \infty$, then the series must diverge. (On the other hand, the fact that $c_k \to 0$ as $k \to \infty$ does *not* guarantee the convergence of the series! See, for example, exercise 4.3.)

2. If L is an integer with $M < L$, then either both $\sum_{k=M}^{\infty} c_k$ and $\sum_{k=L}^{\infty} c_k$ converge or both diverge. That is, the convergence of a series does not depend on its first few terms.

3. (*The triangle inequality*) If $\sum_{k=M}^{\infty} |c_k|$ converges, so does $\sum_{k=M}^{\infty} c_k$. Moreover,

$$\left| \sum_{k=M}^{\infty} c_k \right| \leq \sum_{k=M}^{\infty} |c_k|$$

(see, also, page 28).

!▶**Example 4.1 (the geometric series):**   *Let X be any nonzero real or complex number, and let M be any integer. The corresponding geometric series is*

$$\sum_{k=M}^{\infty} X^k = X^M + X^{M+1} + X^{M+2} + X^{M+3} + \cdots \quad . \tag{4.10}$$

The Nth partial sum is easily computed for any integer N greater than M. First, if $X = 1$, then

$$S_N = \sum_{k=M}^{N} 1^k = \sum_{k=M}^{N} 1 = N - M + 1 \quad .$$

If $X \neq 1$, then

$$(1 - X)S_N = S_N - XS_N$$

$$= [X^M + X^{M+1} + X^{M+2} + \cdots + X^N]$$

$$\quad - X[X^M + X^{M+1} + X^{M+2} + \cdots + X^N]$$

$$= [X^M + X^{M+1} + X^{M+2} + \cdots + X^N]$$

$$\quad - [X^{M+1} + X^{M+2} + X^{M+3} + \cdots + X^{N+1}]$$

$$= X^M - X^{N+1} \quad .$$

Dividing through by $1 - X$ then gives us

$$\sum_{k=M}^{N} X^k = S_N = \frac{X^M - X^{N+1}}{1 - X} \quad . \tag{4.11}$$

Now recall (or verify for yourself) that

$$\lim_{N \to \infty} \left| X^N \right| = \begin{cases} 0 & \text{if } |X| < 1 \\ 1 & \text{if } |X| = 1 \\ \infty & \text{if } |X| > 1 \end{cases} \quad .$$

Consequently, when $|X| < 1$,

$$\lim_{N \to \infty} \sum_{k=M}^{N} X^k = \lim_{N \to \infty} \frac{X^M - X^{N+1}}{1 - X} = \frac{X^M - 0}{1 - X} = \frac{X^M}{1 - X} \quad .$$

It should also be clear that this limit of partial sums will diverge whenever $|X| \geq 1$. Thus, the geometric series $\sum_{k=M}^{\infty} X^k$ converges if and only if $|X| < 1$. Moreover, when $|X| < 1$,

$$\sum_{k=M}^{\infty} X^k = \lim_{N \to \infty} \sum_{k=M}^{N} X^k = \frac{X^M}{1 - X} \quad .$$

(Note that, by the above, the geometric series $\sum_{k=M}^{\infty} |X|^k$ converges if and only if $|X| < 1$. So, in fact, this series converges absolutely if and only if $|X| < 1$.)

In practice, we can rarely find a simple formula for the partial sums of a given infinite series. So, instead, we often rely on one of the many tests for determining convergence. Some that will be useful to us are given below. Their proofs can be found in any decent calculus text.

Theorem 4.4 (bounded partial sums test)
Let $\sum_{k=M}^{\infty} c_k$ be an infinite series such that, for some finite number B and every integer N greater than M,

$$\sum_{k=M}^{N} |c_k| \leq B \quad .$$

Then $\sum_{k=M}^{\infty} c_k$ converges absolutely and

$$\sum_{k=M}^{\infty} |c_k| \leq B \quad .$$

Theorem 4.5 (comparison test)
Let $\sum_{k=M}^{\infty} a_k$ and $\sum_{k=M}^{\infty} b_k$ be two infinite series. Assume that $\sum_{k=M}^{\infty} b_k$ converges absolutely and that, for some finite value B,

$$|a_k| \leq B |b_k| \qquad \text{for every integer } k \geq M \quad .$$

Then $\sum_{k=M}^{\infty} a_k$ also converges absolutely and

$$\sum_{k=M}^{\infty} |a_k| \leq B \sum_{k=M}^{\infty} |b_k| \quad .$$

Theorem 4.6 (integral test)

Assume $\sum_{k=M}^{\infty} c_k$ is an infinite series with only positive terms, and suppose there is a piecewise continuous function f on (M, ∞) such that

1. $f(k) = c_k$ for each integer $k \geq M$, and

2. $f(a) \geq f(b)$ whenever $M \leq a < b$.

Then the infinite series converges absolutely if

$$\int_M^{\infty} f(x)\,dx \; < \; \infty \quad ,$$

and diverges to infinity if

$$\int_M^{\infty} f(x)\,dx \; = \; \infty \quad .$$

Moreover,

$$\int_M^{\infty} f(x)\,dx \; \leq \; \sum_{k=M}^{\infty} c_k \qquad \text{and} \qquad \sum_{k=M+1}^{\infty} c_k \; \leq \; \int_M^{\infty} f(x)\,dx \quad .$$

Theorem 4.7 (alternating series test)

Let $\sum_{k=M}^{\infty} c_k$ be an alternating series whose terms steadily decrease to zero. In other words, assume all of the following:

1. $c_k \to 0$ as $k \to \infty$.

2. $|c_k| \geq |c_{k+1}|$ for each integer $k \geq M$.

3. Either
    $$c_k \; = \; (-1)^k\, |c_k| \qquad \text{for each integer } k \geq M$$
    or
    $$c_k \; = \; (-1)^{k+1}\, |c_k| \qquad \text{for each integer } k \geq M \quad .$$

Then $\sum_{k=M}^{\infty} c_k$ converges, and, for every integer N greater than M,

$$\left| \sum_{k=M}^{\infty} c_k \; - \; \sum_{k=M}^{N} c_k \right| \; \leq \; |c_{N+1}| \quad .$$

?▶Exercise 4.3 (the harmonic series):   Using the integral test, show that the harmonic series,

$$\sum_{k=1}^{\infty} \frac{1}{k} \; = \; 1 + \frac{1}{2} + \frac{1}{3} + \frac{1}{4} + \frac{1}{5} + \frac{1}{6} + \cdots \quad ,$$

diverges to infinity.

?▶Exercise 4.4 (the alternating harmonic series):   Show that the alternating harmonic series,

$$\sum_{k=1}^{\infty} (-1)^{k+1} \frac{1}{k} \; = \; 1 - \frac{1}{2} + \frac{1}{3} - \frac{1}{4} + \frac{1}{5} - \frac{1}{6} + \cdots \quad ,$$

converges conditionally.

The Schwarz Inequality for Infinite Series

One tool we will find useful in discussing the convergence of some infinite series (but which is often not covered in introductory discussions of series) is the Schwarz inequality. The finite summation version of this inequality was discussed and proven near the end of the previous chapter (see theorem 3.7 on page 30). There we saw that, if N is any positive integer, and $\{a_1, a_2, a_3, \ldots, a_N\}$ and $\{b_1, b_2, b_3, \ldots, b_N\}$ are any two sets of N real or complex numbers, then

$$\left| \sum_{k=1}^{N} a_k b_k \right| \leq \left(\sum_{k=1}^{N} |a_k|^2 \right)^{1/2} \left(\sum_{k=1}^{N} |b_k|^2 \right)^{1/2} \quad .$$

Letting $N \to \infty$ then gives us the *Schwarz inequality* for infinite series.

Theorem 4.8 (Schwarz inequality for infinite series)
Let $\{a_1, a_2, a_3, \ldots\}$ and $\{b_1, b_2, b_3, \ldots\}$ be any two infinite sequences of numbers such that

$$\sum_{k=1}^{\infty} |a_k|^2 \qquad and \qquad \sum_{k=1}^{\infty} |b_k|^2$$

are convergent. Then $\sum_{k=1}^{\infty} a_k b_k$ is absolutely convergent and

$$\left| \sum_{k=1}^{\infty} a_k b_k \right| \leq \left(\sum_{k=1}^{\infty} |a_k|^2 \right)^{1/2} \left(\sum_{k=1}^{\infty} |b_k|^2 \right)^{1/2} \quad . \tag{4.12}$$

Two-Sided Series

In Fourier analysis, we often encounter and use *two-sided* infinite series, that is, series of the form

$$\sum_{k=-\infty}^{\infty} c_k \qquad \text{or} \qquad \cdots + c_{-2} + c_{-1} + c_0 + c_1 + c_2 + c_3 + \cdots \quad .$$

For convenience, we'll refer to the type of infinite series discussed in the previous subsection as one-sided series.

For the most part, the "theory of two-sided infinite series" is the obvious extension of the theory of one-sided infinite series. For example, instead of the Nth partial sum of $\sum_{k=-\infty}^{\infty} c_k$, we have the $(M, N)^{\text{th}}$ partial sum

$$S_{MN} = \sum_{k=M}^{N} c_k \quad ,$$

where (M, N) is any pair of integers with $M \leq N$. We then say that $\sum_{k=-\infty}^{\infty} c_k$ converges and

$$\sum_{k=-\infty}^{\infty} c_k = \lim_{\substack{N \to \infty \\ M \to -\infty}} S_{MN} \tag{4.13}$$

if and only if this double limit exists. This double limit, in turn, exists and is defined by

$$\lim_{\substack{N \to \infty \\ M \to \infty}} S_{MN} = \lim_{M \to -\infty} \left[\lim_{N \to \infty} S_{MN} \right] = \lim_{N \to \infty} \left[\lim_{M \to -\infty} S_{MN} \right] \quad ,$$

if and only if the two iterated limits

$$\lim_{M \to -\infty} \left[\lim_{N \to \infty} S_{MN} \right] \quad \text{and} \quad \lim_{N \to \infty} \left[\lim_{M \to -\infty} S_{MN} \right]$$

exist and are equal. (Also see exercise 4.7.)

Let's make a rather simple observation. For any two integers M and N with $M < 0 < N$,

$$\sum_{k=M}^{N} c_k = c_M + c_{M+1} + \ldots + c_{-1} + c_0 + \ldots + c_N$$

$$= c_0 + \left(c_{-1} + c_{-2} + \cdots + c_{-|M|} \right) + \left(c_1 + c_2 + \cdots + c_N \right)$$

$$= c_0 + \sum_{k=1}^{|M|} c_{-k} + \sum_{k=1}^{N} c_k \quad .$$

From this and the basic definitions of the limits, you should have little difficulty in proving the following lemma, which points out that any convergent two-sided series can be viewed as the sum of two one-sided series.

Lemma 4.9

A two-sided series $\sum_{k=-\infty}^{\infty} c_k$ converges if and only if both $\sum_{k=1}^{\infty} c_{-k}$ and $\sum_{k=1}^{\infty} c_k$ converge. Moreover, so long as the infinite series all converge,

$$\sum_{k=-\infty}^{\infty} c_k = c_0 + \sum_{k=1}^{\infty} c_{-k} + \sum_{k=1}^{\infty} c_k \quad .$$

At this point, it should be clear that all the results previously discussed for one-sided series can easily be extended to corresponding results for two-sided series. Rather than repeat those discussions with the obvious modifications, let us assume that these extensions have been made, and get on with it.

?►Exercise 4.5:   *What is the comparison test for two-sided infinite series?*

Symmetric Summations

On occasion, it is appropriate to use a weaker type of convergence for a two-sided series $\sum_{k=-\infty}^{\infty} c_k$. On these occasions we use the *symmetric partial sum,*

$$S_{-NN} = \sum_{k=-N}^{N} c_k \quad .$$

If the limit, as $N \to \infty$, of the symmetric partial sums exists, then we will say that $\sum_{k=-\infty}^{\infty} c_k$ *converges using the symmetric partial sums* (or, more simply, *converges symmetrically*).

Certainly, if the two-sided series is convergent (using the stronger definition indicated in formula (4.13)), then it will converge symmetrically and

$$\sum_{k=-\infty}^{\infty} c_k = \lim_{N \to \infty} \sum_{k=-N}^{N} c_k \quad .$$

However, it is quite possible for a *divergent* series to converge symmetrically because of cancellations occurring in the symmetric partial sum. This, as well as a danger in using symmetric partial sums, is demonstrated in the next example.

!▶ Example 4.2 (the two-sided harmonic series):   *The two-sided harmonic series is*

$$\sum_{\substack{k=-\infty \\ k \neq 0}}^{\infty} \frac{1}{k} = \cdots - \frac{1}{3} - \frac{1}{2} - 1 + 1 + \frac{1}{2} + \frac{1}{3} + \frac{1}{4} + \cdots$$

Since the one-sided harmonic series, $\sum_{k=1}^{\infty} {}^{1}/_{k}$, diverges to infinity (see exercise 4.3), lemma 4.9 tells us that the two-sided harmonic series diverges. However, the terms in S_{-NN} cancel out for any positive integer N,

$$S_{-NN} = \sum_{\substack{k=-N \\ k \neq 0}}^{N} \frac{1}{k} = -\frac{1}{N} - \cdots - \frac{1}{3} - \frac{1}{2} - 1 + 1 + \frac{1}{2} + \frac{1}{3} + \cdots + \frac{1}{N} = 0 \quad .$$

From this, it follows that the two-sided harmonic series converges symmetrically to zero,

$$\lim_{N \to \infty} \sum_{\substack{k=-N \\ k \neq 0}}^{N} \frac{1}{k} = \lim_{N \to \infty} 0 = 0 \quad . \tag{4.14}$$

    This does not justify a claim that the two-sided harmonic series equals zero! As noted above, the two-sided harmonic series diverges, and thus, does not have a well-defined sum. To see why we don't want to even pretend that such a series adds up to anything, let's naively "evaluate" this series using two other sets of limits.
    One "evaluation" of the two-sided harmonic series is

$$\lim_{M \to -\infty} \left[\lim_{N \to \infty} \sum_{\substack{k=M \\ k \neq 0}}^{N} \frac{1}{k} \right] = \lim_{M \to -\infty} \left[\lim_{N \to \infty} \left[\sum_{k=1}^{|M|} \frac{1}{-k} + \sum_{k=1}^{N} \frac{1}{k} \right] \right]$$

$$= \lim_{M \to -\infty} \left[-\sum_{k=1}^{|M|} \frac{1}{k} + \lim_{N \to \infty} \sum_{k=1}^{N} \frac{1}{k} \right]$$

$$= \lim_{M \to -\infty} \left[-\sum_{k=1}^{|M|} \frac{1}{k} + \infty \right] = \lim_{M \to -\infty} [+\infty] = +\infty \quad .$$

On the other hand,

$$\lim_{N \to \infty} \left[\lim_{M \to -\infty} \sum_{\substack{k=M \\ k \neq 0}}^{N} \frac{1}{k} \right] = \lim_{N \to \infty} \left[\lim_{M \to -\infty} \left[\sum_{k=1}^{|M|} \frac{1}{-k} + \sum_{k=1}^{N} \frac{1}{k} \right] \right]$$

$$= \lim_{N \to \infty} \left[- \lim_{M \to -\infty} \sum_{k=1}^{|M|} \frac{1}{k} + \sum_{k=1}^{N} \frac{1}{k} \right]$$

$$= \lim_{N \to \infty} \left[-\infty + \sum_{k=1}^{N} \frac{1}{k} \right] = \lim_{N \to \infty} [-\infty] = -\infty \quad .$$

Thus, naive applications of the above equations lead to

$$\sum_{\substack{k=-\infty \\ k \neq 0}}^{\infty} \frac{1}{k} = \lim_{N \to \infty} \sum_{\substack{k=-N \\ k \neq 0}}^{N} \frac{1}{k} = 0 \quad ,$$

$$\sum_{\substack{k=-\infty \\ k \neq 0}}^{\infty} \frac{1}{k} = \lim_{M \to -\infty} \left[\lim_{N \to \infty} \sum_{\substack{k=M \\ k \neq 0}}^{N} \frac{1}{k} \right] = +\infty \quad ,$$

and

$$\sum_{\substack{k=-\infty \\ k \neq 0}}^{\infty} \frac{1}{k} = \lim_{N \to \infty} \left[\lim_{M \to -\infty} \sum_{\substack{k=M \\ k \neq 0}}^{N} \frac{1}{k} \right] = -\infty \quad ,$$

implying that

$$-\infty = 0 = \infty \quad !$$

Additional Exercises

4.6. Assume (α, β) is a finite interval and f is a piecewise smooth function on (α, β). Assume, further, that f is continuous everywhere in (α, β) except at one point x_0 where f has a jump discontinuity with jump

$$j_0 = \lim_{x \to x_0^+} f(x) - \lim_{x \to x_0^-} f(x) \quad .$$

a. Show that

$$\int_{\alpha}^{\beta} f'(x)\, dx = f(x)\big|_{\alpha}^{\beta} - j_0 \quad . \tag{4.15}$$

b. Show that, for each continuous and piecewise smooth function g on (α, β),

$$\int_{\alpha}^{\beta} f'(x) g(x)\, dx = f(x)g(x)\big|_{\alpha}^{\beta} - j_0 g(x_0) - \int_{\alpha}^{\beta} f(x) g'(x)\, dx \quad .$$

4.7. Another way of defining the double limit in formula (4.13) is to say that the indicated double limit exists if and only if there is a finite number L such that, for each $\epsilon > 0$, there is a corresponding pair of integers (M_ϵ, N_ϵ) such that

$$|S_{MN} - L| < \epsilon \quad \text{whenever} \quad M \leq M_\epsilon \quad \text{and} \quad N_\epsilon \leq N \quad .$$

If this holds, we define the limit to be L,

$$\lim_{\substack{N \to \infty \\ M \to -\infty}} S_{MN} = L \quad .$$

Show that this definition of the double limit is equivalent to the one given in the text.

5

Symmetry and Periodicity

In this chapter, we will review some basic facts regarding functions whose graphs exhibit some sort of repetitive pattern. Either the graphs are symmetric or antisymmetric about the origin (even and odd functions), or they continually repeat themselves at regular intervals along the real line (periodic functions). We are interested in even and odd functions because, on occasion, we will exploit the properties discussed here to simplify our work. Our main interest, however, will be with periodic functions because of the central role these functions will play in our work.

5.1 Even and Odd Functions

Let f be a function defined on a symmetric interval $(-\alpha, \alpha)$ for some $\alpha > 0$. The function is said to be an *even function* on $(-\alpha, \alpha)$ if and only if

$$f(-x) = f(x) \qquad \text{on} \quad (-\alpha, \alpha) \quad .$$

On the other hand, if

$$f(-x) = -f(x) \qquad \text{on} \quad (-\alpha, \alpha) \quad ,$$

then f is said to be an *odd function* on $(-\alpha, \alpha)$. As usual, if no interval is explicitly given, then you should assume $(-\alpha, \alpha)$ is the entire domain of f. Some well-known examples of even functions on $\mathbb{R}$ are

$$1 \quad , \quad x^2 \quad , \quad x^4 \quad , \quad \cos(x) \quad \text{and} \quad \ln|x| \quad .$$

Some well-known examples of odd functions on $\mathbb{R}$ are

$$x \quad , \quad x^3 \quad , \quad \sin(x) \quad \text{and} \quad \tan(x) \quad .$$

Recall that the graph of an even function is symmetric about the line $x = 0$, while the graph of an odd function is antisymmetric about the line $x = 0$. This is illustrated in figure 5.1.

Not all functions are even or odd, but, given such functions, we can use well-known properties to simplify computations. Here are some of those properties we will use later:

1. *The product of two even functions is an even function.*

2. *The product of two odd functions is an even function.*

3. *The product of an even function with an odd function is an odd function.*

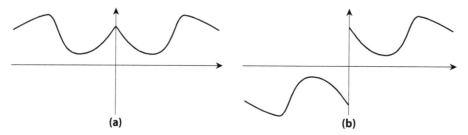

Figure 5.1: Graphs of (a) an even function and (b) an odd function.

4. If f is an even, piecewise continuous function on a finite interval $(-\alpha, \alpha)$, then

$$\int_{-\alpha}^{\alpha} f(x)\,dx = 2 \int_{0}^{\alpha} f(x)\,dx \quad .$$

5. If f is an odd, piecewise continuous function on a finite interval $(-\alpha, \alpha)$, then

$$\int_{-\alpha}^{\alpha} f(x)\,dx = 0 \quad .$$

Each of these properties is easily verified. For example, if both f and g are even functions on $(-\alpha, \alpha)$, then, for each x in $(-\alpha, \alpha)$,

$$fg(-x) = f(-x)g(-x) = f(x)g(x) = fg(x) \quad ,$$

verifying the first property in the above list.

The second and third properties are verified in the same manner. To prove the last two, note that

$$\int_{-\alpha}^{\alpha} f(x)\,dx = I_- + I_+$$

where

$$I_+ = \int_{0}^{\alpha} f(x)\,dx \qquad \text{and} \qquad I_- = \int_{-\alpha}^{0} f(x)\,dx \quad .$$

But, using the substitution $s = -x$,

$$I_- = -\int_{\alpha}^{0} f(-s)\,ds = \begin{cases} \displaystyle\int_{0}^{\alpha} f(s)\,ds & \text{if } f \text{ is even} \\[2ex] -\displaystyle\int_{0}^{\alpha} f(s)\,ds & \text{if } f \text{ is odd} \end{cases}$$

$$= \begin{cases} I_+ & \text{if } f \text{ is even} \\ -I_+ & \text{if } f \text{ is odd} \end{cases} \quad .$$

So, if f is even on $(-\alpha, \alpha)$,

$$\int_{-\alpha}^{\alpha} f(x)\,dx = I_- + I_+ = I_+ + I_+ = 2 \int_{0}^{\alpha} f(x)\,dx \quad ;$$

while, if f is odd on $(-\alpha, \alpha)$,

$$\int_{-\alpha}^{\alpha} f(x)\,dx = I_- + I_+ = -I_+ + I_+ = 0 \quad .$$

Some other properties of even and odd functions are described in the following exercises.

?►Exercise 5.1: Let f be a piecewise smooth and even (odd) function on $(-\alpha, \alpha)$. Show that its derivative, f', is an odd (even) function on $(-\alpha, \alpha)$.

?►Exercise 5.2: Show that an even and piecewise continuous function cannot have a nontrivial jump discontinuity at $x = 0$.

?►Exercise 5.3: Let f be an odd and piecewise smooth function on $(-\alpha, \alpha)$. Show that f' cannot have a nontrivial jump discontinuity at $x = 0$.

5.2 Periodic Functions

Terminology

A function f, defined on the entire real line, is *periodic* if and only if there is a fixed positive value p such that

$$f(x - p) = f(x) \tag{5.1}$$

(as functions of x on $\mathbb{R}$). The value p is called a *period* of f. The corresponding *frequency* ω is related to the period by $\omega = {}^1\!/_p$.[1]

Note that, if f is a periodic function with period p, then, for any integer m,

$$f(x + mp) = f(x + mp - p) = f(x + (m - 1)p) \quad.$$

Thus, if n is any positive integer, then applying the above successively (using $m = n, n - 1, n - 2, \ldots, 1, 0, -1, \ldots, -n$) gives

$$f(x + np) = f(x + (n - 1)p) = \cdots = f(x + 0p)$$

and

$$f(x + 0p) = f(x - 1p) = \cdots = f(x - np) \quad.$$

This tells us equation (5.1) is equivalent to

$$f(x \pm np) = f(x) \qquad \text{for every integer } n \quad. \tag{5.2}$$

We will often use this, implicitly, when defining periodic functions.

!►Example 5.1 (the saw function): Let $p > 0$. The basic saw function with period p is defined by

$$\text{saw}_p(x) = \begin{cases} x & \text{if } 0 < x < p \\ \text{saw}_p(x - p) & \text{in general} \end{cases} \quad.$$

The first line of this formula tells us that the graph of this function is the straight line $y = x$ over the interval $(0, p)$. The second line, which is equivalent to

$$\text{saw}_p(x \pm np) = \text{saw}_p(x) \qquad \text{for any integer } n \quad,$$

tells us that the function is periodic with period p, and that the rest of the graph of $y = \text{saw}(x)$ is generated by shifting that straight line over $(0, p)$ to the left and right by integral multiples of p. That is how the graph in figure 5.2 was sketched.

---

[1] Some texts refer to $\omega = {}^1\!/_p$ as the *circular frequency*. You may also be familiar with the *angular frequency* $\nu = {}^{2\pi}\!/_p$. In this text the term "frequency" will always mean "circular frequency".

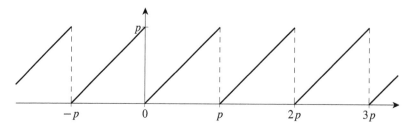

Figure 5.2: The graph of saw$_p$.

Equation (5.2) also points out that the period of a periodic function is not unique. Any positive integral multiple of any given period is another period for that function.

If it exists, the smallest period for a given periodic function is called the *fundamental period* for that function. The corresponding frequency (which, of course, must be the largest frequency for the function) is called the *fundamental frequency*.

There are periodic functions that do not have fundamental periods.

?►Exercise 5.4:   *Verify that any constant function $f(x) = c$ (where c is a constant) is a periodic function with no fundamental period.*

On the other hand, if f is not a constant function but is periodic and at least piecewise continuous, then it should be intuitively obvious that f does have a fundamental period and that every other period of f is an integral multiple of the fundamental period. We'll leave the proof as an exercise (exercise 5.12).

Let's end this discussion on terminology for periodic functions by noting that, in practice, the term "period" is often used for two different, but closely related, entities. First, as we have already seen, any positive number p is referred to as a period for a periodic function f if

$$f(x - p) \ = \ f(x) \quad .$$

In addition, any finite interval whose length equals a period of f (as just defined) is also called a period. Thus, if f is a periodic function with period p, then $(0, p)$, $(-p/2, p/2)$ and $(2, 2 + p)$ are all considered to be periods for f. In practice, it should be clear from the context whether a reference to "a period" is a reference to a length or an interval.

Calculus with Periodic Functions

It is easy to see that any shifting or scaling of a periodic function results in another periodic function, and that any linear combination of periodic functions with a common period is another periodic function. It should also be clear that the derivative of a piecewise smooth periodic function is, itself, periodic. Let us also observe that any periodic function that is piecewise continuous (or piecewise smooth) over any given period must be piecewise continuous (or piecewise smooth) over the entire real line.

?►Exercise 5.5:   *Convince yourself that the claims made in the previous paragraph are true.*

?►Exercise 5.6:   *Give an example showing that a periodic function can be uniformly continuous on a given period without being continuous on the entire real line.*

?► **Exercise 5.7:** *Assume g is a uniformly continuous function on the finite interval $(0, p)$. Let f be the periodic function*

$$f(x) = \begin{cases} g(x) & \text{if } 0 < x < p \\ g(x - p) & \text{in general} \end{cases} .$$

What additional condition(s) must g satisfy for f to be continuous?

We will often need to integrate a periodic function over some given period. The next lemma, which is easily verified (see exercises 5.8 and 5.9), assures us that we can use whatever period is most convenient.

Lemma 5.1
Let f be a periodic, piecewise continuous function with period p. Then, for any real a,

$$\int_a^{a+p} f(x)\, dx = \int_0^p f(x)\, dx \quad .$$

Since there is no need to specify the period, we'll often simply write

$$\int_{\text{period}} f(x)\, dx \qquad \text{for} \qquad \int_a^{a+p} f(x)\, dx$$

where a is an arbitrary real number. Use of this notation assumes, naturally, that f is periodic and that the particular period p has been agreed upon.

?► **Exercise 5.8:** *Sketch the graph of a "generic" real-valued, periodic function f. Let p be any period for the function sketched, and let a be any real number. Demonstrate graphically that "the net area between the graph of f and the X–axis over $(a, a + p)$" will always be the same as "the net area between the graph of f and the X–axis over $(0, p)$".*

?► **Exercise 5.9:** *Prove lemma 5.1. You might start by showing that*

$$\frac{d}{da} \int_a^{a+p} f(x)\, dx = 0 \quad .$$

5.3 Sines and Cosines

We will see that, one way or another, sines and cosines are involved in most of the formulas of Fourier analysis. So it seems prudent to make sure we are quite familiar with these particular trigonometric functions. The graphs of $\sin(x)$ and $\cos(x)$ are sketched in figure 5.3 (in case you forgot what they look like!). These sketches should remind you that, for any integer n,

$$\sin(n\pi) = 0 \qquad \text{and} \qquad \cos(n\pi) = (-1)^n \quad .$$

Do recall that the sine function is an odd function, while the cosine function is an even function. They are related to each other by the formula

$$\sin(x) = \cos\left(x - \frac{\pi}{2}\right) \quad .$$

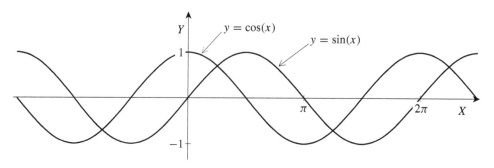

Figure 5.3: The sine and cosine functions.

Also recall that each is a periodic function with fundamental period 2π. Thus, if p is any constant such that either

$$\sin(x - p) = \sin(x) \qquad \text{for every real value } x$$

or

$$\cos(x - p) = \cos(x) \qquad \text{for every real value } x \quad ,$$

then p must be an integral multiple of 2π.

We will often encounter expressions of the form $\sin(2\pi at)$ and $\cos(2\pi at)$ where a is some fixed real number. Clearly, these functions are also periodic functions of t. To determine the possible periods for these functions, observe that, if p is a period for $\sin(2\pi at)$, then

$$\sin(2\pi at - 2\pi |a| p) = \sin(2\pi a(t \pm p)) = \sin(2\pi at)$$

for every real value t. Thus, $2\pi |a| p$ must be a period for the basic sine function, $\sin(x)$; that is,

$$2\pi |a| p = k2\pi \qquad \text{for some integer } k \quad .$$

Solving for the period gives

$$p = \frac{k}{|a|} \qquad \text{where} \quad k = 0, 1, 2, 3, \ldots \quad .$$

This tells us that the fundamental period for $\sin(2\pi at)$ (and $\cos(2\pi at)$) is $p = 1/|a|$ and, hence, the corresponding fundamental frequency must be $\omega = |a|$. This assumes, of course, that $a \neq 0$. If $a = 0$, then, for all t,

$$\sin(2\pi at) = \sin(0) = 0 \qquad \text{and} \qquad \cos(2\pi at) = \cos(0) = 1 \quad .$$

Certain integrals of products of sines and cosines will be particularly important in the development of the Fourier series. The values of these integrals are given in the next theorem.

Theorem 5.2 (orthogonality relations for sines and cosines)
Let $0 < p < \infty$, and let k and n be any pair of positive integers. Then

$$\int_0^p \cos\left(\frac{2\pi k}{p}x\right) dx = \int_0^p \sin\left(\frac{2\pi k}{p}x\right) dx = 0 \quad , \tag{5.3a}$$

$$\int_0^p \cos\left(\frac{2\pi k}{p}x\right) \sin\left(\frac{2\pi n}{p}x\right) dx = 0 \quad , \tag{5.3b}$$

$$\int_0^p \cos\left(\frac{2\pi k}{p}x\right)\cos\left(\frac{2\pi n}{p}x\right)\,dx \;=\; \begin{cases} 0 & \text{if } k \neq n \\[2mm] \dfrac{p}{2} & \text{if } k = n \end{cases} \qquad (5.3c)$$

and

$$\int_0^p \sin\left(\frac{2\pi k}{p}x\right)\sin\left(\frac{2\pi n}{p}x\right)\,dx \;=\; \begin{cases} 0 & \text{if } k \neq n \\[2mm] \dfrac{p}{2} & \text{if } k = n \end{cases} \;. \qquad (5.3d)$$

All the equations in theorem 5.2 can be verified by computing the integrals using basic calculus and trigonometric identities. We'll verify equation (5.3d) and leave the rest as exercises.

PROOF (of equation (5.3d)):   Using the trigonometric identity

$$2\sin(A)\sin(B) \;=\; \cos(A - B) \;-\; \cos(A + B) \quad,$$

we see that

$$\sin\left(\frac{2\pi k}{p}x\right)\sin\left(\frac{2\pi n}{p}x\right) = \frac{1}{2}\left[\cos\left(\frac{2\pi k}{p}x - \frac{2\pi n}{p}x\right) - \cos\left(\frac{2\pi k}{p}x + \frac{2\pi n}{p}x\right)\right]$$

$$= \frac{1}{2}\left[\cos\left(\frac{2\pi(k - n)}{p}x\right) - \cos\left(\frac{2\pi(k + n)}{p}x\right)\right] \quad. \qquad (5.4)$$

Thus, if $k \neq n$,

$$\int_0^p \sin\left(\frac{2\pi k}{p}x\right)\sin\left(\frac{2\pi n}{p}x\right)\,dx$$

$$= \frac{1}{2}\int_0^p \cos\left(\frac{2\pi(k - n)}{p}x\right)\,dx \;-\; \frac{1}{2}\int_0^p \cos\left(\frac{2\pi(k + n)}{p}x\right)\,dx$$

$$= \frac{p}{4\pi(k - n)}\sin\left(\frac{2\pi(k - n)}{p}x\right)\Big|_0^p \;-\; \frac{p}{4\pi(k + n)}\sin\left(\frac{2\pi(k + n)}{p}x\right)\Big|_0^p$$

$$= \frac{p}{4\pi(k - n)}\left[\sin(2\pi(k - n)) - \sin(0)\right] \;-\; \frac{p}{4\pi(k + n)}\left[\sin(2\pi(k + n)) - \sin(0)\right]$$

$$= 0 \quad. \qquad (5.5)$$

On the other hand, the computations in (5.5) are not valid if $k = n$ since they involve division by $k - n$ (which is 0 when $k = n$). Instead, if $k = n$, equation (5.4) reduces to

$$\sin\left(\frac{2\pi k}{p}x\right)\sin\left(\frac{2\pi n}{p}x\right) = \frac{1}{2}\left[\cos(0) - \cos\left(\frac{2\pi(2k)}{p}x\right)\right] = \frac{1}{2} - \frac{1}{2}\cos\left(\frac{4\pi k}{p}x\right) \quad.$$

Hence, when $k = n$,

$$\int_0^p \sin\left(\frac{2\pi k}{p}x\right)\sin\left(\frac{2\pi n}{p}x\right)\,dx = \int_0^p \left[\frac{1}{2} - \frac{1}{2}\cos\left(\frac{4\pi k}{p}x\right)\right]\,dx$$

$$= \left[\frac{1}{2}x - \frac{p}{8\pi k}\sin\left(\frac{4\pi k}{p}x\right)\right]\Big|_0^p$$

$$= \frac{p}{2} - \frac{p}{8\pi k}\left[\sin(4\pi k) - \sin(0)\right] = \frac{p}{2} \quad. \qquad \blacksquare$$

?▶ Exercise 5.10:   Verify equations (5.3a) through (5.3c) in theorem 5.2. (In verifying equations (5.3b) and (5.3c), be sure to consider the cases where $k \neq n$ and $k = n$ separately.)

Additional Exercises

5.11. *Sketch the graph, and identify the fundamental period and frequency for each of the periodic functions given below. In each case, a denotes a positive constant.*

   a. $\sin(x)$ **b.** *the rectified sine function,* $|\sin(x)|$ **c.** $\sin^2(x)$

   d. *The odd saw function,*

$$\text{oddsaw}_a(x) = \begin{cases} x & \text{if } -a/2 < x < a/2 \\ \text{oddsaw}_a(x-a) & \text{in general} \end{cases}$$

   e. *The even saw function,*

$$\text{evensaw}_a(x) = \begin{cases} |x| & \text{if } -a/2 < x < a/2 \\ \text{evensaw}_a(x-a) & \text{in general} \end{cases}$$

   f. *A pulse train,*

$$f(x) = \begin{cases} 0 & \text{if } -a < x < 0 \\ 1 & \text{if } 0 < x < a \\ f(x-2a) & \text{otherwise} \end{cases}$$

5.12. *Let f be a periodic function.*

   a. *Assume p and q are two periods for f with $p < q$. Verify that their difference, $q - p$, is also a period for f.*

   b. *Show that all periods of f are integral multiples of the fundamental period provided f has a fundamental period.*

   c. *Prove that f must have a fundamental period if, in addition to being a periodic, nonconstant function, f is piecewise continuous.*

5.13 a. *Using a computer math package such as Maple or Mathematica, write a "program" or "worksheet" for graphing a periodic function with period p over the interval $(-p/2, 2p)$. Have the function's period and a formula for the function over one period, say, $(0, p)$ or $(-p/2, p/2)$ as the inputs to your program/worksheet.*

   b. *Use your program/worksheet to graph each of the following periodic functions (the first three are from the previous exercise):*

   i. $|\sin(x)|$ **ii.** $\text{evensaw}_6(x)$ **iii.** $\text{oddsaw}_6(x)$

$$\textbf{iv. } f(t) = \begin{cases} 0 & \text{if } -1 < t < 0 \\ 1 & \text{if } 0 < t < 1 \\ f(t-2) & \text{in general} \end{cases}$$

$$\textbf{v. } g(t) = \begin{cases} t^2 & \text{if } -1 < t < 1 \\ g(t-2) & \text{in general} \end{cases}$$

$$\textbf{vi. } h(t) = \begin{cases} 0 & \text{if } -2\pi < t < 0 \\ 1 - \cos(t) & \text{if } 0 < t < 2\pi \\ h(t-4\pi) & \text{in general} \end{cases}$$

6

Elementary Complex Analysis

Fourier analysis could be done without complex-valued functions, but it would be very, very awkward.

6.1 Complex Numbers

Recall that z is a *complex number* if and only if it can be written as

$$z = x + iy$$

where x and y are real numbers and i is a "complex constant" satisfying $i^2 = -1$. The *real part* of z, denoted by $\text{Re}[z]$, is the real number x, while the *imaginary part* of z, denoted by $\text{Im}[z]$, is the real number y. If $\text{Im}[z] = 0$ (equivalently, $z = \text{Re}[z]$), then z is said to be real. Conversely, if $\text{Re}[z] = 0$ (equivalently, $z = i\,\text{Im}[z]$), then z is said to be imaginary.

The *complex conjugate* of $z = x + iy$, which we will denote by z^*, is the complex number $z^* = x - iy$.

In the future, given any statement like "the complex number $z = x+iy$", it should automatically be assumed (unless otherwise indicated) that x and y are real numbers.

The algebra of complex numbers can be viewed as simply being the algebra of real numbers with the addition of a number i whose square is negative one. Thus, choosing some computations that will be of particular interest,

$$zz^* = (x+iy)(x-iy) = x^2 - (iy)^2 = x^2 - i^2y^2 = x^2 - \left(-y^2\right) = x^2 + y^2$$

and

$$\frac{1}{z} = \frac{1}{x+iy} = \frac{1}{x+iy} \cdot \frac{x-iy}{x-iy} = \frac{x-iy}{x^2+y^2} = \frac{z^*}{zz^*} \quad .$$

We will often use the easily verified facts that, for any pair of complex numbers z and w,

$$z^*z = zz^* \quad , \quad (z+w)^* = z^* + w^* \quad \text{and} \quad (zw)^* = (z^*)(w^*) \quad .$$

The set of all complex numbers is denoted by $\mathbb{C}$. By associating the real and imaginary parts of the complex numbers with the coordinates of a two-dimensional Cartesian system, we can identify $\mathbb{C}$ with a plane (called, unsurprisingly, the complex plane). This is illustrated in figure 6.1. Also indicated in this figure are the corresponding polar coordinates r and θ for $z = x + iy$. The value r, which we will also denote by $|z|$, is commonly referred to as either the *magnitude*, the *absolute*

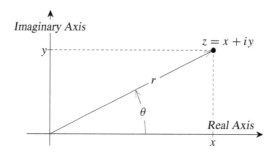

Figure 6.1: Coordinates in the complex plane for $z = x + iy$, where $x > 0$ and $y > 0$.

value, or the *modulus* of z, while θ is commonly called either the *argument*, the *polar angle*, or the *phase* of z. It is easily verified that

$$r = |z| = \sqrt{x^2 + y^2} = \sqrt{z^* z} \ ,$$

$$x = r \cos(\theta) \quad \text{and} \quad y = r \sin(\theta) \ .$$

From this, it follows that the complex number $z = x + iy$ can be written in *polar form*,

$$z = x + iy = r\left[\cos(\theta) + i \sin(\theta)\right] \ .$$

It should also be pretty obvious that

$$|x| \leq |z| \quad \text{and} \quad |y| \leq |z| \ . \tag{6.1}$$

Recall how trivial it is to verify the triangle inequality

$$|z + w| \leq |z| + |w| \tag{6.2}$$

whenever z and w are real numbers. This inequality also holds if z and w are any two complex numbers. Basically, this inequality is an observation about the triangle in the complex plane whose vertices are the points 0, z and $z + w$. Sketch this triangle and you will see that the sides have lengths $|z|$, $|w|$ and $|z + w|$. The observation expressed by inequality (6.2) is that no side of the triangle can be any longer than the sum of the lengths of the other two sides. That is why inequality (6.2) is called the (basic) *triangle inequality* (for complex numbers).

Observe that the polar angle for a complex number is not unique. If

$$z = |z|\left[\cos(\theta_0) + i \sin(\theta_0)\right] \ ,$$

then any θ differing from θ_0 by an integral multiple of 2π is another polar angle for z. This is readily seen by considering how little figure 6.1 changes if the θ there is increased by an integral multiple of 2π. It is also clear that these are the only polar angles for z. We will refer to the polar angle θ with $0 \leq \theta < 2\pi$ as the *principal argument* (or *principal polar angle*) and denote it by Arg[z].[1]

It is instructive to look at the polar form of the product of two complex numbers. So let z and w be two complex numbers with polar forms

$$z = r\left[\cos(\theta) + i \sin(\theta)\right] \quad \text{and} \quad w = \rho\left[\cos(\phi) + i \sin(\phi)\right] \ .$$

---

[1] Warning: Some authors define the principal argument so that $-\pi < \text{Arg}[z] \leq \pi$.

Multiplying z and w together gives

$$zw = (r\,[\cos(\theta) + i\sin(\theta)])\,(\rho\,[\cos(\phi) + i\sin(\phi)])$$
$$= r\rho\,([\cos(\theta)\cos(\phi) - \sin(\theta)\sin(\phi)] + i\,[\cos(\theta)\sin(\phi) + \sin(\theta)\cos(\phi)]) \quad,$$

which, using well-known trigonometric identities, simplifies to

$$zw = r\rho\,[\cos(\theta + \phi) + i\sin(\theta + \phi)] \quad. \tag{6.3}$$

From this, it immediately follows that $|zw| = |z|\,|w|$ and that a polar angle of zw can be found by adding the polar angles of z and w.

?►Exercise 6.1:   *Let N be any positive integer and let z be any complex number with polar angle θ. Show that $\left|z^N\right| = |z|^N$ and that $N\theta$ is a polar angle for z^N.*

6.2   Complex-Valued Functions

Much of our work will involve complex-valued functions defined over subintervals of the real line. If f is such a function on the interval (α, β), then it can be written as

$$f = u + iv$$

where u and v are the real-valued functions on (α, β) given by

$$u(t) = \mathrm{Re}[f(t)] \qquad \text{and} \qquad v(t) = \mathrm{Im}[f(t)] \quad.$$

Naturally, u is called the *real part* of f and is denoted by $\mathrm{Re}[f]$, while v is called the *imaginary part* of f and is denoted by $\mathrm{Im}[f]$. Likewise, the *complex conjugate* of f is

$$f^* = u - iv \quad,$$

and the *magnitude* (or *modulus* or *absolute value*) of f is

$$|f| = \sqrt{u^2 + v^2} = \sqrt{f^*f} \quad.$$

Graphing a complex function f presents a slight difficulty. The values of $f(t)$ correspond to points $(u(t), v(t))$ on the complex plane. Thus, the graph of f would actually be a curve in a three-dimensional TUV–space. Sadly, few of us have the artistic ability to sketch such a graph by hand. And, even if we had a good three-dimensional graphing package for our computer, the medium of this text is, for all practical purposes, two dimensional. So rather than attempt to draw three-dimensional graphs for complex-valued functions, we will simply graph the real and imaginary parts separately.

!►Example 6.1:   *Let us graph $f(t) = \frac{1}{4}(2 + it)^2$ for $-\infty < t < \infty$. Multiplying through,*

$$f(t) = \tfrac{1}{4}(2 + it)^2 = \tfrac{1}{4}(4 + 4it - t^2) = 1 - \tfrac{1}{4}t^2 + it \quad.$$

So, the graph of the real part of $f(t)$ is that of the parabola

$$u(t) = 1 - \frac{1}{4}t^2 \quad,$$

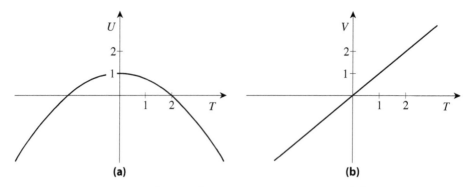

Figure 6.2: Graphing $f(t) = \frac{1}{4}(2 + it)^2$: (a) the real part, $u(t) = 1 - \frac{1}{4}t^2$, and
(b) the imaginary part, $v(t) = t$.

while the graph of the imaginary part of $f(t)$ corresponds to the straight line

$$v(t) = t \quad .$$

These graphs are sketched in figure 6.2.

The reader should realize that, except where it was explicitly stated as otherwise, all the discussion in the previous chapters applied to complex-valued functions as well as real-valued functions. In addition, the following facts concerning an arbitrary complex-valued function $f = u + iv$ should be readily apparent:

1. f is continuous at a point t_0 if and only if u and v are both continuous at t_0.

2. f is continuous on an interval if and only if u and v are both continuous on that interval.

3. The previous statement remains true if the word "continuous" is replaced by any of the conditions — bounded, piecewise continuous, uniformly continuous, smooth, even, periodic, etc. — discussed in the previous chapters.

4. The derivative of f exists on an interval if and only if the derivatives of u and v exist on the interval. Moreover, if the derivatives exist,

$$f' = u' + iv' \quad .$$

5. The integral of f over an interval (α, β) exists if and only if the corresponding integrals of u and v exist. Moreover, if the integrals exist,

$$\int_{\alpha}^{\beta} f(t)\, dt = \int_{\alpha}^{\beta} u(t)\, dt + i \int_{\alpha}^{\beta} v(t)\, dt \quad .$$

Here are two more facts concerning $\int_{\alpha}^{\beta} f(t)\, dt$ that will be useful later on in our work:

$$\left(\int_{\alpha}^{\beta} f(t)\, dt \right)^{*} = \int_{\alpha}^{\beta} f^{*}(t)\, dt \tag{6.4}$$

and

$$\left| \int_{\alpha}^{\beta} f(t)\, dt \right| \le \int_{\alpha}^{\beta} |f(t)|\, dt \quad . \tag{6.5}$$

The first is easily verified:

$$\left(\int_\alpha^\beta f(t)\,dt \right)^* = \left(\int_\alpha^\beta u(t)\,dt + i\int_\alpha^\beta v(t)\,dt \right)^*$$

$$= \int_\alpha^\beta u(t)\,dt - i\int_\alpha^\beta v(t)\,dt$$

$$= \int_\alpha^\beta [u(t) - iv(t)]\,dt = \int_\alpha^\beta f^*(t)\,dt \quad .$$

The second is obviously true when f is real valued. To see why it holds more generally, consider the case where f is a complex-valued, piecewise continuous function on a finite interval (α, β). For each integer N, construct a corresponding N^{th} Riemann sum

$$\mathcal{R}_N = \sum_{k=1}^N f(\bar{x}_k)\,\Delta x$$

for the integral on the left-hand side of inequality (6.5) (see page 38). Then, using the triangle inequality,

$$\left| \int_\alpha^\beta f(t)\,dt \right| = \left| \lim_{N\to\infty} \sum_{k=1}^N f(\bar{x}_k)\,\Delta x \right| \le \lim_{N\to\infty} \sum_{k=1}^N |f(\bar{x}_k)|\,\Delta x = \int_\alpha^\beta |f(t)|\,dt \quad .$$

Extending these computations to cases where f is not piecewise continuous or (α, β) is not finite — but the integrals exist — is easy and will either be left to the interested reader or discussed as the need arises.

6.3 The Complex Exponential

The basic complex exponential, denoted either by e^z or, especially when z is given by a formula that is hard to read as a superscript, by $\exp(z)$, is a complex-valued function of a complex variable. You are probably already acquainted with this function, but it will be so important to our work that it is worthwhile to review its derivation as well as some of its properties and applications.

Derivation

Our goal is to derive a meaningful formula for e^z that extends our notion of the exponential to the case where z is complex. We will derive this formula by requiring that the complex exponential satisfies some of the same basic properties as the well-known real exponential function, and that it reduces to the real exponential function when z is real.

First, let us insist that the law of exponents (i.e., that $e^{A+B} = e^A e^B$) holds. Thus,

$$e^z = e^{x+iy} = e^x e^{iy} \quad . \tag{6.6}$$

We know the first factor, e^x. It's the real exponential from elementary calculus (a function you should be able to graph in your sleep).

To determine the second factor, consider the yet undefined function

$$f(t) = e^{it} \quad .$$

Since we insist that the complex exponential reduces to the real exponential when the exponent is real, we must have

$$f(0) = e^{i0} = e^0 = 1 \quad .$$

Recall, also, that $\frac{d}{dt}e^{at} = ae^{at}$ whenever a is a real constant. Requiring that this formula be true for imaginary constants gives

$$f'(t) = \frac{d}{dt}e^{it} = ie^{it} \tag{6.7}$$

$$\hookrightarrow \qquad f''(t) = \frac{d}{dt}f'(t) = \frac{d}{dt}ie^{it} = i^2 e^{it} = -f(t)$$

$$\hookrightarrow \qquad f''(t) + f(t) = 0 \quad .$$

But the last is a simple differential equation, the general solution of which is easily verified to be $A\cos(t) + B\sin(t)$ where A and B are arbitrary constants. So

$$e^{it} = f(t) = A\cos(t) + B\sin(t) \quad .$$

The constant A is easily determined from the requirement that $e^{i0} = 1$:

$$1 = e^{i0} = A\cos(0) + B\sin(0) = A \cdot 1 + B \cdot 0 = A \quad .$$

From equation (6.7) and the observation that

$$f'(t) = \frac{d}{dt}[A\cos(t) + B\sin(t)] = -A\sin(t) + B\cos(t) \quad ,$$

we see that

$$i = ie^{i0} = f'(0) = -A\sin(0) + B\cos(0) = -A \cdot 0 + B \cdot 1 = B \quad .$$

Thus $A = 1$, $B = i$, and

$$e^{it} = \cos(t) + i\sin(t) \quad . \tag{6.8}$$

Formula (6.8) is Euler's (famous) formula for e^{it}. It and equation (6.6) yield the formula

$$e^{x+iy} = e^x e^{iy} = e^x [\cos(y) + i\sin(y)] \tag{6.9}$$

for all real values of x and y. We will take this formula as the *definition* of the complex exponential.

Properties and Formulas

Using formula (6.9), it can be easily verified that the complex exponential satisfies those properties we assumed in the derivation of that formula. That is, given any two complex numbers A and B,

1. $e^{A+B} = e^A e^B$,

2. e^A is the real exponential of A whenever A is real, and

3. $\frac{d}{dt}e^{At} = Ae^{At}$.

It is also useful to observe that

$$
\begin{aligned}
e^{x-iy} &= e^x e^{i(-y)} \\
&= e^x \left[\cos(-y) + i \sin(-y) \right] \\
&= e^x \left[\cos(y) - i \sin(y) \right] = \left(e^{x+iy} \right)^* \quad .
\end{aligned}
$$

Not only does this give us

$$
\left(e^z \right)^* = e^{(z^*)} \quad ,
$$

but it also provides the second of the following pair of identities (the first being Euler's formula, itself):

$$
e^{x+iy} = e^x \left[\cos(y) + i \sin(y) \right] \tag{6.10a}
$$

and

$$
e^{x-iy} = e^x \left[\cos(y) - i \sin(y) \right] \quad . \tag{6.10b}
$$

Letting $x = 0$ and $y = \theta$, these identities become the pair

$$
e^{i\theta} = \cos(\theta) + i \sin(\theta) \tag{6.11a}
$$

and

$$
e^{-i\theta} = \cos(\theta) - i \sin(\theta) \quad . \tag{6.11b}
$$

We can then solve for $\cos(\theta)$ and $\sin(\theta)$, obtaining

$$
\cos(\theta) = \frac{e^{i\theta} + e^{-i\theta}}{2} \tag{6.12a}
$$

and

$$
\sin(\theta) = \frac{e^{i\theta} - e^{-i\theta}}{2i} \quad . \tag{6.12b}
$$

All of the above pairs of identities will be very useful in our work.

On a number of occasions, we will need to compute the value of $e^{i\theta}$ and $\left| e^{\pm i\theta} \right|$ for specific real values of θ. Computing $\left| e^{\pm i\theta} \right|$ is easy. For any *real value* θ,

$$
\left| e^{\pm i\theta} \right| = \sqrt{\cos^2(\theta) + \sin^2(\theta)} = \sqrt{1} = 1 \quad . \tag{6.13}
$$

This also tells us that $e^{i\theta}$ is a point on the unit circle in the complex plane. In fact, comparing formula (6.11a) with the polar form for the complex number $e^{i\theta}$, we find that θ is, in fact, a polar angle for $e^{i\theta}$. The point $e^{i\theta}$ has been plotted in figure 6.3 for some unspecified θ between 0 and $\pi/2$. The real and imaginary parts of $e^{i\theta}$ can be computed either by using formula (6.11a) (or (6.11b)) or, at least for some values of θ, by inspection of figure 6.3. Clearly, for example,

$$
e^{i\frac{1}{2}\pi} = i \quad , \qquad e^{i\pi} = -1
$$

and

$$
e^{i\frac{3}{2}\pi} = e^{-i\frac{1}{2}\pi} = -i \quad .
$$

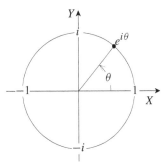

Figure 6.3: Plot of $e^{i\theta}$.

?►Exercise 6.2: *Verify that*

$$e^{i2\pi n} = 1 \quad \text{and} \quad e^{i\pi n} = (-1)^n \quad \text{for} \quad n = 0, \pm 1, \pm 2, \pm 3, \dots \quad .$$

?►Exercise 6.3: *Let z be any complex number and let θ be its polar angle. Verify that*

$$z = |z|\, e^{i\theta} \quad .$$

Much of our work will involve functions of the form $e^{i2\pi\alpha x}$ where α is some fixed real value. From formula (6.11a), it should be clear that $e^{i2\pi\alpha x}$ is a smooth periodic function of x on the entire real line. If $\alpha = 0$, then $e^{i2\pi\alpha x}$ is just the constant function 1. Otherwise, it is a nontrivial periodic function with fundamental frequency $|\alpha|$ and fundamental period $|\alpha|^{-1}$. For future reference, let us note that

$$e^{i2\pi\alpha x} = \cos(2\pi\alpha x) + i\sin(2\pi\alpha x) \quad , \tag{6.14}$$

$$e^{-i2\pi\alpha x} = \cos(2\pi\alpha x) - i\sin(2\pi\alpha x) \quad , \tag{6.15}$$

$$\cos(2\pi\alpha x) = \frac{1}{2}\left[e^{i2\pi\alpha x} + e^{-i2\pi\alpha x} \right] \quad , \tag{6.16}$$

$$\sin(2\pi\alpha x) = \frac{1}{2i}\left[e^{i2\pi\alpha x} - e^{-i2\pi\alpha x} \right] \quad , \tag{6.17}$$

and

$$\left| e^{\pm i2\pi\alpha x} \right| = 1 \quad . \tag{6.18}$$

Complex Exponentials in Trigonometric Computations

Any expression involving sines and cosines can be rewritten in terms of complex exponentials using the above formulas. For many people, these resulting expressions are much easier to manipulate than the original formulas, especially if a table of trigonometric identities is not readily available.

!►Example 6.2: *Let k and n be any pair of positive integers with $k \neq n$ and consider evaluating the integral*

$$\int_0^1 \sin(2k\pi x)\sin(2n\pi x)\, dx \quad .$$

Using formulas (6.12a) and (6.12b),

$$\sin(2k\pi x)\sin(2n\pi x)$$

$$= \left(\frac{e^{i2k\pi x} - e^{-i2k\pi x}}{2i} \right)\left(\frac{e^{i2n\pi x} - e^{-i2n\pi x}}{2i} \right)$$

$$= -\frac{1}{4}\left(e^{i2k\pi x}e^{i2n\pi x} - e^{i2k\pi x}e^{-i2n\pi x} - e^{-i2k\pi x}e^{i2n\pi x} + e^{-i2k\pi x}e^{-i2n\pi x} \right)$$

$$= -\frac{1}{4}\left(e^{i2(k+n)\pi x} - e^{i2(k-n)\pi x} - e^{-i2(k-n)\pi x} + e^{-i2(k+n)\pi x} \right) \quad .$$

Evaluating the integral of each term over $(0, 1)$ is easy. Since k and n are two different positive integers, $k \pm n$ is a nonzero integer and

$$\int_0^1 e^{\pm i2(k\pm n)\pi x}\, dx = \pm\frac{1}{i2(k \pm n)\pi}\, e^{\pm i2(k\pm n)\pi x}\Big|_0^1$$

$$= \pm\frac{1}{i2(k \pm n)\pi}\left(e^{\pm i2(k\pm n)\pi} - e^0 \right) = \pm\frac{1}{i2(k \pm n)\pi}(1 - 1) = 0 \quad .$$

Thus, since each of the integrals vanishes,

$$\int_0^1 \sin(2k\pi x)\sin(2n\pi x)\,dx \;=\; -\frac{1}{4}\left[\int_0^1 e^{i2(k+n)\pi x}\,dx \;-\; \int_0^1 e^{i2(k-n)\pi x}\,dx\right.$$

$$\left. -\int_0^1 e^{-i2(k-n)\pi x}\,dx \;+\; \int_0^1 e^{-i2(k+n)\pi x}\,dx\right] \;=\; 0 \quad.$$

(You should compare these calculations with those used to prove equation (5.3d) in the orthogonality relations for sines and cosines (theorem (5.2) on page 54).)

Complex exponentials can also be used to derive trigonometric identities.

!▶Example 6.3: *Let A and B be any two real values. Then*

$$\sin(A)\sin(B) \;=\; \left(\frac{e^{iA}-e^{-iA}}{2i}\right)\left(\frac{e^{iB}-e^{-iB}}{2i}\right)$$

$$=\; \left(\frac{1}{2i}\right)^2\left(e^{iA}e^{iB} \;-\; e^{iA}e^{-iB} \;-\; e^{-iA}e^{iB} \;+\; e^{-iA}e^{-iB}\right)$$

$$=\; -\frac{1}{4}\left(e^{i(A+B)} \;-\; e^{i(A-B)} \;-\; e^{-i(A-B)} \;+\; e^{-i(A+B)}\right)$$

$$=\; -\frac{1}{2}\left(\frac{e^{i(A+B)}+e^{-i(A+B)}}{2} \;-\; \frac{e^{i(A-B)}+e^{-i(A-B)}}{2}\right)$$

$$=\; \frac{1}{2}\bigl(-\cos(A+B) \;+\; \cos(A-B)\bigr) \quad.$$

With a little rearranging, this becomes the trigonometric identity,

$$2\sin(A)\sin(B) \;=\; \cos(A-B) \;-\; \cos(A+B) \quad.$$

Complex Exponentials, Roots and Factoring

Let γ be any nonzero number (real or complex) and let N be any positive integer. After rewriting the polar form of α in terms of complex exponentials, it is easy to verify (see exercise 6.15) that γ has exactly N N^{th} roots. That is, there are exactly N different solutions to $z^N = \gamma$. In particular, the two square roots of $\gamma = 1$ are 1 and -1 ; the two square roots of $\gamma = -1$ are i and $-i$, and, as you'll compute in exercise 6.16, the two square roots of $\gamma = i$ are σ and $-\sigma$ where

$$\sigma \;=\; \frac{1+i}{\sqrt{2}} \quad.$$

From this, it is easily verified that the classic factoring formula

$$a^2 \;-\; b^2 \;=\; (a+b)(a-b)$$

has the following variations:

$$a^2 \;+\; b^2 \;=\; (a+ib)(a-ib) \quad,$$

$$a^2 \;-\; ib^2 \;=\; (a+\sigma b)(a-\sigma b)$$

and

$$a^2 \;+\; ib^2 \;=\; (a+i\sigma b)(a-i\sigma b) \quad.$$

We leave the few details of confirming this to you, in exercise 6.17.

6.4   Functions of a Complex Variable[*]

The exponential is a function whose variable is not limited to some interval, but can range over the entire complex plane, $\mathbb{C}$. Eventually, we will deal with many other such functions. So let us suppose f is some function for which $f(z)$ is somehow defined for every complex value $z = x + iy$, and let us briefly describe some things concerning f that will later be relevant to our work.

Continuity and Derivatives

Since f is complex valued, it has real and imaginary parts u and v, which can be viewed as functions of two real variables,

$$u(x, y) \ = \ \text{Re}[f(x + iy)] \qquad \text{and} \qquad v(x, y) \ = \ \text{Im}[f(x + iy)] \quad .$$

So,

$$f(z) \ = \ f(x + iy) \ = \ u(x, y) \ + \ iv(x, y) \quad , \tag{6.19}$$

and we can view f as a function of two real variables or as a function of a single complex variable, as convenience dictates.

!▶ **Example 6.4:**   *The real and imaginary parts of the basic complex exponential*

$$e^{x+iy} \ = \ e^x \cos(y) \ + \ ie^x \sin(y)$$

    are, respectively,

$$u(x, y) \ = \ e^x \cos(y) \qquad \text{and} \qquad v(x, y) \ = \ e^x \sin(y) \quad .$$

Let $z_0 = x_0 + iy_0$ be some point on the complex plane. Naturally, we will say that f is *continuous* at z_0 if and only if

$$\lim_{z \to z_0} f(z) \ = \ f(z_0) \quad , \tag{6.20}$$

and we will say that f is continuous on the entire complex plane if and only if it is continuous at every point in $\mathbb{C}$. Keep in mind that this is a two-dimensional limit. Saying that $z = x + iy$ approaches $z_0 = x_0 + iy_0$ means both that x approaches x_0 and that y approaches y_0.

We will continue our convention of removing removable discontinuities. So, if the limit in equation (6.20) exists (as a finite complex value), then we will automatically take $f(z_0)$ as defined and equal to that limit.

In terms of the representation given in equation (6.19), the partial derivatives of f are given by

$$\frac{\partial f}{\partial x} \ = \ \frac{\partial u}{\partial x} \ + \ i\frac{\partial v}{\partial x} \qquad \text{and} \qquad \frac{\partial f}{\partial y} \ = \ \frac{\partial u}{\partial y} \ + \ i\frac{\partial v}{\partial y}$$

provided the corresponding partials of u and v exist. Assuming those partials do exist, you can easily verify that this is completely equivalent to defining the partial derivatives of f at z_0 by

$$\left.\frac{\partial f}{\partial x}\right|_{z_0} = \lim_{\Delta x \to 0} \frac{f(z_0 + \Delta x) - f(z_0)}{\Delta x} \qquad \text{and} \qquad \left.\frac{\partial f}{\partial y}\right|_{z_0} = \lim_{\Delta y \to 0} \frac{f(z_0 + i\Delta y) - f(z_0)}{\Delta y} \quad .$$

---

[*] This section will not be a review for the reader, unless the reader has had an introductory course on complex analysis. It should also be mentioned that, except for one rather optional exercise in chapter 20, this material will not be needed until part IV of this text.

In addition, we define the *(complex) derivative* f' of f at z_0 by

$$f'(z_0) = \left. \frac{df}{dz} \right|_{z_0} = \lim_{\Delta z \to 0} \frac{f(z_0 + \Delta z) - f(z_0)}{\Delta z} \quad , \tag{6.21}$$

provided this limit exists. Naturally, we'll say f is *differentiable* at z_0 if and only if this limit exists. Higher order derivatives are defined in the obvious way:

$$f'' = (f')' \quad , \quad f^{(3)} = (f'')' \quad , \quad \cdots \quad .$$

We should observe that, if f is differentiable at a point z_0, then the partial derivatives of f must also exist at that point. Moreover,

$$f'(z_0) = \lim_{\Delta z \to 0} \frac{f(z_0 + \Delta z) - f(z_0)}{\Delta z}$$

$$= \lim_{\Delta x \to 0} \lim_{\Delta y \to 0} \frac{f(z_0 + \Delta x + i \Delta y) - f(z_0)}{\Delta x + i \Delta y}$$

$$= \lim_{\Delta x \to 0} \frac{f(z_0 + \Delta x) - f(z_0)}{\Delta x} = \left. \frac{\partial f}{\partial x} \right|_{z_0} \quad .$$

Switching the order in which the limits are computed gives

$$f'(z_0) = \lim_{\Delta y \to 0} \lim_{\Delta x \to 0} \frac{f(z_0 + \Delta x + i \Delta y) - f(z_0)}{\Delta x + i \Delta y}$$

$$= \lim_{\Delta y \to 0} \frac{f(z_0 + i \Delta y) - f(z_0)}{i \Delta y} = \frac{1}{i} \left. \frac{\partial f}{\partial y} \right|_{z_0} \quad .$$

So,

$$\left. \frac{\partial f}{\partial x} \right|_{z_0} = f'(z_0) = -i \left. \frac{\partial f}{\partial y} \right|_{z_0} \tag{6.22}$$

whenever f is differentiable at z_0. Thus, not only do the partial derivatives exist at each point where f is differentiable, they also satisfy[2]

$$i \frac{\partial f}{\partial x} = \frac{\partial f}{\partial y} \quad . \tag{6.23}$$

Analyticity
Basic Facts

A function that is differentiable everywhere on the complex plane, $\mathbb{C}$, is said to be *analytic* (on the complex plane). On occasion, we will find it useful to recognize that certain functions are analytic. To this end, let us quote an important theorem from complex analysis:

Theorem 6.1 *(test for analyticity)*
Let f be a function on the complex plane. Then f is analytic on $\mathbb{C}$ if and only if, at each point in $\mathbb{C}$, the partial derivatives of $f(x + iy)$ exist, are continuous and satisfy

$$i \frac{\partial f}{\partial x} = \frac{\partial f}{\partial y} \quad . \tag{6.24}$$

---

[2] Rewritten in terms of the real and imaginary parts of f, equation (6.23) becomes a (famous) pair of equations known as the Cauchy–Riemann equations.

This is one theorem we will not attempt to completely prove. We did part of the proof with our derivation of equation (6.23). Other parts (such as showing the partials are continuous wherever f is analytic), however, require techniques we do not have space to develop here. You will have to trust the author that this is a well-known theorem whose proof is a standard part of any reasonable course in complex analysis.

!▶**Example 6.5:**   Let n be any positive integer, and consider the following two functions on the complex plane:

$$f(z) = z^n \quad \text{and} \quad g(z) = e^z \ .$$

The partial derivatives of $f(x+iy)$ and $g(x+iy)$ are easily computed,

$$\frac{\partial f}{\partial x} = \frac{\partial}{\partial x}(x+iy)^n = n(x+iy)^{n-1} \ ,$$

$$\frac{\partial f}{\partial y} = \frac{\partial}{\partial y}(x+iy)^n = n(x+iy)^{n-1}i \ ,$$

$$\frac{\partial g}{\partial x} = \frac{\partial}{\partial x}e^{x+iy} = e^{x+iy} \quad \text{and} \quad \frac{\partial g}{\partial y} = \frac{\partial}{\partial y}e^{x+iy} = e^{x+iy}i \ .$$

Clearly, these are continuous functions on the complex plane and satisfy

$$i\frac{\partial f}{\partial x} = \frac{\partial f}{\partial y} \quad \text{and} \quad i\frac{\partial g}{\partial x} = \frac{\partial g}{\partial y}$$

everywhere. So z^n and e^z are analytic on the complex plane.
    On the other hand, if we define a function h by

$$h(x+iy) = x^2 + iy^2 \ ,$$

then, whenever $x \neq y$,

$$i\frac{\partial h}{\partial x} = i2x \neq i2y = \frac{\partial h}{\partial y} \ .$$

So h, as defined above, is not analytic on the complex plane.

?▶**Exercise 6.4:**   Verify that any constant function is analytic on the complex plane.

The next theorem lists some results analogous to well-known results from elementary calculus. The validity of this theorem should be obvious from theorem 6.1, above, and equation (6.22).

Theorem 6.2
Let f and g be any two functions analytic on the complex plane. Then

1. the product fg,

2. the linear combination $af + bg$ where a and b are any two complex numbers, and

3. the composition $h(z) = f(g(z))$

are all analytic on the complex plane. Moreover,

1. $(fg)' = f'g + fg'$,

2. $(af + bg)' = af' + bg'$, and

3. $h'(z) = f'(g(z)) \cdot g'(z)$.

It follows from this theorem and our previous example that all polynomials are analytic on the complex plane. So are all linear combinations of complex exponentials, including the sine and cosine functions, which are defined for all complex values by

$$\sin(z) = \frac{e^{iz} - e^{-iz}}{2i} \quad \text{and} \quad \cos(z) = \frac{e^{iz} + e^{-iz}}{2} \quad .$$

?►Exercise 6.5: *Using the above theorem and results from example 6.5, verify that each of the following is analytic on the complex plane:*

$$e^{-3z^2} \quad , \quad \sin\left(z^2\right) \quad and \quad \left(2 + z^3\right) e^{-3z^2} \quad .$$

Properties of Analytic Functions

As anyone who has taken a course in complex variables can attest, much more can be said about a function analytic on the complex plane than can be said about a function that is differentiable on the real line. To illustrate this, let us quote (without proof) two standard theorems that can be found in any introductory text on complex variables.

Theorem 6.3
If f is analytic on the complex plane, then f is infinitely differentiable on the complex plane. That is, for every positive integer n, the nth complex derivative of f exists and is, itself, analytic on the complex plane.

Theorem 6.4
Let f be analytic on the complex plane, and let z_0 be any fixed point on the complex plane. For each nonnegative integer k let

$$a_k = \frac{f^{(k)}(z_0)}{k!} \quad .$$

Then $\sum_{k=0}^{\infty} a_k (z - z_0)^k$ converges absolutely for each complex value z, and

$$f(z) = \sum_{k=0}^{\infty} a_k (z - z_0)^k \quad . \tag{6.25}$$

Conversely, if z_0 is any fixed point on the complex plane and $\sum_{k=0}^{\infty} a_k (z - z_0)^k$ is any power series that converges absolutely for every complex value z, then the function

$$f(z) = \sum_{k=0}^{\infty} a_k (z - z_0)^k$$

is analytic on the complex plane.

If you think about it, these two theorems are remarkable. The first basically assures us that, if a function is complex differentiable everywhere on the complex plane, then *all* of its derivatives — up to any order — exist. The second goes even further and assures us that every such function can be represented by its Taylor series about any point on the plane.

Two results that will be of some value in future work can be quickly derived from the above two theorems (which, of course, is why those theorems were mentioned here).

The first comes from taking equation (6.25) in the last theorem, subtracting a_0 from both sides and then dividing by $z - z_0$. Since $a_0 = f(z_0)$, the result is

$$\frac{f(z) - f(z_0)}{z - z_0} = \sum_{k=0}^{\infty} b_k (z - z_0)^k$$

where $b_k = a_{k+1}$ for each k. After verifying that this last series is also absolutely convergent for each z (which is easy and left to you) and applying theorem 6.4 once again, we have the next corollary.

Corollary 6.5
Let f be analytic on the complex plane and define g by

$$g(z) = \frac{f(z) - f(z_0)}{z - z_0}$$

where z_0 is any fixed complex value. Then g is also analytic on the complex plane.

?▶Exercise 6.6:   *Verify that the sinc function,*

$$\text{sinc}(z) = \frac{\sin(z)}{z} \quad ,$$

is analytic on the entire complex plane.

Next, consider the case where f and g are two analytic functions on the complex plane that are identical on some nontrivial interval (α, β) of the real line; that is, $f(x) = g(x)$ whenever x is a real value with $\alpha < x < \beta$. For any point x_0 in this interval, theorem 6.4 tells us that both f and g can be expressed as Taylor series about x_0,

$$f(z) = \sum_{k=0}^{\infty} a_k (z - x_0)^k \qquad \text{and} \qquad g(z) = \sum_{k=0}^{\infty} b_k (z - x_0)^k \quad .$$

But then, for all real values of x with $\alpha < x < \beta$,

$$0 = f(x) - g(x) = \sum_{k=0}^{\infty} a_k (x - x_0)^k - \sum_{k=0}^{\infty} b_k (x - x_0)^k = \sum_{k=0}^{\infty} (a_k - b_k) (x - x_0)^k \quad .$$

From this, it is obvious (or, if not obvious, very easy to verify) that $a_k = b_k$ for each of the k's. Thus, for every complex value z,

$$f(z) = \sum_{k=0}^{\infty} a_k z^k = \sum_{k=0}^{\infty} b_k z^k = g(z) \quad .$$

This gives us the next corollary of theorem 6.4.

Corollary 6.6
Let f and g be two analytic functions on the complex plane. If $f = g$ on a nontrivial subinterval (α, β) of the real line, then $f = g$ on the entire complex plane.

On occasion, we will find ourselves with a function ϕ defined just on the real line and another function f defined and analytic on the entire complex plane that equals ϕ on the real line (i.e., $f(x) = \phi(x)$ for every real value x). We will refer to any such f as an *analytic extension* of ϕ (to a function on $\mathbb{C}$). Now if g is any "other" such analytic extension of ϕ (i.e., g is defined and analytic on the entire complex plane and equals ϕ on the real line), then, obviously, $f = \phi = g$ on the real line and our last corollary assures us that $f(z) = g(z)$ for each and every complex value z. So there cannot be two different analytic extensions of any function on the real line. This fact will be important enough to be called a theorem.

Theorem 6.7
Let ϕ be a function defined on the real line. If ϕ has an analytic extension to a function on $\mathbb{C}$, then there is exactly one analytic extension of ϕ to a function on $\mathbb{C}$.

In the future, to conserve symbols, we will indicate the analytic extension of a function ϕ by either adding an "E" subscript, ϕ_E, or we will simply use the same symbol for both the original function and its analytic extension.

Additional Exercises

6.7. Let $z = 2 + 3i$ and compute each of the following:

   a. $\text{Re}[z]$ **b.** $\text{Im}[z]$ **c.** $|z|$ **d.** $\text{Arg}[z]$

   e. z^2 **f.** $\text{Re}\left[\dfrac{1}{z}\right]$ **g.** $\text{Im}\left[\dfrac{1}{z}\right]$

6.8. Show that, for any complex number z,

$$\text{Re}[z] = \frac{z + z^*}{2} \qquad \text{and} \qquad \text{Im}[z] = \frac{z - z^*}{2i} \quad .$$

6.9. Show that, if θ is a polar angle for z, then $-\theta$ is a polar angle for z^*.

6.10. For the following, let $f(t) = \dfrac{1}{1 - it}$.

   a. Find and graph the real and imaginary parts of f.

   b. Find and graph $|f(t)|$.

6.11. Let α and ω be two real values (with $\omega > 0$), and sketch graphs for $f(t) = e^{(\alpha + i\omega)t}$ for the cases where $\alpha > 0$, $\alpha < 0$ and $\alpha = 0$.

6.12. Evaluate (i.e., find the real and imaginary parts) of each of the following and plot each on the unit circle:

   a. $\exp\left(i\dfrac{\pi}{4}\right)$ **b.** $\exp\left(i\dfrac{\pi}{3}\right)$ **c.** $\exp\left(i\dfrac{2\pi}{3}\right)$

   d. $\exp\left(-i\dfrac{\pi}{3}\right)$ **e.** $e^{i9\pi}$

6.13. *Let k and n be any two nonzero integers, and let a, b and p be any three nonzero real numbers with $p > 0$. Evaluate each of the following using complex exponentials:*

a. $\displaystyle\int_0^1 \sin^2(2k\pi x)\, dx$

b. $\displaystyle\int_0^x e^{at} \cos(bt)\, dt$

c. $\displaystyle\int_0^p \cos\left(\frac{2\pi k}{p}x\right) \cos\left(\frac{2\pi n}{p}x\right) dx$ (assuming $k \neq n$)

d. $\displaystyle\int_0^p \cos\left(\frac{2\pi k}{p}x\right) \cos\left(\frac{2\pi n}{p}x\right) dx$ (assuming $k = n$)

6.14. *Let A and B be real numbers. Using the complex exponential, derive each of the following trigonometric identities:*

a. $\sin^2(A) = \frac{1}{2} - \frac{1}{2}\cos(2A)$

b. $\cos^2(A) = \frac{1}{2} + \frac{1}{2}\cos(2A)$

c. $\cos(A)\cos(B) = \frac{1}{2}\cos(A+B) + \frac{1}{2}\cos(A-B)$

d. $\sin(A)\cos(B) = \frac{1}{2}\sin(A+B) + \frac{1}{2}\sin(A-B)$

e. $\cos(A+B) = \cos(A)\cos(B) - \sin(A)\sin(B)$

6.15. *Let N be a positive integer and c an arbitrary nonzero number (real or complex). Show that there are exactly N distinct values of z satisfying $z^N = c$, and that they are given by*

$$z_k = re^{i\theta_k} = r\cos(\theta_k) + ir\sin(\theta_k) \qquad \text{for} \quad k = 0, 1, 2, \ldots, N-1 \quad,$$

where, letting ϕ be any single polar angle for c,

$$r = \sqrt[N]{|c|} \qquad \text{and} \qquad \theta_k = \frac{\phi + k2\pi}{N} \quad.$$

6.16. *Using the results from the last exercise, find all distinct solutions to the following equations:*

a. $z^4 = 1$

b. $z^3 = 1$

c. $z^3 = -1$

d. $z^3 = -8$

e. $z^2 = i$

f. $z^2 = -i$

g. $z^3 = i$

6.17. *Let*

$$\sigma = \frac{1+i}{\sqrt{2}} \quad.$$

From the previous problem, we know σ is a square root of i ; that is, $\sigma^2 = i$. Use this to verify the following factoring formulas:

a. $a^2 - ib^2 = (a + \sigma b)(a - \sigma b)$

b. $a^2 + ib^2 = (a + i\sigma b)(a - i\sigma b)$

6.18. *Let f and g be two analytic functions on the complex plane, and let z_0 be some fixed complex value. Show that the function*

$$h(z) = \frac{f(z) - g(z)}{z - z_0}$$

is analytic on the complex plane.

7

*Functions of Several Variables**

Most of the ideas discussed in the previous chapters can be extended to cases where the functions of interest have more than one variable. In this chapter we will briefly review some extensions that will be particularly relevant, and we will develop some fairly deep results concerning integrals of functions of several variables. (In fact, the primary reason this chapter was written was to discuss those "deep" results, and to prevent us from having to prove two to four special cases of each of these results at various widely scattered spots in this text.)

For convenience, we will limit ourselves to discussing functions of two variables. That will suffice for most of our needs. Also, it covers the hard part of extending one-dimensional results to multi-dimensional results, at least for the results we will be needing. Once you've seen the basic ideas expressed here, you should have no trouble extending the definitions and results described in this chapter to corresponding definitions and results for functions of more than two variables.

7.1 Basic Extensions

Presumably, you are already familiar with partial derivatives and double integrals, and can see how the discussion in previous chapters regarding derivatives and integrals can apply to suitably nice functions of two variables. Less clear, perhaps, is how we should extend our notion of a "suitably nice" function of one variable to a useful notion of a "suitably nice" function of two variables.

Regions in the Plane

The first extension is pretty obvious. A function of two variables $f(x, y)$ will normally be defined over a *region* $\mathcal{R}$ in the XY–plane instead of an interval (α, β). One of the simplest types of regions is a rectangle, and, given any two intervals on the real line (a, b) and (c, d), we will let $(a, b) \times (c, d)$ denote the rectangle

$$(a, b) \times (c, d) = \{(x, y) : a < x < b \text{ and } c < y < d\} \quad .$$

This is a *finite* or *bounded* rectangle if a, b, c and d are all finite, and *infinite* or *unbounded* otherwise.

Much more exotic subsets of the plane can be created. To keep us within the realm of practicality, let us agree that the statement "$\mathcal{R}$ is a region in the plane" implies both of the following:

---

* This may be a good chapter to skip at first. Some of the material is a bit more advanced (and technical) than that in the previous chapters, and none of it will be needed for a while.

1. $\mathcal{R}$ is an open set of points in the plane (i.e., if (x, y) is in $\mathcal{R}$, then so is every other point within r of (x, y) for some positive distance r).

2. The boundary of the intersection of the region with any bounded rectangle consists of a finite number of smooth curves each having finite length along with (possibly) a finite number of isolated points.

An arbitrary region will be said to be *bounded* if it is contained in a bounded rectangle. Otherwise, it will be said to be *unbounded*. The entire plane, itself, is the infinite rectangle $(-\infty, \infty) \times (-\infty, \infty)$, which is often denoted in the abbreviated form $\mathbb{R}^2$.

The statement that $\mathcal{R}_0$ *is a subregion of the region* $\mathcal{R}$ simply means that $\mathcal{R}_0$ is a region with every point in $\mathcal{R}_0$ also being in $\mathcal{R}$. It does not exclude the possibility that $\mathcal{R}_0$ and $\mathcal{R}$ are the same region.

Uniform Continuity on Regions

Let $f(x, y)$ be a function of two variables and $\mathcal{R}$ a region in the plane. We will say that $f(x, y)$ is *continuous* on $\mathcal{R}$ if and only if it is continuous at every point in $\mathcal{R}$, that is, if we can write

$$\lim_{(x,y) \to (x_0, y_0)} f(x, y) = f(x_0, y_0) \tag{7.1}$$

for every (x_0, y_0) in $\mathcal{R}$. Additionally, we will say that f is *uniformly continuous* on a bounded region $\mathcal{R}$ if and only if it is continuous on the region and

$$\lim_{\substack{(x,y) \to (x_0, y_0) \\ (x,y) \in \mathcal{R}}} f(x, y) \tag{7.2}$$

exists and is finite for every (x_0, y_0) in the boundary of $\mathcal{R}$.

It should be fairly obvious that any product or linear combination of uniformly continuous functions over a given bounded region will also be uniformly continuous over that region. Showing that other facts regarding uniformly continuous functions of one variable also hold, suitably modified, for uniformly continuous functions of two variables is fairly straightforward and will be left to the interested reader. In particular, we should note the following two-dimensional analogs of lemmas 3.1 through 3.3 (see page 19).

Lemma 7.1
Let f be continuous on a region $\mathcal{R}$ in the plane, and let $\mathcal{R}_0$ be any bounded subregion of $\mathcal{R}$ whose boundary is also contained in $\mathcal{R}$. Then f is uniformly continuous on $\mathcal{R}_0$.

Lemma 7.2
Any function that is uniformly continuous on a given bounded region is also a bounded function on that region.

Lemma 7.3
A function f is uniformly continuous on a bounded region $\mathcal{R}$ if and only if, for every $\epsilon > 0$, there is a corresponding $\Delta r_\epsilon > 0$ such that

$$|f(x, y) - f(\bar{x}, \bar{y})| < \epsilon$$

whenever (x, y) and $(\bar{x}, \bar{y})$ is any pair of points in $\mathcal{R}$ satisfying

$$|(x, y) - (\bar{x}, \bar{y})| < \Delta r_\epsilon \quad .$$

There are two reasons for stating this last lemma. One is that its statement can be viewed as an alternate definition of uniform continuity that can be applied even when $\mathcal{R}$ is an unbounded region. More importantly for us, it provides a way of verifying uniform continuity without having to explicitly verify the existence of limit (7.2) for every different curve making up the boundary of the region. We will illustrate this in the next example by rigorously verifying the unsurprising fact that uniformly continuous functions of one variable also define uniformly continuous functions of two variables.

!▶ Example 7.1:   Let (a, b) and (c, d) be two finite intervals, and assume g is a uniformly continuous function of one variable on (a, b). Let us verify that

$$f(x, y) = g(x)$$

is a uniformly continuous functions of two variables on the rectangle $\mathcal{R} = (a, b) \times (c, d)$.
    Let $\epsilon > 0$. By lemma 3.3 on page 20 we know there is a corresponding distance $\Delta x_\epsilon > 0$ such that

$$|g(x) - g(\bar{x})| < \epsilon$$

whenever x and $\bar{x}$ is any pair of points in (a, b) with

$$|x - \bar{x}| < \Delta x_\epsilon \quad .$$

Let $\Delta r_\epsilon = \Delta x_\epsilon$, and observe that, if (x, y) and $(\bar{x}, \bar{y})$ are any two points in $\mathcal{R}$ with

$$|(x, y) - (\bar{x}, \bar{y})| < \Delta r_\epsilon \quad ,$$

then x and $\bar{x}$ are in (a, b) and,

$$|x - \bar{x}| \leq \sqrt{(x - \bar{x})^2 + (y - \bar{y})^2} = |(x, y) - (\bar{x}, \bar{y})| < \Delta r_\epsilon = \Delta x_\epsilon \quad .$$

So,

$$|f(x, y) - f(\bar{x}, \bar{y})| = |g(x) - g(\bar{x})| < \epsilon \quad ,$$

verifying, according to lemma 7.3, that f is uniformly continuous on $\mathcal{R}$.

    If (a, b), (c, d) and g are as in our last example, and h is a uniformly continuous function of one variable on (c, d), then it should be clear from the example that both

$$f_1(x, y) = g(x) \qquad \text{and} \qquad f_2(x, y) = h(y)$$

are uniformly continuous functions of two variables on the rectangle $(a, b) \times (c, d)$. They must also be uniformly continuous on any subregion of this rectangle (see exercise 7.7). This and the fact that products of uniformly continuous functions on a region are uniformly continuous on that region gives us the following lemma, which we will often use (usually without comment) throughout the rest of this chapter.

Lemma 7.4
Let (a, b) and (c, d) be any two finite intervals, and let $\mathcal{R}_0$ be any subregion of the rectangle $(a, b) \times (c, d)$. Assume that

1. g is a uniformly continuous function of one variable on (a, b);

2. h is a uniformly continuous function of one variable on (c, d), and

3. ϕ is a uniformly continuous function of two variables on $\mathcal{R}_0$.

Then

$$f(x, y) = g(x)h(y)\phi(x, y)$$

is a uniformly continuous function of two variables on $\mathcal{R}_0$.

Piecewise Continuity on Regions

If $\mathcal{R}$ is a bounded region, then the statement "$f(x, y)$ is *piecewise continuous* on $\mathcal{R}$" means that $\mathcal{R}$ can be partitioned into a finite number of subregions over each of which f is uniformly continuous. Consequently, all the discontinuities of a piecewise continuous function on a bounded region must be in some finite collection of smooth curves of finite length (the boundaries of the subregions over which f is uniformly continuous). Let's call these curves the *curves of discontinuity* for f on $\mathcal{R}$.

When the region $\mathcal{R}$ is unbounded, we will refer to a function on $\mathcal{R}$ as being *piecewise continuous* (on $\mathcal{R}$) if and only if the function is piecewise continuous over every bounded subregion of $\mathcal{R}$. It should be obvious that any product or linear combination of piecewise continuous functions over a region will also be a piecewise continuous function over that region.

Continuity of Products

Many of our functions of two variables will be constructed by multiplying two or more simpler functions together. Very often, for example, we will be concerned with functions of the form

$$f(x, y) = g(x)h(y)\phi(x, y)$$

where

1. $g(x)$ is a piecewise continuous function of one variable on the interval (a, b);

2. $h(y)$ is a piecewise continuous function of one variable on the interval (c, d), and

3. $\phi(x, y)$ is a continuous function of two variables on the rectangle $(a, b) \times (c, d)$.

The continuity of such a function can easily be determined from the continuity of its factors. We have already seen this, to some extent, in lemma 7.4. To further illustrate this fact, let f be as above, and let (x_0, y_0) be any point in the rectangle $(a, b) \times (c, d)$. If $g(x)$ is continuous at x_0 and $h(y)$ is continuous at y_0, then

$$\lim_{(x,y)\to(x_0,y_0)} f(x, y) = \lim_{\substack{x\to x_0 \\ y\to y_0}} g(x)h(y)\phi(x, y) = g(x_0)h(y_0)\phi(x_0, y_0) = f(x_0, y_0) \quad ,$$

confirming that the product $f(x, y) = g(x)h(y)\phi(x, y)$ is continuous at (x_0, y_0).

This also tells us that, if this $f(x, y)$ is not continuous at (x_0, y_0), then either $g(x)$ is not continuous at $x = x_0$ or $h(y)$ is not continuous at $y = y_0$. In other words, each point (x_0, y_0) at which $f(x, y)$ is discontinuous must be contained in either

1.   a straight (vertical) line $x = x_0$ on the XY–plane where x_0 is a point at which $g(x)$ has a jump discontinuity,

or

2.   a straight (horizontal) line $y = y_0$ on the XY–plane where y_0 is a point at which $h(y)$ has a jump discontinuity.

These observations (along with lemma 7.4 and the definition of piecewise continuity over intervals) give us the next lemma. We will be referring to this lemma often in the third part (classical Fourier transforms) of this text.

Lemma 7.5
Let f be a function of two variables given by

$$f(x, y) = g(x)h(y)\phi(x, y)$$

where

1. $g(x)$ *is a piecewise continuous function of one variable on an interval* (a, b) ;

2. $h(y)$ *is a piecewise continuous function of one variable on an interval* (c, d) , *and*

3. $\phi(x, y)$ *is a continuous function of two variables on the rectangle* $\mathcal{R} = (a, b) \times (c, d)$ *and is uniformly continuous on every bounded subregion of* $\mathcal{R}$.

Then $f(x, y)$ *is piecewise continuous on* $\mathcal{R}$ *and all the discontinuities of* f *in* $\mathcal{R}$ *are contained in the straight lines*

$$\ldots \quad , \quad x = x_1 \quad , \quad x = x_2 \quad , \quad x = x_3 \quad , \quad \ldots$$
$$\ldots \quad , \quad y = y_1 \quad , \quad y = y_2 \quad , \quad y = y_3 \quad , \quad \ldots$$

where the x_k *'s are the points in the interval* (a, b) *at which* $g(x)$ *is discontinuous, and the* y_k *'s are the points in the interval* (c, d) *at which* $h(y)$ *is discontinuous.*

Moreover, any bounded subregion of $\mathcal{R}$ *intersects only a finite number of these straight lines.*

As an exercise, you should verify the following lemma. It will be used when we discuss convolution (chapter 24).

Lemma 7.6
Let f be a function of two variables given by

$$f(x, y) = g(x)h(y)v(Ax + By)$$

where

1. $g(x)$ *is a piecewise continuous function of one variable on an interval* (a, b) ;

2. $h(y)$ *is a piecewise continuous function of one variable on an interval* (c, d) , *and*

3. $v(s)$ *is a piecewise continuous function on the entire real line with* A *and* B *being two nonzero real constants.*

Then $f(x, y)$ *is piecewise continuous on* $\mathcal{R} = (a, b) \times (c, d)$, *with all the discontinuities of* f *in* $\mathcal{R}$ *being contained in the straight lines*

$$\ldots \quad , \quad x = x_1 \quad , \quad x = x_2 \quad , \quad x = x_3 \quad , \quad \ldots \quad ,$$
$$\ldots \quad , \quad y = y_1 \quad , \quad y = y_2 \quad , \quad y = y_3 \quad , \quad \ldots \quad ,$$
$$\ldots \quad , \quad Ax + By = s_1 \quad , \quad Ax + By = s_2 \quad , \quad Ax + By = s_3 \quad , \quad \ldots$$

where the x_k *'s are the points in the interval* (a, b) *at which* $g(x)$ *is discontinuous, the* y_k *'s are the points in the interval* (c, d) *at which* $h(y)$ *is discontinuous, and the* s_k *'s are the points on* $(-\infty, \infty)$ *at which* $v(s)$ *is discontinuous.*

Moreover, any bounded subregion of $\mathcal{R}$ *intersects only a finite number of these straight lines.*

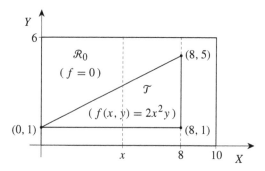

Figure 7.1: Figure for example 7.2.

?▶Exercise 7.1:   *Prove lemma 7.6.*

Finally, let us note that, if the intervals (a, b) and (c, d) are both finite in the two lemmas above, then the discontinuities of $f(x, y)$ in R will all be contained in a finite number of straight lines on the plane. This will be relevant in a few pages.

7.2   Single Integrals of Functions with Two Variables
Functions Defined by Definite Integrals

Much of Fourier analysis involves manipulating functions of the form

$$\psi(x) = \int_c^d f(x, y) \, dy$$

where f is some piecewise continuous function on some rectangle $R = (a, b) \times (c, d)$. Let us assume R is bounded and try to answer three questions that will be particularly important in later work:

1.   *Does this integral unambiguously define the function ψ on the interval (a, b)?*

2.   *Assuming ψ is well defined, what can we say about the continuity of ψ on (a, b)?*

and

3.   *Assuming ψ is well defined, what can we say about differentiating ψ over (a, b)?*

To gather some insight, let's first look at a particular example.

!▶Example 7.2:   *Consider the triangle with vertices $(0, 1)$, $(8, 1)$ and $(8, 5)$ in the rectangle $R = (0, 10) \times (0, 6)$ (see figure 7.1). Let T be the region inside the triangle, R_0 the subregion of R outside the triangle, and define f on R by*

$$f(x, y) = \begin{cases} 2x^2 y & \text{if } (x, y) \text{ is in } T \\ 0 & \text{otherwise} \end{cases} .$$

Clearly, $f(x, y)$ is piecewise continuous on R and is uniformly continuous on both T and R_0.

Now let

$$\psi(x) = \int_0^6 f(x, y)\, dy \quad .$$

$\mathcal{T}$ can be described as the region between $x = 0$ and $x = 8$ bounded by the lines $y = 1$ and $y = {}^x/_2 + 1$. So, when $0 < x < 8$, the above formula for f can be written more explicitly as

$$f(x, y) = \begin{cases} 2x^2 y & \text{if } 1 < y < \frac{1}{2}x + 1 \\ 0 & \text{otherwise} \end{cases} \quad ,$$

and the above formula for ψ reduces to

$$\psi(x) = \int_1^{\frac{1}{2}x+1} 2x^2 y\, dy = x^2 y^2 \Big|_{y=1}^{\frac{1}{2}x+1} = \cdots = \frac{1}{4}x^4 + x^3 \quad .$$

Since $f(x, y) = 0$ for $8 < x < 10$ and all y in $(0, 6)$,

$$\psi(x) = \int_0^6 f(x, y)\, dy = \int_0^6 0\, dy = 0 \quad \text{for } 8 < x < 10 \quad .$$

Combining the above yields

$$\psi(x) = \begin{cases} \frac{1}{4}x^4 + x^3 & \text{if } 0 < x < 8 \\ 0 & \text{if } 8 < x < 10 \end{cases} \quad .$$

Notice that the jump in ψ is at $x = 8$ and that the line $x = 8$ intersects the boundary between $\mathcal{T}$ (where $f(x, y) = 2x^2 y$) and $\mathcal{R}_0$ (where $f(x, y) = 0$) at infinitely many points. Along this boundary, $f(x, y)$ is not unambiguously defined (should it be $2x^2 y$ or 0 ?). So all we have is

$$f(8, y) = \begin{cases} ? & \text{if } 1 < y < 5 \\ 0 & \text{otherwise} \end{cases} \quad ,$$

giving

$$\psi(8) = \int_0^6 f(8, y)\, dy = \int_1^5 ?\, dy = ? \quad .$$

Still, this isn't much of a problem. The formula obtained above for ψ elsewhere on $(0, 10)$ clearly shows that ψ is a well defined, piecewise continuous function on the interval. In fact, it's piecewise smooth, with

$$\psi'(x) = \begin{cases} x^3 + 3x^2 & \text{if } 0 < x < 8 \\ 0 & \text{if } 8 < x < 10 \end{cases} \quad .$$

?►**Exercise 7.2:** Let f be as in the previous example, and let

$$\phi(y) = \int_0^{10} f(x, y)\, dx \quad \text{for } 0 < y < 6 \quad .$$

Show that ϕ is the piecewise continuous function on $(0, 6)$ given by

$$\phi(y) = \begin{cases} \frac{16}{3}\left[65y - 3y^2 + 3y^3 - y^4 \right] & \text{if } 1 < y < 5 \\ 0 & \text{otherwise} \end{cases} \quad .$$

Now consider the general case where $f(x, y)$ is some piecewise continuous function on a bounded rectangle $\mathcal{R} = (a, b) \times (c, d)$, and

$$\psi(x) = \int_c^d f(x, y)\,dy \quad .$$

As the above example and exercise illustrate, the integral defining $\psi(x_0)$ is certainly well defined for a given x_0 in (a, b) so long as the line $x = x_0$ contains only a finite number of points in $\mathcal{R}$ at which $f(x, y)$ is not continuous. However, because $f(x, y)$ is merely piecewise continuous on $\mathcal{R}$, there may be curves along which $f(x, y)$ is not continuous. If one of these curves intersects the line $x = x_0$ at an infinite number of points, then we have a problem defining

$$\psi(x_0) = \int_c^d f(x_0, y)\,dy \quad .$$

However, this did not turn out to be much of a problem in the example because the one point at which that ψ was not well defined was the only point in $(0, 5)$ where that ψ was not continuous.

For convenience, let's define a *line of discontinuity* for $f(x, y)$ (over a region $\mathcal{R}$) to be any straight line in the plane that contains an infinite number of points in $\mathcal{R}$ at which $f(x, y)$ is discontinuous.

!▶ **Example 7.3:**   *For the function $f(x, y)$ defined above in example 7.2, the lines of discontinuity over $(0, 10) \times (0, 6)$ are the lines*

$$x = 8 \quad , \quad y = 1 \quad \text{and} \quad y = \frac{1}{2}x + 1 \quad .$$

Continuity of Functions Defined by Integrals

From our example, it seems reasonable to expect

$$\psi(x) = \int_c^d f(x, y)\,dy$$

to be a piecewise continuous function on (a, b) as long as $f(x, y)$ is piecewise continuous with only a finite number of lines of discontinuity over $\mathcal{R} = (a, b) \times (c, d)$. Furthermore, if $\psi(x)$ is discontinuous at a point $x = x_0$, then we should expect the vertical line $x = x_0$ to be one of those lines of discontinuity for f.

Confirming these expectations is usually fairly simple when given a particular choice for f (as in our example and exercise above). Confirming that we can trust our expectations to hold for every possible f of interest is less easy and will be relegated to an addendum at the end of this chapter (starting on page 84). Part of the difficulty is that some of the arguments available to us depend on the geometry of the curves along which f is discontinuous. In our addendum, we will mainly consider the case where all the discontinuities of f in $\mathcal{R}$ are contained in a finite collection of straight lines. The resulting lemma is given below. Fortunately, it (or its corollary) is exactly what will be needed several times in future discussions.

Theorem 7.7

Let $f(x, y)$ be a piecewise continuous function on a bounded rectangle $(a, b) \times (c, d)$, and assume all the discontinuities of f in this rectangle are contained in a finite number of straight lines. Then

$$\psi(x) = \int_c^d f(x, y)\,dy$$

is a piecewise continuous function on (a, b). Moreover, if $a < \bar{x} < b$ and $x = \bar{x}$ is not a line of discontinuity for f, then ψ is continuous at $\bar{x}$ and

$$\lim_{x \to \bar{x}} \psi(x) = \int_c^d \lim_{x \to \bar{x}} f(x, y) \, dy = \int_c^d f(\bar{x}, y) \, dy \quad .$$

As an immediate corollary, we have:

Corollary 7.8

Let $f(x, y)$ be a piecewise continuous function on a bounded rectangle $(a, b) \times (c, d)$, and assume that all the discontinuities of f in this rectangle are contained in a finite number of straight lines on the plane. If none of these lines of discontinuity are of the form $x = \text{constant}$, then

$$\psi(x) = \int_c^d f(x, y) \, dy$$

is uniformly continuous on (a, b).

Let me mention two things regarding the results just described:

1. In the above theorem and corollary, we required all the discontinuities of f to be contained in a finite number of straight lines. That will suffice for our needs, and it simplifies the proofs in the addendum. In fact, though, it will be pretty obvious from the discussion in the addendum that

$$\psi(x) = \int_c^d f(x, y) \, dy$$

   is piecewise continuous on (a, b) whenever f is a "reasonable" piecewise continuous function on $(a, b) \times (c, d)$ with only a finite number of vertical lines of discontinuity. Crudely speaking, if you can draw all the curves along which f is not continuous, then you are very likely to be able to show that the corresponding ψ is piecewise continuous. Moreover, if none of these curves contain any nontrivial vertical segments, then you should also be able to show that ψ is continuous.

2. On the other hand, the requirement that f be piecewise continuous (i.e., *uniformly* continuous on subregions of $(a, b) \times (c, d)$) is vital in the above theorem and corollary. You cannot derive same sort of results for ψ simply by assuming f is merely continuous on $(a, b) \times (c, d)$. In fact, it's not too difficult to construct a function $f(x, y)$ continuous on a given $(a, b) \times (c, d)$ such that, for some $\bar{x}$ in (a, b),

$$\lim_{x \to \bar{x}} \psi(x) \neq \int_c^d \lim_{x \to \bar{x}} f(x, y) \, dy \quad .$$

   One example is given in exercise 7.8 at the end of this chapter.

Differentiating Functions Defined by Integrals

Again, let

$$\psi(x) = \int_c^d f(x, y) \, dy$$

where (c, d) is a finite interval, and consider computing the derivative of such a function,

$$\psi'(x) = \frac{d}{dx} \int_c^d f(x, y) \, dy \quad .$$

The naive approach would be to just "bring the derivative into the integral" (changing it to a corresponding partial derivative since the integrand is a function of two variables),

$$\frac{d}{dx} \int_c^d f(x, y)\, dy \;=\; \int_c^d \frac{\partial}{\partial x} f(x, y)\, dy \quad .$$

However, as you can easily verify in the next exercise, this naive approach can lead to serious errors.

?►Exercise 7.3: *Let $f(x, y)$ be as in example 7.2 on page 78. Verify that*

$$\frac{d}{dx} \int_0^6 f(x, y)\, dy \;\neq\; \int_0^6 \frac{\partial}{\partial x} f(x, y)\, dy \quad .$$

Using the results from the previous subsection, we can determine conditions under which the naive approach can be safely applied. The result is the next theorem and corollary, which, again, will be just what we will need at various points later on.

Theorem 7.9
Let f be a piecewise continuous function on some bounded rectangle $\mathcal{R} = (a, b) \times (c, d)$, and assume that both of the following hold:

1. *All the discontinuities in $\mathcal{R}$ of f are contained in a finite number of horizontal straight lines (i.e., lines of the form $y = constant$).*

2. *$\partial f / \partial x$ is also a well-defined, piecewise continuous function on $\mathcal{R}$ with all of its discontinuities in $\mathcal{R}$ contained in a finite collection of straight lines, none of which are of the form $x = constant$.*

Then

$$\psi(x) \;=\; \int_c^d f(x, y)\, dy$$

is differentiable and has a uniformly continuous derivative on (a, b). Furthermore, on this interval,

$$\psi'(x) \;=\; \frac{d}{dx} \int_c^d f(x, y)\, dy \;=\; \int_c^d \frac{\partial}{\partial x} f(x, y)\, dy \quad .$$

Details of the above theorem's proof will be discussed in the addendum.

7.3   Double Integrals

Extending the notion of a single integral to that of a double integral (and other multiple integrals) is straightforward and is discussed in any reasonable elementary calculus sequence. I'll assume it's clear that, if $f(x, y)$ is any piecewise continuous function on a bounded region $\mathcal{R}$, then the double integrals

$$\iint_{\mathcal{R}} f(x, y)\, dA \qquad \text{and} \qquad \iint_{\mathcal{R}} |f(x, y)|\, dA$$

are well defined, with the second giving the total volume of the solid region over $\mathcal{R}$ between the XY–plane and the "surface" $z = |f(x, y)|$. Moreover,

$$\left| \iint_{\mathcal{R}} f(x, y) \, dA \right| \le \iint_{\mathcal{R}} |f(x, y)| \, dA \quad .$$

Recall also, that if the region is a rectangle, say, $\mathcal{R} = (a, b) \times (c, d)$, then we actually have three corresponding double integrals:

$$\iint_{\mathcal{R}} f(x, y) \, dA \quad , \quad \int_c^d \int_a^b f(x, y) \, dx \, dy \quad \text{and} \quad \int_a^b \int_c^d f(x, y) \, dy \, dx \quad .$$

Strictly speaking, these three double integrals represent three different things:

1. The first denotes *"the" double integral of f over $\mathcal{R}$* (i.e., the "net volume" under the surface $z = f(x, y)$ if f is real valued).

2. The second tells us to *first integrate with respect to x to get the formula for*

$$\phi(y) = \int_a^b f(x, y) \, dx \quad ,$$

   and then compute

$$\int_c^d \phi(y) \, dy \quad .$$

3. The third says to *first integrate with respect to y to get the formula for*

$$\psi(x) = \int_c^d f(x, y) \, dy \quad ,$$

   and then compute

$$\int_a^b \psi(x) \, dx \quad .$$

In practice, the distinction between these three double integrals is usually ignored because of the following well-known theorem.

Theorem 7.10
Let $f(x, y)$ *be a piecewise continuous function on a bounded rectangle* $\mathcal{R} = (a, b) \times (c, d)$. *Then*

$$\int_c^d \int_a^b f(x, y) \, dx \, dy = \iint_{\mathcal{R}} f(x, y) \, dA = \int_a^b \int_c^d f(x, y) \, dy \, dx$$

provided the integrals

$$\int_a^b f(x, y) \, dx \quad \text{and} \quad \int_c^d f(x, y) \, dy$$

define piecewise continuous functions on (c, d) *and* (a, b), *respectively.*

You may not recall the ending proviso of the two single integrals defining piecewise continuous functions. It's a technicality omitted in most elementary discussions because, in practice, you almost never encounter a case where this requirement is not satisfied. Still, counterexamples do exist (see exercise 7.9 at the end of this chapter), so we will include this requirement simply to ensure that these single integrals can, themselves, be integrated using the standard elementary theories of integration.[1]

?▶ Exercise 7.4:   *Verify the above theorem using Riemann sums. Note where you used the requirement that "integrals ... define piecewise continuous functions ..."*

Combining the above result with theorem 7.7 on the continuity of certain integrals from the previous section gives us the next theorem.

Theorem 7.11
Let $f(x, y)$ be a piecewise continuous function on a bounded rectangle $\mathcal{R} = (a, b) \times (c, d)$, and assume all the discontinuities of f in $\mathcal{R}$ are contained in a finite number of straight lines on the plane. Then the integrals

$$\int_a^b f(x, y)\, dx \qquad and \qquad \int_c^d f(x, y)\, dy$$

define piecewise continuous functions on (c, d) and (a, b), respectively, and

$$\int_c^d \int_a^b f(x, y)\, dx\, dy \;=\; \iint_{\mathcal{R}} f(x, y)\, dA \;=\; \int_a^b \int_c^d f(x, y)\, dy\, dx \quad .$$

7.4   Addenda: Proving Theorems 7.7 and 7.9
Proving Theorem 7.7 on Continuity

Some of the more significant ideas behind the proof of theorem 7.7 can be found in the proof of the first lemma below.

Lemma 7.12
Let $a \leq x_1 < x_2 < b$ and $c \leq y_1 < y_2 \leq d$, and consider the right triangle with vertices (x_1, y_1), (x_2, y_1) and (x_2, y_2). Let $\mathcal{T}$ denote the region inside the triangle, and let $\mathcal{R}_0$ be the subregion of $(a, b) \times (c, d)$ outside the triangle (see figure 7.2). Assume $f(x, y)$ is a function on $(a, b) \times (c, d)$ satisfying both of the following:

1. *f is uniformly continuous on $\mathcal{T}$.*

2. *$f(x, y) = 0$ for every (x, y) in $\mathcal{R}_0$.*

Then

$$\psi(x) \;=\; \int_c^d f(x, y)\, dy$$

is a piecewise continuous function on (a, b). Moreover, if $\bar{x}$ is any point in (a, b) other than x_2,

$$\lim_{x \to \bar{x}} \psi(x) \;=\; \int_c^d f(\bar{x}, y)\, dy \quad . \tag{7.3}$$

---

[1] The readers who are acquainted with Lebesgue's definition of the integral and Fubini's theorem, however, should know how to relax this requirement.

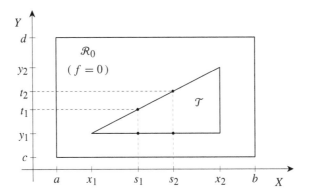

Figure 7.2: Figure for lemma 7.12.

The hard part of proving this lemma is showing that ψ is uniformly continuous on the interval (x_1, x_2). We will prove that part, leaving the rest as an exercise.

PROOF (uniform continuity of ψ on (x_1, x_2)): According to lemma 3.3 on page 20, it suffices to verify that, for any $\epsilon > 0$, there is a corresponding $\Delta x > 0$ such that

$$|\psi(s_2) - \psi(s_1)| < \epsilon$$

whenever s_1 and s_2 are two points in (x_1, x_2) with $|s_2 - s_1| < \Delta x$.

So let s_1 and s_2 be two points in (x_1, x_2). For convenience, assume these two points are labeled so that $s_1 \le s_2$, and let t_1 and t_2 be the values such that (s_1, t_1) and (s_2, t_2) are points on the hypotenuse of the triangle illustrated in figure 7.2. Since f vanishes outside the triangle,

$$
\begin{aligned}
|\psi(s_2) - \psi(s_1)| &= \left| \int_{y_0}^{t_2} f(s_2, y)\, dy - \int_{y_0}^{t_1} f(s_1, y)\, dy \right| \\
&= \left| \int_{y_0}^{t_1} f(s_2, y)\, dy + \int_{t_1}^{t_2} f(s_2, y)\, dy - \int_{y_0}^{t_1} f(s_1, y)\, dy \right| \\
&= \left| \int_{y_0}^{t_1} [f(s_2, y) - f(s_1, y)]\, dy + \int_{t_1}^{t_2} f(s_2, y)\, dy \right| \\
&\le \int_{y_0}^{t_1} |f(s_2, y) - f(s_1, y)|\, dy + \int_{t_1}^{t_2} |f(s_2, y)|\, dy \quad . \qquad (7.4)
\end{aligned}
$$

Now let ϵ be some fixed positive value, and, for reasons soon to be obvious, let

$$\epsilon_1 = \frac{\epsilon}{2(y_2 - y_1)} \quad .$$

As noted in lemma 7.3 on page 74, because f is uniformly continuous on $\mathcal{T}$, there is a $\delta_1 > 0$ such that

$$\left| f(s, t) - f(\bar{s}, \bar{t}) \right| < \epsilon_1 \qquad \text{whenever} \qquad \left| (s, t) - (\bar{s}, \bar{t}) \right| < \delta_1 \quad .$$

Obviously, though, if $|s_2 - s_1| < \delta_1$, then

$$|(s_2, y) - (s_1, y)| < \delta_1 \quad ,$$

and thus,

$$\int_{y_0}^{t_1} |f(s_2, y) - f(s_1, y)| \, dy \; < \; \int_{y_0}^{t_1} \epsilon_1 \, dy \; = \; \epsilon_1(t_1 - y_0) \; < \; \epsilon_1(y_1 - y_0) \quad,$$

which, by our choice of ϵ_1, reduces to

$$\int_{y_0}^{t_1} |f(s_2, y) - f(s_1, y)| \, dy \; < \; \frac{\epsilon}{2} \quad. \tag{7.5}$$

Also remember that any uniformly continuous function on a finite region is a bounded function. That is, for some finite number B,

$$|f(x, y)| \; < \; B \qquad \text{for all } (x, y) \text{ in } \mathcal{T} \quad.$$

Letting m be the slope of the line containing (s_1, t_1) and (s_2, t_2), we find that

$$t_2 - t_1 \; = \; m(s_2 - s_1)$$

and

$$\int_{t_1}^{t_2} |f(s_2, y)| \, dy \; < \; \int_{t_1}^{t_2} B \, dy \; = \; B(t_2 - t_1) \; = \; Bm(s_2 - s_1) \quad.$$

Consequently, choosing

$$\delta_2 \; = \; \frac{\epsilon}{2Bm} \quad,$$

we have

$$\int_{t_1}^{t_2} |f(s_2, y)| \, dy \; < \; \frac{\epsilon}{2} \qquad \text{whenever} \quad |s_2 - s_1| \; < \; \delta_2 \quad. \tag{7.6}$$

Finally, choose Δx to be the smaller of δ_1 and δ_2, and observe that inequalities (7.4), (7.5) and (7.6) all hold whenever $|s_2 - s_1| < \Delta x$. Combining them gives

$$|\psi(s_2) - \psi(s_1)| \; < \; \frac{\epsilon}{2} + \frac{\epsilon}{2} \; = \; \epsilon \tag{7.7}$$

whenever $|s_2 - s_1| < \Delta x$, verifying the uniform continuity of ψ on (x_1, x_2).   ∎

To complete the proof, complete the next exercise.

?▶Exercise 7.5:   Let f, ψ, x_1, x_2 and $\mathcal{T}$ be as in lemma 7.12.

   a: *From the derivation of inequality (7.7), show that*

$$\lim_{x \to \bar{x}} \psi(x) \; = \; \int_c^d f(\bar{x}, y) \, dy \qquad \text{whenever} \quad x_1 < \bar{x} < x_2 \quad.$$

   b: *Using the boundedness of f on $\mathcal{T}$, show that*

$$\lim_{x \to x_1^+} \psi(x) \; = \; 0 \quad.$$

   c: *Finish verifying equation (7.3) when $a < \bar{x} \leq x_1$ and $x_2 < \bar{x} < b$.*

   d: *Finish verifying that ψ is piecewise continuous on (a, b).*

Several things should be obvious if you followed the above proof. First of all, the triangular region $\mathcal{T}$ was chosen mainly so we could present a relatively simple proof illustrating some basic ideas. Using similar arguments, we can prove the following lemma in which f is uniformly continuous on the region inside any given polygon in $\mathcal{R}$.

Lemma 7.13
Let (x_1, y_1), (x_2, y_2), ... and (x_N, y_N) be the vertices of some polygon with all the x's in the closed interval $[a, b]$ and all the y's in the closed interval $[c, d]$. Let $\mathcal{R}_1$ be the region inside this polygon, and let $\mathcal{R}_0$ be the set of all points in $(a, b) \times (c, d)$ located outside this polygon. Assume $f(x, y)$ is a function on $(a, b) \times (c, d)$ satisfying the following:

1. *f is uniformly continuous on $\mathcal{R}_1$.*

2. *$f(x, y) = 0$ for every (x, y) in $\mathcal{R}_0$.*

Then

$$\psi(x) = \int_c^d f(x, y)\, dy$$

is a piecewise continuous function on (a, b). Moreover, if $a < \bar{x} < b$ and $x = \bar{x}$ is not a line of discontinuity for f (i.e., no side of the polygon is contained in the line $x = \bar{x}$), then ψ is continuous at $\bar{x}$ and

$$\lim_{x \to \bar{x}} \psi(x) = \int_c^d f(\bar{x}, y)\, dy \quad .$$

Proving theorem 7.7 is now simple. Remember, that was the theorem stating:

Let $f(x, y)$ be a piecewise continuous function on a bounded rectangle $(a, b) \times (c, d)$, and assume all the discontinuities of f in this rectangle are contained in a finite number of straight lines. Then

$$\psi(x) = \int_c^d f(x, y)\, dy$$

is a piecewise continuous functions on (a, b). Moreover, if $a < \bar{x} < b$ and $x = \bar{x}$ is not a line of discontinuity for f, then ψ is continuous at $\bar{x}$ and

$$\lim_{x \to \bar{x}} \psi(x) = \int_c^d \lim_{x \to \bar{x}} f(x, y)\, dy = \int_c^d f(\bar{x}, y)\, dy \quad .$$

PROOF *(of theorem 7.7, briefly):*   Since all the discontinuities of f in $\mathcal{R} = (a, b) \times (c, d)$ are contained in a finite set of straight lines, these lines partition $\mathcal{R}$ into a finite collection of polygonal regions $\mathcal{R}_1$, $\mathcal{R}_2$, ... and $\mathcal{R}_M$ over each of which f is uniformly continuous. With this partitioning,

$$f = \sum_{m=1}^{M} f_m \quad \text{where} \quad f_m = \begin{cases} f(x, y) & \text{if } (x, y) \text{ is in } \mathcal{R}_m \\ 0 & \text{otherwise} \end{cases} \quad .$$

For each of these f_m's, lemma 7.13 applies and assures us that

$$\psi_m(x) = \int_c^d f_m(x, y)\, dy$$

is piecewise continuous and has discontinuities only at points corresponding to vertical lines of discontinuity for f. The claims of the theorem now immediately follow from the above and the fact that

$$\psi(x) \; = \; \int_c^d f(x, y)\, dy \; = \; \sum_{m=1}^M \int_c^d f_m(x, y)\, dy \; = \; \sum_{m=1}^M \psi_m(x) \quad . \qquad \blacksquare$$

The verification of the results described in this addendum were somewhat simplified by the fact that all the regions considered had boundaries consisting of straight line segments. It's not that difficult, however, to extend the ideas illustrated in the proof of lemma 7.13 so as to show the piecewise continuity of

$$\psi(x) \; = \; \int_a^b f(x, y)\, dx$$

when f is assumed to be uniformly continuous over particular "nonpolygonal" regions. To see this yourself, try doing the next exercise.

?▶**Exercise 7.6:**   Let $\mathcal{D}$ be the unit disk (i.e., $\mathcal{D}$ is the set of all (x, y) where $x^2 + y^2 < 1$), and let $\mathcal{R} = (a, b) \times (c, d)$ be any rectangle containing $\mathcal{D}$. Assume $f(x, y)$ is a piecewise continuous function on $\mathcal{R}$ satisfying both of the following:

   1. f is uniformly continuous on $\mathcal{D}$.

   2. $f(x, y) = 0$ whenever (x, y) is a point in $\mathcal{R}$ with $x^2 + y^2 > 1$.

Verify that

$$\psi(x) \; = \; \int_c^d f(x, y)\, dy$$

is a uniformly continuous function on (a, b).

Proof of Theorem 7.9 on Differentiating Integrals

The "hard part" of proving theorem 7.9 is in proving the next lemma.

Lemma 7.14

Let f be a uniformly continuous function on some bounded rectangle $\mathcal{R} = (a, b) \times (\bar{c}, \bar{d})$. Assume, further, that $\partial f/\partial x$ is a well-defined, piecewise continuous function on $\mathcal{R}$ with all of its discontinuities in $\mathcal{R}$ contained in a finite collection of straight lines, none of which are of the form $x = constant$. Then

$$\psi(x) \; = \; \int_c^d f(x, y)\, dy$$

is uniformly smooth on (a, b). Furthermore, on this interval,

$$\psi'(x) \; = \; \frac{d}{dx} \int_c^d f(x, y)\, dy \; = \; \int_c^d \frac{\partial}{\partial x} f(x, y)\, dy \quad .$$

PROOF: Let x_0 be any point in (a, b). The basic definitions give us

$$\psi'(x_0) = \lim_{\Delta x \to 0} \frac{\psi(x_0 + \Delta x) - \psi(x_0)}{\Delta x}$$

$$= \lim_{\Delta x \to 0} \frac{\int_c^d f(x_0 + \Delta x, y)\, dy - \int_c^d f(x_0, y)\, dy}{\Delta x}$$

$$= \lim_{\Delta x \to 0} \frac{1}{\Delta x} \int_c^d [f(x_0 + \Delta x, y) - f(x_0, y)]\, dy \quad . \tag{7.8}$$

So consider

$$\int_c^d [f(x_0 + \Delta x, y) - f(x_0, y)]\, dy$$

where Δx is any value small enough that the values $x_0 \pm |\Delta x|$ are also in the interval (a, b).
First of all, since f is continuous and $\partial f / \partial x$ is piecewise continuous,

$$f(x_0 + \Delta x, y) - f(x_0, y) = \int_{x_0}^{x_0 + \Delta x} \frac{\partial f}{\partial x}(x, y)\, dx$$

(see theorem 4.1 on page 39), and so,

$$\int_c^d [f(x_0 + \Delta x, y) - f(x_0, y)]\, dy = \int_c^d \int_{x_0}^{x_0 + \Delta x} \frac{\partial f}{\partial x}(x, y)\, dx\, dy \quad .$$

With the assumptions made on f and $\partial f / \partial x$, theorem 7.11 on page 84 assures us that the order of integration can be switched. Doing so, we get

$$\int_c^d [f(x_0 + \Delta x, y) - f(x_0, y)]\, dy = \int_{x_0}^{x_0 + \Delta x} \int_c^d \frac{\partial f}{\partial x}(x, y)\, dy\, dx$$

$$= \int_{x_0}^{x_0 + \Delta x} G(x)\, dx \quad ,$$

where, to simplify subsequent discussion, we've set

$$G(x) = \int_c^d \frac{\partial f}{\partial x}(x, y)\, dy \quad .$$

By corollary 7.8 we know $G(x)$ is a uniformly continuous function over the interval (a, b). This allows us to invoke the mean value theorem for integrals (theorem 4.3 on page 40) to conclude that

$$\int_c^d [f(x_0 + \Delta x, y) - f(x_0, y)]\, dy = \int_{x_0}^{x_0 + \Delta x} G(x)\, dx = G(\bar{x}_{\Delta x})\Delta x$$

for some $\bar{x}_{\Delta x}$ with $|x_0 - \bar{x}_{\Delta x}| \leq |\Delta x|$. Combining this with equation (7.8) gives

$$\psi'(x_0) = \lim_{\Delta x \to 0} \frac{1}{\Delta x} \int_c^d [f(x_0 + \Delta x, y) - f(x_0, y)]\, dy$$

$$= \lim_{\Delta x \to 0} \frac{1}{\Delta x} G(\bar{x}_{\Delta x})\Delta x$$

$$= \lim_{\Delta x \to 0} G(\bar{x}_{\Delta x}) \quad .$$

But G is continuous at x_0 and, clearly, $\bar{x}_{\Delta x} \to x_0$ as $\Delta x \to 0$. So, after recalling the definition of G, we find that

$$\psi'(x_0) = \lim_{\Delta x \to 0} G(\bar{x}_{\Delta x}) = G(x_0) = \int_c^d \frac{\partial f}{\partial x}(x_0, y)\, dy \quad ,$$

which, since x_0 is any point in (a, b) and G is uniformly continuous on (a, b), completes the proof. ∎

The only difference between the lemma just proven and theorem 7.9 on page 82 is that f is not assumed to be uniformly continuous on $(a, b) \times (c, d)$ in theorem 7.9 but is, instead, assumed to be piecewise continuous with all of its discontinuities in some finite set of horizontal lines, say, $y = y_1$, $y = y_2$, ... and $y = y_N$. But this means f is uniformly continuous on each rectangle $(a, b) \times (y_n, y_{n+1})$ (where $y_0 = c$, $y_{N+1} = d$ and $n = 0, 1, ..., N$). Consequently, for each of these n's, the above lemma assures us that

$$\psi_n(x) = \int_{y_n}^{y_{n+1}} f(x, y)\, dy$$

is uniformly smooth on (a, b) with

$$\psi_n'(x) = \int_{y_n}^{y_{n+1}} \frac{\partial f}{\partial x}(x, y)\, dy \quad .$$

From this it follows that

$$\psi = \sum_{n=0}^N \psi_n$$

is also uniformly smooth on (a, b) and, for each x in (a, b),

$$\psi'(x) = \sum_{n=0}^N \psi_n'(x) = \sum_{n=0}^N \int_{y_n}^{y_{n+1}} \frac{\partial f}{\partial x}(x, y)\, dy = \int_c^d \frac{\partial f}{\partial x}(x, y)\, dy \quad .$$

Checking back, we see that this proves theorem 7.9, as desired.

Additional Exercises

7.7. Let $f(x, y)$ be a uniformly continuous function on a bounded region $\mathcal{R}$. Use lemma 7.3 to verify that $f(x, y)$ is also uniformly continuous on any subregion $\mathcal{R}_0$ of $\mathcal{R}$.

7.8. For each (x, y) in the rectangle $(-1, 1) \times (0, 2)$, let

$$f(x, y) = \begin{cases} x^{-2}y & \text{if } 0 < y < |x| \\ x^{-2}(2\,|x| - y) & \text{if } |x| < y < 2\,|x| \\ 0 & \text{otherwise} \end{cases} \quad .$$

a. Sketch the graph of $f(x_0, y)$ as a function of y for an arbitrary nonzero value of x_0 in $(-1, 1)$.

b. Using your graph, verify that, for each nonzero value of x_0 in $(-1, 1)$,

$$\int_0^2 f(x_0, y)\, dy = 1 \quad.$$

c. Let y be any fixed point on the Y–axis of your graph and, by considering what happens to your graph as $x \to 0$, confirm that

$$\lim_{x \to 0} f(x, y) = 0 \quad.$$

d. Using the above results, show that

$$\lim_{x \to 0} \int_0^2 f(x, y)\, dy \neq \int_0^2 \lim_{x \to 0} f(x, y)\, dy \quad.$$

7.9. In the following, we will see one way to construct a piecewise continuous function $f(x, y)$ on $\mathcal{R} = (0, 1) \times (0, 1)$ such that

$$\psi(x) = \int_0^1 f(x, y)\, dy$$

is not piecewise continuous on $(0, 1)$.

a. For $n = 0, 1, 2, \ldots$, let $\mathbf{a}_n$ and $\mathbf{c}_n$ be the points on the plane

$$\mathbf{a}_n = \left(\frac{1}{2^n}, \frac{1}{2^n} \right) \qquad \text{and} \qquad \mathbf{c}_n = \left(\frac{1}{2^{n+1}}, \frac{1}{2^n} \right) \quad.$$

Using $\overline{\mathbf{ac}}$ to denote the straight line segment between points $\mathbf{a}$ and $\mathbf{c}$, let $\mathcal{C}$ be the curve in the plane from $(0, 0)$ to $(1, 1)$ consisting of all line segments of the form $\overline{\mathbf{a}_n \mathbf{c}_n}$ and $\overline{\mathbf{a}_n \mathbf{c}_{n+1}}$. Sketch this curve and note that it contains an infinite number of vertical line segments. Also, in your sketch, label as $\mathcal{R}_0$ and $\mathcal{R}_1$, respectively, the subregions of the rectangle $(0, 1) \times (0, 1)$ above and below $\mathcal{C}$.

b. Find the length of $\mathcal{C}$.

c. For $0 < x < 1$, let

$$\psi(x) = \int_0^1 f(x, y)\, dy$$

where

$$f(x, y) = \begin{cases} 0 & \text{if } (x, y) \text{ is in } \mathcal{R}_0 \\ 1 & \text{if } (x, y) \text{ is in } \mathcal{R}_1 \end{cases} \quad.$$

Sketch the graph of ψ, and convince yourself that ψ has an infinite number of discontinuities in $(0, 1)$ and, hence, is not piecewise continuous on $(0, 1)$.

d. Evaluate $\iint_{\mathcal{R}} f(x, y)\, dA$.

(Note: Because $\mathcal{C}$ contains infinitely many corners, we have violated the second agreement concerning "regions" (see page 74). So, strictly speaking, the above $f(x, y)$ is not piecewise continuous on $\mathcal{R}$. However, you can "round-off" the corners of $\mathcal{C}$ to get a smooth curve $\mathcal{C}_0$ of finite length but still containing infinitely many vertical and horizontal line segments. Using $\mathcal{C}_0$ instead of $\mathcal{C}$ to define $\mathcal{R}_0$, $\mathcal{R}_1$ and f, as above, then gives us a piecewise continuous $f(x, y)$ on $(0, 1) \times (0, 1)$ such that $\psi(x) = \int_0^1 f(x, y)\, dy$ is not piecewise continuous on $(0, 1)$.)

Part II

Fourier Series

8

Heuristic Derivation of the Fourier Series Formulas

Suppose we have a "reasonable" (whatever that means) periodic function f with period p. If Fourier's bold conjecture is true, then this function can be expressed as a (possibly infinite) linear combination of sines and cosines. Let us naively accept Fourier's bold conjecture as true and see if we can derive precise formulas for this linear combination. That is, we will assume there are ω's and corresponding constants A_ω's and B_ω's such that

$$f(t) = \overset{?}{\underset{\omega=?}{\sum}} A_\omega \cos(2\pi\omega t) + \overset{?}{\underset{\omega=?}{\sum}} B_\omega \sin(2\pi\omega t) \qquad \text{for all } t \text{ in } \mathbb{R} \quad . \tag{8.1}$$

Then we will derive (without too much concern for rigor) formulas for the ω's and corresponding A_ω's and B_ω's. Later, we'll investigate the validity of our naively derived formulas.

8.1 The Frequencies

Since f is periodic with period p, it seems reasonable to expect each term of expression (8.1) to also be periodic with period p. Assuming this, we must determine the values of ω such that each of these terms, $A_\omega \cos(2\pi\omega t)$ and $B_\omega \sin(2\pi\omega t)$, has period p.

Certainly, one possible value for ω is 0. After all, if $\omega = 0$, then $\cos(2\pi\omega t)$ and $\sin(2\pi\omega t)$ are the constant functions 1 and 0, which (trivially) have period p, no matter what p is. On the other hand (as noted in our review of the sine and cosine functions), when $\omega \neq 0$, the fundamental period of both $\cos(2\pi\omega t)$ and $\sin(2\pi\omega t)$ is $1/|\omega|$. Thus, for any of these functions to have period p, p must be an integral multiple of $1/|\omega|$. So each ω must satisfy either

$$\omega = 0 \quad \text{or} \quad p = \frac{k}{|\omega|} \quad \text{for any positive integer } k \quad .$$

Consequently, the possible values of ω are given by

$$\omega = \frac{k}{p} \quad \text{where} \quad k = 0, \pm 1, \pm 2, \pm 3, \dots \quad .$$

Using these values for the ω's, equation (8.1) becomes

$$f(t) = \sum_{k=-\infty}^{\infty} A_k \cos\left(\frac{2\pi k}{p}t\right) + \sum_{k=-\infty}^{\infty} B_k \sin\left(\frac{2\pi k}{p}t\right) \quad . \tag{8.2}$$

These last two summations can be further simplified. Observe that, because the cosine function is even and $\cos(0) = 1$,

$$\sum_{k=-\infty}^{\infty} A_k \cos\left(\frac{2\pi k}{p}t\right) = \ldots + A_{-2}\cos\left(\frac{2\pi(-2)}{p}t\right) + A_{-1}\cos\left(\frac{2\pi(-1)}{p}t\right) + A_0\cos\left(\frac{2\pi(0)}{p}t\right)$$

$$+ A_1\cos\left(\frac{2\pi(1)}{p}t\right) + A_2\cos\left(\frac{2\pi(2)}{p}t\right) + \ldots$$

$$= \ldots + A_{-2}\cos\left(\frac{2\pi(2)}{p}t\right) + A_{-1}\cos\left(\frac{2\pi(1)}{p}t\right) + A_0$$

$$+ A_1\cos\left(\frac{2\pi(1)}{p}t\right) + A_2\cos\left(\frac{2\pi(2)}{p}t\right) + \ldots$$

$$= A_0 + [A_{-1}+A_1]\cos\left(\frac{2\pi(1)}{p}t\right) + [A_{-2}+A_2]\cos\left(\frac{2\pi(2)}{p}t\right) + \ldots$$

$$= A_0 + \sum_{k=1}^{\infty} a_k \cos\left(\frac{2\pi k}{p}t\right)$$

where $a_k = A_{-k} + A_k$.

In a very similar manner, you can show

$$\sum_{k=-\infty}^{\infty} B_k \sin\left(\frac{2\pi k}{p}t\right) = \sum_{k=1}^{\infty} b_k \sin\left(\frac{2\pi k}{p}t\right) \qquad \text{where} \quad b_k = B_k - B_{-k} \quad.$$

?▶ Exercise 8.1:   *Verify the last statement.*

With these simplifications equation (8.2) reduces to

$$f(t) = A_0 + \sum_{k=1}^{\infty} a_k \cos\left(\frac{2\pi k}{p}t\right) + \sum_{k=1}^{\infty} b_k \sin\left(\frac{2\pi k}{p}t\right) \quad . \tag{8.3}$$

8.2 The Coefficients

Notice what happens when we integrate both sides of equation (8.3) over the interval $(0, p)$,

$$\int_{t=0}^{p} f(t)\,dt = \int_{t=0}^{p}\left[A_0 + \sum_{k=1}^{\infty} a_k \cos\left(\frac{2\pi k}{p}t\right) + \sum_{k=1}^{\infty} b_k \sin\left(\frac{2\pi k}{p}t\right)\right] dt \tag{8.4a}$$

$$= \int_{t=0}^{p} A_0\,dt + \int_{t=0}^{p}\sum_{k=1}^{\infty} a_k \cos\left(\frac{2\pi k}{p}t\right) dt + \int_{t=0}^{p}\sum_{k=1}^{\infty} b_k \sin\left(\frac{2\pi k}{p}t\right) dt \tag{8.4b}$$

$$= \int_{t=0}^{p} A_0 \, dt \;+\; \sum_{k=1}^{\infty} \int_{t=0}^{p} a_k \cos\!\left(\frac{2\pi k}{p}t\right) dt \;+\; \sum_{k=1}^{\infty} \int_{t=0}^{p} b_k \sin\!\left(\frac{2\pi k}{p}t\right) dt \qquad (8.4c)$$

$$= A_0 \int_{t=0}^{p} dt \;+\; \sum_{k=1}^{\infty} a_k \int_{t=0}^{p} \cos\!\left(\frac{2\pi k}{p}t\right) dt \;+\; \sum_{k=1}^{\infty} b_k \int_{t=0}^{p} \sin\!\left(\frac{2\pi k}{p}t\right) dt \quad . \qquad (8.4d)$$

The integrals in the last line are easily evaluated. All except one turn out to be zero. That exception is

$$\int_{t=0}^{p} dt \;=\; p \quad .$$

Thus, equation sequence (8.4) reduces to

$$\int_{t=0}^{p} f(t) \, dt \;=\; A_0 \cdot p \;+\; \sum_{k=1}^{\infty} a_k \cdot 0 \;+\; \sum_{k=1}^{\infty} b_k \cdot 0 \;=\; pA_0 \quad ,$$

from which it immediately follows that

$$A_0 \;=\; \frac{1}{p} \int_{t=0}^{p} f(t) \, dt \quad . \qquad (8.5)$$

Look at what we just did. We found a formula for A_0 by, first, integrating both sides of equation (8.3) over the interval $(0, p)$ and, then, noting that all but one of the terms in the resulting series vanished. They vanished because, for every nonzero integer k,

$$\int_{t=0}^{p} \sin\!\left(\frac{2\pi k}{p}t\right) dt \;=\; \int_{t=0}^{p} \cos\!\left(\frac{2\pi k}{p}t\right) dt \;=\; 0 \quad ,$$

which, you should note, is the same as equation (5.3a) in the orthogonality relations for sines and cosines (theorem 5.2 on page 54). This is significant because, using the other equations from that theorem, we can derive formulas for the other coefficients in a manner very similar to how we derived equation (8.5).

For example, to find a_3, the coefficient for the $\cos(2\pi\omega_3 t)$ term, multiply both sides of equation (8.3) by that cosine function and then integrate from 0 to p,

$$\int_{t=0}^{p} f(t) \cos\!\left(\frac{2\pi 3}{p}t\right) dt$$

$$= \int_{t=0}^{p} \left[A_0 \;+\; \sum_{k=1}^{\infty} a_k \cos\!\left(\frac{2\pi k}{p}t\right) \;+\; \sum_{k=1}^{\infty} b_k \sin\!\left(\frac{2\pi k}{p}t\right) \right] \cos\!\left(\frac{2\pi 3}{p}t\right) dt \qquad (8.6a)$$

$$= \int_{t=0}^{p} A_0 \cos\!\left(\frac{2\pi 3}{p}t\right) dt \;+\; \int_{t=0}^{p} \sum_{k=1}^{\infty} a_k \cos\!\left(\frac{2\pi k}{p}t\right) \cos\!\left(\frac{2\pi 3}{p}t\right) dt$$

$$+\; \int_{t=0}^{p} \sum_{k=1}^{\infty} b_k \sin\!\left(\frac{2\pi k}{p}t\right) \cos\!\left(\frac{2\pi 3}{p}t\right) dt \qquad (8.6b)$$

$$= \int_{t=0}^{p} A_0 \cos\!\left(\frac{2\pi 3}{p}t\right) dt \;+\; \sum_{k=1}^{\infty} \int_{t=0}^{p} a_k \cos\!\left(\frac{2\pi k}{p}t\right) \cos\!\left(\frac{2\pi 3}{p}t\right) dt$$

$$+\; \sum_{k=1}^{\infty} \int_{t=0}^{p} b_k \sin\!\left(\frac{2\pi k}{p}t\right) \cos\!\left(\frac{2\pi 3}{p}t\right) dt \qquad (8.6c)$$

$$= A_0 \int_{t=0}^{P} \cos\left(\frac{2\pi 3}{p}t\right) dt + \sum_{k=1}^{\infty} a_k \int_{t=0}^{P} \cos\left(\frac{2\pi k}{p}t\right) \cos\left(\frac{2\pi 3}{p}t\right) dt$$

$$+ \sum_{k=1}^{\infty} b_k \int_{t=0}^{P} \sin\left(\frac{2\pi k}{p}t\right) \cos\left(\frac{2\pi 3}{p}t\right) dt \quad .$$

(8.6d)

From theorem 5.2 (the orthogonality relations for sines and cosines), we know that

$$\int_{t=0}^{P} \cos\left(\frac{2\pi 3}{p}t\right) dt = 0$$

and that, for each positive integer k,

$$\int_{t=0}^{P} \cos\left(\frac{2\pi k}{p}t\right) \cos\left(\frac{2\pi 3}{p}t\right) dt = \begin{cases} 0 & \text{if } k \neq 3 \\ \frac{p}{2} & \text{if } k = 3 \end{cases}$$

and

$$\int_{t=0}^{P} \sin\left(\frac{2\pi k}{p}t\right) \cos\left(\frac{2\pi 3}{p}t\right) dt = 0 \quad .$$

Plugging these values back into line (8.6d) gives

$$\int_{t=0}^{P} f(t)\cos\left(\frac{2\pi 3}{p}t\right) dt = A_0 \cdot 0 + \sum_{k=1}^{\infty} a_k \begin{cases} 0 & \text{if } k \neq 3 \\ \frac{p}{2} & \text{if } k = 3 \end{cases} + \sum_{k=1}^{\infty} b_k \cdot 0 = a_3 \cdot \frac{p}{2} \quad .$$

Thus,

$$a_3 = \frac{2}{p} \int_{t=0}^{P} f(t)\cos\left(\frac{2\pi 3}{p}t\right) dt \quad .$$

Clearly, this derivation can be repeated for all the a_k's, yielding

$$a_k = \frac{2}{p} \int_{t=0}^{P} f(t)\cos\left(\frac{2\pi k}{p}t\right) dt \qquad \text{for} \quad k = 1, 2, 3, \ldots \quad .$$

It should also come as no surprise that very similar computations lead to

$$b_k = \frac{2}{p} \int_{t=0}^{P} f(t)\sin\left(\frac{2\pi k}{p}t\right) dt \qquad \text{for} \quad k = 1, 2, 3, \ldots \quad .$$

?► Exercise 8.2:   *Using the above derivation for a_3 as a guide, derive*

$$b_3 = \frac{2}{p} \int_{t=0}^{P} f(t)\sin\left(\frac{2\pi 3}{p}t\right) dt \quad .$$

8.3 Summary

Here is what we just derived: If f is a periodic function with period p, then,

$$f(t) = A_0 + \sum_{k=1}^{\infty} a_k \cos\left(\frac{2\pi k}{p}t\right) + \sum_{k=1}^{\infty} b_k \sin\left(\frac{2\pi k}{p}t\right) \tag{8.7a}$$

where

$$A_0 = \frac{1}{p} \int_{t=0}^{p} f(t)\,dt \tag{8.7b}$$

and, for $k = 1, 2, 3, \ldots$,

$$a_k = \frac{2}{p} \int_{t=0}^{p} f(t) \cos\left(\frac{2\pi k}{p}t\right) dt \tag{8.7c}$$

and

$$b_k = \frac{2}{p} \int_{t=0}^{p} f(t) \sin\left(\frac{2\pi k}{p}t\right) dt \quad . \tag{8.7d}$$

There were, however, a number of "holes" in our derivation. Let's note a few of them:

1.   The entire derivation was based on the assumption that Fourier's bold conjecture is true. How do we know this assumption is true and not just wishful thinking?

2.   At several points we interchanged the order in which an integration and a summation were performed. For example, in going from expression (8.4b) to expression (8.6c), we assumed

$$\int_{t=0}^{p} \sum_{k=1}^{\infty} a_k \cos\left(\frac{2\pi k}{p}t\right) dt = \sum_{k=1}^{\infty} \int_{t=0}^{p} a_k \cos\left(\frac{2\pi k}{p}t\right) dt \quad .$$

Unfortunately, while it is certainly true that an integral of a *finite* sum of functions equals the sum of the integrals of the individual functions, it is not always true that the integral of an *infinite* summation of functions equals the corresponding summation of the integrals of the individual functions (see exercise 8.3).

3.   Indeed, the fact we have infinite summations (i.e., infinite series) of functions should give us pause. How can we, at this point, be sure that any of these infinite series converges?

4.   Finally, since we made no assumptions regarding the integrability of f, we really cannot be sure that the integrals in the formulas we derived for the coefficients are well defined.

Because of these problems with our derivation, we cannot claim to have shown that f can be expressed as indicated by formulas (8.7). In fact, these formulas are valid for many periodic functions and are "essentially valid" for many others. But there are also functions for which these formulas yield nonsense (see exercise 8.4). Consequently, determining when f can be expressed as indicated by formula set (8.7) will be an important part of our future discussions.

Additional Exercises

8.3. *For each positive integer* k, *let*

$$f_k(t) = \begin{cases} k^2 & \text{if } 0 < t < \frac{1}{k} \\ 0 & \text{otherwise} \end{cases} .$$

Also, let $g_1(t) = f_1(t)$ *and, for* $k = 2, 3, 4, \ldots$, *let*

$$g_k(t) = f_k(t) - f_{k-1}(t) .$$

Show that

$$\int_{t=0}^{1} \left[\sum_{k=1}^{\infty} g_k(t) \right] dt \neq \sum_{k=1}^{\infty} \left[\int_{t=0}^{1} g_k(t)\, dt \right]$$

by showing that

$$\int_{t=0}^{1} \left[\sum_{k=1}^{\infty} g_k(t) \right] dt = 0 ,$$

while

$$\sum_{k=1}^{\infty} \left[\int_{t=0}^{1} g_k(t)\, dt \right] = \infty .$$

8.4. *Let* f *be the periodic function*

$$f(t) = \begin{cases} \dfrac{1}{t} & \text{if } 0 < t < 1 \\ f(t-1) & \text{in general} \end{cases} .$$

Show that A_0, *as defined by (8.7b), is infinite when computed using this function.*

8.5. *Recall the trigonometric identity*

$$\sin^2(t) = \frac{1}{2} - \frac{1}{2}\cos(2t) .$$

This means that $f(t) = \sin^2(t)$ *is a periodic function that can be expressed as a finite linear combination of sines and cosines. Thus, Fourier's conjecture is valid for this function. Compute* A_0 *and the* a_k*'s and* b_k*'s (as defined by formulas (8.7b) through (8.7d)) for* $f(t) = \sin^2(t)$ *and show that, in this case, equation (8.7a) reduces to the above trigonometric identity.*

9

The Trigonometric Fourier Series

In the previous chapter we obtained a set of formulas that we suspect will allow us to describe any "reasonable" periodic function as a (possibly infinite) linear combination of sines and cosines. Let us now see about actually *computing* with these formulas.

First, though, a little terminology and notation so that we can conveniently refer to this important set of formulas.

9.1 Defining the Trigonometric Fourier Series
Terminology and Notation

Let f be a periodic function with period p where p is some positive number. The *(trigonometric) Fourier series* for f is the infinite series

$$A_0 + \sum_{k=1}^{\infty} [a_k \cos(2\pi\omega_k t) + b_k \sin(2\pi\omega_k t)] \tag{9.1a}$$

where, for $k = 1, 2, 3, \dots$,

$$\omega_k = \frac{k}{p} \quad , \tag{9.1b}$$

$$A_0 = \frac{1}{p} \int_0^p f(t)\, dt \quad , \tag{9.1c}$$

$$a_k = \frac{2}{p} \int_0^p f(t) \cos(2\pi\omega_k t)\, dt \tag{9.1d}$$

and

$$b_k = \frac{2}{p} \int_0^p f(t) \sin(2\pi\omega_k t)\, dt \quad . \tag{9.1e}$$

The coefficients in expression (9.1a) (the A_0 and the a_k's and b_k's) are called the *(trigonometric) Fourier coefficients* for f. They are well defined as long as the integrals in formulas (9.1c), (9.1d) and (9.1e) are well defined. To ensure this we will usually assume f is at least piecewise continuous on $\mathbb{R}$ (see page 37).

For brevity, we will denote the Fourier series for f by $F.S.\,[f]|_t$. Let us agree that, whenever we encounter an expression like

$$F.S.\,[f]|_t \;=\; A_0 \;+\; \sum_{k=1}^{\infty} [a_k \cos(2\pi\,\omega_k t) + b_k \sin(2\pi\,\omega_k t)] \quad,$$

then it is understood that A_0 and the ω_k's, a_k's and b_k's are given by formulas (9.1b) through (9.1e). While we are at it, we should also note that formulas (9.1d) and (9.1e) could just as well have been written as

$$a_k \;=\; \frac{2}{p}\int_0^p f(t)\cos\!\left(\frac{2\pi k}{p}t\right) dt \qquad \text{and} \qquad b_k \;=\; \frac{2}{p}\int_0^p f(t)\sin\!\left(\frac{2\pi k}{p}t\right) dt \quad.$$

!►Example 9.1:   Let f be the saw function from example 5.1 on page 51,

$$f(t) \;=\; \mathrm{saw}_3(t) \;=\; \begin{cases} t & \text{if } \; 0 < t < 3 \\ f(t-3) & \text{in general} \end{cases}.$$

Here $p = 3$ and

$$\omega_k \;=\; \frac{k}{3} \qquad \text{for } \; k = 1,\,2,\,3,\,\ldots \quad.$$

Formula (9.1c) becomes

$$A_0 \;=\; \frac{1}{p}\int_0^p f(t)\,dt \;=\; \frac{1}{3}\int_0^3 t\,dt \;=\; \frac{1}{3}\left[\frac{1}{2}t^2\Big|_0^3\right] \;=\; \frac{3}{2} \quad.$$

Using formulas (9.1d) and (9.1e) (and integration by parts) we have

$$a_k \;=\; \frac{2}{p}\int_0^p f(t)\cos(2\pi\,\omega_k t)\,dt$$

$$=\; \frac{2}{3}\int_0^3 t\cos\!\left(\frac{2\pi k}{3}t\right) dt$$

$$=\; \frac{2}{3}\left[\frac{3t}{2\pi k}\sin\!\left(\frac{2\pi k}{3}t\right)\Big|_{t=0}^{3} \;-\; \frac{3}{2\pi k}\int_0^3 \sin\!\left(\frac{2\pi k}{3}t\right) dt\right]$$

$$=\; \frac{2}{3}\left[[0-0] \;+\; \left(\frac{3}{2\pi k}\right)^2 [\cos(2\pi k) - \cos(0)]\right] \;=\; 0 \quad,$$

while

$$b_k \;=\; \frac{2}{p}\int_0^p f(t)\sin(2\pi\,\omega_k t)\,dt$$

$$=\; \frac{2}{3}\int_0^3 t\sin\!\left(\frac{2\pi k}{3}t\right) dt$$

$$=\; \frac{2}{3}\left[\frac{-3t}{2\pi k}\cos\!\left(\frac{2\pi k}{3}t\right)\Big|_{t=0}^{3} \;+\; \frac{3}{2\pi k}\int_0^3 \cos\!\left(\frac{2\pi k}{3}t\right) dt\right]$$

$$=\; \frac{2}{3}\left[\left[\frac{-3\cdot 3}{2\pi k} - 0\right] \;+\; \left(\frac{3}{2\pi k}\right)^2 [\sin(2\pi k) - \sin(0)]\right] \;=\; -\frac{3}{k\pi} \quad.$$

The trigonometric Fourier series for $saw_3(t)$ is then obtained by plugging the above values into formula (9.1a),

$$F.S.\,[saw_3]|_t = A_0 + \sum_{k=1}^{\infty} [a_k \cos(2\pi\omega_k t) + b_k \sin(2\pi\omega_k t)]$$

$$= \frac{3}{2} + \sum_{k=1}^{\infty} \left[0 \cos\left(\frac{2\pi k}{3}t\right) - \frac{3}{k\pi} \sin\left(\frac{2\pi k}{3}t\right) \right]$$

$$= \frac{3}{2} - \sum_{k=1}^{\infty} \frac{3}{k\pi} \sin\left(\frac{2\pi k}{3}t\right)$$

$$= \frac{3}{2} - \frac{3}{1\pi} \sin\left(\frac{2\pi 1}{3}t\right) - \frac{3}{2\pi} \sin\left(\frac{2\pi 2}{3}t\right) - \frac{3}{3\pi} \sin\left(\frac{2\pi 3}{3}t\right) - \cdots \quad .$$

Formulas (9.1a) through (9.1e) are just the formulas naively derived in the previous chapter for expressing $f(t)$ as a linear combination of sines and cosines. In that derivation we *assumed*, but did *not* verify, that

$$f(t) = F.S.\,[f]|_t \quad .$$

So, until we prove this equality and determine which functions are reasonable, we can*not* say that we know this equality holds.[1]

?► Exercise 9.1: *Convince yourself that the first term in the trigonometric Fourier series for f, A_0 in (9.1), is the mean (or average) value of $f(t)$ over the interval $(0, p)$.*

Dependence on the Period

The formulas defining the trigonometric Fourier series all involve the period p. But any integral multiple of the fundamental period of a periodic function is a legitimate period for that function. Does this mean we have a different Fourier series for each possible period? The answer, fortunately, is no. The trigonometric Fourier series for a periodic function does not depend on the choice of periods used, even though the computations to find the Fourier series do depend on the actual period chosen. To illustrate this, let us redo example 9.1 using a different choice for p.

!► Example 9.2: *Again, let f be the "saw function" with fundamental period 3,*

$$f(t) = saw_3(t) = \begin{cases} t & \text{if } 0 < t < 3 \\ f(t-3) & \text{in general} \end{cases} \quad .$$

This time let $p = 6$, twice the fundamental period, and let

$$\widehat{A_0} + \sum_{n=1}^{\infty} \left[\widehat{a}_n \cos(2\pi\widehat{\omega}_n t) + \widehat{b}_n \sin(2\pi\widehat{\omega}_n t) \right] \qquad (9.2)$$

be the trigonometric Fourier series for f using $p = 6$. Here then,

$$\widehat{\omega}_n = \frac{n}{6} \qquad \text{for} \quad n = 1, 2, 3, \ldots \quad .$$

---

[1] Of course, if the equality didn't hold for many functions of interest, then this book would be much shorter.

Care must be taken with the computation of the coefficients. If $0 < t < 3$, then $f(t) = t$. On the other hand, if $3 < t < 6$, then, because of the periodicity of f and the fact that here $0 < t - 3 < 3$,

$$f(t) = f(t - 3) = t - 3 \quad .$$

So the integrals in formulas (9.1c), (9.1d) and (9.1e) must be split as follows:

$$\widehat{A}_0 = \frac{1}{6} \int_0^6 f(t)\, dt = \frac{1}{6} \left[\int_0^3 t\, dt + \int_3^6 (t - 3)\, dt \right] \quad , \tag{9.3a}$$

$$\widehat{a}_n = \frac{2}{6} \int_0^6 f(t) \cos\left(\frac{2\pi n}{6}t\right) dt$$

$$= \frac{2}{6} \left[\int_0^3 t \cos\left(\frac{2\pi n}{6}t\right) dt + \int_3^6 (t - 3) \cos\left(\frac{2\pi n}{6}t\right) dt \right] \tag{9.3b}$$

and

$$\widehat{b}_n = \frac{2}{6} \int_0^6 f(t) \sin\left(\frac{2\pi n}{6}t\right) dt$$

$$= \frac{2}{6} \left[\int_0^3 t \sin\left(\frac{2\pi n}{6}t\right) dt + \int_3^6 (t - 3) \sin\left(\frac{2\pi n}{6}t\right) dt \right] \quad . \tag{9.3c}$$

The integrals over $(3, 6)$ can be related to the integrals over $(0, 3)$ through the substitution $\tau = t - 3$ and well-known trigonometric identities,

$$\int_3^6 (t - 3)\, dt = \int_{\tau=0}^3 \tau\, d\tau \quad ,$$

$$\int_3^6 (t - 3) \cos\left(\frac{2\pi n}{6}t\right) dt = \int_{\tau=0}^3 \tau \cos\left(\frac{2\pi n}{6}(\tau + 3)\right) d\tau$$

$$= \int_{\tau=0}^3 \tau \cos\left(\frac{2\pi n}{6}\tau + n\pi\right) d\tau$$

$$= (-1)^n \int_{\tau=0}^3 \tau \cos\left(\frac{2\pi n}{6}\tau\right) d\tau$$

and

$$\int_3^6 (t - 3) \sin\left(\frac{2\pi n}{6}t\right) dt = \int_{\tau=0}^3 \tau \sin\left(\frac{2\pi n}{6}(\tau + 3)\right) d\tau$$

$$= \int_{\tau=0}^3 \tau \sin\left(\frac{2\pi n}{6}\tau + n\pi\right) d\tau$$

$$= (-1)^n \int_{\tau=0}^3 \tau \sin\left(\frac{2\pi n}{6}\tau\right) d\tau \quad .$$

Replacing the τ *with* t *and inserting the above back into equations (9.3a), (9.3b) and (9.3c) gives*

$$\widehat{A}_0 = \frac{1}{6}\left[\int_0^3 t\, dt + \int_0^3 t\, dt\right] = \frac{2}{6}\int_0^3 t\, dt \quad,$$

$$\widehat{a}_n = \frac{2}{6}\left[\int_0^3 t\cos\left(\frac{2\pi n}{6}t\right) dt + (-1)^n \int_0^3 t\cos\left(\frac{2\pi n}{6}t\right) dt\right]$$

$$= \left[1 + (-1)^n\right]\frac{2}{6}\int_0^3 t\cos\left(\frac{2\pi n}{6}t\right) dt$$

and

$$\widehat{b}_n = \frac{2}{6}\left[\int_0^3 t\sin\left(\frac{2\pi n}{6}t\right) dt + (-1)^n \int_0^3 t\sin\left(\frac{2\pi n}{6}t\right) dt\right]$$

$$= \left[1 + (-1)^n\right]\frac{2}{6}\int_0^3 t\sin\left(\frac{2\pi n}{6}t\right) dt \quad.$$

These integrals are very similar to those evaluated in example 9.1. Skipping the details of the computations, we find that

$$\widehat{A}_0 = \frac{3}{2} \quad,$$

$$\widehat{a}_n = 0$$

and

$$\widehat{b}_n = \left[1 + (-1)^n\right]\left[-\frac{3}{n\pi}(-1)^n\right] = \begin{cases} -\dfrac{2\cdot 3}{n\pi} & \text{if } n \text{ is even} \\ 0 & \text{if } n \text{ is odd} \end{cases} \quad.$$

Thus, using $p = 6$,

$$F.S.\,[saw_3]|_t = \widehat{A}_0 + \sum_{n=1}^{\infty}\left[\widehat{a}_n \cos(2\pi\widehat{\omega}_n t) + \widehat{b}_n \sin(2\pi\widehat{\omega}_n t)\right]$$

$$= \frac{3}{2} + \sum_{n=1}^{\infty}\left[0\cdot\cos\left(\frac{2\pi n}{6}t\right) + \begin{cases} -\dfrac{2\cdot 3}{n\pi} & \text{if } n \text{ is even} \\ 0 & \text{if } n \text{ is odd} \end{cases}\sin\left(\frac{2\pi n}{6}t\right)\right]$$

$$= \frac{3}{2} - \sum_{\substack{n=1 \\ n \text{ is even}}}^{\infty} \frac{2\cdot 3}{n\pi}\sin\left(\frac{2\pi n}{6}t\right) \quad.$$

Since the last summation only involves even values of n *, we can simplify it using the substitution* $n = 2k$ *with* $k = 1, 2, 3, \ldots$ *to obtain*

$$F.S.\,[saw_3]|_t = \frac{3}{2} - \sum_{k=1}^{\infty} \frac{2\cdot 3}{2k\pi}\sin\left(\frac{2\pi 2k}{6}t\right)$$

$$= \frac{3}{2} - \frac{3}{1\pi}\sin\left(\frac{2\pi 1}{3}t\right) - \frac{3}{2\pi}\sin\left(\frac{2\pi 2}{3}t\right) - \frac{3}{3\pi}\sin\left(\frac{2\pi 3}{3}t\right) - \cdots \quad,$$

which is exactly the same series as obtained in example 9.1.

What happened above happens in general. No matter what period you use in computing the trigonometric Fourier series, once you simplify your results, you will find that you have the same series you would have obtained using the fundamental period.[2] Because this is an important fact, we'll state it as a theorem.

Theorem 9.1
Suppose f is a periodic, piecewise continuous function other than a constant function. Let p_1 be the fundamental period for f and let $\widehat{p}$ be any other period for f. Then the trigonometric Fourier series for f computed using formulas (9.1a) through (9.1e) with $p = \widehat{p}$ is identical, after simplification, to the corresponding Fourier series computed using $p = p_1$.

The proof will be left as an exercise.

?▶ Exercise 9.2:   *Prove theorem 9.1*

    a: *with the added assumption that $\widehat{p}$ is twice the fundamental period. (Try redoing example 9.2 using an arbitrary periodic, piecewise continuous f instead of $f(t) = \text{saw}_3(t)$.)*

    b: *assuming $\widehat{p}$ is any integral multiple of the fundamental period.*

A Minor Notational Issue

It is standard practice to use the same symbol for the variable in the formula for f, $f(t)$, as for the variable in the Fourier series for f, $F.S.[f]|_t$. We, for example, have been using t for both variables. This should seem reasonable since we anticipate being able to show that periodic functions can be represented by their Fourier series. However, it can lead to somewhat awkward notation. If we replace the "f" in formula (9.1a) with "$f(t)$" we have

$$F.S.[f(t)]|_t = A_0 + \sum_{k=1}^{\infty} [a_k \cos(2\pi \omega_k t) + b_k \sin(2\pi \omega_k t)] \quad . \tag{9.4}$$

The problem is that the symbol t is now being used for two completely different variables in the same equation.[3] In the "$[f(t)]$", t is a dummy variable (i.e., an internal variable) helping to describe the function for which the right-hand side is the Fourier series. Elsewhere in equation (9.4), t denotes a true variable that can be assigned specific values. Using t for these two different types of variables is not necessarily wrong, but it can be a little confusing to the unwary. Just remember, "letting $t = 3$ in expression (9.4)" means

$$F.S.[f(t)]|_3 = A_0 + \sum_{k=1}^{\infty} [a_k \cos(2\pi \omega_k 3) + b_k \sin(2\pi \omega_k 3)]$$

and not

$$F.S.[f(3)]|_3 = A_0 + \sum_{k=1}^{\infty} [a_k \cos(2\pi \omega_k 3) + b_k \sin(2\pi \omega_k 3)] \quad .$$

---

[2] Remember, the only periodic, piecewise continuous functions without fundamental periods are constant functions (see exercise 5.12 on page 56). Constant functions will be discussed later in this chapter.

[3] This may be a good time to re-review the discussion of variables, functions and operators in chapter 2.

9.2 Computing the Fourier Coefficients

As example 9.1 shows, the process of computing the Fourier coefficients for a given function is a fairly straightforward process (provided the integrals are relatively simple). Even so, it can still be a fairly tedious process, especially when the integrals are not so simple. Let us look at a few ways to reduce the amount of work required to compute these coefficients.

In all of the following discussion, f denotes some periodic and piecewise continuous function on $\mathbb{R}$ with period p and with

$$F.S.[f]|_t \; = \; A_0 \; + \; \sum_{k=1}^{\infty} [a_k \cos(2\pi \omega_k t) + b_k \sin(2\pi \omega_k t)] \quad .$$

Very Simple Cases

The computation of the Fourier series for a constant function, a sine function, or a cosine function is particularly easy, especially if we remember the orthogonality relations for sines and cosines (theorem 5.2 on page 54).

If f is a constant function,

$$f(t) \; = \; c \quad \text{(a constant)} \qquad \text{for all} \quad t \quad ,$$

then f is automatically a periodic function with period p for any positive value of p. Computing formula (9.1c) gives

$$A_0 \; = \; \frac{1}{p} \int_0^p f(t)\,dt \; = \; \frac{1}{p} \int_0^p c\,dt \; = \; c \quad .$$

For $k = 1, 2, 3, \ldots$, formulas (9.1d) and (9.1e) are also easy to compute. Or we can use orthogonality relation (5.3a). Either way we get

$$a_k \; = \; \frac{2}{p} \int_0^p f(t) \cos(2\pi \omega_k t)\,dt \; = \; \frac{2}{p} \int_0^p c \cdot \cos\left(\frac{2\pi k}{p} t\right)\,dt \; = \; 0$$

and

$$b_k \; = \; \frac{2}{p} \int_0^p f(t) \sin(2\pi \omega_k t)\,dt \; = \; \frac{2}{p} \int_0^p c \cdot \sin\left(\frac{2\pi k}{p} t\right)\,dt \; = \; 0 \quad .$$

Thus, if $f(t) = c$ for all t, then

$$F.S.[f]|_t \; = \; A_0 \; + \; \sum_{k=1}^{\infty} [a_k \cos(2\pi \omega_k t) + b_k \sin(2\pi \omega_k t)] \; = \; c \quad .$$

Now suppose f is a sine function, say,

$$f(t) \; = \; \sin(2\pi \gamma t)$$

where γ is some fixed, positive value. Since the fundamental period of f is $p = 1/\gamma$, $f(t)$ can be written

$$f(t) \; = \; \sin\left(\frac{2\pi \cdot 1}{p} t\right) \quad .$$

In this case, formulas (9.1c), (9.1d) and (9.1e) for the Fourier coefficients are

$$A_0 \; = \; \frac{1}{p} \int_0^p f(t)\,dt \; = \; \frac{1}{p} \int_0^p \sin\left(\frac{2\pi}{p} t\right)\,dt \quad ,$$

$$a_k = \frac{2}{p} \int_0^p f(t) \cos(2\pi \omega_k t)\, dt = \frac{2}{p} \int_0^p \sin\left(\frac{2\pi \cdot 1}{p}t\right) \cos\left(\frac{2\pi k}{p}t\right) dt$$

and

$$b_k = \frac{2}{p} \int_0^p f(t) \sin(2\pi \omega_k t)\, dt = \frac{2}{p} \int_0^p \sin\left(\frac{2\pi \cdot 1}{p}t\right) \sin\left(\frac{2\pi k}{p}t\right) dt \quad,$$

which, according to the orthogonality relations for sines and cosines, reduce to

$$A_0 = 0 \quad,$$
$$a_k = 0$$

and

$$b_k = \frac{2}{p} \begin{cases} 0 & \text{if } k \neq 1 \\ \frac{p}{2} & \text{if } k = 1 \end{cases} = \begin{cases} 0 & \text{if } k \neq 1 \\ 1 & \text{if } k = 1 \end{cases} \quad.$$

Thus, if $f(t) = \sin(2\pi \gamma t)$, then

$$F.S.\,[f]|_t = A_0 + \sum_{k=1}^{\infty} [a_k \cos(2\pi \omega_k t) + b_k \sin(2\pi \omega_k t)]$$

$$= \sum_{k=1}^{\infty} \begin{cases} 0 & \text{if } k \neq 1 \\ 1 & \text{if } k = 1 \end{cases} \sin\left(\frac{2\pi k}{p}t\right) = \sin\left(\frac{2\pi \cdot 1}{p}t\right) \quad.$$

In other words,

$$F.S.\,\big[\sin(2\pi \gamma t)\big]\big|_t = \sin(2\pi \gamma t) \quad.$$

Very similar computations yield

$$F.S.\,\big[\cos(2\pi \gamma t)\big]\big|_t = \cos(2\pi \gamma t) \quad.$$

We have just shown that, whenever f is either a constant function, a sine function, or a cosine function, then its trigonometric Fourier series is simply $f(t)$, itself. This should not be at all surprising considering how we derived the formulas for the Fourier series in chapter 8. In the future, of course, there will be no real need to explicitly compute the trigonometric Fourier coefficients for such functions. After all, we have just proven the following lemma:

Lemma 9.2
If f is either a constant function, a sine function, or a cosine function, then

$$F.S.\,[f]|_t = f(t) \quad.$$

In particular, we should note that

$$F.S.\,[0]|_t = 0 \quad.$$

Alternative Intervals for Integration

The integrands in formulas (9.1c), (9.1d) and (9.1e) are all periodic functions with period p. Thus, as we saw in lemma 5.1 (on page 53), the values of these integrals remain unchanged if we replace the interval of integration, $(0, p)$, with any other interval of length p. This means that formulas

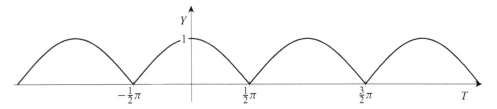

Figure 9.1: The rectified cosine function $|\cos(t)|$.

(9.1c), (9.1d) and (9.1e) are completely equivalent to

$$A_0 = \frac{1}{p} \int_{\text{period}} f(t)\,dt \quad, \tag{9.1c'}$$

$$a_k = \frac{2}{p} \int_{\text{period}} f(t) \cos(2\pi \omega_k t)\,dt \quad, \tag{9.1d'}$$

and

$$b_k = \frac{2}{p} \int_{\text{period}} f(t) \sin(2\pi \omega_k t)\,dt \tag{9.1e'}$$

where it is understood that the integration is done over any convenient interval of length p. In particular, it will often be convenient to evaluate our integrals over the symmetric interval $(-p/2, p/2)$.

!►Example 9.3: *Consider the rectified cosine function[4]*

$$f(t) = |\cos(t)| \quad.$$

This function, graphed in figure 9.1, has fundamental period $p = \pi$. If $-\pi/2 < t < \pi/2$, then

$$f(t) = |\cos(t)| = \cos(t) \quad,$$

while if $\pi/2 < t < \pi$,

$$f(t) = |\cos(t)| = -\cos(t) \quad.$$

To evaluate formula (9.1c) we would have to split the integral,

$$A_0 = \frac{1}{p} \int_0^p f(t)\,dt = \frac{1}{\pi} \int_0^\pi |\cos(t)|\,dt$$

$$= \frac{1}{\pi} \left[\int_0^{\pi/2} \cos(t)\,dt + \int_{\pi/2}^\pi (-\cos(t))\,dt \right] \quad.$$

But why evaluate two integrals, simple though they may be, when only one integral is required using formula (9.1c') with the interval $(-\pi/2, \pi/2)$,

$$A_0 = \frac{1}{p} \int_{\text{period}} f(t)\,dt = \frac{1}{\pi} \int_{-\pi/2}^{\pi/2} |\cos(t)|\,dt$$

$$= \frac{1}{\pi} \int_{-\pi/2}^{\pi/2} \cos(t)\,dt = \frac{1}{\pi} \sin(t)\Big|_{-\pi/2}^{\pi/2} = \frac{2}{\pi} \quad.$$

(We'll compute the other Fourier coefficients later, in example 9.4.)

---

[4] Any function that can be written as $|\cos(\gamma t)|$ (or $|\sin(\gamma t)|$), with γ being some positive constant, will be called a *rectified cosine* (or *rectified sine*) function.

Symmetry

If f is either an even or an odd periodic function, then the computation of the Fourier coefficients can be considerably simplified by making use of some of the observations made in the section on even and odd functions (pages 49 to 51).

Suppose, for example, that f is an *odd* function. Then, as was discussed in that earlier section, the integral of f over any symmetric interval must vanish. In particular,

$$A_0 = \frac{1}{p} \int_{-P/2}^{P/2} f(t)\,dt = 0 \quad .$$

Moreover, because $f(t)\cos(2\pi\omega_k t)$ is also an odd function (being the product of an odd function (the f) with an even function (the cosine function)), we also have

$$a_k = \frac{2}{p} \int_{-P/2}^{P/2} f(t)\cos(2\pi\omega_k t)\,dt = 0 \qquad \text{for} \quad k = 1,\,2,\,3,\,\ldots \quad .$$

On the other hand, $f(t)\sin(2\pi\omega_k t)$ is an even function since it is the product of an odd function (the f) with another odd function (the sine function). So,

$$b_k = \frac{2}{p} \int_{-P/2}^{P/2} f(t)\sin(2\pi\omega_k t)\,dt$$

$$= \frac{2}{p}\left[2\int_{0}^{P/2} f(t)\sin(2\pi\omega_k t)\,dt \right] = \frac{4}{p} \int_{0}^{P/2} f(t)\sin\left(\frac{2\pi k}{p}t\right)\,dt \quad .$$

For convenience, let us summarize the results just obtained.

Theorem 9.3 (Fourier series for odd functions)
Let f be a periodic, piecewise continuous function with period p. If f is an odd function on $\mathbb{R}$, then its trigonometric Fourier series is given by

$$F.S.\,[f]|_t = \sum_{k=1}^{\infty} b_k \sin(2\pi\omega_k t) \quad ,$$

where, for $k = 1, 2, 3, \ldots,$

$$\omega_k = \frac{k}{p} \qquad \text{and} \qquad b_k = \frac{4}{p} \int_{0}^{P/2} f(t)\sin\left(\frac{2\pi k}{p}t\right)\,dt \quad .$$

If, instead, f had been an even function, then we would have obtained the following theorem.

Theorem 9.4 (Fourier series for even functions)
Let f be a periodic, piecewise continuous function with period p. If f is an even function on $\mathbb{R}$, then its trigonometric Fourier series is given by

$$F.S.\,[f]|_t = A_0 + \sum_{k=1}^{\infty} a_k \cos(2\pi\omega_k t) \quad ,$$

where

$$A_0 = \frac{2}{p} \int_{0}^{P/2} f(t)\,dt$$

and, for $k = 1, 2, 3, \ldots$,

$$\omega_k = \frac{k}{p} \quad \text{and} \quad a_k = \frac{4}{p} \int_0^{P/2} f(t) \cos\left(\frac{2\pi k}{p} t\right) dt \quad.$$

?▶ Exercise 9.3: Prove theorem 9.4.

!▶ Example 9.4: Let us complete the computation, begun in example 9.3, of the trigonometric Fourier series for the rectified cosine function

$$f(t) = |\cos(t)|.$$

This function is clearly an even periodic function with period $p = \pi$ (see figure 9.1). Using the formulas from theorem 9.4, we have, for $k = 1, 2, 3, \ldots$,

$$\omega_k = \frac{k}{\pi}$$

and, using a well-known trigonometric identity,

$$
\begin{aligned}
a_k &= \frac{4}{\pi} \int_0^{\pi/2} |\cos(t)| \cos\left(\frac{2\pi k}{\pi} t\right) dt \\
&= \frac{4}{\pi} \int_0^{\pi/2} \cos(t) \cos(2kt) \, dt \\
&= \frac{4}{\pi} \int_0^{\pi/2} \frac{1}{2} [\cos([1+2k]t) + \cos([1-2k]t)] \, dt \\
&= \frac{2}{\pi} \left[\frac{1}{1+2k} \sin([1+2k]t) + \frac{1}{1-2k} \sin([1-2k]t) \right]\Big|_0^{\pi/2} \\
&= \frac{2}{\pi} \left[\frac{1}{1+2k} \sin\left([1+2k]\frac{\pi}{2}\right) + \frac{1}{1-2k} \sin\left([1-2k]\frac{\pi}{2}\right) \right] \quad.
\end{aligned}
$$

The last line can be simplified by observing that

$$\sin\left([1\pm 2k]\frac{\pi}{2}\right) = \sin\left(\frac{\pi}{2} \pm k\pi\right) = \cos(\pm k\pi) = (-1)^k \quad.$$

Thus,

$$a_k = \frac{2}{\pi} \left[\frac{1}{1+2k}(-1)^k + \frac{1}{1-2k}(-1)^k \right] = (-1)^k \frac{4}{\pi(1-4k^2)} \quad.$$

Theorem 9.4 assures us that there are no sine terms in this Fourier series. So, using the above and the value of A_0 computed in example 9.3, we see that the complete trigonometric Fourier series for $f(t) = |\cos(t)|$ is

$$F.S.[f]|_t = A_0 + \sum_{k=1}^{\infty} a_k \cos(2\pi \omega_k t)$$

$$= \frac{2}{\pi} + \sum_{k=1}^{\infty} (-1)^k \frac{4}{\pi(1-4k^2)} \cos(2kt) \quad.$$

Linearity

Sometimes the function of interest can be expressed as a finite linear combination of other periodic, piecewise continuous functions whose Fourier series are already known. When this happens, we can use a simple relation between the Fourier coefficients of the function of interest and the corresponding Fourier coefficients for the functions in the linear combination. That formula is described for the case where f is a linear combination of two functions in the next lemma.

Lemma 9.5 (linearity)
Let f, g and h be periodic, piecewise continuous functions, all with period p. Assume that, for some pair of constants γ and λ,

$$f = \gamma g + \lambda h \quad .$$

Let the corresponding trigonometric Fourier series for f, g and h be as follows:

$$F.S.\,[f]\big|_t = A_0^f + \sum_{k=1}^{\infty} \left[a_k^f \cos(2\pi\omega_k t) + b_k^f \sin(2\pi\omega_k t) \right] \quad ,$$

$$F.S.\,[g]\big|_t = A_0^g + \sum_{k=1}^{\infty} \left[a_k^g \cos(2\pi\omega_k t) + b_k^g \sin(2\pi\omega_k t) \right]$$

and

$$F.S.\,[h]\big|_t = A_0^h + \sum_{k=1}^{\infty} \left[a_k^h \cos(2\pi\omega_k t) + b_k^h \sin(2\pi\omega_k t) \right]$$

with $\omega_k = {k}/{p}$ in each. Then

$$A_0^f = \gamma A_0^g + \lambda A_0^h \quad , \tag{9.5a}$$

and, for $k = 1, 2, 3, \dots$,

$$a_k^f = \gamma a_k^g + \lambda a_k^h \tag{9.5b}$$

and

$$b_k^f = \gamma b_k^g + \lambda b_k^h \quad . \tag{9.5c}$$

PROOF: The formulas in the theorem are a direct result of the linearity of the integrals in formulas (9.1c), (9.1d) and (9.1e). For example, to verify equation (9.5a) we simply observe that

$$A_0^f = \frac{1}{p} \int_0^p f(t)\,dt = \frac{1}{p} \int_0^p \left[\gamma g(t) + \lambda h(t) \right] dt$$

$$= \gamma \frac{1}{p} \int_0^p g(t)\,dt + \lambda \frac{1}{p} \int_0^p h(t)\,dt = \gamma A_0^g + \lambda A_0^h \quad .$$

Verifying equations (9.5b) and (9.5c) is just as easy and is left to the interested reader.   ∎

This lemma tells us that each term in the Fourier series of a linear combination of two suitable functions is simply the corresponding linear combination of the terms of the individual functions. So, under the assumptions of the above lemma,

$$F.S.\,\big[\gamma g + \lambda h\big]\big|_t$$

$$= (\gamma A_0^g + \lambda A_0^h) + \sum_{k=1}^{\infty} \left[(\gamma a_k^g + \lambda a_k^h) \cos(2\pi\omega_k t) + (\gamma b_k^g + \lambda b_k^h) \sin(2\pi\omega_k t) \right]$$

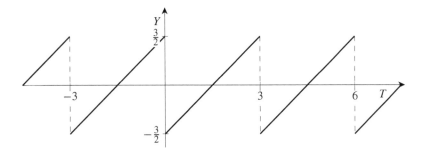

Figure 9.2: The saw function for example 9.5.

$$= \gamma \left[A_0^g + \sum_{k=1}^{\infty} \left[a_k^g \cos(2\pi \omega_k t) + b_k^g \sin(2\pi \omega_k t) \right] \right]$$

$$+ \lambda \left[A_0^h + \sum_{k=1}^{\infty} \left[a_k^h \cos(2\pi \omega_k t) + b_k^h \sin(2\pi \omega_k t) \right] \right]$$

$$= \gamma \, F.S.\,[g]|_t + \lambda \, F.S.\,[h]|_t \quad .$$

The next example illustrates a fairly common (and simple) application of this lemma.

!▶ **Example 9.5:** *Consider the function f sketched in figure 9.2. Clearly, f is just the saw_3 function from example 9.1 "lowered by $3/2$": That is,*

$$f(t) = \mathrm{saw}_3(t) - \frac{3}{2} \quad .$$

So, using the results from example 9.1 and lemma 9.2

$$F.S.\,[f]|_t = F.S.\left[\mathrm{saw}_3 - \frac{3}{2}\right]\Big|_t$$

$$= F.S.\,[\mathrm{saw}_3]|_t - F.S.\left[\frac{3}{2}\right]\Big|_t$$

$$= \frac{3}{2} - \sum_{k=1}^{\infty} \frac{3}{k\pi} \sin\left(\frac{2\pi k}{3}t\right) - \frac{3}{2} = -\sum_{k=1}^{\infty} \frac{3}{k\pi} \sin\left(\frac{2\pi k}{3}t\right) \quad .$$

Obviously, the results of lemma 9.5 can be extended to arbitrary finite linear combinations.

Theorem 9.6 (linearity)
Let N be a finite positive integer; let $f_1, f_2, f_3, \ldots$ and f_N all be periodic, piecewise continuous functions with a common period, and let $\alpha_1, \alpha_2, \alpha_3, \ldots$ and α_N all be constants. Then

$$F.S.\left[\sum_{n=1}^{N} \alpha_n f_n\right]\Bigg|_t = \sum_{n=1}^{N} \alpha_n \, F.S.\,[f_n]|_t$$

where it is understood that $\sum_{n=1}^{N} \alpha_n \, F.S.\,[f_n]|_t$ denotes the series constructed by adding the corresponding terms of the individual series.

As an immediate consequence of this theorem and lemma 9.2 we have:

Corollary 9.7

If f can be expressed as a finite linear combination of a constant function with sine and cosine functions (all with some common period), then that linear combination is the trigonometric Fourier series for f .

!▶**Example 9.6:**   *Let $f(t) = \sin^2(t)$. By a well-known trigonometric identity, we know*

$$\sin^2(t) \;=\; \tfrac{1}{2} \,-\, \tfrac{1}{2}\cos 2t \quad .$$

So

$$F.S.\,[f]|_t \;=\; \tfrac{1}{2} \,-\, \tfrac{1}{2}\cos 2t \quad .$$

Scaling and Shifting

The formulas given in the following theorems are occasionally of value. Their derivations will be left as exercises.

Theorem 9.8 (scaling)

Let f be a periodic, piecewise continuous function with period p and Fourier series

$$F.S.\,[f]|_t \;=\; A_0 \,+\, \sum_{k=1}^{\infty} [a_k \cos(2\pi\omega_k t) + b_k \sin(2\pi\omega_k t)] \quad .$$

If $g(t) = f(\alpha t)$ for some $\alpha > 0$, then g is a periodic, piecewise continuous function with period $\widehat{p} = {p}/{\alpha}$. Moreover, letting $\widehat{\omega}_k = {k}/{\widehat{p}} = \alpha\omega_k$,

$$F.S.\,[g]|_t \;=\; A_0 \,+\, \sum_{k=1}^{\infty} [a_k \cos(2\pi\widehat{\omega}_k t) + b_k \sin(2\pi\widehat{\omega}_k t)] \quad .$$

Theorem 9.9

Let f be a periodic, piecewise continuous function with period p and Fourier series

$$F.S.\,[f]|_t \;=\; A_0 \,+\, \sum_{k=1}^{\infty} [a_k \cos(2\pi\omega_k t) + b_k \sin(2\pi\omega_k t)] \quad .$$

If $g(t) = f(-t)$, then g is a periodic, piecewise continuous function with period p and trigonometric Fourier series

$$F.S.\,[g]|_t \;=\; A_0 \,+\, \sum_{k=1}^{\infty} [a_k \cos(2\pi\omega_k t) - b_k \sin(2\pi\omega_k t)] \quad .$$

Theorem 9.10 (half-period shift)

Let f be a periodic, piecewise continuous function with period p and Fourier series

$$F.S.\,[f]|_t \;=\; A_0 \,+\, \sum_{k=1}^{\infty} [a_k \cos(2\pi\omega_k t) + b_k \sin(2\pi\omega_k t)] \quad .$$

If $g(t) = f(t - {}^p\!/_2)$, then g is a periodic, piecewise continuous function with period p and

$$F.S.\,[g]|_t \;=\; A_0 \;+\; \sum_{k=1}^{\infty} \left[(-1)^k a_k \cos(2\pi\,\omega_k t) + (-1)^k b_k \sin(2\pi\,\omega_k t) \right] \quad .$$

?▶Exercise 9.4   a: *Prove theorem 9.8.*

   b: *Prove theorem 9.9.*

   c: *Prove theorem 9.10.*

Using Computers and Math Packages

Much of the drudgery in computing Fourier coefficients can be eliminated by letting a computer do the computations. The use of a good computer "math package" is especially recommended.[5] Some of these packages can symbolically evaluate many integrals that few of us would care to do by hand. When this is possible, we can actually obtain explicit and exact formulas for all the Fourier coefficients for our function. Even if the integrals are not simple enough to be done symbolically, these packages can often numerically compute very close approximations to as many of the coefficients as we need. Of course, some care must be taken with numerical calculations to ensure the computed answers are within the desired degree of accuracy. In particular, you should be warned that, when numerically computing integrals involving $\sin(\gamma t)$ and $\cos(\gamma t)$, the accuracy of the computations tends to decrease as the value of γ increases.[6]

9.3   Partial Sums and Graphing

Let f be a periodic, piecewise continuous function with trigonometric Fourier series

$$F.S.\,[f]|_t \;=\; A_0 \;+\; \sum_{k=1}^{\infty} [a_k \cos(2\pi\,\omega_k t) + b_k \sin(2\pi\,\omega_k t)] \quad .$$

It would be nice to verify our suspicion that $f(t)$ can be represented by its Fourier series by explicitly summing up the series for all values of t and comparing these values to the corresponding values of $f(t)$. Unfortunately, it is rarely practical to explicitly sum up an infinite series.

What is practical is to compare the function $f(t)$ to various partial sums of the Fourier series. For each positive integer N, the Nth *partial sum* (of $F.S.\,[f]$) is the function

$$F.S._N[f]|_t \;=\; A_0 \;+\; \sum_{k=1}^{N} [a_k \cos(2\pi\,\omega_k t) + b_k \sin(2\pi\,\omega_k t)] \quad .$$

---

[5] Maple and Mathematica were two good math packages available when this was written.

[6] Take a good course in numerical analysis to learn the reasons for this loss of accuracy. One is that the theoretical "worst possible error" resulting from using some integration algorithms is related to the maximum of one of the derivatives of the function being integrated (and, as you can easily check, the larger γ is, the larger too is the maximum of any nontrivial derivative of $\sin(\gamma t)$). Bad luck can also contribute to this loss of accuracy — when γ is large, there is a greater likelihood that two successive iterations of certain algorithms will just use values of t where γt is an integral multiple of π, and this tricks some algorithms into thinking they have found a good approximation for the integral.

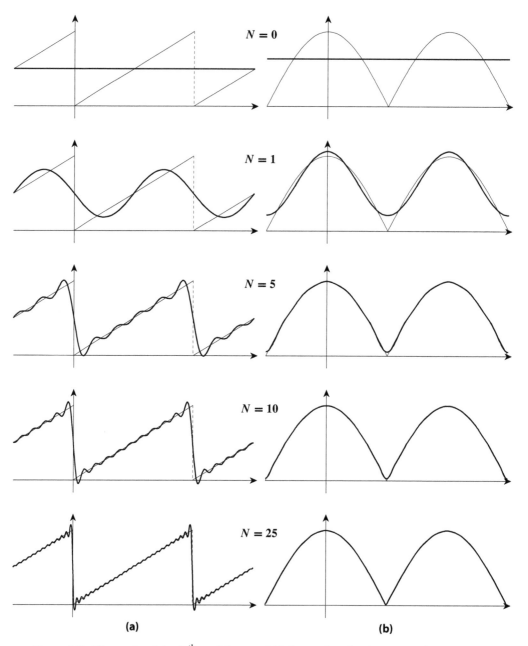

Figure 9.3: The graphs of the Nth partial sums of (a) the saw function from example 9.1 on page
102, $\text{saw}_3(t)$, and (b) the rectified cosine function from examples 9.3 on page 109
and 9.4 on page 111, $|\cos(t)|$. for $N = 0, 1, 5, 10$ and 25. Each partial sum graph
has been superimposed on more faintly drawn graphs of the corresponding saw and
rectified cosine functions.

The 0th partial sum is defined to be

$$F.S._0[f]\|_t = A_0 \quad .$$

Using a computer math package, it is quite easy to compute and graph the Nth partial sum, even for quite large values of N. By looking at these graphs and comparing them to the graph of the original function, we can get an intuitive feel for whether a given function can be represented by its Fourier series, and, when a function can be so represented, for the number of terms needed to get a good approximation to the function. These graphs may also give us some indication of possible problems that may arise.

!▶ *Example 9.7:*   *From the computations in examples 9.1, 9.3 and 9.4 (see pages 102, 109 and 111), we know that the Nth partial sum of the Fourier series for the saw function $saw_3(t)$ and the rectified cosine function $|\cos(t)|$ are*

$$F.S._N[saw_3(t)]|_t = \frac{3}{2} - \sum_{k=1}^{N} \frac{3}{k\pi} \sin\left(\frac{2\pi k}{3}t\right)$$

and

$$F.S._N[|\cos(t)|]|_t = \frac{2}{\pi} + \sum_{k=1}^{\infty} (-1)^k \frac{4}{\pi(1-4k^2)} \cos(2kt) \quad .$$

The graphs of these partial sums have been sketched in figure 9.3 using $N = 0, 1, 5, 10$ and 25.
    Looking at figure 9.3, you can see that the partial sums graphed are fairly good approximations to the saw and the rectified cosine functions, at least when $N \geq 5$. The approximations to the rectified cosine function are especially good — the graph of 25th partial sum approximation to $|\cos(t)|$ is almost indistinguishable from the graph of $|\cos(t)|$. The approximations to the saw function are less good — there are discernable "wiggles", even in the graph of the 25th partial sum approximation, and "strange things" seem to be occurring around the points where the saw function is discontinuous.

We will spend more time later (in chapters 13 and 14) investigating the convergence of the Fourier series. There we will see just how good (and how bad) we can expect the partial sum approximations to be, and just why "strange things" should occur in the graphs of some of the partial sums.

Additional Exercises

9.5. *Determine the fundamental frequency and fundamental period for each of the functions below. Then sketch each function's graph, and find its trigonometric Fourier series.*

a. $f(t) = \begin{cases} 0 & \text{if } -1 < t < 0 \\ 1 & \text{if } 0 < t < 1 \\ f(t-2) & \text{in general} \end{cases}$

b. $g(t) = \begin{cases} t & \text{if } -3 < t < 3 \\ g(t-6) & \text{in general} \end{cases}$

c. $h(t) = \begin{cases} e^t & \text{if } 0 < t < 1 \\ h(t-1) & \text{in general} \end{cases}$

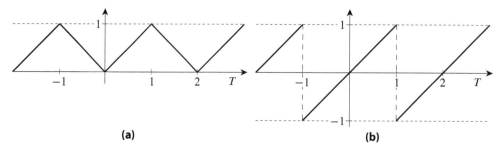

Figure 9.4: Two and one half periods of (a) evensaw(t), the even sawtooth function for exercise
9.6 c, and (b) oddsaw(t), the odd sawtooth function for exercise 9.6 d.

d. $k(t) = \begin{cases} t^2 & \text{if } -1 < t < 1 \\ k(t-2) & \text{in general} \end{cases}$

e. $\cos^2(t)$

f. $|\sin(t)|$ *(a rectified sine function)*

9.6. *Determine whether each of the following functions is even or odd. Then graph each function
(if it has not already been graphed), and find its trigonometric Fourier series. Where
appropriate, use the even- or oddness of the function to reduce the number of integrals you
need to compute.*

a. $f(t) = \begin{cases} -1 & \text{if } -1 < t < 0 \\ +1 & \text{if } 0 < t < 1 \\ f(t-2) & \text{in general} \end{cases}$

b. $g(t) = \begin{cases} 1 & \text{if } |t| < 1 \\ 0 & \text{if } 1 < |t| < 2 \\ g(t-4) & \text{in general} \end{cases}$

c. evensaw(t), the even sawtooth function sketched in figure 9.4a

d. oddsaw(t), the odd sawtooth function sketched in figure 9.4b

e. $h(t) = \begin{cases} t^2 & \text{if } -1 < t < 1 \\ h(t-2) & \text{in general} \end{cases}$

f. $k(t) = \begin{cases} -t^2 & \text{if } -1 < t < 0 \\ +t^2 & \text{if } 0 < t < 1 \\ k(t-2) & \text{in general} \end{cases}$

9.7. *Express each of the following functions in terms of functions from the previous exercise
using scaling, shifting, linear combinations, etc. Then, using your answers from the previous
exercise, determine the trigonometric Fourier series for each. You should not compute any
integrals for these.*

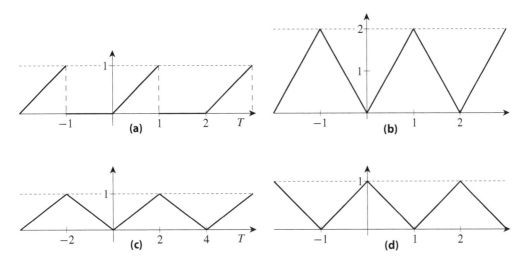

Figure 9.5: Two and one half periods of
(a) $\Phi(t)$, the broken sawtooth function for exercise 9.7 d,
(b) $\Psi_1(t)$, the scaled even sawtooth function, for exercise 9.7 e,
(c) $\Psi_2(t)$, the scaled even sawtooth function for exercise 9.7 f, and
(d) $\Psi_3(t)$, the shifted even sawtooth function for exercise 9.7 g.

a. $G(t) = \begin{cases} +1 & \text{if } |t| < 1 \\ -1 & \text{if } 1 < |t| < 2 \\ G(t-4) & \text{in general} \end{cases}$

b. $H(t) = \begin{cases} 1 - t^2 & \text{if } -1 < t < 1 \\ H(t-2) & \text{in general} \end{cases}$

c. $K(t) = \begin{cases} t^2 - 2t - 3 & \text{if } -1 < t < 1 \\ K(t-2) & \text{in general} \end{cases}$

d. $\Phi(t)$, the broken sawtooth function sketched in figure 9.5a

e. $\Psi_1(t)$, the scaled even sawtooth function sketched in figure 9.5b

f. $\Psi_2(t)$, the scaled even sawtooth function sketched in figure 9.5c

g. $\Psi_3(t)$, the shifted even sawtooth function sketched in figure 9.5d

9.8 a. Using a computer math package (such as Maple or Mathematica) write a "program" or "worksheet" for graphing a periodic function f along with the Nth partial sum of its trigonometric Fourier series,

$$F.S._N[f]|_t = A_0 + \sum_{k=1}^{N} [a_k \cos(2\pi \omega_k t) + b_k \sin(2\pi \omega_k t)] \quad,$$

over the interval $(-p/2, 2p)$. Have the following as the inputs to your program/worksheet: the function's period p, a formula for the function over one period (say, $(0, p)$ or $(-p/2, p/2)$), the value of A_0, the formulas for the a_k's and b_k's, and the value of N. (Also, see exercise 5.13.)

b. *Use your program/worksheet to graph each of the following functions along with the* N^{th} *partial sums of its Fourier series for* $N = 0, 1, 2, 10$ *and* 25. *Examine your graphs and answer these two questions:*

    1. *Are the graphs of the* N^{th} *partial sums converging to the graph of the function as* N *gets larger?*

    2. *Are there points at which strange things are happening in the graphs of the* N^{th} *partial sums, especially for the larger values of* N ?

  i. *The even sawtooth function,* evensaw(t), *from exercise 9.6*

  ii. *The odd sawtooth function,* oddsaw(t), *from exercise 9.6*

  iii. *The broken sawtooth function from exercise 9.7*

  iv. *The function* $G(t)$ *from exercise 9.7*

  v. *The function* $H(t)$ *from exercise 9.7*

  vi. *The function* $K(t)$ *from exercise 9.7*

9.9 a. *Modify your program/worksheet from exercise 9.8 so that it also numerically evaluates the "first* N*" Fourier coefficients (i.e.,* A_0 *and the* a_k*'s and* b_k*'s for* $k = 1$ *to* N *), lists those coefficients, and then graphs both the function and, using the coefficients just computed, the corresponding* N^{th} *partial sum of the trigonometric Fourier series.*

    Here, the inputs should just be the function's period p, *a formula for the function over one period (say,* $(0, p)$ *or* $(-P/2, P/2)$ *) and the value of* N .

b. *Use your program/worksheet to graph each of the following functions and the* N^{th} *partial sums of its Fourier series for* $N = 0, 1, 10$ *and* 25.

  i. $f(t) = \begin{cases} t^2(1-t)^2 & \text{if } 0 < t < 1 \\ f(t-1) & \text{in general} \end{cases}$

  ii. $g(t) = \begin{cases} \sqrt{|t|} & \text{if } -1 < t < 1 \\ g(t-2) & \text{in general} \end{cases}$

  iii. $h(t) = \begin{cases} 0 & \text{if } -1 < t < 0 \\ \sqrt{t} & \text{if } 0 < t < 1 \\ h(t-2) & \text{in general} \end{cases}$

  iv. $k(t) = \begin{cases} \sqrt{1-t^2} & \text{if } |t| < 1 \\ k(t-2) & \text{in general} \end{cases}$

  v. $l(t) = \begin{cases} \sqrt{1-t^2} & \text{if } |t| < 1 \\ 0 & \text{if } 1 < |t| < 3/2 \\ l(t-3) & \text{in general} \end{cases}$

  vi. $\phi(t) = \begin{cases} 0 & \text{if } -2\pi < t < 0 \\ 1 - \cos(t) & \text{if } 0 < t < 2\pi \\ \phi(t-4\pi) & \text{in general} \end{cases}$

10

Fourier Series over Finite Intervals (Sine and Cosine Series)

In practice, the functions of interest are often only defined over finite intervals, not the entire real line. For example, if $f(t)$ is the temperature at time t of the coffee in a person's cup, then $f(t)$ is only defined for $\alpha < t < \beta$ where α is the time the coffee is poured into the cup, and β is the time the cup is finally emptied.

For the rest of this chapter, L will denote some positive real number, and f will denote some piecewise continuous function that is defined only on the interval $(0, L)$ (as illustrated in figure 10.1). As in chapter 8, our interest is in deriving an expression for f in terms of sines and cosines. This time, however, we are only interested in this expression describing the given function over the interval $(0, L)$.

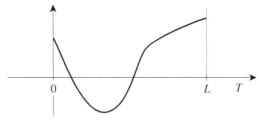

Figure 10.1: A function $f(t)$ defined only on the interval $(0, L)$.

The basic idea behind all the derivations in this chapter is simple. Take any periodic function $\widehat{f}$ (defined on the entire real line) that equals the given function f on $(0, L)$. If, as we suspect, the trigonometric Fourier series for $\widehat{f}$ describes $\widehat{f}$ over the entire real line, then this series must also describe f over the interval $(0, L)$ (where $f(t) = \widehat{f}(t)$). Thus, we should be able to use the trigonometric Fourier series for $\widehat{f}$ as a "Fourier series" for f over the interval $(0, L)$.

Any periodic function $\widehat{f}$ that equals f over the original domain of f is called a *periodic extension* of f. In fact, many different periodic extensions can be constructed for any given f originally defined on the interval $(0, L)$. Consequently, we can derive many different "Fourier series" for any such function. In what follows, we will derive three of the more important Fourier series for functions on a finite interval.

10.1 The Basic Fourier Series

The simplest approach is to simply let $\widehat{f}$ be the periodic extension of f having period $p = L$,

$$
\widehat{f}(t) = \begin{cases} f(t) & \text{if } 0 < t < L \\ \widehat{f}(t - L) & \text{in general} \end{cases}.
$$

We'll call this the *basic periodic extension* of f (see figure 10.2).

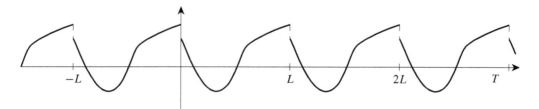

Figure 10.2: The basic periodic extension $\widehat{f}(t)$ of the function graphed in figure 10.1.

For convenience, let us denote by $T.F.S.[f]$ the "Fourier series" for f over $(0, L)$ obtained by using the trigonometric Fourier series for $\widehat{f}$. In other words, for $0 < t < L$,

$$T.F.S.[f]|_t = F.S.[\widehat{f}]|_t \quad .$$

Remember,

$$F.S.[\widehat{f}]|_t = A_0 + \sum_{k=1}^{\infty} [a_k \cos(2\pi \omega_k t) + b_k \sin(2\pi \omega_k t)]$$

where

$$\omega_k = \frac{k}{p} \quad ,$$

$$A_0 = \frac{1}{p} \int_0^p \widehat{f}(t)\, dt \quad ,$$

$$a_k = \frac{2}{p} \int_0^p \widehat{f}(t) \cos\left(\frac{2\pi k}{p} t\right) dt \quad ,$$

and

$$b_k = \frac{2}{p} \int_0^p \widehat{f}(t) \sin\left(\frac{2\pi k}{p} t\right) dt \quad .$$

Since $p = L$ and $f(t) = \widehat{f}(t)$ when $0 < t < L$, the above can be rewritten as

$$T.F.S.[f]|_t = A_0 + \sum_{k=1}^{\infty} [a_k \cos(2\pi \omega_k t) + b_k \sin(2\pi \omega_k t)] \tag{10.1a}$$

where

$$\omega_k = \frac{k}{L} \quad , \tag{10.1b}$$

$$A_0 = \frac{1}{L} \int_0^L f(t)\, dt \quad , \tag{10.1c}$$

$$a_k = \frac{2}{L} \int_0^L f(t) \cos\left(\frac{2\pi k}{L} t\right) dt \quad , \tag{10.1d}$$

and

$$b_k = \frac{2}{L} \int_0^L f(t) \sin\left(\frac{2\pi k}{L} t\right) dt \quad . \tag{10.1e}$$

Observe that, although $\widehat{f}$ was used to derive formulas (10.1a) through (10.1e), it does not explicitly appear in them. So, if we are simply computing $T.F.S.[f]$, there is no need to actually determine the extension, $\widehat{f}$. On the other hand, we will find knowing $\widehat{f}$ is useful when we finally deal with questions concerning the convergence of these series.

!▶**Example 10.1:** Let $L = 2$ and $f(t) = t^2$. Using formula set (10.1), we have (omitting some details of the computations)

$$\omega_k = \frac{k}{L} = \frac{k}{2} \ ,$$

$$A_0 = \frac{1}{L} \int_0^L f(t)\, dt = \frac{1}{2} \int_0^2 t^2\, dt = \frac{4}{3} \ ,$$

$$a_k = \frac{2}{L} \int_0^L f(t) \cos\left(\frac{2\pi k}{L}t\right) dt = \frac{2}{2} \int_0^2 t^2 \cos\left(\frac{2\pi k}{2}t\right) dt = \cdots = \frac{4}{k^2\pi^2} \ ,$$

and

$$b_k = \frac{2}{L} \int_0^L f(t) \sin\left(\frac{2\pi k}{L}t\right) dt = \frac{2}{2} \int_0^2 t^2 \sin\left(\frac{2\pi k}{2}t\right) dt = \cdots = -\frac{4}{k\pi} \ .$$

So, for this function,

$$T.F.S.\,[f]\big|_t = A_0 + \sum_{k=1}^{\infty} [a_k \cos(2\pi\omega_k t) + b_k \sin(2\pi\omega_k t)]$$

$$= \frac{4}{3} + \sum_{k=1}^{\infty} \left[\frac{4}{k^2\pi^2} \cos(k\pi t) - \frac{4}{k\pi} \sin(k\pi t)\right] \ .$$

10.2 The Fourier Sine Series

A somewhat simpler Fourier series for f over $(0, L)$ can be derived by first generating the *odd* extension of f (see figure 10.3) and then taking the simplest periodic extension of this odd function (see figure 10.4).

To be more precise, define f_o to be the odd function on $(-L, L)$ equaling f over $(0, L)$,

$$f_o(t) = \begin{cases} f(t) & \text{if } 0 < t < L \\ -f(-t) & \text{if } -L < t < 0 \end{cases} \ .$$

Since f_o is defined on $(-L, L)$, the simplest periodic extension of f_o must have a period $p = 2L$.

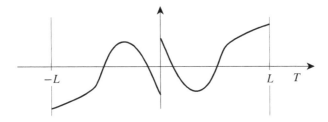

Figure 10.3: The odd extension $f_o(t)$ of the function graphed in figure 10.1.

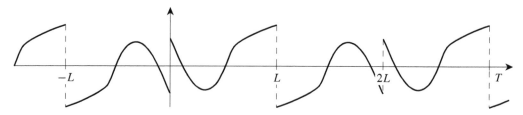

Figure 10.4: The odd periodic extension $\widehat{f}_o(t)$ of the function graphed in figure 10.1 (see also figure 10.3).

Accordingly, here we define our periodic extension $\widehat{f}_o$ by

$$\widehat{f}_o(t) = \begin{cases} f_o(t) & \text{if } -L < t < L \\ \widehat{f}_o(t - 2L) & \text{in general} \end{cases} = \begin{cases} f(t) & \text{if } 0 < t < L \\ -f(-t) & \text{if } -L < t < 0 \\ \widehat{f}_o(t - 2L) & \text{in general} \end{cases} .$$

Clearly, $\widehat{f}_o$ is an odd periodic function. According to theorem 9.3 on page 110,

$$F.S.\left[\widehat{f}_o\right]\big|_t = \sum_{k=1}^{\infty} b_k \sin(2\pi\omega_k t) \tag{10.2a}$$

where $\omega_k = {}^k/_p$ and

$$b_k = \frac{4}{p}\int_0^{p/2} \widehat{f}_o(t) \sin\left(\frac{2\pi k}{p}t\right) dt . \tag{10.2b}$$

The above Fourier series is generally referred to as the *(Fourier) sine series* for f on $(0, L)$, and the b_k's are generally called the *(Fourier) sine coefficients* for f.

Let us denote the series just derived by $F.S.S.[f]$ rather than $F.S.\left[\widehat{f}_o\right]$. Note that, because $p = 2L$ and $\widehat{f}_o(t) = f(t)$ when $0 < t < L$,

$$2\pi\omega_k = 2\pi\frac{k}{2L} = \frac{k\pi}{L}$$

and

$$\frac{4}{p}\int_0^{p/2} \widehat{f}_o(t) \sin\left(\frac{2\pi k}{p}t\right) dt = \frac{2}{L}\int_0^{L} f(t) \sin\left(\frac{k\pi}{L}t\right) dt .$$

Thus, formula set (10.2) can be rewritten as

$$F.S.S.[f]\big|_t = \sum_{k=1}^{\infty} b_k \sin\left(\frac{k\pi}{L}t\right) \tag{10.3a}$$

where

$$b_k = \frac{2}{L}\int_0^{L} f(t) \sin\left(\frac{k\pi}{L}t\right) dt . \tag{10.3b}$$

As in the previous section, the extension $\widehat{f}_o$ was needed to derive the formulas for the sine series and will be used when we discuss convergence, but $\widehat{f}_o$ is not needed for simply computing the sine series of f over $(0, L)$. For that, formulas (10.3a) and (10.3b) suffice.

!▶ **Example 10.2:**   Let $L = 2$ and $f(t) = t^2$. *Plugging this into formula (10.3b), we have (again, omitting some computational details)*

$$b_k = \frac{2}{L} \int_0^L f(t) \sin\left(\frac{k\pi}{L}t\right) dt$$

$$= \frac{2}{2} \int_0^2 t^2 \sin\left(\frac{k\pi}{2}t\right) dt = \frac{16}{k^3\pi^3}\left[(-1)^k - 1\right] - (-1)^k \frac{8}{k\pi} \quad .$$

This gives us the Fourier sine coefficients for t^2 over $(0, 2)$. For the corresponding sine series we then have

$$F.S.S.\,[f]|_t = \sum_{k=1}^{\infty} b_k \sin\left(\frac{k\pi}{L}t\right) = \sum_{k=1}^{\infty} \left\{\frac{16}{k^3\pi^3}\left[(-1)^k - 1\right] - (-1)^k \frac{8}{k\pi}\right\} \sin\left(\frac{k\pi}{2}t\right) \quad .$$

10.3   The Fourier Cosine Series

The *(Fourier) cosine series* for f over $(0, L)$ is given by

$$F.C.S.\,[f]|_t = A_0 + \sum_{k=1}^{\infty} a_k \cos\left(\frac{k\pi}{L}t\right) \tag{10.4a}$$

where

$$A_0 = \frac{1}{L} \int_0^L f(t)\, dt \tag{10.4b}$$

and

$$a_k = \frac{2}{L} \int_0^L f(t) \cos\left(\frac{k\pi}{L}t\right) dt \quad . \tag{10.4c}$$

The A_0 and the a_k's are called the *(Fourier) cosine coefficients* for f.

The cosine series for f is just the trigonometric Fourier series for this function's *even* periodic extension $\widehat{f_e}$ (see figure 10.5). Its derivation is very similar to the derivation of the sine series in the previous section. Of course, instead of using the odd extension, f_o, we use the even extension,

$$f_e(t) = \begin{cases} f(t) & \text{if} \quad 0 < t < L \\ f(-t) & \text{if} \quad -L < t < 0 \end{cases} \quad .$$

The details are left as an exercise.

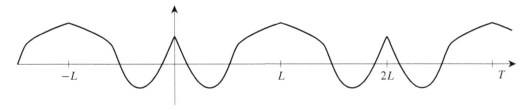

Figure 10.5: The even periodic extension $\widehat{f_e}(t)$ of the function graphed in figure 10.1.

?▶Exercise 10.1:   *Using the derivation of the sine series from the previous section as a guide, derive the cosine series for* f *over* $(0, L)$.

!▶Example 10.3:   Let $L = 2$ and $f(t) = t^2$. *Plugging this into formulas (10.4b) and (10.4c) yields*

$$A_0 = \frac{1}{L} \int_0^L f(t)\,dt = \frac{1}{2} \int_0^2 t^2\,dt = \cdots = \frac{4}{3}$$

and

$$a_k = \frac{2}{L} \int_0^L f(t)\cos\left(\frac{k\pi}{L}t\right) dt = \frac{2}{2} \int_0^2 t^2 \cos\left(\frac{k\pi}{2}t\right) dt = \cdots = (-1)^k \frac{16}{k^2 \pi^2} \quad .$$

This gives us the Fourier cosine coefficients for t^2 over $(0, 2)$. For the corresponding cosine series we then have

$$F.C.S.\,[f]|_t = A_0 + \sum_{k=1}^{\infty} a_k \cos\left(\frac{k\pi}{L}t\right) = \frac{4}{3} + \sum_{k=1}^{\infty} (-1)^k \frac{16}{k^2\pi^2} \cos\left(\frac{k\pi}{2}t\right) \quad .$$

10.4   Using These Series
Computing with These Series

Keep in mind that the three "Fourier series" derived in this chapter are all trigonometric Fourier series for some periodic extension of the function originally given on just the interval $(0, L)$. Consequently, much of the discussion in the previous chapter concerning trigonometric Fourier series for periodic functions also applies to the series derived in this chapter (as well as to any other similarly derived Fourier series). For example, it is an immediate consequence of lemma 9.5 on page 112 on linearity that, if f and g are two piecewise continuous functions on the interval $(0, L)$ with sine series

$$F.S.S.\,[f]|_t = \sum_{k=1}^{\infty} b_k^f \sin\left(\frac{k\pi}{L}t\right) \qquad \text{and} \qquad F.S.S.\,[g]|_t = \sum_{k=1}^{\infty} b_k^g \sin\left(\frac{k\pi}{L}t\right) \quad ,$$

then, for any pair of constants γ and λ,

$$F.S.S.\,[\gamma f + \lambda g]|_t = \sum_{k=1}^{\infty} \left[\gamma b_k^f + \lambda b_k^g\right] \sin\left(\frac{k\pi}{L}t\right) \quad .$$

On the other hand, there are portions of the commentary in the previous chapter that have little or no relevance to the computation of the series being discussed here. Consider, for example, the discussion of trigonometric Fourier series for odd or even periodic functions. That discussion was very relevant to the derivation of the formulas for the sine and cosine series. However, it makes no sense to refer to a function defined on just $(0, L)$ as being either even or odd. So that discussion of Fourier series for odd and even functions gives us no real advice on computing the sine and cosine series using the formulas already derived (i.e., formula sets (10.3) and (10.4)).

The reader should have little difficulty in identifying, as the need arises, which portions of the previous chapter are relevant in computing the coefficients for the sort of Fourier series we are discussing now. That being the case, let's go to another topic.

Which Series Do I Use?

Because each of the Fourier series derived in this chapter is the trigonometric Fourier series for a periodic function equaling the original function f over the interval $(0, L)$, Fourier's bold conjecture suggests that

$$f(t) = T.F.S.[f]|_t = F.S.S.[f]|_t = F.C.S.[f]|_t \quad \text{for} \quad 0 < t < L \ .$$

So, according to Fourier's bold conjecture, f can be represented by each of these three different series (as well as by any other similarly derived series). But if this is true, we are left with a significant practical question: *Which of these series should be used in any given application?* Not too surprisingly, the answer depends on the application at hand and the sort of additional conditions f must satisfy. In fact, determining which is the "best Fourier series" can be a significant part of the mathematics required for solving a given problem. We'll look more into this issue in chapter 16 where we will use sine and cosine series to solve some problems in thermodynamics and mechanics. Meanwhile, some idea of criteria that might be useful in choosing the appropriate series can be gleaned from the next exercise.

?▶Exercise 10.2: *Suppose f is uniformly continuous on $(0, L)$ and $f(0) = f(L) = 0$. Which of the three series discussed in this chapter would be the "obvious" choice to represent $f(t)$ on the interval $(0, L)$? (Hint: For which of these series are all the terms automatically zero when $t = 0$ and $t = L$?)*

Additional Exercises

10.3. For this problem, let $f(t) = 1$ if $0 < t < 1$, and be undefined otherwise.

    a. Sketch the graph of f (defined only on $(0, 1)$!), and, on separate coordinate systems, sketch the graph of each of the following extensions of f :

      i. the basic periodic extension, $\widehat{f}$ **ii.** the odd periodic extension, $\widehat{f}_o$

      iii. the even periodic extension, $\widehat{f}_e$

    b. Find $T.F.S.[f]|_t$, the trigonometric Fourier series for f over $(0, 1)$.

    c. Find $F.S.S.[f]|_t$, the Fourier sine series for f over $(0, 1)$.

    d. Find $F.C.S.[f]|_t$, the Fourier cosine series for f over $(0, 1)$.

10.4. Repeat problem 10.3 using the function $f(t) = t$ when $0 < t < 1$ (with $f(t)$ undefined otherwise).

10.5. Find the Fourier sine series for each of the following functions over the indicated interval:

    a. $g(t) = t^2$ over $(0, 3)$ **b.** $h(t) = \sin(2t)$ over $(0, \pi)$

10.6. Find the Fourier cosine series for each of the following functions over the indicated interval:

    a. $g(t) = t^2$ over $(0, 3)$ **b.** $h(t) = \sin(2t)$ over $(0, \pi)$

10.7 a. *Using a computer math package (such as Maple or Mathematica), write a "program" or "worksheet" for graphing a function f over an interval $(0, L)$ along with the Nth partial sum of its Fourier sine series over that interval,*

$$F.S.S._N[f]\big|_t \;=\; \sum_{k=1}^{N} b_k \sin\!\left(\frac{k\pi}{L}t\right) \quad . \tag{10.5}$$

The inputs to this program/worksheet should be the interval's length L, a formula for the function over the interval, the formulas for the b_k's and the value of N. (Also, see exercise 9.8 on page 119.)

b. *Use your program/worksheet to graph each function from problem 10.5, above, over the given interval along with the Nth partial sums of its Fourier sine series for $N = 1, 2, 10$ and 25. Examine your graphs and answer the following two questions:*

    1. *Do the graphs of the Nth partial sums appear to be converging to the graph of the function as N gets larger?*

    2. *Are there points at which strange things are happening in the graphs of the Nth partial sums, especially for the larger values of N?*

10.8 a. *Modify your program/worksheet from exercise 10.7, above, so that it also numerically evaluates the "first N" Fourier sine coefficients (i.e., the b_k's for $k = 1$ to N), lists those coefficients, and then graphs the function and the corresponding N^{th} partial sum of the Fourier sine series. Here, your inputs should be the interval's length L, a formula for the function over the interval, and the value of N. (See, also, problem 9.9 on page 120.)*

b. *Use your program/worksheet to graph each of the following functions and the Nth partial sum of its Fourier sine series over the given interval for $N = 1, 5, 10$ and 25. Examine the graphs and answer the two questions given in exercise 10.7 b.*

  i. $f(t) = t - t^2$ *over* $(0, 1)$

  ii. $g(t) = 1$ *over* $(0, 1)$

  iii. $h(t) = \sin(t)$ *over* $(0, \pi/2)$

  iv. $k(t) = \begin{cases} t & \text{if } 0 < t < 1 \\ 0 & \text{if } 1 \le t < 2 \end{cases}$ *over* $(0, 2)$

10.9 a. *Redo all of exercise 10.7, above, with the words "sine series" replaced by "cosine series" and with formula (10.5) replaced by*

$$F.C.S._N[f]\big|_t \;=\; A_0 + \sum_{k=1}^{N} a_k \cos\!\left(\frac{k\pi}{L}t\right) \quad .$$

b. *Then modify your cosine series program/worksheet so that it also numerically evaluates the "first N" Fourier cosine coefficients (i.e., A_0 and the a_k's for $k = 1$ to N), lists the coefficients and then graphs the function and the corresponding N^{th} partial sum of the Fourier cosine series.*

c. *Using your cosine series program, redo problem 10.8 b (computing the partial sums of the cosine series instead of the sine series).*

11

Inner Products, Norms and Orthogonality

We've derived several "Fourier series" (including the trigonometric Fourier series, the Fourier cosine series and the Fourier sine series), and we will be discussing yet another Fourier series (the exponential Fourier series) in the next chapter. That's a lot of series, with a lot of different formulas to learn. Fortunately, there is a fairly general framework for describing all of these (and other) Fourier series. This framework is based on an operation with pairs of functions analogous to the dot product operation in vector analysis. Using this, and other ideas from vector analysis, we can then easily describe all Fourier series and derive a single simple formula for computing "the components of a function" relative to any suitable set of base functions.

Throughout this chapter, we will be considering functions defined on some interval (α, β). The functions may be complex valued. To ensure that all the integrals in this chapter are well defined, let us go ahead and assume that (α, β) is a finite interval and that all the functions mentioned in this chapter are piecewise continuous.[1]

11.1 Inner Products

Let f and g be two piecewise continuous functions on the finite interval (α, β). The *inner product* of f with g (over (α, β)) is denoted by $\langle f \mid g \rangle$ and defined by

$$\langle f \mid g \rangle = \int_\alpha^\beta f(t) g^*(t) \, dt$$

(where $g^*(t)$ is the complex conjugate of $g(t)$). If g is a real-valued function, then $g^* = g$ and the above is just

$$\langle f \mid g \rangle = \int_\alpha^\beta f(t) g(t) \, dt \quad .$$

!►*Example 11.1:*   *The inner product of* $3t$ *with* $\sin(2\pi t)$ *over* $(0, 1)$ *is*

$$\langle 3t \mid \sin(2\pi t) \rangle = \int_0^1 3t \sin(2\pi t) \, dt = -\frac{3}{2\pi} \quad .$$

---

[1] Actually, any conditions ensuring the existence of the integrals appearing here would suffice.

?► Exercise 11.1:   *Show that the inner product of t^2 with $9 + i8t$ over $(0, 1)$ is*

$$\left\langle t^2 \mid 9 + i8t \right\rangle = 3 - 2i \quad .$$

At this point, the inner product $\langle f \mid g \rangle$ can be viewed as just shorthand for a particular integral involving the two functions f and g over the interval (α, β). On occasion, we may neglect to explicitly state this interval. When this happens, just remember that (α, β) is always "the interval of current interest". In particular:

1.   If we are discussing "functions that are periodic with period p", then (α, β) is any interval of length p (such as $(0, p)$ or $(-p/2, p/2)$).

2.   If we are discussing "functions that are defined over the interval $(0, L)$" where L is some finite length, then (α, β) is $(0, L)$.

Some of the important properties of inner products are summarized in the next theorem. These properties will allow us to use the inner product in much the same way as the dot product is used for vectors in space.

Theorem 11.1 (properties of the inner product)
Suppose a and b are two (possibly complex) constants, and f, g and h are piecewise continuous functions on the finite interval (α, β). Then

1. $\langle af + bg \mid h \rangle = a \langle f \mid h \rangle + b \langle g \mid h \rangle$,

2. $\langle f \mid ag + bh \rangle = a^* \langle f \mid g \rangle + b^* \langle f \mid h \rangle$,

3. $\langle g \mid f \rangle = \langle f \mid g \rangle^*$,

and

4. $\langle f \mid f \rangle \geq 0$, *with* $\langle f \mid f \rangle = 0$ *if and only if f vanishes on (α, β).*

PROOF: Verifying these properties is easy. For the first,

$$\langle af + bg \mid h \rangle = \int_\alpha^\beta [af(t) + bg(t)]h^*(t)\, dt$$

$$= \int_\alpha^\beta [af(t)h^*(t) + bg(t)h^*(t)]\, dt$$

$$= a \int_\alpha^\beta f(t)h^*(t)\, dt + b \int_\alpha^\beta g(t)h^*(t)\, dt$$

$$= a \langle f \mid h \rangle + b \langle g \mid h \rangle \quad .$$

For the second,

$$\langle f \mid ag + bh \rangle = \int_\alpha^\beta f(t)[ag(t) + bh(t)]^*\, dt$$

$$= \int_\alpha^\beta f(t)[a^* g^*(t) + b^* h^*(t)]\, dt$$

$$= a^* \int_\alpha^\beta f(t) g^*(t) t + b^* \int_\alpha^\beta f(t) h^*(t) \, dt$$

$$= a^* \langle f \mid g \rangle + b^* \langle f \mid h \rangle \quad .$$

The third comes from the observation that

$$\int_\alpha^\beta g(t) f^*(t) \, dt = \int_\alpha^\beta [g^*(t) f(t)]^* \, dt$$

$$= \left[\int_\alpha^\beta g^*(t) f(t) \, dt \right]^* = \left[\int_\alpha^\beta f(t) g^*(t) \, dt \right]^* \quad .$$

Cutting out the middle and rewriting the left- and right-hand sides of this in terms of inner products leaves us with

$$\langle g \mid f \rangle = \langle f \mid g \rangle^* \quad .$$

Finally, since $ff^* = |f|^2 \geq 0$,

$$\langle f \mid f \rangle = \int_\alpha^\beta f(t) f^*(t) \, dt = \int_\alpha^\beta |f(t)|^2 \, dt \quad ,$$

which, clearly, is zero if f vanishes on (α, β) and is positive otherwise. ∎

11.2 The Norm of a Function

Recall that the norm of a vector $\mathbf{v}$ is given by $\|\mathbf{v}\| = \sqrt{\mathbf{v} \cdot \mathbf{v}}$. Similarly, the *norm of a function* f (on the interval (α, β)) is given by

$$\|f\| = \sqrt{\langle f \mid f \rangle} \quad ,$$

which is just shorthand for

$$\|f\| = \left[\int_\alpha^\beta f(t) f^*(t) \, dt \right]^{1/2} = \left[\int_\alpha^\beta |f(t)|^2 \, dt \right]^{1/2} \quad .$$

Notice that property 4 for the inner product assures us that $\|f\| \geq 0$ with $\|f\| > 0$ whenever the piecewise continuous function $f(t)$ is nonzero somewhere on (α, β).

!▶**Example 11.2:**   The norm of $f(t) = 3t + i$ over $(0, 1)$ is

$$\|f\| = \|3t + i\| = \left[\int_0^1 (3t + i)(3t - i) \, dt \right]^{1/2}$$

$$= \left[\int_0^1 (9t^2 + 1) \, dt \right]^{1/2} = \left[3t^3 + t \Big|_{t=0}^1 \right]^{1/2} = 2 \quad .$$

Be careful to not confuse $\|f\|$, the norm of f, with $|f|$, the absolute value of f. $\|f\|$ a single number; $|f|$ is a function. In the above example, $\|f\| = 2$, while

$$|f(t)| = |3t + i| = \sqrt{9t^2 + 1} \quad .$$

In fact, you should observe that the norm of a function $\|f\|$ is actually a measure of the average value of the magnitude of the function $|f(t)|$ over the interval (α, β). Thus, in a sense, the statement "$\|f\|$ is small" is equivalent to the statement "f is close to being zero on most, if not all, of (α, β)." We will use this later when comparing two functions.

?▶ Exercise 11.2:   Let $(\alpha, \beta) = (0, p)$ where p is any finite positive value. Show that

a: $\|1\|^2 = p$.

b: $\|\cos(2\pi\omega_k t)\|^2 = {}^p\!/_2$ where $\omega_k = {}^k\!/_p$ and $k = 1, 2, \ldots$.

c: $\|\sin(2\pi\omega_k t)\|^2 = {}^p\!/_2$ where $\omega_k = {}^k\!/_p$ and $k = 1, 2, \ldots$.

(Hint: You can use the orthogonality relations for sines and cosines.)

11.3 Orthogonal Sets of Functions

Recall that a pair of vectors $\mathbf{u}$ and $\mathbf{v}$ is orthogonal if and only if $\mathbf{u} \cdot \mathbf{v} = 0$. Analogously, we will say that a pair of functions f and g is *orthogonal* (over (α, β)) if and only if

$$\langle f \mid g \rangle = 0 \quad .$$

!▶ Example 11.3:   The two functions $f(t) = t$ and $g(t) = 3t - 2$ form an orthogonal pair over $(0, 1)$ since

$$\langle t \mid 3t - 2 \rangle = \int_0^1 t(3t - 2)\, dt = \int_0^1 \left[3t^2 - 2t\right] dt = t^3 - t^2 \Big|_0^1 = 0 \quad .$$

On the other hand, $f(t) = t$ and $h(t) = 6t - 2$ is not an orthogonal pair of functions over $(0, 1)$ since

$$\langle t \mid 6t - 2 \rangle = \int_0^1 t(6t - 2)\, dt = \int_0^1 \left[6t^2 - 2t\right] dt = 2t^3 - t^2 \Big|_0^1 = 1 \neq 0 \quad .$$

More generally, we will say that a *set* of functions

$$\{\phi_1, \phi_2, \phi_3, \ldots\}$$

is *orthogonal* (over (α, β)) if and only if every distinct pair in the set is orthogonal, that is, if and only if

$$\langle \phi_k \mid \phi_n \rangle = 0 \quad \text{whenever} \quad k \neq n \quad .$$

Note that, because of the symmetry in the inner product (property 3), if $\langle \phi_k \mid \phi_n \rangle = 0$ then $\langle \phi_n \mid \phi_k \rangle = 0$. Thus, to show that $\{\phi_1, \phi_2, \phi_3, \ldots\}$ is orthogonal, we need only show that

$$\langle \phi_k \mid \phi_n \rangle = 0 \quad \text{whenever} \quad k < n \quad .$$

One example is of particular interest to us. It is the set of functions used to construct the trigonometric Fourier series for any periodic function with period p,

$$\{1\,,\ \cos(2\pi\omega_1 t)\,,\ \sin(2\pi\omega_1 t)\,,\ \cos(2\pi\omega_2 t)\,,\ \sin(2\pi\omega_2 t)\,,$$
$$\cos(2\pi\omega_3 t)\,,\ \sin(2\pi\omega_3 t)\,,\ \ldots\} \tag{11.1}$$

where $\omega_k = {}^k/_p$. Let us verify that this is an orthogonal set by confirming that the inner product of every distinct pair in this set is zero:

The inner product of 1 *with each cosine in the set:*

$$\langle\,1\mid\cos(2\pi\omega_k t)\,\rangle \;=\; \int_0^p 1\cdot\cos(2\pi\omega_k t)\,dt$$

$$=\; \int_0^p \cos\!\left(\frac{2\pi k}{p}t\right)dt \;=\; 0\quad.$$

(We can skip the details of the integration here since the last integral is one of the integrals in the orthogonality relations for sines and cosines (theorem 5.2 on page 54)!)

The inner product of 1 *with each sine in the set:*

$$\langle\,1\mid\sin(2\pi\omega_k t)\,\rangle \;=\; \int_0^p 1\cdot\sin(2\pi\omega_k t)\,dt$$

$$=\; \int_0^p \sin\!\left(\frac{2\pi k}{p}t\right)dt \;=\; 0\quad.$$

(Again, we can skip the details of the integration here since the last integral is one of the integrals in the orthogonality relations for sines and cosines!)

The inner product of each cosine in the set with every other cosine in the set (i.e.,

$$\langle\,\cos(2\pi\omega_k t)\mid\cos(2\pi\omega_n t)\,\rangle$$

where k *and* n *are two different positive integers):*

$$\langle\,\cos(2\pi\omega_k t)\mid\cos(2\pi\omega_n t)\,\rangle \;=\; \int_0^p \cos(2\pi\omega_k t)\cos(2\pi\omega_n t)\,dt$$

$$=\; \int_0^p \cos\!\left(\frac{2\pi k}{p}t\right)\cos\!\left(\frac{2\pi n}{p}t\right)dt \;=\; 0\quad.$$

(Again, this was one of the integrals in the orthogonality relations for sines and cosines!)

Is it possible that every inner product we need to check above corresponds to an integral from the orthogonality relations for sines and cosines? Absolutely! That's why they were called the orthogonality relations. We'll leave the final confirmation of this as an exercise.

?▶ Exercise 11.3: *Use the orthogonality relations for sines and cosines to show that each of the following inner products is zero when* n *and* k *are positive integers:*

a: $\langle\,\cos(2\pi\omega_k t)\mid\sin(2\pi\omega_n t)\,\rangle$

b: $\langle\,\sin(2\pi\omega_k t)\mid\sin(2\pi\omega_n t)\,\rangle$ *where* $k\neq n$

Also, convince yourself that this, along with the above "computations", confirms that the set in line (11.1) is orthogonal.

For completeness, *orthonormality* should be briefly discussed. A set of functions (or vectors) $\{\phi_1, \phi_2, \phi_3, \ldots\}$ is said to be *orthonormal* if and only if both of the following hold:

1. The set is an orthogonal set.

2. $\|\phi_k\| = 1$ for each ϕ_k in the set.

If $\{\phi_1, \phi_2, \phi_3, \ldots\}$ is any orthogonal set of nonzero functions, then a corresponding orthonormal set $\{\psi_1, \psi_2, \psi_3, \ldots\}$ can be constructed by "normalizing" each ϕ_k. That is, we define each ψ_k by

$$\psi_k(t) \;=\; \frac{\phi_k(t)}{\|\phi_k\|} \quad .$$

In vector analysis it is fairly standard practice to normalize a given orthogonal basis. It does make the formulas for dot products and norms simpler. In Fourier analysis, however, it is fairly standard to *not* normalize an orthogonal set of functions. Frankly, in practice, it is often just as easy (if not easier) to use your original (non-normalized) orthogonal set of functions.

11.4   Orthogonal Function Expansions

Now suppose

$$\{\,\phi_1\,,\ \phi_2\,,\ \phi_3\,,\ \ldots\}$$

is some orthogonal set of nonzero functions on an interval (α, β). Suppose, further, that f is a function on (α, β) that can be expressed as a (possibly infinite) linear combination of the ϕ_k's,

$$f(t) \;=\; \sum_k c_k \, \phi_k(t) \quad . \tag{11.2}$$

We can derive (somewhat naively) a formula for the coefficients, the c_k's, much the same way we derived the formulas for the coefficients in the trigonometric Fourier series. For example, to find c_3, first take the inner product of each side of equation (11.2) with ϕ_3,

$$\langle\, f \mid \phi_3 \,\rangle \;=\; \left\langle\, \sum_k c_k \, \phi_k \,\middle|\, \phi_3 \,\right\rangle \quad . \tag{11.3}$$

Property 1 for inner products tells us that, so long as the summation has a finite number of terms,

$$\left\langle\, \sum_k c_k \, \phi_k \,\middle|\, \phi_3 \,\right\rangle \;=\; \sum_k c_k \,\langle\, \phi_k \mid \phi_3 \,\rangle \quad .$$

Assuming this formula also holds for summations with infinitely many terms[2] and using the orthogonality of the ϕ_k's (and the definition of the norm), equation (11.3) becomes

$$\langle\, f \mid \phi_3 \,\rangle \;=\; \sum_k c_k \,\langle\, \phi_k \mid \phi_3 \,\rangle \;=\; \sum_k c_k \begin{cases} 0 & \text{if}\quad k \neq 3 \\ \|\phi_3\|^2 & \text{if}\quad k = 3 \end{cases} \;=\; c_3 \, \|\phi_3\|^2 \quad .$$

---

[2] This assumption is one reason our derivation is heuristic and not rigorous.

Dividing through by $\|\phi_3\|^2$ gives

$$c_3 = \frac{\langle\, f \mid \phi_3 \,\rangle}{\|\phi_3\|^2} \quad .$$

Of course, there is nothing special about c_3. The above derivation works equally well for any of the c_k's and gives us the following "quasi-theorem":[3]

Quasi-theorem on orthogonal expansions
Let $\{\phi_1, \phi_2, \phi_3, \ldots\}$ be an orthogonal set of nonzero functions on an interval (α, β), and let f be a function on (α, β). If f can be represented as a (possibly infinite) linear combination of the ϕ_k's, that is, if there are constants c_1, c_2, c_3, $\ldots$ such that

$$f(t) = \sum_k c_k\, \phi_k(t) \qquad \text{on} \quad (\alpha, \beta) \quad , \tag{11.4}$$

then, for each k,

$$c_k = \frac{\langle\, f \mid \phi_k \,\rangle}{\|\phi_k\|^2} \quad . \tag{11.5}$$

The summation in expression (11.4) is often called the *(generalized) Fourier series* for f (with respect to the ϕ_k's) and the corresponding c_k's are called the *(generalized) Fourier coefficients* of f (with respect to the ϕ_k's). Don't forget, however, that our derivation was not completely rigorous; so, we cannot be absolutely certain this quasi-theorem is always valid.

11.5 The Schwarz Inequality for Inner Products[*]

Recall that, if $\mathbf{u}$ and $\mathbf{v}$ are a pair of two- or three-dimensional vectors and θ is the angle between them, then

$$|\mathbf{u} \cdot \mathbf{v}| = \|\mathbf{u}\|\, \|\mathbf{v}\|\, |\cos(\theta)| \leq \|\mathbf{u}\|\, \|\mathbf{v}\| \quad .$$

Given the previous sections, you may suspect that we are going to define "the angle between two functions". Well, we aren't. Even if we could define such an "angle", we would find little or no use for it. On the other hand, we will find use for the inner product analog to the inequality $|\mathbf{u} \cdot \mathbf{v}| \leq \|\mathbf{u}\|\, \|\mathbf{v}\|$. That analog is described in the next theorem.

Theorem 11.2 (Schwarz inequality for inner products)
Let f and g be two piecewise continuous functions on the finite interval (α, β). Then

$$|\langle\, f \mid g \,\rangle| \leq \|f\|\, \|g\| \quad . \tag{11.6}$$

Inequality (11.6) is usually referred to as the *Schwarz inequality* (for inner products). It is also commonly known as the Cauchy–Schwarz inequality, and less commonly known as the Cauchy–Buniakowsky–Schwarz inequality. And if this discussion seems familiar, then you are probably recalling the discussion we had on the Schwarz inequality for finite summations starting on page 30 or the discussion of the Schwarz inequality for infinite series starting on page 45. All of these

---

[3] This is being called a quasi-theorem both to distinguish it from results we obtain rigorously, and because we may still be a little unclear on just what " $f(t) = \sum_k c_k\, \phi_k(t)$ on (α, β) " means when we have infinitely many ϕ_k's.

[*] The material in this and the following section will not be needed until near the end of chapter 13.

Schwarz inequalities are, in fact, different manifestations of the same basic mathematical principle. That's why they have virtually the same name, and why we can use virtually the same proof for each.

PROOF (of theorem 11.2):   We start rewriting inequality (11.6) in integral form,

$$\left| \int_\alpha^\beta f(t)g^*(t)\,dt \right| \le \left(\int_\alpha^\beta |f(t)|^2\,dt \right)^{1/2} \left(\int_\alpha^\beta |g(t)|^2\,dt \right)^{1/2} \ . \tag{11.7}$$

Suppose we can show that

$$\int_\alpha^\beta |f(t)|\,|g(t)|\,dt \le \left(\int_\alpha^\beta |f(t)|^2\,dt \right)^{1/2} \left(\int_\alpha^\beta |g(t)|^2\,dt \right)^{1/2} \ . \tag{11.8}$$

Then inequality (11.7) (and, hence, also inequality (11.6)) follows immediately by combining the above inequality with the fact that

$$\left| \int_\alpha^\beta f(t)g^*(t)\,dt \right| \le \int_\alpha^\beta \left| f(t)g^*(t) \right|\,dt = \int_\alpha^\beta |f(t)|\,|g(t)|\,dt \ .$$

So we only need to verify that inequality (11.8) holds.

Consider, first, the trivial case where either $\|f\|$ or $\|g\|$ is zero; that is, either

$$\int_\alpha^\beta |f(t)|^2\,dt = 0 \quad\text{or}\quad \int_\alpha^\beta |g(t)|^2\,dt = 0 \ .$$

Then either f or g vanishes everywhere on the interval (α, β). Thus, for this case,

$$\int_\alpha^\beta |f|\,|g|\,dx = \int_\alpha^\beta 0\,dx = 0 \ ,$$

and inequality (11.8) reduces to the obviously true statement that $0 \le 0$.

Now consider the case where $\|f\|$ and $\|g\|$ are both nonzero. For convenience, let

$$A = \|f\| \quad\text{and}\quad B = \|g\| \ .$$

Then

$$A^2 = \int_\alpha^\beta |f(t)|^2\,dt \quad , \quad B^2 = \int_\alpha^\beta |g(t)|^2\,dt \ ,$$

and

$$0 \le (B\,|f(t)| - A\,|g(t)|)^2 \quad\text{for}\quad \alpha < t < \beta \ .$$

Thus,

$$0 \le \int_\alpha^\beta (B\,|f(t)| - A\,|g(t)|)^2\,dt$$

$$= \int_\alpha^\beta \left[B^2\,|f(t)|^2 - 2AB\,|f(t)|\,|g(t)| + A^2\,|g(t)|^2 \right]\,dt$$

$$= B^2 \int_\alpha^\beta |f(t)|^2\,dt - 2AB \int_\alpha^\beta |f(t)|\,|g(t)|\,dt + A^2 \int_\alpha^\beta |g(t)|^2\,dt$$

$$= B^2 A^2 - 2AB \int_\alpha^\beta |f(t)|\,|g(t)|\,dt + A^2 B^2$$

$$= 2AB \left[AB - \int_\alpha^\beta |f(t)|\,|g(t)|\,dt \right] \ .$$

Dividing through by $2AB$, which is a positive quantity, and slightly rearranging the resulting inequality gives

$$\int_\alpha^\beta |f(t)| \, |g(t)| \, dt \;\leq\; AB \;=\; \left(\int_\alpha^\beta |f(t)|^2 \, dt\right)^{1/2} \left(\int_\alpha^\beta |g(t)|^2 \, dt\right)^{1/2} \;. \qquad \blacksquare$$

11.6 Bessel's Inequality

Recall that, if $\{\mathbf{i}, \mathbf{j}, \mathbf{k}\}$ is the standard basis for the space of three-dimensional vectors, and

$$\mathbf{u} = u_1\mathbf{i} + u_2\mathbf{j} + u_3\mathbf{k} \qquad \text{and} \qquad \mathbf{v} = v_1\mathbf{i} + v_2\mathbf{j} + v_3\mathbf{k} \;,$$

then the formulas for the vector dot product and vector norm can be written as

$$\mathbf{u} \cdot \mathbf{v} = u_1 v_1 + u_2 v_2 + u_3 v_3 \qquad \text{and} \qquad \|\mathbf{v}\|^2 = |v_1|^2 + |v_2|^2 + |v_3|^2 \;.$$

We can easily derive similar formulas for inner products and norms of functions which are finite linear combinations of functions from an orthogonal set of functions. Let $\{\phi_1, \phi_2, \phi_3, \ldots\}$ be any orthogonal set of piecewise continuous functions on an interval (α, β), and let f and g be two *finite* linear combination of these ϕ_k's, say,

$$f(t) = \sum_{k=1}^N c_k \, \phi_k(t) \qquad \text{and} \qquad g(t) = \sum_{k=1}^N d_k \, \phi_k(t) \;,$$

where N is some finite positive integer. Then, using the basic properties of inner products and the orthogonality of the ϕ_k's,

$$\begin{aligned}
\langle\, f \mid g \,\rangle &= \left\langle\, \sum_{k=1}^N c_k \, \phi_k \;\middle|\; \sum_{n=1}^N d_n \, \phi_n \,\right\rangle \\
&= \sum_{k=1}^N c_k \left\langle\, \phi_k \;\middle|\; \sum_{n=1}^N d_n \, \phi_n \,\right\rangle \\
&= \sum_{k=1}^N c_k \left(\sum_{n=1}^N d_n^* \, \langle\, \phi_k \mid \phi_n \,\rangle\right) \\
&= \sum_{k=1}^N \left(\sum_{n=1}^N c_k d_n^* \begin{cases} \|\phi_k\|^2 & \text{if} \quad n = k \\ 0 & \text{if} \quad n \neq k \end{cases}\right) \;,
\end{aligned}$$

which is more simply written as

$$\langle\, f \mid g \,\rangle = \sum_{k=1}^N c_k d_k^* \, \|\phi_k\|^2 \;. \tag{11.9}$$

From this we also see that

$$\|f\|^2 = \langle\, f \mid f \,\rangle = \sum_{k=1}^N c_k c_k^* \, \|\phi_k\|^2 = \sum_{k=1}^N |c_k|^2 \, \|\phi_k\|^2 \;. \tag{11.10}$$

(Note that, if the set $\{\phi_1, \phi_2, \phi_3, \ldots\}$ is an ortho*normal* set, then the above formulas reduce to

$$\langle\, f \mid g \,\rangle = \sum_{k=1}^{N} c_k d_k^* \quad \text{and} \quad \|f\|^2 = \sum_{k=1}^{N} |c_k|^2 \quad .$$

In practice, however, our sets will be orthogonal but not orthonormal.)

Before we try to extend these formulas to cases where f and g are infinite linear combinations of the ϕ_k's, we should investigate the convergence of the Fourier series more closely. The following formula for the "norm of the error" will be helpful.

Lemma 11.3
Assume $\{\phi_1, \phi_2, \ldots, \phi_N\}$ is a finite orthogonal set of piecewise continuous functions on a finite interval (α, β). Let f be any piecewise continuous function on (α, β), and let

$$c_k = \frac{\langle\, f \mid \phi_k \,\rangle}{\|\phi_k\|^2} \quad \text{for} \quad k = 1, 2, \ldots, N$$

be the corresponding generalized Fourier coefficients. Then,

$$\left\| f - \sum_{k=1}^{N} c_k \phi_k \right\|^2 = \|f\|^2 - \sum_{k=1}^{N} |c_k|^2 \|\phi_k\|^2 \quad .$$

PROOF: For convenience, let S_N be the partial sum

$$S_N(t) = \sum_{k=1}^{N} c_k \phi_k(t) \quad .$$

Observe that

$$\left\| f - \sum_{k=1}^{N} c_k \phi_k \right\|^2 = \|f - S_N\|^2$$

$$= \langle\, f - S_N \mid f - S_N \,\rangle$$

$$= \langle\, f \mid f \,\rangle - \langle\, f \mid S_N \,\rangle - \langle\, S_N \mid f \,\rangle + \langle\, S_N \mid S_N \,\rangle \quad . \quad (11.11)$$

Now

$$\langle\, f \mid f \,\rangle = \|f\|^2 \quad ,$$

$$\langle\, f \mid S_N \,\rangle = \left\langle\, f \mid \sum_{k=1}^{N} c_k \phi_k \,\right\rangle = \sum_{k=1}^{N} c_k^* \langle\, f \mid \phi_k \,\rangle$$

$$= \sum_{k=1}^{N} c_k^* \left(c_k \|\phi_k\|^2 \right) = \sum_{k=1}^{N} |c_k|^2 \|\phi_k\|^2 \quad ,$$

$$\langle\, S_N \mid f \,\rangle = \langle\, f \mid S_N \,\rangle^* = \left(\sum_{k=1}^{N} |c_k|^2 \|\phi_k\|^2 \right)^* = \sum_{k=1}^{N} |c_k|^2 \|\phi_k\|^2$$

and, using equation (11.10) (with S_N replacing f),

$$\langle\, S_N \mid S_N \,\rangle \;=\; \left\langle\, \sum_{k=1}^{N} c_k\, \phi_k \;\middle|\; \sum_{n=1}^{N} c_n\, \phi_n \,\right\rangle \;=\; \sum_{k=1}^{N} |c_k|^2\, \|\phi_k\|^2 \quad.$$

So equation (11.11) becomes

$$\left\| f - \sum_{k=1}^{N} c_k\, \phi_k \right\|^2 \;=\; \|f\|^2 \;-\; \sum_{k=1}^{N} |c_k|^2\, \|\phi_k\|^2 \;-\; \sum_{k=1}^{N} |c_k|^2\, \|\phi_k\|^2 \;+\; \sum_{k=1}^{N} |c_k|^2\, \|\phi_k\|^2$$

$$=\; \|f\|^2 \;-\; \sum_{k=1}^{N} |c_k|^2\, \|\phi_k\|^2 \quad. \qquad\blacksquare$$

An immediate, and useful, consequence is Bessel's inequality.

Theorem 11.4 (Bessel's inequality, general series version)

Assume that $\{\phi_1, \phi_2, \phi_3, \ldots\}$ is an infinite, orthogonal set of piecewise continuous functions on a finite interval (α, β). Let f be any piecewise continuous function on (α, β), and, for each positive integer k, let c_k be the corresponding generalized Fourier coefficient,

$$c_k \;=\; \frac{\langle\, f \mid \phi_k \,\rangle}{\|\phi_k\|^2} \quad.$$

Then

$$\sum_{k=1}^{N} |c_k|^2\, \|\phi_k\|^2 \;\le\; \|f\|^2$$

for every positive integer N. Moreover, the infinite series $\sum_{k=1}^{\infty} |c_k|^2\, \|\phi_k\|^2$ converges and

$$\sum_{k=1}^{\infty} |c_k|^2\, \|\phi_k\|^2 \;\le\; \|f\|^2 \quad.$$

PROOF: Let N be any positive integer. Observe that, using the identity from the previous lemma,

$$0 \;\le\; \left\| f - \sum_{k=1}^{N} c_k\, \phi_k \right\|^2 \;=\; \|f\|^2 \;-\; \sum_{k=1}^{N} |c_k|^2\, \|\phi_k\|^2 \quad.$$

Subtracting the summation from the left- and right-hand sides gives

$$\sum_{k=1}^{N} |c_k|^2\, \|\phi_k\|^2 \;\le\; \|f\|^2 \quad,$$

proving the first claim of the theorem.

This also tells us that every partial sum of $\sum_{k=1}^{\infty} |c_k|^2\, \|\phi_k\|^2$ is bounded by the finite value $\|f\|^2$. Since this series has only nonnegative terms, we know (see theorem 4.4 on page 43) this series must converge and that

$$\sum_{k=1}^{\infty} |c_k|^2\, \|\phi_k\|^2 \;=\; \lim_{N\to\infty} \sum_{k=1}^{N} |c_k|^2\, \|\phi_k\|^2 \;\le\; \|f\|^2 \quad. \qquad\blacksquare$$

Let us see what Bessel's inequality tells us about the trigonometric Fourier series:

!▶ **Example 11.4 (Fourier series and Bessel's inequality):**   *Let* f *be any periodic, piecewise continuous function with period* p *and*

$$F.S.[f]|_t = A_0 + \sum_{k=1}^{\infty} [a_k \cos(2\pi \omega_k t) + b_k \sin(2\pi \omega_k t)]$$

(where $\omega_k = {}^k/_p$ *). Earlier, we saw that*

$$\{1, \cos(2\pi \omega_1 t), \sin(2\pi \omega_1 t), \cos(2\pi \omega_2 t), \sin(2\pi \omega_2 t), \ldots\}$$

is an orthogonal set of functions on any interval of length p *. By the computations from Exercise 11.2 we have*

$$|A_0|^2 \|1\|^2 + \sum_{k=1}^{\infty} \left[|a_k|^2 \|\cos(2\pi \omega_k t)\|^2 + |b_k|^2 \|\sin(2\pi \omega_k t)\|^2 \right]$$

$$= |A_0|^2 p + \sum_{k=1}^{\infty} \left[|a_k|^2 \frac{p}{2} + |b_k|^2 \frac{p}{2} \right] .$$

Bessel's inequality (theorem 11.4) assures us that this series converges and that

$$|A_0|^2 p + \sum_{k=1}^{\infty} \left[|a_k|^2 \frac{p}{2} + |b_k|^2 \frac{p}{2} \right] \leq \|f\|^2 .$$

It is worth noting that, since this series converges, the terms of this series must approach zero as $k \to \infty$ *. That is, we must have*

$$\lim_{k \to \infty} |a_k|^2 \frac{p}{2} = 0 \quad \text{and} \quad \lim_{k \to \infty} |b_k|^2 \frac{p}{2} = 0 ,$$

which, of course, means that the Fourier coefficients, themselves, must vanish as $k \to \infty$ *,*

$$\lim_{k \to \infty} a_k = 0 \quad \text{and} \quad \lim_{k \to \infty} b_k = 0 . \qquad (11.12)$$

You should realize that neither the convergence of

$$|A_0|^2 p + \sum_{k=1}^{\infty} \left[|a_k|^2 \frac{p}{2} + |b_k|^2 \frac{p}{2} \right]$$

in the previous example nor the vanishing of the coefficients as $k \to \infty$ allows us to decisively conclude that the Fourier series

$$A_0 + \sum_{k=1}^{\infty} [a_k \cos(2\pi \omega_k t) + b_k \sin(2\pi \omega_k t)]$$

converges for any given value of t. But they should, at least, strengthen our suspicion that the Fourier series will converge.

Take another look at the last set of limits in the last example. Recalling the definition of the trigonometric Fourier coefficients, we see that equation set (11.12) immediately implies that[4]

$$\lim_{\substack{k \to \infty \\ k \in \mathbb{Z}}} \int_{\text{period}} f(t) \cos\left(\frac{2\pi k}{p} t\right) dt = 0 \quad \text{and} \quad \lim_{\substack{k \to \infty \\ k \in \mathbb{Z}}} \int_{\text{period}} f(t) \sin\left(\frac{2\pi k}{p} t\right) dt = 0$$

whenever f is a periodic, piecewise continuous function with period p. This is a particular case of a famous result called the Riemann–Lebesgue lemma. A fairly general version is given in the next theorem.

Theorem 11.5 (a general Riemann–Lebesgue lemma)
Let f be any piecewise continuous function on a finite interval (α, β), and let $\{\phi_1, \phi_2, \phi_3, \ldots\}$ be any infinite orthogonal set of piecewise continuous functions on (α, β). Assume that, for some finite constant C and every positive integer k,

$$\|\phi_k\| < C .$$

Then

$$\lim_{k \to \infty} \int_\alpha^\beta f(t) \phi_k^*(t) \, dt = 0 .$$

The proof, which is quite easy, is left as an exercise.

?▶ Exercise 11.4: *Prove theorem 11.5 using theorem 11.4, the ideas indicated in example 11.4, and the above discussion.*

Additional Exercises

11.5. *The interval for this exercise is $(0, L)$ for some arbitrary positive number L.*

    a. *Show that the following set of functions is orthogonal over $(0, L)$:*

$$\left\{ \sin\left(\frac{\pi}{L} t\right), \, \sin\left(\frac{2\pi}{L} t\right), \, \sin\left(\frac{3\pi}{L} t\right), \, \ldots \right\} .$$

    b. *Compute $\left\| \sin\left(\frac{k\pi}{L} t\right) \right\|$ where k is any positive integer.*

    c. *Let f be any piecewise continuous function on $(0, L)$, and let*

$$\sum_{k=1}^{\infty} b_k \sin\left(\frac{k\pi}{L} t\right)$$

be the Fourier sine series for f over $(0, L)$ as defined in chapter 10.

---

[4] $\mathbb{Z}$ denotes the set of integers. The "$k \in \mathbb{Z}$" in the limits simply emphasizes that we are only using integer values for k in this limit.

i. *Verify that*

$$b_k = \frac{\left\langle f(t) \mid \sin\left(\frac{k\pi}{L}t\right) \right\rangle}{\left\| \sin\left(\frac{k\pi}{L}t\right) \right\|^2} \qquad \text{for} \quad k = 1, 2, 3, \ldots \quad .$$

ii. *Verify Bessel's inequality for the sine series,*

$$\frac{L}{2} \sum_{k=1}^{\infty} |b_k|^2 \leq \int_0^L |f(t)|^2 \, dt \quad .$$

11.6. *The interval for this exercise is* $(0, L)$ *for some arbitrary positive number* L .

a. *Show that the following set of functions is orthogonal over* $(0, L)$:

$$\left\{ 1, \cos\left(\frac{\pi}{L}t\right), \cos\left(\frac{2\pi}{L}t\right), \cos\left(\frac{3\pi}{L}t\right), \ldots \right\} \quad .$$

b. *Compute* $\|1\|$.

c. *Compute* $\left\| \cos\left(\frac{k\pi}{L}t\right) \right\|$ *where* k *is any positive integer.*

d. *Let* f *be any piecewise continuous function on* $(0, L)$, *and let*

$$A_0 + \sum_{k=1}^{\infty} a_k \cos\left(\frac{k\pi}{L}t\right)$$

be the Fourier cosine series for f *over* $(0, L)$ *as defined in chapter 10.*

i. *Verify that*

$$A_0 = \frac{\langle f(t) \mid 1 \rangle}{\|1\|^2}$$

and

$$a_k = \frac{\left\langle f(t) \mid \cos\left(\frac{k\pi}{L}t\right) \right\rangle}{\left\| \cos\left(\frac{k\pi}{L}t\right) \right\|^2} \qquad \text{for} \quad k = 1, 2, 3, \ldots \quad .$$

ii. *Verify Bessel's inequality for the cosine series,*

$$L |A_0|^2 + \frac{L}{2} \sum_{k=1}^{\infty} |a_k|^2 \leq \int_0^L |f(t)|^2 \, dt \quad .$$

11.7. *For the following, assume* $\omega_k = \frac{k}{p}$ *where* p *is some finite positive number.*

a. *Show that the following is an orthogonal set on any interval of length* p :

$$\left\{ \ldots, e^{i2\pi\omega_{-2}t}, e^{i2\pi\omega_{-1}t}, e^{i2\pi\omega_0 t}, e^{i2\pi\omega_1 t}, e^{i2\pi\omega_2 t}, e^{i2\pi\omega_3 t}, \ldots \right\} \quad .$$

b. *Compute* $\left\| e^{i2\pi\omega_k t} \right\|$ *for* $k = 0, \pm 1, \pm 2, \pm 3, \ldots$.

(Much more will be done with this orthogonal set in the next chapter.)

11.8. *For this problem, let* L *be the length of the interval* (α, β) , *and let* f *be a piecewise continuous function on* (α, β) . *Using the Schwarz inequality and the simple observation that* $f = f \cdot 1$, *derive the inequality*

$$\int_{\alpha}^{\beta} |f(t)| \, dt \leq \sqrt{L} \, \|f\| \quad . \tag{11.13}$$

(This inequality may be of use on later occasions.)

11.9. *In the following, we will see one way to construct an orthogonal set of polynomials on the interval* $(0, 1)$.[5]

   a. *Find all possible real values* a *and* b *so that*

$$\phi_0(t) = 1 \quad \text{and} \quad \phi_1(t) = a + bt$$

   is an orthogonal pair of functions on $(0, 1)$.

   b. *From the answer to the above, it follows that*

$$\phi_0(t) = 1 \quad \text{and} \quad \phi_1(t) = 1 - 2t$$

   is an orthogonal pair of functions on $(0, 1)$. *Now find all possible real values* a , b *and* c *such that* $\{\phi_1, \phi_2, \phi_3\}$ *is an orthogonal set of functions on* $(0, 1)$ *when*

$$\phi_3(t) = a + bt + ct^2 \quad .$$

   (By continuing the sort of computations described above, we can construct an orthogonal set of functions $\{\phi_0, \phi_1, \phi_2, \phi_3, \ldots\}$ *on* $(0, 1)$ *where each* $\phi_k(t)$ *is a polynomial of degree* k .)

11.10. *Assume we have continued the computations indicated in the last exercise, and now have an orthogonal set of polynomials on* $(0, 1)$ $\{\phi_0, \phi_1, \phi_2, \phi_3, \ldots\}$ *with*

$$\phi_0(t) = 1 \quad \text{and} \quad \phi_1(t) = 1 - 2t \quad .$$

   a. *Find the norms of* ϕ_0 *and* ϕ_1 .

   b. *Assume that* $f(t)$ *is a function on* $(0, 1)$ *and that, on this interval,*

$$f(t) = \sum_{k=0}^{\infty} c_k \phi_k(t) \quad .$$

   According to the quasi-theorem on orthogonal expansions on page 135), what are the values of c_0 *and* c_1 *when*

   i. $f(t) = t$?   **ii.** $f(t) = \sqrt{t}$?   **iii.** $f(t) = \sin(\pi t)$?

11.11. *The following concern the* Haar wavelet set *on the interval* $(0, 1)$.[6] *This is the set of functions*

$$\left\{ \psi_0, \ \psi_{1,0}, \ \psi_{2,0}, \ \psi_{2,1}, \ \psi_{3,0}, \ \psi_{3,1}, \ \psi_{3,2}, \ \psi_{3,3}, \ \ldots \right\}$$

   given by

$$\psi_0(t) = 1 \quad \text{for} \quad 0 < t < 1 \quad ,$$

$$\psi_{1,0}(t) = \begin{cases} +1 & \text{if } \ 0 < t < \frac{1}{2} \\ -1 & \text{if } \ \frac{1}{2} < t < 1 \end{cases} \quad ,$$

   and, for $j = 2, 3, \ldots$,

$$\psi_{j,0}(t) = \psi_{1,0}\left(2^{j-1}t\right) = \begin{cases} +1 & \text{if } \ 0 < t < \dfrac{1}{2^j} \\ -1 & \text{if } \ \dfrac{1}{2^j} < t < \dfrac{2}{2^j} \\ 0 & \text{otherwise} \end{cases}$$

---

[5] But a better approach may be to use the "Gram–Schmidt" orthogonalization procedure from linear algebra.

[6] This set of functions is important in "digital multi-resolution analysis".

and

$$\psi_{j,k} = \psi_{j,0}\left(t - \frac{2k}{2^j}\right) \qquad for \quad k = 0, 1, 2, \ldots, 2^{j-1} - 1 \quad .$$

a. Sketch the graphs of the following over the interval $(0, 1)$:

$$\psi_0 \quad , \quad \psi_{1,0} \quad , \quad \psi_{2,0} \quad , \quad \psi_{2,1} \quad , \quad \psi_{3,0} \quad , \quad \psi_{3,1} \quad , \quad \psi_{3,2} \quad and \quad \psi_{3,3} \quad .$$

b. Compute the norms of

$$\psi_0 \quad , \quad \psi_{1,0} \quad , \quad \psi_{2,0} \quad , \quad \psi_{2,1} \quad , \quad \psi_{3,0} \quad , \quad \psi_{3,1} \quad , \quad \psi_{3,2} \quad and \quad \psi_{3,3} \quad .$$

(*Suggestion: Use your sketches from the first part of this exercise to quickly compute these norms.*)

c. Verify that the set

$$\left\{ \psi_0, \psi_{1,0}, \psi_{2,0}, \psi_{2,1}, \psi_{3,0}, \psi_{3,1}, \psi_{3,2}, \psi_{3,3} \right\}$$

is an orthogonal set on $(0, 1)$. (*Suggestion: Use your graphs of these functions to reduce each inner product to an extremely simple integral.*)

11.12. By extending the computations described in the last exercise, it is fairly easy to show that the entire Haar wavelet set on $(0, 1)$ is an orthogonal set on $(0, 1)$. It can also be shown (*though not so easily*) that any piecewise continuous function f on $(0, 1)$ can be written as a (*possibly infinite*) linear combination of the functions in the Haar wavelet set,

$$f(t) = c_0\psi_0 + \sum_{j=1}^{\infty} \sum_{k=0}^{2^{j-1}-1} c_{j,k}\psi_{j,k}(t) \quad .$$

Assuming all this, as well as the quasi-theorem on orthogonal expansions on page 135, compute the corresponding generalized Fourier coefficients

$$c_0 \quad , \quad c_{1,0} \quad , \quad c_{2,0} \quad , \quad c_{2,1} \quad , \quad c_{3,0} \quad , \quad c_{3,1} \quad and \quad c_{3,2}$$

for each of the following choices of f :

a. $f(t) = 1$ for all t **b.** $f(t) = \begin{cases} 1 & if \ 0 < t < \frac{1}{2} \\ 0 & if \ \frac{1}{2} < t < 1 \end{cases}$

c. $f(t) = t$ for all t **d.** $f(t) = \begin{cases} t & if \ 0 < t < \frac{1}{2} \\ 0 & if \ \frac{1}{2} < t < 1 \end{cases}$

11.13. Show that the Schwarz inequality, inequality (11.6), becomes an equality whenever one function is a constant multiple of the other. (*Later — see theorem 25.7 on page 404 — it will be shown that the converse holds; that is, if both sides of the Schwarz inequality equal each other, then one function is a constant multiple of the other.*)

11.14. Let f and g be any two piecewise continuous functions on the interval (α, β), and show that we have the "generalized Pythagorean equation"

$$\|f + g\|^2 = \|f\|^2 + \|g\|^2$$

if and only if f and g is an orthogonal pair on (α, β).

12

The Complex Exponential Fourier Series

In chapter 9 we defined the trigonometric Fourier series for a periodic, piecewise continuous function. That was an infinite series of the form

$$A_0 + \sum_{k=1}^{\infty} [a_k \cos(2\pi \omega_k t) + b_k \sin(2\pi \omega_k t)] \quad .$$

Dealing with this series can be somewhat tedious. Typically, for example, the constant term, the cosine terms and the sine terms must be computed separately.

In this chapter we will derive an alternative — the complex exponential Fourier series — which, basically, is just the trigonometric series rewritten in terms of complex exponentials. This may not seem to be much of an improvement, especially since it will require complex-valued functions in computations that, up to this point, have only involved real-valued functions. In the long run, however, we will find that the advantages of using complex exponentials instead of sines and cosines greatly outweigh the disadvantages of having to deal with complex-valued functions.[1]

12.1 Derivation

Let f be a periodic, piecewise continuous function with period p and trigonometric Fourier series

$$F.S.\,[f]|_t = A_0 + \sum_{k=1}^{\infty} [a_k \cos(2\pi \omega_k t) + b_k \sin(2\pi \omega_k t)] \quad . \tag{12.1}$$

Using

$$\cos(2\pi \omega_k t) = \frac{e^{i2\pi \omega_k t} + e^{-i2\pi \omega_k t}}{2} \quad \text{and} \quad \sin(2\pi \omega_k t) = \frac{e^{i2\pi \omega_k t} - e^{-i2\pi \omega_k t}}{2i} \quad ,$$

let us rewrite formula (12.1) in terms of complex exponentials:

$$F.S.\,[f]|_t = A_0 + \sum_{k=1}^{\infty} \left[a_k \frac{e^{i2\pi \omega_k t} + e^{-i2\pi \omega_k t}}{2} + b_k \frac{e^{i2\pi \omega_k t} - e^{-i2\pi \omega_k t}}{2i} \right]$$

$$= A_0 + \sum_{k=1}^{\infty} \left[C_k\, e^{i2\pi \omega_k t} + D_k\, e^{-i2\pi \omega_k t} \right]$$

---

[1] This may be a good time to review chapter 6, *Elementary Complex Analysis*, especially the section on the complex exponential.

$$= A_0 + \sum_{k=1}^{\infty} C_k \, e^{i 2\pi \omega_k t} + \sum_{k=1}^{\infty} D_k \, e^{-i 2\pi \omega_k t}$$

$$= \sum_{k=1}^{\infty} D_k \, e^{-i 2\pi \omega_k t} + A_0 + \sum_{k=1}^{\infty} C_k \, e^{i 2\pi \omega_k t} \qquad (12.2)$$

where the C_k's and D_k's are constants that could easily (but won't) be computed from the a_k's and b_k's. At this point, remember that

$$\omega_k = \frac{k}{p} \qquad \text{for} \quad k = 1, 2, 3, \dots \quad .$$

Let's take a not-so-bold step and agree that

$$\omega_k = \frac{k}{p} \qquad \text{for} \quad k = 0, \pm 1, \pm 2, \pm 3, \dots \quad .$$

For even more convenience, we can rename our coefficients by letting

$$c_k = \begin{cases} C_k & \text{if} \quad k = 1, 2, 3, \dots \\ A_0 & \text{if} \quad k = 0 \\ D_{-k} & \text{if} \quad k = -1, -2, -3, \dots \end{cases} \quad .$$

Observe that, because $\omega_0 = 0$,

$$A_0 = A_0 \, e^0 = c_0 \, e^{i 2\pi \omega_0 t} \quad . \qquad (12.3)$$

Also, for $k = 1, 2, 3, \dots$,

$$-2\pi \omega_k t = -2\pi \frac{k}{p} t = 2\pi \left(\frac{-k}{p} \right) t = 2\pi \omega_{-k} t \quad .$$

So, using the index substitution $n = -k$ followed by a renaming of the internal variable n as k again,

$$\sum_{k=1}^{\infty} D_k \, e^{-i 2\pi \omega_k t} = \sum_{k=1}^{\infty} D_k \, e^{i 2\pi \omega_{-k} t} = \sum_{n=-1}^{-\infty} D_{-n} \, e^{i 2\pi \omega_n t} = \sum_{k=-1}^{-\infty} c_k \, e^{i 2\pi \omega_k t} \quad . \qquad (12.4)$$

Thus, we can rewrite formula (12.2) as

$$F.S. [f]|_t = \sum_{k=-\infty}^{-1} c_k \, e^{i 2\pi \omega_k t} + c_0 \, e^{i 2\pi \omega_0 t} + \sum_{k=1}^{\infty} c_k \, e^{i 2\pi \omega_k t} \quad ,$$

or, even more concisely, as

$$F.S. [f]|_t = \sum_{k=-\infty}^{\infty} c_k \, e^{i 2\pi \omega_k t} \quad . \qquad (12.5)$$

The formulas for the c_k's can be rigorously derived by first finding the relation between them and the corresponding trigonometric Fourier coefficients and then using the formulas for computing those coefficients (formulas (9.1c), (9.1d) and (9.1e) on page 101). On the other hand, if the set of $e^{i 2\pi \omega_k t}$'s is an orthogonal set of functions, then we should be able to derive formulas for the c_k's more easily using the more general formula from the quasi-theorem on orthogonal function expansions on page 135. This is the approach we will take.

First, observe that, if g is any periodic, piecewise continuous function with period p, and if n is any integer, then the inner product of g with $e^{i2\pi\omega_n t}$ is

$$\left\langle g(t) \mid e^{i2\pi\omega_n t} \right\rangle = \int_0^p g(t) \left(e^{i2\pi\omega_n t} \right)^* dt = \int_0^p g(t) e^{-i2\pi\omega_n t} dt .$$

In particular, if k and n are two different integers, then

$$\left\langle e^{i2\pi\omega_k t} \mid e^{i2\pi\omega_n t} \right\rangle = \int_0^p e^{i2\pi\omega_k t} e^{-i2\pi\omega_n t} dt$$

$$= \int_0^p \exp\left(\frac{i2\pi(k-n)}{p} t \right) dt$$

$$= \frac{p}{i2\pi(k-n)} \exp\left(\frac{i2\pi(k-n)}{p} t \right) \Big|_{t=0}^p$$

$$= \frac{p}{i2\pi(k-n)} [1 - 1]$$

$$= 0 .$$

This verifies that the set of $e^{i2\pi\omega_k t}$'s is an orthogonal set. Thus, according to the quasi-theorem on orthogonal function expansions on page 135, each c_k is given by

$$c_k = \frac{\left\langle f \mid e^{i2\pi\omega_k t} \right\rangle}{\left\| e^{i2\pi\omega_k t} \right\|^2} . \tag{12.6}$$

Equivalently,

$$c_k = \frac{1}{p} \int_0^p f(t) e^{-i2\pi\omega_k t} dt \tag{12.7}$$

since

$$\left\langle f \mid e^{i2\pi\omega_k t} \right\rangle = \int_0^p f(t) e^{-i2\pi\omega_k t} dt$$

and

$$\left\| e^{i2\pi\omega_k t} \right\|^2 = \int_0^p e^{i2\pi\omega_k t} e^{-i2\pi\omega_k t} dt = \int_0^p 1 \, dt = p .$$

Because the derivation of the quasi-theorem on orthogonal function expansions was not completely rigorous, there should be some concern that formulas (12.6) and (12.7) may not be correct. They are correct. The reader can either trust the author or, even better, do exercise 12.2 on page 151.

12.2 Notation and Terminology

Using the formulas just derived in the previous section, we can formally define the complex exponential Fourier series.

Let f be a periodic function with period p. The *(complex exponential) Fourier series* for f, denoted by $F.S.[f]$, is the infinite series

$$F.S.[f]|_t = \sum_{k=-\infty}^{\infty} c_k e^{i2\pi\omega_k t} \tag{12.8a}$$

where, for $k = 0, \pm 1, \pm 2, \pm 3, \ldots,$

$$\omega_k = \frac{k}{p} \tag{12.8b}$$

and

$$c_k = \frac{1}{p} \int_0^p f(t)\, e^{-i2\pi\omega_k t}\, dt \quad . \tag{12.8c}$$

The c_k's are called the *(complex exponential) Fourier coefficients* of f.

As before, we will usually require f to be piecewise continuous on $\mathbb{R}$ just to ensure that the integrals are all well defined.

Compare the next example to example 9.1. It illustrates that fewer computations are often needed to find the complex exponential Fourier series of a function than are needed to find the corresponding trigonometric Fourier series.

!▶ *Example 12.1 (the saw function, again):*   *Let's compute the complex exponential Fourier series for the saw function with fundamental period* 3,

$$f(t) = \text{saw}_3(t) = \begin{cases} t & \text{if } 0 < t < 3 \\ f(t-3) & \text{in general} \end{cases} \quad .$$

Here $p = 3$. *For* $k = 0, \pm 1, \pm 2, \pm 3, \ldots$, *formulas (12.8b) and (12.8c) become*

$$\omega_k = \frac{k}{3}$$

and

$$c_k = \frac{1}{3} \int_0^3 t\, e^{-i2\pi\omega_k t}\, dt = \frac{1}{3} \int_0^3 t \exp\left(-\frac{i2\pi k}{3}t\right) dt \quad .$$

Using integration by parts, we see that, for $k \neq 0$,

$$c_k = \frac{1}{3}\left[\frac{-3t}{i2\pi k}\exp\left(-\frac{i2\pi k}{3}t\right)\Big|_{t=0}^3 + \frac{3}{i2\pi k}\int_0^3 \exp\left(-\frac{i2\pi k}{3}t\right) dt \right]$$

$$= \frac{1}{3}\left[\frac{-3\cdot 3}{i2\pi k} - 0 - \left(\frac{3}{i2\pi k}\right)^2 \left[e^{-i2\pi k} - e^0 \right] \right]$$

$$= \frac{3i}{2\pi k} \quad .$$

Because the above formula for c_k *involves division by* k, *it is not valid when* $k = 0$. *So* c_0 *must be computed separately. Since* $\omega_0 = 0$ *and* $e^0 = 1$,

$$c_0 = \frac{1}{3}\int_0^3 t\, e^{-i2\pi\omega_0 t}\, dt = \frac{1}{3}\int_0^3 t\, dt = \frac{3}{2} \quad .$$

Thus, with $\omega_k = {}^k\!/_3$,

$$F.S.[f]|_t = \frac{3}{2} + \sum_{\substack{k=-\infty \\ k\neq 0}}^{\infty} \frac{3i}{2\pi k} e^{i2\pi\omega_k t} \quad .$$

12.3 Computing the Coefficients

All the comments made in chapter 9 regarding the computation of the trigonometric Fourier coefficients apply, suitably rephrased, to the computation of the complex exponential Fourier coefficients. In particular, if f is any periodic, piecewise continuous function with period p and Fourier series

$$F.S.[f]|_t = \sum_{k=-\infty}^{\infty} c_k e^{i2\pi \omega_k t} \quad ,$$

then:

1. *(independence of period) If p_1 and p_2 are any two periods for f, then the complex exponential Fourier series for f computed using $p = p_1$ is identical, after simplification, to the complex exponential Fourier series computed using $p = p_2$.*

2. *(alternate intervals of integration) Formula (12.8c) is completely equivalent to*

$$c_k = \frac{1}{p} \int_{period} f(t) e^{-i2\pi \omega_k t} dt \quad , \qquad (12.8c')$$

   where it is understood that the integration can be done over any interval of length p.

3. *(symmetry) If f is an even function, then*

$$c_k = c_{-k} \qquad for \quad k = 1, 2, 3, \ldots \quad ;$$

   while if f is an odd function, then $c_0 = 0$ and

$$c_k = -c_{-k} \qquad for \quad k = 1, 2, 3, \ldots \quad .$$

   (Notes: (1) See exercise 12.2 c. (2) To be honest, these particular formulas are seldom of much value in computing coefficients.)

4. *(linearity) If $f = \alpha g + \beta h$ where α and β are constants, and g and h are periodic, piecewise continuous functions each with period p and having Fourier series*

$$F.S.[g]|_t = \sum_{k=-\infty}^{\infty} \widehat{g}_k e^{i2\pi \omega_k t} \qquad and \qquad F.S.[h]|_t = \sum_{k=-\infty}^{\infty} \widehat{h}_k e^{i2\pi \omega_k t}$$

   (with $\omega_k = {}^k/_p$ in each), then

$$c_k = \alpha \widehat{g}_k + \beta \widehat{h}_k \qquad for \quad k = 1, 2, 3, \ldots \quad .$$

5. *If f is a finite linear combination of complex exponential functions having a common period, then that linear combination is the complex exponential Fourier series of f.*

6. *(scaling) If $g(t) = f(\alpha t)$ for some $\alpha > 0$, then*

$$F.S.[g]|_t = \sum_{k=-\infty}^{\infty} c_k e^{i2\pi (\alpha \omega_k) t} \quad .$$

7. *If $g(t) = f(-t)$, then*

$$F.S.[g]|_t = \sum_{k=-\infty}^{\infty} c_{-k} e^{i2\pi \omega_k t} \quad . \qquad (12.9)$$

12.4 Partial Sums

Let f be a periodic function with

$$F.S.[f]|_t = \sum_{k=-\infty}^{\infty} c_k e^{i2\pi\omega_k t} \quad .$$

Since this is a two-sided infinite series (see page 45), we will be interested in the general $(M, N)^{\text{th}}$ partial sum

$$F.S._{MN}[f]|_t = \sum_{k=M}^{N} c_k e^{i2\pi\omega_k t}$$

where M and N are any two integers with $M < N$. Now, if

$$A_0 + \sum_{k=1}^{\infty} [a_k \cos(2\pi\omega_k t) + b_k \sin(2\pi\omega_k t)]$$

is the corresponding trigonometric Fourier series for f, then it is easily verified that $c_0 = A_0$ and that, for each positive integer k,

$$c_{-k} e^{i2\pi\omega_{-k} t} + c_k e^{i2\pi\omega_k t} = a_k \cos(2\pi\omega_k t) + b_k \sin(2\pi\omega_k t) \quad .$$

"Summing these equalities up", with k going from 0 to any finite positive integer N, gives

$$\sum_{k=-N}^{N} c_k e^{i2\pi\omega_k t} = A_0 + \sum_{k=1}^{N} [a_k \cos(2\pi\omega_k t) + b_k \sin(2\pi\omega_k t)] = F.S._N[f]|_t \quad .$$

Thus, the Nth partial sum for the trigonometric Fourier series is identical to the Nth *symmetric* partial sum for the complex exponential series. For this reason, we will occasionally have a particular interest in the Nth symmetric partial sum for the complex exponential series.

Formally, any complex exponential Fourier series can be converted to the corresponding trigonometric series by expressing the complex exponentials in terms of sines and cosines. Likewise, as indicated in the derivation at the beginning of this chapter, any trigonometric series can be converted into the corresponding complex exponential series. So it certainly looks as if these two types of Fourier series are really the same series written in slightly different forms (a fact that we've already indicated by using the same notation, $F.S.[f]$, for both series). The only possible difference between the two lies in the slightly different partial sums used to find the sum of each series. For the complex exponential series,

$$\sum_{k=-\infty}^{\infty} c_k e^{i2\pi\omega_k t} = \lim_{\substack{N \to \infty \\ M \to -\infty}} \sum_{k=M}^{N} c_k e^{i2\pi\omega_k t} \quad ,$$

while for the trigonometric series,

$$A_0 + \sum_{k=1}^{\infty} [a_k \cos(2\pi\omega_k t) + b_k \sin(2\pi\omega_k t)]$$

$$= \lim_{N \to \infty} \left[A_0 + \sum_{k=1}^{N} [a_k \cos(2\pi\omega_k t) + b_k \sin(2\pi\omega_k t)] \right] \quad .$$

In the next two chapters, we will confirm that this difference between the two series is not of practical significance.

Additional Exercises

12.1 a. *By rewriting the sines and cosines in terms of exponentials, convert*

$$1 + \sum_{k=1}^{\infty}\left[\frac{2}{k^2+1}\cos(2\pi\omega_k t) - \frac{2k}{k^2+1}\sin(2\pi\omega_k t)\right]$$

to the corresponding complex exponential Fourier series.

b. *By rewriting the complex exponentials in terms of sines and cosines, convert*

$$\sum_{k=-\infty}^{\infty}\frac{ik}{k^2+4}e^{i2\pi\omega_k t}$$

to the corresponding trigonometric Fourier series.

12.2. *Let f be a periodic, piecewise continuous function with period p, trigonometric Fourier series*

$$A_0 + \sum_{k=1}^{\infty}[a_k\cos(2\pi\omega_k t) + b_k\sin(2\pi\omega_k t)]$$

and complex exponential Fourier series

$$\sum_{k=-\infty}^{\infty}c_k e^{i2\pi\omega_k t}$$

(with $\omega_k = {}^k/_p$ in both series).

a. *Using the formulas for the coefficients (and not the results from chapter 11), show that*

$$A_0 = c_0$$

and that, for $k = 1, 2, 3, \ldots,$

$$c_{-k}e^{i2\pi\omega_{-k}t} + c_k e^{i2\pi\omega_k t} = a_k\cos(2\pi\omega_k t) + b_k\sin(2\pi\omega_k t) \quad .$$

(Suggestion: Use some symbol other than t as the variable of integration in the integrals defining the coefficients.)

b. *Show that, for every nonnegative integer N,*

$$A_0 + \sum_{k=1}^{N}[a_k\cos(2\pi\omega_k t) + b_k\sin(2\pi\omega_k t)] = \sum_{k=-N}^{N}c_k e^{i2\pi\omega_k t} \quad .$$

c. *Show that, if f is an even function, then*

$$c_k = c_{-k} \quad \text{for} \quad k = 1, 2, 3, \ldots \quad ;$$

while, if f is an odd function, then $c_0 = 0$ and

$$c_k = -c_{-k} \quad \text{for} \quad k = 1, 2, 3, \ldots \quad .$$

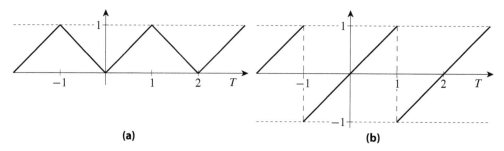

Figure 12.1: Two and one half periods of (a) evensaw(t), the even sawtooth function for exercise 12.3 c, and (b) oddsaw(t), the odd sawtooth function for exercise 12.3 d.

12.3. *Compute the complex exponential Fourier series for each of the following functions. (Most of these functions are being recycled from exercises in chapter 9. Do not, however, convert the answers for those problems to obtain the Fourier series for the following. Instead, compute the coefficients using equation set (12.8) on page 147.)*

a. $f(t) = \begin{cases} 0 & \text{if } -1 < t < 0 \\ 1 & \text{if } 0 < t < 1 \\ f(t-2) & \text{in general} \end{cases}$

b. $g(t) = \begin{cases} e^t & \text{if } 0 < t < 1 \\ g(t-1) & \text{in general} \end{cases}$

c. evensaw(t), *the even sawtooth function sketched in figure 12.1a*

d. oddsaw(t), *the odd sawtooth function, sketched in figure 12.1b*

e. $\sin^2(t)$

f. $|\sin(2\pi t)|$

g. $f(t) = \begin{cases} +1 & \text{if } 0 < |t| < 1 \\ -1 & \text{if } 1 < |t| < 2 \\ f(t-4) & \text{in general} \end{cases}$

h. $f(t) = \begin{cases} t^2 & \text{if } -1 < t < 1 \\ f(t-2) & \text{in general} \end{cases}$

12.4. *(Bessel's inequality for complex exponential series) Let f be a periodic, piecewise continuous function with period p and complex exponential Fourier series*

$$\sum_{k=-\infty}^{\infty} c_k \, e^{i2\pi\omega_k t} \quad .$$

Using the general version of Bessel's inequality (theorem 11.4, page 139), show that $\sum_{k=-\infty}^{\infty} |c_k|^2$ is a convergent series and that

$$\sum_{k=-\infty}^{\infty} |c_k|^2 \leq \frac{1}{p} \int_{period} |f(t)|^2 \, dt \quad .$$

12.5. Let f and g be piecewise continuous, periodic functions with period p, and let

$$F.S.[f]|_t = \sum_{k=-\infty}^{\infty} f_k \, e^{i2\pi \omega_k t} \qquad \text{and} \qquad F.S.[g]|_t = \sum_{k=-\infty}^{\infty} g_k \, e^{i2\pi \omega_k t} \quad .$$

a. Assume that $f = F.S.[f]$, and derive the relation

$$F.S.[fg]|_t = \sum_{k=-\infty}^{\infty} c_k \, e^{i2\pi \omega_k t}$$

where, for each integer k,

$$c_k = \sum_{N=-\infty}^{\infty} f_n g_{k-n} \quad . \tag{12.10}$$

Your derivation need not be completely rigorous. Go ahead and assume that any "integrals of summations" equal the corresponding "summations of integrals", and don't worry about the convergence of the series. (We'll make the derivation rigorous in exercise 13.15 on page 177.)

b. Using the Bessel's inequality from problem 12.4 and the Schwarz inequality for infinite series (theorem 4.8 on page 45), verify that the infinite series in equation (12.10) converges absolutely.

13

Convergence and Fourier's Conjecture

Our initial derivation of the formulas for the Fourier series was based on the conjecture ("Fourier's bold conjecture") that any "reasonable" periodic function can be represented by an infinite linear combination of sines and cosines. However, the only evidence thus far given of this conjecture's validity has been a few pictures and the fact that, if it weren't true, then this book probably would not have been written.

It is time we look at our conjecture more carefully. After all, if we plan to use Fourier series in real applications, we really should know how well any "reasonable" function of interest can be represented by its Fourier series. How accurately, for example, will any particular partial sum approximate the function? Where can problems arise? And just what does it mean for a function to be "reasonable"?

To help answer these questions, we will discuss three types of convergence for infinite series — pointwise, uniform and norm. All three types are important in Fourier analysis and play significant roles in applications.

Unfortunately, some of the important results to be discussed here are not so easily and simply derived. Their proofs and derivations are somewhat lengthy and require much more cleverness than has been needed thus far. Including such proofs and derivations here would make for a very long chapter and may, frankly, hamper the flow of our discussions. Omitting them from the book, however, would be unforgivable. They are important to fully understanding the results presented; they contain truly interesting analysis, and besides, they aren't really that difficult. So, as a compromise, we'll devote the next chapter to these particular proofs and derivations.[1]

13.1 Pointwise Convergence
The Basic Theorems on Pointwise Convergence

Suppose we have a periodic, piecewise continuous function f and its Fourier series

$$F.S.[f]|_t = \sum_{k=-\infty}^{\infty} c_k \, e^{i2\pi\omega_k t} \quad . \tag{13.1}$$

Remember, the sum of such an infinite series is actually the double limit of the partial sums,

$$\sum_{k=-\infty}^{\infty} c_k \, e^{i2\pi\omega_k t} = \lim_{\substack{M\to-\infty \\ N\to\infty}} \sum_{k=M}^{N} c_k \, e^{i2\pi\omega_k t} \quad .$$

---

[1] Before continuing, you may want to quickly skim through the review material on infinite series starting on page 41.

Two questions immediately arise. The first is *Does the above infinite series even make sense for each possible value of t ?* In other words, if t_0 is any given value of t, then can we be sure that the series in formula (13.1) *converges* for $t = t_0$; that is, can we be sure that

$$\lim_{\substack{M \to -\infty \\ N \to \infty}} \sum_{k=M}^{N} c_k e^{i2\pi \omega_k t_0}$$

exists (as a finite number)? If so, then the Fourier series, $F.S.[f]$, is truly a function, and we can ask our second (and more interesting) question: *Are f and $F.S.[f]$ the same function?* More precisely, is it true that

$$\lim_{\substack{M \to -\infty \\ N \to \infty}} \sum_{k=M}^{N} c_k e^{i2\pi \omega_k t_0} = f(t_0)$$

for all values of t_0 (or at least for all values of t_0 at which f is continuous)?

Both of these questions are addressed in the following theorem. Its proof is one of those relegated to the next chapter (pages 179 to 185).

Theorem 13.1 (basic theorem on pointwise convergence)
Let f be a periodic, piecewise continuous function with

$$F.S.[f]|_t = \sum_{k=-\infty}^{\infty} c_k e^{i2\pi \omega_k t} \quad .$$

Assume further that f is piecewise smooth on an interval (a, b), and let t_0 be any point in that interval. Then:

1. *If $f(t)$ is continuous at $t = t_0$, then $F.S.[f]|_{t_0}$ converges and*

    $$\sum_{k=-\infty}^{\infty} c_k e^{i2\pi \omega_k t_0} = f(t_0) \quad .$$

2. *If $f(t)$ has a jump discontinuity at $t = t_0$, then*

    $$\lim_{N \to \infty} \sum_{k=-N}^{N} c_k e^{i2\pi \omega_k t_0} = \frac{1}{2} \left[\lim_{\tau \to t_0^-} f(\tau) + \lim_{\tau \to t_0^+} f(\tau) \right] \quad .$$

Theorem 13.1 assures us that the complex exponential Fourier series for a periodic, piecewise continuous function does pretty well what we expected it to do, at least over intervals where the function is continuous and piecewise smooth. At each point in such an interval, the series converges exactly to the value of the function at that point (so we say that the Fourier series converges *pointwise* to the function over such intervals). Nor does this series behave that badly at those points where f has jump discontinuities. At these points we at least have symmetric convergence of the series to the average of the left- and right-hand limits of the function at that point. Graphically, this is the midpoint of the jump.

Similar results can be derived for the trigonometric Fourier series of f,

$$A_0 + \sum_{k=1}^{\infty} [a_k \cos(2\pi \omega_k t) + b_k \sin(2\pi \omega_k t)] \quad . \tag{13.2}$$

In particular, the next theorem is an immediate consequence of theorem 13.1 and the fact that

$$\sum_{k=-N}^{N} c_k \, e^{i2\pi\omega_k t_0} = A_0 + \sum_{k=1}^{N} [a_k \cos(2\pi\omega_k t) + b_k \sin(2\pi\omega_k t)]$$

for every positive integer N (see the discussion of partial sums starting on page 150).

Theorem 13.2 (pointwise convergence for trigonometric series)
Let f be a periodic, piecewise continuous function. Assume further that f is piecewise smooth on an interval (a, b), and let t_0 be any point in that interval. Then the trigonometric Fourier series for f,

$$A_0 + \sum_{k=1}^{\infty} [a_k \cos(2\pi\omega_k t) + b_k \sin(2\pi\omega_k t)] \quad , \tag{13.3}$$

converges for $t = t_0$. Moreover:

1. *If f is continuous at t_0, then*

$$A_0 + \sum_{k=1}^{\infty} [a_k \cos(2\pi\omega_k t_0) + b_k \sin(2\pi\omega_k t_0)] = f(t_0) \quad .$$

2. *If f is not continuous at t_0, then*

$$A_0 + \sum_{k=1}^{\infty} [a_k \cos(2\pi\omega_k t_0) + b_k \sin(2\pi\omega_k t_0)] = \frac{1}{2}\left[\lim_{\tau \to t_0^-} f(\tau) + \lim_{\tau \to t_0^+} f(\tau) \right] \quad .$$

(A slight refinement of this theorem is given in exercise 13.7 at the end of this chapter.)

Pointwise Convergence and Fourier's Conjecture

As long as f is a piecewise smooth, periodic function, theorems 13.1 and 13.2 assure us that the trigonometric Fourier series and complex exponential Fourier series of f both converge at each point where f is continuous. These theorems further assure us that, at each such point t, the sums of both series equal the value $f(t)$. Consequently, we should view the function f, its trigonometric Fourier series and its complex exponential Fourier series as being the same piecewise continuous function on the entire real line (see page 10), confirming Fourier's bold conjecture for the case where the function is piecewise smooth and periodic. This is an important (and famous) observation, which we might as well state as a theorem.

Theorem 13.3 (on Fourier's bold conjecture, version 1)
Let f be a periodic, piecewise smooth function on $\mathbb{R}$, and let $F.S.[f]$ be either the trigonometric or complex exponential Fourier series for f. Then $F.S.[f]$ converges at every point where f is continuous, and

$$f = F.S.[f]$$

as piecewise continuous functions.[2]

---

[2] That is, $f(t) = F.S.[f]|_t$ at every t where f is continuous (see lemma 3.4 on page 21).

!▶ Example 13.1 (the saw function):   *In examples 9.1 and 12.1, we found that the saw function*

$$\text{saw}_3(t) \;=\; \begin{cases} t & \text{if } \; 0 < t < 3 \\ f(t-3) & \text{in general} \end{cases}$$

has trigonometric Fourier series

$$\frac{3}{2} \;-\; \sum_{k=1}^{\infty} \frac{3}{k\pi} \sin\!\left(\frac{2\pi k}{3}t\right) \tag{13.4}$$

and complex exponential Fourier series

$$\frac{3}{2} \;+\; \sum_{\substack{k=-\infty \\ k\neq 0}}^{\infty} \frac{3i}{2\pi k}\, e^{i 2\pi \omega_k t} \tag{13.5}$$

where $\omega_k = {}^{k}\!/_3$. *This function is certainly piecewise smooth on the entire real line and is continuous at every* t *except where* t *is an integral multiple of* 3 . *So, theorem 13.3 assures us that these two series converge for every* t *not equal to an integral multiple of* 3 , *and that, as piecewise continuous functions,*

$$\text{saw}_3(t) \;=\; \frac{3}{2} \;-\; \sum_{k=1}^{\infty} \frac{3}{k\pi} \sin\!\left(\frac{2\pi k}{3}t\right) \;=\; \frac{3}{2} \;+\; \sum_{\substack{k=-\infty \\ k\neq 0}}^{\infty} \frac{3i}{2\pi k}\, e^{i 2\pi \omega_k t} \quad .$$

In particular, since 5 *is not an integral multiple of* 3 *and*

$$\text{saw}_3(5) \;=\; \text{saw}_3(5-3) \;=\; \text{saw}_3(2) \;=\; 2 \quad,$$

we have

$$\frac{3}{2} \;-\; \sum_{k=1}^{\infty} \frac{3}{k\pi} \sin\!\left(\frac{2\pi k}{3}\cdot 5\right) \;=\; \text{saw}_3(5) \;=\; 2$$

and

$$\frac{3}{2} \;+\; \sum_{\substack{k=-\infty \\ k\neq 0}}^{\infty} \frac{3i}{2\pi k}\, e^{i 2\pi \omega_k 5} \;=\; \text{saw}_3(5) \;=\; 2 \quad .$$

On the other hand, $\text{saw}_3(t)$ *has a jump discontinuity at* $t = 0$ *and*

$$\frac{1}{2}\left[\lim_{\tau \to 0^-} \text{saw}_3(\tau) \;+\; \lim_{\tau \to 0^+} \text{saw}_3(\tau) \right] \;=\; \frac{1}{2}[3+0] \;=\; \frac{3}{2} \quad .$$

According to theorem 13.2, the trigonometric Fourier series for $\text{saw}_3(t)$ *does converge at* $t = 0$, *and*

$$\frac{3}{2} \;-\; \sum_{k=1}^{\infty} \frac{3}{k\pi} \sin\!\left(\frac{2\pi k}{3}0\right) \;=\; \frac{1}{2}\left[\lim_{\tau \to 0^-} \text{saw}_3(\tau) \;+\; \lim_{\tau \to 0^+} \text{saw}_3(\tau) \right] \;=\; \frac{3}{2}$$

(which, in this case, is pretty obvious). We also know from theorem 13.1 that the complex exponential Fourier series converges symmetrically at $t = 0$, *and*

$$\frac{3}{2} \;+\; \lim_{N \to \infty} \sum_{\substack{k=-N \\ k\neq 0}}^{N} \frac{3i}{2\pi k}\, e^{i 2\pi \omega_k 0} \;=\; \frac{1}{2}\left[\lim_{\tau \to 0^-} \text{saw}_3(\tau) \;+\; \lim_{\tau \to 0^+} \text{saw}_3(\tau) \right] \;=\; \frac{3}{2}$$

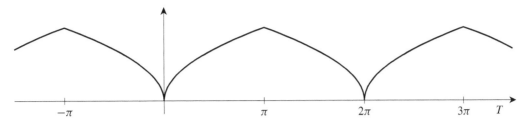

Figure 13.1: Graph of the "periodic square root function" of example 13.2.

(which is also pretty obvious once you look a little more closely at these symmetric partial sums). However, there is no assurance that the complex exponential Fourier series converges in a more general sense at $t = 0$. In fact, plugging $t = 0$ into the complex exponential Fourier series gives

$$\frac{3}{2} + \sum_{\substack{k=-\infty \\ k \neq 0}}^{\infty} \frac{3i}{2\pi k} e^{i2\pi \omega_k 0} = \frac{3}{2} + \sum_{\substack{k=-\infty \\ k \neq 0}}^{\infty} \frac{3i}{2\pi k} = \frac{3}{2} + \frac{3i}{2\pi} \sum_{\substack{k=-\infty \\ k \neq 0}}^{\infty} \frac{1}{k} \quad .$$

But

$$\sum_{\substack{k=-\infty \\ k \neq 0}}^{\infty} \frac{1}{k}$$

is the two-side harmonic series and does not *converge in the more general sense (see example 4.2 on page 47).*

Clearly then, periodic functions that are piecewise smooth on the entire real line can be considered as "reasonable" functions for Fourier analysis. As the next example illustrates, many functions which are "nearly" piecewise smooth can also be represented by their Fourier series.

!►Example 13.2: *Let*

$$f(t) = \begin{cases} \sqrt{|t|} & \text{if } -\pi < t < \pi \\ f(t - 2\pi) & \text{in general} \end{cases} .$$

The graph of this function is sketched in figure 13.1. This is clearly an even, continuous, periodic function with period 2π. Its trigonometric Fourier series and complex exponential Fourier series are

$$A_0 + \sum_{k=1}^{\infty} a_k \cos(kt) \qquad \text{and} \qquad \sum_{k=-\infty}^{\infty} c_k e^{ikt} \qquad (13.6)$$

where

$$A_0 = c_0 = \frac{1}{\pi} \int_0^{\pi} \sqrt{t}\, dt = \frac{2}{3}\sqrt{\pi} \quad .$$

The other coefficients are given by

$$a_k = \frac{2}{\pi} \int_0^{\pi} \sqrt{t} \cos(kt)\, dt \qquad \text{and} \qquad c_k = \frac{1}{\pi} \int_0^{\pi} \sqrt{t} e^{-ikt}\, dt \quad ,$$

which we'll not attempt to explicitly compute. At $t = 0$ (and, by periodicity, at $t = n2\pi$ for $n = \pm 1, \pm 2, \pm 3, \dots$), the derivative of f blows up,

$$\lim_{t \to 0^+} f'(t) = \lim_{t \to 0^+} \frac{1}{2\sqrt{t}} = \infty \quad .$$

So this function is not piecewise smooth on any interval containing an integral multiple of 2π, and theorem 13.3 does not apply.

But look at what happens when t is not an integral multiple of 2π, say $t = 2$. On the interval $(1, 3)$, this function is piecewise smooth (in fact, $f(t) = \sqrt{t}$, which is uniformly smooth on $(1, 3)$). So, theorems 13.1 and 13.2 tell us that both of the series given in line (13.6) converge for $t = 2$ and that

$$A_0 + \sum_{k=1}^{\infty} a_k \cos(k2) = f(2) = \sqrt{2}$$

and

$$\sum_{k=-\infty}^{\infty} c_k e^{ik2} = f(2) = \sqrt{2} \ .$$

More generally, if t_0 is any point other than an integral multiple of 2π, and α is the distance between t_0 and the closest integral multiple of 2π, then f will be piecewise smooth on the interval $(t_0 - {}^\alpha/_2, t_0 + {}^\alpha/_2)$. Theorems 13.1 and 13.2 then assure us that the two series given in line (13.6) converge for $t = t_0$ and equal $f(t_0)$. Since there are only a finite number of integral multiples of 2π on any finite interval, we can view this f, its trigonometric Fourier series, and its complex exponential Fourier series as all being the same function, and we can still write

$$f(t) = A_0 + \sum_{k=1}^{\infty} a_k \cos(kt) \qquad \text{and} \qquad f(t) = \sum_{k=-\infty}^{\infty} c_k e^{ikt}$$

with the understanding that these equalities hold explicitly for all values of t other than integral multiples of 2π.

Using the ideas illustrated in this last example, it is fairly easy to prove the following generalization of theorem 13.3.

Theorem 13.4 (on Fourier's bold conjecture, version 2)
Let f be a piecewise continuous, periodic function on $\mathbb{R}$, and let $F.S.[f]$ be either the trigonometric or complex exponential Fourier series for f. Assume further that, on each finite interval, f is smooth at all but a finite number (possibly zero) of points. Then, on each finite interval, $F.S.[f]|_t$ converges to $f(t)$ at all but a finite number of points, and so, $f = F.S.[f]$ as piecewise continuous functions on $\mathbb{R}$.

For the rest of our discussion of classical Fourier series (part II of this text), we will usually restrict ourselves to periodic functions that satisfy the conditions stated in theorem 13.3 or, at worst, in theorem 13.4. By these theorems then, we know that the functions of interest to us can be represented by both the trigonometric Fourier series and the complex exponential Fourier series, and that these two representations are essentially the same. In view of this we will, henceforth, treat the two series as simply being two different ways of describing the same series, and we will use whichever version — trigonometric or complex exponential — seems most convenient at the time.

The above restriction is not much of a restriction in most applications. To see this, just try sketching the graph of a periodic, piecewise continuous function that does not satisfy the smoothness conditions stated in theorem 13.4. It is possible to construct such a function. We'll do this, ourselves, in section 14.4, where we will even show that, for some of these functions, the associated trigonometric Fourier series diverges at certain points. Perhaps most surprising of all is the fact that even these strange functions can, in a sense, still be represented by their Fourier series. We will discuss this further in section 13.3 (and, in a more generalized setting, in chapter 37).

13.2 Uniform and Nonuniform Approximations
The Error in a Partial Sum Approximation

Knowing that a given periodic function f can be represented by its Fourier series allows us, in theory, to replace f with either its complex exponential or trigonometric Fourier series,

$$\sum_{k=-\infty}^{\infty} c_k e^{i2\pi\omega_k t} \quad \text{or} \quad A_0 + \sum_{k=1}^{\infty} [a_k \cos(2\pi\omega_k t) + b_k \sin(2\pi\omega_k t)] \quad ,$$

respectively. This can be a very powerful tool in certain applications.

On the other hand, adding up the infinitely many terms of a Fourier series is rarely practical (even with the best computers), and so, in practice, we may have to approximate $f(t)$ using a partial sum of its Fourier series,

$$\sum_{k=M}^{N} c_k e^{i2\pi\omega_k t} \quad \text{or} \quad A_0 + \sum_{k=1}^{N} [a_k \cos(2\pi\omega_k t) + b_k \sin(2\pi\omega_k t)] \quad .$$

For these partial sums to be useful, the limits in these summations — N, M, N_c and N_s — must be chosen so that the error in using the partial sum in place of the original function is tolerably small. Of course, the question is now

How do we determine these limits so that the resulting errors are as small as desired?

This is the question we now will address.

For convenience, we will concentrate on the error in using partial sums for complex exponential series. Corresponding results for trigonometric series can then be derived using the relations between the two versions of Fourier series.

In what follows, f denotes some periodic function,

$$F.S.[f]|_t = \sum_{k=-\infty}^{\infty} c_k e^{i2\pi\omega_k t} \quad ,$$

and M and N are a pair of integers with $M < N$. Remember, the corresponding $(M, N)^{\text{th}}$ partial sum is

$$F.S._{MN}[f]|_t = \sum_{k=M}^{N} c_k e^{i2\pi\omega_k t} \quad .$$

Let $E_{MN}(t)$ denote the magnitude of the error in using this partial sum in place of $f(t)$,

$$E_{MN}(t) = \left| f(t) - F.S._{MN}[f]|_t \right| = \left| f(t) - \sum_{k=M}^{N} c_k e^{i2\pi\omega_k t} \right| \quad .$$

Observe that $E_{MN}(t)$ varies as t varies. If f is, say, piecewise smooth, then theorem 13.1 assures us that, for each *individual* t at which f is continuous,

$$\lim_{\substack{N \to \infty \\ M \to -\infty}} E_{MN}(t) = 0 \quad .$$

Thus, if $\epsilon > 0$ is the maximum error we will tolerate and t_0 is a point at which f is continuous, then there is an M_ϵ and an N_ϵ such that $E_{MN}(t_0) < \epsilon$ whenever both $M \leq M_\epsilon$ and $N \geq N_\epsilon$. This does not mean, however, that the error will be less than ϵ at *other* points.

Ideally, of course, we would like to know both of the following:

1.   There is a pair of integers M_ϵ and N_ϵ for each and every $\epsilon > 0$ such that $E_{MN}(t) < \epsilon$ for *every* real value t whenever $M \leq M_\epsilon$ and $N_\epsilon \leq N$.

2.   How to determine that M_ϵ and N_ϵ for any given $\epsilon > 0$.

If the first of these two statements is true for every $\epsilon > 0$, then (and only then) we will say that the $F.S._{MN}[f]$'s *uniformly approximate* f (or, equivalently, that $F.S.[f]$ *converges uniformly* to f). Thus, if f is uniformly approximated by the $F.S._{MN}[f]$'s, then, no matter how small we choose $\epsilon > 0$, we can always find a partial sum $F.S._{MN}[f]$ which differs from f by less than ϵ at every point on the real line.

Note that saying "$F.S.[f]$ uniformly converges to f" is completely equivalent to saying that there is a doubly indexed set of numbers, call them ϵ_{MN}'s, such that

$$E_{MN}(t) \leq \epsilon_{MN} \qquad \text{for all } t \text{ in } \mathbb{R}$$

and satisfying

$$\lim_{\substack{N \to \infty \\ M \to -\infty}} \epsilon_{MN} = 0 \quad .$$

Think of each ϵ_{MN} as describing the largest possible error in using $F.S._{MN}[f]|_t$ to compute the value $f(t)$.[3] Where practical, we will confirm uniform convergence by constructing such a set of ϵ_{MN}'s.

Finally, let me emphasize something implicit in our terminology. If the Fourier series for a function converges uniformly to that function, then that series converges pointwise to that function on the entire real line. That is,

$$\lim_{\substack{N \to \infty \\ M \to -\infty}} \sum_{k=M}^{N} c_k e^{i2\pi \omega_k t} = f(t) \qquad \text{for each } t \text{ in } \mathbb{R} \quad .$$

Moreover, by knowing that the convergence is uniform, we also know that the maximum error in using

$$\sum_{k=M}^{N} c_k e^{i2\pi \omega_k t} \qquad \text{to compute} \qquad f(t)$$

must decrease to zero as M and N approach $-\infty$ and ∞, respectively. While this is certainly the preferred situation, it is not, as we will soon see, always possible.

Continuity and Uniform Approximations

Notice that each partial sum,

$$F.S._{MN}[f]|_t = \sum_{k=M}^{N} c_k e^{i2\pi \omega_k t} \quad ,$$

being a finite linear combination of continuous functions, must itself be a continuous function. Because of this, it is easy to show that these partial sums can*not* uniformly approximate f if f is *not* a continuous function. In fact, if f has a jump discontinuity, then, for each partial sum $F.S._{MN}[f]$, there must be an interval (a_N, b_N) on which the error $E_{MN}(t)$ is nearly half the magnitude of the jump or greater.

To see why, consider the problem of approximating a discontinuous function f with any continuous function S, as illustrated (with real-valued functions) in figure 13.2. In the figure, you can see that, if t_0 is a point at which f has a discontinuity with jump j_0, and if S closely

---

[3] More precisely, each ϵ_{MN} is a computable upper bound on the largest possible error.

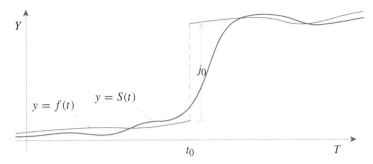

Figure 13.2: A continuous approximation $S(t)$ to a discontinuous function $f(t)$ having a jump j_0 at t_0.

approximates f on, say, the left side of the discontinuity, then S, being continuous, would require a nontrivial interval on the right side of t_0 to move up (or down) by the amount which f "jumped". Over that interval S would no longer be close to f. In particular, if $S(t)$ is within $j_0/2$ of $f(t)$ for every t less than t_0, then there must be a nonzero interval to the right of the jump over which the values of $S(t)$ will not yet be within, say, $j_0/4$ of $f(t)$.[4]

These observations lead to the following little lemma, whose complete proof will be left as an exercise for those who need further convincing.

Lemma 13.5
Let f and S be two functions on the real line with f being piecewise continuous and S being continuous. Assume f has a nontrivial discontinuity with a jump of j_0 at some point t_0. Then there is a nontrivial interval (a, b) such that

    1.   *f is continuous over (a, b), and*

    2.   *$|f(t) - S(t)| > \frac{1}{4}|j_0|$ for every t in (a, b).*

?►Exercise 13.1:   *Rigorously prove lemma 13.5.*

The case of greatest interest to us is where $S = F.S._{MN}[f]$. If f has a nontrivial discontinuity with jump j_0, then this little lemma tells us that, for any choice of M and N, there is an interval over which $E_{MN}(t) > j_0/4$. Thus, the $F.S._{MN}[f]$'s do not uniformly approximate f. Conversely, if the $F.S._{MN}[f]$'s do approximate f uniformly, then f *must be continuous on the entire real line* (otherwise, according to the above, the $F.S._{MN}[f]$'s could not approximate f uniformly!).

These observations are important enough to formalize as a theorem.

Theorem 13.6
Let f be a periodic, piecewise continuous function. If the $F.S._{MN}[f]$'s uniformly approximate f, then f must be a continuous function on the entire real line. Conversely, if f is not a continuous function on the entire real line, then the $F.S._{MN}[f]$'s do not uniformly approximate f. Moreover, if f has a jump of j_0 at t_0, then, for each pair of integers M and N with $M < N$, there is an interval containing t_0 or with t_0 as an endpoint over which

$$\left| f(t) - F.S._{MN}[f]|_t \right| > \frac{1}{4}|j_0| \quad .$$

---

[4] There is nothing magic about $1/4$. Any positive number below $1/2$ can be used.

Along these lines, here is part of a theorem regarding uniform convergence that will be proven after we discuss Fourier series of derivatives (see theorem 15.6 on page 206). It confirms that Fourier series of *continuous*, periodic functions can be expected to converge uniformly.

Theorem 13.7 (continuity and uniform convergence for exponential series)
Let f be a piecewise smooth and periodic function with period p. If f is also continuous, then its Fourier series

$$\sum_{k=-\infty}^{\infty} c_k e^{i2\pi \omega_k t}$$

converges uniformly to f. Moreover, for any real value t and any pair of integers M and N with $M < 0 < N$,

$$\left| f(t) - \sum_{k=M}^{N} c_k e^{i2\pi \omega_k t} \right| \leq \left[\frac{1}{\sqrt{|M|}} + \frac{1}{\sqrt{N}} \right] B$$

where

$$B = \frac{1}{2\pi} \left(p \int_{period} |f'(t)|^2 \, dt \right)^{1/2} \quad .$$

These theorems do not say $F.S.[f]$ uniformly converges to f whenever f is simply a continuous (but not piecewise smooth) periodic function. In fact, as we will see in section 14.4, there are continuous, periodic functions that are not uniformly approximated by their Fourier partial sums. Fortunately, such functions are difficult to construct and do not commonly arise in applications.

The analogs to theorems 13.6 and 13.7 for trigonometric Fourier series are:

Theorem 13.8
Let f be a periodic, piecewise continuous function with trigonometric Fourier series

$$A_0 + \sum_{k=1}^{\infty} [a_k \cos(2\pi \omega_k t) + b_k \sin(2\pi \omega_k t)] \quad .$$

If there is a finite integer N_ϵ for each $\epsilon > 0$ such that

$$\left| f(t) - A_0 - \sum_{k=1}^{N} [a_k \cos(2\pi \omega_k t) + b_k \sin(2\pi \omega_k t)] \right| \leq \epsilon$$

for every real value t and every integer $N \geq N_\epsilon$, then f is continuous on the entire real line. Conversely, if f has a nonzero jump of j_0 at t_0, then, for any positive integer N, there is an interval containing t_0 or with t_0 as an endpoint over which

$$\left| f(t) - A_0 - \sum_{k=1}^{N} [a_k \cos(2\pi \omega_k t) + b_k \sin(2\pi \omega_k t)] \right| > \frac{1}{4} |j_0| \quad .$$

Theorem 13.9 (continuity and uniform convergence for trigonometric series)
Let f be a continuous and piecewise smooth periodic function with period p. Then its trigonometric Fourier series

$$A_0 + \sum_{k=1}^{\infty} [a_k \cos(2\pi \omega_k t) + b_k \sin(2\pi \omega_k t)]$$

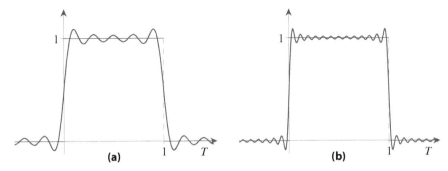

Figure 13.3: Gibbs phenomenon in the graphs of the (a) 10th and (b) 25th partial sum approximation to the Fourier series for a square wave function (sketched faintly).

converges uniformly to f *. Moreover, for any real value of* t *and any positive integer* N *,*

$$\left| f(t) - A_0 - \sum_{k=1}^{N} [a_k \cos(2\pi \omega_k t) + b_k \sin(2\pi \omega_k t)] \right| \leq \frac{B}{\sqrt{N}}$$

where

$$B = \frac{1}{\pi} \left(p \int_{period} |f'(t)|^2 \, dt \right)^{1/2} \quad .$$

?▶ Exercise 13.2: *Assume theorem 13.7 holds and prove theorem 13.9.*

Approximations for Discontinuous Functions

Let us now consider a *dis*continuous, piecewise smooth, periodic function f. We now know that, while f can be represented by its Fourier series, it cannot be uniformly approximated by the partial sums of its Fourier series. Since such functions are often used in applications, it seems prudent to further discuss the behavior of their partial sum approximations both in the neighborhoods of the discontinuities and over intervals not containing discontinuities.

Behavior Near Discontinuities
Gibbs Phenomenon

If we look closely at the graph of a Fourier partial sum approximation to a discontinuous (but piecewise smooth) function f, we see something strange occurring: Not only is the graph of the partial sum approximation not uniformly close to the graph of f, it "oscillates wildly" about the graph of f in the neighborhood of any discontinuity. Looking more closely, we can further see that, on either side of the discontinuity, there is a "hump" in the graph of the partial sum approximation that goes above or below the graph of f by roughly 9% of the magnitude of the jump at the discontinuity. This phenomenon is known as the *Gibbs phenomenon* (or *ringing*).

The Gibbs phenomenon is particularly well illustrated in figure 13.3 by the graphs of the square wave function

$$f(t) = \begin{cases} 0 & \text{if} \quad -\pi < t < 0 \\ 1 & \text{if} \quad 0 < t < \pi \\ f(t - 2\pi) & \text{in general} \end{cases}$$

and the corresponding partial sums $F.S._{10}[f]$ and $F.S._{25}[f]$. Figure 13.3 also illustrates the fact that, as N gets larger, the interval over which the Gibbs phenomenon is significant becomes smaller. Still, the magnitude of the oscillations remains fairly constant.

A rather detailed analysis of the Gibbs phenomenon can be carried out for the shifted and scaled saw function

$$h_0(t) = \frac{J_0}{p} \operatorname{saw}_p(t - t_0) = \begin{cases} \dfrac{J_0}{p}(t - t_0) & \text{if} \quad t_0 < t < t_0 + p \\ h_0(t - p) & \text{in general} \end{cases}$$

where J_0, p and t_0 are any constants with t_0 real and $p > 0$. This function, along with an Nth symmetric partial sum of its Fourier series,

$$F.S._N[h_0]\big|_t = \sum_{k=-N}^{N} c_k e^{i2\pi \omega_k t} \quad ,$$

is sketched in figure 13.4. Note that h_0 is continuous everywhere except at $t = t_0 + Kp$ where K is any integer, and that at these discontinuities the function has a jump of $-J_0$.

The details of the analysis of the Gibbs phenomenon for this function (along with some additional discussion of the Gibbs phenomenon for this function) are given in the next chapter. It is shown there (in proving lemma 14.8 on page 195) that the relative maximums and minimums of $F.S._N[h_0]\big|_t$ (i.e., the peaks and valleys of the wiggles in figure 13.4) occur at the points $t_0 + t_{N,m}$ where

$$t_{N,m} = \begin{cases} \dfrac{mp}{2N} & \text{if } m \text{ is even} \\ \dfrac{mp}{2N+2} & \text{if } m \text{ is odd} \end{cases} \quad , \tag{13.7a}$$

$$m = 0, \pm 1, \pm 2, \pm 3, \ldots, \pm M_N \quad , \tag{13.7b}$$

and

$$M_N = \begin{cases} N - 1 & \text{if } N \text{ is even} \\ N & \text{if } N \text{ is odd} \end{cases} \quad . \tag{13.7c}$$

Moreover, letting

$$\gamma_m = \frac{1}{2} - \frac{1}{\pi} \int_0^{m\pi} \frac{\sin(\tau)}{\tau} d\tau \quad , \tag{13.7d}$$

then, for each nonzero integer m,

$$\lim_{N \to \infty} \left[F.S._N[h_0]\big|_{t_0+t_{N,m}} - h_0(t_0 + t_{N,m}) \right] = \begin{cases} -\gamma_m J_0 & \text{if } m < 0 \\ \gamma_m J_0 & \text{if } 0 < m \end{cases} \quad . \tag{13.7e}$$

The largest of these γ_m's is γ_1, which is approximately 0.09. It is equation (13.7e) that tells us, when N is relatively large, that the "wiggle" in the graph of $F.S._N[h_0]$ closest to each discontinuity will either overshoot or undershoot the graph of h_0 by roughly 9% of the height of the jump of the discontinuity.

Similar behavior occurs with the corresponding partial sum approximations of any periodic, piecewise smooth function. In fact, given the above and previous results discussed in this chapter, it is surprisingly easy to prove the next theorem.

Theorem 13.10

Let f be a periodic, piecewise smooth function with period p and

$$F.S.[f]\big|_t = \sum_{k=-\infty}^{\infty} c_k e^{i2\pi \omega_k t} \quad .$$

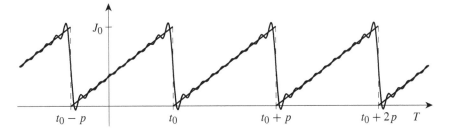

Figure 13.4: Graphs of a shifted and scaled saw function and a partial sum approximation to its Fourier series.

If f has a jump of j_0 at t_0, then

$$\lim_{N \to \infty} \left[\sum_{k=-N}^{N} c_k \, e^{i 2\pi \omega_k (t_0 + t_{N,m})} - f(t_0 + t_{N,m}) \right] = \begin{cases} -\gamma_m \, j_0 & \text{if} \quad m < 0 \\ \gamma_m \, j_0 & \text{if} \quad 0 < m \end{cases}$$

where the $t_{N,m}$'s and γ_m's are as defined in equations (13.7a) and (13.7d).

?▶ Exercise 13.3 (proof of theorem 13.10)   a: Assume that f has only one discontinuity in the interval $0 \le t < p$. Let t_0 be the point in that interval at which f is discontinuous, J_0 the corresponding jump in f and h_0 the shifted and scaled saw function described above. Show that $g = f + h_0$ is a continuous, piecewise smooth, periodic function.

b: Assume theorem 13.7 on page 164 holds as well as equation (13.7e). Show that the claim in theorem 13.10 holds for the case where f has only one discontinuity in the interval $0 \le t < p$.[5]

c: Now prove theorem 13.10.

Behavior Away from Discontinuities
Limited Uniform Approximation

While the partial Fourier sums cannot uniformly approximate discontinuous functions, you may have noticed that the graphs of partial sum approximations do closely approximate the given discontinuous functions over intervals away from any discontinuities (see, for example, figures 13.3 and 13.4). For the scaled and shifted saw function mentioned above,

$$h_0(t) = \frac{J_0}{p} \, \text{saw}_p(t - t_0) \quad ,$$

this will be confirmed in the next chapter, where we will show (lemma 14.7 on page 195) that, if

$$F.S. [h_0]|_t = \sum_{k=-\infty}^{\infty} c_k \, e^{i 2\pi \omega_k t} \quad ,$$

---

[5] One should always be suspicious of those who use a theorem to derive a result before proving that theorem. There is a distinct danger that they will later prove that theorem using the result derived assuming the theorem holds — thus verifying only the vacuous claim that something is true if that "something" is true. Fortunately, the only result from this chapter used in the proof of theorem 13.7 (what we are assuming) is theorem 13.1, the basic theorem on pointwise convergence.

and M and N are two integers with $M < 0 < N$, then

$$\left| h_0(t) - \sum_{k=M}^{N} c_k \, e^{i2\pi\omega_k t} \right| \leq \left[\frac{1}{1-2M} + \frac{1}{1+2N} \right] B(D)$$

where D is the distance between t and the point closest to t at which h_0 is discontinuous, and

$$B(D) = |J_0| \left[\frac{1}{\pi} - \frac{1}{\pi^2} + \frac{1}{\pi \sin\left(\frac{\pi}{p} D\right)} \right] .$$

Note that, as t approaches a point at which h_0 is discontinuous, $D \to 0$ and $B(D) \to \infty$.

The computations leading to the above results are, admittedly, somewhat tedious. That is why they are in the next chapter and not here. But, as with the Gibbs phenomenon, it is fairly easy to take the results derived for the saw function and results discussed (but not yet proven) earlier in this chapter (notably, theorem 13.7 on page 164) to derive a much more general theorem.

Theorem 13.11

Suppose f is a periodic, piecewise smooth function with period p, and t is any fixed real value. Let K be the number of points in the half-closed interval $[t - p/2, t + p/2)$ at which f is discontinuous, and let $t_1, t_2, \ldots, t_K$ be those points of discontinuity with j_k denoting the jump in f at t_k. Then, for any pair of integers M and N with $M < 0 < N$,

$$E_{MN}(t) \leq \left[\frac{1}{\sqrt{|M|}} + \frac{1}{\sqrt{N}} \right] B_0 + \left[\frac{1}{1-2M} + \frac{1}{1+2N} \right] \sum_{k=1}^{K} B_K(t)$$

where

$$B_0 = \frac{1}{2\pi} \left(p \int_{\text{period}} \left| f'(t) + \frac{1}{p} \sum_{k=1}^{K} j_k \right|^2 dt \right)^{1/2}$$

and, for $k = 1, 2, \ldots, K$,

$$B_k(t) = |j_k| \left[\frac{1}{\pi} - \frac{1}{\pi^2} + \frac{1}{\pi \sin\left(\frac{\pi}{p} |t - t_k|\right)} \right] .$$

?▶Exercise 13.4:   Prove theorem 13.11 assuming theorem 13.7 holds. (Hint: See exercise 13.3 on page 167.)

?▶Exercise 13.5:   In the following, let

$$F.S. [\text{saw}_1]|_t = \sum_{k=-\infty}^{\infty} c_k \, e^{i2\pi\omega_k t} .$$

a: Verify that

$$c_k = \begin{cases} \dfrac{i}{2\pi k} & \text{if } k \neq 0 \\[2mm] \dfrac{1}{2} & \text{if } k = 0 \end{cases} .$$

b: Use the error estimate in theorem 13.11 to show that, if N is any positive integer and $1/4 \leq t \leq 3/4$, then

$$\left| \text{saw}_1(t) - \sum_{k=-N}^{N} c_k \, e^{i2\pi\omega_k t} \right| < \frac{2}{1+2N} .$$

c: *What should N be to ensure that*

$$\sum_{k=-N}^{N} c_k \, e^{i2\pi\omega_k t} \quad approximates \quad saw_1(t)$$

with an error of less than $^1/_{100}$ on the interval $\left(^1/_4, \, ^3/_4\right)$?

13.3 Convergence in Norm[*]

Let f be a periodic, piecewise continuous function with period p and Fourier series

$$F.S.\,[f]|_t \;=\; \sum_{k=-\infty}^{\infty} c_k \, e^{i2\pi\omega_k t} \quad .$$

We will say this Fourier series *converges in norm* (or *converges in energy*) to f if and only if

$$\lim_{\substack{N\to\infty \\ M\to-\infty}} \left\| f(t) \;-\; \sum_{k=M}^{N} c_k \, e^{i2\pi\omega_k t} \right\| \;=\; 0 \quad . \tag{13.8}$$

This last equation, of course, can be written as

$$\lim_{\substack{N\to\infty \\ M\to-\infty}} \|E_{MN}(t)\| \;=\; 0 \tag{13.9}$$

where E_{MN} is the corresponding pointwise error,

$$E_{MN}(t) \;=\; f(t) \;-\; \sum_{k=M}^{N} c_k \, e^{i2\pi\omega_k t} \quad .$$

Recalling the definition of the norm, we can see that equations (13.8) and (13.9) are completely equivalent to

$$\lim_{\substack{N\to\infty \\ M\to-\infty}} \int_{\text{period}} \left| f(t) \;-\; \sum_{k=M}^{N} c_k \, e^{i2\pi\omega_k t} \right|^2 dt \;=\; 0 \tag{13.10}$$

and to

$$\lim_{\substack{N\to\infty \\ M\to-\infty}} \int_{\text{period}} |E_{MN}(t)|^2 \, dt \;=\; 0 \quad . \tag{13.11}$$

In words, these equations are saying:

The average value of the square of the error in replacing $f(t)$ by

$$\sum_{k=M}^{N} c_k \, e^{i2\pi\omega_k t}$$

approaches 0 as M and N approach $-\infty$ and ∞, respectively.

---

[*] This continues the discussion on norms begun in chapter 11. It may be a good idea to quickly review that material.

It should now be apparent as to why convergence in the norm is often referred to as *mean-squared convergence*.

Let's consider the particular case where f is continuous and piecewise smooth, as well as periodic. Theorem 13.7 tells us (or will tell us once we prove it) that there is a finite value B such that

$$|E_{MN}(t)| \leq \left[\frac{1}{\sqrt{|M|}} + \frac{1}{\sqrt{N}} \right] B$$

for every real value t and all integers M and N with $M < 0 < N$. Thus,

$$\lim_{\substack{N \to \infty \\ M \to -\infty}} \int_{\text{period}} |E_{MN}(t)|^2 \, dt \leq \lim_{\substack{N \to \infty \\ M \to -\infty}} \int_{\text{period}} \left| \left[\frac{1}{\sqrt{|M|}} + \frac{1}{\sqrt{N}} \right] B \right|^2 \, dt$$

$$= \lim_{\substack{N \to \infty \\ M \to -\infty}} \left| \left[\frac{1}{\sqrt{|M|}} + \frac{1}{\sqrt{N}} \right] B \right|^2 p = 0 \quad,$$

proving the following lemma.

Lemma 13.12
The (complex exponential) Fourier series for a continuous, piecewise smooth, periodic function converges in norm to that function.

That we can show convergence in the norm when we already have uniform convergence should not surprise you. You may even suspect that the results discussed above concerning the Gibbs phenomenon and "almost uniform convergence" can be used to show that the statement of lemma 13.12 remains true if the word "continuous" is removed. You would be correct. In fact, we can go even further and show that the Fourier series of any piecewise continuous, periodic function converges in norm to that function, even when that function is not piecewise smooth.

To see how we might prove the more general claim, go back to the "norm of the error" equation in lemma 11.3 on page 138. In this situation, with M and N being any two integers satisfying $M < N$, that equation becomes

$$\left\| f(t) - \sum_{k=M}^{N} c_k e^{i2\pi\omega_k t} \right\|^2 = \|f\|^2 - \sum_{k=M}^{N} |c_k|^2 \left\| e^{i2\pi\omega_k t} \right\|^2$$

$$= \|f\|^2 - p \sum_{k=M}^{N} |c_k|^2 \quad . \tag{13.12}$$

Thus, the Fourier series of f converges in norm to f,

$$\lim_{\substack{N \to \infty \\ M \to -\infty}} \left\| f(t) - \sum_{k=M}^{N} c_k e^{i2\pi\omega_k t} \right\|^2 = 0 \quad,$$

if and only if

$$\lim_{\substack{N \to \infty \\ M \to -\infty}} \left[\|f\|^2 - p \sum_{k=M}^{N} |c_k|^2 \right] = 0 \quad .$$

But

$$\lim_{\substack{N \to \infty \\ M \to -\infty}} \left[\|f\|^2 - p \sum_{k=M}^{N} |c_k|^2 \right] = \|f\|^2 - p \sum_{k=-\infty}^{\infty} |c_k|^2 \quad .$$

Combining the last two statements gives us the following lemma.

Lemma 13.13
Let f be a periodic, piecewise continuous function with

$$F.S.[f]|_t = \sum_{k=-\infty}^{\infty} c_k e^{i 2\pi \omega_k t} \quad .$$

Then the Fourier series for f converges in norm to f if and only if

$$p \sum_{k=-\infty}^{\infty} |c_k|^2 = \|f\|^2 \quad . \tag{13.13}$$

Equation (13.13) is known as *Bessel's equality* (for Fourier series).[6] To show that this equality holds under very general circumstances, it helps to first show that "Parseval's equality", described below, holds under very general circumstances.

Lemma 13.14 (Parseval's equality)
Let f and g be two piecewise continuous, periodic functions with the same period p and with Fourier series

$$F.S.[f]|_t = \sum_{k=-\infty}^{\infty} f_k e^{i 2\pi \omega_k t} \quad \text{and} \quad F.S.[g]|_t = \sum_{k=-\infty}^{\infty} g_k e^{i 2\pi \omega_k t} \quad .$$

Assume, in addition, that g is continuous and piecewise smooth. Then

1. $\displaystyle\sum_{k=-\infty}^{\infty} f_k g_k^*$ converges absolutely, and

2. $\displaystyle\langle f \mid g \rangle = p \sum_{k=-\infty}^{\infty} f_k g_k^* \quad .$

The last equality is known as *Parseval's equality* (for Fourier series).

PROOF: From Bessel's inequality (see exercise 12.4 on page 152) we already know that

$$\sum_{k=-\infty}^{\infty} |f_k|^2 \leq \frac{1}{p} \|f\|^2 \quad \text{and} \quad \sum_{k=-\infty}^{\infty} |g_k|^2 \leq \frac{1}{p} \|g\|^2 \quad .$$

This and the Schwarz inequality for summations (theorem 4.8 on page 45) give us

$$\sum_{k=-\infty}^{\infty} |f_k g_k^*| = \sum_{k=-\infty}^{\infty} |f_k| |g_k|$$

$$\leq \left(\sum_{k=-\infty}^{\infty} |f_k|^2 \right)^{1/2} \left(\sum_{k=-\infty}^{\infty} |g_k|^2 \right)^{1/2} \leq \frac{1}{p} \|f\| \|g\| \quad ,$$

---

[6] Do *not* call it Bessel's equation; that is something quite different.

which verifies the absolute convergence claimed in the lemma.

Now, for each pair of integers M and N with $M < 0 < N$, let

$$E_{MN}(t) = g(t) - \sum_{k=M}^{N} g_k \, e^{i2\pi\omega_k t} \quad .$$

From lemma 13.12, we know that $\|E_{MN}\| \to 0$ as $(M, N) \to (-\infty, \infty)$. From that and the Schwarz inequality for inner products (theorem 11.2 on page 135), we have

$$\lim_{\substack{N\to\infty \\ M\to-\infty}} |\langle f \mid E_{MN} \rangle| \leq \lim_{\substack{N\to\infty \\ M\to-\infty}} \|f\| \, \|E_{MN}\| = 0 \quad . \tag{13.14}$$

But

$$\langle f \mid E_{MN} \rangle = \int_{\text{period}} f(t) \left(g(t) - \sum_{k=M}^{N} g_k \, e^{i2\pi\omega_k t} \right)^* dt$$

$$= \int_{\text{period}} f(t) g^*(t) \, dt - \sum_{k=M}^{N} g_k^* \int_{\text{period}} f(t) \, e^{-i2\pi\omega_k t} \, dt$$

$$= \langle f \mid g \rangle - \sum_{k=M}^{N} g_k^* \, p f_k \quad .$$

Thus,

$$\langle f \mid g \rangle = \langle f \mid E_{MN} \rangle + p \sum_{k=M}^{N} f_k g_k^*$$

and, using equation (13.14),

$$\langle f \mid g \rangle = \lim_{\substack{N\to\infty \\ M\to-\infty}} \left[\langle f \mid E_{MN} \rangle + p \sum_{k=M}^{N} f_k g_k^* \right] = 0 + p \sum_{k=-\infty}^{\infty} f_k g_k^* \quad . \qquad ∎$$

Using tools that we will develop independently for Fourier transforms, we will be able to show that the additional assumptions made on g in the above lemma are totally unnecessary. That will give us the following theorem (proven in a set of exercises in section 29.5).

Theorem 13.15 (Parseval's equality)
Let f and g be two piecewise continuous, periodic functions with the same period p, and let

$$F.S.[f]|_t = \sum_{k=-\infty}^{\infty} f_k \, e^{i2\pi\omega_k t} \qquad \text{and} \qquad F.S.[g]|_t = \sum_{k=-\infty}^{\infty} g_k \, e^{i2\pi\omega_k t} \quad .$$

Then

1. $\displaystyle\sum_{k=-\infty}^{\infty} f_k g_k^*$ *converges absolutely, and*

2. $\displaystyle\langle f \mid g \rangle = p \sum_{k=-\infty}^{\infty} f_k g_k^* \quad .$

Letting $g = f$ gives us Bessel's equality.

Corollary 13.16 (Bessel's equality)
Let f be a piecewise continuous, periodic function with Fourier series

$$F.S.[f]|_t = \sum_{k=-\infty}^{\infty} c_k e^{i2\pi\omega_k t} \quad .$$

Then

$$\|f\|^2 = p \sum_{k=-\infty}^{\infty} |c_k|^2 \quad .$$

As an immediate corollary to this and lemma 13.13, we have the following major theorem for Fourier analysis.

Theorem 13.17 (norm convergence)
The (complex exponential) Fourier series for a piecewise continuous, periodic function converges in norm to that function.

13.4 The Sine and Cosine Series

After making the obvious modifications, we can apply all the results discussed thus far in this chapter to the various "Fourier series" discussed in chapter 10 for functions just defined on finite intervals. Consider, for example, a function f on a finite interval $(0, L)$ and its Fourier sine series

$$F.S.S.[f]|_t = \sum_{k=1}^{\infty} b_k \sin\left(\frac{k\pi}{L}t\right) \tag{13.15}$$

(see section 10.2 starting on page 123). Remember, this series is the trigonometric Fourier series of the odd periodic extension of f,

$$\widehat{f}_o(t) = \begin{cases} f(t) & \text{if } 0 < t < L \\ -f(-t) & \text{if } -L < t < 0 \\ \widehat{f}_o(t - 2L) & \text{in general} \end{cases} \quad .$$

Assuming f is piecewise smooth on $(0, L)$ and continuous at a point t_0 in $(0, L)$, then it certainly follows that $\widehat{f}_o$ is piecewise smooth on the entire real line and is continuous at t_0. The basic theorem on pointwise convergence for trigonometric series, theorem 13.2 on page 157, then assures us that trigonometric Fourier series for $\widehat{f}_o$ — which is the series given in line (13.15) — converges and equals $\widehat{f}_o(t_0)$. Thus, by the definition of $\widehat{f}_o$,

$$f(t_0) = \widehat{f}_o(t_0) = \sum_{k=1}^{\infty} b_k \sin\left(\frac{k\pi}{L}t_0\right) \quad .$$

Continuing along these lines, we can easily obtain the next theorem.

Theorem 13.18 *(pointwise convergence of sine series)*
Let f be a piecewise smooth function on a finite interval $(0, L)$. Then the sine series of f on $(0, L)$

$$\sum_{k=1}^{\infty} b_k \sin\left(\frac{k\pi}{L}t\right)$$

converges for each t in $(0, L)$. Moreover:

1. If t_0 is a point in $(0, L)$ at which f is continuous, then

$$\sum_{k=1}^{\infty} b_k \sin\left(\frac{k\pi}{L}t_0\right) = f(t_0) \quad .$$

2. If t_0 is a point in $(0, L)$ at which f is not continuous, then

$$\sum_{k=1}^{\infty} b_k \sin\left(\frac{k\pi}{L}t_0\right) = \frac{1}{2}\left[\lim_{t\to t_0^+} f(t) + \lim_{t\to t_0^-} f(t)\right] \quad .$$

The analogous theorem for the cosine series, below, is just as easily verified.

Theorem 13.19 *(pointwise convergence of cosine series)*
Let f be a piecewise smooth function on a finite interval $(0, L)$. Then the cosine series of f on $(0, L)$

$$A_0 + \sum_{k=1}^{\infty} a_k \cos\left(\frac{k\pi}{L}t\right)$$

converges for each t in $(0, L)$. Moreover:

1. If t_0 is a point in $(0, L)$ at which f is continuous, then

$$A_0 + \sum_{k=1}^{\infty} a_k \cos\left(\frac{k\pi}{L}t_0\right) = f(t_0) \quad .$$

2. If t_0 is a point in $(0, L)$ at which f is not continuous, then

$$A_0 + \sum_{k=1}^{\infty} a_k \cos\left(\frac{k\pi}{L}t_0\right) = \frac{1}{2}\left[\lim_{t\to t_0^+} f(t) + \lim_{t\to t_0^-} f(t)\right] \quad .$$

In a similar fashion, we can obtain analogs to the other results discussed in this chapter for the various "Fourier series" of functions on a finite interval. Two that will be of some interest, the sine and cosine series versions of theorem 13.9 on page 164, will be described later in chapter 15 (see page 215). One other of interest is stated below. It follows from the results concerning norm convergence and Bessel's equality (corollary 13.16 and theorem 13.17 in the previous section). I'll leave the details of its verification to you.

Theorem 13.20 *(Bessel's equality for sine and cosine series)*
Let f be a piecewise continuous function on a finite interval $(0, L)$ with sine and cosine series

$$\sum_{k=1}^{\infty} b_k \sin\left(\frac{k\pi}{L}t\right) \quad \text{and} \quad A_0 + \sum_{k=1}^{\infty} a_k \cos\left(\frac{k\pi}{L}t\right) \quad ,$$

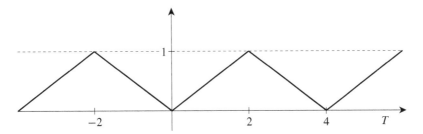

Figure 13.5: Two and one half periods of the even saw function for exercise 13.6 c.

respectively. Then each of these series converges in norm to f. Moreover

$$\| f \|^2 \; = \; \frac{1}{2} L \sum_{k=1}^{\infty} |b_k|^2 \; = \; L \, |A_0|^2 \; + \; \frac{1}{2} L \sum_{k=1}^{\infty} |a_k|^2 \quad .$$

Additional Exercises

13.6. *Consider the Fourier series for each of the functions below at each the following points:*

$$t = 0 \quad , \quad t = \tfrac{1}{2} \quad , \quad t = 1 \quad , \quad t = 2 \quad \text{and} \quad t = \tfrac{5}{2} \quad .$$

In each case:

1. *If the function is continuous at the given point, find the value of its Fourier series at that point.*

2. *If the function is not continuous at the given point, find the value of its trigonometric Fourier series at that point. (Equivalently, find the value to which its complex exponential Fourier series symmetrically converges.)*

Use the results from this chapter! Do not attempt to actually "add up" the infinite series!

a. $f(t) = \begin{cases} 0 & \text{if} \;\; -1 < t < 0 \\ 1 & \text{if} \;\; 0 < t < 1 \\ f(t-2) & \text{in general} \end{cases}$

b. $g(t) = \begin{cases} e^t & \text{if} \;\; 0 < t < 1 \\ g(t-1) & \text{in general} \end{cases}$

c. *The even sawtooth function sketched in figure 13.5*

d. $|\sin(2\pi t)|$

e. $h(t) = \begin{cases} +1 & \text{if} \;\; 0 < |t| < 1 \\ -1 & \text{if} \;\; 1 < |t| < 2 \\ h(t-4) & \text{in general} \end{cases}$

f. $k(t) = \begin{cases} t^2 & \text{if } -1 < t < 1 \\ k(t-2) & \text{in general} \end{cases}$

g. $l(t) = \begin{cases} 0 & \text{if } -1 < t < 0 \\ t^2 & \text{if } 0 < t < 1 \\ l(t-2) & \text{in general} \end{cases}$

13.7. Let f be a periodic, piecewise smooth function with trigonometric Fourier series

$$A_0 + \sum_{k=1}^{\infty} [a_k \cos(2\pi \omega_k t) + b_k \sin(2\pi \omega_k t)] \quad .$$

Assume further that f is piecewise smooth on an interval (a, b), and let t_0 be any point in (a, b). Using either of the theorems on pointwise convergence (theorem 13.1 or 13.2) verify that

$$\sum_{k=1}^{\infty} a_k \cos(2\pi \omega_k t_0) \qquad \text{and} \qquad \sum_{k=1}^{\infty} b_k \sin(2\pi \omega_k t_0)$$

both converge, and that

$$\sum_{k=1}^{\infty} [a_k \cos(2\pi \omega_k t_0) + b_k \sin(2\pi \omega_k t_0)] = \sum_{k=1}^{\infty} a_k \cos(2\pi \omega_k t_0) + \sum_{k=1}^{\infty} b_k \sin(2\pi \omega_k t_0) \quad .$$

(You might start by considering the functions

$$f_e(t) = \tfrac{1}{2}[f(t) + f(-t)] \qquad \text{and} \qquad f_o(t) = \tfrac{1}{2}[f(t) - f(-t)]$$

and their Fourier series.)

13.8. For each function listed in exercise 13.6, decide whether the corresponding Fourier series does or does not converge uniformly to the function.

13.9. In exercises 9.8 and 9.9 you generated the graphs of several partial sums of the trigonometric Fourier series for various functions. Re-examine those graphs and do the following:

  1. Visually identify those functions whose partial sums are uniformly converging to the function.

  2. Identify graphs exhibiting the Gibbs phenomenon. In particular, locate the "over- and undershoots" closest to the discontinuities.

13.10. Let f be a periodic and piecewise continuous function with period p and

$$F.S.[f]|_t = \sum_{k=-\infty}^{\infty} c_k e^{i2\pi \omega_k t} \quad .$$

The mean error in using the $(M, N)^{th}$ partial sum is

$$\frac{1}{p} \int_{period} |E_{MN}(t)| \, dt \qquad \text{where} \qquad E_{MN}(t) = f(t) - \sum_{k=M}^{N} c_k e^{i2\pi \omega_k t} \quad .$$

Assume that theorem 13.17 on norm convergence holds, and, using that theorem and in-equality (11.13) on page 142, verify that the mean error in using the $(M, N)^{th}$ partial sum goes to zero as M and N approach $-\infty$ and ∞, respectively.

13.11. For this problem, if $0 < t < 2$, then $f(t) = t$. Otherwise, $f(t)$ is not defined.

    a. Consider the sine series for f over $(0, 2)$.

        i. To what values does this sine series converge when $t = 0$ and when $t = 2$?

        ii. Does this sine series converge pointwise to $f(t)$ when $0 < t < 2$?

        iii. Does this sine series converge uniformly to f over the interval $(0, 2)$?

    b. Consider the cosine series for f over $(0, 2)$.

        i. To what values does this cosine series converge when $t = 0$ and when $t = 2$?

        ii. Does this cosine series converge pointwise to $f(t)$ when $0 < t < 2$?

        iii. Does this cosine series converge uniformly to f over the interval $(0, 2)$?

13.12. Repeat problem 13.11, above, using the function $f(t) = t(2 - t)$ if $0 < t < 2$ (and undefined otherwise).

13.13. Using results discussed in this chapter for periodic functions:

    a. Finish proving theorem 13.18 on page 174.

    b. Prove theorem 13.20 on page 174.

13.14. Let $f(t)$ be a piecewise smooth function on a finite interval $(0, L)$.

    a. When will we have the Gibbs phenomenon occurring at $t = 0$ in the partial sums for the sine series of f?

    b. When will we have the Gibbs phenomenon occurring at $t = 0$ in the partial sums for the cosine series of f?

13.15. In problem 12.5 on page 153 you derived (somewhat naively) that, if f and g are piecewise continuous, periodic functions with period p, and

$$F.S.[f]|_t = \sum_{k=-\infty}^{\infty} f_k\, e^{i2\pi\omega_k t} \quad \text{and} \quad F.S.[g]|_t = \sum_{k=-\infty}^{\infty} g_k\, e^{i2\pi\omega_k t} \quad,$$

then

$$F.S.[fg]|_t = \sum_{k=-\infty}^{\infty} c_k\, e^{i2\pi\omega_k t} \tag{13.16a}$$

where, for each integer k,

$$c_k = \sum_{n=-\infty}^{\infty} f_n g_{k-n} \quad . \tag{13.16b}$$

    a. Rigorously prove that equation set (13.16) holds assuming that f is continuous and piecewise smooth.

    b. Use theorem 13.17 on norm convergence to show that equation set (13.16) holds even when f is not continuous and piecewise smooth.

14

Convergence and Fourier's Conjecture: The Proofs

As promised, here we will go into the details of verifying the basic theorem on pointwise convergence. In addition, as also promised, we will carefully examine the behavior of the partial sums of the Fourier series of certain saw functions both on intervals away from the discontinuities (to verify "nearly uniform convergence") and on intervals containing points of discontinuity (to study Gibbs phenomenon). And, finally, we will construct a periodic function that, though continuous, has a divergent Fourier series at one point.

14.1 Basic Theorem on Pointwise Convergence

Our first big goal is to prove the following theorem (which is the same as theorem 13.1 on page 156).

Theorem 14.1 *(basic theorem on pointwise convergence)*
Let f be a periodic, piecewise continuous function with

$$F.S.[f]|_t = \sum_{k=-\infty}^{\infty} c_k e^{i2\pi\omega_k t} \quad .$$

Assume, further, that f is piecewise smooth on an interval (α, β), and let t_0 be any point in that interval. Then:

1. *If $f(t)$ is continuous at $t = t_0$, then*

$$\lim_{\substack{N\to\infty \\ M\to-\infty}} \sum_{k=M}^{N} c_k e^{i2\pi\omega_k t_0} = f(t_0) \quad .$$

2. *If $f(t)$ has a jump discontinuity at $t = t_0$, then*

$$\lim_{N\to\infty} \sum_{k=-N}^{N} c_k e^{i2\pi\omega_k t_0} = \frac{1}{2}\left[\lim_{\tau\to t_0^-} f(\tau) + \lim_{\tau\to t_0^+} f(\tau) \right] \quad .$$

Some of the lemmas that we will develop to help prove this theorem will also be used later to more closely examine the convergence of the Fourier series for a simple saw function.

Preliminary Lemmas

A number of "little facts" will be needed. To avoid having to stop in the middle of the main proof to develop them, we will describe these little facts in the following sequence of lemmas and corollaries.

The first is based on the partial sum formula for the geometric series (equation (4.11) on page 43),

$$\sum_{k=M}^{N} X^k = \frac{X^M - X^{N+1}}{1 - X} \tag{14.1}$$

where M and N are any two integers with $M < N$ and X is any complex number other than 1 or 0. Do observe, however, that since the left-hand side is continuous at $X = 1$, the apparent discontinuity at $X = 1$ on the right-hand side of equation (14.1) is clearly trivial.

Replacing X with $e^{-i\gamma x}$ in equation (14.1) yields the following lemma.

Lemma 14.2
If γ is any nonzero number, and M and N are any two integers with $M < N$, then

$$\sum_{k=M}^{N} e^{-ik\gamma x} = \frac{e^{-iM\gamma x} - e^{-i(N+1)\gamma x}}{1 - e^{-i\gamma x}} \tag{14.2}$$

whenever γx is not an integral multiple of 2π.

Again, because the left-hand side of equation (14.2) is clearly continuous, any apparent discontinuity in the right-hand side when γx is any integral multiple of 2π is trivial.

?▶Exercise 14.1:   *Let N be a positive integer and γ and α real values. Derive the following sequence of formulas:*

$$\sum_{k=-N}^{N} e^{-ik\gamma t} = \frac{\sin\left(\gamma[N + \frac{1}{2}]t\right)}{\sin\left(\frac{1}{2}\gamma t\right)} \quad , \tag{14.3a}$$

$$\sum_{k=1}^{N} \cos(2\alpha kt) = \frac{\sin([2N + 1]\alpha t) - \sin(\alpha t)}{2\sin(\alpha t)} \quad , \tag{14.3b}$$

and

$$\sum_{k=1}^{N} \sin(2\alpha kt) = \frac{\cos(\alpha t) - \cos([2N + 1]\alpha t)}{2\sin(\alpha t)} \quad . \tag{14.3c}$$

(Start by letting $M = -N$ in equation (14.2) and then multiplying both the numerator and denominator by $\exp\left(i\frac{\gamma}{2}t\right)$.)

The claims in the next lemma are easily verified by simply evaluating the indicated integrals term by term.

Lemma 14.3
Let $p > 0$, and let M and N be integers with $M < 0 < N$. Then, letting $\gamma = {}^{2\pi}/_{p}$,

$$\int_{-p/2}^{p/2} \left[\sum_{k=M}^{N} e^{-ik\gamma x} \right] dx = p$$

and

$$\int_{-P/2}^{0} \left[\sum_{k=-N}^{N} e^{ik\gamma x} \right] dx = \int_{0}^{P/2} \left[\sum_{k=-N}^{N} e^{ik\gamma x} \right] dx = \frac{p}{2} \quad .$$

?►Exercise 14.2: Prove lemma 14.3.

The next lemma is yet another version of the Riemann–Lebesgue lemma. This particular version is a simple corollary of the more general version on page 141.

Lemma 14.4 (Riemann–Lebesgue lemma)
Let $p > 0$, and assume g is a piecewise continuous function on the interval $(-P/2, P/2)$. Then, letting $\gamma = {}^{2\pi}/_p$,

$$\lim_{\substack{K \to \pm\infty \\ K \in \mathbb{Z}}} \int_{-P/2}^{P/2} g(x) e^{-iK\gamma x} dx = 0 \quad .$$

The Dirichlet Kernel

Let $p > 0$, and let M and N be any two integers with $M < N$. The corresponding *Dirichlet kernel* is the function $D_{M,N}$ given by

$$D_{M,N}(x) = \frac{1}{p} \sum_{k=M}^{N} e^{-i2\pi\omega_k x} \tag{14.4}$$

where, as usual, $\omega_k = {}^k/_p$. Letting $\gamma = {}^{2\pi}/_p$, the above can be rewritten as

$$D_{M,N}(x) = \frac{1}{p} \sum_{k=M}^{N} e^{-ik\gamma x} \quad .$$

From lemma 14.2, we know that

$$D_{M,N}(x) = \frac{e^{-iM\gamma x} - e^{-i(N+1)\gamma x}}{p \left[1 - e^{-i\gamma x}\right]} = \frac{e^{-i2\pi\omega_M x} - e^{-i2\pi\omega_{N+1} x}}{p \left[1 - e^{-i2\pi\omega_1 x}\right]} \quad , \tag{14.5}$$

and from lemma 14.3 it follows that, provided $M < 0 < N$,

$$\int_{-P/2}^{P/2} D_{M,N}(x) \, dx = 1 \tag{14.6}$$

and

$$\int_{-P/2}^{0} D_{-N,N}(x) \, dx = \int_{0}^{P/2} D_{-N,N}(x) \, dx = \frac{1}{2} \quad . \tag{14.7}$$

Our interest in the Dirichlet kernel comes from the fact that, if f is any periodic, piecewise continuous function with

$$F.S.[f]|_t = \sum_{k=-\infty}^{\infty} c_k e^{i2\pi\omega_k t} \quad ,$$

then each partial sum of this series can be expressed as an integral of a translation of f multiplied by the corresponding Dirichlet kernel. To see this, first observe that, for any integer k and real value t_0,

$$
c_k\, e^{i2\pi\omega_k t_0} = \left(\frac{1}{p} \int_{t_0-P/2}^{t_0+P/2} f(\tau) e^{-i2\pi\omega_k \tau}\, d\tau \right) e^{i2\pi\omega_k t_0}
$$

$$
= \frac{1}{p} \int_{t_0-P/2}^{t_0+P/2} f(\tau) e^{-i2\pi\omega_k(\tau-t_0)}\, d\tau = \frac{1}{p} \int_{-P/2}^{P/2} f(t_0+x) e^{-i2\pi\omega_k x}\, dx \quad .
$$

So,

$$
\sum_{k=M}^{N} c_k\, e^{i2\pi\omega_k t_0} = \sum_{k=M}^{N} \frac{1}{p} \int_{-P/2}^{P/2} f(t_0+x) e^{-i2\pi\omega_k x}\, dx
$$

$$
= \int_{-P/2}^{P/2} \frac{1}{p} \sum_{k=M}^{N} f(t_0+x) e^{-i2\pi\omega_k x}\, dx
$$

$$
= \int_{-P/2}^{P/2} f(t_0+x) \left(\frac{1}{p} \sum_{k=M}^{N} e^{-i2\pi\omega_k x} \right) dx \quad .
$$

In other words,

$$
\sum_{k=M}^{N} c_k\, e^{i2\pi\omega_k t_0} = \int_{-P/2}^{P/2} f(t_0+x) D_{M,N}(x)\, dx \quad . \tag{14.8}
$$

The next two lemmas will help reduce the analysis of the pointwise convergence of a Fourier series to a corresponding analysis of a particular integral.

Lemma 14.5
Let f be a periodic, piecewise continuous function with

$$
F.S.[f]|_t = \sum_{k=-\infty}^{\infty} c_k\, e^{i2\pi\omega_k t} \quad .
$$

Let t_0 be any point on the real line at which $f(t)$ is continuous. Then for any pair of integers M and N with $M < 0 < N$,

$$
\sum_{k=M}^{N} c_k\, e^{i2\pi\omega_k t_0} = \int_{-P/2}^{P/2} [f(t_0+x) - f(t_0)] D_{M,N}(x)\, dx + f(t_0) \quad .
$$

Lemma 14.6
Let f be a periodic, piecewise continuous function with

$$
F.S.[f]|_t = \sum_{k=-\infty}^{\infty} c_k\, e^{i2\pi\omega_k t} \quad .
$$

Let t_0 be any point on the real line and let f_0^- and f_0^+ be the values

$$
f_0^- = \lim_{x\to 0^-} f(t_0+x) \quad \text{and} \quad f_0^+ = \lim_{x\to 0^+} f(t_0+x) \quad .
$$

Then, for any positive integer N,

$$\sum_{k=-N}^{N} c_k e^{i2\pi\omega_k t_0} = \int_{-P/2}^{P/2} [f(t_0 + x) - f_0(x)] D_{-N,N}(x)\, dx + \tfrac{1}{2}[f_0^- + f_0^+]$$

where (see figure 14.1)

$$f_0(x) = \begin{cases} f_0^- & \text{if } x < 0 \\ f_0^+ & \text{if } x > 0 \end{cases}.$$

The proofs of these two lemmas are similar and will be combined.

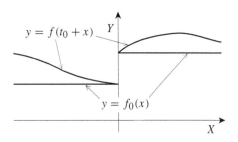

$y = f(t_0 + x)$

$y = f_0(x)$

Figure 14.1: f and f_0.

PROOF (of lemmas 14.5 and 14.6): Let f_0^+, f_0^- and $f_0(x)$ be as in lemma 14.6. Then

$$\int_{-P/2}^{P/2} f(t_0 + x) D_{M,N}(x)\, dx = \int_{-P/2}^{P/2} [f(t_0 + x) - f_0(x) + f_0(x)] D_{M,N}(x)\, dx$$

$$= \int_{-P/2}^{P/2} [f(t_0 + x) - f_0(x)] D_{M,N}(x)\, dx \tag{14.9}$$

$$+ \int_{-P/2}^{P/2} f_0(x) D_{M,N}(x)\, dx \quad.$$

If f is continuous at t_0, then $f_0^- = f(t_0) = f_0^+$ and $f_0(x) = f(t_0)$ for all x. With this and equality (14.6), equation (14.9) becomes

$$\int_{-P/2}^{P/2} f(t_0 + x) D_{M,N}(x)\, dx = \int_{-P/2}^{P/2} [f(t_0 + x) - f(t_0)] D_{M,N}(x)\, dx$$

$$+ f(t_0) \int_{-P/2}^{P/2} D_{M,N}(x)\, dx$$

$$= \int_{-P/2}^{P/2} [f(t_0 + x) - f(t_0)] D_{M,N}(x)\, dx + f(t_0) \quad.$$

This proves lemma 14.5.

Whether or not f is continuous at t_0, if $M = -N$, then, by the definition of $f_0(x)$ and equality (14.6), equation (14.9) simplifies as follows:

$$\int_{-P/2}^{P/2} f(t_0 + x) D_{-N,N}(x)\, dx = \int_{-P/2}^{P/2} [f(t_0 + x) - f_0(x)] D_{-N,N}(x)\, dx$$

$$+ \int_{-P/2}^{0} f_0^- D_{-N,N}(x)\, dx + \int_{0}^{P/2} f_0^+ D_{-N,N}(x)\, dx$$

$$= \int_{-P/2}^{P/2} [f(t_0 + x) - f_0(x)] D_{-N,N}(x)\, dx$$

$$+ \tfrac{1}{2}[f_0^- + f_0^+] \quad.$$

Proof of the Main Theorem

Let f and t_0 be as in the main theorem (theorem 14.1 on page 179). That is, f is a periodic, piecewise continuous function on $\mathbb{R}$ that is piecewise smooth on some interval (α, β) containing the point t_0. For convenience, let us use the notation introduced in lemma 14.6,

$$f_0^- = \lim_{x \to 0^-} f(t_0 + x) \quad , \quad f_0^+ = \lim_{x \to 0^+} f(t_0 + x)$$

and

$$f_0(x) = \begin{cases} f_0^- & \text{if} \quad x < 0 \\ f_0^+ & \text{if} \quad x > 0 \end{cases} .$$

(Again, note that f_0^-, f_0^+ and $f_0(x)$ all reduce to $f(t_0)$ if f is continuous at t_0.) Let M and N be integers with $M < 0 < N$, $D_{M,N}$ the corresponding Dirichlet's kernel for f, and

$$F.S.[f]|_t = \sum_{k=-\infty}^{\infty} c_k e^{i2\pi \omega_k t} \quad .$$

Lemma 14.5 assures us that, if f is continuous at t_0, then

$$\sum_{k=M}^{N} c_k e^{i2\pi \omega_k t_0} = \int_{-P/2}^{P/2} [f(t_0 + x) - f_0(x)] D_{M,N}(x)\, dx \; + \; f(t_0) \quad , \tag{14.10}$$

and lemma 14.6 assures us that, whether or not f is continuous at t_0,

$$\sum_{k=-N}^{N} c_k e^{i2\pi \omega_k t_0} = \int_{-P/2}^{P/2} [f(t_0 + x) - f_0(x)] D_{-N,N}(x)\, dx \; + \; \frac{1}{2}[f_0^- + f_0^+] \quad . \tag{14.11}$$

Comparing these equations to the claims in theorem 14.1, we find that the proof of theorem 14.1 will be complete once we have shown that

$$\lim_{\substack{N \to \infty \\ M \to -\infty}} \int_{-P/2}^{P/2} [f(t_0 + x) - f_0(x)] D_{M,N}(x)\, dx \; = \; 0 \quad . \tag{14.12}$$

We can quickly simplify our problem. First, observe that, using equation (14.5),

$$\int_{-P/2}^{P/2} [f(t_0 + x) - f_0(x)] D_{M,N}(x)\, dx$$

$$= \int_{-P/2}^{P/2} [f(t_0 + x) - f_0(x)] \left[\frac{e^{-iM\gamma x} - e^{-i(N+1)\gamma x}}{p[1 - e^{-i\gamma x}]} \right] dx$$

$$= \frac{1}{p} \int_{-P/2}^{P/2} g(x) e^{-iM\gamma x}\, dx \; - \; \frac{1}{p} \int_{-P/2}^{P/2} g(x) e^{-i(N+1)\gamma x}\, dx \quad ,$$

where

$$\gamma = \frac{2\pi}{p} \quad \text{and} \quad g(x) = \frac{f(t_0 + x) - f_0(x)}{1 - e^{-i\gamma x}} \quad .$$

Thus, to show equation (14.12) holds, it will suffice to show that

$$\lim_{K \to \pm\infty} \int_{-P/2}^{P/2} g(x) e^{-iK\gamma x}\, dx \; = \; 0 \quad .$$

But this last equality follows immediately from the Riemann–Lebesgue lemma (lemma 14.4) provided g is piecewise continuous on $(-P/2, P/2)$.

So all that we now need to show is that

$$g(x) = \frac{f(t_0 + x) - f_0(x)}{1 - e^{-i\gamma x}}$$

is piecewise continuous on $(-P/2, P/2)$.

Since the denominator just above,

$$1 - e^{-i\gamma x} = 1 - \cos\left(\frac{2\pi}{p}x\right) + i \sin\left(\frac{2\pi}{p}x\right) \quad,$$

is nonzero and continuous at every x with $-P/2 \le x < 0$ or $0 < x \le P/2$, it should be clear that

$$\lim_{x \to -P/2^+} g(x) \qquad \text{and} \qquad \lim_{x \to P/2^-} g(x)$$

exist and are finite, and that the only discontinuities in $g(x)$ can be either at $x = 0$ or at one of the finite number of points at which $f(t_0 + x)$ has a jump discontinuity. Further, except possibly at $x = 0$, the resulting discontinuities in g must clearly be jump discontinuities.

All that remains to verifying the piecewise continuity of g on $(-P/2, P/2)$ is showing that the possible discontinuity at $x = 0$ is no worse than a jump discontinuity. In other words, we merely need to verify that the left- and right-hand limits of g exist as finite numbers.

Naively taking the right-hand limit gives

$$\lim_{x \to 0^+} g(x) = \lim_{x \to 0^+} \frac{f(t_0 + x) - f_0^+}{1 - e^{-i\gamma x}} = \frac{f_0^+ - f_0^+}{1 - e^0} \quad,$$

which is indeterminate. Fortunately, because f is piecewise smooth on an interval containing t_0, l'Hôpital's rule can be applied. Doing so,

$$\lim_{x \to 0^+} g(x) = \lim_{x \to 0^+} \frac{f(t_0 + x) - f_0^+}{1 - e^{-i\gamma x}}$$

$$= \lim_{x \to 0^+} \frac{\frac{d}{dx}[f(t_0 + x) - f_0^+]}{\frac{d}{dx}[1 - e^{-i\gamma x}]}$$

$$= \lim_{x \to 0^+} \frac{f'(t_0 + x)}{i\gamma e^{-i\gamma x}} = \frac{1}{i\gamma} \lim_{x \to 0^+} f'(t_0 + x) \quad.$$

Likewise,

$$\lim_{x \to 0^-} g(x) = \frac{1}{i\gamma} \lim_{x \to 0^-} f'(t_0 + x) \quad.$$

Since $f(t)$ is piecewise smooth on an interval about t_0, the left- and right-hand limits of $f'(t)$ at $t = t_0$ exist as finite numbers. Hence, by the above, so do the left- and right-hand limits of $g(x)$ at $x = 0$.

And that completes our proof of theorem 14.1. ∎

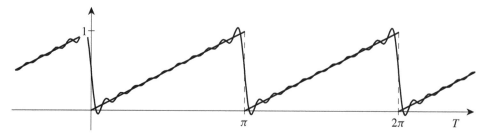

Figure 14.2: Graph of $\frac{1}{\pi} \text{saw}_\pi(t)$ superimposed on a partial sum for its Fourier series.

14.2   Convergence for a Particular Saw Function

Let us examine the error in using the partial sum approximation for a particular discontinuous function, namely,

$$h(t) \;=\; \frac{1}{\pi}\,\text{saw}_\pi(t) \;=\; \begin{cases} \dfrac{t}{\pi} & \text{if } \ 0 < t < \pi \\[2mm] h(t-\pi) & \text{in general} \end{cases} . \tag{14.13}$$

(There are two reasons to choose this function: First, the convergence of its Fourier series is relatively easy to analyze. Second, the results of this analysis can be applied to describing the errors arising when more general functions are approximated by corresponding partial sums.)

The function h, sketched in figure 14.2, is clearly piecewise smooth and periodic with period π. It is continuous everywhere except at integral multiples of π where it has a jump of -1. Its complex exponential and trigonometric Fourier series are easily computed, and are, respectively,

$$\frac{1}{2} + \sum_{\substack{k=-\infty \\ k \neq 0}}^{\infty} \frac{i}{2\pi k} e^{2\pi k t} \quad\quad \text{and} \quad\quad \frac{1}{2} - \sum_{k=1}^{\infty} \frac{1}{k\pi} \sin(2\pi k t) \quad .$$

There are two parts to this study. The first is to show that, while the partial sums cannot uniformly approximate h over the entire real line, they do uniformly approximate h over certain intervals not containing points at which h is discontinuous. The second part is to closely examine the Gibbs phenomenon around the discontinuities. In both parts we will need to derive, as accurately as practical, a usable formula for the error in using the $(M, N)^{\text{th}}$ partial sum for the exponential Fourier sum in place of h,

$$E_{M,N}(t) \;=\; \left| h(t) - F.S._{M,N}[h]\big|_t \right| \tag{14.14}$$

where M and N are any two integers with $M < 0 < N$ and t is a point on the real line other than an integral multiple of π. We will then use the derived formula to see how this error depends on M, N and t.

Limited Uniform Approximation

Let M and N be any two integers with $M < 0 < N$; let t be any point on the real line other than an integral multiple of π, and let $E_{M,N}(t)$ be as above. From lemma 14.5 it easily follows that

$$E_{M,N}(t) \;=\; \left| \int_{-\pi/2}^{\pi/2} [h(t+x) - h(t)] D_{M,N}(x)\,dx \right|$$

where $D_{M,N}$ is the corresponding Dirichlet kernel. By formula (14.5) and the fact that $\omega_k = {}^k/_\pi$, this is the same as

$$E_{M,N}(t) = \left| \int_{-\pi/2}^{\pi/2} [h(t+x) - h(t)] \left[\frac{e^{-i2Mx} - e^{-i2(N+1)x}}{\pi \left[1 - e^{-i2x}\right]} \right] dx \right|$$

$$\leq |\mathit{l}_M(t)| + |\mathit{l}_{N+1}(t)| \tag{14.15}$$

where, for any integer K ,

$$\mathit{l}_K(t) = \int_{-\pi/2}^{\pi/2} \frac{h(t+x) - h(t)}{\pi \left[1 - e^{-i2x}\right]} e^{-i2Kx} \, dx \quad .$$

To further simplify our computations, let us observe that

$$\frac{h(t+x) - h(t)}{\pi \left[1 - e^{-i2x}\right]} \cdot \frac{e^{ix}}{e^{ix}} = \frac{h(t+x) - h(t)}{i2\pi \sin(x)} e^{ix} \quad .$$

Thus,

$$\mathit{l}_K(t) = \frac{1}{i2\pi} \int_{-\pi/2}^{\pi/2} \frac{h(t+x) - h(t)}{\sin(x)} e^{i(1-2K)x} \, dx \quad . \tag{14.16}$$

Some of the details in the following computations will depend on the interval in which t lies. We will first assume

$$0 < t \leq \frac{\pi}{2} \quad . \tag{14.17}$$

With these assumptions on t you can easily verify that

$$h(t+x) - h(t) = \frac{1}{\pi}x + \begin{cases} 1 & \text{if } -\pi/2 < x < -t \\ 0 & \text{if } -t < x < \pi/2 \end{cases} \quad ,$$

and that formula (14.16) for l_K can be rewritten as the sum of two relatively simple integrals,

$$\mathit{l}_K(t) = \frac{1}{i2\pi^2} \int_{-\pi/2}^{\pi/2} \frac{x}{\sin(x)} e^{i(1-2K)x} \, dx + \frac{1}{i2\pi} \int_{-\pi/2}^{-t} \frac{1}{\sin(x)} e^{i(1-2K)x} \, dx \quad . \tag{14.18}$$

Though "relatively simple", these are still not integrals we can easily evaluate. But observe that each is of the form

$$\int_a^b u(x) e^{i(1-2K)x} \, dx \quad .$$

So if u is a uniformly smooth function on (a, b) , then the integration by parts formula can be used in the following "clever" way:

$$\left| \int_a^b u(x) e^{i(1-2K)x} \, dx \right| = \left| \frac{i}{1-2K} u(x) e^{i(1-2K)x} \Big|_a^b - \frac{i}{1-2K} \int_a^b u'(x) e^{i(1-2K)x} \, dx \right|$$

$$\leq \frac{1}{|1-2K|} \left(\left| u(x) e^{i(1-2K)x} \Big|_a^b \right| + \int_a^b |u'(x)| \, dx \right) \quad . \tag{14.19}$$

From this, it should be easy to derive an upper bound on the integral which goes to zero as $K \to \pm\infty$.
     For the first integral on the right-hand side of equation (14.18) we have

$$u(x) = \frac{x}{\sin(x)} \quad .$$

Clearly, this function is continuous and has a continuous derivative everywhere the sine function is nonzero. For $-\pi/2 \le x \le \pi/2$, the sine function vanishes only at $x = 0$. However, using l'Hôpital's rule,

$$\lim_{x \to 0^{\pm}} u(x) = \lim_{x \to 0^{\pm}} \frac{x}{\sin(x)} = \lim_{x \to 0^{\pm}} \frac{1}{\cos(x)} = 1$$

and

$$\lim_{x \to 0^{\pm}} u'(x) = \lim_{x \to 0^{\pm}} \frac{\sin(x) - x\cos(x)}{\sin^2(x)} = \lim_{x \to 0^{\pm}} \frac{x}{2\cos(x)} = 0 \ .$$

So, for this choice of u, any discontinuity in u or u' at 0 is removable. Thus, u is uniformly smooth on $(-\pi/2, \pi/2)$, and the computations indicated in (14.19) are valid when $(a, b) = (-\pi/2, \pi/2)$. Since

$$u(x)e^{i(1-2K)x}\Big|_a^b = \frac{x}{\sin(x)}e^{i(1-2K)x}\Big|_{-\pi/2}^{\pi/2} = \frac{\pi}{2}i^{2K-1} - \frac{\pi}{2}i^{1-2K} = i(-1)^{K+1}\pi \ ,$$

inequality (14.19) becomes

$$\left| \int_{-\pi/2}^{\pi/2} \frac{x}{\sin(x)}e^{i(1-2K)x}\,dx \right| \le \frac{1}{|1-2K|}\left(\pi + \int_{-\pi/2}^{\pi/2} |u'(x)|\,dx \right) \ . \tag{14.20}$$

After a few observations, we will be able to explicitly evaluate the above integral of $|u'(x)|$. The first observation is that, for $0 < x < \pi/2$,

$$\frac{d}{dx}[\sin(x) - x\cos(x)] = x\sin(x) > 0 \ .$$

This tells us that $\sin(x) - x\cos(x)$ is an increasing function on $(0, \pi/2)$. Thus, for $0 < x < \pi/2$,

$$\sin(x) - x\cos(x) > \sin(0) - 0\cos(0) = 0 \ ,$$

which, in turn, assures us that, for $0 < x < \pi/2$,

$$u'(x) = \frac{\sin(x) - x\cos(x)}{\sin^2(x)} > 0 \ .$$

In other words, $u' = |u'|$ on $(0, \pi/2)$. Also, observe that $|u'|$ is an even function,

$$|u'(-x)| = \left| \frac{\sin(-x) - (-x)\cos(-x)}{\sin^2(-x)} \right| = \left| -\frac{\sin(x) - x\cos(x)}{\sin^2(x)} \right| = |u'(x)| \ .$$

So

$$\int_{-\pi/2}^{\pi/2} |u'(x)|\,dx = 2\int_0^{\pi/2} u'(x)\,dx = 2\int_0^{\pi/2} \frac{d}{dx}\left[\frac{x}{\sin(x)} \right]\,dx = 2\left(\frac{\pi}{2} - 1 \right) \ .$$

Plugging this into inequality (14.20) gives

$$\left| \int_{-\pi/2}^{\pi/2} \frac{x}{\sin(x)}e^{i(1-2K)x}\,dx \right| \le \frac{2\pi - 2}{|1-2K|} \ . \tag{14.21}$$

A useful bound for the other integral on the right side of formula (14.18) is more easily derived. This derivation starts with the observation that 0 is not in $(-\pi/2, -t)$ (because we are assuming $t > 0$). So there should be no question that inequality (14.19) holds when

$$u(x) = \frac{1}{\sin(x)} \qquad \text{and} \qquad (a, b) = (-\pi/2, -t) \ .$$

For this case inequality (14.19) becomes

$$\left| \int_{-\pi/2}^{-t} \frac{1}{\sin(x)} e^{i(1-2K)x} \, dx \right| \leq \frac{1}{|1-2K|} \left(\left| \frac{e^{i(1-2K)x}}{\sin(x)} \right|_{-\pi/2}^{-t} + \int_{-\pi/2}^{-t} \frac{\cos(s)}{\sin^2(x)} \, dx \right)$$

$$= \frac{1}{|1-2K|} \left(\left| \frac{e^{i(2K-1)t}}{\sin(t)} - \frac{e^{i(2K-1)\frac{\pi}{2}}}{-1} \right| + \left| \frac{-1}{\sin(x)} \right|_{-\pi/2}^{-t} \right)$$

$$\leq \frac{1}{|1-2K|} \left(\frac{1}{\sin(t)} + 1 + \frac{1}{\sin(t)} - 1 \right)$$

$$= \frac{2}{|1-2K|\sin(t)} \quad . \tag{14.22}$$

Combining formula (14.18) with inequalities (14.21) and (14.22) gives

$$|\mathcal{I}_K(t)| \leq \frac{\pi - 1}{\pi^2 |1-2K|} + \frac{1}{\pi |1-2K|\sin(t)} \quad .$$

Equivalently,

$$|\mathcal{I}_K(t)| \leq \frac{B(t)}{|1-2K|} \tag{14.23a}$$

where

$$B(t) = \frac{1}{\pi} - \frac{1}{\pi^2} + \frac{1}{\pi \sin(t)} \quad . \tag{14.23b}$$

From this and inequality (14.15) it follows that

$$E_{M,N}(t) \leq \left[\frac{1}{1-2M} + \frac{1}{1+2N} \right] B(t) \tag{14.24}$$

at least when $0 < t < \pi/2$ and $M < 0 < N$.

We'll leave the derivation of the error bounds for other values of t as exercises.

?▶ **Exercise 14.3:** *"Redo" the above computations under the assumption that $-\pi/2 \leq t < 0$ (and $M < 0 < N$), and show that, in this case,*

$$E_{M,N}(t) \leq \left[\frac{1}{1-2M} + \frac{1}{1+2N} \right] B(|t|)$$

where B is as given by formula (14.23b). (Suggestion: Start by deriving the corresponding formula for $h(t+x) - h(t)$.)

?▶ **Exercise 14.4:** *Let t be any real value other than an integral multiple of π, and let $E_{M,N}$ be as above (i.e., as defined in equation (14.14) with $M < 0 < N$). Using periodicity, inequality (14.24) and the results of the previous exercise, show that*

$$E_{M,N}(t) \leq \left[\frac{1}{1-2M} + \frac{1}{1+2N} \right] B(D)$$

where B is as given by formula (14.23b) and D is the distance from t to the nearest integral multiple of π.

Gibbs Phenomenon

Let N be any positive integer and, for convenience, let's use S_N instead of $F.S._N[h]$ to denote the Nth partial sum of the trigonometric series for h,

$$S_N(t) = F.S._N[h]|_t = \frac{1}{2} - \sum_{k=1}^{N} \frac{1}{k\pi} \sin(2kt) \quad .$$

The graph of S_N (see figure 14.2) contains very distinctive "wiggles" that are particularly large near the points of discontinuity. Our goals here are to determine

1. the locations of the peaks and valleys of these wiggles (more precisely, the locations of the local maximums and minimums of S_N),

2. how these locations vary as N gets large,

and

3. the difference between $h(t)$ and $S_N(t)$ at these locations, at least for large values of N .

Locating the Wiggles

Since S_N is a finite sum of smooth functions on the entire real line, all of its local maximums and minimums occur at points where its derivative,

$$S_N'(t) = \frac{d}{dt}\left[\frac{1}{2} - \sum_{k=1}^{N} \frac{1}{k\pi} \sin(2kt) \right] = -\frac{2}{\pi} \sum_{k=1}^{N} \cos(2kt) \quad ,$$

is zero. Clearly, none of these local maximums or minimums occur at an integral multiple of π .

Thanks to one of the formulas from exercise 14.1 on page 180, we know that, if t is not an integral multiple of π ,

$$S_N'(t) = \frac{\sin(t) - \sin([2N+1]t)}{\pi \sin(t)} \quad . \tag{14.25}$$

So, to find all values of t for which $S_N'(t) = 0$, it suffices to find all values of t , other than integral multiples of π , satisfying

$$\sin(t) - \sin([2N+1]t) = 0 \quad .$$

This last equation is easily solved if we view it as

$$\sin(t) = \sin(x) \tag{14.26a}$$

where

$$x = (2N+1)t \quad . \tag{14.26b}$$

From figure 14.3, it should be clear that, for each t in the interval $(0, {}^\pi\!/_2)$, the values of x satisfying equation (14.26a) are x_0 , $x_{\pm 1}$, $x_{\pm 2}$, ... where

$$x_m = \begin{cases} m\pi + t & \text{if } m \text{ is even} \\ m\pi - t & \text{if } m \text{ is odd} \end{cases} \quad .$$

With these values for x , equation (14.26b) becomes

$$(2N+1)t = \begin{cases} m\pi + t & \text{if } m \text{ is even} \\ m\pi - t & \text{if } m \text{ is odd} \end{cases} \quad .$$

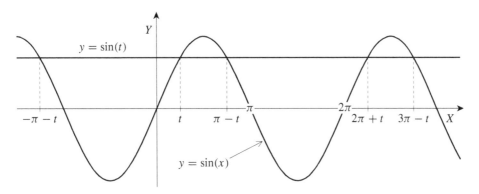

Figure 14.3: Values of x where $\sin(t) = \sin(x)$ for a given t in $(0, \pi/2)$.

Solving for t (and recalling our current assumption that $0 < t < \pi/2$), we find that

$$
t = t_{N,m} = \begin{cases} \dfrac{m\pi}{2N} & \text{if } m \text{ is even} \\[2mm] \dfrac{m\pi}{2N+2} & \text{if } m \text{ is odd} \end{cases} \tag{14.27}
$$

where $m = 1, 2, \ldots, M_N$ and

$$
M_N = \begin{cases} N-1 & \text{if } N \text{ is even} \\ N & \text{if } N \text{ is odd} \end{cases}.
$$

These points (the $t_{N,m}$'s) are the locations of the local maximums and minimums of h on the interval $(0, \pi/2)$.

?► Exercise 14.5:   *Let $t_{N,m}$ be as above. Using the second derivative test and equation (14.25), show that S_N has a local minimum at $t_{N,m}$ when m is odd, and has a local maximum at $t_{N,m}$ when m is even.*

?► Exercise 14.6:   *Show that formula (14.27) with $m = -M_N, \ldots, -2, -1$ gives the locations of the local maximums and minimum on the interval $(-\pi/2, 0)$.*

Because S_N is periodic with period π, the above results assure us that, for any integer K, the local maximums and minimums of S_N on $(K\pi - \pi/2, \ K\pi + \pi/2)$ occur at $t = K\pi + t_{N,m}$ where $m = \pm 1, \pm 2, \pm 3, \ldots, \pm M_N$, and the $t_{N,m}$'s are as given by formula (14.27). Note that

1. of these points, $K\pi + t_{N,1}$ and $K\pi + t_{N,-1}$ are the two points which are closest to the discontinuity,

and

2. for any fixed choice of m, $K\pi + t_{N,m} \to K\pi$ as $m \to \infty$.

Limiting Height of the Wiggles

Again, let us first consider the wiggles in just the interval $(0, \pi/2)$. At each of the $t_{N,m}$'s, the difference between h and the Nth partial sum approximation S_N is

$$
S_N(t_{N,m}) - h(t_{N,m}) = \frac{1}{2} - \sum_{k=1}^{N} \frac{1}{k\pi} \sin(2k t_{N,m}) - \frac{1}{\pi} t_{N,m}.
$$

Since we are interested in the value of this expression for particular choices of m and large values of N, let us see what happens as N goes to infinity (while holding m fixed). Because $t_{N,m} \to 0$ as $N \to \infty$, we see that

$$\lim_{N \to \infty} \left[S_N(t_{N,m}) - h(t_{N,m}) \right] = \frac{1}{2} - \lim_{N \to \infty} \sum_{k=1}^{N} \frac{1}{k\pi} \sin(2kt_{N,m}) \quad . \tag{14.28}$$

Now, for every integer N larger than m, and every positive integer k, let $\tau_k = k\Delta\tau$ where

$$\Delta\tau = 2t_{N,m} = \begin{cases} \dfrac{m\pi}{N} & \text{if } m \text{ is even} \\[2mm] \dfrac{m\pi}{N+1} & \text{if } m \text{ is odd} \end{cases} \quad .$$

Observe that

$$\sum_{k=1}^{N} \frac{1}{k\pi} \sin(2kt_{N,m}) = \frac{1}{\pi} \sum_{k=1}^{N} \frac{1}{k\Delta\tau} \sin(k\Delta\tau)\,\Delta\tau = \frac{1}{\pi} R_N \tag{14.29}$$

where

$$R_N = \sum_{k=1}^{N} \frac{\sin(\tau_k)}{\tau_k} \Delta\tau \quad .$$

But R_N is just a Riemann sum for

$$\int_{\tau_0}^{\tau_N} \frac{\sin(\tau)}{\tau}\, d\tau \quad .$$

Since $\tau_0 = 0 \cdot \Delta\tau = 0$ and

$$\tau_N = N\Delta\tau = \begin{cases} m\pi & \text{if } m \text{ is even} \\[2mm] \dfrac{N}{N+1} m\pi & \text{if } m \text{ is odd} \end{cases} \quad ,$$

we clearly have

$$\lim_{N \to \infty} R_N = \int_0^{m\pi} \frac{\sin(\tau)}{\tau}\, d\tau \quad . \tag{14.30}$$

Thus, after combining equations (14.28), (14.29) and (14.30), we have

$$\lim_{N \to \infty} \left[S_N(t_{N,m}) - h(t_{N,m}) \right] = \frac{1}{2} - \frac{1}{\pi} \int_0^{m\pi} \frac{\sin(\tau)}{\tau}\, d\tau \quad ,$$

which, for future reference, we will write as

$$\lim_{N \to \infty} \left[S_N(t_{N,m}) - h(t_{N,m}) \right] = \gamma_m \tag{14.31a}$$

where

$$\gamma_m = \frac{1}{2} - \frac{1}{\pi} \int_0^{m\pi} \frac{\sin(\tau)}{\tau}\, d\tau \quad . \tag{14.31b}$$

This last integral is not one we can explicitly evaluate by elementary means, but approximations to this integral can easily be found for specific values of m using standard numerical integration methods (such as found in many computer math packages). Using any of these methods, it can be shown that

$$\gamma_1 = \lim_{N \to \infty} \left[S_N(t_{N,1}) - h(t_{N,1}) \right] \approx -0.0895 \quad ,$$

$$\gamma_2 = \lim_{N \to \infty} \left[S_N(t_{N,2}) - h(t_{N,2}) \right] \approx 0.0486 \quad ,$$

$$\gamma_3 = \lim_{N\to\infty} \left[S_N(t_{N,3}) - h(t_{N,3}) \right] \approx -0.0331$$

and

$$\gamma_4 = \lim_{N\to\infty} \left[S_N(t_{N,4}) - h(t_{N,4}) \right] \approx 0.0250 \quad,$$

each with an absolute error of less than 0.00005. Thus, over $(0, \pi/2)$ with N large, the first wiggle in the graph of S_N undershoots the graph of h by about $.09$ units; the second wiggle undershoots the graph of h by about $.05$ units; the third wiggle undershoots the graph of h by about $.03$, and the fourth wiggle overshoots the graph of h by about $.025$ units.

More generally, you can show (in the next exercise) that the γ_N's form an alternating sequence that "steadily approaches" zero. More precisely,

$$\gamma_m = (-1)^m |\gamma_m|$$

with

$$|\gamma_1| > |\gamma_2| > |\gamma_3| > \ldots > 0$$

and

$$\lim_{m\to\infty} \gamma_m = 0 \quad.$$

?▶ Exercise 14.7:   *In the following, γ_m is as given in formula (14.31b). Also, let $a_0 = \frac{1}{2}$ and, for each positive integer k,*

$$a_k = \frac{1}{\pi} \int_{(k-1)\pi}^{k\pi} \frac{|\sin(\tau)|}{\tau} \, d\tau \quad.$$

a: *Show that*

$$\gamma_m = \sum_{k=0}^{m} (-1)^k a_k \qquad \text{for} \quad m = 1, 2, 3, \ldots \quad.$$

(Thus, each γ_m is a partial sum of an alternating series.)

b: *Verify that*

$$a_k > a_{k+1} \geq 0 \qquad \text{for each} \quad k > 1 \qquad \text{and} \qquad \lim_{k\to\infty} a_k = 0 \quad.$$

(This guarantees the conditional convergence of the alternating series $\sum_{k=0}^{\infty} (-1)^k a_k$ — see the alternating series test on page 44.)

c: *Using methods from complex analysis (or methods we will develop later — see exercise 29.18 on page 490), it can be shown that*

$$\lim_{m\to\infty} \int_0^{m\pi} \frac{\sin(\tau)}{\tau} \, d\tau = \frac{\pi}{2} \quad.$$

Using this, confirm that

$$\lim_{m\to\infty} \gamma_m = \sum_{k=0}^{\infty} (-1)^k a_k = 0 \quad.$$

d: *Now, using the results discussed in the above exercises, properties of "alternating series with decreasing terms", and, possibly, induction, verify that*

$$\gamma_m = (-1)^m |\gamma_m| \qquad \text{and} \qquad |\gamma_1| > |\gamma_2| > |\gamma_3| > \ldots > 0 \quad.$$

It should come as no surprise that similar results can be derived concerning the limiting heights of the other wiggles of S_N. We'll leave the derivations of these results as exercises.

?▶ Exercise 14.8:   Let m be a negative integer, and let $t_{N,m}$ and γ_m be as in formulas (14.27) and (14.31b), respectively. Show that

$$\lim_{N \to \infty} \left[S_N(t_{N,m}) - h(t_{N,m}) \right] = \gamma_m \quad .$$

?▶ Exercise 14.9:   Let K and m be any two integers with $m \neq 0$. Show that

$$\lim_{N \to \infty} \left[S_N(K\pi + t_{N,m}) - h(K\pi + t_{N,m}) \right] = \begin{cases} \gamma_m & \text{if} \quad m > 0 \\ -\gamma_m & \text{if} \quad m < 0 \end{cases} \quad .$$

Local Error in the Partial Sum Approximation

We should note that the points where S_N has local maximums and minimums are not quite the same as the points where the difference between S_N and h is locally a maximum or minimum. These points are found by determining where the derivative of

$$\mathcal{E}_N(t) = S_N(t) - h(t)$$

is zero or does not exist.

It turns out that an analysis similar to that just carried out for S_N can be carried out for $\mathcal{E}_N$. In fact, in some ways, the corresponding analysis for $\mathcal{E}_N$ is simpler. We will leave this analysis as an exercise.

?▶ Exercise 14.10:   Let $\mathcal{E}_N$ be as above.

a: Show that, for each positive integer N, the maximum and minimum values of $\mathcal{E}_N$ on $(0, \pi/2)$ occur at the points

$$\tau_{N,m} = \frac{m\pi}{2N+1} \qquad \text{where} \quad m = 1, 2, 3, \dots, N \quad .$$

b: Verify that, on $(-\pi/2, 0)$ and for each positive integer N, the maximum and minimum values of $\mathcal{E}_N$ occur at the points

$$\tau_{N,m} = \frac{m\pi}{2N+1} \qquad \text{where} \quad m = 1, 2, 3, \dots, N \quad .$$

c: Confirm that, for every nonzero integer m,

$$\lim_{N \to \infty} \left[S_N(\tau_{N,m}) - h(\tau_{N,m}) \right] = \gamma_m$$

where γ_m is given by formula (14.31b).

d: Show that $|\gamma_1|$ does not give the maximum error in using S_N for h on $(0, \pi/2)$.

14.3 Convergence for Arbitrary Saw Functions

The results obtained in the previous section for

$$h(t) = \frac{1}{\pi} \text{saw}_\pi(t)$$

can be easily be converted to analogous results for similar functions by scaling and translation. These results (which were referred to in the previous chapter) are summarized in the following lemmas. In each case p, t_0 and J_0 are constants with t_0 real and $0 < p$. The corresponding shifted and scaled saw function h_0 is given by

$$h_0(t) = \frac{J_0}{p} \text{saw}_p(t - t_0) \quad .$$

Lemma 14.7

For any pair of integers M and N with $M < 0 < N$, and any real value t such that $t - t_0$ is not an integral multiple of p,

$$\left| h_0(t) - F.S._{M,N}[h_0]\big|_t \right| \le \left[\frac{1}{1-2M} + \frac{1}{1+2N} \right] B(D)$$

where D is the distance from $t - t_0$ to the integral multiple of p closest to $t - t_0$, and

$$B(D) = |J_0| \left[\frac{1}{\pi} - \frac{1}{\pi^2} + \frac{1}{\pi \sin\left(\frac{\pi}{p} D\right)} \right] \quad .$$

?►Exercise 14.11: *Prove lemma 14.7 using the results from exercise 14.4.*

Lemma 14.8

Let K and N be any two integers with N positive. The local maximums and minimums of $F.S._N[h_0]$ on $(Kp - p/2, Kp + p/2)$ all occur at the points $Kp + t_{N,m}$ where

$$t_{N,m} = \begin{cases} \dfrac{mp}{2N} & \text{if } m \text{ is even} \\[2ex] \dfrac{mp}{2N+2} & \text{if } m \text{ is odd} \end{cases} \quad ,$$

$$m = \pm 1, \pm 2, \pm 3, \ldots, M_N$$

and

$$M_N = \begin{cases} N - 1 & \text{if } N \text{ is even} \\ N & \text{if } N \text{ is odd} \end{cases} \quad .$$

Moreover, for each of these m's,

$$\lim_{N \to \infty} \left[F.S._N[h_0]\big|_{Kp+t_{N,m}} - h_0(Kp + t_{N,m}) \right] = \begin{cases} -\gamma_m J_0 & \text{if } m < 0 \\ \gamma_m J_0 & \text{if } 0 < m \end{cases}$$

where

$$\gamma_m = \frac{1}{2} - \frac{1}{\pi} \int_0^{m\pi} \frac{\sin(\tau)}{\tau} d\tau \quad .$$

The formula for γ_m's in lemma 14.8 is the same as formula (14.31b). Thus, if N is relatively large, the wiggle in the graph of $F.S._N[h_0]$ closest to each discontinuity will either overshoot or undershoot the graph of h_0 by roughly 9% of the height of the jump at the discontinuity.

Lemma 14.9
Let K and N be any two integers with N positive. The local maximums and minimums of $F.S._N[h_0] - h_0$ on $(Kp - p/2, Kp)$ and $(Kp, Kp + p/2)$ all occur at the points $Kp + \tau_{N,m}$ where $m = \pm 1, \pm 2, \pm 3, \ldots$ and N, and

$$\tau_{N,m} = \frac{mp}{2N + 1} \quad .$$

Moreover, for each of these m's,

$$\lim_{N \to \infty} \left[F.S._N[h_0]|_{Kp + \tau_{N,m}} - h_0(Kp + \tau_{N,m}) \right] = \begin{cases} -\gamma_m J_0 & \text{if } m < 0 \\ \gamma_m J_0 & \text{if } 0 < m \end{cases}$$

where γ_m is as in lemma 14.8.

14.4 A Divergent Fourier Series

Our goal is to construct a continuous periodic function f on the real line whose Fourier series does *not* converge at $t = 0$, not even if we only use the symmetric partial sums.

The Basic Pieces of the Construction

Let us start with the Fourier series for the imaginary vertically shifted saw function

$$h(t) = i [\text{saw}_{2\pi}(t) - \pi] \quad .$$

As you can easily verify,

$$F.S.[h]|_t = \sum_{\substack{k=-\infty \\ k \neq 0}}^{\infty} \frac{-1}{k} e^{ikt} \quad .$$

Now, for each positive integer N, let S_N be the Nth symmetric partial sum of this Fourier series,

$$S_N(t) = F.S._{-N,N}[h]|_t = \sum_{\substack{k=-N \\ k \neq 0}}^{N} \frac{-1}{k} e^{ikt} = \sum_{k=-N}^{-1} \frac{-1}{k} e^{ikt} + \sum_{k=1}^{N} \frac{-1}{k} e^{ikt} \quad ,$$

and let g_N be this partial sum multiplied by e^{i2Nt},

$$g_N(t) = e^{i2Nt} S_N(t) = e^{i2Nt} \sum_{\substack{k=-N \\ k \neq 0}}^{N} \frac{-1}{k} e^{ikt} = \sum_{k=-N}^{-1} \frac{-1}{k} e^{i(k+2N)t} + \sum_{k=1}^{N} \frac{-1}{k} e^{i(k+2N)t} \quad .$$

Observe that, after rewriting the exponentials in the formula for S_N in terms of sines and cosines, and letting $m = k + 2N$, the last formula for g_N yields, respectively,

$$S_N(t) = -2i \sum_{k=1}^{N} \frac{1}{k} \sin(kt) \tag{14.32}$$

and

$$g_N(t) = \sum_{m=N}^{2N-1} \frac{1}{2N-m} e^{imt} + \sum_{m=2N+1}^{3N} \frac{1}{2N-m} e^{imt} \quad . \tag{14.33}$$

Some more observations:

1. Since they are finite linear combinations of complex exponentials having period 2π, each S_N and g_N is a continuous (in fact, smooth) periodic function on $\mathbb{R}$ with period 2π.

2. From formula (14.32) and the definition of g_N, it immediately follows that

$$S_N(0) = 0 \quad \text{and} \quad g_N(0) = e^{i2N0}S_N(0) = 0 \quad .$$

3. The Fourier series for each g_N is simply the above finite linear combination of exponentials given in formula (14.33), and the Mth symmetric partial sum of this series is simply the sum of the terms in formula (14.33) with $m \leq M$. In particular, if $M < N$,

$$F.S._{-M,M}[g_N]\big|_t = 0 \quad ;$$

if $M > 3N$,

$$F.S._{-M,M}[g_N]\big|_t = \sum_{m=N}^{2N-1} \frac{1}{2N-m} e^{imt} + \sum_{m=2N+1}^{3N} \frac{1}{2N-n} e^{imt} = g_N(t) \quad ,$$

and, if $M = 2N$,

$$F.S._{-2N,2N}[g_N]\big|_t = \sum_{m=N}^{2N-1} \frac{1}{2N-m} e^{imt} \quad .$$

It will also be worth noting that, using the integral test (theorem 4.6 on page 44), we get

$$F.S._{-2N,2N}[g_N]\big|_0 = \sum_{m=N}^{2N-1} \frac{1}{2N-m} > \int_{x=N}^{2N-1} \frac{1}{2N-x} dx = \ln N \quad .$$

4. From the work in the previous section, it should be clear that the error in using the S_N's to approximate h is bounded. Consequently, there is a finite positive constant B such that, for every positive integer N and every real number t,

$$|S_N(t)| < B \quad .$$

From this and the definition of g_N, it also follows that, for every positive integer N and every real number t,

$$|g_N(t)| < B \quad .$$

The Construction

For each positive integer K, let f_K be the Kth partial sum of

$$f(t) = \sum_{k=1}^{\infty} k^{-2} g_{N_k}(t) \quad \text{where} \quad N_k = 2^{k^3} \quad .$$

That is,

$$f_K(t) = \sum_{k=1}^{K} k^{-2} g_{N_k}(t) \quad \text{where} \quad N_k = 2^{k^3} \quad .$$

Since each f_K is a finite linear combination of continuous functions on $\mathbb{R}$ that are periodic with period p, each f_K is also a continuous, periodic function on $\mathbb{R}$ with period p. Applying the above bound for the g_N's, we see that, for each k and t

$$\left| k^{-2} g_{N_k}(t) \right| < B k^{-2} \quad .$$

But $\sum_{k=1}^{\infty} k^{-2}$ is a well-known convergent series. The comparison test then applies and tells us that the above series f converges absolutely for every real value t. Hence,

$$f(t) = \sum_{k=1}^{\infty} k^{-2} g_{N_k}(t) \qquad \text{where} \quad N_k = 2^{k^3} \tag{14.34}$$

is a well-defined function on the real line, which, like its partial sums, must clearly be periodic.

To verify that f is actually continuous on $\mathbb{R}$, we first find an upper bound on the error in using $f_K(t)$ for $f(t)$. Applying the integral test, again, we have

$$|f(t) - f_K(t)| = \left| \sum_{k=1}^{\infty} k^{-2} g_{N_k}(t) - \sum_{k=1}^{K} k^{-2} g_{N_k}(t) \right|$$

$$\leq \sum_{k=K+1}^{\infty} k^{-2} \left| g_{N_k}(t) \right| \leq \sum_{k=K+1}^{\infty} k^{-2} \leq \int_K^{\infty} x^{-2} \, dx = \frac{1}{K} \quad .$$

So now let t_0 be a point on $\mathbb{R}$ and let $\epsilon > 0$. Choose a positive integer K greater than $4/\epsilon$. Then, for any real value τ,

$$|f(\tau) - f_K(\tau)| < \frac{1}{K} < \frac{\epsilon}{4} \quad .$$

Moreover, since f_K is continuous on $\mathbb{R}$ we can find a $\Delta t > 0$ such that

$$|f_K(t_0) - f_K(t)| < \frac{\epsilon}{2} \qquad \text{whenever} \quad |t_0 - t| < \Delta t \quad .$$

Thus, whenever $|t_0 - t| < \Delta t$,

$$|f(t_0) - f(t)| = |[f(t_0) - f_K(t_0) + f_K(t_0)] - [f(t) - f_K(t) + f_K(t)]|$$

$$\leq |f(t_0) - f_K(t_0)| + |f(t) - f_K(t)| + |f_K(t_0) - f_K(t)|$$

$$\leq \frac{\epsilon}{4} + \frac{\epsilon}{4} + \frac{\epsilon}{2} = \epsilon \quad .$$

This shows that f is continuous at every point of the real line.

The Fourier series for f can be obtained by simply expanding each g_N in formula (14.34) into its corresponding Fourier series:

$$F.S.[f]|_t = \sum_{k=1}^{\infty} k^{-2} F.S.\left[g_{N_k} \right]\big|_t$$

$$= \sum_{k=1}^{\infty} k^{-2} \left[\sum_{m=N_k}^{2N_k-1} \frac{1}{2N_k - m} e^{imt} + \sum_{m=2N_k+1}^{3N_k} \frac{1}{2N_k - n} e^{imt} \right] \quad .$$

You can easily verify for yourself that, for each k, $3N_k < N_{k+1}$. This means that the Fourier series of the g_{N_k}'s do not "overlap", that is, if an exponential of one frequency appears in the Fourier series of a particular g_{N_k}, say, g_{N_Γ}, then that exponential does not appear in the Fourier series of any other g_{N_k} . Moreover, that frequency is less than the frequency of any exponential in any g_{N_k} with $\Gamma < k$. Using this and observations made above, we have, for any positive integer γ ,

$$F.S._{-2N_\gamma,2N_\gamma}[f]\big|_t = \sum_{k=1}^{\gamma} k^{-2} \, F.S._{-2N_\gamma,2N_\gamma}[g_{N_k}]\big|_t$$

$$= \left[\sum_{k=1}^{\gamma-1} k^{-2} \, F.S._{-2N_\gamma,2N_\gamma}[g_{N_k}]\big|_t\right] + \gamma^{-2} \, F.S._{-2N_\gamma,2N_\gamma}[g_{N_\gamma}]\big|_t$$

$$= \left[\sum_{k=1}^{\gamma-1} k^{-2} g_{N_k}(t)\right] + \gamma^{-2} \, F.S._{-N_\gamma,N_\gamma}[g_{N_\gamma}]\big|_t \quad .$$

In particular, then,

$$F.S._{-2N_\gamma,2N_\gamma}[f]\big|_0 = \left[\sum_{k=1}^{\gamma-1} k^{-2} \underbrace{g_{N_k}(0)}_{=0}\right] + \gamma^{-2} \underbrace{F.S._{-N_\gamma,N_\gamma}[g_{N_\gamma}]\big|_0}_{>\ln N_\gamma}$$

$$> \gamma^{-2} \ln 2^{\gamma^3} = \gamma \ln 2 \quad .$$

Thus,

$$\lim_{\gamma\to\infty} F.S._{-N_\gamma,N_\gamma}[f]\big|_0 > \lim_{\gamma\to\infty} \gamma \ln 2 = \infty \quad , \tag{14.35}$$

confirming that, even though $f(t)$ is continuous at $t = 0$, its Fourier series at $t = 0$ does not converge.

Some final observations:

1. It must be noted that choosing a different sequence of integers can lead to different values for the limit of partial sums. For example, if you go back over our development, you will see that

    $$\lim_{\gamma\to\infty} F.S._{-3N_\gamma,3N_\gamma}[f]\big|_0 > \lim_{\gamma\to\infty} 0 = 0 \quad .$$

    So limit (14.35) doesn't even show us that the Fourier series "diverges to infinity" at $t = 0$. Instead, we simply have that

    $$\lim_{N\to\infty} F.S._{-N,N}[f]\big|_0$$

    does not exist.

2. Because of the relation between the symmetric partial sums of the complex exponential Fourier series and the partial sums of the trigonometric series, the above function f also serves as an example of a continuous periodic function whose trigonometric Fourier series fails to converge at $t = 0$.

3. Theorems 13.1 on page 156 and 13.2 on page 157 assure us that the Fourier series of a continuous, periodic function will converge to that function at each point of continuity, provided that function is at least piecewise smooth on an interval about the point in question. Since this does not occur with the above example, this function must not be piecewise smooth on any open interval about 0.

4. By forming suitable linear combinations of translations of f, you can obtain continuous, periodic functions whose Fourier series diverge on any given finite set of points in, say, $(-\pi, \pi)$. And if you are especially clever with your choice of the translations and the

coefficients in these linear combinations, you may be able to generate a convergent series that defines a continuous, periodic function whose Fourier series diverges at, say, every rational point on the real line.

5.   By the way, the construction of f in this section is, basically, that originally given by the mathematician Leopold Fejér in 1911.

?►Exercise 14.12:   *Use a computer mathematics package such as Maple or Mathematica to construct reasonably accurate graphs of the real and imaginary parts of the function f defined by formula (14.34).*

15

Derivatives and Integrals
of Fourier Series

There are two good reasons for considering the differentiation and integration of Fourier series. One is that derivatives and integrals of Fourier series often arise in applications. In the next chapter, for example, we will use Fourier series to solve differential equations describing flow of heat in a rod and vibrations in a string. The other reason is that some results from this chapter have already been quoted (see page 164). So we had better develop those results.

15.1   Differentiation of Fourier Series

Let f be a piecewise smooth, periodic function. Since, as piecewise continuous functions,

$$f(t) \; = \; F.S.\,[f]|_t \; = \; \sum_{k=-\infty}^{\infty} c_k \, e^{i2\pi \omega_k t} \quad ,$$

we certainly have

$$f'(t) \; = \; \frac{d}{dt} \sum_{k=-\infty}^{\infty} c_k \, e^{i2\pi \omega_k t}$$

at all points at which f is differentiable. Now, it is very tempting to assume

$$\frac{d}{dt} \sum_{k=-\infty}^{\infty} c_k \, e^{i2\pi \omega_k t} \; = \; \sum_{k=-\infty}^{\infty} \frac{d}{dt} c_k \, e^{i2\pi \omega_k t} \quad ,$$

but, as our next example shows, *this is not always true.*

!►*Example 15.1 (what can go wrong):*   *Let f be the simple saw function with period 1,*

$$f(t) \; = \; \mathrm{saw}_1(t) \; = \; \begin{cases} t & \text{if } \; 0 < t < 1 \\ f(t-1) & \text{in general} \end{cases} \quad .$$

The Fourier series for this function is easily found to be

$$\frac{1}{2} \; + \; \sum_{\substack{k=-\infty \\ k \neq 0}}^{\infty} \frac{i}{2\pi k} e^{i2\pi kt} \quad .$$

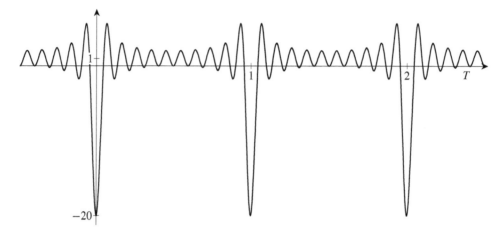

Figure 15.1: Graph of the 10th symmetric partial sum of the series in line (15.2) (with vertical axis compressed).

Clearly,

$$f'(t) \;=\; \left\{ \begin{array}{ll} 1 & \text{if} \quad 0 < t < 1 \\ f'(t-1) & \text{in general} \end{array} \right\} \;=\; 1 \quad.$$

Since the Fourier series of a constant function is simply that constant,

$$\frac{d}{dt}\left[\frac{1}{2} + \sum_{\substack{k=-\infty \\ k \neq 0}}^{\infty} \frac{1}{i2\pi k} e^{i2\pi k t} \right] \;=\; f'(t) \;=\; F.S.\!\left[f' \right] \;=\; 1 \quad. \tag{15.1}$$

On the other hand, differentiating each term of the Fourier series for f gives

$$\frac{d}{dt}\left[\frac{1}{2} \right] + \sum_{\substack{k=-\infty \\ k \neq 0}}^{\infty} \frac{d}{dt}\left[\frac{i}{2\pi k} e^{i2\pi k t} \right] \;=\; -\sum_{\substack{k=-\infty \\ k \neq 0}}^{\infty} e^{i2\pi k t} \quad, \tag{15.2}$$

which certainly does not look like the same series we ended up with in equation (15.1) (i.e., the one-term series "1").

The properly suspicious may wonder if the series in line (15.2) is just a very complicated expression for 1. We will show that this is not the case in exercise 15.5. Meanwhile, to see just how strange this series is, take a look at the graph of its 10th symmetric partial sum sketched in figure 15.1.

Let us look more carefully at the problem of computing the Fourier series of the derivative of a fairly arbitrary periodic function f with period p. To ensure that $F.S.[f]$, f' and $F.S.[f']$ are all well defined, let us assume, at the very least, that f is piecewise smooth. Thus, f' is at least piecewise continuous. Now,

$$F.S.[f]\big|_t \;=\; \sum_{k=-\infty}^{\infty} c_k\, e^{i2\pi \omega_k t} \qquad \text{and} \qquad F.S.\!\left[f' \right]\big|_t \;=\; \sum_{k=-\infty}^{\infty} d_k\, e^{i2\pi \omega_k t}$$

where $\omega_k = {}^{k}\!/_{p}$,

$$c_k \;=\; \frac{1}{p} \int_0^p f(t)\, e^{-i2\pi \omega_k t}\, dt \qquad \text{and} \qquad d_k \;=\; \frac{1}{p} \int_0^p f'(t)\, e^{-i2\pi \omega_k t}\, dt \quad.$$

If f is also a *continuous* function on the real line, then we can use integration by parts (theorem 4.2 on page 40) as follows:

$$d_k = \frac{1}{p} \int_0^p f'(t)\, e^{-i2\pi\omega_k t}\, dt$$

$$= f(t) e^{-i2\pi\omega_k t} \Big|_0^p - \int_0^p f(t) \left[-i2\pi\omega_k e^{-i2\pi\omega_k t} \right] dt$$

$$= \left[f(p) e^{-i2\pi\omega_k p} - f(0) e^0 \right] + i2\pi\omega_k c_k \quad . \tag{15.3}$$

Because of the periodicity (and continuity) of f,

$$f(p) e^{-i2\pi\omega_k p} - f(0) e^0 = f(0) e^{-i2\pi k} - f(0) = 0 \quad .$$

So, in this case, equation (15.3) simplifies to

$$d_k = i2\pi\omega_k c_k \quad , \tag{15.4a}$$

or, equivalently,

$$d_k = \frac{i2\pi k}{p} c_k \quad . \tag{15.4b}$$

This result is important enough to restate as a lemma.

Lemma 15.1 (Fourier series of a derivative)
Assume f is a periodic, continuous, piecewise smooth function with period p and

$$F.S.\,[f]|_t = \sum_{k=-\infty}^{\infty} c_k\, e^{i2\pi\omega_k t} \quad .$$

Then

$$F.S.\,[f']|_t = \sum_{k=-\infty}^{\infty} i2\pi\omega_k c_k\, e^{i2\pi\omega_k t} = \frac{i2\pi}{p} \sum_{k=-\infty}^{\infty} k c_k\, e^{i2\pi\omega_k t} \quad .$$

The assumption in lemma 15.1 that f is piecewise smooth ensures the piecewise continuity of f'. From this we know the terms of the Fourier series for f' are well defined. However, this still does not ensure the convergence of this series nor that it equals f'. Glancing back at the main theorem on convergence (theorem 13.1, page 156), we find we can ensure this convergence and equality by requiring f' to be piecewise smooth. Doing so and making the observation that

$$\frac{d}{dt} \left[c_k\, e^{i2\pi\omega_k t} \right] = i2\pi\omega_k c_k\, e^{i2\pi\omega_k t}$$

gives our main theorem on the differentiation of Fourier series.

Theorem 15.2 (differentiation of Fourier series)
Let f be a periodic, continuous, piecewise smooth function with period p and

$$F.S.\,[f]|_t = \sum_{k=-\infty}^{\infty} c_k\, e^{i2\pi\omega_k t} \quad .$$

If f' is also piecewise smooth, then

$$\sum_{k=-\infty}^{\infty} k c_k\, e^{i2\pi\omega_k t}$$

converges for each t at which f' is continuous, and, as piecewise continuous functions,

$$f'(t) = \sum_{k=-\infty}^{\infty} \frac{d}{dt}\left[c_k e^{i2\pi\omega_k t}\right] = \frac{i2\pi}{p} \sum_{k=-\infty}^{\infty} k c_k e^{i2\pi\omega_k t} \quad .$$

Compare the next example with the example that started this section (example 15.1).

!►Example 15.2:   Let f be the even sawtooth function from exercise 12.3 on page 152,

$$f(t) = \begin{cases} t & \text{if } 0 < t < 1 \\ -t & \text{if } -1 < t < 0 \\ f(t-2) & \text{in general} \end{cases}$$

(see figure 12.1a on page 151). From that exercise we know

$$F.S.[f]|_t = \frac{1}{2} + \sum_{\substack{k=-\infty \\ k\neq 0}}^{\infty} \frac{(-1)^k - 1}{k^2\pi^2} e^{ik\pi t} \quad .$$

The even sawtooth function is certainly continuous and piecewise smooth, and its derivative,

$$f'(t) = \begin{cases} 1 & \text{if } 0 < t < 1 \\ -1 & \text{if } -1 < t < 0 \\ f'(t-2) & \text{in general} \end{cases} ,$$

is piecewise smooth. So theorem 15.2 can be invoked and tells us

$$f'(t) = \frac{d}{dt}\left[\frac{1}{2}\right] + \sum_{\substack{k=-\infty \\ k\neq 0}}^{\infty} \frac{d}{dt}\left[\frac{(-1)^k - 1}{k^2\pi^2} e^{ik\pi t}\right]$$

$$= \sum_{\substack{k=-\infty \\ k\neq 0}}^{\infty} \frac{(-1)^k - 1}{k^2\pi^2} ik\pi\, e^{ik\pi t} = \sum_{\substack{k=-\infty \\ k\neq 0}}^{\infty} i\frac{(-1)^k - 1}{k\pi} e^{ik\pi t} \quad .$$

In terms of trigonometric Fourier series, lemma 15.1 and theorem 15.2 become:

Lemma 15.3 (Fourier series of a derivative)
Assume f is a periodic, continuous, piecewise smooth function with period p and

$$F.S.[f]|_t = A_0 + \sum_{k=1}^{\infty} [a_k \cos(2\pi\omega_k t) + b_k \sin(2\pi\omega_k t)] \quad .$$

Then

$$F.S.[f']|_t = \sum_{k=1}^{\infty} [-2\pi\omega_k a_k \sin(2\pi\omega_k t) + 2\pi\omega_k b_k \cos(2\pi\omega_k t)]$$

$$= \frac{2\pi}{p} \sum_{k=1}^{\infty} [-k a_k \sin(2\pi\omega_k t) + k b_k \cos(2\pi\omega_k t)] \quad .$$

Theorem 15.4 (differentiation of trigonometric series)
Let f be a periodic, continuous, piecewise smooth function with period p and

$$F.S.[f]|_t = A_0 + \sum_{k=1}^{\infty} [a_k \cos(2\pi\omega_k t) + b_k \sin(2\pi\omega_k t)] \quad .$$

If f' is also piecewise smooth, then

$$\sum_{k=1}^{\infty} [-ka_k \sin(2\pi\omega_k t) + kb_k \cos(2\pi\omega_k t)]$$

converges for each t at which f' is continuous, and, as piecewise continuous functions,

$$f'(t) = \sum_{k=1}^{\infty} [-2\pi\omega_k a_k \sin(2\pi\omega_k t) + 2\pi\omega_k b_k \cos(2\pi\omega_k t)]$$

$$= \frac{2\pi}{p} \sum_{k=1}^{\infty} [-ka_k \sin(2\pi\omega_k t) + kb_k \cos(2\pi\omega_k t)] \quad .$$

Of course, as long as our function is sufficiently smooth, we can use the above to find the Fourier series for higher order derivatives. Here is what we obtain when we can repeat theorem 15.2 "$m-1$ times".

Corollary 15.5 (higher order differentiation of the exponential series)
Let f be a periodic, continuous function with period p and Fourier series

$$\sum_{k=-\infty}^{\infty} c_k e^{i2\pi\omega_k t} \quad .$$

Assume further that, for some positive integer m, f is $(m-1)$-times differentiable and $f^{(m-1)}$ is continuous and piecewise smooth. Then

$$\sum_{k=-\infty}^{\infty} k^m c_k e^{i2\pi\omega_k t}$$

converges for each t at which $f^{(m)}$ is continuous, and, as piecewise continuous functions,

$$f^{(m)}(t) = \sum_{k=-\infty}^{\infty} \frac{d^m}{dt^m} \left[c_k e^{i2\pi\omega_k t} \right] = \left(\frac{i2\pi}{p} \right)^m \sum_{k=-\infty}^{\infty} k^m c_k e^{i2\pi\omega_k t} \quad .$$

15.2 Differentiability and Convergence

Equation (15.4b) leads to some useful observations concerning the Fourier coefficients of suitably smooth periodic functions and their derivatives. So, again, let f be periodic, continuous and piecewise smooth (as in lemma 15.1) with

$$F.S.\,[f]|_t \;=\; \sum_{k=-\infty}^{\infty} c_k\, e^{i2\pi\omega_k t} \qquad \text{and} \qquad F.S.\,[f']|_t \;=\; \sum_{k=-\infty}^{\infty} d_k\, e^{i2\pi\omega_k t} \quad .$$

Remember, $\omega_k = {}^k\!/_p$.

The first observation is that

$$d_0 \;=\; \frac{i2\pi \cdot 0}{p} c_0 \;=\; 0 \quad .$$

So the constant term in the Fourier series for f' (which is also the mean value of f' over any period) is zero.

Taking the magnitude of each side of equation (15.4b) and solving for c_k gives

$$|c_k| \;=\; \frac{p}{|k|\,2\pi}\,|d_k| \quad , \tag{15.5}$$

telling us that the c_k's must shrink to zero faster than the d_k's as $k \to \pm\infty$.

Combining this last equation with the observation that

$$|d_k| \;=\; \left| \frac{1}{p} \int_{\text{period}} f'(t)\, e^{-i2\pi\omega_k t}\, dt \right|$$

$$\leq\; \frac{1}{p} \int_{\text{period}} \left| f'(t)\, e^{-i2\pi\omega_k t} \right| dt \;=\; \frac{1}{p} \int_{\text{period}} \left| f'(t) \right| dt \quad ,$$

gives the crude estimate

$$|c_k| \;\leq\; \frac{1}{2\pi\,|k|} \int_{\text{period}} \left| f'(t) \right| dt \quad .$$

More refined arguments involving equation (15.5) lead to the more useful bound on the error described in the next theorem.

Theorem 15.6 (continuity and uniform convergence for exponential series)
Let f be a periodic function with period p and

$$F.S.\,[f]|_t \;=\; \sum_{k=-\infty}^{\infty} c_k\, e^{i2\pi\omega_k t} \quad .$$

Assume, further, that f is piecewise smooth and continuous, and for convenience, let

$$B \;=\; \frac{1}{2\pi} \left(p \int_{\text{period}} \left| f'(t) \right|^2 dt \right)^{1/2} \quad .$$

Then:

1. The series $\sum_{k=-\infty}^{\infty} c_k$ converges absolutely with

$$\sum_{k=-\infty}^{-N-1} |c_k| \leq \frac{B}{\sqrt{N}} \quad \text{and} \quad \sum_{k=N+1}^{\infty} |c_k| \leq \frac{B}{\sqrt{N}} \quad \text{for} \quad N = 1, 2, 3, \ldots \quad .$$

2. The Fourier series for f converges uniformly to f. Moreover, for any real value t and any pair of positive integers M and N,

$$\left| f(t) - \sum_{k=-M}^{N} c_k \, e^{i 2\pi \omega_k t} \right| \leq \left[\frac{1}{\sqrt{M}} + \frac{1}{\sqrt{N}} \right] B \quad .$$

Observe that the claimed bounds in this theorem automatically imply the claimed absolute and uniform convergence of the series. So all we need to verify are those upper bounds. For simplicity, we'll break the proof into two pieces: the proof of the bounds in the first part (in which we'll use equation (15.5), Bessel's inequality and the Schwarz inequality for summations), and the proof of the bounds in the second piece (which uses the basic theorem on convergence and the bounds from the first part).

PROOF (first part of theorem 15.6): Because the two bounds in this part can be obtained by virtually identical arguments, it will suffice to verify just the first bound.

Let N be any positive integer. As noted in the discussion just before the theorem,

$$|c_k| = \frac{p}{2\pi |k|} |d_k| \quad \text{for} \quad k = \pm 1, \pm 2, \pm 3, \ldots$$

where d_k is the kth Fourier coefficient for f'. From this and the Schwarz inequality (see theorem 4.8 on page 45) we get

$$\sum_{k=N+1}^{\infty} |c_k| = \sum_{k=N+1}^{\infty} \frac{p}{2\pi k} |d_k|$$

$$= \frac{p}{2\pi} \sum_{k=N+1}^{\infty} \frac{1}{k} |d_k| = \frac{p}{2\pi} \left(\sum_{k=N+1}^{\infty} \frac{1}{k^2} \right)^{1/2} \left(\sum_{k=N+1}^{\infty} |d_k|^2 \right)^{1/2} \quad . \tag{15.6}$$

Now, by the integral test (theorem 4.6 on page 44),

$$\sum_{k=N+1}^{\infty} \frac{1}{k^2} \leq \int_{N}^{\infty} \frac{1}{x^2} \, dx = \frac{1}{N} \quad ,$$

and, by Bessel's inequality (see exercise 12.4 on page 152),

$$\sum_{k=N+1}^{\infty} |d_k|^2 \leq \sum_{k=-\infty}^{\infty} |d_k|^2 \leq \frac{1}{p} \int_{\text{period}} |f'(t)|^2 \, dt \quad .$$

Plugging these inequalities into equation (15.6) gives

$$\sum_{k=N+1}^{\infty} |c_k| \leq \frac{p}{2\pi} \left(\frac{1}{N} \right)^{1/2} \left(\frac{1}{p} \int_{\text{period}} |f'(t)|^2 \, dt \right)^{1/2} \quad ,$$

which, by the definition of B and a little algebra, can (as claimed in the first part of the theorem) be
written as

$$\sum_{k=N+1}^{\infty} |c_k| \le \frac{B}{\sqrt{N}} \quad . \qquad \blacksquare$$

PROOF (second part of theorem 15.6):   Because f is continuous and piecewise smooth, the
basic theorem on convergence (theorem 13.1 on page 156) assures us that $f(t)$ equals its Fourier
series for every choice of t. So,

$$\left| f(t) - \sum_{k=-M}^{N} c_k e^{i2\pi \omega_k t} \right| = \left| \sum_{k=-\infty}^{\infty} c_k e^{i2\pi \omega_k t} - \sum_{k=-M}^{N} c_k e^{i2\pi \omega_k t} \right|$$

$$= \left| \sum_{k=-\infty}^{-M-1} c_k e^{i2\pi \omega_k t} + \sum_{k=N+1}^{\infty} c_k e^{i2\pi \omega_k t} \right|$$

$$\le \sum_{k=-\infty}^{-M-1} \left| c_k e^{i2\pi \omega_k t} \right| + \sum_{k=N+1}^{\infty} \left| c_k e^{i2\pi \omega_k t} \right|$$

$$= \sum_{k=-\infty}^{-M-1} |c_k| + \sum_{k=N+1}^{\infty} |c_k| \quad .$$

Replacing the two summations in the last line with the corresponding upper bounds described in the
first part of the theorem then yields the claimed bound,

$$\left| f(t) - \sum_{k=-M}^{N} c_k e^{i2\pi \omega_k t} \right| \le \frac{B}{\sqrt{M}} + \frac{B}{\sqrt{N}} = \left[\frac{1}{\sqrt{M}} + \frac{1}{\sqrt{N}} \right] B \quad . \qquad \blacksquare$$

Stronger results can be obtained when the function is known to be even smoother. As an exercise,
you should verify the following by using corollary 15.5 and the ideas described in the above proof.

Theorem 15.7 (smoothness and uniform convergence for exponential series)
Let f be a periodic function with period p and

$$F.S.[f]|_t = \sum_{k=-\infty}^{\infty} c_k e^{i2\pi \omega_k t} \quad .$$

Let m be a positive integer, and assume f is m-times differentiable and $f^{(m)}$ is piecewise continuous. Let

$$B = \left(\frac{p}{2\pi} \right)^m \left(\frac{1}{p(2m-1)} \int_{period} \left| f^{(m)}(t) \right|^2 dt \right)^{1/2} \quad .$$

Then:

1.   The series $\sum_{k=-\infty}^{\infty} c_k$ converges absolutely, and, for each positive integer N,

$$\sum_{k=-\infty}^{-N-1} |c_k| \le \frac{B}{\sqrt{N^{2m-1}}} \qquad and \qquad \sum_{k=N+1}^{\infty} |c_k| \le \frac{B}{\sqrt{N^{2m-1}}} \quad .$$

2. The Fourier series for f converges uniformly to f. Moreover, for any real value t and any pair of positive integers M and N,

$$\left| f(t) - \sum_{k=-M}^{N} c_k\, e^{i2\pi\omega_k t} \right| \leq \left[\frac{1}{\sqrt{M^{2m-1}}} + \frac{1}{\sqrt{N^{2m-1}}} \right] B \quad .$$

?▶ Exercise 15.1:   Prove theorem 15.7.

The corresponding theorem for the trigonometric Fourier series is given below. It, of course, follows immediately from the above theorem and the fact that, for each positive integer N, the trigonometric and complex exponential Fourier series of a given function satisfy

$$A_0 + \sum_{k=1}^{N} [a_k \cos(2\pi\omega_k t) + b_k \sin(2\pi\omega_k t)] = \sum_{k=-N}^{N} c_k\, e^{i2\pi\omega_k t} \quad .$$

Theorem 15.8 (uniform convergence for trigonometric series)
Let f be a periodic function with period p and

$$F.S.[f]|_t = A_0 + \sum_{k=1}^{\infty} [a_k \cos(2\pi\omega_k t) + b_k \sin(2\pi\omega_k t)] \quad .$$

Let m be a positive integer, and assume f is m-times differentiable and $f^{(m)}$ is piecewise continuous. Let

$$B = \left(\frac{p}{2\pi}\right)^m \left(\frac{1}{p(2m-1)} \int_{period} \left| f^{(m)}(t) \right|^2 dt \right)^{1/2} \quad .$$

Then:

1. The series $\sum_{k=1}^{\infty} a_k$ and $\sum_{k=1}^{\infty} b_k$ both converge absolutely with

$$\sum_{k=N+1}^{\infty} |a_k| \leq \frac{2B}{\sqrt{N^{2m-1}}} \quad \text{and} \quad \sum_{k=N+1}^{\infty} |b_k| \leq \frac{2B}{\sqrt{N^{2m-1}}}$$

    for each positive integer N.

2. The Fourier series for f converges uniformly to f. Moreover, for any real value t and any positive integer N,

$$\left| f(t) - A_0 - \sum_{k=1}^{N} [a_k \cos(2\pi\omega_k t) + b_k \sin(2\pi\omega_k t)] \right| \leq \frac{2B}{\sqrt{N^{2m-1}}} \quad .$$

The last few theorems have described how a very smooth function must have a "rapidly converging" Fourier series. Conversely, it can be shown that a function given by a rapidly converging Fourier series must be a very smooth function. We will discuss "rapidly converging series" in general in the next chapter. One immediate consequence of that discussion will be the following theorem.

Theorem 15.9
Assume $\sum_{k=-\infty}^{\infty} c_k$ is an absolutely convergent infinite series of complex numbers, and, for each real value t, let

$$f(t) = \sum_{k=-\infty}^{\infty} c_k\, e^{i2\pi\omega_k t}$$

where $\omega_k = {}^k/_p$ and p is some fixed positive number. Then f is a continuous and periodic function with period p whose complex exponential Fourier series is given by the above series. Moreover, if

$$\sum_{k=-\infty}^{\infty} \left| k^n c_k \right| < \infty$$

for some positive integer n, then f is n-times differentiable, $f^{(n)}$ is continuous, and

$$f^{(m)}(t) = \sum_{k=-\infty}^{\infty} c_k (i2\pi\omega_k)^m e^{i2\pi\omega_k t} \qquad \text{for} \quad m = 1, 2, \ldots, n \quad .$$

15.3   Integrating Periodic Functions and Fourier Series

What if we wanted to integrate the Fourier series of some periodic function f? Say, for the sake of discussion, we wished to integrate

$$F.S.[f]|_\tau = \sum_{k=-\infty}^{\infty} c_k e^{i2\pi\omega_k \tau}$$

(with $\omega_k = {}^k/_p$) from $\tau = a$ to $\tau = t$. We may be tempted to assume the integration can be done term by term,

$$\int_a^t \left[\sum_{k=-\infty}^{\infty} c_k e^{i2\pi\omega_k \tau} \right] d\tau = \sum_{k=-\infty}^{\infty} \int_a^t c_k e^{i2\pi\omega_k \tau} d\tau \quad . \qquad (15.7)$$

Now if $k \neq 0$,

$$\int_a^t c_k e^{i2\pi\omega_k \tau} d\tau = \left. \frac{c_k}{i2\pi\omega_k} e^{i2\pi\omega_k \tau} \right|_a^t = \frac{c_k p}{i2\pi k} \left[e^{i2\pi\omega_k t} - e^{i2\pi\omega_k a} \right] \quad ,$$

while if $k = 0$,

$$\int_a^t c_k e^{i2\pi\omega_k \tau} d\tau = \int_a^t c_0 \, d\tau = c_0[t - a] \quad .$$

So equation (15.7) expands to

$$\int_a^t \left[\sum_{k=-\infty}^{\infty} c_k e^{i2\pi\omega_k \tau} \right] d\tau = c_0[t - a] + \sum_{\substack{k=-\infty \\ k \neq 0}}^{\infty} \frac{c_k p}{i2\pi k} \left[e^{i2\pi\omega_k t} - e^{i2\pi\omega_k a} \right]$$

$$= c_0[t - a] + \sum_{\substack{k=-\infty \\ k \neq 0}}^{\infty} \frac{c_k p}{i2\pi k} e^{i2\pi\omega_k t} - \sum_{\substack{k=-\infty \\ k \neq 0}}^{\infty} \frac{c_k p}{i2\pi k} e^{i2\pi\omega_k a} \quad .$$

Take a look at our last equation. Letting

$$\Gamma_0 = -\sum_{\substack{k=-\infty \\ k \neq 0}}^{\infty} \frac{c_k p}{i2\pi k} e^{i2\pi\omega_k a} \qquad \text{and} \qquad \Gamma_k = \frac{c_k p}{i2\pi k} \qquad \text{for} \quad k = \pm 1, \pm 2, \pm 3, \ldots \quad ,$$

the above equation becomes

$$\int_a^t \left[\sum_{k=-\infty}^{\infty} c_k \, e^{i2\pi\omega_k\tau} \right] d\tau \;=\; c_0[t-a] + \sum_{k=-\infty}^{\infty} \Gamma_k \, e^{i2\pi\omega_k t} \quad . \tag{15.8}$$

which is *NOT* a Fourier series unless $c_0 = 0$. This should be expected; integrals of periodic functions are not necessarily periodic. If this is not obvious, consider

$$g(t) \;=\; \int_a^t f(\tau)\,d\tau \quad .$$

Instead of having $g(t+p) = g(t)$, we have

$$g(t+p) \;=\; \int_a^{t+p} f(\tau)\,d\tau$$

$$=\; \int_a^t f(\tau)\,d\tau \;+\; \int_t^{t+p} f(\tau)\,d\tau \;=\; g(t) + \int_{\text{period}} f(\tau)\,d\tau \quad .$$

So g is periodic if and only if

$$\int_{\text{period}} f(\tau)\,d\tau \;=\; 0 \quad .$$

Recalling that our c_0 is the above integral divided by p, we also see that this previous sentence can be rephrased as "So g is periodic if and only if $c_0 = 0$." That is why we should not expect the right-hand side of equation (15.8) to be a Fourier series unless $c_0 = 0$.

The main question now is whether an integral of a Fourier series can be computed by simply integrating its terms (as we naively assumed in equation (15.7)). The convergence of the term-by-term integrated series (the right-hand side of equation (15.8)) may also be a concern, though the observant reader may have already realized that the integration adds a $1/k$ factor to each term, helping to ensure the convergence of this series.

Answering our questions is fairly easy. Yes, we can safely integrate the Fourier series term by term provided f is at least piecewise continuous. (This is in marked contrast to the situation with differentiating Fourier series.) To be a little more explicit, we have the following:

Theorem 15.10 (integration of Fourier series)
Assume f is a periodic, piecewise continuous function with period p and

$$F.S.[f]|_t \;=\; \sum_{k=-\infty}^{\infty} c_k \, e^{i2\pi\omega_k t} \quad .$$

Then, for each real value τ,

$$\sum_{\substack{k=-\infty \\ k\neq 0}}^{\infty} \frac{c_k\,p}{i2\pi k} \, e^{i2\pi\omega_k\tau}$$

converges absolutely, and, for each pair of real numbers a and t,

$$\int_a^t f(\tau)\,d\tau \;=\; \sum_{k=-\infty}^{\infty} \int_a^t c_k \, e^{i2\pi\omega_k\tau}\,d\tau \;=\; c_0[t-a] + \sum_{k=-\infty}^{\infty} \Gamma_k \, e^{i2\pi\omega_k t}$$

where

$$\Gamma_0 = -\sum_{\substack{k=-\infty \\ k \neq 0}}^{\infty} \frac{c_k P}{i 2\pi k} e^{i 2\pi \omega_k a} \qquad \text{and} \qquad \Gamma_k = \frac{c_k P}{i 2\pi k} \quad \text{for} \quad k = \pm 1, \pm 2, \pm 3, \ldots \quad .$$

Furthermore,

$$\int_a^t f(\tau) \, d\tau \; - \; c_0[t - a]$$

is a continuous, piecewise smooth, periodic function of t with Fourier series

$$\sum_{k=-\infty}^{\infty} \Gamma_k e^{i 2\pi \omega_k t} \quad .$$

PROOF: Verifying this theorem is fairly straightforward if we start with the function

$$h(t) = \int_a^t f(\tau) \, d\tau \; - \; c_0[t - a] \quad .$$

Because h is the sum of an simple polynomial and the integral of a piecewise continuous function, h must be continuous and piecewise smooth with

$$\int_a^t f(\tau) \, d\tau = h(t) + c_0[t - a] \qquad \text{and} \qquad h'(t) = f(t) - c_0 \quad .$$

Its periodicity is also easily verified.

?▶ Exercise 15.2:   *Verify that $h(t + p) = h(t)$.*

Now let $\widehat{\Gamma}_k$ be the kth Fourier coefficient of h. Theorem 15.6 tells us that, for each real t,

$$h(t) = F.S.[h]\big|_t = \sum_{k=-\infty}^{\infty} \widehat{\Gamma}_k e^{i 2\pi \omega_k t}$$

and that this series converges absolutely.

To finish the proof, we need to confirm that each $\widehat{\Gamma}_k$ equals Γ_k. That is left for you.

?▶ Exercise 15.3:   *Let f, h, the c_k's, the Γ_k's and the $\widehat{\Gamma}_k$'s be as above.*

   a: *Verify that $h(a) = 0$, and, using this, show that*

$$\widehat{\Gamma}_0 = -\sum_{\substack{k=-\infty \\ k \neq 0}}^{\infty} \widehat{\Gamma}_k e^{i 2\pi \omega_k a} \quad .$$

   b: *Using the integral formula for the Fourier coefficients of f and h, the relation between h' and f, and integration by parts, verify that*

$$c_k = i 2\pi \omega_k \widehat{\Gamma}_k \qquad \text{for} \quad k = \pm 1, \pm 2, \pm 3, \ldots \quad .$$

   c: *Using the above, finish verifying the claims of theorem 15.10.*

Let's consider integrating the saw function that we tried (unsuccessfully) to differentiate in example 15.1 at the beginning of this chapter.

!▶ **Example 15.3:**   *Recall that*

$$F.S.\,[\text{saw}_1]|_t = \frac{1}{2} + \sum_{\substack{k=-\infty \\ k\neq 0}}^{\infty} \frac{i}{2\pi k}\, e^{i2\pi kt} \quad.$$

Theorem 15.10 assures us that, for every real value t,

$$\int_0^t \text{saw}_1(\tau)\,d\tau = \int_0^t \frac{1}{2}\,d\tau + \sum_{\substack{k=-\infty \\ k\neq 0}}^{\infty} \int_0^t \frac{i}{2\pi k}\, e^{i2\pi k\tau}\,d\tau$$

$$= \frac{1}{2}t + \sum_{\substack{k=-\infty \\ k\neq 0}}^{\infty} \frac{1}{(2\pi k)^2}\left[e^{i2\pi kt} - e^{i2\pi k\cdot 0} \right]$$

$$= \frac{1}{2}t + \sum_{\substack{k=-\infty \\ k\neq 0}}^{\infty} \frac{1}{(2\pi k)^2}\, e^{i2\pi kt} - \sum_{\substack{k=-\infty \\ k\neq 0}}^{\infty} \frac{1}{(2\pi k)^2} \quad.$$

Note that Theorem 15.10 also tells us that the two series in the last line above form the Fourier series for the periodic function

$$h(t) = \int_0^t \text{saw}_1(\tau)\,d\tau - \frac{1}{2}t \quad.$$

Let's look at this function. Its period is $p = 1$, and, for t between 0 and 1,

$$h(t) = \int_0^t \text{saw}_1(\tau)\,d\tau - \frac{1}{2}t = \int_0^t \tau\,d\tau - \frac{1}{2}t = \frac{1}{2}t^2 - \frac{1}{2}t \quad.$$

So

$$h(t) = \begin{cases} \frac{1}{2}t^2 - \frac{1}{2}t & \text{if } 0 \leq t \leq 1 \\ h(t-1) & \text{in general} \end{cases} \quad,$$

which can also be written as

$$h(t) = \frac{1}{2}g(t) - \frac{1}{2}\text{saw}_1(t)$$

where

$$g(t) = \begin{cases} t^2 & \text{if } 0 \leq t \leq 1 \\ g(t-1) & \text{in general} \end{cases} \quad.$$

What is of particular note here is that we can now easily find the Fourier series for g from the Fourier series for h and saw_1:

$$F.S.\,[g]|_t = F.S.\,[2h + \text{saw}_1]|_t$$

$$= 2\,F.S.\,[h]|_t + F.S.\,[\text{saw}_1]|_t$$

$$= 2 \left[\sum_{\substack{k=-\infty \\ k \neq 0}}^{\infty} \frac{1}{(2\pi k)^2} e^{i2\pi kt} - \sum_{\substack{k=-\infty \\ k \neq 0}}^{\infty} \frac{1}{(2\pi k)^2} \right] + \left[\frac{1}{2} + \sum_{\substack{k=-\infty \\ k \neq 0}}^{\infty} \frac{i}{2\pi k} e^{i2\pi kt} \right]$$

$$= \left[\frac{1}{2} - \frac{1}{2\pi^2} \sum_{\substack{k=-\infty \\ k \neq 0}}^{\infty} \frac{1}{k^2} \right] + \sum_{\substack{k=-\infty \\ k \neq 0}}^{\infty} \frac{1 + i\pi k}{2(\pi k)^2} e^{i2\pi kt} \quad .$$

The trigonometric series version of theorem 15.10 is:

Theorem 15.11 (integration of trigonometric Fourier series)
Let f be a periodic, piecewise continuous function with period p and

$$F.S.\,[f]|_t = A_0 + \sum_{k=1}^{\infty} [a_k \cos(2\pi \omega_k t) + b_k \sin(2\pi \omega_k t)] \quad .$$

Then, for any real number τ ,

$$\sum_{k=1}^{\infty} \frac{1}{k} a_k \sin(2\pi \omega_k \tau) \qquad and \qquad \sum_{k=1}^{\infty} \frac{1}{k} b_k \cos(2\pi \omega_k \tau)$$

converge absolutely, and, given any pair of real numbers t and a ,

$$\int_a^t f(\tau)\, d\tau = \int_a^t A_0\, d\tau + \sum_{k=1}^{\infty} \int_a^t [a_k \cos(2\pi \omega_k \tau) + b_k \sin(2\pi \omega_k \tau)]\, d\tau$$

$$= A_0[t - a] + \Gamma_0 + \frac{p}{2\pi} \sum_{k=1}^{\infty} \frac{1}{k} [a_k \sin(2\pi \omega_k t) - b_k \cos(2\pi \omega_k t)]$$

where

$$\Gamma_0 = \frac{p}{2\pi} \sum_{k=1}^{\infty} \frac{1}{k} [-a_k \sin(2\pi \omega_k a) + b_k \cos(2\pi \omega_k a)] \quad .$$

15.4   Sine and Cosine Series

Results similar to those already derived in this chapter can be obtained for the various "Fourier series" discussed in chapter 10 for functions over a finite interval $(0, L)$. You simply apply theorems already derived here to the corresponding periodic extensions.

Suppose, for example, we have the sine series

$$F.S.S.\,[f]|_t = \sum_{k=1}^{\infty} b_k \sin\left(\frac{k\pi}{L} t\right)$$

for some function f which is continuous, piecewise smooth, and has a piecewise smooth derivative on $(0, L)$. Remember, on $(0, L)$,

$$f(t) = f_o(t) = F.S.[f_o]|_t = \sum_{k=1}^{\infty} b_k \sin\left(\frac{k\pi}{L}t\right)$$

where f_o is the odd periodic extension of f.

Clearly, since f and f' are piecewise smooth on $(0, L)$, f_o must be piecewise smooth on the entire real line and have a piecewise smooth derivative. Furthermore, since f is continuous on $(0, L)$, f_o can have discontinuities only at integral multiples of L. So if (and only if)

$$\lim_{t \to 0^+} f(t) = 0 \quad \text{and} \quad \lim_{t \to L^-} f(t) = 0 \quad,$$

then f_o must be continuous on the real line (see figure 10.4 on page 123), and theorem 15.4 assures us that

$$\sum_{k=1}^{\infty} \frac{d}{dt} b_k \sin\left(\frac{k\pi}{L}t\right)$$

converges to $f_o'(t)$ for every real value t. Consequently,

$$f'(t) = f_o'(t) = \sum_{k=1}^{\infty} \frac{d}{dt} b_k \sin\left(\frac{k\pi}{L}t\right) = \sum_{k=1}^{\infty} \frac{k\pi}{L} b_k \sin\left(\frac{k\pi}{L}t\right) \quad \text{for} \quad 0 < t < L \quad,$$

and we have proven the theorem which follows.

Theorem 15.12 (differentiation of the sine series)
Let f be a continuous, piecewise smooth function on a finite interval $(0, L)$ with sine series

$$F.S.S.[f]|_t = \sum_{k=1}^{\infty} b_k \sin\left(\frac{k\pi}{L}t\right) \quad.$$

If f' is also piecewise smooth on $(0, L)$, and

$$\lim_{t \to 0^+} f(t) = 0 \quad \text{and} \quad \lim_{t \to L^-} f(t) = 0 \quad,$$

then $\sum_{k=1}^{\infty} k b_k \cos(2\pi \omega_k t)$ converges for each t at which f' is continuous, and, as piecewise continuous functions,

$$f'(t) = \sum_{k=1}^{\infty} \frac{d}{dt} b_k \sin\left(\frac{k\pi}{L}t\right) = \sum_{k=1}^{\infty} \frac{k\pi}{L} b_k \cos\left(\frac{k\pi}{L}t\right) \quad.$$

Next is the analogous theorem for the cosine series. It's slightly simpler because jump discontinuities are not introduced at integral multiples of L when we extend f in an even periodic manner (see figure 10.5 on page 125).

Theorem 15.13 (differentiation of the cosine series)
Let f be a continuous, piecewise smooth function on a finite interval $(0, L)$ with cosine series

$$F.C.S.[f]|_t = A_0 + \sum_{k=1}^{\infty} a_k \cos\left(\frac{k\pi}{L}t\right) \quad.$$

If f' is also piecewise smooth on $(0, L)$, then $\sum_{k=1}^{\infty} kb_k \sin(2\pi \omega_k t)$ converges for each t at which f' is continuous, and, as piecewise continuous functions,

$$f'(t) = \frac{d}{dt}A_0 + \sum_{k=1}^{\infty} \frac{d}{dt}a_k \cos\left(\frac{k\pi}{L}t\right) = -\sum_{k=1}^{\infty} \frac{k\pi}{L}a_k \sin\left(\frac{k\pi}{L}t\right) \quad .$$

The next two theorems are derived from theorem 15.8 on page 209.

Theorem 15.14 (uniform convergence of the sine series)
Let f be a continuous, piecewise smooth function on a finite interval $(0, L)$ with sine series

$$F.S.S.\,[f]|_t = \sum_{k=1}^{\infty} b_k \sin(2\pi \omega_k t) \quad .$$

If

$$\lim_{t \to 0^+} f(t) = 0 \qquad \text{and} \qquad \lim_{t \to L^-} f(t) = 0 \quad ,$$

then $\sum_{k=1}^{\infty} b_k$ converges absolutely, and, for each $0 < t < L$ and positive integer N,

$$\left| f(t) - F.S.S._N[f]|_t \right| \leq \frac{B}{\sqrt{N}} \qquad \text{where} \qquad B = \frac{2}{\pi}\left(L\int_0^L |f'(t)|^2\,dt\right)^{1/2} \quad .$$

Theorem 15.15 (uniform convergence of the cosine series)
Let f be a continuous, piecewise smooth function on a finite interval $(0, L)$ with cosine series

$$F.C.S.\,[f]|_t = A_0 + \sum_{k=1}^{\infty} a_k \cos(2\pi \omega_k t) \quad .$$

Then $\sum_{k=1}^{\infty} a_k$ converges absolutely, and, for each $0 < t < L$ and positive integer N,

$$\left| f(t) - F.C.S._N[f]|_t \right| \leq \frac{B}{\sqrt{N}} \qquad \text{where} \qquad B = \frac{2}{\pi}\left(L\int_0^L |f'(t)|^2\,dt\right)^{1/2} \quad .$$

?▶Exercise 15.4:   Derive the two theorems above using theorem 15.8.

Additional Exercises

15.5. Here we will look more closely at the partial sums of the series in line (15.2) from example 15.1 on page 201, and we will verify that this series does not converge pointwise to 1 on the real line. For convenience, let S_N denote the Nth symmetric partial sum of the series in line (15.2)

$$S_N(t) = -\sum_{\substack{k=-N \\ k \neq 0}}^{N} e^{i2\pi kt} \quad .$$

The following problems can be simplified by the observation that, using formula (14.3a) on page 180,

$$S_N(t) = 1 - \sum_{k=-N}^{N} e^{i2\pi kt} = 1 - \frac{\sin\left(2\pi\left[N + \frac{1}{2}\right]t\right)}{\sin(\pi t)} .$$

a. Graph $S_N(t)$ over $\left(-\frac{1}{2}, \frac{3}{2}\right)$ for $N = 5$, 10, 20 and 50 with the vertical scale adjusted so that you can clearly see the downward pointing "spikes" at $t = 0$ and $t = 1$. (Use a computer math package. You may use the results of the next exercise to work around the trivial discontinuities at $t = 0$ and $t = 1$.)

b. Show that $S_N(0) = -2N$. This shows that, instead of converging to 1 as $N \to \infty$, the series in (15.2) diverges to $-\infty$ for $t = 0$. (Remember to remove any trivial discontinuities!)

c. Graph $S_N(t)$ over $\left(-\frac{1}{2}, \frac{3}{2}\right)$ for $N = 5$, 10, 20 and 50. Adjust the vertical scale so that you can clearly see the smaller wiggles between the spikes (you may have to "cut off" the spikes). Note that the wiggles in the graph of S_N over this interval do not decrease in size. (Again, use a computer math package.)

d. Show that

$$S_N\left(\frac{1}{2}\right) = \begin{cases} 0 & \text{if } N \text{ is even} \\ 2 & \text{if } N \text{ is odd} \end{cases} .$$

Thus, the series in formula (15.2) on page 202 does not even converge at $t = \frac{1}{2}$. Instead, its partial sums oscillate between 0 and 2.

15.6. The complex exponential Fourier series for each of the functions below was computed in exercise 12.3 (see page 152). For each of these functions:

1. Find a formula (or set of formulas) for the derivative of the given function, and sketch the graph of the derivative.

2. Find the complex exponential Fourier series for the derivative. Use theorem 15.2 when possible.

a. $f(t) = \begin{cases} 0 & \text{if } -1 < t < 0 \\ 1 & \text{if } 0 < t < 1 \\ f(t-2) & \text{in general} \end{cases}$

b. $g(t) = \begin{cases} e^t & \text{if } 0 < t < 1 \\ g(t-1) & \text{in general} \end{cases}$

c. evensaw(t), the even sawtooth function sketched in figure 15.2a

d. oddsaw(t), the odd sawtooth function sketched in figure 15.2b

e. $|\sin(2\pi t)|$

f. $f(t) = \begin{cases} t^2 & \text{if } -1 < t < 1 \\ f(t-2) & \text{in general} \end{cases}$

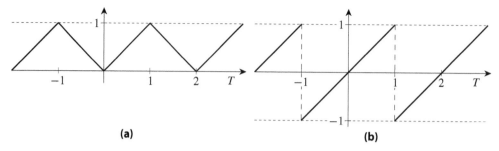

Figure 15.2: Two and one half periods of (a) evensaw(t), the even sawtooth function for
exercises 15.6 c and 15.9 b, and (b) oddsaw(t), the odd sawtooth function for
exercises 15.6 d and 15.9 c.

15.7. *Assume f is a periodic, piecewise smooth function with period p. Let*

$$F.S.\,[f]|_t \;=\; \sum_{k=-\infty}^{\infty} c_k\, e^{i2\pi\omega_k t} \quad and \quad F.S.\,[f']|_t \;=\; \sum_{k=-\infty}^{\infty} d_k\, e^{i2\pi\omega_k t} \;\;.$$

*Let $-p/2 < t_0 < p/2$, and suppose f has a jump discontinuity at t_0 with jump j_0. Suppose,
further, that $f(t)$ is continuous at every other point between $-p/2$ and $p/2$, including the
endpoints $-p/2$ and $p/2$. Derive a formula involving j_0 which relate each d_k to the
corresponding c_k.*

15.8. *Let g be as in exercise 15.3 on page 213, and let Γ_0 denote the constant term in the Fourier
series for g.*

    a. *What is the series formula for Γ_0 obtained in exercise 15.3?*

    b. *Compute Γ_0 using the integral formula for the Fourier coefficients.*

    c. *Compare the answers to the previous two parts of this exercise and show that*

$$\sum_{k=1}^{\infty} \frac{1}{k^2} = \frac{\pi^2}{6} \;\;.$$

15.9. *Find the "Fourier series-like" formula of*

$$g(t) \;=\; \int_0^t f(\tau)\,d\tau$$

*for each of the following choices of f. For which of these f's is the corresponding g
periodic and the series formula obtained the actual Fourier series for g? (Note: The Fourier
series for each f was computed in exercise 12.3.)*

    a. $f(t) = \begin{cases} 0 & \text{if } -1 < t < 0 \\ 1 & \text{if } 0 < t < 1 \\ f(t-2) & \text{in general} \end{cases}$

    b. $f(t) = \text{evensaw}(t)$ *(see figure 15.2a.)*

    c. $f(t) = \text{oddsaw}(t)$ *(see figure 15.2b.)*

    d. $f(t) = \begin{cases} +1 & \text{if } 0 < |t| < 1 \\ -1 & \text{if } 1 < |t| < 2 \\ f(t-4) & \text{in general} \end{cases}$

16

Applications

The use of Fourier series in applications and some of the issues that may arise in such use can be nicely illustrated by solving any of a number of classical problems. We will consider two: a heat flow problem and a vibrating string problem. The first, determining the temperature distribution throughout some heat conducting rod, is basically the same problem Fourier first solved using the series later named for him. The second is the problem of modeling the motion of an elastic string stretched between two points, a problem undoubtedly of interest to all guitar and banjo players.

Since the main goal is to illustrate the use of Fourier series and to examine some of the problems in their use, we will limit ourselves to relatively simple versions of these two problems. You should be aware, however, that the two problems examined here are just elementary prototypes for much wider classes of problems in a wide variety of subjects, including thermodynamics, diffusion processes, vibrational analysis, acoustics, electromagnetics and optics.

16.1 The Heat Flow Problem
Setting Up the Problem

Here is the problem: We have a heat conducting rod of length L, and we want to know how the temperature at different points in the rod varies with time. To keep our discussion relatively simple, we assume the rod is one-dimensional, uniform, positioned along the X–axis with endpoints at $x = 0$ and $x = L$, and with each endpoint being kept at a temperature of 0 degrees (Fahrenheit, Celsius, Kelvin — the actual scale is irrelevant for us). Let

$$u(x, t) \ = \ \text{temperature of the rod's material at horizontal position } x \text{ and time } t .$$

The endpoint conditions can then be written as

$$u(0, t) \ = \ 0 \qquad \text{and} \qquad u(L, t) \ = \ 0 \qquad \text{for all } t \ .$$

Let's also assume the rod's initial temperature distribution is known; that is, we assume

$$u(x, 0) \ = \ f(x)$$

where f is some known function on $(0, L)$. Let us further assume that f is at least piecewise smooth on the interval.

If this were a text on thermodynamics or partial differential equations, we would now derive the heat equation. But this isn't such a text, so we'll simply assume that, at every point in the rod,

$$\frac{\partial u}{\partial t} \ = \ \kappa \frac{\partial^2 u}{\partial x^2}$$

where κ is some positive constant describing the thermal properties of the rod's material. This is the famous heat equation derived by Fourier.[1]

Our goal is to find a usable formula for $u(x, t)$. Since it only makes sense to talk about the temperature where the rod exists, x must be between 0 and L. Gathering all the assumptions from above, we find that $u(x, t)$ must satisfy the following system of equations:

$$\frac{\partial u}{\partial t} - \kappa \frac{\partial^2 u}{\partial x^2} = 0 \qquad \text{for} \quad 0 < x < L \tag{16.1a}$$

$$u(0, t) = 0 \qquad \text{and} \qquad u(L, t) = 0 \tag{16.1b}$$

$$u(x, 0) = f(x) \qquad \text{for} \quad 0 < x < L \tag{16.1c}$$

Implicit in this is the requirement that $u(x, t)$ be a sufficiently smooth function of x and t for the above equations to make sense. Remember, κ is a positive constant, and f is a known piecewise smooth function on $(0, L)$. At this point we have no reason to place any limits on t other than it must be real valued. So, for now, we will assume no other limits on t. Later, however, we will need to modify that assumption.[2]

A Formal Solution

Solving this problem starts with the rather bold assumption that it has a solution. Supposing this, let us try to find a suitable "Fourier series" representation for this solution $u(x, t)$. There doesn't seem to be any periodicity in this problem, but the values of x are limited to the finite interval from $x = 0$ to $x = L$. This suggests that, for each fixed value of t, we represent $u(x, t)$ using one of the "Fourier series" from chapter 10, say, the sine series, letting

$$u(x, t) = \sum_{k=1}^{\infty} b_k \sin\left(\frac{k\pi}{L} x\right) \quad,$$

or the cosine series, letting

$$u(x, t) = A_0 + \sum_{k=1}^{\infty} a_k \cos\left(\frac{k\pi}{L} x\right) \quad.$$

In each case, the representation must change as t changes. This means that the b_k's in the sine series and the A_0 and a_k's in the cosine series will have to be treated as functions of t, and not as constants.

To further narrow our choices, let's note what happens when we plug $x = 0$ and $x = L$ into the sine series representation for $u(x, t)$:

$$u(0, t) = \sum_{k=1}^{\infty} b_k \sin\left(\frac{k\pi}{L} 0\right) = \sum_{k=1}^{\infty} b_k \sin(0) = \sum_{k=1}^{\infty} b_k \cdot 0 = 0$$

and

$$u(L, t) = \sum_{k=1}^{\infty} b_k \sin\left(\frac{k\pi}{L} L\right) = \sum_{k=1}^{\infty} b_k \sin(k\pi) = \sum_{k=1}^{\infty} b_k \cdot 0 = 0 \quad.$$

These equations match the endpoint conditions in equation set (16.1b). Also, for equations (16.1a) and (16.1b) to make sense, $u(x, t)$ should be at least a continuous and piecewise smooth function

---

[1] Each person reading this should go through the derivation of the heat equation at least once in their life. Reasonable derivations can be found in most introductory texts on partial differential equations.

[2] Part of "solving" many a problem is determining just what the problem is, and what can or should be considered as "known" at the onset. Here, for example, we "know" we can find $u(x, t)$ for all time t. We are wrong.

of x on $(0, L)$ with

$$\lim_{x \to 0^+} u(x, t) = 0 \qquad \text{and} \qquad \lim_{x \to L^-} u(x, t) = 0 \quad .$$

Under these conditions, theorem 15.14 on page 216 assures us that the partial sums of the above sine series will *uniformly* approximate $u(x, t)$ for $0 < x < L$ and a fixed value of t.

So let us choose the sine series to represent $u(x, t)$. Also, to emphasize their dependence on t and to avoid some confusion later, let's denote the kth coefficient by $\phi_k(t)$, instead of b_k. With these choices, we have

$$u(x, t) = \sum_{k=1}^{\infty} \phi_k(t) \sin\left(\frac{k\pi}{L}x\right) \tag{16.2}$$

where the ϕ_k's are functions to be determined.

The next step is to plug this representation for u into the heat equation. For now, we will ignore the warnings given at the beginning of chapter 15 and naively compute the derivatives in the heat equation by differentiating the terms in the series,

$$\frac{\partial u}{\partial t} = \frac{\partial}{\partial t} \sum_{k=1}^{\infty} \phi_k(t) \sin\left(\frac{k\pi}{L}x\right)$$

$$= \sum_{k=1}^{\infty} \frac{\partial}{\partial t}\left[\phi_k(t) \sin\left(\frac{k\pi}{L}x\right)\right] = \sum_{k=1}^{\infty} \phi_k'(t) \sin\left(\frac{k\pi}{L}x\right)$$

and

$$\frac{\partial^2 u}{\partial x^2} = \frac{\partial^2}{\partial x^2} \sum_{k=1}^{\infty} \phi_k(t) \sin\left(\frac{k\pi}{L}x\right)$$

$$= \sum_{k=1}^{\infty} \frac{\partial^2}{\partial x^2}\left[\phi_k(t) \sin\left(\frac{k\pi}{L}x\right)\right]$$

$$= \sum_{k=1}^{\infty} \phi_k(t)\left[-\left(\frac{k\pi}{L}\right)^2 \sin\left(\frac{k\pi}{L}x\right)\right] = -\sum_{k=1}^{\infty} \left(\frac{k\pi}{L}\right)^2 \phi_k(t) \sin\left(\frac{k\pi}{L}x\right) \quad .$$

With these expressions for the derivatives, equation (16.1a) becomes

$$\sum_{k=1}^{\infty} \phi_k'(t) \sin\left(\frac{k\pi}{L}x\right) + \kappa \sum_{k=1}^{\infty} \left(\frac{k\pi}{L}\right)^2 \phi_k(t) \sin\left(\frac{k\pi}{L}x\right) = 0 \quad .$$

Letting

$$\lambda = \kappa \left(\frac{\pi}{L}\right)^2 \quad ,$$

this can be written more concisely as

$$\sum_{k=1}^{\infty} \left[\phi_k'(t) + k^2 \lambda \phi_k(t)\right] \sin\left(\frac{k\pi}{L}x\right) = 0 \quad .$$

Look at this last equation. For each value of t, the left-hand side looks like a sine series which, according to the equation, equals 0 for all x in $(0, L)$. Surely, this is only possible if each coefficient is 0. Here, though, the coefficients are expressions involving the ϕ_k's. So each of these expressions must equal 0. This gives us a bunch of differential equations,

$$\frac{d\phi_k}{dt} + k^2\lambda\phi_k = 0 \qquad \text{for} \quad k = 1, 2, 3, \dots \quad . \tag{16.3}$$

These differential equations are easy to solve. Each is nothing more than

$$\frac{dy}{dt} + \gamma y = 0$$

with $y = \phi_k$ and $\gamma = k^2\lambda$ — one of the simplest first order linear equations around. You should have no problem confirming that its general solution is $y = Be^{-\gamma t}$ where B is an arbitrary constant. Hence,

$$\phi_k(t) = B_k e^{-k^2\lambda t} \qquad \text{for} \quad k = 1, 2, 3, \dots$$

where the B_k's are yet unknown constants.

With these formulas for the ϕ_k's, formula (16.2) becomes

$$u(x, t) = \sum_{k=1}^{\infty} B_k e^{-k^2\lambda t} \sin\left(\frac{k\pi}{L}x\right) \quad . \tag{16.4}$$

In deriving this expression for $u(x, t)$, we assumed $u(x, t)$ exists and satisfies the heat equation (equation (16.1a)) and the endpoint conditions in equation set (16.1b). All that remains is to further refine our expression so it also satisfies the initial condition of equation (16.1c), $u(x, 0) = f(x)$. Using the above formula for $u(x, t)$ in this equation, we get

$$f(x) = u(x, 0) = \sum_{k=1}^{\infty} B_k e^{-k^2\lambda \cdot 0} \sin\left(\frac{k\pi}{L}x\right) = \sum_{k=1}^{\infty} B_k \sin\left(\frac{k\pi}{L}x\right)$$

for $0 < x < L$. Cutting out the middle yields

$$f(x) = \sum_{k=1}^{\infty} B_k \sin\left(\frac{k\pi}{L}x\right) \qquad \text{for} \quad 0 < x < L \quad ,$$

which, by an amazing coincidence, looks exactly as if we are representing our known function f by its Fourier sine series. Surely then, each B_k must be the corresponding Fourier sine coefficient for f,

$$B_k = \frac{2}{L} \int_0^L f(x) \sin\left(\frac{k\pi}{L}x\right) dx \quad .$$

That finishes our derivation. *If* the solution exists and our (occasionally naive) suppositions are valid, then our heat flow problem (equation set (16.1)) is solved by

$$u(x, t) = \sum_{k=1}^{\infty} B_k e^{-k^2\lambda t} \sin\left(\frac{k\pi}{L}x\right) \tag{16.5a}$$

where

$$\lambda = \kappa \left(\frac{\pi}{L}\right)^2 \tag{16.5b}$$

and

$$B_k = \frac{2}{L} \int_0^L f(x) \sin\left(\frac{k\pi}{L}x\right) dx \qquad \text{for} \quad k = 1, 2, 3, \dots \quad . \tag{16.5c}$$

This set of formulas is often called a *formal solution* to the heat equation problem because we obtained it through a process of formal manipulations which seemed reasonable, but were not all rigorously justified.

!► Example 16.1:   *Consider solving our heat flow problem when $L = \pi$, $\kappa = \ln 2$ and the rod is initially a constant temperature throughout, say,*

$$f(x) = 100 \quad.$$

Here: $\pi/L = 1$, *formula (16.5b) simplifies to*

$$\lambda = \kappa \left(\frac{\pi}{L}\right)^2 = \ln 2 \quad,$$

and

$$e^{-k^2 \lambda t} = e^{-k^2 (\ln 2) t} = \left(e^{\ln 2}\right)^{-k^2 t} = 2^{-k^2 t} \qquad for \quad k = 1, 2, 3, \dots \quad.$$

Formula (16.5c) yields

$$B_k = \frac{2}{L} \int_0^L f(x) \sin\left(\frac{k\pi}{L} x\right) dx$$

$$= \frac{2}{\pi} \int_0^\pi 100 \sin(kx) \, dx = \frac{200}{k\pi} \left[1 - (-1)^k\right] \quad.$$

Hence, according to formula (16.5a), the formal solution to this heat flow problem is

$$u(x,t) = \sum_{k=1}^{\infty} B_k e^{-k^2 \lambda t} \sin\left(\frac{k\pi}{L} x\right)$$

$$= \sum_{k=1}^{\infty} \frac{200}{k\pi} \left[1 - (-1)^k\right] 2^{-k^2 t} \sin(kx)$$

$$= \frac{400}{\pi} \left(\frac{1}{2}\right)^t \sin(x) + \frac{400}{3\pi} \left(\frac{1}{2}\right)^{3^2 t} \sin(3x) + \frac{400}{5\pi} \left(\frac{1}{2}\right)^{5^2 t} \sin(5x)$$

$$+ \frac{400}{7\pi} \left(\frac{1}{2}\right)^{7^2 t} \sin(7x) + \frac{400}{9\pi} \left(\frac{1}{2}\right)^{9^2 t} \sin(9x) + \cdots \quad.$$

Validity and Properties of the Formal Solution

The question remains as to whether formula set (16.5) is a valid solution to our heat flow problem. There are several parts to this question: Does the series converge for all values of x and t of interest? If so, is the resulting function suitably smooth for the expressions in equation set (16.1) to make sense, and if so, does this function satisfy those equations?

Unfortunately, we cannot completely address these questions using the theory developed thus far. Until now, all of our infinite series have been generated from "known" functions. Here though, the function of interest (our solution) is given by an infinite series; so we need to develop some material regarding whether such a function is well defined, continuous, differentiable, etc. We also need to confirm that, if it is differentiable, then it can be differentiated by differentiating the series term by term. We will develop this material rigorously in section 16.3 and apply it to validating the above solution formula in section 16.4.

Partial answers to these questions, along with some insight, can be gained by examining the terms of our series,

$$B_k \, e^{-k^2 \lambda t} \sin\left(\frac{k\pi}{L} x\right) \qquad for \quad k = 1, 2, 3, \dots \quad.$$

Remember $\lambda > 0$ and

$$|B_k| = \left| \frac{2}{L} \int_0^L f(x) \sin\left(\frac{k\pi}{L}x\right) dx \right|$$

$$\leq \frac{2}{L} \int_0^L |f(x)| \left|\sin\left(\frac{k\pi}{L}x\right)\right| dx \leq \frac{2}{L} \int_0^L |f(x)| \, dx \quad .$$

So, letting

$$A = \frac{2}{L} \int_0^L |f(x)| \, dx \quad ,$$

we see that

$$\left| B_k \, e^{-k^2\lambda t} \sin\left(\frac{k\pi}{L}x\right) \right| \leq A \, e^{-k^2\lambda t} \quad \text{for} \quad k = 1, 2, 3, \ldots \quad .$$

If t is also positive, then each $e^{-k^2\lambda t}$ shrinks to 0 very rapidly as $k \to \infty$. This ensures that the series formula for $u(x,t)$ converges absolutely. Consequently, we are assured that $u(x,t)$, as defined by formula set (16.5), is well defined when $0 \leq x \leq L$ and $0 < t$.

In section 16.3 we will also see that these exponentially decreasing terms ensure that, as long as $t > 0$, $u(x,t)$ is an infinitely smooth function of both x and t whose partial derivatives can all be computed by differentiating the series term by term. This will allow us to rigorously confirm our formal solution to be a valid solution to our heat flow problem (and a very nice one, at that) when $t > 0$.

On the other hand, if $t < 0$, then $e^{-k^2\lambda t} = e^{k^2\lambda|t|} \to \infty$ as $k \to \infty$. Thus, unless the B_k's shrink to 0 extremely rapidly as $k \to \infty$, the terms of our series solution will blow up, and the series itself diverges whenever $t < 0$.

In short:

> The series formula given by formula set (16.5) succeeds beautifully as a solution to our heat flow problem for $t > 0$ and, typically, fails miserably for $t < 0$.

There is something else worth noting about our series solution: Each term in that series,

$$u(x,t) = \sum_{k=1}^{\infty} B_k \, e^{-k^2\lambda t} \sin\left(\frac{k\pi}{L}x\right) \quad ,$$

rapidly shrinks to 0 as $t \to \infty$. From this it can readily be shown that the maximum and minimum temperatures in the rod must be approaching 0 degrees fairly quickly as t gets large.

?▶Exercise 16.1: Let $u(x,t)$ be the infinite series solution found in above example 16.1.

a: Verify that

$$|u(x,t)| < \frac{400}{\pi} \sum_{k=1}^{\infty} \left(\frac{1}{2}\right)^{tk} \quad \text{when} \quad t > 0 \quad . \tag{16.6}$$

b: Using the above and the formula for computing the sum of a geometric series, show that

$$|u(x,t)| < \frac{800}{\pi} \left(\frac{1}{2}\right)^t \quad \text{when} \quad t \geq 1 \quad .$$

c: Assuming $u(x,t)$ is the temperature distribution in a rod, what does the above tell you about the maximum temperature in the rod when $t = 1$? when $t = 2$? when $t = 10$?

?▶ Exercise 16.2:   *Again, let $u(x, t)$ be the infinite series solution found in example 16.1, above. This time, consider this infinite series when $t = -1$.*

  a: *Write out this series.*

  b: *Verify that this series does not converge to anything when $x = {}^{\pi}/_2$.*

  c: *Show that this series cannot be the sine series for any piecewise continuous function on $(0, \pi)$. (Remember, if it were the sine series for such a function, then the coefficients would be bounded.)*

Uniqueness of the Solution

There is one more question we should ask regarding our series solution: *If it is a solution, is it the only possible solution, or have we just found one possible way the temperature might vary?*

To a great extent, we can answer this question by redoing our derivation a little more carefully and by applying some of the results concerning integrals of functions with two variables from chapter 7. To see this, suppose $u(x, t)$ is any solution to our heat flow problem that it is valid for $0 < x < L$ and $a < t < b$ (with $a \le 0 \le b$ so that the initial condition makes sense). Because part of being a solution means that $u(x, t)$ is a sufficiently smooth function of x and t, for the equations in the heat flow problem (equation set (16.1)) to make sense, it is reasonable to assume $u(x, t)$ is twice differentiable with respect to x, differentiable with respect to t, and that

$$u(x, t) \quad , \quad \frac{\partial u}{\partial x} \quad , \quad \frac{\partial^2 u}{\partial x^2} \quad \text{and} \quad \frac{\partial u}{\partial t}$$

are all uniformly continuous functions of x and t. The pointwise convergence theorem for sine series (theorem 13.18 on page 174) then assures us that, for each x in $(0, L)$ and t in (a, b),

$$u(x, t) = \sum_{k=1}^{\infty} \phi_k(t) \sin\left(\frac{k\pi}{L}x\right)$$

where

$$\phi_k(t) = \frac{2}{L} \int_0^L u(x, t) \sin\left(\frac{k\pi}{L}x\right) dx \quad .$$

Using some of the results from chapter 7, the fact that $u(x, t)$ satisfies equations (16.1a) and (16.1b), and integration by parts, we find that each ϕ_k is a smooth function, and

$$\phi_k{}'(t) = \frac{d}{dt}\left[\frac{2}{L} \int_0^L u(x, t) \sin\left(\frac{k\pi}{L}x\right) dx\right] = \cdots = k^2 \lambda \phi_k(t) \tag{16.7}$$

where, as before,

$$\lambda = \kappa \left(\frac{\pi}{L}\right)^2$$

(see exercise 16.3, below). Solving this differential equation gives us

$$\phi_k(t) = B_k e^{-k^2 \lambda t} \quad \text{for} \quad k = 1, 2, 3, \ldots$$

with the B_k's being undetermined constants. Then using the above two formulas for ϕ_k,

$$B_k = \lim_{t \to 0} \phi_k(t) = \lim_{t \to 0} \frac{2}{L} \int_0^L u(x, t) \sin\left(\frac{k\pi}{L} x\right) dx$$

$$= \frac{2}{L} \int_0^L \lim_{t \to 0} u(x, t) \sin\left(\frac{k\pi}{L} x\right) dx$$

$$= \frac{2}{L} \int_0^L u(x, 0) \sin\left(\frac{k\pi}{L} x\right) dx = \frac{2}{L} \int_0^L f(x) \sin\left(\frac{k\pi}{L} x\right) dx \quad .$$

Putting this all together, we find that we have rigorously rederived the formulas in set (16.5) as formulas describing any given solution to our heat flow problem. Consequently, any solution to our heat flow problem that satisfies the smoothness conditions assumed above *must be given by formula set (16.5)*. (This doesn't mean there might not be other formulas describing this function, only that no formula can describe a different function satisfying our problem.)

?►Exercise 16.3: *Verify equation (16.7) by doing all the computations indicated by the "···".*

16.2 The Vibrating String Problem
Setting Up the Problem

Envision an elastic string (such as you might find on any guitar or banjo) stretched between two fixed points on the X–axis, say, from $x = 0$ to $x = L$ (with $L > 0$). For simplicity, we'll assume the string only moves vertically, and we will let

$u(x, t) = $ vertical position at time t of the portion of string located at horizontal position x.

Because the ends of the string are fixed at $x = 0$ and $x = L$, $u(x, t)$ is only defined for $0 \leq x \leq L$, and we have the endpoint conditions

$$u(0, t) = 0 \quad \text{and} \quad u(L, t) = 0 \quad .$$

After making a few idealizations and applying a little physics, it can be shown that

$$\frac{\partial^2 u}{\partial t^2} = c^2 \frac{\partial^2 u}{\partial x^2}$$

where c is some positive constant (the reason for using c^2 instead of c will be clear later).[3] This is the basic (one-dimensional) *wave equation*.[4]

We will assume the initial shape of the string is given by the graph of some known function f on $(0, L)$,

$$u(x, 0) = f(x) \quad \text{for} \quad 0 < x < L \quad .$$

---

[3] More precisely, $c = \sqrt{\tau/\rho}$ where τ and ρ are, respectively, the tension in and the linear density of the string when the stretched string is at rest.

[4] Another famous equation whose derivation we are skipping. Look up the derivation in any decent introductory book on partial differential equations.

For most (unbroken) strings we would expect f to be continuous and piecewise smooth. In addition, since the string is fastened at the endpoints, we should have $f(0) = 0$ and $f(L) = 0$.

As it turns out, this is not quite enough to completely specify $u(x, t)$. An additional initial condition is necessary. We will take that condition to be

$$\left. \frac{\partial u}{\partial t} \right|_{(x,0)} = 0 \quad \text{for} \quad 0 < x < L \quad .$$

In other words, we assume the string is not moving at time $t = 0$. This would be the case, for example, if we held the string in some fixed shape until releasing it at $t = 0$.

Gathering all the above equations together, we find that $u(x, t)$ must satisfy the following system of equations:

$$\frac{\partial^2 u}{\partial t^2} - c^2 \frac{\partial^2 u}{\partial x^2} = 0 \quad \text{for} \quad 0 < x < L \tag{16.8a}$$

$$u(0, t) = 0 \quad \text{and} \quad u(L, t) = 0 \tag{16.8b}$$

$$u(x, 0) = f(x) \quad \text{for} \quad 0 < x < L \tag{16.8c}$$

$$\left. \frac{\partial u}{\partial t} \right|_{(x,0)} = 0 \quad \text{for} \quad 0 < x < L \tag{16.8d}$$

Again, there is an implicit requirement that $u(x, t)$ be a sufficiently smooth function for the above equations to make sense. Keep in mind that c is a positive constant and f is a known uniformly continuous and piecewise smooth function on $(0, L)$ satisfying $f(0) = 0 = f(L)$. (Later we will realize that f' must also be piecewise smooth.) While it is reasonable to be interested in solving this problem just for $t > 0$, such a restriction on t turns out to be mathematically unnecessary. So we will assume the above equations are valid for $-\infty < t < \infty$.

A Formal Solution

The process of finding a solution to our vibrating string problem is very similar to the process we went through to solve our heat flow problem. As then, we begin by supposing a solution $u(x, t)$ exists, and, as with our heat flow problem, the end conditions (equation set (16.8b)) suggest that $u(x, t)$ should be represented by a Fourier sine series on $0 < x < L$ with the coefficients being functions of time,

$$u(x, t) = \sum_{k=1}^{\infty} \phi_k(t) \sin\left(\frac{k\pi}{L}x\right) \quad . \tag{16.9}$$

As we noted with the heat flow problem, this formula equals 0 when $x = 0$ or $x = L$.

Naively differentiating, we get

$$\frac{\partial^2 u}{\partial t^2} = \frac{\partial^2}{\partial t^2} \sum_{k=1}^{\infty} \phi_k(t) \sin\left(\frac{k\pi}{L}x\right) = \sum_{k=1}^{\infty} \phi_k''(t) \sin\left(\frac{k\pi}{L}x\right)$$

and

$$\frac{\partial^2 u}{\partial x^2} = \frac{\partial^2}{\partial x^2} \sum_{k=1}^{\infty} \phi_k(t) \sin\left(\frac{k\pi}{L}x\right) = -\sum_{k=1}^{\infty} \phi_k(t) \left(\frac{k\pi}{L}\right)^2 \sin\left(\frac{k\pi}{L}x\right) \quad .$$

With these expressions for the derivatives, equation (16.8a) becomes

$$\sum_{k=1}^{\infty} \phi_k''(t) \sin\left(\frac{k\pi}{L}x\right) + c^2 \sum_{k=1}^{\infty} \phi_k(t) \left(\frac{k\pi}{L}\right)^2 \sin\left(\frac{k\pi}{L}x\right) = 0 \quad ,$$

which is written more concisely as

$$\sum_{k=1}^{\infty} \left[\phi_k''(t) + (k\nu)^2 \phi_k(t) \right] \sin\left(\frac{k\pi}{L}x\right) = 0$$

using, for lexicographic convenience,

$$\nu = \frac{c\pi}{L} \quad .$$

This time, each ϕ_k must satisfy the second order linear differential equation

$$\frac{d^2\phi_k}{dt^2} + (k\nu)^2 \phi_k = 0 \quad .$$

Again, we have a simple differential equation that should be familiar to anyone who has had an elementary course in differential equations. Its solution is

$$\phi_k(t) = A_k \sin(k\nu t) + B_k \cos(k\nu t)$$

where A_k and B_k are constants yet to be determined.

With this formula for the ϕ_k's, equation (16.9) becomes

$$u(x,t) = \sum_{k=1}^{\infty} [A_k \sin(k\nu t) + B_k \cos(k\nu t)] \sin\left(\frac{k\pi}{L}x\right) \quad . \tag{16.10}$$

The A_k's and B_k's will be determined by the initial conditions, equations (16.8c) and (16.8d). For the second initial condition, we will need the partial of u with respect to t, which we might as well (naively) compute here:

$$\frac{\partial u}{\partial t} = \frac{\partial}{\partial t} \sum_{k=1}^{\infty} [A_k \sin(k\nu t) + B_k \cos(k\nu t)] \sin\left(\frac{k\pi}{L}x\right)$$

$$= \sum_{k=1}^{\infty} [A_k k\nu \cos(k\nu t) - B_k k\nu \sin(k\nu t)] \sin\left(\frac{k\pi}{L}x\right) \quad . \tag{16.11}$$

Combining formula (16.10) for $u(x,t)$ with the first initial condition gives us

$$f(x) = u(x,0) = \sum_{k=1}^{\infty} [A_k \sin(k\nu 0) + B_k \cos(k\nu 0)] \sin\left(\frac{k\pi}{L}x\right)$$

$$= \sum_{k=1}^{\infty} [A_k \cdot 0 + B_k \cdot 1] \sin\left(\frac{k\pi}{L}x\right)$$

$$= \frac{2}{L} \int_0^L f(x) \sin\left(\frac{k\pi}{L}x\right) dx \qquad \text{for} \quad k = 1, 2, 3, \dots \quad ,$$

which looks remarkably like an equation we obtained while solving our heat flow problem. As before, we are compelled to conclude that the B_k's are the Fourier sine coefficients for f. That is,

$$B_k = \frac{2}{L} \int_0^L f(x) \sin\left(\frac{k\pi}{L}x\right) dx \qquad \text{for} \quad k = 1, 2, 3, \dots \quad .$$

The second initial condition, along with formula (16.11), yields

$$0 = \left.\frac{\partial u}{\partial t}\right|_{(x,0)} = \sum_{k=1}^{\infty} [A_k k v \cos(kv0) - B_k k v \sin(kv0)] \sin\left(\frac{k\pi}{L}x\right)$$

$$= \sum_{k=1}^{\infty} [A_k k v \cdot 1 - B_k k v \cdot 0] \sin\left(\frac{k\pi}{L}x\right)$$

$$= \sum_{k=1}^{\infty} A_k k v \sin\left(\frac{k\pi}{L}x\right) \qquad \text{for} \quad 0 < x < L \quad,$$

strongly suggesting that

$$A_k = 0 \qquad \text{for} \quad k = 1, 2, 3, \ldots \quad.$$

Our derivation is complete. If our vibrating string problem (equation set (16.8)) has a solution and our (occasionally naive) computations are valid, then that solution is given by

$$u(x, t) = \sum_{k=1}^{\infty} B_k \cos\left(\frac{kc\pi}{L}t\right) \sin\left(\frac{k\pi}{L}x\right) \tag{16.12a}$$

where

$$B_k = \frac{2}{L}\int_0^L f(x) \sin\left(\frac{k\pi}{L}x\right) dx \qquad \text{for} \quad k = 1, 2, 3, \ldots \quad. \tag{16.12b}$$

Once again, we have derived a "formal solution", that is, a formula obtained through formal (naive) manipulations which we hope can be rigorously verified later.

!▶ **Example 16.2:** *Consider solving our vibrating string problem assuming $L = 1$ and $c = 3$, and starting with the middle point of the string pulled up a distance of $\frac{1}{2}$ (see figure 16.1). That is, $u(x, 0) = f(x)$ with*

$$f(x) = \begin{cases} x & \text{if } 0 \leq x \leq \frac{1}{2} \\ 1 - x & \text{if } \frac{1}{2} \leq x \leq 1 \end{cases} .$$

With these choices, equation (16.12b) is

$$B_k = \frac{2}{L}\int_0^L f(x) \sin\left(\frac{k\pi}{L}x\right) dx$$

$$= 2\int_0^{1/2} x \sin(k\pi x)\, dx + 2\int_{1/2}^1 (1-x) \sin(k\pi x)\, dx = \cdots = \sin\left(\frac{k\pi}{2}\right)\left(\frac{2}{k\pi}\right)^2 .$$

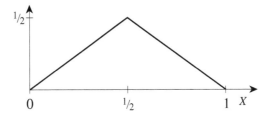

Figure 16.1: The initial shape of the string in example 16.2.

Thus, according to formula (16.12a), the solution to this vibrating string problem is

$$u(x, t) = \sum_{k=1}^{\infty} \sin\left(\frac{k\pi}{2}\right) \left(\frac{2}{k\pi}\right)^2 \cos(k3\pi t) \sin(k\pi x)$$

$$= 1 \left(\frac{2}{\pi}\right)^2 \cos(3\pi t) \sin(\pi x) + 0 \left(\frac{2}{2\pi}\right)^2 \cos(2 \cdot 3\pi t) \sin(2\pi x)$$

$$+ (-1) \left(\frac{2}{3\pi}\right)^2 \cos(3 \cdot 3\pi t) \sin(3\pi x) + 0 \left(\frac{2}{4\pi}\right)^2 \cos(4 \cdot 3\pi t) \sin(4\pi x)$$

$$+ \cdots \quad .$$

An Alternate Solution Formula and Validating Our Solution

Formula set (16.12) can be converted to a much simpler form once we recall a well-known trigonometric identity for the product of the sine and cosine functions. That identity with formula (16.12a) yields

$$u(x, t) = \sum_{k=1}^{\infty} B_k \sin\left(\frac{k\pi}{L}x\right) \cos\left(\frac{kc\pi}{L}t\right)$$

$$= \sum_{k=1}^{\infty} B_k \frac{1}{2} \left[\sin\left(\frac{k\pi}{L}x + \frac{kc\pi}{L}t\right) + \sin\left(\frac{k\pi}{L}x - \frac{kc\pi}{L}t\right) \right] \quad .$$

This, of course, is the same as

$$u(x, t) = \frac{1}{2} \left[\sum_{k=1}^{\infty} B_k \sin\left(\frac{k\pi}{L}(x + ct)\right) + \sum_{k=1}^{\infty} B_k \sin\left(\frac{k\pi}{L}(x - ct)\right) \right] \tag{16.13}$$

provided the two infinite series converge.

   Recall now, both that the B_k's are the Fourier sine coefficients for f, and that the sine series for f is just the trigonometric series for the odd periodic extension of f,

$$f_o(s) = \begin{cases} f(s) & \text{if } 0 < s < L \\ -f(s) & \text{if } -L < s < 0 \\ f(s - 2L) & \text{in general} \end{cases} \quad .$$

If you check, you'll find that our assumptions for f guarantee that its odd periodic extension is continuous and piecewise smooth on the entire real line. The basic theorem on pointwise convergence (page 156) assures us that the Fourier series for f_o converges to $f_o(s)$ for every s on the real line. In other words, we can safely write

$$f_o(s) = F.S.[f_o]|_s = \sum_{k=1}^{\infty} B_k \sin\left(\frac{k\pi}{L}s\right) \qquad \text{for all} \quad -\infty < s < \infty \quad .$$

This means equation (16.13) simplifies to

$$u(x, t) = \frac{1}{2}[f_o(x + ct) + f_o(x - ct)] \tag{16.14}$$

for every $0 < x < L$ and real value t.

Equation (16.14) holds whenever $u(x, t)$ is given by equation set (16.12) and f is a continuous, piecewise smooth function on $(0, L)$ with $f(0) = 0 = f(L)$. Conversely, by reversing the above steps, we can clearly derive equation set (16.12) from formula (16.14). Thus, formula (16.14) and formula set (16.12) are equivalent, and we can verify that both describe a valid solution to our vibrating string problem by confirming that either one is a valid solution. Naturally, it is formula (16.14) that we will verify.

The confirmation is straightforward. Since f_o is continuous, $u(x, t)$, as defined by formula (16.14), is clearly a well-defined, continuous function of x and t. Letting $t = 0$ and $0 < x < L$, we get

$$u(x, 0) = \frac{1}{2}[f_o(x + 0) + f_o(x - 0)] = \frac{1}{2}[f(x) + f(x)] = f(x) \quad,$$

verifying that the initial condition of equation (16.8d) is satisfied. To verify the other initial condition in equation (16.8d), we first observe that, by the chain rule,

$$\frac{\partial u}{\partial t} = \frac{1}{2}\left[\frac{\partial}{\partial t} f_o(x + ct) + \frac{\partial}{\partial t} f_o(x - ct)\right]$$

$$= \frac{1}{2}\left[\left(f_o''(x + ct)\right)(+c) + \left(f_o''(x - ct)\right)(-c)\right]$$

$$= \frac{c}{2}\left[f_o''(x + ct) - f_o''(x - ct)\right] \quad.$$

Plugging $t = 0$ then gives

$$\left.\frac{\partial u}{\partial t}\right|_{(x,0)} = \frac{c}{2}\left[f_o''(x) - f_o''(x)\right] = 0 \quad.$$

To see that the required endpoint conditions are satisfied (i.e., that $u(0, t) = 0 = u(L, t)$), first observe that, because f_o is an odd function and is periodic with period $2L$,

$$f_o(-ct) = -f_o(ct)$$

and

$$f(L - ct) = f(L - ct - 2L) = f(-L - ct) = -f(L + ct) \quad.$$

Thus,

$$u(0, t) = \frac{1}{2}[f_o(0 + ct) + f_o(0 - ct)] = \frac{1}{2}[f_o(ct) - f_o(ct)] = 0$$

and

$$u(L, t) = \frac{1}{2}[f_o(L + ct) + f_o(L - ct)] = \frac{1}{2}[f_o(L + ct) - f_o(L + ct)] = 0 \quad.$$

Confirming that $u(x, t)$ satisfies the wave equation (equation (16.8a)) is a simple matter of computing the appropriate derivatives. Assuming $f_o(s)$ is twice differentiable at $s = x + ct$ and $s = x - ct$, and using the chain rule, we find that

$$\frac{\partial^2}{\partial x^2} f_o(x \pm ct) = f_o''(x \pm ct) \quad,$$

while

$$\frac{\partial^2}{\partial t^2} f_o(x \pm ct) = \left[f_o''(x \pm ct)\right](\pm c)^2 = c^2 f_o''(x \pm ct) \quad.$$

Hence,

$$\frac{\partial^2 u}{\partial t^2} = c^2 \frac{1}{2}\left[f_o''(x + ct) + f_o''(x - ct)\right] = c^2 \frac{\partial^2 u}{\partial x^2} \quad,$$

verifying that the wave equation is satisfied at every (x, t) such that $f_o(s)$ is twice differentiable at $s = x + ct$ and $s = x - ct$.

There is a slight technical difficulty here. When we set up our problem, we saw no reason to assume anything about the second derivative of f. On the other hand, requiring f to be, say, "piecewise twice differentiable" on $(0, L)$ would hardly be much of a practical restriction. Besides, assuming the wave equation and initial condition both hold, we see that

$$\left. \frac{\partial^2 u}{\partial t^2} \right|_{(x,0)} = c^2 \left. \frac{\partial^2 u}{\partial x^2} \right|_{(x,0)} = c^2 \frac{d^2}{dx^2} u(x, 0) = c^2 f''(x) \quad .$$

So if f'' is not reasonably well defined on $(0, L)$, neither is the initial acceleration throughout the string.

All this suggests that we modify our requirements on our initial condition to

f is a continuous, piecewise smooth function on $(0, L)$ *with a piecewise smooth first derivative* and satisfying $f(0) = 0 = f(L)$.

Consider it done.

With these modified requirements, the second derivative of the odd periodic extension of f, f_o'', is certainly a well-defined, piecewise continuous function on the real line, and the computations done two paragraphs or so ago confirm that, as piecewise continuous functions,

$$\frac{\partial^2 u}{\partial t^2} - c^2 \frac{\partial^2 u}{\partial x^2} = 0$$

for $0 < x < L$ and $-\infty < t < \infty$. This completes our verification that $u(x, t)$, as given by either formula set (16.12) or formula (16.14), is a solution to our vibrating string problem.

Do these formulas describe the only solution? Yes. It can be shown (using methods outside the scope of this text) that the general solution to the wave equation is given by

$$u(x, t) = g(x + ct) + h(x - ct) \tag{16.15}$$

where g and h are arbitrary piecewise smooth functions on $\mathbb{R}$ with piecewise smooth derivatives. If you impose the endpoint conditions and the initial conditions of our vibrating string problem and solve for g and h you get

$$g(s) = h(s) = \frac{1}{2} f_o(s) \qquad \text{for all } s \text{ in } \mathbb{R} \quad ,$$

from whence then follows formula (16.14) for the solution.

?▶Exercise 16.4:   *Using equation (16.15), convince yourself that the motion of a vibrating string can be described as the superposition of two fixed shapes, one traveling to the left with speed c and the other traveling to the right with speed c. (For obvious reasons, these two "traveling shapes" are more commonly referred to as traveling waves.)*

Harmonics of a Vibrating String

One advantage of the Fourier series solution to our vibrating string problem is that it allows us to analyze the sound produced by such a string by looking at the components of the series solution. For convenience, let's rewrite that solution as

$$u(x, t) = \sum_{k=1}^{\infty} B_k \, u_k(x, t)$$

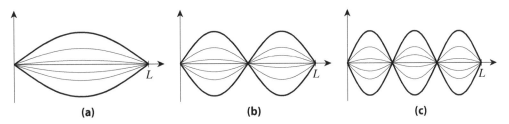

Figure 16.2: The (a) first harmonic, (b) second harmonic and (c) third harmonic for a vibrating string of length L sketched at various times as functions of x.

where

$$u_k(x,t) = \sin\left(\frac{k\pi}{L}x\right)\cos(2\pi v_k t) \qquad \text{and} \qquad v_k = \frac{kc}{2L} \ .$$

The individual u_k's are often referred to as the *modes of vibration* or the *harmonics*, with u_1 being the first or "fundamental" mode/harmonic. The graphs of the first three harmonics — $u_1(x,t)$, $u_2(x,t)$ and $u_3(x,t)$ — have been sketched as functions of x for various values of t in figure 16.2. Notice that $u_k(x,t)$ is nothing more than a sine function of x being scaled by a sinusoid function of time with frequency v_k. It is that v_k which determines the pitch of the sound resulting from that mode of vibration. The magnitude of B_k, of course, helps determine the "loudness" of the sound due to the kth harmonic, with the perceived loudness increasing as B_k increases. (However, the relation between B_k and the apparent loudness is not linear and is strongly influenced by the ability of the ear to perceive different pitches.)

In theory, one can produce a "pure tone" corresponding to any one of these frequencies (say v_3) by imposing just the right initial condition (namely, $u(x,0) = u_3(x,0)$). In practice, this is very difficult, and the sound heard is usually a combination of the sounds corresponding to many of the harmonics. Typically, most of the sound heard is due to the fundamental harmonic, because, typically, people pluck strings in such a manner that the first harmonic is the dominant term in the series solution. For example, whether in a violin or a banjo, v_1 is approximately 440 cycles/second for a string tuned to A above middle C. The other harmonics provide the "overtones" that modify the sound we hear and help us distinguish between a vibrating violin string and a vibrating banjo string.

!▶Example 16.3: *In exercise 16.2 we obtained*

$$u(x,t) = 1\left(\frac{2}{\pi}\right)^2 \cos(3\pi t)\sin(\pi x) + 0\left(\frac{2}{2\pi}\right)^2 \cos(2\cdot 3\pi t)\sin(2\pi x)$$

$$+ (-1)\left(\frac{2}{3\pi}\right)^2 \cos(3\cdot 3\pi t)\sin(3\pi x) + 0\left(\frac{2}{4\pi}\right)^2 \cos(4\cdot 3\pi t)\sin(4\pi x)$$

$$+ \cdots$$

as a solution to a vibrating string problem. From this we see that the first four harmonics for this string are

$$u_1(x,t) = \sin(\pi x)\cos(2\pi v_1 t) \qquad , \qquad u_2(x,t) = \sin(2\pi x)\cos(2\pi v_2 t) \quad ,$$

$$u_3(x,t) = \sin(3\pi x)\cos(2\pi v_3 t) \qquad and \qquad u_4(x,t) = \sin(4\pi x)\cos(2\pi v_4 t)$$

where

$$v_1 = \frac{3}{2} \quad , \quad v_2 = 3 \quad , \quad v_3 = \frac{9}{2} \quad and \quad v_4 = 6 \ .$$

The fundamental harmonic frequency is $v_1 = {}^3/_2$, and the other harmonic frequencies are integral multiples of the fundamental. In this case, the first harmonic is certainly the dominant component *of the above solution. However, if the units on t are seconds, then the first harmonic frequency of $^3/_2$ cycles per second is somewhat below what most people can hear, and so, as far as most people are concerned, the first harmonic will not contribute significantly to the sound heard from this vibrating string.*

16.3 Functions Defined by Infinite Series

In previous chapters, we mainly discussed infinite series that were generated (as Fourier series) from known functions. In this chapter we suddenly find ourselves interested in defining functions using known series of functions. Let's consider this situation for a little bit.

Strongly Convergent Series

Often we are fortunate enough to be dealing with infinite series of functions in which the terms of the series are bounded by numbers that, themselves, are terms of an absolutely convergent series. We will call such infinite series of functions "strongly convergent".[5] To be more precise and, perhaps, a little more clear, suppose we have an interval (a, b) and a sequence of functions on this interval, say, $\psi_1, \psi_2, \psi_3, \dots$. The corresponding infinite series of functions

$$\sum_{k=1}^{\infty} \psi_k(s)$$

will be called *strongly convergent* on (a, b) if and only if there is a sequence of nonnegative real numbers $\Gamma_1, \Gamma_2, \Gamma_3, \dots$ such that both of the following hold:

1. $|\psi_k(s)| \leq \Gamma_k$ for all s in (a, b) and $k = 1, 2, 3, \dots$.

2. $\displaystyle\sum_{k=1}^{\infty} \Gamma_k < \infty$.

!▶Example 16.4: *Plugging in $t = 1$ into the series obtained in example 16.1 on page 223 gives the following infinite series of functions on $(0, \pi)$:*

$$\sum_{k=1}^{\infty} \frac{200}{k\pi} \left[1 - (-1)^k \right] 2^{-k^2} \sin(kx) \quad .$$

This is a strongly convergent series of functions on $(0, \pi)$. To see that, let

$$\psi_k(x) = \frac{200}{k\pi} \left[1 - (-1)^k \right] 2^{-k^2} \sin(kx) \qquad for \quad k = 1, 2, 3, \dots \quad .$$

For each of these ψ_k's and every x in $(0, \pi)$, we clearly have

$$|\psi_k(x)| = \left| \frac{200}{k\pi} \left[1 - (-1)^k \right] 2^{-k^2} \sin(kx) \right| \leq \Gamma_k \qquad with \qquad \Gamma_k = \frac{400}{\pi} \left(\frac{1}{2} \right)^k \quad .$$

---

[5] We could also refer to these series as being "uniformly absolutely convergent". Though linguistically awkward, it's certainly a more accurate description.

Since $\left|{}^1\!/_2\right| < 1$, we can use the geometric series formula (see example 4.1 on page 42) to "add up" all these Γ_k's, obtaining

$$\sum_{k=1}^{\infty} \Gamma_k = \sum_{k=1}^{\infty} \frac{400}{\pi} \left(\frac{1}{2}\right)^k = \frac{400}{\pi} < \infty \quad .$$

Thus,

$$\sum_{k=1}^{\infty} \frac{200}{k\pi} \left[1 - (-1)^k\right] 2^{-k^2} \sin(kx)$$

satisfies the requirements for being strongly convergent on $(0, \pi)$.

?▶Exercise 16.5: *Letting $x = {}^\pi\!/_2$ in the series obtained in example 16.1 on page 223 gives the following infinite series of functions on $(0, \pi)$:*

$$\sum_{k=1}^{\infty} \frac{200}{k\pi} \left[1 - (-1)^k\right] \sin\left(\frac{k\pi}{2}\right) 2^{-k^2 t} \quad .$$

a: *Show this is a strongly convergent series of functions on $(1, \infty)$.*

b: *Show this is a strongly convergent series of functions on any interval (T, ∞) where T is a fixed positive number.*

Certainly, if $\sum_{k=1}^{\infty} \psi_k(s)$ is strongly convergent on some interval, then $\sum_{k=1}^{\infty} \psi_k(s)$ is an absolutely convergent series of numbers for each s in the interval. Consequently,

$$h(s) = \sum_{k=1}^{\infty} \psi_k(s)$$

is a well-defined function on the interval. It also turns out that when the ψ_k's are "sufficiently nice", so is h, and h can be integrated and differentiated by integrating and differentiating the ψ_k's. That is the gist of the following theorem.

Theorem 16.1 (strongly convergent series)
Suppose h is given by a strongly convergent series of continuous functions on the interval (a, b)

$$h(s) = \sum_{k=1}^{\infty} \psi_k(s) \quad . \tag{16.16}$$

Then all of the following hold:

1. *h is continuous on (a, b) with*

$$\lim_{s \to s_0} h(s) = \sum_{k=1}^{\infty} \psi_k(s_0) \qquad \text{for each } s \text{ in } (a, b) \quad .$$

2. *Let (α, β) be any finite subinterval of (a, b) (including, possibly, (α, β), itself). If each ψ_k is uniformly continuous on (α, β), then so is h. Moreover,*

$$\lim_{s \to \alpha^+} h(s) = \sum_{k=1}^{\infty} \psi_k(\alpha) \qquad \text{and} \qquad \lim_{s \to \beta^-} h(s) = \sum_{k=1}^{\infty} \psi_k(\beta) \quad .$$

3.   If g is a bounded, piecewise continuous function on (a, b), and s_0 and s are any two points in (a, b), then

$$\sum_{k=1}^{\infty} \int_{s_0}^{s} g(\sigma) \psi_k(\sigma) \, d\sigma$$

converges, and

$$\int_{s_0}^{s} g(\sigma) h(\sigma) \, d\sigma \; = \; \sum_{k=1}^{\infty} \int_{s_0}^{s} g(\sigma) \psi_k(\sigma) \, d\sigma \quad . \tag{16.17}$$

4.   If each ψ_k is a smooth function on (a, b), and $\sum_{k=1}^{\infty} \psi_k{}'(s)$ is also strongly convergent on (a, b), then h is a smooth function on (a, b) with

$$h'(s) \; = \; \sum_{k=1}^{\infty} \psi_k{}'(s) \quad .$$

5.   Suppose each ψ_k is an nth order differentiable function on (a, b) with $\psi_k{}^{(n)}$ being continuous for each integer k and some fixed integer n. If

$$\sum_{k=1}^{\infty} \psi_k{}'(s) \quad , \quad \sum_{k=1}^{\infty} \psi_k{}''(s) \quad , \quad \dots \quad \text{and} \quad \sum_{k=1}^{\infty} \psi_k{}^{(n)}(s)$$

are all strongly convergent on (a, b), then h is n-times differentiable on (a, b), $h^{(n)}$ is continuous on (a, b) with

$$h^{(m)}(s) \; = \; \sum_{k=1}^{\infty} \psi_k{}^{(m)}(s) \qquad \text{for} \quad m = 1, \, 2, \, \dots, \, n \quad .$$

The proof of this theorem is relatively straightforward, though a little tedious. For convenience, we'll break it into several parts starting with a "part 0" in which we derive some results that will be useful throughout the rest of the proof.

PROOF (theorem 16.1, part 0):   By the definition of strong convergence we can assume there are nonnegative real numbers Γ_1, Γ_2, Γ_3, $\dots$ such that

$$|\psi_k(s)| \; \leq \; \Gamma_k \quad \text{for each positive integer } k \text{ and each } s \text{ in } (a, b) \tag{16.18}$$

and

$$\sum_{k=1}^{\infty} \Gamma_k \; < \; \infty \quad . \tag{16.19}$$

Now, for any s in (a, b) and any positive integer N,

$$\left| h(s) - \sum_{k=1}^{N} \psi_k(s) \right| = \left| \sum_{k=1}^{\infty} \psi_k(s) - \sum_{k=1}^{N} \psi_k(s) \right| = \left| \sum_{k=N+1}^{\infty} \psi_k(s) \right| \leq \sum_{k=N+1}^{\infty} |\psi_k(s)| \quad .$$

Combining this with inequality (16.18) gives us

$$\left| h(s) - \sum_{k=1}^{N} \psi_k(s) \right| \leq \sum_{k=N+1}^{\infty} \Gamma_k \quad .$$

We will find it more convenient to write this last inequality as

$$\left| h(s) - \sum_{k=1}^{N} \psi_k(s) \right| \leq E(N) \tag{16.20}$$

where

$$E(N) = \sum_{k=N+1}^{\infty} \Gamma_k \qquad \text{for} \quad N = 1, 2, 3, \ldots \quad .$$

Keep in mind that, because of the convergence of the series in line (16.19), the infinite series on the right-hand side of our last formula converges and approaches 0 as $N \to \infty$. Thus,

$$\lim_{N \to \infty} E(N) = 0 \quad . \tag{16.21}$$

PROOF (theorem 16.1, parts 1 and 2):   To prove part 1, it suffices to show that, for each s_0 in (a, b) and each $\epsilon > 0$, there is a corresponding Δs such that

$$|h(s_0) - h(s)| < \epsilon \qquad \text{whenever} \quad |s_0 - s| < \Delta s$$

and where $h(s_0)$ and $h(s)$ are computed using formula (16.16).

So let s_0 and $\epsilon > 0$ be chosen. Because equation (16.21) holds, we can choose an integer N_ϵ so that $E(N_\epsilon) < \epsilon/3$. For convenience, let

$$h_\epsilon(s) = \sum_{k=1}^{N_\epsilon} \psi_k(s)$$

and observe that, by inequality (16.20) and our choice of N_ϵ,

$$|h(s) - h_\epsilon(s)| = \left| h(s) - \sum_{k=1}^{N_\epsilon} \psi_k(s) \right| \leq E(N_\epsilon) < \frac{\epsilon}{3}$$

for each s in (a, b). Observe also, that h_ϵ, being a finite sum of continuous functions, is a continuous function. So there is a $\Delta s > 0$ such that

$$|h_\epsilon(s_0) - h_\epsilon(s)| < \frac{\epsilon}{3} \qquad \text{whenever} \quad |s_0 - s| < \Delta s \quad .$$

Consequently, whenever $|s_0 - s| \leq \Delta s$,

$$|h(s_0) - h(s)| = |h(s_0) - h_\epsilon(s_0) + h_\epsilon(s_0) - h(s) + h_\epsilon(s) - h_\epsilon(s)|$$

$$\leq |h(s_0) - h_\epsilon(s_0)| + |h_\epsilon(s_0) - h_\epsilon(s)| + |h(s) - h_\epsilon(s)|$$

$$< \frac{\epsilon}{3} + \frac{\epsilon}{3} + \frac{\epsilon}{3}$$

$$= \epsilon \quad ,$$

completing the proof of the first part.

To prove part 2 we simply repeat the above, replacing s_0 with α and β, and restricting s to being between α and β. ∎

PROOF (theorem 16.1, part 3):   Since we've just shown h to be continuous, we know gh is piecewise continuous and the integral on the left-hand side of equation (16.17) is well defined. Also, since "an integral of a finite sum is the corresponding sum of the integrals",

$$\int_{s_0}^{s} g(\sigma)h(\sigma)\,d\sigma = \int_{s_0}^{s} g(\sigma)\left[\sum_{k=1}^{N}\psi_k(\sigma) + h(\sigma) - \sum_{k=1}^{N}\psi_k(\sigma)\right]d\sigma$$

$$= \sum_{k=1}^{N}\int_{s_0}^{s} g(\sigma)\psi_k(\sigma)\,d\sigma + \int_{s_0}^{s} g(\sigma)\left[h(\sigma) - \sum_{k=1}^{N}\psi_k(\sigma)\right]d\sigma$$

for every positive integer N. So

$$\int_{s_0}^{s} g(\sigma)h(\sigma)\,d\sigma = \lim_{N\to\infty}\sum_{k=1}^{N}\int_{s_0}^{s} g(\sigma)\psi_k(\sigma)\,d\sigma$$

$$+ \lim_{N\to\infty}\int_{s_0}^{s} g(\sigma)\left[h(\sigma) - \sum_{k=1}^{N}\psi_k(\sigma)\right]d\sigma \quad . \tag{16.22}$$

Consider the last limit in the last line above. Since g is bounded, we can let B denote some finite value such that

$$|g(\sigma)| \leq B \quad \text{whenever} \quad a < \sigma < b \quad .$$

Using this and inequality (16.20), we have

$$\left|\int_{s_0}^{s} g(\sigma)\left[h(\sigma) - \sum_{k=1}^{N}\psi_k(\sigma)\right]d\sigma\right| \leq \int_{s_0}^{s}|g(\sigma)|\left|h(\sigma) - \sum_{k=1}^{N}\psi_k(\sigma)\right|d\sigma$$

$$\leq \int_{s_0}^{s} BE(N)\,d\sigma$$

$$\leq BE(N)[s - s_0] \quad ,$$

which, because $E(N) \to 0$ as $N \to \infty$, means that

$$\lim_{N\to\infty}\int_{s_0}^{s} g(\sigma)\left[h(\sigma) - \sum_{k=1}^{N}\psi_k(\sigma)\right]d\sigma = 0 \quad .$$

Plugging this back into equation (16.22) we finally get

$$\int_{s_0}^{s} g(\sigma)h(\sigma)\,d\sigma = \lim_{N\to\infty}\sum_{k=1}^{N}\int_{s_0}^{s} g(\sigma)\psi_k(\sigma)\,d\sigma + 0 \quad ,$$

confirming both the convergence of the series of integrals and equation (16.17). ∎

PROOF (theorem 16.1, parts 4 and 5): Because part 5 clearly follows by applying the results from part 16.1 n times, we will just prove part 4.

The first part of the theorem tells us that

$$g(s) = \sum_{k=1}^{\infty}\psi_k'(s)$$

is a continuous function on (a, b). From part 3 we know

$$\int_{s_0}^{s} g(\sigma)\, d\sigma = \sum_{k=1}^{\infty} \int_{s_0}^{s} \psi_k'(\sigma)\, d\sigma$$

$$= \sum_{k=1}^{\infty} [\psi_k(s) - \psi_k(s_0)]$$

$$= \sum_{k=1}^{\infty} \psi_k(s) - \sum_{k=1}^{\infty} \psi_k(s_0) = h(s) - h(s_0) \quad .$$

So,

$$h(s) = h(s_0) + \int_{s_0}^{s} g(\sigma)\, d\sigma$$

and

$$h'(s) = \frac{d}{ds} h(s_0) + \frac{d}{ds} \int_{s_0}^{s} g(\sigma)\, d\sigma = g(s) = \sum_{k=1}^{\infty} \psi_k'(s) \quad . \qquad \blacksquare$$

Applications to Fourier Series

The results described in the previous subsection can be applied to any Fourier series whose coefficients are terms in an absolutely convergent series. Theorem 15.9 on page 209, for example, is an immediate corollary of the theorem on strongly convergent series in the previous subsection. So is the following, which will be of particular interest to us in validating the series solution obtained in section 16.1.

Theorem 16.2 (strongly convergent sine series)
Suppose $\sum_{k=1}^{\infty} b_k$ is an absolutely convergent infinite series of complex numbers and L is some positive value. Let f be defined on $(0, L)$ by

$$f(x) = \sum_{k=1}^{\infty} b_k \sin\left(\frac{k\pi}{L} x\right) \quad .$$

Then f is a uniformly continuous function on $(0, L)$ with

$$\lim_{x \to 0^+} f(x) = 0 \qquad and \qquad \lim_{x \to L^-} f(x) = 0 \quad .$$

Moreover:

1. *f is a uniformly continuous function on $(0, L)$ with*

$$\lim_{x \to 0^+} f(x) = 0 \qquad and \qquad \lim_{x \to L^-} f(x) = 0 \quad .$$

2. *The above series is the Fourier sine series for f. That is,*

$$b_k = \frac{2}{L} \int_0^L f(t) \sin\left(\frac{k\pi}{L} x\right) dx \qquad for \quad k = 1, 2, 3, \ldots \quad .$$

3.   If

$$\sum_{k=1}^{\infty} \left| k^n b_k \right| < \infty$$

for some positive integer n, then f is n-times differentiable, $f^{(n)}$ is continuous, and

$$f^{(m)}(x) = \sum_{k=1}^{\infty} b_k \frac{d^m}{dx^m} \sin\left(\frac{k\pi}{L}x\right) \qquad for \quad m = 1, 2, \ldots, n \quad .$$

?►Exercise 16.6:   Confirm the claims of the above theorem using the theorem on strongly convergent series (theorem 16.1 on page 235).

?►Exercise 16.7:   Write out the corresponding "strongly convergent cosine series" theorem.

Convergence of a Parameterized Series

The next result will be useful in computing the limit of a function given by a series that is not strongly convergent. It is a subtle result, and we will prove it by employing a remarkably clever construction usually attributed to the early nineteenth-century mathematician Niels Abel.

Lemma 16.3
Let ϕ_1, ϕ_2, ϕ_3, ... be a sequence of functions on $[0, 1)$ such that, for $k = 1, 2, 3, \ldots$,

$$\lim_{t \to 0^+} \phi_k(t) = 1$$

and

$$1 \leq \phi_k(t) \leq \phi_{k+1}(t) \qquad for\ all\ t\ in\ (0, 1) \quad .$$

Further assume that a_1, a_2, a_3, ... is a sequence of numbers such that $\sum_{k=1}^{\infty} a_k$ converges, as does[6]

$$\sum_{k=1}^{\infty} a_k \phi_k(t) \qquad for\ each\ t\ in\ (0, 1) \quad .$$

Then

$$\lim_{t \to 0^+} \sum_{k=1}^{\infty} a_k \phi_k(t) = \sum_{k=1}^{\infty} a_k \quad .$$

PROOF: We need to show that, for any given $\epsilon > 0$, there is a corresponding $\Delta t > 0$ such that

$$\left| \sum_{k=1}^{\infty} a_k - \sum_{k=1}^{\infty} a_k \phi_k(t) \right| \leq \epsilon \qquad whenever \quad 0 < t < \Delta t \quad .$$

So let $\epsilon > 0$ be chosen.

---

[6] In fact, we could show that this series converges for each t in $(0, 1)$ if $\sum_{k=1}^{\infty} a_k$ converges, but we won't need that fact and it would make the proof longer.

Since $\sum_{k=1}^{\infty} a_k$ is convergent, there is an integer $N = N_\epsilon$ such that

$$\left| \sum_{k=M+1}^{\infty} a_k \right| = \left| \sum_{k=1}^{\infty} a_k - \sum_{k=1}^{M} a_k \right| < \frac{\epsilon}{4} \qquad \text{whenever} \quad M \geq N \quad . \tag{16.23}$$

Now, for each integer k greater than N, let

$$A_k = \sum_{j=N+1}^{k} a_j \quad .$$

Note that $A_{N+1} = a_{N+1}$ and, for $k = N + 2, N + 3, N + 4, \ldots,$

$$a_k + A_{k-1} = A_k$$

and

$$|A_k| = \left| \sum_{j=N+1}^{k} a_j \right| = \left| \sum_{j=N+1}^{\infty} a_j - \sum_{j=k+1}^{\infty} a_j \right|$$

$$\leq \left| \sum_{j=N+1}^{\infty} a_j \right| + \left| \sum_{j=k+1}^{\infty} a_j \right| \leq \frac{\epsilon}{4} + \frac{\epsilon}{4} = \frac{\epsilon}{2} \quad . \tag{16.24}$$

Here is the clever bit: Let $0 < t < 1$ and observe that, for $M > N$,

$$\sum_{k=N+1}^{M} a_k \, \phi_k(t) = a_{N+1} \, \phi_{N+1}(t) + \sum_{k=N+2}^{M} (a_k + A_{k-1} - A_{k-1}) \phi_k(t)$$

$$= A_{N+1} \, \phi_{N+1}(t) + \sum_{k=N+2}^{M} (A_k - A_{k-1}) \phi_k(t) \quad .$$

For the sake of brevity, let ψ_k denote $\phi_k(t)$. Then, expanding out the last formula and rearranging a few terms, we find that

$$\sum_{k=N+1}^{M} a_k \psi_k = A_{N+1} \psi_{N+1} + \sum_{k=N+2}^{M} (A_k - A_{k-1}) \psi_k$$

$$= A_{N+1} \psi_{N+1} + (A_{N+2} - A_{N+1}) \psi_{N+2}$$
$$\quad + (A_{N+3} - A_{N+2}) \psi_{N+3} + \cdots + (A_M - A_{M-1}) \psi_M$$

$$= A_{N+1}(\psi_{N+1} - \psi_{N+2}) + A_{N+2}(\psi_{N+2} - \psi_{N+3})$$
$$\quad + A_{N+3}(\psi_{N+3} - \psi_{N+4}) + \cdots + A_{M-1}(\psi_{M-1} - \psi_M) + A_M \psi_M$$

$$= A_M \psi_M + \sum_{k=N+1}^{M-1} A_k (\psi_k - \psi_{k+1}) \quad .$$

This, along with inequality (16.24), gives

$$\left| \sum_{k=N+1}^{M} a_k \psi_k \right| \leq |A_M| |\psi_M| + \sum_{k=N+1}^{M-1} |A_k| |\psi_k - \psi_{k+1}|$$

$$\leq |\psi_M| + \frac{\epsilon}{2} \left[\sum_{k=N+1}^{M-1} |\psi_k - \psi_{k+1}| \right] . \tag{16.25}$$

Remember, $0 \leq \phi_{k+1}(t) \leq \phi_k(t) \leq 1$ for each positive integer k, and ψ_k is just shorthand for $\phi_k(t)$. So $|\psi_M| = \psi_M$ and

$$|\psi_k - \psi_{k+1}| = |\phi_k(t) - \phi_{k+1}(t)| = \phi_k(t) - \phi_{k+1}(t) = \psi_k - \psi_{k+1} .$$

Plugging this into inequality (16.25) gives us

$$\left| \sum_{k=N+1}^{M} a_k \psi_k \right| \leq \frac{\epsilon}{2} \left[\psi_M + \sum_{k=N+1}^{M-1} (\psi_k - \psi_{k+1}) \right] .$$

But, since

$$\sum_{k=1}^{N-1} (\psi_k - \psi_{k+1}) = (\psi_{N+1} - \psi_{N+2}) + (\psi_{N+2} - \psi_{N+3})$$

$$+ (\psi_{N+3} - \psi_{N+4}) + \cdots + (\psi_{M-1} - \psi_M)]$$

$$= \psi_{N+1} - \psi_M ,$$

and $\psi_{N+1} = \phi_{N+1}(t) \leq 1$, our last inequality reduces to

$$\left| \sum_{k=N+1}^{M} a_k \psi_k \right| \leq \frac{\epsilon}{2} \left[(\psi_{N+1} - \psi_M) + \psi_M \right] \leq \frac{\epsilon}{2} \psi_{N+1} \leq \frac{\epsilon}{2} .$$

Thus, after letting $M \to \infty$, we have

$$\left| \sum_{k=N+1}^{\infty} a_k \phi_k(t) \right| \leq \frac{\epsilon}{2} \qquad \text{for each } t \text{ in } (0,1) . \tag{16.26}$$

Finally, consider

$$\sum_{k=1}^{N} a_k - \sum_{k=1}^{N} a_k \phi_k(t) = \sum_{k=1}^{N} a_k [1 - \phi_k(t)] .$$

Since this is a finite sum and $\phi_k(t) \to 1$ as $t \to 0^+$ for each positive integer k,

$$\lim_{t \to 0^+} \sum_{k=1}^{N} a_k [1 - \phi_k(t)] = \sum_{k=1}^{N} a_k \lim_{t \to 0^+} [1 - \phi_k(t)] = 0 .$$

This means there is a $\Delta t > 0$ such that

$$\left| \sum_{k=1}^{N} a_k [1 - \phi_k(t)] \right| < \frac{\epsilon}{4} \qquad \text{whenever} \quad 0 < t < \Delta t . \tag{16.27}$$

Combining this with inequalities (16.23) and (16.26), we discover that, whenever $0 < t < \Delta t$,

$$\left| \sum_{k=1}^{\infty} a_k - \sum_{k=1}^{\infty} a_k \, \phi_k(t) \right| = \left| \sum_{k=1}^{N} a_k[1 - \phi_k(t)] + \sum_{k=N+1}^{\infty} a_k - \sum_{k=N+1}^{\infty} a_k \, \phi_k(t) \right|$$

$$\leq \left| \sum_{k=1}^{N} a_k[1 - \phi_k(t)] \right| + \left| \sum_{k=N+1}^{\infty} a_k \right| + \left| \sum_{k=N+1}^{\infty} a_k \, \phi_k(t) \right|$$

$$\leq \frac{\epsilon}{4} + \frac{\epsilon}{2} + \frac{\epsilon}{4} = \epsilon \quad .$$ ∎

16.4   Verifying the Heat Flow Problem Solution

Let's now verify that a solution to the heat flow problem of the first section for positive time is given by

$$u(x, t) = \sum_{k=1}^{\infty} B_k \, e^{-k^2 \lambda t} \sin\left(\frac{k\pi}{L} x\right) \tag{16.28a}$$

where

$$\lambda = \kappa \left(\frac{\pi}{L}\right)^2 \tag{16.28b}$$

and

$$B_k = \frac{2}{L} \int_0^L f(x) \sin\left(\frac{k\pi}{L} x\right) dx \quad \text{for} \quad k = 1, 2, 3, \ldots \quad . \tag{16.28c}$$

Remember, L and κ are positive constants and f is a known piecewise continuous function on the interval $(0, L)$.

We start by deriving some convenient upper bounds on the terms of the above series. For reasons that will soon be obvious, we will restrict ourselves to values of t greater than T where T is some positive value (however, because T can be *any* positive value, our results will still hold for any $t > 0$). It will also be convenient to let

$$R = \exp\left(-\frac{1}{2}\lambda T\right) \quad . \tag{16.29}$$

Observe that, because T and λ are positive, we know $0 < R < 1$ and (see example 4.1 on page 42)

$$\sum_{k=1}^{\infty} R^k = \frac{R}{1 - R} < \infty \quad . \tag{16.30}$$

Consider the B_k's. These are the Fourier sine coefficients of f. Consequently, letting

$$A = \frac{2}{L} \int_0^L |f(x)| \, dx$$

we must clearly have

$$|B_k| = \left| \frac{2}{L} \int_0^L f(x) \sin\left(\frac{k\pi}{L} x\right) dx \right| \leq A \quad \text{for} \quad k = 1, 2, 3, \ldots \quad .$$

Next, consider the exponential factor in each term of formula (16.28a). In fact, anticipating future needs, let's consider the expression

$$k^n e^{-k^2 \lambda t} \qquad \text{for} \quad k = 1, 2, 3, \dots$$

assuming n is some fixed nonnegative integer (and $t \geq T$). If γ is any positive number and k is any positive integer, then $k^2 \gamma t \geq k \gamma T$, and so,

$$e^{-k^2 \gamma t} \leq e^{-k \gamma T} \quad .$$

Also, the maximum of $x^n e^{-x \gamma T}$ on the interval $[1, \infty)$ is easily found using elementary calculus to be

$$C_n = \begin{cases} \left(\dfrac{n}{\gamma T} \right)^n e^{-n} & \text{if} \quad 1 < \dfrac{n}{\gamma T} \\[2ex] e^{-\gamma T} & \text{if} \quad \dfrac{n}{\gamma T} \leq 1 \end{cases} \quad .$$

?►Exercise 16.8:   *Using elementary calculus, verify the above claim that $x^n e^{-x \gamma T} \leq C_n$.*

In particular now, let $\gamma = \lambda/2$. Then

$$k^n e^{-k^2 \lambda t} \leq k^n e^{-k \lambda T} = k^n e^{-k \gamma T} e^{-k \gamma T} \leq C_n e^{-k \gamma T} = C_n \left[e^{-\gamma T} \right]^k \quad ,$$

which can be written more simply as

$$k^n e^{-k^2 \lambda t} \leq C_n R^k$$

using the R defined above in line (16.29).

So that we can apply the results from the previous sections, we will now consider the series solution assuming t is fixed. That is, we will let h be formula (16.28a) with $t \geq T$ treated as a constant. For convenience, let's write this as

$$h(x) = \sum_{k=1}^{\infty} \psi_k(x) \qquad \text{where} \qquad \psi_k(x) = B_k e^{-k^2 \lambda t} \sin\left(\frac{k\pi}{L} x \right) \quad .$$

By the bounds derived above for the B_k's and the exponentials, we see that

$$|\psi_k(x)| = \left| B_k e^{-k^2 \lambda t} \sin\left(\frac{k\pi}{L} x \right) \right| \leq A C_0 R^k \quad .$$

But, as noted in inequality (16.30),

$$\sum_{k=1}^{\infty} A C_n R^k = A C_0 \sum_{k=1}^{\infty} R^k < \infty \quad .$$

This tells us that the series defining h is a strongly convergent series of uniformly continuous functions (of x) on $(0, L)$. Moreover, each ψ_k is clearly infinitely differentiable and, for each positive integer n, we have

$$\left| \psi_k^{(n)}(x) \right| = \left| \frac{\partial^n}{\partial x^n} B_k e^{-k^2 \lambda t} \sin\left(\frac{k\pi}{L} x \right) \right|$$

$$\leq |B_k| e^{-k^2 \lambda t} \left(\frac{k\pi}{L} \right)^n$$

$$\leq A \left(\frac{\pi}{L} \right)^n k^n e^{-k^2 \lambda t} \leq A \left(\frac{\pi}{L} \right)^n C_n R^k \quad .$$

From this it follows that

$$\sum_{k=1}^{\infty} \psi_k(x) \quad , \quad \sum_{k=1}^{\infty} \psi_k{}'(x) \quad , \quad \sum_{k=1}^{\infty} \psi_k{}''(x) \quad , \quad \sum_{k=1}^{\infty} \psi_k{}^{(3)}(x) \quad , \quad \dots$$

are all strongly convergent series of uniformly continuous, differentiable functions on $(0, L)$. The theorem on strongly convergent series (theorem 16.1 on page 235) informs us that h must then be uniformly continuous and infinitely differentiable on $(0, L)$. But $h(x) = u(x, t)$ for an arbitrary positive t. So we've just verified that $u(x, t)$ is a uniformly continuous and infinitely differentiable function of x on $(0, L)$ for each $t > 0$. Moreover, from part 2 of theorem 16.1, we get

$$u(0, t) = h(0) = \lim_{x \to 0^+} h(x) = \sum_{k=1}^{\infty} B_k e^{-k^2 \lambda t} \sin\left(\frac{k\pi}{L} 0\right) = \sum_{k=1}^{\infty} B_k e^{-k^2 \lambda t} \cdot 0 = 0$$

and

$$u(L, t) = h(L) = \lim_{x \to L^-} h(x) = \sum_{k=1}^{\infty} B_k e^{-k^2 \lambda t} \sin\left(\frac{k\pi}{L} L\right) = \sum_{k=1}^{\infty} B_k e^{-k^2 \lambda t} \cdot 0 = 0 \quad .$$

Thus, the endpoint conditions of equations (16.1b) are satisfied. In addition, from part 5 of theorem 16.1, we know

$$\frac{\partial^2}{\partial x^2} u(x, t) = h''(x) = \sum_{k=1}^{\infty} \psi_k{}''(x)$$

$$= \sum_{k=1}^{\infty} \frac{\partial^2}{\partial x^2} B_k e^{-k^2 \lambda t} \sin\left(\frac{k\pi}{L} x\right)$$

$$= -\sum_{k=1}^{\infty} \left(\frac{k\pi}{L}\right)^2 B_k e^{-k^2 \lambda t} \sin\left(\frac{k\pi}{L} x\right) \quad . \tag{16.31}$$

If we hold x fixed and let t vary over (T, ∞), then arguments very similar to those described above lead us to conclude that $u(x, t)$, as defined by formula (16.28a), is an infinitely smooth function of t on (T, ∞) for each x in $(0, L)$ with

$$\frac{\partial}{\partial t} u(x, t) = \sum_{k=1}^{\infty} \frac{\partial}{\partial t} B_k e^{-k^2 \lambda t} \sin\left(\frac{k\pi}{L} x\right) = -\sum_{k=1}^{\infty} B_k k^2 \lambda e^{-k^2 \lambda t} \sin\left(\frac{k\pi}{L} x\right) \tag{16.32}$$

for each $0 < x < L$ and $T < t < \infty$. Since this holds for any $T > 0$, $u(x, t)$ must be an infinitely smooth function of t on $(0, \infty)$ for each $0 < x < L$. Consequently, $u(x, t)$ is a "smooth enough" function of x and t for the derivatives in the heat equation to make sense when $0 < x < L$ and $0 < t$. Furthermore, after recalling the formula for λ and using formula (16.31), we find that (16.32) can be written as

$$\frac{\partial}{\partial t} u(x, t) = -\sum_{k=1}^{\infty} B_k k^2 \kappa \left(\frac{\pi}{L}\right)^2 e^{-k^2 \lambda t} \sin\left(\frac{k\pi}{L} x\right)$$

$$= -\kappa \sum_{k=1}^{\infty} B_k \left(\frac{k\pi}{L}\right)^2 e^{-k^2 \lambda t} \sin\left(\frac{k\pi}{L} x\right) = \kappa \frac{\partial^2}{\partial x^2} u(x, t)$$

or, equivalently, as

$$\frac{\partial u}{\partial t} - \kappa \frac{\partial^2 u}{\partial x^2} = 0 \quad ,$$

confirming that our formula for $u(x, t)$ satisfies the heat equation whenever $0 < x < L$ and $0 < t$.

Finally, to verify that the initial condition given in equation (16.1c) holds, let x be any fixed point in $(0, L)$ and consider

$$\sum_{k=1}^{\infty} a_k \, \phi_k(t)$$

where

$$a_k \;=\; B_k \sin\!\left(\frac{k\pi}{L}x\right) \qquad \text{and} \qquad \phi_k(t) \;=\; e^{-k^2\lambda t} \quad .$$

This summation is just the formula for $u(x, t)$ written to match the notation in lemma 16.3 on page 240. It is easily verified that the lemma applies and assures us that

$$\lim_{t \to 0^+} u(x, t) \;=\; \lim_{t \to 0^+} \sum_{k=1}^{\infty} a_k \, \phi_k(t) \;=\; \sum_{k=1}^{\infty} a_k \, \phi_k(0) \;=\; \sum_{k=1}^{\infty} B_k \sin\!\left(\frac{k\pi}{L}x\right) \quad .$$

And since the B_k's are the Fourier sine coefficients of f, and f is piecewise smooth on $(0, L)$, this last equation is just

$$u(x, 0) \;=\; \lim_{t \to 0^+} u(x, t) \;=\; f(x)$$

for every x at which f is continuous, confirming that (16.1c) holds and, thus, completing our verification that formula set (16.5) on page 222 satisfies our heat flow problem for all positive t.

There are several points worth noting:

1. Strictly speaking, the initial condition we verified was that $u(x_0, 0) = f(x_0)$ for each x_0 at which f is continuous. By being a little more careful with the analysis and using the "nearly uniform" convergence of the Fourier series discussed in chapter 13, you can actually verify that

$$\lim_{\substack{t \to 0^+ \\ x \to x_0}} u(x, t) \;=\; f(x_0)$$

for each x_0 at which f is continuous. (Whether you care to verify this, of course, is another matter.)

2. A lot of work went into proving lemma 16.3 simply so we could verify that $u(x, 0) = f(x)$. We could have avoided all that labor if we had assumed f was also uniformly continuous on $(0, L)$ with $f(0) = 0 = f(L)$. Theorem 15.7 on page 208 would have then guaranteed the absolute convergence of $\sum_{k=1}^{\infty} B_k$. That alone would have ensured the strong convergence of our series solution as a series of functions of t on $(0, \infty)$, and would have allowed us to conclude that $u(x, 0) = f(x)$ without recourse to lemma 16.3. The disadvantage would have been that our results would not have applied to perfectly reasonable cases such as in example 16.1 on page 223.

3. On the other hand, if you are satisfied with "$u(x, 0) = f(x)$ in norm", then the requirement that f be piecewise smooth can be relaxed to "f is piecewise continuous". Then, using Bessel's equality and lemma 16.3, you get

$$\lim_{t \to 0^+} \int_0^L |u(x, t) - f(x)|^2 \, dx \;=\; \frac{L}{2} \lim_{t \to 0^+} \sum_{k=1}^{\infty} |B_k|^2 \left(1 - e^{-k^2\lambda t}\right)^2 \;=\; 0 \quad .$$

Additional Exercises

16.9. Consider the series solution to the heat flow problem of exercise 16.1 on page 223. Using the first 25 terms of this solution, sketch the temperature distribution throughout the rod at $t = 0$, $t = \frac{1}{10}$, $t = 1$ and $t = 10$. (Use the computer math package you used for sketching partial sums to sine series in exercise 10.7 on page 128.)

16.10. Using the formal solution derived in the first section of this chapter, find the solution to the heat flow problem described in equation set (16.1) on page 220 assuming $L = 2$, $\kappa = 3$ and

   a. $f(x) = 5\sin(\pi x)$ **b.** $f(x) = x$

Which of these solutions will be valid for all t and which will just be valid for $t > 0$?

16.11. If the endpoints of our heat conducting rod are insulated instead of being kept at 0 degrees, then the temperature distribution $u(x, t)$ satisfies the following set of equations:

$$\frac{\partial u}{\partial t} - \kappa \frac{\partial^2 u}{\partial x^2} = 0 \qquad \text{for} \quad 0 < x < L, \; 0 < t$$

$$\left.\frac{\partial u}{\partial x}\right|_{(0,t)} = 0 \quad \text{and} \quad \left.\frac{\partial u}{\partial x}\right|_{(L,t)} = 0 \qquad \text{for} \quad 0 < t$$

$$u(x, 0) = f(x) \qquad \text{for} \quad 0 < x < L$$

where κ and L are positive constants, and f is piecewise smooth on $(0, L)$.

   a. Why, in this case, would it be better to represent $u(x, t)$ using a cosine series,

$$u(x, t) = \phi_0(t) + \sum_{k=1}^{\infty} \phi_k(t) \cos\left(\frac{k\pi}{L}x\right) \quad,$$

instead of the sine series used for the problem in the first section of this chapter?

   b. Derive the formal series solution for this heat flow problem.

   c. Find the solution to this problem assuming $\kappa = 2$, $L = 3$ and

$$f(x) = \begin{cases} 1 & \text{if} \quad 0 < x < \frac{3}{2} \\ 0 & \text{if} \quad \frac{3}{2} < x < 3 \end{cases} \quad,$$

and sketch the temperature distribution (using the first 25 terms of your series solution) for $t = 0$, $t = \frac{1}{10}$, $t = 1$ and $t = 10$.

   d. What happens to the solution found in the last part as $t \to \infty$? Sketch the temperature distribution "at $t = \infty$".

   e. What can be said about the differentiability of the solution derived above in the first part of this exercise?

16.12. If our heat conducting rod contains sources of heat, and we start with the rod at 0 degrees and keep the endpoints at 0 degrees, then the temperature distribution $u(x,t)$ satisfies the following set of equations:

$$\frac{\partial u}{\partial t} - \kappa \frac{\partial^2 u}{\partial x^2} = f(x) \qquad \text{for } 0 < x < L , 0 < t$$

$$u(0,t) = 0 \quad \text{and} \quad u(L,t) = 0 \qquad \text{for } 0 < t$$

$$u(x,0) = 0 \qquad \text{for } 0 < x < L$$

Again, κ and L are positive constants, and f is piecewise smooth on $(0,L)$.

a. Derive the formal series solution to this problem assuming that a solution exists. (Hint: Start with formula (16.2).)

b. Find the solution to this problem assuming $\kappa = 4$, $L = 3$ and

$$f(x) = \begin{cases} 1 & \text{if } 1 < x < 2 \\ 0 & \text{otherwise} \end{cases},$$

and sketch the temperature distribution (using the first 25 terms of your series solution) for $t = 0$, $t = \frac{1}{10}$, $t = 1$ and $t = 10$.

c. What happens to the solution found in the last part as $t \to \infty$? Sketch the temperature distribution "at $t = \infty$".

d. What can be said about the differentiability of the solution derived above in the first part of this exercise?

16.13. Find the formal solution $u(x,t)$ to the following "vibrating string" problem:

$$\frac{\partial^2 u}{\partial t^2} - c^2 \frac{\partial^2 u}{\partial x^2} = 0 \qquad \text{for } 0 < x < L$$

$$u(0,t) = 0 \quad \text{and} \quad u(L,t) = 0$$

$$u(x,0) = 0 \qquad \text{for } 0 < x < L$$

$$\left.\frac{\partial u}{\partial t}\right|_{(x,0)} = f(x) \qquad \text{for } 0 < x < L$$

where L and c are positive constants and f is piecewise smooth on $(0,L)$.

16.14. A more realistic model for the vibrating string that takes into account the dampening of the vibrations due to air resistance is partially given by the equations

$$\frac{\partial^2 u}{\partial t^2} + 2\beta \frac{\partial u}{\partial t} - c^2 \frac{\partial^2 u}{\partial x^2} = 0 \qquad \text{for } 0 < x < L ,$$

$$u(0,t) = 0 \quad \text{and} \quad u(L,t) = 0$$

where L, β and c are positive constants with β being much smaller than c (assume $\beta L < c\pi$ for the following).

a. Derive, as completely as possible, the formal series solution to the above system of equations. Since no initial conditions are given, your answer will contain arbitrary constants.

b. How rapidly do the vibrations die out?

c. How does the "β" term modify the frequencies at which the individual terms of the solution vibrate?

Part III

Classical Fourier Transforms

17

Heuristic Derivation of the Classical Fourier Transform

In the previous chapters, we developed some useful tools for dealing with periodic functions on the real line. The question now is whether we can extend the basic concepts already developed and obtain comparable tools for dealing with *nonperiodic* functions on the real line. Obviously, the answer is yes (otherwise, this would be a much shorter text), and, judging from the above title, this must be the chapter where that extension is done.

What we will actually derive here (with limited concern for rigor) are the two integral formulas on which the Fourier transforms are based, along with a fundamental relation between these two formulas. The basic idea behind the derivation is straightforward. We will take our nonperiodic function f and, for each $p > 0$, compute the Fourier series for a periodic function f_p having period p and equaling f over the interval $(-p/2, p/2)$. Then we will see what happens as $p \to \infty$.

Part of our derivation requires that we recognize a certain limit of a summation as being an integral over the real line. To prepare for that, we will first look at Riemann sums over the entire real line.

17.1 Riemann Sums over the Entire Real Line

In chapter 4 we discussed computing the integral of a function over a finite interval using a sequence of Riemann sums.[1] Though rarely done, a similar approach can be used to evaluate

$$\int_{-\infty}^{\infty} g(x)\,dx$$

where $g(x)$ is, say, a continuous function on $\mathbb{R}$. To ensure that the areas and infinite series arising in the following discussion are well defined and finite, we will assume $g(x)$ vanishes "sufficiently rapidly" as $x \to \pm\infty$. (Just what is "sufficiently rapid", however, will not concern us at this time.)

Let us first assume g is a real-valued function. That way, $\int_{-\infty}^{\infty} g(x)\,dx$ can be viewed as the net area between the graph of g and the X–axis. For each $\Delta x > 0$, we should be able to approximate this net area using the net area enclosed by the rectangles indicated in figure 17.1. We construct this approximation by

1.   first partitioning the entire real line into an infinite number of subintervals

$$\ldots \quad , \quad (x_{-1}, x_0) \quad , \quad (x_0, x_1) \quad , \quad (x_1, x_2) \quad , \quad (x_2, x_3) \quad , \quad \ldots$$

---

[1] This may be a good time to quickly review the subsection on well-defined integrals starting on page 37.

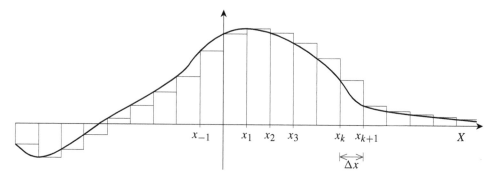

Figure 17.1: The graph of a real-valued function g and the rectangles for a Riemann sum
approximation of $\displaystyle\int_{-\infty}^{\infty} g(x)\,dx$.

where

$$x_k = k\Delta x \qquad \text{for} \quad k = 0, \pm 1, \pm 2, \ldots \quad ,$$

2.   then observing that

$$\int_{x_k}^{x_{k+1}} g(x)\,dx \approx g(x_k)\Delta x \qquad \text{for} \quad k = 0, \pm 1, \pm 2, \ldots \quad ,$$

3.   and, finally, adding up these approximations, getting

$$\int_{-\infty}^{\infty} g(x)\,dx = \sum_{k=-\infty}^{\infty} \int_{x_k}^{x_{k+1}} g(x)\,dx \approx \sum_{k=-\infty}^{\infty} g(x_k)\,\Delta x \quad .$$

This gives us the infinite series $\sum_{k=-\infty}^{\infty} g(x_k)\Delta x$ as a "Riemann sum" approximation for the integral
$\int_{-\infty}^{\infty} g(x)\,dx$. It is certainly reasonable to expect this approximation to improve as $\Delta x \to 0$. More
precisely, we should expect

$$\int_{-\infty}^{\infty} g(x)\,dx = \lim_{\Delta x \to 0^+} \sum_{k=-\infty}^{\infty} g(x_k)\,\Delta x \quad . \tag{17.1}$$

This assumes, of course, that the infinite series $\sum_{k=-\infty}^{\infty} g(x_k)\,\Delta x$ converges for each $\Delta x > 0$.
If, instead, g is complex valued with real and imaginary parts u and v,

$$g(x) = u(x) + iv(x) \quad ,$$

then, since u and v are real-valued functions, equation (17.1) can be used to find the corresponding
integrals of u and v. Thus,

$$\int_{-\infty}^{\infty} g(x)\,dx = \int_{-\infty}^{\infty} u(x)\,dx + i\int_{-\infty}^{\infty} v(x)\,dx$$

$$= \lim_{\Delta x \to 0^+} \sum_{k=-\infty}^{\infty} u(x_k)\,\Delta x + i\lim_{\Delta x \to 0^+} \sum_{k=-\infty}^{\infty} v(x_k)\,\Delta x$$

$$= \lim_{\Delta x \to 0^+} \sum_{k=-\infty}^{\infty} [u(x_k) + i v(x_k)]\Delta x$$

$$= \lim_{\Delta x \to 0^+} \sum_{k=-\infty}^{\infty} g(x_k)\, \Delta x \quad .$$

So we should also expect equation (17.1) to be valid whenever g is a complex-valued function on $\mathbb{R}$.

Keep in mind that we have *not* proven the validity of equation (17.1). To do that, we need to verify that all the various infinite series involved converge, and that the error between the approximations and the integral goes to zero as $\Delta x \to 0$. All we have done is to derive an equation (equation (17.1)) that we can reasonably suspect as being valid when the function g is "nice enough". Consequently, in the next section, when we get an expression of the form

$$\lim_{\Delta x \to 0^+} \sum_{k=-\infty}^{\infty} g(x_k)\Delta x \quad ,$$

where $x_k = k\Delta x$, we will feel reasonably confident — but not absolutely certain — that this expression can be replaced with

$$\int_{-\infty}^{\infty} g(x)\, dx \quad .$$

Naturally, whatever results we derive using this substitution will have to be rigorously justified, eventually.

17.2 The Derivation

Let f be some "sufficiently nice" function defined on the entire real line. We will not assume f is periodic, but, to simplify our derivation, we will assume the following:

1. $f(t)$ is smooth on the entire real line.

2.   All the following integrals and infinite series involving f are well defined and finite.

For each $p > 0$, let f_p be the corresponding periodic function with period p and which equals f over the interval $(-p/2, p/2)$,

$$f_p(t) = \begin{cases} f(t) & \text{if } -p/2 < t < p/2 \\ f_p(t-p) & \text{in general} \end{cases}$$

(see figure 17.2). Clearly, f_p is a periodic, piecewise smooth function that is continuous at every point between $-p/2$ and $p/2$. Thus, for all $-p/2 < t < p/2$,

$$f(t) = f_p(t) = \text{F.S.}\left[f_p\right]\big|_t = \sum_{k=-\infty}^{\infty} c_k\, e^{i2\pi \omega_k t} \tag{17.2}$$

where, for each integer k,

$$\omega_k = \frac{k}{p} \quad \text{and} \quad c_k = \frac{1}{p} \int_{-p/2}^{p/2} f_p(t)\, e^{-i2\pi \omega_k t}\, dt \quad .$$

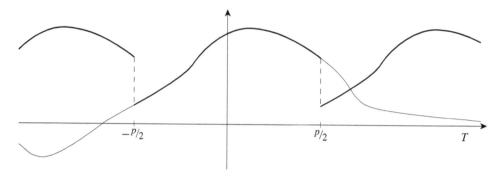

Figure 17.2: Graphs of a smooth function f on $\mathbb{R}$ (thin curve) and a periodic approximation f_p (thicker curve).

Letting

$$\Delta\omega = \frac{1}{p} \quad ,$$

and using the fact that $f(t) = f_p(t)$ when $-p/2 < t < p/2$, we can rewrite the formulas for ω_k and c_k as

$$\omega_k = k\,\Delta\omega \qquad \text{and} \qquad c_k = \Delta\omega \int_{-p/2}^{p/2} f(t)\,e^{-i2\pi\omega_k t}\,dt \quad .$$

Let us now define a function F_p by

$$F_p(\omega) = \int_{-p/2}^{p/2} f(t)\,e^{-i2\pi\omega t}\,dt \quad ,$$

and observe that the above formula for c_k can be written as

$$c_k = \Delta\omega\,F_p(\omega_k) \quad . \tag{17.3}$$

Combining equations (17.3) and (17.2) gives us

$$f(t) = \sum_{k=-\infty}^{\infty} F_p(\omega_k)\,e^{i2\pi\omega_k t}\,\Delta\omega \qquad \text{for} \quad -\frac{p}{2} < t < \frac{p}{2} \quad . \tag{17.4}$$

Looking back over our definitions, it should be clear that, as $p \to \infty$,

$$\text{``}-\frac{p}{2} < t < \frac{p}{2}\text{''} \;\to\; \text{``}-\infty < t < \infty\text{''} \quad ,$$

$$f_p(t) \;\to\; f(t) \quad ,$$

$$\Delta\omega \;\to\; 0$$

and

$$F_p(\omega) \;\to\; F(\omega)$$

where

$$F(\omega) = \int_{-\infty}^{\infty} f(t)\,e^{-i2\pi\omega t}\,dt \quad . \tag{17.5}$$

In taking the limit of equation (17.4), let us use the fact that $\Delta\omega \to 0$ as $p \to \infty$, and take the limits as $p \to \infty$ and $\Delta\omega \to 0$ separately:

$$f(t) = \lim_{p \to \infty} \sum_{k=-\infty}^{\infty} F_p(\omega_k) \, e^{i2\pi\omega_k t} \, \Delta\omega$$

$$= \lim_{\substack{p \to \infty \\ \Delta\omega \to 0}} \sum_{k=-\infty}^{\infty} F_p(\omega_k) \, e^{i2\pi\omega_k t} \, \Delta\omega$$

$$= \lim_{\Delta\omega \to 0} \sum_{k=-\infty}^{\infty} \lim_{p \to \infty} F_p(\omega_k) \, e^{i2\pi\omega_k t} \, \Delta\omega$$

$$= \lim_{\Delta\omega \to 0} \sum_{k=-\infty}^{\infty} F(\omega_k) \, e^{i2\pi\omega_k t} \, \Delta\omega \quad . \tag{17.6}$$

Comparing this last limit to the limit in equation (17.1) on page 252 (with $x = \omega$ and $g(x) = F(x) \, e^{i2\pi xt}$), we see that this last limit is simply a Riemann sum formula for the integral

$$\int_{-\infty}^{\infty} F(\omega) \, e^{i2\pi\omega t} \, d\omega \quad .$$

Thus, according to equation (17.1), equation (17.6) can be written as

$$f(t) = \int_{-\infty}^{\infty} F(\omega) \, e^{i2\pi\omega t} \, d\omega \qquad \text{for} \quad -\infty < t < \infty \quad . \tag{17.7}$$

17.3 Summary

Our goal was to extend the basic formulas for Fourier series to cases where the functions of interest are not periodic. What we obtained were formulas (17.5) and (17.7). And if you consider how formula (17.7) is related to formula (17.5), you will realize that we have actually derived (provided our many assumptions are valid) the following:

> If f is a "reasonably nice" function on $\mathbb{R}$, and if F is the function constructed from f by
>
> $$F(\omega) = \int_{-\infty}^{\infty} f(t) \, e^{-i2\pi\omega t} \, dt \quad , \tag{17.8}$$
>
> then the original function f can be recovered from F through the formula
>
> $$f(t) = \int_{-\infty}^{\infty} F(\omega) \, e^{i2\pi\omega t} \, d\omega \quad . \tag{17.9}$$

The two integrals in these formulas are called the *Fourier integrals* and will be the basis for much of the rest of our study. It is worth remembering that the first integral came directly from the formula for the Fourier coefficients for a periodic function, while the second came directly from the formula for reconstructing a periodic function from its Fourier coefficients (i.e., its Fourier series representation).

Also, don't forget that our derivation was not rigorous. We made many assumptions, including the nebulous " f is reasonably nice". To help pin down the meaning of "reasonably nice", we will introduce the concept of "absolute integrability" in the next chapter. After that, we will be able to properly start our development of the Fourier transform.

By the way, one of the things we will discover is that the derivation in this chapter is not only nonrigorous — it is misleading. Relatively few functions of interest are as "nice" as this derivation requires. Certainly, we do not want to restrict ourselves to only smooth functions that vanish "sufficiently rapidly" on the real line! Determining how to deal with these "less than sufficiently nice" functions will be one of our big challenges.

18

Integrals on Infinite Intervals

Throughout the rest of this book, a large part of our work will involve integrals over infinite intervals (usually the entire real line). While we could treat such integrals as limits of "infinite Riemann summations", as in the previous chapter, it is much more natural (and easier) to view them as limits of integrals over finite subintervals. For example, if our interval is $(-\infty, \infty)$, then

$$\int_{-\infty}^{\infty} f(x)\, dx \;=\; \lim_{\substack{b \to \infty \\ a \to -\infty}} \int_{a}^{b} f(x)\, dx \quad .$$

This requires, of course, that $\int_{a}^{b} f(x)\, dx$ exists for every finite interval (a, b) and that the above double limit exists.

Since these integrals will be so fundamental to our work, we had better discuss a few issues that could cause problems if we are not careful. The most pressing of these is determining when we can safely assume our integrals "make sense".

?► Exercise 18.1:   *Why does $\int_{-\infty}^{\infty} \cos(x)\, dx$ not make sense?*

18.1 Absolutely Integrable Functions
Definition

A function f is said to be *absolutely integrable* over an interval (α, β) if and only if we can legitimately write

$$\int_{\alpha}^{\beta} |f(x)|\, dx \;<\; \infty \quad . \tag{18.1}$$

This inequality certainly holds if (α, β) is a finite interval and f is piecewise continuous on the interval. On the other hand, if (α, β) is, say, $(-\infty, \infty)$ and the continuity of f is unknown, then, for inequality (18.1) to hold, we must have both

1. $\displaystyle \int_{a}^{b} |f(x)|\, dx$ being a well-defined integral for each finite subinterval (a, b), and

2.   the double limit

$$\lim_{\substack{b \to \infty \\ a \to -\infty}} \int_{a}^{b} |f(x)|\, dx$$

existing as a finite value.

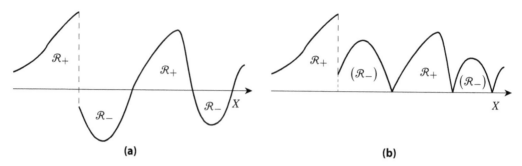

Figure 18.1: (a) The graph of some real-valued function f and (b) the graph of its absolute value.

For the next several chapters we will mainly be considering piecewise continuous functions on the real line; so the "well definition" of $\int_a^b |f(x)|\, dx$ will not be an issue. The finiteness of the above double limit, however, will be an important consideration.

Geometric Significance

It may be helpful to briefly review what absolute integrability means geometrically when f is a real-valued, piecewise continuous function over an interval (α, β). Recall that

$$\int_\alpha^\beta f(x)\, dx = \text{"net area" of the region enclosed by } x = \alpha,\ x = \beta,\ \text{the graph of } f(x)$$
$$\text{and the } X\text{-axis}$$

$$= \text{Area of region } \mathcal{R}_+ \ - \ \text{Area of region } \mathcal{R}_-$$

where (see figure 18.1a)

$$\mathcal{R}_+ = \text{the region with } \alpha < x < \beta \text{ bounded above by the graph of } f(x) \text{ and below by the } X\text{-axis}$$

and

$$\mathcal{R}_- = \text{the region with } \alpha < x < \beta \text{ bounded above by the } X\text{-axis and below by the graph of } f(x) \quad.$$

Since

$$|f(x)| = \begin{cases} f(x) & \text{if } f(x) \geq 0 \\ -f(x) & \text{if } f(x) < 0 \end{cases},$$

it should be clear (see figure 18.1b) that $|f|$ is also piecewise continuous on (α, β) and that

$$\int_\alpha^\beta |f(x)|\, dx = \text{Area of region } \mathcal{R}_+ \ + \ \text{Area of region } \mathcal{R}_-$$

$$= \text{total area of the region enclosed by } x = \alpha,\ x = \beta,\ \text{the graph of } f(x) \text{ and the } X\text{-axis} \quad.$$

So f being absolutely integrable over (α, β) is equivalent to the above total area being finite.

The situation is especially simple if (α, β) is a *finite* interval. From the above it should be clear that, whenever f is a real-valued, piecewise continuous function on a finite interval (α, β):

1. The integrals $\int_\alpha^\beta f(x)\, dx$ and $\int_\alpha^\beta |f(x)|\, dx$, being integrals of piecewise continuous functions over a finite interval, automatically exist and are finite.

2. Hence, f is automatically absolutely integrable on (α, β).

3. Moreover,

$$\left| \int_{\alpha}^{\beta} f(x)\,dx \right| = |\text{Area of region } \mathcal{R}_+ \; - \; \text{Area of region } \mathcal{R}_-|$$

$$\leq \; \text{Area of region } \mathcal{R}_+ \; + \; \text{Area of region } \mathcal{R}_-$$

$$= \int_{\alpha}^{\beta} |f(x)|\,dx \quad .$$

The situation is less simple if (α, β) is an infinite interval (i.e., $\alpha = -\infty$ and/or $\beta = \infty$). Then the above total area can be infinite, and thus, f might not be absolutely integrable over the interval.

For f to be absolutely integrable on an infinite interval (α, β), the areas of regions $\mathcal{R}_+$ and $\mathcal{R}_-$ must be well defined and finite. Think about this for a moment; it means that, in some sense, $f(x)$ must "shrink to zero fairly rapidly as x gets large". This is illustrated, for example, in exercise 18.10 at the end of this chapter.[1] Of course, if the areas of regions $\mathcal{R}_+$ and $\mathcal{R}_-$ are well defined and finite, then the integral of f over (α, β), being

$$\int_{\alpha}^{\beta} f(x)\,dx = \text{Area of region } \mathcal{R}_+ \; - \; \text{Area of region } \mathcal{R}_- \quad ,$$

is well defined and finite. Moreover, we still have

$$\left| \int_{\alpha}^{\beta} f(x)\,dx \right| = |\text{Area of region } \mathcal{R}_+ \; - \; \text{Area of region } \mathcal{R}_-|$$

$$\leq \; \text{Area of region } \mathcal{R}_+ \; + \; \text{Area of region } \mathcal{R}_-$$

$$= \int_{\alpha}^{\beta} |f(x)|\,dx \quad .$$

On the other hand, if this real-valued, piecewise continuous function f is not absolutely integrable over an infinite interval (α, β), then the area of $\mathcal{R}_+$ or $\mathcal{R}_-$ (or both) must be infinite. In this case, the only way

$$\int_{\alpha}^{\beta} f(x)\,dx = \lim_{\substack{a \to \alpha \\ b \to \beta}} \int_{a}^{b} f(x)\,dx$$

can converge to a finite number is for the areas of the regions above and below the X–axis to just happen to "cancel out" each other as the limits are computed. This is a very unstable type of convergence and can be grossly affected by any manipulation that affects how these cancellations occur.[2] This is illustrated in exercises 18.14 and 18.15.

Some Examples

!▶ *Example 18.1 (the rectangle function):*   The rectangle function *over the interval* (a, b), denoted by $\text{rect}_{(a,b)}$, is given by

$$\text{rect}_{(a,b)}(x) = \begin{cases} 1 & \text{if } \; a < x < b \\ 0 & \text{otherwise} \end{cases} \quad .$$

---

[1] However, $f(x)$ does not have to *steadily* shrink to zero or even be bounded! That is illustrated in exercise 18.16.

[2] We could say that such integrals are *conditionally integrable*. Recall the distinction between absolutely convergent and conditionally convergent infinite series, as well as the difficulties with conditionally convergent series. The situation here with integrals is completely analogous.

Since this is a nonnegative function, $\left|\text{rect}_{(a,b)}(x)\right| = \text{rect}_{(a,b)}(x)$. If a and b are finite, then

$$\int_{-\infty}^{\infty} \left|\text{rect}_{(a,b)}(x)\right| \, dx = \int_{-\infty}^{a} 0 \, dx + \int_{a}^{b} 1 \, dx + \int_{b}^{\infty} 0 \, dx = b - a < \infty \quad .$$

This shows that $\text{rect}_{(a,b)}$ is absolutely integrable over $\mathbb{R}$ whenever a and b are finite. On the other hand, if $b = \infty$, then

$$\int_{-\infty}^{\infty} \left|\text{rect}_{(a,\infty)}(x)\right| \, dx = \int_{-\infty}^{a} 0 \, dx + \int_{a}^{\infty} 1 \, dx = \infty \quad .$$

So $\text{rect}_{(a,\infty)}$ is not absolutely integrable on $\mathbb{R}$.

!► **Example 18.2:**   Consider $e^{(-a+ib)x} \, \text{step}(x)$, where a and b are two real numbers with $a > 0$ and step is the step function,

$$\text{step}(x) = \text{rect}_{(0,\infty)}(x) = \begin{cases} 1 & \text{if } 0 < x \\ 0 & \text{if } x < 0 \end{cases} \quad .$$

Noting that

$$\left|e^{(-a+ib)x}\right| = \left|e^{-ax} e^{ibx}\right| = \left|e^{-ax}\right| \left|e^{ibx}\right| = e^{-ax} \cdot 1 \quad ,$$

we see that

$$\int_{-\infty}^{\infty} \left|e^{(-a+ib)x} \, \text{step}(x)\right| \, dx = \int_{-\infty}^{0} 0 \, dx + \int_{0}^{\infty} \left|e^{(-a+ib)x}\right| \, dx$$

$$= \int_{0}^{\infty} e^{-ax} \, dx$$

$$= -\frac{1}{a} e^{-ax} \Big|_{0}^{\infty}$$

$$= \lim_{x \to \infty} \frac{-1}{a} e^{-ax} - \frac{-1}{a} e^{-a \cdot 0}$$

$$= 0 + \frac{1}{a} < \infty \quad .$$

So $e^{(-a+ib)x} \, \text{step}(x)$, with $a > 0$ and b real, is absolutely integrable on the real line.

?► **Exercise 18.2:**   Show that $e^{(a+ib)x} \, \text{step}(x)$, with $a > 0$ and b real, is not absolutely integrable on the real line.

!► **Example 18.3:**   Consider the sine function over the entire real line. Rather than trying to compute

$$\int_{-\infty}^{\infty} |\sin(x)| \, dx \quad ,$$

just look at the graph of $|\sin(x)|$ (figure 18.2). Since the total area under this graph is certainly not finite, it is clear that the sine function is not absolutely integrable over the entire real line.

?► **Exercise 18.3:**   Convince yourself that no periodic, piecewise continuous function (other than the zero function) can be absolutely integrable over $\mathbb{R}$.

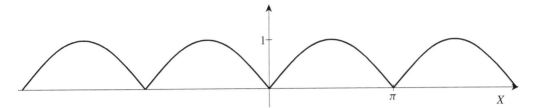

Figure 18.2: Graph of $|\sin(x)|$.

18.2   The Set of Absolutely Integrable Functions

We will often need to assume that our functions are both piecewise continuous and absolutely integrable on an interval (α, β).[3] To help avoid constantly rewriting "piecewise continuous and absolutely integrable on the interval (α, β)", let us agree to denote by $\mathcal{A}[(\alpha, \beta)]$ the set of all functions that are both piecewise continuous and absolutely integrable on the interval (α, β),

$$\mathcal{A}[(\alpha, \beta)] \;=\; \left\{ f : f \text{ is piecewise continuous on } (\alpha, \beta) \text{ and } \int_{\alpha}^{\beta} |f(x)|\, dx < \infty \right\} \quad .$$

This will allow us to use the phrase " f is in $\mathcal{A}[(\alpha, \beta)]$ " as shorthand for the phrase " f is piecewise continuous and absolutely integrable on the interval (α, β)."

If no interval (α, β) is explicitly given, then (α, β) should be assumed to be the entire real line. That is,

$$\mathcal{A} \;=\; \mathcal{A}[\mathbb{R}] \;=\; \mathcal{A}[(-\infty, \infty)] \quad .$$

18.3   Many Useful Facts

The following lemmas give a number of useful little facts concerning absolutely integrable functions. All of them will be used one way or another later on.

By the way, we are not going to rederive all those elementary formulas that follow immediately from treating an integral over an infinite interval as a limit of integrals over finite subintervals. For example, if a and b are any pair of constants and

$$\int_{-\infty}^{\infty} f(x)\, dx \qquad \text{and} \qquad \int_{-\infty}^{\infty} g(x)\, dx$$

are known to be well-defined finite integrals (i.e., the limits

$$\lim_{\substack{b \to \infty \\ a \to -\infty}} \int_{a}^{b} f(x)\, dx \qquad \text{and} \qquad \lim_{\substack{b \to \infty \\ a \to -\infty}} \int_{a}^{b} g(x)\, dx$$

---

[3] It is the assumption of absolute integrability that is most important. The assumption of piecewise continuity in most of the following lemmas can be replaced by just about any other assumption ensuring the existence of the necessary integrals over finite intervals. In particular, those acquainted with the Lebesgue theory of integration should try replacing any assumption of a function being "piecewise continuous" with the more general assumption of the function being "bounded and measurable on each finite subinterval".

exist and are finite), then it will be assumed that you realize $\int_{-\infty}^{\infty} [af(x)+bg(x)] \, dx$ is a well-defined finite integral with

$$\int_{-\infty}^{\infty} [af(x) + bg(x)] \, dx \; = \; a \int_{-\infty}^{\infty} f(x) \, dx \; + \; b \int_{-\infty}^{\infty} g(x) \, dx \quad .$$

Tests for Absolute Integrability

The basic way of testing whether a given function f is absolutely integrable on (α, β) is to simply evaluate

$$\int_{\alpha}^{\beta} |f(x)| \, dx \quad ,$$

and see if you get a finite number. Often, however, it is easier to use one of the following lemmas.

The first lemma is simply the reiteration of the fact that, if f is piecewise continuous on a finite interval (α, β), then so is $|f|$, and thus, the integral of $|f|$ over (α, β) exists and is finite.

Lemma 18.1
If a function is piecewise continuous on a finite interval, then that function is absolutely integrable on that interval.

The next two lemmas give the integral analogs of two tests for the convergence of infinite series: the bounded partial sums test on page 43 and the comparison test on page 43. Since the following can be proven in much the same way as the corresponding infinite series versions, and since the proofs of the infinite series versions can be found in most calculus texts, we'll leave the proofs as exercises for the interested reader.

Lemma 18.2 (bounded integrals test)
Let f be a function defined on an interval (α, β), and suppose there is a finite constant M such that, for every interval (a, b) with $\alpha < a < b < \beta$, $\int_{a}^{b} |f(x)| \, dx$ exists and

$$\int_{a}^{b} |f(x)| \, dx \; < \; M \quad .$$

Then f is absolutely integrable on (α, β).

Lemma 18.3 (comparison test)
Let f and g be two piecewise continuous functions on the interval (α, β), and assume that, on this interval,

$$|f(x)| \; \leq \; |g(x)| \quad .$$

Then

$$\int_{\alpha}^{\beta} |f(x)| \, dx \; \leq \; \int_{\alpha}^{\beta} |g(x)| \, dx \quad ,$$

and consequently:

1. *If g is in $\mathcal{A}[(\alpha, \beta)]$, then so is f.*

2. *If f is not in $\mathcal{A}[(\alpha, \beta)]$, then neither is g.*

?►Exercise 18.4:   *Prove*

  a: *lemma 18.2.*   **b:** *lemma 18.3.*

!►Example 18.4:   *Consider the piecewise continuous function*

$$f(x) = \sin(bx)\, e^{-ax}\, \text{step}(x)$$

where a *and* b *are two positive real numbers. From example 18.2, we know*

$$\int_{-\infty}^{\infty} \left| e^{-ax}\, \text{step}(x) \right| dx < \infty \quad .$$

This and the fact that $g(x) = e^{-ax}\, \text{step}(x)$ *is piecewise continuous means that* g *is in* $\mathcal{A}$. *Clearly, also,*

$$\left| \sin(bx)\, e^{-ax}\, \text{step}(x) \right| \leq \left| e^{-ax}\, \text{step}(x) \right| \qquad for \quad -\infty < x < \infty \quad .$$

So the comparison test (lemma 18.3) assures us that

$$\int_{-\infty}^{\infty} \left| \sin(bx)\, e^{-ax}\, \text{step}(x) \right| dx \leq \int_{-\infty}^{\infty} \left| e^{-ax}\, \text{step}(x) \right| dx < \infty \quad .$$

Thus $\sin(bx)\, e^{-ax}\, \text{step}(x)$ *is in* $\mathcal{A}$.

   The next can be thought of as a "limit comparison" test, and uses the observation that, when $\alpha > 1$ and $X > 0$,

$$\int_{X}^{\infty} \frac{1}{x^{\alpha}}\, dx = \lim_{x \to \infty} \frac{1}{(1-\alpha)x^{\alpha-1}} - \frac{1}{(1-\alpha)X^{\alpha-1}} = 0 + \frac{1}{(\alpha-1)X^{\alpha-1}} < \infty \quad .$$

Lemma 18.4 (a limit comparison test)
Let f *be a piecewise continuous function on* $\mathbb{R}$. *If there is a real constant* $\alpha > 1$ *such that*

$$\lim_{x \to \pm\infty} |x|^{\alpha}\, f(x) = 0 \quad ,$$

then f *is in* $\mathcal{A}$.

PROOF: Because $|x|^{\alpha}\, f(x) \to 0$ as $x \to \pm\infty$, there must be a finite positive X such that

$$|x|^{\alpha}\, f(x) < 1 \qquad whenever \quad X \leq |x| \quad .$$

Now define

$$g(x) = \begin{cases} f(x) & \text{if} \quad |x| < X \\ |x|^{-\alpha} & \text{if} \quad X \leq |x| \end{cases} \quad .$$

Observe that $|f(x)| \leq |g(x)|$ for every x in $\mathbb{R}$. So, using the comparison test,

$$\int_{-\infty}^{\infty} |f(x)|\, dx \leq \int_{-\infty}^{\infty} |g(x)|\, dx$$

$$= \int_{-\infty}^{-X} |x|^{-\alpha}\, dx + \int_{-X}^{X} |f(x)|\, dx + \int_{X}^{\infty} x^{-\alpha}\, dx$$

$$= \frac{1}{(\alpha-1)X^{\alpha-1}} + \int_{-X}^{X} |f(x)|\, dx + \frac{1}{(\alpha-1)X^{\alpha-1}} < \infty \quad .$$

!▶ Example 18.5 (Gaussian functions):   Consider $f(x) = e^{-\gamma x^2}$ where γ is any positive real number. Using L'Hôpital's rule and the fact that $e^{-s} \to 0$ as $s \to \infty$, we see that

$$\lim_{x \to \pm\infty} |x|^2 |f(x)| = \lim_{x \to \pm\infty} \frac{x^2}{e^{\gamma x^2}} = \lim_{x \to \pm\infty} \frac{2x}{2\gamma x e^{\gamma x^2}} = \lim_{x \to \pm\infty} \frac{e^{-\gamma x^2}}{\gamma} = 0 \ .$$

Thus, according to lemma 18.4 (with $\alpha = 2$), $e^{-\gamma x^2}$ is in $\mathcal{A}$.

?▶ Exercise 18.5:   Show that $x^n e^{-\gamma x^2}$ is in $\mathcal{A}$ if $\gamma > 0$ and n is a nonnegative integer.

?▶ Exercise 18.6:   Verify that

$$\frac{1}{a^2 + 4\pi^2 x^2}$$

is absolutely integrable whenever a is a nonzero real number.

The next lemma can often simplify the task of verifying whether a given complex-valued function is or is not absolutely integrable.

Lemma 18.5
Let u and v be, respectively, the real and imaginary parts of a complex-valued function f on an interval (α, β). Then f is in $\mathcal{A}[(\alpha, \beta)]$ if and only if both u and v are in $\mathcal{A}[(\alpha, \beta)]$.

PROOF: First of all, we already know that f is piecewise continuous if and only if both u and v are piecewise continuous. So all we need to show is that f is absolutely integrable if and only if both u and v are absolutely integrable.

Suppose f is absolutely integrable on (α, β). Then, since u and v are the real and imaginary parts of f,
$$|u(x)| \leq |f(x)| \qquad \text{and} \qquad |v(x)| \leq |f(x)|$$
for every x in (α, β) (see inequality set (6.1) on page 58). Hence,

$$\int_\alpha^\beta |u(x)|\, dx \leq \int_\alpha^\beta |f(x)|\, dx < \infty$$

and

$$\int_\alpha^\beta |v(x)|\, dx \leq \int_\alpha^\beta |f(x)|\, dx < \infty \ .$$

On the other hand, the triangle inequality assures us that
$$|f(x)| = |u(x) + iv(x)| \leq |u(x)| + |v(x)|$$
for every x in (α, β). So, if u and v are absolutely integrable on (α, β), then

$$\int_\alpha^\beta |f(x)|\, dx \leq \int_\alpha^\beta [|u(x)| + |v(x)|]\, dx$$

$$= \int_\alpha^\beta |u(x)|\, dx + \int_\alpha^\beta |v(x)|\, dx < \infty \ . \qquad \blacksquare$$

?▶ Exercise 18.7:   Show that

$$\frac{1}{1 + i2\pi x}$$

is not absolutely integrable on the real line.

Absolute Integrability and the Integral

The next lemma extends the observations made in the previous section concerning the geometric significance of absolute integrability when f is real valued.

Lemma 18.6
If f is in $\mathcal{A}[(\alpha, \beta)]$, then $\int_\alpha^\beta f(x)\,dx$ exists and is finite. Moreover,

$$\left| \int_\alpha^\beta f(x)\,dx \right| \le \int_\alpha^\beta |f(x)|\,dx \quad . \tag{18.2}$$

PROOF: We've already seen in the previous section that this lemma's claim holds when f is real valued.

Suppose, now, that f is complex valued with real and imaginary parts u and v, respectively. Lemma 18.5 assures us that these two real-valued functions (u and v) are absolutely integrable on (α, β). Hence, since u and v are real valued, we know

$$\int_\alpha^\beta u(x)\,dx \qquad \text{and} \qquad \int_\alpha^\beta v(x)\,dx$$

exist and are finite real values. Clearly then, so is the corresponding integral of f. In fact,

$$\int_\alpha^\beta f(x)\,dx \;=\; \int_\alpha^\beta [u(x) + iv(x)]\,dx \;=\; \int_\alpha^\beta u(x)\,dx \;+\; i \int_\alpha^\beta v(x)\,dx \quad .$$

Finally, recall that inequality (18.2) has already been verified for the case where (α, β) is a finite interval (in section 6.2 starting on page 59). Thus, if (α, β) is, say, $(-\infty, \infty)$, then

$$\left| \int_{-\infty}^{\infty} f(x)\,dx \right| \;=\; \lim_{\substack{b\to\infty \\ a\to-\infty}} \left| \int_a^b f(x)\,dx \right|$$

$$\le\; \lim_{\substack{b\to\infty \\ a\to-\infty}} \int_a^b |f(x)|\,dx \;=\; \int_{-\infty}^{\infty} |f(x)|\,dx \quad ,$$

confirming inequality (18.2) when (α, β) is $(-\infty, \infty)$. Obviously, similar computations will confirm the inequality when (α, β) is any other infinite interval. ∎

Constructing Absolutely Integrable Functions

In our work, we will find ourselves manipulating absolutely integrable functions. The following lemmas will assure us that the results of many of our manipulations will also be absolutely integrable.

Lemma 18.7
Suppose f is in $\mathcal{A}$, and let γ be any fixed nonzero real number. Then the functions given by $f(x - \gamma)$ and $f(\gamma x)$ are in $\mathcal{A}$.

PROOF: As noted in chapter 3, $f(x - \gamma)$ and $f(\gamma x)$ are piecewise continuous functions of x whenever $f(x)$ is. Hence we need only show the absolute integrability, which is easily verified using the well-known substitutions $\sigma = x - \gamma$ and $\tau = \gamma x$, and the fact that f is absolutely integrable on $\mathbb{R}$:

$$\int_{-\infty}^{\infty} |f(x - \gamma)|\, dx = \int_{-\infty}^{\infty} |f(\sigma)|\, d\sigma < \infty$$

and

$$\int_{-\infty}^{\infty} |f(\gamma x)|\, dx = \frac{1}{|\gamma|} \int_{-\infty}^{\infty} |f(\tau)|\, d\tau < \infty \quad . \qquad \blacksquare$$

!▶Example 18.6:   Let

$$g(x) = e^{(a+ib)x} \, \text{step}(-x) \qquad \text{and} \qquad f(x) = e^{(-a-ib)x} \, \text{step}(x)$$

where a and b are two real numbers with $a > 0$ and step is the step function (from exercise 18.2 on page 260). Observe that, with $\gamma = -1$,

$$g(x) = e^{(-a-ib)(-x)} \, \text{step}(-x) = f(-x) = f(\gamma x) \quad .$$

Since f was shown to be in $\mathcal{A}$ in example 18.2 on page 260, lemma 18.7 assures us that g is also in $\mathcal{A}$.

Lemma 18.8
Any finite linear combination of functions in $\mathcal{A}[(\alpha, \beta)]$ is also in $\mathcal{A}[(\alpha, \beta)]$ (i.e., $\mathcal{A}[(\alpha, \beta)]$ is a linear space of functions).

PROOF: Let f be any linear combination of functions in $\mathcal{A}[(\alpha, \beta)]$, say,

$$f = c_1 f_1 + c_2 f_2 + \cdots + c_N f_N$$

where N is some positive integer, the c_k's are constants, and the f_k's are functions in $\mathcal{A}[(\alpha, \beta)]$. Being a linear combination of piecewise continuous functions on (α, β), f must also be piecewise continuous on (α, β). And, using the triangle inequality, we see that

$$\int_{\alpha}^{\beta} |f(x)|\, dx \leq \int_{\alpha}^{\beta} \left[|c_1 f_1(x)| + |c_2 f_2(x)| + \cdots + |c_N f_N(x)| \right] dx$$

$$= |c_1| \int_{\alpha}^{\beta} |f_1(x)|\, dx + |c_2| \int_{\alpha}^{\beta} |f_2(x)|\, dx + \cdots + |c_N| \int_{\alpha}^{\beta} |f_N(x)|\, dx$$

$$< \infty \quad . \qquad \blacksquare$$

!▶Example 18.7:   Let $\alpha > 0$. Observe that

$$e^{-\alpha|x|} = \begin{cases} e^{\alpha x} & \text{if } x < 0 \\ e^{-\alpha x} & \text{if } 0 < x \end{cases} = e^{\alpha x} \, \text{step}(-x) + e^{-\alpha x} \, \text{step}(x) \quad .$$

From examples 18.2 and 18.6 we know that $e^{\alpha x} \, \text{step}(-x)$ and $e^{-\alpha x} \, \text{step}(x)$ are in $\mathcal{A}$. Lemma 18.8 then assures us that their sum, $e^{-\alpha|x|}$, is also in $\mathcal{A}$.

Lemma 18.9
Let f be in $\mathcal{A}[(\alpha, \beta)]$, and assume g is a bounded, piecewise continuous function on (α, β). Then the product fg is in $\mathcal{A}[(\alpha, \beta)]$.

PROOF: Since g is bounded, there is a finite value M such that

$$|g(x)| \leq M \qquad \text{for} \quad \alpha < x < \beta \quad .$$

Thus, because f is assumed to be absolutely integrable,

$$\int_\alpha^\beta |f(x)g(x)| \, dx \leq M \int_\alpha^\beta |f(x)| \, dx < \infty \quad .$$

This, and the fact that products of piecewise continuous functions are piecewise continuous, tells us that fg is in $\mathcal{A}[(\alpha, \beta)]$. ∎

The following corollary will be of special interest to us.

Corollary 18.10
Let α be any real number. If f is in $\mathcal{A}$, then so are the functions

$$f(x) \, e^{i2\pi\alpha x} \qquad \text{and} \qquad f(x) \, e^{-i2\pi\alpha x} \quad .$$

A Limit Lemma

The last lemma will be used on occasion in some proofs. It helps describe how an absolutely integrable function $f(x)$ on $\mathbb{R}$ must "shrink to zero" as $x \to \pm\infty$.

Lemma 18.11
Suppose f is in $\mathcal{A}$. For each $\epsilon > 0$, there is then a finite positive length ℓ_ϵ such that

$$\int_b^\infty |f(x)| \, dx \leq \tfrac{1}{2}\epsilon \quad , \qquad \int_{-\infty}^a |f(x)| \, dx \leq \tfrac{1}{2}\epsilon$$

and

$$0 \leq \int_{-\infty}^\infty |f(x)| \, dx - \int_a^b |f(x)| \, dx \leq \epsilon$$

whenever $a \leq -\ell_\epsilon$ and $\ell_\epsilon \leq b$.

PROOF: Because of the way we define integrals on infinite intervals,

$$\lim_{b \to \infty} \int_b^\infty |f(x)| \, dx = \lim_{b \to \infty} \left[\int_0^\infty |f(x)| \, dx - \int_0^b |f(x)| \, dx \right] = 0 \quad .$$

This means that, for each positive value ρ, there is a finite positive number B_ρ such that

$$\int_b^\infty |f(x)| \, dx \leq \rho \qquad \text{whenever} \quad B_\rho \leq b \quad .$$

Likewise, for each $\rho > 0$, there is a finite positive number A_ρ such that

$$\int_{-\infty}^a |f(x)| \, dx \leq \rho \qquad \text{whenever} \quad a \leq -A_\rho \quad .$$

Consequently,

$$0 \leq \int_{-\infty}^{\infty} |f(x)| \, dx \; - \; \int_{a}^{b} |f(x)| \, dx$$

$$= \int_{-\infty}^{a} |f(x)| \, dx \; + \; \int_{b}^{\infty} |f(x)| \, dx \; \leq \; \rho + \rho \; = \; 2\rho$$

whenever $a \leq -A_\rho$ and $B_\rho \leq b$. These inequalities then immediately give the inequalities of the lemma after taking ℓ_ϵ to be the larger of A_ρ and B_ρ with $\rho = \epsilon/2$. ∎

18.4 Functions with Two Variables[*]

With the obvious modifications, the basic ideas and results just developed can be extended to apply to functions of two (or more) variables. This, in turn, will allow us to extend many of the results involving integrals of functions over bounded intervals and rectangles from chapter 7 to corresponding results involving integrals of functions over unbounded intervals and rectangles.[4] These results are mainly concerned with the continuity and differentiation of certain integrals, and with the interchanging of the order of integration in double integrals on unbounded rectangles. They will be of special interest to us because many of the most useful formulas and properties in the theory and application of Fourier transforms can be derived as special cases of the more general results discussed here. Proving them here, in fairly general form, will save us from proving several variations of each later on.

Basic Extensions

If $f(x, y)$ is a function of two variables on an unbounded region $\mathcal{R}$, then the double integral of f over $\mathcal{R}$ is defined by

$$\iint_{\mathcal{R}} f(x, y) \, dA \; = \; \lim_{\substack{a \to -\infty \\ b \to \infty \\ c \to -\infty \\ d \to \infty}} \iint_{\mathcal{R}_{abcd}} f(x, y) \, dA$$

where $\mathcal{R}_{abcd}$ denotes the intersection of $\mathcal{R}$ with the rectangle $(a, b) \times (c, d)$. This requires, of course, that the above integral over $\mathcal{R}_{abcd}$ exists for all intervals (a, b) and (c, d), and that the quadruple limit exists.

A function of two variables $f(x, y)$ is *absolutely integrable* over a region $\mathcal{R}$ of the plane if and only if

$$\iint_{\mathcal{R}} |f(x, y)| \, dA$$

exists and is finite. Geometrically, f is absolutely integrable if and only if the total volume of the solid region above $\mathcal{R}$ in the plane and below the surface $z = |f(x, y)|$ is finite.

The set of all piecewise continuous, absolutely integrable functions over $\mathcal{R}$ will be denoted by $\mathcal{A}[\mathcal{R}]$.

---

[*] The material in this section, as in chapter 7, will not be needed for a while. It probably won't hurt if you delay reading it until we start referring to it.

[4] Before starting this section, you may want to review at least the first part of chapter 7.

Analogs to all the lemmas previously developed in this chapter for functions of one variable can also be derived for functions of two variables. We'll list a few, and let the reader convince him- or herself of their validity.

Lemma 18.12
Any piecewise continuous function on a bounded region is also absolutely integrable on that region.

Lemma 18.13
Let f be a piecewise continuous function on some region $\mathcal{R}$, and suppose there is a finite constant M such that, for every bounded subregion $\mathcal{R}_0$ of $\mathcal{R}$,

$$\iint_{\mathcal{R}_0} |f(x, y)| \, dA \leq M \quad .$$

Then f is absolutely integrable on $\mathcal{R}$.

Lemma 18.14
Let f and g be two piecewise continuous functions on a region $\mathcal{R}$, and assume that, on this region,

$$|f(x, y)| \leq |g(x, y)| \quad .$$

Then

$$\iint_{\mathcal{R}} |f(x, y)| \, dA \leq \iint_{\mathcal{R}} |g(x, y)| \, dA \quad ,$$

and thus:

1. *If g is in $\mathcal{A}[\mathcal{R}]$, so is f.*

2. *If f is not in $\mathcal{A}[\mathcal{R}]$, neither is g.*

Lemma 18.15
Any linear combination of functions in $\mathcal{A}[\mathcal{R}]$ is a function in $\mathcal{A}[\mathcal{R}]$, as is the product of any function in $\mathcal{A}[\mathcal{R}]$ with any bounded, piecewise continuous function on $\mathcal{R}$.

Lemma 18.16
If f is in $\mathcal{A}[\mathcal{R}]$ for some region $\mathcal{R}$, then $\iint_{\mathcal{R}} f(x, y) \, dA$ exists and is finite. Moreover,

$$\left| \iint_{\mathcal{R}} f(x, y) \, dA \right| \leq \iint_{\mathcal{R}} |f(x, y)| \, dA \quad .$$

Lemma 18.17
Let f be in $\mathcal{A}[\mathcal{R}]$ for some region $\mathcal{R}$. For each $\epsilon > 0$, there is a finite positive length ℓ_ϵ such that, whenever $a \leq -\ell_\epsilon$, $c \leq -\ell_\epsilon$, $\ell_\epsilon \leq b$ and $\ell_\epsilon \leq d$,

$$0 \leq \iint_{\mathcal{R}} |f(x, y)| \, dA - \iint_{\mathcal{R}_{abcd}} |f(x, y)| \, dA \leq \epsilon \quad ,$$

where $\mathcal{R}_{abcd}$ is the intersection of $\mathcal{R}$ with the rectangle $(a, b) \times (c, d)$.

Functions on Unbounded Rectangles

Most, if not all, of our functions of two variables will be defined over rectangles in the plane. Since we've already discussed piecewise continuous functions on bounded rectangles in chapter 7, we will spend the rest of this chapter seeing how the discussion and results from that chapter extend when the rectangles are unbounded. In particular, our development of Fourier transforms will be greatly simplified by using the results developed here concerning the continuity and differentiation of functions of the form

$$\psi(x) = \int_{-\infty}^{\infty} f(x, y)\, dy \quad,$$

along with the results developed here concerning the interchanging of the order of integration.

Unfortunately, while our discussion in the next section will parallel that in sections 7.2 and 7.3, the conditions we will have to impose on $f(x, y)$ will not be as simple as imposed in those earlier sections (mainly, piecewise continuity). This is because "infinities" can easily be introduced when integrating piecewise continuous functions over infinite intervals. This can even happen when the function being integrated is absolutely integrable on $\mathbb{R}^2$ (see exercise 18.18). To help ensure this does not happen, we will often insist that our functions satisfy some sort of "uniform absolute integrability" requirement.

So let's see what "uniform absolute integrability" is.

Uniform Absolute Integrability on Strips

A rectangle $\mathcal{R} = (a, b) \times (c, d)$ will be called a (*thin*) *strip* if one of these intervals is finite and the other is infinite. Since it will simplify the exposition, we will limit the following discussion to strips of the form $(a, b) \times (-\infty, \infty)$, although it should be obvious that similar results apply for functions defined on other thin strips.

It should be noted that, when $\mathcal{R}$ is the strip $(a, b) \times (-\infty, \infty)$, the definition of the integral of f over $\mathcal{R}$ reduces to

$$\iint_{\mathcal{R}} f(x, y)\, dA = \lim_{\substack{c \to -\infty \\ d \to \infty}} \iint_{\mathcal{R}_{cd}} f(x, y)\, dA$$

where $\mathcal{R}_{cd}$ denotes the bounded rectangle $\mathcal{R}_{cd} = (a, b) \times (c, d)$.

A slightly stronger version of "absolute integrability" will be needed to ensure that

$$\int_{-\infty}^{\infty} f(x, y)\, dy$$

is "well behaved" as a function of x. Accordingly, we define $f(x, y)$ to be *uniformly absolutely integrable* on the strip $\mathcal{R} = (a, b) \times (-\infty, \infty)$ if and only if there is a piecewise continuous and absolutely integrable function f_0 of one variable on $(-\infty, \infty)$ such that, on $\mathcal{R}$,

$$|f(x, y)| \leq |f_0(y)| \quad.$$

If f_0 is such a function, then, from the discussion in chapter 7 (see, specifically, theorem 7.11 on page 84), we know that, for every finite interval (c, d),

$$\iint_{\mathcal{R}_{cd}} |f_0(y)|\, dA = \int_c^d \int_a^b |f_0(y)|\, dx\, dy = (b - a) \int_c^d |f_0(y)|\, dy \quad.$$

So, if f is a piecewise continuous function on the strip and $|f(x, y)| \leq |f_0(y)|$, then

$$\iint_{\mathcal{R}} |f(x, y)| \, dA = \lim_{\substack{c \to -\infty \\ d \to \infty}} \iint_{\mathcal{R}_{cd}} |f(x, y)| \, dA$$

$$\leq \lim_{\substack{c \to -\infty \\ d \to \infty}} \iint_{\mathcal{R}_{cd}} |f_0(y)| \, dA$$

$$= \lim_{\substack{c \to -\infty \\ d \to \infty}} (b - a) \int_c^d |f_0(y)| \, dy$$

$$= (b - a) \int_{-\infty}^{\infty} |f_0(y)| \, dy < \infty \quad .$$

Thus, any uniformly absolutely integrable function on a strip is also just plain absolutely integrable on the strip.

Single Integrals of Functions with Two Variables
Continuity of Functions Defined by Integrals

Our first theorem requiring uniform absolute integrability is an analog to theorem 7.7 on the continuity of a single integral of a function with two variables (see page 80). The necessity of this requirement (or something similar) is illustrated in exercise 18.17 at the end of this chapter.

Theorem 18.18
Let (a, b) be a finite interval, and let $f(x, y)$ be piecewise continuous and uniformly absolutely integrable on the strip $(a, b) \times (-\infty, \infty)$. Assume further that, on each bounded subrectangle $\mathcal{R}$ of this strip, all the discontinuities of f in $\mathcal{R}$ are contained in a finite number of straight lines. Then

$$\psi(x) = \int_{-\infty}^{\infty} f(x, y) \, dy$$

is a piecewise continuous function on (a, b). Moreover, if $a < x_0 < b$ and $x = x_0$ is not a line of discontinuity for f, then ψ is continuous at x_0 and

$$\lim_{x \to x_0} \psi(x) = \int_{-\infty}^{\infty} \lim_{x \to x_0} f(x, y) \, dy = \int_{-\infty}^{\infty} f(x_0, y) \, dy \quad .$$

As an immediate corollary we have:

Corollary 18.19
Let

$$f(x, y) = g(x)h(y)v(Ax + By)\phi(x, y)$$

where g, h and v are all piecewise continuous functions on the real line, ϕ is a continuous function on the entire plane, and A and B are any two nonzero real numbers. Assume further that, for each point x_0 at which g is continuous, there is an interval (a, b) containing x_0 such that f is uniformly absolutely integrable on the strip $(a, b) \times (-\infty, \infty)$. Then

$$\psi(x) = \int_{-\infty}^{\infty} f(x, y) \, dy$$

is a well-defined and piecewise continuous function on the entire real line. Furthermore, if x_0 is any real value at which g is continuous, then

$$\lim_{x \to x_0} \int_{-\infty}^{\infty} f(x, y)\, dy \;=\; \int_{-\infty}^{\infty} f(x_0, y)\, dy \quad .$$

PROOF (of theorem 18.18):   Because f is assumed to be uniformly absolutely integrable, there is an absolutely integrable function f_0 on $\mathbb{R}$ such that

$$|f(x, y)| \;\leq\; |f_0(y)|$$

for all (x, y) in the strip on which f is continuous. In particular, if x_0 is any fixed point in (a, b), then

$$|f(x_0, y)| \;\leq\; |f_0(y)| \quad .$$

This, along with lemma 18.3 and the fact that, by our assumptions, f can only have a finite number of discontinuities on the line $x = x_0$ in any bounded rectangle, assures us that $f(x_0, y)$ is a piecewise continuous, absolutely integrable function of y on the real line. So

$$\psi(x_0) \;=\; \int_{-\infty}^{\infty} f(x_0, y)\, dy$$

is well defined and finite at any point x_0 in (a, b).

   To show the claimed continuity of ψ, let x_0 be any point in (a, b) not on a line of discontinuity for f. Pick any finite positive value ϵ, and let f_0 be as above. It will suffice to show there is a corresponding Δx such that

$$|\psi(x) - \psi(x_0)| \;\leq\; \epsilon \qquad \text{whenever} \qquad |x - x_0| \;\leq\; \Delta x \quad .$$

   Since f_0 is in $\mathcal{A}$, lemma 18.11 on page 267 assures us that there is a finite positive value ℓ such that

$$\int_{-\infty}^{-\ell} |f_0(y)|\, dy \;\leq\; \frac{1}{6}\epsilon \qquad \text{and} \qquad \int_{\ell}^{\infty} |f_0(y)|\, dy \;\leq\; \frac{1}{6}\epsilon \quad .$$

Let

$$\psi_\ell(x) \;=\; \int_{-\ell}^{\ell} f(x, y)\, dy \quad .$$

From lemma 7.7 on page 80 we know ψ_ℓ is continuous at x_0, and that

$$\lim_{x \to x_0} \psi_\ell(x) \;=\; \psi_\ell(x_0) \;=\; \int_{-\ell}^{\ell} f(x_0, y)\, dy \quad .$$

Thus, there is a $\Delta x > 0$ such that, whenever x is within Δx of x_0,

$$|\psi_\ell(x) - \psi_\ell(x_0)| \;<\; \frac{1}{3}\epsilon \quad .$$

But also, for each x in (a, b),

$$|\psi(x) - \psi_\ell(x)| \;=\; \left| \int_{-\infty}^{\infty} f(x, y)\, dy - \int_{-\ell}^{\ell} f(x, y)\, dy \right|$$

$$=\; \left| \int_{-\infty}^{-\ell} f(x, y)\, dy + \int_{\ell}^{\infty} f(x, y)\, dy \right|$$

$$\leq \int_{-\infty}^{-\ell} |f(x,y)|\, dy + \int_{\ell}^{\infty} |f(x,y)|\, dy$$

$$\leq \int_{-\infty}^{-\ell} |f_0(y)|\, dy + \int_{\ell}^{\infty} |f_0(y)|\, dy \leq \frac{1}{6} + \frac{1}{6} = \frac{1}{3}\epsilon \quad.$$

Hence, whenever $|x - x_0| \leq \Delta x$,

$$|\psi(x) - \psi(x_0)| = |\psi(x) - \psi_\ell(x) + \psi_\ell(x) - \psi_\ell(x_0) + \psi_\ell(x_0) - \psi(x_0)|$$

$$\leq |\psi(x) - \psi_\ell(x)| + |\psi_\ell(x) - \psi_\ell(x_0)| + |\psi_\ell(x_0) - \psi(x_0)|$$

$$\leq \frac{1}{3}\epsilon + \frac{1}{3}\epsilon + \frac{1}{3}\epsilon = \epsilon \quad.\qquad \blacksquare$$

Differentiating Functions Defined by Integrals

The next theorem is easy to prove. Simply take the proof of theorem 7.9 on page 82 and replace the finite interval (c, d) with the infinite interval $(-\infty, \infty)$ and replace all references to theorem 7.11 and corollary 7.8 with references to theorem 18.22 and theorem 18.18, respectively.[5]

Theorem 18.20
Assume $f(x, y)$ and $\partial f/\partial x$ are both well-defined, piecewise continuous functions on a strip $\mathcal{R} = (a, b) \times (-\infty, \infty)$. Suppose, further, all of the following:

1. *For every point x_0 in (a, b), there is a finite subinterval $(\bar{a}, \bar{b})$ of (a, b) such that*
   (a) *$\bar{a} < x_0 < \bar{b}$, and*
   (b) *$f(x, y)$ and $\partial f/\partial x$ are both uniformly absolutely integrable on the strip $(\bar{a}, \bar{b}) \times (-\infty, \infty)$.*

2. *For each bounded rectangle $\mathcal{R}_0$ of $\mathcal{R}$,*
   (a) *all the discontinuities of f over $\mathcal{R}_0$ are contained in a finite number of lines of the form $y = constant$, and*
   (b) *all the discontinuities of $\partial f/\partial x$ over $\mathcal{R}_0$ are contained in a finite number of straight lines, none of which are of the form $x = constant$.*

Then

$$\psi(x) = \int_{-\infty}^{\infty} f(x, y)\, dy$$

is a smooth function on (a, b). Furthermore, on this interval,

$$\psi'(x) = \frac{d}{dx} \int_{-\infty}^{\infty} f(x, y)\, dy = \int_{-\infty}^{\infty} \frac{\partial}{\partial x} f(x, y)\, dy \quad.$$

The following is an immediate corollary of the last theorem. It will be especially useful when discussing derivatives of Fourier transforms and convolutions.

---

[5] As far as proving the results in these few sections, the logical order would be to prove theorem 18.18 first, then theorem 18.22 and finally theorem 18.20. For purposes of exposition, however, it seems more reasonable to present theorem 18.20 before theorem 18.22.

Corollary 18.21
Let h be a piecewise continuous function on the real line, v a continuous and piecewise smooth function on $(-\infty, \infty)$, and $\phi(x, y)$ a continuous function on $\mathbb{R}^2$ whose partial derivative with respect to x is also a well-defined continuous function on $\mathbb{R}^2$. Let A and B be any two nonzero real values, and define f by

$$f(x, y) = h(y)v(Ax + By)\phi(x, y) \quad .$$

Further assume that, for each value of x, there is a corresponding interval (a, b) containing x such that both f and $\partial f / \partial x$ are uniformly absolutely integrable on the strip $(a, b) \times (-\infty, \infty)$. Then

$$\psi(x) = \int_{-\infty}^{\infty} f(x, y) \, dy$$

is a smooth function on the real line and

$$\psi'(x) = \frac{d}{dx} \int_{-\infty}^{\infty} f(x, y) \, dy = \int_{-\infty}^{\infty} \frac{\partial}{\partial x} f(x, y) \, dy \quad .$$

Double Integrals

A little more care needs to be taken with double integrals over $\mathbb{R}^2$ than is necessary with double integrals over bounded rectangles. For example, it is quite possible to have

$$\int_{-\infty}^{\infty} \int_{-\infty}^{\infty} f(x, y) \, dx \, dy \neq \int_{-\infty}^{\infty} \int_{-\infty}^{\infty} f(x, y) \, dy \, dx$$

even though both iterated double integrals are well defined and finite (an example is given in exercise 18.19). Of course, if this is the case, then we cannot say the double integral of f over $\mathbb{R}^2$ is well defined.

General conditions ensuring that the above does not happen are given in the next two theorems.[6]

Theorem 18.22
Let $f(x, y)$ be a piecewise continuous and uniformly absolutely integrable function on a strip $\mathcal{R} = (a, b) \times (-\infty, \infty)$. Assume further that, for every finite interval (c, d), all the discontinuities of $f(x, y)$ on the rectangle $(a, b) \times (c, d)$ are contained in a finite number of straight lines, none of which are of the form $x = x_0$. Then

$$\psi(x) = \int_{-\infty}^{\infty} f(x, y) \, dy$$

is a well-defined, uniformly continuous function on the interval (a, b),

$$\phi(y) = \int_{a}^{b} f(x, y) \, dx$$

is a well-defined, piecewise continuous and absolutely integrable function on the entire real line, and

$$\int_{-\infty}^{\infty} \int_{a}^{b} f(x, y) \, dx \, dy = \iint_{\mathcal{R}} f(x, y) \, dA = \int_{a}^{b} \int_{-\infty}^{\infty} f(x, y) \, dy \, dx \quad .$$

---

[6] Those acquainted with the Lebesgue theory should compare these theorems to Fubini's theorem.

This theorem can be proven via a relatively straightforward application of the corresponding theorem for double integrals over bounded rectangles (theorem 7.11 on page 84) along with theorem 18.18 and lemma 18.17. We'll leave the details for the interested reader (exercise 18.8, below).

Using corollary 18.19 instead of theorem 18.18 proves:

Theorem 18.23
Let
$$f(x, y) = g(x)h(y)v(Ax + By)\phi(x, y)$$

where g, h and v are all piecewise continuous functions on the real line, ϕ is a continuous function on the entire plane, and A and B are any two nonzero real numbers. Assume, further, that f is absolutely integrable on $\mathbb{R}^2$ and is uniformly absolutely integrable on every strip of the form $(a, b) \times (-\infty, \infty)$ and on every strip of the form $(-\infty, \infty) \times (c, d)$. Then, the integrals

$$\int_{-\infty}^{\infty} f(x, y)\, dx \qquad \text{and} \qquad \int_{-\infty}^{\infty} f(x, y)\, dy$$

define piecewise continuous, absolutely integrable functions on the real line, and

$$\int_{-\infty}^{\infty} \int_{-\infty}^{\infty} f(x, y)\, dx\, dy = \iint_{\mathcal{R}} f(x, y)\, dA = \int_{-\infty}^{\infty} \int_{-\infty}^{\infty} f(x, y)\, dy\, dx \quad .$$

Less general, but more easily recognized, conditions ensuring the validity of the last equation are described in the following corollary.

Corollary 18.24
Let
$$f(x, y) = g(x)h(y)v(Ax + By)\phi(x, y)$$

where g, h and v are all piecewise continuous functions on the real line, ϕ is a continuous and bounded function on the entire plane, and A and B are any two nonzero real numbers. Assume, further, that any one of the following sets of conditions holds:

1. g and h are in $\mathcal{A}$, and v is bounded.

2. g and v are in $\mathcal{A}$, and both v and h are bounded.

3. h and v are in $\mathcal{A}$, and both v and g are bounded.

4. There is a bounded region $\mathcal{R}_0$ such that $f(x, y) = 0$ whenever (x, y) is not in $\mathcal{R}_0$.

Then f is absolutely integrable on $\mathbb{R}^2$, the integrals

$$\int_{-\infty}^{\infty} f(x, y)\, dx \qquad \text{and} \qquad \int_{-\infty}^{\infty} f(x, y)\, dy$$

define piecewise continuous, absolutely integrable functions on the real line, and

$$\int_{-\infty}^{\infty} \int_{-\infty}^{\infty} f(x, y)\, dx\, dy = \iint_{\mathcal{R}} f(x, y)\, dA = \int_{-\infty}^{\infty} \int_{-\infty}^{\infty} f(x, y)\, dy\, dx \quad .$$

?►Exercise 18.8   a: Prove theorem 18.22.

   b: Prove theorem 18.23.

   c: Prove corollary 18.24 using theorem 18.23.

Additional Exercises

18.9. By computing the appropriate integrals, determine which of the following is absolutely integrable over $\mathbb{R}$ and which is not.

    a. $\dfrac{1}{1+x^2}$ **b.** $\dfrac{x}{1+x^2}$ **c.** $\dfrac{1}{\sqrt{1+x^2}}$

18.10. For each real constant γ, let $f_\gamma(x) = x^{-\gamma} \operatorname{step}(x-1)$ where step is the step function (see example 18.2 on page 260).

    a. Sketch the graphs of $f_\gamma(x)$ for $\gamma = 2$, $\gamma = 1$, $\gamma = \frac{1}{2}$, $\gamma = 0$, $\gamma = -\frac{1}{2}$, $\gamma = -1$ and $\gamma = -2$.

    b. Determine all the values of γ for which $f_\gamma(x)$ is absolutely integrable on the real line.

18.11. For each real constant γ, let $g_\gamma(x) = x^{-\gamma} \operatorname{rect}_{(0,1)}(x)$ where $\operatorname{rect}_{(0,1)}(x)$ is the rectangle function over $(0, 1)$ (see example 18.1 on page 259).

    a. Sketch the graphs of $g_\gamma(x)$ for $\gamma = 2$, $\gamma = 1$, $\gamma = \frac{1}{2}$, $\gamma = 0$, $\gamma = -\frac{1}{2}$, $\gamma = -1$ and $\gamma = -2$.

    b. Determine all the values of γ for which $g_\gamma(x)$ is absolutely integrable on the real line.

    (Notice that $f(x) = x^{-\gamma} \operatorname{rect}_{(0,1)}(x)$ is not bounded — and hence, is not piecewise continuous — on the interval $(0, 1)$ if $\gamma > 0$. This exercise demonstrates that a function does not have to be piecewise continuous to be absolutely integrable.)

18.12. Let $\alpha > 0$. For each of the following functions, determine all the real values of γ for which the given function is in $\mathcal{A}$.

    a. $x^\gamma e^{-\alpha x} \operatorname{step}(x)$ **b.** $x^\gamma e^{\alpha x} \operatorname{step}(-x)$

    c. $x^\gamma e^{-\alpha|x|}$ **d.** $x^\gamma e^{-\alpha x^2}$

18.13. Using the lemmas and work already done, determine which of the following functions are absolutely integrable over the real line and which are not.

    a. $\dfrac{\sin(x^2)}{1+x^2}$ **b.** $\dfrac{1+e^{|x|}}{1+x^2}$

    c. $\dfrac{1}{1+i2\pi x}$ **d.** $\operatorname{sinc}^2(2\pi x)$

18.14. Let

$$ f(x) = \sum_{k=0}^{\infty} (-1)^k \frac{1}{k+1} \operatorname{rect}_{(k,k+1)} \qquad \text{and} \qquad g(x) = \sum_{k=0}^{\infty} (-1)^k \operatorname{rect}_{(k,k+1)} \quad . $$

    a. Sketch the graphs of f, $|f|$ and g.

    b. Show that f is not absolutely integrable over the real line. (Suggestion: Evaluate $\int_{-\infty}^{\infty} |f(x)|\, dx$, and compare the result to the harmonic series — see exercise 4.3 on page 44.)

c. Show that, even though it is not absolutely integrable, f is "integrable" over the real line in the sense that

$$\int_{-\infty}^{\infty} f(x)\,dx = \lim_{\substack{b\to\infty\\a\to-\infty}} \int_a^b f(x)\,dx$$

exists and is finite. (Again, you might start by evaluating the integral. Then compare the result to the alternating harmonic series — see exercise 4.4 on page 44.)

d. Note that g is a bounded, piecewise continuous function on $\mathbb{R}$. Show that, even though g is bounded and f is "integrable" (as described above), their product, fg, is not "integrable". That is, show that

$$\int_{-\infty}^{\infty} f(x)g(x)\,dx = \lim_{\substack{b\to\infty\\a\to-\infty}} \int_a^b f(x)g(x)\,dx = \infty \quad.$$

Why does this not contradict lemma 18.9?

18.15. Repeat the previous problem using

$$f(x) = \operatorname{sinc}(\pi x) = \frac{\sin(\pi x)}{\pi x} \quad\text{and}\quad g(x) = e^{i\pi x} \quad.$$

(Notes: (1) For some, this may be a challenging problem. (2) Remember, to show fg is not integrable, it suffices to show that the imaginary part of fg is not integrable.)

18.16. Let

$$f(x) = \sum_{k=1}^{\infty} 2^k \operatorname{rect}_{(0,2^{-2k})}(x-k) \quad.$$

Sketch the graph of f and verify that this function is not bounded but is absolutely integrable on the real line. (So a function f can be absolutely integrable on the real line even though $f(x)$ does not steadily shrink to zero as $x \to \pm\infty$.)

18.17. Let $\mathcal{R}$ be the region in the XY–plane bounded by the curves

$$y = \frac{1}{|x|} \quad\text{and}\quad y = 1 + \frac{1}{|x|} \quad,$$

and let

$$f(x,y) = \begin{cases} 6\left(y - \frac{1}{|x|}\right)\left(y - \frac{1}{|x|} - 1\right) & \text{if } (x,y) \text{ is in } \mathcal{R} \\ 0 & \text{otherwise} \end{cases}.$$

a. Sketch the region $\mathcal{R}$.

b. Sketch, as a function of y, the graph of $z = f(x,y)$ assuming

i. $x > 0$.   **ii.** $x < 0$.   **iii.** $x = 0$.

(These graphs should convince you that $f(x, y)$ is a continuous and absolutely integrable function of y for each real x.)

c. Verify that f is continuous and bounded on $\mathbb{R}^2$.

d. By computing the appropriate integrals and limits, show that

$$\lim_{x \to 0} \int_{-\infty}^{\infty} f(x, y) \, dy \neq \int_{-\infty}^{\infty} \lim_{x \to 0} f(x, y) \, dy$$

even though all the limits and integrals in this expression are well defined and finite.

e. Why does this inequality not violate theorem 18.18 on page 271?

18.18. Let $\mathcal{R}$ be the region in the XY–plane bounded by the curves

$$x = 1 \quad , \quad x = -1 \quad , \quad y = \frac{1}{\sqrt{|x|}} \quad \text{and} \quad y = 0 \ ,$$

and let ψ be the function on the real line given by

$$\psi(x) = \int_{-\infty}^{\infty} f(x, y) \, dy$$

where

$$f(x, y) = \begin{cases} 1 & \text{if } (x, y) \text{ is in } \mathcal{R} \\ 0 & \text{otherwise} \end{cases} .$$

a. Sketch the region $\mathcal{R}$.

b. Show that $f(x, y)$ is absolutely integrable on $\mathbb{R}^2$. (*Thus, since f is also obviously piecewise continuous on $\mathbb{R}^2$, we know f is in $\mathcal{A}[\mathbb{R}^2]$.*)

c. Evaluate $\int_{-\infty}^{\infty} f(x, y) \, dy$ to obtain a formula for $\psi(x)$.

d. What happens to $\psi(x)$ when "$x = 0$"? (*This shows that ψ is not piecewise continuous and, hence, is not in $\mathcal{A}[\mathbb{R}]$.*)

18.19 a. By explicitly computing the integrals, verify that

$$\int_{-\infty}^{\infty} \int_{-\infty}^{\infty} f(x, y) \, dx \, dy \neq \int_{-\infty}^{\infty} \int_{-\infty}^{\infty} f(x, y) \, dy \, dx$$

when

$$f(x, y) = (x - y)e^{-(x-y)^2} \text{step}(y) \ .$$

You may use the fact that

$$\int_{-\infty}^{\infty} e^{-s^2} \, ds = \sqrt{\pi} \ .$$

(*For a derivation of this last equation, either look in your old calculus book or look ahead to the first few pages of chapter 23.*)

b. Why does this inequality not violate either theorem 18.23 or corollary 18.24?

c. Is this $f(x, y)$ absolutely integrable on $\mathbb{R}^2$?

19

The Fourier Integral Transforms

We are now ready for the *first* official set of definitions for the Fourier transforms. These definitions are directly inspired by the integral formulas (17.8) and (17.9) on page 255, and will, accordingly, be called the Fourier *integral* transforms. To ensure our integrals are well defined, *we will only use these definitions for the Fourier transforms of functions in $\mathcal{A}$, the set of piecewise continuous, absolutely integrable functions on the entire real line,* $\mathbb{R}$.[1] This will not be completely satisfactory. Many functions of interest are not absolutely integrable. Consequently, one of our goals will later be to intelligently extend the basic definitions given in this chapter so that we can deal with interesting functions that are not absolutely integrable.

Words of warning to those who have already seen the Fourier transform in applications: Different disciplines have different conventions and notation for the Fourier transforms. Don't be surprised if the formulas we are about to give for the Fourier transforms look a little strange, and if one theorem (the principle of near-equivalence) appears to disagree with your interpretation of the transforms. In fact, there is no real conflict, and we will later discuss some of the standard conventions and notation used in applications. For now, however, it may be best just to forget everything you thought you knew about Fourier transforms.

19.1 Definitions, Notation and Terminology

Let ϕ be any function in $\mathcal{A}$, the set of piecewise continuous, absolutely integrable functions on the real line. The *(direct) Fourier integral transform* of ϕ, denoted by $\mathcal{F}_I[\phi]$, is the function on the real line given by

$$\mathcal{F}_I[\phi]|_x = \mathcal{F}_I[\phi(y)]|_x = \int_{-\infty}^{\infty} \phi(y)\, e^{-i2\pi xy}\, dy \quad . \tag{19.1}$$

The *Fourier inverse integral transform* of ϕ, denoted by $\mathcal{F}_I^{-1}[\phi]$, is the function on the real line given by

$$\mathcal{F}_I^{-1}[\phi]|_x = \mathcal{F}_I^{-1}[\phi(y)]|_x = \int_{-\infty}^{\infty} \phi(y)\, e^{i2\pi xy}\, dy \quad . \tag{19.2}$$

(Remember, corollary 18.10 on page 267 assures us that the product of any function $\phi(y)$ in $\mathcal{A}$ with $e^{\pm i2\pi xy}$ is a piecewise continuous, absolutely integrable function of y for each real value x. So $\mathcal{F}_I[\phi]$ and $\mathcal{F}_I^{-1}[\phi]$, as defined by the above integrals, are well-defined functions on the real line.)

---

[1] Those acquainted with the Lebesgue theory of integration may want to try replacing "$\mathcal{A}$" with "$\mathcal{L}$, the set of measurable and absolutely integrable functions on $\mathbb{R}$", in what follows.

Together, $\mathcal{F}_I[\phi]$ and $\mathcal{F}_I^{-1}[\phi]$ are called the Fourier integral transforms of ϕ, though, in common practice, $\mathcal{F}_I[\phi]$ is usually "the" Fourier integral transform of ϕ. The integrals on the right-hand side of formulas (19.1) and (19.2) are called the *Fourier integrals* and the formulas, themselves, are often referred to as the *integral formulas for the Fourier (integral) transforms.*

!▶ **Example 19.1 (transform of the pulse function):** Let $a > 0$ (i.e., a is a positive real number). The (symmetric) pulse function of half-width a, denoted by pulse_a and graphed in figure 19.1a, is given by

$$\text{pulse}_a(x) = \text{rect}_{(-a,a)}(x) = \begin{cases} 1 & \text{if } -a < x < a \\ 0 & \text{otherwise} \end{cases} \ .$$

From example 18.1 on page 259, we know pulse_a is in $\mathcal{A}$. Its Fourier transform is easily computed:

$$\mathcal{F}_I[\text{pulse}_a]\big|_x = \int_{-\infty}^{\infty} \text{pulse}_a(y)\, e^{-i2\pi xy}\, dy$$

$$= \int_{-\infty}^{-a} 0 \cdot e^{-i2\pi xy}\, dy + \int_{-a}^{a} 1 \cdot e^{-i2\pi xy}\, dy + \int_{a}^{\infty} 0 \cdot e^{-i2\pi xy}\, dy$$

$$= \frac{1}{-i2\pi x}\left[e^{-i2\pi ax} - e^{i2\pi ax} \right]$$

$$= \frac{1}{i2\pi x}\left[e^{i2\pi ax} - e^{-i2\pi ax} \right] \ .$$

We can rewrite this in a somewhat more convenient form after recalling that

$$\sin(A) = \frac{e^{iA} - e^{-iA}}{2i} \qquad \text{and} \qquad \text{sinc}(A) = \frac{\sin(A)}{A} \ .$$

So,

$$\mathcal{F}_I[\text{pulse}_a]\big|_x = \frac{1}{i2\pi x}\left[e^{i2\pi ax} - e^{-i2\pi ax} \right]$$

$$= \frac{2a}{2a\pi x}\left[\frac{e^{i2\pi ax} - e^{-i2\pi ax}}{2i} \right] = 2a\frac{\sin(2\pi ax)}{2\pi ax} \ .$$

That is,

$$\mathcal{F}_I[\text{pulse}_a]\big|_x = 2a\,\text{sinc}(2\pi ax) \ .$$

?▶ **Exercise 19.1:** Let $a > 0$. From example 18.2 on page 260, we know that $e^{-ay}\,\text{step}(y)$ is in $\mathcal{A}$. Show that

$$\mathcal{F}_I\left[e^{-ay}\,\text{step}(y)\right]\big|_x = \frac{1}{a + i2\pi x} \ .$$

Also, sketch both $e^{-ay}\,\text{step}(y)$ and the real and imaginary parts of its Fourier integral transform.

The *process* of changing ϕ to $\mathcal{F}_I[\phi]$ is also referred to as the *(direct) Fourier integral transform* and is denoted by $\mathcal{F}_I$. Thus, "the Fourier integral transform" can refer to either a particular function $\mathcal{F}_I[\phi]$ or the process of obtaining $\mathcal{F}_I[\phi]$ from any given ϕ in $\mathcal{A}$.

Likewise, the process of changing ϕ to $\mathcal{F}_I^{-1}[\phi]$ is called the *Fourier inverse integral transform* and is denoted by $\mathcal{F}_I^{-1}$.

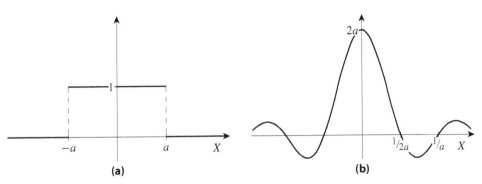

Figure 19.1: The graphs of (a) $\text{pulse}_a(x)$ and (b) its Fourier integral transform, $2a\,\text{sinc}(2\pi a x)$. (These graphs correspond to $a \approx 1$.)

Collectively, $\mathcal{F}_I$ and $\mathcal{F}_I^{-1}$ are often referred to as the Fourier integral transforms, though $\mathcal{F}_I$ is usually viewed as "the" Fourier integral transform and $\mathcal{F}_I^{-1}$ as "the" Fourier inverse integral transform. Both are transforms as discussed in the section in chapter 2 on operators and transforms, and the domain of each is $\mathcal{A}$, the set of piecewise continuous, absolutely integrable functions on the real line.[2]

19.2   Near-Equivalence

There is a striking similarity between integrals on the right-hand side of formulas (19.1) and (19.2). Between them, only the sign in the exponential differs. Let us formalize this observation and some of its more obvious consequences as the *principle of near-equivalence*.[3]

Theorem 19.1 (principle of near-equivalence)
Let ϕ be an absolutely integrable and piecewise continuous function on $\mathbb{R}$ (i.e., ϕ is in $\mathcal{A}$). Then the function $\phi(-y)$ is also in $\mathcal{A}$. Moreover,

$$\mathcal{F}_I[\phi(y)]|_x \;=\; \mathcal{F}_I^{-1}[\phi(y)]|_{-x} \;=\; \mathcal{F}_I^{-1}[\phi(-y)]|_x \tag{19.3}$$

and

$$\mathcal{F}_I^{-1}[\phi(y)]|_x \;=\; \mathcal{F}_I[\phi(y)]|_{-x} \;=\; \mathcal{F}_I[\phi(-y)]|_x \quad . \tag{19.4}$$

PROOF: That $\phi(-y)$ is in $\mathcal{A}$ was verified with the proof of lemma 18.7 on page 265. The first equality in line (19.3) comes from the observation that

$$\mathcal{F}_I[\phi(y)]|_x \;=\; \int_{-\infty}^{\infty} \phi(y)\,e^{-i2\pi xy}\,dy$$

$$=\; \int_{-\infty}^{\infty} \phi(y)\,e^{i2\pi(-x)y}\,dy \;=\; \mathcal{F}_I^{-1}[\phi(y)]|_{-x} \quad .$$

---

[2] This may be a good time to review that section on operators and transforms starting on page 12. In particular, the discussion concerning the use of dummy variables in formulas for transforms is especially relevant to the next section.

[3] In some texts this is called *symmetry*.

Next, using the substitution $y = -s$ and the fact that, in the computations below, the y and the s are dummy variables, we see that

$$\mathcal{F}_I^{-1}[\phi(y)]|_{-x} = \int_{y=-\infty}^{\infty} \phi(y)\, e^{i2\pi(-x)y}\, dy$$

$$= \int_{s=+\infty}^{-\infty} \phi(-s)\, e^{i2\pi(-x)(-s)}(-1)\, ds$$

$$= \int_{s=-\infty}^{\infty} \phi(-s)\, e^{i2\pi xs}\, ds$$

$$= \int_{y=-\infty}^{\infty} \phi(-y)\, e^{i2\pi xy}\, dy = \mathcal{F}_I^{-1}[\phi(-y)]|_x \quad .$$

This proves the second equality in line (19.3).

The rest of the proof is left as an exercise.   ∎

?►Exercise 19.2:   *Prove the equalities in line (19.4) of theorem 19.1.*

!►Example 19.2:   *Let $a > 0$. From exercise 19.1 we know*

$$\mathcal{F}_I\big[e^{-ay}\,\text{step}(y)\big]\big|_x = \frac{1}{a+i2\pi x} \quad .$$

By this and the principle of near-equivalence,

$$\mathcal{F}_I^{-1}\big[e^{-ay}\,\text{step}(y)\big]\big|_x = \mathcal{F}_I\big[e^{-ay}\,\text{step}(y)\big]\big|_{-x} = \frac{1}{a+i2\pi(-x)} = \frac{1}{a-i2\pi x} \quad .$$

?►Exercise 19.3:   *Let $a > 0$, and consider the function*

$$f(y) = e^{ay}\,\text{step}(-y) \quad .$$

Sketch the graph of f, and confirm that $f(y) = g(-y)$ where

$$g(y) = e^{-ay}\,\text{step}(y) \quad .$$

Using this, the principle of near-equivalence and the results from either of the last two exercises, show that

$$\mathcal{F}_I\big[e^{ay}\,\text{step}(-y)\big]\big|_x = \frac{1}{a-i2\pi x} \quad .$$

Using the principle of near-equivalence, it is easy to derive and prove some simple, but useful, facts about the transforms of even and odd functions. Suppose, for example, ϕ is an even function (i.e., $\phi(-y) = \phi(y)$) in $\mathcal{A}$. The principle of near-equivalence then tells us that

$$\mathcal{F}_I[\phi(y)]|_{-x} = \mathcal{F}_I[\phi(-y)]|_x = \mathcal{F}_I[\phi(y)]|_x \quad ,$$

$$\mathcal{F}_I^{-1}[\phi(y)]|_{-x} = \mathcal{F}_I^{-1}[\phi(-y)]|_x = \mathcal{F}_I^{-1}[\phi(y)]|_x$$

and

$$\mathcal{F}_I^{-1}[\phi(y)]|_x = \mathcal{F}_I[\phi(-y)]|_x = \mathcal{F}_I[\phi(y)]|_x \quad .$$

This gives us the following corollary.

Corollary 19.2 (transforms of even functions)
Let ϕ be an even function in $\mathcal{A}$. Then both $\mathcal{F}_I[\phi]$ and $\mathcal{F}_I^{-1}[\phi]$ are even functions. Moreover,

$$\mathcal{F}_I^{-1}[\phi] \;=\; \mathcal{F}_I[\phi] \quad .$$

Similar arguments lead to the corresponding corollary for odd functions.

Corollary 19.3 (transforms of odd functions)
Let ϕ be an odd function in $\mathcal{A}$. Then both $\mathcal{F}_I[\phi]$ and $\mathcal{F}_I^{-1}[\phi]$ are odd functions. Moreover,

$$\mathcal{F}_I^{-1}[\phi] \;=\; -\mathcal{F}_I[\phi] \quad .$$

?►Exercise 19.4:   *Prove corollary 19.3 using the principle of near-equivalence.*

?►Exercise 19.5:   *Let $a > 0$. Is the pulse function from example 19.1 even or odd? Use the result of example 19.1 and one of the above corollaries to quickly find the Fourier inverse integral transform of pulse_a.*

19.3 Linearity

In chapter 18 we saw that any linear combination of functions from $\mathcal{A}$ is another function in $\mathcal{A}$; that is, $\mathcal{A}$ is a linear space of functions. We will now show that $\mathcal{F}_I$ and $\mathcal{F}_I^{-1}$ are linear transforms on this linear space.

Theorem 19.4 (linearity)
Let ϕ and ψ be any two functions in $\mathcal{A}$, and let α and β be any two (possibly complex) constants. Then the linear combination $\alpha\phi + \beta\psi$ is in $\mathcal{A}$. Moreover,

$$\mathcal{F}_I[\alpha\phi + \beta\psi] \;=\; \alpha\mathcal{F}_I[\phi] + \beta\mathcal{F}_I[\psi]$$

and

$$\mathcal{F}_I^{-1}[\alpha\phi + \beta\psi] \;=\; \alpha\mathcal{F}_I^{-1}[\phi] + \beta\mathcal{F}_I^{-1}[\psi] \quad .$$

PROOF: Since the proofs of these two equations are virtually identical (and almost trivial), we will just confirm the first, which is easily done using the definition of $\mathcal{F}_I$ and the linearity of integration,

$$\mathcal{F}_I[\alpha\phi + \beta\psi] \;=\; \int_{-\infty}^{\infty} [\alpha\phi(y) + \beta\psi(y)]\, e^{-i2\pi xy}\, dy$$

$$=\; \int_{-\infty}^{\infty} \left[\alpha\phi(y)\, e^{-i2\pi xy} + \beta\psi(y)\, e^{-i2\pi xy}\right] dy$$

$$=\; \alpha\int_{-\infty}^{\infty} \phi(y)\, e^{-i2\pi xy}\, dy + \beta\int_{-\infty}^{\infty} \psi(y)\, e^{-i2\pi xy}\, dy$$

$$=\; \alpha\mathcal{F}_I[\phi] + \beta\mathcal{F}_I[\psi] \quad .$$

!▶**Example 19.3:**   Let $a > 0$. *From previous examples and exercises we know*

$$\mathcal{F}_I\left[e^{ay}\,\text{step}(-y)\right]\big|_x = \frac{1}{a - i2\pi x} \quad , \quad \mathcal{F}_I\left[e^{-ay}\,\text{step}(y)\right]\big|_x = \frac{1}{a + i2\pi x} \quad ,$$

and

$$e^{-a|y|} = e^{ay}\,\text{step}(-y) + e^{-ay}\,\text{step}(y) \quad .$$

Using these equations and the linearity of the transform,

$$\begin{aligned}
\mathcal{F}_I\left[e^{-a|y|}\right]\big|_x &= \mathcal{F}_I\left[e^{ay}\,\text{step}(-y) + e^{-ay}\,\text{step}(y)\right]\big|_x \\
&= \mathcal{F}_I\left[e^{ay}\,\text{step}(-y)\right]\big|_x + \mathcal{F}_I\left[e^{-ay}\,\text{step}(y)\right]\big|_x \\
&= \frac{1}{a - i2\pi x} + \frac{1}{a + i2\pi x} \\
&= \frac{a + i2\pi x}{(a - i2\pi x)(a + i2\pi x)} + \frac{a - i2\pi x}{(a + i2\pi x)(a - i2\pi x)} \\
&= \frac{2a}{a^2 + 4\pi^2 x^2} \quad .
\end{aligned}$$

?▶**Exercise 19.6:**   Let

$$f(y) = \begin{cases} -e^{ay} & \text{if } y < 0 \\ e^{-ay} & \text{if } 0 < y \end{cases}$$

where a is any positive number. Show that

$$\mathcal{F}_I[f]\big|_x = \frac{-i4\pi x}{a^2 + 4\pi^2 x^2} \quad .$$

Also, sketch the graphs of f and its transform.

19.4 Invertibility

The astute reader has probably noticed that our notation and terminology suggest that the Fourier inverse integral transform, $\mathcal{F}_I^{-1}$, is the inverse transform of the Fourier integral transform, $\mathcal{F}_I$. That reader even may have recalled something suggesting this relation in the summary at the end of our derivation of the formulas which inspired the integral transform formulas of this chapter. Let us quote that summary, with a certain phrase emphasized:

... we have actually derived (*provided our many assumptions are valid*) the following:

If f is a "reasonably nice" function on $\mathbb{R}$, and if F is the function constructed from f by

$$F(\omega) = \int_{-\infty}^{\infty} f(t)\,e^{-i2\pi\omega t}\,dt \quad ,$$

then the original function f can be recovered from F through the formula

$$f(t) = \int_{-\infty}^{\infty} F(\omega)\,e^{i2\pi\omega t}\,d\omega \quad .$$

Condensing this statement and using the notation developed in this chapter give us:

> *Provided the many assumptions made in chapter 17 are valid:*
>
> If f and F are two "reasonably nice" functions on $\mathbb{R}$ with
>
> $$ F = \mathcal{F}_I[f] \quad , $$
>
> then
>
> $$ f = \mathcal{F}_I^{-1}[F] \quad . $$

Unfortunately, not only was the above derived with limited concern for rigor, it turns out that the "many assumptions" made in its derivation are, in general, *not* valid. In particular, it is quite possible to have a piecewise continuous, absolutely integrable function f whose Fourier integral transform, $F = \mathcal{F}_I[f]$, is not absolutely integrable. In that case, we don't even have $\mathcal{F}_I^{-1}[F]$ defined. For example, from exercise 19.1 we know that the Fourier integral transform of

$$ f(t) = e^{-t} \operatorname{step}(t) $$

is

$$ F(\omega) = \frac{1}{1 + i2\pi\omega} \quad , $$

which is easily shown *not* to be in $\mathcal{A}$ (see exercise 18.7 on page 264). So this function is not even in the domain of the Fourier inverse integral transform.

Fortunately, there are some very important functions in $\mathcal{A}$ whose integral transforms are also in $\mathcal{A}$. For these functions we have the following theorem, which is so important that we will henceforth refer to it as the *fundamental theorem on invertibility*.

Theorem 19.5 (fundamental theorem on invertibility)
Let f and F be two piecewise continuous, absolutely integrable functions on the real line. Then[4]

$$ F = \mathcal{F}_I[f] \quad \Longleftrightarrow \quad \mathcal{F}_I^{-1}[F] = f \quad . $$

This theorem assures us that our non-rigorously derived claim that

$$ f(t) = \int_{-\infty}^{\infty} F(\omega)\, e^{i2\pi\omega t}\, d\omega $$

whenever

$$ F(\omega) = \int_{-\infty}^{\infty} f(t)\, e^{-i2\pi\omega t}\, dt $$

is true provided *both* f and F are in $\mathcal{A}$.

Another useful way to state the fundamental theorem on invertibility is:

Theorem 19.5′ (fundamental theorem on invertibility, alternative version)
Let ϕ be in $\mathcal{A}$ and assume that $\mathcal{F}_I[\phi]$ and $\mathcal{F}_I^{-1}[\phi]$ are in $\mathcal{A}$. Then

$$ \mathcal{F}_I^{-1}\big[\mathcal{F}_I[\phi]\big] = \phi \quad \text{and} \quad \mathcal{F}_I\big[\mathcal{F}_I^{-1}[\phi]\big] = \phi \quad . $$

?▶Exercise 19.7: *Convince yourself that theorems 19.5 and 19.5′ are equivalent.*

---

[4] The " $\Longleftrightarrow$ " is a graphic shorthand for "if and only if."

A good proof of the fundamental theorem on invertibility is nontrivial. Our proof (a good proof) will require mathematical machinery that will be developed over the next several chapters. For that reason, we will wait to prove this theorem. We will not wait, however, to use it. The fundamental theorem on invertibility is just too useful in applications and too important in developing the theory of Fourier transforms.[5]

!▶**Example 19.4:**   *Consider the two functions*

$$f(x) = e^{-|x|} \quad and \quad F(x) = \frac{2}{1 + 4\pi^2 x^2} \quad .$$

Both are continuous and easily shown to be absolutely integrable on the real line (see exercise 18.6 on page 264 and example 18.7). Furthermore, from example 19.3 we know that

$$\mathcal{F}_I\left[e^{-|y|}\right]\Big|_x = \frac{2}{1 + 4\pi^2 x^2} \quad .$$

The fundamental theorem on invertibility then tells us that

$$e^{-|y|} = \mathcal{F}_I^{-1}\left[\frac{2}{1 + 4\pi^2 x^2}\right]\Big|_y \quad .$$

This is certainly an easy way to find this inverse integral transform, much easier than directly computing the integral in the integral formula,

$$\mathcal{F}_I^{-1}\left[\frac{2}{1 + 4\pi^2 x^2}\right]\Big|_y = \int_{-\infty}^{\infty} \frac{2}{1 + 4\pi^2 x^2} e^{i2\pi xy} \, dx \quad .$$

19.5   Other Integral Formulas (A Warning)

Warning!
Not everyone uses the same set of integral formulas for the Fourier integral transforms.

Our choice, formulas (19.1) and (19.2), is one of the more commonly used sets of integral formulas for defining $\mathcal{F}_I$ and $\mathcal{F}_I^{-1}$, but it is not the only possible set. For example, many engineers prefer to define the direct Fourier integral transform by

$$\mathcal{F}_I[\phi]|_x = \int_{-\infty}^{\infty} \phi(y) e^{-ixy} \, dy \quad .$$

To ensure that the corresponding theorem on invertibility holds, they then define the corresponding inverse integral transform by

$$\mathcal{F}_I^{-1}[\phi]|_x = \frac{1}{2\pi} \int_{-\infty}^{\infty} \phi(y) e^{ixy} \, dy \quad .$$

---

[5] Using an unproven theorem to develop a mathematical theory and then using elements of that theory to prove the theorem is somewhat risky. There is a danger of both wasting time on something that may not be true, and of falsely verifying the theorem in question by using results based on assuming the theorem is true.
  To avoid the circular argument that the fundamental theorem on invertibility is true because we are pretending it is true, we will carefully avoid using this theorem when deriving results we will later use to prove the theorem. Fortunately, the author already knows which results we are going to use.

Among mathematicians,

$$\frac{1}{\sqrt{2\pi}} \int_{-\infty}^{\infty} \phi(y)\, e^{-ixy}\, dy$$

is often favored as the formula for the Fourier integral transform with

$$\frac{1}{\sqrt{2\pi}} \int_{-\infty}^{\infty} \phi(y)\, e^{ixy}\, dy$$

defining the corresponding Fourier inverse integral transform.

You will also find definitions with the signs in the exponents switched, say,

$$\mathcal{F}_I[\phi]|_x = \int_{-\infty}^{\infty} \phi(y)\, e^{i2\pi xy}\, dy \qquad \text{and} \qquad \mathcal{F}_I^{-1}[\phi]|_x = \int_{-\infty}^{\infty} \phi(y)\, e^{-i2\pi xy}\, dy \quad .$$

And other variations have surely been used. In general, you can obtain a perfectly valid "theory of Fourier transforms" by starting with the defining formula

$$\mathcal{F}_I[\phi]|_x = B \int_{-\infty}^{\infty} \phi(y)\, e^{-iAxy}\, dy$$

where A and B are any two nonzero real numbers. The corresponding inverse integral transform is then

$$\mathcal{F}_I^{-1}[\phi]|_x = \frac{A}{2\pi B} \int_{-\infty}^{\infty} \phi(y)\, e^{iAxy}\, dy \quad .$$

Whatever your choices of A and B, the basic ideas and manipulations remain the same. The resulting formulas, of course, are slightly different for different choices of A and B. As you can imagine, this can cause some difficulties, and care should be exercised when using formulas, tables, or software from various sources. Be sure either to check that both you and the other source are basing calculations on the same integral formula for the direct Fourier integral transform, or that you know how to convert the other source's formulas to your theory.

?▶Exercise 19.8: *Suppose we had used the definitions*

$$\mathcal{F}_I[\phi]|_x = \int_{-\infty}^{\infty} \phi(y)\, e^{-ixy}\, dy \qquad \text{and} \qquad \mathcal{F}_I^{-1}[\phi]|_x = \frac{1}{2\pi} \int_{-\infty}^{\infty} \phi(y)\, e^{ixy}\, dy \quad .$$

What then would be the formulas for

a: $\mathcal{F}_I\!\left[e^{-ay}\, \text{step}(y) \right]\big|_x$? *(Compare with the results obtained in exercise 19.1.)*

b: *the principle of near-equivalence?*

19.6   Some Properties of the Transformed Functions

As noted in the section on invertibility, a Fourier integral transform of an absolutely integrable function might not, itself, be absolutely integrable. In other ways, however, these transforms turn out to be fairly "nice". Understanding just how these functions are "nice" will help simplify some of our discussions later. It will also help us understand some of the limitations of the classical theory of Fourier transforms.

The Properties

Let ϕ be a piecewise continuous, absolutely integrable function on the real line (i.e., ϕ is in $\mathcal{A}$), and let Ψ be either the direct Fourier integral transform of ϕ, $\mathcal{F}_I[\phi]$, or the Fourier inverse integral transform of ϕ, $\mathcal{F}_I^{-1}[\phi]$. We can write this in a shorthand form as

$$\Psi(x) = \mathcal{F}_I^{\mp}[\phi]|_x = \int_{-\infty}^{\infty} \phi(y) e^{\pm i2\pi xy} \, dy \quad . \tag{19.5}$$

It turns out that the integrability of ϕ forces Ψ to satisfy a number of "pointwise" properties. Four that will be of interest to us are described in the following theorem.

Theorem 19.6
Let $\Psi = \mathcal{F}_I^{\mp}[\phi]$ where ϕ is in $\mathcal{A}$. Then:

1. $\Psi(0) = \displaystyle\int_{-\infty}^{\infty} \phi(y) \, dy$.

2. Ψ is a bounded function. In fact, for each x on the real line,

$$|\Psi(x)| \leq \int_{-\infty}^{\infty} |\phi(y)| \, dy \quad .$$

3. Ψ is a continuous function; that is, for every point x_0 on the real line,

$$\lim_{x \to x_0} \Psi(x) = \Psi(x_0) \quad . \tag{19.6}$$

4. $\Psi(x)$ "vanishes at infinity". More precisely,

$$\lim_{x \to \pm\infty} \Psi(x) = 0 \quad .$$

The first two properties are easily verified. The first is simply formula (19.5) with $x = 0$. Verifying the second property is almost as easy:

$$|\Psi(x)| = \left| \int_{-\infty}^{\infty} \phi(y) e^{\pm i2\pi xy} \, dy \right| \leq \int_{-\infty}^{\infty} \left| \phi(y) e^{\pm i2\pi xy} \right| \, dy$$

$$= \int_{-\infty}^{\infty} |\phi(y)| \left| e^{\pm i2\pi xy} \right| \, dy = \int_{-\infty}^{\infty} |\phi(y)| \, dy \quad .$$

The third property, the continuity of Ψ, is worth a bit more discussion. When we rewrite equation (19.6) using the definition of Ψ, we see that the claim of Ψ being continuous at x_0 is equivalent to the claim of

$$\lim_{x \to x_0} \int_{-\infty}^{\infty} \phi(y) e^{\pm i2\pi xy} \, dy = \int_{-\infty}^{\infty} \phi(y) e^{\pm i2\pi x_0 y} \, dy \quad , \tag{19.7}$$

or, equivalently, since the exponential is continuous,

$$\lim_{x \to x_0} \int_{-\infty}^{\infty} \phi(y) e^{\pm i2\pi xy} \, dy = \int_{-\infty}^{\infty} \lim_{x \to x_0} \phi(y) e^{\pm i2\pi xy} \, dy \quad . \tag{19.8}$$

These equations may appear reasonable. This is, however, somewhat deceptive. It is quite possible to have a continuous function $f(x, y)$ such that

$$\lim_{x \to x_0} \int_{\alpha}^{\beta} f(x, y) \, dy \neq \int_{\alpha}^{\beta} \lim_{x \to x_0} f(x, y) \, dy$$

(see exercise 18.17 on page 277). So the third property, above, assures us that this unfortunate situation does not happen if $f(x, y) = \phi(y)e^{\pm i2\pi xy}$ and ϕ is in $\mathcal{A}$.

The proof of Ψ's continuity is a bit more involved than were the proofs of the previous two properties. It will be discussed a little later in this section.

The statement of the fourth property is known as the Riemann–Lebesgue lemma.[6] We will also delay the proof of this property until later in this section, both because it is somewhat detailed, and because we will need an inequality derived while proving the continuity of Ψ. (For a geometric interpretation of the Riemann–Lebesgue lemma, see exercise 19.19 at the end of this chapter.)

Look at what these properties tell us: If a given function is a Fourier integral transform of an absolutely integrable function, then that given function must be continuous, bounded and must vanish at infinity. Conversely, *if a given function is not continuous or is not bounded or does not vanish at infinity, then it cannot be a Fourier transform of an absolutely integrable function!* This fact will be important in determining which functions are transformable under the classical theory of Fourier transforms.

!▶ Example 19.5: The pulse function $\text{pulse}_a(x)$ is not continuous at $x = \pm a$. Thus, $\text{pulse}_a(x)$ *cannot be the Fourier integral transform (or Fourier inverse integral transform) of any absolutely integrable, piecewise continuous function. (Combined with the fundamental theorem on invertibility, this also tells us that the sinc function is not absolutely integrable on the real line.)*

!▶ Example 19.6: The function x^{-2} is not bounded — it blows up at $x = 0$. Thus, x^{-2} *cannot be the Fourier integral transform (or Fourier inverse integral transform) of any absolutely integrable, piecewise continuous function.*

!▶ Example 19.7: The constant function $f(x) = 1$ does not vanish as $x \to \infty$. Thus, it *cannot be the Fourier integral transform (or Fourier inverse integral transform) of any absolutely integrable, piecewise continuous function.*

?▶ Exercise 19.9: Verify that all the transforms computed previously in this chapter are bounded, continuous and vanish at infinity.

?▶ Exercise 19.10: Which of the following functions cannot be the Fourier integral transform (or Fourier inverse integral transform) of an absolutely integrable, piecewise continuous function:

$$\sin(x) \quad , \quad x^2 \quad , \quad e^{-x^2} \quad , \quad \ln|x| \quad \text{and} \quad \text{step}(x) \quad ?$$

?▶ Exercise 19.11: To see how transforms of functions from $\mathcal{A}$ can truly be "nicer" than the original functions, come up with

a: an example of a function from $\mathcal{A}$ that is not bounded,

b: an example of a function from $\mathcal{A}$ that is not continuous, and

c: an example of a function from $\mathcal{A}$ that does not vanish at infinity.

(Hint: Consider the functions in some of the examples and exercises in the previous chapter.)

---

[6] There are several versions of the Riemann–Lebesgue lemma, including versions that arise in the study of Fourier series. See, for example, theorem 11.5 on page 141 and lemma 14.4 on page 181.

Verifying the Continuity

Our goal here is to verify property 3 in theorem 19.6. That is, assuming ϕ is in $\mathcal{A}$, we want to show that

$$\Psi(x) = \int_{-\infty}^{\infty} \phi(y) \, e^{\pm i 2\pi x y} \, dy$$

is continuous at each point on the real line.

We should start by observing that if ϕ is zero everywhere on the real line, then Ψ also vanishes on the real line and, so, is obviously continuous. Accordingly, for the rest of our discussion we may (and will) assume ϕ is nonzero over some interval. This will ensure that we don't divide by zero at one point.

The quickest (legitimate) way to verify that Ψ is continuous would probably be to apply the theorem on the continuity of an integral with a parameter (corollary 18.19 on page 271). Instead, we'll undertake a slightly more detailed analysis that will also give us a bound useful in proving the Riemann–Lebesgue lemma. That analysis requires an equality and an inequality that you can easily verify.

?▶Exercise 19.12:   *Verify that*

$$\left| e^{\pm i 2\Theta} - 1 \right| = 2 \, |\sin(\Theta)| \qquad \text{whenever } \Theta \text{ is a real number} \quad .$$

?▶Exercise 19.13:   *Show that*

$$|\sin(\Theta)| \leq |\Theta| \qquad \text{whenever } \Theta \text{ is a real number} \quad .$$

To confirm the continuity of Ψ on the real line, we need to show that

$$\lim_{s \to x} \Psi(s) = \Psi(x) \qquad \text{for every } x \text{ in } \mathbb{R} \quad .$$

Letting $s = x + \Delta x$ this becomes

$$\lim_{\Delta x \to 0} \Psi(x + \Delta x) = \Psi(x) \quad ,$$

which is the same as

$$\lim_{\Delta x \to 0} |\Psi(x + \Delta x) - \Psi(x)| = 0 \quad .$$

Recall[7] that, to confirm this last limit, it suffices to show there is a $\delta_\epsilon > 0$ for each $\epsilon > 0$ such that

$$|\Psi(x + \Delta x) - \Psi(x)| \leq \epsilon \qquad \text{whenever} \qquad |\Delta x| < \delta_\epsilon \quad .$$

We will show this (and, hence, the continuity of Ψ) by deriving, via several strings of inequalities, a fairly explicit formula for δ_ϵ.

So let ϵ be some arbitrary positive value. Using the identity from the first exercise above, we see that

$$|\Psi(x + \Delta x) - \Psi(x)| = \left| \int_{-\infty}^{\infty} \phi(y) \, e^{\pm i 2\pi (x + \Delta x) y} \, dy - \int_{-\infty}^{\infty} \phi(y) \, e^{\pm i 2\pi x y} \, dy \right|$$

$$= \left| \int_{-\infty}^{\infty} \phi(y) \, e^{\pm i 2\pi x y} \left[e^{\pm i 2\pi \Delta x \, y} - 1 \right] dy \right|$$

---

[7] This may be a good time to review *A Refresher on Limits* starting on page 27.

$$\leq \int_{-\infty}^{\infty} \left| \phi(y) \, e^{\pm i 2\pi xy} \left[e^{\pm i 2\pi \Delta x \, y} - 1 \right] \right| dy$$

$$= 2 \int_{-\infty}^{\infty} |\phi(y)| \, |\sin(\pi \Delta x \, y)| \, dy \quad . \tag{19.9}$$

From lemma 18.11 on page 267 we know there is a finite distance ℓ_ϵ such that

$$\int_{-\infty}^{-\ell_\epsilon} |\phi(y)| \, dy \leq \tfrac{1}{8}\epsilon \qquad \text{and} \qquad \int_{\ell_\epsilon}^{\infty} |\phi(y)| \, dy \leq \tfrac{1}{8}\epsilon \quad .$$

Using this, the inequality from exercise 19.13 and the fact that the sine function is bounded by 1, we get

$$\int_{-\infty}^{\infty} |\phi(y)| \, |\sin(\pi \Delta x \, y)| \, dy$$

$$= \int_{-\infty}^{-\ell_\epsilon} |\phi(y)| \, |\sin(\pi \Delta x \, y)| \, dy + \int_{-\ell_\epsilon}^{\ell_\epsilon} |\phi(y)| \, |\sin(\pi \Delta x \, y)| \, dy$$

$$+ \int_{\ell_\epsilon}^{\infty} |\phi(y)| \, |\sin(\pi \Delta x \, y)| \, dy$$

$$\leq \int_{-\infty}^{-\ell_\epsilon} |\phi(y)| \, dy + \int_{-\ell_\epsilon}^{\ell_\epsilon} |\phi(y)| \, |\pi \Delta x \, y| \, dy + \int_{\ell_\epsilon}^{\infty} |\phi(y)| \, dy$$

$$\leq \tfrac{1}{4}\epsilon + |\Delta x| \, \pi \int_{-\ell_\epsilon}^{\ell_\epsilon} |\phi(y)| \, |y| \, dy \quad . \tag{19.10}$$

But,

$$\int_{-\ell_\epsilon}^{\ell_\epsilon} |\phi(y)| \, |y| \, dy \leq \int_{-\ell_\epsilon}^{\ell_\epsilon} |\phi(y)| \, \ell_\epsilon \, dy \leq \ell_\epsilon \int_{-\infty}^{\infty} |\phi(y)| \, dy \quad .$$

Combining this with inequalities (19.9) and (19.10) gives us

$$|\Psi(x + \Delta x) - \Psi(x)| \leq 2 \int_{-\infty}^{\infty} |\phi(y)| \, |\sin(\pi \Delta x \, y)| \, dy$$

$$\leq \tfrac{1}{2}\epsilon + |\Delta x| \, 2\pi \ell_\epsilon \int_{-\infty}^{\infty} |\phi(y)| \, dy \quad .$$

Thus, setting

$$\delta_\epsilon = \tfrac{1}{2}\epsilon \left(2\pi \ell_\epsilon \int_{-\infty}^{\infty} |\phi(y)| \, dy \right)^{-1} \quad ,$$

we have that, whenever $|\Delta x| < \delta_\epsilon$,

$$|\Psi(x + \Delta x) - \Psi(x)| \leq \tfrac{1}{2}\epsilon + |\Delta x| \, 2\pi \ell_\epsilon \int_{-\infty}^{\infty} |\phi(y)| \, dy$$

$$\leq \tfrac{1}{2}\epsilon + \tfrac{1}{2}\epsilon \left(2\pi \ell_\epsilon \int_{-\infty}^{\infty} |\phi(y)| \, dy \right)^{-1} 2\pi \ell_\epsilon \int_{-\infty}^{\infty} |\phi(y)| \, dy$$

$$= \tfrac{1}{2}\epsilon + \tfrac{1}{2}\epsilon = \epsilon \quad . \qquad \blacksquare$$

By deriving a formula for δ_ϵ that does not depend on x, we have actually shown that Ψ is not merely continuous on the real line — it is uniformly continuous. For reference in our proof of the Riemann–Lebesgue lemma, let us formally re-state what we have just derived.

Lemma 19.7
Assume ϕ is a nontrivial function $\mathcal{A}$, and let

$$\Psi(x) = \int_{-\infty}^{\infty} \phi(y)\, e^{\pm i 2\pi x y}\, dy \quad .$$

For each $\epsilon > 0$, let ℓ_ϵ be any positive value such that

$$\int_{-\infty}^{-\ell_\epsilon} |\phi(y)|\, dy \leq \frac{1}{8}\epsilon \quad \text{and} \quad \int_{\ell_\epsilon}^{\infty} |\phi(y)|\, dy \leq \frac{1}{8}\epsilon \quad .$$

Also, let

$$\delta_\epsilon = \frac{1}{2}\epsilon \left(2\pi \ell_\epsilon \int_{-\infty}^{\infty} |\phi(y)|\, dy \right)^{-1} \quad .$$

Then

$$|\Psi(\bar{x}) - \Psi(x)| \leq \epsilon \quad \text{whenever} \quad |\bar{x} - x| < \delta_\epsilon \quad .$$

Verifying the Riemann–Lebesgue Lemma

Since the fourth property described in theorem 19.6 is, itself, a famous theorem in integration theory (although traditionally called a lemma), let us state it as such:

Theorem 19.8 (Riemann–Lebesgue lemma)
Let ϕ be absolutely integrable and piecewise continuous on the real line, and let

$$\Psi(x) = \int_{-\infty}^{\infty} \phi(y)\, e^{\pm i 2\pi x y}\, dy \quad .$$

Then

$$\lim_{x \to \pm\infty} \Psi(x) = 0 \quad .$$

PROOF: Again, the claim of this theorem is clearly true if ϕ is zero everywhere on the real line. So, in what follows we may (and will) make the additional assumption that ϕ is nonzero over some interval.

By the basic definition, it will suffice to show that, for any $\epsilon > 0$, there is a corresponding distance X_ϵ such that

$$|\Psi(x)| \leq \epsilon \quad \text{whenever} \quad X_\epsilon \leq |x| \quad .$$

We will show this by using the uniform continuity of Ψ derived in the previous section along with the version of the Riemann–Lebesgue lemma obtained for Fourier series in chapter 14.

We start by letting ϵ be any positive value. Set $\gamma = \epsilon/3$, choose ℓ_γ to be any positive real number such that

$$\int_{-\infty}^{-\ell_\gamma} |\phi(y)|\, dy \leq \frac{1}{8}\gamma \quad \text{and} \quad \int_{\ell_\gamma}^{\infty} |\phi(y)|\, dy \leq \frac{1}{8}\gamma \quad ,$$

and set

$$\delta_\epsilon = \frac{1}{2}\gamma \left(2\pi\,\ell_\gamma \int_{-\infty}^{\infty} |\phi(y)|\, dy \right)^{-1} \quad .$$

Remember, from lemma 19.7, we know

$$|\Psi(x) - \Psi(\bar{x})| \le \gamma \qquad \text{whenever} \qquad |x - \bar{x}| \le \delta_\epsilon \quad . \tag{19.11}$$

Now (this is the clever part) choose any finite real value p large enough that

$$\frac{1}{p} < \delta_\epsilon \qquad \text{and} \qquad 2\ell_\gamma < p \quad . \tag{19.12}$$

For convenience, let $v = \frac{1}{p}$. Observe that, by inequality (19.11) and the first inequality in set (19.12),

$$|\Psi(x) - \Psi(\bar{x})| \le \gamma \qquad \text{whenever} \qquad |x - \bar{x}| \le v \quad . \tag{19.13}$$

Next consider $\Psi(kv)$ for $k = 0, \pm 1, \pm 2, \dots$. From the second of the two inequalities in set (19.12),

$$\left| \int_{-\infty}^{-p/2} \phi(y)\, e^{\pm i 2\pi k v y}\, dy \right| \le \int_{-\infty}^{-p/2} \left| \phi(y)\, e^{\pm i 2\pi k v y} \right| dy$$

$$= \int_{-\infty}^{-p/2} |\phi(y)|\, dy$$

$$\le \int_{-\infty}^{-\ell_\gamma} |\phi(y)|\, dy \le \frac{1}{8}\gamma \quad .$$

Likewise,

$$\left| \int_{p/2}^{\infty} \phi(y)\, e^{\pm i 2\pi k v y}\, dy \right| \le \frac{1}{8}\gamma \quad .$$

So,

$$|\Psi(kv)| = \left| \int_{-\infty}^{\infty} \phi(y)\, e^{\pm i 2\pi k v y}\, dy \right|$$

$$\le \left| \int_{-\infty}^{-p/2} \phi(y)\, e^{\pm i 2\pi k v y}\, dy \right| + \left| \int_{-p/2}^{p/2} \phi(y)\, e^{\pm i 2\pi k v y}\, dy \right|$$

$$+ \left| \int_{p/2}^{\infty} \phi(y)\, e^{\pm i 2\pi k v y}\, dy \right|$$

$$\le \frac{1}{4}\gamma + \left| \int_{-p/2}^{p/2} \phi(y)\, e^{\pm i 2\pi k v y}\, dy \right| \quad . \tag{19.14}$$

But, from the Riemann–Lebesgue lemma for Fourier series (lemma 14.4 on page 181), we also know that

$$\lim_{k \to \pm\infty} \int_{-p/2}^{p/2} \phi(y)\, e^{\pm i 2\pi k v y}\, dy = 0 \quad ,$$

which means that there must be a positive integer N_γ such that

$$\left| \int_{-p/2}^{p/2} \phi(y)\, e^{\pm i 2\pi k v y}\, dy \right| \le \gamma \qquad \text{whenever} \qquad N_\gamma \le |k| \quad .$$

With this and inequality (19.14) we then have

$$|\Psi(kv)| \leq \tfrac{1}{4}\gamma + \gamma < 2\gamma \qquad \text{whenever} \qquad N_\gamma \leq |k| \quad . \tag{19.15}$$

Finally, set $X_\epsilon = N_\gamma v$, and let x be any real value with $|x| \geq X_\epsilon$. Clearly, if x is positive, then it must be within v of one of the following points:

$$N_\gamma v \quad , \quad (1 + N_\gamma)v \quad , \quad (2 + N_\gamma)v \quad , \quad (3 + N_\gamma)v \quad , \quad \ldots \quad ;$$

while if x is negative, then it must be within v of one of the following points:

$$-N_\gamma v \quad , \quad -(1 + N_\gamma)v \quad , \quad -(2 + N_\gamma)v \quad , \quad (-3 + N_\gamma)v \quad , \quad \ldots \quad .$$

In other words, there is an integer k with $N_\gamma \leq |k|$ such that $|x - kv| \leq v$. Thus, we can apply both inequalities (19.13) and (19.15), obtaining

$$|\Psi(x)| = |\Psi(kv) + \Psi(x) - \Psi(kv)|$$
$$\leq |\Psi(kv)| + |\Psi(x) - \Psi(kv)| < 2\gamma + \gamma \quad ,$$

which, because $\gamma = \epsilon/3$ and x is any real value with $|x| \geq X_\epsilon$, means that

$$|\Psi(x)| \leq \epsilon \qquad \text{whenever} \qquad X_\epsilon \leq |x| \quad . \qquad \blacksquare$$

Additional Exercises

19.14. *In the following, a and b denote real numbers with $a > 0$.*

    a. Find $\mathcal{F}_I\left[e^{(-a+ib)y}\, \text{step}(y)\right]\big|_x$ *by computing the appropriate integral.*

    b. *Using your answer to the above, find each of the following:*

        i. $\mathcal{F}_I\left[e^{(-2+i3)y}\, \text{step}(y)\right]\big|_x$ **ii.** $\mathcal{F}_I\left[e^{(-2+i3)y}\, \text{step}(y)\right]\big|_5$

        iii. $\mathcal{F}_I\left[e^{(-2-i4\pi)y}\, \text{step}(y)\right]\big|_5$ **iv.** $\mathcal{F}_I\left[e^{(-a-ib)y}\, \text{step}(y)\right]\big|_x$

    c. *Find each of the following. Do not evaluate any integrals. Instead, use your answers to the above and near-equivalence.*

        i. $\mathcal{F}_I^{-1}\left[e^{(-a+ib)y}\, \text{step}(y)\right]\big|_x$ **ii.** $\mathcal{F}_I\left[e^{(a+ib)y}\, \text{step}(-y)\right]\big|_x$

        iii. $\mathcal{F}_I^{-1}\left[e^{(a+ib)y}\, \text{step}(-y)\right]\big|_x$

    d. *Find each of the following using your answers to the above and linearity.*

        i. $\mathcal{F}_I\left[e^{-a|y|}e^{iby}\right]\big|_x$

        ii. $\mathcal{F}_I\left[e^{-ay}\cos(by)\, \text{step}(y)\right]\big|_x$ *(Express the cosine in terms of complex exponentials.)*

19.15 a. *Find each of the following by computing the integral in the integral formula.*

    i. $\mathcal{F}_I\left[\text{rect}_{(0,1)}(y)\right]\big|_x$ **ii.** $\mathcal{F}_I\left[y\,\text{rect}_{(0,1)}(y)\right]\big|_x$

  b. *Verify that* $\text{rect}_{(-1,0)}(y) = \text{rect}_{(0,1)}(-y)$ *and find the following using your answers to the above, near-equivalence and linearity.*

    i. $\mathcal{F}_I\left[\text{rect}_{(-1,0)}(y)\right]\big|_x$ **ii.** $\mathcal{F}_I\left[y\,\text{rect}_{(-1,0)}(y)\right]\big|_x$

    iii. $\mathcal{F}_I^{-1}\left[\text{rect}_{(-1,0)}(y)\right]\big|_x$ **iv.** $\mathcal{F}_I\left[(1+y)\,\text{rect}_{(-1,0)}(y)\right]\big|_x$

    v. $\mathcal{F}_I\left[(1-y)\,\text{rect}_{(0,1)}(y)\right]\big|_x$

  c. *The basic triangle function* $\text{tri}(y)$ *is given by*

$$\text{tri}(y) = \begin{cases} 1+y & \text{if } -1 < y < 0 \\ 1-y & \text{if } 0 < y < 1 \\ 0 & \text{otherwise} \end{cases} \quad .$$

    i. *Sketch the graph of this function.*

    ii. *Express this function in terms of rectangle functions.*

    iii. *Using results from previous problems and properties of the transforms, show that*

$$\mathcal{F}_I[\text{tri}]\big|_x = \text{sinc}^2(\pi x) \quad .$$

19.16. *Let* a *and* b *be nonzero real numbers with* $a > 0$.

  a. *By computing the integral in the integral formula (as in example 19.1 on page 280), show that*[8]

$$\mathcal{F}_I\left[e^{i2\pi by}\,\text{pulse}_a(y)\right]\bigg|_x = 2a\,\text{sinc}(2\pi a[x-b])$$

  b. *Using near-equivalence, linearity and the above, find:*

    i. $\mathcal{F}_I^{-1}\left[e^{i2\pi by}\,\text{pulse}_a(y)\right]\big|_x$ **ii.** $\mathcal{F}_I\left[e^{-i2\pi by}\,\text{pulse}_a(y)\right]\big|_x$

    iii. $\mathcal{F}_I\left[\sin(2\pi by)\,\text{pulse}_a(y)\right]\big|_x$ **iv.** $\mathcal{F}_I\left[\cos(2\pi by)\,\text{pulse}_a(y)\right]\big|_x$

    v. $\mathcal{F}_I^{-1}\left[\sin(2\pi by)\,\text{pulse}_a(y)\right]\big|_x$ **vi.** $\mathcal{F}_I^{-1}\left[\cos(2\pi by)\,\text{pulse}_a(y)\right]\big|_x$

19.17. *In the following,* ϕ *denotes a piecewise continuous, absolutely integrable function on the real line.*

  a. *Assume* ϕ *is real valued and even. Show that* $\mathcal{F}_I[\phi]$ *is also real valued and even, and that*

$$\mathcal{F}_I[\phi]\big|_x = 2\int_0^\infty \phi(y)\cos(2\pi xy)\,dy \quad .$$

  b. *Assume* ϕ *is real valued and odd. Show that* $\mathcal{F}_I[\phi]$ *is imaginary valued and odd, and that*

$$\mathcal{F}_I[\phi]\big|_x = -2i\int_0^\infty \phi(y)\sin(2\pi xy)\,dy \quad .$$

[8] A simpler approach to computing this will be discussed in chapter 21.

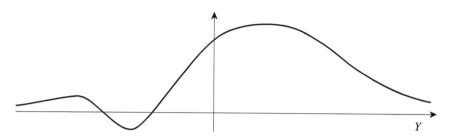

Figure 19.2: The graph of the function $f(y)$ for exercise 19.19.

19.18. Let $a > 0$,

$$h(y) = e^{-a|y|} \qquad \text{and} \qquad H(x) = \frac{2a}{a^2 + 4\pi^2 x^2} \quad .$$

From the work in this and the previous chapter, we know that both of these functions are in $\mathcal{A}$ and that $H = \mathcal{F}_I[h]$. Thus, by the fundamental theorem on invertibility,

$$h(y) = \mathcal{F}_I^{-1}[H]|_y = \int_{-\infty}^{\infty} \frac{2a}{a^2 + 4\pi^2 x^2} e^{i2\pi xy} dx \quad .$$

Use this fact in doing the following exercises.

a. Find the following transforms:

   i. $\mathcal{F}_I^{-1}\left[\dfrac{1}{a^2 + 4\pi^2 x^2}\right]\Big|_y$ **ii.** $\mathcal{F}_I\left[\dfrac{1}{a^2 + 4\pi^2 x^2}\right]\Big|_y$

   iii. $\mathcal{F}_I\left[\dfrac{1}{a^2 + x^2}\right]\Big|_y$ *(Hint: Multiply the numerator and denominator by $4\pi^2$.)*

b. Evaluate the following integrals:

   i. $\displaystyle\int_{-\infty}^{\infty} \frac{e^{i2\pi x}}{1 + 4\pi^2 x^2} dx$ **ii.** $\displaystyle\int_{-\infty}^{\infty} \frac{e^{i4\pi x}}{9 + 4\pi^2 x^2} dx$

   iii. $\displaystyle\int_{-\infty}^{\infty} \frac{1}{1 + 4\pi^2 x^2} dx$ **iv.** $\displaystyle\int_{-\infty}^{\infty} \frac{\sin(2\pi x)}{1 + 4\pi^2 x^2} dx$

   v. $\displaystyle\int_{-\infty}^{\infty} \frac{\cos(2\pi x)}{1 + 4\pi^2 x^2} dx$

19.19. Let $F(x) = \mathcal{F}_I[f]|_x$ where f is the function sketched in figure 19.2. Go ahead and assume f is real valued, continuous and absolutely integrable.

a. Sketch, as a function of y, the real and imaginary parts of $f(y)e^{-i2\pi xy}$ for some fixed positive value x (choose x large enough that your graphs contain several "humps").

b. What happens to the graphs of the real and imaginary parts of $f(y)e^{-i2\pi xy}$ as x gets larger? In particular, what about the areas contained in adjacent "humps" above and below the Y–axis?

c. Develop a geometric (and non-rigorous) argument that $F(x) \to 0$ as $x \to \infty$ based on the "near-cancellation of areas in adjacent 'humps' in the graphs of the real and imaginary parts of $f(y)e^{-i2\pi xy}$". (This is the geometric interpretation of the Riemann–Lebesgue lemma.)

20

Classical Fourier Transforms and Classically Transformable Functions

We know that

$$\mathcal{F}_I\big[e^{-y}\,\text{step}(y)\big]\big|_x \;=\; \int_{-\infty}^{\infty} e^{-y}\,\text{step}(y)\,e^{-i2\pi xy}\,dy \;=\; \frac{1}{1 + i2\pi x} \quad .$$

Now if $(1 + i2\pi x)^{-1}$, the function on the right-hand side of these equations, were absolutely integrable on the real line, then its integral inverse Fourier transform would be defined by the integral formula for $\mathcal{F}_I^{-1}$, and the fundamental theorem on invertibility would assure us that

$$\mathcal{F}_I^{-1}\Big[\frac{1}{1 + i2\pi x}\Big]\Big|_y \;=\; \int_{-\infty}^{\infty} \frac{1}{1 + i2\pi x}\,e^{i2\pi xy}\,dx \;=\; e^{-y}\,\text{step}(y) \quad .$$

But, as you verified in exercise 18.7 on page 264, $(1 + i2\pi x)^{-1}$ is not absolutely integrable. So, we cannot invoke the fundamental theorem on invertibility to evaluate its Fourier inverse integral transform.

In fact, since $(1 + i2\pi x)^{-1}$ is not absolutely integrable, its Fourier inverse integral transform is not even defined.

So why don't we just

$$\text{define} \qquad \mathcal{F}_I^{-1}\Big[\frac{1}{1 + i2\pi x}\Big]\Big|_y \qquad \text{to be} \qquad e^{-y}\,\text{step}(y) \quad ?$$

Basically, that is just what we will do in this chapter. We will extend our definitions for the Fourier transforms in this and one other rather obvious manner, and we will verify

1. that these extensions are legitimate extensions (i.e., give the same results as the integral transforms of chapter 19 when used to compute the transforms of functions in $\mathcal{A}$),

and

2. that the properties of linearity, near-equivalence and invertibility hold using the extended definitions.

On occasion, we will refer to some formulas derived in chapter 19. To simplify matters, these formulas are summarized in table 20.1.

Table 20.1: Selected Integral Transforms from Chapter 19. (a and b are real constants.)

$f(y)$	$F(x) = \mathcal{F}_I[f(y)]\|_x$	Restrictions	See
$f(y)$	$\displaystyle\int_{-\infty}^{\infty} f(y)\,e^{-i2\pi xy}\,dy$	$f \in \mathcal{A}$	formula 19.1, page 279
$\text{pulse}_a(y)$	$2a\,\text{sinc}(2\pi ax)$	$0 < a$	example 19.1, page 280
$e^{-ay}\,\text{step}(y)$	$\dfrac{1}{a + i2\pi x}$	$0 < a$	exercise 19.1, page 280
$e^{ay}\,\text{step}(-y)$	$\dfrac{1}{a - i2\pi x}$	$0 < a$	exercise 19.3, page 282
$e^{-a|y|}$	$\dfrac{2a}{a^2 + 4\pi^2 x^2}$	$0 < a$	example 19.3, page 284
$\dfrac{1}{a^2 + y^2}$	$\dfrac{\pi}{a}e^{-2\pi a|x|}$	$0 < a$	exercise 19.18, page 296
$e^{i2\pi by}\,\text{pulse}_a(y)$	$2a\,\text{sinc}(2\pi a[x - b])$	$0 < a$	exercise 19.16, page 295

20.1 The First Extension
The Set of Integral Transforms

Recall again that $\mathcal{A}$ denotes the set of all functions that are piecewise continuous and absolutely integrable on the entire real line. For convenience, $\mathcal{T}$ will denote the set of all functions on $\mathbb{R}$ that are Fourier integral transforms of functions from $\mathcal{A}$. That is,

$$\mathcal{T} = \left\{ \Psi : \Psi = \mathcal{F}_I[\phi] \text{ for some } \phi \text{ in } \mathcal{A} \right\} \quad .$$

This will allow us to say "Ψ is in $\mathcal{T}$" as a shorthand for "Ψ is the Fourier integral transform of some piecewise continuous, absolutely integrable function on the real line".

The astute reader probably already realizes that, because of the near-equivalence of the transforms, $\mathcal{T}$ is also the set of all Fourier inverse integral transforms of functions in $\mathcal{A}$,

$$\mathcal{T} = \left\{ \Psi : \Psi = \mathcal{F}_I^{-1}[\phi] \text{ for some } \phi \text{ in } \mathcal{A} \right\} \quad .$$

Thus, we can also use the phrase "Ψ is in $\mathcal{T}$" as shorthand for "Ψ is the Fourier inverse integral transform of some piecewise continuous, absolutely integrable function on the real line".

Our basic plan is to define, say, the inverse transform of any given Ψ in $\mathcal{T}$ to be the function ϕ in $\mathcal{A}$ such that $\Psi = \mathcal{F}_I[\phi]$. This plan will work fine so long as there is only one function ϕ for which $\Psi = \mathcal{F}_I[\phi]$. If there is a second function ψ with $\Psi = \mathcal{F}_I[\psi]$, then we have a problem. Just which function, ϕ or ψ, do we use as the inverse transform Ψ? Fortunately, as our next lemma states, this difficulty does not arise.

Lemma 20.1 *(uniqueness of the integral transforms)*
Let Ψ be a function in $\mathcal{T}$. Then there is exactly one function f in $\mathcal{A}$ such that $\Psi = \mathcal{F}_I[f]$, and there is exactly one function g in $\mathcal{A}$ such that $\Psi = \mathcal{F}_I^{-1}[g]$. Moreover, $g(x) = f(-x)$.

PROOF: Because of what it means for Ψ to be in $\mathcal{T}$, there must be at least one function f in $\mathcal{A}$ with $\Psi = \mathcal{F}_I[f]$. Let g be the corresponding function given by $g(x) = f(-x)$. By the principle of near-equivalence (theorem 19.1 on page 281) g is also in $\mathcal{A}$ and

$$\Psi = \mathcal{F}_I[f] = \mathcal{F}_I^{-1}[g] \quad .$$

Thus, there is at least one f in $\mathcal{A}$ and one g in $\mathcal{A}$ such that $\Psi = \mathcal{F}_I[f]$, $\Psi = \mathcal{F}_I^{-1}[g]$ and $g(x) = f(-x)$.

What remains is to verify that f is the only function whose direct Fourier integral transform is Ψ. To do this, let ϕ be any function in $\mathcal{A}$ satisfying $\Psi = \mathcal{F}_I[\phi]$. By linearity (theorem 19.4), we know that $\phi - f$ is in $\mathcal{A}$ and that

$$\mathcal{F}_I[\phi - f] = \mathcal{F}_I[\phi] - \mathcal{F}_I[f] = \Psi - \Psi = 0 \quad .$$

But the zero function is certainly piecewise continuous and absolutely integrable on the real line. So the fundamental theorem on invertibility (theorem 19.5 on page 285) assures us that

$$\phi - f = \mathcal{F}_I^{-1}[0] = \int_{-\infty}^{\infty} 0 \cdot e^{i2\pi xy} \, dy = 0 \quad .$$

Therefore,

$$\phi = f \quad .$$

Virtually identical arguments can be used to show that there are no functions in $\mathcal{A}$ other than g whose Fourier inverse integral transform equals Ψ. (Or you can use near-equivalence and the fact that f is the only function in $\mathcal{A}$ whose Fourier integral transform equals Ψ.) ∎

Transforms of Integral Transforms

In light of our last lemma, we can now define the Fourier transforms of functions in $\mathcal{T}$. Since integrals are not directly used in these definitions, we will not refer to them as integral transforms, and we will not include that irritating subscript I in the notation.

Let Ψ be a function in $\mathcal{T}$. The *(direct) Fourier transform* of Ψ, denoted by $\mathcal{F}[\Psi]$, is defined to be the function in $\mathcal{A}$ whose Fourier inverse integral transform equals Ψ. In other words, we define $\mathcal{F}[\Psi]$ to be the absolutely integrable, piecewise continuous function g that makes the following mathematical statement true:

$$\mathcal{F}[\Psi] = g \quad \Longleftrightarrow \quad \Psi(x) = \mathcal{F}_I^{-1}[g]\big|_x = \int_{-\infty}^{\infty} g(y) \, e^{i2\pi xy} \, dy \quad .$$

Likewise, the *inverse Fourier transform* of Ψ, denoted by $\mathcal{F}^{-1}[\Psi]$, is defined to be the function in $\mathcal{A}$ whose direct Fourier integral transform equals Ψ. That is, $\mathcal{F}^{-1}[\Psi]$ is defined to be the absolutely integrable, piecewise continuous function f that makes the following mathematical statement true:

$$\mathcal{F}^{-1}[\Psi] = f \quad \Longleftrightarrow \quad \Psi(x) = \mathcal{F}_I[f]\big|_x = \int_{-\infty}^{\infty} f(y) \, e^{-i2\pi xy} \, dy \quad .$$

!►Example 20.1:   Because $e^{-y}\,\text{step}(y)$ *is in* $\mathcal{A}$, *and*

$$\mathcal{F}_I^{-1}\!\left[e^{-y}\,\text{step}(y)\right]\Big|_x = \int_{-\infty}^{\infty} e^{-y}\,\text{step}(y)\,e^{i2\pi xy}\,dy = \frac{1}{1 - i2\pi x} \quad ,$$

the function $(1 - i2\pi x)^{-1}$ *is in* $\mathcal{T}$ *and*

$$\mathcal{F}\!\left[\frac{1}{1 - i2\pi x}\right]\Big|_y = e^{-y}\,\text{step}(y) \quad .$$

?►Exercise 20.1:   *Verify that*

$$\mathcal{F}^{-1}\!\left[\frac{1}{1 + i2\pi x}\right]\Big|_y = e^{-y}\,\text{step}(y) \quad .$$

There are functions that are in both $\mathcal{A}$ and $\mathcal{T}$. For each such function Ψ we have two definitions for a direct Fourier transform: the integral transform definition,

$$\mathcal{F}_I[\Psi]\big|_y = \int_{-\infty}^{\infty} \Psi(x)\,e^{-i2\pi xy}\,dx \quad ,$$

and the one developed in this section,

$$\mathcal{F}[\Psi] = \phi \qquad \text{where } \phi \text{ is the function in } \mathcal{A} \text{ such that } \Psi = \mathcal{F}_I^{-1}[\phi] \quad .$$

Since Ψ and ϕ are both absolutely integrable, and $\Psi = \mathcal{F}_I^{-1}[\phi]$, the fundamental theorem on invertibility holds and tells us that $\mathcal{F}_I[\Psi] = \phi$. Thus,

$$\mathcal{F}[\Psi] = \phi = \mathcal{F}_I[\Psi] \quad .$$

Likewise, in this case, we can easily verify that

$$\mathcal{F}^{-1}[\Psi] = \mathcal{F}_I^{-1}[\Psi] \quad .$$

For future reference, let us state this little observation as a lemma.

Lemma 20.2
Let Ψ *be a function in both* $\mathcal{A}$ *and* $\mathcal{T}$. *Then*

$$\mathcal{F}[\Psi] = \mathcal{F}_I[\Psi] \qquad \text{and} \qquad \mathcal{F}^{-1}[\Psi] = \mathcal{F}_I^{-1}[\Psi] \quad .$$

!►Example 20.2:   *From previous work we know that the functions*

$$\frac{2}{1 + 4\pi^2 x^2} \qquad \text{and} \qquad e^{-|y|}$$

are both absolutely integrable and that

$$\frac{2}{1 + 4\pi^2 x^2} = \mathcal{F}_I\!\left[e^{-|y|}\right]\Big|_x \quad .$$

By the definition in this section

$$\mathcal{F}^{-1}\!\left[\frac{2}{1 + 4\pi^2 x^2}\right]\Big|_y = e^{-|y|} \quad ,$$

which is exactly the same as was obtained for

$$\mathcal{F}_I^{-1}\!\left[\frac{2}{1 + 4\pi^2 x^2}\right]\Big|_y$$

in example 19.4 on page 286.

Basic Properties

The last lemma above tells us that our new definitions of the Fourier transforms are equivalent to the old definitions when both definitions can be applied. This is reassuring. Now let's see if linearity, near-equivalence and invertibility hold with our new definition.

First of all, observe that, by the definition of the Fourier transforms on $\mathcal{T}$, we automatically have the following invertibility lemma.

Lemma 20.3
Let ϕ be in $\mathcal{A}$, and let Ψ be in $\mathcal{T}$. Then

$$\Psi = \mathcal{F}_I[\phi] \quad \Longleftrightarrow \quad \mathcal{F}^{-1}[\Psi] = \phi$$

and

$$\phi = \mathcal{F}[\Psi] \quad \Longleftrightarrow \quad \mathcal{F}_I^{-1}[\phi] = \Psi \quad .$$

Linearity and near-equivalence follow fairly directly from the linearity and near-equivalence of the integral transforms.

Lemma 20.4
Let ϕ and ψ be any two functions in $\mathcal{T}$, and let a and b be any two complex numbers. Then the linear combination $a\phi + b\psi$ is in $\mathcal{T}$. Moreover,

$$\mathcal{F}[a\phi + b\psi] = a\mathcal{F}[\phi] + b\mathcal{F}[\psi] \tag{20.1}$$

and

$$\mathcal{F}^{-1}[a\phi + b\psi] = a\mathcal{F}^{-1}[\phi] + b\mathcal{F}^{-1}[\psi] \quad . \tag{20.2}$$

PROOF: Let $f = \mathcal{F}[\phi]$ and $g = \mathcal{F}[\psi]$. By the definition of transforms of functions in $\mathcal{T}$, f and g must be the two functions in $\mathcal{A}$ such that $\phi = \mathcal{F}_I^{-1}[f]$ and $\psi = \mathcal{F}_I^{-1}[g]$. Since we know the integral transforms of functions in $\mathcal{A}$ are linear, we know that $af + bg$ is in $\mathcal{A}$ and that

$$a\phi + b\psi = a\mathcal{F}_I^{-1}[f] + b\mathcal{F}_I^{-1}[g] = \mathcal{F}_I^{-1}[af + bg] \quad ,$$

showing that the linear combination $a\phi + b\psi$, being a Fourier inverse integral transform of a function in $\mathcal{A}$, is in $\mathcal{T}$. By the definition, the direct Fourier transform of this linear combination is obtained by "inverting" this last equality. Doing this inversion and using the definitions of f and g give

$$\mathcal{F}[a\phi + b\psi] = af + bg = a\mathcal{F}[\phi] + b\mathcal{F}[\psi] \quad ,$$

confirming equation (20.1). Equation (20.2) can then be confirmed by virtually identical arguments using $f = \mathcal{F}^{-1}[\phi]$ and $g = \mathcal{F}^{-1}[\psi]$. The details will be left as an exercise. ∎

?► Exercise 20.2: *Prove that equation (20.2) holds in the above lemma.*

Lemma 20.5
Let $\phi(y)$ be a function in $\mathcal{T}$. Then the function $\phi(-y)$ is also in $\mathcal{T}$. Moreover,

$$\mathcal{F}[\phi(y)]|_x = \mathcal{F}^{-1}[\phi(y)]|_{-x} = \mathcal{F}^{-1}[\phi(-y)]|_x \tag{20.3}$$

and

$$\mathcal{F}^{-1}[\phi(y)]|_x = \mathcal{F}[\phi(y)]|_{-x} = \mathcal{F}[\phi(-y)]|_x \quad . \tag{20.4}$$

PROOF: Let $g = \mathcal{F}[\phi]$ and $f = \mathcal{F}^{-1}[\phi]$. From the definition of the Fourier transforms of ϕ and lemma 20.1, we know f and g are in $\mathcal{A}$, and that

$$\phi(y) = \mathcal{F}_I[f(x)]|_y \qquad , \qquad \phi(y) = \mathcal{F}_I^{-1}[g(x)]|_y$$

and

$$g(x) = f(-x) \quad .$$

By the definitions of f and g, this last equality can be written as

$$\mathcal{F}[\phi(y)]|_x = \mathcal{F}^{-1}[\phi(y)]|_{-x} \quad ,$$

verifying the first equality in equation set (20.3).

Using the first equality in the above list and the fact that near-equivalence holds for the integral transforms, we see that

$$\phi(-y) = \mathcal{F}_I[f(x)]|_{-y} = \mathcal{F}_I[f(-x)]|_y \quad ,$$

verifying that $\phi(-y)$ is in $\mathcal{T}$. Moreover, inverting this last line and using the definition of f gives

$$\mathcal{F}^{-1}[\phi(-y)]|_x = f(-x) = \mathcal{F}^{-1}[\phi(y)]|_{-x} \quad ,$$

which verifies the second equality in equation set (20.3).

As was probably expected, verifying equation set (20.4) is left as an exercise. ∎

?►Exercise 20.3:   *Verify the equations in line (20.4) of the above lemma.*

20.2   The Set of Classically Transformable Functions

The two function sets $\mathcal{A}$ and $\mathcal{T}$ can be viewed as the two components of the set of all functions for which the "classical" Fourier transforms can be defined. Accordingly, we will say that a function ψ is *classically transformable* if and only if ψ can be written

$$\psi = \psi_\mathcal{A} + \psi_\mathcal{T}$$

where $\psi_\mathcal{A}$ is some function in $\mathcal{A}$ and $\psi_\mathcal{T}$ is some function in $\mathcal{T}$.

Several simple observations are worth making at this point. The first is that, by this definition, any function in $\mathcal{A}$ or in $\mathcal{T}$ is automatically classically transformable. This is because the zero function is in both $\mathcal{A}$ and $\mathcal{T}$, and thus, can always serve as either $\psi_\mathcal{A}$ or $\psi_\mathcal{T}$. For example, we know that the function $\psi(x) = e^{-x} \operatorname{step}(x)$ is in $\mathcal{A}$. This function also satisfies our definition for being classically transformable since

$$e^{-x} \operatorname{step}(x) = \psi_\mathcal{A} + \psi_\mathcal{T}$$

where

$$\psi_\mathcal{A}(x) = e^{-x} \operatorname{step}(x) \qquad \text{and} \qquad \psi_\mathcal{T}(x) = 0 \quad .$$

A second observation is that the choice of $\psi_\mathcal{A}$ and $\psi_\mathcal{T}$ is not unique. Consider, for example, the function $\psi(x) = e^{-|x|}$. This function is in both $\mathcal{A}$ and $\mathcal{T}$. This means that we could use the pair

$$\psi_\mathcal{A}(x) = e^{-|x|} \qquad \text{and} \qquad \psi_\mathcal{T}(x) = 0 \quad ,$$

or the pair

$$\psi_A(x) = 0 \quad \text{and} \quad \psi_T(x) = e^{-|x|} \quad .$$

We could also use the pair

$$\psi_A(x) = \frac{1}{2}e^{-|x|} \quad \text{and} \quad \psi_T(x) = \frac{1}{2}e^{-|x|} \quad .$$

or even

$$\psi_A(x) = \frac{1}{3}e^{-|x|} + \frac{1}{1+x^2} \quad \text{and} \quad \psi_T(x) = \frac{2}{3}e^{-|x|} - \frac{1}{1+x^2} \quad ,$$

since $(1 + x^2)^{-1}$ is also in both $\mathcal{A}$ and $\mathcal{T}$.

For the final observation, consider finding the (direct) Fourier transform of any classically transformable function

$$\psi = \psi_A + \psi_T \quad .$$

Since ψ_A is in $\mathcal{A}$, its Fourier transform is given by the integral transform $\mathcal{F}_I[\psi_A]$. Since ψ_T is in $\mathcal{T}$, its Fourier transform, $\mathcal{F}[\psi_T]$, is as defined in the previous section. Technically, we have not yet defined $\mathcal{F}[\psi]$, the Fourier transform of the sum of ψ_A and ψ_T, but really, is there any question as to how we should define $\mathcal{F}[\psi]$? We should, naturally, assume the Fourier transform is linear, and define $\mathcal{F}[\psi]$ by

$$\mathcal{F}[\psi] = \text{Fourier transform of } \psi_A + \psi_T$$

$$= \text{Fourier transform of } \psi_A + \text{Fourier transform of } \psi_T$$

$$= \mathcal{F}_I[\psi_A] + \mathcal{F}[\psi_T] \quad . \tag{20.5}$$

That is our last simple observation of this section.

A few readers may feel uneasy about equation (20.5) because of the many possible choices for ψ_A and ψ_T.[1] Can we be sure that the above computation for $\mathcal{F}[\psi]$ will give the same result using a different choice for ψ_A and ψ_T? That is, if

$$\psi = \psi_A + \psi_T = \phi_A + \phi_T$$

where ψ_A, ψ_T and ϕ_A, ϕ_T are two different pairs of functions with ψ_A and ϕ_A in $\mathcal{A}$, and ψ_T and ϕ_T in $\mathcal{T}$, are we then certain that computing $\mathcal{F}[\psi]$ by

$$\mathcal{F}[\psi] = \mathcal{F}[\psi_A + \psi_T] = \mathcal{F}_I[\psi_A] + \mathcal{F}[\psi_T]$$

gives the same result as computing $\mathcal{F}[\psi]$ by

$$\mathcal{F}[\psi] = \mathcal{F}[\phi_A + \phi_T] = \mathcal{F}_I[\phi_A] + \mathcal{F}[\phi_T] \quad ?$$

Or is there a danger that

$$\mathcal{F}_I[\psi_A] + \mathcal{F}[\psi_T] \neq \mathcal{F}_I[\phi_A] + \mathcal{F}[\phi_T] \quad ? \tag{20.6}$$

To allay fears in this regard, note that, since

$$\psi_A + \psi_T = \psi = \phi_A + \phi_T \quad ,$$

we must have

$$\psi_A - \phi_A = \phi_T - \psi_T \quad . \tag{20.7}$$

---

[1] If you trust equation (20.5), you can skip to the next section.

Since the right-hand side of this equation is a linear combination of functions in $\mathcal{T}$, we have, using the definition for transforms of functions in $\mathcal{T}$,

$$\mathcal{F}[\phi_r - \psi_r] = \mathcal{F}[\phi_r] - \mathcal{F}[\psi_r] \quad .\tag{20.8}$$

On the other hand, the left side of equation (20.7), being a linear combination of functions in $\mathcal{A}$, is another function in $\mathcal{A}$. Thus, each side of this equation is a function in both $\mathcal{A}$ and $\mathcal{T}$, and we have

$$\mathcal{F}[\phi_r - \psi_r] = \mathcal{F}[\phi_A - \psi_A] = \mathcal{F}_I[\psi_A - \phi_A] = \mathcal{F}_I[\psi_A] - \mathcal{F}_I[\phi_A] \quad .\tag{20.9}$$

Combining equations (20.8) and (20.9) gives

$$\mathcal{F}_I[\psi_A] - \mathcal{F}_I[\phi_A] = \mathcal{F}[\phi_r - \psi_r] = \mathcal{F}[\phi_r] - \mathcal{F}[\psi_r] \quad .$$

After cutting out the middle and doing some elementary algebra, this becomes

$$\mathcal{F}_I[\psi_A] + \mathcal{F}[\psi_r] = \mathcal{F}_I[\phi_A] + \mathcal{F}[\phi_r] \quad .$$

So inequality (20.6) is not possible, and we are assured that computing $\mathcal{F}[\psi]$ by

$$\mathcal{F}[\psi] = \mathcal{F}[\psi_A + \psi_r] = \mathcal{F}_I[\psi_A] + \mathcal{F}[\psi_r]$$

gives the same result as computing $\mathcal{F}[\psi]$ by

$$\mathcal{F}[\psi] = \mathcal{F}[\phi_A + \phi_r] = \mathcal{F}_I[\phi_A] + \mathcal{F}[\phi_r] \quad .$$

20.3 The Complete Classical Fourier Transforms Definition

Let ψ be any classically transformable function. To define the *(direct) (classical) Fourier transform* of ψ and the *(classical) Fourier inverse transform* of ψ, denoted, respectively, by $\mathcal{F}[\psi]$ and $\mathcal{F}^{-1}[\psi]$, let ψ_A and ψ_r be any pair of functions with ψ_A in $\mathcal{A}$, ψ_r in $\mathcal{T}$ and

$$\psi = \psi_A + \psi_r \quad .$$

We then define $\mathcal{F}[\psi]$ and $\mathcal{F}^{-1}[\psi]$ by

$$\mathcal{F}[\psi] = \mathcal{F}_I[\psi_A] + \mathcal{F}[\psi_r]\tag{20.10}$$

and

$$\mathcal{F}^{-1}[\psi] = \mathcal{F}_I^{-1}[\psi_A] + \mathcal{F}^{-1}[\psi_r] \quad .\tag{20.11}$$

Remember that $\mathcal{F}_I[\psi_A]$ and $\mathcal{F}_I^{-1}[\psi_A]$ are given by the integral formulas from chapter 19,

$$\mathcal{F}_I[\psi_A]\big|_x = \int_{-\infty}^{\infty} \psi_A(y)\, e^{-i2\pi xy}\, dy$$

and

$$\mathcal{F}_I^{-1}[\psi_A]\big|_x = \int_{-\infty}^{\infty} \psi_A(y)\, e^{i2\pi xy}\, dy \quad ;$$

while $\mathcal{F}[\psi_\tau]$ and $\mathcal{F}^{-1}[\psi_\tau]$ are as defined in the first section of this chapter,

$$\mathcal{F}[\psi_\tau] \;=\; g \qquad \text{where } g \text{ is the function in } \mathcal{A} \text{ such that } \psi_\tau \;=\; \mathcal{F}_I^{-1}[g]$$

and

$$\mathcal{F}^{-1}[\psi_\tau] \;=\; f \qquad \text{where } f \text{ is the function in } \mathcal{A} \text{ such that } \psi_\tau \;=\; \mathcal{F}_I[f] \quad .$$

Together, $\mathcal{F}[\psi]$ and $\mathcal{F}^{-1}[\psi]$ are called the *(classical) Fourier transforms of* ψ, though $\mathcal{F}[\psi]$ is commonly referred to as "the" (classical) Fourier transform of ψ. The processes of changing a classically transformable function ψ to $\mathcal{F}[\psi]$ and to $\mathcal{F}^{-1}[\psi]$ are also referred to as the (classical) Fourier transforms. Naturally, of course, the process of converting ψ to $\mathcal{F}[\psi]$ is called the *(classical) (direct) Fourier transform* and is denoted by $\mathcal{F}$, while the process of converting ψ to $\mathcal{F}^{-1}[\psi]$ is called the *(classical) Fourier inverse transform* and is denoted by $\mathcal{F}^{-1}$.

It should be clear from our definitions and earlier discussions that the above definitions for $\mathcal{F}[\psi]$ and $\mathcal{F}^{-1}[\psi]$ reduce to earlier definitions when ψ is in either $\mathcal{A}$ or $\mathcal{T}$. In particular, if ψ is piecewise continuous and absolutely integrable, then, using $\psi_\mathcal{A} = \psi$ and $\psi_\tau = 0$, equations (20.10) and (20.11) become

$$\mathcal{F}[\psi]|_x \;=\; \mathcal{F}_I[\psi]|_x \;=\; \int_{-\infty}^{\infty} \psi(y)\, e^{-i 2\pi x y}\, dy$$

and

$$\mathcal{F}^{-1}[\psi]|_x \;=\; \mathcal{F}_I^{-1}[\psi]|_x \;=\; \int_{-\infty}^{\infty} \psi(y)\, e^{i 2\pi x y}\, dy \quad .$$

Something else worth noticing is that the right sides of formulas (20.10) and (20.11) are, themselves, sums of functions in $\mathcal{A}$ and $\mathcal{T}$. Thus, they are classically transformable functions. This is a fact which we will use so much (and usually without thinking) that we should state it as a theorem.

Theorem 20.6
Let ψ be a classically transformable function. Then its classical Fourier transforms, $\mathcal{F}[\psi]$ and $\mathcal{F}^{-1}[\psi]$, are also classically transformable functions.

Some Fundamental Properties

Based on past sections, you probably now expect some theorems on linearity, near-equivalence and invertibility.

Here they are:

Theorem 20.7 (linearity)
Let ϕ and ψ be any two classically transformable functions, and let a and b be any two complex numbers. Then the linear combination $a\phi + b\psi$ is classically transformable. Moreover,

$$\mathcal{F}[a\phi + b\psi] \;=\; a\mathcal{F}[\phi] + b\mathcal{F}[\psi] \tag{20.12}$$

and

$$\mathcal{F}^{-1}[a\phi + b\psi] \;=\; a\mathcal{F}^{-1}[\phi] + b\mathcal{F}^{-1}[\psi] \quad . \tag{20.13}$$

PROOF: Since ϕ and ψ are classically transformable, there are functions $\phi_\mathcal{A}$ and $\psi_\mathcal{A}$ in $\mathcal{A}$, and functions ϕ_τ and ψ_τ in $\mathcal{T}$ such that

$$\phi \;=\; \phi_\mathcal{A} + \phi_\tau \qquad \text{and} \qquad \psi \;=\; \psi_\mathcal{A} + \psi_\tau \quad .$$

Clearly then,

$$a\phi + b\psi = \Psi_A + \Psi_T$$

where

$$\Psi_A = a\phi_A + b\psi_A \quad \text{and} \quad \Psi_T = a\phi_T + b\psi_T \quad,$$

which we know to be in A and T, respectively, because of the linearity results already discussed for the transforms of these types of functions (see theorem 19.4 on page 283 and lemma 20.4 on page 301).

Also by these lemmas and the definition of the classical Fourier transform,

$$
\begin{aligned}
\mathcal{F}[a\phi + b\psi] &= \mathcal{F}[\Psi_A + \Psi_T] \\
&= \mathcal{F}_I[\Psi_A] + \mathcal{F}[\Psi_T] \\
&= \mathcal{F}_I[a\phi_A + b\psi_A] + \mathcal{F}[a\phi_T + b\psi_T] \\
&= a\mathcal{F}_I[\phi_A] + b\mathcal{F}_I[\psi_A] + a\mathcal{F}[\phi_T] + b\mathcal{F}[\psi_T] \\
&= a\left(\mathcal{F}_I[\phi_A] + \mathcal{F}[\phi_T]\right) + b\left(\mathcal{F}_I[\psi_A] + \mathcal{F}[\psi_T]\right) \\
&= a\mathcal{F}[\phi] + b\mathcal{F}[\psi] \quad,
\end{aligned}
$$

proving that equation (20.12) holds.

Virtually identical computations (with $\mathcal{F}^{-1}$ replacing $\mathcal{F}$) show that equation (20.13) holds.

■

As the above proof illustrates, these theorems follow pretty directly from the definitions of the classical transforms and the corresponding results already proven for the transforms of functions in A and T. We will leave the proofs of the next two as exercises.

Theorem 20.8 (principle of near-equivalence)
Let ψ be a classically transformable function. Then the function $\psi(-y)$ is also classically transformable. Moreover,

$$\mathcal{F}[\psi(y)]|_x = \mathcal{F}^{-1}[\psi(y)]|_{-x} = \mathcal{F}^{-1}[\psi(-y)]|_x \tag{20.14}$$

and

$$\mathcal{F}^{-1}[\psi(y)]|_x = \mathcal{F}[\psi(y)]|_{-x} = \mathcal{F}[\psi(-y)]|_x \quad. \tag{20.15}$$

?►Exercise 20.4:   Prove theorem 20.8.

Theorem 20.9 (invertibility)
Let ϕ and ψ be classically transformable functions. Then

$$\psi = \mathcal{F}[\phi] \quad \Longleftrightarrow \quad \mathcal{F}^{-1}[\psi] = \phi \quad.$$

Equivalently,

$$\mathcal{F}^{-1}[\mathcal{F}[\phi]] = \phi = \mathcal{F}\left[\mathcal{F}^{-1}[\phi]\right] \quad.$$

?►Exercise 20.5:   Prove theorem 20.9.

Just as with the integral transforms, the principle of near-equivalence for the classical transforms leads directly to some simple, but occasionally useful, observations concerning even and odd functions. Suppose, for example, ψ is a classically transformable function that is even, $\psi(-x) = \psi(x)$.

By the principle of near-equivalence (theorem 20.8):

$$\mathcal{F}[\psi(y)]|_{-x} \; = \; \mathcal{F}[\psi(-y)]|_{x} \; = \; \mathcal{F}[\psi(y)]|_{x} \quad,$$

$$\mathcal{F}^{-1}[\psi(y)]|_{-x} \; = \; \mathcal{F}^{-1}[\psi(-y)]|_{x} \; = \; \mathcal{F}^{-1}[\psi(y)]|_{x}$$

and

$$\mathcal{F}^{-1}[\psi(y)]|_{x} \; = \; \mathcal{F}[\psi(-y)]|_{x} \; = \; \mathcal{F}[\psi(y)]|_{-x} \quad.$$

This proves the first of the following two corollaries.

Corollary 20.10 (transforms of even functions)
Let ψ be an even, classically transformable function. Then both $\mathcal{F}[\psi]$ and $\mathcal{F}^{-1}[\psi]$ are even functions. Moreover, $\mathcal{F}^{-1}[\psi] = \mathcal{F}[\psi]$.

Corollary 20.11 (transforms of odd functions)
Let ψ be an odd, classically transformable function. Then both $\mathcal{F}[\psi]$ and $\mathcal{F}^{-1}[\psi]$ are odd functions. Moreover, $\mathcal{F}^{-1}[\psi] = -\mathcal{F}[\psi]$.

?►Exercise 20.6:   *Prove corollary 20.11.*

The theorem on invertibility (theorem 20.9) tells us what we have been expecting all along, namely, that the two Fourier transforms (properly defined) are both invertible and each is the inverse transform of the other. A minor consequence of this is that we can now use the phrases "Fourier inverse transform" and "inverse Fourier transform" interchangeably.

For the record, we should also mention the following corollary. It is an immediate consequence of the above theorem on invertibility and the lemma on the uniqueness of the integral transforms (lemma 20.1 on page 299). Though hardly worth much more discussion in itself, it will be used implicitly in much of what follows.

Corollary 20.12
Assume that either f is classically transformable and that $F = \mathcal{F}[f]$ or that F is classically transformable and that $f = \mathcal{F}^{-1}[F]$. Then all of the following hold:

1. Both f and F are classically transformable.

2. $F = \mathcal{F}[f]$.

3. $f = \mathcal{F}^{-1}[F]$.

Moreover,

$$f \text{ is in } \mathcal{A} \quad \Longleftrightarrow \quad F \text{ is in } \mathcal{T}$$

and

$$f \text{ is in } \mathcal{T} \quad \Longleftrightarrow \quad F \text{ is in } \mathcal{A} \quad.$$

20.4 What Is and Is Not Classically Transformable?

The extent to which we can invoke the classical theory for the Fourier transforms is largely determined by the set of classically transformable functions. So, maybe, we should get some idea as to which functions are classically transformable and which are not.

Let's start by looking at the set of all classically transformable functions and reviewing some of the manipulations that can be done with them. From the theorems in the previous section we know the following:

1. The set of all classically transformable functions is a *linear space*; that is, every (finite) linear combination of classically transformable functions is classically transformable.[2]

2. The set of classically transformable functions is "closed under the Fourier transforms", by which we mean that, whenever ψ is classically transformable, so are its Fourier transforms, $\mathcal{F}[\psi]$ and $\mathcal{F}^{-1}[\psi]$.

3. The set of classically transformable functions is "closed under variable reflection" — if $\psi(x)$ is classically transformable, then so is $\psi(-x)$.

In the next chapter, we will also verify the following:

4. The set of classically transformable functions is "closed under translation" — if $\psi(x)$ is classically transformable and a is any fixed real number, then the corresponding translation of ψ by a, $\psi(x-a)$, is also classically transformable.

5. The set of classically transformable functions is "closed under any nontrivial scaling of the variable" — if $\psi(x)$ is classically transformable and a is any nonzero real number, then the function $\psi(ax)$, is also classically transformable.

6. The set of classically transformable functions is "closed under multiplication by sines and cosines" — if $\psi(x)$ is classically transformable and a is any real number, then the functions $\psi(x)\sin(ax)$ and $\psi(x)\cos(ax)$ (and, hence $\psi(x)e^{iax}$) are all classically transformable.

This list gives us some idea of the sort of manipulations that can be done safely with classically transformable functions within the classical theory of Fourier analysis. On the other hand, we will see that the set of classically transformable functions is not closed under either multiplication or differentiation — the product of two classically transformable functions might not be classically transformable, and the derivative of a classically transformable function might not be classically transformable. So, when we attempt to perform these operations, we will need to take some extra precautions.

Let's look a little more closely at the functions in the set of classically transformable functions. Suppose ψ is any classically transformable function with ψ_A and ψ_T being functions in $\mathcal{A}$ and $\mathcal{T}$, respectively, such that

$$\psi = \psi_A + \psi_T \quad .$$

Recall that, since ψ_A is in $\mathcal{A}$, we know ψ_A is piecewise continuous and absolutely integrable on the entire real line. From theorem 19.6 on page 288 we also know that ψ_T, being the Fourier integral transform of some function in $\mathcal{A}$, must be a bounded and continuous function on the entire real line which vanishes at $\pm\infty$. What about the sum, ψ, of these two functions?

First of all, since both ψ_A and ψ_T are at least piecewise continuous, we know that ψ, their sum, must be piecewise continuous.

---

[2] So we could refer to this set as the *space of classically transformable functions.*

We can also see that both $\psi_A(x)$ and $\psi_T(x)$ must, in some sense, get small as $x \to \pm\infty$. In the case of ψ_A, we have

$$\int_{-\infty}^{\infty} \left| \psi_A(s) \right| ds < \infty \quad,$$

which, clearly, can only happen if, for any finite positive length ℓ,

$$\lim_{x \to \pm\infty} \int_{x}^{x+\ell} \left| \psi_A(s) \right| ds = 0 \quad.$$

For ψ_T we explicitly have

$$\lim_{x \to \pm\infty} \psi_T(x) = 0 \quad,$$

which also clearly implies that, for every finite positive length ℓ,

$$\lim_{x \to \pm\infty} \int_{x}^{x+\ell} \left| \psi_T(s) \right| ds = 0 \quad.$$

Thus, for any finite positive length ℓ,

$$\lim_{x \to \pm\infty} \int_{x}^{x+\ell} \left| \psi(s) \right| ds = \lim_{x \to \pm\infty} \int_{x}^{x+\ell} \left| \psi_A(s) + \psi_T(s) \right| ds$$

$$\leq \lim_{x \to \pm\infty} \int_{x}^{x+\ell} \left| \psi_A(s) \right| ds + \lim_{x \to \pm\infty} \int_{x}^{x+\ell} \left| \psi_T(s) \right| ds = 0 \quad.$$

This gives us the following little lemma.

Lemma 20.13
If ψ is a classically transformable function, then ψ is piecewise continuous on the entire real line and, for any finite positive value ℓ,

$$\lim_{x \to \pm\infty} \int_{x}^{x+\ell} \left| \psi(s) \right| ds = 0 \quad. \tag{20.16}$$

This lemma partially characterizes classically transformable functions by describing conditions every classically transformable function must satisfy. It does not tell us, however, that every piecewise continuous function ψ on $\mathbb{R}$ satisfying equation (20.16) is classically transformable. On the other hand, it does give us a way of showing that many functions are not classically transformable.

Corollary 20.14 (test for non-transformability)
Let ψ be a function on the real line. If ψ is not piecewise continuous, or if there is a finite positive length ℓ such that

$$\lim_{x \to \pm\infty} \int_{x}^{x+\ell} \left| \psi(s) \right| ds$$

either does not exist or is not zero, then ψ is not classically transformable.

!▶ **Example 20.3:** *Since*

$$\lim_{x \to \infty} \int_{x}^{x+\pi} |\sin(s)| \, ds = \lim_{x \to \infty} \text{``area under } |\sin(s)| \text{ between } s = x \text{ and } s = x + \pi \text{''}$$

$$= \text{``area under } \sin(s) \text{ between } s = 0 \text{ and } s = \pi \text{''}$$

$$= \int_0^\pi \sin(s)\, ds$$

$$= 2 \neq 0 \quad,$$

we know that $\sin(x)$ *is not classically transformable.*

?▶Exercise 20.7:   *Show* e^{iax} *is not classically transformable for any real or complex value* a .

As a special case of the last corollary, we have

Corollary 20.15
Let ψ *be a function on the real line. If either*

$$\lim_{x \to \infty} |\psi(x)| \qquad \text{or} \qquad \lim_{x \to -\infty} |\psi(x)|$$

exists (as a finite or infinite number) and is nonzero, then ψ *is not classically transformable.*

!▶Example 20.4:   *Since*

$$\lim_{x \to \infty} 1 = 1 \neq 0 \quad,$$

we know that the constant function 1 *is not classically transformable.*

?▶Exercise 20.8:   *Show that* $\arctan(x)$ *is not classically transformable.*

The fact that exponentials and constant functions are not classically transformable will later prompt us to further generalize our definitions of the Fourier transform.

20.5   Finite Duration and Finite Bandwidth Functions

The two sets of functions we are about to describe — "the functions with finite duration" and "the functions with finite bandwidth" — play significant roles in many applications.

Duration and Functions with Finite Duration

In everyday English, the "duration" of something is the length of time something effectively exists (i.e., is nonzero in some sense). When this something is a function on the real line, we get the "duration" of that function being the length of the smallest interval over which the function is nonzero. More precisely, the *duration* of a function f on the real line is the value $T_0 = b - a$ where (a, b) is the smallest interval such that

$$f(x) = 0 \qquad \text{whenever} \quad x < a \quad \text{or} \quad b < x \quad .$$

The interval (a, b) will be called the *interval of duration* for f .

This, of course, assumes there is an interval over which f is nontrivial. If there isn't (i.e., if f is zero everywhere), we'll just say the duration is zero.

The interval of duration for a function can be infinite. If it is finite and the function is piecewise continuous, then it is easily verified that the function is absolutely integrable on the real line and, hence, is classically transformable. As you can well imagine, the set of all piecewise continuous functions with finite durations is an important set of classically integrable functions. For one thing, such functions correspond to measurements of processes that are, themselves, of finite duration.

Often we will not know (or need) the precise interval of duration for a function f, only that the interval is contained in some other interval of the form $[-T, T]$ where T is some positive real number. (We'll also allow T to be 0 if the duration is 0.) Any such value T will be referred to as a *bound on the interval of duration* for f. Note, then, that T is a bound on the interval of duration for f if and only if

$$f(x) = 0 \qquad \text{whenever} \quad T < |x| \quad .$$

For brevity, we may say "T is a *duration bound* for f" instead of "T is a bound on the duration interval for f".

?▶ Exercise 20.9: *Give an example of a function with finite duration.*

Bandwidth and Finite Bandwidth Functions

The *bandwidth* of a function is just the duration of the Fourier transform of the function. Naturally, we can only speak of the bandwidth of a transformable function. These functions are important because, in many applications, there are good reasons to believe that the functions describing the processes occurring have finite bandwidths. Indeed, in some applications these functions are more important than the finite duration functions.

Let f be a transformable function with finite bandwidth. We will refer to any nonnegative real number Ω as being a *bandwidth bound* for f if and only if Ω is a duration bound for the Fourier transform of f. In other words, Ω is a bandwidth bound for f if and only if, letting $F = \mathcal{F}[f]$,

$$F(\omega) = 0 \qquad \text{whenever} \quad \Omega < |\omega| \quad .$$

?▶ Exercise 20.10: *Give an example of a function with finite bandwidth.*

Since finite-bandwidth functions are clearly absolutely integrable, theorem 19.6 on page 288 applies telling us that finite-bandwidth functions are automatically continuous on the real line. Another important property of finite-bandwidth functions is described in the famous sampling theorem:

Theorem 20.16 (sampling theorem)
Assume f is a function with finite bandwidth, and let Ω be a bandwidth bound for f. Then, for any t in $\mathbb{R}$,

$$f(t) = \sum_{k=-\infty}^{\infty} f(t_k) \, \text{sinc}(2\pi\Omega \, [t - t_k]) \tag{20.17a}$$

where

$$t_k = k\Delta t \qquad \text{for} \quad k = 0, \pm 1, \pm 2, \pm 3, \ldots \tag{20.17b}$$

with

$$\Delta t = \frac{1}{2\Omega} \quad . \tag{20.17c}$$

This theorem tells us that a finite-bandwidth function can be completely reconstructed from a "sampling" of the values of the function (the sampling being the set of values of $f(t)$ at the t_k's). Observe that the choice of Δt is not completely arbitrary. We must have

$$\Delta t \leq \frac{1}{2\Omega_0}$$

where Ω_0 is the smallest possible bandwidth bound for f. (The value $2\Omega_0$ is often called the *Nyquist frequency*.)

Using "Fourier analysis", we (i.e., you) will derive and verify the sampling theorem in exercises 20.20 and 20.21 at the end of this chapter.

Effective Duration and Bandwidth, and Bandwidth Theorems

You may wonder about those functions having both finite duration and finite bandwidth. Don't bother. Except for the zero function, there are none. If the duration of a nonzero function is finite, then its bandwidth must be infinite. Conversely, any nonzero function with finite bandwidth must have infinite duration. We'll prove this in exercise 20.22.

What is often possible is to use "effective bounds" on the durations and bandwidths. That is, instead of attempting to describe absolute bounds on the duration and bandwidth, we define values T_{eff} and Ω_{eff}, called, respectively, an effective duration bound and an effective bandwidth bound, for our function f so that

$$|f(x)| \quad \text{is negligibly small whenever} \quad T_{\text{eff}} < |x| \quad ,$$

and, letting $F = \mathcal{F}[f]$,

$$|F(\omega)| \quad \text{is negligibly small whenever} \quad \Omega_{\text{eff}} < |\omega| \quad .$$

Precisely what is meant by "negligibly small" depends on the application and on the needs (and ability) of those interested in that application. We won't discuss possible criteria for "negligibly small" at this time (one example is given in exercise 20.19 on page 316). What should be noted, however, is that there are invariably mathematical restrictions on your ability to choose T_{eff} and Ω_{eff}. It turns out that, if you want one of these values to be small, then you must allow the other to be relatively large. Typically, the relation between these two values can be described by an inequality of the form

$$T_{\text{eff}}\,\Omega_{\text{eff}} \;\geq\; C$$

where C is some constant that depends on your precise definition of T_{eff} and Ω_{eff}. The inequality, itself, is often referred to as an *uncertainty principle* and the statement of its validity is often called a *bandwidth theorem*. The importance of this inequality, naturally, depends on the application. In quantum mechanics, for example, it is very important and is the basis for the Heisenberg uncertainty principle. We'll try to return to this subject and verify a couple of important versions of the uncertainty principle after developing sufficient tools.

A Little More on Terminology

In other texts, you will find a number of other terms for functions with finite duration. These include *duration limited* and, with particular types of applications, *time limited* and *spatially limited*. Many mathematicians will also refer to these functions as having *bounded support*.[3] Functions with finite bandwidth are also commonly referred to as *bandwidth limited* functions. You should also be aware that the terminology usage is not consistent throughout the literature. In particular, the words "duration" and "bandwidth" are often used by others where we will use the terms "duration bound" and "bandwidth bound". Usually, though, it is fairly clear from the context when a particular author is using, say, "duration" to mean what we defined it to mean, or to mean what we defined as a "bound on the interval of duration".

---

[3] The *support* of a function is the smallest closed set containing all points at which the function is nonzero.

20.6   More on Terminology, Notation and Conventions*

Classical?

The "theory of Fourier transforms" being developed in this part of the text is basically the same theory presented in most traditional introductions to Fourier analysis. It is being referred to as "classical" both because it is fairly close to what is traditionally presented and, more importantly, to distinguish it from a more general theory we will develop later. Since it is the only theory we will be discussing for the next several chapters, and since so much of the discussion will apply to the more general theory as well, we might as well stop overusing the word "classical" and use it only when there is a good reason to emphasize that we are discussing the classical theory.

Denoting the Transforms

Different workers in different disciplines have different ways of denoting Fourier transforms, and each one has its disadvantages. In many applications it is convenient to use lower-case Latin letters — f, g, h, etc. — for the "untransformed" functions (i.e., the functions to be plugged into $\mathcal{F}$ or which pop out of $\mathcal{F}^{-1}$), and the corresponding upper-case Latin letters — F, G, H, ETC. — for the corresponding transformed functions. Also, in these applications, it is often convenient to use "x" or "t" as the variable in the "untransformed" functions, and "ω" (the lower-case Greek letter "omega") as the variable in the corresponding "transformed" functions. The lower-case Greek letter "nu", written "ν", is also often used as a variable, especially as a substitute for "$2\pi\omega$". These conventions arose naturally in applications because, in applications, it often makes sense to distinguish between functions of position or time (often representing quantities that can be directly measured, such as voltage or illumination intensity) and corresponding functions of frequencies that are related to the functions of position or time through the Fourier transform.[4]

In other situations, the convention of using $f(t)$ and $F(\omega)$ to denote a function and its corresponding direct Fourier transform can be awkward or even misleading. This is especially true when developing the mathematics of Fourier analysis. Imagine the difficulty in describing the principle of near-equivalence using this convention! We have had, and will have, many occasions where a single function can be viewed as both an "untransformed" function and a "transformed" function. How should such a function be denoted? And which symbol — x, t, or ω — should denote the variable? Because of these difficulties many people eschew the aforementioned convention of distinguishing between "untransformed" functions and "transformed" functions and, as much as possible, avoid direct reference to the variables being used, especially when they are dummy variables. These folks prefer the notation $\mathcal{F}[f]$ and $\mathcal{F}^{-1}[f]$ (or even $\hat{f}$ and $\check{f}$) to denote the Fourier transforms of f. They may even go so far as to use Greek symbols such as ϕ and ψ to denote functions rather than letters from the Latin alphabet.

This last set of conventions and notation can lead to very elegant writing. Unfortunately, especially when carried to excess, it does not lend itself well to describing many of the more mundane formulas we use. As a result, additional notation has to be developed and a good part of the reader's time is spent learning this new (but elegant) notation.

?▶Exercise 20.11:   *Describe the principle of near-equivalence without using dummy variables.*

We will adopt a pragmatic approach regarding notation. We will use whichever of the above

---

* Warning: The author shamelessly expresses personal opinions in this section.

[4] The author tries to be tolerant of those who use f instead of ω for frequency, but those who go so far as to use f to denote both a variable and a function in a single expression (e.g., "$F(f) = \mathcal{F}[f(x)]|_f$") are guilty of abusing notation and their readers.

sets of conventions and notation is convenient at the time. This will *not* include using $\hat{f}$ and $\check{f}$ to denote the Fourier transforms of f. We will also feel free to combine and modify these systems of notation, keeping in mind our discussion of variables, formulas, functions and operators in chapter 2. Sometimes, for clarity, we may even express results (or do computations) twice, using a different set of conventions for each.

Time Domains and Such

In some of the literature you will find references to the "time domain" (or "spatial domain") and the "frequency domain" of a function.[5] Strictly speaking, this terminology is nonsense. A function has one and only one domain; namely, the set of all numbers that can be plugged into the function. That's it. If, for example, the function is any of the classically transformable functions discussed in this chapter, then its domain is the set of all real numbers. Period.

That said, even this author is guilty of using these questionable terms in informal conversation — but always with the understanding that we are not talking about a single function. We are really talking about something which could be called a *signal*, and which corresponds to an ordered pair of functions (f, F) with $F = \mathcal{F}[f]$. In practice, both of these functions describe the same process or phenomenon. The first, f, describes how the process or phenomenon varies as either time or position varies. It is likely to correspond to something that can be measured, such as the changing voltage at some point in a circuit. As such, it is appropriate to refer to f as the "time (or spatial) description of the signal". For brevity, we might even abuse the terminology a little and refer to f as the "time (or spatial) component of the signal", though this incorrectly suggests that f is describing a time or position instead of being described in terms of time or position. We might even, in a moment of weakness, further abuse the terminology and refer to f as "the function in the time (or spatial) domain".

The other function, F, describes the same process, but in terms of another variable, which, because of the physics involved, often corresponds to some sort of frequency. Consequently, it is often appropriate to refer to F as the "frequency description of the signal". Abuse the terminology a little, and this becomes " F is the frequency component of the signal". Abuse it further, and we have " $F(\omega)$ is the function in the frequency domain."

In this text, any further usage of the phrases "time domain" or "frequency domain" of a function is hereby forbidden. Your employment of these terms in private conversations will be left as a matter for your own conscience.

Additional Exercises

20.12. *Let a and b denote real constants with $a > 0$. Using the results summarized in table 20.1 on page 298, find the following inverse Fourier transforms:*

a. $\mathcal{F}^{-1}[\operatorname{sinc}(2\pi a\omega)]|_t$

b. $\mathcal{F}^{-1}\left[\dfrac{1}{a + i2\pi\omega}\right]\Big|_t$

c. $\mathcal{F}^{-1}\left[\dfrac{1}{a - i2\pi\omega}\right]\Big|_t$

d. $\mathcal{F}^{-1}\left[\dfrac{1}{a + ib + i2\pi\omega}\right]\Big|_t$

e. $\mathcal{F}^{-1}\left[\dfrac{1}{a + ib - i2\pi\omega}\right]\Big|_t$

f. $\mathcal{F}^{-1}[2a\operatorname{sinc}(2\pi a[\omega - b])]|_t$

---

[5] If you haven't seen such references, stop reading this. This discussion is only for those who have seen these phrases.

20.13. Let a and b denote real constants with $a > 0$. Using your answers to the previous exercise and the principle of near-equivalence, find the following Fourier transforms:

a. $\mathcal{F}[\operatorname{sinc}(2\pi a t)]|_\omega$

b. $\mathcal{F}\left[\dfrac{1}{a + i2\pi t}\right]\Big|_\omega$

c. $\mathcal{F}\left[\dfrac{1}{a - i2\pi t}\right]\Big|_\omega$

d. $\mathcal{F}\left[\dfrac{1}{a + ib + i2\pi t}\right]\Big|_\omega$

e. $\mathcal{F}\left[\dfrac{1}{a + ib - i2\pi t}\right]\Big|_\omega$

f. $\mathcal{F}[2a\operatorname{sinc}(2\pi a[\omega - b])]|_t$

20.14. Find the following transforms by factoring out appropriate constants from some of your answers to the previous exercise. In each case a, b and c are real with $a > 0$ and $c > 0$.

a. $\mathcal{F}\left[\dfrac{1}{a + it}\right]\Big|_\omega$

b. $\mathcal{F}\left[\dfrac{1}{a + ib + ict}\right]\Big|_\omega$

20.15. Find the following transforms using linearity and near-equivalence along with the entries in table 20.1 and your answers to the above exercises. In each case a, b and c are real constants with $a > 0$ and $c > 0$.

a. $\mathcal{F}\left[\dfrac{1}{a - it}\right]\Big|_\omega$

b. $\mathcal{F}\left[\dfrac{1}{a + ib - ict}\right]\Big|_\omega$

c. $\mathcal{F}^{-1}\left[\dfrac{1}{a + i\omega}\right]\Big|_t$

d. $\mathcal{F}^{-1}\left[\dfrac{1}{a + ib + ic\omega}\right]\Big|_t$

e. $\mathcal{F}^{-1}\left[\dfrac{1}{a - i\omega}\right]\Big|_t$

f. $\mathcal{F}^{-1}\left[\dfrac{1}{a + ib - ic\omega}\right]\Big|_t$

g. $\mathcal{F}\left[\dfrac{1}{a^2 + c^2 t^2}\right]\Big|_\omega$

h. $\mathcal{F}^{-1}\left[\dfrac{1}{a^2 + c^2 \omega^2}\right]\Big|_t$

20.16. Compute the following transforms:

a. $\mathcal{F}[\operatorname{sinc}(10\pi t)]|_\omega$

b. $\mathcal{F}\left[\dfrac{1}{3 - i2\pi t}\right]\Big|_\omega$

c. $\mathcal{F}\left[\dfrac{1}{2 + i2\pi t}\right]\Big|_\omega$

d. $\mathcal{F}^{-1}\left[\dfrac{1}{3 - i2\pi\omega}\right]\Big|_t$

20.17. Factor the denominator in each of the following functions and find the partial fraction expansion for the function.[6] Then find the Fourier transform of the function using linearity (possibly factoring out a few constants) and some of your answers to previous exercises in this set. Where they appear, a and c are both positive. (Note: For some of these, you may want to review the discussion of "complex factoring" on page 65.)

a. $\dfrac{1}{6 + i2\pi t + 4\pi^2 t^2}$ $\left(\text{Hint: } 6 + i2\pi t + 4\pi^2 t^2 = (3 - i2\pi t)(2 + i2\pi t)\right)$

b. $\dfrac{1}{6 + i5\pi t + 6\pi^2 t^2}$ $\left(\text{Hint: } 6 + i5\pi t + 6\pi^2 t^2 = (3 - i2\pi t)(2 + i3\pi t)\right)$

c. $\dfrac{t}{a^2 + 4\pi^2 t^2}$

d. $\dfrac{t}{a^2 + c^2 t^2}$

e. $\dfrac{1}{a^2 - it^2}$

f. $\dfrac{1}{a^2 - ic^2 t^2}$

g. $\dfrac{1}{a^2 + it^2}$

h. $\dfrac{1}{a^2 + ic^2 t^2}$

---

[6] If necessary, review "partial fractions" in your old calculus text!

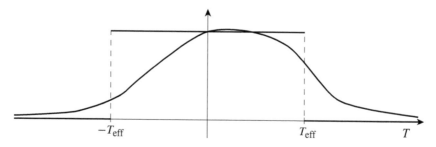

$-T_{\text{eff}}$ T_{eff} T

Figure 20.1: A function f and its effective duration for problem 20.19.

20.18. *Identify each of the following functions as being either classically transformable or not being classically transformable. (In these formulas, assume that $a > 0$.)*

$$\cos(ax) \quad , \quad x^2 \quad , \quad x^{-2} \quad , \quad \text{step}(x) \quad , \quad x^{-2}\,\text{step}(x-1) \quad ,$$

$$e^{-ax}\,\text{step}(x) \quad , \quad e^{ax}\,\text{step}(x) \quad , \quad \frac{1}{1-ix} \quad , \quad \frac{1}{1-x} \quad \text{and} \quad e^{-ax^2} \quad .$$

20.19. *Let f and F be two absolutely integrable and nonnegative real-valued functions on the real line with $F = \mathcal{F}[f]$. Assume further that $f(0) \neq 0$ and $F(0) \neq 0$. For such an f we can define the effective duration bound T_{eff} and the effective bandwidth bound Ω_{eff} by the equations*

$$f(0)\,T_{\text{eff}} = \int_{-\infty}^{\infty} f(t)\,dt \quad \text{and} \quad F(0)\,\Omega_{\text{eff}} = \int_{-\infty}^{\infty} F(\omega)\,d\omega \quad . \tag{20.18}$$

(Basically, T_{eff} and Ω_{eff} are being defined, respectively, as the half widths of the pulse functions having heights $f(0)$ and $F(0)$ and enclosing the same areas as the graphs of $f(t)$ and $F(\omega)$. See figure 20.1. Admittedly, this approach is of limited practical value.)

a. *Verify that the equations in line (20.18) can be written as*

$$f(0)\,T_{\text{eff}} = F(0) \quad \text{and} \quad F(0)\,\Omega_{\text{eff}} = f(0) \quad .$$

b. *Using the results of the previous exercise, verify that*

$$T_{\text{eff}}\,\Omega_{\text{eff}} \geq 1 \quad .$$

(Actually, you should derive $T_{\text{eff}}\Omega_{\text{eff}} = 1$.)

c. *Using the above definitions, find the effective duration and effective bandwidth for $f(t) = e^{-|t|}$.*

20.20. *In this exercise set, we will derive formula set (20.17) in the sampling theorem (theorem 20.16 on page 311). Accordingly, in the following, f is a finite-bandwidth function with bandwidth bound Ω, and, for each integer k,*

$$t_k = k\Delta t \quad \text{with} \quad \Delta t = \frac{1}{2\Omega} \quad .$$

Also, let $p = 2\Omega$, and let F_p be the periodic function given by

$$F_p(\omega) = \begin{cases} F(\omega) & \text{if } -\Omega < \omega < \Omega \\ F_p(\omega - p) & \text{in general} \end{cases} \quad .$$

a. Convince yourself (pictures will do) that, because f has finite bandwidth with Ω as a bandwidth bound,

$$F(\omega) = F(\omega)\,\text{pulse}_\Omega(\omega) = F_p(\omega)\,\text{pulse}_\Omega(\omega) \quad,$$

and, taking the inverse transform of this, derive the fact that

$$f(t) = \mathcal{F}^{-1}\big[F_p(\omega)\,\text{pulse}_\Omega(\omega)\big]\big|_t \tag{20.19}$$

along with the integral formulas

$$f(t) = \int_{-\Omega}^{\Omega} F(\omega)\,e^{i2\pi\omega t}\,d\omega = \int_{-\Omega}^{\Omega} F_p(\omega)\,e^{i2\pi\omega t}\,d\omega \quad. \tag{20.20}$$

b. Because $F_p(\omega)$ is a piecewise continuous periodic function of ω with period p, it can be expressed as a Fourier series

$$F_p(\omega) = \sum_{n=-\infty}^{\infty} C_n e^{i2\pi t_n \omega} \quad.$$

(Note that the traditional roles of t and ω have been switched.) Using the above, verify that, for each n,

$$t_n = n\Delta t \quad\text{and}\quad C_n = f(-t_n)\Delta t \quad.$$

c. Now, combine results from the previous two parts of this exercise set (and use other material developed in this and the previous chapter) to finish deriving formula (20.17a) in the sampling theorem. Go ahead and assume that the linearity of the transform holds even when the summation is an infinite series. At the end, you will probably want to use the change of index $n = -k$.

20.21. Because we assumed the infinite sum could be treated like an ordinary finite sum and did not worry about the convergence of the series, the derivation in the previous exercise is not completely rigorous. Let us address this weakness by considering the pointwise error in using the $(M, N)^{\text{th}}$ partial sum of the series in formula (20.16) for $f(t)$,

$$E_{MN}(t) = f(t) - \sum_{k=M}^{N} f(t_k)\,\text{sinc}(2\pi\Omega\,[t - t_k])$$

(where f and everything else are as in the previous exercise). Our goal is to show that, for each real t, $|E_{MN}(t)| \to 0$ as $(M, N) \to (-\infty, \infty)$. This will confirm that the series in question really does converge, and that it converges to $f(t)$ for each t on the real line.

a. Make use of relations rigorously obtained in the last exercise set to show that

$$E_{MN}(t) = \int_{-\Omega}^{\Omega} \mathcal{E}_{M'N'}(\omega)\,e^{i2\pi t\omega}\,d\omega$$

where $M' = -N$, $N' = -M$ and

$$\mathcal{E}_{M'N'}(\omega) = F_p(\omega) - \sum_{n=M'}^{N'} C_n e^{i2\pi t_n \omega}$$

is the pointwise error in using the $(M', N')^{\text{th}}$ partial sum of the Fourier series for F_p in place of F_p.

b. Next, apply appropriate inequalities (including inequality (11.13) on page 142) to show that

$$|E_{MN}(t)| \leq \sqrt{2\Omega} \, \|\mathcal{E}_{M'N'}\| \quad ,$$

and observe that the theorem on the norm convergence of Fourier series (theorem 13.17 on page 173) assures us that, indeed,

$$|E_{MN}(t)| \rightarrow 0 \qquad \text{as} \qquad (M, N) \rightarrow (-\infty, \infty) \quad .$$

20.22. In this exercise set, you will show that the only function having both finite duration and finite bandwidth is the zero function (i.e., the function $f(t) = 0$ for all t). Accordingly, let us assume that we do have a function f having both finite duration (with duration bound T) and finite bandwidth (with bandwidth bound Ω). (Warning: You will need to use some of the more advanced material about analytic functions summarized in section 6.4.)

a. Show that there is a finite integer M such that, for every t in $\mathbb{R}$,

$$f(t) = \sum_{k=-M}^{M} f(t_k) \, \text{sinc}(2\pi\Omega \, [t - t_k])$$

where

$$t_k = k\Delta t \qquad \text{for} \quad k = 0, \pm 1, \pm 2, \pm 3, \ldots$$

with

$$\Delta t = \frac{1}{2\Omega} \quad .$$

b. Using the above formula for f, material from section 6.4 and the fact that f has finite duration, show that f must be the zero function.

21

Some Elementary Identities: Translation, Scaling and Conjugation

There are several easily derived identities that can simplify the computation of many transforms and play significant roles both in applications and in further development of our theory. Some of these, such as those identities associated with the linearity of the transforms and the principle of near-equivalence, have already been discussed. In this chapter, we will discuss those identities involving translation, "modulation", scaling and complex conjugation. We will also discuss a few topics relating to these identities (such as the intelligent use of tables).

For convenience, many of the formulas for transforms we have already computed are listed in table 21.1.

21.1 Translation
The Translation Identities

The translation identities (also known as the shifting identities) relate the translation $\phi(x - \gamma)$ of any classically transformable function $\phi(x)$ with a product of the transform of the function and a corresponding complex exponential.

Theorem 21.1 (the translation identities)
Let f and F be any two classically transformable functions with $F(\omega) = \mathcal{F}[f(t)]|_\omega$, and let γ be any fixed real value. Then

$$f(t - \gamma) \quad , \quad F(\omega - \gamma) \quad , \quad e^{i2\pi\gamma t} f(t) \quad and \quad e^{-i2\pi\gamma\omega} F(\omega)$$

are all classically transformable. Moreover,

$$\mathcal{F}\big[f(t - \gamma)\big]\big|_\omega = e^{-i2\pi\gamma\omega} F(\omega) \tag{21.1a}$$

and

$$\mathcal{F}^{-1}\big[F(\omega - \gamma)\big]\big|_t = e^{i2\pi\gamma t} f(t) \quad . \tag{21.1b}$$

Equivalently,

$$\mathcal{F}^{-1}\big[e^{-i2\pi\gamma\omega} F(\omega)\big]\big|_t = f(t - \gamma) \tag{21.1c}$$

and

$$\mathcal{F}\big[e^{i2\pi\gamma t} f(t)\big]\big|_\omega = F(\omega - \gamma) \quad . \tag{21.1d}$$

Table 21.1: Selected Fourier Transforms from Previous Work

$f(t) = \mathcal{F}^{-1}[F(\omega)]\|_t$	$F(\omega) = \mathcal{F}[f(t)]\|_\omega$	Restrictions	See
$\text{pulse}_\alpha(t)$	$2\alpha\,\text{sinc}(2\pi\alpha\omega)$	$0 < \alpha$	example 19.1, page 280
$e^{-\alpha t}\,\text{step}(t)$	$\dfrac{1}{\alpha + i2\pi\omega}$	$0 < \alpha$	exercise 19.1, page 280
$e^{\alpha t}\,\text{step}(-t)$	$\dfrac{1}{\alpha - i2\pi\omega}$	$0 < \alpha$	exercise 19.3, page 282
$\dfrac{1}{\alpha + it}$	$2\pi\, e^{2\pi\alpha\omega}\,\text{step}(-\omega)$	$0 < \alpha$	exercise 20.14, page 315
$\dfrac{1}{\alpha - it}$	$2\pi\, e^{-2\pi\alpha\omega}\,\text{step}(\omega)$	$0 < \alpha$	exercise 20.15, page 315
$e^{-\alpha|t|}$	$\dfrac{2\alpha}{\alpha^2 + 4\pi^2\omega^2}$	$0 < \alpha$	example 19.3, page 284
$\dfrac{1}{\alpha^2 + t^2}$	$\dfrac{\pi}{\alpha}e^{-2\pi\alpha|\omega|}$	$0 < \alpha$	exercise 19.18, page 296

Before we prove this theorem, you should look at the claimed equivalence of, say, identities (21.1a) and (21.1c). This equivalence comes directly from the invertibility of the Fourier transforms. If this is not obvious, let

$$g(t) \ = \ f(t - \gamma) \qquad \text{and} \qquad G(\omega) \ = \ e^{-i2\pi\gamma\omega}F(\omega) \quad .$$

Assuming these functions are classically transformable, the theorem on invertibility (theorem 20.9 on page 306) assures us that

$$\mathcal{F}[g(t)]\big|_\omega \ = \ G(\omega) \quad \Longleftrightarrow \quad g(t) \ = \ \mathcal{F}^{-1}[G(\omega)]\big|_t \quad .$$

Replacing $g(t)$ and $G(\omega)$ with their formulas in terms of f and F, we see that

$$\mathcal{F}\big[f(t - \gamma)\big]\big|_\omega \ = \ e^{-i2\pi\gamma\omega}F(\omega) \quad \Longleftrightarrow \quad f(t - \gamma) \ = \ \mathcal{F}^{-1}\big[e^{-i2\pi\gamma\omega}F(\omega)\big]\big|_t \quad . \qquad (21.2)$$

In other words, if we can show one of the equations in (21.2) is true, then we automatically know that the other one is also true. Since these equations are the equations in identities (21.1a) and (21.1c), this tells us that *both* (21.1a) and (21.1c) must be true if either *one* is true. In fact, we really should view (21.1a) and (21.1c) as being the same identity, just written two different ways.

Likewise, (21.1b) and (21.1d) are really the same identity, written two different ways.

As you can imagine, this sort of situation will occur several times again in this text. When it does, it will be assumed that you, the reader, can recognize why "invertibility" implies the equivalence of two given equations.

PROOF (of theorem 21.1): We will limit our proof to explicitly showing that the indicated functions are classically transformable and that identities (21.1a) and (21.1c) hold. Verifying identities (21.1b) and (21.1d) will be left as an exercise.

For our part of the proof, we need to consider three cases: the case where f is in $\mathcal{A}$, the case where f is in $\mathcal{T}$ and the general case where f is any classically transformable function.

First, assume f is in $\mathcal{A}$. We've already noted (in lemma 18.7 on page 265) the fact that any translation of f by a real value is also in $\mathcal{A}$. Also, since f is absolutely integrable, we can use the integral formula for its transform,

$$F(y) = \mathcal{F}[f(x)]|_y = \int_{-\infty}^{\infty} f(x) e^{-i2\pi xy} dx \quad .$$

Using this formula and the substitution $x = t - \gamma$ (so $t = x + \gamma$ and $dt = dx$), we have

$$\mathcal{F}[f(t-\gamma)]|_\omega = \int_{t=-\infty}^{\infty} f(t-\gamma) e^{-i2\pi\omega t} dt$$

$$= \int_{x=-\infty}^{\infty} f(x) e^{-i2\pi\omega(x+\gamma)} dx$$

$$= \int_{x=-\infty}^{\infty} f(x) e^{-i2\pi\omega x} e^{-i2\pi\gamma\omega} dx$$

$$= e^{-i2\pi\gamma\omega} \int_{x=-\infty}^{\infty} f(x) e^{-i2\pi\omega x} dx = e^{-i2\pi\gamma\omega} F(\omega) \quad ,$$

verifying, for this case, that $e^{-i2\pi\gamma\omega} F(\omega)$ is classically transformable (in fact, it is in $\mathcal{T}$) and that identity (21.1a) holds. By "invertibility" (as discussed just before this proof), we also know identity (21.1c) holds.

Now assume f is in $\mathcal{T}$. Then F must be in $\mathcal{A}$ (if this is not obvious, see corollary 20.12 on page 307), and so,

$$f(y) = \mathcal{F}^{-1}[F(x)]|_y = \int_{-\infty}^{\infty} F(x) e^{i2\pi xy} dx \quad .$$

We already know (corollary 18.10 on page 267) that, since F is in $\mathcal{A}$, so is the product of $F(\omega)$ with $e^{-i2\pi\gamma\omega}$. So we can use the integral formula to find the inverse transform of this product. Doing so, we obtain

$$\mathcal{F}^{-1}\left[e^{-i2\pi\gamma\omega} F(\omega)\right]\Big|_t = \int_{-\infty}^{\infty} e^{-i2\pi\gamma\omega} F(\omega) e^{i2\pi\omega t} d\omega$$

$$= \int_{-\infty}^{\infty} F(\omega) e^{-i2\pi\gamma\omega} e^{i2\pi\omega t} d\omega$$

$$= \int_{-\infty}^{\infty} F(\omega) e^{i2\pi\omega(t-\gamma)} d\omega = f(t-\gamma) \quad ,$$

verifying, for this case, that $f(t-\gamma)$ must be classically transformable (in fact, it is in $\mathcal{T}$) and that identity (21.1c) holds. By "invertibility", identity (21.1a) must also hold.

Finally, consider the case where f is any classically transformable function; that is,

$$f = f_{\mathcal{A}} + f_{\mathcal{T}}$$

where $f_{\mathcal{A}}$ is some function in $\mathcal{A}$ and $f_{\mathcal{T}}$ is some function in $\mathcal{T}$. Let $F_{\mathcal{A}} = \mathcal{F}[f_{\mathcal{A}}]$ and $F_{\mathcal{T}} = \mathcal{F}[f_{\mathcal{T}}]$. Note that

$$f(t-\gamma) = f_{\mathcal{A}}(t-\gamma) + f_{\mathcal{T}}(t-\gamma)$$

and that

$$F = \mathcal{F}[f] = \mathcal{F}[f_A + f_T] = \mathcal{F}[f_A] + \mathcal{F}[f_T] = F_A + F_T \quad .$$

Since f_A and f_T are in A and T, respectively, the previous parts of this proof assure us that $f_A(t - \gamma)$ and $f_T(t - \gamma)$ are classically transformable and that

$$\mathcal{F}\big[f_A(t - \gamma)\big]\big|_{\omega} = e^{-i2\pi\gamma\omega} F_A(\omega) \qquad \text{and} \qquad \mathcal{F}\big[f_T(t - \gamma)\big]\big|_{\omega} = e^{-i2\pi\gamma\omega} F_T(\omega) \quad .$$

Thus, being the sum of two classically transformable functions, $f(t - \gamma)$ must be classically transformable. Furthermore,

$$\begin{aligned}
\mathcal{F}\big[f(t - \gamma)\big]\big|_{\omega} &= \mathcal{F}\big[f_A(t - \gamma) + f_T(t - \gamma)\big]\big|_{\omega} \\
&= \mathcal{F}\big[f_A(t - \gamma)\big]\big|_{\omega} + \mathcal{F}\big[f_T(t - \gamma)\big]\big|_{\omega} \\
&= e^{-i2\pi\gamma\omega} F_A(\omega) + e^{-i2\pi\gamma\omega} F_T(\omega) \\
&= e^{-i2\pi\gamma\omega}\big[F_A(\omega) + F_T(\omega)\big] \\
&= e^{-i2\pi\gamma\omega} F(\omega) \quad .
\end{aligned}$$

So, whenever f is classically transformable, we have both that $e^{-i2\pi\gamma\omega} F(\omega)$, being the transform of the classically transformable function $f(t - \gamma)$, is classically transformable and that identity (21.1a) holds. Moreover, by invertibility, so does identity (21.1c).

Since f and F are arbitrary classically transformable functions and γ is an arbitrary real value, we have, in fact, shown that $\psi(t - \gamma)$ and $e^{\pm i2\pi\gamma\omega}\psi(\omega)$ are classically transformable for any classically transformable function ψ. Thus, in particular, $F(\omega - \gamma)$ and $e^{i2\pi\gamma t} f(t)$ must be classically transformable.

This completes our part of the proof.   ∎

?▶ Exercise 21.1:   *Verify identities (21.1b) and (21.1d) two ways:*

    a: *Show identities (21.1b) and (21.1d) hold by simply repeating, with suitable modifications, the computations done in the above proof. Be sure to consider the case where F is in A, the case where F is in T and the case where F is the sum of functions from A and T.*

    b: *Show that identities (21.1b) and (21.1d) hold by using identities (21.1a) and (21.1c) and the principle of near-equivalence.*

We will refer to equations (21.1a) through (21.1d) as the *translation* (or *shifting*) *identities*. Observe that the first one can be written as

$$\mathcal{F}\big[f(t - \gamma)\big]\big|_{\omega} = e^{-i2\pi\gamma\omega}\mathcal{F}[f(t)]\big|_{\omega} \quad .$$

Similar observations can be made for each of the identities. Changing our notation slightly, then, we can see that the translation identities can also be written as

$$\mathcal{F}\big[\psi(t - \gamma)\big]\big|_{\omega} = e^{-i2\pi\gamma\omega}\mathcal{F}[\psi]\big|_{\omega} \quad , \tag{21.1a$'$}$$

$$\mathcal{F}^{-1}\big[\psi(\omega - \gamma)\big]\big|_{t} = e^{i2\pi\gamma t}\mathcal{F}^{-1}[\psi(\omega)]\big|_{t} \quad , \tag{21.1b$'$}$$

$$\mathcal{F}^{-1}\big[e^{-i2\pi\gamma\omega}\psi(\omega)\big]\big|_{t} = \mathcal{F}^{-1}[\psi(\omega)]\big|_{t-\gamma} \tag{21.1c$'$}$$

and

$$\mathcal{F}\left[e^{i2\pi\gamma t}\psi(t)\right]\Big|_{\omega} = \mathcal{F}[\psi(t)]|_{\omega-\gamma} \tag{21.1d$'$}$$

where ψ is any classically transformable function and γ is any fixed real value. (Again, it should be noted that this list contains redundant information with identities (21.1c$'$) and (21.1d$'$) being completely equivalent to identities (21.1a$'$) and (21.1b$'$), respectively.)

Which version of the translation identities you use is matter of preference. We'll illustrate the use of both.

!► **Example 21.1:** *Consider computing* $\mathcal{F}\left[e^{-2t}\,\text{step}(t-3)\right]$. *It should be clear that we will want to use identity (21.1a) with* $\gamma = 3$,

$$\begin{aligned}\mathcal{F}\left[e^{-2t}\,\text{step}(t-3)\right]\Big|_{\omega} &= \mathcal{F}[f(t-3)]|_{\omega} \\ &= e^{-i2\pi 3\omega}F(\omega) = e^{-i6\pi\omega}F(\omega) \quad .\end{aligned} \tag{21.3}$$

To use this formula, we must find the correct formula for the function f and its corresponding transform. We have

$$f(t-3) = e^{-2t}\,\text{step}(t-3) \quad,$$

which is not the formula for f but the formula for the translation of f by 3. To recover the formula for f from this, we use the substitution $x = t - 3$,

$$f(x) = e^{-2(x+3)}\,\text{step}(x) = e^{-6}e^{-2x}\,\text{step}(x) \quad .$$

Now we can find the formula for $F = \mathcal{F}[f]$. Factoring out the constant and using table 21.1 on page 320,

$$F(\omega) = \mathcal{F}\left[e^{-6}e^{-2x}\,\text{step}(x)\right]\Big|_{\omega} = e^{-6}\mathcal{F}\left[e^{-2x}\,\text{step}(x)\right]\Big|_{\omega} = e^{-6}\cdot\frac{1}{2+i2\pi\omega} \quad .$$

Plugging this into equation (21.3) completes our computations:

$$\begin{aligned}\mathcal{F}\left[e^{-2t}\,\text{step}(t-3)\right]\Big|_{\omega} &= e^{-i6\pi\omega}F(\omega) \\ &= e^{-i6\pi\omega}\left(e^{-6}\cdot\frac{1}{2+i2\pi\omega}\right) = \frac{1}{2+i2\pi\omega}e^{-6-i6\pi\omega} \quad .\end{aligned}$$

!► **Example 21.2:** *Again, consider the problem of computing the Fourier transform of* $e^{-2t}\,\text{step}(t-3)$. *If we had recognized that*

$$e^{-2t}\,\text{step}(t-3) = e^{-6}e^{-2(t-3)}\,\text{step}(t-3) \quad,$$

then we could have computed the transform using identity (21.1a$'$) as follows:

$$\begin{aligned}\mathcal{F}\left[e^{-2t}\,\text{step}(t-3)\right]\Big|_{\omega} &= \mathcal{F}\left[e^{-6}e^{-2(t-3)}\,\text{step}(t-3)\right]\Big|_{\omega} \\ &= e^{-6}\mathcal{F}\left[e^{-2(t-3)}\,\text{step}(t-3)\right]\Big|_{\omega} \\ &= e^{-6}\left(e^{-i2\pi 3\omega}\mathcal{F}\left[e^{-2t}\,\text{step}(t)\right]\Big|_{\omega}\right) \\ &= e^{-6}\left(e^{-i2\pi 3\omega}\cdot\frac{1}{2+i2\pi\omega}\right) = \frac{1}{2+i2\pi\omega}e^{-6-i6\pi\omega} \quad .\end{aligned}$$

?▶Exercise 21.2: Let α and γ be two real numbers with $\alpha > 0$. Using the appropriate translation identity and the fact that

$$\mathcal{F}\left[e^{\alpha t}\,\text{step}(-t)\right]\Big|_{\omega} = \frac{1}{\alpha - i2\pi\omega}$$

(see table 21.1 on page 320), show that

$$\mathcal{F}\left[e^{\alpha t}\,\text{step}(\gamma - t)\right]\Big|_{\omega} = \frac{1}{\alpha - i2\pi\omega}\,e^{\alpha\gamma - i2\gamma\pi\omega} \quad .$$

!▶Example 21.3: Consider the problem of finding the transform of the function

$$e^{(-2+i8\pi)t}\,\text{step}(t) \quad .$$

Observing that we can rewrite this as the product of a function whose transform we know with a complex exponential, we can try to use identity (21.1d'),

$$\mathcal{F}\left[e^{(-2+i8\pi)t}\,\text{step}(t)\right]\Big|_{\omega} = \mathcal{F}\left[e^{i2\pi 4t}e^{-2t}\,\text{step}(t)\right]\Big|_{\omega} = \mathcal{F}\left[e^{-2t}\,\text{step}(t)\right]\Big|_{\omega-4} \quad . \tag{21.4}$$

From table 21.1 on page 320,

$$\mathcal{F}\left[e^{-2t}\,\text{step}(t)\right]\Big|_{y} = \frac{1}{2 + i2\pi y} \quad .$$

This and the sequence of equalities in (21.4) give us

$$\mathcal{F}\left[e^{(-2+i8\pi)t}\,\text{step}(t)\right]\Big|_{\omega} = \mathcal{F}\left[e^{-2t}\,\text{step}(t)\right]\Big|_{\omega-4} = \frac{1}{2 + i2\pi(\omega - 4)} \quad .$$

(Note: By not multiplying out the denominator, we have left our answer in the form of "a simple translation of a relatively simple function". In practice, this tends to be the preferred way to express such functions. It certainly simplifies the graphing of these functions.)

?▶Exercise 21.3: Let α and γ be two real numbers with $\alpha > 0$. Using the appropriate translation identity and the fact that

$$\mathcal{F}\left[e^{\alpha t}\,\text{step}(-t)\right]\Big|_{\omega} = \frac{1}{\alpha - i2\pi\omega}$$

(see table 21.1 on page 320), show that

$$\mathcal{F}\left[e^{(\alpha + i2\pi\gamma)t}\,\text{step}(-t)\right]\Big|_{\omega} = \frac{1}{\alpha - i2\pi(\omega - \gamma)} \quad .$$

Also, sketch the real and imaginary parts of this transform.

The Modulation Identities

It is not at all uncommon to encounter the product of a sine or cosine function with a function whose transform is already known. Finding the transforms of such products is easy using the translation identities and the complex exponential formulas for the sine and cosine,

$$\sin(2\pi\gamma x) = \frac{e^{i2\pi\gamma x} - e^{-i2\pi\gamma x}}{2i} \quad \text{and} \quad \cos(2\pi\gamma x) = \frac{e^{i2\pi\gamma x} + e^{-i2\pi\gamma x}}{2} \quad .$$

!► Example 21.4: Consider finding the Fourier transform of

$$\sin(6\pi t)\, e^{-2t}\, \text{step}(t) \quad.$$

For convenience, let

$$f(t) \;=\; e^{-2t}\, \text{step}(t) \quad.$$

Then, rewriting the sine function in complex exponential form and using both the linearity of the transform and translation identity (21.1d), we have

$$\mathcal{F}\!\left[\sin(6\pi t)\, e^{-2t}\, \text{step}(t)\right]\Big|_{\omega} \;=\; \mathcal{F}[\sin(2\pi 3t)\, f(t)]\big|_{\omega}$$

$$=\; \mathcal{F}\!\left[\frac{e^{i2\pi 3t} - e^{-i2\pi 3t}}{2i}\, f(t)\right]\Big|_{\omega}$$

$$=\; \frac{1}{2i}\left(\mathcal{F}\!\left[e^{i2\pi 3t} f(t)\right]\Big|_{\omega} - \mathcal{F}\!\left[e^{i2\pi(-3)t} f(t)\right]\Big|_{\omega}\right)$$

$$=\; \frac{1}{2i}\left[F(\omega - 3) - F(\omega - (-3))\right]$$

$$=\; \frac{i}{2}\left[F(\omega + 3) - F(\omega - 3)\right]$$

where $F = \mathcal{F}[f]$. From table 21.1 on page 320 we find that

$$F(y) \;=\; \mathcal{F}\!\left[e^{-2t}\, \text{step}(t)\right]\Big|_{y} \;=\; \frac{1}{2 + i2\pi y} \quad.$$

So

$$\mathcal{F}\!\left[\sin(2\pi 3t)\, e^{-2t}\, \text{step}(t)\right]\Big|_{\omega} \;=\; \frac{i}{2}\left[F(\omega + 3) - F(\omega - 3)\right]$$

$$=\; \frac{i}{2}\left[\frac{1}{2 + i2\pi(\omega + 3)} - \frac{1}{2 + i2\pi(\omega - 3)}\right] \quad.$$

Look back over the last example. Cleverly embedded is a derivation of the following: For any classically transformable function f, the product $\sin(2\pi 3t)\, f(t)$ is also transformable and

$$\mathcal{F}[\sin(2\pi 3t)\, f(t)]|_{\omega} \;=\; \frac{i}{2}\left[F(\omega + 3) - F(\omega - 3)\right] \qquad \text{where} \quad F = \mathcal{F}[f(t)] \quad.$$

Replacing 3 with γ then gives the first identity listed in the following theorem.

Theorem 21.2 (modulation identities)
The product of any classically transformable function with a sine or cosine function is another classically transformable function. Moreover, if f and F is any pair of classically transformable functions with $F = \mathcal{F}[f]$, and, if γ is any fixed real number, then

$$\mathcal{F}\!\left[\sin(2\pi \gamma t)\, f(t)\right]\Big|_{\omega} \;=\; \frac{i}{2}\left[F(\omega + \gamma) - F(\omega - \gamma)\right] \quad, \tag{21.5a}$$

$$\mathcal{F}\!\left[\cos(2\pi \gamma t)\, f(t)\right]\Big|_{\omega} \;=\; \frac{1}{2}\left[F(\omega + \gamma) + F(\omega - \gamma)\right] \quad, \tag{21.5b}$$

$$\mathcal{F}^{-1}\!\left[\sin(2\pi \gamma \omega)\, F(\omega)\right]\Big|_{t} \;=\; \frac{i}{2}\left[f(t - \gamma) - f(t + \gamma)\right] \tag{21.5c}$$

and

$$\mathcal{F}^{-1}\!\left[\cos(2\pi \gamma \omega)\, F(\omega)\right]\Big|_{t} \;=\; \frac{1}{2}\left[f(t - \gamma) + f(t + \gamma)\right] \quad. \tag{21.5d}$$

Identities (21.5a) through (21.5d) are called the *modulation identities*, and the reader can readily verify that all can be derived directly from the translation identities.

?►Exercise 21.4:   *Derive identity (21.5b) from the appropriate translation identity.*

?►Exercise 21.5:   *Show that*

$$\mathcal{F}\left[\cos(2\pi 3t)\, e^{-2t}\, \text{step}(t)\right]\Big|_{\omega} = \frac{1}{2}\left[\frac{1}{2 + i2\pi(\omega+3)} + \frac{1}{2 + i2\pi(\omega-3)}\right]$$

a: *first, by using the appropriate translation identity (as was done in example 21.4),*

b: *and then by using the appropriate modulation identity.*

By the way, these are called the modulation identities because of the forms of the functions appearing in the left sides of identities (21.5). They are all expressed as "amplitude modulations" of sine and cosine functions. That is, each is written as a fixed sine or cosine function multiplied by some function, and that function is viewed as modulating (i.e., adjusting or varying) the amplitude of that sine or cosine function. Such expressions arise naturally in many applications. For example,

$$f(t)\cos(2\pi\omega_c t)$$

could well describe the signal transmitted over time by an old-time AM radio station (remember, "AM" stands for "amplitude modulation"). The function f contains the information the station wishes to communicate — music, news, commercials, etc. — and the value ω_c, called the carrier frequency, is the frequency to which you tune your radio to hear the station. However, before you can hear the station, your radio must extract the function $f(t)$ from the signal actually transmitted. This extraction is actually done (in some radios, at least) by electronic analogs of the procedures described in the next exercise.

?►Exercise 21.6 (de-modulation):   *Let*

$$g(t) = f(t)\cos(2\pi\omega_c t)$$

where ω_c is some fixed positive value and f is some function with finite bandwidth (see page 311). Let Ω be a bandwidth bound for f, and assume $\Omega < \omega_c$.

a: *Sketch possible graphs of $F(\omega)$, $F(\omega + 2\omega_c)$ and $F(\omega - 2\omega_c)$ assuming $F = \mathcal{F}[f]$. Using these graphs, convince yourself that*

$$\text{pulse}_\Omega(\omega)\, F(\omega) = F(\omega) \qquad \text{for} \quad -\infty < \omega < \infty \quad,$$

while

$$\text{pulse}_\Omega(\omega)\, F(\omega \pm 2\omega_c) = 0 \qquad \text{for} \quad -\infty < \omega < \infty \quad.$$

b: *Let*

$$h(t) = g(t)\cos(2\pi\omega_c t) \qquad \text{and} \qquad H(\omega) = \mathcal{F}[h]\big|_{\omega} \quad.$$

Using a standard trigonometric identity, the modulation identities and the observations made above, show that

$$f = 2\,\mathcal{F}^{-1}\left[\text{pulse}_\Omega(\omega)\, H(\omega)\right] \quad.$$

21.2 Scaling

The scaling identities tell us that the Fourier transform of a function with a "scaled variable" is an appropriately scaled version of the corresponding transform of the original function. They will occasionally be helpful in computing transforms, and they are important in understanding how transforms of functions vary as we make certain changes in the original functions.

Theorem 21.3 (scaling identities)
Any classically transformable function with its variable scaled by a nonzero real number is another classically transformable function. Moreover, if f and F is any pair of classically transformable functions with $F = \mathcal{F}[f]$, and if γ is any real nonzero constant, then

$$\mathcal{F}[f(\gamma t)]\big|_\omega = \frac{1}{|\gamma|} F\left(\frac{\omega}{\gamma}\right) \tag{21.6a}$$

and

$$\mathcal{F}^{-1}[F(\gamma\omega)]\big|_t = \frac{1}{|\gamma|} f\left(\frac{t}{\gamma}\right) \quad . \tag{21.6b}$$

Equivalently

$$\mathcal{F}^{-1}\left[F\left(\frac{\omega}{\gamma}\right)\right]\bigg|_t = |\gamma| \, f(\gamma t) \tag{21.6c}$$

and

$$\mathcal{F}\left[f\left(\frac{t}{\gamma}\right)\right]\bigg|_\omega = |\gamma| \, F(\gamma\omega) \quad . \tag{21.6d}$$

Do note that the last two identities are redundant given the first two. If that's not obvious, let $\alpha = 1/\gamma$. The last two identities then become

$$\mathcal{F}^{-1}[F(\alpha\omega)]\big|_t = \frac{1}{|\alpha|} f\left(\frac{t}{\alpha}\right) \qquad \text{and} \qquad \mathcal{F}[f(\alpha t)]\big|_\omega = \frac{1}{|\alpha|} F\left(\frac{\omega}{\alpha}\right) \quad ,$$

which are identical, save for the symbol used, to the first two identities. In fact, the very same arguments show that the first two identities are, themselves, completely equivalent. For this reason many texts refer to equation (21.6a) as "the" scaling identity.

We might also observe that identities (21.6a) and (21.6b) can be written as

$$\mathcal{F}[\psi(\gamma t)]\big|_\omega = \frac{1}{|\gamma|} \mathcal{F}[\psi(t)]|_{\omega/\gamma} \tag{21.6a$'$}$$

and

$$\mathcal{F}^{-1}[\psi(\gamma\omega)]\big|_t = \frac{1}{|\gamma|} \mathcal{F}^{-1}[\psi(\omega)]|_{t/\gamma} \quad . \tag{21.6b$'$}$$

The proof of theorem 21.3 is both straightforward and a good exercise for the reader.

?►Exercise 21.7   a: *Derive identity (21.6a) assuming f is absolutely integrable.*

   b: *Show that identity (21.6a) also holds when f is in $\mathcal{T}$.*

   c: *Finish proving theorem 21.3.*

!►Example 21.5:   *Consider finding the Fourier transform of $(2 - 3it)^{-1}$.*
       By the scaling identity,

$$\mathcal{F}\left[\frac{1}{2 - 3it}\right]\bigg|_\omega = \mathcal{F}\left[\frac{1}{2 - i(3t)}\right]\bigg|_\omega = \mathcal{F}[f(3t)]\big|_\omega = \frac{1}{|3|} F\left(\frac{\omega}{3}\right) \tag{21.7}$$

where

$$f(3t) \;=\; \frac{1}{2 - i(3t)} \qquad \text{and} \qquad F(y) \;=\; \mathcal{F}[f(x)]|_y \quad .$$

We can find the "unscaled" formula for f by using the substitution $x = 3t$,

$$f(x) \;=\; \frac{1}{2 - ix} \quad .$$

This is a function listed in table 21.1 on page 320. From that table we get

$$F(y) \;=\; \mathcal{F}\!\left[\frac{1}{2 - ix}\right]\Big|_y \;=\; 2\pi\, e^{-2\pi 2y}\,\text{step}(y) \;=\; 2\pi\, e^{-4\pi y}\,\text{step}(y) \quad .$$

So equation (21.7) becomes

$$\mathcal{F}\!\left[\frac{1}{2 - i3t}\right]\Big|_\omega \;=\; \frac{1}{|3|} F\!\left(\frac{\omega}{3}\right) \;=\; \frac{2\pi}{3} e^{-4\pi(\omega/3)}\,\text{step}\!\left(\frac{\omega}{3}\right) \quad . \tag{21.8}$$

At this point we should observe that

$$\text{step}\!\left(\frac{\omega}{3}\right) \;=\; \begin{cases} 1 & \text{if } 0 < {}^{\omega}\!/_3 \\ 0 & \text{otherwise} \end{cases} \;=\; \begin{cases} 1 & \text{if } 0 < \omega \\ 0 & \text{otherwise} \end{cases} \;=\; \text{step}(\omega) \quad .$$

So, the equalities in line (21.8) simplify to

$$\mathcal{F}\!\left[\frac{1}{2 - i3t}\right]\Big|_\omega \;=\; \frac{2\pi}{3} e^{-4\pi\omega/3}\,\text{step}(\omega) \quad .$$

?▶Exercise 21.8:   Show that

$$\mathcal{F}\!\left[\frac{1}{\alpha - i\beta t}\right]\Big|_\omega \;=\; \frac{2\pi}{\beta} e^{-2\pi\alpha\omega/\beta}\,\text{step}(\omega)$$

and

$$\mathcal{F}\!\left[\frac{1}{\alpha + i\beta t}\right]\Big|_\omega \;=\; \frac{2\pi}{\beta} e^{2\pi\alpha\omega/\beta}\,\text{step}(-\omega)$$

whenever α and β are positive real numbers.

21.3 Practical Transform Computing

In practice, few people compute formulas for Fourier transforms from first principles. Typically, you have a table of known transforms (such as table 21.1) and a list of identities (we hardly have enough identities, yet, to make a decent list), and you "cleverly" use the identities to convert the formulas at hand to formulas involving functions appearing on your table of transforms. This is not a process requiring deep knowledge of Fourier analysis. Instead, what you really need (aside from reasonable tables) are *(1)* some competency in "pattern recognition", *(2)* moderate bookkeeping skills and *(3)* the wits to avoid inexcusable blunders.

"Competency in 'pattern recognition'", means the ability to look at a formula and recognize which identities are likely to help reduce the formula to something involving functions you know are in your table of transforms. This competency requires practice to develop and some — but not that much — understanding of the theory of Fourier transforms. It also helps to have tables with which you are reasonably familiar.

The bookkeeping skills are needed simply to keep from getting lost or confused in your calculations. This becomes especially important in calculations involving more than one identity. Use different symbols for different functions and variables, and write down the formulas describing how these functions and variables are related to each other. Beginning students, especially, should avoid reusing symbols in a series of calculations. If you want to use an identity containing, say, the symbol "f" but you've already used "f" to denote something else and there is the slightest danger of later confusing the two different quantities being represented by "f", then rewrite that identity with "f" replaced with some other symbol, say, "g". Between the Latin and Greek alphabets, and your own imagination, there are plenty of symbols available.

In spite of good bookkeeping, errors, sometimes, do hoppen. But some errors are particularly hard to excuse. They proclaim "This person is so ignorant of the basic concepts that he or she cannot even read a simple table!" What sort of errors proclaim this? Here are a few with which this author is sadly familiar:

1. *Misuse of dummy variables*: Be careful about dummy variable substitutions. Remember, the expression

$$\mathcal{F}[f(t)]|_\omega$$

   is really shorthand for

   > the formula for $\mathcal{F}[f]$ using the ω as the variable and with f being the function whose formula is $f(t)$.

   So, for example, both sides of the equation

$$\mathcal{F}\left[e^{-|t|}\right]\bigg|_\omega = \frac{2}{1 + 4\pi^2\omega^2}$$

   are formulas of ω, and we can use the substitution $\omega = y - 2$ to get the equation

$$\mathcal{F}\left[e^{-|t|}\right]\bigg|_{y-2} = \frac{2}{1 + 4\pi^2(y-2)^2} .$$

   Neither side, however, is truly a function of t, and we can *NOT* use the substitution $t = x - 2$ to claim

$$\mathcal{F}\left[e^{-|x-2|}\right]\bigg|_\omega = \mathcal{F}\left[e^{-|t|}\right]\bigg|_\omega = \frac{2}{1 + 4\pi^2\omega^2} .$$

   In fact, identity (21.1a$'$) tells us that, instead,

$$\mathcal{F}\left[e^{-|x-2|}\right]\bigg|_\omega = e^{-i2\pi 2\omega}\,\mathcal{F}\left[e^{-|x|}\right]\bigg|_\omega = e^{-i4\pi\omega}\,\frac{2}{1 + 4\pi^2\omega^2} .$$

2. *Ignoring the stated restrictions*: Any reasonable table of transforms or identities will indicate when each transform formula or identity can be used. Ignoring these restrictions will usually (but not always) result in errors. Consider, for example, finding the transform of $(-3+it)^{-1}$. According to table 21.1 on page 320,

$$\mathcal{F}\left[\frac{1}{\alpha + it}\right]\bigg|_\omega = 2\pi\,e^{2\pi\alpha\omega}\,\text{step}(-\omega) \qquad \textit{provided} \qquad 0 < \alpha$$

   and

$$\mathcal{F}\left[\frac{1}{\alpha - it}\right]\bigg|_\omega = 2\pi\,e^{-2\pi\alpha\omega}\,\text{step}(\omega) \qquad \textit{provided} \qquad 0 < \alpha .$$

(Mis)using the first identity (with $\alpha = -3$, in defiance of the restriction that $0 < \alpha$) gives

$$\mathcal{F}\left[\frac{1}{-3 + it}\right]\bigg|_{\omega} = 2\pi\, e^{2\pi(-3)\omega}\, \text{step}(-\omega) = 2\pi\, e^{-6\pi\omega}\, \text{step}(-\omega) \quad.$$

But these computations cannot be trusted since the restriction that $0 < \alpha$ was violated. In fact, the function obtained is wrong. Doing the computation correctly, that is, using the other identity with $\alpha = 3$, so the restriction that $0 < \alpha$ is satisfied, we have

$$\mathcal{F}\left[\frac{1}{-3 + it}\right]\bigg|_{\omega} = -\mathcal{F}\left[\frac{1}{3 - it}\right]\bigg|_{\omega}$$

$$= -2\pi\, e^{-2\pi 3\omega}\, \text{step}(\omega) = -2\pi\, e^{-6\pi\omega}\, \text{step}(\omega) \quad.$$

This result is very different from the one obtained by misusing the first identity. (Sketch the two functions to see just how big a difference results from the two sign differences.)

3. *Using formulas and functions that are not understood*: When you first begin using tables, there is a good chance that you will encounter functions and formulas that you do not yet understand. For example, using the tables in the appendix you can easily derive

$$\mathcal{F}\left[e^{i2\pi t}2\,\text{sinc}(2\pi t)\right]\bigg|_{\omega} = \delta_1 * \text{pulse}_1(\omega) \quad.$$

The pulse function we know, but, since we have not yet discussed the delta function or convolution, it is likely that the "δ_1" and "$*$" are meaningless symbols to you. If so, then "$\delta_1 * \text{pulse}_1(\omega)$" just stands for "a group of meaningless symbols", and you can hardly say that you know what $\mathcal{F}\left[e^{i2\pi t}2\,\text{sinc}(2\pi t)\right]\big|_{\omega}$ is.

!▶ **Example 21.6:**   *Let's find the Fourier transform of*

$$\frac{e^{i6\pi t}}{2 - 5i + it} \quad.$$

Since this can be written as

$$e^{i2\pi 3t} \cdot \frac{1}{2 + i(t - 5)} \quad,$$

translation identities (21.1a),

$$\mathcal{F}\left[f(t - \gamma)\right]\big|_{\omega} = e^{-i2\pi\gamma\omega}F(\omega) \quad, \tag{21.9}$$

and (21.1d),

$$\mathcal{F}\left[e^{i2\pi\gamma t}f(t)\right]\big|_{\omega} = F(\omega - \gamma) \quad, \tag{21.10}$$

seem worth trying. (Here, γ is any real value.)
    Applying identity (21.10) with $\gamma = 3$ yields

$$\mathcal{F}\left[\frac{e^{i6\pi t}}{2 - 5i + it}\right]\bigg|_{\omega} = \mathcal{F}\left[e^{i2\pi 3t} \cdot \frac{1}{2 - 5i + it}\right]\bigg|_{\omega}$$

$$= \mathcal{F}\left[e^{i2\pi 3t}f(t)\right]\bigg|_{\omega} \tag{21.11}$$

$$= F(\omega - 3)$$

where

$$f(t) = \frac{1}{2 - 5i + it}$$

and

$$F(y) = \mathcal{F}[f(t)]|_y = \mathcal{F}\left[\frac{1}{2 - 5i + it}\right]\Bigg|_y = \mathcal{F}\left[\frac{1}{2 + i(t - 5)}\right]\Bigg|_y \quad . \tag{21.12}$$

(Observe that we used y rather than ω for the variable in the formula for F . That way, once we finally find the formula for F(y), we can find the formula for equation (21.13) by just replacing y with ω − 3 .)

To avoid possible confusion with some of the other equations in these computations, let us rewrite equation (21.9) with the symbols f , F and γ replaced with g , G and β , respectively,

$$\mathcal{F}[g(t - \beta)]|_y = e^{-i2\pi\beta y} G(y) \quad .$$

(So, in our rewritten identity, G = \mathcal{F}[g] and β is any real number.) We have also replaced the symbol ω with the symbol y since that is the symbol for the variable in line (21.12). Using our rewritten identity (with β = 5), the equations in line (21.12) become

$$\begin{aligned} F(y) &= \mathcal{F}\left[\frac{1}{2 + i(t - 5)}\right]\Bigg|_y \\ &= \mathcal{F}[g(t - 5)]|_y \\ &= e^{-i2\pi 5y} G(y) = e^{-i10\pi y} G(y) \quad , \end{aligned} \tag{21.13}$$

where

$$g(t - 5) = \frac{1}{2 + i(t - 5)} \quad \text{and} \quad G(y) = \mathcal{F}[g(x)]|_y \quad .$$

Substituting x for t − 5 in the above formula for g(t − 5) gives us our formula for g ,

$$g(x) = \frac{1}{2 + ix} \quad .$$

The transform of this function can be found in table 21.1 on page 320. Keeping in mind that, here, the symbols for the variables are x and y instead of t and ω , table 21.1 tells us that

$$\begin{aligned} G(y) &= \mathcal{F}[g(x)]|_y = \mathcal{F}\left[\frac{1}{2 + ix}\right]\Bigg|_y \\ &= 2\pi e^{2\pi 2y} \text{step}(-y) = 2\pi e^{4\pi y} \text{step}(-y) \quad . \end{aligned}$$

This gives us the formula for G(y). Plugging this into equation (21.13) then gives us the formula for F(y),

$$\begin{aligned} F(y) &= e^{-i10\pi y} G(y) \\ &= e^{-i10\pi y}\left[2\pi e^{4\pi y} \text{step}(-y)\right] = 2\pi e^{(4\pi - i10\pi)y} \text{step}(-y) \quad . \end{aligned}$$

Finally, looking back at equation (21.11), we see that the desired transform is simply F(y) with y = ω − 3 . So

$$\mathcal{F}\left[\frac{e^{i6\pi t}}{2 - 5i + it}\right]\Bigg|_\omega = F(\omega - 3) = 2\pi e^{(4\pi - i10\pi)(\omega - 3)} \text{step}(3 - \omega) \quad .$$

21.4 Complex Conjugation and Related Symmetries
Complex Conjugation Identities

The complex conjugation identities are very similar to the identities making up the principle of near-equivalence. We can derive them easily by observing that

$$\int_{-\infty}^{\infty} \phi^*(y)\, e^{-i2\pi xy}\, dy \;=\; \int_{-\infty}^{\infty} \left[\phi(y)\, e^{i2\pi xy}\right]^*\, dy \;=\; \left[\int_{-\infty}^{\infty} \phi(y)\, e^{i2\pi xy}\, dy\right]^*$$

and

$$\int_{-\infty}^{\infty} \phi^*(y)\, e^{i2\pi xy}\, dy \;=\; \int_{-\infty}^{\infty} \left[\phi(y)\, e^{-i2\pi xy}\right]^*\, dy \;=\; \left[\int_{-\infty}^{\infty} \phi(y)\, e^{-i2\pi xy}\, dy\right]^*$$

whenever ϕ is a function in $\mathcal{A}$. Recognizing that the integrals on the left and right sides of these equations are the integral formulas for the Fourier transforms, we see that these equations can be rewritten as

$$\mathcal{F}[\phi^*] \;=\; \left(\mathcal{F}^{-1}[\phi]\right)^* \qquad \text{and} \qquad \mathcal{F}^{-1}[\phi^*] \;=\; \left(\mathcal{F}[\phi]\right)^* \quad .$$

These equations are the complex conjugation identities, and we have just gone through the first part of the proof of our next theorem.

Theorem 21.4 (complex conjugation identities)
For any classically transformable function ψ,

$$\mathcal{F}[\psi^*] \;=\; \left(\mathcal{F}^{-1}[\psi]\right)^* \qquad \text{and} \qquad \mathcal{F}^{-1}[\psi^*] \;=\; \left(\mathcal{F}[\psi]\right)^* \quad . \tag{21.14}$$

PROOF: The brief computations done just above (but with $\psi = \phi$) confirmed that both of these equations hold when ψ is in $\mathcal{A}$.

To show that the first equation in line (21.14) holds when ψ is in $\mathcal{T}$, let $\phi = \mathcal{F}^{-1}[\psi]$. That is, ϕ is the function in $\mathcal{A}$ such that $\psi = \mathcal{F}[\phi]$. Since we know the conjugation identities hold for functions in $\mathcal{A}$, we know that

$$\psi^* \;=\; \left(\mathcal{F}[\phi]\right)^* \;=\; \mathcal{F}^{-1}[\phi^*] \quad .$$

Thus,

$$\mathcal{F}[\psi^*] \;=\; \mathcal{F}\left[\mathcal{F}^{-1}[\phi^*]\right] \;=\; \phi^* \;=\; \left(\mathcal{F}^{-1}[\psi]\right)^* \quad ,$$

verifying the validity of the first equation in line (21.14) when ψ is in $\mathcal{T}$.

So we now know the first equation in line (21.14) holds whenever ψ is in either $\mathcal{A}$ or in $\mathcal{T}$. But, by definition, if ψ is any classically transformable function, then $\psi = \psi_{\mathcal{A}} + \psi_{\mathcal{T}}$ where $\psi_{\mathcal{A}}$ and $\psi_{\mathcal{T}}$ are in $\mathcal{A}$ and $\mathcal{T}$, respectively. By this, the above and the linearity of the transforms and conjugation, we have

$$\begin{aligned}
\mathcal{F}[\psi^*] &= \mathcal{F}\left[\psi_{\mathcal{A}}^* + \psi_{\mathcal{T}}^*\right] \\
&= \mathcal{F}\left[\psi_{\mathcal{A}}^*\right] + \mathcal{F}\left[\psi_{\mathcal{T}}^*\right] \\
&= \left(\mathcal{F}^{-1}[\psi_{\mathcal{A}}]\right)^* + \left(\mathcal{F}^{-1}[\psi_{\mathcal{T}}]\right)^* \\
&= \left(\mathcal{F}^{-1}[\psi_{\mathcal{A}} + \psi_{\mathcal{T}}]\right)^* = \left(\mathcal{F}^{-1}[\psi]\right)^* \quad .
\end{aligned}$$

Thus, the first equation in line (21.14) holds when ψ is any classically transformable function.

The verification of the second equation in line (21.14) is very similar and will be left as an exercise.   ∎

?►Exercise 21.9:   *Complete the proof of the above theorem by verifying that*

$$\mathcal{F}^{-1}\left[\psi^*\right] \;=\; \left(\mathcal{F}[\psi]\right)^*$$

whenever:

   a: ψ *is in* $\mathcal{T}$.

   b: ψ *is classically transformable.*

Some Related Symmetries*

We've already observed that the Fourier transform of an even transformable function is even, and that the Fourier transform of an odd transformable function is another odd function. We can expand on these observations by also looking at the real and imaginary parts of our functions.

Let's start with an absolutely integrable, piecewise continuous function f having real and imaginary parts u and v, respectively. Then

$$\mathcal{F}[f]|_x \;=\; \int_{-\infty}^{\infty} f(y)\, e^{-i2\pi xy}\, dy$$

$$=\; \int_{-\infty}^{\infty} [u(y) + iv(y)]\, [\cos(2\pi xy) - i\sin(2\pi xy)]\, dy$$

$$=\; \int_{-\infty}^{\infty} [u(y)\cos(2\pi xy) + v(y)\sin(2\pi xy)]\, dy$$

$$+\; i \int_{-\infty}^{\infty} [-u(y)\sin(2\pi xy) + v(y)\cos(2\pi xy)]\, dy \quad .$$

From this, we see that the real and imaginary parts of $F = \mathcal{F}[f]$ are given by

$$\mathrm{Re}[F(x)] \;=\; \int_{-\infty}^{\infty} [u(y)\cos(2\pi xy) + v(y)\sin(2\pi xy)]\, dy$$

and

$$\mathrm{Im}[F(x)] \;=\; \int_{-\infty}^{\infty} [-u(y)\sin(2\pi xy) + v(y)\cos(2\pi xy)]\, dy \quad .$$

Observe that, if f is real valued (i.e., $u = f$ and $v = 0$), then these equations reduce to

$$\mathrm{Re}[F(x)] \;=\; \int_{-\infty}^{\infty} u(y)\cos(2\pi xy)\, dy$$

and

$$\mathrm{Im}[F(x)] \;=\; -\int_{-\infty}^{\infty} u(y)\sin(2\pi xy)\, dy \quad .$$

With additional assumptions regarding the symmetry of f, these equations reduce further. Consider, for example, what happens when f is also an even function. Then $u(y)\cos(2\pi xy)$ is an even function of y, $u(y)\sin(2\pi xy)$ is an odd function of y, and, as we noted in our discussion of integrals of even and odd functions in chapter 5,

$$\mathrm{Re}[F(x)] \;=\; \int_{-\infty}^{\infty} u(y)\cos(2\pi xy)\, dy \;=\; 2\int_{0}^{\infty} u(y)\cos(2\pi xy)\, dy$$

---

* To be honest, we will have little future need for the results derived in this subsection, but they are occasionally useful in certain applications.

Table 21.2: Symmetry Relations between f and F with $F = \mathcal{F}[f]$

f is even	$\Longleftrightarrow$	F is even
f is real valued and even	$\Longleftrightarrow$	F is real valued and even
f is imaginary valued and even	$\Longleftrightarrow$	F is imaginary valued and even
f is odd	$\Longleftrightarrow$	F is odd
f is real valued and odd	$\Longleftrightarrow$	F is imaginary valued and odd
f is imaginary valued and odd	$\Longleftrightarrow$	F is real valued and odd
f is real valued	$\Longleftrightarrow$	$F(-\omega) = [F(\omega)]^*$

and

$$\mathrm{Im}[F(x)] = -\int_{-\infty}^{\infty} u(y) \sin(2\pi xy) \, dy = 0 \quad .$$

Since the imaginary part of F is 0,

$$F(x) = \mathrm{Re}[F(x)] = 2\int_{0}^{\infty} u(y) \cos(2\pi xy) \, dy \quad .$$

So F must be real valued. Furthermore, because the cosine is an even function, we can easily verify (again) that F is an even function,

$$F(-x) = 2\int_{0}^{\infty} u(y) \cos(2\pi(-x)y) \, dy = 2\int_{0}^{\infty} u(y) \cos(2\pi xy) \, dy = F(x) \quad .$$

In summary, here is what we've just shown:

> If f is a classically transformable function that is also even, real valued and absolutely integrable, then its Fourier transform is an even and real-valued function.

Looking back over our derivation of this statement, it should be clear that a number of other similarly worded statements, such as

> If f is a classically transformable function that is also odd, real valued and absolutely integrable, then its Fourier transform is an odd and imaginary-valued function

are also true. You may even suspect these statements remain true with the phrase "absolutely integrable" removed. They do, and you should have little trouble confirming this. So let us just state a brief theorem and provide a short table summarizing some statements you can easily prove.

Theorem 21.5
The implications given in table 21.2 are true whenever f and F are classically transformable functions with $F = \mathcal{F}[f]$.

?▶ Exercise 21.10 (symmetries):   *We showed that $F = \mathcal{F}[f]$ is even and real valued when f is an even, real-valued function in $\mathcal{A}$. Now show F is even and real valued when*

    a: *f is an even, real-valued function in $\mathcal{T}$.*

    b: *f is an even, real-valued, classically transformable function.*

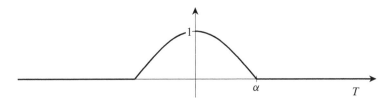

Figure 21.1: The truncated cosine function, $\cos\left(\frac{\pi}{2\alpha}t\right)\text{pulse}_\alpha(t)$ with $\alpha > 0$.

Additional Exercises

Unless otherwise indicated, all of the following exercises are to be done using table 21.1 on page 320 and the identities and properties of the classical Fourier transform developed in this and the previous chapter. Do not compute these transforms by directly evaluating integrals.

21.11. *Using the translation and modulation identities, compute the following:*

a. $\mathcal{F}\left[\text{pulse}_3(t-4)\right]\big|_\omega$

b. $\mathcal{F}\left[\text{pulse}_3(t+4)\right]\big|_\omega$

c. $\mathcal{F}\left[e^{i8\pi t}\,\text{pulse}_3(t)\right]\big|_\omega$

d. $\mathcal{F}\left[e^{-i8\pi t}\,\text{pulse}_3(t)\right]\big|_\omega$

e. $\mathcal{F}\left[\dfrac{1}{3+4i-it}\right]\big|_\omega$

f. $\mathcal{F}^{-1}\left[\dfrac{1}{3+i4\pi-i2\pi\omega}\right]\big|_t$

g. $\mathcal{F}\left[e^{-5|t-2|}\right]\big|_\omega$

h. $\mathcal{F}\left[e^{i10\pi t}e^{-3|t|}\right]\big|_\omega$

i. $\mathcal{F}^{-1}\left[e^{i10\pi\omega}e^{-3|\omega|}\right]\big|_t$

j. $\mathcal{F}^{-1}\left[\dfrac{e^{i\pi\omega}}{3-i2\pi\omega}\right]\big|_t$

k. $\mathcal{F}\left[e^{-3t}\,\text{step}(t-4)\right]\big|_\omega$

l. $\mathcal{F}\left[\dfrac{1}{16+(t-3)^2}\right]\big|_\omega$

m. $\mathcal{F}\left[\dfrac{1}{34-10t+t^2}\right]\big|_\omega$ *(Hint: See the previous problem.)*

n. $\mathcal{F}\left[\dfrac{1}{29+4t+t^2}\right]\big|_\omega$

o. $\mathcal{F}\left[e^{-3t}\cos(2\pi t)\,\text{step}(t)\right]\big|_\omega$

p. $\mathcal{F}\left[\cos(6\pi t)\,e^{-5|t|}\right]\big|_\omega$

q. $\mathcal{F}\left[\sin(6\pi t)\,e^{-5|t|}\right]\big|_\omega$

r. $\mathcal{F}\left[\dfrac{\sin^2(2\pi t)}{2\pi t}\right]\big|_\omega$

s. $\mathcal{F}\left[\cos\left(\frac{\pi}{2\alpha}t\right)\text{pulse}_\alpha(t)\right]\big|_\omega$ *with $\alpha > 0$*   *(the truncated cosine in figure 21.1)*

21.12. *Using scaling (and table 21.1), find the following transforms:*

a. $\mathcal{F}\left[\dfrac{1}{12+i3t}\right]\big|_\omega$

b. $\mathcal{F}\left[\dfrac{1}{\alpha^2+\gamma^2t^2}\right]\big|_\omega$ *where $\alpha > 0$ and $\gamma > 0$*

21.13. *Find the following transforms using table 21.1, the scaling and/or shifting identities, and the fundamental properties of Fourier transforms. Where they appear, assume α, β and γ are real numbers with $\alpha > 0$ and $\gamma > 0$. Plan on using at least two identities for each.*

a. $\mathcal{F}\left[e^{-i4\pi t}\operatorname{sinc}(t - 3)\right]\Big|_{\omega}$

b. $\mathcal{F}\left[e^{i\pi t}\operatorname{pulse}_3(t - 1)\right]\Big|_{\omega}$

c. $\mathcal{F}\left[\dfrac{\sin(2\pi t)}{6 + 3i + it}\right]\Big|_{\omega}$

d. $\mathcal{F}\left[\dfrac{e^{i6\pi t}}{4 + 12i + 3it}\right]\Big|_{\omega}$

e. $\mathcal{F}\left[\dfrac{1}{\alpha + i\beta + i\gamma t}\right]\Big|_{\omega}$

f. $\mathcal{F}\left[\dfrac{1}{\alpha + i\beta - i\gamma t}\right]\Big|_{\omega}$

21.14. *In the following, let $\alpha > 0$. You will also use the fact, derived in exercise 19.15 on page 295, that*

$$\mathcal{F}\left[\operatorname{rect}_{(0,1)}(t)\right]\Big|_{\omega} = i\frac{e^{-i2\pi\omega} - 1}{2\pi\omega} \quad .$$

a. *Find the values of β and b (in terms of α) such that*

$$\operatorname{rect}_{(0,\alpha)}(t) = \operatorname{rect}_{(0,1)}(\beta t) \quad \text{and} \quad \operatorname{rect}_{(-\alpha,0)}(t) = \operatorname{rect}_{(0,1)}(bt) \quad .$$

b. *Using the scaling identity and the above, find the following:*

i. $\mathcal{F}\left[\operatorname{rect}_{(0,\alpha)}(t)\right]\Big|_{\omega}$

ii. $\mathcal{F}\left[\operatorname{rect}_{(-\alpha,0)}(t)\right]\Big|_{\omega}$

21.15. *Let $-\infty < A < B < \infty$.*

a. *Find the values of α and t_0 (in terms of A and B) such that*

$$\operatorname{rect}_{(A,B)}(t) = \operatorname{pulse}_{\alpha}(t - t_0) \quad .$$

b. *Using the above and a translation identity, find the direct Fourier transform of $\operatorname{rect}_{(A,B)}(t)$.*

c. *Use the formula just obtained to compute $\mathcal{F}\left[\operatorname{rect}_{(0,1)}(t)\right]\Big|_{\omega}$. Show that the result is equivalent to the formula given in exercise 21.14, above.*

d. *Use the formula just obtained to compute $\mathcal{F}\left[\operatorname{rect}_{(5,6)}(t)\right]\Big|_{\omega}$.*

21.16. *The translation identities were derived assuming real-valued translations. Occasionally, it may be tempting to use the translation identities with imaginary translations. The following shows that this is generally a BAD idea.*

a. *What is the correct formula for $\mathcal{F}^{-1}\left[\dfrac{1}{2\pi + i2\pi\omega}\right]\Big|_{t}$?*

b. *Verify that*

$$\frac{1}{2\pi + i2\pi\omega} = \frac{-1}{2\pi - i2\pi(\omega - 2i)} \quad ,$$

and "evaluate" the inverse Fourier transform of this using translation identity 21.1b on page 319 with — in violation of the stated restriction — $\gamma = 2i$.

c. *Confirm that the result just obtained by (mis)using a translation identity is wrong by comparing it with the correct formula.[1]*

---

[1] The classical translation identities can *sometimes* be safely used with complex-valued translations. This will be discussed further in chapter 35 where we will develop a "generalized translation" operator.

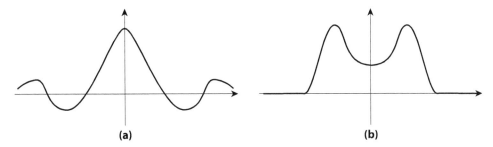

(a)   (b)

Figure 21.2: Graphs for exercise 21.17 with (a) being the graph of f and (b) being the graph of $F = \mathcal{F}[f]$.

21.17. Let f and F be the two real-valued functions sketched in figure 21.2, and assume they are classically transformable with $F = \mathcal{F}[f]$.

  a. On the same coordinate system, sketch the graphs of $f(at)$ and its Fourier transform for $a = 1$, $a = \frac{1}{2}$ and $a = \frac{1}{4}$. (For $a = 1$, simply copy the graphs sketched in figure 21.2. For the other values of a, sketch the appropriately "scaled" function using the scaling identity to properly scale the graph of the transform.)

  b. What happens to the graphs of $f(at)$ and its transform as $a \to 0$?

  c. On the same coordinate system, sketch the graphs of $f(at)$ and its Fourier transform for $a = 1$, $a = 2$ and $a = 4$.

  d. What happens to the graphs of $f(at)$ and its transform as $a \to +\infty$?

21.18. Let ϕ be any function on the real line, and denote by ϕ_E and ϕ_O the corresponding functions given by

$$\phi_E(x) = \frac{1}{2}[\phi(x) + \phi(-x)] \quad \text{and} \quad \phi_O(x) = \frac{1}{2}[\phi(x) - \phi(-x)] \quad .$$

These two functions, ϕ_E and ϕ_O, are called the even part and the odd part of ϕ, respectively.

      Verify that the terminology is appropriate by showing that

  a. ϕ_E is an even function.

  b. ϕ_O is an odd function.

  c. $\phi = \phi_E + \phi_O$.

21.19. Let f and F be two classically transformable functions with real parts u and R, and imaginary parts v and I,

$$f = u + iv \quad \text{and} \quad F = R + iI \quad .$$

Let u_E, v_E, R_E and I_E be the even parts, respectively, of f and F, and let u_O, v_O, R_O and I_O be the odd parts, respectively, of f and F (as defined in the previous exercise).

  a. Assume f is in $\mathcal{A}$.

    i. Show that $R_E = \mathcal{F}[u_E]$ and $R_0 = i\mathcal{F}[v_0]$.

    ii. What are the corresponding formulas for I_E and I_O *(in terms of the Fourier transforms of u_E , u_O , v_E and/or v_O)?*

  b. Verify that the relations derived in the previous exercises between the parts of f — u_E , u_O , v_E and v_O — and the parts of F — R_E , R_O , I_E and I_O — also hold

    i. when f is in $\mathcal{T}$, and

    ii. when f is any classically transformable function.

21.20 a. Let f be an even function, and show that

$$\mathrm{Re}[\mathcal{F}[f]] \ = \ \mathcal{F}[\mathrm{Re}[f]] \qquad and \qquad \mathrm{Im}[\mathcal{F}[f]] \ = \ \mathcal{F}[\mathrm{Im}[f]]$$

    when

    i. f is in $\mathcal{A}$.

    ii. f is any classically transformable function.

  b. Let f be an odd function, and show that

$$\mathrm{Re}[\mathcal{F}[f]] \ = \ i\mathcal{F}[\mathrm{Im}[f]] \qquad and \qquad \mathrm{Im}[\mathcal{F}[f]] \ = \ -i\mathcal{F}[\mathrm{Re}[f]]$$

    when

    i. f is in $\mathcal{A}$.

    ii. f is any classically transformable function.

22

Differentiation and Fourier Transforms

One reason the Fourier transform is so important in many applications is that it can convert expressions involving derivatives of unknown functions to simpler algebraic expressions. We will derive the main formulas for this conversion immediately below and then spend most of the rest of this chapter discussing some of the results that follow directly from these formulas.

22.1 The Differentiation Identities
Initial Derivations

Deriving the main identities of this chapter is fairly straightforward if we use the integral formulas and make a few "reasonable" assumptions. Later, we will look a little more closely at these assumptions and derivations.

Two derivations will be presented. In both, f and F are classically transformable functions with $F = \mathcal{F}[f]$.

The First Derivation

Since we have indicated that the transform should simplify expressions involving the derivative of f, let's start by trying to convert $\mathcal{F}[f']$ to a simple expression involving F. Right off, we are making an assumption, namely, that f' exists and is classically transformable. By the properties of classically transformable functions, this means f' must be piecewise continuous; hence, f, itself, must be piecewise smooth. Assuming f' is also absolutely integrable, we have

$$\mathcal{F}\big[f'(t)\big]\big|_\omega = \int_{-\infty}^{\infty} f'(t)\, e^{-i2\pi\omega t}\, dt \quad .$$

Seeing the integral of a derivative with another function suggests that we try using the integration by parts formula (see theorem 4.2 on page 40)

$$\int_\alpha^\beta f'(t)g(t)\, dt = f(t)g(t)\big|_\alpha^\beta - \int_\alpha^\beta f(t)g'(t)\, dt$$

with g being the exponential and (α, β) being the entire real line, $(-\infty, \infty)$. Glancing back at theorem 4.2, we see that two more assumptions are being made: that f is continuous and that the integration by parts formula holds when the interval is not finite. Making these assumptions, we try integration by parts,

$$\mathcal{F}\left[f'(t)\right]\big|_{\omega} = f(t)\,e^{-i2\pi\omega t}\Big|_{t=-\infty}^{\infty} - \int_{-\infty}^{\infty} f(t)\left[\frac{\partial}{\partial t}e^{-i2\pi\omega t}\right]dt$$

$$= f(t)\,e^{-i2\pi\omega t}\Big|_{t=-\infty}^{\infty} - \int_{-\infty}^{\infty} f(t)\,(-i2\pi\omega)e^{-i2\pi\omega t}\,dt \quad .$$

Since the integration is with respect to t, the $(-i2\pi\omega)$ can be factored out. Noting, also, that the resulting integral is the integral formula for the Fourier transform of f (assuming f is absolutely integrable) gives

$$\mathcal{F}\left[f'(t)\right]\big|_{\omega} = f(t)\,e^{-i2\pi\omega t}\Big|_{t=-\infty}^{\infty} + i2\pi\omega \int_{-\infty}^{\infty} f(t)\,e^{-i2\pi\omega t}\,dt$$

$$= f(t)\,e^{-i2\pi\omega t}\Big|_{t=-\infty}^{\infty} + i2\pi\omega F(\omega) \quad .$$

With a little thought it should be clear that, as long as f is "reasonably nice", then $f(t)$ should vanish as $t \to \pm\infty$. Assuming this (and recalling that the magnitude of the complex exponential is always 1), we then have

$$f(t)\,e^{-i2\pi\omega t}\Big|_{t=-\infty}^{\infty} = \lim_{t\to\infty} f(t)\,e^{-i2\pi\omega t} - \lim_{t\to-\infty} f(t)\,e^{-i2\pi\omega t} = 0 - 0 = 0 \quad ,$$

and the last formula above for $\mathcal{F}[f']$ becomes

$$\mathcal{F}\left[f'(t)\right]\big|_{\omega} = i2\pi\omega F(\omega) \quad . \tag{22.1}$$

In any situation where all of the above assumptions hold, the above derivation can be accepted as a rigorous proof of equation (22.1). For future reference, let us list all the assumptions made in deriving this equation:

1. f is piecewise smooth and absolutely integrable.

2. f' is absolutely integrable.

3. f is continuous.

4. The integration by parts formula is valid.

5. $\displaystyle\lim_{t\to\pm\infty} f(t) = 0$.

We did not list the assumptions that f and f' are classically transformable, since these assumptions are automatically satisfied if the first two assumptions in the list hold. The astute reader may also complain that the assumption of f being continuous is really part of the assumption that the integration by parts formula is valid. However, the continuity of f will turn out to be such an important assumption that it should be explicitly stated.

The Second Derivation

Inverting equation (22.1) and factoring out the constant gives

$$f'(t) = \mathcal{F}^{-1}[i2\pi\omega F(\omega)]\big|_t = i2\pi\,\mathcal{F}^{-1}[\omega F(\omega)]\big|_t \quad . \tag{22.2}$$

It would be nice if we could derive this making a few assumptions concerning F instead of f, and viewing f as $\mathcal{F}^{-1}[F]$. It turns out that this derivation is even easier than the derivation of equation (22.1):

$$f'(t) \;=\; \frac{d}{dt}\mathcal{F}^{-1}[F(\omega)]\big|_t$$

$$=\; \frac{d}{dt}\int_{-\infty}^{\infty} F(\omega)\,e^{i2\pi\omega t}\,d\omega \tag{22.3}$$

$$=\; \int_{-\infty}^{\infty} \frac{\partial}{\partial t}\left[F(\omega)\,e^{i2\pi\omega t}\right] d\omega \tag{22.4}$$

$$=\; \int_{-\infty}^{\infty} F(\omega)\,(i2\pi\omega)e^{i2\pi\omega t}\,d\omega \tag{22.5}$$

$$=\; i2\pi \int_{-\infty}^{\infty} \omega F(\omega)\,e^{i2\pi\omega t}\,d\omega \tag{22.6}$$

$$=\; i2\pi\,\mathcal{F}^{-1}[\omega F(\omega)]\big|_t \quad . \tag{22.7}$$

What assumptions were made in this sequence of equations? Obviously, we assumed that the functions $F(\omega)$ and $\omega F(\omega)$ are absolutely integrable, and that $f = \mathcal{F}^{-1}[F]$ is differentiable.[1] Also, in going from line (22.3) to (22.4), we assumed that we could switch the order in which we perform "integration with respect to ω" and "differentiation with respect to t". Other than those assumptions, we simply used well-known facts from calculus. Thus, this derivation can be accepted as a rigorous proof of equation (22.2) under the following assumptions:

1. F is classically transformable.

2. $F(\omega)$ and $\omega F(\omega)$ are absolutely integrable functions on $\mathbb{R}$.

3. $f(t) \;=\; \displaystyle\int_{-\infty}^{\infty} F(\omega)\,e^{i2\pi\omega t}\,dt$ is differentiable.

4. $\displaystyle\frac{d}{dt}\int_{-\infty}^{\infty} F(\omega)\,e^{i2\pi\omega t}\,d\omega \;=\; \int_{-\infty}^{\infty}\frac{\partial}{\partial t}\left[F(\omega)\,e^{i2\pi\omega t}\right] d\omega \quad .$

The Basic Identities

Equations (22.1) and (22.2) are two of the differentiation identities. The following theorem describes the "most general" situations (within the classical theory) in which they may be used.

Theorem 22.1 (differentiation identities, part I)
Let f and F be two classically transformable functions with $F = \mathcal{F}[f]$. Assume either that

1. *f is continuous and piecewise smooth with f' being classically transformable,*

or that

2. *$\omega F(\omega)$ is classically transformable.*

---

[1] We also assumed $\omega F(\omega)$ is classically transformable. This, however, is easily shown to be true if F is classically transformable and $\omega F(\omega)$ is absolutely integrable.

Then both of the above sets of conditions hold as do the following two equivalent equations:

$$\mathcal{F}\left[f'(t)\right]\big|_{\omega} = i2\pi\omega F(\omega) \tag{22.8a}$$

and

$$\mathcal{F}^{-1}[\omega F(\omega)]\big|_{t} = \frac{1}{i2\pi} f'(t) \quad . \tag{22.8b}$$

A very similar pair of differentiation identities are given in the following theorem, which follows from the above by "near equivalenc".

Theorem 22.2 (differentiation identities, part II)
Let f and F be two classically transformable functions with $F = \mathcal{F}[f]$. Assume either that

    1. F is continuous and piecewise smooth with F' being classically transformable,

or that

    2. $tf(t)$ is classically transformable.

Then both of the above sets of conditions hold as do the following two equivalent equations:

$$\mathcal{F}^{-1}\left[F'(\omega)\right]\big|_{t} = -i2\pi t f(t) \tag{22.8c}$$

and

$$\mathcal{F}[tf(t)]\big|_{\omega} = -\frac{1}{i2\pi} F'(\omega) \quad . \tag{22.8d}$$

Equations (22.8a) through (22.8d) are the *differentiation identities*. They can also be written as

$$\mathcal{F}\left[f'(t)\right]\big|_{\omega} = i2\pi\omega \mathcal{F}[f(t)]\big|_{\omega} \quad , \tag{22.8a$'$}$$

$$\mathcal{F}^{-1}[\omega F(\omega)]\big|_{t} = \frac{1}{i2\pi}\frac{d}{dt}\mathcal{F}^{-1}[F(\omega)]\big|_{t} \quad , \tag{22.8b$'$}$$

$$\mathcal{F}^{-1}\left[F'(\omega)\right]\big|_{t} = -i2\pi t \mathcal{F}^{-1}[F(\omega)]\big|_{t} \tag{22.8c$'$}$$

and

$$\mathcal{F}[tf(t)]\big|_{\omega} = -\frac{1}{i2\pi}\frac{d}{d\omega}\mathcal{F}[f(t)]\big|_{\omega} \quad . \tag{22.8d$'$}$$

Whichever way you write the identities, it is important to remember that they are *NOT* always applicable and that theorems 22.1 and 22.2 state when each identity is valid.

The assumptions given in theorems 22.1 and 22.2 are much more general than those indicated in our derivations of the first two differentiation identities. In fact, these theorems are more general than we can completely prove at this time. What we can rigorously prove now are the four lemmas on the differential identities given in the next section. Later, in chapter 30, we will confirm that these lemmas generalize to the theorems above.

We will go ahead and use theorems 22.1 and 22.2 with the understanding that their proofs will be carefully completed later on. Where convenient or necessary, we will also note when our computations can be justified by one of the lemmas from the next section.

Before stating the next section, however, let's look at a few examples of these identities in action.

!▶**Example 22.1:**   *Consider finding the Fourier transform of the function*

$$g(t) = \begin{cases} t & \text{if } -1 < t < 1 \\ 0 & \text{otherwise} \end{cases} \quad .$$

This function is clearly classically transformable since it is clearly piecewise continuous and absolutely integrable. Observe that

$$g(t) = t \begin{cases} 1 & \text{if} \quad -1 < t < 1 \\ 0 & \text{otherwise} \end{cases} = t \, \text{pulse}_1(t) \quad .$$

We already know about the pulse function; its Fourier transform is $2 \, \text{sinc}(2\pi\omega)$ (see table 21.1 on page 320). Since both $\text{pulse}_1(t)$ and $t \, \text{pulse}_1(t)$ are absolutely integrable, theorem 22.2 applies and assures us that identity (22.8d) holds (with $f = \text{pulse}_1$). Using the alternate version of this identity (equation (22.8d′)),

$$\mathcal{F}[g(t)]|_\omega = \mathcal{F}\left[t \, \text{pulse}_1(t)\right]\big|_\omega$$

$$= -\frac{1}{i2\pi}\frac{d}{d\omega}\mathcal{F}\left[\text{pulse}_1(t)\right]\big|_\omega$$

$$= -\frac{1}{i2\pi}\frac{d}{d\omega}2\,\text{sinc}(2\pi\omega)$$

$$= -\frac{1}{i2\pi}\frac{d}{d\omega}\frac{\sin(2\pi\omega)}{\pi\omega}$$

$$= -\frac{1}{i2\pi}\left[\frac{(\pi\omega)(2\pi)\cos(2\pi\omega) - \pi\sin(2\pi\omega)}{(\pi\omega)^2}\right]$$

$$= i\frac{2\pi\omega\cos(2\pi\omega) - \sin(2\pi\omega)}{2\pi^2\omega^2} \quad .$$

(Note that, because $g(t) = t \, \text{pulse}_1(t)$ is absolutely integrable, we could also have found $\mathcal{F}[g]$ by directly evaluating the integral formula for this transform. You may want to compute $\mathcal{F}[g]$ this way and compare the results and work done with the results and work done in this example. In particular, ask yourself "Which method is easier?")

?► Exercise 22.1: *Using identity (22.8d) (or (22.8d′)), show that*

$$\mathcal{F}\left[\frac{t}{1+t^2}\right]\bigg|_\omega = \begin{cases} i\pi e^{2\pi\omega} & \text{if} \quad \omega < 0 \\ -i\pi e^{-2\pi\omega} & \text{if} \quad 0 < \omega \end{cases} \quad .$$

In this case the function being transformed is not absolutely integrable, so we must trust theorem 22.2 to justify the use of this identity here.

!► Example 22.2 (a differential equation): Let us try to solve the differential equation

$$\frac{dy}{dt} + 3y = \text{pulse}_1(t) \quad . \tag{22.9}$$

We will start by assuming a classically transformable solution $y(t)$ exists, and seeking its Fourier transform $Y = \mathcal{F}[y]$.

Taking the Fourier transform of both sides of the differential equation and using differential identity (22.8a) along with other properties of the transform, we obtain

$$\mathcal{F}\left[\frac{dy}{dt} + 3y\right]\bigg|_\omega = \mathcal{F}\left[\text{pulse}_1(t)\right]\big|_\omega \tag{22.10}$$

$$\longrightarrow \qquad \mathcal{F}\left[\frac{dy}{dt}\right]\bigg|_\omega + 3\mathcal{F}[y]|_\omega = 2\,\text{sinc}(2\pi\omega) \tag{22.11}$$

$$\longrightarrow \qquad i2\pi\omega Y(\omega) + 3Y(\omega) = 2\,\text{sinc}(2\pi\omega) \quad . \tag{22.12}$$

The last equation is a simple algebraic equation for Y *. By simple algebra*

$$[i2\pi\omega + 3]\, Y(\omega) \;=\; 2\sin c(2\pi\omega) \quad . \tag{22.13}$$

So,

$$Y(\omega) \;=\; \frac{2\sin c(2\pi\omega)}{3 + i2\pi\omega} \quad . \tag{22.14}$$

Finally, provided Y *really is classically transformable, we can obtain the corresponding solution to the differential equation by taking the inverse transform,*

$$y(t) \;=\; \mathcal{F}^{-1}[Y]|_t \;=\; \mathcal{F}^{-1}\!\left[\frac{2\sin c(2\pi\omega)}{3 + i2\pi\omega}\right]\Bigg|_t \quad . \tag{22.15}$$

Unfortunately, the evaluation of this transform is currently beyond us and will have to wait until we discuss convolution (chapter 24).

Do observe that we did not actually show there is a classically transformable solution to the differential equation in the previous example. Nor did we verify there that our use of differential identity (22.8a) was actually valid (i.e., that the appropriate assumptions given in theorem 22.1 hold). All that was shown was that, *if the differential equation has a classically transformable solution* y and *if this solution is "sufficiently nice"* (say, continuous, piecewise smooth and with a classically transformable derivative), then this solution is

$$y(t) \;=\; \mathcal{F}^{-1}[Y]|_t \;=\; \mathcal{F}^{-1}\!\left[\frac{2\sin c(2\pi\omega)}{3 + i2\pi\omega}\right]\Bigg|_t \quad . $$

The next step should be to verify that this does give a solution to the differential equation. One way to do this is to simply check that the final formula for $y(t)$ gives a suitable continuous and differentiable function that satisfies the given differential equation. Another approach is indicated in exercise 22.8 on page 357.

The next example illustrates the importance of verifying the applicability of any differential identity used, and how the misuse of these identities can lead to nonsense.

!▶ *Example 22.3 (misusing a differentiation identity):*   *Let us try to find the Fourier transform of*

$$\frac{t^2}{1 + t^2} \quad . $$

Blindly using identity (22.8d′),

$$\mathcal{F}\!\left[\frac{t^2}{1+t^2}\right]\Bigg|_\omega \;=\; \mathcal{F}\!\left[t\,\frac{t}{1+t^2}\right]\Bigg|_\omega \;=\; -\frac{1}{i2\pi}\frac{d}{d\omega}\mathcal{F}\!\left[\frac{t}{1+t^2}\right]\Bigg|_\omega \quad . \tag{22.16}$$

Plugging in the results of exercise 22.1 and computing the derivatives,

$$\mathcal{F}\!\left[\frac{t^2}{1+t^2}\right]\Bigg|_\omega \;=\; -\frac{1}{i2\pi}\frac{d}{d\omega}\begin{cases} i\pi e^{2\pi\omega} & \text{if } \omega < 0 \\ -i\pi e^{-2\pi\omega} & \text{if } 0 < \omega \end{cases}$$

$$\;=\; -\frac{1}{i2\pi}\begin{cases} i2\pi^2 e^{2\pi\omega} & \text{if } \omega < 0 \\ i2\pi^2 e^{-2\pi\omega} & \text{if } 0 < \omega \end{cases}$$

$$\;=\; -\pi e^{-2\pi|\omega|} \quad . $$

Taking the inverse transform of both sides gives

$$\frac{t^2}{1+t^2} \;=\; -\mathcal{F}^{-1}\!\left[\pi e^{-2\pi\,|\omega|}\right]\Big|_t \quad . \tag{22.17}$$

But from table 21.1 on page 320 we find that

$$\mathcal{F}^{-1}\!\left[\pi e^{-2\pi\,|\omega|}\right]\Big|_t \;=\; \frac{1}{1+t^2} \quad .$$

Thus, if we can believe both of the last two equations, then we must agree that

$$\frac{t^2}{1+t^2} \;=\; -\mathcal{F}^{-1}\!\left[\pi e^{-2\pi\,|\omega|}\right]\Big|_t \;=\; -\frac{1}{1+t^2}$$

for all real values of t. In particular, using $t = 0$, we get the dubious equation

$$0 \;=\; -1 \quad !$$

So what went wrong?

What went wrong was that, in line (22.16), we used the identity

$$\mathcal{F}[tf(t)]|_\omega \;=\; -\frac{1}{i2\pi}\frac{d}{d\omega}\mathcal{F}[f(t)]|_\omega$$

with

$$f(t) \;=\; \frac{t}{1+t^2} \qquad \text{and} \qquad F(\omega) \;=\; \begin{cases} i\pi e^{2\pi\omega} & \text{if } \omega < 0 \\[2mm] -i\pi e^{-2\pi\omega} & \text{if } 0 < \omega \end{cases}$$

but without checking that this was a situation in which that identity could be applied. Now, theorem 22.2 does assure us that this identity is applicable if either

1. *F is continuous and piecewise smooth, and F' is classically transformable,*

or

2. *$tf(t)$ is classically transformable.*

But we can easily verify that neither set of conditions holds here: F is not continuous — it has a jump discontinuity at $\omega = 0$. Moreover,

$$\lim_{t\to\infty} tf(t) \;=\; \lim_{t\to\infty}\frac{t^2}{1+t^2} \;=\; 1 \quad ,$$

which, according to corollary 20.15 on page 310, means that $tf(t)$ is not classically transformable. Consequently, we have no assurance that the identity used is valid here.[2] In fact, we have shown rather conclusively that its use was a serious mistake.

?▶ Exercise 22.2: Verify that

$$\sin 2\pi\omega \;=\; 2\pi\omega\,\mathrm{sinc}(2\pi\omega) \qquad \text{and} \qquad \mathrm{pulse}_1{}'(t) \;=\; 0 \quad .$$

From this, known facts concerning Fourier transforms and the misuse of one of the differentiation identities, derive the obviously false conclusion that

$$\sin(2\pi\omega) \;=\; 0 \qquad \text{for every real value } \omega \quad .$$

---

[2] Actually, we only need to verify that one of the two sets of conditions does not hold. After all, the theorem states that if either one of these two sets of conditions holds, then both must hold. Consequently, if one set does not hold, then the other cannot hold either.

22.2   Rigorous Derivation of the Differential Identities

Let us start with some general results that will be of interest beyond just proving the lemmas given in this section.

Absolutely Integrable and Piecewise Smooth Functions

In a sense, the next theorem can be thought of as another Riemann–Lebesgue lemma.

Theorem 22.3
Suppose f is a continuous, piecewise smooth and classically transformable function on the real line. Suppose, also, that its derivative, f', is absolutely integrable on the real line. Then

$$\lim_{x \to \pm\infty} f(x) = 0 \quad .$$

PROOF: Since f is both continuous and piecewise smooth, we know (see theorem 4.1 on page 39) that, for any real value of x,

$$f(x) - f(0) = \int_0^x f'(s)\, ds \quad .$$

Rearranging slightly and letting $x \to \infty$ gives

$$\lim_{x \to \infty} f(x) = \lim_{x \to \infty} \int_0^x f'(s)\, ds + f(0) = \int_0^\infty f'(s)\, ds + f(0) \quad ,$$

which is a well-defined finite value since f' is absolutely integrable. Thus, the limit of $f(x)$ as $x \to \infty$ exists and is finite. But, from our discussion on "transformability" (see corollary 20.15 on page 310) we know that, if this limit exists and is nonzero, then f cannot be classically transformable, contradicting our assumption that f is classically transformable. Hence, not only does $\lim_{x \to \infty} f(x)$ exist, it must be zero.

Virtually identical arguments also confirm that

$$\lim_{x \to -\infty} f(x) = 0 \quad . \qquad\blacksquare$$

Extending the theorem on integration by parts over a finite interval (α, β) (theorem 4.2 on page 40), to a corresponding theorem on integration by parts over $(-\infty, \infty)$ just requires the addition of conditions ensuring the existence of the appropriate limits as $(\alpha, \beta) \to (-\infty, \infty)$. The proof of the resulting theorem, stated below, is trivial and left to the interested reader.

Theorem 22.4 (integration by parts)
Let f and g be two continuous and piecewise smooth functions on the real line, and assume either that

  1.   *the products $f'g$ and fg' are both absolutely integrable on the real line,*

or that

  2.   *both limits*

$$\lim_{x \to -\infty} f(x)g(x) \qquad and \qquad \lim_{x \to \infty} f(x)g(x)$$

exist and are finite, and either the product $f'g$ or the product fg' is absolutely integrable on the real line.

Then

$$\int_{-\infty}^{\infty} f'(x)g(x)\,dx \quad , \quad \int_{-\infty}^{\infty} f(x)g'(x)\,dx \quad \text{and} \quad f(x)g(x)\Big|_{-\infty}^{\infty}$$

all exist and are finite. Moreover,

$$\int_{-\infty}^{\infty} f'(x)g(x)\,dx \;=\; f(x)g(x)\Big|_{-\infty}^{\infty} \;-\; \int_{-\infty}^{\infty} f(x)g'(x)\,dx \quad .$$

The Lemmas on the Differential Identities

Here are the four lemmas on the differential identities that we can rigorously verify at this time. The first two describe conditions under which our initial derivations of identities (22.8a) and (22.8b) are valid. The other two describe conditions under which similar computations lead to identities (22.8c) and (22.8d). Together, they form the starting point for the complete proofs of theorems 22.1 and 22.2.

By the way, when the fundamental theorem on invertibility is finally proven, that proof will involve formulas derived using the lemmas below. So we must be careful here not to use the invertibility of the Fourier transforms in the statements or the proofs of these lemmas.

Lemma 22.5
Let f and F be two classically transformable functions with $F = \mathcal{F}[f]$. In addition, assume that f is continuous and piecewise smooth, and that both f and f' are absolutely integrable. Then $\omega F(\omega)$ is classically transformable and

$$\mathcal{F}\big[f'(t)\big]\big|_{\omega} \;=\; i2\pi\omega F(\omega) \quad .$$

PROOF: First of all, by the assumptions, f' is classically transformable. In fact, it is in $\mathcal{A}$. So if the equation in the lemma is valid, then $\omega F(\omega)$ must be classically transformable since it is the Fourier transform of a classically transformable function (in fact, $\omega F(\omega)$ will be a function in $\mathcal{T}$). So it will suffice to confirm that the above equation is valid. But this equation is exactly the same as equation (22.1) on page 340, and we have already observed that equation (22.1) is valid provided

1. f is piecewise smooth and absolutely integrable,

2. f' is absolutely integrable,

3. f is continuous,

4. the "integration by parts" formula

$$\int_{-\infty}^{\infty} f'(t)\,e^{-i2\pi\omega t}\,dt \;=\; f(t)\,e^{-i2\pi\omega t}\Big|_{t=-\infty}^{\infty} \;-\; \int_{-\infty}^{\infty} f(t)\left[\frac{\partial}{\partial t}e^{-i2\pi\omega t}\right]dt$$

   is valid, and

5. $\displaystyle\lim_{t\to\pm\infty} f(t) = 0 \quad .$

Verifying that the first three conditions hold is trivial. They are assumptions in the lemma!
Also, from theorem 22.3, we know that, because the first three conditions hold,

$$\lim_{t \to \pm\infty} f(t) = 0 \quad .$$

Hence, the fifth of the above conditions holds.

Finally, it is trivial to verify that the assumptions in theorem 22.4 are satisfied when f is as in our lemma and $g(t) = e^{-i2\pi\omega t}$, assuring us that

$$\int_{-\infty}^{\infty} f'(t) e^{-i2\pi\omega t} \, dt = \int_{-\infty}^{\infty} f'(t) g(t) \, dt$$

$$= f(t)g(t) \Big|_{-\infty}^{\infty} - \int_{-\infty}^{\infty} f(t)g'(t) \, dt$$

$$= f(t) e^{-i2\pi\omega t} \Big|_{t=-\infty}^{\infty} - \int_{-\infty}^{\infty} f(t) \left[\frac{d}{dt} e^{-i2\pi\omega t} \right] dt \quad .$$

So the fourth condition holds.

Thus, since all the needed conditions hold, the equation in lemma 22.5 must be valid. ∎

Lemma 22.6
Let f and F be two classically transformable functions with $f = \mathcal{F}^{-1}[F]$. Assume, further, that both $F(\omega)$ and $\omega F(\omega)$ are absolutely integrable. Then f is smooth, f' is classically transformable, and

$$\mathcal{F}^{-1}[\omega F(\omega)]|_t = \frac{1}{i2\pi} f'(t) \quad .$$

PROOF: If we can verify that the equation in this lemma is valid, then we will automatically know that both f and f' are continuous because they are Fourier inverse transforms of functions in $\mathcal{A}$. So we merely need to verify the equation, and, from our derivation of equation (22.2) on page 340, we know the equation in the lemma is valid if

1. $f(t) = \displaystyle\int_{-\infty}^{\infty} F(\omega) e^{i2\pi\omega t} \, d\omega$ is differentiable, and

2. $\displaystyle\frac{d}{dt} \int_{-\infty}^{\infty} F(\omega) e^{i2\pi\omega t} \, d\omega = \int_{-\infty}^{\infty} \frac{\partial}{\partial t} \left[F(\omega) e^{i2\pi\omega t} \right] d\omega \quad .$

Verifying that these conditions hold is easy because a more general case has already been discussed in corollary 18.21 on page 274. Reviewing that corollary (and the material leading up to that corollary) it should be clear that the above two conditions hold provided $F(\omega) e^{i2\pi\omega t}$ and $\frac{\partial}{\partial t} \left[F(\omega) e^{i2\pi\omega t} \right]$ are bounded by an absolutely integrable function of ω. But, for all ω and t,

$$\left| F(\omega) e^{i2\pi\omega t} \right| = |F(\omega)| \left| e^{i2\pi\omega t} \right| \leq |F(\omega)|$$

and

$$\left| \frac{\partial}{\partial t} \left[F(\omega) e^{i2\pi\omega t} \right] \right| = \left| F(\omega) 2\pi\omega e^{i2\pi\omega t} \right| \leq 2\pi |\omega F(\omega)| \quad .$$

Fortunately, for the case being considered, $F(\omega)$ and $\omega F(\omega)$ are absolutely integrable functions of ω. So corollary 18.21 applies and assures us that

$$f(t) = \int_{-\infty}^{\infty} F(\omega) e^{i2\pi\omega t} \, d\omega$$

is differentiable, and

$$\frac{d}{dt}\int_{-\infty}^{\infty} F(\omega)\, e^{i2\pi\omega t}\, d\omega \;=\; \int_{-\infty}^{\infty}\frac{\partial}{\partial t}\left[F(\omega)\, e^{i2\pi\omega t}\right] d\omega \quad. \qquad\blacksquare$$

The next two lemmas can easily be confirmed using lemmas 22.5 and 22.6 and the principle of near-equivalence. Those who prefer not to employ near-equivalence can, instead, repeat the proofs of the previous two lemmas with the roles of f and F switched.

Lemma 22.7
Let f and F be two classically transformable functions with $f = \mathcal{F}^{-1}[F]$. Assume, additionally, that F is continuous and piecewise smooth, and that both F and F' are absolutely integrable. Then $tf(t)$ is classically transformable and

$$\mathcal{F}^{-1}\left[F'(\omega)\right]\big|_{t} \;=\; -i2\pi t f(t) \quad.$$

Lemma 22.8
Let f and F be two classically transformable functions with $F = \mathcal{F}[f]$. Assume, further, that both $f(t)$ and $tf(t)$ are absolutely integrable. Then F is smooth, F' is classically transformable, and

$$\mathcal{F}[tf(t)]\big|_{\omega} \;=\; -\frac{1}{i2\pi}F'(\omega) \quad.$$

22.3 Higher Order Differential Identities

Applying the theorems and lemmas of the previous section to problems involving higher order derivatives is completely straightforward. For example, if $F = \mathcal{F}[f]$ where f, f' and f'' are all classically transformable with both f and f' being continuous and piecewise smooth, then theorem 22.1 can be invoked twice to obtain the second order version of the first differential identity (identity (22.8a)),

$$\mathcal{F}\left[f''\right]\big|_{\omega} \;=\; \mathcal{F}\left[(f')'\right]\big|_{\omega} \;=\; i2\pi\omega\left(\mathcal{F}\left[f'\right]\big|_{\omega}\right)$$
$$=\; i2\pi\omega\left(i2\pi\omega\mathcal{F}[f]\big|_{\omega}\right) \;=\; (i2\pi\omega)^{2}F(\omega) \quad.$$

These computations can be repeated as many times as we wish provided the function and its derivatives are "suitably nice". We will go ahead and state the theorems for the higher order differential identities, but don't think too much of them. They are simply the statements that theorems 22.1 and 22.2 can be applied repeatedly.

Theorem 22.9 (higher order differentiation identities, part I)
Let n be some positive integer, and let f and F be two classically transformable functions with $F = \mathcal{F}[f]$. Assume either that

1. f, f', f'', $\ldots$ and $f^{(n-1)}$ are all continuous, piecewise smooth, classically transformable functions with $f^{(n)}$ being classically transformable,

or that

2. $F(\omega)$, $\omega F(\omega)$, $\omega^2 F(\omega)$, $\ldots$ and $\omega^n F(\omega)$ are all classically transformable.

Then both of the above sets of conditions hold as do the following two equivalent equations:

$$\mathcal{F}\left[f^{(n)}(t)\right]\Big|_{\omega} = (i2\pi\omega)^n F(\omega)$$

and

$$\mathcal{F}^{-1}\left[\omega^n F(\omega)\right]\Big|_{t} = \left(\frac{1}{i2\pi}\right)^n f^{(n)}(t) \quad .$$

Theorem 22.10 (higher order differentiation identities, part II)
Let f and F be two classically transformable functions with $F = \mathcal{F}[f]$. Assume either that

1. F, F', F'', $\dots$ *and $F^{(n-1)}$ are all continuous, piecewise smooth, classically transformable functions with $F^{(n)}$ being classically transformable,*

or that

2. $f(t)$, $tf(t)$, $t^2 f(t)$, $\dots$ *and $t^n f(t)$ are all classically transformable.*

Then both of the above sets of conditions hold as do the following two equivalent equations:

$$\mathcal{F}^{-1}\left[F^{(n)}(\omega)\right]\Big|_{t} = (-i2\pi t)^n f(t)$$

and

$$\mathcal{F}\left[t^n f(t)\right]\Big|_{\omega} = \left(-\frac{1}{i2\pi}\right)^n F^{(n)}(\omega) \quad .$$

!▶Example 22.4: *Let's try to find the Fourier transform of $t^2 e^{-t}$ step(t). Obviously, it would help if we can use the identity from theorem 22.10*

$$\mathcal{F}\left[t^n f(t)\right]\Big|_{\omega} = \left(-\frac{1}{i2\pi}\right)^n F^{(n)}(\omega)$$

with $n = 2$, $f(t) = e^{-t}$ step(t) and (from, say, table 21.1 on page 320)

$$F(\omega) = \mathcal{F}\left[e^{-t} \text{ step}(t)\right]\Big|_{\omega} = \frac{1}{1 + i2\pi\omega} \quad .$$

Verifying that $f(t)$, $tf(t)$ and $t^2 f(t)$ are classically transformable, here, is easy (see exercise 18.12 on page 276), so theorem 22.10 assures us that the desired identity holds. Thus,

$$\mathcal{F}\left[t^2 e^{-t} \text{ step}(t)\right]\Big|_{\omega} = \mathcal{F}\left[t^2 f(t)\right]\Big|_{\omega}$$

$$= \left(-\frac{1}{i2\pi}\right)^2 F''(\omega) = -\frac{1}{4\pi^2} \frac{d^2}{d\omega^2}\left[\frac{1}{1 + i2\pi\omega}\right] \quad .$$

By elementary calculus,

$$\frac{d^2}{d\omega^2}\left[\frac{1}{1 + i2\pi\omega}\right] = \frac{d}{d\omega}\left(\frac{d}{d\omega}\left[\frac{1}{1 + i2\pi\omega}\right]\right)$$

$$= \frac{d}{d\omega}\left(\frac{-i2\pi}{(1 + i2\pi\omega)^2}\right) = \frac{2(i2\pi)^2}{(1 + i2\pi\omega)^3} \quad .$$

Therefore,

$$\mathcal{F}\left[t^2 e^{-t} \text{ step}(t)\right]\Big|_{\omega} = -\frac{1}{4\pi^2}\frac{2(i2\pi)^2}{(1 + i2\pi\omega)^3} = \frac{2}{(1 + i2\pi\omega)^3} \quad .$$

22.4 Anti-Differentiation and Integral Identities

We have used the differentiation identities to compute transforms of derivatives. Conversely, we can use these same identities to find transforms of functions when the transforms of their derivatives are already known. After all, identity (22.8a),

$$\mathcal{F}\left[f'(t)\right]\big|_{\omega} = i2\pi\omega F(\omega) \quad ,$$

can just as easily be written as

$$F(\omega) = \frac{1}{i2\pi\omega}\mathcal{F}\left[f'(t)\right]\big|_{\omega} \quad . \tag{22.18}$$

!►Example 22.5: *Consider finding* $F = \mathcal{F}[f]$ *where* f *is the triangle function from problem 19.15c on page 295,*

$$f(t) = \text{tri}_1(t) = \begin{cases} 1+t & \text{if } -1 < t < 0 \\ 1-t & \text{if } 0 < t < 1 \\ 0 & \text{otherwise} \end{cases} \quad .$$

This function is certainly absolutely integrable, continuous and piecewise smooth. While its derivative,

$$f'(t) = \begin{cases} +1 & \text{if } -1 < t < 0 \\ -1 & \text{if } 0 < t < 1 \\ 0 & \text{otherwise} \end{cases} \quad ,$$

is not continuous, it is absolutely integrable. What's more, the transform of f' *is very easily computed using the integral formula,*

$$\mathcal{F}\left[f'(t)\right]\big|_{\omega} = \int_{-\infty}^{\infty} f'(t)\,e^{-i2\pi\omega t}\,dt$$

$$= \int_{-1}^{0} e^{-i2\pi\omega t}\,dt \;-\; \int_{0}^{1} e^{-i2\pi\omega t}\,dt$$

$$= \frac{e^{i2\pi\omega} - 1}{i2\pi\omega} - \frac{1 - e^{-i2\pi\omega}}{i2\pi\omega} = \frac{\cos(2\pi\omega) - 1}{i\pi\omega} \quad .$$

So, using equation (22.18),

$$F(\omega) = \frac{1}{i2\pi\omega}\mathcal{F}\left[f'(t)\right]\big|_{\omega} = \frac{1}{i2\pi\omega}\cdot\frac{\cos(2\pi\omega) - 1}{i\pi\omega} = \frac{1 - \cos(2\pi\omega)}{2\pi^2\omega^2} \quad .$$

(For comparison, you should compute $F(\omega)$ *directly, using the integral formula. See, also, exercise 19.15c on page 295.)*

Since f is the anti-derivative of f', equation (22.18) can be viewed as a formula for computing the Fourier transform of an anti-derivative of a function using the presumably known transform of the function. But remember, anti-derivatives can be expressed as integrals with appropriate limits. So we should be able to derive identities involving integrals from corresponding differential identities.

Let's derive one such identity. Assume g is any classically transformable function with transform $G = \mathcal{F}[g]$, and define f by

$$f(t) = \int_a^t g(x)\,dx$$

where a is a real constant or $-\infty$ or $+\infty$. (If $a = \pm\infty$, then we must also assume g is absolutely integrable.) From elementary calculus, we know both that f, being the integral of a piecewise continuous function, must be continuous and piecewise smooth, and that

$$f'(t) = g(t) \quad .$$

Notice, this means f' is classically transformable. If f is also classically transformable, then theorem 22.1 assures us that the first differential identity,

$$\mathcal{F}\left[f'(t)\right]\big|_\omega = i2\pi\omega F(\omega) \quad ,$$

holds for this f. Combining our last two equations gives

$$F(\omega) = \frac{1}{i2\pi\omega}\mathcal{F}\left[f'\right]\big|_\omega = \frac{1}{i2\pi\omega}\mathcal{F}[g]\big|_\omega = \frac{1}{i2\pi\omega}G(\omega) \quad ,$$

further assuring us that $\omega^{-1}G(\omega)$ is classically transformable. Moreover, since

$$F(\omega) = \mathcal{F}[f(t)]\big|_\omega = \mathcal{F}\left[\int_a^t g(x)\,dx\right]\bigg|_\omega \quad ,$$

our previous equation can be written as

$$\mathcal{F}\left[\int_a^t g(x)\,dx\right]\bigg|_\omega = \frac{G(\omega)}{i2\pi\omega} \quad .$$

This gives us the following theorem.

Theorem 22.11 (transform of an integral)
Let a be either some real constant or $\pm\infty$, and let g and G be classically transformable functions with $G = \mathcal{F}[g]$. If $a = \pm\infty$, also assume g is absolutely integrable. Assume, further, that the integral

$$\int_a^t g(x)\,dx$$

is a classically transformable function of t. Then $\omega^{-1}G(\omega)$ is classically transformable and

$$\mathcal{F}\left[\int_a^t g(x)\,dx\right]\bigg|_\omega = \frac{G(\omega)}{i2\pi\omega} \quad . \tag{22.19}$$

Starting with the assumption that $G(\omega)$ and $\omega^{-1}G(\omega)$ are classically transformable leads to the next theorem, which does not require the integral to be classically transformable.

Theorem 22.12 (integral of an inverse transform)
Let g and G be two classically transformable functions with $g = \mathcal{F}^{-1}[G]$. Assume, further, that $\omega^{-1}G(\omega)$ is classically transformable. Then, for any real value a,

$$\mathcal{F}^{-1}\left[\frac{G(\omega)}{\omega}\right]\bigg|_t - \mathcal{F}^{-1}\left[\frac{G(\omega)}{\omega}\right]\bigg|_a = i2\pi\int_a^t g(x)\,dx \quad . \tag{22.20}$$

In general, equation (22.20) is of limited value in finding $\mathcal{F}^{-1}\left[\omega^{-1}G(\omega)\right]\big|_t$ because it requires *a priori* knowledge of $\mathcal{F}^{-1}\left[\omega^{-1}G(\omega)\right]\big|_a$. Sometimes, though, this requirement can be replaced with other conditions on G.

Corollary 22.13
Let g and G be two classically transformable functions with $g = \mathcal{F}^{-1}[G]$. Assume, further, that $\omega^{-1}G(\omega)$ is piecewise continuous and absolutely integrable. Then

$$\mathcal{F}^{-1}\left[\frac{G(\omega)}{\omega}\right]\bigg|_t = i2\pi \int_{-\infty}^{t} g(x)\,dx \tag{22.21a}$$

and

$$\mathcal{F}^{-1}\left[\frac{G(\omega)}{\omega}\right]\bigg|_t = -i2\pi \int_{t}^{\infty} g(x)\,dx \quad . \tag{22.21b}$$

If, in addition, G is an even function, then

$$\mathcal{F}^{-1}\left[\frac{G(\omega)}{\omega}\right]\bigg|_t = i2\pi \int_{0}^{t} g(x)\,dx \quad . \tag{22.21c}$$

The proofs of theorem 22.12 and its corollary will be left as exercises (exercise 22.12 at the end of this chapter).

The two theorems above really should be viewed as corollaries of theorem 22.1 on page 341, the first theorem on differential identities. From the second theorem on differential identities, theorem 22.2 on page 342, we can obtain the following near-equivalent versions of the above.

Theorem 22.14 (inverse transform of an integral)
Let a be either some real constant or $\pm\infty$, and let g and G be classically transformable functions with $g = \mathcal{F}[G]$. If $a = \pm\infty$, also assume G is absolutely integrable. Assume, further, that the integral

$$\int_{a}^{\omega} G(x)\,dx$$

is a classically transformable function of ω. Then $t^{-1}g(t)$ is classically transformable and

$$\mathcal{F}^{-1}\left[\int_{a}^{\omega} G(x)\,dx\right]\bigg|_t = -\frac{g(t)}{i2\pi t} \quad . \tag{22.22}$$

Theorem 22.15 (integral of a transform)
Let g and G be two classically transformable functions with $G = \mathcal{F}[g]$. Assume, further, that $t^{-1}g(t)$ is classically transformable. Then, for any real value a,

$$\mathcal{F}\left[\frac{g(t)}{t}\right]\bigg|_\omega - \mathcal{F}\left[\frac{g(t)}{t}\right]\bigg|_a = -i2\pi \int_{a}^{\omega} G(x)\,dx \quad . \tag{22.23}$$

Corollary 22.16
Let g and G be two classically transformable functions with $G = \mathcal{F}[g]$. Assume, further, that $t^{-1}g(t)$ is piecewise continuous and absolutely integrable. Then

$$\mathcal{F}\left[\frac{g(t)}{t}\right]\bigg|_\omega = -i2\pi \int_{-\infty}^{\omega} G(x)\,dx \tag{22.24a}$$

and

$$\mathcal{F}\left[\frac{g(t)}{t}\right]\bigg|_\omega = i2\pi \int_\omega^\infty G(x)\,dx \quad . \tag{22.24b}$$

If, in addition, g is an even function, then

$$\mathcal{F}\left[\frac{g(t)}{t}\right]\bigg|_\omega = -i2\pi \int_0^\omega G(x)\,dx \quad . \tag{22.24c}$$

We will end this section with an application of one of the identities from the above theorems and corollaries. Before that, though, a couple of points should be noted:

1. All the above integral identities were derived from differential identities in theorems 22.1 and 22.2. In fact, any results derived using these integral identities could also be derived, with just a little more work, using the corresponding differential identities. (So the reader with limited time should concentrate more on the differential identities than on these integral identities.)

2. The requirements assumed for these identities are pretty stringent and are not satisfied by a great many classically transformable functions. Consider, for example, theorem 22.11 when $a = -\infty$. Then, not only do we need g to be absolutely integrable, but

$$f(t) = \int_a^t g(x)\,dx$$

    must be classically transformable. With a little thought, you should realize that this requires g to satisfy

$$\int_{-\infty}^\infty g(x)\,dx = 0 \quad ,$$

    which is not generally the case.

    Notice, also, the requirement in theorem 22.12 that both $G(\omega)$ and $\omega^{-1}G(\omega)$ be classically transformable. But, typically, $\omega^{-1}G(\omega)$ is not classically transformable because it "blows up" at $\omega = 0$ — at least, it "blows up" if, say, $G(\omega)$ is continuous and nonzero at $\omega = 0$. So, requiring $\omega^{-1}G(\omega)$ to be classically transformable also imposes the requirement that $G(\omega)$ be continuous at $\omega = 0$ with $G(0) = 0$.[3]

!▶**Example 22.6:**   *Let α and β be any two positive values, and consider the function*

$$f(t) = \frac{e^{-\alpha|t|} - e^{-\beta|t|}}{t} \quad ,$$

which can be written as

$$f(t) = \frac{g(t)}{t} \quad \text{with} \quad g(t) = e^{-\alpha|t|} - e^{-\beta|t|} \quad .$$

Since g is the difference of two continuous and absolutely integrable functions, it should be clear that $t^{-1}g(t)$ will be piecewise continuous and absolutely integrable so long as it has, at worst, a jump discontinuity at $t = 0$. Now,

$$g(0) = e^{-\alpha|0|} - e^{-\beta|0|} = 1 - 1 = 0$$

---

[3] The reader should recall that, if g is absolutely integrable, then $G = \mathcal{F}[g]$ is continuous and $G(0) = \int_{-\infty}^\infty g(x)\,dx$. So the two implicit requirements discussed here on g and G are complementary.

and

$$\lim_{t \to 0^+} \frac{g(t)}{t} = \lim_{t \to 0^+} \frac{e^{-\alpha|t|} - e^{-\beta|t|}}{t} = \lim_{t \to 0} \frac{e^{-\alpha t} - e^{-\beta t}}{t} \quad .$$

Using L'Hôpital's rule, then,

$$\lim_{t \to 0^+} \frac{g(t)}{t} = \lim_{t \to 0} \frac{-\alpha e^{-\alpha t} + \beta e^{-\beta t}}{1} = \beta - \alpha \quad .$$

Similarly,

$$\lim_{t \to 0^-} \frac{g(t)}{t} = \alpha - \beta \quad .$$

Thus, $t^{-1}g(t)$ does not "blow up" at $t = 0$; it merely has a jump discontinuity there.

Since $g(t)$ and $t^{-1}g(t)$ are both piecewise continuous and absolutely integrable, we can use the identities in corollary 22.16. Noting that

$$g(t) = e^{-\alpha|t|} - e^{-\beta|t|}$$

is an even function, we might as well use identity (22.24c), which tells us that

$$\mathcal{F}\left[\frac{e^{-\alpha|t|} - e^{-\beta|t|}}{t}\right]\Bigg|_\omega = \mathcal{F}\left[\frac{g(t)}{t}\right]\Bigg|_\omega = -i2\pi \int_0^\omega G(x)\, dx \quad ,$$

where (see table 21.1 on page 320)

$$G(\omega) = \mathcal{F}[g(t)]|_\omega = \mathcal{F}\left[e^{-\alpha|t|} - e^{-\beta|t|}\right]\Big|_\omega = \frac{2\alpha}{\alpha^2 + 4\pi^2\omega^2} - \frac{2\beta}{\beta^2 + 4\pi^2\omega^2} \quad .$$

Now, from elementary calculus, we know that

$$\int_0^\omega \frac{2\gamma}{\gamma^2 + 4\pi^2 x^2}\, dx = \frac{1}{\pi}\arctan\left(\frac{2\pi}{\gamma}\omega\right) \qquad \text{for} \quad \gamma > 0 \quad .$$

Thus,

$$\mathcal{F}\left[\frac{e^{-\alpha|t|} - e^{-\beta|t|}}{t}\right]\Bigg|_\omega = -i2\pi \int_0^\omega \left[\frac{2\alpha}{\alpha^2 + 4\pi^2\omega^2} - \frac{2\beta}{\beta^2 + 4\pi^2\omega^2}\right] dx$$

$$= -i2\pi \left[\frac{1}{\pi}\arctan\left(\frac{2\pi}{\alpha}\omega\right) - \frac{1}{\pi}\arctan\left(\frac{2\pi}{\beta}\omega\right)\right]$$

$$= i2 \left[\arctan\left(\frac{2\pi}{\beta}\omega\right) - \arctan\left(\frac{2\pi}{\alpha}\omega\right)\right] \quad .$$

(Note: This shows that the difference between two arctangent functions is classically transformable. However, an individual arctangent is not. We know this is so because $\arctan(x)$ approaches $^\pi/_2$ — not 0 — as $x \to \infty$.)

?▶ Exercise 22.3: *Let α and β be any two positive values. Verify that*

$$f(t) = \frac{e^{-\alpha t} - e^{-\beta t}}{t}\, \text{step}(t)$$

is piecewise continuous and absolutely integrable, and that

$$\mathcal{F}[f(t)]|_\omega = \frac{1}{2}\ln\left(\frac{\beta^2 + 4\pi^2\omega^2}{\alpha^2 + 4\pi^2\omega^2}\right) + i\left[\arctan\left(\frac{2\pi}{\beta}\omega\right) - \arctan\left(\frac{2\pi}{\alpha}\omega\right)\right] \quad .$$

Additional Exercises

22.4. Let $\alpha > 0$ and, using table 21.1 on page 320, the appropriate differentiation identities and other basic identities as needed, compute each of the following transforms.

a. $\mathcal{F}\left[te^{-\alpha t}\,\text{step}(t)\right]\big|_{\omega}$

b. $\mathcal{F}\left[t^2 e^{-\alpha t}\,\text{step}(t)\right]\big|_{\omega}$

c. $\mathcal{F}\left[t^3 e^{-\alpha t}\,\text{step}(t)\right]\big|_{\omega}$

d. $\mathcal{F}^{-1}\left[\omega\,e^{-\alpha\omega}\,\text{step}(\omega)\right]\big|_{t}$

e. $\mathcal{F}\left[te^{\alpha t}\,\text{step}(-t)\right]\big|_{\omega}$

f. $\mathcal{F}\left[t^2 e^{\alpha t}\,\text{step}(-t)\right]\big|_{\omega}$

g. $\mathcal{F}\left[te^{-\alpha|t|}\right]\big|_{\omega}$

h. $\mathcal{F}^{-1}\left[\omega\,e^{-\alpha|\omega|}\right]\big|_{t}$

i. $\mathcal{F}\left[t^2\,\text{pulse}_1(t)\right]\big|_{\omega}$

22.5. Let $\alpha > 0$. Compute each of the following transforms by expressing the function being transformed as the derivative of a known function, and then using the appropriate differentiation identities, table 21.1 on page 320 and other basic identities as needed.

a. $\mathcal{F}\left[\dfrac{1}{(\alpha + it)^2}\right]\bigg|_{\omega}$

b. $\mathcal{F}\left[\dfrac{1}{(\alpha + it)^3}\right]\bigg|_{\omega}$

c. $\mathcal{F}^{-1}\left[\dfrac{1}{(\alpha + i\omega)^2}\right]\bigg|_{t}$

d. $\mathcal{F}^{-1}\left[\dfrac{1}{(\alpha + i\omega)^3}\right]\bigg|_{t}$

e. $\mathcal{F}\left[\dfrac{1}{(\alpha - it)^2}\right]\bigg|_{\omega}$

f. $\mathcal{F}\left[\dfrac{1}{(\alpha - it)^3}\right]\bigg|_{\omega}$

g. $\mathcal{F}\left[\dfrac{t}{(\alpha^2 + t^2)^2}\right]\bigg|_{\omega}$

h. $\mathcal{F}^{-1}\left[\dfrac{\omega}{(\alpha^2 + \omega^2)^2}\right]\bigg|_{t}$

22.6. In the following, assume $\alpha > 0$.

a. By now you have computed the Fourier transform of $t^n e^{-\alpha t}\,\text{step}(t)$ for $n = 0$, $n = 1$, $n = 2$ and $n = 3$ (see the above exercises). Go ahead and compute the Fourier transform of this function for $n = 4$ and $n = 5$.

b. Show that

$$\mathcal{F}\left[t^n e^{-\alpha t}\,\text{step}(t)\right]\big|_{\omega} = \frac{n!}{(\alpha + i2\pi\omega)^{n+1}} \qquad \text{for} \quad n = 1, 2, 3, \ldots \quad .$$

c. Let n denote an arbitrary positive integer. Using appropriate identities along with the answer to the previous part of this exercise, determine the Fourier transform of each of the following:

i. $t^n e^{\alpha t}\,\text{step}(-t)$

ii. $\dfrac{1}{(\alpha + it)^n}$

iii. $\dfrac{1}{(\alpha - it)^n}$

22.7. There is a classically transformable function $y(t)$ satisfying the differential equation

$$\frac{dy}{dt} + 3y = e^{-|t|} \quad .$$

    a. Take the Fourier transform of both sides of this equation and find $Y = \mathcal{F}[y]$.

    b. Use your answer from the previous exercise (and partial fractions) to find $y(t)$.

22.8. Let us reconsider example 22.2 on page 343. We saw that, if a classically transformable solution y to the given differential equation exists, then $y = \mathcal{F}^{-1}[Y]$ where

$$Y(\omega) = \frac{2\operatorname{sinc}(2\pi\omega)}{3 + i2\pi\omega} \quad .$$

    a. Since the sinc function is continuous and $3 + i2\pi\omega$ is never zero for real values of ω, it should be clear that Y is continuous. Show that $Y(\omega)$ and $\omega Y(\omega)$ are classically transformable by showing that

      i. $Y(\omega)$ is absolutely integrable, and

      ii. $\omega Y(\omega)$ is the product of a known classically transformable function with a sine function.

      (Hint: It may help to recall the definition of the sinc function.)

    b. Now use theorem 22.1 to show all the following (without attempting to compute the formula for $y(t)$!):

      i. y is continuous and piecewise smooth.

      ii. y and y' are classically transformable.

      iii. $y(t)$ satisfies the differential equation in example 22.2.

      (Note: For the last one, you simply need to verify that the steps done in example 22.2 to find y can be done in reverse order, i.e., that $(22.15) \rightarrow (22.14) \rightarrow (22.13) \rightarrow \cdots \rightarrow (22.9)$.)

22.9. Assume that each of the following differential equations has a classically transformable solution $y(t)$, and find the Fourier transform of that solution, $Y(\omega) = \mathcal{F}[y(t)]|_\omega$. Do not attempt to find $y(t)$.

    a. $\dfrac{d^2y}{dt^2} - 9y = \operatorname{pulse}_1(t)$

    b. $\dfrac{d^2y}{dt^2} - 4\dfrac{dy}{dt} + 3y = e^{-t}\operatorname{step}(t)$

22.10. Let n be some positive integer, and let f and F be classically transformable functions with $F = \mathcal{F}[f]$. Assume, further, that f, f', f'', $\ldots$ and $f^{(n-1)}$ are all continuous, piecewise smooth, classically transformable functions, and that $f^{(n)}$ is absolutely integrable.

    a. Verify that there is a finite constant M such that

$$|F(\omega)| \le \frac{M}{\omega^n} \quad \text{for every real value } \omega \quad .$$

    b. Verify that F is in $\mathcal{A}$ if $n > 1$.

22.11. *(A differential identity for discontinuous functions) Let f be a piecewise smooth and absolutely integrable function whose derivative, f', is also absolutely integrable. Assume, further, that f is continuous everywhere on the real line except at one point t_0, at which f has a jump discontinuity with jump j_0. Let $F = \mathcal{F}[f]$, and derive an equation (similar to identity (22.1) on page 340) relating $\mathcal{F}\left[f'(t)\right]\big|_\omega$ to $F(\omega)$ and j_0.*

22.12. *Let g and G be as in theorem 22.12 on page 352. Let*

$$F(\omega) \;=\; \frac{G(\omega)}{\omega} \qquad \text{and} \qquad f \;=\; \mathcal{F}^{-1}[F] \quad .$$

    a. *Using theorem 22.1, show that*

$$f'(t) \;=\; i2\pi \, g(t) \quad .$$

    b. *Now prove theorem 22.12.*

    c. *Using properties of transforms of absolutely integrable functions (see theorem 19.6 on page 288), verify the claims of corollary 22.13 on page 353.*

23

Gaussians and Gaussian-Like Functions

Most of this chapter is devoted to examining those functions commonly known as Gaussians. These are functions that naturally arise in many applications. In part, this is because they describe some of the most commonly expected probability distributions, and consequently, are used to model such diverse phenomena as the "noise" in electronic and optical devices, the likelihood of a missile hitting its target, and the distribution of grades in a large class. In addition, they arise as fundamental solutions to the differential equations describing heat flow and diffusion problems. We will also find Gaussian functions invaluable in further developing the mathematics used in everyday applications in science and engineering (and mathematics). In particular, Gaussian functions make up the "identity sequences" that will play a major role in confirming the fundamental theorem on invertibility and the more general theorems on the differential identities in chapter 22, and, in part IV of this text, they will serve as the basic "test functions" on which the generalized theory of functions and transforms will be developed.

Some of the more significant formulas we will derive for Gaussians are also valid, slightly modified, for other functions that are similar to Gaussians in certain ways. So, after thoroughly discussing Gaussian functions, we will broaden our discussion to these "Gaussian-like functions" .

23.1 Basic Gaussians
Definition and Some Basics

We will refer to a function g as a *Gaussian function* (or, more simply, as a *Gaussian*) if and only if it can be written as

$$g(x) \ = \ Ae^{-\gamma(x-\zeta)^2}$$

where A, γ and ζ are constants with $\gamma > 0$.[1] Both A and ζ (the "shift") may be complex.

To simplify matters, let's first look at Gaussians having zero shift. We will refer to these as the "basic Gaussians". That is, g is a *basic Gaussian function* (more simply, a *basic Gaussian*) if and only if it can be given by

$$g(x) \ = \ Ae^{-\gamma x^2}$$

where A is some (complex) constant and $\gamma > 0$. The graph of a basic Gaussian — which you should recognize as the (in)famous "bell curve" — is sketched in figure 23.1.

---

[1] Throughout this chapter, γ will *always* denote a positive constant, whether or not the author remembers to say so.

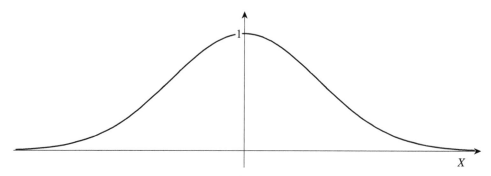

Figure 23.1: Graph of a basic Gaussian, $e^{-\gamma x^2}$ for some $\gamma > 0$.

Basic Gaussian functions satisfy a number of properties that we will find useful. Here are some that you can easily verify:

1. Any basic Gaussian is an even function.

2. If $g(x)$ is a basic Gaussian and a is any nonzero real number, then g with the variable scaled by a, $g(ax)$, is also a basic Gaussian.

3. Basic Gaussian functions are infinitely smooth. In fact, $g^{(n)}(x)$, the nth derivative of a basic Gaussian $g(x)$, is simply the product of $g(x)$ with an nth degree polynomial. (See exercise 23.1 if this is not obvious.)

4. Basic Gaussian functions and their derivatives shrink to zero very rapidly "near infinity". More precisely, if g is a basic Gaussian, and n and m are any pair of nonnegative integers, then

$$\lim_{x \to \pm\infty} x^m g(x) = 0 \qquad \text{and} \qquad \lim_{x \to \pm\infty} x^m g^{(n)}(x) = 0 \quad .$$

    (See example 18.5 and exercise 18.5 starting on page 264.)

5. Basic Gaussian functions and their derivatives are absolutely integrable. In fact, if g is a basic Gaussian, and n and m are any two nonnegative integers, then $x^m g^{(n)}(x)$ is an absolutely integrable function on the entire real line. (Again, see example 18.5 and exercise 18.5.)

?▶Exercise 23.1:   Assuming $\gamma > 0$ and $g(x) = e^{-\gamma x^2}$, verify that $g^{(n)}(x)$, the nth derivative of g, is the product of $g(x)$ with an nth degree polynomial when

a: $n = 0, 1, 2$ and 3.   **b:** n is any nonnegative integer.

Integral of the Basic Gaussian

You will not find a simple formula for the indefinite integral of $e^{-\gamma x^2}$. None is known. The value of the definite integral

$$\int_{-\infty}^{\infty} e^{-\gamma s^2} \, ds \quad ,$$

however, is easily computed via a clever trick. Since this value will be needed, let us compute it now (assuming, of course, that $\gamma > 0$).

For convenience, let I denote the value we are seeking. Clearly,

$$I = \int_{-\infty}^{\infty} e^{-\gamma s^2} \, ds = \int_{-\infty}^{\infty} e^{-\gamma x^2} \, dx = \int_{-\infty}^{\infty} e^{-\gamma y^2} \, dy \quad .$$

The "clever trick" is based on the observation that I^2, the product of I with itself, can be expressed as a double integral over the entire XY–plane,

$$\begin{aligned} I^2 &= \left(\int_{-\infty}^{\infty} e^{-\gamma x^2} \, dx \right) \left(\int_{-\infty}^{\infty} e^{-\gamma y^2} \, dy \right) \\ &= \int_{-\infty}^{\infty} \left(\int_{-\infty}^{\infty} e^{-\gamma x^2} \, dx \right) e^{-\gamma y^2} \, dy \\ &= \int_{-\infty}^{\infty} \left(\int_{-\infty}^{\infty} e^{-\gamma x^2} e^{-\gamma y^2} \, dx \right) dy = \int_{-\infty}^{\infty} \int_{-\infty}^{\infty} e^{-\gamma (x^2 + y^2)} \, dx \, dy \quad . \end{aligned}$$

This double integral is easily computed using polar coordinates (r, θ) where

$$x = r \cos(\theta) \qquad \text{and} \qquad y = r \sin(\theta) \quad .$$

Recall that

$$x^2 + y^2 = r^2 \qquad \text{and} \qquad dx \, dy = r \, dr \, d\theta \quad .$$

So, converting to polar coordinates and using elementary integration techniques,

$$I^2 = \int_0^{2\pi} \int_0^{\infty} e^{-\gamma r^2} r \, dr \, d\theta = \int_0^{2\pi} \frac{1}{2\gamma} \, d\theta = \frac{\pi}{\gamma} \quad . \tag{23.1}$$

Taking the square root gives

$$I = \sqrt{\frac{\pi}{\gamma}} \qquad \text{or} \qquad I = -\sqrt{\frac{\pi}{\gamma}} \quad .$$

But $e^{-\gamma s^2}$ is a positive function on $(-\infty, \infty)$; so we must take the positive square root for I,

$$\int_{-\infty}^{\infty} e^{-\gamma s^2} \, ds = I = \sqrt{\frac{\pi}{\gamma}} \quad . \tag{23.2}$$

Fourier Transforms of Basic Gaussians

Since the basic Gaussian functions are absolutely integrable, we can write

$$\mathcal{F}\left[e^{-\gamma t^2} \right]\Big|_{\omega} = \int_{-\infty}^{\infty} e^{-\gamma t^2} e^{-i 2\pi \omega t} \, dt \quad .$$

Unfortunately, computing this integral directly is not a trivial task.[2] We will compute this integral indirectly using differential identities from the previous chapter. Some care, however, must be taken because we will later employ some of the results obtained here to complete the proofs of theorems 22.1 and 22.2 (on the differential identities). For this reason, we will only use the lemmas actually proven in the previous chapter (lemmas 22.5 through 22.8, starting on page 347) to justify our use of the differential identities.

---

[2] One standard approach uses contour integration in the complex plane.

We start our computations by stating our notation:

$$g(t) = e^{-\gamma t^2} \quad \text{and} \quad G(\omega) = \mathcal{F}\left[e^{-\gamma t^2}\right]\Big|_\omega$$

where γ is an arbitrary fixed positive real number. Observe that

$$g'(t) = \frac{d}{dt}\left[e^{-\gamma t^2}\right] = -2\gamma t e^{-\gamma t^2} = -2\gamma t g(t) \quad .$$

Since $g'(t)$ and $tg(t)$ are continuous and absolutely integrable, we can take the Fourier transform of each side of this equation, obtaining

$$\mathcal{F}\left[g'(t)\right]\Big|_\omega = -2\gamma \mathcal{F}\left[tg(t)\right]\Big|_\omega \quad . \tag{23.3}$$

Moreover, because $g(t)$, $g'(t)$ and $tg(t)$ are all smooth and absolutely integrable, lemma 22.5 on page 347 assures us that $\omega\, G(\omega)$ is classically transformable and

$$\mathcal{F}\left[g'(t)\right]\Big|_\omega = i2\pi\omega\, G(\omega) \quad ,$$

while lemma 22.8 on page 349 assures us both that $G(\omega)$ is a smooth function with a classically transformable derivative, G', and that

$$\mathcal{F}\left[tg(t)\right]\Big|_\omega = -\frac{1}{i2\pi}G'(\omega) \quad .$$

So equation (23.3) becomes

$$i2\pi\omega\, G(\omega) = -2\gamma\left[-\frac{1}{i2\pi}G'(\omega)\right] \quad .$$

Simplifying this last equation and rewriting G' more explicitly gives

$$\frac{dG}{d\omega} = -\beta 2\omega\, G(\omega) \tag{23.4}$$

where, for convenience, we've let $\beta = \pi^2/\gamma$. This is a simple first order ordinary differential equation that you can easily solve using either separation of variables or integrating factors. We will just observe that, by the product rule and equation (23.4),

$$\frac{d}{d\omega}\left[e^{\beta\omega^2} G(\omega)\right] = \beta 2\omega\, e^{\beta\omega^2} G(\omega) + e^{\beta\omega^2}\frac{dG}{d\omega}$$

$$= \beta 2\omega\, e^{\beta\omega^2} G(\omega) + e^{\beta\omega^2}\left[-\beta 2\omega G(\omega)\right] = 0 \quad .$$

But the only way a smooth function can have zero derivative is for that function to be a constant, say, A. Thus,

$$e^{\beta\omega^2} G(\omega) = A \quad ,$$

which, of course, means that

$$G(\omega) = A e^{-\beta\omega^2} \quad . \tag{23.5}$$

All that remains to computing $G(\omega)$ is determining the value of A. Letting $\omega = 0$ in equation (23.5) and recalling what g denotes, we obtain

$$A = G(0) = \int_{-\infty}^{\infty} g(t)\, e^{-i2\pi 0 t}\, dt = \int_{-\infty}^{\infty} e^{-\gamma t^2}\, dt \quad .$$

By an amazing stroke of luck, this is the same integral we evaluated a page or two ago. From that calculation, we know

$$A = \sqrt{\frac{\pi}{\gamma}} \quad .$$

With this, equation (23.5) becomes

$$G(\omega) = \sqrt{\frac{\pi}{\gamma}} e^{-\beta\omega^2} \quad .$$

After recalling what G and β denote (and using the alternative notation for exponentials), we see that the last equation is

$$\mathcal{F}\left[e^{-\gamma t^2}\right]\Big|_{\omega} = \sqrt{\frac{\pi}{\gamma}} \exp\left(-\frac{\pi^2}{\gamma}\omega^2\right) \qquad \text{for every} \quad \gamma > 0 \quad . \tag{23.6}$$

While the inverse Fourier transform of g can be found through similar computations, it is easier to use the fact that, because g is an even function, we know $\mathcal{F}^{-1}[g] = \mathcal{F}[g]$. Applying this with equation (23.6) yields

$$\mathcal{F}^{-1}\left[e^{-\gamma\omega^2}\right]\Big|_{t} = \sqrt{\frac{\pi}{\gamma}} \exp\left(-\frac{\pi^2}{\gamma}t^2\right) \qquad \text{for each} \quad \gamma > 0 \quad . \tag{23.7}$$

It is certainly worth noting that the Fourier transform and Fourier inverse transform of a basic Gaussian function are, themselves, basic Gaussian functions. In particular, letting $\gamma = \pi$ in equation (23.6) gives us

$$\mathcal{F}\left[e^{-\pi t^2}\right]\Big|_{\omega} = e^{-\pi\omega^2} \quad ,$$

a most aesthetically pleasing formula!

?▶Exercise 23.2:   *Let $\alpha > 0$. Using the above, verify that*

$$\mathcal{F}\left[e^{-\alpha\pi t^2}\right]\Big|_{\omega} = \frac{1}{\sqrt{\alpha}} \exp\left(-\frac{\pi}{\alpha}\omega^2\right)$$

and

$$\mathcal{F}^{-1}\left[e^{-\alpha\pi\omega^2}\right]\Big|_{t} = \frac{1}{\sqrt{\alpha}} \exp\left(-\frac{\pi}{\alpha}t^2\right) \quad .$$

(On occasion, these are more convenient to use than identities (23.6) and (23.7).)

Notes on the Derivations

Some of the formulas derived in this and the next section will figure in the final proofs of three theorems already discussed: the fundamental theorem on invertibility and the two general theorems on the differential identities (theorems 19.5, 22.1 and 22.2). Because of this, we carefully avoided using those theorems or results derived from those theorems to obtain formulas (23.6) and (23.7). Instead, we employed identities already shown to be valid when the functions being transformed are all absolutely integrable (as are the basic Gaussians, their derivatives and their products with polynomials). You may want to look back over the above derivation to confirm this. Alternatively, you may want to verify that formulas (23.6) and (23.7) can be obtained by directly using the integral formulas for the Fourier transforms, some basic calculus and some of the results discussed in section 18.4 regarding the differentiation of fairly general integrals.

23.2 General Gaussians
Formulas for Gaussians

By our definition, a (general) Gaussian function g is just a basic Gaussian with the variable shifted by some quantity ζ,

$$g(x) = Ae^{-\gamma(x-\zeta)^2} \quad . \tag{23.8}$$

Remember, while γ must be positive, the constants A and ζ may be complex.[3]

A fair amount of information can be derived from a little algebra. Let a and b be the real and imaginary components of ζ. Then, keeping the real part of the shift in the Gaussian,

$$e^{-\gamma(x-\zeta)^2} = e^{-\gamma(x-a-ib)^2} = \exp\left(-\gamma\left[(x-a)^2 - i2(x-a)b + (ib)^2\right]\right)$$

$$= \exp\left(-\gamma(x-a)^2 + i2\gamma bx - i2\gamma ab + \gamma b^2\right)$$

$$= e^{-i2\gamma ab + \gamma b^2} e^{i2\gamma bx} e^{-\gamma(x-a)^2} \quad .$$

That is,

$$Ae^{-\gamma(x-a-ib)^2} = Be^{i\lambda x} e^{-\gamma(x-a)^2} \tag{23.9}$$

where

$$B = Ae^{-i2\gamma ab + \gamma b^2} \quad \text{and} \quad \lambda = 2b\gamma \quad .$$

This tells us that any Gaussian function is simply a basic Gaussian shifted along the real axis and multiplied by a constant and a complex exponential. It also tells us that the formula on the right-hand side of equation (23.9) can serve just as well as formula (23.8) to describe any Gaussian. So these two formulas are equivalent, and either can be used in any computations involving Gaussians.

Other equivalent formulas for Gaussians can easily be derived using the same sort of algebra that led to equation (23.9). Since we will be needing them off and on for the rest of this text, we'll summarize these formulas in the next lemma.

Lemma 23.1 (formulas for a Gaussian function)
If any one of the following statements are true, then all are true.

1. *g is a Gaussian function.*

2. *There is a positive constant γ and complex constants A and ζ such that*

$$g(x) = Ae^{-\gamma(x-\zeta)^2} \quad \text{for} \quad -\infty < x < \infty \quad . \tag{23.10a}$$

3. *There is a positive constant γ, a complex constant B, and real constants a and λ such that*

$$g(x) = Be^{i\lambda x}e^{-\gamma(x-a)^2} \quad \text{for} \quad -\infty < x < \infty \quad . \tag{23.10b}$$

4. *There is a positive constant γ, a complex constant C, and real constants b and μ such that*

$$g(x) = Ce^{\mu x}e^{-\gamma(x-ib)^2} \quad \text{for} \quad -\infty < x < \infty \quad . \tag{23.10c}$$

---

[3] We may be stretching the traditional definitions of Gaussians by allowing complex shifts. Be aware that the graph of $e^{-\gamma(x-\zeta)^2}$ is not the standard bell curve if ζ is not real.

5. There is a positive constant γ and complex constants D and σ, such that

$$g(x) = De^{\sigma x}e^{-\gamma x^2} \qquad \text{for} \quad -\infty < x < \infty \quad . \tag{23.10d}$$

Moreover, the above constants are related by

$$\zeta = a + ib \quad , \qquad \lambda = 2b\gamma \quad , \qquad \mu = 2a\gamma \quad , \qquad \sigma = \mu + i\lambda \quad ,$$

$$B = Ae^{\gamma b^2 - i2ab\gamma} \quad , \qquad C = Ae^{-\gamma a^2 - i2ab\gamma} \quad \text{and} \quad D = Ae^{-\gamma \zeta^2} \quad .$$

?►Exercise 23.3: *Derive formulas (23.10c) and (23.10d) from formula (23.10a), and verify the relations between the constants claimed in the above lemma.*

From these formulas and our previous discussions regarding basic Gaussians, it should be obvious that, if g is any Gaussian function, and if n and m are any two nonnegative integers, then all of the following hold:

1. $g(x)$ with the variable scaled by any real, nonzero number a, $g(ax)$, is another Gaussian.

2. The translation of $g(x)$ by any complex number a, $g(x - a)$, is another Gaussian.

3. g is infinitely smooth, and the nth derivative of g is the product of $g(x)$ with some n^{th} degree polynomial.

4. $g(x)$ and each of its derivatives shrink to zero very rapidly as $x \to \pm\infty$. That is,

$$\lim_{x \to \pm\infty} x^m g(x) = 0 \qquad \text{and} \qquad \lim_{x \to \pm\infty} x^m g^{(n)}(x) = 0 \quad .$$

5. $g(x)$ and all of its derivatives are absolutely integrable on the real line. In fact, $x^m g^{(n)}(x)$ is absolutely integrable on the real line.

6. The product of $g(x)$ with any other Gaussian function is another Gaussian.

7. The product of $g(x)$ with any exponential function — real or complex — is another Gaussian.

Transforms of Arbitrary Gaussians

Thanks to some of the equations just derived, the formulas for the Fourier transforms of basic Gaussian functions can be extended to formulas for the transforms of all Gaussian functions through straightforward applications of the translation identities.

So let γ, a and b be fixed real values with $\gamma > 0$. For the Fourier transform of a basic Gaussian translated along the real axis by a we have, by equation (23.6) and one of the translation identities,

$$\mathcal{F}\left[e^{-\gamma(t-a)^2}\right]\Big|_\omega = e^{-i2\pi a\omega}\mathcal{F}\left[e^{-\gamma t^2}\right]\Big|_\omega$$

$$= e^{-i2\pi a\omega}\sqrt{\frac{\pi}{\gamma}}\exp\left(-\frac{\pi^2}{\gamma}\omega^2\right) = \sqrt{\frac{\pi}{\gamma}}\exp\left(-i2\pi a\omega - \frac{\pi^2}{\gamma}\omega^2\right) \quad .$$

Using this, equation (23.9) and another translation identity, we find that

$$\mathcal{F}\left[e^{-\gamma(t-a-ib)^2}\right]\Big|_\omega = \mathcal{F}\left[Be^{i2b\gamma t}e^{-\gamma(t-a)^2}\right]\Big|_\omega$$

$$= B\,\mathcal{F}\left[e^{i2\pi \frac{b\gamma}{\pi}t}e^{-\gamma(t-a)^2}\right]\Big|_\omega$$

$$= B \, \mathcal{F}\left[e^{-\gamma(t-a)^2} \right]\Big|_{\omega - \frac{b\gamma}{\pi}}$$

$$= B \sqrt{\frac{\pi}{\gamma}} \exp\left(-i2\pi a \left(\omega - \frac{b\gamma}{\pi} \right) - \frac{\pi^2}{\gamma}\left(\omega - \frac{b\gamma}{\pi} \right)^2 \right) \qquad (23.11)$$

where

$$B = e^{-i2\gamma ab + \gamma b^2} \quad .$$

Equation (23.11) can be simplified a bit after noting that

$$-i2\pi a \left(\omega - \frac{b\gamma}{\pi} \right) - \frac{\pi^2}{\gamma}\left(\omega - \frac{b\gamma}{\pi} \right)^2 = -i2\pi a\omega + i2ab\gamma - \frac{\pi^2}{\gamma}\left(\omega^2 - 2\omega\frac{b\gamma}{\pi} + \frac{b^2\gamma^2}{\pi^2} \right)$$

$$= i2ab\gamma - \gamma b^2 - i2\pi(a + ib)\omega - \frac{\pi^2}{\gamma}\omega^2 \quad .$$

So,

$$\exp\left(-i2\pi a \left(\omega - \frac{b\gamma}{\pi} \right) - \frac{\pi^2}{\gamma}\left(\omega - \frac{b\gamma}{\pi} \right)^2 \right) = e^{i2ab\gamma - \gamma b^2} \, e^{-i2\pi(a+ib)\omega} \exp\left(-\frac{\pi^2}{\gamma}\omega^2 \right)$$

$$= \frac{1}{B} e^{-i2\pi(a+ib)\omega} \exp\left(-\frac{\pi^2}{\gamma}\omega^2 \right) \quad ,$$

and equation (23.11) reduces to

$$\mathcal{F}\left[e^{-\gamma(t-a-ib)^2} \right]\Big|_{\omega} = \sqrt{\frac{\pi}{\gamma}} e^{-i2\pi(a+ib)\omega} \exp\left(-\frac{\pi^2}{\gamma}\omega^2 \right) \quad .$$

Letting $\xi = a + ib$, this further simplifies to

$$\mathcal{F}\left[e^{-\gamma(t-\xi)^2} \right]\Big|_{\omega} = \sqrt{\frac{\pi}{\gamma}} e^{-i2\pi\xi\omega} \exp\left(-\frac{\pi^2}{\gamma}\omega^2 \right) \qquad (23.12)$$

where ξ is any complex constant.

By the principle of near-equivalence, we also have

$$\mathcal{F}^{-1}\left[e^{-\gamma(\omega-\xi)^2} \right]\Big|_{t} = \sqrt{\frac{\pi}{\gamma}} e^{-i2\pi\xi(-t)} \exp\left(-\frac{\pi^2}{\gamma}(-t)^2 \right)$$

$$= \sqrt{\frac{\pi}{\gamma}} e^{i2\pi\xi t} \exp\left(-\frac{\pi^2}{\gamma}t^2 \right) \qquad (23.13)$$

for any complex constant ξ.

Observe that each of the two transforms above is a basic Gaussian multiplied by a complex exponential, which, as we saw earlier, means that these transforms are also Gaussian functions. To repeat: *The Fourier transforms of all Gaussian functions are, themselves, Gaussians.* This is not simply an interesting bit of mathematical trivia. It has consequences in many of the applications in which Gaussians naturally arise. It is also one of the reasons that the set of Gaussian functions will play a major role later, both in some of our more advanced discussions of the classical theory (such as our proof of the fundamental theorem on invertibility), and in our work extending Fourier analysis beyond the confines of the classical theory.

Let us end this subsection on computing transforms of Gaussians by reconsidering the translation identities in chapter 21. Those identities were derived assuming real translations only. Indeed, they are not generally valid when the shift is complex (recall exercise 21.16 on page 336). But notice that, letting $g(t) = e^{-\gamma t^2}$ and $G = \mathcal{F}[g]$, equation (23.12) is

$$\mathcal{F}[g(t - \xi)]|_\omega = e^{-i2\pi\xi\omega}G(\omega) \quad,$$

which certainly looks like a translation identity. Here, though, we are allowing the shift, ξ, to be a complex value. So there are cases where the translation identities do hold using complex shifts. We will expand on this in part IV of this text.

?▶Exercise 23.4: *Let $\gamma > 0$, and let ξ be any (possibly complex) value. Confirm that*

$$\mathcal{F}\left[e^{i2\pi\xi t}e^{-\gamma t^2}\right]\bigg|_\omega = \sqrt{\frac{\pi}{\gamma}}\exp\left(-\frac{\pi^2}{\gamma}(\omega - \xi)^2\right) \tag{23.14}$$

and

$$\mathcal{F}^{-1}\left[e^{-i2\pi\xi\omega}e^{-\gamma\omega^2}\right]\bigg|_t = \sqrt{\frac{\pi}{\gamma}}\exp\left(-\frac{\pi^2}{\gamma}(t - \xi)^2\right) \tag{23.15}$$

a: *using invertibility and equations (23.12) and (23.13).*

b: *without using invertibility. (Use basic algebra and translation identities from theorem 21.1 on page 319 in a manner similar to our derivation of equations (23.12) and (23.13).)*

More Notes on the Derivations (and Proving Invertibility)

As already mentioned, some of the results in this section will play a role in finally proving the fundamental theorem on invertibility and the general theorems on the differential identities. So, again, let us briefly review our derivations.

Equations (23.12) and (23.13) were derived using results from the previous section, elementary algebra, the translation identities from theorem 21.1 and the principle of near-equivalence. However, the functions being transformed here are all absolutely integrable. So the Fourier transforms here can be written in integral form, and the translation identities and near-equivalence identities used can be verified easily via elementary calculus. Consequently, we can write equations (23.12) and (23.13) in integral form,

$$\int_{-\infty}^{\infty} e^{-\gamma(t-\xi)^2} e^{-i2\pi\omega t} \, dt = \sqrt{\frac{\pi}{\gamma}} e^{-i2\pi\xi\omega}\exp\left(-\frac{\pi^2}{\gamma}\omega^2\right) \tag{23.16}$$

and

$$\int_{-\infty}^{\infty} e^{-\gamma(\omega-\xi)^2} e^{i2\pi\omega t} \, d\omega = \sqrt{\frac{\pi}{\gamma}} e^{i2\pi\xi t}\exp\left(-\frac{\pi^2}{\gamma}t^2\right) \quad. \tag{23.17}$$

Moreover, we are certain these equations hold even though some of our "Fourier analysis" theorems (such as the fundamental theorem on invertibility) remain to be proven.

Likewise, by exercise 23.4 b above, we know that equations (23.14) and (23.15) can be written in integral form,

$$\int_{-\infty}^{\infty} e^{i2\pi\xi t}e^{-\gamma t^2} e^{-i2\pi\omega t} \, dt = \sqrt{\frac{\pi}{\gamma}}\exp\left(-\frac{\pi^2}{\gamma}(\omega - \xi)^2\right) \tag{23.18}$$

and

$$\int_{-\infty}^{\infty} e^{-i2\pi\xi\omega}e^{-\gamma\omega^2} e^{i2\pi\omega t} \, d\omega = \sqrt{\frac{\pi}{\gamma}}\exp\left(-\frac{\pi^2}{\gamma}(t - \xi)^2\right) \quad, \tag{23.19}$$

and that we can verify these equations directly using calculus and algebra.

It is important to realize that we now know the above integral equations hold (for suitable choices of γ, t and ω) even though we have not yet proven the fundamental theorem on invertibility or the two theorems on the differential identities. This means that we can use these equations in proving those theorems. Since that is the plan, let us formally state the above observations so we do not have to rethink all this when we finally get around to proving those theorems.

Theorem 23.2
Let $\gamma > 0$, and let ξ be any complex number. Then equations (23.16) and (23.18) hold for every real ω, and equations (23.17) and (23.19) hold for every real t.

If it is still unclear why we know this theorem is valid, do the next exercise.

?► Exercise 23.5:   *Let $\gamma > 0$. Using basic algebra, calculus and results from section 18.4 on differentiating integrals, verify that equation (23.16) holds*

   a: *when $\xi = 0$.*

   b: *when ξ is any complex number.*

23.3 Gaussian-Like Functions
Basic Gaussian-Like Functions

Thus far in this chapter, all of our computations have basically involved functions of the form

$$g(x) = e^{-\gamma x^2} \qquad \text{where} \quad \gamma > 0 \quad .$$

Many of those computations, however, did not truly require that γ be a positive real number. So let us reconsider those computations using a function of the form

$$f(x) = e^{-\lambda x^2}$$

where λ is a *complex* number. We will let $\gamma = \mathrm{Re}[\lambda]$ and $\kappa = \mathrm{Im}[\lambda]$, so that $\lambda = \gamma + i\kappa$ and

$$f(x) = e^{-(\gamma+i\kappa)x^2} = e^{-\gamma x^2}\left[\cos(\kappa x^2) - i\sin(\kappa x^2)\right] \quad .$$

To ensure that f is absolutely integrable, we will continue to assume γ is positive. The graph of such a function, with $\kappa < 0$, has been sketched in figure 23.2. Since the formula of f is so similar to that of a true Gaussian function, we might call f a basic *Gaussian-like* function.

While we won't refer to the above f as a Gaussian (unless $\kappa = 0$), it should be obvious that this function satisfies properties very similar to those satisfied by the basic Gaussians. For example, f, as defined above, is certainly an even, infinitely differentiable function on the real line. Noting that

$$|f(x)| = \left|e^{-(\gamma+i\kappa)x^2}\right| = \left|e^{-\gamma x^2}\right|\left|e^{-i\kappa x^2}\right| = e^{-\gamma x^2} \quad ,$$

it is also clear that f is absolutely integrable (assuming $\gamma > 0$). And with a little thought, you should also realize that every derivative of f, as well as every product of f with a polynomial, is also absolutely integrable.

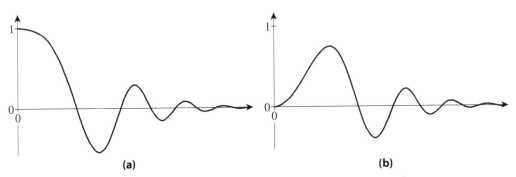

Figure 23.2: Graph of (a) the real part and (b) the imaginary part of $e^{-(\gamma + i\kappa)x^2}$ with $\gamma > 0$ and $\kappa < 0$.

Integrals of Basic Gaussian-Like Functions

The evaluation of the integral of f over the real line,

$$ I = \int_{-\infty}^{\infty} f(x)\,dx = \int_{-\infty}^{\infty} e^{-\lambda x^2}\,dx \quad , $$

will, doubtlessly, be the most difficult part of our work in extending the results of the previous sections to Gaussian-like functions. True, the same "clever trick" that led to equation (23.1) on page 361 can be used here just as well to show that

$$ I^2 = \frac{\pi}{\lambda} = \frac{\pi}{\gamma + i\kappa} \quad , \tag{23.20} $$

assuring us that I is one of the two square roots of π/λ. Unfortunately, the argument that I must be the positive square root, which we used when $\lambda = \gamma$, is no longer valid. So, instead, let's look at the real and imaginary parts of both I and each of the possible square roots of π/λ when $\lambda = \gamma + i\kappa$.

Rewriting f in terms of its real and imaginary parts yields

$$ I = \int_{-\infty}^{\infty} e^{-\gamma x^2} \left[\cos(\kappa x^2) - i\,\sin(\kappa x^2) \right] dx $$

$$ = \int_{-\infty}^{\infty} e^{-\gamma x^2} \cos(\kappa x^2)\,dx \; - \; i \int_{-\infty}^{\infty} e^{-\gamma x^2} \sin(\kappa x^2)\,dx \quad . $$

So

$$ \mathrm{Re}[I] = \int_{-\infty}^{\infty} e^{-\gamma x^2} \cos(\kappa x^2)\,dx $$

and

$$ \mathrm{Im}[I] = - \int_{-\infty}^{\infty} e^{-\gamma x^2} \sin(\kappa x^2)\,dx \quad . $$

Take another look at the graphs of f in figure 23.2. In particular, look at the graph of the imaginary part and notice how each "hump" above the X–axis is larger than the adjacent hump on the right and below the X–axis. This means (assuming our graph is reasonably accurate) that the "net area" between the graph of the imaginary part of f and the X–axis must be positive whenever $\kappa < 0$. On the other hand, since $\sin(\kappa x^2)$ is an odd function of κ, this net area will be negative when $\kappa > 0$. Thus,

$$ \mathrm{Im}[I] < 0 \qquad \text{when} \quad \kappa > 0 \tag{23.21a} $$

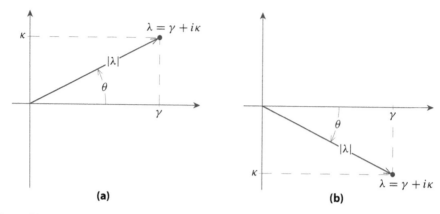

Figure 23.3: Relation between θ and $\lambda = \gamma + i\kappa$ where (a) $\kappa > 0$ and (b) $\kappa < 0$. In both cases $\gamma > 0$.

and

$$\text{Im}[I] > 0 \qquad \text{when} \quad \kappa < 0 \quad . \tag{23.21b}$$

(A rigorous verification of these inequalities will be left as an exercise, exercise 23.9.)

To find useful expressions for the square roots of π/λ, it helps to observe that, because $\gamma > 0$, λ can be written in polar form,

$$\lambda = |\lambda| e^{i\theta} \qquad \text{where} \quad \theta = \arctan\left(\frac{\kappa}{\gamma}\right) \quad .$$

This is illustrated in figure 23.3. Also, while we have that figure at hand, we might as well observe the following relations between the sign of κ and θ :

$$\kappa > 0 \quad \Longleftrightarrow \quad 0 < \theta < \frac{\pi}{2} \quad ,$$

$$\kappa = 0 \quad \Longleftrightarrow \quad \theta = 0 \quad ,$$

and

$$\kappa < 0 \quad \Longleftrightarrow \quad -\frac{\pi}{2} < \theta < 0 \quad .$$

We will use these relations in a little bit as well as the observation that

$$\cos(\theta) = \frac{\gamma}{|\lambda|} \quad .$$

Using the polar representation for λ, equation (23.20) can be written as

$$I^2 = \frac{\pi}{\lambda} = \frac{\pi}{|\lambda|} e^{-i\theta} \quad .$$

Solving for I, we get

$$I = \pm\left(\frac{\pi}{|\lambda|} e^{-i\theta}\right)^{1/2} = \pm\sqrt{\frac{\pi}{|\lambda|}} \exp\left(-i\frac{1}{2}\theta\right) \quad .$$

That is,

$$I = I_1 \qquad \text{or} \qquad I = I_2$$

where

$$I_1 = \sqrt{\frac{\pi}{|\lambda|}} \exp\left(-i\frac{1}{2}\theta\right) = \sqrt{\frac{\pi}{|\lambda|}} \left[\cos\left(\frac{1}{2}\theta\right) - i\sin\left(\frac{1}{2}\theta\right)\right]$$

and

$$I_2 = -\sqrt{\frac{\pi}{|\lambda|}} \exp\left(-i\frac{1}{2}\theta\right) = \sqrt{\frac{\pi}{|\lambda|}}\left[-\cos\left(\frac{1}{2}\theta\right) + i\sin\left(\frac{1}{2}\theta\right)\right] \quad .$$

Therefore,

$$\mathrm{Im}[I_1] = -\sqrt{\frac{\pi}{|\lambda|}}\sin\left(\frac{1}{2}\theta\right) \qquad \text{and} \qquad \mathrm{Im}[I_2] = \sqrt{\frac{\pi}{|\lambda|}}\sin\left(\frac{1}{2}\theta\right) \quad .$$

From these formulas and the relations between κ and θ, we see that

$$\mathrm{Im}[I_1] < 0 \quad \text{and} \quad \mathrm{Im}[I_2] > 0 \qquad \text{when} \quad 0 < \kappa \quad ,$$

while

$$\mathrm{Im}[I_1] > 0 \quad \text{and} \quad \mathrm{Im}[I_2] < 0 \qquad \text{when} \quad \kappa < 0 \quad .$$

Comparing these inequalities with inequalities (23.21) originally derived for the imaginary part of I, we finally see that I must be the first of the two possible square roots, I_1.

Thus $I = I_1$. Replacing I and I_1 with the formulas they represent then gives us

$$\int_{-\infty}^{\infty} e^{-\lambda x^2}\,dx = \sqrt{\frac{\pi}{|\lambda|}}\exp\left(-i\frac{1}{2}\theta\right) = \sqrt{\frac{\pi}{|\lambda|}}\left[\cos\left(\frac{1}{2}\theta\right) - i\sin\left(\frac{1}{2}\theta\right)\right] \qquad (23.22)$$

where

$$\theta = \arctan\left(\frac{\kappa}{\gamma}\right) \quad .$$

Don't forget, $\lambda = \gamma + i\kappa$ and $\gamma > 0$.

An alternative version of equation (23.22) not explicitly involving θ can be derived easily. Using a little basic trigonometry, we see that

$$\cos\left(\frac{1}{2}\theta\right) = \sqrt{\frac{1}{2} + \frac{1}{2}\cos(\theta)} = \sqrt{\frac{1}{2} + \frac{\gamma}{2|\lambda|}}$$

and

$$\sin\left(\frac{1}{2}\theta\right) = \mathrm{sgn}(\kappa)\sqrt{\frac{1}{2} - \frac{1}{2}\cos(\theta)} = \mathrm{sgn}(\kappa)\sqrt{\frac{1}{2} - \frac{\gamma}{2|\lambda|}}$$

where

$$\mathrm{sgn}(\kappa) = \begin{cases} -1 & \text{if } \kappa < 0 \\ 0 & \text{if } \kappa = 0 \\ +1 & \text{if } 0 < \kappa \end{cases} \quad .$$

With these equations and a little algebra, equation (23.22) "reduces" to

$$\int_{-\infty}^{\infty} e^{-\lambda x^2}\,dx = \sqrt{\frac{\pi}{2|\lambda|}}\left(\sqrt{1 + \frac{\gamma}{|\lambda|}} - i\,\mathrm{sgn}(\kappa)\sqrt{1 - \frac{\gamma}{|\lambda|}}\right) \qquad (23.23)$$

still assuming, of course, that $\lambda = \gamma + i\kappa$ and $\gamma > 0$.

Finally, it may be of some interest to observe that, after splitting equation (23.23) into its real and imaginary parts, we have

$$\int_{-\infty}^{\infty} e^{-\gamma x^2}\cos(\kappa x^2)\,dx = \sqrt{\frac{\pi}{2|\lambda|}}\left[1 + \frac{\gamma}{|\lambda|}\right]$$

and

$$\int_{-\infty}^{\infty} e^{-\gamma x^2}\sin(\kappa x^2)\,dx = \sqrt{\frac{\pi}{2|\lambda|}}\left[1 - \frac{\gamma}{|\lambda|}\right]$$

whenever $\gamma > 0$ and $\kappa \geq 0$. (In fact, these equations can be justified when $\gamma = 0$, even though the integrands are no longer absolutely integrable.)

Fourier Transforms of Gaussian-Like Functions

Again, let $f(x) = e^{-\lambda x^2}$ where $\lambda = \gamma + i\kappa$ and $\gamma > 0$. You can easily verify that the Fourier transform of f can be obtained through the same sort of computations we used to find the Fourier transform of the basic Gaussian functions. The result of those computations is

$$\mathcal{F}\left[e^{-\lambda t^2}\right]\Big|_\omega = I_\lambda \exp\left(-\frac{\pi^2}{\lambda}\omega^2\right) \tag{23.24}$$

where I_λ is the integral over the real line of $e^{-\lambda x^2}$ — the same integral evaluated just a few paragraphs ago. From that discussion, we have three equivalent formulas for this constant:

$$I_\lambda = \sqrt{\frac{\pi}{|\lambda|}} \exp\left(-i\frac{1}{2}\theta\right) \quad,$$

$$I_\lambda = \sqrt{\frac{\pi}{|\lambda|}} \left[\cos\left(\frac{1}{2}\theta\right) - i\sin\left(\frac{1}{2}\theta\right)\right] \quad,$$

and

$$I_\lambda = \sqrt{\frac{\pi}{2|\lambda|}} \left(\sqrt{1 + \frac{\gamma}{|\lambda|}} - i\,\mathrm{sgn}(\kappa)\sqrt{1 - \frac{\gamma}{|\lambda|}}\,\right)$$

where, in the first two formulas,

$$\theta = \arctan\left(\frac{\kappa}{\gamma}\right) \quad.$$

Likewise,

$$\mathcal{F}^{-1}\left[e^{-\lambda\omega^2}\right]\Big|_t = I_\lambda \exp\left(-\frac{\pi^2}{\lambda}t^2\right) \quad.$$

You can also verify that the discussion regarding general Gaussian functions in section 23.2 remains valid if the word "Gaussian" is replaced by "Gaussian-like", and the values γ and $\sqrt{\pi/\gamma}$ are replaced, respectively, by λ and the above defined I_λ. In particular, for any complex number α,

$$\mathcal{F}\left[e^{-\lambda(t-\alpha)^2}\right]\Big|_\omega = e^{-i2\pi\alpha\omega} I_\lambda \exp\left(-\frac{\pi^2}{\lambda}\omega^2\right)$$

and

$$\mathcal{F}\left[e^{i2\pi\alpha t}e^{-\lambda t^2}\right]\Big|_\omega = I_\lambda \exp\left(-\frac{\pi^2}{\lambda}(\omega-\alpha)^2\right) \quad.$$

Finally, let us observe that, because $e^{-\lambda t^2}$ is an even function, the transforms

$$\mathcal{F}\left[e^{-\gamma t^2}\cos\left(\kappa t^2\right)\right]\Big|_\omega \quad \text{and} \quad \mathcal{F}\left[e^{-\gamma t^2}\sin\left(\kappa t^2\right)\right]\Big|_\omega$$

can be found by taking the real and imaginary parts of equation (23.24) (see exercise 21.20 a on page 338). The computations are straightforward (though a little tedious) and are left as an exercise (exercise 23.10).

Additional Exercises

23.6. *Find each of the following transforms. Wherever it appears, assume that $\gamma > 0$.*

a. $\mathcal{F}\left[e^{-4t^2}\right]\Big|_\omega$

b. $\mathcal{F}^{-1}\left[e^{-4\omega^2}\right]\Big|_t$

c. $\mathcal{F}\left[e^{-4\pi t^2}\right]\Big|_\omega$

d. $\mathcal{F}\left[e^{-\gamma(t-3)^2}\right]\Big|_\omega$

e. $\mathcal{F}^{-1}\left[e^{-\gamma(\omega-3)^2}\right]\Big|_t$

f. $\mathcal{F}\left[e^{i6\pi t}e^{-9t^2}\right]\Big|_\omega$

g. $\mathcal{F}\left[\sin(12\pi t)\,e^{-9t^2}\right]\Big|_\omega$

h. $\mathcal{F}\left[t\,e^{-\gamma t^2}\right]\Big|_\omega$

i. $\mathcal{F}\left[t^2\,e^{-\gamma t^2}\right]\Big|_\omega$

j. $\mathcal{F}\left[(t+3)e^{-4\pi(t-5)^2}\right]\Big|_\omega$

k. $\mathcal{F}\left[e^{i6\pi t}\,e^{-4\pi(t-5)^2}\right]\Big|_\omega$

23.7. *Find the Fourier transform of each of the following using identities (23.6) and (23.7) (see page 363), elementary algebra and the Fourier translation identities. Do not use the more general Gaussian identities derived in section 23.2.*

a. $e^{6\gamma t}\,e^{-\gamma t^2}$ *where* $\gamma > 0$
    (Hint: Write this as a single exponential and then "complete the square".)

b. $e^{10t}\,e^{-t^2}$

c. $e^{-3(t-2-6i)^2}$

d. $e^{-9\pi(t-6i)^2}$

23.8. *Redo the previous exercise using the more general identities from section 23.2.*

23.9 a. *For the following, assume γ and κ are both positive real numbers.*

   i. *For each nonnegative integer n, rigorously verify that*

$$\int_{x_n}^{x_{n+1}} e^{-\gamma x^2}\sin\left(\kappa x^2\right)dx \;>\; 0 \qquad \text{where} \quad x_n = \sqrt{\frac{2n\pi}{\kappa}} \quad .$$

     (Suggestion: Simplify the problem using the substitution $\tau = \kappa x^2$.)

   ii. *Now show that*

$$\int_{-\infty}^{\infty} e^{-\gamma x^2}\sin\left(\kappa x^2\right)dx \;>\; 0 \quad .$$

   b. *Using the result of the last part, show that, if $\gamma > 0$ and $\kappa < 0$, then*

$$\int_{-\infty}^{\infty} e^{-\gamma x^2}\sin\left(\kappa x^2\right)dx \;<\; 0 \quad .$$

23.10. *Let γ and κ be two positive numbers, and, by applying the results of exercise 21.20 a on page 338 to equation (23.24), find the following:*

a. $\mathcal{F}\left[e^{-\gamma t^2}\cos\left(\kappa t^2\right)\right]\Big|_\omega$

b. $\mathcal{F}\left[e^{-\gamma t^2}\sin\left(\kappa t^2\right)\right]\Big|_\omega$

24

Convolution and Transforms of Products

Earlier, we obtained

$$y(t) = \mathcal{F}^{-1}\left[\frac{2\operatorname{sinc}(2\pi\omega)}{3 + i2\pi\omega}\right]\Big|_t$$

as a solution to a differential equation (see example 22.2 on page 343). At the time we did not attempt to further evaluate it, because, well, it just looked too darned hard.

Let us reconsider this formula. It is the product of two relatively simple functions,

$$F(\omega) = 2\operatorname{sinc}(2\pi\omega) \qquad \text{and} \qquad G(\omega) = \frac{1}{3 + i2\pi\omega} \quad ,$$

whose inverse transforms,

$$f(t) = \mathcal{F}^{-1}[F]\big|_t = \operatorname{pulse}_1(t) \qquad \text{and} \qquad g(t) = \mathcal{F}^{-1}[G]\big|_t = e^{-3t}\operatorname{step}(t) \quad ,$$

can be found by such elementary means as looking them up in table 21.1 on page 320. An obvious question now arises: *Is there a relatively simple formula of $f(t)$ and $g(t)$ that can be relied on to give the inverse transform of the product $F(\omega)G(\omega)$?*

The answer to this question is yes, at least for "most practical cases". The formula is the convolution formula, and this chapter is a study of that formula.

24.1 Derivation of the Convolution Formula

We will derive the convolution formula by attempting to evaluate $\mathcal{F}^{-1}[FG]\big|_t$ in terms of f and g (where, as usual, $f = \mathcal{F}^{-1}[F]$ and $g = \mathcal{F}^{-1}[G]$). For this derivation, we will assume both f and G are in $\mathcal{A}$ so that we can use the integral formulas for their transforms,

$$F(\omega) = \int_{-\infty}^{\infty} f(s)\,e^{-i2\pi s\omega}\,ds \qquad \text{and} \qquad g(\tau) = \int_{-\infty}^{\infty} G(\omega)\,e^{i2\pi\omega\tau}\,d\omega \quad .$$

Assuming that the product FG is also absolutely integrable, we then have

$$\mathcal{F}^{-1}[F(\omega)G(\omega)]\big|_t = \int_{-\infty}^{\infty} F(\omega)G(\omega)\,e^{i2\pi\omega t}\,d\omega$$

$$= \int_{-\infty}^{\infty}\left[\int_{-\infty}^{\infty} f(s)\,e^{-i2\pi s\omega}\,ds\right]G(\omega)\,e^{i2\pi\omega t}\,d\omega \quad .$$

Thus,

$$\mathcal{F}^{-1}[F(\omega)G(\omega)]|_t \;=\; \int_{-\infty}^{\infty}\int_{-\infty}^{\infty} f(s)G(\omega)\,e^{i2\pi\omega(t-s)}\,ds\,d\omega \quad .$$

Let us further assume that the order of integration of this last double integral can be interchanged. Then

$$\mathcal{F}^{-1}[F(\omega)G(\omega)]|_t \;=\; \int_{-\infty}^{\infty}\int_{-\infty}^{\infty} f(s)G(\omega)\,e^{i2\pi\omega(t-s)}\,d\omega\,ds$$

$$=\; \int_{-\infty}^{\infty} f(s)\left[\int_{-\infty}^{\infty} G(\omega)\,e^{i2\pi\omega(t-s)}\,d\omega\right]ds \quad .$$

But the inner integral in the last line is just the integral for $g(\tau)$ with $\tau = t - s$. So the above reduces to

$$\mathcal{F}^{-1}[F(\omega)G(\omega)]|_t \;=\; \int_{-\infty}^{\infty} f(s)g(t-s)\,ds \quad . \tag{24.1}$$

The integral formula on the right-hand side of this last equation is the convolution formula. While we've derived it as a formula for computing the inverse transform of a product, we will later discover that it and some other closely related formulas are important tools both for the general development of the mathematics of Fourier analysis and for solving many specific problems in engineering and science. Because of this, we are going to take a short break from discussing "Fourier transforms" and focus our efforts on understanding this important formula.

However, before turning our attention completely away from Fourier transforms, let us briefly look back at the assumptions made above in deriving equation (24.1). In addition to assuming that both f and G are classically transformable and absolutely integrable (i.e., that both f and G are in $\mathcal{A}$), we assumed that

1. the product FG is absolutely integrable, and

2. the order of integration in

$$\int_{-\infty}^{\infty}\int_{-\infty}^{\infty} f(s)G(\omega)\,e^{i2\pi\omega(t-s)}\,ds\,d\omega$$

   can be interchanged.

In fact, neither of these two additional assumptions was necessary. For one thing, if f is in $\mathcal{A}$, then $F = \mathcal{F}[f]$ is bounded (see theorem 19.6 on page 288), and thus, being the product of an absolutely integrable function with a bounded function, FG is automatically absolutely integrable — there was no need to assume it.

The issue of interchanging the order of integration in double integrals such as above was discussed in more general terms at the end of chapter 18.[1] The above double integral is just a special example of the double integrals considered in corollary 18.24 on page 275. Because f and G are assumed to be absolutely integrable functions on the real line and $e^{i2\pi\omega t}$ is a bounded continuous function, corollary 18.24 assures us that the product $f(s)G(\omega)\,e^{i2\pi\omega(t-s)}$ is absolutely integrable on the $S\Omega$-plane and that

$$\int_{-\infty}^{\infty}\int_{-\infty}^{\infty} f(s)G(\omega)\,e^{i2\pi\omega(t-s)}\,ds\,d\omega \;=\; \int_{-\infty}^{\infty}\int_{-\infty}^{\infty} f(s)G(\omega)\,e^{i2\pi\omega(t-s)}\,d\omega\,ds \quad .$$

Again, there was no need to assume this.

---

[1] We'll be referring to that discussion several times in this chapter. If you haven't yet done so, you might want to review that discussion now.

So all our derivation of equation (24.1) really required was that both f and G be in $\mathcal{A}$.

As an exercise, you should verify that the roles of f and g can be switched in the derivation of equation (24.1).

?►Exercise 24.1: *Re-derive equation (24.1) assuming that both F and g are in $\mathcal{A}$.*

24.2 Basic Formulas and Properties of Convolution

Definition

Let f and g be two functions defined on the real line. The *convolution* of f with g, denoted by either $f * g$ or $f(x) * g(x)$, is the function given by

$$f * g(x) \; = \; \int_{-\infty}^{\infty} f(s)g(x - s)\,ds$$

provided this integral exists and is finite for all real values of x. This integral is often called the *convolution integral*.

!►Example 24.1: If

$$f(t) \; = \; e^{-2t}\,\text{step}(t) \qquad \text{and} \qquad g(t) \; = \; e^{5t} \quad ,$$

then

$$f(s) \; = \; e^{-2s}\,\text{step}(s) \qquad \text{and} \qquad g(x - s) \; = \; e^{5(x-s)} \quad ,$$

and

$$f * g(x) \; = \; \int_{s=-\infty}^{\infty} f(s)g(x - s)\,ds$$

$$= \; \int_{-\infty}^{\infty} e^{-2s}\,\text{step}(s)\, e^{5(x-s)}\,ds$$

$$= \; e^{5x} \int_{0}^{\infty} e^{-7s}\,ds \; = \; e^{5x}\left[-\frac{1}{7}e^{-7s}\Big|_{s=0}^{\infty}\right] \; = \; \frac{1}{7}e^{5x} \quad .$$

!►Example 24.2:

$$\text{pulse}_1(x) * \sin(x) \; = \; \int_{-\infty}^{\infty} \text{pulse}_1(s)\sin(x - s)\,ds$$

$$= \; \int_{-1}^{1} \sin(x - s)\,ds \; = \; \cos(x - 1) \; - \; \cos(x + 1) \quad .$$

?►Exercise 24.2: *Show that, if $f(x) = \text{pulse}_2(x)$ and $g(x) = e^{3x}$, then*

$$f * g(x) \; = \; \frac{1}{3}\left[e^{3(x+2)} \; - \; e^{3(x-2)}\right] \quad .$$

As the next example shows, it is quite possible to have a convolution integral that is not finite. In such cases we will simply say that the corresponding convolution does not exist.

!▶ *Example 24.3:*   *Attempting to find the convolution of*

$$f(t) = e^{2t} \quad \text{with} \quad g(t) = e^{-3t}$$

yields

$$e^{2x} * e^{-3x} = \int_{-\infty}^{\infty} e^{2s} e^{-3(x-s)} \, ds = e^{-3x} \int_{-\infty}^{\infty} e^{5s} \, ds \quad,$$

which is not a finite integral. So $e^{2x} * e^{-3x}$ *does not exist.*

We will discuss conditions guaranteeing the existence of a given convolution later. You might be interested to know that it is possible (but not common) to even have functions f and g such that

$$f * g(x) = \int_{-\infty}^{\infty} f(s) g(x - s) \, ds$$

is well defined for some values of x and undefined (or infinite) for other values of x (an example is given in exercise 24.22 at the end of this chapter). Our interests, however, will mainly be with the cases in which the integral is well defined and finite for all real values of x. This will mean that, for every real value of x, we will want the product $f(s)g(x - s)$ to be a piecewise continuous and absolutely integrable function of s over the real line.

About the Notation

A few words must be said about notation and how, in our desire to use the "standard notation" commonly found in the literature, we are also perpetuating some rather bad (yet, convenient) notation.

Consider the two ways we have for indicating the formula of the convolution of $\text{pulse}_2(x)$ with e^{3x}. The first way is to write

$$f * g(x) = \frac{1}{3} \left[e^{3(x+2)} - e^{3(x-2)} \right]$$

where

$$f(x) = \text{pulse}_2(x) \quad \text{and} \quad g(x) = e^{3x} \quad .$$

This is notationally correct, but not nearly as convenient as simply saying

$$\text{pulse}_2(x) * e^{3x} = \frac{1}{3} \left[e^{3(x+2)} - e^{3(x-2)} \right] \quad , \tag{24.2}$$

which illustrates how convolution formulas are often described. Observe, however, that the symbol "x" is being used for two *different* things. On the left side of this equation x is a dummy variable used to describe the functions being convolved, while the x on the right-hand side is a true variable in the final formula for the convolution. In practice, equation (24.2) is understood to mean that, for example, when $x = 1$, the convolution of $\text{pulse}_2(x)$ with e^{3x} is

$$\frac{1}{3} \left[e^{3(1+2)} - e^{3(1-2)} \right] = \frac{1}{3} \left[e^9 - e^{-3} \right] \quad .$$

On the other hand, naively replacing x with 1 in the left-hand side gives

$$\text{pulse}_2(1) * e^{3 \cdot 1} = 1 * e^3 = \int_{-\infty}^{\infty} 1 \cdot e^3 \, ds = \infty \quad ,$$

which is certainly not what equation (24.2) was intended to imply!

The questionable double use of one symbol in equation (24.2) can be easily avoided by writing the equation as

$$\text{pulse}_2(s) * e^{3s}\bigg|_x = \frac{1}{3}\left[e^{3(x+2)} - e^{3(x-2)}\right] \quad .$$

Unfortunately, this does not seem to be common practice. So, innocent reader, be aware that, unless you are explicitly told otherwise, the phrase "the value of $f(x) * g(x)$ when $x = x_0$", as well as the notations

$$f(x) * g(x)\big|_{x_0} \qquad \text{and} \qquad f(x) * g(x)\big|_{x=x_0} \quad ,$$

should all be taken to mean

the value obtained by first finding the formula for $f * g(x)$, and then evaluating this formula with x replaced by x_0

and *NOT* $f(x_0) * g(x_0)$.

Alternate Definitions

Be warned that texts using one of the other definitions for the Fourier integral transforms will often define convolution by

$$f * g(x) = \frac{1}{\sqrt{2\pi}} \int_{-\infty}^{\infty} f(s)g(x - s)\,ds \quad .$$

As with the variations in the Fourier integral formulas, this variation in the definition of convolution affects specific formulas, but not the basic concepts.

You should also be aware that the convolution formula given here is the one appropriate when using the Fourier transform. When using the Laplace transform, the appropriate definition is

$$f * g(x) = \int_0^x f(s)g(x - s)\,ds \quad .$$

The difference between this and our convolution formula — the different limits in the integral — is significant. For example, the existence of $f * g$ for a particular choice of functions may depend on whether we are using the Laplace transform version or the Fourier transform version of convolution.

24.3 Algebraic Properties
Basic Algebraic Properties

In computing convolutions, you may wonder if there is a simple relationship between $f * g$ and $g * f$ that we could use to simplify our computation of, say, $f * g$ given that we already know $g * f$. Well, by definition,

$$f * g(x) = \int_{-\infty}^{\infty} f(s)g(x - s)\,ds \qquad \text{and} \qquad g * f(x) = \int_{-\infty}^{\infty} g(s)f(x - s)\,ds \quad .$$

If, in the second integral, we use the change of variables $\sigma = x - s$ (so that $s = x - \sigma$ and $ds = -d\sigma$), then, keeping in mind that s and σ are dummy variables,

$$g * f(x) = \int_{s=-\infty}^{\infty} g(s)f(x-s)\,ds$$

$$= \int_{\sigma=\infty}^{-\infty} g(x-\sigma)f(\sigma)(-1)\,d\sigma$$

$$= -\int_{\sigma=\infty}^{-\infty} f(\sigma)g(x-\sigma)\,d\sigma$$

$$= \int_{-\infty}^{\infty} f(\sigma)g(x-\sigma)\,d\sigma = f * g(x) \quad .$$

So there is a simple relation between $f * g$ and $g * f$ — they are the same. To use a phrase from algebra, "convolution is commutative".

Commutativity is one of the basic algebraic properties of convolution. For convenience, this and the other basic algebraic properties of convolution are listed below (using f, g and h to denote arbitrary functions on the real line).

1. **Convolution is commutative:** *If $f * g$ exists, so does $g * f$. Moreover,*

    $$f * g = g * f \quad .$$

2. **Convolution distributes over addition:** *If $f * g$ and $f * h$ exist, so does $f * (g + h)$. Moreover,*
    $$f * (g + h) = (f * g) + (f * h) \quad .$$

3. **Constants factor out of convolution:** *For any constant α,*

    $$(\alpha f) * g = f * (\alpha g) = \alpha(f * g)$$

    provided any one of these convolutions exist.

4. **For convolution, the zero function is zero:** $f * 0 = 0 \quad .$

?▶Exercise 24.3:   *Verify that the second, third and fourth properties listed above always hold.*

The properties listed above suggest a similarity between convolution and multiplication. Indeed, $f * g$ is often referred to as "the convolution product". The analogy between convolution and multiplication is further reinforced by the fact that the convolution formula was derived as a Fourier transform of the product of two functions. Some care, however, must be exercised in treating convolution as a type of multiplication. For example, $f * 1 \neq f$. Instead,

$$f * 1 = \int_{-\infty}^{\infty} f(s)\,ds \quad .$$

You may have noted that "associativity" is not in the above list. There is a reason for that.

The Myth of Associativity

Many authors (and instructors) claim that convolution is associative; that is, they claim that given any three functions f, g and h, then, as long as the convolutions exist,

$$(f * g) * h = f * (g * h) \quad .$$

They are wrong.

?▶ Exercise 24.4: *Verify that*
$$(f * g) * h \neq f * (g * h)$$

when

$$f(x) = 1 \quad , \quad g(x) = x \, e^{-x^2} \quad \text{and} \quad h(x) = \text{step}(x)$$

by computing the appropriate convolutions. (Don't forget to compute the convolutions in the parentheses first.)

The myth that "convolution is associative" is widely believed because it is *often* true that

$$(f * g) * h = f * (g * h) \quad .$$

To see why this is so, and to determine when associativity can be assumed, let us examine the integrals defining $(f * g) * h$ and $f * (g * h)$.[2]

For any given real value x,

$$(f * g) * h(x) = \int_{s=-\infty}^{\infty} [f * g(s)] h(x - s) \, ds$$

$$= \int_{s=-\infty}^{\infty} \left[\int_{t=-\infty}^{\infty} f(t) g(s - t) \, dt \right] h(x - s) \, ds$$

$$= \int_{s=-\infty}^{\infty} \int_{t=-\infty}^{\infty} f(t) g(s - t) h(x - s) \, dt \, ds \quad .$$

Similar computations (and a simple change of variables) yields

$$f * (g * h)(x) = \int_{t=-\infty}^{\infty} \int_{s=-\infty}^{\infty} f(t) g(s - t) h(x - s) \, ds \, dt \quad .$$

Thus, the statement that
$$(f * g) * h(x) = f * (g * h)(x)$$

is completely equivalent to the statement that

$$\int_{s=-\infty}^{\infty} \int_{t=-\infty}^{\infty} \Psi_x(s, t) \, dt \, ds = \int_{t=-\infty}^{\infty} \int_{s=-\infty}^{\infty} \Psi_x(s, t) \, ds \, dt$$

where

$$\Psi_x(s, t) = f(t) g(s - t) h(x - s) \quad .$$

In other words, the question of whether associativity holds here has the same answer as the question of whether we can interchange the order of integration of a related double integral. That issue was discussed in somewhat greater generality at the end of chapter 18. It is not difficult to show (see

---

[2] See also exercise 24.21 on page 398 for additional conditions ensuring associativity.

exercises 24.13 and 24.14) that theorem 18.23 on page 275 and its corollary apply directly to the case at hand, and lead to the following theorem and corollary.[3]

Theorem 24.1

Let f, g and h be piecewise continuous functions on the real line. If, for each real value x,

$$\Psi_x(s,t) = f(t)g(s-t)h(x-s)$$

is an absolutely integrable function on the ST–plane and is uniformly absolutely integrable on every strip of the form $(a,b) \times (-\infty, \infty)$ or $(-\infty, \infty) \times (c,d)$, then

$$(f * g) * h = f * (g * h) \quad .$$

Corollary 24.2

Let f, g and h be piecewise continuous functions on the real line. Then

$$(f * g) * h = f * (g * h)$$

if any one of the following sets of conditions holds:

1. g is bounded, and both f and h are absolutely integrable.

2. f and g are bounded, and both g and h are absolutely integrable.

3. g and h are bounded, and both f and g are absolutely integrable.

4. There is a real number c such that $f(\sigma)$, $g(\sigma)$ and $h(\sigma)$ are all zero whenever $\sigma < c$.[4]

5. There is a real number c such that $f(\sigma)$, $g(\sigma)$ and $h(\sigma)$ are all zero for $c < \sigma$.

6. For two of these functions, say f and g, there is a positive constant c such that $f(\sigma)$ and $g(\sigma)$ are both zero whenever $|\sigma| > c$.

24.4   Computing Convolutions

"Computing a convolution" can mean one of two things. It may mean "evaluating $f * g(x)$ for a specific value of x". In practice, it is more likely to mean "finding a formula (or set of formulas) for $f * g(x)$ that is valid for all real values of x".

Either way, the basic algebraic properties should be used to reduce the work required to compute the necessary integrals. Where appropriate, use the distributive property to break the one integral into two or more simpler integrals, and, for each case, decide whether it would be easier to set up and compute

$$\int_{-\infty}^{\infty} f(s)g(x-s)\,ds \qquad \text{or} \qquad \int_{-\infty}^{\infty} g(s)f(x-s)\,ds \quad .$$

---

[3] Here, and at a few other places in this chapter, the material from the end of chapter 18 leads to a very general — but perhaps obscure — theorem, followed by a corollary that is less general, but more easily applied.

[4] Such functions are said to be *causal* and play an important role in many applications.

Both integrals give the same result (remember, $f * g = g * f$), but one might occasionally be easier to set up or evaluate.

Setting up the integral for the actual computation of $f * g$ is usually fairly straightforward when at least one of the functions is not "piecewise defined" (i.e., when either $f(x)$ or $g(x)$ is described by a single manageable formula for all real values of x). If both functions are piecewise defined, however, then $f * g$ can be expected to also be piecewise defined, and a certain amount of bookkeeping becomes necessary to keep track of all the various pieces to the final formulas describing the convolution.

One approach to computing the convolution of two piecewise-defined functions f and g is outlined below. In this approach, we simplify the bookkeeping by roughly sketching the graphs of $f(s)g(x-s)$ for certain choices of x. The making of these sketches is partially based on the observation that

$$g(x-s) = g(-(s-x)) \quad .$$

This means that the graph of $g(x-s)$, as a function of s, is the *translation by x* of the graph of $g(-s)$, which, itself, is the *reflection (across $s = 0$)* of the graph of the original function $g(s)$. It is the relative positions of the various "pieces" of the graphs of $g(x-s)$ and $f(s)$ that determine the intervals for the various formulas describing $f * g$.

In the following, we refer to the *singularities* in the functions or their graphs. In practice, a function "has a singularity at x_0" if there is something "unusual" about the function around x_0, with the precise meaning of "unusual" dependent on the interests of those involved. Here, we are interested in manipulating formulas, so here "$f(x)$ has a singularity at x_0" means that the formula we are using to compute $f(x)$ changes at $x = x_0$.

The Procedure (Illustrated)

The procedure consists of a series of steps. We will illustrate these steps by finding the formula for $f * g$ where

$$f(s) = \begin{cases} 0 & \text{if} \quad s < 1 \\ s & \text{if} \quad 1 < s \end{cases}$$

and

$$g(s) = \begin{cases} 2s^2 & \text{if} \quad 0 < s < 1 \\ 0 & \text{otherwise} \end{cases} \quad .$$

(Note: This procedure is for computing $f * g$ when f and g each has, at most, a finite number of singularities. Modifications are needed if f or g has an infinite number of singularities.)

Step 1: *Sketch the graph of each function (as a function of s), and note where the singularities of each occur. Also, determine whether you would prefer to compute $f * g(x)$*

$$\text{as} \quad \int_{-\infty}^{\infty} f(s)g(x-s) \, ds \quad \text{or as} \quad \int_{-\infty}^{\infty} g(s)f(x-s) \, ds \quad .$$

(The choice here is largely a matter of individual preference, which, in turn, is developed from experience. The author's experience is that it rarely makes much difference which is chosen.)

The graphs of the functions in our example have been sketched in figure 24.1. The only singularity of $f(s)$ is at $s = 1$, while $g(s)$ has singularities at $s = 0$ and $s = 1$.

We will proceed computing $f * g$ as $\int_{-\infty}^{\infty} f(s)g(x-s) \, ds$.

(If, instead, $\int_{-\infty}^{\infty} g(s)f(x-s) \, ds$ is chosen, then the roles of f and g in the following steps should be interchanged.)

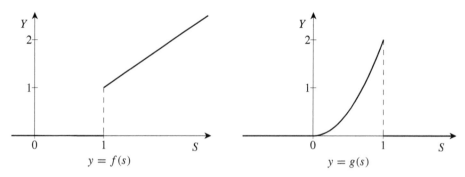

Figure 24.1: Step 1 in computing $f * g(x)$: graphing $f(s)$ and $g(s)$ separately.

Step 2:   Sketch the graph of $g(-s)$ (the reflection of $g(s)$), and note where this graph has singularities.

   The graph of $g(-s)$ from our example has been sketched in figure 24.2. The only singularities in $g(-s)$ are at $s = -1$ and $s = 0$.

Step 3:   Sketch the translation of the previous graph by an arbitrary x (this will be the graph of $g(x - s)$), and note where this graph has singularities. Also, at this point, determine the general set of formulas describing $g(x - s)$.

   The graph of $g(x - s)$ from our example has also been sketched in figure 24.2. The only singularities in $g(x - s)$ are at $s = x - 1$ and $s = x$.

   Replacing the s with $x - s$ in the set of formulas for $g(s)$ gives us

$$g(x - s) \;=\; \begin{cases} 2(x - s)^2 & \text{if } \; 0 < x - s < 1 \\ 0 & \text{otherwise} \end{cases}$$

$$=\; \begin{cases} 2(x - s)^2 & \text{if } \; x - 1 < s < x \\ 0 & \text{otherwise} \end{cases} \;.$$

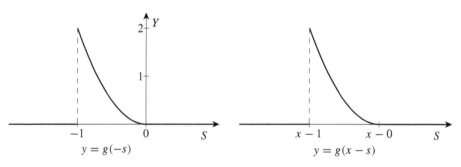

Figure 24.2: Steps 2 and 3 in computing $f * g(x)$: graphing $g(-s)$ and $g(x - s)$.

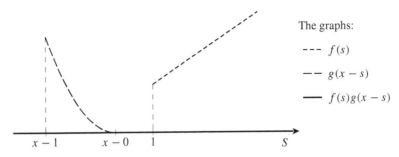

The graphs:

--- $f(s)$

-- $g(x - s)$

— $f(s)g(x - s)$

Figure 24.3: Step 4 in computing $f * g(x)$: superimposing $g(x - s)$ on $f(s)$ to help obtain $f(s)g(x - s)$ when all singularities of $g(x - s)$ are to the left of all singularities of $f(s)$.

Step 4: *Superimpose the graphs of $f(s)$ and $g(x - s)$ with x chosen so that all the singularities of $g(x - s)$ are to the left of all the singularities of $f(s)$. Do not choose a specific value for x, but do determine the maximum possible value for x here. Then:*

a. *Sketch the graph of the product $f(s)g(x - s)$, and note where this product has singularities. (This graph can be rather crude; just be sure you can identify the singularities.)*

b. *Using this sketch and the formulas for $f(s)$ and $g(x - s)$, determine the formula for $f(s)g(x - s)$ over each of the intervals bounded by the singularities just found.*

c. *Finally, compute the formula for*

$$f * g(x) = \int_{-\infty}^{\infty} f(s)g(x - s)\, ds$$

for this case. Write this formula down some place safe along with the values of x for which it is valid.

The graphs of $f(s)$, $g(x - s)$ and $f(s)g(x - s)$ from our example have been sketched in figure 24.3 for the case where $x < 1$. Clearly, $f(s)g(x - s) = 0$ for all values of s. Thus, when $x < 1$,

$$f * g(x) = \int_{-\infty}^{\infty} f(s)g(x - s)\, ds = \int_{-\infty}^{\infty} 0\, ds = 0 \quad .$$

Step Next: *"Slide" the graph of $g(x - s)$ to the right (i.e., increase the value of x) until a singularity in the graph of this function "passes" a singularity in the graph of $f(s)$. Sketch the resulting graph of $g(x - s)$ with the graph of $f(s)$, and determine the largest x can be before another pair of singularities "pass each other". Then:*

a. *Sketch the graph of the product $f(s)g(x - s)$, noting where this product has singularities. (Again, this graph can be rather crude provided you can identify the singularities.)*

b. *Using this sketch and the formulas for $f(s)$ and $g(x - s)$, determine the formula for $f(s)g(x - s)$ over each of the intervals bounded by the singularities just found.*

c. *Finally, compute the formula for*

$$f * g(x) = \int_{-\infty}^{\infty} f(s)g(x - s)\, ds$$

for this case. Write this formula down some place safe along with the values of x for which it is valid.

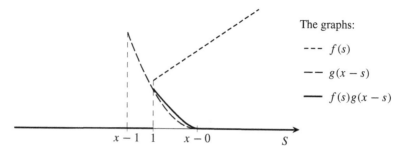

The graphs:

--- $f(s)$

-- $g(x-s)$

— $f(s)g(x-s)$

$x-1$ 1 $x-0$ S

Figure 24.4: First application of "Step Next" in computing $f * g(x)$: superimposing $g(x-s)$ on $f(s)$ to help obtain $f(s)g(x-s)$ when one singularity of $g(s)$ is to the right of one singularity of $f(s)$.

The graphs of $f(s)$, $g(x-s)$ and $f(s)g(x-s)$ from our example have been sketched in figure 24.4 for the case where $1 < x < 2$. The singularities in $f(s)g(x-s)$ are at $s = 1$ and $s = x$. Using the graph and the above formulas for $f(s)$ and $g(x-s)$,

$$f(s)g(x-s) = \begin{cases} 0 & \text{if } s < 1 \\ 2s(x-s)^2 & \text{if } 1 < s < x \\ 0 & \text{if } x < s \end{cases} .$$

Thus, for $1 < x < 2$,

$$f * g(x) = \int_{-\infty}^{\infty} f(s)g(x-s)\,ds$$

$$= \int_{-\infty}^{1} 0\,ds + \int_{1}^{x} 2s(x-s)^2\,ds + \int_{x}^{\infty} 0\,ds$$

$$= \cdots = \left(\frac{1}{6}x + \frac{1}{2}\right)(x-1)^3 .$$

Subsequent Steps: *Repeat "Step Next" until all possible cases have been accounted for (i.e., until all the singularities in the graph of $g(x-s)$ are to the right of all the singularities in the graph of $f(s)$).*

The graphs of $f(s)$, $g(x-s)$ and $f(s)g(x-s)$ from our example have been sketched in figure 24.5 for the case where $2 < x$. The only singularities in $f(s)g(x-s)$ are at $s = x-1$ and $s = x$. Using the graph and the above formulas for $f(s)$ and $g(x-s)$,

$$f(s)g(x-s) = \begin{cases} 0 & \text{if } s < x-1 \\ 2s(x-s)^2 & \text{if } x-1 < s < x \\ 0 & \text{if } x < s \end{cases} .$$

Thus, for $2 < x$,

$$f * g(x) = \int_{-\infty}^{\infty} f(s)g(x-s)\,ds$$

$$= \int_{-\infty}^{x-1} 0\,ds + \int_{x-1}^{x} 2s(x-s)^2\,ds + \int_{x}^{\infty} 0\,ds$$

$$= \cdots = \frac{2}{3}x - \frac{1}{2} .$$

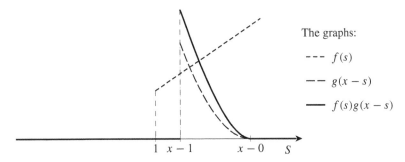

Figure 24.5: Last application of "Step Next" in computing $f * g(x)$: superimposing $g(x - s)$ on $f(s)$ to help obtain $f(s)g(x - s)$ when all singularities of $g(s)$ are to the right of all singularities of $f(s)$.

Repeating *Step Next* is no longer possible since both singularities in the graph of $g(x-s)$ are now to the right of the one singularity in the graph of $f(s)$.

Last Step: *Combining the results of the previous steps, write out the complete set of formulas for* $f * g(x)$.

By the above steps, the complete set of formulas for $f * g(x)$ is

$$f * g(x) = \begin{cases} 0 & \text{if} \quad x < 1 \\ \left(\frac{1}{6}x + \frac{1}{2}\right)(x - 1)^3 & \text{if} \quad 1 < x < 2 \\ \frac{2}{3}x - \frac{1}{2} & \text{if} \quad 2 < x \end{cases} \quad . \qquad (24.3)$$

This convolution is sketched in figure 24.6.

?▶Exercise 24.5: *Show that the convolution of the step function with itself,* step $*$ step, *is the* ramp function,

$$\text{ramp}(x) = \begin{cases} 0 & \text{if} \quad x < 0 \\ x & \text{if} \quad 0 < x \end{cases} \quad .$$

Also, sketch the ramp function and explain how it got its name.

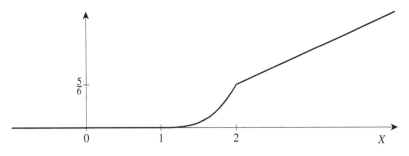

Figure 24.6: Last step in computing $f * g(x)$: sketching the graph (using formula (24.3)).

24.5 Existence, Smoothness and Derivatives of Convolutions
Existence and Continuity

If you compare the graph of $f * g$ in figure 24.6 to the graphs of f and g in figure 24.1, you will probably notice that, while the graphs of f and g clearly contain jumps, the graph of $f * g$ appears to be continuous. This illustrates something you should come to expect: Typically, the convolution of any two piecewise continuous functions will be continuous, provided the convolution exists.

To get an intuitive feeling for why this is so, let's consider the case where both f and g are nonnegative real-valued functions. Then, for each x,

$$f * g(x) = \int_{-\infty}^{\infty} f(s)g(x - s)\,ds$$

$$= \text{``area enclosed by the graph of } f(s)g(x - s) \text{ and the } S\text{-axis ''}.$$

So, for any two real values x_0 and x_1, the difference between $f * g(x_0)$ and $f * g(x_1)$ will be the difference in the areas corresponding to the graphs of $f(s)g(x_0 - s)$ and $f(s)g(x_1 - s)$. But, if you think about it, you should normally expect these two graphs to be rather similar if x_0 and x_1 are close to each other. After all, the graph of $g(x_1 - s)$ will just be the graph of $g(x_0 - s)$ shifted to the left or right by a distance of $|x_1 - x_0|$. So the corresponding net areas — and hence, the convolutions $f * g(x_0)$ and $f * g(x_1)$ — should also be about the same whenever x_0 and x_1 are about the same.

!▶**Example 24.4:**   *In particular, consider* step * step(x), *the convolution of the step function with itself. Let* $0 < x_0 < x_1$ *with* x_1 *being just a little bit larger than* x_0. *The graphs of* step(s) step$(x - s)$ *for* $x = x_0$ *and* $x = x_1$ *have been sketched in figure 24.7. Observe that these two graphs differ only over the interval* (x_0, x_1). *Though the difference in the graphs is significant over this interval, the corresponding area is only that of a rectangle of height one and width* $x_1 - x_0$. *Thus,*

$$\text{step} * \text{step}(x_1) = \int_0^{x_1} 1\,ds$$

$$= \int_0^{x_0} 1\,ds + \int_{x_0}^{x_1} 1\,ds$$

$$= \text{step} * \text{step}(x_0)$$

$$\quad + \left[\text{area of rectangle of height } 1 \text{ and width } x_1 - x_0\right]$$

$$= \text{step} * \text{step}(x_0) + x_1 - x_0 \quad .$$

So clearly,

$$\lim_{x_1 \to x_0^+} \text{step} * \text{step}(x_1) = \text{step} * \text{step}(x_0)$$

and

$$\lim_{x_0 \to x_1^-} \text{step} * \text{step}(x_0) = \text{step} * \text{step}(x_1) \quad .$$

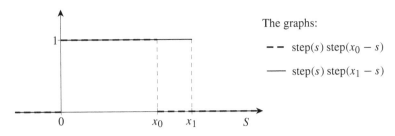

Figure 24.7: Graph of $\text{step}(s)\,\text{step}(x_0 - s)$ and $\text{step}(s)\,\text{step}(x_1 - s)$ when $0 < x_0 < x_1$.

Of course, not only are the intuitive arguments given above that $f * g(x_0)$ is close to $f * g(x_1)$ when x_0 is close to x_1 rather nonrigorous, they are meaningless if the enclosed areas are not finite.

These two issues — existence and continuity — were previously (and more rigorously) discussed in chapter 18 for functions more generally defined by

$$\psi(x) = \int_{-\infty}^{\infty} h(x, s)\, ds \quad .$$

A convolution

$$f * g(x) = \int_{-\infty}^{\infty} f(s) g(x - s)\, ds \quad ,$$

is just a special case of this. Employing the results of that discussion, we can easily obtain (see exercise 24.17 at the end of this chapter) the following theorem describing fairly general conditions ensuring that a convolution exists and is continuous.

Theorem 24.3

Let f and g be two piecewise continuous functions on $\mathbb{R}$. Then $f * g$ is a well-defined and continuous function on the real line if, for each real value x_0, there is a corresponding interval (a, b) and a corresponding function h_0 such that

1. $a < x_0 < b$,

2. h_0 *is in* $\mathcal{A}$, *and*

3. *for every (x, s) in the strip $(a, b) \times (-\infty, \infty)$,*

$$|f(s) g(x - s)| \leq |h_0(s)| \quad .$$

A few common situations where the conditions in the above theorem can be verified (see exercise 24.18 at the end of this chapter) are given in the next corollary.

Corollary 24.4

Suppose f and g are two piecewise continuous functions on the real line. Then $f * g$ exists and is a continuous function on $\mathbb{R}$ whenever any one of the following sets of conditions holds:

1. *One of the two functions is absolutely integrable, and the other is bounded.*

2. *One of the two functions vanishes outside of some finite interval.*

3. *There is a real number c such that $f(\sigma)$ and $g(\sigma)$ are both zero whenever $c < \sigma$.*

4. *There is a real number c such that $f(\sigma)$ and $g(\sigma)$ are both zero whenever $c > \sigma$.*

You may be somewhat surprised, however, to learn that f and g both being in $\mathcal{A}$ is not enough to guarantee that $f * g$ exists or is continuous (see exercise 24.22 at the end of this chapter).

Smoothness and Differentiation

The fact that, typically (if it exists), the convolution of two piecewise continuous functions is continuous is a particular case of a broader observation that, typically (if it exists), the convolution of any two functions is "smoother" than either of the two functions. For example, if f is piecewise continuous and g is continuous and piecewise smooth, then $f * g$, if it exists, can be expected to be smooth. The formula relating the derivative of $f * g$ to corresponding convolution of the derivative is easy to derive. Ignoring, for the moment, questions about integrability, existence, etc.,

$$(f * g)'(x) = \frac{d}{dx} \int_{-\infty}^{\infty} f(s) g(x - s) \, ds$$

$$= \int_{-\infty}^{\infty} \frac{\partial}{\partial x} [f(s) g(x - s)] \, ds$$

$$= \int_{-\infty}^{\infty} f(s) g'(x - s) \, ds = f * g'(x) \quad.$$

By "commutativity", we might then expect that

$$f * g' = (f * g)' = (g * f)' = g * f' = f' * g \quad,$$

provided, of course, that f is "suitably differentiable".

To determine when the above derivations are valid, let us again turn back to the end of chapter 18. There we discussed differentiating

$$\psi(x) = \int_{-\infty}^{\infty} h(x, s) \, ds$$

where h was a relatively arbitrary function of two variables. A convolution,

$$f * g(x) = \int_{-\infty}^{\infty} f(s) g(x - s) \, ds \quad,$$

is just a special case. In particular, corollary 18.21 on page 274 applies and yields the following theorem.

Theorem 24.5
Let f and g be two functions on $\mathbb{R}$ with f being piecewise continuous, and with g being both continuous and piecewise smooth. Assume that, for each x_0 on the real line, there is a corresponding finite interval (a, b), and two functions h_0 and h_1 such that

1. *$a < x_0 < b$,*

2. *both h_0 and h_1 are in $\mathcal{A}$, and*

3. *for every (x, s) in the strip $(a, b) \times (-\infty, \infty)$,*

$$|f(s) g(x - s)| \leq |h_0(s)| \quad \text{and} \quad |f(s) g'(x - s)| \leq |h_1(s)| \quad.$$

Then $f * g$ is a well-defined and smooth function on the real line. Moreover,

$$(f * g)' = f * (g') = (g') * f \quad.$$

A few cases where the conditions given in this last theorem are easily verified are given in the next corollary.

Corollary 24.6

Let f and g be piecewise continuous functions on $\mathbb{R}$ with g also being continuous and piecewise smooth. Then $f * g$ is a well-defined and smooth function on the real line, and

$$(f * g)' = f * (g') = (g') * f$$

whenever any one of the following sets of conditions holds:

1. f is absolutely integrable, and both g and g' are bounded.

2. f is bounded, and both g and g' are absolutely integrable.

3. One of the two functions vanishes outside of some finite interval.

4. There is a real constant c such that $f(\sigma)$ and $g(\sigma)$ are both zero whenever $c < \sigma$.

5. There is a real constant c such that $f(\sigma)$ and $g(\sigma)$ are both zero whenever $c > \sigma$.

It is true (though we won't prove it) that, under reasonable "integrability conditions", the convolution of two piecewise continuous and piecewise smooth functions is continuous and piecewise smooth. Without the continuity of g, however, we do *NOT* have $(f * g)' = f * (g')$! This is illustrated in the next two exercises.

?▶Exercise 24.6:   *Compute*

$$e^x * \text{step}(x) \quad , \quad (e^x * \text{step}(x))' \quad \text{and} \quad e^x * \text{step}'(x) \quad ,$$

and verify that

$$(e^x * \text{step}(x))' \neq e^x * \text{step}'(x) \quad .$$

?▶Exercise 24.7:   *Show that, for every function f in $\mathcal{A}$, the convolution $f * \text{step}$ is continuous and piecewise smooth, and that*

$$(f * \text{step})' = f \quad .$$

Extending the above discussion to cases involving second and higher order derivatives is just a matter of repeatedly applying the above results. For future reference, here are two of the results that can be derived by repeated applications of theorem 24.5.

Theorem 24.7

Let f be a piecewise continuous function on $\mathbb{R}$, and let g be a smooth function on $\mathbb{R}$ whose derivative is piecewise smooth. Assume that, for each x_0 on the real line, there is a corresponding finite interval (a, b) and three functions h_0, h_1 and h_2 such that

1. $a < x_0 < b$,

2. h_0, h_1 and h_2 are all in $\mathcal{A}$, and

3. for every (x, s) in the strip $(a, b) \times (-\infty, \infty)$,

$$|f(s)g(x - s)| \leq |h_0(s)| \quad , \quad \left| f(s)g'(x - s) \right| \leq |h_1(s)|$$

and

$$\left| f(s)g''(x-s) \right| \le |h_2(s)| \quad .$$

Then $f * g$ is a well-defined and smooth function on the real line whose derivative is smooth. Moreover,

$$(f * g)'' = f * (g'') = (g'') * f \quad .$$

Theorem 24.8
Let f be a piecewise continuous function on $\mathbb{R}$ and g an infinitely smooth function on $\mathbb{R}$. Assume that, for each x_0 on the real line, there is a corresponding finite interval (a, b) and functions h_0, h_1, h_2, ... such that

1. $a < x_0 < b$,

2. h_0, h_1, h_2, ... are all in $\mathcal{A}$, and

3. for each nonnegative integer n and every (x, s) in the strip $(a, b) \times (-\infty, \infty)$,

$$\left| f(s)g^{(n)}(x-s) \right| \le |h_n(s)| \quad .$$

Then $f * g$ is a well-defined, infinitely smooth function on the real line. Moreover,

$$(f * g)^{(n)} = f * (g^{(n)}) = (g^{(n)}) * f \qquad \text{for} \quad n = 1, 2, 3, \ldots \quad .$$

24.6 Convolution and Fourier Analysis
Fourier Transforms

By now you may have forgotten just what led to our discussion of convolution. It was our derivation of equation (24.1) on page 376, which we can now write as

$$\mathcal{F}^{-1}[FG] = f * g \quad \text{where} \quad f = \mathcal{F}^{-1}[F] \quad \text{and} \quad g = \mathcal{F}^{-1}[G] \quad . \tag{24.4}$$

From our discussion there and exercise 24.1, we know this equation is valid (and the product FG is absolutely integrable) whenever both F and g are in $\mathcal{A}$ or both f and G are in $\mathcal{A}$. Though we didn't note it then, we should now observe that the same assumptions ensure that, for each real value t, $f(s)g(t-s)$ is the product of a bounded function with an absolutely integrable function of s. Thus, $f(s)g(t-s)$ is also an absolutely integrable function of s, assuring us that

$$f * g(x) = \int_{-\infty}^{\infty} f(s)g(x-s)\, ds$$

is a well-defined, continuous function of x on the entire real line.

Notice, too, that these same assumptions (both F and g being in $\mathcal{A}$ or both f and G being in $\mathcal{A}$) also ensure that the product fg is absolutely integrable and that, for each real value of ω, $F(s)G(\omega-s)$ is an absolutely integrable function of s. This, in turn, assures us that $F * G$ is well defined and continuous on the real line. Moreover, after looking back at our derivation of equation

(24.1), it should be clear that a very similar derivation leads to

$$\mathcal{F}[fg] = F * G \qquad \text{where} \quad F = \mathcal{F}[f] \quad \text{and} \quad G = \mathcal{F}[g] \qquad (24.5)$$

under the same assumptions.

While we're at it, we should also consider the case where, say, both g and G are in $\mathcal{A}$ (such as, for example, when g is a Gaussian function). Recall that each classically transformable function f can be written as $f = f_1 + f_2$ where f_1 is in $\mathcal{A}$ and f_2 is in $\mathcal{T}$. This, of course, means that $F = F_1 + F_2$ where $F_1 = \mathcal{F}[f_1]$ is in $\mathcal{T}$ and $F_2 = \mathcal{F}[f_2]$ is in $\mathcal{A}$. By the observations just made, you can easily verify that the product of any one of the functions

$$F_1 \quad , \quad F_2 \quad , \quad f_1 \quad \text{or} \quad f_2$$

with any real translation of either

$$G \quad \text{or} \quad g$$

is absolutely integrable. Moreover,

$$\mathcal{F}^{-1}[FG] = \mathcal{F}^{-1}[(F_1 + F_2)\,G] = \mathcal{F}^{-1}[F_1 G] + \mathcal{F}^{-1}[F_2 G]$$
$$= f_1 * g + f_2 * g = (f_1 + f_2) * g = f * g$$

and

$$\mathcal{F}[fg] = \mathcal{F}[(f_1 + f_2)\,g] = \mathcal{F}[f_1 g] + \mathcal{F}[f_2 g]$$
$$= F_1 * G + F_2 * G = (F_1 + F_2) * G = F * G \quad .$$

Obviously, similar results can be derived when both f and F are in $\mathcal{A}$.

For future reference, let us officially summarize everything we've just verified.

Theorem 24.9 (first theorem on convolution identities)
Let f, g, F and G be classically transformable functions with $F = \mathcal{F}[f]$ and $G = \mathcal{F}[g]$. Assume that any one of the following holds:

1. *Both F and g are in $\mathcal{A}$.*

2. *Both f and G are in $\mathcal{A}$.*

3. *Both g and G are in $\mathcal{A}$.*

4. *Both f and F are in $\mathcal{A}$.*

Then all of the following hold:

1. *The products fg and FG are absolutely integrable.*

2. *The convolutions $f * g$ and $F * G$ exist and are classically transformable.*

3. $\mathcal{F}[fg] = F * G$ *and* $\mathcal{F}^{-1}[FG] = f * g$.

We will refer to equations $\mathcal{F}[fg] = F * G$ and $\mathcal{F}^{-1}[FG] = f * g$ as the *convolution identities*. It can also be shown that these identities are valid as long as the functions involved are simply bounded (and classically transformable). The full proof of this will have to wait until chapter 30 after we develop a little more machinery (namely, the "fundamental identity" and "identity sequences"), but we will go ahead and state (and use) this important theorem.

Theorem 24.10 (second theorem on convolution identities)
Let f, g, F and G all be bounded, classically transformable functions with $F = \mathcal{F}[f]$ and $G = \mathcal{F}[g]$. Then

1. the products fg and FG are absolutely integrable;

2. the convolutions $f * g$ and $F * G$ exist and are classically transformable, and

3. $\mathcal{F}[fg] = F * G$ and $\mathcal{F}^{-1}[FG] = f * g$.

It is possible to have two classically transformable functions whose product or convolution is not classically transformable (again, see exercise 24.22 at the end of this chapter). In practice, though, such situations rarely arise.

?► Exercise 24.8:   *Using convolution, show that*

$$\mathcal{F}^{-1}\left[\frac{2\operatorname{sinc}(2\pi\omega)}{3+i2\pi\omega}\right]\bigg|_t = \begin{cases} 0 & \text{if } t < -1 \\ \frac{1}{3}e^{-3t}\left[e^{3t} - e^{-3}\right] & \text{if } -1 < t < 1 \\ \frac{1}{3}e^{-3t}\left[e^{3} - e^{-3}\right] & \text{if } 1 < t \end{cases}.$$

Fourier Series

Suppose we have two piecewise continuous functions f and ψ with ψ being absolutely integrable on $\mathbb{R}$ and f being periodic with period p. Remember, f can then be expressed as its Fourier series,

$$F.S.[f]\big|_t = \sum_{k=-\infty}^{\infty} f_k\, e^{i2\pi\omega_k t}$$

where

$$\omega_k = \frac{k}{p} \quad \text{and} \quad f_k = \frac{1}{p}\int_0^p f(t)\, e^{-i2\pi\omega_k t}\, dt \quad.$$

Since f is bounded and ψ is absolutely integrable, the convolution of the two, $f * \psi$, is a well-defined function on the real line. In fact, since both f and ψ are piecewise continuous, we know $f * \psi$ is continuous. Notice also that, because f has period p,

$$f * \psi(x - p) = \int_{-\infty}^{\infty} f((x - p) - s)\psi(s)\, ds$$

$$= \int_{-\infty}^{\infty} f(x - s - p)\psi(s)\, ds$$

$$= \int_{-\infty}^{\infty} f(x - s)\psi(s)\, ds = f * \psi(x)$$

for all real x. So $f * \psi$ is also periodic with period p and is, itself, describable in terms of its Fourier series,

$$f * \psi(t) = F.S.[f * \psi]\big|_t = \sum_{k=-\infty}^{\infty} c_k\, e^{i2\pi\omega_k t}$$

where

$$c_k = \frac{1}{p} \int_0^p f * \psi(t) \, e^{-i2\pi\omega_k t} \, dt = \frac{1}{p} \int_0^p \left(\int_{-\infty}^\infty f(t-s)\psi(s) \, ds \right) e^{-i2\pi\omega_k t} \, dt \quad .$$

The order of integration in this last double integral can be switched (see theorem 18.22 on page 274). Doing so, and doing a few other simple manipulations, gives us

$$c_k = \frac{1}{p} \int_{-\infty}^\infty \psi(s) \int_0^p f(t-s) \, e^{-i2\pi\omega_k t} \, dt \, ds$$

$$= \int_{-\infty}^\infty \psi(s) \, e^{-i2\pi\omega_k s} \left(\frac{1}{p} \int_0^p f(t-s) \, e^{-i2\pi\omega_k(t-s)} \, dt \right) ds \quad . \tag{24.6}$$

But the inner integral is easily seen to be the kth Fourier coefficient for f,

$$\frac{1}{p} \int_0^p f(t-s) \, e^{-i2\pi\omega_k(t-s)} \, dt = \frac{1}{p} \int_{-s}^{-s+p} f(\tau) \, e^{-i2\pi\omega_k \tau} \, d\tau = f_k \quad .$$

Plugging this back into equation (24.6) and letting $\Psi = \mathcal{F}[\psi]$, we get a particularly simple formula for the c_k's,

$$c_k = \int_{-\infty}^\infty \psi(s) \, e^{-i2\pi\omega_k s} f_k \, ds = f_k \int_{-\infty}^\infty \psi(s) \, e^{-i2\pi\omega_k s} \, ds = f_k \Psi(\omega_k) \quad .$$

To summarize:

Theorem 24.11
Let f be a periodic, piecewise continuous function with period p and Fourier series

$$F.S.[f]|_t = \sum_{k=-\infty}^\infty f_k \, e^{i2\pi\omega_k t} \quad ,$$

and let ψ be an absolutely integrable, piecewise continuous function on the real line. Then the convolution $f * \psi$ is a continuous, periodic function with period p. Moreover, letting $\Psi = \mathcal{F}[\psi]$,

$$F.S.[f * \psi]|_t = \sum_{k=-\infty}^\infty f_k \Psi(\omega_k) \, e^{i2\pi\omega_k t} \quad .$$

Additional Exercises

24.9. Find $f * g(x)$ for each of the following choices of $f(x)$ and $g(x)$, if the convolution exists. If it does not exist, say so.

    a. $f(x) = \text{step}(x)e^{-3x}$, $g(x) = e^{-2x}$

b. $f(x) = \text{step}(x)e^{-3x}$, $\quad g(x) = e^{-5x}$

c. $f(x) = \text{step}(x)e^{-3x}$, $\quad g(x) = \sin(x)$

d. $f(x) = x^2$, $\quad g(x) = x^2 - 4$

e. $f(x) = \text{pulse}_3(x)$, $\quad g(x) = x^2 - 4$

f. $f(x) = x\,\text{pulse}_3(x)$, $\quad g(x) = x^3$

g. $f(x) = \text{step}(x)$, $\quad g(x) = \dfrac{1}{1+x^2}$

h. $f(x) = 3x + 4$, $\quad g(x) = e^{-x^2}$

i. $f(x) = e^{4x}$, $\quad g(x) = e^{-x^2}$

j. $f(x) = e^{-\alpha|x|}$, $\quad g(x) = e^{\beta x}$ with $0 < \beta < \alpha$

24.10 a. Compute $\text{pulse}_\alpha * \sin(x)$ for arbitrary $\alpha > 0$.

b. Suppose g is not the zero function, but $f * g$ is the zero function. Is it necessarily true that f must be the zero function? (Give a reason for your answer!)

24.11. Assume the convolution $f * g$ exists, and let α be any real number. Show that

$$f(s) * g(s - \alpha)\big|_x = f * g(x - \alpha) \quad .$$

24.12 a. Verify that

$$f(x) * e^{i2\pi\alpha x} = F(\alpha)e^{i2\pi\alpha x}$$

whenever α is a real number, f is in $\mathcal{A}$, and $F = \mathcal{F}[f]$.

b. Let α, f and F be as in the previous part of this exercise. What additional condition on f will ensure that $f(x) * \cos(2\pi\alpha x) = F(\alpha)\cos(2\pi\alpha x)$?

c. Using results just derived, evaluate the following:

    i. $e^{-x^2} * e^{i2\pi x}$ **ii.** $e^{-x^2} * \cos(2\pi x)$

    iii. $e^{-x^2} * \sin(2\pi x)$ **iv.** $e^{-3|x|} * e^{i\pi x}$

    v. $\left[e^{-x}\,\text{step}(x)\right] * \cos(6\pi x)$

24.13. Verify that theorem 24.1 on the associativity of convolution follows from theorem 18.23 on page 275.

24.14. Using corollary 18.24 on page 275, prove corollary 24.2 on the associativity of convolution for each of the following cases:

a. f and g are in $\mathcal{A}$, and h is bounded.

b. There is a real constant c such that $f(\sigma)$, $g(\sigma)$ and $h(\sigma)$ are all zero whenever $c < \sigma$. (Hint: Find a triangle outside of which $f(t)g(s - t)h(x - s)$ vanishes.)

c. For two of these functions, say f and g, there is a positive constant c such that $f(\sigma)$ and $g(\sigma)$ are both zero whenever $|\sigma| > c$.

24.15. Let α be a fixed positive value and β some other real value. For each of the following, determine the values of β (relative to α) for which the indicated convolution exists, and evaluate the convolution for those values of β.

    a. $\left[e^{-\alpha x} \operatorname{step}(x)\right] * \left[e^{-\beta x} \operatorname{step}(x)\right]$ **b.** $\left[e^{-\alpha x} \operatorname{step}(x)\right] * \left[e^{\beta x} \operatorname{step}(-x)\right]$

24.16. Evaluate each of the following convolutions if it exists. If it does not exist, say so.

    a. $\operatorname{pulse}_2 * \operatorname{step}(x)$

    b. $\operatorname{ramp} * \operatorname{ramp}(x)$

    c. $\left[e^{2x} \operatorname{step}(x)\right] * \left[e^{5x} \operatorname{rect}_{(0,2)}(x)\right]$

    d. $\left[e^{2x} \operatorname{step}(x)\right] * \left[e^{8x} \operatorname{rect}_{(2,3)}(x)\right]$

    e. $\left[e^{2x} \operatorname{step}(x)\right] * \left[\left(3e^{5x} \operatorname{rect}_{(0,2)}(x)\right) - \left(6e^{8x} \operatorname{rect}_{(2,3)}(x)\right)\right]$

    f. $\left[x^2 \operatorname{pulse}_1(x)\right] * \left[x \operatorname{rect}_{(1,4)}(x)\right]$

24.17. Use corollary 18.19 on page 271 to derive theorem 24.3 on the existence and continuity of convolution.

24.18. Let f and g be two piecewise continuous functions on the real line. For each case below, confirm that $f * g$ exists and is continuous by applying theorem 24.3 after first verifying that, for each point x_0 of the real line, there is a corresponding interval (a, b) and a corresponding function h_0 such that

    1. $a < x_0 < b$,

    2. h_0 is in A, and

    3. for every (x, s) in the strip $(a, b) \times (-\infty, \infty)$,

$$|f(s)g(x - s)| \leq |h_0(s)| \quad .$$

    a. One of the two functions is absolutely integrable, and the other is bounded.

    b. One of the two functions vanishes outside of some finite interval.

    c. There is a real constant c such that $f(\sigma)$ and $g(\sigma)$ are both zero whenever $c < \sigma$.

24.19. Using convolution, find each of the following:

    a. $\mathcal{F}\left[\dfrac{1}{(2 + it)(4 + it)}\right]\Big|_\omega$ **b.** $\mathcal{F}^{-1}\left[\dfrac{1}{(3 + i2\pi\omega)(5 - i2\pi\omega)}\right]\Big|_t$

    c. $\mathcal{F}^{-1}\left[\dfrac{\operatorname{sinc}(6\pi\omega)}{6 + i2\pi\omega}\right]\Big|_t$ **d.** $\mathcal{F}\left[\operatorname{sinc}^2(2\pi t)\right]\Big|_\omega$

    e. $\mathcal{F}\left[\dfrac{1}{(9 + 4\pi^2 t^2)(4 - i2\pi t)}\right]\Big|_\omega$ **f.** $\mathcal{F}\left[\dfrac{1}{(\alpha - i2\pi t)^2}\right]\Big|_\omega$ where $0 < \alpha$

    g. $\mathcal{F}\left[\dfrac{1}{(\alpha^2 + 4\pi^2 t^2)^2}\right]\Big|_\omega$ where $0 < \alpha$ **h.** $\mathcal{F}^{-1}\left[e^{-2\pi|\omega|} \operatorname{sinc}(2\pi\omega)\right]\Big|_t$

24.20. *Using convolution, find a classically transformable solution to each of the following differential equations:*

   a. $\dfrac{dy}{dt} - 4y = \text{pulse}_1(t)$

   b. $\dfrac{d^2y}{dt^2} - 9y = e^{-3t}\,\text{step}(t)$

24.21. *Let f, g, h, F, G and H be bounded, classically transformable functions with $F = \mathcal{F}[f]$, $G = \mathcal{F}[g]$ and $H = \mathcal{F}[h]$. Using theorem 24.10 on page 394, show that the triple convolutions $f * (g * h)$ and $(f * g) * h$ are well defined and that*

$$ f * (g * h) \;=\; (f * g) * h \quad . $$

24.22. *For each positive integer n, let f_n be the function*

$$ f_n(x) = \begin{cases} 2^n & \text{if } n < x < n + 2^{-2n} \\ 0 & \text{otherwise} \end{cases} , $$

   and define the functions f and g by

$$ f(x) = \sum_{n=1}^{\infty} f_n(x) \quad \text{and} \quad g(x) = f(-x) \quad . $$

   a. *Sketch the graph of f_n for $n = 1$, $n = 2$ and 3.*

   b. *Sketch the graphs of f and g.*

   c. *From the sketches just done, it should be clear that f and g are piecewise continuous. Verify that they are also absolutely integrable by verifying that*

$$ \int_{-\infty}^{\infty} |g(x)|\, dx \;=\; \int_{-\infty}^{\infty} |f(x)|\, dx \;=\; 1 \quad . $$

   (Hint: You may want to review the discussion of geometric series in example 4.1 on page 42.)

   d. *Show that, even though f and g are both in $\mathcal{A}$, their convolution $f * g$ is not a well-defined function in $\mathcal{A}$. In particular, show that*

$$ f * g(0) = \infty \quad . $$

   Are there any other points at which $f * g$ is infinite?

   e. *Show that $f * g\left(\frac{1}{2}\right) = 0$.*

   f. *Show that $f * g(x)$ is finite whenever $0 < x < \frac{1}{2}$. What happens to the value of $f * g(x)$ as $x \to 0^+$?*

25

Correlation, Square-Integrable Functions and the Fundamental Identity

The three main topics in this chapter involve integral formulas that are very similar to the convolution formula. So you could view this chapter as a continuation of the one on convolution.

The first integral formula will define correlation, an operation often used in applications to measure similarities between two functions. This operation is so much like convolution that we'll be able to prove some of the major results regarding correlation by simply referring to analogous results already proven for convolution.

One thing we will discover is that applications of correlation often involve integrals of squares of functions. This will naturally lead to a brief discussion of functions whose squares are absolutely integrable and the derivation of equations analogous to the Parseval and Bessel equalities derived late in chapter 13 for Fourier series.

Finally, we will discuss an identity that will play a fundamental role in further developing the mathematics of Fourier analysis. It, too, is closely related to convolution, and will be a major tool in confirming the one yet unproven theorem on the convolution identities, theorem 24.10.

25.1 Correlation
Cross-Correlation

Let f and g be two functions on the real line. If it exists, the *correlation* (also called the *cross-correlation*) of f with g, denoted by either $f \star g$ or $f(x) \star g(x)$, is the function given by[1]

$$f \star g(x) \;=\; \int_{-\infty}^{\infty} f^*(s)g(s+x)\,ds \quad . \tag{25.1a}$$

Letting $\sigma = s + x$, we obtain the equivalent formula

$$f \star g(x) \;=\; \int_{-\infty}^{\infty} f^*(\sigma - x)g(\sigma)\,d\sigma \quad . \tag{25.1b}$$

For $f \star g(x)$ to exist for every real value x, the above integrals must be well defined (and finite) for every such x. Thus, verifying that $f \star g$ exists for a given choice of f and g will usually mean verifying that the product $f(s)g(s+x)$ is a piecewise continuous and absolutely integrable function of s for each real value x.

---

[1] Warning: The notation used for correlation varies widely from author to author. Some, for example, prefer using ρ_{fg} where we are using $f \star g$.

Two facts should be readily apparent from the above definitions. One is that correlation does *not* commute. In general, $f \star g \neq g \star f$. Instead,

$$g \star f(x) = [f \star g(-x)]^* \qquad (25.2)$$

provided the correlations exist.

?▶Exercise 25.1:   *Verify equation (25.2).*

Another obvious fact is that the operations of correlation and convolution are very similar. Indeed, comparing the definitions of the two leads to the following lemma.

Lemma 25.1 (relation between correlation and convolution)
Let f and g be two functions on the real line. If either $f(x) \star g(x)$ or $f^*(-x) * g(x)$ exists, then both exist and

$$f(x) \star g(x) = f^*(-x) * g(x) \quad . \qquad (25.3)$$

?▶Exercise 25.2:   *Prove lemma 25.1.*

Because of the relation between correlation and convolution, every result described in chapter 24 for convolutions can be rephrased as an analogous result for correlations. To keep this chapter relatively short, we'll leave to the interested reader the task of compiling a complete list of all the correlation analogs of results from chapter 24.

Still, we should at least determine the analogs to the Fourier convolution identities (see page 392). As usual, we start by letting f and g be classically transformable functions with $F = \mathcal{F}[f]$ and $G = \mathcal{F}[g]$. Assuming the appropriate Fourier convolution identity holds, we see that

$$\mathcal{F}[f \star g(t)]|_\omega = \mathcal{F}\left[f^*(-t) * g(t)\right]\big|_\omega = H(\omega)G(\omega) \qquad (25.4)$$

where $H(\omega) = \mathcal{F}\left[f^*(-t)\right]\big|_\omega$. Using near-equivalence and the complex conjugation identities (theorem 21.4 on page 332), we then find that

$$H(\omega) = \mathcal{F}\left[f^*(-t)\right]\big|_\omega = \mathcal{F}^{-1}\left[f^*(t)\right]\big|_\omega = \left(\mathcal{F}[f(t)]|_\omega\right)^* = F^*(\omega) \quad .$$

With this, equation (25.4) becomes

$$\mathcal{F}[f \star g(t)]|_\omega = F^*(\omega)G(\omega) \quad .$$

This equation is one of the *correlation identities*. The other correlation identities can be obtained in a similar fashion or using near-equivalence. For reference, here is the complete list of the correlation identities:

$$\mathcal{F}[f \star g] = F^*G \quad , \qquad (25.5a)$$

$$\mathcal{F}^{-1}\left[F^*G\right] = f \star g \quad , \qquad (25.5b)$$

$$\mathcal{F}\left[f^*g\right] = F \star G \qquad (25.5c)$$

and

$$\mathcal{F}^{-1}[F \star G] = f^*g \quad . \qquad (25.5d)$$

Explicit conditions under which these identities hold can be derived from the corresponding theorems for convolution (theorems 24.9 and 24.10, see page 393).[2] Since we will be using it soon, we'll state the analog to the second theorem on the convolution identities, theorem 24.10.

Theorem 25.2 (second theorem on correlation identities)
Let f, g, F and G all be bounded, classically transformable functions with $F = \mathcal{F}[f]$ and $G = \mathcal{F}[g]$. Then

1. *the products f^*g and F^*G are absolutely integrable;*

2. *the correlations $f \star g$ and $F \star G$ exist and are classically transformable, and*

3. *all the correlation identities in equation set (25.5) hold.*

?▶Exercise 25.3: *What's the first theorem on correlation identities?*

Auto-Correlation

It is somewhat interesting to see what happens to the correlation identities when f and g are the same function. When this is the case,

$$f^*g = f^*f = |f|^2 \quad \text{and} \quad F^*G = F^*F = |F|^2 \quad ,$$

and the correlation identities become

$$\mathcal{F}[f \star f] = |F|^2 \quad , \tag{25.6a}$$

$$\mathcal{F}^{-1}\left[|F|^2\right] = f \star f \quad , \tag{25.6b}$$

$$\mathcal{F}\left[|f|^2\right] = F \star F \tag{25.6c}$$

and

$$\mathcal{F}^{-1}[F \star F] = |f|^2 \tag{25.6d}$$

where, naturally, $F = \mathcal{F}[f]$.

It is common to refer to the correlation of a function f with itself, $f \star f$, as the *auto-correlation* of f. Accordingly, we will refer to the above identities as the *auto-correlation identities*. Explicit conditions under which they are valid are given in the next theorem, which is simply theorem 25.2 with $f = g$.

Theorem 25.3 (auto-correlation identities)
Let f and F be bounded, classically transformable functions with $F = \mathcal{F}[f]$. Then

1. *$|f|^2$ and $|F|^2$ are absolutely integrable functions;*

2. *the auto-correlations $f \star f$ and $F \star F$ exist and are classically transformable, and*

3. *all the auto-correlation identities of equation set (25.6) are valid.*

---

[2] Many authors refer to a statement that identities (25.5a) through (25.5d) hold as a *Wiener–Khintchine theorem*.

Now suppose f and F are as in the last theorem, and recall that the Fourier transform of an absolutely integrable function is bounded by the integral of the absolute value of that function (theorem 19.6 on page 288). Then, for every real value t,

$$\left|\mathcal{F}^{-1}\left[|F|^2\right]\big|_t\right| \leq \int_{-\infty}^{\infty} |F(\omega)|^2 \, d\omega = \mathcal{F}^{-1}\left[|F|^2\right]\big|_0 \quad .$$

With two applications of identity (25.6b), this yields,

$$|f \star f(t)| = \left|\mathcal{F}^{-1}\left[|F|^2\right]\big|_t\right| \leq \mathcal{F}^{-1}\left[|F|^2\right]\big|_0 = f \star f(0) \quad ,$$

telling us that the maximum value of $|f \star f(t)|$ must occur at $t = 0$. In fact, somewhat stronger results can be derived using a version of the Schwarz inequality that we will discuss in the next section. Using that inequality, you can easily prove the following theorem (see exercise 25.12).

Theorem 25.4
The auto-correlation of a piecewise continuous function f exists if and only if

$$\int_{-\infty}^{\infty} |f(x)|^2 \, dx < \infty \quad .$$

Moreover, if the auto-correlation exists (and f is not the zero function), then

$$|f \star f(x)| < f \star f(0) \qquad \text{whenever} \quad x \neq 0 \quad .$$

So any auto-correlation of a nontrivial function has its maximum at the origin and only at the origin. It is this property that makes correlation so useful in many applications. To illustrate this, here is a brief and highly simplified discussion of the mathematics of radar detection.

!▶**Example 25.1 (a very simplified model of radar range detection):**   *To detect the distance between a radar installation and some target (an incoming missile, a speeding car, etc.), the radar device sends out an electromagnetic pulse that, as a function of time t, can be described as a real-valued, piecewise continuous and absolutely integrable function $f(t)$. (This function actually describes what would be detected at the radar site. At any other location the pulse detected would be given by $Af(t - \Delta t)$ with A being the attenuation factor (a constant between 0 and 1) and Δt being the length of time it takes the radar pulse to travel to that location.)*
     At the same time, the radar installation is also measuring the corresponding electromagnetic radiation it receives. Some of this radiation is "noise" (ambient radiation from the environment, reflections of the radar pulse off nearby trees, etc.) and some is the reflection of the transmitted pulse off the target. The resulting signal detected is given by

$$g(t) = Af(t - 2\Delta t) + \eta(t)$$

where A is some attenuation factor, Δt is the length of time it takes the radar pulse to travel between the radar site and the target, and η is a real-valued function describing the ambient noise. Initially, the value of Δt is not known. That is the information you want to extract from your measurements.
     Part of the difficulty is that the noise, $\eta(t)$, may be masking the reflected radar pulse, $Af(t - 2\Delta t)$. Typically, the exact formula for $\eta(t)$ cannot be predicted. However, it is also typical for the random fluctuations in the noise to "average out" to some fixed value η_0 over any reasonable time period. This means that, if we take the correlation of the noise with our pulse, we should expect to get

$$f \star \eta(t) = \int_{-\infty}^{\infty} f^*(s)\eta(s + t) \, ds = \alpha\eta_0$$

where

$$\alpha = \int_{-\infty}^{\infty} f^*(s) \, ds \quad .$$

This assumes the effective duration of f is large enough for the fluctuations in η to "average out". Assuming so, then taking the correlation of g with f yields

$$f \star g(t) = f(t) \star [Af(t - 2\Delta t) + \eta(t)]$$
$$= A[f(t) \star f(t - 2\Delta t)] + f \star \eta(t) = A[f(t) \star f(t - 2\Delta t)] + \alpha\eta_0 \quad .$$

Now observe that

$$f(t) \star f(t - 2\Delta t) = \int_{-\infty}^{\infty} f^*(s) f(s - 2\Delta t + t) \, ds$$
$$= \int_{-\infty}^{\infty} f^*(s) f(s + [t - 2\Delta t]) \, ds = f \star f(t - 2\Delta t) \quad .$$

Consequently, the cross-correlation of f with g can be expressed in terms of the auto-correlation of f via

$$f \star g(t) = A f \star f(t - 2\Delta t) + \alpha\eta_0 \quad .$$

Since A, α, η_0 and Δt are all constants, the maximum value of $|f \star g(t)|$ must be at the same t where $f \star f(t - 2\Delta t)$ is maximum. According to theorem 25.4 this is only when $t - 2\Delta t = 0$ — that is, when $t = 2\Delta t$. So, if we observe that $f \star g(t)$ achieves its maximum value at some time t_0 (this can be done electronically), then we know $t_0 = 2\Delta t$. Dividing through by 2 gives us Δt, the time it takes the radar pulse to travel between the radar site and the target, and multiplying this time by the velocity of the radar pulse (i.e., the speed of light) gives us the distance from the site to the target.

25.2 Square-Integrable/Finite Energy Functions
Definitions and Basic Facts

A function f on the real line is said to be *square integrable* (on $\mathbb{R}$) if and only if

$$\int_{-\infty}^{\infty} |f(x)|^2 \, dx \tag{25.7}$$

exists and is finite. After, recalling that the norm of f over the interval $(-\infty, \infty)$ is given by

$$\|f\| = \left[\int_{-\infty}^{\infty} |f(x)|^2 \, dx \right]^{1/2}$$

(provided the integral is well defined — see chapter 11), we see that it is also appropriate to refer to square-integrable functions as *finite normed* functions. Additionally, in some applications, the energy in the process described by f is given by expression (25.7). So square-integrable functions are also called *finite energy* functions.

In chapter 11 we saw that

$$\left| \int_a^b f(x)g^*(x)\, dx \right| \leq \int_a^b |f(x)g(x)|\, dx$$

$$\leq \left(\int_a^b |f(x)|^2\, dx \right)^{1/2} \left(\int_a^b |g(x)|^2\, dx \right)^{1/2}$$

whenever f and g are piecewise continuous functions on a finite interval (a, b) (see the Schwarz inequality, theorem 11.2 on page 135, and its proof). Letting $a \to -\infty$ and $b \to \infty$ gives us yet another version of the Schwarz inequality.

Lemma 25.5 (Schwarz inequality)
Let f and g be two piecewise continuous, square-integrable functions on the real line. Then the products fg and fg^* are absolutely integrable functions on the real line, and

$$\left| \int_{-\infty}^{\infty} f(x)g^*(x)\, dx \right| \leq \int_{-\infty}^{\infty} |f(x)g(x)|\, dx$$

$$\leq \left(\int_{-\infty}^{\infty} |f(x)|^2\, dx \right)^{1/2} \left(\int_{-\infty}^{\infty} |g(x)|^2\, dx \right)^{1/2} \quad .$$

With this version of the Schwarz inequality, it is easy to confirm that any linear combination of square-integrable functions is also square integrable.

Corollary 25.6
The set of all piecewise continuous, square-integrable functions on the real line is a linear space; that is, any linear combination of two piecewise continuous, square-integrable functions on $\mathbb{R}$ is another piecewise continuous, square-integrable function on the real line.

?▶Exercise 25.4:   *Prove corollary 25.6.*

Recall that the Schwarz inequality can be viewed as a generalization of the inequality $|\mathbf{u} \cdot \mathbf{v}| \leq \|\mathbf{u}\| \|\mathbf{v}\|$ where $\mathbf{u}$ and $\mathbf{v}$ are any two three-dimensional vectors. You may also recall that $|\mathbf{u} \cdot \mathbf{v}| = \|\mathbf{u}\| \|\mathbf{v}\|$ if and only if one of the two vectors is a constant multiple of the other (i.e., the angle between the two is 0 or π). An analogous fact, important in many applications involving correlation, also holds for square-integrable functions (and would have been proven in earlier discussions of the Schwarz inequality had there been any need for it then).

Theorem 25.7 (Schwarz inequality)
Let f and g be two piecewise continuous, square-integrable functions on the real line. Then the products fg and fg^* are absolutely integrable functions on the real line, and

$$\left| \int_{-\infty}^{\infty} f(x)g^*(x)\, dx \right| \leq \left(\int_{-\infty}^{\infty} |f(x)|^2\, dx \right)^{1/2} \left(\int_{-\infty}^{\infty} |g(x)|^2\, dx \right)^{1/2} \tag{25.8}$$

with equality holding if and only if one of the functions is a constant multiple of the other.

PROOF: From lemma 25.5, we already know inequality (25.8) holds. Also, it readily follows from exercise 11.13 on page 144 that

$$\left| \int_{-\infty}^{\infty} f(x)g^*(x)\, dx \right| = \left(\int_{-\infty}^{\infty} |f(x)|^2\, dx \right)^{1/2} \left(\int_{-\infty}^{\infty} |g(x)|^2\, dx \right)^{1/2} \tag{25.9}$$

whenever either function is a constant multiple of the other. Consequently, all we need to verify is that one function is a constant multiple of the other, say $g = \lambda f$ for some constant λ, whenever equation (25.9) holds.

So assume equation (25.9) holds. For convenience, let

$$A = \left(\int_{-\infty}^{\infty} |f(x)|^2 \, dx \right)^{1/2} \qquad \text{and} \qquad B = \left(\int_{-\infty}^{\infty} |g(x)|^2 \, dx \right)^{1/2} \quad ,$$

and let R and θ be the magnitude and principal argument for the integral of fg^*,

$$\int_{-\infty}^{\infty} f(x)g^*(x) \, dx = Re^{i\theta} \quad .$$

Observe that we only need to consider the case where neither A nor B is the zero function (otherwise either f or g is automatically "0 times the other"). Also note that, in terms of A, B and R, equation (25.9) is

$$R = AB \quad .$$

Now, using the foresight of the author, let $\lambda = \alpha e^{i\theta}$ where $\alpha = A/B$. Then

$$\begin{aligned}
|f - \lambda g|^2 &= (f - \lambda g)(f^* - \lambda^* g^*) \\
&= |f|^2 + |\lambda|^2 |g|^2 - \lambda^* fg^* - \lambda f^* g \\
&= |f|^2 + \alpha^2 |g|^2 - \alpha \left[e^{-i\theta} fg^* + e^{i\theta} f^* g \right] \quad .
\end{aligned}$$

Integrating $|f - \lambda g|^2$ and applying the above, we get

$$\begin{aligned}
\int_{-\infty}^{\infty} |f(x) - \lambda g(x)|^2 \, dx &= \int_{-\infty}^{\infty} |f(x)|^2 \, dx + \alpha^2 \int_{-\infty}^{\infty} |g(x)|^2 \, dx \\
&\quad - \alpha \left[e^{-i\theta} \int_{-\infty}^{\infty} f(x)g^*(x) \, dx + e^{i\theta} \int_{-\infty}^{\infty} f^*(x)g(x) \, dx \right] \\
&= A^2 + \alpha^2 B^2 - \alpha \left[e^{-i\theta} \left(e^{i\theta} R \right) + e^{i\theta} \left(e^{i\theta} R \right)^* \right] \\
&= A^2 + \alpha^2 B^2 - 2\alpha R \\
&= A^2 + \left(\frac{A}{B} \right)^2 B^2 - 2 \left(\frac{A}{B} \right) AB \\
&= 0 \quad ,
\end{aligned}$$

which is only possible if $f - \lambda g = 0$. Thus, f is a constant multiple of g, namely, $f = \lambda g$.   ∎

Fourier Transforms and Square-Integrable Functions

You should realize that most, if not all, classically transformable functions that commonly arise in practice are bounded, as are their Fourier transforms. After all, if f is a classically transformable function, then it is the sum of an absolutely integrable function f_A and the Fourier transform of an absolutely integrable function. Since Fourier transforms of absolutely integrable functions are automatically bounded, f can be unbounded only if its absolutely integrable part, f_A, is unbounded. But functions that are both unbounded and absolutely integrable are rather uncommon in "real-world" applications. Indeed, the only such functions we've seen were pathologies to help illustrate where naive intuition might lead us astray (as in exercise 18.16 on page 277 and exercise 24.22 on page 398).

So what has this to do with square-integrable functions? Well, suppose f is some function arising from some "real-world" application, and you have reason to believe it is classically transformable. You may not know the formula for f (that happens in the real world), but it is a pretty safe bet that it and its Fourier transform F are both bounded functions. Now look back at theorem 25.3 on page 401. It tells us $|f|^2$ and $|F|^2$ are absolutely integrable, which is just one way of saying both f and F are square integrable. Consequently, classically transformable functions that arise in practice can usually be assumed to be square integrable.[3]

Take another look at theorem 25.3. Not only does it tell us that f and F are square integrable, it tells us that $\mathcal{F}\left[|f|^2\right] = F \star F$. Since the left-hand side of this equation is the Fourier transform of an absolutely integrable function, both sides describe a function that is well defined and continuous everywhere. In particular, we can plug in the point 0. But

$$\mathcal{F}\left[|f|^2\right]\Big|_0 = \int_{-\infty}^{\infty} |f(t)|^2 \, e^{-i2\pi \cdot 0 \cdot t} \, dt = \int_{-\infty}^{\infty} |f(t)|^2 \, dt$$

and

$$F \star F(0) = \int_{-\infty}^{\infty} F^*(\omega) F(\omega + 0) \, d\omega = \int_{-\infty}^{\infty} |F(\omega)|^2 \, d\omega \quad .$$

Thus, since $\mathcal{F}\left[|f|^2\right] = F \star F$,

$$\int_{-\infty}^{\infty} |f(t)|^2 \, dt = \mathcal{F}\left[|f|^2\right]\Big|_0 = F \star F(0) = \int_{-\infty}^{\infty} |F(\omega)|^2 \, d\omega \quad .$$

This gives us:

Theorem 25.8 (Bessel's equality)
If f and F are bounded, classically transformable functions with $F = \mathcal{F}[f]$, then both f and F are square integrable, and

$$\int_{-\infty}^{\infty} |f(t)|^2 \, dt = \int_{-\infty}^{\infty} |F(\omega)|^2 \, d\omega \quad . \tag{25.10}$$

Equation (25.10) is known as *Bessel's equality* (for square-integrable functions on the real line).[4] In terms of norms it can also be written as

$$\|f\| = \|F\| \quad . \tag{25.10'}$$

Just as the auto-correlation identities are special cases of the more general correlation identities, Bessel's equality is a special case of a more general identity involving two classically transformable functions f and g and their Fourier transforms F and G. Assuming all these functions are bounded, theorem 25.2 on correlation assures us that

$$\int_{-\infty}^{\infty} f^*(t) g(t) \, dt = \mathcal{F}\left[f^*g\right]\Big|_0$$

$$= F \star G(0) = \int_{-\infty}^{\infty} F^*(\omega) G(\omega) \, d\omega \quad ,$$

---

[3] It also turns out that every square-integrable function is "Fourier transformable" using the more general theory of Fourier transforms discussed in part IV of this text. That, along with the fact that square-integrable functions have finite "energy" (equivalently, "norms"), makes the class of square-integrable functions particularly important in real-world applications (and abstract mathematics).

[4] Don't call it Bessel *equation*. Bessel's equation is something altogether different.

giving us:

Theorem 25.9 (Parseval's equality)
Suppose f, g, F *and* G *are all bounded, classically transformable functions with* $F = \mathcal{F}[f]$
and $G = \mathcal{F}[g]$. *Then the products* f^*g *and* F^*G *are absolutely integrable, and*

$$\int_{-\infty}^{\infty} f^*(t)g(t)\,dt = \int_{-\infty}^{\infty} F^*(\omega)G(\omega)\,d\omega \quad . \tag{25.11}$$

Equation (25.11) is known as *Parseval's equality* (for square-integrable functions on the real line). Using the inner product notation from chapter 11, Parseval's equality can also be written as

$$\langle\, g \mid f \,\rangle = \langle\, G \mid F \,\rangle \quad . \tag{25.11'}$$

Keep in mind that the general validity of both Parseval's and Bessel's equalities follows from theorem 25.2, and that theorem, in turn, is based on the corresponding convolution theorem, theorem 24.10, which will not be proven until chapter 30. In the meantime, we will optimistically assume the necessary theorems supporting Parseval's and Bessel's equalities can be proven, and we will see what can be done using these equalities.

Integration Using Parseval's and Bessel's Equalities

As the next examples illustrate, Parseval's and Bessel's equalities can simplify the computation of some integrals.

!▶**Example 25.2:** *Using Bessel's equality,*

$$\int_{-\infty}^{\infty} \text{sinc}^2(2\pi t)\,dt = \int_{-\infty}^{\infty} |\text{sinc}(2\pi t)|^2\,dt$$

$$= \int_{-\infty}^{\infty} |\mathcal{F}[\text{sinc}(2\pi t)]|_\omega|^2\,d\omega = \int_{-\infty}^{\infty} \left|\frac{1}{2}\text{pulse}_1(\omega)\right|^2\,d\omega \quad .$$

But,

$$\left|\frac{1}{2}\text{pulse}_1(\omega)\right|^2 = \begin{cases} \dfrac{1}{4} & \text{if } -1 < \omega < 1 \\ 0 & \text{otherwise} \end{cases} = \frac{1}{4}\text{pulse}_1(\omega) \quad .$$

So,

$$\int_{-\infty}^{\infty} \text{sinc}^2(2\pi t)\,dt = \int_{-\infty}^{\infty} \frac{1}{4}\text{pulse}_1(\omega)\,d\omega = \frac{1}{4}\int_{-1}^{1} d\omega = \frac{1}{4}\cdot 2 = \frac{1}{2} \quad .$$

!▶**Example 25.3:** *Using Parseval's identity and tables of transforms, we have*

$$\int_{-\infty}^{\infty} \frac{1}{(2+i2\pi t)(3-i2\pi t)}\,dt = \int_{-\infty}^{\infty} \left(\frac{1}{2-i2\pi t}\right)^* \left(\frac{1}{3-i2\pi t}\right)\,dt$$

$$= \int_{-\infty}^{\infty} \left(e^{-2\omega}\text{step}(\omega)\right)^* \left(e^{-3\omega}\text{step}(\omega)\right)\,d\omega \quad .$$

But, as is easily verified,

$$\left(e^{-2\omega}\text{step}(\omega)\right)^* \left(e^{-3\omega}\text{step}(\omega)\right) = e^{-5\omega}\text{step}(\omega) \quad .$$

So,

$$\int_{-\infty}^{\infty} \frac{1}{(2+i2\pi t)(3-i2\pi t)}\, dt \;=\; \int_{-\infty}^{\infty} e^{-5\omega}\, \text{step}(\omega)\, d\omega \;=\; \int_{0}^{\infty} e^{-5\omega}\, d\omega \;=\; \frac{1}{5} \;.$$

A somewhat deeper application of Bessel's identity is given in the next subsection.

Duration and Bandwidth for Square-Integrable Functions

Given a square-integrable function $f(x)$ it is common to define the quantities E and Δx by

$$E \;=\; \|f\|^2 \;=\; \int_{-\infty}^{\infty} |f(x)|^2\, dx$$

and

$$\Delta x \;=\; \left[\frac{1}{E} \int_{-\infty}^{\infty} x^2\, |f(x)|^2\, dx \right]^{1/2} \;.$$

As already noted, in many applications E corresponds to the "energy" in the process described by f. The value Δx (which can be infinite) gives a measure of both the width of the graph of f and, as demonstrated in the following theorem and example, can be used to determine an effective duration bound for f.[5]

Theorem 25.10

Suppose f is a square-integrable function on $\mathbb{R}$. Let

$$E \;=\; \int_{-\infty}^{\infty} |f(x)|^2\, dx \qquad \text{and} \qquad \Delta x \;=\; \left[\frac{1}{E} \int_{-\infty}^{\infty} x^2\, |f(x)|^2\, dx \right]^{1/2} \;.$$

Then, for any $\alpha > 1$,

$$\int_{-\alpha\Delta x}^{\alpha\Delta x} |f(x)|^2\, dx \;\geq\; \frac{\alpha^2 - 1}{\alpha^2} E$$

and

$$\int_{-\infty}^{-\alpha\Delta x} |f(x)|^2\, dx \;+\; \int_{\alpha\Delta x}^{\infty} |f(x)|^2\, dx \;\leq\; \frac{1}{\alpha^2} E \;.$$

For a rather simple proof of this theorem, see exercise 25.14 at the end of this chapter.

▶ Example 25.4:   *Remember that an "effective duration bound" X_{eff} for a function f is a length such that the values of $f(x)$ are "significant" only when x is in the interval $(-X_{\text{eff}}, X_{\text{eff}})$. For a square-integrable function f, it is often appropriate to choose X_{eff} so that the energy in f over $(-X_{\text{eff}}, X_{\text{eff}})$ is some significant percentage of the total energy in f. Thus, if by significant percentage we mean, say, 99%, then we want to choose X_{eff} so that*

$$\int_{-X_{\text{eff}}}^{X_{\text{eff}}} |f(x)|^2\, dx \;\geq\; \frac{99}{100} E \qquad \text{where} \qquad E \;=\; \int_{-\infty}^{\infty} |f(x)|^2\, dx \;.$$

According to theorem 25.10, we can do this by choosing $X_{\text{eff}} = \alpha\Delta x$ where

$$\Delta x \;=\; \left[\frac{1}{E} \int_{-\infty}^{\infty} x^2\, |f(x)|^2\, dx \right]^{1/2} \qquad \text{and} \qquad \frac{\alpha^2 - 1}{\alpha^2} \;=\; \frac{99}{100} \;,$$

---

[5] Those acquainted with probability theory should observe that we can view $|f(x)|^2 / E$ as a probability density function, and that, using this probability density function, Δx is the standard deviation of $|x|$ from 0.

which, as you can easily determine, means choosing X_{eff} to be at least $10\Delta x$. Notice that this also means that

$$\left\| f - f \text{ pulse}_{10\Delta x} \right\|^2 = \int_{-\infty}^{\infty} \left| f(x) - f(x) \text{ pulse}_{10\Delta x}(x) \right|^2 \, dx$$

$$= \int_{-\infty}^{-10\Delta x} |f(x)|^2 \, dx + \int_{10\Delta x}^{\infty} |f(x)|^2 \, dx \leq \frac{1}{10^2} E \quad .$$

?► Exercise 25.5: *What should* α *be to ensure that*

$$\left\| f - f \text{ pulse}_{\alpha\Delta x} \right\|^2 \leq \frac{1}{10} E$$

where f *is a square-integrable function and* E *and* Δx *are as above?*

Relation between Duration and Bandwidth Measures

Since the bandwidth of a function is simply the duration of the Fourier transform of that function, the formulas above provide means for measuring both the effective duration and the effective bandwidth of any square-integrable, classically transformable function. We'll give the explicit formulas for these measures in the next theorem, the bandwidth theorem for square-integrable functions (also called the uncertainty principle).

Theorem 25.11 (bandwidth theorem / uncertainty principle)
Suppose f *and* F *are bounded, classically transformable functions with* $F = \mathcal{F}[f]$, *and let*

$$\Delta t = \left[\frac{1}{E} \int_{-\infty}^{\infty} t^2 |f(t)|^2 \, dt \right]^{1/2} \quad \text{and} \quad \Delta\omega = \left[\frac{1}{E} \int_{-\infty}^{\infty} \omega^2 |F(\omega)|^2 \, d\omega \right]^{1/2}$$

where

$$E = \int_{-\infty}^{\infty} |f(t)|^2 \, dt = \int_{-\infty}^{\infty} |F(\omega)|^2 \, d\omega \quad .$$

Further assume that neither f *nor* F *is the zero function, and that either*

1. Δt *or* $\Delta\omega$ *is infinite,*

or that

2. $\omega F(\omega)$ *is classically transformable, and both* f' *and* $\omega F(\omega)$ *are bounded,*

or that

3. $tf(t)$ *is classically transformable, and both* F' *and* $tf(t)$ *are bounded.*

Then

$$\Delta t \, \Delta\omega \geq \frac{1}{4\pi} \tag{25.12}$$

with equality holding if and only if

$$f(t) = Ae^{-\gamma t^2}$$

for some pair of constants A *and* γ *with* $\gamma > 0$.

Most of the details of the proof will be developed shortly in a series of exercises. A few observations are in order, however, before starting those exercises. The first is that, since f and F are bounded, classically transformable functions with $F = \mathcal{F}[f]$, theorem 25.8 assures us that they are square-integrable functions and that the formula(s) given for E are valid and finite. The second observation is that inequality (25.12) cannot hold if f or F is the zero function. After all, if f or its transform, F, is the zero function, then both must be the zero function and we would obviously have $\Delta t\, \Delta \omega = 0$ instead of inequality (25.12). That is why we insisted that neither f nor F be the zero function. It also ensures that the integrals defining Δt and $\Delta \omega$ are either positive or are infinite. A third observation is that, if either Δt and $\Delta \omega$ is infinite, then so is their product, and the claim of the theorem, inequality (25.12), follows trivially.

Proving the bandwidth theorem under either of the other two sets of assumptions is a bit trickier. In the following exercises, we'll assume the second set holds. (To prove the theorem assuming the third set holds, just repeat what follows with the roles of f and F reversed.) Before starting these exercises, however, let's make one more observation that will be useful for these exercises: According to the second set of assumptions, $\omega F(\omega)$ is classically transformable. Thus, by theorem 22.1 on page 341, we know f is continuous and piecewise smooth, and that f' is classically transformable.

?▶ Exercise 25.6 (proof of the bandwidth theorem):   *In the following, let f, F, E, Δt and $\Delta \omega$ be as in theorem 25.11, and assume that condition set (2) in that theorem holds. Since we have already considered the cases where either Δt or $\Delta \omega$ is infinite or zero, go ahead and assume both are finite and positive. For convenience, let*

$$g(t) \;=\; |f(t)|^2 \qquad \text{and} \qquad G \;=\; \mathcal{F}[g] \quad .$$

(Don't forget to verify that any identity used is valid under the given circumstances.)

a: *Verify that*

$$E^2 \;=\; \left| \mathcal{F}\big[t g'(t)\big]\big|_0 \right|^2$$

by first using the appropriate differential identities to obtain

$$\mathcal{F}\big[t g'(t)\big]\big|_0 \;=\; -E \quad .$$

b: *Using the above result and the Schwarz inequality, show that*

$$E \;\le\; 4(\Delta t)^2 \int_{-\infty}^{\infty} |f'(t)|^2 \, dt$$

with equality holding if and only if

$$f(t) \;=\; A e^{-\gamma t^2}$$

for some pair of constants A and γ with $\gamma > 0$. (To show γ must be real, it may help to first show that

$$\frac{d}{dt}\left| e^{-\gamma t^2} \right|^2 \;=\; 2\left| e^{-\gamma t^2} 2\gamma t e^{-\gamma t^2} \right|$$

if and only if γ is real.)

c: *Using a differential identity and an appropriate identity from this chapter, show that*

$$\int_{-\infty}^{\infty} |f'(t)|^2 \, dt \;=\; (2\pi \, \Delta \omega)^2 E \quad .$$

d: *Finally, using the results just derived, confirm that*

$$\Delta t\, \Delta \omega \;\ge\; \frac{1}{4\pi}$$

*with equality holding if and only if f is the Gaussian described in part **b** of this exercise set.*

Refinements

Our measures of duration and bandwidth can be refined by taking into account the fact that the actual intervals of effective durations might be centered about points other than the origin. We do this by defining the center $\bar{x}$ of the (effective) interval of duration for $f(x)$ by

$$\bar{x} = \frac{1}{E} \int_{-\infty}^{\infty} x \, |f(x)|^2 \, dx \quad .$$

The corresponding measure of duration is then given by

$$\Delta x = \left[\frac{1}{E} \int_{-\infty}^{\infty} (x - \bar{x})^2 \, |f(x)|^2 \, dx \right]^{1/2} \quad .$$

As before, E is the square of the norm of f .[6]

By straightforward use of translation we can easily "refine" the theorems just developed to take into account the possibly off-centered intervals of effective duration. Those refined versions are given below, with the proofs being left to the reader.

Theorem 25.12
Suppose f is a square-integrable function on $\mathbb{R}$ and let

$$E = \int_{-\infty}^{\infty} |f(x)|^2 \, dx \quad .$$

If

$$\bar{x} = \frac{1}{E} \int_{-\infty}^{\infty} x \, |f(x)|^2 \, dx \quad \text{and} \quad \Delta x = \left[\frac{1}{E} \int_{-\infty}^{\infty} (x - \bar{x})^2 \, |f(x)|^2 \, dx \right]^{1/2}$$

are finite, then, for any $\alpha > 1$,

$$\int_{\bar{x}-\alpha\Delta x}^{\bar{x}+\alpha\Delta x} |f(x)|^2 \, dx \geq \frac{\alpha^2 - 1}{\alpha^2} E$$

and

$$\int_{-\infty}^{\bar{x}-\alpha\Delta x} |f(x)|^2 \, dx + \int_{\bar{x}+\alpha\Delta x}^{\infty} |f(x)|^2 \, dx \leq \frac{1}{\alpha^2} E \quad .$$

Theorem 25.13 (bandwidth theorem / uncertainty principle)
Suppose f and F are bounded, classically transformable functions with $F = \mathcal{F}[f]$, and let

$$\bar{t} = \frac{1}{E} \int_{-\infty}^{\infty} t \, |f(t)|^2 \, dt \quad , \quad \Delta t = \left[\frac{1}{E} \int_{-\infty}^{\infty} (t - \bar{t})^2 \, |f(t)|^2 \, dt \right]^{1/2} \quad ,$$

$$\bar{\omega} = \frac{1}{E} \int_{-\infty}^{\infty} \omega \, |f(\omega)|^2 \, d\omega \quad \text{and} \quad \Delta\omega = \left[\frac{1}{E} \int_{-\infty}^{\infty} (\omega - \bar{\omega})^2 \, |F(\omega)|^2 \, d\omega \right]^{1/2}$$

---

[6] Again, those acquainted with probability should verify for themselves that $|f(x)|^2 / E$ can be treated as a probability density function for a real random variable x, and that $\bar{x}$ and Δx are then, respectively, the expected value and standard deviation of x .

where

$$E = \int_{-\infty}^{\infty} |f(t)|^2 \, dt = \int_{-\infty}^{\infty} |F(\omega)|^2 \, d\omega \quad .$$

Assume these integrals exist and are finite. Assume further that neither f nor F is the zero function and either that

1. *$\omega F(\omega)$ is classically transformable, and both f' and $\omega F(\omega)$ are bounded*

or that

2. *$t f(t)$ is classically transformable, and both F' and $t f(t)$ are bounded.*

Then

$$\Delta t \, \Delta \omega \geq \frac{1}{4\pi}$$

with equality holding if and only if

$$f(t) = A e^{i2\pi \bar{\omega} t} e^{-\gamma (t - \bar{t})^2}$$

for some pair of constants A and γ with $\gamma > 0$.

?▶Exercise 25.7   **a:** *Prove theorem 25.12 using translation and theorem 25.10.*

   b: *Prove theorem 25.13 using translation and theorem 25.11.*

25.3   The Fundamental Identity

The main identity of this section may seem rather innocuous and less useful than the other equations we've derived thus far. Truth is, it is not an identity you commonly see in applications, at least not in the form given here. However, it will play a fundamental role in further developing the basic mathematics of Fourier analysis. In particular, it will be useful in verifying several major theorems we have delayed proving, and, along with certain facts concerning Gaussian functions, it will lead to a new definition for Fourier transforms that will allow us to view many more functions of interest as being "Fourier transformable".

We'll first state the fundamental identity and then, in a little bit, discuss when this identity is valid.

Let f, g, F and G be classically transformable functions with, as usual, $F = \mathcal{F}[f]$ and $G = \mathcal{F}[g]$. The *fundamental identity* (of Fourier analysis) is

$$\int_{-\infty}^{\infty} F(x) g(x) \, dx = \int_{-\infty}^{\infty} f(y) G(y) \, dy \quad . \tag{25.13a}$$

Observe that this identity can also be expressed as

$$\int_{-\infty}^{\infty} \mathcal{F}[f]|_x \, g(x) \, dx = \int_{-\infty}^{\infty} f(y) \mathcal{F}[g]|_y \, dy \tag{25.13b}$$

or as

$$\int_{-\infty}^{\infty} F(x) \mathcal{F}^{-1}[G]|_x \, dx = \int_{-\infty}^{\infty} \mathcal{F}^{-1}[F]|_y \, G(y) \, dy \quad . \tag{25.13c}$$

You should also notice that here it is impossible to follow the convention of t denoting the variable in the "untransformed" functions and ω denoting the variable in the "transformed" functions. While we are on the subject, it may be wise to observe that the variables appearing in the fundamental identity are all variables internal to the integrals (i.e., they are dummy variables). So the variables appearing on one side of the identity have no direct relationship with the variables appearing on the other side. We can, just as correctly, write the fundamental identity as

$$\int_{-\infty}^{\infty} F(s)g(s)\,ds = \int_{-\infty}^{\infty} f(s)G(s)\,ds \quad .$$

Using different symbols for the internal variables, however, will simplify keeping track of the formulas involved in some of the manipulations we will be doing later.

The astute reader may also notice a similarity between the fundamental identity and Parseval's equality. In fact, each can be derived from the other. However, because of the importance of the fundamental identity to the basic development of the advanced theory of Fourier analysis, and because we will use the fundamental identity to verify theorems from which we obtained Parseval's equality, we will derive our results for the fundamental identity independently.

Some conditions under which the fundamental identity does hold are described in the next theorem.

Theorem 25.14 (first theorem on the fundamental identity of Fourier analysis)
Let f, g, F and G be classically transformable functions with $F = \mathcal{F}[f]$ and $G = \mathcal{F}[g]$. Then the products fG and Fg are absolutely integrable, and

$$\int_{-\infty}^{\infty} F(x)g(x)\,dx = \int_{-\infty}^{\infty} f(y)G(y)\,dy$$

whenever any one of the following sets of conditions holds:

1. *Both f and g are absolutely integrable.*

2. *Both F and G are absolutely integrable.*

3. *Both g and G are absolutely integrable.*

4. *Both f and F are absolutely integrable.*

We will confirm this theorem's claims assuming that either the first or third set of conditions holds. The proofs assuming the second or fourth sets of conditions are very similar and will be left as exercises.

PROOF *(theorem 25.14, assuming f and g are absolutely integrable)*:   Since f and g are absolutely integrable, F and G are the bounded functions given by

$$F(x) = \int_{-\infty}^{\infty} f(y)\,e^{-i2\pi xy}\,dy \quad \text{and} \quad G(y) = \int_{-\infty}^{\infty} g(x)\,e^{-i2\pi xy}\,dx \quad .$$

Thus, Fg and fG, being products of absolutely integrable functions with bounded functions, are each absolutely integrable. Moreover

$$\int_{-\infty}^{\infty} F(x)g(x)\,dx = \int_{-\infty}^{\infty} \left(\int_{-\infty}^{\infty} f(y)\,e^{-i2\pi xy}\,dy \right) g(x)\,dx$$

$$= \int_{-\infty}^{\infty} \int_{-\infty}^{\infty} f(y)g(x)\,e^{-i2\pi xy}\,dy\,dx$$

and

$$\int_{-\infty}^{\infty} f(y)G(y)\,dy = \int_{-\infty}^{\infty} f(y)\left(\int_{-\infty}^{\infty} g(x)\,e^{-i2\pi xy}\,dx\right)dy$$

$$= \int_{-\infty}^{\infty}\int_{-\infty}^{\infty} f(y)g(x)\,e^{-i2\pi xy}\,dx\,dy \quad.$$

Reviewing corollary 18.24 on page 275 (concerning interchanging the order of integration of double integrals over the XY–plane), we see that this is one of the cases where the order of integration can be interchanged (in that corollary, let $g(x)$, $h(y)$, $v(s)$ and $\phi(x,y)$ be $g(x)$, $f(y)$, 1 and $e^{-i2\pi xy}$, respectively). So,

$$\int_{-\infty}^{\infty} F(x)g(x)\,dx = \int_{-\infty}^{\infty}\int_{-\infty}^{\infty} f(y)g(x)\,e^{-i2\pi xy}\,dy\,dx$$

$$= \int_{-\infty}^{\infty}\int_{-\infty}^{\infty} f(y)g(x)\,e^{-i2\pi xy}\,dx\,dy = \int_{-\infty}^{\infty} f(y)G(y)\,dy \quad. \qquad \blacksquare$$

?►Exercise 25.8: *Prove theorem 25.14 assuming both F and G are absolutely integrable.*

PROOF (theorem 25.14, assuming g and G are absolutely integrable):   In this case we use the fact that, because f is classically transformable, it and its transform, F, can be written as

$$f = f_A + f_T \qquad \text{and} \qquad F = F_T + F_A$$

where f_A and F_A are absolutely integrable functions and

$$F_T = \mathcal{F}[f_A] \qquad \text{and} \qquad F_A = \mathcal{F}[f_T] \quad.$$

From the first part of this theorem (proven above), we already know that, since f_A and g are absolutely integrable,

$$\int_{-\infty}^{\infty} F_T(x)g(x)\,dx = \int_{-\infty}^{\infty} f_A(y)G(y)\,dy \quad.$$

Likewise, from the second part (proven in exercise 25.8), we know that, since F_A and G are absolutely integrable,

$$\int_{-\infty}^{\infty} F_A(x)g(x)\,dx = \int_{-\infty}^{\infty} f_T(y)G(y)\,dy \quad.$$

So,

$$\int_{-\infty}^{\infty} F(x)g(x)\,dx = \int_{-\infty}^{\infty} [F_T(x) + F_A(x)]\,g(x)\,dx$$

$$= \int_{-\infty}^{\infty} F_T(x)g(x)\,dx + \int_{-\infty}^{\infty} F_A(x)g(x)\,dx$$

$$= \int_{-\infty}^{\infty} f_A(y)G(y)\,dy + \int_{-\infty}^{\infty} f_T(y)G(y)\,dy$$

$$= \int_{-\infty}^{\infty} [f_A(y) + f_T(y)]\,G(y)\,dy = \int_{-\infty}^{\infty} f(y)G(y)\,dy \quad. \qquad \blacksquare$$

?►Exercise 25.9: *Prove theorem 25.14 assuming both f and F are absolutely integrable.*

Some other conditions under which the fundamental identity holds are given in the next theorem. It is closely related to the second theorem on convolutions (theorem 24.10 on page 394) and will be proven, along with theorem 24.10, in chapter 30.

Theorem 25.15 (second theorem on the fundamental identity of Fourier analysis)
Let f, g, F and G be bounded, classically transformable functions with $F = \mathcal{F}[f]$ and $G = \mathcal{F}[g]$. Then the products fG and Fg are absolutely integrable, and

$$\int_{-\infty}^{\infty} F(x)g(x)\,dx = \int_{-\infty}^{\infty} f(y)G(y)\,dy \quad .$$

The really important applications of the fundamental identity will have to wait, but in the following example we will show how this identity can be used to reduce the work in evaluating certain integrals.

!▶ Example 25.5: *Consider evaluating*

$$\int_{-\infty}^{\infty} \frac{2\operatorname{sinc}(2\pi x)}{3 - ix}\,dx \quad .$$

We know

$$2\operatorname{sinc}(2\pi x) = \mathcal{F}\left[\operatorname{pulse}_1(y)\right]\big|_x \quad and \quad \mathcal{F}\left[\frac{1}{3 - ix}\right]\bigg|_y = 2\pi e^{-6\pi y}\operatorname{step}(y) \quad .$$

So, by the fundamental identity,

$$\int_{-\infty}^{\infty} \frac{2\operatorname{sinc}(2\pi x)}{3 - ix}\,dx = \int_{-\infty}^{\infty} 2\operatorname{sinc}(2\pi x)\frac{1}{3 - ix}\,dx$$

$$= \int_{-\infty}^{\infty} \mathcal{F}\left[\operatorname{pulse}_1(y)\right]\big|_x \frac{1}{3 - ix}\,dx$$

$$= \int_{-\infty}^{\infty} \operatorname{pulse}_1(y)\,\mathcal{F}\left[\frac{1}{3 - ix}\right]\bigg|_y dy$$

$$= \int_{-1}^{1} 2\pi e^{-6\pi y}\operatorname{step}(y)\,dy = 2\pi \int_{0}^{1} e^{-6\pi y}\,dy \quad .$$

This last integral, unlike the integral we began with, is easily evaluated. Evaluating it (and cutting out the middle of the above sequence of equalities) leaves us with

$$\int_{-\infty}^{\infty} \frac{2\operatorname{sinc}(2\pi x)}{3 - ix}\,dx = \frac{1}{3}\left[1 - e^{-6\pi}\right] \quad .$$

?▶ Exercise 25.10: *Show that*

$$\int_{-\infty}^{\infty} [\operatorname{sinc}(2\pi x)]^2\,dx = \frac{1}{2} \quad .$$

Additional Exercises

25.11. Using either the integral formula for correlation or its relation with convolutions, find $f \star g(x)$ for each of the following choices of $f(x)$ and $g(x)$:

   a. $f(x) = e^{-3x}\,\text{step}(x)$, $g(x) = e^{i2\pi x}$

   b. $f(x) = e^{i2\pi x}$, $g(x) = e^{-3x}\,\text{step}(x)$

   c. $f(x) = e^{-3x}\,\text{step}(x)$, $g(x) = e^{-3x}\,\text{step}(x)$

25.12. Prove theorem 25.4 on page 402 using the version of the Schwarz inequality given in theorem 25.7 on page 404.

25.13. Evaluate each of the following integrals using either Bessel's or Parseval's equality:

   a. $\displaystyle\int_{-\infty}^{\infty} \text{sinc}(2\pi\alpha x)\,\text{sinc}(2\pi\beta x)\,dx$ where $0 < \alpha \le \beta$

   b. $\displaystyle\int_{-\infty}^{\infty} \frac{\text{sinc}(2\pi x)}{1+x^2}\,dx$ **c.** $\displaystyle\int_{-\infty}^{\infty} \frac{1}{(6+i2\pi x)^2}\,dx$

   d. $\displaystyle\int_{-\infty}^{\infty} \frac{1}{\left(\pi^2 + x^2\right)^2}\,dx$ **e.** $\displaystyle\int_{-\infty}^{\infty} \frac{x^2}{\left(\pi^2 + x^2\right)^2}\,dx$

   f. $\displaystyle\int_{-\infty}^{\infty} \frac{1}{(3+i2\pi x)(5-i2\pi x)}\,dx$ **g.** $\displaystyle\int_{-\infty}^{\infty} \frac{1}{(3+i2\pi x)(5+i2\pi x)}\,dx$

25.14. Suppose f is a square-integrable function, and E and Δx are as in theorem 25.10 on page 408. Assume Δx is finite and nonzero. In addition, let $\alpha > 1$ and, for convenience, let

$$E_\alpha = \int_{-\alpha\Delta x}^{\alpha\Delta x} |f(x)|^2\,dx \quad , \quad X_\alpha = \left[\frac{1}{E_\alpha}\int_{-\alpha\Delta x}^{\alpha\Delta x} x^2\,|f(x)|^2\,dx\right]^{1/2} \quad ,$$

$$\mathcal{E}_\alpha = \int_{-\infty}^{-\alpha\Delta x} |f(x)|^2\,dx + \int_{\alpha\Delta x}^{\infty} |f(x)|^2\,dx \quad ,$$

and

$$\xi_\alpha = \left[\frac{1}{\mathcal{E}_\alpha}\int_{-\infty}^{-\alpha\Delta x} x^2\,|f(x)|^2\,dx + \frac{1}{\mathcal{E}_\alpha}\int_{\alpha\Delta x}^{\infty} x^2\,|f(x)|^2\,dx\right]^{1/2} \quad .$$

   a. Verify the following system of equalities and inequalities:

$$E_\alpha + \mathcal{E}_\alpha = E \quad ,$$

$$(X_\alpha)^2\,E_\alpha + (\xi_\alpha)^2\,\mathcal{E}_\alpha = (\Delta x)^2 E \quad ,$$

$$X_\alpha \le \alpha\Delta x \le \xi_\alpha \quad .$$

   b. Using the above system, show that

$$\mathcal{E}_\alpha \le \frac{1}{\alpha^2} E \quad \text{and} \quad E_\alpha \ge \frac{\alpha^2 - 1}{\alpha^2} E \quad .$$

c. *Now verify the claims in theorem 25.10.*

25.15. *Let ϵ be any (small) positive value and consider the function*

$$f(x) = \begin{cases} 0 & \text{if } -1 < x < 1 \\ |x|^{-(3+\epsilon)/2} & \text{otherwise} \end{cases}.$$

a. *Verify that f is square integrable and compute E and Δx (where E and Δx are as in theorem 25.10).*

b. *Show that, for all $\alpha > 1$,*

$$\int_{-\infty}^{-\alpha \Delta x} |f(x)|^2 \, dx + \int_{\alpha \Delta x}^{\infty} |f(x)|^2 \, dx = \frac{1}{\alpha^{2+\epsilon}} EM$$

where M is some constant between 0 and 1. Compare this expression to the bounds in theorem 25.10.

25.16. *For each of the functions given below,*

1. *verify that the function and its Fourier transform are square integrable;*

2. *compute the measures of duration and bandwidth (Δt and $\Delta \omega$ as defined in theorem 25.11 on page 409), and*

3. *show that $\Delta t \, \Delta \omega \geq (4\pi)^{-1}$, as claimed in the bandwidth theorem.*

a. $\text{pulse}_1(t)$ b. $e^{-4\pi t^2}$ c. $\dfrac{1}{\pi^2 + t^2}$

25.17. *Let f and h be two classically transformable functions, and consider the equation*

$$\int_{-\infty}^{\infty} f(x)h(x) \, dx = \int_{-\infty}^{\infty} \mathcal{F}[f]|_y \, \mathcal{F}^{-1}[h]|_y \, dy \quad .$$

Verify that this equation is equivalent to the fundamental identity of Fourier analysis by

a. *first deriving this equation from the fundamental identity, and then*

b. *deriving the fundamental identity from this equation.*

25.18. *Determine some conditions under which the equation in the previous problem is valid.*

25.19. *Let f, h, F and H be classically transformable functions with $F = \mathcal{F}[f]$ and $H = \mathcal{F}[h]$. Verify that the equation*

$$\int_{-\infty}^{\infty} f(x)h(x) \, dx = \int_{-\infty}^{\infty} F(y)H(-y) \, dy$$

is equivalent to the fundamental identity of Fourier analysis.

25.20. *Derive the fundamental identity of Fourier analysis from the identity*

$$\mathcal{F}^{-1}[FH] = f * h$$

where $F = \mathcal{F}[f]$ and $H = \mathcal{F}[h]$. (Start by letting $G = h$.)

25.21. Evaluate the following integrals using the fundamental identity of Fourier analysis. (*Yes, they are the very same integrals as in exercise 25.13.*)

a. $\displaystyle\int_{-\infty}^{\infty} \mathrm{sinc}(2\pi\alpha x)\,\mathrm{sinc}(2\pi\beta x)\,dx$ where $0 < \alpha \le \beta$

b. $\displaystyle\int_{-\infty}^{\infty} \frac{\mathrm{sinc}(2\pi x)}{1 + x^2}\,dx$

c. $\displaystyle\int_{-\infty}^{\infty} \frac{1}{(6 + i2\pi x)^2}\,dx$

d. $\displaystyle\int_{-\infty}^{\infty} \frac{1}{(\pi^2 + x^2)^2}\,dx$

e. $\displaystyle\int_{-\infty}^{\infty} \frac{x^2}{(\pi^2 + x^2)^2}\,dx$

f. $\displaystyle\int_{-\infty}^{\infty} \frac{1}{(3 + i2\pi x)(5 - i2\pi x)}\,dx$

g. $\displaystyle\int_{-\infty}^{\infty} \frac{1}{(3 + i2\pi x)(5 + i2\pi x)}\,dx$

26

Generalizing the Classical Theory: A Naive Approach

Many functions encountered in everyday applications — including all nonzero constant functions periodic functions, exponentials and polynomials — are *not* classically transformable. As a result, the purely classical theory for Fourier transforms is too limited for many applications. Fortunately, there is a more general theory under which many more functions, including all of those mentioned above, are "Fourier transformable". This theory, which we will refer to as the "generalized theory" since it generalizes the classical theory, will be developed, as rigorously and completely as we can, in part IV of this text.

Admittedly, however, a reasonably rigorous and complete development of the generalized theory is a significant undertaking and may require more time or effort than some readers may wish to spend at this point in their careers. This chapter is for those readers. Here we will develop elements of the generalized theory by using mathematical entities called "delta functions". Although not truly functions in the classical sense, delta functions can, with some precautions, be treated as functions and can be used to derive a generalized theory in which constants and periodic functions are "Fourier transformable". They will also allow us to obtain a useful generalization of the classical concept of differentiation.

Of course, with a naive theory, some of the basic concepts must be naively defined. One of those is that of "Fourier transformability". Basically, something is *Fourier transformable* if its Fourier transform can be defined in some legitimate manner. Ultimately, that legitimate manner will be described in part IV. Until then, you can view the set of Fourier transformable functions as consisting of all classically transformable functions along with everything we add in this chapter.

26.1 Delta Functions
Derivation of the Working Definition

Let a be some real number and suppose (i.e., pretend) that the Fourier transform of $e^{i2\pi at}$ can be defined, somehow. Denote this transform, $\mathcal{F}\left[e^{i2\pi at}\right]$, by δ_a. Observe that, if g is any other transformable function with $G = \mathcal{F}[g]$, then, naively using a convolution identity,

$$\mathcal{F}\left[g(t)e^{i2\pi at}\right]\Big|_{\omega} = G * \delta_a(\omega) = \int_{-\infty}^{\infty} G(\omega - s)\delta_a(s)\,ds \quad .$$

On the other hand, using a translation identity,

$$\mathcal{F}\left[g(t)e^{i2\pi at}\right]\Big|_{\omega} = G(\omega - a) \quad .$$

Combining these two computations gives us

$$\int_{-\infty}^{\infty} G(\omega - s)\delta_a(s)\,ds = G(\omega - a) \quad .$$

For convenience, let ω be fixed and define ϕ by $\phi(s) = G(\omega - s)$. The last equation can then be written as

$$\int_{-\infty}^{\infty} \phi(s)\delta_a(s)\,ds = \phi(a) \quad . \tag{26.1}$$

Since G and ω are fairly arbitrary, so is ϕ (though we should limit ourselves to ϕ's that are continuous at a to ensure that $\phi(a)$ is well defined).

Equation (26.1) was derived assuming δ_a exists and is the Fourier transform of $e^{i2\pi a t}$. Conversely, as it turns out, equation (26.1) describes a property that can define δ_a. Since there are advantages to defining δ_a via this property, we will do so.

The Working Definition

Let a be some real number. The *delta function* at a, denoted by δ_a, is the "function" such that, if ϕ is any function on the real line continuous at a, then

$$\int_{-\infty}^{\infty} \phi(s)\delta_a(s)\,ds = \phi(a) \quad . \tag{26.2}$$

For convenience, the delta function at 0 will often be referred to as *the* delta function and will be denoted by δ (i.e., $\delta = \delta_0$).

You may wonder why we define a delta function δ_a this way instead of just giving a formula for computing the values of $\delta_a(x)$ for different values of x. Here's why: No such formula exists. Indeed, strictly speaking, there is no function satisfying the above definition for δ_a. We will verify this in a little bit when we try to visualize delta functions.

Delta Functions as Generalized Functions

It turns out that a delta function corresponds to something we will be calling a generalized function in the next part of this book. For now, think of a generalized function f as something for which the integral

$$\int_{-\infty}^{\infty} \phi(s)f(s)\,ds$$

somehow "makes sense" whenever ϕ is a suitably continuous and integrable function on the real line. When dealing with delta functions, "suitably continuous and integrable" can be interpreted to simply mean "continuous at whatever points we have delta functions".[1]

Since generalized functions might only be defined in terms of integrals, we must use those integrals to determine when two expressions describe the same generalized function. Accordingly, we say that two generalized functions f and g are equal (and write $f = g$) if and only if

$$\int_{-\infty}^{\infty} \phi(s)f(s)\,ds = \int_{-\infty}^{\infty} \phi(s)g(s)\,ds \tag{26.3}$$

for every suitably integrable and continuous function ϕ.

---

[1] Yes, this is all pretty vague. If you really want to know what it means for the integral to "make sense" and when a given ϕ is "suitably continuous and integrable", then you will just have to read chapters 32 and 33.

!▶ *Example 26.1:* *Let's compare δ_a with the translation of δ by a where a is some fixed real number. Assume ϕ is any function on the real line that is continuous at a. Then, by the definition of δ_a,*

$$\int_{-\infty}^{\infty} \phi(s)\delta_a(s)\,ds \;=\; \phi(a) \quad .$$

Using the change of variable $x = s - a$, we also have

$$\int_{-\infty}^{\infty} \phi(s)\delta(s-a)\,ds \;=\; \int_{-\infty}^{\infty} \phi(x+a)\delta(x)\,dx \;=\; \phi(0+a) \;=\; \phi(a) \quad .$$

Thus,

$$\int_{-\infty}^{\infty} \phi(s)\delta(s-a)\,ds \;=\; \int_{-\infty}^{\infty} \phi(s)\delta_a(s)\,ds$$

for every function ϕ on the real line that is continuous at a. This tells us that, as generalized functions, $\delta(s-a) = \delta_a(s)$.

The fact that every delta function is a translation of the delta function will be used often and without comment in many of our later discussions.

?▶ *Exercise 26.1* *a: Show that δ is even. That is, show that $\delta(-s) = \delta(s)$ as a generalized function. (Remember, this means you must show that*

$$\int_{-\infty}^{\infty} \phi(s)\delta(-s)\,ds \;=\; \int_{-\infty}^{\infty} \phi(s)\delta(s)\,ds$$

for every function ϕ on the real line that is continuous at 0.)

b: Is δ_a even if $a \neq 0$?

Visualizing Delta Functions

The delta functions can also be viewed as representing certain limiting processes, and that, in turn, leads to a naive "graphical visualization" that can be insightful. Since any delta function δ_a is just a translation of the delta function δ, we'll just describe this limiting process for δ. You, yourself, can do the "translation by a" to get the corresponding limiting process and visualization for δ_a.

The Delta Function as a Limiting Process

There can be several versions of this limiting process. Our version involves a set of functions $\{\psi_\epsilon : \epsilon > 0\}$ with[2]

$$\psi_\epsilon(s) \;=\; \frac{1}{2\epsilon}\,\mathrm{pulse}_\epsilon(s) \;=\; \begin{cases} \dfrac{1}{2\epsilon} & \text{if } -\epsilon < s < \epsilon \\[2mm] 0 & \text{otherwise} \end{cases} \quad .$$

Now let be ϕ any piecewise continuous, real-valued function continuous at 0. This means ϕ is uniformly continuous on $(-\epsilon, \epsilon)$ whenever ϵ is less than the distance between 0 and the closest

---

[2] This set of functions is a simple example of an "identity sequence". We will discuss identity sequences in greater detail and generality in chapter 29.

discontinuity of ϕ, and the mean value theorem for integrals (theorem 4.3 on page 40) then tells us that, for some $\bar{s}_\epsilon$ with $-\epsilon \leq \bar{s}_\epsilon \leq \epsilon$,

$$\int_{-\infty}^{\infty} \phi(s)\psi_\epsilon(s)\,ds = \int_{-\infty}^{\infty} \phi(s)\frac{1}{2\epsilon}\,\text{pulse}_\epsilon(s)\,ds$$

$$= \frac{1}{2\epsilon}\int_{-\epsilon}^{\epsilon} \phi(s)\,ds$$

$$= \frac{1}{2\epsilon} \times [\epsilon - (-\epsilon)]\phi(\bar{s}_\epsilon) = \phi(\bar{s}_\epsilon) \quad .$$

Since $-\epsilon \leq \bar{s}_\epsilon \leq \epsilon$ and ϕ is continuous at 0, it follows that

$$\lim_{\epsilon \to 0^+} \int_{-\infty}^{\infty} \phi(s)\psi_\epsilon(s)\,ds = \lim_{\epsilon \to 0^+} \phi(\bar{s}_\epsilon) = \phi(0) \quad .$$

Of course, we assumed ϕ was real valued just so we could apply the mean value theorem. If ϕ is a complex-valued, piecewise continuous function continuous at 0, then its real part u and imaginary part v are, themselves, real-valued, piecewise continuous functions continuous at 0, and, by the above,

$$\lim_{\epsilon \to 0^+} \int_{-\infty}^{\infty} \phi(s)\,\psi_\epsilon(s)\,ds = \lim_{\epsilon \to 0^+} \int_{-\infty}^{\infty} [u(s)+iv(s)]\,\psi_\epsilon(s)\,ds$$

$$= \lim_{\epsilon \to 0^+} \int_{-\infty}^{\infty} u(s)\,\psi_\epsilon(s)\,ds$$

$$+ i\lim_{\epsilon \to 0^+} \int_{-\infty}^{\infty} v(s)\,\psi_\epsilon(s)\,ds$$

$$= u(0) + iv(0)$$

$$= \phi(0) \quad .$$

But, by our original working definition of the delta function,

$$\phi(0) = \int_{-\infty}^{\infty} \phi(s)\delta(s)\,ds \quad .$$

Thus, for every piecewise continuous function on $\mathbb{R}$ continuous at 0,

$$\lim_{\epsilon \to 0^+} \int_{-\infty}^{\infty} \phi(s)\,\psi_\epsilon(s)\,ds = \int_{-\infty}^{\infty} \phi(s)\delta(s)\,ds \quad . \tag{26.4}$$

Because of equation (26.4), it is often claimed that the delta function is the limit of these ψ_ϵ's. We'll soon see that, within the confines of the classical theory, there are problems with this claim. But there is little harm in viewing these ψ_ϵ's as *approximations* to the delta function, at least when ϵ is small.

Graphically Visualizing Delta Functions

For a little more visual insight, let's look at the graph of ψ_ϵ when ϵ is close to 0. For such values of ϵ the corresponding values of $\psi_\epsilon(x)$ are large when x is inside the small interval $(-\epsilon, \epsilon)$ and are zero outside this interval. Moreover, each of these graphs encloses a region of unit area (see figure 26.1a) and looks like a very tall "spike" above $s = 0$.[3] Because of this, the delta function is

---

[3] Similar graphs are obtained using other identity sequences.

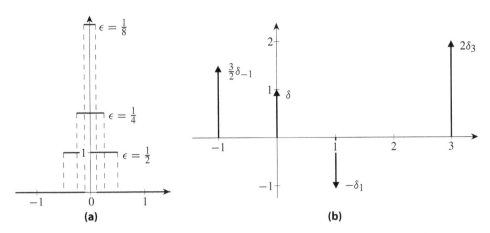

Figure 26.1: Delta functions — (a) viewed as a limit of pulse functions, and (b) graphically represented using vertical arrows.

often visualized as being zero everywhere except at the origin, at which its graph has "an infinitely high, infinitesimally narrow spike of unit area".

Aside from being mathematically questionable, an infinitely high, infinitesimally narrow spike of unit area is very difficult to graph accurately. As an alternative, it is standard practice to *graphically represent* the delta function by an upward pointing unit arrow starting on the X–axis at $x = 0$. More generally, the graphical representation of $\beta\delta_a(x)$ is a vertical arrow of length $|\beta|$ starting on the X–axis at $x = a$. It points up if $\beta > 0$ and down if $\beta < 0$ provided β is real (see figure 26.1b). If β is complex, then we graphically represent the real and imaginary parts of $\beta\delta_a(x)$, $\text{Re}[\beta]\delta_a(x)$ and $\text{Im}[\beta]\delta_a(x)$, on separate coordinate systems.

You can now see why we did not attempt to give a formula for computing the values of $\delta(x)$ for various values of x. If such a formula did exist, and if our above "graphs" are accurate, then that formula would have to be $\delta(x) = 0$ whenever $x \neq 0$ (that would also be the formula if our graphs are not accurate — see exercise 29.20). But then

$$\int_{-\infty}^{\infty} \phi(s)\delta_a(s)\,ds = \int_{-\infty}^{0} \phi(s) \cdot 0\,ds + \int_{0}^{\infty} \phi(s) \cdot 0\,ds = 0$$

for any function ϕ on the real line. In particular, using $\phi(s) = \cos(s)$ and our working definition of δ (equation (26.2)), this last equation implies that

$$1 = \cos(0) = \int_{-\infty}^{\infty} \cos(s)\delta(s)\,ds = 0 \quad !$$

Obviously then, the claim that "$\delta(s)$ is a function that vanishes whenever $x \neq 0$" should not be taken too seriously.

Analysis with the Delta Function

For the rest of this chapter we will adopt the naive view that delta functions exist and, at least within integrals, can be treated as fairly ordinary "integrable" functions (with extraordinary properties). We will also view them as being Fourier transformable, and will assume that the identities we showed to be valid for all classically transformable functions remain valid for all Fourier transformable functions. Naturally, all the results naively derived here will be justified in our more complete discussions in part IV.

By the way, in all of the following, a is some real number.

Fourier Transforms

By the integral formulas for the Fourier transforms and the working definition for the delta functions,

$$\mathcal{F}[\delta_a]|_\omega = \int_{-\infty}^{\infty} \delta_a(t)\, e^{-i2\pi t\omega}\, dt = e^{-i2\pi a\omega}$$

and

$$\mathcal{F}^{-1}[\delta_a]|_t = \int_{-\infty}^{\infty} \delta_a(\omega)\, e^{i2\pi \omega t}\, d\omega = e^{i2\pi a t} \quad .$$

Inverting these equations gives

$$\mathcal{F}^{-1}\!\left[e^{-i2\pi a\omega}\right] = \delta_a \qquad \text{and} \qquad \mathcal{F}\!\left[e^{i2\pi a t}\right] = \delta_a \quad .$$

(The first, of course, is equivalent to $\mathcal{F}^{-1}\!\left[e^{i2\pi a\omega}\right] = \delta_{-a}$.)

In particular, since $\delta = \delta_0$,

$$\mathcal{F}[\delta] = 1 = \mathcal{F}^{-1}[\delta] \qquad \text{and} \qquad \mathcal{F}[1] = \delta = \mathcal{F}^{-1}[1] \quad .$$

It is now a simple matter to find the Fourier transforms of all constants, sines and cosines.

!▶Example 26.2:   *For any real value* a *,*

$$\mathcal{F}[a] = \mathcal{F}[a \cdot 1] = a\mathcal{F}[1] = a\delta \quad .$$

!▶Example 26.3:   *For any real value* a *,*

$$\mathcal{F}[\sin(2\pi a t)] = \mathcal{F}\!\left[\frac{1}{2i}\left(e^{i2\pi a t} - e^{-i2\pi a t}\right)\right]$$

$$= \frac{1}{2i}\left(\mathcal{F}\!\left[e^{i2\pi a t}\right] - \mathcal{F}\!\left[e^{i2\pi(-a)t}\right]\right) = \frac{1}{2i}[\delta_a - \delta_{-a}] \quad .$$

?▶Exercise 26.2:   *Show that, for any real value* a *,*

$$\mathcal{F}[\cos(2\pi a t)] = \frac{1}{2}[\delta_a + \delta_{-a}] \quad .$$

Convolution

If f is any piecewise continuous function on the real line and a is any point at which f is continuous, then, by the definitions of convolution and the delta function,

$$f * \delta_a(x) = \int_{-\infty}^{\infty} f(x - s)\delta_a(s)\, ds = f(x - a) \quad .$$

In other words, the convolution of f with δ_a is just f translated by a . In particular then,

$$f * \delta = f * \delta_0 = f \quad .$$

Multiplication

Let f be a function continuous at a, and consider the two products $f\delta_a$ and $f(a)\delta_a$. (Note the difference. The first is a product of the *function* f with the delta function. The second is the product of the *value* of f at a with the delta function.) If ϕ is any other function continuous at a, then

$$\int_{-\infty}^{\infty} \phi(s)[f(s)\delta_a(s)]\,ds \;=\; \int_{-\infty}^{\infty} [\phi(s)f(s)]\delta_a(s)\,ds \;=\; \phi(a)f(a)$$

and

$$\int_{-\infty}^{\infty} \phi(s)[f(a)\delta_a(s)]\,ds \;=\; \int_{-\infty}^{\infty} [\phi(s)f(a)]\delta_a(s)\,ds \;=\; \phi(a)f(a) \quad .$$

Thus,

$$\int_{-\infty}^{\infty} \phi(s)[f(s)\delta_a(s)]\,ds \;=\; \int_{-\infty}^{\infty} \phi(s)[f(a)\delta_a(s)]\,ds$$

for every function ϕ continuous at a. This means

$$f\delta_a \;=\; f(a)\delta_a \quad .$$

In other words, the product of a delta function at a with some function is the same as the product of that delta function with the value of that function at a.

!▶ **Example 26.4:** *Two simple examples:*

$$(x^2 - 2)\delta_3(x) \;=\; (3^2 - 2)\delta_3(x) \;=\; 7\,\delta_3(x)$$

and

$$x\delta(x) \;=\; 0 \cdot \delta(x) \;=\; 0 \quad .$$

Various Integrals

Assume f is continuous at a. Then, by definition,

$$\int_{-\infty}^{\infty} f(x)\delta_a(x)\,dx \;=\; f(a) \quad .$$

In particular, using the constant function $f(x) = 1$, we have

$$\int_{-\infty}^{\infty} \delta_a(x)\,dx \;=\; \int_{-\infty}^{\infty} f(x)\delta_a(x)\,dx \;=\; f(a) \;=\; 1 \quad .$$

Now let (r, s) be any subinterval of the real line with neither r nor s equaling a, and consider computing

$$\int_{r}^{s} f(x)\delta_a(x)\,dx \quad .$$

To apply our definition of the delta function we need to convert this to an integral over the entire real line, say, by using the rectangle function over (r, s),

$$\mathrm{rect}_{(r,s)}(x) \;=\; \begin{cases} 1 & \text{if } r < a < s \\ 0 & \text{otherwise} \end{cases} \quad .$$

Doing so, we find that

$$\int_r^s f(x)\delta_a(x)\,dx = \int_{-\infty}^{\infty} \text{rect}_{(r,s)}(x)\,f(x)\delta_a(x)\,dx$$

$$= \text{rect}_{(r,s)}(a)\,f(a) = \begin{cases} 1\cdot f(a) & \text{if } r < a < s \\ 0\cdot f(a) & \text{otherwise} \end{cases} \; .$$

Thus, provided $a \neq r$ and $a \neq s$,

$$\int_r^s f(x)\delta_a(x)\,dx = \begin{cases} f(a) & \text{if } r < a < s \\ 0 & \text{otherwise} \end{cases} \; .$$

!▶ Example 26.5:   *Consider the integral of just the delta function over the interval* $(-\infty, x)$ *where* x *is any nonzero real number. By the above, with* $a = 0$,

$$\int_{-\infty}^{x} \delta(s)\,ds = \int_{-\infty}^{x} 1\cdot \delta(s)\,ds = \begin{cases} 1 & \text{if } -\infty < 0 < x \\ 0 & \text{otherwise} \end{cases} = \text{step}(x) \quad .$$

Continuing the computations started in the last example leads to some equations (the equations in the next exercise) suggesting that the delta function is, in some sense, the derivative of the step function. This, in turn, suggests that we should re-examine our concept of differentiation. We'll carry out this re-examination later in this chapter.

?▶ Exercise 26.3:   *Let* γ *be any real number, and let* (a, b) *be a finite interval with neither* a *nor* b *equaling* γ. *Verify the following:*

a: $\displaystyle\int_{-\infty}^{x} \delta_\gamma(s)\,ds = \text{step}(x - \gamma)$

b: $\displaystyle\int_{a}^{b} \delta_\gamma(s)\,ds = \text{step}(b - \gamma) - \text{step}(a - \gamma) = \text{rect}_{(a,b)}(\gamma)$

26.2   Transforms of Periodic Functions

In example 26.3 and exercise 26.2, we found the Fourier transforms of sine and cosine functions by rewriting them in terms of complex exponentials and using the fact that $\mathcal{F}\left[e^{i2\pi a t}\right] = \delta_a$ whenever a is a real number. Obviously, the same approach can be used to find the Fourier transform of any function that can be written as a linear combination of corresponding complex exponentials.

But remember, any "reasonable" periodic function can be represented by its Fourier series, and this Fourier series is just a (possibly infinite) linear combination of complex exponentials.

So let p be some positive value, and let f be any "reasonable" periodic function with period p and Fourier series

$$F.S.\,[f]\big|_t = \sum_{k=-\infty}^{\infty} c_k\, e^{i2\pi \omega_k t} \quad .$$

As you doubtlessly recall, the corresponding frequencies and Fourier coefficients are given by

$$\omega_k = \frac{k}{p} \quad \text{and} \quad c_k = \frac{1}{p} \int_{\text{period}} f(t) \, e^{-i2\pi \omega_k t} \, dt \quad .$$

Since we are assuming f is "reasonable", we have

$$f(t) = F.S.[f]|_t = \sum_{k=-\infty}^{\infty} c_k \, e^{i2\pi \omega_k t} \quad .$$

Taking the Fourier transform of both sides, and computing the transform of the series in a somewhat naive manner, we have

$$\mathcal{F}[f] = \mathcal{F}\left[\sum_{k=-\infty}^{\infty} c_k \, e^{i2\pi \omega_k t} \right]$$

$$= \sum_{k=-\infty}^{\infty} \mathcal{F}\left[c_k \, e^{i2\pi \omega_k t} \right]$$

$$= \sum_{k=-\infty}^{\infty} c_k \mathcal{F}\left[e^{i2\pi \omega_k t} \right] = \sum_{k=-\infty}^{\infty} c_k \, \delta_{\omega_k} \quad .$$

The naivety is in assuming that the transform of the infinite sum is the infinite sum of the transforms. Eventually (in chapter 37) we will verify that this method for finding the Fourier transform (or inverse transform) is valid whenever f is anything that can be represented by its complex exponential Fourier series. And, in the spirit of the chapter we are currently in, we will go ahead and assume our naive computations are valid.

Before working an example, it's worth noting that, if we let $\Delta\omega = {}^{1}\!/_{p}$, then

$$\omega_k = \frac{k}{p} = k\,\Delta\omega \quad \text{for} \quad k = 0,\, \pm 1,\, \pm 2,\, \pm 3,\, \ldots \quad ,$$

and the above formula for $\mathcal{F}[f]$ becomes

$$\mathcal{F}[f] = \sum_{k=-\infty}^{\infty} c_k \, \delta_{\omega_k} = \sum_{k=-\infty}^{\infty} c_k \, \delta_{k\Delta\omega} \quad .$$

Equivalently,

$$\mathcal{F}[f]|_{\omega} = \sum_{k=-\infty}^{\infty} c_k \, \delta(\omega - \omega_k) = \sum_{k=-\infty}^{\infty} c_k \, \delta(\omega - k\,\Delta\omega) \quad .$$

!▶ Example 26.6:   *Consider the saw function from example 12.1 on page 148,*

$$\text{saw}_3(t) = \begin{cases} t & \text{if } 0 < t < 3 \\ \text{saw}_3(t-3) & \text{in general} \end{cases} \quad .$$

Since saw_3 *is piecewise smooth, we know it equals its Fourier series (as piecewise continuous functions). Thus, using the Fourier series computed in example 12.1,*

$$\text{saw}_3(t) = \frac{3}{2} + \sum_{\substack{k=-\infty \\ k \neq 0}}^{\infty} \frac{3i}{2\pi k} e^{i2\pi \omega_k t} \quad \text{where} \quad \omega_k = \frac{k}{3} \quad ,$$

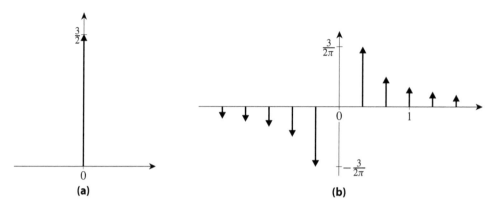

Figure 26.2: The (a) real part and (b) imaginary part of the Fourier transform of saw_3.

and

$$\mathcal{F}[saw_3(t)] = \mathcal{F}\left[\frac{3}{2} + \sum_{\substack{k=-\infty \\ k \neq 0}}^{\infty} \frac{3i}{2\pi k} e^{i2\pi \omega_k t}\right]$$

$$= \frac{3}{2}\mathcal{F}[1] + \sum_{\substack{k=-\infty \\ k \neq 0}}^{\infty} \frac{3i}{2\pi k}\mathcal{F}\left[e^{i2\pi \omega_k t}\right] = \frac{3}{2}\delta + \sum_{\substack{k=-\infty \\ k \neq 0}}^{\infty} \frac{3i}{2\pi k}\delta_{\omega_k} \quad .$$

Equivalently,

$$\mathcal{F}[saw_3]\big|_\omega = \frac{3}{2}\delta(\omega) + \sum_{\substack{k=-\infty \\ k \neq 0}}^{\infty} \frac{3i}{2\pi k}\delta_{k/3}(\omega) = \frac{3}{2}\delta(\omega) + \sum_{\substack{k=-\infty \\ k \neq 0}}^{\infty} \frac{3i}{2\pi k}\delta\left(\omega - \frac{k}{3}\right) \quad .$$

The graphical representation of this transform has been sketched in figure 26.2.

?▶ Exercise 26.4:   *Show that*

$$\mathcal{F}^{-1}[saw_3] = \frac{3}{2}\delta + \sum_{\substack{k=-\infty \\ k \neq 0}}^{\infty} \frac{3i}{2\pi k}\delta_{-k/3} = \frac{3}{2}\delta - \sum_{\substack{n=-\infty \\ n \neq 0}}^{\infty} \frac{3i}{2\pi n}\delta_{n/3} \quad .$$

The third or fourth time you compute a Fourier transform of a periodic function, you will probably find yourself skipping a few steps and simply "converting" the Fourier series for the given function directly to the corresponding series of delta functions for the Fourier transform or inverse Fourier transform of the given function. That's fine, provided you continue to realize what you are skipping. In fact, for a couple of reasons, it's worthwhile to explicitly describe the appropriate conversions in the following "quasi-theorem".

Quasi-Theorem on Transforms of Periodic Functions
Let f be periodic with period p and Fourier series

$$F.S.[f]\big|_x = \sum_{k=-\infty}^{\infty} c_k e^{i2\pi \omega_k x} \quad .$$

Then, letting $\Delta y = {}^1\!/_p$,

$$\mathcal{F}[f] = \sum_{k=-\infty}^{\infty} c_k \,\delta_{k\Delta y} \qquad \text{and} \qquad \mathcal{F}^{-1}[f] = \sum_{k=-\infty}^{\infty} c_{-k} \,\delta_{k\Delta y} \quad .$$

?▶ Exercise 26.5: *Derive (naively) the formulas for $\mathcal{F}[f]$ and $\mathcal{F}^{-1}[f]$ in the above quasi-theorem starting with the assumption that*

$$f(x) = \sum_{k=-\infty}^{\infty} c_k \, e^{i2\pi \omega_k x} \quad .$$

You may have noticed that we are being a little vague concerning f in our quasi-theorem. We are not insisting that it be piecewise continuous. We are not even explicitly stating that f is a function! This vagueness reflects the fact that, eventually (in chapter 37), the claims of this quasi-theorem will be verified assuming very general assumptions concerning the nature of f.

26.3 Arrays of Delta Functions
Regular Arrays

As we've just seen, the Fourier transforms of periodic functions can all be expressed as infinite series of delta functions. Such expressions are traditionally referred to as "arrays of delta functions".

To be a little more precise, a *regular array of delta functions* with spacing Δx is any expression of the form

$$\sum_{k=-\infty}^{\infty} c_k \,\delta_{k\Delta x}$$

where the c_k's (the *coefficients* of the array) are constants and the Δx (the *spacing* of the array) is some positive value. For example, the Fourier transform of the saw function in example 26.6,

$$\frac{3}{2}\delta + \sum_{\substack{k=-\infty \\ k\neq 0}}^{\infty} \frac{3i}{2\pi k}\delta_{k/3} \quad ,$$

is a regular array of delta functions with spacing ${}^1\!/_3$. Similarly, the Fourier transform of $\cos(2\pi t)$,

$$\frac{1}{2}\left[\delta_1 + \delta_{-1}\right]$$

(see exercise 26.2), is a regular array of delta functions with spacing 1 since

$$\frac{1}{2}\left[\delta_1 + \delta_{-1}\right] = \sum_{k=-\infty}^{\infty} c_k \,\delta_{k\Delta x}$$

where

$$\Delta x = 1 \quad \text{and} \quad c_k = \begin{cases} \dfrac{1}{2} & \text{if } k = \pm 1 \\[2mm] 0 & \text{otherwise} \end{cases} \quad .$$

Two other arrays are sketched in figure 26.3.

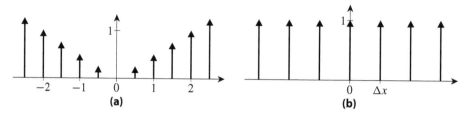

Figure 26.3: Two arrays of delta functions — (a) the array for example 26.7 and (b) the comb function with spacing Δx.

At this time, we are interested in regular arrays of delta functions simply because they arise when we compute the Fourier transforms of periodic functions. Later, in part V, we will also find them playing a major role in the Fourier analysis of "sampled data".

Computing the Fourier transforms of regular arrays of delta functions is easy. Simply compute the transform term by term (again, the validity of this will be confirmed in chapter 37).

!► **Example 26.7:** *The array sketched in figure 26.3a is*

$$ f = \sum_{k=-\infty}^{\infty} \frac{1}{4} |k| \delta_{k/2} \quad . $$

Its Fourier transform is

$$ \mathcal{F}[f]|_y = \mathcal{F}\left[\sum_{k=-\infty}^{\infty} \frac{1}{4} |k| \delta_{k/2} \right]\bigg|_y $$

$$ = \sum_{k=-\infty}^{\infty} \mathcal{F}\left[\frac{1}{4} |k| \delta_{k/2} \right]\bigg|_y $$

$$ = \sum_{k=-\infty}^{\infty} \frac{1}{4} |k| \, \mathcal{F}\left[\delta_{k/2} \right]\bigg|_y = \sum_{k=-\infty}^{\infty} \frac{1}{4} |k| \, e^{-i2\pi(k/2)y} \quad . $$

Via the substitution $n = -k$, this becomes

$$ \mathcal{F}[f]|_y = \sum_{n=-\infty}^{\infty} \frac{1}{4} |n| \, e^{i2\pi \omega_n y} \qquad where \quad \omega_n = \frac{n}{2} \quad . $$

The results of such computations look like Fourier series for periodic functions. In particular, the series obtained in the last example looks like a Fourier series for a periodic function having period $p = 2$ (since, in general, $\omega_n = {}^n/_p$). With incredible luck, you might even recognize a resulting series as the Fourier series of some well-known function. In practice, of course, this rarely happens. For example, the last series above does not correspond to any Fourier series we derived in our discussions of the classical Fourier series (part II of this book). In fact, because the coefficients are increasing as k increases, you can show that this "Fourier series" cannot correspond to any piecewise continuous function on the real line (see exercise 26.17 at the end of this chapter). Just what this "Fourier series" represents will have to remain a mystery until chapter 37 where we will seriously discuss periodic "generalized" functions.

Comb Functions and Other Periodic, Regular Arrays

It is quite possible for regular arrays to also be periodic. For example, the comb function with spacing Δx, sketched in figure 26.3b and given by

$$\text{comb}_{\Delta x} = \sum_{k=-\infty}^{\infty} \delta_{k \Delta x} \quad ,$$

is a rather simple regular array with spacing Δx. Moreover, if you shift it by Δx, you end up with what you started with (do this visually with the figure). So $\text{comb}_{\Delta x}$ is also periodic with period Δx.

Our main interest in periodic, regular arrays will be in the role they play in "sampling" and the "discrete theory" of Fourier analysis. Since this discussion won't occur until chapter 40, we will delay a more complete discussion of these arrays and their transforms until chapter 39. Still, computing the Fourier transform of a relatively simple periodic, regular array

$$f = \sum_{k=-\infty}^{\infty} f_k \, \delta_{k \Delta t}$$

can easily be done with the material we've already developed. In fact, we have two ways to compute this transform:

1. We can compute $F = \mathcal{F}[f]$ the same way we can compute the Fourier transform of any regular array:

$$F(\omega) = \mathcal{F}\left[\sum_{k=-\infty}^{\infty} f_k \, \delta_{k \Delta t} \right]\bigg|_{\omega} = \sum_{k=-\infty}^{\infty} f_k \, \mathcal{F}[\delta_{k \Delta t}]|_{\omega} = \sum_{k=-\infty}^{\infty} f_k \, e^{-i 2\pi k \Delta t \, \omega} \quad .$$

   This, after the substitution $n = -k$, gives us the Fourier series for F.

2. On the other hand, because it is periodic, f can be represented by its Fourier series,

$$f(t) = \sum_{n=-\infty}^{\infty} c_n \, e^{i 2\pi \omega_n t} \quad .$$

   We can then compute $F = \mathcal{F}[f]$ by transforming this series:

$$F = \mathcal{F}\left[\sum_{n=-\infty}^{\infty} c_n \, e^{i 2\pi \omega_n t} \right] = \sum_{n=-\infty}^{\infty} c_n \, \mathcal{F}\left[e^{i 2\pi \omega_n t} \right] = \sum_{n=-\infty}^{\infty} c_n \, \delta_{\omega_n} \quad .$$

   This gives an explicit formula for F as an array of delta functions. Since such a formula is usually preferred over the Fourier series representation, this approach to computing F is the one normally recommended. Admittedly, a little more work is required — the period p must be determined, and the Fourier coefficients of f must be computed. But p is usually obvious once you've sketched the array f, and the computation of each Fourier coefficient,

$$c_n = \frac{1}{p} \int_{\text{period}} f(t) \, e^{-i 2\pi \omega_n t} \, dt \quad ,$$

   is easily done using the original delta function array formula for f (you may assume the integration can be done term by term).

Do observe that, since both of the above approaches are valid, the Fourier transform (and, similarly, the inverse Fourier transform) of any periodic, regular array will be another periodic, regular array of delta functions.

?▶Exercise 26.6:   *Let Δt be a positive constant. Using the second method described above, show that the Fourier series of $\text{comb}_{\Delta t}$ is*

$$\sum_{n=-\infty}^{\infty} \Delta\omega\, e^{i2\pi n\Delta\omega t} \qquad where \quad \Delta\omega = \frac{1}{\Delta t} \quad .$$

Then show that

$$\mathcal{F}[\text{comb}_{\Delta t}] = \Delta\omega\, \text{comb}_{\Delta\omega} \quad .$$

26.4   The Generalized Derivative
Re-Examining Differentiation

For reasons soon to be obvious, we will begin referring to the derivative as defined in chapter 3 (and elementary calculus) as the classical derivative. That is, the *classical derivative* of a piecewise differentiable function f is the function f' given by

$$f'(x) = \lim_{\Delta x \to 0} \frac{f(x+\Delta x) - f(x)}{\Delta x} \quad .$$

Since we will be especially interested in integrals involving derivatives, let's also recall that, as long as f is also *continuous* on the real line,

$$f(b) - f(a) = \int_a^b f'(x)\,dx \qquad whenever \quad -\infty < a < b < \infty \quad . \tag{26.5}$$

So important is this fact that we often call its statement "the fundamental theorem of calculus". It is also important to recall that this equation fails to hold if f has a discontinuity between a and b. This was dramatically illustrated in example 3.6 on page 23 where we saw that, because the classical derivative of the step function is 0,

$$\text{step}(b) - \text{step}(a) = 1 \neq 0 = \int_a^b \text{step}'(x)\,dx \qquad whenever \quad a < 0 < b \quad .$$

We are recalling all this because, thanks to the delta function, we can now refine our concept of differentiation so as to account for discontinuities in the functions being differentiated. To see this, look back at exercise 26.3 b on page 426. There, letting $\gamma = 0$, you found that

$$\text{step}(b) - \text{step}(a) = \int_a^b \delta(x)\,dx \qquad whenever \quad -\infty < a < b < \infty \quad ,$$

provided neither a nor b is 0. The similarity between this equation and equation (26.5) suggests that the delta function can, in a sense, be viewed as a "derivative" of the step function. Of course, it's not the classical derivative — the classical derivative of the step function is still 0 — so let's call this new type of derivative something else, say, the "generalized" derivative (since it involves the delta function, which is a generalized function) and, to distinguish it from the classical derivative, let's denote the generalized derivative of step by D step.

To be a little more general, if we go back to exercise 26.3 b on page 426, we see that, for any real value γ and any finite interval (a, b) with neither a nor b equaling γ,

$$\text{step}(b - \gamma) \; - \; \text{step}(a - \gamma) \; = \; \int_a^b D[\text{step}(x - \gamma)] \, dx$$

provided we define "the generalized derivative of $\text{step}(x - \gamma)$" by

$$D[\text{step}(x - \gamma)] \; = \; \delta_\gamma(x) \quad .$$

Let us do so.

To be even more general, here is our goal: For each piecewise smooth function f we want to find another "function" Df satisfying

$$f(b) \; - \; f(a) \; = \; \int_a^b Df(x) \, dx \tag{26.6}$$

whenever (a, b) is a finite interval with endpoints at which f is continuous (so $f(a)$ and $f(b)$ are well defined). This Df will be referred to as the "generalized derivative" of f .

Defining the Generalized Derivative

There are two classes of functions for which we already know the corresponding Df's satisfying equation (26.6). One is the set of all piecewise smooth functions that are also *continuous* on the real line. After all, if f is such a function, then equation (26.5) holds, and that is the same as equation (26.6) with $Df = f'$. This then is how we will define the generalized derivative for a continuous, piecewise smooth function. It's just that function's classical derivative.

The other set is the set of step functions. As we just saw, if f is a step function, then equation (26.6) holds if we choose Df to be the delta function at the point where f has its jump discontinuity. And it is not difficult to see that, if f is a constant multiple of a step function, say, $f(x) = c \, \text{step}(x - \gamma)$, then equation (26.6) holds if Df is $c\delta_\gamma$, which, we should note, is just the delta function at the point where f has its jump multiplied by the value of that jump in f .

All of this suggests that the generalized derivative of any piecewise smooth function will account for any jump in that function if that derivative has a term consisting of a delta function at each point where the function has a jump, with that delta function being multiplied by the value of the jump. So it seems reasonable for the generalized derivative of any piecewise smooth function to be the classical derivative modified by the addition of the suggested linear combination of delta functions. Let's try that.

To be precise: Suppose f is any piecewise smooth function on the real line, and let $\{\ldots, x_1, x_2, x_3, \ldots\}$ be the set of all points at which f has discontinuities (this set may be empty). The *generalized derivative* of f , denoted by Df , is defined by

$$Df \; = \; f' + \sum_k j_k \, \delta_{x_k}$$

where f' is the classical derivative of f , the summation is taken over all points at which f is discontinuous, and, for each k in the summation, j_k is the jump in f at x_k .

Do note that the above summation "over all points at which f is discontinuous" has no terms if f is continuous on the real line. Thus, if f is a continuous and piecewise smooth function on the real line, then the above formula for Df reduces to $Df = f'$.

!▶ Example 26.8:   Let $f(x) = x^3$. Its classical derivative is $f'(x) = 3x^2$. Since f is obviously continuous on the real line

$$Df(x) \;=\; f'(x) \;=\; 3x^2 \quad .$$

!▶ Example 26.9:   Let $f(x) = \text{step}(x - 4)$. The classical derivative is simply

$$\text{step}'(x - 4) \;=\; \begin{cases} \dfrac{d}{dx}0 & \text{if } x < 4 \\[2mm] \dfrac{d}{dx}1 & \text{if } 4 < x \end{cases} \;=\; \begin{cases} 0 & \text{if } x < 4 \\ 0 & \text{if } 4 < x \end{cases} \;=\; 0 \quad .$$

There is one jump discontinuity, a jump of 1 at $x = 4$. So,

$$Df \;=\; f' + \sum_k j_k\,\delta_{x_k} \;=\; 0 + 1\delta_4 \;=\; \delta_4 \quad .$$

!▶ Example 26.10:   Let

$$f(x) \;=\; \begin{cases} 1 + x^2 & \text{if } x < 0 \\ 4 - x^2 & \text{if } 0 < x \end{cases} \quad .$$

This is a piecewise smooth function with one discontinuity, at $x = 0$. The jump in the function at the discontinuity is

$$j_0 \;=\; \lim_{x\to 0^+} f(x) \;-\; \lim_{x\to 0^-} f(x)$$

$$=\; \lim_{x\to 0^+}\left[4 - x^2\right] \;-\; \lim_{x\to 0^-}\left[1 + x^2\right] \;=\; 3 \quad .$$

The classical derivative is easily found,

$$f'(x) \;=\; \begin{cases} \dfrac{d}{dx}[x^2 + 1] & \text{if } x < 0 \\[2mm] \dfrac{d}{dx}[4 - x^2] & \text{if } 0 < x \end{cases} \;=\; \begin{cases} 2x & \text{if } x < 0 \\ -2x & \text{if } 0 < x \end{cases} \quad .$$

So

$$Df(x) \;=\; f'(x) + \sum_k j_k\,\delta_{x_k}(x) \;=\; \begin{cases} 2x & \text{if } x < 0 \\ -2x & \text{if } 0 < x \end{cases} + 3\delta(x) \quad .$$

?▶ Exercise 26.7:   Let a and b be two real numbers and verify the following:

a: $D\,\text{rect}_{(a,b)} \;=\; \delta_a - \delta_b$ (assuming $a < b$)

b: $D\,\text{pulse}_a \;=\; \delta_{-a} - \delta_a$ (assuming $0 < a$)

?▶ Exercise 26.8:   Confirm the following:

a: $D[x^3\,\text{step}(x - 2)] \;=\; 3x^2\,\text{step}(x - 2) + 8\delta_2(x)$

b: $D[x^3\,\text{rect}_{(0,2)}(x)] \;=\; 3x^2\,\text{rect}_{(0,2)}(x) - 8\delta_2(x)$

Basic Identities for the Generalized Derivative
The Identities

Not too surprisingly, many of those identities and rules for derivatives from elementary calculus are also valid, suitably reinterpreted, for the generalized derivative. Some that will be particularly useful to us are described in the next theorem.

Theorem 26.1 *(elementary identities for the generalized derivative)*
Let f and g be two piecewise smooth functions on $\mathbb{R}$, and let α and β be two constants. All of the following then hold:

1. *(linearity)* $\quad D[\alpha f + \beta g] = \alpha Df + \beta Dg$.

2. *(generalized product rule)* *If g is continuous, then* $D[fg] = [Df]g + f[g']$.

3. *(generalized fundamental theorem of calculus)* *If $-\infty < \alpha \leq \beta < \infty$, and if f is continuous at both α and β, then*
$$\int_\alpha^\beta Df(x)\,dx = f(\beta) - f(\alpha) \quad .$$

4. *(generalized integration by parts)* *If $-\infty < \alpha \leq \beta < \infty$, f is continuous at both α and β, and g is continuous on the real line, then*
$$\int_\alpha^\beta Df(x)\,g(x)\,dx = f(x)g(x)\Big|_\alpha^\beta - \int_\alpha^\beta f(x)\,g'(x)\,dx \quad .$$

5. *(limited chain rule)* *If $g(x) = f(\alpha x - \beta)$ where α and β are real and α is nonzero, then $Dg(x) = \alpha Df(\alpha x - \beta)$.*

The first, second and last identities in this theorem can often simplify the computation of generalized derivatives. The third identity, the generalized fundamental theorem of calculus, is the identity we hoped would be satisfied when we defined the generalized derivative. In addition, that identity combined with the generalized product rule leads to the integration by parts identity in the theorem, and that, in turn, will play a big role when we discuss the generalized differential identities for Fourier transforms.

Before discussing the proof of our theorem, let's see how some of these identities, especially the product rule, can simplify some computations (provided we remember the generalized derivatives of certain basic discontinuous functions — the rectangle, step and pulse functions — and that the product of a continuous function g with a delta function δ_a simplifies to $g(a)\delta_a$).

!▶ Example 26.11: Let
$$f(x) = \begin{cases} x^2 & \text{if} \ \ -2 < x < 2 \\ 4 & \text{if} \ \ 2 < x < 3 \\ 0 & \text{otherwise} \end{cases} \quad .$$

Observe that $f(x)$ can also be written as
$$f(x) = x^2 \operatorname{rect}_{(-2,2)}(x) + 4 \operatorname{rect}_{(2,3)}(x) \quad .$$

Using linearity, the product rule and the results of exercise 26.7, we have
$$Df(x) = D\left[x^2 \operatorname{rect}_{(-2,2)}(x) + 4 \operatorname{rect}_{(2,3)}(x)\right]$$
$$= D\left[x^2 \operatorname{rect}_{(-2,2)}(x)\right] + 4D \operatorname{rect}_{(2,3)}(x)$$

$$= 2x \operatorname{rect}_{(-2,2)}(x) + x^2 \, D \operatorname{rect}_{(-2,2)}(x) + 4[\delta_2(x) - \delta_3(x)]$$

$$= 2x \operatorname{rect}_{(-2,2)}(x) + x^2[\delta_{-2}(x) - \delta_2(x)] + 4\delta_2(x) - 4\delta_3(x) \quad .$$

But,

$$x^2[\delta_{-2}(x) - \delta_2(x)] = x^2\delta_{-2}(x) - x^2\delta_2(x)$$

$$= (-2)^2\delta_{-2}(x) - (2)^2\delta_2(x) = 4\delta_{-2}(x) - 4\delta_2(x) \quad .$$

So,

$$Df(x) = 2x \operatorname{rect}_{(-2,2)}(x) + 4\delta_{-2}(x) - 4\delta_2(x) + 4\delta_2(x) - 4\delta_3(x)$$

$$= 2x \operatorname{rect}_{(-2,2)}(x) + 4\delta_{-2}(x) - 4\delta_3(x) \quad .$$

?▶ Exercise 26.9:   *Using linearity and the product rule, verify that*

$$D\left[x^3 \operatorname{rect}_{(0,2)}(x)\right] = 3x^2 \operatorname{rect}_{(0,2)}(x) - 8\,\delta_2(x) \quad .$$

Proving Theorem 26.1

PROOF (linearity claim of theorem 26.1):   Let $\{\ldots, x_0, x_1, x_2, \ldots\}$ be the set of all points at which either f or g has discontinuities, and, for each of these x_k's, let j_k and J_k denote, respectively, the jump in f and the jump in g at x_k (with the jump being 0 if the corresponding function is continuous at x_k). We then have

$$Df = f' + \sum_k j_k \delta_{x_k} \qquad \text{and} \qquad Dg = g' + \sum_k J_k \delta_{x_k} \quad .$$

Obviously, $\{\ldots, x_0, x_1, x_2, \ldots\}$ is also the set of all points at which $\alpha f + \beta g$ can have discontinuities. Letting $\mathcal{J}_k$ be the jump in $\alpha f + \beta g$, we see that

$$\mathcal{J}_k = \lim_{x \to x_k^+} [\alpha f(x) + \beta g(x)] - \lim_{x \to x_k^-} [\alpha f(x) + \beta g(x)]$$

$$= \alpha\left[\lim_{x \to x_k^+} f(x) - \lim_{x \to x_k^-} f(x)\right] + \beta\left[\lim_{x \to x_k^+} g(x) - \lim_{x \to x_k^-} g(x)\right] = \alpha j_k + \beta J_k \quad .$$

So, applying the definition of the generalized derivative to $\alpha f + \beta g$ along with the above computations and the linearity of the classical derivative, we have

$$D[\alpha f + \beta g] = [\alpha f + \beta g]' + \sum_k \mathcal{J}_k \delta_{x_k}$$

$$= \alpha f' + \beta g' + \sum_k [\alpha j_k + \beta J_k]\delta_{x_k}$$

$$= \alpha\left[f' + \sum_k j_k \delta_{x_k}\right] + \beta\left[g' + \sum_k J_k \delta_{x_k}\right] = \alpha Df + \beta Dg \quad . \qquad ∎$$

For the rest of the proof of theorem 26.1, we only need to be concerned with the discontinuities of f . So, to avoid needless repetition, for the rest of this section, $\{\ldots, x_0, x_1, x_2, \ldots\}$ will denote the set of all distinct points at which f has discontinuities, and, for each of these x_k's, j_k will be the corresponding jump in f.

PROOF (generalized product rule): Since g is continuous, the x_k's are also the only points at which the product fg can have discontinuities. Letting $\mathcal{J}_k$ be the jump in fg at x_k, observe that

$$\mathcal{J}_k = \lim_{x \to x_k^+} f(x)g(x) - \lim_{x \to x_k^-} f(x)g(x)$$

$$= \left[\lim_{x \to x_k^+} f(x) - \lim_{x \to x_k^-} f(x) \right] g(x_k) = j_k\, g(x_k) \quad.$$

Thus,

$$g \sum_k j_k\, \delta_{x_k} = \sum_k j_k\, g\, \delta_{x_k} = \sum_k j_k\, g(x_k)\, \delta_{x_k} = \sum_k \mathcal{J}_k\, \delta_{x_k} \quad.$$

Using this, the relation between classical and generalized derivatives, and the classical product rule, we get

$$D[fg] = [fg]' + \sum_k \mathcal{J}_k\, \delta_{x_k}$$

$$= [f']g + f[g'] + g \sum_k j_k\, \delta_{x_k}$$

$$= \left[[f']g + g \sum_k j_k\, \delta_{x_k} \right] + f[g']$$

$$= \left[f' + \sum_k j_k\, \delta_{x_k} \right] g + f[g'] = [Df]g + f[g'] \quad. \qquad \blacksquare$$

For the rest of our proof, we will restrict ourselves to cases where f has only a finite number of discontinuities, and leave the cases where f has an infinite number of discontinuities as exercises (exercises 26.21 and 26.22). This will allow us to simplify our discussion slightly, while still illustrating the fundamental ideas, one of which is described in the next lemma.

Lemma 26.2
Suppose f is a piecewise smooth function on $\mathbb{R}$ with only a finite number N of discontinuities, say, at $x_1, x_2, \ldots$ and x_N. Then

$$f(x) = h(x) + \sum_{k=1}^{N} j_k\, \text{step}(x - x_k) \tag{26.7}$$

where h is a piecewise smooth function that is continuous on the entire real line. Moreover,

$$f' = h' \quad \text{and} \quad Df = h' + \sum_{k=1}^{N} j_k\, \delta_{x_k} \quad. \tag{26.8}$$

PROOF: Let h be the piecewise smooth function given by

$$h(x) = f(x) - \sum_{k=1}^{N} j_k\, \text{step}(x - x_k) \quad. \tag{26.9}$$

Since the x_k's are the only points at which f or the above step functions are discontinuous, they are also the only points at which h can be discontinuous. It should also be pretty clear that the jumps in h due to jumps in f will be canceled out by the corresponding jumps in the step functions. After all,

$$\text{the jump in } h \text{ at } x_n = \lim_{x \to x_n^+} h(x) - \lim_{x \to x_n^-} h(x)$$

$$= \lim_{x \to x_n^+} \left[f(x) - \sum_{k=1}^{N} j_k \, \text{step}(x - x_k) \right]$$

$$- \lim_{x \to x_n^-} \left[f(x) - \sum_{k=1}^{N} j_k \, \text{step}(x - x_k) \right]$$

$$= \lim_{x \to x_n^+} f(x) - \lim_{x \to x_n^-} f(x)$$

$$- \sum_{k=1}^{N} j_k \left[\lim_{x \to x_n^+} \text{step}(x - x_k) - \lim_{x \to x_n^-} \text{step}(x - x_k) \right]$$

$$= j_n - \sum_{k=1}^{N} j_k \begin{cases} 1 & \text{if } k = n \\ 0 & \text{if } k \neq n \end{cases}$$

$$= j_n - j_n$$

$$= 0 \quad .$$

Thus, h is continuous (which, we'll note for use in the near future, also means that $Dh = h'$).
  Solving equation (26.9) for f gives us

$$f(x) = h(x) + \sum_{k=1}^{N} j_k \, \text{step}(x - x_k) \quad .$$

Moreover, because of the linearity of the derivatives and the known formulas for the derivatives of the step functions,

$$f'(x) = h'(x) + \sum_{k=1}^{N} j_k \, \text{step}'(x - x_k) = h'(x)$$

and

$$Df(x) = Dh(x) + \sum_{k=1}^{N} j_k D \, \text{step}(x - x_k) = h'(x) + \sum_{k=1}^{N} j_k \, \delta_{x_k}(x) \quad . \qquad \blacksquare$$

The main reason for assuming, above, that f only has a finite number of discontinuities is to avoid dealing with the possibility that $\sum_k j_k \, \text{step}(x - x_k)$ might be an infinite summation that diverges. This is not a serious problem. One way to deal with it is described at the end of this chapter in exercise 26.21. Using the results of that exercise, you can then redo the following proofs even when f has infinitely many discontinuities (see exercise 26.22).
  Now, utilizing the lemma just proven, let's see about verifying the rest of theorem 26.1.

PROOF (generalized fundamental theorem of calculus when f has finitely many discontinuities): Using equation (26.7) and the relations in line (26.8), we see that

$$\int_a^b Df(x)\,dx = \int_a^b \left[h'(x) + \sum_{k=1}^N j_k \delta_{x_k}(x) \right] dx$$

$$= \int_a^b h'(x)\,dx + \sum_{k=1}^N j_k \int_a^b \delta_{x_k}(x)\,dx \quad .$$

But h is continuous. So the classical fundamental theorem of calculus assures us that

$$\int_a^b h'(x)\,dx = h(b) - h(a) \quad ,$$

while exercise 26.3 b tells us that

$$\int_a^b \delta_{x_k}(x)\,dx = \text{step}(b - x_k) - \text{step}(a - x_k) \qquad \text{for} \quad k = 1, 2, \ldots, N \quad .$$

Combining these three equations, and recalling the relation between f and h (equation (26.7)), we find that

$$\int_a^b Df(x)\,dx = h(b) - h(a) + \sum_{k=1}^N j_k[\text{step}(b - x_k) - \text{step}(a - x_k)]$$

$$= \left[h(b) + \sum_{k=1}^N j_k\,\text{step}(b - x_k) \right] - \left[h(a) + \sum_{k=1}^N j_k\,\text{step}(a - x_k) \right]$$

$$= f(b) - f(a) \quad . \qquad \blacksquare$$

Confirming the generalized integration by parts formula is now a snap.

PROOF (generalized integration by parts when f has finitely many discontinuities): Using the generalizations of the fundamental theorem of calculus and product rule just obtained, we see that

$$f(x)g(x)\Big|_a^b = \int_a^b D[f(x)g(x)]\,dx = \int_a^b [Df(x)\,g(x) + f(x)\,g'(x)]\,dx \quad .$$

So,

$$f(x)g(x)\Big|_a^b = \int_a^b Df(x)\,g(x)\,dx + \int_a^b f(x)\,g'(x)\,dx \quad .$$

Solving for the first integral in the last equation then yields

$$\int_a^b Df(x)\,g(x)\,dx = f(x)g(x)\Big|_a^b - \int_a^b f(x)\,g'(x)\,dx \quad . \qquad \blacksquare$$

The proof of the chain rule formula in theorem 26.1 is left as a relatively simple exercise.

?▶Exercise 26.10 (limited chain rule when f has finitely many discontinuities):   Let α and β be real numbers with α being nonzero.

  a: Assume $\alpha > 0$, and let $g(x) = \text{step}(\alpha x + \beta - \gamma)$ where γ is any real number. Letting $v = \gamma/\alpha - \beta/\alpha$, verify that

  i: $g(x) = \text{step}(x - v)$,

  ii: $Dg = \delta_v$, and

  iii: $Dg(x) = \alpha h(\alpha x + \beta - \gamma)$ where $h = D\,\text{step}$. *(Use the results from exercise 26.13 c at the end of this chapter.)*

  b: Verify that $Dg(x) = \alpha h(\alpha x + \beta - \gamma)$ where $h = D\,\text{step}$, assuming that $\alpha < 0$, γ is any real number and $g(x) = \text{step}(\alpha x + \beta - \gamma)$.

  c: Finish verifying the chain rule formula in theorem 26.1 assuming f has only a finite number of discontinuities.

Extending the Generalized Derivative
Higher Order Derivatives

Higher order generalized derivatives of a function f — $D^2 f$, $D^3 f$, etc. — are defined in the obvious manner:

$$D^2 f = D[Df] \quad , \quad D^3 f = D[D^2 f] = D[D[Df]] \quad , \quad \cdots \quad .$$

!▶Example 26.12:   *In exercise 26.23 a, you showed that $D\,\text{ramp} = \text{step}$. So,*

$$D^2 \text{ramp} = D[D\,\text{ramp}] = D\,\text{step} = \delta \quad .$$

?▶Exercise 26.11:   Show that $D^3[x^2 \text{step}(x)] = 2\delta(x)$.

More General Generalized Derivatives

Thus far, we have said nothing about "the generalized derivative of the delta function, $D\delta$". We could not. Our definition for generalized differentiation presupposes that the function being differentiated is piecewise smooth. And the delta function is not piecewise smooth. It's not even a true function.

   Nonetheless, "$D\delta$" can be reasonably defined, as can the generalized derivative of any other Fourier transformable function. We will see this in chapter 35. Naively, you can think of $D\delta$ as that "thing" such that, using the generalized integration by parts formula,

$$\int_{-\infty}^{\infty} \phi(x) D\delta(x)\, dx = -\int_{-\infty}^{\infty} \phi'(x)\delta(x)\, dx = -\phi'(0) \tag{26.10}$$

whenever ϕ is a smooth function on the real line. But don't bother trying to visualize this as we did the delta function. That pushes our naive theory beyond its reasonable limits. And if you find yourself dealing with things like $D\delta$ on a regular basis, then it is time to break down and learn the generalized theory in part IV of this text. Otherwise the math just starts looking too mystical.

The Generalized Fourier Differential Identities

Let's now redo our first derivation of a Fourier differential identity using the generalized derivative instead of the classical derivative.

We start by assuming f is an absolutely integrable, piecewise smooth function on the real line. For simplicity, let's further assume that, taking the limit over the set of the points where f is continuous,

$$\lim_{t \to \pm\infty} f(t) = 0 \quad .$$

For each positive integer k, choose two real numbers $a_k \leq -k$ and $b_k \geq k$ at which f is continuous. The generalized integration by parts formula tells us that, for each of these k's,

$$\int_{a_k}^{b_k} Df(t) \, e^{-i2\pi\omega t} \, dt = f(t) \, e^{-i2\pi\omega t} \Big|_{a_k}^{b_k} - \int_{a_k}^{b_k} f(t) \frac{\partial}{\partial t} e^{-i2\pi\omega t} \, dt$$

$$= f(b_k) \, e^{-i2\pi\omega b_k} - f(a_k) \, e^{-i2\pi\omega a_k} + i2\pi\omega \int_{a_k}^{b_k} f(t) \, e^{-i2\pi\omega t} \, dt \quad .$$

Clearly though, as $k \to \infty$,

$$a_k \to -\infty \quad , \quad f(a_k) \to 0 \quad , \quad b_k \to \infty \quad \text{and} \quad f(b_k) \to 0 \quad .$$

So, if we let $k \to \infty$, the above equation of integrals reduces to

$$\int_{-\infty}^{\infty} Df(t) \, e^{-i2\pi\omega t} \, dt = 0 - 0 + i2\pi\omega \int_{-\infty}^{\infty} f(t) \, e^{-i2\pi\omega t} \, dt \quad ,$$

which we immediately recognize as being

$$\mathcal{F}[Df]|_\omega = i2\pi\omega \, F(\omega) \qquad \text{where} \quad F = \mathcal{F}[f] \quad . \tag{26.11}$$

This formula looks just like the first differential formula derived for the Fourier transform (equation (22.1) on page 340), only with Df replacing f'. In deriving equation (26.11), however, we only assumed the following:

1. f is piecewise smooth.

2. f is absolutely integrable.

3. $f(t) \to 0$ as $t \to \pm\infty$.

There is another similarity between equations (26.11) and (22.1) on page 340. In both cases, a more careful analysis can show that the identities hold under much weaker assumptions. In fact, we will discover in chapter 35 that all the assumptions listed above can be replaced with the single assumption that f is Fourier transformable. Not even piecewise smoothness is required, at least, not once we've figured out how to define the generalized derivative for Fourier transformable functions (such as the delta function) that are not piecewise smooth.

Here is the complete theorem describing the generalized Fourier differential identities (including the one just derived above). Except for minor differences in terminology, it is the same as theorem 35.12 on page 611, which we will later prove using the more advanced theory of generalized functions.

Theorem 26.3 (generalized Fourier differential identities)
Let f and F be Fourier transformable with $F = \mathcal{F}[f]$, and let n be any positive integer. Then

$$\mathcal{F}\left[D^n f\right]\big|_\omega = (i2\pi\omega)^n F(\omega) \quad , \tag{26.12a}$$

$$\mathcal{F}^{-1}\left[D^n F\right]\big|_t = (-i2\pi t)^n f(t) \quad , \tag{26.12b}$$

$$\mathcal{F}^{-1}\left[\omega^n F(\omega)\right] = \left(\frac{1}{i2\pi}\right)^n D^n f \quad , \tag{26.12c}$$

and

$$\mathcal{F}\left[t^n f(t)\right] = \left(-\frac{1}{i2\pi}\right)^n D^n F \quad . \tag{26.12d}$$

!▶**Example 26.13:**   In example 22.3 on page 344, we unsuccessfully attempted to find the Fourier transform of $t^2\left(1+t^2\right)^{-1}$ using the observation that

$$\frac{t^2}{1+t^2} = t\,f(t) \qquad \text{where} \quad f(t) = \frac{t}{1+t^2}$$

and the fact, derived earlier, that the Fourier transform of this f is

$$F(\omega) = \begin{cases} i\pi e^{2\pi\omega} & \text{if } \omega < 0 \\[2mm] -i\pi e^{-2\pi\omega} & \text{if } 0 < \omega \end{cases} \quad .$$

Since F is piecewise smooth and is clearly Fourier transformable (in fact F is clearly in $\mathcal{A}$), theorem 26.3 assures us that

$$\mathcal{F}\left[\frac{t^2}{1+t^2}\right]\bigg|_\omega = \mathcal{F}[t\,f(t)]|_\omega = -\frac{1}{i2\pi}DF(\omega) \quad .$$

Moreover, because F is piecewise smooth and has its only discontinuity at $\omega = 0$, we know that

$$DF(\omega) = F'(\omega) + j_0\,\delta(\omega)$$

where $F'(\omega)$ is the classical derivative of F and j_0 is the jump in F at $\omega = 0$.
    Using the above formula for $F(\omega)$ we see that

$$F'(\omega) = \begin{cases} i2\pi^2 e^{2\pi\omega} & \text{if } \omega < 0 \\[2mm] i2\pi^2 e^{-2\pi\omega} & \text{if } 0 < \omega \end{cases} = i2\pi^2 e^{-2\pi|\omega|}$$

and

$$j_0 = \lim_{\omega\to 0^+} F(\omega) - \lim_{\omega\to 0^-} F(\omega)$$

$$= \lim_{\omega\to 0^+} -i\pi e^{-2\pi\omega} - \lim_{\omega\to 0^-} i\pi e^{2\pi\omega} = -i2\pi \quad .$$

So

$$DF(\omega) = F'(\omega) + j_0\,\delta(\omega) = i2\pi^2 e^{-2\pi|\omega|} - i2\pi\,\delta(\omega) \quad ,$$

and thus,

$$\mathcal{F}\left[\frac{t^2}{1+t^2}\right]\bigg|_\omega = -\frac{1}{i2\pi}DF(\omega)$$

$$= -\frac{1}{i2\pi}\left[i2\pi^2 e^{-2\pi|\omega|} - i2\pi\,\delta(\omega)\right] = -\pi e^{-2\pi|\omega|} + \delta(\omega) \quad .$$

?►Exercise 26.12:   Redo the last example using the product rule and the observation that

$$F(\omega) = i\pi e^{2\pi\omega} \text{step}(-\omega) - i\pi e^{-2\pi\omega} \text{step}(\omega) .$$

Oddly enough, while we cannot yet adequately describe (or even define) the generalized derivatives of the delta function, we can easily describe their transforms. They are, as the next example shows, simple polynomials.

!►Example 26.14:   Letting $f = \delta$ (so $F = \mathcal{F}[\delta] = 1$), identity (26.12a) becomes

$$\mathcal{F}\left[D^n\delta\right]\Big|_\omega = (i2\pi\omega)^n \cdot 1 = (i2\pi\omega)^n .$$

Convolution and Generalized Derivatives

In chapter 24, we saw that the convolution of two suitably integrable, piecewise smooth functions f and g is a smooth function, and that

$$(f * g)' = f * (g')   \text{provided } g \text{ is continuous} .$$

You probably now suspect that $D(f * g) = f * Dg$ whether or not g is continuous. For particular choices of f and g this is easily verified.

!►Example 26.15:   Let f be any function in $\mathcal{A}$. In exercise 24.7 on page 391 we saw that $f * \text{step}$ is continuous and piecewise smooth (so, $D(f * \text{step}) = (f * \text{step})'$). We also saw that

$$(f * \text{step})' = f .$$

But we now know

$$f = f * \delta = f * D \text{step} .$$

So,

$$D(f * \text{step}) = (f * \text{step})' = f * D \text{step} .$$

More generally, the following theorem can be proven.

Theorem 26.4
Let f and g be any two Fourier transformable functions. If $f * g$ and $f * Dg$ exist, then $D(f * g) = f * Dg$.

In the spirit of this chapter, here is a slightly naive proof.

PROOF (naive):   Let $F = \mathcal{F}[f]$ and $G = \mathcal{F}[g]$. Using one of the generalized Fourier differentiation identities, we see that

$$\mathcal{F}[D(f * g)]\Big|_\omega = i2\pi\omega \,\mathcal{F}[f * g]\Big|_\omega = i2\pi\omega[F(\omega)G(\omega)] = F(\omega)[i2\pi\omega G(\omega)] .$$

Taking the inverse transform of the left and right sides of this string of equalities then gives

$$D(f * g) = \mathcal{F}^{-1}\left[F(\omega)[i2\pi\omega G(\omega)]\right]$$
$$= \mathcal{F}^{-1}[F(\omega)] * \mathcal{F}^{-1}[i2\pi\omega G(\omega)] = f * Dg .$$

Additional Exercises

26.13 a. *What value should c be so that $\delta(3x) = c\delta(x)$?*

    b. *In general, what should c be so that $\delta(bx) = c\delta(x)$ if b is a nonzero real value?*

    c. *Assume a and b are fixed real values with $b \neq 0$. What values should c and γ be so that $\delta_a(bx) = c\delta_\gamma(x)$?*

26.14. *For each of the following, find the Fourier transform, and sketch the graphical representation of the transform found:*

    a. $e^{i8\pi t}$ **b.** $\cos(6\pi t)$ **c.** $\sin(6\pi t)$

    d. $\sin^2(6\pi t)$ **e.** $e^{i6\pi t}\cos(6\pi t)$ **f.** $1 + 2\delta(t) - e^{-|t|}$

26.15. *Evaluate/simplify each of the following expressions:*

    a. $\displaystyle\int_{-\infty}^{\infty} x^2\delta(x)\,dx$ **b.** $\displaystyle\int_{-\infty}^{\infty} x^2\delta_3(x)\,dx$ **c.** $\displaystyle\int_{-\infty}^{\infty} x^2\delta_3(2x-5)\,dx$

    d. $x^2 * \delta(x)$ **e.** $x^2 * \delta_3(x)$ **f.** $x^2\delta(x)$

    g. $x^2\delta_3(x)$ **h.** $\displaystyle\int_0^5 x^2\delta_3(x)\,dx$ **i.** $\displaystyle\int_5^{\infty} x^2\delta_3(x)\,dx$

26.16. *Find the Fourier transform for each of the following periodic functions, and sketch the real and imaginary parts of the resulting array of delta functions. (Note: You found the Fourier series for these functions in exercise 12.3 on page 152.)*

    a. $f(t) = \begin{cases} 0 & \text{if } -1 < t < 0 \\ 1 & \text{if } 0 < t < 1 \\ f(t-2) & \text{in general} \end{cases}$

    b. $g(t) = \begin{cases} e^t & \text{if } 0 < t < 1 \\ g(t-1) & \text{in general} \end{cases}$

    c. $\text{evensaw}(t)$ *the even sawtooth function sketched in figure 26.4a*

    d. $\text{oddsaw}(t)$, *the odd sawtooth function, sketched in figure 26.4b*

    e. $|\sin(2\pi t)|$

26.17. *Using the Bessel's inequality derived in exercise 12.4 on page 152, show that neither of the two series obtained in example 26.7 on page 430 can be the Fourier series for any piecewise continuous, periodic function.*

26.18. *Sketch each of the following regular arrays of delta functions, and compute their Fourier transforms:*

    a. $\displaystyle\sum_{k=-\infty}^{\infty} k^3\,\delta_{3k}$ **b.** $\displaystyle\sum_{k=-\infty}^{\infty} \frac{4k^2}{1+k^2}\,\delta_{k/2}$

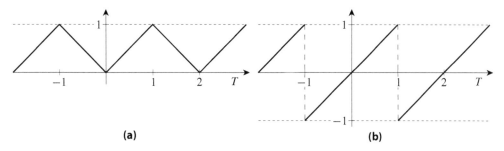

(a)   **(b)**

Figure 26.4: Two and a half periods of (a) evensaw(t), the even sawtooth function, and (b) oddsaw(t), the odd sawtooth function for exercises 26.16 c and 26.16 d.

26.19. *Based on the methods discussed in this chapter for computing the Fourier transforms of periodic functions and regular arrays, justify the following two claims:*

    a. *If f is periodic with period p, then $\mathcal{F}[f]$ is a regular array of delta functions with spacing $1/p$. (Compare this statement with the quasi-theorem on page 428.)*

    b. *If f is a regular array of delta functions with spacing Δt, then $\mathcal{F}[f]$ is periodic with period $1/\Delta t$.*

26.20. *Sketch*

$$f = \sum_{k=-\infty}^{\infty} (-1)^k \, \delta_k$$

and convince yourself that f is a periodic and regular array of delta functions. Then find the Fourier series of f, and compute $\mathcal{F}[f]$ using this Fourier series.

26.21. *Assume f is a piecewise smooth function on the real line with infinitely many disconti-nuities. Let $\{\ldots, x_1, x_2, x_3, \ldots\}$ be the set of all points at which f is discontinuous. For each of these x_k's, let j_k be the jump in f at x_k and let*

$$h_k(x) = \begin{cases} \text{step}(x - x_k) - 1 & \text{if } x_k < 0 \\ \text{step}(x - x_k) & \text{if } 0 \le x_k \end{cases} .$$

    (In the following, the summations are over "all discontinuities of f".)

    a. *Sketch the graph of h_k for each of the following cases:*

        i. $x_k < 0$ **ii.** $0 \ge x_k$

    b. *Verify that only a finite number of terms in $\sum_k j_k h_k(x)$ are nonzero for each real value x. (Hence, this infinite summation converges and defines a piecewise continuous function on the real line.)*

    c. *Verify that the following function g is piecewise smooth and continuous on the real line:*

$$g = f - \sum_k j_k h_k \quad .$$

26.22. *Some of the claims made in theorem 26.1 were confirmed only for the cases where the discontinuous function, f, had only finitely many discontinuities. Using the results of the previous exercise and naively assuming*

$$D \sum_k j_k h_k = \sum_k j_k D h_k \quad ,$$

verify the following parts of theorem 26.1 for the cases where f has an infinite number of discontinuities.

a. The generalized fundamental theorem of calculus.

b. The generalized integration by parts formula.

c. The limited chain rule formula.

26.23. Sketch the graph of each of the following functions. Then find both the classical and the generalized derivative for the function just graphed.

a. $\mathrm{ramp}(x)$

b. $g(x) = \begin{cases} x^3 & \text{if } -1 < x < 2 \\ 0 & \text{otherwise} \end{cases}$

c. $h(x) = \begin{cases} x & \text{if } x < -1 \\ x^2 & \text{if } -1 < x < 2 \\ x^3 & \text{if } 2 < x \end{cases}$

d. $f(x) = \begin{cases} 0 & \text{if } -1 < x < 0 \\ 1 & \text{if } 0 < x < 1 \\ f(x-2) & \text{in general} \end{cases}$

26.24. Using either the limited chain rule or the generalized product rule (or both), find the generalized derivatives of the following:

a. $e^{-3x}\,\mathrm{step}(x)$ **b.** $e^{3x}\,\mathrm{step}(2-x)$

c. $\mathrm{ramp}(x)\,\mathrm{rect}_{(1,3)}(x)$ **d.** $\dfrac{1}{3+i2\pi x}\,\mathrm{pulse}_3(x)$

e. $\displaystyle\sum_{k=-\infty}^{\infty} \mathrm{rect}_{(0,1)}(x-2k)$ *(compare with problem 26.23 d, above)*

26.25. Compute the following second derivatives:

a. $D^2\!\left[x^3\,\mathrm{step}(2-x)\right]$ **b.** $D^2\!\left[(x^2-4)\,\mathrm{pulse}_2(x)\right]$

26.26. Using the generalized Fourier differential identities listed in theorem 26.3, find the Fourier transforms of:

a. $\dfrac{t}{3-i2\pi t}$ **b.** $t\sin(6\pi t)$

c. $t^2 + 3t - 4$ **d.** $D^2\delta_3(t)$

26.27. Each of the following differential equations has a classically transformable solution (assuming the derivatives are generalized derivatives). Find those solutions.

a. $\dfrac{dy}{dt} + 3y = \delta(t)$ **b.** $\dfrac{d^2y}{dt^2} - 9y = \delta(t)$

c. $\dfrac{d^2y}{dt^2} - 4\dfrac{dy}{dt} + 3y = \delta(t)$

27

Fourier Analysis in the Analysis of Systems

Let us pause in our development of the theory of Fourier transforms to see how these transforms might be useful in applications. Rather than closely examine any particular application, however, we will discuss the analysis appropriate for a large class of problems in which the goal is to predict the outputs of various physical or mathematical devices based on the expected inputs. Consequently, our discussion will remain somewhat general.

27.1 Linear, Shift-Invariant Systems
Systems, Linearity and Shift Invariance

Many physical systems (such as telescopes and radios) convert some type of input into some sort of output (for example, an optical telescope will convert a given optical image into an enlarged optical image, and a radio will convert radio waves into sound waves). Mathematically, these various inputs and outputs can often be described by functions on the real line, which, naturally, we will refer to as *input functions* and *output functions*, respectively. The action of the physical system can then be modeled by the operator that converts the various input functions into the corresponding output functions. Following standard practice, we will refer to this operator as a "system". Thus, as far as we are concerned in this chapter, a *system* is an operator. Its domain is the set of all possible input functions, and its range is the set of all possible output functions.

Given a system S, we will indicate that y is the output function corresponding to input function x (we will also refer to y as the response of the system to x) via

$$ y = S[x] \qquad \text{or} \qquad S: x \mapsto y \quad . $$

Occasionally, it will be convenient to denote both the input and corresponding output functions by the same basic symbol with subscripts distinguishing between the input and output (I for input and O for output),

$$ f_O = S[f_I] \qquad \text{or} \qquad S: g_I \mapsto g_O \quad . $$

Since a system is just an operator, we can view all known operators as "systems". In particular, the Fourier transform and the generalized derivative from chapter 26,

$$ f_O = \mathcal{F}[f_I] \qquad \text{and} \qquad f_O = D[f_I] = Df_I \quad , $$

are systems. Other basic operators give us other systems.

!▶ Example 27.1:   Let a be some fixed, positive value. The corresponding "amplification by a factor of a" system $\mathcal{A}$ is simply "multiply by a",

$$\mathcal{A}: f(t) \longmapsto af(t) \quad .$$

So, for each input f_I, the corresponding output is $f_O = \mathcal{A}[f_I] = af_I$.
   In particular, if $a = 4$ and $f(t) = e^{-3t} \text{step}(t)$, then

$$\mathcal{A}[f]|_t = 4e^{-3t} \text{step}(t) \quad .$$

For future use, here are a few more systems:

$$\mathcal{Q}: f(t) \longmapsto [f(t)]^2 \tag{27.1}$$

$$\mathcal{I}: f(t) \longmapsto \int_0^t f(s)\,ds \tag{27.2}$$

$$\mathcal{K}: f(t) \longmapsto f * \text{pulse}_1(t) \tag{27.3}$$

?▶ Exercise 27.1:   Let $f(t) = e^{-3t} \text{step}(t)$ and, using the above definitions, compute

$$\mathcal{F}[f]|_t \quad , \quad D[f]|_t \quad , \quad \mathcal{Q}[f]|_t \quad , \quad \mathcal{I}[f]|_t \quad \text{and} \quad \mathcal{K}[f]|_t \quad .$$

Not all systems are amiable to the sort of analysis we are about to develop. To help identify those that are amiable, we will introduce the ideas of system "linearity" and "shift invariance".
   Linearity we know about. A system (operator) S is linear if and only if

$$S[\alpha f + \beta g] = \alpha S[f] + \beta S[g]$$

for every pair of input functions f and g, and every pair of constants α and β.
   The system S is *shift invariant* if, whenever f_I and g_I are input functions related by

$$g_I(t) = f_I(t - a)$$

for some real constant a, then the corresponding outputs $f_O = S[f_I]$ and $g_O = S[g_I]$ are also related by

$$g_O(t) = f_O(t - a) \quad .$$

If the inputs and outputs are viewed as functions of time, then shift invariance simply means that the only effect of delaying the input by a certain length of time is to delay the output by the same length of time.[1]
   Sometimes the above formulas defining the shift-invariance of S are abbreviated as

$$f_O(t) = S[f_I(t)] \quad \Longrightarrow \quad f_O(t - a) = S[f_I(t - a)]$$

for every real value a and input function f_I. This is fine so long as we keep in mind that the "t" appearing with f_O is a true variable, while the "t" appearing with f_I is a dummy (i.e., internal) variable. On occasion, to avoid possible confusion and the stupid mistakes that could result, we will use another symbol for the internal variable, say, τ, and express the above as

$$S[f_I(\tau - a)]|_t = S[f_I(\tau)]|_{t-a} \quad .$$

---

[1] Warning: While the term "shift invariant" is in widespread use for the property just described, so are other terms, including "translation invariant", "time invariant", "stationary" and "fixed".

For the rest of this section, we will mainly consider systems that are linear and shift invariant. For obvious reasons, the systems satisfying these properties are commonly referred to as *linear, shift-invariant systems*. For equally obvious reasons, it is standard practice to further abbreviate our terminology and refer to these systems as *LSI systems*.

!▶ Example 27.2: Let Q be the system given by formula (27.1). This system is not linear. We can easily verify this by taking, say, $\alpha = 2$, $\beta = 0$, $f(t) = t$ and $g(t) = 0$. Then

$$Q[\alpha f + \beta g] = Q[2t] = (2t)^2 = 4t^2 ,$$

while

$$\alpha Q[f] + \beta Q[g] = 2Q[t] = 2t^2 .$$

Clearly then,

$$Q[\alpha f + \beta g] \neq \alpha Q[f] + \beta Q[g] .$$

However, this system is shift invariant. To verify this, suppose f_I and g_I are any two input functions related by

$$g_I(t) = f_I(t - a)$$

for some real constant a. Then

$$f_O(t) = Q[f_I]|_t = [f_I(t)]^2$$

and so,

$$f_O(t - a) = [f_I(t - a)]^2 .$$

But also,

$$g_O(t) = Q[g_I]|_t = [g_I(t)]^2 = [f_I(t - a)]^2 .$$

Comparing the above formulas for $g_O(t)$ and $f_O(t - a)$, we see that

$$g_O(t) = [f_I(t - a)]^2 = f_O(t - a) .$$

!▶ Example 27.3: Let I be the integral system with formula (27.2). Given any two functions f and g, and constants α and β, the linearity of integration gives us

$$I[\alpha f + \beta g]|_t = \int_0^t [\alpha f(s) + \beta g(s)]\, ds$$

$$= \alpha \int_0^t f(s)\, ds + \beta \int_0^t g(s)\, ds = \alpha I[f]|_t + \beta I[g]|_t .$$

So I is a linear system.

On the other hand, if we take, say, $f(t) = 3t^2$ and $a = 1$, we see that

$$I[f(\tau - a)]|_t = \int_0^t f(s - a)\, ds$$

$$= \int_0^t 3(s - 1)^2\, ds$$

$$= (s - 1)^3 \Big|_{s=0}^t = (t - 1)^3 - (-1)^3 ,$$

while

$$\mathcal{I}[f(\tau)]|_{t-a} = \int_0^{t-a} f(s)\,ds$$

$$= \int_0^{t-1} 3s^2\,ds$$

$$= s^3\Big|_{s=0}^{t-1} = (t-1)^3 \quad .$$

Thus, for this function f and constant a,

$$\mathcal{I}[f(\tau - a)]|_t \neq \mathcal{I}[f(\tau)]|_{t-a} \quad ,$$

showing that $\mathcal{I}$ is not shift invariant.

!▶ Example 27.4:   The system $\mathcal{K}$ defined by formula (27.3) is an LSI system. To verify this, let f and g be any two input functions, and let α and β be any two constants. By the linearity of convolution,

$$\mathcal{K}[\alpha f + \beta g] = [\alpha f + \beta g] * \text{pulse}_1$$

$$= \alpha f * \text{pulse}_1 + \beta g * \text{pulse}_1 = \alpha \mathcal{K}[f] + \beta \mathcal{K}[g] \quad ,$$

confirming that $\mathcal{K}$ is linear. Moreover, if a is any fixed real value,

$$\mathcal{K}[f(\tau - a)]|_t = f(t-a) * \text{pulse}_1(t)$$

$$= \int_{-\infty}^{\infty} f((t-s) - a)\,\text{pulse}_1(s)\,ds$$

$$= \int_{-\infty}^{\infty} f((t-a) - s)\,\text{pulse}_1(s)\,ds$$

$$= f * \text{pulse}_1(t-a)$$

$$= \mathcal{K}[f(\tau)]|_{t-a} \quad .$$

So this system is also shift invariant.

For reasons that will become obvious in a few pages, systems for which the output is the convolution of the input with some fixed function — such as in the last example — can often be expected in applications. Given this last example, you should have no trouble verifying that such systems are both linear and shift invariant.

?▶ Exercise 27.2:   Let h be some fixed function on the real line, and let S be the system whose output is given by the convolution of the input with h,

$$S[f] = f * h \quad .$$

Show that S is linear and shift invariant. (Also, what is the domain of such a system?)

Operational Continuity

In much of what follows, we will need whatever system S is at hand to satisfy one more property, namely, that if $\{f_\gamma\}_{\gamma=1}^{\gamma_0}$ is any sequence of input functions converging to some input function, then

$$S\left[\lim_{\gamma \to \gamma_0} f_\gamma\right] = \lim_{\gamma \to \gamma_0} S[f_\gamma] \quad . \tag{27.4}$$

We'll refer to this property as *operational continuity* and refer to any system S for which the above statement holds as being *operationally continuous*.

In particular, suppose we have a linear, operationally continuous system S, and a convergent infinite series

$$\sum_{k=1}^{\infty} c_k \, \phi_k(t) \tag{27.5}$$

where each c_k is a constant and each ϕ_k is an input function for S. Strictly speaking, the linearity of S only assures us that

$$S\left[\sum_{k=1}^{N} c_k \, \phi_k(t)\right] = \sum_{k=1}^{N} c_k S[\phi_k(t)] \qquad \text{for} \quad N = 1, 2, 3, \ldots \quad .$$

Combine this with the operational continuity of S, however, and we get

$$S\left[\sum_{k=1}^{\infty} c_k \, \phi_k(t)\right] = S\left[\lim_{N \to \infty} \sum_{k=1}^{N} c_k \, \phi_k(t)\right]$$

$$= \lim_{N \to \infty} S\left[\sum_{k=1}^{N} c_k \, \phi_k(t)\right]$$

$$= \lim_{N \to \infty} \sum_{k=1}^{N} c_k S[\phi_k(t)] = \sum_{k=1}^{\infty} c_k S[\phi_k(t)] \quad .$$

So linearity along with operational continuity allows us to write

$$S\left[\sum_{k=1}^{\infty} c_k \, \phi_k(t)\right] = \sum_{k=1}^{\infty} c_k S[\phi_k(t)] \quad .$$

We will find this important on at least a couple of occasions.

Though important, operational continuity is not something we will test for in this chapter. One reason for this is that the above definition failed to include an explanation of just what

$$\lim_{\gamma \to \gamma_0} f_\gamma$$

means when the f_γ's are functions instead of numbers. As you may recall, this was an issue when we discussed the convergence of partial sums of Fourier series in chapter 13. There we found it necessary to develop three different notions for convergence: pointwise convergence, uniform convergence and convergence in the norm. Unfortunately, our sequences now may involve such things as delta functions, and developing a rigorous notion of convergence for these sequences would take us deep into part IV of this text.

Fortunately, LSI systems that are not operationally continuous rarely, if ever, arise in practice. That is probably why so many other authors assume operational continuity without comment.[2] So, for the rest of this chapter, let us pretend that all LSI systems of interest are operationally continuous, that all classically convergent limits and series are convergent in whatever sense we need, and that we may naively use equation (27.4) whenever convenient. After all, in this chapter we are illustrating applications of Fourier analysis, not rigorously developing the theory.

Differential Equations and Systems, Part I

Consider the differential equation

$$\frac{dy}{dt} + 3y = f \quad . \tag{27.6}$$

This is a simple first order, linear, nonhomogeneous differential equation relating two functions y and f. As you can easily verify (exercise 27.4 on page 461), the process of computing f from any given function y can be viewed as a linear, shift-invariant system. In applications, however, you are rarely interested in computing f from y. Instead, you are more likely to know f and need to find a corresponding function y satisfying the above equation.

One such solution can easily be found using Fourier transforms whenever f is Fourier transformable. Letting $Y = \mathcal{F}[y]$ and $F = \mathcal{F}[f]$, and using the Fourier differential identities, we can quickly convert the above differential equation relating y and f to an algebraic equation relating Y and F:

$$\mathcal{F}\left[\frac{dy}{dt} + 3y\right]\bigg|_\omega = \mathcal{F}[f]|_\omega$$

$\hookrightarrow \qquad\qquad i2\pi\omega Y(\omega) + 3Y(\omega) = F(\omega)$

$\hookrightarrow \qquad\qquad (i2\pi\omega + 3)Y(\omega) = F(\omega)$

$\hookrightarrow \qquad\qquad Y(\omega) = \dfrac{1}{3 + i2\pi\omega} F(\omega) \quad .$

Therefore,

$$y = \mathcal{F}^{-1}[Y] = \mathcal{F}^{-1}\left[\frac{1}{3 + i2\pi\omega} F(\omega)\right] \quad ,$$

which, after application of a well-known convolution identity, can be written as

$$y = \mathcal{F}^{-1}\left[\frac{1}{3 + i2\pi\omega}\right] * \mathcal{F}^{-1}[F(\omega)] = h * f$$

where

$$h(t) = \mathcal{F}^{-1}\left[\frac{1}{3 + i2\pi\omega}\right]\bigg|_t = e^{-3t} \operatorname{step}(t) \quad .$$

Rewriting $h * f$ in integral form, we obtain the following two equivalent formulas for $y(t)$:

$$y(t) = \int_{-\infty}^{\infty} f(t - s)e^{-3s} \operatorname{step}(s)\, ds = \int_0^{\infty} f(t - s)e^{-3s}\, ds$$

and

$$y(t) = \int_{-\infty}^{\infty} f(s)e^{-3(t-s)} \operatorname{step}(t - s)\, ds = e^{-3t} \int_{-\infty}^{t} f(s)e^{3s}\, ds \quad .$$

---

[2] Or else they can't imagine systems not being operationally continuous. Cynical mathematicians, however, can.

Observe that the above process for computing y from f can be viewed as a system with f being the input function and y the output function. Further observe that this process reduces to computing a convolution,

$$y = h * f \quad \text{where} \quad h(t) = e^{-3t} \text{step}(t) \quad . \tag{27.7}$$

So, as you verified in exercise 27.2, this system is linear and shift invariant.

You should also note that there was nothing special about the differential equation we started with. Clearly, if we have any other n^{th} order, linear differential equation with constant coefficients,

$$a_n \frac{d^n y}{dt^n} + a_{n-1} \frac{d^{n-1} y}{dt^{n-1}} + \cdots + a_1 \frac{dy}{dt} + a_0 y = f \quad , \tag{27.8}$$

and take the Fourier transform of both sides, we get

$$P(\omega)Y(\omega) = F(\omega)$$

where $Y = \mathcal{F}[y]$, $F = \mathcal{F}[f]$ and P is some polynomial (in the above, $P(\omega) = i2\pi\omega + 3$). As long as $1/P(\omega)$ is classically transformable, we can continue the process described above (with $P(\omega)$ replacing $i2\pi\omega + 3$) to obtain

$$y = h * f \quad \text{where} \quad h = \mathcal{F}^{-1}\left[\frac{1}{P(\omega)}\right] \quad .$$

Consequently, whenever $1/P(\omega)$ is classically transformable, this process for finding a solution y to differential equation (27.8) can be treated as an LSI system with input f and output y. We will refer to this system as the *LSI system corresponding to the differential equation*. We will have more to say about solving these systems after discussing impulse response and transfer functions.

Before going on, let us note that the above process only gives one particular solution to the given differential equation. The general solution to equation (27.8), as you should recall from a course in differential equations, is given by

$$y_{\text{gen}} = y_{\text{p}} + y_{\text{h}}$$

where y_{p} is any particular solution and y_{h} is the general solution (containing arbitrary constants) to the corresponding homogeneous equation,

$$a_n \frac{d^n y}{dt^n} + a_{n-1} \frac{d^{n-1} y}{dt^{n-1}} + \cdots + a_1 \frac{dy}{dt} + a_0 y = 0 \quad .$$

For the differential equation we started with, equation (27.6), the corresponding homogeneous equation is

$$\frac{dy}{dt} + 3y = 0 \quad .$$

As you can easily verify, the general solution to this equation is

$$y_{\text{h}}(t) = ce^{-3t}$$

where c is an arbitrary constant. So the general solution to equation (27.6) is

$$y(t) = h * f(t) + ce^{-3t} \quad \text{where} \quad h(t) = e^{-3t} \text{step}(t)$$

and c is an arbitrary constant.

27.2 Computing Outputs for LSI Systems
Input Response and Transfer Functions

As noted a few pages ago, one can often expect the output of an LSI system to be the convolution of the input with some fixed function. To help see why, let us first derive a formula for the output of an LSI system when the input is a convolution.

Lemma 27.1

Let S be an LSI system with a piecewise continuous input function g. Suppose f is another piecewise continuous function for which the convolution $f * g$ exists and is an input function for S. Then

$$S[f * g] = f * S[g] .$$ (27.9)

Before starting our proof/derivation of this lemma, do recall that we are implicitly assuming our systems are operationally continuous, with the definition of operational continuity based on a limit we only naively understand. Consequently, our proof/derivation is not completely rigorous (we'll call it naive). However, it does illustrate how this lemma would be proven if a complete rigorous definition of operational continuity had been given.

PROOF (naive):   Recalling that an integral on $(-\infty, \infty)$ can be written as the limit of Riemann sums (see page 252), we have, for each real value τ,

$$f * g(\tau) = \int_{-\infty}^{\infty} f(s)g(\tau - s)\,ds = \lim_{\Delta s \to 0} \sum_{k=-\infty}^{\infty} f(s_k)g(\tau - s_k)\Delta s$$

where $s_k = k\Delta s$. Plugging this into the system and using the operational continuity, linearity and shift invariance of S, we see that

$$S[f * g(\tau)]\big|_t = S\left[\lim_{\Delta s \to 0} \sum_{k=-\infty}^{\infty} f(s_k)g(\tau - s_k)\Delta s \right]\Bigg|_t$$

$$= \lim_{\Delta s \to 0} S\left[\sum_{k=-\infty}^{\infty} f(s_k)g(\tau - s_k)\Delta s \right]\Bigg|_t$$

$$= \lim_{\Delta s \to 0} \sum_{k=-\infty}^{\infty} f(s_k)S[g(\tau - s_k)]\big|_t \,\Delta s$$

$$= \lim_{\Delta s \to 0} \sum_{k=-\infty}^{\infty} f(s_k)S[g(\tau)]\big|_{t-s_k} \,\Delta s ,$$

for each real value t. However, the last formula above is just another limit of Riemann sums and, thus, yields another integral,

$$\lim_{\Delta s \to 0} \sum_{k=-\infty}^{\infty} f(s_k)S[g(\tau)]\big|_{t-s_k} \,\Delta s = \int_{-\infty}^{\infty} f(s)S[g(\tau)]\big|_{t-s} \,ds ,$$

which just happens to be the convolution integral for f and $S[g]$. Thus, the above sequence of equalities reduces to

$$S[f * g(\tau)]|_t = \int_{-\infty}^{\infty} f(s)S[g(\tau)]|_{t-s} \, ds = f * S[g](t) \quad . \qquad \blacksquare$$

The main reason for assuming f and g are piecewise continuous in the lemma above was so we could use the classical notion of an integral being the limit of a Riemann sum in the proof. In fact, identity (27.9) can be shown to hold for more general choices for f and g. Of particular interest is the case where one of these functions is the delta function. We'll state the result and leave the (naive) proof as an exercise (exercise 27.5 at the end of this chapter).

Lemma 27.2
Suppose S is an LSI system for which the delta function, δ, is an input function. Then, for any piecewise continuous input function f,

$$S[\delta * f] = S[\delta] * f$$

provided the convolution $S[\delta] * f$ exists.

Now suppose we know the output of an LSI system when the input is the delta function. Following standard conventions, we will call this output the *impulse response function* (for the system S) and denote it by h. That is,

$$h = \text{the impulse response function for } S = S[\delta] \quad .$$

Recall, also, that $f = \delta * f$ for any piecewise continuous function on $\mathbb{R}$. Combining this with our last lemma, we see that, for any piecewise continuous input function f,

$$S[f] = S[\delta * f] = S[\delta] * f = h * f$$

(provided the convolution $h * f$ exists). In practice, this often means we can compute the outputs corresponding to all inputs of interest by simply convolving the input functions with the impulse response function. Thus, in a sense, the impulse response function tells us everything we need to know about the system.

Up to now in this chapter, we've hardly said anything about Fourier analysis. We can rectify that by defining the *transfer function* H for an LSI system S to be the Fourier transform of the system's impulse response function. That is,

$$H = \text{the transfer function for } S = \mathcal{F}[h]$$

where

$$h = S[\delta] \quad .$$

Now let $F = \mathcal{F}[f]$. By the above, we know $S[f] = h * f$. Taking the Fourier transform of this, we get

$$\mathcal{F}[S[f]] = \mathcal{F}[h * f] = \mathcal{F}[h]\,\mathcal{F}[f] = HF \quad .$$

Thus,

$$S[f] = \mathcal{F}^{-1}[HF] \quad .$$

This gives us another way to compute the output of an LSI system, provided, of course, the functions involved are Fourier transformable and the product HF exists and is transformable.

We've just derived some important relations. Let us preserve them in a theorem.

Theorem 27.3

Suppose S is an LSI system with impulse response function h and transfer function H. Let f be any (piecewise continuous) input function.[3] Then the corresponding output is given by

$$S[f] = h * f \tag{27.10}$$

provided this convolution exists. Alternatively, if f is Fourier transformable and the product HF is Fourier transformable, then

$$S[f] = \mathcal{F}^{-1}[HF] \qquad \text{where} \quad F = \mathcal{F}[f] \quad . \tag{27.11}$$

It may be tempting to assume every LSI system has both an impulse response function and a transfer function. In practice, this often seems to be the case simply because so many of the LSI systems we commonly encounter have both impulse response and transfer functions. In fact, though, it is easy to define an LSI system not having one or the other of these two functions. We'll demonstrate this in the next two examples. For both we will use the function $\exp(\pi x^2)$, which is *not* Fourier transformable. Certainly, it is not classically transformable (you can verify this), nor is it even Fourier transformable using the more general theory developed in part IV of this text (you'll have to trust the author on this).

!▶ **Example 27.5:** *Let S be the system given by*

$$S[f]|_t = e^{\pi t^2} * f(t) \quad .$$

Since this system is given by a convolution, we know it is linear and shift invariant. The domain of this system is the set of all functions on $\mathbb{R}$ for which the convolution with $\exp(\pi t^2)$ exists. This includes the delta function. So this system has an impulse response function, namely,

$$h(t) = S[\delta]|_t = e^{\pi t^2} * \delta(t) = e^{\pi t^2} \quad .$$

However, because this h is not Fourier transformable, it makes no sense to talk about this system's transfer function, $H = \mathcal{F}[h]$.

!▶ **Example 27.6:** *Now define another system S by the formula*

$$S[f]|_t = \mathcal{F}^{-1}\left[e^{\pi \omega^2} F(\omega)\right]\Big|_t \qquad \text{where} \quad F = \mathcal{F}[f] \quad . \tag{27.12}$$

The domain of this system (i.e., the set of all valid input functions) consists of all Fourier transformable functions such that the products of their Fourier transforms with $\exp(\pi \omega^2)$ are also Fourier transformable. In particular, the function $2\operatorname{sinc}(2\pi t)$ is a valid input function because the product of its Fourier transform, $\operatorname{pulse}_1(\omega)$, with $\exp(\pi \omega^2)$ is the absolutely integrable function

$$e^{\pi \omega^2} \operatorname{pulse}_1(\omega) = \begin{cases} e^{\pi \omega^2} & \text{if } -1 < \omega < 1 \\ 0 & \text{otherwise} \end{cases} \quad .$$

So,

$$S[2\operatorname{sinc}(2\pi \tau)]|_t = \mathcal{F}^{-1}\left[e^{\pi \omega^2} \operatorname{pulse}_1(\omega)\right]\Big|_t = \int_{-1}^{1} e^{\pi \omega^2} e^{i2\pi \omega t} \, d\omega \quad ,$$

---

[3] Using a more general theory, it can be verified that piecewise continuity is not necessary.

which, though not easily evaluated, is a well-defined integral for each real value t.

On the other hand, if $f(t) = \delta(t)$, then $F(\omega) = 1$ and

$$e^{\pi\omega^2} F(\omega) = e^{\pi\omega^2} \quad ,$$

which is not Fourier transformable. So the delta function is not a valid input for this system, and it makes no sense to refer to an impulse response function, $h = S[\delta]$, for this system.

Strictly speaking, this system does not have a transfer function either. After all, it hardly makes sense to have the transform of the impulse response function when there is no impulse response function. However, if you compare equations (27.11) and (27.12), you will see that $\exp(\pi\omega^2)$ plays the role of the transfer function in computing the output from this system. (If systems such as this played a significant role in applications, we would broaden our definition of transfer functions to include such role players. But they don't, so we won't.)

Differential Equations and Systems, Part II

Let us reconsider the problem of solving

$$\frac{dy}{dt} + 3y = f \quad ,$$

only this time let us first find the impulse response function and transfer function for the corresponding LSI system $S: f \mapsto y$.

Since the impulse response function h is the output (denoted above by y) corresponding to a delta function input (denoted above by f), we can find h by solving

$$\frac{dh}{dt} + 3h = \delta$$

using the process described in the previous subsection on differential equations (see page 452). First we take the Fourier transform of both sides of this equation,

$$\mathcal{F}\left[\frac{dh}{dt} + 3h\right]\bigg|_{\omega} = \mathcal{F}[\delta]\big|_{\omega}$$

$$\hookrightarrow \qquad i2\pi\omega H(\omega) + 3H(\omega) = 1$$

$$\hookrightarrow \qquad (i2\pi\omega + 3)H(\omega) = 1 \quad .$$

Dividing through by $i2\pi\omega + 3$ gives us the transfer function,

$$H(\omega) = \frac{1}{i2\pi\omega + 3} = \frac{1}{3 + i2\pi\omega} \quad .$$

Taking the inverse Fourier transform of this then gives us the impulse response function,

$$h(t) = \mathcal{F}^{-1}[H]\big|_t = \mathcal{F}^{-1}\left[\frac{1}{3 + i2\pi\omega}\right]\bigg|_t = e^{-3t}\,\text{step}(t) \quad .$$

Now suppose we want a solution to

$$\frac{dy}{dt} + 3y = f$$

for some given f. Since we've just found the impulse response function h for the corresponding LSI system $S: f \mapsto y$, we can compute a solution using

$$y = h * f \qquad \text{where} \quad h(t) = e^{-3t} \, \text{step}(t) \quad,$$

provided the convolution exists. (Notice that this is the same solution, formula (27.7), as obtained on page 453.) On the other hand, if we don't like this particular formula and f is Fourier transformable and the product exists and is transformable, then we can compute our solution via

$$y = \mathcal{F}^{-1}[HF] \qquad \text{where} \quad F = \mathcal{F}[f] \quad.$$

In particular, one solution to

$$\frac{dy}{dt} + 3y = e^{4t}$$

is

$$y(t) = h * f(t) = \int_{-\infty}^{\infty} h(s) f(t-s) \, ds$$

$$= \int_{-\infty}^{\infty} e^{-3s} \, \text{step}(s) e^{4(t-s)} \, ds$$

$$= \int_{0}^{\infty} e^{-3s} e^{4t} e^{-4s} \, ds = e^{4t} \int_{0}^{\infty} e^{-7s} \, ds = \frac{1}{7} e^{4t} \quad.$$

On the other hand, using $y = h * f$ to solve

$$\frac{dy}{dt} + 3y = \text{sinc}(2\pi t)$$

yields

$$y(t) = \left[e^{-3t} \, \text{step}(t) \right] * \text{sinc}(2\pi t) = \cdots = \int_{0}^{\infty} e^{-3s} \, \text{sinc}(2\pi(t-s)) \, ds \quad, \qquad (27.13)$$

which, though valid, is not a particularly nice integral for computation. So, let's try using the transfer function. Here,

$$F(\omega) = \mathcal{F}[\text{sinc}(2\pi t)]|_{\omega} = \frac{1}{2} \, \text{pulse}_1(\omega) \quad.$$

So,

$$y(t) = \mathcal{F}^{-1}[HF]|_t = \mathcal{F}^{-1}\left[\frac{1}{3 + i2\pi\omega} \cdot \frac{1}{2} \, \text{pulse}_1(\omega) \right]\Big|_t \quad.$$

Fortunately for us, this product is absolutely integrable. Consequently, we can compute this inverse transform via the integral formula,

$$y(t) = \frac{1}{2} \int_{-\infty}^{\infty} \frac{1}{3 + i2\pi\omega} \, \text{pulse}_1(\omega) \, e^{i2\pi\omega t} \, d\omega = \frac{1}{2} \int_{-1}^{1} \frac{1}{3 + i2\pi\omega} \, e^{i2\pi\omega t} \, d\omega \quad. \qquad (27.14)$$

Again, we have an integral we cannot evaluate by elementary means. So, if we want to compute values for $y(t)$ for specific choices of t, we will have to numerically evaluate either the integral in formula (27.13) or in formula (27.14). Of the two, the latter may be a better choice for numerical integration since it is an integral over a finite interval.

Frequency Response and Periodic Inputs
Complex Exponential Inputs and Outputs

There is often a close relation between the transfer function of an LSI system and the outputs corresponding to complex exponential inputs. This relation is easily derived using output relation (27.11) (i.e., $S[f] = \mathcal{F}^{-1}[HF]$) with

$$f(t) = e^{i2\pi\overline{\omega}t}$$

where $\overline{\omega}$ denotes any fixed real value. Assuming the transfer function H is continuous at $\overline{\omega}$ and recalling that

$$\mathcal{F}\left[e^{i2\pi\overline{\omega}t}\right] = \delta_{\overline{\omega}} \quad\text{and}\quad H \cdot \delta_{\overline{\omega}} = H(\overline{\omega})\,\delta_{\overline{\omega}} \quad,$$

we see that

$$S\left[e^{i2\pi\overline{\omega}\tau}\right]\Big|_t = \mathcal{F}^{-1}[HF]\big|_t$$

$$= \mathcal{F}^{-1}[H \cdot \delta_{\overline{\omega}}]\big|_t$$

$$= \mathcal{F}^{-1}[H(\overline{\omega})\,\delta_{\overline{\omega}}]\big|_t = H(\overline{\omega})\mathcal{F}^{-1}[\delta_{\overline{\omega}}]\big|_t \quad.$$

But $\mathcal{F}^{-1}[\delta_{\overline{\omega}}]\big|_t = e^{i2\pi\overline{\omega}t}$. Plugging that into the above gives the following theorem.

Theorem 27.4
Let S be an LSI system with transfer function H . Then, for each real value $\overline{\omega}$ at which H is continuous,

$$S\left[e^{i2\pi\overline{\omega}\tau}\right]\Big|_t = H(\overline{\omega})\,e^{i2\pi\overline{\omega}t} \quad. \tag{27.15}$$

Equation (27.15) says that the only effect the system has on an input consisting of a complex exponential with frequency ω is to scale the amplitude by a factor of $H(\omega)$.[4] Using this, we can quickly compute the response of the system to a complex exponential input of any frequency, provided we have the transfer function H . For this reason, many authors also refer to the transfer function as the *frequency response* function.

There are LSI systems whose transfer functions are not sufficiently continuous for us to apply theorem 27.4 (see exercise 27.10 at the end of this chapter), but these tend to be systems for which equation (27.15) would be of little value anyway (again, see exercise 27.10).

Periodic Inputs

One nice application of equation (27.15) is the computation of the response of a system to a periodic input. To see this, let f be any reasonable periodic function with period p . Remember, such a function can be represented by its Fourier series,

$$f(t) = F.S.[f]\big|_t = \sum_{k=-\infty}^{\infty} c_k\, e^{i2\pi\omega_k t}$$

where

$$\omega_k = \frac{k}{p} \quad\text{and}\quad c_k = \frac{1}{p}\int_{\text{period}} f(t)\, e^{-i2\pi\omega_k t}\, dt \quad.$$

---

[4] For those acquainted with eigenfunctions and eigenvalues: The complex exponentials are eigenfunctions for the system and, for each frequency ω, $H(\omega)$ is the eigenvalue corresponding to the eigenfunction $e^{i2\pi\omega t}$.

Now let S be any LSI system for which periodic functions are valid inputs, and let H, as usual, denote the corresponding transfer function. Then, if f is as above,

$$S[f]|_t = S\left[\sum_{k=-\infty}^{\infty} c_k e^{i2\pi\omega_k\tau}\right]\Bigg|_t = \sum_{k=-\infty}^{\infty} c_k S\left[e^{i2\pi\omega_k\tau}\right]\Bigg|_t \quad .$$

According to equation (27.15), this reduces to

$$S[f]|_t = \sum_{k=-\infty}^{\infty} c_k H(\omega_k) e^{i2\pi\omega_k t}$$

provided H is continuous at each ω_k.

!►*Example 27.7:*   *Consider finding a solution y to*

$$\frac{dy}{dt} + 3y = \text{saw}_3(t) \quad .$$

A few pages ago, we saw that the transfer function for the LSI system S corresponding to this differential equation is

$$H(\omega) = \frac{1}{3 + i2\pi\omega} \quad .$$

We also know

$$\text{saw}_3(t) = F.S.[\text{saw}_3]|_t = \frac{3}{2} + \sum_{\substack{k=-\infty \\ k\neq 0}}^{\infty} \frac{3i}{2\pi k} e^{i2\pi\omega_k t} \quad \text{with} \quad \omega_k = \frac{k}{3} \quad .$$

Applying the above, we get

$$y(t) = S[\text{saw}_t]|_t = \frac{3}{2}H(0)e^{i2\pi0t} + \sum_{\substack{k=-\infty \\ k\neq 0}}^{\infty} \frac{3i}{2\pi k} H\left(\frac{k}{3}\right) e^{i2\pi\omega_k t}$$

$$= \frac{3}{2} \cdot \frac{1}{3} + \sum_{\substack{k=-\infty \\ k\neq 0}}^{\infty} \frac{3i}{2\pi k} \cdot \frac{1}{3 + i2\pi \cdot \frac{k}{3}} e^{i2\pi\omega_k t}$$

$$= \frac{1}{2} + \sum_{\substack{k=-\infty \\ k\neq 0}}^{\infty} \frac{9i}{2\pi k[9 + i2\pi]} e^{i2\pi\omega_k t} \quad .$$

Experimentally Determining Transfer Functions

Suppose we have some mysterious physical system and the means of measuring the output corresponding to any applied input. For example, this system may be an unfamiliar electric circuit, and we can measure the voltage at some point corresponding to an applied voltage at some other point in the circuit. Suppose, further, that we are fairly certain this physical system can be described mathematically as an LSI system having some, yet unknown, impulse response function h and transfer function H.

Given all this, it is natural to ask *if we can use our measurements to determine h or H?*

Well, since h is the output function corresponding to a delta function input, we could try to determine the function $h(t)$ by generating an input to our physical system that can be described by the delta function and then measuring the corresponding output. Unfortunately, such "delta function inputs" are very difficult, if not impossible, to generate in the real world.[5] Even generating a good approximation to the delta function may be problematic. On the other hand, sinusoidal inputs of fixed frequencies can often be generated fairly easily (and accurately). With the help of equation (27.15) it then becomes a simple exercise to compute H from the measured outputs to these sinusoidal inputs.

?► Exercise 27.3: *For the following, assume S is an LSI system for which we have determined, via measurements, that*

$$S[\cos(2\pi\omega\tau)]|_t = e^{-\omega}\cos(2\pi\omega t) \qquad \text{for each} \quad \omega \geq 0 \quad.$$

a: *Derive*

$$S[\sin(2\pi\omega\tau)]|_t = e^{-\omega}\sin(2\pi\omega t) \qquad \text{for each} \quad \omega \geq 0$$

using the shift invariance of S. (Remember, $\sin(2\pi\omega t) = \cos(2\pi\omega t - {}^{\pi}/_2)$.)

b: *Using the above, the linearity of S and equation (27.15), show that the transfer function for this system is*

$$H(\omega) = e^{-|\omega|} \quad.$$

(Consider the cases $\omega \geq 0$ and $\omega < 0$ separately.)

Additional Exercises

27.4. *Let A and B be any two constants. Verify that the following is an LSI system:*

$$S: y \longmapsto B\frac{dy}{dt} + Ay \quad.$$

27.5. *Using the properties of the delta function, equation (26.4) on page 422 and the naive notion of operational continuity, show that lemma 27.2 on page 455 follows from lemma 27.1.*

27.6. *Let $S: f \mapsto y$ be the system corresponding to the differential equation*

$$\frac{dy}{dt} - 6y = f \quad.$$

a. *Find the transfer function, $H(\omega)$, and the impulse response function, $h(t)$, for this system.*

b. *Using either H or h, find a solution to each of the following differential equations. (One or more of your answers may have to be left in integral form.)*

i. $\dfrac{dy}{dt} - 6y = 24e^{2t}$ **ii.** $\dfrac{dy}{dt} - 6y = 20\sin(2t)$

iii. $\dfrac{dy}{dt} - 6y = \text{sinc}(t)$ **iv.** $\dfrac{dy}{dt} - 6y = -6\,\text{step}(t)$

---

[5] Actually, this may be rather fortunate.

27.7. *Three second order differential equations are given below. For each, find the transfer function H and impulse response function h for the corresponding LSI system $S: f \mapsto y$. Also, for each equation, write out two expressions for a solution y — one in terms of h and one in terms of H.*

a. $\dfrac{d^2 y}{dt^2} + 7\dfrac{dy}{dt} + 10y = f$ **b.** $\dfrac{d^2 y}{dt^2} - 10\dfrac{dy}{dt} + 29y = f$

c. $\dfrac{d^2 y}{dt^2} + 10\dfrac{dy}{dt} + 25y = f$

27.8. *Let α be some positive constant. Determine the impulse response function h and the transfer function H for the corresponding "time delay" system*

$$S: f(t) \longmapsto f(t - \alpha) \quad .$$

27.9. *Let S be a system with transfer function*

$$H(\omega) = A\,\mathrm{pulse}_\Omega(\omega)$$

where A and Ω are positive constants. (Such a system is called an ideal low-pass filter.)

a. *Describe what this system does to each input of the form $f(t) = e^{i2\pi\omega t}$. In particular, why might Ω be called the "cut-off frequency", and why might A be called the "amplification"?*

b. *Find the corresponding impulse response function.*

c. *Assume $A = 6$ and $\Omega = {}^9\!/\!_2$. Compute the following outputs for this low-pass filter.*

  i. $S[\sin(4\pi t)]$ **ii.** $S[\sin(10\pi t)]$

  iii. $S[10\,\mathrm{sinc}(10\pi t)]$ **iv.** $S[\mathrm{saw}_1(t)]$

27.10. *Let S be the system*

$$S: f(t) \longmapsto \int_{-\infty}^{\infty} f(s)\, ds \quad .$$

a. *Verify that this is an LSI system.*

b. *Find both the impulse response function and the transfer function for this system.*

c. *Does the domain of this system include nonzero periodic functions? (Why?)*

d. *Why would it not be appropriate to refer to the transfer function of this system as the "frequency response" function for this system?*

27.11. *Let S be an LSI system with transfer function H. For each nonnegative real number $\overline{\omega}$, let*

$$f_{\overline{\omega}}(t) = S[\cos(2\pi\overline{\omega}t)] \qquad \text{and} \qquad g_{\overline{\omega}}(t) = S[\sin(2\pi\overline{\omega}t)] \quad .$$

a. *Verify that*

$$g_{\overline{\omega}}(t) = f_{\overline{\omega}}\!\left(t - \frac{1}{4\overline{\omega}}\right) \quad .$$

b. *Show that, for each $\overline{\omega} \geq 0$,*

$$H(\overline{\omega}) = f_{\overline{\omega}}(0) + i g_{\overline{\omega}}(0)$$

and

$$H(-\overline{\omega}) = f_{\overline{\omega}}(0) - i g_{\overline{\omega}}(0) \quad .$$

28

Multi-Dimensional Fourier Transforms

Throughout this text, we are mainly concerned with the "one-dimensional" theory of Fourier analysis; that is, Fourier analysis involving functions of just one variable. You should be aware, however, that we can extend the theory to include functions of two, three or more variables. In fact, if you are ultimately interested in such areas as optics and image processing, then you will invariably find yourself dealing with the two-dimensional theory.

Fortunately, extending the one-dimensional theory of Fourier analysis to a higher-dimensional theory is relatively straightforward and natural. We will illustrate this in this chapter by briefly discussing Fourier transforms for functions of two or more variables. To keep discussion brief, many details will be omitted, especially when the results are direct analogs of results already discussed.

28.1 Basic Definitions

Assuming the functions are sufficiently integrable, the two-dimensional Fourier integral transform of $f(x, y)$ is

$$F(\omega, v) = \mathcal{F}[f(x, y)]|_{(\omega, v)} = \int_{-\infty}^{\infty} \int_{-\infty}^{\infty} f(x, y) e^{-i2\pi(\omega x + v y)} \, dx \, dy \quad ,$$

the three-dimensional Fourier integral transform of $f(x, y, z)$ is

$$F(\omega, v, \mu) = \mathcal{F}[f(x, y, z)]|_{(\omega, v, \mu)} = \int_{-\infty}^{\infty} \int_{-\infty}^{\infty} \int_{-\infty}^{\infty} f(x, y, z) e^{-i2\pi(\omega x + v y + \mu z)} \, dx \, dy \, dz \quad ,$$

and so on.

For more concise definitions, it is convenient to use vector notation $\mathbf{x} = (x_1, x_2, \ldots, x_N)$ and $\boldsymbol{\omega} = (\omega_1, \omega_2, \ldots, \omega_N)$ with $\boldsymbol{\omega} \cdot \mathbf{x}$ denoting the classic Cartesian dot product formula

$$\boldsymbol{\omega} \cdot \mathbf{x} = \omega_1 x_1 + \omega_2 x_2 + \cdots + \omega_N x_N \quad .$$

The N-dimensional Fourier integral transform and inverse Fourier integral transform are then defined, respectively, by

$$F(\boldsymbol{\omega}) = \mathcal{F}[f(\mathbf{x})]|_{\boldsymbol{\omega}} = \int_{-\infty}^{\infty} \cdots \int_{-\infty}^{\infty} f(\mathbf{x}) e^{-i2\pi \boldsymbol{\omega} \cdot \mathbf{x}} \, dx_1 \cdots dx_N$$

and

$$f(\mathbf{x}) = \mathcal{F}^{-1}[F(\boldsymbol{\omega})]|_{\mathbf{x}} = \int_{-\infty}^{\infty} \cdots \int_{-\infty}^{\infty} F(\boldsymbol{\omega}) e^{i2\pi \boldsymbol{\omega} \cdot \mathbf{x}} \, d\omega_1 \cdots d\omega_N \quad .$$

These formulas, of course, require that the integrands be "sufficiently integrable"; that is, the f in the first and the F in the second must satisfy the N-dimensional analogs of being absolutely integrable and piecewise continuous.

For more general definitions of the transforms, we define the *partial Fourier transforms with respect to the jth variables*

$$\mathcal{F}\left[f(\mathbf{x}); x_j, \omega_j\right] \qquad \text{and} \qquad \mathcal{F}^{-1}\left[F(\boldsymbol{\omega}); \omega_j, x_j\right]$$

as the standard one-dimensional transforms treating all but the jth components of $\mathbf{x}$ and $\boldsymbol{\omega}$ as constants. These definitions only require that the functions involved be Fourier transformable "with respect to the jth variable". "Sufficient integrability" is not required. However, when the functions are sufficiently integrable, these partial transforms are given by

$$\mathcal{F}\left[f(\mathbf{x}); x_j, \omega_j\right] = \int_{-\infty}^{\infty} f(\mathbf{x})\, e^{-i2\pi\omega_j x_j}\, dx_j$$

and

$$\mathcal{F}^{-1}\left[F(\boldsymbol{\omega}); \omega_j, x_j\right] = \int_{-\infty}^{\infty} F(\boldsymbol{\omega})\, e^{i2\pi\omega_j x_j}\, d\omega_j \quad .$$

Note that, in particular,

$$\mathcal{F}[f(x, y)]|_{(\omega, v)} = \int_{-\infty}^{\infty} \int_{-\infty}^{\infty} f(x, y)\, e^{-i2\pi(\omega x + vy)}\, dx\, dy$$

$$= \int_{-\infty}^{\infty} \left[\int_{-\infty}^{\infty} f(x, y)\, e^{-i2\pi\omega x}\, dx\right] e^{-i2\pi vy}\, dy$$

$$= \mathcal{F}[\mathcal{F}[f(x, y); x, \omega]; y, v] \quad .$$

It is easy to expand on this and show that, at least when f is a sufficiently integrable function of N variables,

$$\mathcal{F}[f(\mathbf{x})]|_{\boldsymbol{\omega}} = \mathcal{F}[\cdots \mathcal{F}[\mathcal{F}[f(\mathbf{x}); x_1, \omega_1]; x_2, \omega_2]\ldots; x_N, \omega_N] \quad . \tag{28.1}$$

That is, the N-dimensional integral Fourier transform can be viewed as N iterations of the one-dimensional transform applied successively to the variables. Likewise, at least when F is a sufficiently integrable function of N variables,

$$\mathcal{F}^{-1}[F(\boldsymbol{\omega})]|_{\mathbf{x}} = \mathcal{F}^{-1}\left[\cdots \mathcal{F}^{-1}\left[\mathcal{F}^{-1}[F(\boldsymbol{\omega}); \omega_1, x_1]; \omega_2, x_2\right]\ldots; \omega_N, x_N\right] \quad . \tag{28.2}$$

Formulas (28.1) and (28.2) were derived assuming the functions were sufficiently integrable. However, the partial transforms in these formulas only require that the functions be Fourier transformable with respect to the individual variables. So let us use formulas (28.1) and (28.2) to extend our definitions of the N-dimensional Fourier transforms. That is, $\mathcal{F}[f(\mathbf{x})]$ is given by formula (28.1) whenever that formula yields a well-defined result, and $\mathcal{F}^{-1}[F(\mathbf{x})]$ is given by formula (28.2) whenever that formula yields a well-defined result.

It should be clear that just about all the classical one-dimensional theory developed in chapters 18 through 25 (as well as the theory and results that will be developed later) can be redeveloped to obtain N-dimensional analogs. This includes the expected analogs of the principle of near-equivalence,

$$\mathcal{F}^{-1}[\phi(\mathbf{x})]|_{\mathbf{y}} = \mathcal{F}[\phi(\mathbf{x})]|_{-\mathbf{y}} = \mathcal{F}[\phi(-\mathbf{x})]|_{\mathbf{y}} \quad ,$$

linearity,

$$\mathcal{F}[a\phi + b\psi] = a\mathcal{F}[\phi] + b\mathcal{F}[\psi] \quad ,$$

invertibility,

$$F = \mathcal{F}[f] \quad \Longleftrightarrow \quad \mathcal{F}^{-1}[F] = f \quad ,$$

and the N-dimensional versions of all the standard identities. Here is a short list of some of these identities (in each, f and F are functions of N variables with $F(\boldsymbol{\omega}) = \mathcal{F}[f(\mathbf{x})]|_{\boldsymbol{\omega}}$):

1. *(Translation)* If $\mathbf{a} = (a_1, a_2, \ldots, a_N)$ is a constant vector with real components, then

$$\mathcal{F}[f(\mathbf{x} - \mathbf{a})]|_{\boldsymbol{\omega}} = e^{-i2\pi \mathbf{a}\cdot\boldsymbol{\omega}} F(\boldsymbol{\omega})$$

and

$$\mathcal{F}\left[e^{i2\pi \mathbf{a}\cdot\mathbf{x}} f(\mathbf{x})\right]\bigg|_{\boldsymbol{\omega}} = F(\boldsymbol{\omega} - \mathbf{a}) \quad .$$

2. *(Scaling)* If γ is a nonzero real number, then

$$\mathcal{F}[f(\gamma \mathbf{x})]|_{\boldsymbol{\omega}} = \frac{1}{|\gamma|^N} F\left(\frac{\boldsymbol{\omega}}{\gamma}\right) \quad .$$

3. *(Differentiation)* Let j be an integer with $1 \leq j \leq N$. Assume either that

    (a) f is continuous and piecewise smooth, and $\partial f / \partial x_j$ is classically transformable,

   or that

    (b) $\omega_j F(\boldsymbol{\omega})$ is classically transformable.

   Then both of the above sets of conditions hold as does the following equation:

$$\mathcal{F}\left[\frac{\partial f}{\partial x_j}\right]\bigg|_{\boldsymbol{\omega}} = i2\pi \omega_j F(\boldsymbol{\omega}) \quad .$$

4. *(Convolution)* Let f, g, F and G all be bounded, classically transformable functions with $F = \mathcal{F}[f]$ and $G = \mathcal{F}[g]$. Then

    (a)   the products fg and FG are absolutely integrable,

    (b)   the convolutions $f * g$ and $F * G$ exist and are classically transformable, and

    (c) $\mathcal{F}[fg] = F * G$ and $\mathcal{F}^{-1}[FG] = f * g$,

   where $f * g$ is the N-dimensional convolution given by

$$f * g(\mathbf{x}) = \int_{-\infty}^{\infty} \cdots \int_{-\infty}^{\infty} f(\mathbf{s})g(\mathbf{x} - \mathbf{s}) \, ds_1 \ldots ds_N \quad .$$

It should be further noted that the one-dimensional generalized theory developed in chapter 26 also extends in a natural fashion, with the delta function of N variables at $\mathbf{a} = (a_1, \cdots, a_N)$, $\delta_{\mathbf{a}}(\mathbf{x})$, being the generalized function such that

$$\int_{-\infty}^{\infty} \cdots \int_{-\infty}^{\infty} \phi(\mathbf{x}) \, \delta_{\mathbf{a}}(\mathbf{x}) \, dx_1 \cdots dx_N = \phi(\mathbf{a})$$

whenever ϕ is a function of N variables continuous at $\mathbf{a}$. In addition (as we will soon see), some of our transforms will involve delta functions of a single variable, with that one variable being a linear combination of two or more variables. In particular, we will see things like

$$f(x, y) = \delta_y(x) = \delta(x - y) \quad .$$

As you can easily verify, this is the generalized function such that

$$\int_{-\infty}^{\infty} \int_{-\infty}^{\infty} \phi(x, y)\delta_y(x) \, dx \, dy = \int_{-\infty}^{\infty} \phi(y, y) \, dy \quad .$$

28.2 Computing Multi-Dimensional Transforms

The computation of a multi-dimensional Fourier transform basically starts with either the integral formula (provided the function being transformed is sufficiently integrable), or with formula (28.1) to break the computation into a succession of one-dimensional transforms. We'll usually find that using formula (28.1) is more convenient.

!▶**Example 28.1:**   *Let us find the two-dimensional Fourier transform of* $f(x, y) = e^{-2\pi|x-y|}$,

$$F(\omega, v) = \mathcal{F}\left[e^{-2\pi|x-y|}\right]\Big|_{(\omega,v)} = \mathcal{F}\left[\mathcal{F}\left[e^{-2\pi|x-y|}; x, \omega\right]; y, v\right] \quad.$$

In the inner partial transform, y is treated as a constant, allowing us to use a translation identity,

$$\mathcal{F}\left[e^{-2\pi|x-y|}; x, \omega\right] = e^{-i2\pi y\omega}\mathcal{F}\left[e^{-2\pi|x|}; x, \omega\right] = e^{-i2\pi y\omega}\frac{1}{\pi + \pi\omega^2} \quad.$$

In the outer partial transform, y is a variable, but ω is treated as a constant. Thus,

$$F(\omega, v) = \mathcal{F}\left[\mathcal{F}\left[e^{-2\pi|x-y|}; x, \omega\right]; y, v\right]$$

$$= \mathcal{F}\left[e^{-i2\pi\omega y}\frac{1}{\pi + \pi\omega^2}; y, v\right]$$

$$= \frac{1}{\pi + \pi\omega^2}\mathcal{F}\left[e^{i2\pi(-\omega)y}; y, v\right] = \frac{1}{\pi + \pi\omega^2}\delta_{-\omega}(v) \quad.$$

Equivalently,

$$F(\omega, v) = \frac{1}{\pi + \pi\omega^2}\delta(\omega + v) \quad.$$

Transforms of Separable Functions

We will say that a function f of N variables is *separable* if and only if its formula can be written as

$$f(x_1, x_2, \ldots, x_N) = f_1(x_1)f_2(x_2)\cdots f_N(x_N)$$

where each f_j is a function of a single variable.

If the individual f_j's are Fourier transformable and their one-dimensional transforms are known, then the N-dimensional Fourier transform of f is easily computed. To see why, let's consider a separable function of two variables

$$f(x, y) = f_1(x)f_2(y) \quad,$$

and assume we already know the one-dimensional transforms $F_1 = \mathcal{F}[f_1]$ and $F_2 = \mathcal{F}[f_2]$. Then, by definition,

$$\mathcal{F}[f_1(x); x, \omega] = F_1(\omega) \quad \text{and} \quad \mathcal{F}[f_2(y); y, v] = F_2(v) \quad.$$

Moreover, because the partial transform with respect to the first variables treats y as a constant, and the partial transform with respect to the second variables treats ω as a constant,

$$\mathcal{F}[f_1(x)f_2(y)]|_{(\omega,v)} = \mathcal{F}[\mathcal{F}[f_1(x)f_2(y); x, \omega]; y, v]$$

$$= \mathcal{F}[f_2(y)\mathcal{F}[f_1(x); x, \omega]; y, v]$$

$$= \mathcal{F}[f_2(y)F_1(\omega); y, \nu]$$

$$= F_1(\omega)\mathcal{F}[f_2(y); y, \nu] = F_1(\omega)F_2(\nu) \quad.$$

In other words,
$$\mathcal{F}[f_1(x)f_2(y)]|_{(\omega,\nu)} = \mathcal{F}[f_1(x)]|_\omega \times \mathcal{F}[f_2(y)]|_\nu \quad.$$

   Expanding this to functions of more than two variables and to the inverse transform is straight-forward and yields the next theorem.

Theorem 28.1
Let $\{f_1, f_2, \ldots, f_N\}$ and $\{F_1, F_2, \ldots, F_N\}$ be two sets of Fourier transformable functions of one variable, with $F_j = \mathcal{F}[f_j]$ for each j. If f is a separable function of N variables given by

$$f(x_1, x_2, \ldots, x_N) = f_1(x_1)f_2(x_2)\cdots f_N(x_N) \quad,$$

then its N-dimensional Fourier transform is given by

$$\mathcal{F}[f(\mathbf{x})]|_{\boldsymbol{\omega}} = F_1(\omega_1)F_2(\omega_2)\cdots F_N(\omega_N) \quad.$$

If, on the other hand, F is a separable function of N variables given by

$$F(\omega_1, \omega_2, \ldots, \omega_N) = F_1(\omega_1)F_2(\omega_2)\cdots F_N(\omega_N) \quad,$$

then its N-dimensional inverse Fourier transform is given by

$$\mathcal{F}^{-1}[F(\boldsymbol{\omega})]|_{\mathbf{x}} = f_1(x_1)f_2(x_2)\cdots f_N(x_N) \quad.$$

!▶ Example 28.2:   *Let α and β be two positive values, and let*

$$f(x, y) = \begin{cases} 1 & \text{if } |x| < \alpha \text{ and } |y| < \beta \\ 0 & \text{otherwise} \end{cases} \quad.$$

Observe that, in fact,
$$f(x, y) = \text{pulse}_\alpha(x)\,\text{pulse}_\beta(y) \quad.$$

Applying theorem 28.1 and what we know of the transforms of pulse functions, we get

$$\mathcal{F}[f(x, y)]|_{(\omega,\nu)} = \mathcal{F}\left[\text{pulse}_\alpha(x)\,\text{pulse}_\beta(y)\right]\Big|_{(\omega,\nu)}$$

$$= \mathcal{F}\left[\text{pulse}_\alpha(x)\right]\Big|_\omega \times \mathcal{F}\left[\text{pulse}_\beta(y)\right]\Big|_\nu$$

$$= 2\alpha\,\text{sinc}(2\pi\alpha\omega) \times 2\beta\,\text{sinc}(2\pi\beta\omega)$$

$$= 4\alpha\beta\,\text{sinc}(2\pi\alpha\omega)\,\text{sinc}(2\pi\beta\omega) \quad.$$

   Sometimes, a particular variable x_j does not explicitly appear in a given formula for $f(\mathbf{x})$, and, because of this, you may sometimes overlook the transform of $f_j(x_j)$ in using theorem 28.1. To avoid this error, explicitly write out any "missing" f_j's as the constant function 1. That is, if x_j does not explicitly appear in the formula for $f(\mathbf{x})$, then set

$$f_j(x_j) = 1(x_j) = 1 \qquad \text{for all} \quad x_j \quad.$$

!► Example 28.3:   *The basic vertical slit function is the function of two variables given by*

$$f(x, y) = \begin{cases} 1 & if \quad |x| < 1 \\ 0 & otherwise \end{cases} = \text{pulse}_1(x) \quad,$$

which we can also write as

$$f(x, y) = \text{pulse}_1(x) \cdot 1 = \text{pulse}_1(x) 1(y) \quad.$$

So,

$$\mathcal{F}[f(x, y)]|_{(\omega, \nu)} = \mathcal{F}\left[\text{pulse}_1(x) 1(y)\right]\big|_{(\omega, \nu)}$$

$$= \mathcal{F}\left[\text{pulse}_1(x)\right]\big|_{\omega} \times \mathcal{F}[1(y)]|_{\nu} = 2 \operatorname{sinc}(2\pi\omega) \delta(\nu) \quad.$$

Using Polar Coordinates

Let us now limit ourselves to two-dimensional transforms, treating (x, y) and (ω, ν) as Cartesian coordinates of points on the plane. Let us also assume our functions are sufficiently integrable so that we may use the basic integral definition of the two-dimensional Fourier transform,

$$F(\omega, \nu) = \mathcal{F}[f(x, y)]|_{(\omega, \nu)} = \int_{-\infty}^{\infty} \int_{-\infty}^{\infty} f(x, y) e^{-i2\pi(x\omega + y\nu)} \, dx \, dy \quad. \tag{28.3}$$

It may well happen that the problem at hand involves some sort of circular symmetry, prompting us to use the polar coordinates (r, θ) and (ρ, ϕ) where

$$x = r\cos(\theta) \quad, \quad y = r\sin(\theta) \quad, \quad \omega = \rho\cos(\phi) \quad \text{and} \quad \nu = \rho\sin(\phi) \quad.$$

Recall that

$$r = \sqrt{x^2 + y^2} \quad, \quad \rho = \sqrt{\omega^2 + \nu^2}$$

and

$$\int_{-\infty}^{\infty} \int_{-\infty}^{\infty} [\cdots] \, dx \, dy = \int_{r=0}^{\infty} \int_{\theta=\tau}^{\tau+2\pi} [\cdots] r \, d\theta \, dr$$

where τ can be any real value. In addition, using a well-known trigonometric identity, we have

$$x\omega + y\nu = r\rho \left[\cos(\theta)\cos(\phi) + \sin(\theta)\sin(\phi)\right] = r\rho\cos(\theta - \phi) \quad.$$

Thus, the polar coordinate version of formula (28.3) is

$$\widehat{F}(\rho, \phi) = \int_{r=0}^{\infty} \int_{\theta=\tau}^{\tau+2\pi} \widehat{f}(r, \theta) e^{i2\pi r\rho\cos(\theta - \phi)} r \, d\theta \, dr \tag{28.4}$$

where $\widehat{f}$ and $\widehat{F}$ are simply the polar coordinate versions of f and F,

$$\widehat{f}(r, \theta) = f(r\cos(\theta), r\sin(\theta)) \quad \text{and} \quad \widehat{F}(\rho, \phi) = F(\rho\cos(\phi), \rho\sin(\phi)) \quad.$$

   In practice, if you are considering using polar coordinates, then you are likely to have started with the polar coordinate formula $\widehat{f}(r, \theta)$. It is even quite possible that your function is circularly symmetric, meaning that $\widehat{f}(r, \theta)$ is really a function of r only. So let us assume so and continue, letting

$$\widehat{f}(r) = \widehat{f}(r, \theta) \quad.$$

Using this in formula (28.4), letting $\tau = \phi - \pi$, and then applying a simple change of variables, we get

$$\widehat{F}(\rho, \phi) = \int_{r=0}^{\infty} \int_{\theta=\phi-\pi}^{\phi+\pi} \widehat{f}(r) e^{i2\pi r\rho \cos(\theta-\phi)} r \, d\theta \, dr$$

$$= \int_{r=0}^{\infty} \widehat{f}(r) \left[\int_{\theta=\phi-\pi}^{\phi+\pi} e^{i2\pi r\rho \cos(\theta-\phi)} \, d\theta \right] r \, dr$$

$$= \int_{r=0}^{\infty} \widehat{f}(r) \left[\int_{\sigma=-\pi}^{\pi} e^{i2\pi r\rho \cos(\sigma)} \, d\sigma \right] r \, dr \quad .$$

That is,

$$\widehat{F}(\rho, \phi) = \int_{r=0}^{\infty} \widehat{f}(r) K(2\pi r\rho) r \, dr \qquad \text{where} \quad K(\zeta) = \int_{\sigma=-\pi}^{\pi} e^{i\zeta \cos(\sigma)} \, d\sigma \quad .$$

Note that the right side does not depend on ϕ. So, $\widehat{F}(\rho, \phi)$ also does not depend on ϕ. Just like $\widehat{f}$, $\widehat{F}$ is circularly symmetric, and we can write $\widehat{F}(\rho) = \widehat{F}(\rho, \phi)$.

Let's continue just a little further with K. For any real value ζ,

$$K(\zeta) = \int_{\sigma=-\pi}^{\pi} e^{i\zeta \cos(\sigma)} \, d\sigma$$

$$= \int_{\sigma=-\pi}^{\pi} [\cos(\zeta \cos(\sigma)) + i \sin(\zeta \cos(\sigma))] \, d\sigma$$

$$= \int_{\sigma=-\pi}^{\pi} \cos(\zeta \cos(\sigma)) \, d\sigma + i \int_{\sigma=-\pi}^{\pi} \sin(\zeta \cos(\sigma)) \, d\sigma \quad . \tag{28.5}$$

The second integral in the last line can be greatly simplified. Using the symmetry properties of the cosine and sine functions, you can easily verify that

$$\int_{\sigma=-\pi}^{\pi} \sin(\zeta \cos(\sigma)) \, d\sigma = 2 \left[\int_{\sigma=0}^{\pi/2} \sin(\zeta \cos(\sigma)) \, d\sigma + \int_{\sigma=\pi/2}^{\pi} \sin(\zeta \cos(\sigma)) \, d\sigma \right]$$

$$= 2 \left[\int_{\sigma=0}^{\pi/2} \sin(\zeta \cos(\sigma)) \, d\sigma - \int_{\sigma=0}^{\pi/2} \sin(\zeta \cos(\sigma)) \, d\sigma \right] = 0 \quad .$$

The other integral in line (28.5) is not so easily simplified. However, if you are acquainted with Bessel functions, then you may recall that one formula for the zero[th] order Bessel function J_0 is

$$J_0(\zeta) = \frac{1}{2\pi} \int_{\sigma=-\pi}^{\pi} \cos(\zeta \cos(\sigma)) \, d\sigma \quad .$$

Combining all the above, we see that the polar coordinate formula for the two-dimensional Fourier transform of a (suitably integrable) circularly symmetric function $\widehat{f}$ is

$$\widehat{F}(\rho) = 2\pi \int_{r=0}^{\infty} \widehat{f}(r) J_0(2\pi r\rho) r \, dr \quad . \tag{28.6a}$$

Repeating the above to find the polar coordinate formula for the two-dimensional inverse Fourier transform of a (suitably integrable) circularly symmetric function $\widehat{F}$ yields the almost identical formula

$$\widehat{f}(r) = 2\pi \int_{\rho=0}^{\infty} \widehat{F}(\rho) J_0(2\pi r\rho) \rho \, d\rho \quad . \tag{28.6b}$$

The advantage in using the Bessel function J_0 instead of the integral it replaces is that Bessel functions are "well known". Their values are well-tabulated, they are standard functions in standard computer math packages, and there are numerous identities for dealing with them. The disadvantage, of course, is that you will have to learn about them from some other text. (The classic text by Watson (reference [13]), originally published in 1922 and still in print, is highly recommended.)

By the way, the integral formulas in equation set (28.6) are examples of *Hankel transforms*. To be precise, they are zero[th] order Hankel transforms. (If you want to learn more about Hankel transforms, you can start with chapter 9 of reference [10].)

_____

Additional Exercises

28.1. Find the two-dimensional Fourier transform $F(\omega, \nu)$ for each of the following:

a. $f(x, y) = \text{pulse}_1(x - y)$

b. $f(x, y) = \dfrac{1}{1 + (x - y)^2}$

c. $f(x, y) = \delta(x - 2y)$

d. $f(x, y) = e^{i2\pi xy}$

e. $f(x, y) = e^{i2\pi xy} e^{-3|x|}$

f. $f(x, y) = \dfrac{e^{i2\pi xy}}{3 - i2\pi x}$

g. $f(x, y) = \dfrac{e^{-i2\pi y}}{3 - i2\pi x}$

h. $f(x, y) = xe^{(-3+i2\pi y)x} \text{step}(x)$

i. $f(x, y) = ye^{(-3+i2\pi y)x} \text{step}(x)$

j. $f(x, y) = e^{-(x^2+y^2)}$

k. $f(x, y) = \dfrac{e^{-i2\pi x}}{3 - i2\pi x}$

l. $f(x, y) = \text{pulse}_3(y)$

28.2. Find the three-dimensional Fourier transform $F(\omega, \nu, \mu)$ of each of the following:

a. $f(x, y, z) = \text{pulse}_3(x - y + z)$

b. $f(x, y, z) = e^{i2\pi x(y-z)} \text{pulse}_3(x)$

c. $f(x, y, z) = e^{4\pi(x^2+y^2+z^2)}$

d. $f(x, y, z) = e^{-2\pi|x-z|} e^{-2\pi|y-z|}$

28.3. For $a > 0$, the corresponding circular pulse function is

$$f(x, y) = \begin{cases} 1 & \text{if } x^2 + y^2 < a^2 \\ 0 & \text{otherwise} \end{cases}.$$

Find the polar coordinate form of the two-dimensional Fourier transform of this function in terms of Bessel functions. In doing your computations, you may use the fact that the Bessel functions of orders one and zero, J_1 and J_0, respectively, are related by

$$\frac{d}{ds} [s J_1(s)] = s J_0(s) \quad,$$

along with the fact that $J_1(0) = 0$.

29

Identity Sequences

Before going much further in our study of Fourier analysis, we need to develop a set of analytical tools called "identity sequences". Except for providing approximations to the delta function, these are not things often used in day-to-day applications. Their importance, rather, are as tools for further developing and justifying the Fourier analysis that is used in day-to-day applications. They will be essential in chapter 30 where we finally prove the few major theorems, such as the fundamental theorem on invertibility, that we could not prove earlier. In particular, using identity sequences we will derive a subtle "test for equality" that (1) will help us validate those major theorems, and (2) will provide, along with the fundamental identity from chapter 25, the starting point for the development of a much more general theory of Fourier analysis.

The basic ideas behind identity sequences are so simple that some people have trouble grasping them. So we will first illustrate these ideas employing a very simple identity sequence. Following that, we'll define and discuss identity sequences in general. Finally, we will examine the identity sequences that will be of greatest value to us in later chapters, the Gaussian identity sequences. By the way, while developing Gaussian identity sequences, we will also describe the set of "exponentially integrable" functions. This set will ultimately replace the set of classically transformable functions as the set of functions that can be "Fourier transformed" using the more general theory of part IV.

29.1   An Elementary Identity Sequence

For each $\epsilon > 0$, let ϕ_ϵ be the following scaled pulse function:

$$\phi_\epsilon(s) = \frac{1}{2\epsilon} \, \text{pulse}_\epsilon(s) \quad .$$

The set of these ϕ_ϵ's form what we will be calling an "identity sequence".

You may remember these functions from the chapter on delta functions (see the discussion starting on page 421) where, using slightly different notation, we saw that, for any piecewise continuous function f continuous at 0,

$$\lim_{\epsilon \to 0^+} \int_{-\infty}^{\infty} f(s) \, \phi_\epsilon(s) \, ds = f(0) \quad .$$

Observe, now, that if f is piecewise continuous on $\mathbb{R}$ and is continuous at a point x, then, by the above and a simple change of variable,

$$\lim_{\epsilon \to 0^+} \int_{-\infty}^{\infty} f(s) \, \phi_\epsilon(s - x) \, ds = \lim_{\epsilon \to 0^+} \int_{-\infty}^{\infty} f(\sigma + x) \, \phi_\epsilon(\sigma) \, d\sigma$$

$$= f(0 + x) = f(x) \quad .$$

This gives us the following theorem:

Theorem 29.1
For each $\epsilon > 0$, let ϕ_ϵ be the scaled pulse function

$$\phi_\epsilon(s) = \frac{1}{2\epsilon} \text{pulse}_\epsilon(s) \quad .$$

Then, for any piecewise continuous function f on the real line,

$$\lim_{\epsilon \to 0^+} \int_{-\infty}^{\infty} f(s)\, \phi_\epsilon(s - x)\, ds = f(x)$$

whenever x is a point at which f is continuous.

As the next example shows, we can (in theory at least) use this theorem to recover the formula for a function $f(s)$ if we know the formulas for certain integrals of f. In practice, we'll rarely have need to do such a thing, but the example illustrates an important principle, namely, that *any piecewise continuous function can be completely defined by a suitable set of integrals involving that function.*

!▶ **Example 29.1:** Let's find the formula $f(x)$ for the function satisfying

$$\int_{x-\epsilon}^{x+\epsilon} f(s)\, ds = 6x^2\epsilon + 2\epsilon^3 \quad \text{whenever} \quad -\infty < x < \infty \quad \text{and} \quad \epsilon > 0 \quad .$$

Theorem 29.1 tells us that, for each real value x,

$$f(x) = \lim_{\epsilon \to 0^+} \int_{-\infty}^{\infty} f(s)\, \phi_\epsilon(s - x)\, ds$$

$$= \lim_{\epsilon \to 0^+} \int_{-\infty}^{\infty} f(s)\, \frac{1}{2\epsilon} \text{pulse}_\epsilon(s - x)\, ds = \lim_{\epsilon \to 0^+} \frac{1}{2\epsilon} \int_{x-\epsilon}^{x+\epsilon} f(s)\, ds \quad .$$

Plugging in the given formula for this last integral yields

$$f(x) = \lim_{\epsilon \to 0^+} \frac{1}{2\epsilon}[6x^2\epsilon + 2\epsilon^3] = \lim_{\epsilon \to 0^+} [3x^2 + \epsilon^2] \quad ,$$

which, of course, means that $f(x) = 3x^2$.

A corollary of theorem 29.1 follows. It describes a test for determining when two functions, which may have been defined by different means, are really the same. Similar tests will play important roles in verifying the fundamental theorem on invertibility and in developing a more general theory for Fourier analysis.

Corollary 29.2
Let f and g be two piecewise continuous functions on the real line. Then $f = g$ if and only if

$$\int_{-\infty}^{\infty} f(s)\, \text{pulse}_\epsilon(s - x)\, ds = \int_{-\infty}^{\infty} g(s)\, \text{pulse}_\epsilon(s - x)\, ds \qquad (29.1)$$

for every $\epsilon > 0$ and every point x on the real line.

PROOF: Obviously, if $f = g$, then equation (29.1) holds for every real x and $\epsilon > 0$.

Now assume equation (29.1) holds for all real x and positive ϵ. We need to show that $f(x) = g(x)$ for every point x at which f and g are continuous. So let x be such a point. By our assumption and theorem 29.1, we have

$$f(x) = \lim_{\epsilon \to 0} \int_{-\infty}^{\infty} f(s) \frac{1}{2\epsilon} \text{pulse}_\epsilon (s - x) \, ds$$

$$= \lim_{\epsilon \to 0} \frac{1}{2\epsilon} \int_{-\infty}^{\infty} f(s) \, \text{pulse}_\epsilon (s - x) \, ds$$

$$= \lim_{\epsilon \to 0} \frac{1}{2\epsilon} \int_{-\infty}^{\infty} g(s) \, \text{pulse}_\epsilon (s - x) \, ds$$

$$= \lim_{\epsilon \to 0} \int_{-\infty}^{\infty} g(s) \frac{1}{2\epsilon} \text{pulse}_\epsilon (s - x) \, ds = g(x) \quad . \quad \blacksquare$$

29.2 General Identity Sequences
Definitions

For convenience, we'll let $\{\psi_\gamma\}_{\gamma=\alpha}^{\gamma_0}$ denote any indexed set of functions — with the ψ_γ's being the functions and the indexing parameter γ starting at α and going to, but not including, the limit point γ_0 (which is usually 0 or ∞). For example, $\{\text{pulse}_\epsilon\}_{\epsilon=1}^{0}$ denotes the set of all pulse functions with half-widths (the ϵ's) less than or equal to 1.

Now suppose we have some set S of functions of interest. (For example, S might be the set of all piecewise continuous functions on the real line.) We will say that a given $\{\psi_\gamma\}_{\gamma=\alpha}^{\gamma_0}$ is an *identity sequence*[1] (for S) if and only if

$$\lim_{\gamma \to \gamma_0} \int_{-\infty}^{\infty} f(s)\psi_\gamma (s - x) \, ds = f(x) \tag{29.2}$$

whenever f is a function in S and x is a point at which f is continuous. Implicit in this definition is that the above limit is taken "over the indexing set". In particular, if the starting point is $\alpha = 1$ and the limit point is $\gamma_0 = 0$, then the above limit is the one-sided limit

$$\lim_{\gamma \to 0^+} \int_{-\infty}^{\infty} f(s)\psi_\gamma (s - x) \, ds = f(x) \quad .$$

!▶ *Example 29.2:* In the previous section, we saw that, for any piecewise continuous function f and any real point x at which f is continuous,

$$\lim_{\epsilon \to 0^+} \int_{-\infty}^{\infty} f(s) \frac{1}{2\epsilon} \text{pulse}_\epsilon (s - x) \, ds = f(x) \quad .$$

Thus, $\left\{ \frac{1}{2\epsilon} \text{pulse}_\epsilon \right\}_{\epsilon=1}^{0}$ *is an identity sequence for the set of all piecewise continuous functions on the real line.*

---

[1] We are using the term "sequence" somewhat incorrectly since we are not requiring the set of γ's be a true sequence. It would be more proper to call these sets of functions something like "parameterized identity sets", but we won't.

A few comments should be made about S, the set of functions "of interest". While all sorts of esoteric sets can be considered as interesting, our interest will usually be with mundane sets such as "the set of all bounded, piecewise continuous functions on $\mathbb{R}$" or "the set of all classically transformable functions". In every case where we specify S, the following will hold:

1. S is a linear space of functions. That is, if f and g are two functions in S, then so is any linear combination of f and g.

2. Each function in S is continuous over some partitioning of the real line. That is, if f is any function in S and (a, b) is any finite interval, then f is well defined and continuous at all but, at most, a finite number of points in (a, b).

3. All nontrivial real translations and scalings of functions in S are also in S. That is, if f is any function in S and α is any nonzero real number, then the translation of f by α, $f(x - \alpha)$, and the variable scaling of f by α, $f(\alpha x)$, are also "functions of interest".

Often, it will be so obvious that these conditions hold that we won't explicitly note the fact. Also, when we are discussing identity sequences in general without explicitly specifying the set S — as in the next paragraph — then we will implicitly assume the above conditions hold.

Let's get back to equation (29.2),

$$ f(x) = \lim_{\gamma \to \gamma_0} \int_{-\infty}^{\infty} f(s) \psi_\gamma (s - x) \, ds \quad . $$

Yes, the above integral looks much like a convolution integral (or the closely related correlation integral). We also want to observe that, with $x = 0$, this equation becomes

$$ f(0) = \lim_{\gamma \to \gamma_0} \int_{-\infty}^{\infty} f(s) \psi_\gamma (s) \, ds \quad . $$

Now consider the limit of $f * \psi_\gamma$ at a point x where f is continuous. By our last observation (using $f(x - s)$ in place of $f(s)$)

$$ \lim_{\gamma \to \gamma_0} f * \psi_\gamma (x) = \lim_{\gamma \to \gamma_0} \int_{-\infty}^{\infty} f(x - s) \psi_\gamma (s) \, ds = f(x - 0) = f(x) \quad . $$

Conversely (assuming f is continuous at x), if

$$ \lim_{\gamma \to \gamma_0} f * \psi_\gamma (x) = f(x) \quad , $$

then, with $x = 0$, we have

$$ f(0) = \lim_{\gamma \to \gamma_0} f * \psi_\gamma (0) = \lim_{\gamma \to \gamma_0} \int_{-\infty}^{\infty} f(s) \psi_\gamma (s) \, ds \quad , $$

from which we see (after a simple change of variables) that, for any other x at which f is continuous,

$$ \lim_{\gamma \to \gamma_0} \int_{-\infty}^{\infty} f(s) \psi_\gamma (s - x) \, ds = \lim_{\gamma \to \gamma_0} \int_{-\infty}^{\infty} f(\sigma + x) \psi_\gamma (\sigma) \, d\sigma = f(0 + x) = f(x) \quad . $$

What we have just done is to show that two other equations could have been used in place of equation (29.2) in our definition of identity sequences. Using these equations instead of equation (29.2) will simplify some of our work later on. For reference let us summarize what we've just shown in the following theorem.

Theorem 29.3 *(definitions for identity sequences)*
An indexed set of functions on the real line $\{\psi_\gamma\}_{\gamma=\alpha}^{\gamma_0}$ is an identity sequence for some set of functions S if and only if one of the following conditions holds:

1. *For each f in S and each real value x at which f is continuous,*

$$\lim_{\gamma \to \gamma_0} \int_{-\infty}^{\infty} f(s)\psi_\gamma(s-x)\,ds \ = \ f(x) \quad .$$

2. *For each f in S that is continuous at 0,*

$$\lim_{\gamma \to \gamma_0} \int_{-\infty}^{\infty} f(s)\psi_\gamma(s)\,ds \ = \ f(0) \quad .$$

3. *For each f in S and each real value x at which f is continuous,*

$$\lim_{\gamma \to \gamma_0} f * \psi_\gamma(x) \ = \ f(x) \quad .$$

Moreover, if any one of these conditions holds, then they all hold.

Generating Identity Sequences

To help us figure out how to generate additional (and more useful) identity sequences, let us look at some of the geometric aspects of the one identity sequence we already know about, namely, $\{\phi_\epsilon\}_{\epsilon=1}^{0}$ where

$$\phi_\epsilon(s) \ = \ \frac{1}{2\epsilon}\,\text{pulse}_\epsilon(s) \quad .$$

To begin, note that each ϕ_ϵ can be viewed as a scaling by $1/\epsilon$ of ϕ_1 and its variable,

$$\phi_\epsilon(s) \ = \ \frac{1}{2\epsilon}\,\text{pulse}_\epsilon(s) \ = \ \begin{cases} \frac{1}{2\epsilon} & \text{if} \quad -\epsilon < s < \epsilon \\ 0 & \text{otherwise} \end{cases}$$

$$= \ \frac{1}{\epsilon} \begin{cases} \frac{1}{2} & \text{if} \quad -1 < \frac{1}{\epsilon}s < 1 \\ 0 & \text{otherwise} \end{cases} \ = \ \frac{1}{\epsilon}\phi_1\left(\frac{1}{\epsilon}s\right) \quad .$$

Moreover, the graph of ϕ_1 encloses a region of unit area,

$$\int_{-\infty}^{\infty} \phi_1(s)\,ds \ = \ \frac{1}{2}\int_{-\infty}^{\infty} \text{pulse}_1(s)\,ds \ = \ \frac{1}{2}\int_{-1}^{1} ds \ = \ 1 \quad .$$

Finally, notice (see figure 29.1a) what the graphs of these ϕ_ϵ's look like as ϵ shrinks to 0. For each ϵ, the graph of ϕ_ϵ also encloses a region of unit area, but as $\epsilon \to 0$, the graphs become thinner and taller, and appear to be approaching an "infinitely tall and infinitesimally narrow spike" at $s = 0$.

You may suspect that other identity sequences can be generated by suitably scaling a single "base" function whose graph encloses a region of unit area. Pursuing that suspicion (and replacing $1/\epsilon$ with γ) leads to the following theorem. It also turns out that, whenever you generate an identity sequence via the process described in the theorem, then, as $\gamma \to \infty$, the graphs of the functions in the sequence appear to approach an "infinitely tall and narrow spike enclosing unit area" at the origin.

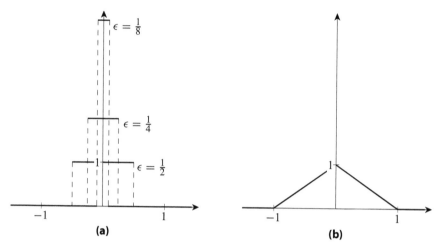

Figure 29.1: Some functions in two identity sequences — (a) $\frac{1}{2\epsilon}$ pulse$_\epsilon$ for various values of ϵ, and (b) the basic triangle function, tri, from example 29.3.

Theorem 29.4

Let Ψ be any piecewise continuous function that vanishes outside of some finite interval and satisfies

$$\int_{-\infty}^{\infty} \Psi(s)\, ds \; = \; 1 \quad .$$

For each $\gamma > 0$ let

$$\psi_\gamma(s) \; = \; \gamma \Psi(\gamma s) \quad .$$

Then $\{\psi_\gamma\}_{\gamma=1}^{\infty}$ is an identity sequence for the set of all piecewise continuous functions on $\mathbb{R}$.

!▶ Example 29.3:   For the base function Ψ, use the triangle function,

$$\text{tri}(s) \; = \; \begin{cases} 1+s & \text{if } \; -1 < s < 0 \\ 1-s & \text{if } \; 0 < s < 1 \\ 0 & \text{otherwise} \end{cases} \quad .$$

The graph of tri is sketched in figure 29.1b. Clearly, tri is piecewise continuous and vanishes outside the interval $(-1, 1)$. Moreover,

$$\int_{-\infty}^{\infty} \text{tri}(s)\, ds \; = \; \text{"area under the triangle"} \; = \; 1 \quad .$$

So, theorem 29.4 assures us that $\{\gamma\,\text{tri}(\gamma s)\}_{\gamma=1}^{\infty}$ is an identity sequence for the set of all piecewise continuous functions on the real line. That is, if f is any piecewise continuous function on $\mathbb{R}$ and x is any point at which f is continuous, then

$$\lim_{\gamma \to \infty} \int_{-\infty}^{\infty} f(s)\, \gamma\, \text{tri}(\gamma[s-x])\, ds \; = \; f(x) \quad .$$

?▶ Exercise 29.1:   Sketch the graphs of $\gamma\,\text{tri}(\gamma s)$ for $\gamma = 2$, $\gamma = 4$ and $\gamma = 8$ where tri is the triangle function from example 29.3.

Theorem 29.4 gives a method for generating an identity sequence $\{\psi_\gamma\}_{\gamma=1}^\infty$ from a suitable "base function" Ψ. It turns out that, if we are willing to restrict our set of "functions of interest" somewhat, then we can relax the requirement that Ψ vanish outside some finite interval. That will be the case with the "Gaussian identity sequences" discussed in the next section. As a result, we can generate many, many theorems similar to theorem 29.4. What's more, a great many of these theorems can be proven following the same basic approach. Since this is the case, and since Gaussian identity sequences will play a more significant role in the rest of this book than the identity sequences of theorem 29.4, the proof of theorem 29.4 will be left as an exercise to be done after a more general discussion of similar theorems and the proof of the main theorem in the next section.

29.3 Gaussian Identity Sequences

Gaussian identity sequences are simply identity sequences consisting of suitable scalings of Gaussian functions (we'll give a formal definition later). Currently, we are interested in such identity sequences because of the role they will play in developing the classical theory. Later, we will also find that the results discussed here are important in developing a much more general theory for the Fourier transform.

We will need to restrict the set of "functions of interest" somewhat to use Gaussian identity sequences. Because of that, we will now introduce the notions of "exponential boundedness" and "exponential integrability" extending the notions of a function being bounded and being absolutely integrable. This will enable us to identify the functions of interest for much of the rest of this text.

Exponentially Bounded and Integrable Functions

Let f be a function defined on the real line. We will say f is *exponentially bounded* (on the real line) if and only if there are real constants B and α such that

$$|f(x)| \leq Be^{\alpha|x|} \qquad \text{for all real values of } x \quad . \tag{29.3}$$

Of course, B must be nonnegative (positive, actually, unless f is the zero function), but α can be negative. We should also note the following:

1. *If inequality (29.3) holds for one given value of α, then it holds for all greater values of α (because $e^{\alpha_1|x|} \leq e^{\alpha_2|x|}$ whenever $\alpha_1 \leq \alpha_2$).*

2. If f is merely bounded on the real line, then it is automatically exponentially bounded (because inequality (29.3) then holds with $\alpha = 0$).

3. If f is piecewise continuous and inequality (29.3) holds for some $\alpha > 0$, then, for any $\beta > \alpha$,

$$\int_{-\infty}^\infty |f(x)| e^{-\beta|x|}\, dx \leq \int_{-\infty}^\infty Be^{\alpha|x|} e^{-\beta|x|}\, dx$$

$$= 2B \int_0^\infty e^{-(\beta-\alpha)x}\, dx = \frac{2B}{\beta-\alpha} < \infty \quad .$$

So $f(x)e^{-\beta|x|}$ is absolutely integrable on the real line whenever $\beta > \alpha$.

The last observation is especially significant to us. It describes "exponential integrability", a property that will be more important than mere exponential boundedness. To be precise, we will say a function f is *exponentially integrable* (on the real line) if and only if it is continuous over some partitioning of $\mathbb{R}$ and there is a real value β such that

$$f(x)e^{-\beta|x|}$$

is absolutely integrable on the real line. From the above observations, we know that any exponentially bounded, piecewise continuous function is automatically exponentially integrable. We should also observe that every piecewise continuous, absolutely integrable function f on the real line is automatically exponentially integrable (because, then, $f(x)e^{-\beta|x|}$ with $\beta = 0$ is absolutely integrable).

?▶ Exercise 29.2:   *Which of the following are exponentially bounded on the real line:*

$$\sin(x) \quad , \quad e^{3x} \quad , \quad x^2 \quad , \quad e^{-x^2} \quad , \quad e^{x^2} \quad ?$$

?▶ Exercise 29.3:   *Verify that $|x|^{-1/2}$ is not exponentially bounded, but is exponentially integrable on the real line.*

As was the case with absolutely integrable functions, there are a number of simple but useful facts regarding exponentially integrable functions. Some that we will be using soon (and without comment) are described in the following lemmas. Most are direct analogs of lemmas for absolutely integrable functions from chapter 18.

Lemma 29.5
If f is any exponentially integrable function and α is any nonzero real number, then the functions $f(\alpha x)$ and $f(x - \alpha)$ are also exponentially integrable.

Lemma 29.6
Any finite linear combination of exponentially integrable functions is exponentially integrable.

Lemma 29.7
If f is exponentially integrable and g is exponentially bounded and piecewise continuous, then the product fg is exponentially integrable.

Lemma 29.8
If f is any exponentially integrable and piecewise continuous function, and γ and α are any two real numbers with $\gamma > 0$, then the function

$$f(x)e^{-\gamma(x-\alpha)^2}$$

is in $\mathcal{A}$, the set of absolutely integrable, piecewise continuous functions on $\mathbb{R}$. Moreover, if f is also exponentially bounded, then

$$\lim_{x \to \pm\infty} f(x)e^{-\gamma(x-\alpha)^2} = 0 \quad .$$

We'll prove the first of these lemmas and leave the confirmation of the rest as exercises.

PROOF (of lemma 29.5):   Since f is exponentially integrable, there is a $\beta > 0$ such that $f(x)e^{-\beta|x|}$ is absolutely integrable. So, using $\widehat{\beta} = \beta|\alpha|$ and an obvious change of variables,

$$\int_{-\infty}^{\infty} |f(\alpha x)| \, e^{-\widehat{\beta}|x|} \, dx \;=\; \int_{-\infty}^{\infty} |f(\alpha x)| \, e^{-\beta|\alpha x|} \, dx \;=\; \frac{1}{|\alpha|} \int_{-\infty}^{\infty} |f(s)| \, e^{-\beta|s|} \, ds \quad,$$

which is finite. Hence, $f(\alpha x)e^{-\widehat{\beta}|x|}$ is absolutely integrable, and $f(\alpha x)$ is exponentially integrable.

Verifying that $f(x - \alpha)$ is exponentially integrable begins with an obvious change of variables,

$$\int_{-\infty}^{\infty} |f(x - \alpha)| \, e^{-\beta|x|} \, dx \;\leq\; \int_{-\infty}^{\infty} |f(s)| \, e^{-\beta|s+\alpha|} \, ds \quad. \tag{29.4}$$

Now, by basic arithmetic and the triangle inequality,

$$|s| \;=\; |s + \alpha - \alpha| \;\leq\; |s + \alpha| + |\alpha| \quad.$$

Consequently, $|s + \alpha| \geq |s| - |\alpha|$. From this and the fact that $\beta > 0$, we have

$$-\beta|s + \alpha| \;\leq\; -\beta|s| + \beta|\alpha| \qquad \text{and} \qquad e^{-\beta|s+\alpha|} \;\leq\; e^{-\beta|s|}e^{\beta|\alpha|} \quad.$$

Using this last inequality in the right-hand side of equation (29.4) gives us

$$\int_{-\infty}^{\infty} |f(x - \alpha)| \, e^{-\beta|x|} \, dx \;\leq\; e^{\beta|\alpha|} \int_{-\infty}^{\infty} |f(s)| \, e^{-\beta|s|} \, ds \quad.$$

Because of our choice for β, this last integral is finite. Hence, $f(x - \alpha)e^{-\beta|x|}$ is absolutely integrable, which, in turn, means that $f(x - \alpha)$ is exponentially integrable. ∎

?►Exercise 29.4:   Prove

    a: *lemma 29.6.*   **b:** *lemma 29.7.*   **c:** *lemma 29.8.*

Let us end this discussion of exponentially integrable functions with an observation and a comment. The observation is that every classically transformable function is exponentially integrable (see the next exercise). So any result we obtain concerning exponentially integrable functions also applies to classically transformable functions.[2] The comment is that, eventually, we will discover that the set of exponentially integrable functions is also, essentially, the set of all "Fourier transformable" functions using the more general theory we will develop (with the aid of Gaussian identity sequences) in part IV of this text.

?►Exercise 29.5:   Show that every classically transformable function is also exponentially integrable.

Gaussian Identity Sequences

We can now state the main theorem of this section. Its proof will be given later in this chapter after we discuss the general problem of verifying identity sequences.

---

[2] On the other hand, as you'll see in exercise 29.13 at the end of this chapter, there are classically transformable functions that are not exponentially bounded.

Theorem 29.9

For every $\gamma > 0$, let ψ_γ be the scaled Gaussian

$$\psi_\gamma(s) \;=\; \gamma\, e^{-\pi(\gamma s)^2} \quad .$$

Then $\{\psi_\gamma\}_{\gamma=1}^\infty$ is an identity sequence for the set of all exponentially integrable functions on the real line. That is,

$$\lim_{\gamma \to \infty} \int_{-\infty}^{\infty} f(s)\, \gamma\, e^{-\pi(\gamma(s-x))^2}\, ds \;=\; f(x) \qquad\qquad (29.5)$$

whenever f is an exponentially integrable function on the real line and x is a point on the real line at which f is continuous.

For obvious reasons, the identity sequence described in the above theorem will be referred to as a Gaussian identity sequence. Note that, using the substitution $\sigma = \pi\gamma^2$, we can also write equation (29.5) as

$$\lim_{\sigma \to \infty} \int_{-\infty}^{\infty} f(s)\, \sqrt{\frac{\sigma}{\pi}}\, e^{-\sigma(s-x)^2}\, ds \;=\; f(x) \quad .$$

So theorem 29.9 also tells us that

$$\left\{ \sqrt{\frac{\sigma}{\pi}}\, e^{-\sigma s^2} \right\}_{\sigma=1}^{\infty}$$

is an identity sequence for the set of exponentially integrable functions. Similarly, it should be clear that the substitutions $\alpha = \gamma^2$ and $\epsilon = \gamma^{-2}$ lead to other identity sequences, namely,

$$\left\{ \sqrt{\alpha}\, e^{-\alpha\pi s^2} \right\}_{\alpha=1}^{\infty} \qquad \text{and} \qquad \left\{ \frac{1}{\sqrt{\epsilon}}\, \exp\!\left(-\frac{\pi}{\epsilon} s^2\right) \right\}_{\epsilon=1}^{0} \quad .$$

We'll refer to all of these as Gaussian identity sequences.

At this point you probably have a good notion of what a Gaussian identity sequence is. For the sake of having a formal definition, we will say $\{\phi_\sigma\}_{\sigma=\alpha}^{\sigma_0}$ is a *Gaussian identity sequence* if and only if, for each indexing parameter σ and real value s,

$$\phi_\sigma(s) \;=\; \sqrt{A(\sigma)}\, e^{-\pi A(\sigma) s^2}$$

where $A(\sigma)$ is some positive-valued function that approaches $+\infty$ as the parameter σ approaches σ_0. Using the substitution $\gamma = A(\sigma)$ and theorem 29.9, it is easy enough to verify that all Gaussian identity sequences are, in fact, identity sequences for the set of all exponentially integrable functions on the real line.

Gaussian identity sequences will provide tools for analyzing formulas involving classically transformable functions and for extending the theory so that any exponentially integrable function is "Fourier transformable". In particular, in chapter 30 we will find a "test for equality", analogous to that given in corollary 29.2 on page 472 (with Gaussians instead of pulse functions), that will be especially useful in verifying the fundamental theorem on invertibility and the major theorems on the differential identities from chapter 22. Other identity sequences can be used, but the many nice properties of Gaussian functions will make Gaussian identity sequences the preferred identity sequences for our future work.

But, at this time, we had better verify that Gaussian identity sequences are, in fact, identity sequences.

29.4 Verifying Identity Sequences
The Four-Step Path

Let us consider the problem of verifying that some given indexed set of functions $\{\psi_\gamma\}_{\gamma=1}^\infty$ is an identity sequence for some set S of functions. By theorem 29.3 on page 475, we know it suffices to show that

$$\lim_{\gamma\to\infty}\int_{-\infty}^{\infty} f(s)\psi_\gamma(s)\,ds \ = \ f(0) \tag{29.6}$$

for every f in S that is continuous at 0. This is fairly easily done when the ψ_γ's are pulse functions and f is continuous on an interval containing 0 (as in theorem 29.1 on page 472). For less simple cases it is often (but not always) convenient to remember that this equation is equivalent to

$$\lim_{\gamma\to\infty}\left| f(0) - \int_{-\infty}^{\infty} f(s)\psi_\gamma(s)\,ds \right| = 0 \quad,$$

which, itself, can be verified by showing that, for any given $\epsilon > 0$, there is a corresponding γ_ϵ such that[3]

$$\left| f(0) - \int_{-\infty}^{\infty} f(s)\psi_\gamma(s)\,ds \right| \leq \epsilon \qquad \text{whenever} \quad \gamma > \gamma_\epsilon \quad. \tag{29.7}$$

Proving this, in turn, can often be broken down to four steps:

1.   First simplify the problem by showing that inequality (29.7) is equivalent to

$$\left| \int_{-\infty}^{\infty} g(s)\psi_\gamma(s)\,ds \right| \leq \epsilon \qquad \text{whenever} \quad \gamma > \gamma_\epsilon \tag{29.8}$$

for some suitable function g that is continuous and vanishes at $s = 0$. (Consequently, to show that $\{\psi_\gamma\}_{\gamma=1}^\infty$ is an identity sequence for S, it suffices to show that there is a $\gamma_\epsilon > 0$ such that inequality (29.8) holds.)

2.   Verify that, because of the continuity of g, there is a $\delta > 0$ such that

$$\left| \int_{-\delta}^{\delta} g(s)\psi_\gamma(s)\,ds \right| \leq \frac{\epsilon}{3} \qquad \text{for all} \quad \gamma > 1 \quad.$$

3.   Verify that there is then a $\gamma_\epsilon > 0$ such that, whenever $\gamma > \gamma_\epsilon$,

$$\left| \int_{\delta}^{\infty} g(s)\psi_\gamma(s)\,ds \right| \leq \frac{\epsilon}{3} \quad \text{and} \quad \left| \int_{-\infty}^{-\delta} g(s)\psi_\gamma(s)\,ds \right| \leq \frac{\epsilon}{3} \quad.$$

4.   Finally, observe that, whenever $\gamma > \gamma_\epsilon$,

$$\left| \int_{-\infty}^{\infty} g(s)\psi_\gamma(s)\,ds \right| \leq \left| \int_{-\infty}^{-\delta} g(s)\psi_\gamma(s)\,ds \right| + \left| \int_{-\delta}^{\delta} g(s)\psi_\gamma(s)\,ds \right|$$

$$+ \left| \int_{\delta}^{\infty} g(s)\psi_\gamma(s)\,ds \right|$$

$$\leq \frac{\epsilon}{3} + \frac{\epsilon}{3} + \frac{\epsilon}{3} = \epsilon \quad.$$

This, as noted in step 1, confirms that $\{\psi_\gamma\}_{\gamma=1}^\infty$ is an identity sequence for S.

---

[3] You may want to review the *Refresher on Limits* (see page 27).

The approach to "verifying an identity sequence" outlined above will be a convenient approach to verifying the identity sequences of greatest interest to us. You should realize, however, that it is not always applicable, and that, even when it is, there may be easier ways to prove that a given $\{\psi_\gamma\}_{\gamma=1}^\infty$ is an identity sequence for some set S of functions.[4]

Proof of Theorem 29.9

To illustrate these steps — and to prove theorem 29.9 — let us consider the case where

$$\psi_\gamma(x) = \gamma\, e^{-\pi(\gamma x)^2}$$

and $f(x)$ is any exponentially integrable function on the real line continuous at $x = 0$. As just discussed, we need to show that, for every $\epsilon > 0$, there is a corresponding $\gamma_\epsilon > 0$ such that inequality (29.7) holds. For our choice of ψ_γ, inequality (29.7) is

$$\left| f(0) - \int_{-\infty}^\infty f(s)\, \gamma\, e^{-\pi(\gamma s)^2}\, dx \right| \le \epsilon \qquad \text{whenever} \quad \gamma > \gamma_\epsilon \; . \tag{29.9}$$

So let ϵ be some positive (but small) value, and let's proceed, step by step, to verify that there is a corresponding $\gamma_\epsilon > 0$ such that (29.9) holds.

Step 1: Using, say, equation (23.2) on page 361, we see that

$$\int_{-\infty}^\infty \gamma\, e^{-\pi(\gamma s)^2}\, ds = 1 \qquad \text{for every} \quad \gamma > 0 \; . \tag{29.10}$$

So,

$$f(0) = f(0) \int_{-\infty}^\infty \gamma\, e^{-\pi(\gamma s)^2}\, ds = \int_{-\infty}^\infty f(0)\, \gamma\, e^{-\pi(\gamma s)^2}\, ds$$

and

$$f(0) - \int_{-\infty}^\infty f(s)\, \gamma\, e^{-\pi(\gamma s)^2}\, ds = \int_{-\infty}^\infty f(0)\, \gamma\, e^{-\pi(\gamma s)^2}\, ds - \int_{-\infty}^\infty f(s)\, \gamma\, e^{-\pi(\gamma s)^2}\, ds$$

$$= \int_{-\infty}^\infty [f(0) - f(s)]\, \gamma\, e^{-\pi(\gamma s)^2}\, ds$$

$$= \int_{-\infty}^\infty g(s)\, \gamma\, e^{-\pi(\gamma s)^2}\, ds$$

where $g(s) = f(0) - f(s)$. Thus, verifying that inequality (29.9) holds for some $\gamma_\epsilon > 0$ is equivalent to verifying that

$$\left| \int_{-\infty}^\infty g(s)\, \gamma\, e^{-\pi(\gamma s)^2}\, ds \right| \le \epsilon \qquad \text{whenever} \quad \gamma > \gamma_\epsilon \tag{29.11}$$

for some $\gamma_\epsilon > 0$.

Do note that g, being a linear combination of exponentially integrable functions, is also exponentially integrable. Moreover, because $f(x)$ is continuous at 0, so is $g(x)$, and

$$g(0) = \lim_{s\to 0} g(s) = \lim_{s\to 0} [f(0) - f(s)] = 0 \; .$$

---

[4] In fact, those acquainted with the Lebesgue theory can often verify equation (29.6) using a much simpler two-step approach involving the dominated convergence theorem.

Step 2: We just saw that $g(s) \to 0$ as $s \to 0$. So there must be a positive distance δ such that

$$|g(s)| < \frac{\epsilon}{3} \qquad \text{whenever} \qquad -\delta < s < \delta \quad .$$

This, along with equation (29.10), means that

$$\int_{-\delta}^{\delta} |g(s)| \, \gamma \, e^{-\pi(\gamma s)^2} \, ds \leq \int_{-\delta}^{\delta} \frac{\epsilon}{3} \gamma \, e^{-\pi(\gamma s)^2} \, ds < \frac{\epsilon}{3} \cdot \int_{-\infty}^{\infty} \gamma \, e^{-\pi(\gamma s)^2} \, ds = \frac{\epsilon}{3} \quad .$$

Thus, for all $\gamma > 0$,

$$\left| \int_{-\delta}^{\delta} g(s) \, \gamma \, e^{-\pi(\gamma s)^2} \, ds \right| \leq \int_{-\delta}^{\delta} |g(s)| \, \gamma \, e^{-\pi(\gamma s)^2} \, ds < \frac{\epsilon}{3} \quad . \tag{29.12}$$

Step 3: Since g is exponentially integrable, we can choose a positive value β such that $|g(s)| e^{-\beta|s|}$ is absolutely integrable on the real line. With such a choice for β we know that the integrals

$$\int_{\delta}^{\infty} |g(s)| e^{-\beta s} \, ds \qquad \text{and} \qquad \int_{-\infty}^{-\delta} |g(s)| e^{\beta s} \, ds$$

are finite.

Now, let's consider

$$\int_{\delta}^{\infty} |g(s)| \, \gamma \, e^{-\pi(\gamma s)^2} \, ds \quad ,$$

which we will rewrite as

$$\int_{\delta}^{\infty} |g(s)| e^{-\beta s} h_\gamma(s) \, ds \qquad \text{with} \qquad h_\gamma(s) = \gamma \, e^{-\pi(\gamma s)^2 + \beta s} \quad .$$

By completing the square, we can rewrite h_γ as a shifted Gaussian,

$$h_\gamma(s) = \gamma \exp\left(-\pi \gamma^2 \left(s - \frac{\beta}{2\pi\gamma^2} \right)^2 + \left(\frac{\beta}{2\pi\gamma^2} \right)^2 \right) \quad .$$

That is,

$$h(s) = \gamma A_\gamma e^{-\pi\gamma^2(s - s_\gamma)^2}$$

where

$$A_\gamma = \exp\left(\frac{\beta^2}{4\pi\gamma^2} \right)$$

and

$$s_\gamma = \frac{\beta}{2\pi\gamma^2} \quad .$$

Choosing γ large enough that

$$\frac{\beta}{2\pi\gamma^2} < \delta$$

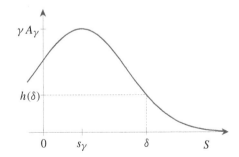

Figure 29.2: The shifted Gaussian $h(s)$ when $s_\gamma < \delta$.

ensures that $s_\gamma < \delta$ and, thus, also ensures that $h(\delta)$ is the maximum value of $h(s)$ on (δ, ∞) (see figure 29.2). Thus, as long as we choose γ as above,

$$\int_{\delta}^{\infty} |g(s)| \, \gamma \, e^{-\pi(\gamma s)^2} \, ds = \int_{\delta}^{\infty} |g(s)| e^{-\beta s} h_\gamma(s) \, ds \leq h_\gamma(\delta) \int_{\delta}^{\infty} |g(s)| e^{-\beta s} \, ds \quad .$$

Combine this with the fact that

$$\lim_{\gamma \to \infty} h_\gamma(\delta) = \lim_{\gamma \to \infty} \gamma e^{-\pi(\gamma\delta)^2} e^{\beta\delta} = e^{\beta\delta} \lim_{\gamma \to \infty} \gamma e^{-\pi\delta^2\gamma^2} = 0 \quad,$$

and we get

$$\lim_{\gamma \to \infty} \int_\delta^\infty |g(s)| \gamma e^{-\pi(\gamma s)^2} \, ds = 0 \quad.$$

This means there is some large value for γ, call it γ_ϵ^+, such that

$$\int_\delta^\infty |g(s)| \gamma e^{-\pi(\gamma s)^2} \, ds < \frac{\epsilon}{3} \qquad \text{whenever} \quad \gamma > \gamma_\epsilon^+ \quad.$$

By virtually identical computations, we can obviously show there is also a positive value γ_ϵ^- such that

$$\int_{-\infty}^{-\delta} |g(s)| \gamma e^{-\pi(\gamma s)^2} \, ds < \frac{\epsilon}{3} \qquad \text{whenever} \quad \gamma > \gamma_\epsilon^- \quad.$$

Taking γ_ϵ to be the larger of γ_ϵ^- and γ_ϵ^+, we then have, for $\gamma > \gamma_\epsilon$,

$$\left| \int_{-\infty}^{-\delta} g(s) \gamma e^{-\pi(\gamma s)^2} \, ds \right| \leq \int_{-\infty}^{-\delta} |g(s)| \gamma e^{-\pi(\gamma s)^2} \, ds < \frac{\epsilon}{3}$$

and

$$\left| \int_\delta^\infty g(s) \gamma e^{-\pi(\gamma s)^2} \, ds \right| \leq \int_\delta^\infty |g(s)| \gamma e^{-\pi(\gamma s)^2} \, ds < \frac{\epsilon}{3} \quad.$$

Step 4: Let $\gamma > \gamma_\epsilon$ where γ_ϵ is as above. From the last two inequalities and inequality (29.12) we know that

$$\left| \int_{-\infty}^\infty g(s) \gamma e^{-\pi(\gamma s)^2} \, ds \right| \leq \left| \int_{-\infty}^{-\delta} g(s) \gamma e^{-\pi(\gamma s)^2} \, ds \right| + \left| \int_{-\delta}^\delta g(s) \gamma e^{-\pi(\gamma s)^2} \, ds \right|$$

$$+ \left| \int_\delta^\infty g(s) \gamma e^{-\pi(\gamma s)^2} \, ds \right|$$

$$< \frac{\epsilon}{3} + \frac{\epsilon}{3} + \frac{\epsilon}{3} = \epsilon \quad.$$

Thus, as noted in step 1,

$$\left| f(0) - \int_{-\infty}^\infty f(s) \gamma e^{-\pi(\gamma s)^2} \, ds \right| \leq \epsilon \qquad \text{whenever} \quad \gamma > \gamma_\epsilon \quad,$$

which was all we needed to confirm that, as claimed in theorem 29.9,

$$\left\{ \gamma e^{-(\gamma x)^2} \right\}_{\gamma=1}^\infty$$

is an identity sequence for the set of exponentially integrable functions.   ∎

Verifying Other Identity Sequences

It is not at all difficult to modify the above proof of theorem 29.9 so as to obtain a proof of, say, theorem 29.4 on page 476. The only significant changes, aside from replacing

$$\psi_\gamma(s) = \gamma\, e^{-\pi(\gamma s)^2} \qquad \text{with} \qquad \psi_\gamma(s) = \gamma\Psi(\gamma s) \quad,$$

are in the second and third steps.

The main change in the second step is to take into account the fact that, because Ψ is not necessarily a positive function, the integral of $|\phi_\gamma|$ is not necessarily 1. So let

$$B = \int_{-\infty}^{\infty} |\Psi(x)|\, dx \quad ;$$

verify that $1 \le B < \infty$, and choose $\delta > 0$ so that

$$|g(s)| < \frac{\epsilon}{3B} \qquad \text{whenever} \qquad -\delta < s < \delta \quad .$$

You should then have no trouble verifying that

$$\left| \int_{-\delta}^{\delta} g(s)\, \psi_\gamma(s)\, ds \right| < \frac{\epsilon}{3} \qquad \text{for all} \quad \gamma > 0 \quad .$$

The necessary modifications to the third step are also simple. You use the fact that Ψ vanishes outside of some interval to show the existence of a γ_ϵ^+ and a γ_ϵ^- such that, in fact,

$$\int_{\delta}^{\infty} |g(s)\, \psi_\gamma(s)|\, ds + \int_{-\infty}^{-\delta} |g(s)\, \psi_\gamma(s)|\, ds = 0 \qquad \text{whenever} \quad \gamma > \gamma_\epsilon^+ \quad .$$

The details are left to you.

?► Exercise 29.6:   *Prove theorem 29.4 on page 476.*

(Also see exercise 29.15 at the end of this chapter.)

29.5 An Application (with Exercises)

We will illustrate here how identity sequences can be used by going way back to the topic of Fourier series and completing the proof of an equality from chapter 13. Other significant applications of Gaussian identity sequences will be described in chapter 30 (and exercises at the end of this chapter).

Parseval's Equality for Fourier Series

Let us finish the proof of theorem 13.15 on page 172 (Parseval's equality for Fourier series). That theorem concerned the inner product of two periodic, piecewise continuous functions f and g with a common period p and Fourier series

$$F.S.[f]|_t = \sum_{k=-\infty}^{\infty} f_k\, e^{i2\pi\omega_k t} \qquad \text{and} \qquad F.S.[g]|_t = \sum_{k=-\infty}^{\infty} g_k\, e^{i2\pi\omega_k t} \quad .$$

Recall that the inner product of the periodic functions f and g is given by

$$\langle f \mid g \rangle = \int_{\text{period}} f(t)g^*(t)\,dt \quad .$$

The claim of the theorem was that

1. $\displaystyle\sum_{k=-\infty}^{\infty} f_k g_k^*$ converges absolutely, and

2. $\displaystyle\langle f \mid g \rangle = p \sum_{k=-\infty}^{\infty} f_k g_k^* \quad .$

We already know this claim is true with the additional assumption that g is continuous and piecewise smooth (see lemma 13.14 on page 171). What's more, you can easily verify that the arguments showing the absolute convergence of $\sum_{k=-\infty}^{\infty} f_k g_k^*$ given in the proof of lemma 13.14 remain valid when g is merely piecewise continuous. So all we really need to verify is Parseval's equality itself,

$$\langle f \mid g \rangle = p \sum_{k=-\infty}^{\infty} f_k g_k^* \quad .$$

To do so, let $\{\psi_\alpha\}_{\alpha=1}^{\infty}$ be the Gaussian identity sequence with

$$\psi_\alpha(x) = \sqrt{\alpha}\, e^{-\pi \alpha x^2} \quad ,$$

and consider $g * \psi_\alpha$ for each $\alpha \geq 1$. From theorem 24.11 on page 395 we know $g * \psi_\alpha$ is a continuous, periodic function with period p and Fourier series

$$F.S.[g * \psi_\alpha]|_t = \sum_{k=-\infty}^{\infty} g_k\, \mathcal{F}[\psi_\alpha]|_{\omega_k}\, e^{i2\pi\omega_k t} = \sum_{k=-\infty}^{\infty} g_k \exp\!\left(-\frac{\pi}{\alpha}\,\omega_k{}^2\right) e^{i2\pi\omega_k t} \quad .$$

But, as we well know, ψ_α is also a smooth function with an absolutely integrable derivative. Hence $g * \psi$ must also be a smooth function (see corollary 24.6 on page 391), and lemma 13.14, mentioned above, assures us that

$$\langle f \mid g * \psi_\alpha \rangle = p \sum_{k=-\infty}^{\infty} f_k \left[g_k \exp\!\left(-\frac{\pi}{\alpha}\,\omega_k{}^2\right)\right]^*$$

$$= p \sum_{k=-\infty}^{\infty} f_k g_k^* \exp\!\left(-\frac{\pi}{\alpha}\,\omega_k{}^2\right) \quad . \tag{29.13}$$

The next major portion of the proof is in going through the details of showing

$$\lim_{\alpha \to \infty} \langle f \mid g * \psi_\alpha \rangle = \langle f \mid g \rangle \tag{29.14}$$

and

$$\lim_{\alpha \to \infty} \sum_{k=-\infty}^{\infty} f_k g_k^* \exp\!\left(-\frac{\pi}{\alpha}\,\omega_k{}^2\right) = \sum_{k=-\infty}^{\infty} f_k g_k^* \quad . \tag{29.15}$$

These details are left to you. (It is in showing that equation (29.14) holds that we use the fact that we have a Gaussian identity sequence.)

?►Exercise 29.7: *Prove equation (29.14). Start by verifying that*

$$\langle f \mid g * \psi_\alpha \rangle = \int_0^p \int_{-\infty}^\infty f(t) g^*(t-s) \psi_\alpha(s) \, ds \, dt \quad .$$

Then apply some of the general integration results discussed in the last sections of chapters 7 and 18 along with the fact that $\{\psi_\alpha\}_{\alpha=1}^\infty$ is a Gaussian identity sequence to show that

$$\lim_{\alpha \to \infty} \langle f \mid g * \psi_\alpha \rangle = \lim_{\alpha \to \infty} \int_{-\infty}^\infty \psi_\alpha(s) \left[\int_0^p f(t) g^*(t-s) \, dt \right] ds$$

$$= \int_0^p f(t) g^*(t) \, dt \quad .$$

?►Exercise 29.8: *Prove equation (29.15).*

Finally, combining equations (29.14), (29.15), and (29.13), we get

$$\langle f \mid g \rangle = \lim_{\alpha \to \infty} \langle f \mid g * \psi_\alpha \rangle$$

$$= p \lim_{\alpha \to \infty} \sum_{k=-\infty}^\infty f_k g_k^* \exp\left(-\frac{\pi}{\alpha} \omega_k^2\right) = p \sum_{k=-\infty}^\infty f_k g_k^* \quad ,$$

confirming that Parseval's equality holds. ∎

29.6 Laplace Transforms as Fourier Transforms

The discussion of exponential boundedness and integrability may have sparked memories of the Laplace transform from a course on differential equations. It turns out that we can easily relate the Laplace and classical Fourier transforms, and, through this relation, derive formulas for inverting the Laplace transform.

Let us give slightly more general definitions of terms you may recall. Let f be a function on $(0, \infty)$ and α some real number. We will say that f has *exponential order* α if and only if $f(t) e^{-\beta t}$ is piecewise continuous and absolutely integrable on $(0, \infty)$ for every $\beta > \alpha$. It is worth noting that if f has exponential order α, then, because $\left| e^{iy} \right| = 1$, we also have $f(t) e^{-st}$ being absolutely integrable on $(0, \infty)$ for every complex number $s = x + iy$ with $x > \alpha$, .

So let f be any function on $(0, \infty)$ of exponential order α. We then define the *Laplace transform* of f, denoted by either $\mathcal{L}[f]$ or $F_\mathcal{L}$, to be the function given by

$$F_\mathcal{L}(s) = \mathcal{L}[f]\big|_s = \int_0^\infty f(t) e^{-st} \, dt$$

for every complex value $s = x + iy$ with $x \geq \alpha$.

To get a useful relation between Laplace and Fourier transforms, let f_0 be the extension of f on $(-\infty, \infty)$ given by

$$f_0(t) = \begin{cases} f(t) & \text{if } t > 0 \\ 0 & \text{if } t \leq 0 \end{cases} ,$$

and observe that, for $x > \alpha$,

$$\int_0^\infty f(t)\, e^{-(x+i2\pi\omega)t}\, dt \;=\; \int_{-\infty}^\infty f_0(t) e^{-xt}\, e^{-i2\pi\omega t}\, dt \quad.$$

In other words, our two transforms are related by

$$F_{\mathcal{L}}(x + i2\pi\omega) \;=\; \mathcal{L}[f]\big|_{x+i2\pi\omega} \;=\; \mathcal{F}\big[f_0(t)e^{-xt}\big]\big|_\omega \quad. \tag{29.16}$$

To get an inversion formula for the Laplace transform, let $\beta > \alpha$ and observe that, for $t > 0$,

$$f(t) \;=\; e^{\beta t} f_0(t) e^{-\beta t} \;=\; e^{\beta t} \mathcal{F}^{-1}\big[\mathcal{F}\big[f_0(\tau)e^{-\beta\tau}\big]\big|_\omega\big]\big|_t \quad.$$

Applying equation (29.16) then gives the following theorem on inverting the Laplace transform (i.e., finding the f such that $F_{\mathcal{L}} = \mathcal{L}[f]$ for a given $F_{\mathcal{L}}$):

Theorem 29.10
If $F_{\mathcal{L}}$ is the Laplace transform of some function of exponential order α, then $F_{\mathcal{L}} = \mathcal{L}[f]$ where f is the function on $(0, \infty)$ given by

$$f(t) \;=\; e^{\beta t} \mathcal{F}^{-1}[F_{\mathcal{L}}(\beta + i2\pi\omega)]\big|_t \tag{29.17}$$

where β can be any real number greater than α.

It is tempting to use the integral formula for the inverse Fourier transform in equation (29.17) to obtain

$$f(t) \;=\; e^{\beta t} \int_{-\infty}^\infty F_{\mathcal{L}}(\beta + i2\pi\omega)\, e^{i2\pi\omega t}\, d\omega$$

$$= \int_{-\infty}^\infty F_{\mathcal{L}}(\beta + i2\pi\omega)\, e^{(\beta + i2\pi\omega)t}\, d\omega \;=\; \frac{1}{2\pi} \int_{-\infty}^\infty F_{\mathcal{L}}(\beta + iy)\, e^{(\beta + iy)t}\, dy \quad.$$

Unfortunately, this requires that $F_{\mathcal{L}}(\beta + i2\pi\omega)$ be an absolutely integrable function of ω, which, in general, cannot be assumed. However, if f is exponentially bounded with $\big|f(t)e^{-\alpha t}\big|$ being bounded, then you can show the next best thing, namely, that for every $t > 0$ and $\beta > \alpha$,

$$f(t) \;=\; \frac{1}{2\pi} \lim_{Y\to\infty} \int_{-Y}^{Y} F_{\mathcal{L}}(\beta + iy)\, e^{(\beta + iy)t}\, dy \quad. \tag{29.18}$$

This is the inversion formula for the Laplace transform commonly found in texts discussing the Laplace transform. Its derivation from equation (29.17) will be left to you (exercise 29.18 c). It involves an identity sequence.

It should be noted that inversion formulas (29.17) and (29.18) are rarely used to compute inverse Laplace transforms. What is important is the following immediate corollary of theorem 29.10:

Corollary 29.11
Let f and g be two functions on $(0, \infty)$ of any exponential orders. If $\mathcal{L}[f] = \mathcal{L}[g]$, then $f = g$.

This corollary assures us that inverse Laplace transforms are well defined and unique, and this assures us that we can safely find inverse transforms using tables of Laplace transforms along with the various properties of the inverse Laplace transforms.

Additional Exercises

29.9. Let f be a piecewise continuous function on the real line, and let x be a point at which f has a jump discontinuity. Show that

$$\lim_{\epsilon \to 0^+} \int_{-\infty}^{\infty} f(s) \frac{1}{2\epsilon} \operatorname{pulse}_\epsilon(s - x) \, ds = \frac{1}{2}\left[\lim_{h \to x^-} f(h) + \lim_{h \to x^+} f(h) \right]$$

$$= \text{"the midpoint of the jump at } x\text{"} \quad .$$

29.10. Using theorem 29.1 on page 472, find the formulas for f, g and h, respectively, assuming the following equations hold for every real value x and every $\epsilon > 0$:

a. $\displaystyle\int_{x-\epsilon}^{x+\epsilon} f(s) \, ds = 6\epsilon x$

b. $\displaystyle\int_{x-\epsilon}^{x+\epsilon} g(s) \, ds = 2\cos(x)\sin(\epsilon)$

c. $\displaystyle\int_{x-\epsilon}^{x+\epsilon} h(s) \, ds = \begin{cases} 0 & \text{if } x \leq -\epsilon \\ x + \epsilon & \text{if } -\epsilon \leq x \leq \epsilon \\ 2\epsilon & \text{if } \epsilon \leq x \end{cases}$

29.11. Verify that every polynomial is exponentially bounded on the real line.

29.12 a. For what real values of α is $|x|^\alpha$ exponentially bounded on the real line?

b. For what real values of α is $|x|^\alpha$ exponentially integrable on the real line?

29.13 Find or construct a piecewise continuous, absolutely integrable function on $\mathbb{R}$ that is not exponentially bounded, thereby showing that there are classically transformable functions that are not exponentially bounded. (You might consider a sequence of increasingly narrow, yet taller, pulses along the positive real line similar to that in exercise 18.16 on page 277.)

29.14. Using a suitable Gaussian identity sequence, find the formulas for f and g, respectively, assuming the following equations hold for each real value x and each $\gamma > 0$:

a. $\displaystyle\int_{-\infty}^{\infty} f(s) \, e^{-\pi\gamma(s-x)^2} \, ds = \frac{4x}{\sqrt{\gamma}}$

b. $\displaystyle\int_{-\infty}^{\infty} g(s) \, e^{-\pi\gamma(s-x)^2} \, ds = \frac{1}{\sqrt{\gamma}} \exp\left(\frac{\pi}{\gamma}(1 - 2\gamma x)\right)$

29.15. Let Ψ be any absolutely integrable function on the real line that satisfies

$$\int_{-\infty}^{\infty} \Psi(x) \, dx = 1 \quad ,$$

and, for each $\gamma > 0$, let

$$\psi_\gamma(x) = \gamma \Psi(\gamma x) \quad .$$

Using the four-step path, show that $\{\psi_\gamma\}_{\gamma=1}^{\infty}$ is an identity sequence for the set of all bounded, piecewise continuous functions.

29.16. Let f be an exponentially integrable function, and let x be a point at which f has a jump discontinuity. Show that

$$\lim_{\gamma \to \infty} \int_{-\infty}^{\infty} f(s)\, \gamma\, e^{-\pi \gamma^2 (s-x)^2}\, dx \;=\; \frac{1}{2}\left[\lim_{h \to x^-} f(h) \;+\; \lim_{h \to x^+} f(h) \right]$$

$$= \text{"the midpoint of the jump at } x\text{"} \quad.$$

29.17. Show that $\{2\gamma\, \mathrm{sinc}(2\gamma \pi s)\}_{\gamma=1}^{\infty}$ is an identity sequence for each set S of functions given below. Do not try using the four-step path. Instead, start by using either Parseval's equality or the fundamental identity of Fourier analysis as indicated.

    a. S is the set of all bounded, classically transformable functions whose Fourier transforms are bounded. (Assume Parseval's equality from chapter 25 is valid here.)

    b. $S = \mathcal{A} \cap \mathcal{T}$, the set of all absolutely integrable functions that are also Fourier transforms of absolutely integrable functions. (Use the fundamental identity from chapter 25.)

29.18 a. Suppose f and F are bounded, classically transformable functions with $F = \mathcal{F}[f]$. Assuming Parseval's equality is valid, and using results from the previous problem, show that

$$f(t) \;=\; \lim_{\gamma \to \infty} \int_{-\gamma}^{\gamma} F(\omega)\, e^{i 2\pi \omega t}\, d\omega$$

for each t in $\mathbb{R}$ at which f is continuous. (Do not assume F is absolutely integrable.)

    b. Using the above with $t = 0$, evaluate

$$\lim_{\gamma \to \infty} \int_{-\gamma}^{\gamma} \mathrm{sinc}(\omega)\, d\omega \quad.$$

    c. Using the above, rigorously verify that formula (29.18) on page 488 for inverting the Laplace transform follows from inversion formula (29.17) under the conditions given for formula (29.18).

29.19 a. Assume f and F are classically transformable functions with $F = \mathcal{F}[f]$ and f in $\mathcal{A}$. Show that

$$\lim_{\epsilon \to 0^+} \int_{-\infty}^{\infty} e^{-\epsilon \omega^2} F(\omega)\, e^{i 2\pi t \omega}\, d\omega \;=\; f(t)$$

without using "invertibility". (Hint: See if the fundamental identity of Fourier analysis applies.)

    b. Using this and theorem 18.18 on page 271, prove the fundamental theorem on invertibility, theorem 19.5 on page 285. (Be careful to not use "invertibility" or any other results that have not yet been proven, such as Parseval's equality from theorem 25.9.)

29.20. Using your favorite identity sequence, verify the claim that, if the delta function, δ, were a function continuous at some nonzero point x_0, then $\delta(x_0) = 0$.

30

Gaussians as Test Functions and Proofs of Important Theorems

There are two major reasons for this chapter. One is to finally confirm those few important theorems that we've been using with only a promise of proof "sometime in the future": the fundamental theorem on invertibility, the main theorems on the differentiation identities, the second theorem on the fundamental identity and the second theorem on the convolution identities. The other is to introduce some basic concepts that will later lead to a much more general and powerful theory of Fourier analysis. In particular, you should note the "Gaussian test for equality" in the next section and the way we use it with the fundamental identity to prove the theorems on invertibility and differential identities. It will be these ideas and procedures that will be expanded upon in part IV of this book to obtain that more general theory.

30.1 Testing for Equality with Gaussians

For each of those big theorems we want to validate, it will be necessary to verify that two different formulas (say, $\mathcal{F}[f']|_\omega$ and $i2\pi\omega\mathcal{F}[f]|_\omega$) describe the same function. While we could explicitly use Gaussian identity sequences to do this, we will find it convenient to hide our use of these identity sequences in the test described in the following theorem.

Theorem 30.1 (Gaussian test for equality)
Let f and g be two exponentially integrable functions. Then $f = g$ if and only if

$$\int_{-\infty}^{\infty} f(x)\phi(x)\,dx = \int_{-\infty}^{\infty} g(x)\phi(x)\,dx \tag{30.1}$$

for every Gaussian function ϕ.

PROOF: Obviously, if $f = g$, then equation (30.1) holds for every Gaussian function ϕ.

Now suppose equation (30.1) holds whenever ϕ is a Gaussian, and let $\{\phi_\gamma\}_{\gamma=1}^{\infty}$ be a Gaussian identity sequence, say, with

$$\phi_\gamma(x) = \gamma\, e^{-\pi(\gamma x)^2} \quad .$$

Then, for each real t and $\gamma \geq 1$, we must also have

$$\int_{-\infty}^{\infty} f(x)\phi_\gamma(x - t)\,dx = \int_{-\infty}^{\infty} g(x)\phi_\gamma(x - t)\,dx \tag{30.2}$$

since each $\phi_\gamma(x-t)$ is a Gaussian function of x. Combining this with the fact that $\{\phi_\gamma\}_{\gamma=1}^\infty$ is an identity sequence for the set of exponentially integrable functions (see theorem 29.9 on page 480) gives us

$$f(t) \;=\; \lim_{\gamma\to\infty} \int_{-\infty}^{\infty} f(x)\phi_\gamma(x-t)\,dx \;=\; \lim_{\gamma\to\infty} \int_{-\infty}^{\infty} g(x)\phi_\gamma(x-t)\,dx \;=\; g(t)$$

for each t at which f and g are continuous. Thus $f = g$.   ∎

Keep in mind that every classically transformable function is exponentially integrable. So we can certainly employ these tests whenever f and g are classically transformable.[1]

30.2 The Fundamental Theorem on Invertibility
The Theorem

Recall that we first defined the Fourier *integral* transforms,

$$\mathcal{F}_I[f]|_x \;=\; \int_{-\infty}^{\infty} f(y)\,e^{-i2\pi xy}\,dy \qquad \text{and} \qquad \mathcal{F}_I^{-1}[G]|_x \;=\; \int_{-\infty}^{\infty} G(y)\,e^{i2\pi xy}\,dy$$

where f and G are any two functions from $\mathcal{A}$, the set of all piecewise continuous, absolutely integrable functions on the real line.[2]

One important theorem in chapter 19 was left unproven. That was the fundamental theorem on invertibility (theorem 19.5 on page 285), which asserted the following:

> If f and F are two functions in $\mathcal{A}$, then
>
> $$F \;=\; \mathcal{F}_I[f] \quad \Longleftrightarrow \quad \mathcal{F}_I^{-1}[F] = f \quad .$$

Though left unproven, this has been a very important and useful theorem. Indeed, our very definitions in chapter 20 of the Fourier transforms, $\mathcal{F}$ and $\mathcal{F}^{-1}$, required this theorem to be true.

As a step towards proving this theorem (and to illustrate a pattern that will be repeated many times later), we will divide this theorem into the following two complementary lemmas.

Lemma 30.2 *(invertibility of the direct transform)*
If f and F are two functions in $\mathcal{A}$, then

$$F \;=\; \mathcal{F}_I[f] \quad \Longrightarrow \quad \mathcal{F}_I^{-1}[F] = f \quad .$$

Lemma 30.3 *(invertibility of the inverse transform)*
If f and F are two functions in $\mathcal{A}$, then

$$F \;=\; \mathcal{F}_I[f] \quad \Longleftarrow \quad \mathcal{F}_I^{-1}[F] = f \quad .$$

---

[1] Because of the role Gaussians play in "testing for equality", we will begin referring to Gaussians as "test functions" in part IV of this book.

[2] You may want to quickly review chapter 19 at this time.

We'll prove the first lemma in detail, and leave the details of the second for you. There are three major elements to the proof of either:

1.   Confirming, without assuming the fundamental theorem on invertibility, that the claim of the complementary lemma holds when the functions are Gaussians. We will refer to the lemma in which we state this claim as the "associated lemma".

2.   Using versions of the fundamental identity to "move" the transform operations from fairly arbitrary functions to Gaussian functions.

3.   Using the Gaussian test for equality.

Since we have already discussed the test for equality, let us briefly discuss that "associated lemma" and the appropriate versions of the fundamental identities.

The Associated Lemma and the Fundamental Identities

Here is the associated lemma. Observe that it is basically the same as lemma 30.3 (the complement to lemma 30.2) with "Gaussians" replacing "functions in $\mathcal{A}$".

Lemma 30.4
If ϕ is a Gaussian, then $\mathcal{F}_I\big[\mathcal{F}_I^{-1}[\phi]\big] = \phi$.

(Before we prove this lemma, you should go back to page 367 and re-read the notes on the derivations of the formulas for the Fourier transforms of Gaussians.)

PROOF: Suppose ϕ is a Gaussian. By definition (i.e., equation (23.8) on page 364 with $\gamma = \alpha\pi$), ϕ can be expressed as

$$\phi(x) = A\,e^{-\alpha\pi(x-\xi)^2}$$

where A and ξ are two complex constants and $\alpha > 0$. Now let $\Phi = \mathcal{F}_I^{-1}[\phi]$. Recalling the definition of the integral transforms, as well as equation (23.13) from theorem 23.2 on page 368 (and the discussion just before that theorem), we see that

$$\Phi(y) = \mathcal{F}_I^{-1}[\phi]\big|_y = \int_{-\infty}^{\infty} A\,e^{-\alpha\pi(x-\xi)^2}\,e^{i2\pi xy}\,dx$$

$$= A\frac{1}{\sqrt{\alpha}}\,e^{i2\pi\xi y}\exp\!\left(-\frac{\pi}{\alpha}y^2\right) = A\sqrt{\beta}\,e^{i2\pi\xi y}e^{-\pi\beta y^2}$$

where, for convenience, we've let $\beta = {}^1\!/_\alpha$. Invoking theorem 23.2 again (using equation (23.18)) and recalling our definitions, we find that, for all x in $\mathbb{R}$,

$$\mathcal{F}_I\big[\mathcal{F}_I^{-1}[\phi]\big]\big|_x = \mathcal{F}_I[\Phi]\big|_x$$

$$= A\sqrt{\beta}\int_{-\infty}^{\infty} e^{i2\pi\xi y}e^{-\beta\pi y^2}\,e^{-i2\pi xy}\,dy$$

$$= A\sqrt{\beta}\cdot\frac{1}{\sqrt{\beta}}\exp\!\left(-\frac{\pi}{\beta}(x-\xi)^2\right)$$

$$= A\,e^{-\alpha\pi(x-\xi)^2}$$

$$= \phi(x) \quad.$$

This is where the fundamental identity of Fourier analysis becomes especially important; so you may want to review our discussion of this identity in chapter 25. In particular, to finish proving the fundamental theorem on invertibility, we will need to know the fundamental identity is applicable using the integral transforms and Gaussians, and that it can be derived without using the fundamental theorem on invertibility. That is the claim of the next lemma.

Lemma 30.5 (fundamental identities with integral transform and Gaussians)
Assume ψ is any function in $\mathcal{A}$, and let ϕ be either a Gaussian or a Fourier integral transform of a Gaussian. Then the product of either of these functions with the direct or inverse Fourier integral transform of the other is absolutely integrable. Moreover,

$$\int_{-\infty}^{\infty} \mathcal{F}_I[\psi]|_x\, \phi(x)\, dx = \int_{-\infty}^{\infty} \psi(y) \mathcal{F}_I[\phi]|_y\, dy \tag{30.3a}$$

and

$$\int_{-\infty}^{\infty} \mathcal{F}_I^{-1}[\psi]|_x\, \phi(x)\, dx = \int_{-\infty}^{\infty} \psi(y) \mathcal{F}_I^{-1}[\phi]|_y\, dy \quad. \tag{30.3b}$$

The details of proving this lemma are left to you. Keep in mind the following:

1. From the previous lemma and the integral formulas for the transforms of Gaussians, we know that the statement

    ϕ *is either a Gaussian or a Fourier integral transform of a Gaussian*

    is completely equivalent to

    ϕ *and its Fourier integral transforms are Gaussians.*

2. The computations in proving the identities in the above lemma are virtually identical to those in the proof of the first part of theorem 25.14 on page 413.

?►Exercise 30.1: *Prove lemma 30.5 without using the invertibility of the transforms.*

Proof of the First Lemma on Invertibility (Lemma 30.2)

Remember: Both f and F are in $\mathcal{A}$ with $F = \mathcal{F}_I[f]$. Our goal is to show that $\mathcal{F}_I^{-1}[F] = f$. To do this, let ϕ be any Gaussian function. By identity (30.3b), the definition of F and then identity (30.3a), we have

$$\int_{-\infty}^{\infty} \mathcal{F}_I^{-1}[F]|_x\, \phi(x)\, dx = \int_{-\infty}^{\infty} F(y) \mathcal{F}_I^{-1}[\phi]|_y\, dy$$

$$= \int_{-\infty}^{\infty} \mathcal{F}_I[f]|_y\, \mathcal{F}_I^{-1}[\phi]|_y\, dy = \int_{-\infty}^{\infty} f(x) \mathcal{F}_I\big[\mathcal{F}_I^{-1}[\phi]\big]\big|_x\, dx \quad.$$

But in the associated lemma, we saw that $\mathcal{F}_I\big[\mathcal{F}_I^{-1}[\phi]\big] = \phi$ whenever ϕ is a Gaussian. So the above reduces to

$$\int_{-\infty}^{\infty} \mathcal{F}_I^{-1}[F]|_x\, \phi(x)\, dx = \int_{-\infty}^{\infty} f(x) \phi(x)\, dx \quad.$$

Since this equation holds for every Gaussian ϕ, our Gaussian test for equality (theorem 30.1) assures us that

$$\mathcal{F}_I^{-1}[F] = f \quad. \qquad\blacksquare$$

?► Exercise 30.2: *Finish the proof of the fundamental theorem on invertibility by proving lemma 30.2.*

30.3 The Fourier Differential Identities

In chapter 22 we derived several differential identities for Fourier transforms, and we verified these identities under various conditions (see lemmas 22.5 through 22.8, starting on page 347). We also stated, but did not prove, two theorems (theorems 22.1 and 22.2, starting on page 341) asserting that these identities held under much more general conditions. Our goal here is to finish the proofs of these two theorems.

If you check back, you will find that either theorem 22.1 or theorem 22.2 can be obtained from the other via near-equivalence. That being the case, let us concentrate on proving just one of these theorems, say, theorem 22.1. Let us further simplify our discussion by splitting that theorem into the following two "complementary sub-theorems".

Lemma 30.6 (first sub-theorem on the differential identities)
Let f and F be two classically transformable functions with $F = \mathcal{F}[f]$. If f is continuous and piecewise smooth, and f' is classically transformable, then $x F(x)$ is classically transformable, and

$$\mathcal{F}[f']\big|_x = i2\pi x\, F(x) \quad .$$

Lemma 30.7 (second sub-theorem on the differential identities)
Let f and F be two classically transformable functions with $F = \mathcal{F}[f]$. If $x F(x)$ is classically transformable, then f is continuous and piecewise smooth, f' is classically transformable, and

$$\mathcal{F}^{-1}[i2\pi y F(y)]\big|_x = f'(x) \quad .$$

There will be noteworthy similarities between our proofs of these two sub-theorems and our proof of the fundamental theorem on invertibility. An "associated lemma" stating that the differential identities hold when the functions are Gaussians will first be verified. Also, we will again employ the fundamental identity to "move" transforms from fairly arbitrary functions to the test functions. On the other hand, there will be some analysis specifically related to the fact that we are dealing with derivatives. In particular, we will look more closely at integration by parts. Once we've discussed these topics, the proofs of the sub-theorems will proceed quickly.

The Associated Lemma

In this case, the associated lemma is just the observation that we already know the differential identities hold when the functions are Gaussians. Since you can easily (re)verify this using the properties of Gaussians and lemmas 22.5 through 22.8, we will simply state the two identities we will use in our final proof.

Lemma 30.8
If ϕ is any Gaussian function, then ϕ, ϕ' and $x\phi(x)$ are smooth, absolutely integrable functions with

$$\frac{d}{dy}\mathcal{F}[\phi]\big|_y = \mathcal{F}[-i2\pi x\, \phi(x)]\big|_y \qquad \text{and} \qquad \mathcal{F}^{-1}[\phi']\big|_y = -i2\pi y\mathcal{F}^{-1}[\phi]\big|_y \quad .$$

From the second part of theorem 25.14 on the fundamental identity (page 413), we know that, if f, F, g and G are classically transformable functions with $F = \mathcal{F}[f]$ and $G = \mathcal{F}[g]$, then Fg and fG are absolutely integrable, and

$$\int_{-\infty}^{\infty} F(x)g(x)\,dx \;=\; \int_{-\infty}^{\infty} f(y)G(y)\,dy$$

provided both g and G are absolutely integrable. This will certainly be the case if either g or G is a Gaussian. Moreover, as you can easily verify from the formulas for the transforms of Gaussians and the previous lemma, both g and G will be absolutely integrable if either is a Gaussian, a derivative of a Gaussian, or a product of a Gaussian with a polynomial. Consequently, we are already assured that the following lemma holds.

Lemma 30.9 *(fundamental identities with Gaussians)*
Let ψ be any classically transformable function, and assume ϕ is either a Gaussian function, the derivative of a Gaussian function, or a Gaussian function multiplied by a polynomial. Then the product of either function with a Fourier transform of the other is absolutely integrable. Moreover,

$$\int_{-\infty}^{\infty} \mathcal{F}[\psi]|_x\, \phi(x)\,dx \;=\; \int_{-\infty}^{\infty} \psi(y)\mathcal{F}[\phi]|_y\,dy$$

and

$$\int_{-\infty}^{\infty} \mathcal{F}^{-1}[\psi]|_x\, \phi(x)\,dx \;=\; \int_{-\infty}^{\infty} \psi(y)\mathcal{F}^{-1}[\phi]|_y\,dy \quad .$$

Some Integral/Differential Results

Here are a few results concerning differentiation we will find useful.

Lemma 30.10
If g is exponentially integrable, then the function

$$f(x) \;=\; \int_{0}^{x} g(s)\,ds$$

is exponentially bounded.

PROOF: First observe that, for any real x and positive β,

$$|f(x)| \;=\; \left| \int_{0}^{x} g(s)\,ds \right|$$

$$\leq \int_{-|x|}^{|x|} |g(s)|\,ds$$

$$= \int_{-|x|}^{|x|} |g(s)|\,e^{-\beta|s|}e^{\beta|s|}\,ds \;\leq\; e^{\beta|x|} \int_{-|x|}^{|x|} |g(s)|\,e^{-\beta|s|}\,ds \quad .$$

Now, because g is exponentially integrable, we can choose β so that

$$A \;=\; \int_{-\infty}^{\infty} |g(s)|\,e^{-\beta|s|}\,ds \;<\; \infty \quad .$$

Consequently, for every real value x, the previous sequence of inequalities can be continued as follows:

$$|f(x)| \leq e^{\beta|x|} \int_{-|x|}^{|x|} |g(s)| e^{-\beta|s|} ds \leq e^{\beta|x|} \int_{-\infty}^{\infty} |g(s)| e^{-\beta|s|} ds = A e^{\beta|x|} \quad . \qquad \blacksquare$$

Lemma 30.11 (integration by parts with Gaussians)
Let f be a continuous, piecewise smooth function on the real line, and assume f' is exponentially integrable. Then f is exponentially bounded, and

$$\int_{-\infty}^{\infty} f'(x)\phi(x) \, dx = - \int_{-\infty}^{\infty} f(x)\phi'(x) \, dx$$

whenever ϕ is a Gaussian.

PROOF: From lemma 30.10 and the fact that

$$f(x) - f(0) = \int_{0}^{x} f'(s) \, ds \quad ,$$

we know f is exponentially bounded.

Now let ϕ be any Gaussian. As noted in lemma 29.8 on page 478, the exponential boundedness of f ensures that the product $f\phi$ is absolutely integrable and that

$$\lim_{x \to \pm\infty} f(x)\phi(x) = 0 \quad .$$

So, applying the integration by parts formula from theorem 4.2 on page 40, we have

$$\int_{-\infty}^{\infty} f'(x)\phi(x) \, dx = \lim_{\substack{b \to \infty \\ a \to -\infty}} \int_{a}^{b} f'(x)\phi(x) \, dx$$

$$= \lim_{\substack{b \to \infty \\ a \to -\infty}} \left[f(x)\phi(x)\Big|_{a}^{b} - \int_{a}^{b} f(x)\phi'(x) \, dx \right]$$

$$= 0 - \int_{-\infty}^{\infty} f(x)\phi'(x) \, dx = - \int_{-\infty}^{\infty} f(x)\phi'(x) \, dx \quad . \qquad \blacksquare$$

To use the above integration by parts formula we need to know that f is continuous and piecewise smooth. Sometimes, though, the real problem is verifying that a given function is continuous and piecewise smooth. That, to indicate a particularly pertinent example, is the difficult part in the proof of the second sub-theorem we are trying to confirm. To simplify our task, we will isolate those "hard bits" in the following lemma and corollary.

Lemma 30.12
Assume h is a piecewise continuous and exponentially integrable function on $\mathbb{R}$. Then h is a constant function if and only if

$$\int_{-\infty}^{\infty} h(x)\phi'(x) \, dx = 0 \qquad (30.4)$$

for every Gaussian function ϕ.

Letting the h in this lemma be given by $h = f - g$ and observing that the equation

$$\int_{-\infty}^{\infty} h(x)\phi'(x)\,dx \;=\; \int_{-\infty}^{\infty} [f(x) - g(x)]\,\phi'(x)\,dx \;=\; 0$$

is equivalent to

$$\int_{-\infty}^{\infty} f(x)\phi'(x)\,dx \;=\; \int_{-\infty}^{\infty} g(x)\phi'(x)\,dx \quad,$$

immediately gives the following corollary .

Corollary 30.13
Let f and g be two piecewise continuous, exponentially integrable functions. Then f and g differ by a constant if and only if

$$\int_{-\infty}^{\infty} f(x)\phi'(x)\,dx \;=\; \int_{-\infty}^{\infty} g(x)\phi'(x)\,dx \tag{30.5}$$

for every Gaussian function ϕ .

The proof of this lemma requires a little wily reasoning.

PROOF *(of lemma 30.12):*   Obviously, if h is a constant function and ϕ is any Gaussian, then h is continuous, $h' = 0$ and the integration by parts formula applies, yielding

$$\int_{-\infty}^{\infty} h(x)\phi'(x)\,dx \;=\; -\int_{-\infty}^{\infty} h'(x)\phi(x)\,dx \;=\; 0 \quad.$$

Now assume equation (30.4) holds for every Gaussian ϕ .

Recall that any shifted Gaussian is another Gaussian. So, if ϕ is any Gaussian and t is any real number, then the function $\phi(x + t)$ is another Gaussian function of x with derivative $\phi'(x + t)$. Consequently, our assumption implies that

$$\int_{-\infty}^{\infty} h(x)\phi'(x + t)\,dx \;=\; 0 \tag{30.6}$$

for every Gaussian function ϕ and every real value t .

Now let ϕ be any fixed Gaussian, and define the function Ψ by

$$\Psi(t) \;=\; \int_{-\infty}^{\infty} h(x - t)\phi(x)\,dx \quad.$$

Of course, by a simple change of variables, this is the same as

$$\Psi(t) \;=\; \int_{-\infty}^{\infty} h(x)\phi(x + t)\,dx \quad.$$

Using corollary 18.21 on page 274 regarding the derivatives of integrals with parameters, you can easily verify that Ψ is a smooth function on $\mathbb{R}$ and that

$$\Psi'(t) \;=\; \int_{-\infty}^{\infty} \frac{\partial}{\partial t} h(x)\phi(x + t)\,dx \;=\; \int_{-\infty}^{\infty} h(x)\phi'(x + t)\,dx \quad.$$

From equation (30.6), we know this last integral vanishes for every real t. So Ψ is a smooth function with $\Psi' = 0$. This means Ψ must be a constant. Hence, $\Psi(t) = \Psi(0)$ for each real value t. Writing "$\Psi(t) = \Psi(0)$" in terms of h and ϕ, and recalling how ϕ was chosen, we see that

$$\int_{-\infty}^{\infty} h(x - t)\phi(x)\,dx = \int_{-\infty}^{\infty} h(x)\phi(x)\,dx$$

for every Gaussian ϕ and every real value t. The Gaussian test for equality now applies and assures us that, as exponentially integrable functions of x,

$$h(x - t) = h(x) \qquad \text{for each } t \text{ in } \mathbb{R} \quad .$$

Now let x_0 be a point at which h is continuous. Choosing t so that $s = x_0 - t$, the above equation tells us that, at any point s where h is continuous,

$$h(s) = h(x_0) \quad .$$

Remember, however, that an exponentially integrable function has at most a finite number of (nontrivial) discontinuities in each finite interval. So, even if we suppose s_0 is a point at which h is not continuous, we still have intervals (a, s_0) and (s_0, b) over which the above equation holds. Thus,

$$\lim_{s \to s_0} h(s) = \lim_{s \to s_0} h(x_0) = h(x_0) \quad ,$$

telling us that the supposed discontinuity is trivial. So h has no nontrivial discontinuities. It is continuous and

$$h(s) = h(x_0) \qquad \text{for all} \quad -\infty < s < \infty \quad . \qquad \blacksquare$$

Proving the Sub-Theorems on Differentiation

We can now, with very little difficulty, prove our sub-theorems on the differential identities, lemmas 30.6 and 30.7. For convenience, we'll restate each theorem just before proving it.

First Sub-Theorem on the Differential Identities (lemma 30.6)
Let f and F be two classically transformable functions with $F = \mathcal{F}[f]$. If f is continuous and piecewise smooth, and f' is classically transformable, then $xF(x)$ is classically transformable, and

$$\mathcal{F}[f']\big|_x = i2\pi x\, F(x) \quad .$$

PROOF: Let ϕ be any Gaussian. Then, using the fundamental identities, integration by parts and a differential identity from the associated lemma,

$$\int_{-\infty}^{\infty} \mathcal{F}[f']\big|_x \phi(x)\,dx = \int_{-\infty}^{\infty} f'(y)\mathcal{F}[\phi]\big|_y\,dy$$

$$= -\int_{-\infty}^{\infty} f(y)\frac{d}{dy}\mathcal{F}[\phi]\big|_y\,dy$$

$$= -\int_{-\infty}^{\infty} f(y)\mathcal{F}[-i2\pi x\,\phi(x)]\big|_y\,dy$$

$$= -\int_{-\infty}^{\infty} \mathcal{F}[f]\big|_x\,(-i2\pi x)\phi(x)\,dx \quad .$$

Thus, for every Gaussian ϕ,

$$\int_{-\infty}^{\infty} \mathcal{F}[f']\big|_x \, \phi(x) \, dx \;=\; \int_{-\infty}^{\infty} [i2\pi x F(x)] \, \phi(x) \, dx \quad ,$$

which, by our Gaussian test for equality (theorem 30.1), implies that $\mathcal{F}[f']|_x = i2\pi x F(x)$. Moreover, since $x F(x)$ is the transform of a classically transformable function, it must, itself, be classically transformable. $\blacksquare$

Second Sub-Theorem on the Differential Identities (lemma 30.7)

Let f and F be two classically transformable functions with $F = \mathcal{F}[f]$. If $xF(x)$ is classically transformable, then f is continuous and piecewise smooth, f' is classically transformable, and

$$\mathcal{F}^{-1}[i2\pi y F(y)]\big|_x \;=\; f'(x) \quad .$$

PROOF: Since $y F(y)$ is classically transformable, so is $i2\pi y F(y)$. Now let

$$h(t) \;=\; \mathcal{F}^{-1}[i2\pi y F(y)]\big|_t \quad .$$

As the transform of a classically transformable function, h is classically transformable and, hence, piecewise continuous. From elementary calculus, we know the function g given by

$$g(x) \;=\; f(0) + \int_0^x h(t) \, dt$$

is a continuous and piecewise smooth function on the real line with

$$g(0) \;=\; f(0) \qquad \text{and} \qquad g'(x) \;=\; h(x) \;=\; \mathcal{F}^{-1}[i2\pi y F(y)]\big|_x \quad .$$

This also shows that g' is the transform of a classically transformable function and, therefore, is classically transformable, itself.

Notice that we would have proven everything we are trying to prove if g can be replaced by f in the last two sentences. Thus, it will now suffice to verify that $f = g$.

Let ϕ be any Gaussian. Since g is continuous and piecewise smooth, we can use the integration by parts formula (along with the above formula for g' and one of the fundamental identities) as follows:

$$\int_{-\infty}^{\infty} g(x)\phi'(x) \, dx \;=\; -\int_{-\infty}^{\infty} g'(x)\phi(x) \, dx$$

$$=\; -\int_{-\infty}^{\infty} \mathcal{F}^{-1}[i2\pi y F(y)]\big|_x \, \phi(x) \, dx$$

$$=\; -\int_{-\infty}^{\infty} F(y) \, i2\pi y \, \mathcal{F}^{-1}[\phi]\big|_y \, dy \quad .$$

But $F = \mathcal{F}[f]$, and, from lemma 30.8, we know $-i2\pi y \mathcal{F}^{-1}[\phi]\big|_y = \mathcal{F}^{-1}[\phi']\big|_y$. Using this in the last integral above, followed by applications of the fundamental identities and the invertibility of the transforms, yields

$$\int_{-\infty}^{\infty} g(x)\phi'(x) \, dx \;=\; \int_{-\infty}^{\infty} \mathcal{F}[f]\big|_y \, \mathcal{F}^{-1}[\phi']\big|_y \, dy$$

$$=\; \int_{-\infty}^{\infty} f(x) \mathcal{F}\Big[\mathcal{F}^{-1}[\phi']\Big]\Big|_x \, dx \;=\; \int_{-\infty}^{\infty} f(x)\phi'(x) \, dx \quad .$$

Thus,

$$\int_{-\infty}^{\infty} g(x)\phi'(x)\,dx \;=\; \int_{-\infty}^{\infty} f(x)\phi'(x)\,dx$$

for every Gaussian function ϕ. Corollary 30.13 then assures us that f and g differ by some constant. However, we have already observed that $f(0) = g(0)$, so the constant by which f and g differ must be zero. Hence, $f = g$. ∎

30.4 The Fundamental and Convolution Identities of Fourier Analysis

For the rest of this chapter we will concentrate on proving the validity of the fundamental identity of Fourier analysis and the Fourier convolution identities when all the functions involved are bounded (theorem 25.15 on page 415 and theorem 24.10 on page 394). Our approach will be to confirm the fundamental identity first, and then derive the convolution identities as a corollary.

The first theorem on the fundamental identity (theorem 25.14) will be an important tool in the analysis that follows. We will also find an identity sequence and the transform of that sequence, described immediately below, very helpful.

A Unitary Sequence and Integrability

Let us take the Gaussian identity sequence $\{\psi_\alpha\}_{\alpha=1}^{\infty}$ where

$$\psi_\alpha(x) \;=\; \sqrt{\alpha}\, e^{-\alpha\pi x^2} \quad,$$

and construct the corresponding sequence $\{u_\alpha\}_{\alpha=1}^{\infty}$ of Fourier transforms of the ψ_α's. That is, for each $\alpha \geq 1$,

$$u_\alpha(x) \;=\; \mathcal{F}[\psi_\alpha]|_x \;=\; \mathcal{F}\left[\sqrt{\alpha}\, e^{-\alpha\pi y^2}\right]\Big|_x \;=\; \exp\!\left(-\frac{\pi}{\alpha}x^2\right) \quad.$$

You might have noticed that each of these u_α's is a Gaussian function. It's also worth noting that, for each real value x,

$$\lim_{\alpha\to\infty} u_\alpha(x) \;=\; \lim_{\alpha\to\infty} \exp\!\left(-\frac{\pi}{\alpha}x^2\right) \;=\; e^0 \;=\; 1 \quad,$$

and that, whenever x is real and $1 \leq \alpha < \sigma < \infty$,

$$0 \;<\; u_\alpha(x) \;=\; \exp\!\left(-\frac{\pi}{\alpha}x^2\right) \;\leq\; \exp\!\left(-\frac{\pi}{\sigma}x^2\right) \;=\; u_\sigma(x) \;<\; 1 \quad.$$

A relatively simple (but surprisingly useful) identity involving the integral of the u_α's with an arbitrary classically transformable function f can be derived from the fact that the transforms of the u_α's form an identity sequence. First of all, since the u_α's are Gaussians, we know the fundamental identity applies (see lemma 30.9), and so

$$\int_{-\infty}^{\infty} u_\alpha(x)f(x)\,dx \;=\; \int_{-\infty}^{\infty} \mathcal{F}[\psi_\alpha]|_x\, f(x)\,dx \;=\; \int_{-\infty}^{\infty} \psi_\alpha(y)F(y)\,dy \qquad (30.7)$$

where, as usual, $F = \mathcal{F}[f]$. Then, because $\{\psi_\alpha\}_{\alpha=1}^{\infty}$ is an identity sequence for the set of exponentially integrable functions (and all classically transformable functions are exponentially integrable), we have

$$\lim_{\alpha \to \infty} \int_{-\infty}^{\infty} u_\alpha(x) f(x) \, dx = \lim_{\alpha \to \infty} \int_{-\infty}^{\infty} \psi_\alpha(y) F(y) \, dy = F(0)$$

provided F is continuous at 0. Notice that this equation holds whether or not f is absolutely integrable. However, if f is absolutely integrable, then

$$F(0) = \int_{-\infty}^{\infty} f(x) e^{-i2\pi 0x} \, dx = \int_{-\infty}^{\infty} f(x) \, dx$$

and equation (30.7) tells us the not-very-surprising fact that

$$\lim_{\alpha \to \infty} \int_{-\infty}^{\infty} u_\alpha(x) f(x) \, dx = \int_{-\infty}^{\infty} f(x) \, dx \quad . \tag{30.8}$$

Now consider

$$\lim_{\alpha \to \infty} \int_{-\infty}^{\infty} u_\alpha(x) |g(x)| \, dx$$

where g is any exponentially integrable, piecewise continuous function. Because each u_α is a Gaussian function, we know each product $u_\alpha |g|$ is absolutely integrable. So each of the integrals in the above limit is well defined and finite. If g, itself, happens to be absolutely integrable, then equation (30.8) holds with $f = |g|$,

$$\lim_{\alpha \to \infty} \int_{-\infty}^{\infty} u_\alpha(x) |g(x)| \, dx = \int_{-\infty}^{\infty} |g(x)| \, dx \quad .$$

On the other hand, if g is not absolutely integrable, then it is fairly easy to verify that

$$\lim_{\alpha \to \infty} \int_{-\infty}^{\infty} u_\alpha(x) |g(x)| \, dx = \infty$$

(see exercise 30.3, below). Consequently, if $\int_{-\infty}^{\infty} u_\alpha(x) |g(x)| \, dx$ converges to a finite value as $\alpha \to \infty$, then g cannot be not absolutely integrable.

Restating these observations and derivations in a more grammatically acceptable manner gives us the following lemma.

Lemma 30.14

Let g be an exponentially integrable function on the real line. Then g is absolutely integrable if and only if there is a finite number B such that

$$\lim_{\alpha \to \infty} \int_{-\infty}^{\infty} \exp\left(-\frac{\pi}{\alpha}x^2\right) |g(x)| \, dx = B \quad .$$

Moreover, if g is absolutely integrable, then

$$\lim_{\alpha \to \infty} \int_{-\infty}^{\infty} \exp\left(-\frac{\pi}{\alpha}x^2\right) g(x) \, dx = \int_{-\infty}^{\infty} g(x) \, dx \quad .$$

?►Exercise 30.3: *Let g be an exponentially integrable function on the real line.*

    a: *Verify that*

$$\int_{-\sqrt{\alpha}}^{\sqrt{\alpha}} \exp\left(-\frac{\pi}{\alpha}x^2\right) |g(x)| \, dx \geq e^{-\pi} \int_{-\sqrt{\alpha}}^{\sqrt{\alpha}} |g(x)| \, dx \qquad \text{whenever} \quad \alpha > 0 \quad .$$

b: *Using the result of the previous exercise, show that, if g is not absolutely integrable, then*

$$\lim_{\alpha \to \infty} \int_{-\infty}^{\infty} \exp\left(-\frac{\pi}{\alpha} x^2\right) |g(x)| \, dx = \infty \quad .$$

By the way, $\{u_\alpha\}_{\alpha=1}^{\infty}$ is an example of a *unitary sequence*. Basically, a unitary sequence is a indexed set of functions that converges to 1, the unit constant function, in such a way that a lemma similar to lemma 30.14 holds using that set.

The Fundamental Identity of Fourier Analysis

The second theorem on the fundamental identity (theorem 25.15 on page 415) asserted that the identity holds when all the functions are bounded. We now have the tools for verifying this. So let us again state the theorem and begin its proof.

Second Theorem on the Fundamental Identity (theorem 25.15)
Let f, g, F and G be bounded, classically transformable functions with $F = \mathcal{F}[f]$ and $G = \mathcal{F}[g]$. Then the products fG and Fg are absolutely integrable, and

$$\int_{-\infty}^{\infty} F(x)g(x) \, dx = \int_{-\infty}^{\infty} f(y)G(y) \, dy \quad .$$

Much of this theorem follows from the first theorem on the fundamental identity (theorem 25.14 on page 413), which stated that the fundamental identity held when certain pairs of functions were in $\mathcal{A}$, the set of absolutely integrable, piecewise continuous functions. Consequently, the first thing we want to do is to figure out exactly what remains to be proven.

To do this, suppose f, g, F and G are bounded, classically transformable functions with $F = \mathcal{F}[f]$ and $G = \mathcal{F}[g]$. Since every classically transformable function can be written as the sum of a function from $\mathcal{A}$ with a function from $\mathcal{T}$, the set of transforms of functions from $\mathcal{A}$, we can express f and g as

$$f = f_A + f_T \quad \text{and} \quad g = g_A + g_T$$

where f_A and g_A are from $\mathcal{A}$ and f_T and g_T are from $\mathcal{T}$.

Observe that f_T and g_T, being transforms of absolutely integrable functions, must be bounded. But, by assumption, f and g are bounded functions. Thus, f_A and g_A, each being the difference of a pair of bounded functions ($f_A = f - f_T$ and $g_A = g - g_T$, respectively), must also be bounded functions. Similar arguments confirm that the functions

$$\mathcal{F}[f_A] \quad , \quad \mathcal{F}[f_T] \quad , \quad \mathcal{F}[g_A] \quad \text{and} \quad \mathcal{F}[g_T]$$

are also bounded functions on the real line.

Now expand the integrals of Fg and fG in terms of f_A, g_A, f_T and g_T. Ignoring the question of integrability for the moment, we naively get

$$\int_{-\infty}^{\infty} F(x)g(x) \, dx = \int_{-\infty}^{\infty} \mathcal{F}[f_A + f_T]|_x \, [g_A(x) + g_T(x)] \, dx$$

$$= \int_{-\infty}^{\infty} \mathcal{F}[f_A]|_x \, g_A(x) \, dx + \int_{-\infty}^{\infty} \mathcal{F}[f_A]|_x \, g_T(x) \, dx$$

$$+ \int_{-\infty}^{\infty} \mathcal{F}[f_T]|_x \, g_A(x) \, dx + \int_{-\infty}^{\infty} \mathcal{F}[f_T]|_x \, g_T(x) \, dx$$

(30.9)

and

$$\int_{-\infty}^{\infty} f(y)G(y)\,dy \;=\; \int_{-\infty}^{\infty} [f_A(y) + f_T(y)]\,\mathcal{F}[g_A + g_T]|_y \,dy$$

$$= \int_{-\infty}^{\infty} f_A(y)\mathcal{F}[g_A]|_y \,dy \;+\; \int_{-\infty}^{\infty} f_A(y)\mathcal{F}[g_T]|_y \,dy$$

$$+ \int_{-\infty}^{\infty} f_T(y)\mathcal{F}[g_A]|_y \,dy \;+\; \int_{-\infty}^{\infty} f_T(y)\mathcal{F}[g_T]|_y \,dy \quad . \tag{30.10}$$

From theorem 25.14, we know the products

$$\mathcal{F}[f_A]\,g_A \quad , \qquad \mathcal{F}[f_T]\,g_T \quad , \qquad f_A\mathcal{F}[g_A] \qquad \text{and} \qquad f_T\mathcal{F}[g_T]$$

are absolutely integrable, and that

$$\int_{-\infty}^{\infty} \mathcal{F}[f_A]|_x\, g_A(x)\,dx \;=\; \int_{-\infty}^{\infty} f_A(y)\mathcal{F}[g_A]|_y \,dy$$

and

$$\int_{-\infty}^{\infty} \mathcal{F}[f_T]|_x\, g_T(x)\,dx \;=\; \int_{-\infty}^{\infty} f_T(y)\mathcal{F}[g_T]|_y \,dy \quad .$$

Comparing the integrals in these equations to the integrals in the expansions of the integrals of Fg and fG in equations (30.9) and (30.10), we find that all we need to show now is that the products

$$\mathcal{F}[f_A]\,g_T \quad , \qquad \mathcal{F}[f_T]\,g_A \quad , \qquad f_T\mathcal{F}[g_A] \qquad \text{and} \qquad f_T\mathcal{F}[g_A]$$

are absolutely integrable, and that

$$\int_{-\infty}^{\infty} \mathcal{F}[f_A]|_x\, g_T(x)\,dx \;=\; \int_{-\infty}^{\infty} f_A(y)\mathcal{F}[g_T]|_y \,dy$$

and

$$\int_{-\infty}^{\infty} \mathcal{F}[f_T]|_x\, g_A(x)\,dx \;=\; \int_{-\infty}^{\infty} f_T(y)\mathcal{F}[g_A]|_y \,dy \quad .$$

This, in turn, is guaranteed by the next lemma.

Lemma 30.15
Suppose $F = \mathcal{F}[f]$ and $g = \mathcal{F}^{-1}[G]$ where f and G are two bounded functions in $\mathcal{A}$. Then the products Fg and fG are absolutely integrable and

$$\int_{-\infty}^{\infty} F(x)g(x)\,dx \;=\; \int_{-\infty}^{\infty} f(y)G(y)\,dy \quad .$$

Before we prove this lemma, you should verify that the second theorem on the fundamental identity does follow from the lemma by doing the next exercise.

?▶Exercise 30.4:   *Assume lemma 30.15 holds.*

a: *Let f_A be a bounded function in $\mathcal{A}$, and let g_T be a function in $\mathcal{T}$ whose Fourier transform is bounded. Show that the products $\mathcal{F}[f_A]\,g_T$ and $f_A\mathcal{F}[g_T]$ are absolutely integrable, and that*

$$\int_{-\infty}^{\infty} \mathcal{F}[f_A]|_x\, g_T(x)\,dx \;=\; \int_{-\infty}^{\infty} f_A(y)\mathcal{F}[g_T]|_y \,dy \quad .$$

b: Let g_A be a bounded function in $\mathcal{A}$, and let f_T be a function in $\mathcal{T}$ whose Fourier transform is bounded. Show that the products $\mathcal{F}[f_T] g_A$ and $f_T \mathcal{F}[g_A]$ are absolutely integrable, and that

$$\int_{-\infty}^{\infty} \mathcal{F}[f_T]\big|_x \, g_A(x) \, dx = \int_{-\infty}^{\infty} f_T(y) \mathcal{F}[g_A]\big|_y \, dy \quad .$$

c: Prove theorem 25.15, the second theorem on the fundamental identity (restated on page 503).

With the proof of theorem 25.15 reduced to proving lemma 30.15, our goal has become the proof of that lemma. To make this proof a little more manageable, we'll prove portions of the lemma in the following two "sub-lemmas". The first describes the results of a sequence of computations that we will need more than once. The second verifies that our functions are square integrable.

Lemma 30.16
Let f, g, F and G be bounded, classically transformable functions with $F = \mathcal{F}[f]$ and $G = \mathcal{F}[g]$. Assume, further, that f is absolutely integrable. Then the product fG is absolutely integrable and

$$\lim_{\alpha \to \infty} \int_{-\infty}^{\infty} F(x)g(x) \exp\left(-\frac{\pi}{\alpha}x^2\right) dx = \int_{-\infty}^{\infty} f(y)G(y) \, dy \quad .$$

PROOF: First of all, we know the product fG is absolutely integrable because it is the product of an absolutely integrable function with a bounded function, as is (we note in passing) the product of g with any Gaussian function.

For convenience and for each $\alpha > 0$, let

$$u_\alpha(x) = \exp\left(-\frac{\pi}{\alpha}x^2\right) \qquad \text{and} \qquad \psi_\alpha(y) = \mathcal{F}[u_\alpha]\big|_y = \sqrt{\alpha}\, e^{-\alpha \pi x^2}$$

(as in the previous subsection), and consider the integral of Fgu_α over the real line. Because both f and gu_α are absolutely integrable, the first theorem on the validity of the fundamental identity (theorem 25.14 on page 413) tells us that the fundamental identity can be applied as follows:

$$\int_{-\infty}^{\infty} F(x)g(x)u_\alpha(x) \, dx = \int_{-\infty}^{\infty} \mathcal{F}[f]\big|_x \, g(x)u_\alpha(x) \, dx = \int_{-\infty}^{\infty} f(y)\mathcal{F}[gu_\alpha]\big|_y \, dy \quad .$$

Moreover, because u_α is a Gaussian, we know the transform of gu_α can be computed (see theorem 24.9 on page 393) by convolution,

$$\mathcal{F}[gu_\alpha]\big|_y = G * \psi_\alpha(y) = \int_{-\infty}^{\infty} G(y-s)\psi_\alpha(s) \, ds \quad .$$

Combining the last two equations, we get

$$\int_{-\infty}^{\infty} F(x)g(x)u_\alpha(x) \, dx = \int_{-\infty}^{\infty} f(y) \int_{-\infty}^{\infty} G(y-s)\psi_\alpha(s) \, ds \, dy$$

$$= \int_{-\infty}^{\infty} \int_{-\infty}^{\infty} f(y)G(y-s)\psi_\alpha(s) \, ds \, dy \quad . \tag{30.11}$$

Since f and ψ_α are absolutely integrable and G is bounded, corollary 18.24 on page 275 applies, assuring us that the order of integration in the last double integral can be switched. Thus

$$\int_{-\infty}^{\infty} F(x)g(x)u_\alpha(x) \, dx = \int_{-\infty}^{\infty} \int_{-\infty}^{\infty} f(y)G(y-s)\psi_\alpha(s) \, dy \, ds = \int_{-\infty}^{\infty} \Phi(s)\psi_\alpha(s) \, ds$$

where

$$\Phi(s) = \int_{-\infty}^{\infty} f(y)G(y-s)\,dy \quad .$$

It is easily verified (using corollary 18.19 on page 271, the boundedness of G and the fact that f is in $\mathcal{A}$) that $\Phi(s)$ is a well-defined, bounded, continuous function with

$$\Phi(0) = \int_{-\infty}^{\infty} f(y)G(y)\,dy \quad .$$

It should also be recalled that $\{\psi_\alpha\}_{\alpha=1}^{\infty}$ is a Gaussian identity sequence. So

$$\lim_{\alpha \to \infty} \int_{-\infty}^{\infty} F(x)g(x)u_\alpha(x)\,dx = \lim_{\alpha \to \infty} \int_{-\infty}^{\infty} \Phi(s)\psi_\alpha(s)\,ds$$

$$= \Phi(0)$$

$$= \int_{-\infty}^{\infty} f(y)G(y)\,dy \quad . \qquad \blacksquare$$

Lemma 30.17
Let h and H be bounded, classically transformable functions with $H = \mathcal{F}[h]$. Also assume that either h or H is absolutely integrable. Then both h and H are square integrable, and

$$\int_{-\infty}^{\infty} |H(x)|^2\,dx = \int_{-\infty}^{\infty} |h(y)|^2\,dy \quad .$$

We'll verify this lemma assuming h is absolutely integrable, and leave the "nearly equivalent" proof of this lemma's validity when H is absolutely integrable to the interested reader.

PROOF *(assuming h is absolutely integrable):*   Applying the previous lemma (with $f = h$ and $G = h^*$), we find both that fG is absolutely integrable and that, letting $F = \mathcal{F}[f]$ and $g = \mathcal{F}^{-1}[G]$,

$$\lim_{\alpha \to \infty} \int_{-\infty}^{\infty} F(x)g(x) \exp\left(-\frac{\pi}{\alpha}x^2\right)\,dx = \int_{-\infty}^{\infty} f(y)G(y)\,dy \quad . \qquad (30.12)$$

But

$$fG = hh^* = |h|^2 \quad , \qquad F = \mathcal{F}[h] = H \quad ,$$

and, by one of the conjugation identities (theorem 21.4 on page 332),

$$g = \mathcal{F}^{-1}[h^*] = (\mathcal{F}[h])^* = H^* \quad .$$

Thus, $|h|^2$, being fG, is absolutely integrable (i.e., h is square integrable). Furthermore, since $Fg = HH^* = |H|^2$, equation (30.12) is equivalent to

$$\lim_{\alpha \to \infty} \int_{-\infty}^{\infty} |H(x)|^2 \exp\left(-\frac{\pi}{\alpha}x^2\right)\,dx = \int_{-\infty}^{\infty} |h(y)|^2\,dy \quad .$$

Now we can apply lemma 30.14 on page 502 (with $g = |H|^2$ and $B = \int_{-\infty}^{\infty} |h(y)|^2\,dy$), confirming immediately that $|H|^2$ is absolutely integrable (so H is square integrable), and that

$$\int_{-\infty}^{\infty} |H(x)|^2\,dx = \lim_{\alpha \to \infty} \int_{-\infty}^{\infty} |H(x)|^2 \exp\left(-\frac{\pi}{\alpha}x^2\right)\,dx = \int_{-\infty}^{\infty} |h(y)|^2\,dy \quad . \qquad \blacksquare$$

We can now finish proving our main lemma, lemma 30.15, which (in case you forgot) is:

Suppose $F = \mathcal{F}[f]$ and $g = \mathcal{F}^{-1}[G]$ where f and G are two bounded functions in $\mathcal{A}$. Then the products Fg and fG are absolutely integrable, and

$$\int_{-\infty}^{\infty} F(x)g(x)\, dx = \int_{-\infty}^{\infty} f(y)G(y)\, dy \quad .$$

PROOF (of lemma 30.15):   From lemma 30.17, we know F and g are square-integrable functions on the real line. According to the Schwarz inequality (lemma 25.7 on page 404) then, the product of these two square integrable functions, Fg, is absolutely integrable. Consequently, as we saw in lemma 30.14 on page 502,

$$\int_{-\infty}^{\infty} F(x)g(x)\, dx = \lim_{\alpha \to \infty} \int_{-\infty}^{\infty} F(x)g(x) \exp\left(-\frac{\pi}{\alpha}x^2\right) dx \quad .$$

But lemma 30.16 tells us that the product fG is absolutely integrable and that

$$\lim_{\alpha \to \infty} \int_{-\infty}^{\infty} F(x)g(x) \exp\left(-\frac{\pi}{\alpha}x^2\right) dx = \int_{-\infty}^{\infty} f(y)G(y)\, dy \quad .$$

So both Fg and fG are absolutely integrable, and

$$\int_{-\infty}^{\infty} F(x)g(x)\, dx = \lim_{\alpha \to \infty} \int_{-\infty}^{\infty} F(x)g(x) \exp\left(-\frac{\pi}{\alpha}x^2\right) dx = \int_{-\infty}^{\infty} f(y)G(y)\, dy \quad . \quad \blacksquare$$

The Convolution Identities

The following corollary of our second theorem on the fundamental identity (theorem 25.15, restated on page 503) is essentially the theorem on convolution we wanted to prove earlier (theorem 24.10 on page 394). Deriving it from theorem 25.15 requires just a little more work. The details of that work, along with the verification that the desired theorem on convolution follows, will be left as exercises.

Corollary 30.18 (Fourier convolution identity)
Assume f, h, F and H are all bounded, classically transformable functions with $F = \mathcal{F}[f]$ and $H = \mathcal{F}[h]$. Then

1. *the product fh is absolutely integrable;*

2. *the convolution $F * H$ exists and is classically transformable,*

and

3. $\mathcal{F}[fh] = F * H$.

?►Exercise 30.5:   *Prove corollary 30.18 by doing the following:*

a: *Use theorem 25.15 to show that the product fh is in $\mathcal{A}$, thus allowing us to write, for each ω in $\mathbb{R}$,*

$$\mathcal{F}[fh]|_{\omega} = \int_{-\infty}^{\infty} f(y)G(y)\, dy \quad \text{where} \quad G(y) = h(y)\, e^{-i2\pi \omega y} \quad .$$

b: *Then confirm that $\mathcal{F}^{-1}[G]|_x = H(\omega - x)$.*

c: Finally, verify that theorem 25.15 assures us both that $F * H$ exists and that

$$\mathcal{F}[fh] = F * H \quad .$$

(This also shows that $F * H$, being the transform of a classically transformable function, is also classically transformable. In fact, since fh is in $\mathcal{A}$, $F * H$ must be in $\mathcal{T}$.)

?▶ Exercise 30.6:   Verify that theorem 24.10 on page 394 follows from the last corollary and near-equivalence.

Part IV

Generalized Functions and Fourier Transforms

31

A Starting Point
for the Generalized Theory

The theory we have developed thus far for the Fourier transform (the "classical theory") provides a number of mathematical tools that could be useful in solving many problems of real interest — provided those problems only involve classically transformable functions. Unfortunately, this is often not the case. Even fairly simple real-world problems are likely to involve constant functions, polynomials and exponential functions, none of which can be "Fourier transformed" using the classical theory. Since we want to deal with such functions, let us now turn our attention to finding a more general way of defining the Fourier transform.

31.1 Starting Points

Suppose f is a classically transformable function. By the fundamental identity, we know

$$\int_{-\infty}^{\infty} \mathcal{F}[f]|_x \, \phi(x) \, dx = \int_{-\infty}^{\infty} f(y) \mathcal{F}[\phi]|_y \, dy$$

for each Gaussian function ϕ. Suppose, further, that F is some other exponentially integrable function and that

$$\int_{-\infty}^{\infty} F(x) \phi(x) \, dx = \int_{-\infty}^{\infty} f(y) \mathcal{F}[\phi]|_y \, dy$$

whenever ϕ is a Gaussian. Since the right-hand sides of these two equations are the same, so are the left-hand sides. Thus, for each Gaussian function ϕ,

$$\int_{-\infty}^{\infty} F(x) \phi(x) \, dx = \int_{-\infty}^{\infty} \mathcal{F}[f]|_x \, \phi(x) \, dx \quad .$$

This, according to our Gaussian test for equality (theorem 30.1 on page 491), tells us that $F = \mathcal{F}[f]$. In other words, we have just proven the following lemma.

Lemma 31.1
Let f and F be two exponentially integrable functions with f being classically transformable. Then $F = \mathcal{F}[f]$ if and only if

$$\int_{-\infty}^{\infty} F(x) \phi(x) \, dx = \int_{-\infty}^{\infty} f(y) \mathcal{F}[\phi]|_y \, dy \qquad \text{for each Gaussian } \phi \quad . \qquad (31.1)$$

This lemma gives us a test for determining if one function is the transform of another. Admittedly, it is not a test you are likely to use much within the classical theory.[1] Look closely, however, at the integral on the right side of equation (31.1). Since ϕ is a Gaussian function, so is $\mathcal{F}[\phi]$. Consequently, this integral is well defined whenever f is *any exponentially integrable function*, not just when f is classically transformable.

This raises a question: Does this test give us a means for defining Fourier transforms of functions that are not classically transformable?

Let us propose a definition based on this test, and then see if it works.

A Proposed Definition for the Fourier Transform

Let f *be any piecewise continuous, exponentially integrable function on the real line. Its Fourier transform* $F = \mathcal{F}[f]$ *is defined to be the function that satisfies*

$$\int_{-\infty}^{\infty} F(x)\phi(x)\,dx = \int_{-\infty}^{\infty} f(y)\mathcal{F}[\phi]|_y\,dy$$

for every Gaussian function ϕ.

Let us see what we obtain as the Fourier transform of the constant function $f = 1$ using this definition. For convenience (and to anticipate notation we'll develop later), let δ denote the Fourier transform of 1, $\delta = \mathcal{F}[1]$, as defined by the proposed definition. That is, $\delta(x)$ is defined as the function that satisfies

$$\int_{-\infty}^{\infty} \delta(x)\phi(x)\,dx = \int_{-\infty}^{\infty} 1 \cdot \mathcal{F}[\phi]|_y\,dy \tag{31.2}$$

for each Gaussian ϕ. We can greatly simplify the integral on the right once we realize that it is just the integral for the inverse Fourier transform of $\mathcal{F}[\phi]$ at 0,

$$\int_{-\infty}^{\infty} 1 \cdot \mathcal{F}[\phi]|_y\,dy = \int_{-\infty}^{\infty} \mathcal{F}[\phi]|_y\, e^{i2\pi 0 \cdot y}\,dy = \mathcal{F}^{-1}[\mathcal{F}[\phi]]|_0 = \phi(0) \quad .$$

With this, equation (31.2) becomes

$$\int_{-\infty}^{\infty} \delta(x)\phi(x)\,dx = \phi(0) \tag{31.3}$$

for each Gaussian function ϕ. In particular then,

$$\int_{-\infty}^{\infty} \delta(x)e^{-x^2}\,dx = e^{-0^2} = 1 \quad . \tag{31.4}$$

Equation (31.3) describes integrals involving δ. To find a more explicit formula for $\delta(x)$, let us try using equation (31.3) along with a Gaussian identity sequence, say

$$\left\{ \gamma e^{-\pi(\gamma s)^2} \right\}_{\gamma=1}^{\infty} \quad .$$

Because this is an identity sequence,

$$\delta(x) = \lim_{\gamma \to \infty} \int_{-\infty}^{\infty} \delta(s)\, \gamma e^{-\pi(\gamma(s-x))^2}\,ds$$

---

[1] Though, if you re-examine the work we did in chapter 30, you will find that this test was surreptitiously incorporated into our proofs of the fundamental theorem on invertibility and the differentiation identities.

for every real value x at which δ is continuous. Combined with equation (31.3), this yields

$$\delta(x) = \lim_{\gamma \to \infty} \int_{-\infty}^{\infty} \delta(s)\, \gamma e^{-\pi(\gamma(s-x))^2}\, ds$$

$$= \lim_{\gamma \to \infty} \gamma e^{-\pi(\gamma(0-x))^2} = \lim_{\gamma \to \infty} \gamma e^{-\pi(\gamma x)^2}$$

for every real value x at which δ is continuous. Even if you don't recognize this last limit, you can easily compute it for $x \neq 0$ via L'Hôpital's rule:

$$\lim_{\gamma \to \infty} \gamma e^{-\pi(\gamma x)^2} = \lim_{\gamma \to \infty} \frac{\gamma}{e^{\pi(\gamma x)^2}} = \lim_{\gamma \to \infty} \frac{1}{2\pi \gamma x^2 e^{\pi(\gamma x)^2}} = 0 \quad !$$

So (after removing the trivial discontinuity at $x = 0$), δ must be zero everywhere on the real line. Consequently,

$$\int_{-\infty}^{\infty} \delta(x) e^{-x^2}\, dx = \int_{-\infty}^{\infty} 0 \cdot e^{-x^2}\, dx = 0 \tag{31.5}$$

in complete disagreement with equation (31.4)!

The disagreement between equations (31.4) and (31.5) should disturb you. It means that, unless we are willing to accept "$0 = 1$", we must conclude that our proposed method for defining the Fourier transform of 1 is flawed. And here is how it is flawed: Our proposed definition requires the existence of a function that, in fact, might not exist. If, for example, there were an exponentially integrable function δ that satisfied equation (31.2) for each Gaussian ϕ, then both equation (31.4) and equation (31.5) would hold. Since equations (31.4) and (31.5) cannot both be true (unless $1 = 0$), the function δ cannot exist.

So why are we wasting time on an idea that fails? Because, as it turns out, it does not completely fail. The proposed definition does work for some functions. It certainly works when f is classically transformable, and it can be shown to work if f is any square-integrable function. More importantly, we will discover that, if we relax our requirement of F being a function in the classical sense, then the proposed definition (suitably modified) does "work" when f is any exponentially integrable function. That is, we can use the proposed definition to define Fourier transforms of arbitrary exponentially integrable functions — but only after suitably generalizing our basic notions of "functions" and "integration" and developing the necessary theory for understanding and manipulating these "generalized functions and integrals". This will be a significant undertaking, but the end result will be an incredibly useful and elegant set of tools for dealing with a wide variety of problems in the sciences and engineering.

Exactly what is meant by a "generalized function" will be described in chapter 33. The starting point leading to that discussion is the realization that any exponentially integrable function f is completely determined by the values of all the integrals of the form

$$\int_{-\infty}^{\infty} f(s)\phi(s)\, ds$$

where ϕ can be any Gaussian. That, essentially, is what our Gaussian test for equality tells us. Moreover, if we know the values of all such integrals, then we can find the value of $f(x)$ for each real x by

$$f(x) = \lim_{\gamma \to \infty} \int_{-\infty}^{\infty} f(s)\phi_\gamma(s - x)\, ds$$

where $\{\phi_\gamma\}_{\gamma=1}^{\infty}$ is any Gaussian identity sequence.

Likewise, a "generalized function f" will be something determined by all the values of some-thing analogous to the integrals of the form

$$\int_{-\infty}^{\infty} f(s)\phi(s)\,ds$$

where ϕ is any Gaussian. In particular, we will discover that the Fourier transform of 1 is defined by an equation analogous to equation (31.3).

Additional Exercises

31.1. *For each of the following choices of f, show that there is no exponentially integrable, piecewise continuous function F such that*

$$\int_{-\infty}^{\infty} F(x)\phi(x)\,dx = \int_{-\infty}^{\infty} f(y)\mathcal{F}[\phi]|_y\,dy$$

for every Gaussian function ϕ :

a. $f(y) = e^{i2\pi y}$

b. $f(y) = y$

32

Gaussian Test Functions

Because of the role of Gaussian functions in the Gaussian test for equality (theorem 30.1 on page 491), it is natural to refer to Gaussians as "test functions" for determining the equality of two formulas. Other functions can be used as test functions (see corollary 29.2 on page 472), but, because Fourier transforms of Gaussians are also Gaussian functions, Gaussians were more convenient for our work in chapter 30. On the other hand, derivatives and linear combinations of Gaussian functions are seldom purely Gaussian. That will make using just Gaussians as our test functions somewhat awkward in the next few chapters. So, in this chapter, we will use Gaussians to generate a larger collection of "test functions" (called, appropriately enough, the "Gaussian test functions"), and we will verify that this expanded set satisfies a number of properties that will be important in our development of the generalized theory.

32.1 The Space of Gaussian Test Functions

There are two stages in developing the complete set of Gaussian test functions. The first is to define the "base" Gaussian test functions. As the name suggests, these functions make up the base set of functions from which we will construct all other Gaussian test functions.

Base Gaussian Test Functions

A function ϕ on the real line will be called a *base Gaussian test function* if and only if it can be given by the formula

$$\phi(x) = Ax^n e^{-\gamma(x-\zeta)^2} \tag{32.1}$$

where A and ζ are fixed complex values, n is any nonnegative integer and $\gamma > 0$.

If n is zero, then the above is just a formula for our beloved Gaussians. So the set of all base Gaussian test functions includes all Gaussian functions. Moreover, since A can be zero, the zero function is a base Gaussian test function (an important, though not very exciting, test function). On the other hand, nonzero constant functions, polynomials and exponentials are not base Gaussian test functions.

Other formulas could just as easily have been used to define base Gaussian test functions. All we need to do is to replace the above formula for the Gaussian with any of the other equivalent formulas listed back in lemma 23.1 on page 364. That gives us the following lemma.

Lemma 32.1

Let ϕ be a function on $\mathbb{R}$. If any one of the following statements is true, then all are true.

1. ϕ is a base Gaussian test function.

2. There is a positive constant γ, a nonnegative integer n, and complex constants A and ζ such that
$$\phi(x) = Ax^n e^{-\gamma(x-\zeta)^2} \qquad \text{for} \quad -\infty < x < \infty \ .$$

3. There is a positive constant γ, a nonnegative integer n, a complex constant B, and real constants a and λ such that
$$\phi(x) = Bx^n e^{i\lambda x} e^{-\gamma(x-a)^2} \qquad \text{for} \quad -\infty < x < \infty \ .$$

4. There is a positive constant γ, a nonnegative integer n, a complex constant C, and real constants b and μ such that
$$\phi(x) = Cx^n e^{\mu x} e^{-\gamma(x-ib)^2} \qquad \text{for} \quad -\infty < x < \infty \ .$$

5. There is a positive constant γ, a nonnegative integer n, and complex constants D and σ such that
$$\phi(x) = Dx^n e^{\sigma x} e^{-\gamma x^2} \qquad \text{for} \quad -\infty < x < \infty \ .$$

Moreover, if the above hold, then

$$\zeta = a + ib \quad , \qquad \lambda = 2b\gamma \quad , \qquad \mu = 2a\gamma \quad , \qquad \sigma = \mu + i\lambda \ ,$$
$$B = Ae^{\gamma b^2 - i2ab\gamma} \quad , \qquad C = Ae^{-\gamma a^2 - i2ab\gamma} \quad \text{and} \quad D = Ae^{-\gamma \zeta^2} \ .$$

The next lemma lists a number of easily verified facts we will later find useful.

Lemma 32.2 (some properties of base Gaussian test functions)
All of the following hold:

1. Every Gaussian is a base Gaussian test function.

2. Every base Gaussian test function is a bounded, smooth and absolutely integrable function on the real line.

3. The product of any base Gaussian test function with any exponentially integrable function is absolutely integrable on the real line.

4. If ϕ is a base Gaussian test function, then so is each of the following:
    (a) $\phi\psi$ where ψ is also any base Gaussian test function
    (b) $Cx^k e^{\sigma x}\phi(x)$ where C and σ are any two complex numbers and k is any nonnegative integer
    (c) $\phi(ax)$ where a is any nonzero real value

5. If ϕ is a base Gaussian test function, then each of the following is a linear combination of base Gaussian test functions:

    (a) $\phi(x - \xi)$ where ξ is any real number

    (b) $\phi^{(m)}$ where m is any positive integer

    (c) $\mathcal{F}[\phi]$

    (d) $\mathcal{F}^{-1}[\phi]$

Little needs to be said about confirming the first three statements above — the first was an observation made just after defining base Gaussian test functions, and the next two statements follow immediately from what we already know about Gaussians. We'll go ahead and verify parts 4a and 5a, and leave the verification of the rest to you.

PROOF *(that $\phi\psi$ is a base Gaussian test function whenever ϕ and ψ are any two base Gaussian test functions)*:   From lemma 32.1, we know there are complex constants A, B, a and b, nonnegative integers k and m, and positive values κ and λ such that

$$\phi(x) = Ax^k e^{ax} e^{-\kappa x^2} \quad \text{and} \quad \psi(x) = Bx^m e^{bx} e^{-\lambda x^2} \quad.$$

So, by basic algebra,

$$\phi(x)\psi(x) = Cx^n e^{cx} e^{-\gamma x^2}$$

where $C = AB$, n is the nonnegative integer $k + m$, $c = a + b$ and γ is the positive value $\kappa + \lambda$. Thus, as noted in lemma 32.1, the product $\phi\psi$ is a base Gaussian test function.   ∎

PROOF *(that $\phi(x - \xi)$ is a base Gaussian test function whenever ϕ is a base Gaussian test function and ξ is a real number)*:   From lemma 32.1, we know there are complex constants A and η, and a nonnegative integer n such that

$$\phi(x) = Ax^n e^{-\gamma(x-\eta)^2} \quad.$$

So

$$\phi(x - \xi) = A(x - \xi)^n e^{-\gamma(x-\xi-\eta)^2} = A(x - \xi)^n e^{-\gamma(x-\zeta)^2}$$

where ζ is the complex constant $\xi + \eta$. But the $(x - \xi)^n$ can be multiplied out,

$$(x - \xi)^n = x^n + c_{n-1}x^{n-1} + \cdots + c_0 = \sum_{k=0}^{n} c_k x^k$$

where the c_k's are computable complex numbers.[1] Thus, letting $A_k = Ac_k$,

$$\phi(x - \xi) = A\sum_{k=0}^{n} c_k x^k e^{-\gamma(x-\zeta)^2} = \sum_{k=0}^{n} A_k x^k e^{-\gamma(x-\zeta)^2} \quad, \tag{32.2}$$

which is a linear combination of base Gaussian test functions.   ∎

?►Exercise 32.1   a: *Verify part 4b of lemma 32.2.*

  b: *Verify part 4c of lemma 32.2.*

---

[1] In fact, the binomial theorem tells us that $c_k = \dfrac{n!}{k!(n - k)!}(-\xi)^{n-k}$.

?▶ Exercise 32.2: *Show that the derivative of any base Gaussian test function is a linear combination of base Gaussian test functions, and then verify part 5b of lemma 32.2.*

?▶ Exercise 32.3: *Finish proving lemma 32.2 by using the results of the above exercises to show that $\mathcal{F}[\phi]$ and $\mathcal{F}^{-1}[\phi]$ are linear combinations of base Gaussian test functions whenever ϕ is a base Gaussian test function.*

The Space of Gaussian Test Functions

Typically, a sum of base Gaussian test functions is not, itself, a base Gaussian test function. This would make it inconvenient to use only base Gaussian test functions as our test functions. So let us expand our set of test functions one last time by also throwing in all finite sums of base Gaussian test functions. To be precise, let us define a function ϕ on the real line to be a *Gaussian test function* if and only if

$$\phi = \sum_{k=1}^{N} \phi_k$$

where N is some positive integer and each ϕ_k is a base Gaussian test function.

The set of all Gaussian test functions will be denoted by $\mathcal{G}$. In anticipation of the next lemma, we will often refer to $\mathcal{G}$ as the *(linear) space of Gaussian test functions.*

Lemma 32.3

Any linear combination of Gaussian test functions is another Gaussian test function. (Hence, $\mathcal{G}$ is a linear space of functions.)

PROOF: It will suffice to show $a\phi + b\psi$ is a Gaussian test function assuming ϕ and ψ are any two Gaussian test functions, and a and b are any two constants. Making these assumptions, we know that (by definition) there are positive integers K and M, and base Gaussian test functions $\phi_1, \ldots, \phi_K, \psi_1, \ldots$ and ψ_M such that

$$\phi = \sum_{k=1}^{K} \phi_k \qquad \text{and} \qquad \psi = \sum_{k=1}^{M} \psi_k \quad .$$

So,

$$a\phi + b\psi = a\sum_{k=1}^{K} \phi_k + b\sum_{k=1}^{M} \psi_k = \sum_{k=1}^{N} \Psi_k$$

where $N = K + M$ and

$$\Psi_k = \begin{cases} a\phi_k & \text{if } k = 1, 2, \ldots, K \\ b\psi_{k-K} & \text{if } k = K+1, K+2, \ldots, N \end{cases} \quad .$$

Since a constant times a base Gaussian test function is another base Gaussian test function, each of the Ψ_k's is a base Gaussian test function, and $a\phi + b\psi$ is a finite sum of base Gaussian test functions. Consequently, $a\phi + b\psi$ satisfies the sole requirement for being a Gaussian test function. ∎

A number of easily verified facts about Gaussian test functions will be important later. Many that follow almost immediately from lemma 32.3 are listed in the next lemma.

Lemma 32.4 *(some properties of Gaussian test functions)*
The following all hold:

1. *The set of Gaussian test functions contains all Gaussians and all base Gaussian test functions.*

2. *Every Gaussian test function is a bounded, smooth and absolutely integrable function on the real line.*

3. *The product of any Gaussian test function with any exponentially integrable function is absolutely integrable on the real line.*

4. *If ϕ is a Gaussian test function, then so is each of the following:*

    (a) *$Cx^k e^{\sigma x}\phi(x)$ where C and σ are any two complex numbers and k is any nonnegative integer*

    (b) *$\phi\psi$ where ψ is also a Gaussian test function*

    (c) *$\phi(ax)$ where a is any nonzero real value*

    (d) *$\phi(x - \zeta)$ where ζ is any real number*

    (e) *$\phi^{(m)}$ where m is any positive integer*

    (f) *$\mathcal{F}[\phi]$*

    (g) *$\mathcal{F}^{-1}[\phi]$*

5. *If ϕ is a Gaussian test function, then there are two Gaussian test functions ψ^+ and ψ^- such that $\phi = \mathcal{F}[\psi^+]$ and $\phi = \mathcal{F}^{-1}[\psi^-]$.*

   You should verify for yourself that all but the last statement in the above list are immediate consequences of lemmas 32.2 and 32.3. And with those statements verified, confirmation of the last statement becomes almost trivial. Simply choose ψ^+ and ψ^- as follows:

$$\psi^+ = \mathcal{F}^{-1}[\phi] \qquad \text{and} \qquad \psi^- = \mathcal{F}[\phi] \quad .$$

Statements 4f and 4g of the lemma and the invertibility of the classical Fourier transforms then assure us that ψ^+ and ψ^- are Gaussian test functions such that

$$\mathcal{F}[\psi^+] = \phi \qquad \text{and} \qquad \mathcal{F}^{-1}[\psi^-] = \phi \quad .$$

32.2   On Using the Space of Gaussian Test Functions

There are two "meta-properties" of the Gaussian test functions that will make them very useful. The first is that, as far as classical analysis is concerned, they are extremely nice functions. Take another look at the second and third statements in the last lemma to see just how nice these functions are. (While you are at it, go ahead and memorize the entire list of properties given in that lemma. These facts, along with the fact that $\mathcal{G}$ is a linear space, will be more useful to us than the actual formulas for the Gaussian test functions. Frankly, most of the generalized theory we will develop can be developed using any set of functions for which a similar list can be given, and it will normally be a good idea to *avoid* using explicit formulas for our test functions, simply to reduce the temptation to complicate simple arguments with unnecessary computations.)

The other meta-property of our test functions is that several tests for equality can be constructed with the help of Gaussian test functions. To see this, let f and g be any two exponentially integrable functions on $\mathbb{R}$. Certainly, if $f = g$, then

$$\int_{-\infty}^{\infty} f(x)\phi(x)\,dx \;=\; \int_{-\infty}^{\infty} g(x)\phi(x)\,dx$$

for all ϕ in $\mathcal{G}$. Conversely, if this last equation holds for each ϕ in $\mathcal{G}$, then it certainly holds whenever ϕ is simply a Gaussian (since $\mathcal{G}$ contains all Gaussian functions), and so, the Gaussian test for equality described in theorem 30.1 on page 491 tells us that $f = g$. This gives us the following version of our test for equality.

Theorem 32.5 (test function test for equality)
Let f and g be two exponentially integrable functions on $\mathbb{R}$. Then $f = g$ if and only if

$$\int_{-\infty}^{\infty} f(x)\phi(x)\,dx \;=\; \int_{-\infty}^{\infty} g(x)\phi(x)\,dx \qquad \text{for each } \phi \text{ in } \mathcal{G} \quad . \tag{32.3}$$

The test just described in theorem 32.5 will be fundamental in our work on generalizing the classical theory of functions. Combining it with the fundamental identity gives another test for equality, this time involving Fourier transforms.

Theorem 32.6 (test function test for Fourier transforms)
Let f and F be two classically transformable functions. If any one of the following statements is true, then all of the statements are true.

1. $F = \mathcal{F}[f]$.

2. *For each ϕ in $\mathcal{G}$,*
$$\int_{-\infty}^{\infty} F(x)\phi(x)\,dx \;=\; \int_{-\infty}^{\infty} f(y)\mathcal{F}[\phi]\big|_y\,dy \quad .$$

3. *For each ψ in $\mathcal{G}$,*
$$\int_{-\infty}^{\infty} F(x)\mathcal{F}^{-1}[\psi]\big|_x\,dx \;=\; \int_{-\infty}^{\infty} f(y)\psi(y)\,dy \quad .$$

PROOF: If the first statement is true, then the second and third statements follow immediately by the fundamental identity of Fourier analysis (theorem 25.14 on page 413) and the fact that Gaussian test functions and their Fourier transforms are absolutely integrable.

Now assume the second statement is true, and let ϕ be any Gaussian test function. Again applying the fundamental identity of Fourier analysis, we have

$$\int_{-\infty}^{\infty} f(y)\mathcal{F}[\phi]\big|_y\,dy \;=\; \int_{-\infty}^{\infty} \mathcal{F}[f]\big|_x\,\phi(x)\,dx \quad ,$$

which, combined with the second statement, tells us that

$$\int_{-\infty}^{\infty} F(x)\phi(x)\,dx \;=\; \int_{-\infty}^{\infty} \mathcal{F}[f]\big|_x\,\phi(x)\,dx$$

for each ϕ in $\mathcal{G}$. This, according to the test function test for equality (theorem 32.5), means that $F = \mathcal{F}[f]$. Thus the first statement holds, and, as we just saw, since the first statement holds, so does the third.

Finally, assume the third statement holds, and let ϕ be any Gaussian test function. Letting $\psi = \mathcal{F}[\phi]$ and applying the third statement and the invertibility of the classical Fourier transform gives

$$\int_{-\infty}^{\infty} F(x)\phi(x)\,dx = \int_{-\infty}^{\infty} F(x)\mathcal{F}^{-1}[\psi]|_x\,dx$$

$$= \int_{-\infty}^{\infty} f(y)\psi(y)\,dy = \int_{-\infty}^{\infty} f(y)\mathcal{F}[\phi]|_y\,dy \quad .$$

So the second statement also holds. And thus, as just proven, so then does the first. ∎

Three more tests that are similar in spirit and which will be relevant later are partially described in the following exercise.

?► Exercise 32.4: *Partial statements for three "tests for equality" are given below. In each, f and g are exponentially integrable functions, ϕ denotes an arbitrary Gaussian test function, and ψ is a Gaussian test function related, somehow, to ϕ. For each partial statement, determine the formula relating ψ to ϕ.*

a: *Let a be any real value. Then $g(x) = f(x - a)$ if and only if*

$$\int_{-\infty}^{\infty} g(x)\phi(x)\,dx = \int_{-\infty}^{\infty} f(s)\psi(s)\,ds \qquad \text{for each } \phi \text{ in } \mathcal{G} \quad .$$

(Answer: $\psi(s) = \phi(s + a)$)

b: *Let a be any nonzero real value. Then $g(x) = f(ax)$ if and only if*

$$\int_{-\infty}^{\infty} g(x)\phi(x)\,dx = \int_{-\infty}^{\infty} f(s)\psi(s)\,ds \qquad \text{for each } \phi \text{ in } \mathcal{G} \quad .$$

c: *Assume f is continuous, piecewise smooth and has an exponentially integrable derivative. Then $g = f'$ if and only if*

$$\int_{-\infty}^{\infty} g(x)\phi(x)\,dx = \int_{-\infty}^{\infty} f(s)\psi(s)\,ds \qquad \text{for each } \phi \text{ in } \mathcal{G} \quad .$$

32.3 Two Other Test Function Spaces and a Confession

Loosely speaking, a "space of test functions" is a linear space of "nice functions" that can be used to construct tests for equality such as in theorem 32.5. Just what is meant by "nice functions" depends on the goal at hand. Our goal is to develop a generalized theory of functions that supports a more useful theory of Fourier transforms. So we will stick to the space of Gaussian test functions.

Other test function spaces can be used to generalize the classical theory of Fourier analysis. One you should be aware of is the *space of rapidly decreasing test functions*, often denoted by $\mathcal{S}$. A function ϕ is in $\mathcal{S}$ if and only if it satisfies the following two conditions:

1. ϕ is an infinitely differentiable function on $\mathbb{R}$.

2. $x^n \phi^{(m)}(x)$ is a bounded function on $\mathbb{R}$ for every pair of nonnegative integers n and m.

This is the space originally developed by Laurent Schwartz in his generalization of the classical theory of Fourier analysis. It is also the test function space most other texts employ in developing a generalized theory for Fourier transforms.[2]

Another important test function space for us is the one we will call the *space of very rapidly decreasing test functions* and will denote by $\mathcal{H}$. A function ϕ is in $\mathcal{H}$ if and only if it satisfies the following two conditions:

1. ϕ is an analytic function on the entire complex plane.[3]

2.   For every $\alpha > 0$ there is a corresponding finite constant M_α such that
$$|\phi(x + iy)| \;\leq\; M_\alpha \, e^{-\alpha|x|}$$
for all real values x and y with $|y| \leq \alpha$.

The "generalized theory of functions" based on using either $\mathcal{G}$ or $\mathcal{H}$ as a test function space is a little more general than that based on $\mathcal{S}$. For example, the Fourier transform of e^x can be defined using either $\mathcal{G}$ or $\mathcal{H}$, but not using just $\mathcal{S}$.

On the other hand, the set of "generalized functions" we will develop in the next chapter using $\mathcal{G}$ is identical to the set we would obtain using $\mathcal{H}$. (For those acquainted with the terminology: This is because $\mathcal{G}$ can be viewed as a dense subspace of $\mathcal{H}$ using the α-norms that will be described in section 32.5.) Frankly, $\mathcal{G}$ is a stripped-down version of $\mathcal{H}$ whose use will allow the author to streamline much of the early development of our generalized theory of functions. For one thing, because we have explicit formulas for our test functions, we have less need for some of the more general results from the theory of analytic functions on the complex plane.[4] It must be admitted, however, that in some ways $\mathcal{H}$ is a better test function space than $\mathcal{G}$. The reasons are technical and not worth getting into here. Later, we will encounter a few difficulties because we are using $\mathcal{G}$ instead of $\mathcal{H}$, but we can wait until then to address those difficulties.[5,6]

32.4   More on Gaussian Test Functions

The topics in this section — simple multipliers, complex analysis with test functions and bounds on test functions — will become important at various points in our development of the algebra and calculus of generalized functions. Since, however, the initial development of generalized functions will not require the following material, you may want to first skim through this section and then return to the necessary topics as the need arises.

Simple Multipliers

As was just noted (part 4 of lemma 32.4), the product $h\phi$ is a Gaussian test function whenever ϕ is a Gaussian test function and h either is a Gaussian test function or is given by
$$h(x) \;=\; Cx^k e^{\sigma x}$$

---

[2] See, for example the texts by Richards and Youn (reference [9]) and by Strichartz (reference [11]).

[3] "Analyticity" is described in section 6.4 starting on page 66. On several occasions we will note that a particular function is "analytic". For the most part, however, you need not worry if you do not find the significance of "analyticity" apparent.

[4] That is why you need not worry overmuch about the significance of "analyticity".

[5] Thus, in streamlining our initial development, we complicate the later development — there's a moral lesson in this.

[6] If you really want to see the development based on $\mathcal{H}$, go to the author's original papers (references [3] – [8]). Be warned, though, these are papers in a professional journal for mathematicians — the notation is different, the development is done in a more general setting, and a greater background in real and complex analysis is assumed than is assumed in this text.

where C and σ are any two complex numbers and k is any nonnegative integer. In particular, then, the products

$$x^2\phi(x) \quad , \quad e^{3x}\phi(x) \quad , \quad 4xe^{i2\pi x}\phi(x) \quad \text{and} \quad e^{-x^2}\phi(x)$$

are all Gaussian test functions. Furthermore, if h is the sum of, say x^2, e^{3x} and $4xe^{i2\pi x}$, then

$$
\begin{aligned}
h(x)\phi(x) &= \left[x^2 + e^{3x} + 4xe^{i2\pi x}\right]\phi(x) \\
&= x^2\phi(x) + e^{3x}\phi(x) + 4xe^{i2\pi x}\phi(x) \\
&= \text{a sum of Gaussian test functions} = \text{a Gaussian test function} \quad .
\end{aligned}
$$

Continuing along these lines quickly leads to the next lemma.

Lemma 32.7
Let h be a linear combination of functions of the form

$$x^n e^{cx} e^{-\gamma(x-\zeta)^2}$$

where n is any nonnegative integer, c and ζ are any two complex constants, and γ is any nonnegative real value. Then the product $h\phi$ is a Gaussian test function whenever ϕ is a Gaussian test function.

?▶Exercise 32.5: *Prove the above lemma.*

In light of the above lemma, let us agree to call a function h a *simple multiplier* for $\mathcal{G}$ if and only if h is a linear combination of functions of the form

$$x^n e^{cx} e^{-\gamma(x-\zeta)^2}$$

where n is any nonnegative integer, c and ζ are any two complex constants, and γ is any nonnegative real value. Using this terminology, the above lemma becomes:

Lemma 32.7′
Let h be a simple multiplier for $\mathcal{G}$. Then $h\phi$ is a Gaussian test function whenever ϕ is a Gaussian test function.

Clearly, all constants, all polynomials, all exponential functions and all Gaussian test functions are simple multipliers. Some other simple multipliers are described in exercise 32.12 on page 535. It should also be fairly easy to see that products and linear combinations of simple multipliers are simple multipliers.

?▶Exercise 32.6: *Let g and h be simple multipliers for $\mathcal{G}$, and let α and β be any two constants. Verify that the product gh and the linear combination $\alpha g + \beta h$ are simple multipliers for $\mathcal{G}$.*

Strictly speaking, the functions just discussed here should be referred to as "simple multipliers *for the Gaussian test functions*" (i.e., "*for $\mathcal{G}$*"), and not merely as "simple multipliers". This is because we will later develop a somewhat broader definition for a "simple multiplier", and we will discover that different test function spaces have different corresponding sets of "simple multipliers". However, this won't occur until chapter 36. Fortunately, all the results we will derive using the "simple multipliers for $\mathcal{G}$" can also be obtained using, say, the "simple multipliers for $\mathcal{H}$". So, at least until chapter 36, it won't hurt to use the term "simple multiplier" when we really mean "simple multiplier for $\mathcal{G}$".

Some Complex Analysis with Gaussian Test Functions
Test Functions as Functions of a Complex Variable

Let's go back and reconsider any one of those formulas describing an arbitrary base Gaussian test function ϕ, say,

$$\phi(s) = As^n e^{-\gamma(s-\zeta)^2} \qquad \text{for} \quad -\infty < s < \infty \tag{32.4}$$

where A and ζ are complex constants, n is a nonnegative integer and $\gamma > 0$.

Up to this point, we have been considering Gaussian test functions as functions on the real line. However, if we replace the real variable s in equation (32.4) with the complex variable $x + iy$, then we have a formula for a function on the complex plane,

$$\phi(x + iy) = A(x + iy)^n e^{-\gamma(x+iy-\zeta)^2} \quad . \tag{32.5}$$

Those who have had a course in complex analysis or have read about analyticity in section 6.4 will even recognize that this is an analytic function on $\mathbb{C}$. Furthermore, you can easily verify that any other formula given in lemma 32.1 for the above ϕ on $\mathbb{R}$ gives exactly the same function on $\mathbb{C}$ when the real variable is replaced by a complex variable. That is, assuming ϕ is as above and, say, D and σ are the complex constants such that

$$\phi(s) = Ds^n e^{\sigma s} e^{-\gamma s^2} \qquad \text{for} \quad -\infty < s < \infty \quad ,$$

then you can easily show via basic algebra that

$$D(x + iy)^n e^{\sigma(x+iy)} e^{-\gamma(x+iy)^2} = A(x + iy)^n e^{-\gamma(x+iy-\zeta)^2}$$

for every complex value $x + iy$.

What this means is that, in practice, we can treat our base Gaussian test functions as analytic functions of a complex variable. And since all other Gaussian test functions are linear combinations of base Gaussian test functions, we can also treat all Gaussian test functions as analytic functions on $\mathbb{C}$ when the need arises.

You should realize that, strictly speaking, formulas (32.4) and (32.5) define two different functions because the corresponding domains, $\mathbb{R}$ and $\mathbb{C}$, respectively, are different. Perhaps we should use a slightly different notation for the second function, say, adding a "subscript E",

$$\phi_E(x + iy) = A(x + iy)^n e^{-\gamma(x+iy-\zeta)^2} \quad , \tag{32.5$'$}$$

to indicate that this is an *extension* of ϕ to a function defined on all of $\mathbb{C}$. Let us do this — denote the extension of a test function ϕ to a function on $\mathbb{C}$ by ϕ_E — when there is a danger in confusing the two. Usually though, this is not necessary, and we can safely use the same symbol to denote both the original function defined on $\mathbb{R}$ and the extension, defined on $\mathbb{C}$, obtained by viewing the variable in the original function's formula as an arbitrary complex value. Just remember, if "ϕ is a test function", then:

1. Unless otherwise indicated, ϕ is a function on the real line.

2. Unless otherwise indicated, the variable in an expression such as "$\phi(s)$" should be treated as an arbitrary real value.

3. In any expression involving ϕ with an explicitly complex variable, say, "$\phi(x + iy)$", the ϕ denotes the function on the complex plane obtained by replacing s by $x + iy$ in the formula for $\phi(s)$ (i.e., $\phi(x + iy) = \phi_E(x + iy)$).

It is worth noting that similar observations can be made regarding the simple multipliers discussed in the previous section. Just look at the formulas defining simple multipliers. Clearly, these formulas also define functions on the complex plane.

Complex Translation of Test Functions

Take another look at our proof, on page 517, that $\phi(x - \xi)$ is a linear combination of base Gaussian test functions whenever ϕ is a base Gaussian test function and ξ is a fixed real number. There we assumed ξ to be real simply because ϕ was only defined as a function on the real line. But, as we just noted, we can treat our test functions as functions on $\mathbb{C}$. So it makes sense to consider $\phi(x - \xi)$ even when ξ is complex. Furthermore, the computations leading to equation (32.2) can be repeated almost verbatim, yielding

$$\phi(x - \xi) = \sum_{k=0}^{n} A_k x^k e^{-\gamma(x-\zeta)^2}$$

for some integer n and some set of complex constants ζ, A_0, ... and A_n. Thus, if ϕ is any base Gaussian test function and ξ is any fixed number, real or complex, then $\phi(x - \xi)$ is a linear combination of base Gaussian test functions (and, hence, is a Gaussian test function). Combine this with the fact that every Gaussian test function is a linear combination of base Gaussian test functions, and we get the following addition to our list of properties of Gaussian test functions (lemma 32.4 on page 519):

Lemma 32.8
If $\phi(x)$ is a Gaussian test function, then so is $\phi(x - \xi)$ when ξ is any complex constant.

Now recall something we observed after deriving the formulas for the Fourier transforms of arbitrary Gaussians in chapter 23, namely, that the Fourier translation identities hold for *complex* translations when the functions are Gaussians. To be precise, we saw that, if $\gamma > 0$, $g(t) = e^{-\gamma t^2}$ and $G = \mathcal{F}[g]$, then

$$\mathcal{F}\left[e^{i2\pi\xi t}g(t)\right]\bigg|_\omega = G(\omega - \xi) \tag{32.6}$$

for any fixed complex value ξ (see exercise 23.4 on page 367 and the discussion preceding it). It is not hard to show that this observation also holds with "Gaussian test functions" replacing "Gaussians".

Theorem 32.9
Let ψ and Ψ be Gaussian test functions with $\Psi = \mathcal{F}[\psi]$, and let a be any complex constant. Then the functions $\psi(x - a)$ and $\Psi(x - a)$ are also in $\mathcal{G}$, and the following identities hold:

$$\mathcal{F}\left[e^{i2\pi a t}\psi(t)\right]\bigg|_\omega = \Psi(\omega - a) \quad , \tag{32.7a}$$

$$\mathcal{F}^{-1}\left[e^{-i2\pi a\omega}\Psi(\omega)\right]\bigg|_t = \psi(t - a) \quad , \tag{32.7b}$$

$$\mathcal{F}^{-1}[\Psi(\omega - a)]|_t = e^{i2\pi a t}\psi(t) \tag{32.7c}$$

and

$$\mathcal{F}[\psi(t - a)]|_\omega = e^{-i2\pi a\omega}\Psi(\omega) \quad . \tag{32.7d}$$

In view of lemma 32.8 and the invertibility of the Fourier transform, we only need to confirm equations (32.7a) and (32.7b). Naturally, we will go through the details of showing the first equation holds, and leave the details of verifying the other as an exercise.

PROOF (of identity (32.7a)):   First, assume ϕ is any base Gaussian test function. Let $\Phi = \mathcal{F}[\phi]$, and choose A, σ, n and γ to be the constants (with A and σ complex, n a nonnegative integer and $\gamma > 0$) such that

$$\phi(t) = At^n e^{\sigma t}e^{-\gamma t^2} \quad .$$

For convenience, let

$$\beta = \frac{\sigma}{i2\pi} \quad , \quad g(t) = e^{-\gamma t^2} \quad \text{and} \quad G(\omega) = \mathcal{F}[g(t)]|_\omega \quad ,$$

so that

$$\phi(t) = At^n e^{i2\pi\beta t} g(t) \quad .$$

Applying equation (32.6), the chain rule and a Fourier differential identity, we then see that

$$\mathcal{F}\left[e^{i2\pi\beta t} g(t)\right]\Big|_\omega = G(\omega - \beta) \quad ,$$

and

$$\Phi(\omega) = \mathcal{F}\left[At^n e^{i2\pi\beta t} g(t)\right]\Big|_\omega$$

$$= A\left(\frac{i}{2\pi}\right)^n \frac{d^n}{d\omega^n} G(\omega - \beta) = A\left(\frac{i}{2\pi}\right)^n G^{(n)}(\omega - \beta) \quad .$$

Now,

$$e^{i2\pi a t} \phi(t) = e^{i2\pi a t} At^n e^{i2\pi\beta t} g(t) = At^n e^{i2\pi(a+\beta)t} g(t) \quad .$$

So, by the same arguments used to compute Φ,

$$\mathcal{F}\left[e^{i2\pi a t} \phi(t)\right]\Big|_\omega = \mathcal{F}\left[At^n e^{i2\pi(a+\beta)t} g(t)\right]\Big|_\omega$$

$$= A\left(\frac{i}{2\pi}\right)^n G^{(n)}(\omega - (a+\beta))$$

$$= A\left(\frac{i}{2\pi}\right)^n G^{(n)}((\omega - a) - \beta) = \Phi(\omega - a) \quad .$$

Now let ψ be any Gaussian test function and let $\Psi = \mathcal{F}[\psi]$. By definition, there is an integer N and base Gaussian test functions ϕ_1, ϕ_2, ... and ϕ_N such that

$$\psi = \sum_{k=1}^N \phi_k \quad .$$

Then, letting $\Phi_k = \mathcal{F}[\phi_k]$,

$$\Psi(\omega) = \mathcal{F}[\psi(t)]|_\omega = \mathcal{F}\left[\sum_{k=1}^N \phi_k(t)\right]\Bigg|_\omega = \sum_{k=1}^N \mathcal{F}[\phi_k(t)]|_\omega = \sum_{k=1}^N \Phi_k(\omega) \quad .$$

Since each ϕ_k is a base Gaussian test function, the results from the previous paragraph assure us that

$$\mathcal{F}\left[e^{i2\pi a t} \phi_k(t)\right]\Big|_\omega = \Phi_k(\omega - a) \quad \text{for} \quad k = 1, 2 \ldots, N \quad .$$

Thus,

$$\mathcal{F}\left[e^{i2\pi a t} \psi(t)\right]\Big|_\omega = \mathcal{F}\left[e^{i2\pi a t} \sum_{k=1}^N \phi_k(t)\right]\Bigg|_\omega$$

$$= \sum_{k=1}^N \mathcal{F}\left[e^{i2\pi a t} \phi_k(t)\right]\Big|_\omega$$

$$= \sum_{k=1}^N \Phi_k(\omega - a) = \Psi(\omega - a) \quad . \qquad \blacksquare$$

?▶Exercise 32.7:   *Verify equation (32.7b).*

Analytic Conjugates

Finding the complex conjugate ϕ^* of a function ϕ defined on just the real line is easy. Simply take the complex conjugate of $\phi(x)$ assuming that x is an arbitrary real number. In practice, this means taking the complex conjugate of every constant in the formula for ϕ. Consider, for example, any base Gaussian test function ϕ (viewed as a function on just the real line). Let γ be the positive constant, n the nonnegative integer, and A and ζ the complex values such that

$$\phi(x) \;=\; A x^n e^{-\gamma(x-\zeta)^2}$$

for all real values of x. The complex conjugate of ϕ, ϕ^*, is then given by

$$\phi^*(x) \;=\; \phi(x)^* \;=\; A^* x^n e^{-\gamma(x-\zeta^*)^2} \quad .$$

Notice that this function, ϕ^*, is also a base Gaussian test function, and that, as an analytic function on the complex plane, ϕ^* is given by

$$\phi^*(z) \;=\; A^* z^n e^{-\gamma(z-\zeta^*)^2}$$

where we now view z as an arbitrary complex value.

There is a subtle and sometimes confusing point here. If z is not a real value, then $\phi^*(z)$ is *not* the same as $[\phi(z)]^*$. To see this, let $z = x + iy$. Then

$$\phi^*(z) \;=\; A^* (x+iy)^n e^{-\gamma(x+iy-\zeta^*)^2} \quad ,$$

while

$$[\phi(z)]^* \;=\; \left[A(x+iy)^n e^{-\gamma(x+iy-\zeta)^2} \right]^* \;=\; A^* (x-iy)^n e^{-\gamma(x-iy-\zeta^*)^2} \quad .$$

In computing $[\phi(z)]^*$ we have actually taken the complex conjugate of both the function's formula and the variable. Comparing the above two equations, we see that, in fact,

$$[\phi(z)]^* \;=\; \phi^*(z^*) \quad .$$

Replacing z with z^* in this formula, and recalling that $z^{**} = z$, leads to the correct relation between $\phi^*(z)$ and $[\phi(z)]^*$,

$$[\phi(z^*)]^* \;=\; \phi^*(z^{**}) \;=\; \phi^*(z) \quad .$$

More generally, for any analytic function f on the complex plane, f^* will denote the function given by

$$f^*(z) \;=\; [f(z^*)]^* \qquad \text{for all } z \text{ in } \mathbb{C} \quad .$$

To distinguish $f^*(z)$ from $[f(z)]^*$, we will refer to f^* as the *analytic conjugate* of f.

The following lemma, which extends the observations made a couple of paragraphs ago, can be verified using the test for analyticity and the uniqueness theorem for analytic functions discussed in section 6.4. The details will be left to the interested reader (exercise 32.14 at the end of this chapter).

Lemma 32.10
Let f be an analytic function on the entire complex plane. Then its analytic conjugate, f^*, is the single analytic function on the complex plane that satisfies $f^*(x) = [f(x)]^*$ for each real x.

It should be clear that

$$[f + g]^* = f^* + g^* \quad , \quad (fg)^* = f^* g^* \quad \text{and} \quad (f^*)^* = f$$

for any pair of functions f and g that are analytic on $\mathbb{C}$. From this and the above discussion we immediately obtain the next lemma.

Lemma 32.11

A function is a Gaussian test function if and only if its analytic conjugate is a Gaussian test function.

Bounds on Test Functions

Because of the way Gaussian test functions are constructed from Gaussians, we know that the value of any test function at a real point x must become very small very quickly as $|x|$ increases. This is described explicitly through a set of bounds that, on occasion, we will find useful.

To derive these bounds, let ϕ be any Gaussian test function, and let α be some real number. Consider the function on $\mathbb{R}$ given by

$$e^{\alpha|x|}\phi(x) \quad .$$

Remember, both $e^{\alpha x}\phi(x)$ and $e^{-\alpha x}\phi(x)$ are Gaussian test functions. And, since Gaussian test functions are bounded functions on the real line, there is a finite positive constant M_α larger than either of these functions at any point on the real line. So, for any real value x,

$$e^{\alpha|x|}|\phi(x)| = \begin{cases} e^{-\alpha x}|\phi(x)| & \text{if } x \leq 0 \\ e^{\alpha x}|\phi(x)| & \text{if } 0 \leq x \end{cases} \leq \begin{cases} M_\alpha & \text{if } x \leq 0 \\ M_\alpha & \text{if } 0 \leq x \end{cases} = M_\alpha \quad .$$

Dividing through by $e^{\alpha|x|}$ gives us the next lemma.

Lemma 32.12 (simple bounds for test functions)

For each Gaussian test function ϕ and each real number α, there is a finite positive value M_α such that

$$|\phi(x)| \leq M_\alpha e^{-\alpha|x|} \quad \text{for all } x \text{ in } \mathbb{R} \quad .$$

Similar bounds can be derived for Gaussian test functions evaluated at points in the complex plane off of the real line. That derivation is left as an exercise.

?▶Exercise 32.8: *Let ϕ be a Gaussian test function, and let α and β be two nonnegative real values. Show there is a finite constant M such that*

$$|\phi(x + iy)| \leq Me^{-\alpha|x|}$$

for each pair of real values x and y where $|y| \leq \beta$. (First consider the case where ϕ is a base Gaussian test function and use one of the formulas for such functions from lemma 32.1 on page 516.)

Notice that, when $\alpha = \beta$, the inequality of this last exercise can be written as

$$|\phi(x + iy)|e^{\alpha|x|} \leq M$$

for each pair of real values x and y where $|y| \leq \alpha$. In other words, $\phi(x + iy)e^{\alpha|x|}$ is a bounded function on the strip in the complex plane of all points that lie within a distance of α of the real axis. In the next section, we will use these bounds to measure the similarities between different test functions.

32.5 Norms and Operational Continuity[*]
Norms

In developing the generalized theory, it is sometimes helpful to have very stringent rules for determining when two test functions are "almost the same". The rules we will use are based on the "α-norms" defined in the next paragraph.

Let ϕ be a Gaussian test function, and let α be any nonnegative real number. The *α-norm* of ϕ, denoted by $\|\phi\|_\alpha$, is the nonnegative real number given by

$$\|\phi\|_\alpha = \max\left\{ |\phi(x+iy)|\, e^{\alpha|x|} : -\infty < x < \infty \text{ and } -\alpha \le y \le \alpha \right\} \quad .$$

In other words, $\|\phi\|_\alpha$ is the maximum value of

$$|\phi(x+iy)|\, e^{\alpha|x|}$$

over the strip in the complex plane of all points that lie within a distance α of the real axis. (That this maximum exists and is finite can be easily verified using the properties of Gaussian test functions and material normally developed in elementary calculus.)

!►*Example 32.1:*   *Let's compute* $\|\phi\|_\alpha$ *when*

$$\phi(x) = e^{-x^2} \quad \text{and} \quad \alpha = 4 \quad .$$

Note that, for any pair of real numbers x *and* y,

$$\left| e^{-(x+iy)^2} \right| = \left| e^{-(x^2-y^2+i2xy)} \right| = e^{-x^2} e^{y^2} \left| e^{-i2xy} \right| = e^{-x^2} e^{y^2} \quad .$$

So,

$$\begin{aligned}
\|\phi\|_4 &= \max\left\{ \left| e^{-(x+iy)^2} \right| e^{4|x|} : -\infty < x < \infty,\ -4 \le y \le 4 \right\} \\
&= \max\left\{ e^{-x^2} e^{4|x|} e^{y^2} : -\infty < x < \infty,\ -4 \le y \le 4 \right\} \\
&= \max\left\{ e^{-x^2} e^{4|x|} : -\infty < x < \infty \right\} \times \max\left\{ e^{y^2} : -4 \le y \le 4 \right\} \quad .
\end{aligned}$$

Of course,

$$\max\left\{ e^{y^2} : -4 \le y \le 4 \right\} = e^{4^2} = e^{16} \quad .$$

Observe, also, that

$$\begin{aligned}
\max\left\{ e^{-x^2} e^{4|x|} : -\infty < x < \infty \right\} &= \max\left\{ e^{-x^2} e^{4x} : 0 \le x \right\} \\
&= \max\left\{ e^4 e^{-(x^2-4x+4)} : 0 \le x \right\} \\
&= e^4 \max\left\{ e^{-(x-2)^2} : 0 \le x \right\} = e^4 \cdot 1 \quad .
\end{aligned}$$

So the last formula for $\|\phi\|_4$ *reduces to*

$$\|\phi\|_4 = e^4 e^{16} = e^{20} \quad .$$

---

[*] This section is for the more advanced readers. "Less advanced" readers can skip it. They will just have to trust the author a little more in a few sections later on.

Keep in mind that $\|\phi\|_\alpha$ is just the maximum value of

$$|\phi(x+iy)|\,e^{\alpha|x|}$$

where x and y are any real numbers with $|y| \leq \alpha$. From this fact you should be able to quickly confirm the claims in the next few lemmas.

Lemma 32.13
Assume ϕ is any Gaussian test function and $\alpha \geq 0$. If M is a real number such that

$$|\phi(x+iy)|\,e^{\alpha|x|} \;\leq\; M$$

whenever x and y are two real numbers with $|y| \leq \alpha$, then $\|\phi\|_\alpha \leq M$.

Lemma 32.14
If ϕ is any Gaussian test function and $\alpha \geq 0$, then

$$|\phi(x+iy)|\,e^{\alpha|x|} \;\leq\; \|\phi\|_\alpha \qquad \text{and} \qquad |\phi(x+iy)| \;\leq\; e^{-\alpha|x|}\,\|\phi\|_\alpha \quad.$$

whenever x and y are two real numbers with $|y| \leq \alpha$.

Lemma 32.15
If $0 \leq \alpha \leq \beta < \infty$ and ϕ is a Gaussian test function, then

$$\|\phi\|_\alpha \;\leq\; \|\phi\|_\beta \quad.$$

?► Exercise 32.9: *Verify each of the above three lemmas.*

You may recall that "norms", in general, are expected to satisfy certain properties. The next lemma asserts that our α-norms satisfy those properties.

Lemma 32.16
Let $\alpha \geq 0$, let ϕ and ψ be any two Gaussian test functions, and let c be any complex number. Then

1. $\|\phi + \psi\|_\alpha \;\leq\; \|\phi\|_\alpha + \|\psi\|_\alpha$,

2. $\|c\phi\|_\alpha \;=\; |c|\,\|\phi\|_\alpha$, and

3. $0 \leq \|\phi\|_\alpha < \infty$ with $\|\phi\|_\alpha = 0$ if and only if ϕ is the zero function.

PROOF: From the triangle inequality and lemma 32.14, above, we know that, if x and y are any two real numbers with $|y| \leq \alpha$, then

$$e^{\alpha|x|}\,|\phi(x+iy) + \psi(x+iy)| \;\leq\; e^{\alpha|x|}\,|\phi(x+iy)| + e^{\alpha|x|}\,|\psi(x+iy)|$$
$$\leq\; \|\phi\|_\alpha + \|\psi\|_\alpha \quad.$$

Thanks to lemma 32.13, this confirms the first claim.

Verification of the second claim is straightforward,

$$
\begin{aligned}
\|c\phi\|_\alpha &= \max\left\{ |c\phi(x+iy)|\, e^{\alpha|x|} : -\infty < x < \infty, -\alpha \le y \le \alpha \right\} \\
&= \max\left\{ |c|\, |\phi(x+iy)|\, e^{\alpha|x|} : -\infty < x < \infty, -\alpha \le y \le \alpha \right\} \\
&= |c| \max\left\{ |\phi(x+iy)|\, e^{\alpha|x|} : -\infty < x < \infty, -\alpha \le y \le \alpha \right\} = |c|\,\|\phi\|_\alpha \quad .
\end{aligned}
$$

Finally, the claim that $0 \le \|\phi\|_\alpha < \infty$ follows immediately from the definition of the α-norm and properties of Gaussian functions, as does the fact that $\|\phi\|_\alpha = 0$ whenever ϕ is the zero function. On the other hand, if $\|\phi\|_\alpha = 0$, then lemma 32.14 (with $y = 0$) tells us that

$$
0 \le |\phi(x)| \le e^{-\alpha|x|}\,\|\phi\|_\alpha = 0 \qquad \text{for each} \quad -\infty < x < \infty \quad ,
$$

confirming that $\phi(x) = 0$ for all real x. ∎

Bounded Operations

Let $\mathcal{O}$ denote some operation, such as differentiation or Fourier transformation, that converts each Gaussian test function ϕ into another Gaussian test function $\psi = \mathcal{O}[\phi]$. We will refer to this operation as being *(operationally) bounded* (on $\mathcal{G}$) if and only if, for each $\alpha \ge 0$, there are corresponding nonnegative constants M_α and $\beta = \beta(\alpha)$ such that

$$
\|\mathcal{O}[\phi]\|_\alpha \le M_\alpha \|\phi\|_\beta \qquad \text{for each } \phi \text{ in } \mathcal{G} \quad .
$$

For reasons that will become evident later, we will also refer to such operations as being *(operationally) continuous*.

The "operational boundedness" of an operator $\mathcal{O}$ assures us that, in some ways at least, $\mathcal{O}$ is well behaved. In particular, it assures us that $\mathcal{O}[\phi]$ will be "small" whenever ϕ is a "small" Gaussian test function (as determined by its α-norms). This, along with linearity, becomes important when we try to generalize classical operations such as differentiation and Fourier transformation. Fortunately, as we'll see in the next lemma and theorem, many operations of interest to us are operationally bounded.

Lemma 32.17
Let $\mathcal{O}_1$ and $\mathcal{O}_2$ be two operationally bounded operations on $\mathcal{G}$, and let a and b be two constants. Then the composition $\mathcal{O}_1 \circ \mathcal{O}_2$ and the linear combination $a\mathcal{O}_1 + b\mathcal{O}_2$, defined by

$$
\mathcal{O}_1 \circ \mathcal{O}_2[\phi] = \mathcal{O}_1\left[\mathcal{O}_2[\phi]\right] \qquad \text{and} \qquad [a\mathcal{O}_1 + b\mathcal{O}_2][\phi] = a\mathcal{O}_1[\phi] + b\mathcal{O}_2[\phi] \quad ,
$$

are operationally bounded.

PROOF: Because $\mathcal{O}_1$ and $\mathcal{O}_2$ are operationally bounded, we know that, for each $\alpha \ge 0$, there are corresponding nonnegative real constants A_α, B_α, $\kappa = \kappa(\alpha)$ and $\lambda = \lambda(\alpha)$ such that

$$
\|\mathcal{O}_1[\phi]\|_\alpha \le A_\alpha \|\phi\|_\kappa \qquad \text{and} \qquad \|\mathcal{O}_2[\phi]\|_\alpha \le B_\alpha \|\phi\|_\lambda
$$

for each Gaussian test function ϕ.

So let α be any nonnegative real number, and let ϕ be any Gaussian test function.

From the operational boundedness of $\mathcal{O}_1$ and $\mathcal{O}_2$, we have

$$
\begin{aligned}
\|\mathcal{O}_1 \circ \mathcal{O}_2[\phi]\|_\alpha &= \|\mathcal{O}_1[\mathcal{O}_2[\phi]]\|_\alpha \\
&\le A_\alpha \|\mathcal{O}_2[\phi]\|_{\kappa(\alpha)} \\
&\le A_\alpha B_{\kappa(\alpha)} \|\phi\|_{\lambda(\kappa(\alpha))} = M_\alpha \|\phi\|_\beta
\end{aligned}
$$

where $M_\alpha = A_\alpha B_{\kappa(\alpha)}$ and $\beta = \lambda(\kappa(\alpha))$. Hence, the composition is operationally bounded.

Using inequalities from lemma 32.16, we see also that

$$
\|a\mathcal{O}_1[\phi] + b\mathcal{O}_2[\phi]\|_\alpha \le |a| \, \|\mathcal{O}_1[\phi]\|_\alpha + |b| \, \|\mathcal{O}_2[\phi]\|_\alpha \le |a| \, A_\alpha \|\phi\|_\kappa + |b| \, B_\alpha \|\phi\|_\lambda
$$

where $\kappa = \kappa(\alpha)$ and $\lambda = \lambda(\alpha)$. Let $M_\alpha = |a| \, A_\alpha + |b| \, B_\alpha$, and let β be the larger of κ and λ. Then, using the inequality from lemma 32.15, we can extend the above string of inequalities to

$$
\|a\mathcal{O}_1[\phi] + b\mathcal{O}_2[\phi]\|_\alpha \le |a| \, A_\alpha \|\phi\|_\beta + |b| \, B_\alpha \|\phi\|_\beta = M_\alpha \|\phi\|_\beta \quad . \qquad \blacksquare
$$

Theorem 32.18

Each of the following operations on $\mathcal{G}$ is operationally bounded:

1. *the Fourier transform*

2. *the inverse Fourier transform*

3. *multiplication by a fixed elementary multiplier*

4. *differentiation*

5. *translation by any fixed real or complex value*

6. *variable scaling by any fixed nonzero real number*

7. *analytic conjugation*

The first and the third parts of this theorem are the most difficult to prove. We'll prove those parts and the fourth part (on differentiation). The proof of the second part is almost identical to that of the first part and, so, will be omitted. The rest will be left as exercises.

Before getting into the details for each part, let us consider just what is needed to verify that an operation $\mathcal{O}$ is operationally bounded. Suppose we can show that, for each $\alpha \ge 0$, there are corresponding constants M_α and $\beta(\alpha)$ such that

$$
e^{\alpha|x|} \left| \mathcal{O}[\phi] \right|_{x+iy} \le M_\alpha \|\phi\|_{\beta(\alpha)}
$$

whenever ϕ is in $\mathcal{G}$, and x and y are real numbers with $|y| \le \alpha$. Lemma 32.13 then assures us that

$$
\|\mathcal{O}[\phi]\|_\alpha \le M_\alpha \|\phi\|_{\beta(\alpha)} \quad .
$$

Consequently, to verify $\mathcal{O}$ is operationally bounded, it suffices to show that, for each $\alpha \ge 0$, there are corresponding nonnegative real values M_α and $\beta(\alpha)$ such that

$$
e^{\alpha|x|} \left| \mathcal{O}[\phi] \right|_{x+iy} \le M_\alpha \|\phi\|_{\beta(\alpha)}
$$

for each Gaussian test function ϕ and each pair of real numbers x and y with $|y| \le \alpha$. That is how we will prove the claimed operational boundedness in the following.

PROOF (theorem 32.18, boundedness of the Fourier transform):   Let ϕ be any Gaussian test function and α any nonnegative real number. For convenience, let $\psi = \mathcal{F}[\phi]$ and consider

$$e^{\alpha|x|} |\psi(x+iy)|$$

where x and y are two real numbers with $|y| \leq \alpha$. Since we will want to use the complex translation identities, which are valid for Gaussian test functions, it is also convenient to let

$$\widehat{\alpha} = \begin{cases} \dfrac{\alpha}{2\pi} & \text{if } x \geq 0 \\[2mm] -\dfrac{\alpha}{2\pi} & \text{if } x < 0 \end{cases} \quad ,$$

and to observe that

$$e^{\alpha|x|} |\psi(x+iy)| = \left| e^{-i2\pi(i\widehat{\alpha})x} \psi(x+iy) \right| \quad .$$

Using the complex translation identities in theorem 32.9 on page 525, we obtain

$$\psi(x+iy) = \mathcal{F}\left[e^{i2\pi(-iy)t} \phi(t) \right]\Big|_x = \mathcal{F}\left[e^{2\pi yt} \phi(t) \right]\Big|_x \quad ,$$

and

$$e^{\alpha|x|} |\psi(x+iy)| = \left| e^{-i2\pi(i\widehat{\alpha})x} \mathcal{F}\left[e^{2\pi yt} \phi(t) \right]\Big|_x \right| = \left| \mathcal{F}\left[e^{2\pi y(t-i\widehat{\alpha})} \phi(t - i\widehat{\alpha}) \right]\Big|_x \right| \quad .$$

Fortunately for us, Fourier transforms of Gaussian test functions are given by the integral formula. Using this, the above and the fact that $|y| \leq \alpha$, we get

$$e^{\alpha|x|} |\psi(x+iy)| = \left| \int_{-\infty}^{\infty} e^{2\pi y(t-i\widehat{\alpha})} \phi(t - i\widehat{\alpha}) e^{-i2\pi tx} \, dt \right|$$

$$= \left| \int_{-\infty}^{\infty} \phi(t - i\widehat{\alpha}) e^{2\pi yt} e^{-i2\pi(y\widehat{\alpha}+tx)} \, dt \right|$$

$$\leq \int_{-\infty}^{\infty} \left| \phi(t - i\widehat{\alpha}) e^{2\pi yt} e^{-i2\pi(y\widehat{\alpha}+tx)} \right| dt$$

$$\leq \int_{-\infty}^{\infty} |\phi(t - i\widehat{\alpha})| \, e^{2\pi \alpha|t|} \, dt \quad . \tag{32.8}$$

After recalling the definition of $\widehat{\alpha}$ and lemma 32.14 on page 530, we see that

$$|\phi(t - i\widehat{\alpha})| = \left| \phi\left(t \pm i\frac{\alpha}{2\pi} \right) \right| \leq e^{-\beta|t|} \|\phi\|_\beta \qquad \text{whenever} \quad \frac{\alpha}{2\pi} < \beta \quad .$$

Combined with inequality (32.8), this gives

$$e^{\alpha|x|} |\psi(x+iy)| \leq \int_{-\infty}^{\infty} e^{-\beta|t|} \|\phi\|_\beta \, e^{2\pi\alpha|t|} \, dt$$

$$= \left(\int_{-\infty}^{\infty} e^{-(\beta-2\pi\alpha)|t|} \, dt \right) \|\phi\|_\beta = \frac{2}{\beta - 2\pi\alpha} \|\phi\|_\beta$$

whenever $\beta > 2\pi\alpha$. In particular, letting $\beta = 2\pi\alpha + 1$, we see that, for each Gaussian test function ϕ and each pair of real numbers x and y with $|y| \leq \alpha$,

$$e^{\alpha|x|} \left| \mathcal{F}[\phi]|_{x+iy} \right| = e^{\alpha|x|} |\psi(x+iy)| \leq 2 \|\phi\|_{2\pi\alpha+1} \quad . \qquad \blacksquare$$

PROOF (theorem 32.18, boundedness of multiplication by simple multipliers): Let h be any simple multiplier. Using the definition of simple multipliers along with computations similar to those done for lemma 32.1 on page 516, we can readily verify that h is a linear combination of functions of the form

$$x^n e^{(c+ib)x} e^{-\gamma x^2}$$

where n is a nonnegative integer, and c, b and γ are real constants with $\gamma \geq 0$. And from lemma 32.17, we know that any linear combination of operationally bounded operations is, itself, operationally bounded. So it will suffice to show "multiplication by h" is operationally bounded when

$$h(x) = x^n e^{(c+ib)x} e^{-\gamma x^2}$$

for some nonnegative integer n and real values c, b and γ with $\gamma \geq 0$.

So let $\alpha \geq 0$, let ϕ be any Gaussian test function, and let x and y be any two real numbers with $|y| \leq \alpha$. Using a little algebra and the triangle inequality, we see that

$$|h(x+iy)| = \left| (x+iy)^n e^{(c+ib)(x+iy)} e^{-\gamma(x+iy)^2} \right|$$

$$= |(x+iy)|^n \left| e^{cx} e^{-by} e^{i(cy+bx)} e^{-\gamma x^2} e^{\gamma y^2} e^{i\gamma 2xy} \right|$$

$$\leq (|x|+|y|)^n e^{cx} e^{-by} e^{-\gamma x^2} e^{\gamma y^2} \quad,$$

which, because $e^{-\gamma x^2} \leq 1$ and $|y| \leq \alpha$, reduces to

$$|h(x+iy)| \leq e^{|b|\alpha + \gamma \alpha^2} (|x|+\alpha)^n e^{|c||x|} \quad. \tag{32.9}$$

Observe now that, if $|x| \leq \alpha$, then

$$(|x|+\alpha)^n \leq (\alpha+\alpha)^n = 2^n \alpha^n \leq 2^n \alpha^n e^{n|x|} \quad,$$

while, if $\alpha \leq |x|$,

$$(|x|+\alpha)^n \leq (|x|+|x|)^n = 2^n |x|^n \leq 2^n e^{n|x|} \quad.$$

Either way,

$$(|x|+\alpha)^n \leq 2^n (1+\alpha^n) e^{n|x|} \quad.$$

This, along with inequality (32.9), gives

$$e^{\alpha|x|} |h(x+iy)\phi(x+iy)| \leq e^{\alpha|x|} e^{|b|\alpha + \gamma \alpha^2} \left[2^n (1+\alpha^n) e^{n|x|} \right] e^{|c||x|} |\phi(x+iy)|$$

$$\leq A_\alpha e^{\beta|x|} |\phi(x+iy)|$$

where

$$A_\alpha = 2^n (1+\alpha^n) e^{|b|\alpha + \gamma \alpha^2} \quad \text{and} \quad \beta = \alpha + n + |c| \quad.$$

But, from lemma 32.14 and the fact that $|y| \leq \alpha \leq \beta$, we know

$$e^{\beta|x|} |\phi(x+iy)| \leq \|\phi\|_\beta \quad.$$

That, combined with the previous inequality, then yields

$$e^{\alpha|x|} |h(x+iy)\phi(x+iy)| \leq A_\alpha \|\phi\|_\beta \quad. \qquad \blacksquare$$

PROOF *(theorem 32.18, boundedness of differentiation):* Let ϕ be any Gaussian test function. By one of the differentiation identities,

$$\mathcal{F}[\phi']\big|_x = i2\pi x \mathcal{F}[\phi]\big|_x \quad .$$

Thus,

$$\phi' = \mathcal{F}^{-1}[h\mathcal{F}[\phi]]$$

where h is the simple multiplier $h(x) = i2\pi x$. Letting $\mathcal{O}_h$ denote the operation of "multiplication by h", we can rewrite this as a composition of three operations,

$$\phi' = \mathcal{F}^{-1} \circ \mathcal{O}_h \circ \mathcal{F}[\phi] \quad .$$

But we already verified that these operations — $\mathcal{F}^{-1}$, $\mathcal{O}_h$ and $\mathcal{F}$ — are operationally bounded, as are the compositions of operationally bounded operators (see lemma 32.17). So it immediately follows that the operation of differentiation, being a composition of operationally bounded operations, is operationally bounded. ∎

?▶Exercise 32.10: *Prove translation is a bounded operation on $\mathcal{G}$ (part 5 of theorem 32.18) by showing that, for any fixed complex value $\zeta = \mu + i\nu$ and any given $\alpha \geq 0$,*

$$\|\psi\|_\alpha \leq e^{\alpha|\mu|} \|\phi\|_{\alpha+|\nu|}$$

whenever ϕ and ψ are Gaussian test functions related by the formula $\psi(s) = \phi(s - \zeta)$.

?▶Exercise 32.11: *Prove variable scaling by any fixed nonzero real value is a bounded operation on $\mathcal{G}$ (part 6 of theorem 32.18).*

Additional Exercises

32.12. *Without using the results from section 32.4, determine which of the following are always Gaussian test functions whenever ϕ is a Gaussian test function. Assume α is an arbitrary positive constant and λ is an arbitrary complex constant.*

a. $e^{\lambda x}\phi(x)$ **b.** $\sin(\alpha x)\phi(x)$

c. $\cos(\alpha x)\phi(x)$ **d.** $\text{sinc}(\alpha x)\phi(x)$

e. $f\phi$ where f is any polynomial   **f.** $x^{-1}\phi(x)$

g. $\sinh(\alpha x)\phi(x)$ **h.** $e^{x^2}\phi(x)$

i. $\phi(3x)$ **j.** $\phi(ix)$

k. $\text{Re}[\phi]$, the real part of ϕ **l.** $\text{Im}[\phi]$, the imaginary part of ϕ

32.13. *Give some examples of functions that are not simple multipliers for the Gaussian test functions. In other words, describe various choices for a function h such that $h\phi$ is not a Gaussian test function for some Gaussian test function ϕ.*

32.14. Suppose f is an analytic function on $\mathbb{C}$. For each pair of real values x and y, let

$$u(x, y) \;=\; \text{Re}[f(x + iy)] \qquad \text{and} \qquad v(x, y) \;=\; \text{Im}[f(x + iy)] \quad .$$

a. Observe that, by the definition of u and v,

$$f(z) \;=\; u(x, y) + iv(x, y) \qquad \text{where} \quad z = x + iy \quad .$$

What are the corresponding formulas for $f(z^*)$, $f^*(z)$ and $f^*(z^*)$?

b. Using these formulas and either the Cauchy–Riemann equations or the test described in theorem 6.1 on page 67, verify that the analytic conjugate of f is analytic.

c. Finish the proof of lemma 32.10 on page 527.

32.15. Let ψ and ϕ be two Gaussian test functions. Using the claims from lemma 32.4 on page 519 and the observations made regarding complex conjugates of test functions (as in lemma 32.11 on page 528), verify that each of the following is a Gaussian test function:

a. $\psi * \phi$ **b.** $\psi \star \phi$ **c.** $\psi^* \star \phi$

32.16. Assume f is a piecewise continuous, absolutely integrable function on $\mathbb{R}$, and ψ and ϕ are two Gaussian test functions. Verify the following:

a. $f * \psi$ is a bounded, continuous function on $\mathbb{R}$.

b. $\displaystyle\int_{-\infty}^{\infty} \left[f * \psi(x)\right]\phi(x)\,dx \;=\; \int_{-\infty}^{\infty} f(x)\left[\psi^* \star \phi(x)\right] dx \,.$

(The next exercise is a more ambitious version of this exercise. The results from either can be applied later in an exercise on generalizing convolution.)

32.17. Assume f is a piecewise continuous, exponentially integrable function on $\mathbb{R}$, and ψ and ϕ are two Gaussian test functions. Using the bounds described in lemma 32.12 on page 528, verify the following:

a. $f * \psi$ is a continuous and exponentially bounded function on $\mathbb{R}$.

b. $\displaystyle\int_{-\infty}^{\infty} \left[f * \psi(x)\right]\phi(x)\,dx \;=\; \int_{-\infty}^{\infty} f(x)\left[\psi^* \star \phi(x)\right] dx \,.$

32.18. Let $\alpha > 0$, and let ϕ be in $\mathcal{G}$. Show that

$$\lim_{\epsilon \to 0^+} \|\phi - u_\epsilon\phi\|_\alpha \;=\; 0$$

assuming:

a. $u_\epsilon(x) = e^{-\epsilon x^2}$.

b. $u_\epsilon(x) = u(\epsilon x)$ where u is any simple multiplier satisfying $u(0) = 1$.

33

Generalized Functions

We are now ready to develop the basic elements of a generalized function theory that will support, in a reasonably natural fashion, a much more general way of defining Fourier transforms. Naturally, we do not want this new theory to be something completely different from what we already know. It should extend the well-known classical theory of functions, and should, in spirit and notation, be similar enough to the classical theory that we can begin using the new theory with a minimum of psychological pain. This gives us the following design parameters for our new theory:

1.   The set of generalized functions should contain all classical functions of interest (in our case, all exponentially integrable functions).

2.   The basic familiar operations on functions — addition, multiplication, translation, differentiation and so forth — should extend in a fairly natural way to corresponding operations with the generalized functions.

3.   The generalized theory must agree with the classical theory whenever the classical theory can be applied. For example, the generalized Fourier transform of $e^{-|x|}$ must be the same as the classical Fourier transform of this function.

It turns out that generalized functions can be viewed as special cases of entities known as "functionals". Accordingly, our exposition begins with a definition and brief discussion of functionals. Then, after modifying our notation slightly and imposing one or two new requirements, we will be able to say exactly what generalized functions are and begin describing the basic manipulations we can perform with them.

33.1   Functionals
Functions of Test Functions

Go back to the observation made at the end of chapter 31. That observation (slightly rephrased in light of our discussion of Gaussian test functions) was that any exponentially integrable function f is completely determined by the values of all integrals of the form

$$\int_{-\infty}^{\infty} f(x)\phi(x)\,dx \tag{33.1}$$

where ϕ is a Gaussian test function. Here is an expression in which we are keeping f fixed and considering values resulting from various choices of ϕ. In other words, ϕ is the "variable" in this expression. So, if we define Γ by

$$\Gamma[\phi] \;=\; \int_{-\infty}^{\infty} f(x)\phi(x)\,dx \qquad \text{for each } \phi \text{ in } \mathscr{G} \quad,$$

then Γ is a "complex-valued function of Gaussian test functions". Mathematicians refer to such things, which treat test functions as input and yield complex values as output, as functionals. That is, a *functional* Γ (on the test function space $\mathscr{G}$) is something that determines a particular complex value for each ϕ in $\mathscr{G}$. For now, we will let $\Gamma[\phi]$ represent the "value of Γ at ϕ". Soon, though, we will adopt an alternative notation that will better serve us.

Let's look at three particularly important sets of functionals on $\mathscr{G}$.

!▶ **Example 33.1 (functionals defined by functions):**   Let f be an exponentially integrable function on $\mathbb{R}$. The corresponding functional Γ_f is defined by

$$\Gamma_f[\phi] \;=\; \int_{-\infty}^{\infty} f(x)\phi(x)\,dx \qquad \text{for each } \phi \text{ in } \mathscr{G} \quad.$$

In particular, for $f(x) = 2x + 3$, we have the functional Γ_{2x+3} where, for each Gaussian test function ϕ,

$$\Gamma_{2x+3}[\phi] \;=\; \int_{-\infty}^{\infty} (2x+3)\phi(x)\,dx \quad.$$

?▶ **Exercise 33.1:**   Let Γ_{2x+3} be as above and verify each of the following:

   a: $\Gamma_{2x+3}[\phi] = 3\sqrt{\pi}$ when $\phi(x) = e^{-x^2}$

   b: $\Gamma_{2x+3}[\phi] = 5\sqrt{\pi}$ when $\phi(x) = 2e^{-4(x-1)^2}$

   c: $\Gamma_{2x+3}[\phi] = 8\sqrt{\pi}$ when $\phi(x) = e^{-x^2} + 2e^{-4(x-1)^2}$

!▶ **Example 33.2 (evaluation functionals):**   Let a be some complex number. The "evaluation at a" functional E_a is given by

$$E_a[\phi] \;=\; \phi(a) \qquad \text{for each } \phi \text{ in } \mathscr{G} \quad.$$

In particular, we have the "evaluation at 0" functional E_0 given by

$$E_0[\phi] \;=\; \phi(0) \qquad \text{for each } \phi \text{ in } \mathscr{G} \quad.$$

?▶ **Exercise 33.2:**   Let E_a be as above and $\phi(x) = e^{-x^2}$. Verify each of the following:

   a: $E_0[\phi] \;=\; 1$ **b:** $E_1[\phi] \;=\; e^{-1}$ **c:** $E_i[\phi] \;=\; e$

!▶ **Example 33.3 (the norm functional):**   The norm functional N is the (square-integral) norm introduced in chapter 11 for functions on the real line. That is,

$$N[\phi] \;=\; \|\phi\| \;=\; \left[\int_{-\infty}^{\infty} |\phi(x)|^2\,dx\right]^{1/2} \qquad \text{for each } \phi \text{ in } \mathscr{G} \quad.$$

?▶ **Exercise 33.3:**   Let N be as above and verify each of the following:

   a: $N[\phi] = \dfrac{1}{\sqrt{2}}$ when $\phi(x) = e^{-2\pi x^2}$ **b:** $N[\phi] = \dfrac{1}{\sqrt{2}}$ when $\phi(x) = -e^{-2\pi x^2}$

In the next section, we will see that our main interest is in those functionals that, in some loose sense, can be thought of as being described by

$$\Gamma[\phi] = \int_{-\infty}^{\infty} f(x)\phi(x)\,dx$$

for "some function-like thing f". (Those "function-like things" will be our "generalized functions".) To exclude those functionals that cannot be viewed in this manner, we will impose two additional conditions: linearity and functional continuity.

Linearity for Functionals

As you probably guessed, a functional Γ on $\mathscr{G}$ is said to be *linear* if and only if

$$\Gamma[\alpha\phi + \beta\psi] = \alpha\Gamma[\phi] + \beta\Gamma[\psi]$$

for every pair of Gaussian test functions ϕ and ψ, and every pair of constants α and β.

Let's check for linearity in the previous examples.

!►Example 33.4:   Let f be any fixed exponentially integrable function on $\mathbb{R}$, and let Γ_f be the corresponding functional as defined in example 33.1. Given any two Gaussian test functions ϕ and ψ, and any two constants α and β, we see that

$$\Gamma_f[\alpha\phi + \beta\psi] = \int_{-\infty}^{\infty} f(x)[\alpha\phi(x) + \beta\psi(x)]\,dx$$

$$= \alpha\int_{-\infty}^{\infty} f(x)\phi(x)\,dx + \beta\int_{-\infty}^{\infty} f(x)\psi(x)\,dx$$

$$= \alpha\Gamma_f[\phi] + \beta\Gamma_f[\psi] \quad.$$

So all the functionals described in example 33.1 are linear.

!►Example 33.5:   Let a be some fixed complex number, and let E_a be the corresponding evaluation at a functional defined in example 33.2. Given any two Gaussian test functions ϕ and ψ, and any two constants α and β, we see that

$$E_a[\alpha\phi + \beta\psi] = \alpha\phi(a) + \beta\psi(a) = \alpha E_a[\phi] + \beta E_a[\psi] \quad.$$

So all the evaluation functionals described in example 33.2 are linear.

!►Example 33.6:   Let N be the norm functional, $N[\phi] = \|\phi\|$. From exercise 33.3, we know

$$N\left[e^{-2\pi x^2}\right] + N\left[-e^{-2\pi x^2}\right] = \frac{1}{\sqrt{2}} + \frac{1}{\sqrt{2}} = \sqrt{2} \quad.$$

However,

$$N\left[e^{-2\pi x^2} + \left(-e^{-2\pi x^2}\right)\right] = N[0] = \left[\int_{-\infty}^{\infty} |0|^2\,dx\right]^{1/2} = 0 \quad.$$

So

$$N\left[e^{-2\pi x^2} + \left(-e^{-2\pi x^2}\right)\right] \neq N\left[e^{-2\pi x^2}\right] + N\left[-e^{-2\pi x^2}\right] \quad,$$

showing us that the norm functional is not linear (and will not correspond to a generalized function).

Continuity for Functionals

The basic idea of functional continuity is simple. A functional Γ is *(functionally) continuous* if and only if the values $\Gamma[\phi]$ and $\Gamma[\psi]$ are almost the same whenever the two test functions ϕ and ψ are "almost the same". What is not so simple is determining just what is meant by "the two test functions are 'almost the same' ". Fortunately, functional continuity need not be a big issue for us, and most of our work can be done using a relatively naive understanding of functional continuity. As long as we take a few simple precautions (to be described later), it is fairly safe to assume that the functionals naturally arising in applications are functionally continuous. Imposing functional continuity will mean that we cannot easily use classical functions that are not exponentially integrable; in practice, though, this is not a serious limitation. On the other hand, requiring functional continuity will allow us to compute derivatives of certain expressions in a natural manner (to be discussed in section 33.5), and will allow us to avoid some of those pathologies dreamt up by mathematicians simply to show that they can dream up strange things (see, for example, exercise 33.25 at the end of this chapter).

For those who are interested, and for those few derivations requiring it, a precise definition of functional continuity is given later in this chapter (section 33.6). For now, though, we will accept the statement

> A functional Γ is (functionally) continuous if and only if $\Gamma[\phi]$ and $\Gamma[\psi]$ are almost the same whenever ϕ and ψ are almost the same Gaussian test functions

as a working definition for functional continuity, and we will not worry about the precise meanings of "almost the same" in this definition.

Along these lines, it should be stated that all the functionals described thus far — excluding the pathology in exercise 33.25 — are functionally continuous. This will be verified in section 33.6.

33.2   Generalized Functions
The Generalized Function Notation for Functionals

It will greatly simplify our work if we adopt a notation for "the value $\Gamma[\phi]$ " that looks like shorthand notation for

$$\int_{-\infty}^{\infty} f(x)\phi(x)\,dx$$

for some fixed f. Accordingly, to each functional Γ on $\mathcal{G}$ we will associate some symbol, say, f, and let f denote that functional with $\langle f, \phi \rangle$ denoting "the value of Γ at ϕ". That is,

$$\langle f, \phi \rangle = \Gamma[\phi] \quad .$$

This notation will be most convenient when Γ is both linear and functionally continuous. If and only if this is the case, then f, as just defined, will be called a *generalized function*.

In other words, a generalized function f is a continuous, linear functional on the space of Gaussian test functions with "the value of f at a test function ϕ " being denoted by $\langle f, \phi \rangle$ instead of $f[\phi]$. This "generalized function notation" will allow us to describe the algebra, calculus and Fourier analysis of continuous, linear functionals so that these theories appear similar to the well-known classical theories. Indeed, to some extent, the generalized function notation will help hide the fact that we are actually dealing with function**als** instead of functions.

Looking back, you will find that we have already described two sets of continuous, linear functionals on $\mathcal{G}$: the "evaluation functionals" defined in example 33.2 and the functionals corresponding to exponentially integrable functions defined in example 33.1. (The linearity of these functionals was confirmed in examples 33.4 and 33.5, and, for now, you are trusting the author that they are functionally continuous.) These will be rather important sets of functionals for us. So, let us take a second look at these functionals and describe the standard generalized function notation for them.

Evaluation Functionals and Delta Functions

Consider E_a, the evaluation at a functional for some point a on the complex plane. As a generalized function, E_a is usually denoted by δ_a and is called the *delta (generalized) function* at a. In other words, the delta function at a is that generalized function δ_a such that

$$\langle \delta_a , \phi \rangle = E_a[\phi] = \phi(a) \qquad \text{for each } \phi \text{ in } \mathscr{G} \quad .$$

Each δ_a is commonly referred to as a delta function. It is also standard practice to distinguish the delta function at zero as being "the" delta function and to write it as δ instead of δ_0. So, for each Gaussian test function ϕ,

$$\langle \delta , \phi \rangle = \phi(0) \quad .$$

?►Exercise 33.4: Let $\phi(x) = e^{-3x^2}$ and verify each of the following:

   a: $\langle \delta_2 , \phi \rangle = e^{-12}$ **b:** $\langle \delta , \phi \rangle = 1$ **c:** $\langle \delta_i , \phi \rangle = e^3$

Classical Functions and Generalized Functions
Classical versus Generalized Functions

Since we will be dealing with both generalized functions and classical functions, it may be prudent to review the difference between these two types of entities.

A generalized function is a continuous, linear functional. If f denotes a generalized function, then its domain is the space of test functions $\mathscr{G}$, and a formula for f is a formula for computing $\langle f , \phi \rangle$ for each Gaussian test function ϕ. For example, a formula for δ_2 is

$$\langle \delta_2 , \phi \rangle = \phi(2) \qquad \text{for each } \phi \text{ in } \mathscr{G} \quad .$$

What you have been calling a function since calculus is what we will start referring to as a *classical function*. Thus, if f denotes a classical function, then its domain is some set of numbers (in this book, $\mathbb{R}$ unless otherwise specified), and a formula for f is the formula for computing $f(x)$ for each number x in the domain.

Where necessary, a given "f" will be identified as being a generalized function or a classical function either by using the adjective "generalized" or "classical", or by reasonable clues as to its nature. For example, if you see something like "let f be a continuous function on $\mathbb{R}$", then you should realize that f is a classical function because you've just been told that its domain is $\mathbb{R}$ and not $\mathscr{G}$. On the other hand, if a certain entity is just referred to as "a function" (e.g., the "delta function"), then it will be safer to assume it is a generalized function.

Fortunately, in practice, it is often unnecessary to carefully distinguish between classical and generalized functions. This is due to the association between certain classical functions and corresponding generalized functions described next.

Classical Functions as Generalized Functions

We've seen that, if f is any exponentially integrable function on the real line, then Γ_f, given by

$$\Gamma_f[\phi] = \int_{-\infty}^{\infty} f(x)\phi(x)\,dx \qquad \text{for each } \phi \text{ in } \mathcal{G} \quad,$$

is a continuous, linear functional on $\mathcal{G}$. When a functional is generated in this fashion, it is standard practice to use the same symbol to denote the corresponding generalized function as denoted the original classical function (in this case, "f"). Thus, using our generalized function notation, we write

$$\langle f, \phi \rangle = \int_{-\infty}^{\infty} f(x)\phi(x)\,dx \qquad \text{for each } \phi \text{ in } \mathcal{G} \tag{33.2}$$

where the "f" on the left denotes a generalized function (i.e., a continuous, linear functional on $\mathcal{G}$) and the "f" on the right denotes the original classical function.

Any formula $f(x)$ for the classical function f will also be used to denote the corresponding generalized function. Keep in mind, however, that the formula for the corresponding generalized function is not the classical formula $f(x)$ but, instead, is the formula for computing $\langle f, \phi \rangle$,

$$\langle f, \phi \rangle = \langle f(x), \phi(x) \rangle = \int_{-\infty}^{\infty} f(x)\phi(x)\,dx \quad .$$

Using our generalized function notation, example 33.1 and exercise 33.1 become:

!▶**Example 33.7:**   *Let f be the classical function given by the formula*

$$f(x) = 2x + 3$$

where x is any real value. The corresponding generalized function, also denoted by either f or $2x + 3$, is given by the formula

$$\langle f, \phi \rangle = \langle 2x + 3, \phi(x) \rangle = \int_{-\infty}^{\infty} (2x + 3)\phi(x)\,dx$$

where ϕ is any Gaussian test function.

?▶**Exercise 33.5:**   *Let f be the function $f(x) = 2x + 3$, and verify the following:*

a: $\langle f, \phi \rangle = 3\sqrt{\pi}$ *when* $\phi(x) = e^{-x^2}$

b: $\langle 2x + 3, 2e^{-4(x-1)^2} \rangle = 5\sqrt{\pi}$

c: $\langle f, \phi \rangle = 8\sqrt{\pi}$ *when* $\phi(x) = e^{-x^2} + 2e^{-4(x-1)^2}$

The use of the same notation for functions and their corresponding generalized functions underscores the fact that our generalized theory should be equivalent to the well-known classical theory whenever the classical theory applies. It also supports the common practice of "identifying" exponentially integrable functions with the generalized functions they generate; that is, in practice, one often does not explicitly distinguish between the classical, exponentially integrable function f and the corresponding generalized function f defined by equation (33.2). Under this identification, the set of all exponentially integrable functions is treated as a subset of the set of all generalized functions. While this is often a convenient practice, it is not without its pitfalls. There are occasions where it is important to distinguish between "f as a classical function" and "f as a generalized function". Still, this practice can simplify discussion and does reduce the need for extra symbols, and so, with some trepidation on the part of the author, we will adopt it.

Convention (identifying exponentially integrable functions with generalized functions)
Exponentially integrable functions on the real line are automatically identified with their correspond-
ing generalized functions. That is, if f is an exponentially integrable function on $\mathbb{R}$, then f and
its formula, $f(x)$, are also viewed as the generalized function given by

$$\langle\, f , \phi \,\rangle \;=\; \langle\, f(x) , \phi(x) \,\rangle \;=\; \int_{-\infty}^{\infty} f(x)\phi(x)\,dx \qquad \text{for each } \phi \text{ in } \mathcal{G}$$

where the integral is computed using the formula for the classical function f.

Constants

Most people would agree that constant functions are the simplest things that can be viewed as
generalized functions. Of course, the two most noteworthy are the zero function, 0, and the unit
function, 1. As generalized functions, they are defined by

$$\langle\, 0 , \phi \,\rangle \;=\; \int_{-\infty}^{\infty} 0 \cdot \phi(x)\,dx \;=\; 0$$

and

$$\langle\, 1 , \phi \,\rangle \;=\; \int_{-\infty}^{\infty} 1 \cdot \phi(x)\,dx \;=\; \int_{-\infty}^{\infty} \phi(x)\,dx$$

where ϕ denotes an arbitrary Gaussian test function. We might as well also observe that, if c is any
constant and ϕ is any Gaussian test function, then

$$\langle\, c , \phi \,\rangle \;=\; \int_{-\infty}^{\infty} c \cdot \phi(x)\,dx \;=\; c\int_{-\infty}^{\infty} \phi(x)\,dx \;=\; c\langle\, 1 , \phi \,\rangle \quad .$$

?▶Exercise 33.6: Let $\phi(x) = e^{-x^2}$, *and verify each of the following:*

    a: $\langle\, 0 , \phi \,\rangle = 0$ **b:** $\langle\, 1 , \phi \,\rangle = \sqrt{\pi}$ **c:** $\langle\, 3 , \phi \,\rangle = 3\sqrt{\pi}$

The Importance of Exponential Integrability

There are good reasons for limiting the above discussion to classical functions that are exponentially
integrable. For one thing, it ensures that the integrals remain finite.

!▶Example 33.8: *Consider trying to define a generalized function f by*

$$\langle\, f , \phi \,\rangle \;=\; \int_{-\infty}^{\infty} f(x)\phi(x)\,dx \tag{33.3}$$

*when f is the classical function $f(x) = x^{-2}$. In particular, consider evaluating $\langle\, f , \phi \,\rangle$ for
the Gaussian test function $\phi(x) = \exp(-x^2)$. Since the functions are all positive and even,*

$$\left\langle\, x^{-2} , e^{-x^2} \,\right\rangle \;=\; \int_{-\infty}^{\infty} x^{-2}e^{-x^2}\,dx \;\geq\; \int_{0}^{1} x^{-2}e^{-x^2}\,dx \;\geq\; e^{-1}\int_{0}^{1} x^{-2}\,dx \quad .$$

*Unfortunately, this last integral is infinite. So equation (33.3) does not define a legitimate func-
tional on $\mathcal{G}$ when $f(x) = x^{-2}$, much less one which is continuous and linear. Consequently,
equation (33.3) does not give us a generalized function corresponding to the classical function
x^{-2}. (There are generalized function analogs to classical functions like x^{-2}. These analogs, the
"pole functions", will be developed in chapter 38.)*

Even if the integrals are all well defined and finite, you can be pretty certain that the formula

$$\langle f , \phi \rangle = \int_{-\infty}^{\infty} f(x)\phi(x)\,dx \tag{33.4}$$

does *not* define f as a generalized function when f is a classical function that is not exponentially integrable. The reason is somewhat subtle — the functional won't be functionally continuous. There will be ϕ's such that the $\langle f , \phi \rangle$'s are relatively large even though the ϕ's are "small" Gaussian test functions. (For an illustration of this, see exercise 33.26 at the end of this chapter.)

Consequently, we will use equation (33.4) to identify a classical function with a generalized function if and only if the classical function is exponentially integrable. This makes it important to be able to distinguish functions that are exponentially integrable from those that are not. Sometimes, this is easy. Other times, we may find the following test helpful.

Lemma 33.1 (test for exponential integrability)
Let f be a classical function which is at least continuous on some partitioning of the real line, and let ℓ be any finite length. Then f is exponentially integrable if and only if there are finite real constants M and β such that

$$\int_{x}^{x+\ell} |f(s)|\,ds \ \leq \ Me^{\beta|x|} \qquad \text{for all} \quad -\infty < x < \infty \quad . \tag{33.5}$$

PROOF: First, assume f is exponentially integrable. Then, by definition, there is a finite positive β such that $f(x)e^{-\beta|x|}$ is absolutely integrable. Noting that $|s| \leq |x|+\ell$ whenever $x \leq s \leq x+\ell$, we see that

$$\int_{x}^{x+\ell} |f(s)|\,ds \ = \ \int_{x}^{x+\ell} |f(s)|\,e^{-\beta|s|}e^{\beta|s|}\,ds$$

$$\leq \ \int_{x}^{x+\ell} |f(s)|\,e^{-\beta|s|}e^{\beta(|x|+\ell)}\,ds \ \leq \ \left[\int_{-\infty}^{\infty} |f(s)|\,e^{-\beta|s|}\,ds\right] e^{\beta\ell}\,e^{\beta|x|} \quad ,$$

which gives inequality (33.5) once we've cut out the middle and let

$$M \ = \ e^{\beta\ell}\int_{-\infty}^{\infty} |f(s)|\,e^{-\beta|s|}\,ds \quad .$$

Now assume inequality (33.5) holds, and observe that

$$\int_{0}^{\infty} |f(s)|\,e^{-(\beta+1)|s|}\,ds \ = \ \sum_{k=0}^{\infty} \int_{k\ell}^{k\ell+\ell} |f(s)|\,e^{-(\beta+1)s}\,ds$$

$$\leq \ \sum_{k=0}^{\infty} \int_{k\ell}^{k\ell+\ell} |f(s)|\,e^{-(\beta+1)k\ell}\,ds$$

$$= \ \sum_{k=0}^{\infty} e^{-(\beta+1)k\ell} \int_{k\ell}^{k\ell+\ell} |f(s)|\,ds \quad .$$

Using inequality (33.5), we then get

$$\int_{0}^{\infty} |f(s)|\,e^{-(\beta+1)|s|}\,ds \ \leq \ \sum_{k=0}^{\infty} e^{-(\beta+1)k\ell}\,Me^{\beta k\ell} \ = \ M\sum_{k=0}^{\infty} e^{-k\ell} \ = \ M\sum_{k=0}^{\infty} \left[e^{-\ell}\right]^{k} \quad .$$

But this last summation is just a geometric series that we know converges since $e^{-\ell} < 1$, (see example 4.1 on page 42). So this last sequence of inequalities reduces to

$$\int_0^\infty |f(s)| \, e^{-(\beta+1)|s|} \, ds \; < \; \infty \; .$$

By similar arguments, we also have

$$\int_{-\infty}^0 |f(s)| \, e^{-(\beta+1)|s|} \, ds \; < \; \infty \; .$$

Thus, with $\alpha = \beta + 1$,

$$\int_{-\infty}^\infty |f(x)| \, e^{-\alpha|x|} \, dx \; = \; \int_{-\infty}^0 |f(s)| \, e^{-\alpha|s|} \, ds \; + \; \int_0^\infty |f(s)| \, e^{-\alpha|s|} \, ds \; < \; \infty \; ,$$

confirming that f is exponentially integrable. ∎

?▶Exercise 33.7: *Let γ denote a real number. Verify that x^γ is exponentially integrable if and only if $\gamma > -1$.*

More on Defining and Denoting Generalized Functions
Defining Generalized Functions

In practice, a generalized function f is rarely defined as "the generalized function representation of a linear functional Γ". Instead, we usually define f directly by describing how to evaluate $\langle f, \phi \rangle$ for each Gaussian test function ϕ. For example, rather than defining the evaluation at zero functional, E_0, and then defining the delta function as the generalized function corresponding to E_0, we can simply define the delta function δ to be the generalized function satisfying

$$\langle \delta, \phi \rangle \; = \; \phi(0) \qquad \text{for each } \phi \text{ in } \mathcal{G} \; .$$

For brevity, we may even say something like "the delta function δ is the generalized function given by

$$\langle \delta, \phi \rangle \; = \; \phi(0) \qquad \text{"};$$

with the understanding that this formula describes how to compute $\langle \delta, \phi \rangle$ for each ϕ in $\mathcal{G}$.

Of course, we should verify that any given formula for $\langle f, \phi \rangle$ truly defines f as a generalized function. Strictly speaking, this means verifying that the functional given by that formula for $\langle f, \phi \rangle$ is both functionally continuous and linear. Those who read section 33.6, however, will realize that each linear functional we encounter outside of exercises 33.25 and 33.26 is functionally continuous. So they do not need to worry about verifying functional continuity. And those who do not read section 33.6 will not have the tools to verify functional continuity. They will just have to trust those who read that section (and he who wrote it) and accept the fact that, in this text and in "real-world applications", there is little need to verify functional continuity; it usually can (and will) be safely assumed.

As a result, as far as we are concerned, verifying that a given f is a generalized function amounts to verifying, with $\Gamma[\phi] = \langle f, \phi \rangle$, that

$$\Gamma[\alpha\phi + \beta\psi] \; = \; \alpha\Gamma[\phi] \; + \; \beta\Gamma[\psi]$$

for each pair of Gaussian test functions ϕ and ψ, and each pair of complex numbers α and β. In terms of f and our generalized function notation, the above equation is

$$\langle f, \alpha\phi + \beta\psi \rangle = \alpha\langle f, \phi \rangle + \beta\langle f, \psi \rangle \quad . \tag{33.6}$$

Thus, when defining a generalized function f, we must assure ourselves one way or another that equation (33.6) holds for each pair of Gaussian test functions ϕ and ψ, and each pair of complex numbers α and β.

!▶*Example 33.9:*   *Let's try to define two generalized functions f and g by stating that*

$$\langle f, \phi \rangle = \phi'(0) \qquad and \qquad \langle g, \phi \rangle = |\phi(0)|^2$$

for each Gaussian test function ϕ. We can verify f is a generalized function by simply noting that, if ϕ and ψ are any two Gaussian test functions, and α and β are any two complex numbers, then

$$\langle f, \alpha\phi + \beta\psi \rangle = \frac{d}{dx}\left[\alpha\phi(x) + \beta\psi(x)\right]\Big|_{x=0}$$

$$= \alpha\phi'(0) + \beta\psi'(0) = \alpha\langle f, \phi \rangle + \beta\langle f, \psi \rangle \quad .$$

On the other hand, the above does not define g as a generalized function. To show this, it suffices to show that

$$\langle g, \alpha\phi + \beta\psi \rangle \neq \alpha\langle g, \phi \rangle + \beta\langle g, \psi \rangle$$

for some choice of α, β, ϕ and ψ. In particular, using $\alpha = 2$, $\beta = 0$, $\phi(x) = e^{-x^2}$ and $\psi(x) = 0$, we see that

$$\langle g, 2\phi \rangle = \left|2e^{0^2}\right|^2 = 4 \qquad while \qquad 2\langle g, \phi \rangle = 2\left|e^{0^2}\right|^2 = 2 \quad .$$

So,

$$\langle g, 2\phi \rangle \neq 2\langle g, \phi \rangle \quad ,$$

and we cannot treat g as a generalized function.

Do keep in mind that, strictly speaking, the formula for a generalized function f is the formula for computing $\langle f, \phi \rangle$ for each ϕ in $\mathcal{G}$. This fact is implicit whenever reference is made to any formula involving a generalized function whether or not the formula for computing $\langle f, \phi \rangle$ is explicitly stated. For example, if f is a generalized function and "$f(x) = \sin(x)$", is stated, then it is to be understood that the statement is shorthand for "f is the generalized function satisfying

$$\langle f, \phi \rangle = \int_{-\infty}^{\infty} \sin(x)\,\phi(x)\,dx \qquad \text{for each } \phi \text{ in } \mathcal{G} \qquad ".$$

More on Notation

Because of our discussion so far, you may be viewing $\langle f, \phi \rangle$ as shorthand for a "generalized integral" over the real line of the product $f\phi$, even when f is not a true function. Heuristically, that is a good way to view $\langle f, \phi \rangle$. In fact, many authors write

$$\int_{-\infty}^{\infty} f(x)\phi(x)\,dx \qquad \text{instead of} \qquad \langle f, \phi \rangle$$

even when the generalized function f does not correspond to a classical function. That is why you often find the delta function defined to be the function δ such that, "for each suitable ϕ",

$$\int_{-\infty}^{\infty} \delta(x)\phi(x)\,dx \; = \; \phi(0) \quad ,$$

even though, as we saw in chapter 31 (and chapter 26), there is no such classical function.

Along these same lines, it is fairly common to "attach" a variable to a generalized function — writing $\delta(x)$ instead of δ, for example — even though there is no formula of x describing that generalized function. The x (or whatever symbol is used) is employed as a dummy variable to help the reader keep track of any manipulations being done and the relations between the various generalized functions present. To give an example (which will make more sense after we discuss products), the phrase "consider $x\delta$" could mean that we want to consider the product of the delta function with some yet unknown function being denoted by x. But if we say "consider $x\delta(x)$", then it is clear that we really mean "consider the product of the function $f(x) = x$ with the delta function."

Be warned, however, that the practices of writing

$$\int_{-\infty}^{\infty} f(x)\phi(x)\,dx \qquad \text{instead of} \qquad \langle\, f\,,\phi\,\rangle$$

and of attaching a variable to a generalized function are not without their dangers. They can lull the unwary into treating something like the delta function as a true function, and that, taken too seriously, can lead to the derivation of such nonsense as $0 = 1$ (as in chapter 31).

Since "attaching variables" will help simplify and clarify some of our discussions, we will do so — but only when it does help, and always keeping in mind that these are dummy variables. However, we will not explicitly write $\langle\, f\,,\phi\,\rangle$ as an integral unless f truly is a classical function on the real line.

33.3 Basic Algebra of Generalized Functions
Equality

Remember, two generalized functions f and g are completely defined as soon as we know how to find the values of $\langle\, f\,,\phi\,\rangle$ and $\langle\, g\,,\phi\,\rangle$ for each Gaussian test function ϕ. That being the case, we naturally say f and g are equal, and write $f = g$, if and only if

$$\langle\, f\,,\phi\,\rangle = \langle\, g\,,\phi\,\rangle \qquad \text{for every } \phi \text{ in } \mathcal{G}\,.$$

!▶ *Example 33.10:* Let f *be a generalized function and suppose*

$$\langle\, f\,,\phi\,\rangle = -\int_{0}^{\infty} \phi'(x)\,dx$$

for each Gaussian test function ϕ. *Note that, for any Gaussian test function* ϕ,

$$\langle\, f\,,\phi\,\rangle = -\int_{0}^{\infty} \phi'(x)\,dx = -\phi(x)\Big|_{0}^{\infty} = \phi(0) = \langle\, \delta\,,\phi\,\rangle \quad .$$

Thus, $f = \delta$.

?▶Exercise 33.8:　Let f be the generalized function satisfying

$$\langle f, \phi \rangle = -\int_0^\infty x\phi'(x)\,dx$$

for each Gaussian test function ϕ. Show that $f = \text{step}$. (Try evaluating the above integral using integration by parts.)

We should observe that, if f and g are two exponentially integrable functions, then the "test function test for equality" in theorem 32.5 on page 520 tells us that f and g are the same classical function if and only if

$$\int_{-\infty}^\infty f(x)\phi(x)\,dx = \int_{-\infty}^\infty g(x)\phi(x)\,dx \qquad \text{for every } \phi \text{ in } \mathscr{G}.$$

This last line is equivalent to

$$\langle f, \phi \rangle = \langle g, \phi \rangle \qquad \text{for every } \phi \text{ in } \mathscr{G},$$

which, by definition of generalized equality, is true if and only if f and g are equal as generalized functions. Thus, whenever f and g are exponentially integrable functions,

$$f = g \quad \text{as classical functions} \qquad \Longleftrightarrow \qquad f = g \quad \text{as generalized functions}　.$$

In other words, our concept of equality for generalized functions reduces to the classical concept whenever the classical concept makes sense.

Addition

Suppose, for the moment, f and g are exponentially integrable functions. From basic calculus we know

$$\int_{-\infty}^\infty [f(x) + g(x)]\,\phi(x)\,dx = \int_{-\infty}^\infty f(x)\phi(x)\,dx + \int_{-\infty}^\infty g(x)\phi(x)\,dx$$

whenever ϕ is a Gaussian test function. Using the generalized function notation, this equation is

$$\langle f + g, \phi \rangle = \langle f, \phi \rangle + \langle g, \phi \rangle \qquad \text{for each } \phi \text{ in } \mathscr{G}　. \tag{33.7}$$

Here, $f + g$ is the classical sum of f and g; that is, $f + g$ is the classical function whose value at each real x is given by $f(x) + g(x)$.

Because it tells us how to compute $\langle f + g, \phi \rangle$ for each Gaussian test function ϕ, equation (33.7) defines the generalized function corresponding to the classical sum of f and g. However, the right-hand side of this equation remains well defined when we replace the classical functions f and g with any two generalized functions f and g. So let's try using this equation to define "$f + g$" for any such pair of generalized functions. In other words, given any pair of generalized functions f and g, let's define the *generalized sum* of f and g, which we will cleverly denote by $f + g$, to be the generalized function satisfying

$$\langle f + g, \phi \rangle = \langle f, \phi \rangle + \langle g, \phi \rangle \qquad \text{for each } \phi \text{ in } \mathscr{G}　. \tag{33.8}$$

Of course, stating that the above defines $f + g$ as a generalized function doesn't make it so. We should verify that this defines a generalized function. As discussed a few pages ago, this means verifying that

$$\langle f + g, \alpha\phi + \beta\psi \rangle = \alpha\langle f + g, \phi \rangle + \beta\langle f + g, \psi \rangle$$

whenever α and β are any two constants, and ϕ and ψ are any two Gaussian test functions. To do so, first observe that, by the definition of the generalized sum and the fact that f and g are, themselves, generalized functions,

$$\langle f+g , \alpha\phi + \beta\psi \rangle = \langle f , \alpha\phi + \beta\psi \rangle + \langle g , \alpha\phi + \beta\psi \rangle$$
$$= [\alpha\langle f , \phi \rangle + \beta\langle f , \psi \rangle] + [\alpha\langle g , \phi \rangle + \beta\langle g , \psi \rangle] \quad .$$

The last line is just the sum of two linear combination of ordinary numbers. By elementary algebra, this line equals

$$\alpha[\langle f , \phi \rangle + \langle g , \phi \rangle] + \beta[\langle f , \psi \rangle + \langle g , \psi \rangle] \quad ,$$

which, according to our definition of the generalized sum, is

$$\alpha\langle f+g , \phi \rangle + \beta\langle f+g , \psi \rangle \quad .$$

So,

$$\langle f+g , \alpha\phi + \beta\psi \rangle = \alpha\langle f+g , \phi \rangle + \beta\langle f+g , \psi \rangle$$

whenever α and β are any two constants, and ϕ and ψ are any two Gaussian test functions. This verifies that $f+g$, as defined above, is a generalized function.

?►Exercise 33.9:   *Verify that, by the above definition,*

$$\left\langle \delta(x) + x^2 , \phi \right\rangle = \phi(0) + \int_{-\infty}^{\infty} x^2 \phi(x)\, dx$$

for each Gaussian test function ϕ.

Simple Multiplication

There are some very nontrivial issues concerning the multiplication of two arbitrary generalized functions. For that reason, we will hold off the general discussion until chapter 36 and concern ourselves here with products in which one of the factors is a simple multiplier, as defined in the previous chapter (see page 522). Recall that the set of simple multipliers includes all polynomials and exponential functions, as well as the Gaussian test functions themselves.

To see how we will define these products (and why one factor will be assumed to be a simple multiplier), consider the integral of the classical product of two exponentially integrable functions f and h multiplied by a Gaussian test function ϕ. Then, of course,

$$\int_{-\infty}^{\infty} [h(x)f(x)] \phi(x)\, dx = \int_{-\infty}^{\infty} [f(x)h(x)] \phi(x)\, dx$$
$$= \int_{-\infty}^{\infty} f(x) [h(x)\phi(x)]\, dx \quad .$$

If h is a simple multiplier, then the product $h\phi$ is also a Gaussian test function, and we can rewrite the above as

$$\langle hf , \phi \rangle = \langle fh , \phi \rangle = \langle f , h\phi \rangle \quad .$$

However, if h is not a simple multiplier, then $h\phi$ will not generally be a Gaussian test function, and we may run into problems in trying to use the last equation (especially if h is not even a classical function).

All of this suggests the following definition for simple products: Let f be any generalized function and h any simple multiplier. The *generalized products* hf and fh are defined, respectively, to be the generalized functions given by

$$\langle\, hf \,,\, \phi \,\rangle = \langle\, f \,,\, h\phi \,\rangle \qquad \text{and} \qquad \langle\, fh \,,\, \phi \,\rangle = \langle\, f \,,\, h\phi \,\rangle$$

for each ϕ in $\mathcal{G}$. (Note that, while we've given separate definitions for hf and fh, these definitions automatically ensure that $hf = fh$.)

The verification that this defines a generalized function is left to you.

?▶ Exercise 33.10: *Let h be any simple multiplier, and let f be any generalized function. Verify that the generalized products hf and fh are generalized functions by showing that*

$$\langle\, hf \,,\, \alpha\phi + \beta\psi \,\rangle = \alpha\langle\, hf \,,\, \phi \,\rangle + \beta\langle\, hf \,,\, \psi \,\rangle$$

and

$$\langle\, fh \,,\, \alpha\phi + \beta\psi \,\rangle = \alpha\langle\, fh \,,\, \phi \,\rangle + \beta\langle\, fh \,,\, \psi \,\rangle$$

for every pair of constants α and β, and every pair of Gaussian test functions ϕ and ψ.

From the discussion leading to our definition of generalized multiplication, it should be clear that the generalized product of h with f is equivalent to the classical product of h with f whenever h is a simple multiplier and f is a exponentially integrable function. This also takes care of a potential ambiguity in our definition when both f and h are simple multipliers. In this case we have two equations defining fh,

$$\langle\, fh \,,\, \phi \,\rangle = \langle\, f \,,\, h\phi \,\rangle \qquad \text{and} \qquad \langle\, fh \,,\, \phi \,\rangle = \langle\, h \,,\, f\phi \,\rangle$$

where ϕ is an arbitrary Gaussian test function. But since simple multipliers are classical functions, both of the above equations reduce to the same equation, namely

$$\langle\, fh \,,\, \phi \,\rangle = \int_{-\infty}^{\infty} f(x)h(x)\phi(x)\,dx \quad ,$$

assuring us that the two definitions for fh are completely equivalent and define the same generalized function as we would have gotten by just multiplying f and h as classical functions.

The simplest simple multipliers, of course, are the constants. Moreover, if A is a constant, f is a generalized function, and ϕ is any Gaussian test function, then, by the definition of the generalized product and the linearity of the functional,

$$\langle\, Af \,,\, \phi \,\rangle = \langle\, f \,,\, A\phi \,\rangle = A\langle\, f \,,\, \phi \,\rangle \quad .$$

In particular,

$$\langle\, 0f \,,\, \phi \,\rangle = 0\langle\, f \,,\, \phi \,\rangle = 0 = \langle\, 0 \,,\, \phi \,\rangle$$

and

$$\langle\, 1f \,,\, \phi \,\rangle = 1\langle\, f \,,\, \phi \,\rangle = \langle\, f \,,\, \phi \,\rangle \quad ,$$

assuring us that the expected equalities

$$0f = 0 \qquad \text{and} \qquad 1f = f$$

hold for any generalized function f.

Let's now look at some products in which the f is not a classical function. In the process, we will also derive an identity that will simplify many computations involving delta functions.

!▶Example 33.11 (products with delta functions):   Assume a is any point on the complex plane, and let δ_a be the delta function at a. If h is any simple multiplier, and ϕ is any Gaussian test function, then, by the definitions of generalized products and delta functions,

$$\left\langle h\delta_a , \phi \right\rangle = \left\langle \delta_a , h\phi \right\rangle = h(a)\phi(a) \quad .$$

But also, since $h(a)$ is a constant,

$$h(a)\phi(a) = h(a)\left\langle \delta_a , \phi \right\rangle = \left\langle h(a)\delta_a , \phi \right\rangle \quad .$$

Thus,

$$\left\langle h\delta_a , \phi \right\rangle = \left\langle h(a)\delta_a , \phi \right\rangle \qquad \text{for each } \phi \text{ in } \mathcal{G} \quad .$$

In other words, the generalized product of any simple multiplier h with a delta function at a point a is simply the value of h at a times the delta function,

$$h\delta_a = h(a)\delta_a \quad . \tag{33.9}$$

For example,

$$(x^2 - 1)\delta_3(x) = (3^2 - 1)\delta_3 = 8\delta_3(x)$$

and

$$e^x \delta_{i\pi}(x) = e^{i\pi}\delta_{i\pi}(x) = -\delta_{i\pi}(x) \quad .$$

?▶Exercise 33.11:   Verify each of the following:

   a: $x\delta(x) = 0$ **b:** $x^2\delta_3(x) = 9\delta_3(x)$ **c:** $\sin\left(\tfrac{1}{2}x\right)\delta_\pi(x) = \delta_\pi(x)$

Properties of Addition and Multiplication

In doing algebra with classical functions, most people freely employ all sorts of well-known properties (such as the associativity of addition) to simplify their work. Naturally, we want to know the extent to which these properties hold when dealing with generalized functions.[1]

Let's first verify that generalized addition is commutative. Let f and g be any two generalized functions. By the definition of $f + g$ and $g + f$, and the fact that the addition of complex numbers is commutative, we have

$$\left\langle f + g , \phi \right\rangle = \left\langle f , \phi \right\rangle + \left\langle g , \phi \right\rangle = \left\langle g , \phi \right\rangle + \left\langle f , \phi \right\rangle = \left\langle g + f , \phi \right\rangle$$

for each Gaussian test function ϕ. Hence, $f + g = g + f$, confirming the commutativity of generalized addition.

?▶Exercise 33.12:   Let f, g and h be any three generalized functions. Verify each of the following:

   a: $f + (g + h) = (f + g) + h$ (associativity of addition)

   b: $f + 0 = f$

---

[1] Admittedly, verifying all these properties for generalized functions is neither difficult, nor particularly exciting. Still, it must be done.

Let's now consider simple generalized multiplication.

We've already seen that the simple product is commutative, and that $0f = 0$ and $1f = f$ whenever f is a generalized function. To check whether the product is associative, let f be any generalized function, and let g and h be two simple multipliers. Since the classical product of any two simple multipliers is also a simple multiplier, we can form the generalized product $f(gh)$. It is given by

$$\langle f(gh), \phi \rangle = \langle f, (gh)\phi \rangle$$

where ϕ is any Gaussian test function. But, because the product $(gh)\phi$ is simply the classical product of the functions g, h and ϕ, we know $(gh)\phi = g(h\phi)$. So,

$$\langle f(gh), \phi \rangle = \langle f, g(h\phi) \rangle = \langle fg, h\phi \rangle = \langle (fg)h, \phi \rangle \qquad \text{for each } \phi \text{ in } \mathcal{G} \quad.$$

Thus, $f(gh) = (fg)h$, at least when g and h are simple multipliers.

?▶ Exercise 33.13:   *Extend the above discussion and show that*

$$f(gh) = (fg)h$$

whenever any two of these factors are simple multipliers.

At this point, you are doubtlessly wondering whether the generalized product distributes over the generalized sum. Well, let g be any simple multiplier, let f and h be any two generalized functions, and let ϕ be any Gaussian test function. Then, by the definitions,

$$\langle g(f+h), \phi \rangle = \langle f+h, g\phi \rangle$$
$$= \langle f, g\phi \rangle + \langle h, g\phi \rangle$$
$$= \langle gf, \phi \rangle + \langle gh, \phi \rangle = \langle (gf)+(gh), \phi \rangle \quad.$$

So yes, we do have $g(f+h) = (gf)+(gh)$ whenever g is a simple multiplier, and f and h are generalized functions.

Does addition distribute over multiplication? Do the next exercise and find out.

?▶ Exercise 33.14:   *Let g and h be two simple multipliers, and let f be any generalized function. Show that*

$$(g+h)f = (gf) + (hf) \quad.$$

Linear Combinations and Subtraction

Suppose f and g are any two generalized functions, and α and β are any two complex numbers. From the above we know that the linear combination $\alpha f + \beta g$, being the generalized sum of two generalized functions each of which is the product of a constant with a generalized function, is another generalized function. Moreover, for each ϕ in $\mathcal{G}$,

$$\langle \alpha f + \beta g, \phi \rangle = \langle \alpha f, \phi \rangle + \langle \beta g, \phi \rangle = \alpha \langle f, \phi \rangle + \beta \langle g, \phi \rangle \quad.$$

Naturally, we will refer to the product $(-1)f$ as the *(generalized) negative* of f and denote it by $-f$. Likewise, the *(generalized) difference* of f and g, denoted by $f - g$, is just the linear combination $f + (-1)g$. Using the fact that addition distributes over multiplication, we see that

$$g - g = 1g + (-1)g = [1 + (-1)]g = 0g = 0 \quad,$$

which should come as no surprise whatsoever.

?▶ Exercise 33.15:   Let f, g and h be generalized functions. Verify that $f = g + h$ if and only if $f - g = h$.

33.4 Generalized Functions Based on Other Test Function Spaces

It should be noted that the precise meaning of "f is a generalized function" depends somewhat on the space of test functions being used (and the precise definition of functional continuity depends on the criteria being used to determine when two test functions are "almost the same"). In general, a "generalized function" is a continuous, linear functional on whatever space of test functions is of interest. Because of our interests, that space of test functions will usually be $\mathcal{G}$, the space of Gaussian test functions. On the few occasions where we need to refer to another test function space, we will use $\mathcal{H}$, the space of very rapidly decreasing test functions (see section 32.3 on page 521). Fortunately, for reasons alluded to in section 32.3, we can treat the generalized function theory based on $\mathcal{G}$ as being the same as that based on $\mathcal{H}$.

33.5 Some Consequences of Functional Continuity*

While a complete understanding of "functional continuity" will not be necessary for most of our work, there will be a few occasions where knowing some consequences of functional continuity will be useful. In this section, we'll go ahead and derive some of those consequences using the intuitive understanding that, if f is a generalized function, then functional continuity assures us that the values of $\langle f, \phi \rangle$ and $\langle f, \psi \rangle$ are "almost the same" whenever ϕ and ψ are "almost the same test functions". Those brave souls wishing for a rigorous definition of functional continuity and a rigorous development of the material in this section can then proceed to the section following this.

A Simple Limit Lemma

Suppose f is some generalized function and ϕ is any Gaussian test function. Observe that

$$\lim_{\epsilon \to 0^+} e^{-\epsilon \pi x^2} \phi(x) = e^0 \phi(x) = \phi(x) \qquad \text{for each } x \text{ in } \mathbb{R} \quad .$$

This suggests that, for small positive values of ϵ, $e^{-\epsilon \pi x^2} \phi(x)$ is "almost the same test function" as $\phi(x)$, with their difference shrinking to 0 as $\epsilon \to 0^+$. The functional continuity of f would then imply that

$$\left\langle f(x), e^{-\epsilon \pi x^2} \phi(x) \right\rangle \qquad \text{and} \qquad \left\langle f(x), \phi(x) \right\rangle$$

---

* This is another section that can be skimmed through initially and read more carefully later as the need arises — and that
  need won't arise in the next few chapters.

are almost the same values, with their difference shrinking to 0 as $\epsilon \to 0^+$. In other words, we should expect the following lemma to be true.

Lemma 33.2
For any generalized function f and any Gaussian test function ϕ,

$$\lim_{\epsilon \to 0^+} \left\langle f(x), e^{-\epsilon \pi x^2} \phi(x) \right\rangle = \left\langle f(x), \phi(x) \right\rangle \quad .$$

This lemma is true and can easily be proven by anyone who reads section 33.6. So its proof will be left as an exercise for those brave souls (exercise 33.29 at the end of this chapter).

Smoothness of Evaluated Functionals with Parameters

Recall how we've occasionally made use of the fact that, if $\Phi(x, \lambda)$ is a "sufficiently reasonable" function of x and λ, then the function h given by

$$h(\lambda) = \int_{-\infty}^{\infty} \Phi(x, \lambda) \, dx$$

is a smooth function of the parameter λ, and

$$h'(\lambda) = \frac{d}{d\lambda} \int_{-\infty}^{\infty} \Phi(x, \lambda) \, dx = \int_{-\infty}^{\infty} \frac{\partial}{\partial \lambda} \Phi(x, \lambda) \, dx \quad .$$

For example, we used these facts in deriving the classical differentiation identities for Fourier transforms in chapter 22. They are also applicable, in a limited sense, to the generalized theory. To see this, suppose f is some fixed generalized function, and define the function h on $\mathbb{R}$ by

$$h(\lambda) = \left\langle f(x), e^{-3(x-\lambda)^2} \right\rangle \qquad \text{for} \quad -\infty < \lambda < \infty \quad .$$

If f happens to be a piecewise continuous and exponentially bounded function on $\mathbb{R}$, then the above formula for h can be written as

$$h(\lambda) = \int_{-\infty}^{\infty} f(x) e^{-3(x-\lambda)^2} \, dx \quad ,$$

and the results from the last section of chapter 18 can be applied to show that h is a smooth function on the real line whose derivative can be computed in a naive manner,

$$h'(\lambda) = \frac{d}{d\lambda} \int_{-\infty}^{\infty} f(x) e^{-3(x-\lambda)^2} \, dx = \int_{-\infty}^{\infty} f(x) \frac{\partial}{\partial \lambda} e^{-3(x-\lambda)^2} \, dx \quad .$$

Using our generalized integral notation this can also be written as

$$h'(\lambda) = \frac{d}{d\lambda} \left\langle f(x), \Psi(x, \lambda) \right\rangle = \left\langle f(x), \frac{\partial}{\partial \lambda} \Psi(x, \lambda) \right\rangle \tag{33.10}$$

where

$$\Psi(x, \lambda) = e^{-3(x-\lambda)^2} \quad . \tag{33.11}$$

(Yes, we could compute the partial derivative here, but that is not particularly relevant to the point being made.)

Now, let us consider any expression of the form

$$h(\lambda) = \left\langle f(x), \Psi(x, \lambda) \right\rangle \tag{33.12}$$

where f is a fixed generalized function and $\Psi(x, \lambda)$ describes some "family of Gaussian test functions continuously parameterized by λ" (such as, for example, given by equation (33.11)). When saying " $\Psi(x, \lambda)$ describes some 'family ... parameterized by λ '," it is meant that all of the following hold:

1. The parameter λ can be any value in some given interval (a, b).

2. For each fixed value of λ in (a, b), $\Psi(x, \lambda)$ is a Gaussian test function of x.

3. If λ_0 is in (a, b) and λ is almost the same value as λ_0, then $\Psi(x, \lambda)$ is "almost the same" Gaussian test function as $\Psi(x, \lambda_0)$.

Remember, we are using an intuitive notion of functional continuity, so use an intuitive notion as to what the last condition means. In particular, you may assume the third condition implies that

$$\lim_{\lambda \to \lambda_0} \Psi(x, \lambda) = \Psi(x, \lambda_0) \qquad \text{for each } x \text{ in } \mathbb{R} \quad . \tag{33.13}$$

Since we are not assuming f is a classical function, we cannot appeal to results from chapter 18 to show that h, as defined in equation (33.12), is continuous or to verify that an equation similar to equation (33.10) holds. Still, if λ_0 is any point in (a, b) and λ is almost the same value as λ_0, then the third condition, above, says that

$$\Psi(x, \lambda) \qquad \text{and} \qquad \Psi(x, \lambda_0)$$

are "almost the same test function", and the functional continuity of f then ensures that

$$h(\lambda) = \langle f(x), \Psi(x, \lambda) \rangle \qquad \text{and} \qquad h(\lambda_0) = \langle f(x), \Psi(x, \lambda_0) \rangle$$

are almost the same values. With a little thought, you'll realize that this can happen for each λ_0 in (a, b) only if h is continuous on (a, b). Thus, the functional continuity of f ensures that the evaluated functional in equation (33.12) is a continuous classical function of λ. That is, for each λ_0 in (a, b),

$$\lim_{\lambda \to \lambda_0} \langle f(x), \Psi(x, \lambda) \rangle = \langle f(x), \Psi(x, \lambda_0) \rangle \quad .$$

Do observe that, according to equation (33.13), our last equation can also be written

$$\lim_{\lambda \to \lambda_0} \langle f(x), \Psi(x, \lambda) \rangle = \left\langle f(x), \lim_{\lambda \to \lambda_0} \Psi(x, \lambda) \right\rangle \quad .$$

Now consider the derivative of h at some point λ_0. By the definition of derivatives and the linearity of generalized functions,

$$h'(\lambda_0) = \lim_{\Delta\lambda \to 0} \frac{1}{\Delta\lambda} [h(\lambda_0 + \Delta\lambda) - h(\lambda_0)]$$

$$= \lim_{\Delta\lambda \to 0} \frac{1}{\Delta\lambda} [\langle f(x), \Psi(x, \lambda_0 + \Delta\lambda) \rangle - \langle f(x), \Psi(x, \lambda_0) \rangle]$$

$$= \lim_{\Delta\lambda \to 0} \left\langle f(x), \frac{1}{\Delta\lambda} [\Psi(x, \lambda_0 + \Delta\lambda) - \Psi(x, \lambda_0)] \right\rangle \quad .$$

If

$$\lim_{\Delta\lambda \to 0} \left\langle f(x), \frac{1}{\Delta\lambda} [\Psi(x, \lambda_0 + \Delta\lambda) - \Psi(x, \lambda_0)] \right\rangle$$

$$= \left\langle f(x), \lim_{\Delta\lambda \to 0} \frac{1}{\Delta\lambda} [\Psi(x, \lambda_0 + \Delta\lambda) - \Psi(x, \lambda_0)] \right\rangle \quad ,$$

and if the resulting partial derivative of Ψ exists and is, itself, a Gaussian test function for $\lambda = \lambda_0$, then

$$h'(\lambda_0) = \left\langle f(x), \lim_{\Delta\lambda \to 0} \frac{1}{\Delta\lambda} [\Psi(x, \lambda_0 + \Delta\lambda) - \Psi(x, \lambda_0)] \right\rangle = \left\langle f(x), \frac{\partial}{\partial\lambda}\Psi(x, \lambda)\Big|_{\lambda_0} \right\rangle ,$$

and thus,

$$\frac{d}{d\lambda} \langle f(x), \Psi(x, \lambda) \rangle = \left\langle f(x), \frac{\partial}{\partial\lambda}\Psi(x, \lambda) \right\rangle .$$

The preceding computations can be rigorously justified for a number of choices of Ψ after a more complete development of functional continuity. Two choices that will be of particular interest to us are given in the next two lemmas.

Lemma 33.3
Let

$$H(\tau) = \langle f(x), e^{\sigma\tau x}\phi(x) \rangle$$

where f is any fixed generalized function, σ is any fixed complex value, and ϕ is any fixed Gaussian test function. Then H is a smooth function on the real line, and

$$H'(\tau) = \left\langle f(x), \frac{\partial}{\partial\tau}e^{\sigma\tau x}\phi(x) \right\rangle = \langle f(x), \sigma x\, e^{\sigma\tau x}\phi(x) \rangle .$$

Lemma 33.4
Let

$$K(\tau) = \left\langle f(x), e^{-\sigma\tau x^2}h(x) \right\rangle$$

where f is any fixed generalized function, σ is any fixed positive value, and h is any fixed simple multiplier. Then K is a smooth function on $(0, \infty)$, and

$$K'(\tau) = \left\langle f(x), \frac{\partial}{\partial\tau}e^{-\sigma\tau x^2}h(x) \right\rangle = \left\langle f(x), -\sigma x^2 e^{-\sigma\tau x^2}h(x) \right\rangle .$$

Rigorous proofs of these two lemmas are developed in section 33.6.

There are two or so observations that should be made regarding the function H defined in lemma 33.3, above.

The first is that its formula is valid even if τ is treated as a complex variable. Those acquainted with the theory of analytic functions on the complex plane and who have read section 6.4 should have no trouble verifying that this formula defines an analytic function on $\mathbb{C}$.

?▶Exercise 33.16:   *For each complex value τ, let $H(\tau)$ be as defined by the formula in lemma 33.3. Verify that H is an analytic function on the complex plane.*

Next, consider the last line in lemma 33.3, which can be written as

$$H'(\tau) = \sigma \langle f(x), e^{\sigma\tau x}[x\phi(x)] \rangle .$$

That same lemma then tells us that H' is a smooth function on $\mathbb{R}$, with

$$H''(\tau) = \sigma \left\langle f(x), \frac{\partial}{\partial\tau}e^{\sigma\tau x}[x\phi(x)] \right\rangle = \sigma^2 \left\langle f(x), x^2 e^{\sigma\tau x}\phi(x) \right\rangle .$$

Repeating this again and again and again ..., leads to the next result.

Corollary 33.5

Let

$$H(\tau) = \langle f(x), e^{\sigma \tau x} \phi(x) \rangle$$

where f is any fixed generalized function, σ is any fixed complex value, and ϕ is any fixed Gaussian test function. Then H is an infinitely differentiable function on the real line. Moreover, for $n = 1, 2, 3, \dots$,

$$H^{(n)}(\tau) = \left\langle f(x), \frac{\partial^n}{\partial \tau^n} e^{\sigma \tau x} \phi(x) \right\rangle = \sigma^n \langle f(x), x^n e^{\sigma \tau x} \phi(x) \rangle \quad .$$

Obviously, a similar statement can be made regarding K from lemma 33.4.

Simple Tests for Equality

When we say two generalized functions f and g are equal, we mean

$$\langle f, \phi \rangle = \langle g, \phi \rangle \tag{33.14}$$

for *every* Gaussian test function ϕ. On occasion, explicitly verifying this for every ϕ in $\mathcal{G}$ is not so straightforward, and, on some of those occasions, we will find one of the following lemmas useful. These lemmas assure us that equation (33.14) holds for *all* ϕ in $\mathcal{G}$ provided it holds for *certain* choices of ϕ in $\mathcal{G}$.

Lemma 33.6

Suppose f and g are two generalized functions satisfying

$$\left\langle f(x), e^{\sigma x} e^{-\gamma x^2} \right\rangle = \left\langle g(x), e^{\sigma x} e^{-\gamma x^2} \right\rangle$$

for every complex number σ and every positive value γ. Then $f = g$.

Lemma 33.7

Suppose f and g are two generalized functions satisfying

$$\left\langle f(x), e^{-\gamma(x-\xi)^2} \right\rangle = \left\langle g(x), e^{-\gamma(x-\xi)^2} \right\rangle$$

for every complex number ξ and every positive value γ. Then $f = g$.

Either lemma can easily be obtained as a corollary of the other, so we will prove one, and leave the proof of the other as an exercise.

PROOF (of lemma 33.6): To prove this lemma, it suffices to confirm that equation (33.14) holds for an arbitrarily chosen Gaussian test function ϕ. Since such functions are sums of base Gaussian test functions, and the formulas for base Gaussian test functions are relatively simple, let's first restrict our choices to base Gaussian test functions.

Assume ψ is any base Gaussian test function. Then the formula for ψ can be written as

$$\psi(x) = A x^n e^{cx} e^{-\gamma x^2}$$

where n is some nonnegative integer, A and c are complex constants, and γ is a positive constant (see lemma 32.1 on page 516). Using these constants, define H to be the function

$$H(\tau) = \left\langle f(x), e^{\tau x} \left[A e^{cx} e^{-\gamma x^2} \right] \right\rangle \quad .$$

By linearity and the basic assumption of the lemma we are proving, we also have

$$H(\tau) = A \left\langle f(x), e^{(\tau+c)x} e^{-\gamma x^2} \right\rangle$$
$$= A \left\langle g(x), e^{(\tau+c)x} e^{-\gamma x^2} \right\rangle = \left\langle g(x), e^{\tau x} \left[A e^{cx} e^{-\gamma x^2} \right] \right\rangle \quad .$$

Now, if $n = 0$, then $\psi(x) = A e^{cx} e^{-\gamma x^2}$, and

$$\langle f, \psi \rangle = \left\langle f(x), e^{0 \cdot x} \left[A e^{cx} e^{-\gamma x^2} \right] \right\rangle$$
$$= H(0)$$
$$= \left\langle g(x), e^{0 \cdot x} \left[A e^{cx} e^{-\gamma x^2} \right] \right\rangle = \langle g, \psi \rangle \quad .$$

On the other hand, if $n > 0$, then we can apply corollary 33.5 and a little algebra, obtaining

$$\langle f, \psi \rangle = \left\langle f(x), A x^n e^{cx} e^{-\gamma x^2} \right\rangle$$
$$= \left\langle f(x), x^n e^{0 \cdot x} \left[A e^{cx} e^{-\gamma x^2} \right] \right\rangle$$
$$= H^{(n)}(0)$$
$$= \left\langle g(x), x^n e^{0 \cdot x} \left[A e^{cx} e^{-\gamma x^2} \right] \right\rangle$$
$$= \left\langle f(x), A x^n e^{cx} e^{-\gamma x^2} \right\rangle = \langle g, \psi \rangle \quad .$$

Thus,

$$\langle f, \psi \rangle = \langle g, \psi \rangle$$

for every base Gaussian test function ψ.

Now, let ϕ be any Gaussian test function. By definition, ϕ is a finite sum of base Gaussian functions, say,

$$\phi = \sum_{k=1}^{N} \psi_k \quad .$$

Using this, linearity and the result from the previous paragraph, we have

$$\langle f, \phi \rangle = \left\langle f, \sum_{k=1}^{N} \psi_k \right\rangle = \sum_{k=1}^{N} \langle f, \psi_k \rangle$$
$$= \sum_{k=1}^{N} \langle g, \psi_k \rangle = \left\langle g, \sum_{k=1}^{N} \psi_k \right\rangle = \langle g, \phi \rangle \quad . \qquad \blacksquare$$

?▶Exercise 33.17: Prove lemma 33.7 using lemma 33.6.

Those acquainted with the theory of analytic functions on the complex plane can go further and prove the following two theorems.

Theorem 33.8
Suppose f and g are two generalized functions satisfying

$$\left\langle f(x) , e^{\alpha x} e^{-\gamma x^2} \right\rangle = \left\langle g(x) , e^{\alpha x} e^{-\gamma x^2} \right\rangle$$

for each pair of real numbers α and γ with $\gamma > 0$. Then $f = g$.

Theorem 33.9
Suppose f and g are two generalized functions satisfying

$$\left\langle f(x) , e^{-\gamma (x-a)^2} \right\rangle = \left\langle g(x) , e^{-\gamma (x-a)^2} \right\rangle$$

for each pair of real numbers a and γ with $\gamma > 0$. Then $f = g$.

This last theorem is especially noteworthy because of its similarity to our classical Gaussian test for equality (theorem 30.1 on page 491).

?►Exercise 33.18   a: *Using lemmas 33.3 and 33.6, prove theorem 33.8. (Start by showing*

$$\Psi(\zeta) = \left\langle h(x) , e^{\zeta x} e^{-\gamma x^2} \right\rangle$$

is an analytic function on the complex plane for any generalized function h and positive value γ. Then use corollary 6.6 on page 70.)

b: *Prove theorem 33.9.*

33.6 The Details of Functional Continuity
Definition and Basic Examples

Recall our working definition of functional continuity: A functional Γ on $\mathcal{G}$ is functionally continuous if and only if the values of $\Gamma[\phi]$ and $\Gamma[\psi]$ are almost the same whenever ϕ and ψ are "almost the same Gaussian test function". There are many ways one might judge whether two test functions are "almost the same", but the one that has been found appropriate for this theory is based on the α-norms developed in section 32.5 (starting on page 529). Basically, two Gaussian test functions are viewed as being "almost the same" if and only if, in some sense, the α-norms of their difference are all relatively small.

Consequently, the criteria for functional continuity must also be based on these α-norms.

Let Γ be a linear functional on the space of Gaussian test functions. We will call Γ *(functionally) continuous* if and only if there are two nonnegative real constants M and a such that

$$|\Gamma[\phi]| \leq M \|\phi\|_a$$

for every Gaussian test function ϕ. Functionals satisfying this inequality are also referred to as being *functionally bounded*. Assuming this inequality holds, we have

$$|\Gamma[\phi] - \Gamma[\psi]| = |\Gamma[\phi - \psi]| \leq M \|\phi - \psi\|_a$$

for every pair of Gaussian test functions ϕ and ψ. So, if the difference between ϕ and ψ is small as measured by the α-norm with $\alpha = a$, so too is the difference between the values $\Gamma[\phi]$ and $\Gamma[\psi]$.

For most cases of interest to us, we will be writing $\langle f, \phi \rangle$ instead of $\Gamma[\phi]$ and trying to verify that f is a generalized function. In terms of this notation, the above criteria for functional continuity is that there are two nonnegative real constants, M and a such that

$$|\langle f, \phi \rangle| \leq M \|\phi\|_a$$

for every Gaussian test function ϕ. This then, along with linearity, is what we should verify to show that a given f is a generalized function.

Verifying functional continuity for the cases we've already seen (classical functions and delta functions) is fairly easy using an inequality from lemma 32.14 on page 530, namely that, given any Gaussian test function ϕ and any nonnegative value a,

$$|\phi(x + iy)| \leq e^{-a|x|} \|\phi\|_a \tag{33.15}$$

whenever x and y are real values with $|y| \leq a$.

!▶ Example 33.12:   *Let f be any exponentially integrable function on the real line. By definition, there is a nonnegative real value a such that the function $e^{-a|x|} f(x)$ is absolutely integrable. So if ϕ is any Gaussian test function,*

$$
\begin{aligned}
|\langle f, \phi \rangle| &= \left| \int_{-\infty}^{\infty} f(x)\phi(x)\,dx \right| \\
&\leq \int_{-\infty}^{\infty} |f(x)|\,|\phi(x)|\,dx \\
&\leq \int_{-\infty}^{\infty} |f(x)|\,e^{-a|x|}\,\|\phi\|_a\,dx = \left(\int_{-\infty}^{\infty} |f(x)|\,e^{-a|x|}\,dx \right) \|\phi\|_a \quad .
\end{aligned}
$$

Thus, letting

$$M = \int_{-\infty}^{\infty} |f(x)|\,e^{-a|x|}\,dx$$

(which we know is finite because of our choice for a), we have

$$|\langle f, \phi \rangle| \leq M \|\phi\|_a$$

for every Gaussian test function ϕ. Since we already know linearity holds, this inequality tells us that functional continuity also holds and, thus, completes our proof that every exponentially integrable function is a generalized function.

?▶ Exercise 33.19:   *Let ζ be any fixed point on the complex plane and verify that functional continuity holds for the delta function at this point, δ_ζ (thus completing the proof that the delta functions are generalized functions).*

So far, all the generalized functions we have seen are linear combinations of the sort of generalized functions described in the above example and exercises. This makes the next lemma rather significant to us.

Lemma 33.10
Functional continuity holds for any linear combination of generalized functions.

PROOF: Let f and g be two generalized functions, and let α and β be any two constants. Since functional continuity holds for f and g, there are real constants M_1, M_2, a_1 and a_2 such that

$$\left| \langle f , \phi \rangle \right| \leq M_1 \|\phi\|_{a_1} \quad \text{and} \quad \left| \langle g , \phi \rangle \right| \leq M_2 \|\phi\|_{a_2}$$

for each ϕ in $\mathcal{G}$. Let a be the larger of a_1 and a_2. Then, as was noted in lemma 32.15 on page 530,

$$\|\phi\|_{a_1} \leq \|\phi\|_a \quad \text{and} \quad \|\phi\|_{a_2} \leq \|\phi\|_a \quad .$$

Thus, with $M = \alpha M_1 + \beta M_2$ and ϕ being any Gaussian test function, we see that

$$\begin{aligned}
\left| \langle \alpha f + \beta g , \phi \rangle \right| &= \left| \alpha \langle f , \phi \rangle + \beta \langle g , \phi \rangle \right| \\
&\leq \alpha \left| \langle f , \phi \rangle \right| + \beta \left| \langle g , \phi \rangle \right| \\
&\leq \alpha M_1 \|\phi\|_{a_1} + \beta M_2 \|\phi\|_{a_2} \\
&\leq \alpha M_1 \|\phi\|_a + \beta M_2 \|\phi\|_a = M \|\phi\|_a \quad .
\end{aligned}$$
∎

Proving Lemmas 33.3 and 33.4

These two lemmas can be proven in much the same manner. We will prove the first one here. As an exercise, you can prove the second one by yourself. To simplify matters slightly, we will start with another lemma describing a bound we will use in proving lemma 33.3, and which you may use it in proving lemma 33.4.

Lemma 33.11
Let λ and θ be complex values with $|\theta| \leq 1$. Then

$$\left| \frac{1}{\theta} \left[e^{\lambda \theta} - 1 \right] - \lambda \right| \leq \frac{1}{2} |\theta| \, |\lambda|^2 \, e^{|\lambda|} \quad .$$

PROOF: Observe that

$$e^{\lambda \theta} - 1 = \int_0^\theta \lambda e^{\lambda s} \, ds \quad \text{and} \quad \lambda = \frac{1}{\theta} \int_0^\theta \lambda \, ds \quad .$$

So,

$$\begin{aligned}
\frac{1}{\theta} \left[e^{\lambda \theta} - 1 \right] - \lambda &= \frac{1}{\theta} \int_0^\theta \lambda e^{\lambda s} \, ds - \frac{1}{\theta} \int_0^\theta \lambda \, ds \\
&= \frac{1}{\theta} \int_0^\theta \lambda \left[e^{\lambda s} - 1 \right] ds \\
&= \frac{1}{\theta} \int_0^\theta \lambda \int_0^s \lambda e^{\lambda t} \, dt \, ds = \frac{\lambda^2}{\theta} \int_0^\theta \int_0^s e^{\lambda t} \, dt \, ds \quad .
\end{aligned} \tag{33.16}$$

By assumption, $|\theta| \leq 1$. So, when $|t| \leq |\theta|$,

$$|\lambda t| \leq |\lambda| \qquad \text{and} \qquad \left| e^{\lambda t} \right| \leq e^{|\lambda t|} \leq e^{|\lambda|} \quad .$$

This and a little basic integration gives us

$$\left| \int_0^\theta \int_0^s e^{\lambda t} \, dt \, ds \right| \leq \int_0^{|\theta|} \int_0^s e^{|\lambda|} \, dt \, ds = \frac{1}{2} |\theta|^2 \, e^{|\lambda|} \quad ,$$

which, combined with equation (33.16), yields

$$\left| \frac{1}{\theta} \left[e^{\lambda \theta} - 1 \right] - \lambda \right| = \left| \frac{\lambda^2}{\theta} \right| \left[\frac{1}{2} |\theta|^2 \, e^{|\lambda|} \right] \leq \frac{1}{2} |\theta| \, |\lambda|^2 \, e^{|\lambda|} \quad . \qquad \blacksquare$$

Before starting on lemma 33.3's proof, let's recall the lemma's statement:

Let

$$H(\tau) = \left\langle \, f(x) \, , \, e^{\sigma \tau x} \phi(x) \, \right\rangle$$

where f is any fixed generalized function, σ is any fixed complex value, and ϕ is any fixed Gaussian test function. Then H is a smooth function on the real line, and

$$H'(\tau) = \left\langle \, f(x) \, , \, \frac{d}{d\tau} e^{\sigma \tau x} \phi(x) \, \right\rangle = \left\langle \, f(x) \, , \, \sigma x \, e^{\sigma \tau x} \phi(x) \, \right\rangle \quad .$$

PROOF (of lemma 33.3):   For the moment, suppose we know that

$$\lim_{\Delta \tau \to 0} \frac{H(\tau + \Delta \tau) - H(\tau)}{\Delta \tau} = \left\langle \, f(x) \, , \, \sigma x e^{\sigma \tau x} \phi(x) \, \right\rangle$$

or, equivalently, that

$$\lim_{\Delta \tau \to 0} \left| \frac{H(\tau + \Delta \tau) - H(\tau)}{\Delta \tau} - \left\langle \, f(x) \, , \, \sigma x e^{\sigma \tau x} \phi(x) \, \right\rangle \right| = 0 \qquad (33.17)$$

for each real value τ (and arbitrary choices of f, ϕ and σ). Either equation tells us that the given H is differentiable (and hence, continuous) everywhere on the real line and that the formula claimed in the lemma for $H'(\tau)$ is true. All that is then left to prove is the claimed smoothness of H (i.e., that H' is continuous). But, since H', itself, is then given by

$$H'(\tau) = \left\langle \, f(x) \, , \, e^{\tau x} \psi(x) \, \right\rangle \qquad \text{where} \qquad \psi(x) = \sigma x \phi(x) \quad ,$$

the arguments that H is smooth also apply to H'. Thus, H' must also be continuous on $\mathbb{R}$.

Consequently, to prove lemma 33.3, it suffices to verify equation (33.17) for each real value τ.

So let τ and $\Delta \tau$ be two real numbers, with τ being viewed as fixed, and $\Delta \tau$ being viewed as a variable parameter. Since we will be taking the limit as $\Delta \tau \to 0$, we may assume $0 < |\Delta \tau| < 1$. By linearity,

$$\frac{H(\tau + \Delta \tau) - H(\tau)}{\Delta \tau} - \left\langle \, f(x) \, , \, \sigma x \phi(x) \, \right\rangle$$

$$= \frac{1}{\Delta \tau} \left[\left\langle \, f(x) \, , \, e^{\sigma(\tau + \Delta \tau)x} \phi(x) \, \right\rangle - \left\langle \, f(x) \, , \, e^{\sigma \tau x} \phi(x) \, \right\rangle \right]$$

$$\qquad - \left\langle \, f(x) \, , \, \sigma x e^{\sigma \tau x} \phi(x) \, \right\rangle$$

$$= \left\langle \, f(x) \, , \, \left(\frac{1}{\Delta \tau} \left[e^{\sigma(\Delta \tau)x} - 1 \right] - \sigma x \right) e^{\sigma \tau x} \phi(x) \, \right\rangle \quad .$$

For convenience, let's rewrite this as

$$\frac{H(\tau + \Delta\tau) - H(\tau)}{\Delta\tau} - \big\langle f(x), \sigma x \phi(x) \big\rangle = \big\langle f, \Psi_{\Delta\tau} \big\rangle \tag{33.18}$$

where $\Psi_{\Delta\tau}$ is the Gaussian test function

$$\Psi_{\Delta\tau}(x) = \left(\frac{1}{\Delta\tau} \left[e^{\sigma(\Delta\tau)x} - 1 \right] - \sigma x \right) e^{\sigma\tau x} \phi(x) \quad.$$

Remember now, the functional continuity of f means there are two real numbers M and a such that

$$\big| \big\langle f, \Psi_{\Delta\tau} \big\rangle \big| \leq M \| \Psi_{\Delta\tau} \|_a \tag{33.19}$$

for each nonzero $\Delta\tau$. Our next major goal is to obtain a useful bound for the right side of this inequality in terms of $\Delta\tau$. To help find that bound, consider the formula for $\Psi_{\Delta\tau}(z)$ when z is any complex value. Applying lemma 33.11 (with $\theta = \Delta\tau$ and $\lambda = \sigma z$), we see that

$$|\Psi_{\Delta\tau}(z)| = \left| \left(\frac{1}{\Delta\tau} \left[e^{\sigma(\Delta\tau)z} - 1 \right] - \sigma z \right) e^{\sigma\tau z} \phi(z) \right|$$

$$\leq \frac{1}{2} |\Delta\tau| \, |\sigma z|^2 \, e^{|\sigma z|} \cdot \left| e^{\sigma\tau z} \phi(z) \right| = |\Delta\tau| \, e^{|\sigma z|} \, |\psi(z)|$$

where ψ is the Gaussian test function

$$\psi(z) = \frac{1}{2} \sigma^2 z^2 e^{\sigma\tau z} \phi(z) \quad.$$

In particular, assume $z = x + iy$ with $|y| \leq a$. Then, letting $A = e^{|\sigma||a|}$,

$$e^{|\sigma x + iy|} \leq e^{|\sigma|(|x| + |y|)} \leq e^{|\sigma|(|x| + |a|)} = A e^{|\sigma||x|} \quad.$$

From this, the above inequality relating $\Psi_{\Delta\tau}$ and ψ, and lemma 32.14 on page 530, we get

$$|\Psi_{\Delta\tau}(x + iy)| \, e^{a|x|} \leq |\Delta\tau| \, e^{|\sigma(x+iy)|} \, |\psi(x + iy)| \, e^{a|x|}$$

$$\leq |\Delta\tau| \, A e^{|\sigma||x|} \, |\psi(x + iy)| \, e^{a|x|}$$

$$= |\Delta\tau| \, A \, |\psi(x + iy)| \, e^{(a+|\sigma|)|x|} \leq |\Delta\tau| \, A \, \| \psi \|_{a+|\sigma|} \quad.$$

Lemma 32.13 on page 530 then assures us that

$$\| \Psi_{\Delta\tau} \|_a \leq |\Delta\tau| \, A \, \| \psi \|_{a+|\sigma|} \quad.$$

Combining this with equation (33.18) and inequality (33.19) yields

$$\left| \frac{H(\tau + \Delta\tau) - H(\tau)}{\Delta\tau} - \big\langle f(x), \sigma x e^{\sigma\tau x} \phi(x) \big\rangle \right| = \big| \big\langle f, \Psi_{\Delta\tau} \big\rangle \big|$$

$$\leq M \| \Psi_{\Delta\tau} \|_a \leq |\Delta\tau| \, A \, \| \psi \|_{a+|\sigma|} \quad.$$

Hence,

$$\lim_{\Delta\tau \to 0} \left| \frac{H(\tau + \Delta\tau) - H(\tau)}{\Delta\tau} - \big\langle f(x), \sigma x e^{\sigma\tau x} \phi(x) \big\rangle \right| \leq \lim_{\Delta\tau \to 0} |\Delta\tau| \, A \, \| \psi \|_{a+|\sigma|} = 0 \quad. \quad \blacksquare$$

?►Exercise 33.20: *Prove lemma 33.4 on page 556.*

Additional Exercises

33.21. *Several formulas for computing* $\langle f, \phi \rangle$ *for an arbitrary Gaussian test function* ϕ *are given below. For each, determine if the given formula defines a generalized function on* $\mathcal{G}$ *. (You may assume functional continuity holds whenever the formula at least defines a functional on* $\mathcal{G}$ *.)*

a. $\langle f, \phi \rangle = \displaystyle\int_{-\infty}^{\infty} \phi(3x)\, dx$

b. $\langle f, \phi \rangle = \displaystyle\int_{-\infty}^{\infty} x^2 \mathcal{F}[\phi]|_x\, dx$

c. $\langle f, \phi \rangle = \displaystyle\int_{-\infty}^{\infty} \phi(ix)\, dx$

d. $\langle f, \phi \rangle = \displaystyle\int_{0}^{1}\int_{0}^{1} \phi(x + iy)\, dx\, dy$

e. $\langle f, \phi \rangle = |\phi(0)|$

f. $\langle f, \phi \rangle = \phi'(0)$

33.22. *Let* f *be a fixed generalized function, let* α *be a nonzero real number, and let* ζ *be some real or complex number. Consider each of the following formulas for computing* $\langle g, \phi \rangle$ *for each Gaussian test function, and verify that each defines a generalized function* g *. (In each, you may assume functional continuity holds.)*

a. $\langle g, \phi \rangle = \langle f, \mathcal{F}[\phi] \rangle$

b. $\langle g, \phi \rangle = \langle f, \mathcal{F}^{-1}[\phi] \rangle$

c. $\langle g, \phi \rangle = \left\langle f(x), \phi\!\left(\frac{1}{\alpha}x\right) \right\rangle$

d. $\langle g, \phi \rangle = \langle f(x), \phi(x + \zeta) \rangle$

e. $\langle g, \phi \rangle = \langle f, \phi' \rangle$

33.23. *Which of the following classical functions are exponentially integrable and, hence, can be directly treated as generalized functions?*

a. *Any classically transformable function*

b. $\sin(x)$

c. $\text{Arctan}(x)$

d. $e^{3|x|}$

e. e^{x^3}

f. $x^3 - 4x^2 + 8$

g. $\sqrt{|x|}$

h. $\dfrac{1}{\sqrt{|x|}}$

i. $\dfrac{1}{\sqrt{|x|^3}}$

j. $\ln|x|$

33.24. *Verify that* $2\pi\delta = f$ *where*

$$\langle f, \phi \rangle = \int_{0}^{2\pi} \phi\!\left(e^{i\theta}\right) d\theta \qquad \text{for each } \phi \text{ in } \mathcal{G} \ .$$

(Doing this problem requires knowledge of Cauchy's integral formula or residue theory from complex variables.)

33.25. *(A pathology) Consider the functional* Γ *defined as follows:*

1. *For each base Gaussian test function* $\phi(x) = Ax^n e^{-\gamma(x-\zeta)^2}$,

$$\Gamma[\phi] = \Gamma\!\left[Ax^n e^{-\gamma(x-\zeta)^2}\right] = \begin{cases} A & \text{if } \gamma = 1 \\ 0 & \text{otherwise} \end{cases} \ .$$

2. For each finite summation of base Gaussian test functions $\phi_1 + \phi_2 + \cdots + \phi_n$,

$$\Gamma[\phi_1 + \phi_2 + \cdots + \phi_n] = \Gamma[\phi_1] + \Gamma[\phi_2] + \cdots + \Gamma[\phi_n] \quad .$$

(Note that, since all Gaussian test functions are finite summations of base Gaussian functions, this does define Γ as a functional on $\mathcal{G}$. In fact, it is a linear functional.)

Now consider the "sequence" of test functions

$$\phi_\gamma(x) = e^{-\gamma x^2} \quad \text{as} \quad \gamma \to 1 \quad .$$

Convince yourself that Γ is not functionally continuous by verifying that, while the test functions ϕ_γ are clearly getting close to the test function ϕ_1 as $\gamma \to 1$, the values of $\Gamma[\phi_\gamma]$ are not getting close to $\Gamma[\phi_1]$ as $\gamma \to 1$.

33.26. In the following, let $f(x) = \exp\left(|x|^{3/2}\right)$ and, for each positive integer n, let ϕ_n be the Gaussian test function

$$\phi_n(x) = A_n\, e^{-(x-n)^2} \quad \text{where} \quad A_n = \exp\left(-n^{3/2}\right) \quad .$$

a. Is f exponentially integrable?

b. Does

$$\langle\, f , \phi \,\rangle = \int_{-\infty}^{\infty} \exp\left(|x|^{3/2}\right) \phi(x)\, dx$$

define a linear functional on $\mathcal{G}$?

c i. Graph ϕ_n for several choices of n and, using these graphs, convince yourself that the ϕ_n's are "shrinking to 0" as $n \to \infty$.

ii. Show that $\|\phi_n\|_\alpha \to 0$ as $n \to \infty$ for any $\alpha \geq 0$.

d. Show that $\langle\, f , \phi_n \,\rangle \not\to 0$ as $n \to \infty$. (*Hint: Split the above integral into integrals over $(-\infty, n)$ and (n, ∞).*)

e. Why does the result in the last part show that f does not define a continuous linear functional on $\mathcal{G}$? (*Hence, f cannot be viewed as a generalized function.*)

33.27. Three ways of defining a generalized function corresponding to x^{-1} (which is a classical function, but is not exponentially integrable) are given below.

a i. Let ϕ be any Gaussian test function and verify that

$$\lim_{\epsilon \to 0+} \left[\int_{-\infty}^{-\epsilon} \frac{1}{x}\phi(x)\, dx + \int_{\epsilon}^{\infty} \frac{1}{x}\phi(x)\, dx \right]$$

exists and is finite. (*This limit is known as the Cauchy finite part of $\int_{-\infty}^{\infty} x^{-1}\phi(x)\, dx$.*)

ii. Verify that f_1, defined by

$$\langle\, f_1 , \phi \,\rangle = \lim_{\epsilon \to 0+} \left[\int_{-\infty}^{-\epsilon} \frac{1}{x}\phi(x)\, dx + \int_{\epsilon}^{\infty} \frac{1}{x}\phi(x)\, dx \right] \quad ,$$

is a generalized function (assume functional continuity holds).

b i. *Show that*

$$\frac{1}{x}\left[\phi(x) - \phi(0)e^{-x^2}\right]$$

is absolutely integrable whenever ϕ is a Gaussian test function.

ii. *Verify that f_2, defined by*

$$\langle f_2, \phi \rangle = \int_{-\infty}^{\infty} \frac{1}{x}\left[\phi(x) - \phi(0)e^{-x^2}\right] dx \quad,$$

is a generalized function (assume functional continuity holds).

c. *Verify that f_3, defined by*

$$\langle f_3, \phi \rangle = -\int_{-\infty}^{\infty} \ln|x|\,\phi'(x)\,dx$$

is a generalized function (assume functional continuity holds).

d. *Let f_1, f_2 and f_3 be as above, and show that $f_1 = f_2 = f_3$. That is, verify that*

$$\langle f_1, \phi \rangle = \langle f_2, \phi \rangle = \langle f_3, \phi \rangle \qquad \text{for each } \phi \text{ in } \mathcal{G} \quad.$$

33.28. *For the following, let f be a fixed, nonzero generalized function.*

a. *Show there must be a Gaussian test function ψ_0 such that $\langle f, \psi_0 \rangle = 1$.*

b. *Let ψ_0 be a fixed Gaussian test function such that $\langle f, \psi_0 \rangle = 1$, and let ψ be any other Gaussian test function. Show there is a value β such that $\langle f, \psi - \beta\psi_0 \rangle = 0$.*

c. *Suppose g is a generalized function such that, for any ϕ in $\mathcal{G}$,*

$$\langle f, \phi \rangle = 0 \quad \Longleftrightarrow \quad \langle g, \phi \rangle = 0 \quad.$$

Verify that g is a constant multiple of f (i.e., $g = cf$ for some constant c).

33.29. *Prove lemma 33.2 on page 554. (Use the results of exercise 32.18 on page 536.)*

33.30. *Let f be a linear functional on $\mathcal{G}$.*

a. *Assume f is also functionally continuous, and show that $\langle f, \phi_n \rangle \to 0$ as $n \to \infty$ whenever $\{\phi_n\}_{n=1}^{\infty}$ is any sequence of Gaussian test functions satisfying*

$$\lim_{n \to \infty} \|\phi_n\|_{\alpha} = 0 \qquad \text{for all } \alpha \geq 0 \quad.$$

b. *Now suppose $\langle f, \phi_n \rangle \to 0$ as $n \to \infty$ whenever $\{\phi_n\}_{n=1}^{\infty}$ is a sequence of Gaussian test functions such that*

$$\lim_{n \to \infty} \|\phi_n\|_{\alpha} = 0 \qquad \text{for all } \alpha \geq 0 \quad.$$

Show that f is functionally continuous.

33.31. *Verify that the functionals in exercises 33.21 d and 33.27 b ii are both functionally continuous.*

34

Sequences and Series
of Generalized Functions

Various sequences and infinite series of generalized functions will be appearing in the next several chapters. We will find them useful in describing many generalized functions that cannot be viewed as classical functions (delta functions, for example), and they can be used to extend some classical formulas to formulas that cannot be derived using the classical theory alone. Furthermore, the elements of one particularly important class of generalized functions are all infinite series of delta functions. These are the "arrays of delta functions", and they will play major roles in at least three subsequent chapters.

34.1 Sequences and Limits
Limits of Sequences of Generalized Functions

Suppose $\{f_\gamma\}_{\gamma=\alpha}^{\gamma_0}$ is a sequence of generalized functions.[1] We will say this sequence is *convergent (in the generalized sense)* if and only if there is a generalized function g such that

$$\lim_{\gamma \to \gamma_0} \langle f_\gamma , \phi \rangle = \langle g , \phi \rangle \qquad \text{for each } \phi \text{ in } \mathcal{G} \quad . \tag{34.1}$$

Naturally, if it exists, then the generalized function g in the above equation is called the *(generalized) limit* of the sequence. And, just as naturally, all the standard alternative ways of saying

the sequence $\{f_\gamma\}_{\gamma=\alpha}^{\gamma_0}$ is convergent with limit g

are at our disposal when discussing generalized limits, including:

the sequence $\{f_\gamma\}_{\gamma=\alpha}^{\gamma_0}$ converges, in the generalized sense, to g ,

$$\lim_{\gamma \to \gamma_0} f_\gamma = g \qquad \text{(in the generalized sense)} \quad ,$$

$$f_\gamma \to g \quad \text{as} \quad \gamma \to \gamma_0 \qquad \text{(in the generalized sense)} \quad ,$$

and

g is the (generalized) limit of the f_γ's as $\gamma \to \gamma_0$.

---

[1] As in chapter 29, we are using the term "sequence" for any indexed set of things whether or not the indices, themselves, can be indexed by the positive integers. All we require for the indices in $\{f_\gamma\}_{\gamma=\alpha}^{\gamma_0}$ is that the statement "as $\gamma \to \gamma_0$" can be readily understood.

It is worth noting that, using one of these alternatives, equation (34.1) becomes

$$\lim_{\gamma \to \gamma_0} \left\langle f_\gamma , \phi \right\rangle = \left\langle \lim_{\gamma \to \gamma_0} f_\gamma , \phi \right\rangle \qquad \text{for each } \phi \text{ in } \mathcal{G} \quad .$$

As will be illustrated in the next example, the qualifying phrase "in the generalized sense" helps prevent confusion when there are different ways of interpreting "the limit". Sometimes, for variety, we might say "as generalized functions" instead of "in the generalized sense". And sometimes, if there is no significant danger of confusion, we may just omit any qualifying phrase.

!▶**Example 34.1:** *For each $\epsilon > 0$, let h_ϵ be the classical function*

$$h_\epsilon(x) = \frac{1}{2\epsilon} \text{pulse}_\epsilon(x) \quad .$$

In chapters 26 and 29, we saw that $\{h_\epsilon\}_{\epsilon=1}^0$ is an identity sequence and that

$$\lim_{\epsilon \to 0} \int_{-\infty}^{\infty} \phi(x) h_\epsilon(x)\, dx = \lim_{\epsilon \to 0} \frac{1}{2\epsilon} \int_{-\epsilon}^{\epsilon} \phi(x)\, dx = \phi(0)$$

whenever ϕ is a piecewise continuous function on $\mathbb{R}$. In particular then, for each Gaussian test function ϕ,

$$\lim_{\epsilon \to 0} \left\langle h_\epsilon , \phi \right\rangle = \lim_{\epsilon \to 0} \int_{-\infty}^{\infty} \phi(x) h_\epsilon(x)\, dx = \phi(0) = \left\langle \delta , \phi \right\rangle \quad .$$

Thus, in the generalized sense,

$$\lim_{\epsilon \to 0} h_\epsilon = \delta \quad .$$

However, as classical, piecewise continuous functions,

$$\lim_{\epsilon \to 0} h_\epsilon = 0$$

because, for each nonzero real value x,

$$\lim_{\epsilon \to 0} h_\epsilon(x) = \lim_{\epsilon \to 0} \frac{1}{2\epsilon} \text{pulse}_\epsilon(x) = 0 \quad .$$

(If this last statement is not obvious, look back at figure 26.1a on page 422, or look ahead at figure 34.1a on page 577.)

?▶**Exercise 34.1:** *Convince yourself that the generalized limit of any Gaussian identity sequence is the delta function.*

The next lemma assures us that at least some of the manipulations commonly performed using classical limits are valid with the generalized limit.

Lemma 34.1
Assume

$$\left\{ f_\gamma \right\}_{\gamma=\alpha}^{\gamma_0} \qquad \text{and} \qquad \left\{ g_\gamma \right\}_{\gamma=\alpha}^{\gamma_0}$$

are two convergent sequences of generalized functions. Then, for any two constants a and b, and any simple multiplier h,

$$\left\{ af_\gamma + bg_\gamma \right\}_{\gamma=\alpha}^{\gamma_0} \qquad \text{and} \qquad \left\{ hf_\gamma \right\}_{\gamma=\alpha}^{\gamma_0}$$

are also convergent. Moreover,

$$\lim_{\gamma \to \gamma_0} \left[a f_\gamma + b g_\gamma \right] = a \lim_{\gamma \to \gamma_0} f_\gamma + b \lim_{\gamma \to \gamma_0} g_\gamma \quad \text{and} \quad \lim_{\gamma \to \gamma_0} \left[h f_\gamma \right] = h \lim_{\gamma \to \gamma_0} f_\gamma \quad .$$

PROOF: Since $\{f_\gamma\}_{\gamma=\alpha}^{\gamma_0}$ and $\{g_\gamma\}_{\gamma=\alpha}^{\gamma_0}$ are convergent, their respective limits,

$$\lim_{\gamma \to \gamma_0} f_\gamma \quad \text{and} \quad \lim_{\gamma \to \gamma_0} g_\gamma \quad ,$$

are well-defined generalized functions. Hence, so are

$$a \lim_{\gamma \to \gamma_0} f_\gamma + b \lim_{\gamma \to \gamma_0} g_\gamma \quad \text{and} \quad h \lim_{\gamma \to \gamma_0} f_\gamma \quad .$$

Now, to prove our lemma, it will suffice to confirm that, for each Gaussian test function ϕ,

$$\lim_{\gamma \to \gamma_0} \left\langle a f_\gamma + b g_\gamma , \phi \right\rangle = \left\langle a \lim_{\gamma \to \gamma_0} f_\gamma + b \lim_{\gamma \to \gamma_0} g_\gamma , \phi \right\rangle \tag{34.2}$$

and

$$\lim_{\gamma \to \gamma_0} \left\langle h f_\gamma , \phi \right\rangle = \left\langle h \lim_{\gamma \to \gamma_0} f_\gamma , \phi \right\rangle \quad . \tag{34.3}$$

So let ϕ be an arbitrary Gaussian test function.

Applying what we know about linear combinations of generalized functions and classical limits, we obtain

$$\lim_{\gamma \to \gamma_0} \left\langle a f_\gamma + b g_\gamma , \phi \right\rangle = \lim_{\gamma \to \gamma_0} \left[a \left\langle f_\gamma , \phi \right\rangle + b \left\langle g_\gamma , \phi \right\rangle \right]$$

$$= a \lim_{\gamma \to \gamma_0} \left\langle f_\gamma , \phi \right\rangle + b \lim_{\gamma \to \gamma_0} \left\langle g_\gamma , \phi \right\rangle$$

$$= a \left\langle \lim_{\gamma \to \gamma_0} f_\gamma , \phi \right\rangle + b \left\langle \lim_{\gamma \to \gamma_0} g_\gamma , \phi \right\rangle$$

$$= \left\langle a \lim_{\gamma \to \gamma_0} f_\gamma + b \lim_{\gamma \to \gamma_0} g_\gamma , \phi \right\rangle \quad ,$$

confirming equation (34.2). To confirm equation (34.3), we simply apply the definition of multiplication by a simple multiplier:

$$\lim_{\gamma \to \gamma_0} \left\langle h f_\gamma , \phi \right\rangle = \lim_{\gamma \to \gamma_0} \left\langle f_\gamma , h\phi \right\rangle = \left\langle \lim_{\gamma \to \gamma_0} f_\gamma , h\phi \right\rangle = \left\langle h \lim_{\gamma \to \gamma_0} f_\gamma , \phi \right\rangle \quad . \quad \blacksquare$$

Generalized Limits

In practice, sequences are often not explicitly given, but are implicitly described by some given "limit" expression. This is something you should already be used to from calculus, and there should be no need for us to go into detail about how to interpret all the different possible "limit expressions" that can be defined.

For example, if we write

$$\lim_{\gamma \to 0^+} f_\gamma \quad ,$$

then we are clearly considering the limit of a sequence $\{f_\gamma\}_{\gamma=\alpha}^{0}$ in which $0 < \gamma$ for each index γ, and rather than saying "the sequence does (or does not) converge", we may say "the (one-sided) limit does (or does not) exist".

Likewise, when we write

$$\lim_{\gamma \to 0^-} f_\gamma \quad ,$$

we are considering the limit of a sequence $\{f_\gamma\}_{\gamma=\alpha}^0$ in which $\gamma < 0$ for each index γ, and, rather than saying "the sequence does (or does not) converge", we may say "the (one-sided) limit does or (does not) exist".

On the other hand, if we have

$$\lim_{\gamma \to 0} f_\gamma \text{ exists}$$

(and the γ's are real numbers), then you should realize that the one-sided limits

$$\lim_{\gamma \to 0^-} f_\gamma \qquad \text{and} \qquad \lim_{\gamma \to 0^+} f_\gamma$$

exist, are equal, and each gives the value of $\lim_{\gamma \to 0} f_\gamma$.

!▶**Example 34.2:**   Let f be any fixed generalized function. Lemma 33.2 on page 554 assures us that

$$\lim_{\epsilon \to 0^+} \left\langle f(x), e^{-\epsilon\pi x^2}\phi(x) \right\rangle = \left\langle f(x), \phi(x) \right\rangle \qquad \text{for each } \phi \text{ in } \mathcal{G} \quad .$$

Combining this with the definition of simple multiplication yields

$$\lim_{\epsilon \to 0^+} \left\langle f(x)e^{-\epsilon\pi x^2}, \phi(x) \right\rangle = \lim_{\epsilon \to 0^+} \left\langle f(x), e^{-\epsilon\pi x^2}\phi(x) \right\rangle = \left\langle f(x), \phi(x) \right\rangle$$

for each ϕ in $\mathcal{G}$. Thus,

$$\lim_{\epsilon \to 0^+} f(x)e^{-\epsilon\pi x^2} = f(x) \qquad \text{(in the generalized sense)} \quad .$$

?▶**Exercise 34.2:**   *Assume our indices may be complex. What does it mean to say*

$$\lim_{\gamma \to i} f_\gamma \text{ exists} \quad ?$$

Caveats on Computing Limits

Some care must be taken in computing generalized limits. For example, suppose we are to compute the generalized limits of

$$e^{\gamma x} \text{step}(x) \quad , \qquad \frac{\gamma}{\gamma^2 + x^2} \quad \text{and} \qquad \frac{x}{\gamma^2 + x^2} \qquad \text{as } \gamma \to 0 \quad .$$

The naive approach would be to simply replace the γ's with 0, giving

$$e^{0x} \text{step}(x) = \text{step}(x) \quad , \qquad \frac{0}{0^2 + x^2} = 0 \quad \text{and} \qquad \frac{x}{0^2 + x^2} = \frac{1}{x} \quad .$$

Unfortunately, a more careful analysis will show that *two of these are wrong!* (Actually, the fact that the last computation yields $1/x$ — which is not an exponentially integrable function and, thus, cannot be viewed as a generalized function — suggests that something is wrong with that computation.)

Remember, to verify

$$\lim_{\gamma \to \gamma_0} f_\gamma = g \qquad \text{(in the generalized sense)} \quad,$$

we need to show

$$\langle g, \phi \rangle = \lim_{\gamma \to \gamma_0} \langle f_\gamma, \phi \rangle \qquad \text{for each } \phi \text{ in } \mathcal{G} \quad.$$

If the f_γ's happen to be classical functions, then we can write this last equation as

$$\langle g, \phi \rangle = \lim_{\gamma \to \gamma_0} \int_{-\infty}^{\infty} f_\gamma(x)\phi(x)\,dx \qquad \text{for each } \phi \text{ in } \mathcal{G} \quad. \tag{34.4}$$

Consequently, a more careful computation of a generalized limit sometimes should involve a careful analysis of a limit of some parameterized sequence of integrals. Some very general results regarding such limits were developed in chapter 18 (see theorem 18.18 on page 271). Using those results and the mean value theorem for derivatives, we can obtain the following theorem, which can be useful for computing the limit in equation (34.4). The details of its proof will be left to the interested reader (exercise 34.4 at the end of this chapter).

Theorem 34.2
Let $h(\gamma, x)$ be a piecewise continuous function on a strip $\mathcal{R} = (a, b) \times (-\infty, \infty)$ where (a, b) is a finite interval containing 0. Assume, further, all of the following:

1. *All discontinuities of h on $\mathcal{R}$ are contained in a finite collection of lines of the form $x = \text{constant}$.*

2. *There is a piecewise continuous and absolutely integrable function of one variable $\widehat{h}$ such that*
$$|h(\gamma, x)| \leq \widehat{h}(x) \qquad \text{for each } (\gamma, x) \text{ in } \mathcal{R} \quad.$$

3. *On $\mathcal{R}$, $h(\gamma, x)$ is a continuous and piecewise smooth function of γ for all but, at most, a finite number of x's.*

4. *For some finite constant C,*
$$\left| \frac{\partial h}{\partial \gamma} \right| \leq C$$
   at every point in $\mathcal{R}$ where this partial derivative exists.

Then $h(\gamma, x)$ is an absolutely integrable function of x for each γ in (a, b) and

$$\lim_{\gamma \to 0} \int_{-\infty}^{\infty} h(\gamma, x)\,dx = \int_{-\infty}^{\infty} h(0, x)\,dx \quad.$$

Typically, we'll want to find the g such that, for each Gaussian test function ϕ,

$$\langle g, \phi \rangle = \lim_{\gamma \to \gamma_0} \int_{-\infty}^{\infty} f(\gamma, x)\phi(x)\,dx$$

(this is equation (34.4) using the notation $f(\gamma, x)$ instead of $f_\gamma(x)$). So we will be applying the above theorem with

$$h(\gamma, x) = f(\gamma, x)\phi(x) \qquad \text{and} \qquad \frac{\partial h}{\partial \gamma} = \frac{\partial f}{\partial \gamma}\phi$$

for an arbitrary ϕ in $\mathcal{G}$. After recalling that any such ϕ is infinitely smooth and "very rapidly vanishes at infinity", you'll find it easy to derive (exercise 34.4 at the end of this chapter) the following corollary to the above theorem. This gives a relatively simple test for determining when generalized limits can be computed naively.

Corollary 34.3

Let $f(\gamma, x)$ be a piecewise continuous function on a strip $\mathcal{R} = (a, b) \times (-\infty, \infty)$ where (a, b) is a finite interval containing 0. Assume, further, all of the following:

1. All discontinuities of f on $\mathcal{R}$ are contained in a finite collection of lines of the form $x = constant$.

2. There are finite constants B and β such that

$$|f(\gamma, x)| \leq Be^{\beta|x|}$$

    at every point in $\mathcal{R}$ where f is continuous.

3. On $\mathcal{R}$, $f(\gamma, x)$ is a continuous and piecewise smooth function of γ for all but, at most, a finite number of x's.

4. There are finite constants C and σ such that

$$\left| \frac{\partial f}{\partial \gamma} \right| \leq Ce^{\sigma|x|}$$

    at every point in $\mathcal{R}$ where this partial derivative exists.

Then $f(\gamma, x)$ is an exponentially integrable function of x for each γ in (a, b) and

$$\lim_{\gamma \to 0} f(\gamma, x) = f(0, x) \qquad \text{(in the generalized sense)} \quad.$$

!▶Example 34.3: Consider the generalized limit

$$\lim_{\gamma \to 0} e^{-\gamma x} \operatorname{step}(x) \quad.$$

To see if the above corollary applies, we'll let

$$f(\gamma, x) = e^{-\gamma x} \operatorname{step}(x) \qquad \text{and} \qquad (a, b) = (-1, 1) \quad.$$

This is certainly a piecewise continuous function on $\mathcal{R} = (-1, 1) \times (-\infty, \infty)$. Checking the other conditions required by the corollary:

1. Since the only discontinuities in $f(\gamma, x)$ are in the line $x = 0$ (where the step function has a jump discontinuity), the first condition holds.

2. Because $-1 < \gamma < 1$ and $|\operatorname{step}(x)| \leq 1$, we have

$$\left| e^{-\gamma x} \operatorname{step}(x) \right| \leq e^{|x|} \qquad \text{for each } (\gamma, x) \text{ in } \mathcal{R} \quad,$$

    confirming the second condition.

3. Since $e^{-\gamma x} \operatorname{step}(x)$ is certainly a smooth function of γ whenever $x \neq 0$, the third condition holds.

4. *Recalling that* $|x| < e^{|x|}$, *we see that, whenever* $x \neq 0$ *and* $-1 < \gamma < 1$,

$$\left|\frac{\partial f}{\partial \gamma}\right| = \left|-xe^{-\gamma x} \,\text{step}(x)\right| \leq |x|\,e^{|x|} < e^{2|x|} \quad,$$

confirming that the fourth condition holds.

So the conditions required by the corollary hold. Consequently, the corollary assures us that, in the generalized sense,

$$\lim_{\gamma \to 0} e^{-\gamma x}\,\text{step}(x) = e^{-0x}\,\text{step}(x) = 1 \cdot \text{step}(x) = \text{step}(x) \quad.$$

Corollary 34.3 gives conditions under which a generalized limit

$$\lim_{\gamma \to 0} f(\gamma, x)$$

can be computed by naively replacing γ with 0 in the formula for $f(\gamma, x)$. If those conditions do not hold, then you had better analyze the limit in equation (34.4) carefully.

In practice, it is typically the second condition in corollary 34.3 which fails to hold when the naive approach to computing the generalized limit fails. So that is a good condition to check first. In particular, see if there is a point x_0 such that $f(\gamma, x_0)$ is continuous at (γ, x_0) when $\gamma \neq 0$, but for which

$$\lim_{\gamma \to 0} (\gamma, x_0) = \infty \quad.$$

If such an x_0 exists, then $f(\gamma, x_0)$ cannot be bounded as described in the second condition, and you are thereby warned that naively replacing γ with 0 in the formula for $f(\gamma, x)$ will probably not give you the generalized limit you seek.

!►*Example 34.4:* *Consider the generalized limit*

$$\lim_{\gamma \to 0} f(\gamma, x) \qquad \text{with} \qquad f(\gamma, x) = \frac{\gamma}{\gamma^2 + x^2} \quad.$$

Clearly, $f(\gamma, x)$ *is continuous, at least wherever* $\gamma \neq 0$. *Thus, if the second condition in corollary 34.3 holds, there are finite constants* B *and* β *such that*

$$\left|\frac{\gamma}{\gamma^2 + x^2}\right| \leq Be^{\beta|x|} \tag{34.5}$$

for every γ *in some nontrivial interval* (a, b) *containing* 0. *In particular, at least for every nonzero* γ *in* (a, b),

$$\left|\frac{\gamma}{\gamma^2 + 0^2}\right| \leq B \quad. \tag{34.6}$$

However,

$$\lim_{\gamma \to 0} \frac{\gamma}{\gamma^2 + 0^2} = \lim_{\gamma \to 0} \frac{1}{\gamma} = \infty \quad.$$

So there cannot be a finite B *such that inequality (34.6) holds. Thus, the second condition in corollary 34.3 does not hold for our sequence, and the corollary does not assure us that the generalized limit*

$$\lim_{\gamma \to 0} \frac{\gamma}{\gamma^2 + x^2}$$

can be computed by simply letting $\gamma = 0$. *Instead, we must consider more carefully the limit*

$$\lim_{\gamma \to 0} \int_{-\infty}^{\infty} \frac{\gamma}{\gamma^2 + x^2} \phi(x) \, dx$$

for an arbitrary Gaussian test function ϕ. *(In fact, this limit does not exist, but the one-sided limits as* $\gamma \to 0^+$ *and* $\gamma \to 0^-$ *do. You will verify that in exercises 34.6 d and 34.6 e.)*

Finally, let me remind you that you can usually convert any limit to a limit of the form considered in corollary 34.3 via some simple substitution. For example,

$$f(\gamma, x) = g(|\gamma|, x) \quad\quad \text{converts} \quad\quad \lim_{\gamma_0 \to 0^+} g(\gamma_0, x) \quad\quad \text{to} \quad\quad \lim_{\gamma \to 0} f(\gamma, x) \quad,$$

and

$$f(\gamma, x) = h\left(\frac{1}{|\gamma|}, x\right) \quad\quad \text{converts} \quad\quad \lim_{\beta \to \infty} h(\beta, x) \quad\quad \text{to} \quad\quad \lim_{\gamma \to 0} f(\gamma, x) \quad.$$

34.2 Infinite Series (Summations)

Now suppose we have

$$\sum_{k=k_0}^{\infty} g_k$$

where each g_k is a generalized function and k_0 is some fixed integer. Since any infinite series is simply the limit of its partial sums,

$$\sum_{k=k_0}^{\infty} g_k = \lim_{N \to \infty} \sum_{k=k_0}^{N} g_k \quad,$$

the basic theory of generalized infinite series follows immediately from the theory of generalized limits just discussed. Consequently, we'll say that our infinite series *converges (in the generalized sense)* if and only if there is a generalized function f such that

$$f = \lim_{N \to \infty} \sum_{k=k_0}^{N} g_k \quad.$$

Remember, this equation means

$$\langle f, \phi \rangle = \lim_{N \to \infty} \left\langle \sum_{k=k_0}^{N} g_k, \phi \right\rangle \quad\quad \text{for each } \phi \text{ in } \mathcal{G} \quad. \tag{34.7}$$

Naturally, if this is the case, then the generalized function f will be referred to as the *sum* of the infinite series, and we will write

$$f = \sum_{k=k_0}^{\infty} g_k \quad.$$

With this notation and the observation that

$$\lim_{N \to \infty} \left\langle \sum_{k=k_0}^{N} g_k , \phi \right\rangle = \lim_{N \to \infty} \sum_{k=k_0}^{N} \left\langle g_k , \phi \right\rangle = \sum_{k=k_0}^{\infty} \left\langle g_k , \phi \right\rangle \quad,$$

we see that, if our series of generalized functions is convergent, then the classical infinite series of numbers

$$\sum_{k=k_0}^{\infty} \left\langle g_k , \phi \right\rangle$$

also converges for each Gaussian test function ϕ, and $\sum_{k=k_0}^{\infty} g_k$ is the generalized function such that

$$\left\langle \sum_{k=k_0}^{\infty} g_k , \phi \right\rangle = \sum_{k=k_0}^{\infty} \left\langle g_k , \phi \right\rangle \qquad \text{for each } \phi \text{ in } \mathscr{G} \quad.$$

As with limits, we will include the qualifying phrase "in the generalized sense" (or some other phrase to that effect) when there is some danger of confusion, and often omit it otherwise.

!► Example 34.5:   *Consider the following infinite series of rectangle functions*

$$\sum_{k=0}^{\infty} \text{rect}_{(k,k+1)} \quad.$$

If ϕ is any Gaussian test function,

$$\left\langle \sum_{k=0}^{N} \text{rect}_{(k,k+1)} , \phi \right\rangle = \sum_{k=0}^{N} \left\langle \text{rect}_{(k,k+1)} , \phi \right\rangle$$

$$= \sum_{k=0}^{N} \int_{-\infty}^{\infty} \text{rect}_{(k,k+1)}(x) \, \phi(x) \, dx$$

$$= \sum_{k=0}^{N} \int_{k}^{k+1} \phi(x) \, dx$$

$$= \left[\int_{0}^{1} \phi(x) \, dx + \int_{1}^{2} \phi(x) \, dx + \cdots + \int_{N}^{N+1} \phi(x) \, dx \right]$$

$$= \int_{0}^{N+1} \phi(x) \, dx \quad.$$

So,

$$\lim_{N \to \infty} \left\langle \sum_{k=0}^{N} \text{rect}_{(k,k+1)} , \phi \right\rangle = \lim_{N \to \infty} \int_{0}^{N+1} \phi(x) \, dx$$

$$= \int_{0}^{\infty} \phi(x) \, dx = \left\langle \text{step} , \phi \right\rangle \quad,$$

and thus, in the generalized sense,

$$\sum_{k=0}^{\infty} \text{rect}_{(k,k+1)} = \text{step} \quad. \tag{34.8}$$

It's worth noting that both sides of the last equation are well-defined, piecewise continuous functions, and, as you can easily confirm (after removing the trivial discontinuities from the left-hand side),

$$\sum_{k=0}^{\infty} \text{rect}_{(k,k+1)}(x) = \text{step}(x)$$

for each real nonzero value x. Thus, equation (34.8) holds in a classical sense, as well as in the generalized sense.

(Later in this chapter, we will discuss some infinite series that cannot be interpreted classically. These will be the "arrays of delta functions".)

The theory for two-sided infinite series of generalized functions is the straightforward extension of that for one-sided series. Such a series,

$$\sum_{k=-\infty}^{\infty} g_k \quad ,$$

is convergent (in the generalized sense) and equals a generalized function f if and only if

$$\langle f, \phi \rangle = \lim_{\substack{N \to \infty \\ M \to -\infty}} \left\langle \sum_{k=M}^{N} g_k, \phi \right\rangle \qquad \text{for each } \phi \text{ in } \mathcal{G} \quad . \tag{34.9}$$

If this series is convergent, then, for each Gaussian test function ϕ, the two-sided infinite series of numbers

$$\sum_{k=-\infty}^{\infty} \langle g_k, \phi \rangle$$

converges, and the generalized function that we can denote by either f or $\sum_{k=-\infty}^{\infty} g_k$ is the generalized function satisfying

$$\left\langle \sum_{k=-\infty}^{\infty} g_k, \phi \right\rangle = \sum_{k=-\infty}^{\infty} \langle g_k, \phi \rangle \qquad \text{for each } \phi \text{ in } \mathcal{G} \quad .$$

The next lemma assures us that at least some of the manipulations commonly performed using classical infinite series are valid with infinite series of generalized functions. It is an analog of lemma 34.1 and can be proven directly from that lemma.

Lemma 34.4

Let k_0 denote either some integer or $-\infty$, and assume

$$\sum_{k=k_0}^{\infty} f_k \qquad \text{and} \qquad \sum_{k=k_0}^{\infty} g_k$$

are two convergent infinite series of generalized functions. Then, for any two constants a and b, and any simple multiplier h,

$$\sum_{k=k_0}^{\infty} [af_k + bg_k] \qquad \text{and} \qquad \sum_{k=k_0}^{\infty} hf_k$$

are also convergent. Moreover,

$$\sum_{k=k_0}^{\infty} [af_k + bg_k] = a \sum_{k=k_0}^{\infty} f_k + b \sum_{k=k_0}^{\infty} g_k \qquad \text{and} \qquad \sum_{k=k_0}^{\infty} hf_k = h \sum_{k=k_0}^{\infty} f_k \quad .$$

?►Exercise 34.3:   *Prove the above lemma using lemma 34.1 on page 568.*

Finally, keep in mind that equations (34.7) and (34.9) are, respectively, equivalent to

$$\lim_{N \to \infty} \left| \left\langle f - \sum_{k=k_0}^{N} g_k , \phi \right\rangle \right| = 0 \qquad \text{for each } \phi \text{ in } \mathscr{G} \qquad (34.7')$$

and

$$\lim_{\substack{N \to \infty \\ M \to -\infty}} \left| \left\langle f - \sum_{k=M}^{N} g_k , \phi \right\rangle \right| = 0 \qquad \text{for each } \phi \text{ in } \mathscr{G} \quad . \qquad (34.9')$$

For a given infinite series, one of these equations may be relatively easy to verify, especially if the series can be viewed as a classical series of functions for which pointwise error bounds have already been determined (as, for example, in exercises 34.8 and 34.9 at the end of this chapter.)

34.3 A Little More on Delta Functions
Delta Functions on the Real Line

Let's extend some of the observations made in example 34.1 and exercise 34.1.

Suppose c is a real number and $\{h_\gamma\}_{\gamma=\alpha}^{\gamma_0}$ is any of our favorite identity sequences for the set of Gaussian test functions. Then, by the nature of identity sequences,

$$\lim_{\gamma \to \gamma_0} \left\langle h_\gamma(x - c) , \phi(x) \right\rangle = \lim_{\gamma \to \gamma_0} \int_{-\infty}^{\infty} h_\gamma(x - c) \phi(x) \, dx = \phi(c) = \left\langle \delta_c(x) , \phi(x) \right\rangle$$

for any Gaussian test function ϕ. This tells us that

$$\lim_{\gamma \to \gamma_0} h_\gamma(x - c) = \delta_c(x) \qquad \text{(in the generalized sense)}$$

no matter which identity sequence we use. It also tells us that, when γ is close to γ_0, we can view the classical function $h_\gamma(x - c)$ as an *approximation* for the generalized function δ_c. In particular, using the pulse function identity sequence from example 34.1, $\{h_\epsilon\}_{\epsilon=1}^{0}$ with

$$h_\epsilon(x) = \frac{1}{2\epsilon} \text{pulse}_\epsilon(x) \quad ,$$

we can view $h_\epsilon(x - c)$ as approximating δ_c whenever ϵ is a small positive value.

Now consider "the graph of $\delta_c(x)$". Strictly speaking, this is a meaningless concept. We cannot plot the "value of $\delta_c(x)$" for each real value x, because there is no formula that can give the "value of $\delta_c(x)$" for each real value x (as we saw in chapter 31). We can, however, visualize δ_c by using certain geometric properties of the graphs of many identity sequences approximating this delta function. For simplicity, we'll use the pulse function identity sequence, though the same results could also be derived using, say, any Gaussian identity sequence.

As we just noted, "$\delta_c(x) \approx h_\epsilon(x - c)$" when ϵ is a small positive value and

$$h_\epsilon(x - c) = \frac{1}{2\epsilon} \text{pulse}(x - c) \quad .$$

The graphs of a few such $h_\epsilon(x - c)$'s have been sketched in figure 34.1a (with $c = 0$). As we saw back in chapter 29, each of these graphs has zero height everywhere except for a small interval

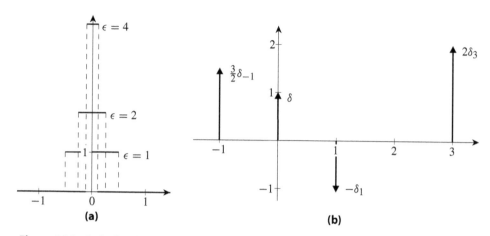

Figure 34.1: Delta functions — (a) viewed as a limit of pulse functions, and (b) graphically represented using vertical arrows.

about the point $x = c$ where the graph becomes a tall and narrow "spike". As $\epsilon \to 0$ the height of this spike becomes infinite while the width shrinks to 0. However, the area enclosed by this spike is always the same,

$$\text{area} = \text{base} \times \text{height} = (2\epsilon) \times \left(\frac{1}{2\epsilon}\right) = 1 \quad .$$

As $\epsilon \to 0$, these graphs seem to approach an "infinitely tall, infinitesimally narrow spike at $x = c$ enclosing unit area". Because of this, some people visualize the graph of δ_c as this idealized spike.

Do not take this visualization too seriously.[2] It is simply a graphical reminder that δ_c can be approximated, in some sense, by a function whose graph is a tall, narrow spike enclosing unit area. And don't trust any computation based on this visualization unless you can verify the computation independently using legitimate techniques. There is too much danger of misusing the "infinities" and deriving a result that appears reasonable but is actually equivalent to " $1 = 0$ ", just as we derived when treating the delta function as a classical function in chapter 31.

Graphing an "infinitely tall, infinitesimally narrow spike" in such a way as to indicate that it, somehow, encloses unit area is a task beyond the graphical talents of most humans and computers. As an alternative, it has become standard practice to graphically represent a delta function at $x = c$ by an upward pointing arrow of unit length starting at $x = c$. Naturally, given any real constant A, the graphical representation of $A\delta_c$ is just that arrow with length multiplied by $|A|$, pointing upward if A is positive and pointing downward if A is negative. And if A is complex, say $A = a + ib$, then we sketch the graphical representations of the real and imaginary parts, $a\delta_c$ and $b\delta_c$, separately. A few examples are sketched in figure 34.1b.

Delta Functions off the Real Line

Much of what we just said about approximating and visualizing δ_c falls apart when c is not a real number. How, for example, could we claim that δ_i is approximated by any function of the form

$$\frac{1}{2\epsilon} \text{pulse}(x - i)$$

if the pulse function is not even defined when the input, $x - i$, is not real?

---

[2] Though it is useful if you are trying to teach "all" of Fourier analysis in two weeks. If your students can accept "boxes of zero width with unit area", they will accept anything else you tell them.

On the other hand, Gaussian functions are defined on the complex plane. So, if $\{h_\gamma\}_{\gamma=\alpha}^{\gamma_0}$ is a Gaussian identity sequence, say, with

$$h_\gamma(x) = \sqrt{\gamma}\, e^{-\gamma\pi x^2} \quad ,$$

then it does make sense to write $h_\gamma(x - c)$ when c is complex and x is real. In fact, eventually, we will verify in exercise 35.46 on page 625) that

$$\lim_{\gamma\to\infty} \sqrt{\gamma}\, e^{-\gamma\pi(x-c)^2} = \delta_c(x) \qquad \text{(in the generalized sense)}$$

when c is any complex value. Thus, for example, if γ is fairly large, we can treat

$$\sqrt{\gamma}\, e^{-\gamma\pi(x-i)^2}$$

as an approximation to δ_i. The difficulty here is that we need to treat this approximation as a complex-valued function on the complex plane. The graphs of such functions are a good deal more complex (and more difficult to sketch) than those used in the previous subsection to justify the visualization given there. Consequently, it probably is not a good idea to try to visualize δ_c as an "infinitely tall and infinitesimally narrow spike at $z = c$" when c is not on the real line.

We could, presumably, still "graphically represent" delta functions at complex points by little arrows "pointing upward" from corresponding points on the complex plane — provided we don't try to attach too much geometric meaning to these arrows. Truth is, though, that the classical functions and the delta functions on the real line are pretty well all of the basic generalized functions for which graphical representations are practical. So don't worry about "graphing" these other generalized functions any more than you would worry about counting from 1 to $\sqrt{2} + 3i$ on your fingers.

34.4 Arrays of Delta Functions

An *array of delta functions* is simply a summation in which each term is a constant times a delta function. When ambiguity is not a danger, we will refer to such a summation as simply an *array*. The array may be *finite* (i.e., have a finite number of terms), as with

$$3\delta_{-1} + 6\delta_0 - 4\delta_2 \quad ,$$

or it may be *infinite* (i.e., have an infinite number of terms), as with

$$\sum_{k=-\infty}^{\infty} \frac{1}{1+k^2}\delta_{2k} \quad .$$

The constants in the summation will, naturally enough, be referred to as the *coefficients* of the array, and the set of points where the array "has delta functions" will be called the *support* of the array. For example, the coefficients of the first array above are 3, 6 and -4, and the corresponding support is the set $\{-1,\, 0,\, 2\}$. For the second array, three of the coefficients are

$$\frac{1}{1+0^2} = 1 \quad , \qquad \frac{1}{1+1^2} = \frac{1}{2} \quad \text{and} \qquad \frac{1}{1+2^2} = \frac{1}{5} \quad ,$$

and the support is the set of all even integers,

$$\{\ldots,\, -4,\, -2,\, 0,\, 2,\, 4,\, 6,\, \ldots\} \quad .$$

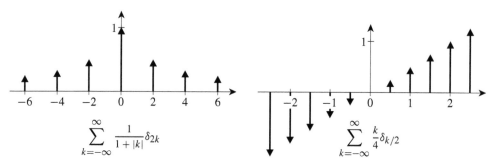

Figure 34.2: Two arrays of delta functions.

Two other arrays have been sketched ("graphically represented") in figure 34.2.

For convenience, let's agree that whatever symbol denotes the array can also be used — with suitable subscripts — to denote the individual coefficients. Thus, if

$$ g = \sum_{k=-\infty}^{\infty} \frac{1}{1+k^2} \delta_{2k} \quad , $$

then it is understood that

$$ g_k = \frac{1}{1+k^2} \quad \text{for} \quad k = 0, \pm1, \pm2, \pm3, \ldots \quad . $$

More commonly, we will say something like "assume f is an arbitrary array", and write

$$ f = \sum_{k=-\infty}^{\infty} f_k \, \delta_{x_k} \quad . $$

Note that this last expression does not necessarily mean that f is an infinite array. After all, any finite array can also be written this way with all but a finite number of the coefficients being zero. For example, the array

$$ 3\delta_{-1} + 6\delta_0 - 4\delta_2 \quad , $$

can be written as

$$ \sum_{k=-\infty}^{\infty} f_k \, \delta_k \quad \text{where} \quad f_k = \begin{cases} 3 & \text{if} \quad k = -1 \\ 6 & \text{if} \quad k = 0 \\ -4 & \text{if} \quad k = 2 \\ 0 & \text{otherwise} \end{cases} \quad . $$

To further simplify future discussions, let us refer to an array f as being *regular* if and only if there is a positive value Δx, called the *spacing* of the array, such that

$$ f = \sum_{k=-\infty}^{\infty} f_k \, \delta_{k\Delta x} \quad . $$

In other words, a regular array of delta functions with spacing Δx is an array whose support is contained in the set of integral multiples of the spacing,

$$ \{\ldots, -2\Delta x, -1\Delta x, 0\Delta x, 1\Delta x, 2\Delta x, 3\Delta x, \ldots\} \quad . $$

Most arrays that will be of interest to us will be regular arrays.

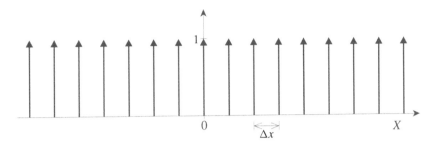

Figure 34.3: The comb function with spacing Δx .

!► Example 34.6 (comb functions): *The "comb functions" make up an especially important class of regular arrays. For any positive value* Δx *, the comb function with spacing* Δx *, denoted by* $\mathrm{comb}_{\Delta x}$ *, is the array*

$$\mathrm{comb}_{\Delta x} \;=\; \sum_{k=-\infty}^{\infty} \delta_{k\Delta x} \quad . \tag{34.10}$$

You can see how it got its name from figure 34.3.[3]

You should realize that, while individual delta functions are generalized functions, it is quite possible to create an array of delta functions that is not a generalized function. Certainly, every finite array is a generalized function since a finite array is just a finite linear combination of generalized functions. For an infinite array to define a generalized function, however, it must converge as an infinite series of generalized functions.

To see the main issue involved, consider computing $\langle f , \phi \rangle$ for some Gaussian test function ϕ and some regular array

$$f \;=\; \sum_{k=-\infty}^{\infty} f_k \, \delta_{k\Delta x} \quad . \tag{34.11}$$

Whether or not we have convergence, we can write

$$\langle f , \phi \rangle = \left\langle \sum_{k=-\infty}^{\infty} f_k \, \delta_{k\Delta x} , \phi \right\rangle$$

$$= \lim_{\substack{N \to \infty \\ M \to -\infty}} \left\langle \sum_{k=M}^{N} f_k \, \delta_{k\Delta x} , \phi \right\rangle$$

$$= \lim_{\substack{N \to \infty \\ M \to -\infty}} \sum_{k=M}^{N} f_k \left\langle \delta_{k\Delta x} , \phi \right\rangle$$

$$= \lim_{\substack{N \to \infty \\ M \to -\infty}} \sum_{k=M}^{N} f_k \, \phi(k\Delta x) = \sum_{k=-\infty}^{\infty} f_k \, \phi(k\Delta x) \quad .$$

Clearly then, if f is truly a generalized function; that is, if the array in formula (34.11) converges in the generalized sense, then the last infinite series above must converge to a finite value for each

---
[3] Provided you can imagine an infinitely long comb whose teeth are widely spaced and barbed. Still, "the comb function" is probably a better name than "the picket fence function".

ϕ in $\mathcal{G}$. Conversely, if this last series does not converge to a finite number for some ϕ in $\mathcal{G}$, then the array in formula (34.11) cannot be convergent in the generalized sense and, thus, does not define a generalized function.

!► **Example 34.7:**   *Consider the array*

$$ f = \sum_{k=-\infty}^{\infty} e^{k^2} \delta_k \quad . $$

Computing $\langle f, \phi \rangle$ *using the Gaussian test function* $\phi(x) = e^{-x^2}$ *gives*

$$ \langle f, \phi \rangle = \left\langle \sum_{k=-\infty}^{\infty} e^{k^2} \delta_k(x), e^{-x^2} \right\rangle $$

$$ = \sum_{k=-\infty}^{\infty} e^{k^2} \left\langle \delta_k(x), e^{-x^2} \right\rangle $$

$$ = \sum_{k=-\infty}^{\infty} e^{k^2} e^{-k^2} $$

$$ = \sum_{k=-\infty}^{\infty} 1 = \infty \quad . $$

Thus, this array of delta functions does not converge in the generalized sense, and we cannot view f as a generalized function.

To help guarantee convergence, we may require our regular array f to have *exponentially bounded coefficients*; that is, we may require all the coefficients to satisfy

$$ |f_k| \leq M e^{\gamma |k|} $$

for some fixed pair of positive constants M and γ.

Lemma 34.5
Any regular array of delta functions having exponentially bounded coefficients is a generalized function.

PROOF: Assume

$$ f = \sum_{k=-\infty}^{\infty} f_k \delta_{k \Delta x} $$

is a regular array, and M and γ are constants such that

$$ |f_k| \leq M e^{\gamma |k|} \qquad \text{for} \quad k = 0, \pm 1, \pm 2, \pm 3, \ldots \quad . $$

Also, for convenience, let $\beta = {}^{(1+\gamma)}/_{\Delta x}$ (so $\gamma - \beta \Delta x = -1$).
As we saw above,

$$ \langle f, \phi \rangle = \sum_{k=-\infty}^{\infty} f_k \phi(k \Delta x) \quad . \tag{34.12} $$

To show this defines a generalized function (i.e., a continuous linear functional on $\mathcal{G}$), we will first confirm that the summation on the right converges for each ϕ in $\mathcal{G}$ (thus verifying that f is a well-defined functional on $\mathcal{G}$). Then we will verify that this functional is linear. Verifying the functional continuity of f will be left as an exercise (exercise 34.13 at the end of this chapter) for those acquainted with the details of functional continuity developed at the end of the last chapter (sections 33.5 and 33.6).

So let ϕ be any Gaussian test function. From lemma 32.12 on page 528, we know there is a corresponding constant B_β such that

$$|\phi(x)| \leq B_\beta e^{-\beta|x|} \qquad \text{whenever} \qquad -\infty < x < \infty \quad .$$

Thus,

$$\sum_{k=-\infty}^{\infty} |f_k|\, |\phi(k\Delta x)| \leq \sum_{k=-\infty}^{\infty} M e^{\gamma|k|} B_\beta\, e^{-\beta|k\Delta x|}$$

$$\leq M B_\beta \sum_{k=-\infty}^{\infty} e^{(\gamma - \beta\Delta x)|k|} = M B_\beta \sum_{k=-\infty}^{\infty} e^{-|k|} \quad .$$

The last infinite series is a geometric series which you can easily verify as being convergent. That, along with the above inequalities, assures us that the infinite series in equation (34.12) is absolutely convergent, hence confirming that it converges for each ϕ in $\mathcal{G}$.

To verify the necessary linearity, let ϕ and ψ be any two Gaussian test functions and let a and b be any two constants. From the previous paragraph, we know

$$\sum_{k=-\infty}^{\infty} f_k\, \phi(k\Delta x) \qquad \text{and} \qquad \sum_{k=-\infty}^{\infty} f_k\, \psi(k\Delta x)$$

are absolutely convergent infinite series of numbers. Thus,

$$\langle f, a\phi + b\psi \rangle = \sum_{k=-\infty}^{\infty} f_k\big[a\phi(k\Delta x) + b\psi(k\Delta x)\big]$$

$$= a \sum_{k=-\infty}^{\infty} f_k\, \phi(k\Delta x) + b \sum_{k=-\infty}^{\infty} f_k\, \psi(k\Delta x)$$

$$= a\langle f, \phi \rangle + b\langle f, \psi \rangle \quad . \qquad \blacksquare$$

Additional Exercises

34.4 a. Prove theorem 34.2 on page 571. In addition to using theorem 18.18 on page 271, you should consider using the mean value theorem for derivatives (see your calculus text) to show that

$$|h(\gamma, x) - h(0, x)| \leq C\,|\gamma|$$

for all γ in (a, b) and all but a finite number of x's.

b. Obtain corollary 34.3 from theorem 34.2. (The bounds on Gaussian test functions discussed in lemma 32.12 on page 528 will be useful here.)

34.5. *For each of the following generalized limits, verify that corollary 34.3 applies, and compute the limit:*

    a. $\displaystyle\lim_{\gamma\to 0} e^{\gamma x}$ **b.** $\displaystyle\lim_{\gamma\to 0} e^{-\gamma|x|}$

    c. $\displaystyle\lim_{\gamma\to\infty} e^{-\gamma|x|}$ **d.** $\displaystyle\lim_{\gamma\to\infty} \frac{1}{\pi} \operatorname{Arctan}(\gamma x)$

34.6. *Compute each of the following generalized limits using the basic definition of generalized limits. That is, find each given $\lim_{\gamma\to\gamma_0} f_\gamma$ by determining the g such that*

$$\lim_{\gamma\to\gamma_0} \langle f_\gamma , \phi \rangle = \langle g , \phi \rangle \qquad \text{for each } \phi \text{ in } \mathcal{G} \quad .$$

In all but the first, you will need to analyze the limit of integrals corresponding to the limit of $\langle f_{\gamma_0} , \phi \rangle$'s.

    a. $\displaystyle\lim_{\gamma\to 0} \delta_\gamma$

    b. $\displaystyle\lim_{\gamma\to\infty} e^{-i2\pi\gamma x}$ *(Don't forget the integral formula for $\mathcal{F}[\phi]$.)*

    c. $\displaystyle\lim_{\gamma\to\infty} \gamma \operatorname{sinc}(2\pi\gamma x)$ *(Take a look at exercise 29.17 on page 490.)*

    d. $\displaystyle\lim_{\gamma\to 0^+} \frac{\gamma}{\gamma^2 + x^2}$ *(Consider theorem 34.2 after using the substitution $x = \gamma s$.)*

    e. $\displaystyle\lim_{\gamma\to 0^-} \frac{\gamma}{\gamma^2 + x^2}$ *(Consider theorem 34.2 after using the substitution $x = \gamma s$.)*

34.7. *Show that*

$$\lim_{\gamma\to\infty} \sin(\gamma x) = 0 \qquad \text{as generalized functions}$$

even though this limit does not exist classically for any nonzero value of x. (Try using integration by parts in computing $\langle \sin(\gamma x) , \phi(x) \rangle$.)

34.8. *Let f be a classical function that is continuous, piecewise smooth and periodic with period p.[4] Remember that its Fourier series is given by*

$$\sum_{k=-\infty}^{\infty} c_k \, e^{i2\pi\omega_k x}$$

where

$$c_k = \frac{1}{p} \int_0^p f(t) e^{-i2\pi\omega_k t}\, dt \qquad \text{and} \qquad \omega_k = \frac{k}{p} \quad .$$

Remember, also (see theorem 15.6 on page 206), that there is a fixed constant B such that

$$\left| f(x) - \sum_{k=-M}^{N} c_k \, e^{i2\pi\omega_k x} \right| \le \left[\frac{1}{\sqrt{M}} + \frac{1}{\sqrt{N}} \right] B$$

for every real value x and every pair of integers M and N with $M < 0 < N$.

    Using the above, confirm that

$$f = \sum_{k=-\infty}^{\infty} c_k \, e^{i2\pi\omega_k x} \qquad \text{(in the generalized sense)} \quad .$$

---

[4] More general periodic functions will be discussed in chapter 37.

34.9. Let $\sum_{k=0}^{\infty} g_k$ be an infinite series of classical functions that converges uniformly to a classical function f on the real line. Assume all the functions are exponentially integrable, and verify that

$$f = \sum_{k=0}^{\infty} g_k \qquad (in\ the\ generalized\ sense) \quad .$$

34.10 a. Show that, if h is any simple multiplier, and

$$f = \sum_{k=-\infty}^{\infty} f_k\ \delta_{k\Delta x}$$

is any regular array of delta functions, then

$$hf = \sum_{k=-\infty}^{\infty} h(k\Delta x) f_k\ \delta_{k\Delta x} \quad .$$

b. For each equation below, find a simple multiplier h satisfying the given equation.

i. $h\ \mathrm{comb}_1 = \sum_{k=-\infty}^{\infty} k^2\ \delta_k \quad .$

ii. $h\ \mathrm{comb}_1 = \sum_{k=-\infty}^{\infty} (-1)^k\ \delta_k \quad .$

iii. $h\ \mathrm{comb}_{\Delta x} = \sum_{k=-\infty}^{\infty} (-1)^k\ \delta_{k\Delta x} \quad .$

34.11. Sketch the graphical representations for the following arrays of delta functions.

a. $x\ \mathrm{comb}_1(x)$

b. $x\ \mathrm{comb}_{\Delta x}(x)$ where $\Delta x = 1/2$

c. $\sin(x)\ \mathrm{comb}_{\Delta x}(x)$ where $\Delta x = \pi/4$

d. $\sum_{k=-\infty}^{\infty} \dfrac{k^2}{10}\ \delta_{2k}$

34.12. Not all arrays are regular. For each of the following:

1. Explain why the given array is not a regular array.

2. Verify that it is a generalized function (assume functional continuity).

3. If the support of the given array is in $\mathbb{R}$, sketch the graphical representation of the array as well as you can.

a. $\sum_{k=-\infty}^{\infty} \delta_{i+k}$

b. $\sum_{k=-\infty}^{\infty} \dfrac{1}{k^2}\delta_{1/k}$

c. $\sum_{k=1}^{\infty} \dfrac{1}{k^3} \sum_{n=-k}^{k} \delta_{n/k}$

d. $\sum_{k=1}^{\infty} \dfrac{1}{k^3}\ \mathrm{comb}_{1/k}$

(The last one may be a real challenge.)

34.13. Assume f is a regular array with exponentially bounded coefficients. From the proof given for lemma 34.5, we know f is a linear functional on $\mathcal{G}$. Verify that it is also functionally continuous; that is, verify that there are positive constants M and a such that

$$\left| \langle f, \phi \rangle \right| \leq M \|\phi\|_a$$

for every Gaussian test function ϕ. (Hint: Use example 33.12 on page 560 as a guide.)

35

Basic Transforms
of Generalized Fourier Analysis

In this chapter, we will develop the generalized analogs of several basic operators from classical Fourier analysis and calculus. Naturally, we will start by generalizing the Fourier transforms, but we will also generalize the operations of variable scaling, translation and differentiation, and we will explore how these different operations interact.

You will probably notice that the opening paragraphs of the first few sections are strikingly (numbingly?) similar to each other. You might even notice that certain technical issues regarding whether the generalized transforms are "well defined" or "unique" are only lightly touched upon in these sections. The reason for all this is that each of these transforms is "adjoint defined". A detailed general discussion of adjoint-defined transforms occupies section 35.6. There we will define "adjoint-defined", lay heavy hands on those issues glossed over in the previous sections, and completely verify that all the transforms defined in this chapter are "well-defined, linear, continuous transforms that uniquely generalize the corresponding classical transforms".

35.1 Fourier Transforms
Definition and Examples

Our first attempt to generalize the classical Fourier transform was based on a test derived from the fundamental identity of Fourier analysis. Recall that identity:

$$\int_{-\infty}^{\infty} \mathcal{F}[f]|_x \, \phi(x) \, dx = \int_{-\infty}^{\infty} f(y) \mathcal{F}[\phi]|_y \, dy \quad .$$

We know this identity is valid whenever f and ϕ are "appropriately chosen" functions. In particular, we saw in chapter 32 that this identity holds whenever f is a classically transformable function and ϕ is a Gaussian test function (theorem 32.6 on page 520). In that case, though, $\mathcal{F}[\phi]$ is also a Gaussian test function, and both f and $\mathcal{F}[f]$ can be treated as generalized functions. Consequently, we can rewrite the above equation as

$$\langle \, \mathcal{F}[f] \, , \phi \, \rangle = \langle \, f \, , \mathcal{F}[\phi] \, \rangle \tag{35.1}$$

or, equivalently, as

$$\langle \, F \, , \phi \, \rangle = \langle \, f \, , \mathcal{F}[\phi] \, \rangle \tag{35.2}$$

where F is the classical transform of f.

The interesting thing about the last equation is that the right-hand side makes sense for *any* generalized function f, whether or not it is classically transformable. Moreover, as we saw in exercise 32.6 on page 520 (and will verify in detail later in section 35.6), for any given generalized function f, we can define a corresponding generalized function, call it $\widehat{f}$ for now, to be the generalized function satisfying

$$\langle\, \widehat{f}, \phi \,\rangle = \langle\, f, \mathcal{F}[\phi] \,\rangle \qquad \text{for each } \phi \text{ in } \mathcal{G} \quad .$$

Because of the relation of this equation to equation (35.2), it seems natural to refer to $\widehat{f}$ as the (generalized) Fourier transform of f.

That is exactly what we will do.

Let f be any generalized function. The *(generalized) Fourier transform* of f, denoted by $\mathcal{F}[f]$, is defined to be the generalized function satisfying

$$\langle\, \mathcal{F}[f], \phi \,\rangle = \langle\, f, \mathcal{F}[\phi] \,\rangle \qquad \text{for each } \phi \text{ in } \mathcal{G} \quad . \tag{35.3}$$

Similarly, the *(generalized) Fourier inverse transform* of a generalized function F, denoted by $\mathcal{F}^{-1}[F]$, is defined to be the generalized function satisfying

$$\langle\, \mathcal{F}^{-1}[F], \phi \,\rangle = \langle\, F, \mathcal{F}^{-1}[\phi] \,\rangle \qquad \text{for each } \phi \text{ in } \mathcal{G} \quad . \tag{35.4}$$

It is important to realize that we have not redefined the transform of any classically transformable function. After all, if f is a classically transformable function with classical transform F and generalized transform $\widehat{f}$, then, by the above definition of generalized transform and the classical fundamental identity (equation (35.2)),

$$\langle\, \widehat{f}, \phi \,\rangle = \langle\, f, \mathcal{F}[\phi] \,\rangle = \langle\, F, \phi \,\rangle \qquad \text{for each } \phi \text{ in } \mathcal{G} \quad .$$

Thus, $\widehat{f} = F$. Likewise, the generalized inverse Fourier transform of a classically transformable function is just the same as the classical inverse Fourier transform of that function.

So, just to be exceedingly clear: *Do not waste effort in finding the generalized Fourier transform (or inverse transform) of a classically transformable function when you already know its classical transform!* The generalized definitions above are only needed when the classical definitions are not valid (or when you don't know if a given function is classically transformable). To see this, let's consider the transform of a classically transformable function.

!▶**Example 35.1:**   *Back in chapter 19, we saw that $e^{-2t}\,\text{step}(t)$ is a classically transformable function with classical Fourier transform $(2+i2\pi\omega)^{-1}$. Thus, the generalized Fourier transform of $e^{-2t}\,\text{step}(t)$ is also $(2+i2\pi\omega)^{-1}$.*

Though totally unnecessary (except to prove a point), let us compute the Fourier transform from the generalized definition. Letting ϕ be any Gaussian test function, then, by the definition of the generalized transform and the fact that $e^{-2t}\,\text{step}(t)$ is exponentially integrable,

$$\left\langle\, \mathcal{F}\!\left[e^{-2t}\,\text{step}(t)\right]\Big|_{\omega}, \phi(\omega) \,\right\rangle = \left\langle\, e^{-2t}\,\text{step}(t), \mathcal{F}[\phi(\omega)]|_{t} \,\right\rangle$$

$$= \int_{-\infty}^{\infty} e^{-2t}\,\text{step}(t)\,\mathcal{F}[\phi(\omega)]|_{t}\,dt \quad .$$

Since ϕ is a Gaussian test function, it is absolutely integrable and the above can be written as

$$\left\langle\, \mathcal{F}\!\left[e^{-2t}\,\text{step}(t)\right]\Big|_{\omega}, \phi(\omega) \,\right\rangle = \int_{-\infty}^{\infty} e^{-2t}\,\text{step}(t)\int_{-\infty}^{\infty} \phi(\omega)\,e^{-i2\pi t\omega}\,d\omega\,dt$$

$$= \int_{-\infty}^{\infty}\int_{-\infty}^{\infty} e^{-2t}\,\text{step}(t)\,\phi(\omega)\,e^{-i2\pi t\omega}\,d\omega\,dt \quad .$$

Because both e^{-2t} step(t) and $\phi(\omega)$ are absolutely integrable, corollary 18.24 on page 275 assures us that the order of integration in the last integral can be interchanged. Doing so, and computing the resulting integral, we obtain

$$\left\langle \mathcal{F}\left[e^{-2t} \text{step}(t) \right] \Big|_{\omega} , \phi(\omega) \right\rangle = \int_{-\infty}^{\infty} \int_{-\infty}^{\infty} e^{-2t} \text{step}(t) \phi(\omega) e^{-i2\pi t\omega} \, dt \, d\omega$$

$$= \int_{-\infty}^{\infty} \phi(\omega) \left(\int_{0}^{\infty} e^{-2t} e^{-i2\pi t\omega} \, dt \right) d\omega$$

$$= \int_{-\infty}^{\infty} \phi(\omega) \frac{1}{2 + i2\pi\omega} \, d\omega \quad .$$

Thus, for each ϕ in $\mathcal{G}$,

$$\left\langle \mathcal{F}\left[e^{-2t} \text{step}(t) \right] \Big|_{\omega} , \phi(\omega) \right\rangle = \left\langle \frac{1}{2 + i2\pi\omega} , \phi(\omega) \right\rangle \quad ,$$

(re)verifying that the generalized Fourier transform of e^{-2t} step(t) is $(2 + i2\pi\omega)^{-1}$.

Now let's try to find the Fourier transforms of some (generalized) functions for which we cannot use the classical theory.

!► Example 35.2 (transform of the step function):   *While the step function is not classically transformable, it clearly is exponentially integrable and, so, can be treated as a generalized function. Using the generalized definition to find its Fourier transform gives*

$$\left\langle \mathcal{F}[\text{step}] , \phi \right\rangle = \left\langle \text{step} , \mathcal{F}[\phi] \right\rangle$$

$$= \int_{-\infty}^{\infty} \text{step}(x) \mathcal{F}[\phi]\big|_{x} \, dx = \int_{0}^{\infty} \mathcal{F}[\phi]\big|_{x} \, dx$$

for each Gaussian test function ϕ. And because Gaussian test functions are absolutely integrable, we can write the last integral as

$$\int_{0}^{\infty} \int_{-\infty}^{\infty} \phi(y) e^{-i2\pi xy} \, dy \, dx \quad .$$

Thus, we can describe $\mathcal{F}[\text{step}]$ as the generalized function such that

$$\left\langle \mathcal{F}[\text{step}] , \phi \right\rangle = \int_{0}^{\infty} \int_{-\infty}^{\infty} \phi(y) e^{-i2\pi xy} \, dy \, dx \qquad \text{for each } \phi \text{ in } \mathcal{G} \quad .$$

You should verify, yourself, that the order of integration cannot be changed in this case. (Other, more useful, ways of describing $\mathcal{F}[\text{step}]$ will be developed later.)

As the last example illustrates, the definition of the generalized transform for f often does not naturally lead to a formula for $\mathcal{F}[f]$ in terms of other generalized functions. This is something we just have to live with. Sometimes, though, with luck and a little work, we can find a formula for the transform in terms of other known generalized functions.

!▶**Example 35.3 (Fourier transform of the unit constant):**   *Letting ϕ be any Gaussian test function, observe that*

$$\langle \mathcal{F}[1] , \phi \rangle = \langle 1 , \mathcal{F}[\phi] \rangle$$

$$= \int_{-\infty}^{\infty} 1 \cdot \mathcal{F}[\phi]|_x \, dx$$

$$= \int_{-\infty}^{\infty} \mathcal{F}[\phi]|_x \, e^{i2\pi \cdot 0 \cdot x} \, dx = \mathcal{F}^{-1}[\mathcal{F}[\phi]]|_0 \quad.$$

Because ϕ is classically transformable, we already know $\mathcal{F}^{-1}[\mathcal{F}[\phi]]|_0 = \phi(0)$. Combining this with the last equation above and the definition of the delta function, we get

$$\langle \mathcal{F}[1] , \phi \rangle = \phi(0) = \langle \delta , \phi \rangle \qquad \text{for each } \phi \text{ in } \mathcal{G} \quad.$$

Hence,

$$\mathcal{F}[1] = \delta \quad.$$

?▶**Exercise 35.1 (Fourier transforms of complex exponentials):**   *Let a be any real number, and show that*

$$\mathcal{F}\left[e^{i2\pi ax}\right] = \delta_a \quad.$$

Finding the Fourier transform of a delta function (on the real line) is just as easy.

!▶**Example 35.4 (Fourier transform of a delta function):**   *For ϕ in $\mathcal{G}$,*

$$\langle \mathcal{F}[\delta_3] , \phi \rangle = \langle \delta_3 , \mathcal{F}[\phi] \rangle = \mathcal{F}[\phi]|_3 \quad.$$

This, along with the fact that

$$\mathcal{F}[\phi]|_3 = \int_{-\infty}^{\infty} \phi(x) \, e^{-i2\pi 3x} \, dx = \int_{-\infty}^{\infty} e^{-i6\pi x} \phi(x) \, dx = \left\langle e^{-i6\pi x} , \phi \right\rangle \quad,$$

gives us

$$\langle \mathcal{F}[\delta_3] , \phi \rangle = \left\langle e^{-i6\pi x} , \phi \right\rangle \qquad \text{for each } \phi \text{ in } \mathcal{G} \quad.$$

Thus,

$$\mathcal{F}[\delta_3] = e^{-i6\pi x} \quad.$$

?▶**Exercise 35.2 (Fourier transform of delta functions):**   *Let a be any fixed real number, and show that*

$$\mathcal{F}[\delta_a]|_x = e^{-i2\pi ax} \quad.$$

In particular,

$$\mathcal{F}[\delta]|_x = e^{-i2\pi \cdot 0 \cdot x} = 1 \quad.$$

(In a few pages we'll discover that the formulas derived in the above exercises also hold when a is complex.)

Inverse transforms of exponentials and delta functions can be computed in a similar fashion, or by using the invertibility described below.

Linearity and Invertibility

Let f and g be any two generalized functions, and let α and β be any two constants. By the definition of the generalized Fourier transform and an observation made while discussing linear combinations of generalized functions (see page 552), we have, for each ϕ in $\mathcal{G}$,

$$\left\langle \mathcal{F}[\alpha f + \beta g] , \phi \right\rangle = \left\langle \alpha f + \beta g , \mathcal{F}[\phi] \right\rangle$$
$$= \alpha \left\langle f , \mathcal{F}[\phi] \right\rangle + \beta \left\langle g , \mathcal{F}[\phi] \right\rangle$$
$$= \alpha \left\langle \mathcal{F}[f] , \phi \right\rangle + \beta \left\langle \mathcal{F}[g] , \phi \right\rangle = \left\langle \alpha \mathcal{F}[f] + \beta \mathcal{F}[g] , \phi \right\rangle \quad .$$

Thus,

$$\mathcal{F}[\alpha f + \beta g] = \alpha \mathcal{F}[f] + \beta \mathcal{F}[g] \quad .$$

This, along with virtually identical computations using the inverse transform, proves the expected theorem on linearity.

Theorem 35.1 (linearity of the generalized transforms)
If f and g are any two generalized functions, and α and β are any two constants, then

$$\mathcal{F}[\alpha f + \beta g] = \alpha \mathcal{F}[f] + \beta \mathcal{F}[g]$$

and

$$\mathcal{F}^{-1}[\alpha f + \beta g] = \alpha \mathcal{F}^{-1}[f] + \beta \mathcal{F}^{-1}[g] \quad .$$

!▶Example 35.5: Let $a > 0$. *Using the linearity of the transform and the results from exercise 35.1, above, we see that*

$$\mathcal{F}[\sin(2\pi a t)] = \frac{1}{2i} \left[\mathcal{F}\left[e^{i2\pi a t} \right] - \mathcal{F}\left[e^{-i2\pi a t} \right] \right]$$
$$= \frac{1}{2i} \left[\mathcal{F}\left[e^{i2\pi a t} \right] - \mathcal{F}\left[e^{i2\pi(-a)t} \right] \right] = \frac{1}{2i} \left[\delta_a - \delta_{-a} \right] \quad .$$

?▶Exercise 35.3: *Letting $a > 0$, verify that*

$$\mathcal{F}[\cos(2\pi a t)] = \frac{1}{2} \left[\delta_a + \delta_{-a} \right] \quad .$$

Showing that the two generalized transforms $\mathcal{F}$ and $\mathcal{F}^{-1}$ are inverses of each other is ludicrously simple. By the definition of the generalized Fourier transforms, we have

$$\left\langle \mathcal{F}^{-1}\left[\mathcal{F}[f] \right] , \phi \right\rangle = \left\langle \mathcal{F}[f] , \mathcal{F}^{-1}[\phi] \right\rangle = \left\langle f , \mathcal{F}\left[\mathcal{F}^{-1}[\phi] \right] \right\rangle$$

for any generalized function f and any Gaussian test function ϕ. Moreover, because ϕ is classically transformable, we know $\mathcal{F}\left[\mathcal{F}^{-1}[\phi] \right] = \phi$. Thus, for any generalized function f,

$$\left\langle \mathcal{F}^{-1}\left[\mathcal{F}[f] \right] , \phi \right\rangle = \left\langle f , \phi \right\rangle \qquad \text{for each } \phi \text{ in } \mathcal{G} \quad .$$

In other words, $\mathcal{F}^{-1}\left[\mathcal{F}[f] \right] = f$ (equivalently: $F = \mathcal{F}[f] \Rightarrow \mathcal{F}^{-1}[F] = f$).
By virtually identical calculations we also get $\mathcal{F}\left[\mathcal{F}^{-1}[F] \right] = F$ for any generalized function F, completing the proof of our expected invertibility theorem.

Theorem 35.2 (*invertibility of the generalized transforms*)
Let f and F be any two generalized functions. Then

$$\mathcal{F}^{-1}\big[\mathcal{F}[f]\big] = f \qquad \text{and} \qquad \mathcal{F}\big[\mathcal{F}^{-1}[F]\big] = F \quad .$$

Equivalently,

$$F = \mathcal{F}[f] \quad \Longleftrightarrow \quad \mathcal{F}^{-1}[F] = f \quad .$$

!►Example 35.6:   Let a be any real constant. In exercises 35.1 and 35.2 you showed that

$$\delta_a = \mathcal{F}\Big[e^{i2\pi a x}\Big] \qquad \text{and} \qquad e^{-i2\pi a x} = \mathcal{F}[\delta_a]|_x \quad .$$

Theorem 35.2 then assures us that

$$\mathcal{F}^{-1}[\delta_a]|_x = e^{i2\pi a x} \qquad \text{and} \qquad \mathcal{F}^{-1}\Big[e^{-i2\pi a x}\Big] = \delta_a \quad .$$

In particular,

$$\mathcal{F}^{-1}[\delta] = 1 \qquad \text{and} \qquad \mathcal{F}^{-1}[1] = \delta \quad .$$

35.2   Generalized Scaling of the Variable
Basic Definition and Notation
Definition

A common way to modify a classical function f is to scale its variable by some fixed nonzero real number σ, thereby transforming the function $f(x)$ to the function $f(\sigma x)$. To determine the generalized version of this operation, let us consider integrating such a scaled, exponentially integrable function $f(\sigma x)$ multiplied by an arbitrary Gaussian test function $\phi(x)$. If $\sigma > 0$, then, using the change of variables $\sigma x = y$ and the fact that y is a dummy variable,

$$\int_{-\infty}^{\infty} f(\sigma x)\phi(x)\,dx = \int_{-\infty}^{\infty} f(y)\phi\Big(\frac{y}{\sigma}\Big)\frac{1}{\sigma}\,dy = \frac{1}{\sigma}\int_{-\infty}^{\infty} f(x)\phi\Big(\frac{x}{\sigma}\Big)\,dx \quad .$$

If $\sigma < 0$, the same change of variables gives

$$\int_{-\infty}^{\infty} f(\sigma x)\phi(x)\,dx = \int_{\infty}^{-\infty} f(y)\phi\Big(\frac{y}{\sigma}\Big)\frac{1}{\sigma}\,dy = -\frac{1}{\sigma}\int_{-\infty}^{\infty} f(x)\phi\Big(\frac{x}{\sigma}\Big)\,dx \quad .$$

Recalling that $|\sigma| = \sigma$ if $\sigma > 0$ and $|\sigma| = -\sigma$ if $\sigma < 0$, we see that the above two equations can be written as

$$\int_{-\infty}^{\infty} f(\sigma x)\phi(x)\,dx = \frac{1}{|\sigma|}\int_{-\infty}^{\infty} f(x)\phi\Big(\frac{x}{\sigma}\Big)\,dx \quad , \qquad (35.5)$$

which, in generalized integral notation, is

$$\big\langle\, f(\sigma x)\,,\,\phi(x) \,\big\rangle = \frac{1}{|\sigma|}\Big\langle\, f(x)\,,\,\phi\Big(\frac{x}{\sigma}\Big) \,\Big\rangle \quad .$$

Observe that the right side of this last equation makes sense for any generalized function f, even if it is not a classical, exponentially integrable function. Moreover, as was seen in exercise 32.6

on page 520 (and as will be carefully verified in section 35.6), this formula defines a new generalized function, which we might as well denote by $f(\sigma x)$. This gives us a generalized definition for scaling the variable, namely, if $f(x)$ is any generalized function and σ is any real, nonzero value, then $f(\sigma x)$ is defined to be the generalized function satisfying

$$\langle\, f(\sigma x)\,,\,\phi(x)\,\rangle \;=\; \frac{1}{|\sigma|}\left\langle\, f(x)\,,\,\phi\!\left(\frac{x}{\sigma}\right)\,\right\rangle \qquad \text{for each } \phi \text{ in } \mathcal{G} \quad. \tag{35.6}$$

The value σ will be referred to as the corresponding scaling factor.

    Of course, if the classical definition of $f(\sigma x)$ applies, then there is no difference between $f(\sigma x)$ defined classically and $f(\sigma x)$ defined by the above generalized definition. We know this because equation (35.6), defining $f(\sigma x)$ as a generalized function, is identical to equation (35.5), which was derived using the classical definition of $f(\sigma x)$. So let's consider a case where f is not a classical function.

!▶ **Example 35.7:**   *Consider the delta function with the variable scaled by 3. For any Gaussian test function ϕ we have, by equation (35.6),*

$$\langle\, \delta(3x)\,,\,\phi(x)\,\rangle \;=\; \frac{1}{3}\left\langle\, \delta(x)\,,\,\phi\!\left(\frac{x}{3}\right)\,\right\rangle \;=\; \frac{1}{3}\phi\!\left(\frac{0}{3}\right) \;=\; \frac{1}{3}\phi(0) \quad.$$

But also,

$$\frac{1}{3}\phi(0) \;=\; \left\langle\, \frac{1}{3}\delta(x)\,,\,\phi(x)\,\right\rangle \quad.$$

So,

$$\langle\, \delta(3x)\,,\,\phi(x)\,\rangle \;=\; \left\langle\, \frac{1}{3}\delta(x)\,,\,\phi(x)\,\right\rangle \qquad \text{for each } \phi \text{ in } \mathcal{G} \quad.$$

In other words,

$$\delta(3x) \;=\; \frac{1}{3}\delta(x) \quad.$$

?▶ **Exercise 35.4:**   *Verify that* $\delta_6(2x) \;=\; \frac{1}{2}\delta_3(x)$.

    It is worth emphasizing that we only allow the scaling factor to be a nonzero real number. The reason for this is something on which you should briefly meditate.

?▶ **Exercise 35.5:**   *What "goes wrong" in equation (35.6) if σ is either zero or an imaginary number?*

The Scaling Operator

The process of converting any $f(x)$ to $f(\sigma x)$ can be viewed as a transformation of generalized functions. We will, on occasion, denote this transformation by S_σ. Thus, for any given generalized function $f(x)$ and any nonzero real number σ,

$$f(\sigma x) \quad,\quad S_\sigma[f]|_x \quad \text{and} \quad S_\sigma[f]$$

all mean the same thing. We may, for example, write $S_3[\delta]$ for $\delta(3x)$. As this shows, the use of the scaling operator notation can eliminate the need to "attach variables" to generalized functions. This can simplify the bookkeeping in some computations, especially when the computation involves a long sequence of computations that would otherwise require the introduction of a number of dummy variables. It also allows us to rewrite the generalized definition of scaling as follows: Let f be any generalized function and σ any nonzero real number. Then $S_\sigma[f]$ is the generalized function given by

$$\langle\, S_\sigma[f]\,,\phi\,\rangle \;=\; \frac{1}{|\sigma|}\langle\, f\,,S_{1/\sigma}[\phi]\,\rangle \qquad \text{for each } \phi \text{ in } \mathcal{G} \quad .$$

Implicit in this definition is that, for each ϕ in $\mathcal{G}$, $S_{1/\sigma}[\phi]$ is the Gaussian test function obtained by scaling the variable of ϕ by $1/\sigma$,

$$S_{1/\sigma}[\phi]|_x \;=\; \phi\!\left(\frac{1}{\sigma}\cdot x\right) \quad .$$

?►Exercise 35.6: *Show that $S_{-1}[\delta_\alpha] = \delta_{-\alpha}$ for any complex value α.*

A few useful (and easily verified) properties of the scaling operator are described in the next exercise.

?►Exercise 35.7: *Verify each of the following, assuming f and g are two generalized functions, h is a simple multiplier, α and β are fixed complex numbers, and σ and τ are nonzero real numbers:*

a: *(linearity)* $S_\sigma[\alpha f + \beta g] \;=\; \alpha S_\sigma[f] + \beta S_\sigma[g]$

b: $S_\sigma[hf] \;=\; S_\sigma[h]S_\sigma[f]$

c: *(commutativity)* $S_\sigma\,[S_\tau[f]] \;=\; S_{\sigma\tau}[f] \;=\; S_\tau\,[S_\sigma[f]]$

Even and Odd Generalized Functions

Given any generalized function $f(x)$, we will abbreviate $f((-1)x)$ by $f(-x)$, just as we do classically. And, since $|-1| = 1$, our generalized definition of $f(-x)$ reduces to $f(-x)$ being the generalized function such that

$$\langle\, f(-x)\,,\phi(x)\,\rangle \;=\; \langle\, f(x)\,,\phi(-x)\,\rangle \qquad \text{for each } \phi \text{ in } \mathcal{G} \quad .$$

Just as with classical functions, we will refer to a generalized function f as being *even* if and only if $f(-x) = f(x)$. From the discussion in the previous paragraph, we see that this is equivalent to saying that f is even if and only if

$$\langle\, f(x)\,,\phi(x)\,\rangle \;=\; \langle\, f(x)\,,\phi(-x)\,\rangle \qquad \text{for each } \phi \text{ in } \mathcal{G} \quad .$$

Likewise, a generalized function f will be called an *odd* generalized function if and only if $f(-x) = -f(x)$. Equivalently, f is odd if and only if

$$\langle\, f(x)\,,\phi(x)\,\rangle \;=\; -\langle\, f(x)\,,\phi(-x)\,\rangle \qquad \text{for each } \phi \text{ in } \mathcal{G} \quad .$$

In terms of the scaling operator,

$$f \text{ is even} \quad\Longleftrightarrow\quad S_{-1}[f] = f \quad ,$$

and

$$f \text{ is odd} \quad\Longleftrightarrow\quad S_{-1}[f] = -f \quad .$$

Earlier, we noted that, if $f(x)$ is also a classical function, then $f(\alpha x)$ defined using the generalized definition of scaling is the same as $f(\alpha x)$ defined classically for any real number α, including $\alpha = -1$. From this, it automatically follows that a classical function is, respectively, even or odd as a generalized function if and only if it is even or odd in the classical sense.

Now consider a couple of generalized functions that are not classical functions.

?►Exercise 35.8: *Verify that the delta function, δ, is even.*

?►Exercise 35.9: *Verify that the delta function at 2, δ_2, is neither even nor odd.*

Scaling and the Fourier Transforms

The scaling identities for the classical Fourier transforms were described in theorem 21.3 on page 327. Unsurprisingly, the scaling identities for the generalized Fourier transforms, described in the next theorem, are very similar.

Theorem 35.3 (scaling identities)
Let f and F be any pair of generalized functions with $F = \mathcal{F}[f]$. Then, for any nonzero real constant σ,

$$\mathcal{F}[f(\sigma x)]|_y = \frac{1}{|\sigma|}F\left(\frac{1}{\sigma}y\right) \tag{35.7a}$$

and

$$\mathcal{F}^{-1}[F(\sigma y)]|_x = \frac{1}{|\sigma|}f\left(\frac{1}{\sigma}x\right) \quad . \tag{35.7b}$$

Equivalently,

$$\mathcal{F}^{-1}\left[F\left(\frac{1}{\sigma}y\right)\right]\Big|_x = |\sigma|\, f(\sigma x) \tag{35.7c}$$

and

$$\mathcal{F}\left[f\left(\frac{1}{\sigma}x\right)\right]\Big|_y = |\sigma|\, F(\sigma y) \quad . \tag{35.7d}$$

We will prove equation (35.7a) and leave the similar proof of equation (35.7b) as an exercise. First, though, let's observe that, in terms of the scaling operator, equations (35.7a) and (35.7b) are

$$\mathcal{F}[S_\sigma[f]] = \frac{1}{|\sigma|}S_{1/\sigma}[F] \quad \text{and} \quad \mathcal{F}^{-1}[S_\sigma[F]] = \frac{1}{|\sigma|}S_{1/\sigma}[f] \quad ,$$

while equations (35.7c) and (35.7d) are

$$\mathcal{F}^{-1}\left[S_{1/\sigma}[F]\right] = |\sigma|\, S_\sigma[f] \quad \text{and} \quad \mathcal{F}\left[S_{1/\sigma}[f]\right] = |\sigma|\, S_\sigma[F] \quad .$$

PROOF (of equation (35.7a)): Let ϕ be any Gaussian test function, and let Φ be its Fourier transform, $\Phi(x) = \mathcal{F}[\phi(y)]|_x$. For convenience, also let $\alpha = 1/\sigma$ (so $\alpha\sigma = 1$). Using the generalized definitions of scaling and the Fourier transform,

$$\left\langle \frac{1}{|\sigma|}F\left(\frac{1}{\sigma}y\right), \phi(y) \right\rangle = |\alpha|\left\langle F(\alpha y), \phi(y) \right\rangle$$

$$= |\alpha|\frac{1}{|\alpha|}\left\langle F(y), \phi\left(\frac{1}{\alpha}y\right) \right\rangle$$

$$= \left\langle \mathcal{F}[f(x)]|_y, \phi(\sigma y) \right\rangle = \left\langle f(x), \mathcal{F}[\phi(\sigma y)]|_x \right\rangle \quad .$$

But ϕ and Φ are classically transformable functions and, by the classical scaling identities,

$$\mathcal{F}[\phi(\sigma y)]|_x = \frac{1}{|\sigma|}\Phi\left(\frac{1}{\sigma}x\right) \quad .$$

So,

$$\left\langle \frac{1}{|\sigma|} F\left(\frac{1}{\sigma}y\right) , \phi(y) \right\rangle = \left\langle f(x) , \frac{1}{|\sigma|}\Phi\left(\frac{1}{\sigma}x\right) \right\rangle$$

$$= \frac{1}{|\sigma|}\left\langle f(x) , \Phi\left(\frac{1}{\sigma}x\right) \right\rangle$$

$$= \left\langle f(\sigma x) , \Phi(x) \right\rangle = \left\langle f(\sigma x) , \mathcal{F}[\phi(y)]|_x \right\rangle .$$

Applying the defining equation for the generalized Fourier transform one last time then yields

$$\left\langle \frac{1}{|\sigma|} F\left(\frac{1}{\sigma}y\right) , \phi(y) \right\rangle = \left\langle \mathcal{F}[f(\sigma x)]|_y , \phi(y) \right\rangle$$

for each Gaussian test function ϕ, verifying equation (35.7a).   ∎

?►Exercise 35.10:   *Verify equation (35.7b) in theorem 35.3.*

Near-Equivalence of the Generalized Fourier Transforms

The near-equivalence identities for the generalized Fourier transforms are straightforward conse-quences of the scaling identities and the near-equivalence identities for the classical transforms. We will state these identities in the next theorem and leave their proofs for exercises.

Theorem 35.4 (principle of near-equivalence)
Let f be any generalized function. Then

$$\mathcal{F}[f(y)]|_x = \mathcal{F}^{-1}[f(y)]|_{-x} = \mathcal{F}^{-1}[f(-y)]|_x \tag{35.8}$$

and

$$\mathcal{F}^{-1}[f(y)]|_x = \mathcal{F}[f(y)]|_{-x} = \mathcal{F}[f(-y)]|_x . \tag{35.9}$$

?►Exercise 35.11:   *Verify the equations in line (35.8) of theorem 35.4.*

In terms of the scaling operator, equations (35.8) and (35.9) can be written as

$$\mathcal{F}[f] = S_{-1}\left[\mathcal{F}^{-1}[f]\right] = \mathcal{F}^{-1}\left[S_{-1}[f]\right]$$

and

$$\mathcal{F}^{-1}[f] = S_{-1}\mathcal{F}[f] = \mathcal{F}\left[S_{-1}[f]\right] .$$

As an immediate consequence of the generalized principle of near-equivalence, we have the following generalizations of the classical results concerning the transforms of even and odd functions (see corollaries 19.2 and 19.3, starting on page 283).

Corollary 35.5
Let f be an even generalized function. Then both $\mathcal{F}[f]$ and $\mathcal{F}^{-1}[f]$ are even. Moreover,

$$\mathcal{F}^{-1}[f] = \mathcal{F}[f] .$$

Corollary 35.6
Let f be an odd generalized function. Then both $\mathcal{F}[f]$ and $\mathcal{F}^{-1}[f]$ are odd. Moreover,

$$\mathcal{F}^{-1}[f] = -\mathcal{F}[f] .$$

?►Exercise 35.12:   Let $a > 0$. From a previous example and exercise, we know

$$\mathcal{F}[\sin(2\pi at)] = \frac{1}{2i}\left[\delta_a - \delta_{-a}\right] \quad and \quad \mathcal{F}[\cos(2\pi at)] = \frac{1}{2}\left[\delta_a + \delta_{-a}\right] \quad.$$

Using the above corollaries, now find

$$\mathcal{F}^{-1}[\sin(2\pi at)] \quad and \quad \mathcal{F}^{-1}[\cos(2\pi at)] \quad.$$

35.3   Generalized Translation/Shifting
Translation along the Real Axis
Definition

Let a be some real value. If f is an exponentially integrable function and ϕ is any Gaussian test function, then, using the change of variables $y = x - a$, we obtain

$$\int_{-\infty}^{\infty} f(x-a)\phi(x)\,dx = \int_{-\infty}^{\infty} f(y)\phi(y+a)\,dy = \int_{-\infty}^{\infty} f(x)\phi(x+a)\,dx \quad.$$

Thus, using our generalized function notation,

$$\left\langle\, f(x-a)\,,\,\phi(x)\,\right\rangle = \left\langle\, f(x)\,,\,\phi(x+a)\,\right\rangle \quad. \tag{35.10}$$

Once again we have an equation involving a classical function f in which the right-hand side is well defined and, itself, defines a generalized function even when f is not a classical function (see exercise 33.22 on page 564 and plan on reading section 35.6). And, again, this inspires a definition generalizing the operation that led to the equation.

Let $f(x)$ be any generalized function and a any real value. The *(generalized) translation* of f by a, which, for now, we will denote by $f(x - a)$, is defined to be the generalized function satisfying

$$\left\langle\, f(x-a)\,,\,\phi(x)\,\right\rangle = \left\langle\, f(x)\,,\,\phi(x+a)\,\right\rangle \qquad \text{for each } \phi \text{ in } \mathcal{G} \quad. \tag{35.11}$$

The generalized function $f(x - a)$ is also known as the *(generalized) shift* of f by a.

Following the pattern laid out in the previous two sections, we first observe that our generalized definition of $f(x - a)$ gives the same thing as the classical notion of $f(x - a)$ when the classical notion applies (compare equation (35.10), which was derived assuming the classical notion of translation to equation (35.11), which defines the corresponding generalized translation). Now that we've made that observation, let's do an example involving a generalized function that is not a classical function.

!►Example 35.8:   Consider the delta function translated by any real number a. If ϕ is any Gaussian test function, then, by our definition of $\delta(x - a)$,

$$\left\langle\, \delta(x-a)\,,\,\phi(x)\,\right\rangle = \left\langle\, \delta(x)\,,\,\phi(x+a)\,\right\rangle$$
$$= \phi(0+a) = \phi(a) = \left\langle\, \delta_a(x)\,,\,\phi(x)\,\right\rangle \quad.$$

Thus,

$$\delta(x-a) = \delta_a(x) \quad.$$

The Translation Operator

When convenient, the process of converting a generalized function $f(x)$ to $f(x - a)$ will be indicated by T_a, and we will refer to T_a as the *translation operator* corresponding to a given (real) value a. Under this notation

$$f(x - a) \quad , \quad T_a[f]|_x \quad \text{and} \quad T_a[f]$$

all mean the same. The advantages (and disadvantages) of using the translation operator are similar to the advantages (and disadvantages) of using the scaling operator. Rewriting the definition for translation in terms of the translation operator, we get that, for any generalized function f and any real number a, $T_a[f]$ is the generalized function such that, for each Gaussian test function ϕ,

$$\langle T_a[f], \phi \rangle = \langle f, T_{-a}[\phi] \rangle \tag{35.12}$$

where $T_{-a}[\phi]$ is the Gaussian test function

$$T_{-a}[\phi]|_x = \phi(x - (-a)) = \phi(x + a) \quad . \tag{35.13}$$

There are two reasons for introducing the translation operator notation. The first is that it is a cleaner way to express the generalized translation. We don't need to "attach variables" to our expressions. The second, and more important, reason is that we are about to expand our definition of the generalized translation, and under this extended definition there are situations where it is necessary to distinguish between the generalized translation and the classical translation of a function.

Complex Translation*

You may wonder why we restricted a to being a real value when defining $T_a[f]$. After all, in a previous chapter we saw that $\phi(x + a)$ is a Gaussian test function whenever ϕ is a Gaussian test function and a is any fixed real *or complex* constant. Truth is, there is no need to insist on a just being a real number in equations (35.12) and (35.13).[1] So let us take the bold step of replacing the adjective "real" with "complex" in our definition of generalized translation: Given any generalized function f and any complex constant a, we define the *generalized translation (or shift)* of f by a, denoted by $T_a[f]$, to be the generalized function such that, for each Gaussian test function ϕ,

$$\langle T_a[f], \phi \rangle = \langle f, T_{-a}[\phi] \rangle \tag{35.14}$$

where $T_{-a}[\phi]$ is the Gaussian test function

$$T_{-a}[\phi]|_x = \phi(x - (-a)) = \phi(x + a) \quad . \tag{35.15}$$

!▶**Example 35.9:** *The generalized translation of the step function by i is the generalized function $T_i[\text{step}]$ given by*

$$\langle T_i[\text{step}], \phi \rangle = \langle \text{step}, T_{-i}[\phi] \rangle = \int_{-\infty}^{\infty} \text{step}(s)\phi(s - (-i)) \, ds = \int_{0}^{\infty} \phi(x + i) \, dx$$

for each Gaussian test function ϕ.

?▶**Exercise 35.13:** *Verify that $T_a[\delta] = \delta_a$ for each complex value a.*

---

* You may want to review the material on test functions as functions of a complex variable (page 524) and complex translation of test functions (page 525) before starting this subsection.
[1] No mathematical reason, that is.

While it is fairly safe to use " $f(x - a)$ " to denote the generalized translation of a function f by a *real* value a, it is decidedly *unsafe* to use this same notation when a is not a real number. It's safe when a is real because, if f is a classical function and a is real, then the generalized translation of f is given by the corresponding classical translation. This often fails to be true when a is not real. Indeed, there are many classical functions that are only defined, classically, as functions of a real-valued variable (what, for example, could we mean by "the step function at i, step(i)"?). For such a function there is no classical analog to the complex shift. Worse yet, even when $f(x - a)$ is, classically, a well-defined function for complex values of a, it is still often the case that $T_a f|_x \neq f(x - a)$ as generalized functions. (We'll discuss this further a little later in this section.)

Since the two are not always equivalent, we will use different notation for the generalized and the classical translations. Let us agree to use " $T_a[f]$ " to indicate the generalized translation of f by a, and to use " $f(x - a)$ " to indicate the corresponding classical translation.

Naturally, there are many cases where a generalized translation is essentially the same as the corresponding classical translation. As we have already seen, for example, " $T_a f|_x = f(x - a)$ " whenever f is an exponentially integrable function and a is a real number. Another rather important case, described in the next theorem, is where the function f is analytic on the complex plane and is "exponentially bounded on horizontal strips (of $\mathbb{C}$)" with the last quoted phrase meaning that, given any horizontal strip of the complex plane

$$S_{(a,b)} = \{x + iy : a < y < b\}$$

with $-\infty < a < b < \infty$, there are corresponding finite positive constants M and c such that

$$|f(x + iy)| \leq M e^{c|x|} \qquad \text{for all } x + iy \text{ in } S_{(a,b)} \quad .$$

Theorem 35.7
Let a be any fixed complex value, and let f be a classical function defined and analytic on the entire complex plane. Suppose, further, that f is exponentially bounded on horizontal strips of $\mathbb{C}$. Then the generalized translation of f by a, $T_a[f]$, is given by the corresponding classical translation, $f(x - a)$. That is,

$$\langle T_a[f], \phi \rangle = \langle f(s - a), \phi(s) \rangle \qquad \text{for each } \phi \text{ in } \mathcal{G} \quad .$$

Our proof of this theorem will involve another theorem we've employed several times before, namely, theorem 18.20 (page 273) on differentiating certain integrals. Use of that theorem, however, requires that certain functions and their derivatives be "sufficiently integrable". To help determine whether the derivatives of our functions are "sufficiently integrable", here is a useful lemma:

Lemma 35.8
Let f be a function analytic on the entire complex plane. If f is exponentially bounded on horizontal strips, then so is its derivative, f'.

Sadly, the author knows of no simple way to prove this last lemma without developing much more material from the theory of complex analysis. So its proof will be left as exercise 35.49 (at the end of this chapter) for those who have had a course in that subject.

PROOF (of theorem 35.7): Let ϕ be any Gaussian test function. Remember, by definition,

$$\langle T_a[f], \phi \rangle = \langle f, T_{-a}[\phi] \rangle = \int_{-\infty}^{\infty} f(s)\phi(s + a)\, ds \quad ,$$

while

$$\left\langle\, f(s-a)\,,\,\phi(s)\,\right\rangle \;=\; \int_{-\infty}^{\infty} f(s-a)\phi(s)\,ds \quad.$$

So, to verify the claim of the theorem, it will suffice to show that

$$\int_{-\infty}^{\infty} f(s)\phi(s+a)\,ds \;=\; \int_{-\infty}^{\infty} f(s-a)\phi(s)\,ds \quad. \tag{35.16}$$

To help confirm this last equality, let

$$h(t) \;=\; \int_{-\infty}^{\infty} f(s-a+ta)\phi(s+ta)\,ds \quad,$$

and observe that

$$h(0) \;=\; \int_{-\infty}^{\infty} f(s-a)\phi(s)\,ds \qquad \text{and} \qquad h(1) \;=\; \int_{-\infty}^{\infty} f(s)\phi(s+a)\,ds \quad.$$

Confirming equation (35.16) can then be accomplished by confirming that

$$h(0) \;=\; h(1) \quad.$$

Using the fact that both f and f' are exponentially bounded on strips (along with results that, by now, should be well known to you), it is a simple exercise to show that h is a smooth function on $\mathbb{R}$ and that

$$h'(t) \;=\; \frac{d}{dt}\int_{-\infty}^{\infty} f(s-a+ta)\phi(s+ta)\,ds \;=\; \int_{-\infty}^{\infty} \frac{\partial}{\partial t}[f(s-a+ta)\phi(s+ta)]\,ds \quad.$$

However, by the product and chain rules,

$$\frac{\partial}{\partial t}[f(s-a+ta)\phi(s+ta)]$$
$$= \; f'(s-a+ta)a\,\phi(s+ta) \;+\; f(s-a+ta)\,\phi'(s+ta)a$$
$$= \; a[f'(s-a+ta)\phi(s+ta) \;+\; f(s-a+ta)\phi'(s+ta)]$$
$$= \; a\frac{\partial}{\partial s}[f(s-a+ta)\phi(s+ta)] \quad.$$

Plugging this back into the above formula for $h'(t)$ and using basic calculus and the facts that f is exponentially bounded and ϕ is a Gaussian test function, we have

$$h'(t) \;=\; a\int_{-\infty}^{\infty} \frac{\partial}{\partial s}[f(s-a+ta)\phi(s+ta)]\,ds$$
$$= \; af(s-a+ta)\phi(s+ta)\Big|_{s=-\infty}^{\infty} \;=\; 0 \quad.$$

This means h is a constant function. Hence, in particular, $h(0) = h(1)$. ∎

It's not too difficult to verify that all simple multipliers and all Gaussian test functions are exponentially bounded on horizontal strips. Consequently (as we anticipated in equation (35.15)), there is no difference between the generalized and the corresponding classical translations of these functions, even when the translations are by complex values.

?►Exercise 35.14:   *Show f is exponentially bounded on horizontal strips of $\mathbb{C}$ whenever*

   a: *f is a Gaussian function.*

   b: *f is a Gaussian test function.*

   c: *f is a simple multiplier.*

Basic Properties and Identities

A few easily verified properties of the translation operator are given in the next exercise.

?►Exercise 35.15:   *Verify each of the following, assuming f and g are two generalized functions, h is a simple multiplier, and a, b, α and β are fixed complex numbers:*

   a: *(linearity) $T_a[\alpha f + \beta g] = \alpha T_a[f] + \beta T_a[g]$*

   b: *(commutativity) $T_a[T_b[f]] = T_{a+b}[f] = T_b[T_a[f]]$*

   c: *$T_a[hf] = T_a[h]\,T_a[f]$*

   d: *$T_0[f] = f$*

Translation and the Fourier Transforms

As with the scaling identities, the classical translation identities for Fourier transforms (described in theorem 21.1 on page 319) generalize to analogous identities for the generalized Fourier transforms.

Theorem 35.9 (the translation identities)
Let f and F be generalized functions with $F(y) = \mathcal{F}[f(x)]|_y$, and let a be any complex number. Then

$$\mathcal{F}[T_a[f]]|_y = e^{-i2\pi a y}F(y) \tag{35.17a}$$

and

$$\mathcal{F}^{-1}[T_a[F]]|_x = e^{i2\pi a x}f(x) \quad . \tag{35.17b}$$

Equivalently,

$$\mathcal{F}^{-1}\left[e^{-i2\pi a y}F(y)\right] = T_a[f] \tag{35.17c}$$

and

$$\mathcal{F}\left[e^{i2\pi a x}f(x)\right] = T_a[F] \quad . \tag{35.17d}$$

PROOF (of identities (35.17a) and (35.17c)):   Let ϕ be any Gaussian test function, and let $\Phi = \mathcal{F}[\phi]$. By the generalized definitions of the Fourier transform and translation

$$\left\langle\, \mathcal{F}[T_a[f]]|_y \,, \phi(y) \,\right\rangle = \left\langle\, \mathcal{F}[T_a[f]] \,, \phi \,\right\rangle$$
$$= \left\langle\, T_a[f] \,, \mathcal{F}[\phi] \,\right\rangle = \left\langle\, f \,, T_{-a}[\mathcal{F}[\phi]] \,\right\rangle \quad . \tag{35.18}$$

Now, $T_{-a}[\mathcal{F}[\phi]]$ is computed classically. Rewriting it in more classical notation and using the (extended) classical translation identities — which we know are valid for Gaussian test functions (see theorem 32.9 on page 525) — we see that

$$T_{-a}[\mathcal{F}[\phi]]|_x = T_{-a}[\Phi]|_x = \Phi(x - (-a)) = \mathcal{F}\left[e^{i2\pi(-a)y}\phi(y)\right]\Big|_x \quad .$$

So

$$\langle f , T_{-a}[\mathcal{F}[\phi]] \rangle = \langle f , \mathcal{F}\left[e^{-i2\pi ay}\phi(y)\right] \rangle$$

$$= \langle \mathcal{F}[f]|_y , e^{-i2\pi ay}\phi(y) \rangle$$

$$= \langle F(y) , e^{-i2\pi ay}\phi(y) \rangle = \langle e^{-i2\pi ay}F(y) , \phi(y) \rangle \quad .$$

Combined with equation (35.18), this gives

$$\langle \mathcal{F}[T_a[f]]|_y , \phi(y) \rangle = \langle e^{-i2\pi ay}F(y) , \phi(y) \rangle$$

for each Gaussian test function ϕ, verifying identity (35.17a).

Identity (35.17c) is then obtained by taking the inverse Fourier transform of both sides of identity (35.17a).   ∎

?►Exercise 35.16:   Verify identities (35.17b) and (35.17d).

Let's look at some applications of these identities, starting with the computations of the transforms of arbitrary delta functions and exponentials.

!►Example 35.10:   Let a be any complex value. As we have already seen (example 35.3 on page 590 and exercise 35.13 on page 598),

$$\mathcal{F}[1] = \delta \quad \text{and} \quad T_a[\delta] = \delta_a \quad .$$

Combined with translation identity 35.17d, these give us

$$\mathcal{F}\left[e^{i2\pi ax}\right] = \mathcal{F}\left[e^{i2\pi ax} \cdot 1\right] = T_a[\mathcal{F}[1]] = T_a[\delta] = \delta_a \quad .$$

?►Exercise 35.17:   Letting a denote an arbitrary complex number, verify the following:

a: $\mathcal{F}^{-1}\left[e^{-i2\pi ax}\right] = \delta_a$ **b:** $\mathcal{F}[\delta_a]|_x = e^{-i2\pi ax}$ **c:** $\mathcal{F}^{-1}[\delta_a]|_x = e^{i2\pi ax}$

Next, consider the problem of finding the transform of a classically transformable function multiplied by a real exponential.

!►Example 35.11:   From the classical theory we know

$$\mathcal{F}[\text{sinc}(2\pi x)] = \frac{1}{2}\text{pulse}_1 \quad .$$

Using this and identity (35.17d), we get

$$\mathcal{F}\left[e^{6\pi x}\text{sinc}(2\pi x)\right]\bigg|_y = \mathcal{F}\left[e^{i2\pi(-3i)x}\text{sinc}(2\pi x)\right]\bigg|_y = \frac{1}{2}T_{-3i}[\text{pulse}_1] \quad .$$

?►Exercise 35.18:   Show that $\mathcal{F}\left[e^{6\pi x}\sin(2\pi x)\right] = \frac{1}{2i}\left[\delta_{1-3i} - \delta_{-1-3i}\right]$.

With a little cleverness, we can use the above identities to get a more convenient representation for the Fourier transform of the step function.

!▶**Example 35.12:**   Let α be any positive value. From the classical theory we know

$$\mathcal{F}\left[e^{-2\pi\alpha x}\,\text{step}(x)\right]\Big|_y = \frac{1}{2\pi\alpha + i2\pi y} \quad .$$

Using this and identity (35.17d), we have

$$\mathcal{F}\left[\text{step}(x)\right]\Big|_y = \mathcal{F}\left[e^{2\pi\alpha x}e^{-2\pi\alpha x}\,\text{step}(x)\right]\Big|_y$$

$$= \mathcal{F}\left[e^{i2\pi(-\alpha i)x}\left(e^{-2\pi\alpha x}\,\text{step}(x)\right)\right]\Big|_y = T_{-\alpha i}\left[\frac{1}{2\pi\alpha + i2\pi y}\right] \quad ,$$

which, after factoring out $(i2\pi)^{-1}$, can be written as

$$\mathcal{F}\left[\text{step}(x)\right]\Big|_y = \frac{1}{i2\pi}T_{-\alpha i}\left[\frac{1}{y - i\alpha}\right] \quad .$$

In particular, taking $\alpha = 1$,

$$\mathcal{F}\left[\text{step}(x)\right]\Big|_y = \frac{1}{i2\pi}T_{-i}\left[\frac{1}{y - i}\right] \quad .$$

?▶**Exercise 35.19:**   Letting α be any positive value, verify that

$$\mathcal{F}\left[\text{step}(-x)\right]\Big|_y = \frac{-1}{i2\pi}T_{\alpha i}\left[\frac{1}{y + i\alpha}\right] \quad .$$

Comparing Some Classical and Generalized Translations

It is instructive to compare the classical translation of

$$f(x) = \frac{1}{x + i}$$

by $i\alpha$, $f(x - i\alpha)$, to the corresponding generalized translation, $T_{i\alpha}[f]$.

For simplicity, assume α is a real number.

By definition,

$$\langle\, T_{i\alpha}[f] , \phi\,\rangle = \langle\, f , T_{-i\alpha}[\phi]\,\rangle = \int_{-\infty}^{\infty} \frac{\phi(s + i\alpha)}{s + i}\,ds \tag{35.19}$$

for each Gaussian test function ϕ. On the other hand, the classical translation of f by $i\alpha$ is given by the formula

$$f(x - i\alpha) = \frac{1}{(x - i\alpha) + i} = \frac{1}{x + i - i\alpha} \quad . \tag{35.20}$$

So, for any Gaussian test function ϕ,

$$\langle\, f(x - i\alpha) , \phi(x)\,\rangle = \int_{-\infty}^{\infty} \frac{\phi(x)}{x + i - i\alpha}\,dx \quad . \tag{35.21}$$

The naive reader may be tempted to point out that the integral in equation (35.21) can be obtained from the integral in equation (35.19) using the substitution $x = s + i\alpha$. Unfortunately, that is a complex change of variables, and complex changes of variables are *not* generally valid.

What we can do is derive formulas for $T_{i\alpha}[f]$ in terms of the classical translation. We start with the observation that

$$T_{i\alpha}[f] = \mathcal{F}\big[\mathcal{F}^{-1}[T_{i\alpha}[f]]\big] \quad . \tag{35.22}$$

From the classical theory we know

$$\mathcal{F}^{-1}[f(x)]\big|_y = \mathcal{F}^{-1}\Big[\frac{1}{x+i}\Big]\Big|_y$$

$$= -i2\pi\,\mathcal{F}^{-1}\Big[\frac{1}{2\pi - i2\pi x}\Big]\Big|_y = -i2\pi e^{2\pi y}\,\text{step}(-y) \quad .$$

So, by the translation identities,

$$\mathcal{F}^{-1}[T_{i\alpha}[f]]\big|_y = e^{i2\pi(i\alpha)y}\Big[-i2\pi e^{2\pi y}\,\text{step}(-y)\Big] = -i2\pi e^{2\pi(1-\alpha)y}\,\text{step}(-y) \quad .$$

Thus, equation (35.22) can be written as

$$T_{i\alpha}[f] = -i2\pi\,\mathcal{F}\Big[e^{2\pi(1-\alpha)y}\,\text{step}(-y)\Big] \quad . \tag{35.23}$$

There are three cases to consider: $\alpha < 1$, $\alpha > 1$ and $\alpha = 1$.

Consider first the case where $\alpha < 1$. Then $1 - \alpha > 0$ and $e^{2\pi(1-\alpha)y}\,\text{step}(-y)$ is classically transformable. From the tables

$$T_{i\alpha}[f]\big|_x = -i2\pi\,\mathcal{F}\Big[e^{2\pi(1-\alpha)y}\,\text{step}(-y)\Big]\Big|_x = \frac{-i2\pi}{2\pi(1-\alpha) - i2\pi x} = \frac{1}{x+i-i\alpha} \quad .$$

Comparing this with formula (35.20), we thus find that

$$T_{i\alpha}[f]\big|_x = f(x - i\alpha) \qquad \text{when} \quad \alpha < 1 \quad .$$

Next, assume $\alpha > 1$. Then $1 - \alpha < 0$ and the function $e^{2\pi(1-\alpha)y}\,\text{step}(-y)$ is not classically transformable. However, after observing that

$$\text{step}(-y) = 1 - \text{step}(y) \quad ,$$

we see that

$$e^{2\pi(1-\alpha)y}\,\text{step}(-y) = e^{2\pi(1-\alpha)y}[1 - \text{step}(y)]$$

$$= e^{-2\pi(\alpha-1)y} - e^{-2\pi(\alpha-1)y}\,\text{step}(y) \quad ,$$

which is the difference between an exponential function and a classically transformable function. Using this with equation (35.23) (and, again, referring to the tables), we obtain

$$T_{i\alpha}[f]\big|_x = -i2\pi\,\mathcal{F}\Big[e^{-2\pi(\alpha-1)y} - e^{-2\pi(\alpha-1)y}\,\text{step}(y)\Big]\Big|_x$$

$$= -i2\pi\,\mathcal{F}\Big[e^{i2\pi[i(\alpha-1)]y}\Big]\Big|_x + i2\pi\,\mathcal{F}\Big[e^{-2\pi(\alpha-1)y}\,\text{step}(y)\Big]\Big|_x$$

$$= -i2\pi\,\delta_{i(\alpha-1)}(x) + i2\pi\cdot\frac{1}{2\pi(\alpha-1) + i2\pi x}$$

$$= -i2\pi\,\delta_{i(\alpha-1)}(x) + \frac{1}{x+i-i\alpha} \quad .$$

Comparing this with formula (35.20) we see that, in this case, the classical and the generalized translations are not the same. Instead,

$$T_{i\alpha}[f]|_x = f(x - i\alpha) - i2\pi \delta_{i(\alpha-1)}(x) \qquad \text{when} \quad \alpha > 0 \quad .$$

Finally, consider what we have when $\alpha = 1$. Then $1 - \alpha = 0$ and the classical translation is

$$f(x - i) = \frac{1}{(x - i) + i} = \frac{1}{x} \quad .$$

This function is not classically transformable. It "blows up" at $x = 0$. In fact, it is not even exponentially integrable, and so, does not define a generalized function. Consequently, as far as we are concerned, there is no equation relating the generalized function $T_i[f]$ to the classical function $f(x - i)$. (However $T_i[f]$ can be related to the "pole function" that will be developed in chapter 38 as an analog to $1/x$.)

35.4 The Generalized Derivative
Definition

Let us begin with a classical function f that is continuous and piecewise smooth on the real line. For now, let's also assume both f and its derivative, f', are exponentially bounded. If ϕ is any Gaussian test function, then, using integration by parts,

$$\int_{-\infty}^{\infty} f'(x)\phi(x)\, dx = f(x)\phi(x)\big|_{-\infty}^{\infty} - \int_{-\infty}^{\infty} f(x)\phi'(x)\, dx \quad .$$

But since f is exponentially bounded and ϕ is a Gaussian test function,

$$f(x)\phi(x)\big|_{-\infty}^{\infty} = \lim_{x\to\infty} f(x)\phi(x) - \lim_{x\to-\infty} f(x)\phi(x) = 0 \quad .$$

So the above integration by parts formula simplifies to

$$\int_{-\infty}^{\infty} f'(x)\phi(x)\, dx = - \int_{-\infty}^{\infty} f(x)\phi'(x)\, dx \quad ,$$

which we can also write as

$$\langle\, f'\,,\phi\,\rangle = -\langle\, f\,,\phi'\,\rangle \quad . \tag{35.24}$$

While this last equation was derived assuming f is a classical function, the equation's right-hand side is well defined for any generalized function f. So, once again, we have an equation that can define a generalized analog of the operation leading to the equation. This time the operation is differentiation, and we will formally define the generalized analog as follows: For each generalized function f, the corresponding *(generalized) derivative*, denoted by Df, is defined to be the generalized function satisfying

$$\langle\, Df\,,\phi\,\rangle = -\langle\, f\,,\phi'\,\rangle \qquad \text{for each } \phi \text{ in } \mathcal{G} \quad . \tag{35.25}$$

Higher order generalized derivatives are defined in the obvious way: $D^2 f = DDf$, $D^3 f = DDDf$ and so forth. Repeatedly applying the above definition we see that, given any generalized function f and any nonnegative integer n, $D^n f$ is the generalized function satisfying

$$\langle\, D^n f\,,\phi\,\rangle = (-1)^n \langle\, f\,,\phi^{(n)}\,\rangle \qquad \text{for each } \phi \text{ in } \mathcal{G} \quad .$$

If f is the sort of classical function assumed at the beginning of this section, then the generalized derivative Df and the classical derivative f' are the same. It is possible, however, to have a classical function f whose generalized derivative Df differs in a nontrivial manner from its classical derivative f'. That is why we will use different notation for the two types of derivatives. We will explore the relation between classical and generalized derivatives more fully after looking at the generalized derivatives of a few specific generalized functions.

!▶Example 35.13:   *The generalized derivative of the delta function, $D\delta$, is the generalized function such that*

$$\langle\, D\delta\,,\,\phi\,\rangle \;=\; -\langle\,\delta\,,\,\phi'\,\rangle \;=\; -\phi'(0) \qquad \text{for each } \phi \text{ in } \mathcal{G} \quad .$$

?▶Exercise 35.20:   *Convince yourself that, for any positive integer n,*

$$\langle\, D^n\delta\,,\,\phi\,\rangle \;=\; (-1)^n\phi^{(n)}(0) \qquad \text{for each } \phi \text{ in } \mathcal{G} \quad .$$

To see how the generalized and classical derivatives can differ, let's find the derivatives of the step function.

!▶Example 35.14:   *Way back in example 3.5 on page 23, we saw that the classical derivative of the step function,*

$$\text{step}(x) \;=\; \begin{cases} 0 & \text{if } x < 0 \\ 1 & \text{if } 0 < x \end{cases} \quad ,$$

is

$$\text{step}' \;=\; 0 \quad .$$

On the other hand, for any Gaussian test function ϕ,

$$\langle\, D\,\text{step}\,,\,\phi\,\rangle \;=\; -\langle\,\text{step}\,,\,\phi'\,\rangle \;=\; -\int_{-\infty}^{\infty} \text{step}(x)\phi'(x)\,dx \;=\; -\int_{0}^{\infty} \phi'(x)\,dx \quad .$$

This last integral is easily evaluated,

$$-\int_{0}^{\infty} \phi'(x)\,dx \;=\; -\left[\lim_{x\to\infty}\phi(x) \;-\; \phi(0)\right] \;=\; \phi(0) \quad .$$

But $\phi(0) = \langle\,\delta\,,\,\phi\,\rangle$. So the above reduces to

$$\langle\, D\,\text{step}\,,\,\phi\,\rangle \;=\; \langle\,\delta\,,\,\phi\,\rangle \qquad \text{for each } \phi \text{ in } \mathcal{G} \quad .$$

In other words,

$$D\,\text{step} \;=\; \delta \quad .$$

Notice the difference between the generalized and the classical derivatives of the step function: The generalized derivative has a delta function at the point where the step function has a discontinuity. This, as we will soon see, is indicative of the general relationship between the classical and the generalized derivative of any piecewise smooth function.

Relation with Classical Derivatives

Let's see how f' is related to Df when f is a piecewise smooth function whose classical derivative, f', is exponentially integrable. For now, assume f has exactly one discontinuity, say, a jump of j_0 at x_0. Remember,

$$j_0 = \lim_{x \to x_0^+} f(x) - \lim_{x \to x_0^-} f(x) \quad .$$

At this point we should observe that $f(x)$ can be computed from its derivative via

$$f(x) - \lim_{s \to x_0^+} f(s) = \int_{x_0}^{x} f'(s) \, ds \qquad \text{when} \quad x_0 < x$$

and

$$\lim_{s \to x_0^-} f(s) - f(x) = \int_{x}^{x_0} f'(s) \, ds \qquad \text{when} \quad x < x_0 \quad .$$

From this and the exponential integrability of f' we can easily deduce that f, itself, must be exponentially bounded on the real line (see lemma 30.10 on page 496).

Now let ϕ be any Gaussian test function. By definition,

$$\langle\, Df \,,\, \phi \,\rangle = -\langle\, f \,,\, \phi' \,\rangle = -\int_{-\infty}^{\infty} f(x)\phi'(x) \, dx \quad . \tag{35.26}$$

Because f is not continuous at x_0, this last integral cannot be integrated by parts as we did in deriving equation (35.24). But we can use integration by parts after splitting the integral into integrals over intervals on which f is continuous. Doing so,

$$\int_{-\infty}^{\infty} f(x)\phi'(x) \, dx = \int_{-\infty}^{x_0} f(x)\phi'(x) \, dx + \int_{x_0}^{\infty} f(x)\phi'(x) \, dx$$

$$= f(x)\phi(x)\big|_{-\infty}^{x_0} - \int_{-\infty}^{x_0} f'(x)\phi(x) \, dx$$

$$+ f(x)\phi(x)\big|_{x_0}^{\infty} - \int_{x_0}^{\infty} f'(x)\phi(x) \, dx \quad . \tag{35.27}$$

Now,

$$\int_{-\infty}^{x_0} f'(x)\phi(x) \, dx + \int_{x_0}^{\infty} f'(x)\phi(x) \, dx = \int_{-\infty}^{\infty} f'(x)\phi(x) \, dx = \langle\, f' \,,\, \phi \,\rangle \quad ,$$

and, because f is exponentially bounded and ϕ is in $\mathcal{G}$,

$$f(x)\phi(x)\big|_{-\infty}^{x_0} + f(x)\phi(x)\big|_{x_0}^{\infty} = \left[\lim_{x \to x_0^-} f(x)\phi(x_0) - 0 \right] + \left[0 - \lim_{x \to x_0^+} f(x)\phi(x_0) \right]$$

$$= \left[\lim_{x \to x_0^-} f(x) - \lim_{x \to x_0^+} f(x) \right]\phi(x_0)$$

$$= -j_0 \, \phi(x_0)$$

$$= -\langle\, j_0 \, \delta_{x_0} \,,\, \phi \,\rangle \quad .$$

So equation (35.27) becomes

$$\int_{-\infty}^{\infty} f(x)\phi'(x) \, dx = -\langle\, f' \,,\, \phi \,\rangle - \langle\, j_0 \, \delta_{x_0} \,,\, \phi \,\rangle = -\langle\, f' + j_0 \, \delta_{x_0} \,,\, \phi \,\rangle \quad .$$

Plugging this back into equation (35.26) then gives us

$$\langle \, Df \, , \, \phi \, \rangle \; = \; \langle \, f' + j_0 \, \delta_{x_0} \, , \, \phi \, \rangle \qquad \text{for each } \phi \text{ in } \mathcal{G} \quad .$$

Thus, under the assumptions assumed here for f,

$$Df \; = \; f' \; + \; j_0 \, \delta_{x_0} \quad .$$

In other words, the generalized derivative is the classical derivative plus a delta function at the point of discontinuity multiplied by the value of the jump at the discontinuity.

What happens when f has more than one discontinuity should also be fairly obvious, given the above computations.

Theorem 35.10

Let f be a piecewise smooth function on the real line whose classical derivative is exponentially integrable. If f has an infinite number of discontinuities, then also assume f is exponentially bounded. Let $\{ \dots, x_0, x_1, x_2, \dots \}$ be the set of all points at which f has discontinuities. Then f is exponentially integrable, and

$$Df \; = \; f' \; + \; \sum_k j_k \, \delta_{x_k}$$

where the summation is taken over all points at which f is discontinuous and, for each k, j_k is the jump in f at x_k.

The details of the proof of this last theorem will be left to the reader (exercise 35.47 at the end of the chapter).

Of course, if f is continuous on the real line, then the above reduces to the reassuring corollary below.

Corollary 35.11

If f is a piecewise smooth and continuous function on the real line with an exponentially integrable derivative, then f is exponentially integrable and its generalized derivative is the same as its classical derivative.

!▶ **Example 35.15:**   Let

$$f(x) \; = \; \begin{cases} 1 + x^2 & \text{if } \; x < 0 \\ 4 - x^2 & \text{if } \; 0 < x \end{cases} \quad .$$

This is a piecewise smooth function with one discontinuity, at $x = 0$. The jump in the function at the discontinuity is

$$j_0 \; = \; \lim_{x \to 0^+} f(x) \; - \; \lim_{x \to 0^-} f(x)$$

$$= \; \lim_{x \to 0^+} \left[4 - x^2 \right] \; - \; \lim_{x \to 0^-} \left[1 + x^2 \right] \; = \; 3 \quad .$$

The classical derivative is easily found,

$$f'(x) \; = \; \begin{cases} \dfrac{d}{dx}[x^2 + 1] & \text{if } \; x < 0 \\ \dfrac{d}{dx}[4 - x^2] & \text{if } \; 0 < x \end{cases} \; = \; \begin{cases} 2x & \text{if } \; x < 0 \\ -2x & \text{if } \; 0 < x \end{cases} \quad .$$

So,

$$Df(x) = f'(x) + \sum_k j_k \, \delta_{x_k}(x) = \begin{cases} 2x & \text{if } x < 0 \\ -2x & \text{if } 0 < x \end{cases} + 3\delta(x) \quad .$$

?►Exercise 35.21: *Let a and b be two real numbers, and verify the following:*

  a: $D \, \text{rect}_{(a,b)} = \delta_a - \delta_b$ *(assuming $a < b$)*

  b: $D \, \text{pulse}_a = \delta_{-a} - \delta_a$ *(assuming $0 < a$)*

?►Exercise 35.22: *Show that*

  a: $D[x^3 \, \text{step}(x - 2)] = 3x^2 \, \text{step}(x - 2) + 8\delta_2(x)$

  b: $D[x^3 \, \text{rect}_{(0,2)}(x)] = 3x^2 \, \text{rect}_{(0,2)}(x) - 8\delta_2(x)$

Elementary Differential Identities

In elementary calculus you learned an number of identities and rules for the classical derivative, and you probably suspect that some of these also hold, suitably reinterpreted, for the generalized derivative. Your suspicions are correct. Here are four (linearity and versions of the chain and product rules) we can verify at this time:

1. (linearity) For any two generalized functions f and g, and any two constants α and β,

$$D[\alpha f + \beta g] = \alpha Df + \beta Dg \quad .$$

2. (differentiation and scaling) If f is any generalized function and σ is any nonzero real number, then

$$D[f(\sigma x)] = \sigma \, Df(\sigma x) \quad .$$

   In terms of the scaling operator, this is

$$D[S_\sigma[f]] = \sigma \, S_\sigma[Df] \quad .$$

3. (differentiation and translation) For any generalized function f and any real number a,

$$D[f(x - a)] = Df(x - a) \quad .$$

   More generally,

$$D[T_a[f]] = T_a[Df]$$

   for any generalized function f and any complex number a.

4. (product rule) The generalized derivative of an elementary multiplier is an elementary multiplier. Moreover, if f and g are two generalized functions and at least one is an elementary multiplier, then

$$D[fg] = [Df]g + f[Dg] \quad .$$

   We will quickly verify the second and fourth identities and leave the verification of the first and third identities to you.

PROOF (Differentiation and scaling):   Let ϕ be any Gaussian test function. By the definitions of generalized differentiation and scaling,

$$\left\langle D\left[S_\sigma[f]\right], \phi \right\rangle = \left\langle S_\sigma[f], \phi' \right\rangle = \frac{1}{|\sigma|} \left\langle f, S_{1/\sigma}[\phi'] \right\rangle \quad . \tag{35.28}$$

On ϕ, however, the scaling and differentiation operators are simply the classical operators and can be manipulated using the rules learned in elementary calculus. So, for each real value x,

$$S_{1/\sigma}[\phi']\big|_x = \phi'\left(\frac{x}{\sigma}\right) = \sigma \frac{1}{\sigma} \phi'\left(\frac{x}{\sigma}\right) = \sigma \frac{d}{dx}\left[\phi\left(\frac{x}{\sigma}\right)\right] = \sigma \frac{d}{dx}\left[S_{1/\sigma}[\phi]\big|_x\right] \quad .$$

Thus,

$$S_{1/\sigma}[\phi'] = \sigma \left[S_{1/\sigma}[\phi]\right]' \quad .$$

Plugging this into equation (35.28) and using the definitions of generalized scaling and differentiation once again yields

$$\left\langle D\left[S_\sigma[f]\right], \phi \right\rangle = \frac{1}{|\sigma|} \left\langle f, \sigma \left[S_{1/\sigma}[\phi]\right]' \right\rangle$$

$$= \sigma \cdot \frac{1}{|\sigma|} \left\langle Df, S_{1/\sigma}[\phi] \right\rangle$$

$$= \sigma \left\langle S_\sigma[Df], \phi \right\rangle = \left\langle \sigma S_\sigma[Df], \phi \right\rangle \quad . \qquad \blacksquare$$

PROOF (Product Rule):   The fact that derivatives of simple multipliers are, themselves, simple multipliers follows directly from the formulas defining simple multipliers (see page 523).

     Assume g is the simple multiplier, and let ϕ be any Gaussian test function. By the definitions of generalized differentiation and simple multiplication,

$$\left\langle D[fg], \phi \right\rangle = -\left\langle fg, \phi' \right\rangle = -\left\langle f, g\phi' \right\rangle \quad . \tag{35.29}$$

Since g is a simple multiplier, it is a smooth, classical function, and the classical product rule applies in computing the derivative of the product $g\phi$,

$$[g\phi]' = g'\phi + g\phi' \quad .$$

Consequently, $g\phi' = [g\phi]' - g'\phi$. When we plug this into the right-hand side of equation (35.29) and then use linearity along with the definitions of generalized differentiation and simple multiplication, we get

$$\left\langle D[fg], \phi \right\rangle = -\left\langle f, [g\phi]' - g'\phi \right\rangle$$

$$= -\left\langle f, [g\phi]' \right\rangle + \left\langle f, g'\phi \right\rangle$$

$$= \left\langle Df, g\phi \right\rangle + \left\langle fg', \phi \right\rangle$$

$$= \left\langle [Df]g, \phi \right\rangle + \left\langle fg', \phi \right\rangle = \left\langle [Df]g + fg', \phi \right\rangle \quad .$$

     Now cut out the middle of the last set of equalities and recall that, because g is a smooth function, $g' = Dg$. This gives us

$$\left\langle D[fg], \phi \right\rangle = \left\langle [Df]g + f[Dg], \phi \right\rangle \quad ,$$

verifying that $D[fg] = [Df]g + f[Dg]$ when g is a simple multiplier. Obviously, repeating the above computations with the roles of f and g interchanged will also give this last equation when f is a simple multiplier. $\blacksquare$

?►Exercise 35.23: *Prove the linearity of the generalized derivative. That is, verify that*

$$D[\alpha f + \beta g] = \alpha Df + \beta Dg$$

whenever f and g are any two generalized functions, and α and β are any two constants.

?►Exercise 35.24: *Show that*
$$D[T_a[f]] = T_a[Df]$$

for any generalized function f and any complex number a.

!►Example 35.16: *Earlier, we discovered that $D\,\text{step} = \delta$. From the chain rule formula $D[f(\sigma x)] = \sigma Df(\sigma x)$ with f being the step function and $\sigma = -1$, we then have*

$$D[\text{step}(-x)] = -D\,\text{step}(-x) = -\delta(-x) \quad,$$

which reduces to
$$D[\text{step}(-x)] = -\delta(x)$$

because δ is an even generalized function.

!►Example 35.17: *To compute the generalized derivative of*

$$f(x) = e^{-3x}\,\text{step}(x) \quad,$$

we can use the product rule along with a few other results already discussed:

$$\begin{aligned}
Df(x) &= \left(De^{-3x}\right)\text{step}(x) + e^{-3x}(D\,\text{step}(x)) \\
&= -3e^{-3x}\,\text{step}(x) + e^{-3x}\delta(x) \\
&= -3e^{-3x}\,\text{step}(x) + e^{-3\cdot 0}\delta(x) \\
&= -3e^{-3x}\,\text{step}(x) + \delta(x) \quad.
\end{aligned}$$

?►Exercise 35.25: *Show that $D[e^{3x}\,\text{step}(-x)] = 3e^{3x}\,\text{step}(-x) - \delta(x)$.*

The Fourier Differential Identities

The generalized versions of the differential identities discussed in chapter 22 are described in the next theorem. This theorem also illustrates how much simpler life can be using the generalized theory. Not only does it generalize four theorems from chapter 22 (theorems 22.1, 22.2, 22.9 and 22.10), but it is a more simply stated theorem.

Theorem 35.12
Let f and F be generalized functions with $F(y) = \mathcal{F}[f(x)]|_y$, and let n be any positive integer. Then
$$\mathcal{F}\left[D^n f\right]\big|_y = (i2\pi y)^n F(y) \tag{35.30a}$$

and

$$\mathcal{F}^{-1}\left[D^n F\right]\big|_x = (-i2\pi x)^n f(x) \quad. \tag{35.30b}$$

Equivalently,

$$\mathcal{F}^{-1}\big[y^n F(y)\big] = \left(\frac{1}{i2\pi}\right)^n D^n f \qquad (35.30\text{c})$$

and

$$\mathcal{F}\big[x^n f(x)\big] = \left(-\frac{1}{i2\pi}\right)^n D^n F \quad . \qquad (35.30\text{d})$$

We will quickly go through proofs of two of the identities and leave proving the rest as an exercise.

PROOF (of identities (35.30a) and (35.30c)):   First observe that identity (35.30c) can be obtained by simply taking the inverse transform of both sides of identity (35.30a). So we only need verify equation (35.30a). That is, we need only show that

$$\Big(\mathcal{F}\big[D^n f\big]\big|_y \,,\, \phi(y)\Big) = \Big((i2\pi y)^n F(y)\,,\, \phi(y)\Big) \qquad \text{for each } \phi \text{ in } \mathcal{G} \quad .$$

So let ϕ be any Gaussian test function, and let $\Phi = \mathcal{F}[\phi]$. By the definitions of the generalized transforms and derivatives,

$$\Big(\mathcal{F}\big[D^n f\big]\big|_y \,,\, \phi(y)\Big) = \Big(D^n f\,,\, \Phi\Big) = (-1)^n \Big(f\,,\, \Phi^{(n)}\Big) = \Big(f\,,\, (-1)^n \Phi^{(n)}\Big) \quad .$$

Fortunately, ϕ and Φ are classically transformable functions that clearly satisfy the requirements for the higher order classical differential identity given in theorem 22.10 on page 350. So we know

$$(-1)^n \Phi^{(n)} = (-1)^n \mathcal{F}\big[(-i2\pi y)^n \phi(y)\big] = \mathcal{F}\big[(i2\pi y)^n \phi(y)\big] \quad .$$

Plugging this into the preceding equation and, again, using the definitions of the generalized transforms, we get

$$\Big(\mathcal{F}\big[D^n f\big]\big|_y \,,\, \phi(y)\Big) = \Big(f\,,\, \mathcal{F}\big[(i2\pi y)^n \phi(y)\big]\Big)$$

$$= \Big(F(y)\,,\, (i2\pi y)^n \phi(y)\Big) = \Big((i2\pi y)^n F(y)\,,\, \phi(y)\Big) \quad . \qquad \blacksquare$$

?▶ Exercise 35.26:   *Verify identities (35.30b) and (35.30d) in theorem 35.12.*

!▶ Example 35.18:   *Consider finding the inverse Fourier transform of $y(3 + i2\pi y)^{-1}$. Using identity (35.30c) with $n = 1$ and $F(y) = (3 + i2\pi y)^{-1}$, we have*

$$\mathcal{F}^{-1}\left[\frac{y}{3 + i2\pi y}\right] = \mathcal{F}^{-1}[yF(y)] = \frac{1}{i2\pi}Df$$

where, as we well know from the classical theory,

$$f(x) = \mathcal{F}^{-1}[F(y)]\big|_x = \mathcal{F}^{-1}\left[\frac{1}{3 + i2\pi y}\right]\Big|_x = e^{-3x}\,\text{step}(x) \quad .$$

From the computations in example 35.17, we also know that

$$Df(x) = -3e^{-3x}\,\text{step}(x) + \delta(x) \quad .$$

So

$$\mathcal{F}^{-1}\left[\frac{y}{3 + i2\pi y}\right]\Big|_x = \frac{1}{i2\pi}Df(x) = \frac{1}{i2\pi}\left[-3e^{-3x}\,\text{step}(x) + \delta(x)\right] \quad .$$

?► Exercise 35.27: *Show that*

$$\mathcal{F}^{-1}\left[\frac{y}{3 - i2\pi y}\right]\Bigg|_x = \frac{1}{i2\pi}[3e^{3x}\,\text{step}(-x) - \delta(x)] \quad.$$

35.5   Transforms of Limits and Series

Many generalized functions can be defined in terms of limits and/or infinite series of generalized functions. For example,

$$\text{comb}_1 = \sum_{k=-\infty}^{\infty} \delta_k = \lim_{\substack{N\to\infty \\ M\to\infty}} \sum_{k=M}^{N} \delta_k \quad.$$

Let us now consider the Fourier transform of some generalized function f that is the generalized limit of some sequence of other generalized functions, say,

$$f = \lim_{k\to\infty} g_k \quad.$$

Remember, this means

$$\langle\, f\,,\phi\,\rangle = \lim_{k\to\infty}\langle\, g_k\,,\phi\,\rangle \qquad \text{for each } \phi \text{ in } \mathcal{G} \quad.$$

By this and the definition of the generalized Fourier transform, we then have

$$\langle\, \mathcal{F}[f]\,,\phi\,\rangle = \langle\, f\,,\mathcal{F}[\phi]\,\rangle = \lim_{k\to\infty}\langle\, g_k\,,\mathcal{F}[\phi]\,\rangle = \lim_{k\to\infty}\langle\, \mathcal{F}[g_k]\,,\phi\,\rangle$$

for each Gaussian test function ϕ. Thus,

$$\mathcal{F}[f] = \lim_{k\to\infty}\mathcal{F}[g_k] \qquad \text{(in the generalized sense)} \quad.$$

Similar computations can be done with any other type of convergent sequence of generalized functions using any other transform we've developed thus far in this chapter. Consequently, we have the following lemma, which we will find useful on several occasions.

Lemma 35.13

Assume $\{g_\gamma\}_{\gamma=\gamma_0}^{\alpha}$ *is a convergent sequence of generalized functions with*

$$f = \lim_{\gamma\to\alpha} g_\gamma \quad.$$

Then, in the generalized sense,

$$\mathcal{F}[f] = \lim_{\gamma\to\alpha}\mathcal{F}[g_\gamma] \quad, \qquad \mathcal{F}^{-1}[f] = \lim_{\gamma\to\alpha}\mathcal{F}^{-1}[g_\gamma] \quad,$$

$$S_\sigma[f] = \lim_{\gamma\to\alpha}S_\sigma[g_\gamma] \quad, \qquad T_a[f] = \lim_{\gamma\to\alpha}T_a[g_\gamma] \qquad \text{and} \qquad Df = \lim_{\gamma\to\alpha}Dg_\gamma$$

where σ is any nonzero real number and a is any complex number.

?► Exercise 35.28: *Verify the claims made in the last lemma that*

$$T_a[f] = \lim_{\gamma\to\alpha}T_a[g_\gamma] \qquad \text{and} \qquad Df = \lim_{\gamma\to\alpha}Dg_\gamma \quad.$$

As an immediate corollary, we have the corresponding lemma for infinite sums of generalized functions.

Lemma 35.14

Suppose

$$f = \sum_{k=-\infty}^{\infty} g_k$$

where the summation is a convergent series of generalized functions. Then, in the generalized sense,

$$\mathcal{F}[f] = \sum_{k=-\infty}^{\infty} \mathcal{F}[g_k] \quad , \quad \mathcal{F}^{-1}[f] = \sum_{k=-\infty}^{\infty} \mathcal{F}^{-1}[g_k] \quad ,$$

$$S_\sigma[f] = \sum_{k=-\infty}^{\infty} S_\sigma[g_k] \quad , \quad T_a[f] = \sum_{k=-\infty}^{\infty} T_a[g_k] \quad \text{and} \quad Df = \sum_{k=-\infty}^{\infty} Dg_k$$

where σ is any nonzero real number and a is any complex number.

!▶**Example 35.19:**

$$\mathcal{F}[\text{comb}_1]|_\omega = \mathcal{F}\left[\sum_{k=-\infty}^{\infty} \delta_k\right]\Bigg|_\omega = \sum_{k=-\infty}^{\infty} \mathcal{F}[\delta_k]|_\omega = \sum_{k=-\infty}^{\infty} e^{-i2\pi k\omega} \quad .$$

35.6 Adjoint-Defined Transforms in General

We've generalized a number of transforms — Fourier transforms, scaling, translation and the derivative — using pretty much the same ideas. Let's now look at these ideas a little more closely.

Adjoints

Let $\mathcal{B}$ be some set of exponentially integrable functions (e.g., the set of all classically transformable functions), and let $\mathcal{L}$ be some transform defined on $\mathcal{B}$ such that $\mathcal{L}[f]$ is an exponentially integrable function for each f in $\mathcal{B}$. An *adjoint* of $\mathcal{L}$, which we will denote by $\mathcal{L}^A$, is a corresponding transformation defined on the space of Gaussian test functions that satisfies all of the following:

1. $\mathcal{L}^A[\phi]$ is in $\mathcal{G}$ whenever ϕ is in $\mathcal{G}$.

2. $\mathcal{L}^A$ is linear and "operationally continuous" on $\mathcal{G}$.

3. For each f in $\mathcal{B}$ and ϕ in $\mathcal{G}$,

$$\int_{-\infty}^{\infty} \mathcal{L}[f]|_x \, \phi(x) \, dx = \int_{-\infty}^{\infty} f(y) \, \mathcal{L}^A[\phi]\big|_y \, dy \quad . \tag{35.31}$$

This equation is called the *adjoint identity* for $\mathcal{L}$.

Requiring that $\mathcal{L}^A$ be linear, of course, means that we must have

$$\mathcal{L}^A[a\phi + b\psi] = a\mathcal{L}^A[\phi] + b\mathcal{L}^A[\psi]$$

for each pair of constants a and b, and each pair of Gaussian test functions ϕ and ψ. The "operational continuity" requirement is, essentially, the requirement that $\mathcal{L}^A[\phi]$ and $\mathcal{L}^A[\psi]$ be "almost the same test function" whenever ϕ and ψ, themselves, are "almost the same test function". For a more complete definition and discussion of operational continuity, turn back to the subsection in chapter 32 on bounded operations starting on page 531. We will, in a few pages, carefully examine the relation between operational continuity and functional continuity, and we will see that, in practice, virtually all adjoints that naturally arise in computations are operationally continuous. Consequently, while operational continuity is a necessary requirement for the theory, it's not a significant issue in most applications.

To get a better feel for all this, let's identify the $\mathcal{B}$, $\mathcal{L}$, $\mathcal{L}^A$ and adjoint identity for some of the transforms already generalized:

For the Fourier transform: $\mathcal{B}$ is the set of all classically transformable functions. $\mathcal{L}$ is the classical Fourier transform, $\mathcal{F}$. $\mathcal{L}^A$ is also the classical Fourier transform, $\mathcal{F}$. The adjoint identity is the fundamental identity of Fourier analysis,

$$\int_{-\infty}^{\infty} \mathcal{F}[f]|_x \, \phi(x) \, dx = \int_{-\infty}^{\infty} f(y) \mathcal{F}[\phi]|_y \, dy \quad .$$

For scaling the variable with a positive scaling factor σ : $\mathcal{B}$ is the set of all classical functions on the real line. $\mathcal{L}$ is "multiply the variable by σ", $\mathcal{L}[f]|_x = f(\sigma x)$. $\mathcal{L}^A$ is given by

$$\mathcal{L}^A[\phi]\big|_x = \frac{1}{\sigma} \phi\left(\frac{x}{\sigma}\right) \quad .$$

The adjoint identity is the change of variables formula

$$\int_{-\infty}^{\infty} f(\sigma x)\phi(x) \, dx = \int_{-\infty}^{\infty} f(y) \frac{1}{\sigma} \phi\left(\frac{y}{\sigma}\right) dy \quad .$$

For translation by a real value a : $\mathcal{B}$ is the set of all exponentially integrable functions on the real line. $\mathcal{L}$ is "subtract a from the variable", $\mathcal{L}[f]|_x = f(x - a)$. $\mathcal{L}^A$ is "add a to the variable", $\mathcal{L}^A[\phi]\big|_x = \phi(x + a)$. The adjoint identity is the change of variables formula

$$\int_{-\infty}^{\infty} f(x - a)\phi(x) \, dx = \int_{-\infty}^{\infty} f(y)\phi(y + a) \, dy \quad .$$

For translation by a complex value a : Though we did not explicitly state it, we can take $\mathcal{B}$ to be the set of all simple multipliers, with $\mathcal{L}$ being "subtract a from the variable", $\mathcal{L}[f]|_x = f(x - a)$. $\mathcal{L}^A$ is "add a to the variable", $\mathcal{L}^A[\phi]\big|_x = \phi(x + a)$. The adjoint identity is the change of variables formula

$$\int_{-\infty}^{\infty} f(x - a)\phi(x) \, dx = \int_{-\infty}^{\infty} f(y)\phi(y + a) \, dy \quad ,$$

which we verified as being valid for simple multipliers in the proof of theorem 35.7 on page 599 (see also exercise 35.14 on page 601).

?▶ Exercise 35.29: *What are $\mathcal{B}$, $\mathcal{L}$, $\mathcal{L}^A$ and the adjoint identity for*

a: *scaling the variable with a negative scaling factor σ ?*

b: *multiplying by a fixed simple multiplier h ?*

c: *differentiation?*

Adjoint-Defined Generalized Transforms

Now assume we have a $\mathcal{B}$, $\mathcal{L}$, $\mathcal{L}^A$ and adjoint identity as described above. For each generalized function f, define $\mathcal{L}^{AA}[f]$ by requiring

$$\left\langle \mathcal{L}^{AA}[f], \phi \right\rangle = \left\langle f, \mathcal{L}^A[\phi] \right\rangle \qquad \text{for each } \phi \text{ in } \mathcal{G} \quad . \tag{35.32}$$

One question immediately arises:

>   *Does this formula define $\mathcal{L}^{AA}[f]$ as a generalized function?*

To answer this, we must determine whether equation (35.32) defines $\mathcal{L}^{AA}[f]$ as a linear, functionally continuous functional on $\mathcal{G}$. Certainly the formula gives a well-defined complex value for $\langle \mathcal{L}^{AA}[f], \phi \rangle$ for each ϕ in $\mathcal{G}$. So it does define a functional on $\mathcal{G}$. Furthermore, whenever a and b are two constants, and ϕ and ψ are two Gaussian test functions, then, by the definition of $\mathcal{L}^{AA}$, the linearity of $\mathcal{L}^A$ and the fact that f is a generalized function,

$$\begin{aligned}
\left\langle \mathcal{L}^{AA}[f], a\phi + b\psi \right\rangle &= \left\langle f, \mathcal{L}^A[a\phi + b\psi] \right\rangle \\
&= \left\langle f, a\mathcal{L}^A[\phi] + b\mathcal{L}^A[\psi] \right\rangle \\
&= a\left\langle f, \mathcal{L}^A[\phi] \right\rangle + b\left\langle f, \mathcal{L}^A[\psi] \right\rangle \\
&= a\left\langle \mathcal{L}^{AA}[f], \phi \right\rangle + b\left\langle \mathcal{L}^{AA}[f], \psi \right\rangle \quad .
\end{aligned}$$

So the required linearity holds.

Functional continuity can also be shown to hold using the assumed operational continuity of $\mathcal{L}^A$. We will verify this rigorously in the last part of this section (see page 620), but you can easily see why this should be true. After all, if

$$\phi \quad \text{and} \quad \psi \quad \text{are "almost the same" test functions} \quad ,$$

then the operational continuity of $\mathcal{L}^A$ ensures that

$$\mathcal{L}^A[\phi] \quad \text{and} \quad \mathcal{L}^A[\psi] \quad \text{are "almost the same" test functions} \quad .$$

This, in turn, implies that

$$\left\langle f, \mathcal{L}^A[\phi] \right\rangle \quad \text{and} \quad \left\langle f, \mathcal{L}^A[\psi] \right\rangle \quad \text{are "almost the same" values}$$

(since functional continuity holds for f). But

$$\left\langle \mathcal{L}^{AA}[f], \phi \right\rangle = \left\langle f, \mathcal{L}^A[\phi] \right\rangle \qquad \text{and} \qquad \left\langle \mathcal{L}^{AA}[f], \psi \right\rangle = \left\langle f, \mathcal{L}^A[\psi] \right\rangle \quad .$$

So

$$\left\langle \mathcal{L}^{AA}[f], \phi \right\rangle \qquad \text{and} \qquad \left\langle \mathcal{L}^{AA}[f], \psi \right\rangle$$

are "almost the same" values whenever ϕ and ψ are "almost the same Gaussian test functions", and this, loosely speaking, is what we mean when we say $\mathcal{L}^{AA}[f]$ is functionally continuous.

So the answer to our question is *yes, $\mathcal{L}^{AA}[f]$ is a generalized function.*

The process of converting each generalized function f to the corresponding generalized function $\mathcal{L}^{AA}[f]$, as defined by equation (35.32), can be viewed as a transformation of generalized functions. For a while, we will continue to denote this transform by $\mathcal{L}^{AA}$ to distinguish it from the original transform $\mathcal{L}$, which was only defined for those functions in $\mathcal{B}$. And, thanks to of a lack of imagination, we will refer to this $\mathcal{L}^{AA}$ as either the *(generalized) transform defined by the adjoint $\mathcal{L}^A$* or, when we don't care to specify the adjoint involved, an *adjoint-defined generalization of $\mathcal{L}$*. (This last bit of terminology anticipates some of the results described below.)

Basic Properties of Adjoint-Defined Transforms

Here they are:

Theorem 35.15
Let B be some set of exponentially integrable functions on $\mathbb{R}$, and let $\mathcal{L}$ be some transform defined on B such that $\mathcal{L}[f]$ is an exponentially integrable function for each f in B. Assume $\mathcal{L}$ has an adjoint $\mathcal{L}^A$, and let $\mathcal{L}^{AA}$ be the generalized transform defined by the adjoint $\mathcal{L}^A$. Then:

1. $\mathcal{L}^{AA}[f]$ *is a well-defined generalized function for each generalized function f.*

2. *(linearity) For any two generalized functions f and g, and any two constants a and b,*

$$\mathcal{L}^{AA}[af + bg] = a\mathcal{L}^{AA}[f] + b\mathcal{L}^{AA}[g] \quad .$$

3. *(continuity) Given any sequence $\{g_\gamma\}_{\gamma=\gamma_0}^{\alpha}$ of generalized functions, if*

$$\lim_{\gamma \to \alpha} g_\gamma = f \qquad \text{(in the generalized sense)} \quad ,$$

   then

$$\lim_{\gamma \to \alpha} \mathcal{L}^{AA}[g_\gamma] = \mathcal{L}^{AA}[f] \qquad \text{(in the generalized sense)} \quad .$$

4. *(equivalence on B) For each function f in B, $\mathcal{L}^{AA}[f] = \mathcal{L}[f]$.*

PROOF: From our discussion in the previous subsection, we already know the first claim is true (with the rigorous proof of functional continuity beginning on page 620).

To see that the second claim holds, simply note that, for any Gaussian test function ϕ,

$$
\begin{aligned}
\langle \mathcal{L}^{AA}[af + bg], \phi \rangle &= \langle af + bg, \mathcal{L}^A[\phi] \rangle \\
&= a\langle f, \mathcal{L}^A[\phi] \rangle + b\langle g, \mathcal{L}^A[\phi] \rangle \\
&= a\langle \mathcal{L}^{AA}[f], \phi \rangle + b\langle \mathcal{L}^{AA}[g], \phi \rangle \\
&= \langle a\mathcal{L}^{AA}[f] + b\mathcal{L}^{AA}[g], \phi \rangle \quad .
\end{aligned}
$$

The third claim follows from the observation that, for each ϕ in $\mathcal{G}$,

$$
\begin{aligned}
\lim_{\gamma \to \alpha} \langle \mathcal{L}^{AA}[g_\gamma], \phi \rangle &= \lim_{\gamma \to \alpha} \langle g_\gamma, \mathcal{L}^A[\phi] \rangle \\
&= \langle f, \mathcal{L}^A[\phi] \rangle = \langle \mathcal{L}^{AA}[f], \phi \rangle \quad .
\end{aligned}
$$

Finally, assume f is in B and ϕ is in $\mathcal{G}$. Using the definitions and the adjoint identity, we find that

$$
\begin{aligned}
\langle \mathcal{L}[f], \phi \rangle &= \int_{-\infty}^{\infty} \mathcal{L}[f]\big|_x \, \phi(x) \, dx \\
&= \int_{-\infty}^{\infty} f(y) \, \mathcal{L}^A[\phi]\big|_y \, dy \\
&= \langle f, \mathcal{L}^A[\phi] \rangle = \langle \mathcal{L}^{AA}[f], \phi \rangle \quad ,
\end{aligned}
$$

verifying the last claim of the theorem.

Uniqueness of Adjoint-Defined Transforms

Again, let $\mathcal{B}$, $\mathcal{L}$ and $\mathcal{L}^A$ be as described at the beginning of theorem 35.15, just above, and let $\mathcal{L}^{AA}$ be the generalized transform defined by the adjoint $\mathcal{L}^A$. Given the results stated in theorem 35.15, it is tempting to refer to $\mathcal{L}^{AA}$ as the generalized transform that generalizes the classical transform $\mathcal{L}$. It is also tempting to just drop the "^{AA}" from "$\mathcal{L}^{AA}$" and let $\mathcal{L}[f]$ denote the generalized function given by

$$\left\langle \mathcal{L}[f], \phi \right\rangle = \left\langle f, \mathcal{L}^A[\phi] \right\rangle$$

for every generalized function f whether or not it is in $\mathcal{B}$. In practice, it's usually safe to fall to this temptation. We did so, ourselves, in defining the generalized Fourier transforms, and, to some extent, in defining the generalized scaling transforms, the generalized translation transforms, and the generalized derivative. In doing so, however, we implicitly assumed that these were the only generalizations of the original transforms and, therefore, could unambiguously be referred to as *the* corresponding generalized transforms. But what if two different generalized adjoint-defined transforms corresponded to the same original transform $\mathcal{L}$? Then, rather than dealing with *the one* generalization of $\mathcal{L}$, we would have to deal with two different generalizations — a rather undesirable situation and one likely to lead to ambiguities in our work.

!▶ **Example 35.20:**   *Let $\mathcal{B}$ consist of all constant functions, and let $\mathcal{L}$ be the "zero" transformation on $\mathcal{B}$. That is, $\mathcal{L}[c] = 0$ for each constant c. Then, of course,*

$$\int_{-\infty}^{\infty} \mathcal{L}[c]\big|_x \, \phi(x) = \int_{-\infty}^{\infty} 0 \cdot \phi(x) = 0$$

for each c in $\mathcal{B}$ and each ϕ in $\mathcal{G}$. For each Gaussian test function ϕ, let $\mathcal{L}_1^A[\phi]$ and $\mathcal{L}_2^A[\phi]$ be given by

$$\mathcal{L}_1^A[\phi] = 0 \qquad \text{and} \qquad \mathcal{L}_2^A[\phi] = \phi' \quad .$$

Both $\mathcal{L}_1^A$ and $\mathcal{L}_2^A$ are easily verified to be linear and operationally continuous transforms on $\mathcal{G}$. Moreover, if c is any constant function and ϕ is any Gaussian test function, then

$$\int_{-\infty}^{\infty} c \, \mathcal{L}_1^A[\phi]\big|_y \, dy = \int_{-\infty}^{\infty} c \cdot 0 \, dy = 0 = \int_{-\infty}^{\infty} \mathcal{L}[c]\big|_x \, \phi(x)$$

and

$$\int_{-\infty}^{\infty} c \, \mathcal{L}_2^A[\phi]\big|_y \, dy = \int_{-\infty}^{\infty} c\phi'(y) \, dy = c\,\phi(y)\big|_{-\infty}^{\infty} = 0 = \int_{-\infty}^{\infty} \mathcal{L}[c]\big|_x \, \phi(x) \quad .$$

Consequently, both $\mathcal{L}_1^A$ and $\mathcal{L}_2^A$ are adjoints for $\mathcal{L}$, and thus, both define corresponding transformations $\mathcal{L}_1^{AA}$ and $\mathcal{L}_2^{AA}$ of generalized functions via equation (35.32). These two generalized transforms, however, are not the same. In particular, using $f(x) = x$, we see that

$$\left\langle \mathcal{L}_1^{AA}[f], \phi \right\rangle = \left\langle f, \mathcal{L}_1^A[\phi] \right\rangle = \left\langle f, 0 \right\rangle = 0 \quad ,$$

while, using integration by parts,

$$\left\langle \mathcal{L}_2^{AA}[f], \phi \right\rangle = \left\langle f, \mathcal{L}_2^A[\phi] \right\rangle$$

$$= \int_{-\infty}^{\infty} x\phi'(x) \, dx$$

$$= x\phi(x)\big|_{-\infty}^{\infty} - \int_{-\infty}^{\infty} 1 \cdot \phi(x) \, dx = 0 + \left\langle -1, \phi \right\rangle$$

for each Gaussian test function ϕ. So,

$$\mathcal{L}_1^{AA}[f] \;=\; 0 \;\neq\; -1 \;=\; \mathcal{L}_2^{AA}[f] \quad .$$

Hence, we have at least two equally valid choices, $\mathcal{L}_1^{AA}$ and $\mathcal{L}_2^{AA}$ (and possibly others we haven't thought of) that can be viewed as generalizations of our original transform of constant functions.

Since the generalized transform $\mathcal{L}^{AA}$ is defined by the corresponding adjoint $\mathcal{L}^A$, we can have multiple generalized transforms only if there are multiple adjoints corresponding to the original transform as defined for functions in the set B. That was the case in the last example. There are ways to ensure that multiple adjoints are not possible. Basically, you must be sure that the original transform is defined on such a wide variety of functions that the adjoint equation cannot be satisfied by more than one choice of "$\mathcal{L}^A$". A number of different conditions can be imposed on B to guarantee this, but only one, the one described in the next theorem, is of great interest to us.

Theorem 35.16
Let B be a set of exponentially integrable functions on $\mathbb{R}$, and assume $\mathcal{L}$ is a transform on B having an adjoint-defined generalization. Assume, further, that B contains all the Gaussian functions. Then $\mathcal{L}$ has exactly one adjoint-defined generalization.

PROOF: As noted above, it will suffice to confirm that $\mathcal{L}$ can have only one adjoint. And to confirm this, it suffices to show that any "two" adjoints of $\mathcal{L}$, $\mathcal{L}_1^A$ and $\mathcal{L}_2^A$, are really the same; that is, that

$$\mathcal{L}_1^A[\phi] \;=\; \mathcal{L}_2^A[\phi] \qquad \text{for each } \phi \text{ in } \mathcal{G} \quad .$$

So assume $\mathcal{L}_1^A$ and $\mathcal{L}_2^A$ are two adjoints for $\mathcal{L}$. Let ϕ be any Gaussian test function, and let ψ be any function in B. By the definition of the adjoint,

$$\int_{-\infty}^{\infty} \mathcal{L}_1^A[\phi]\big|_y \, \psi(y)\, dy \;=\; \int_{-\infty}^{\infty} \psi(y)\mathcal{L}_1^A[\phi]\big|_y \, dy \;=\; \int_{-\infty}^{\infty} \mathcal{L}[\psi]\big|_x \, \phi(x)\, dx$$

and

$$\int_{-\infty}^{\infty} \mathcal{L}_2^A[\phi]\big|_y \, \psi(y)\, dy \;=\; \int_{-\infty}^{\infty} \psi(y)\mathcal{L}_2^A[\phi]\big|_y \, dy \;=\; \int_{-\infty}^{\infty} \mathcal{L}[\psi]\big|_x \, \phi(x)\, dx \quad .$$

Thus, for every ψ in B and ϕ in $\mathcal{G}$,

$$\int_{-\infty}^{\infty} \mathcal{L}_1^A[\phi]\big|_y \, \psi(y)\, dy \;=\; \int_{-\infty}^{\infty} \mathcal{L}_2^A[\phi]\big|_y \, \psi(y)\, dy \quad .$$

But B contains all Gaussians. So our last equation holds whenever ψ is a Gaussian, and we can apply our old Gaussian test for equality (theorem 30.1 on page 491), to conclude that

$$\mathcal{L}_1^A[\phi] \;=\; \mathcal{L}_2^A[\phi] \qquad \text{for each } \phi \text{ in } \mathcal{G} \quad . \qquad\blacksquare$$

If you check, you will find that the set "B" for each of the generalized transforms defined thus far contains all the Gaussians. That, along with our last lemma, assures us that these are the only possible (adjoint-defined) generalizations of the corresponding classical transforms.

On Operational and Functional Continuity

Here is our rigorous discussion of the functional continuity of adjoint-defined transforms. We'll split our discussion in two. We will first complete the proof of the first part of theorem 35.15 on page 617 by showing how operational continuity ensures the functional continuity of adjoint-defined generalized transforms. Then we will consider the transforms defined in this chapter.

Functional Continuity of Adjoint-Defined Transforms

Assume B is some set of exponentially integrable functions and $\mathcal{L}$ is some transform on B such that $\mathcal{L}[f]$ is an exponentially integrable function whenever f is in B. Let $\mathcal{L}^A$ be an adjoint corresponding to $\mathcal{L}$, and, for some given generalized function f, define $\mathcal{L}^{AA}[f]$ as we've been doing since page 616,

$$\left\langle \mathcal{L}^{AA}[f], \phi \right\rangle = \left\langle f, \mathcal{L}^A[\phi] \right\rangle \qquad \text{for each } \phi \text{ in } \mathcal{G} \quad .$$

Our goal is to confirm that this defines $\mathcal{L}^{AA}[f]$ as a generalized function. Since we already know the above equation defines $\mathcal{L}^{AA}[f]$ as a linear functional on $\mathcal{G}$, all that remains is to show that this functional is functionally continuous. And, by the definition of functional continuity, it will suffice to verify the existence of two nonnegative constants B and b such that

$$\left| \left\langle \mathcal{L}^{AA}[f], \phi \right\rangle \right| \leq B \, \|\phi\|_b \qquad \text{for each } \phi \text{ in } \mathcal{G} \quad .$$

Now, by definition, the adjoint transform $\mathcal{L}^A$ is an operationally continuous operator on $\mathcal{G}$. From our discussion of operational continuity in chapter 32 (see the subsection on bounded operations, starting on page 531), we know this means that, for each $\alpha \geq 0$, there are corresponding nonnegative real constants M_α and $\beta(\alpha)$ such that

$$\left\| \mathcal{L}^A[\phi] \right\|_\alpha \leq M_\alpha \, \|\phi\|_{\beta(\alpha)} \qquad \text{for each } \phi \text{ in } \mathcal{G} \quad .$$

In addition, we know f is functionally continuous because f is assumed to be a generalized function and generalized functions are, by definition, functionally continuous. So there are nonnegative real constants C and c such that

$$\left| \left\langle f, \phi \right\rangle \right| \leq C \, \|\phi\|_c \qquad \text{for each } \phi \text{ in } \mathcal{G} \quad .$$

Combining the above relations, we get

$$\left| \left\langle \mathcal{L}^{AA}[f], \phi \right\rangle \right| = \left| \left\langle f, \mathcal{L}^A[\phi] \right\rangle \right| \leq C \left\| \mathcal{L}^A[\phi] \right\|_c \leq C M_a \, \|\phi\|_{\beta(c)}$$

for each Gaussian test function ϕ. Thus, letting $B = C M_c$ and $b = \beta(c)$,

$$\left| \left\langle \mathcal{L}^{AA}[f], \phi \right\rangle \right| \leq B \, \|\phi\|_b \qquad \text{for each } \phi \text{ in } \mathcal{G} \quad . \qquad \blacksquare$$

Particular Transforms

This will be brief because all the work has already been done in the subsection in chapter 32 on bounded operations (starting on page 531). From lemma 32.17 and theorem 32.18 of that subsection, we know all of the following are operationally continuous on $\mathcal{G}$:

1. the Fourier transform

2. the inverse Fourier transform

3. multiplication by any fixed elementary multiplier

4. differentiation

5. translation by any fixed real or complex value

6. variable scaling by any fixed nonzero real number

7. analytic conjugation

8. any linear combination of the above

9. any composition of the above

10. any linear combination of any compositions of the above

11. any composition of any linear combinations of any compositions of the above

12. ...

From this, theorem 35.15, and the rest of the above discussion on adjoint-defined transforms, it immediately follows that all the generalized transforms defined earlier in this chapter (as well as all the generalized transforms we are likely to construct from these transforms) are linear operations transforming generalized functions into generalized functions and satisfying all the other properties described in theorem 35.15.

35.7 Generalized Complex Conjugation

If f is any exponentially integrable function and ϕ is any Gaussian test function, then we know that

$$\int_{-\infty}^{\infty} f^*(x)\phi(x)\,dx = \left(\int_{-\infty}^{\infty} f(x)\phi^*(x)\,dx \right)^* , \tag{35.33}$$

which we can also write as

$$\left\langle f^*, \phi \right\rangle = \left\langle f, \phi^* \right\rangle^* \tag{35.34}$$

since the analytic complex conjugate of any Gaussian test function is another Gaussian test function. Now look at the right side of this last equation. Unsurprisingly (considering the number of times we've done something like this in this chapter), the right side of this equation is well defined for any generalized function f, whether or not f is a classical function. So equation (35.34) can be used to define f^* as a functional on $\mathcal{G}$ for any generalized function f. However, if you check carefully, you'll discover that equation (35.33) is not quite an adjoint identity, so we cannot immediately appeal to the work in the previous section to justify a claim that this equation defines f^* as a generalized function (i.e., as a continuous linear functional on $\mathcal{G}$). The linearity of the functional must be confirmed the old-fashioned way:

$$\begin{aligned}
\left\langle f^*, [a\phi + b\psi] \right\rangle &= \left\langle f, [a\phi + b\psi]^* \right\rangle^* \\
&= \left\langle f, a^*\phi^* + b^*\psi^* \right\rangle^* \\
&= \left(a^* \left\langle f, \phi^* \right\rangle + b^* \left\langle f, \psi^* \right\rangle \right)^* \\
&= a \left\langle f, \phi^* \right\rangle^* + b \left\langle f, \psi^* \right\rangle^* = a \left\langle f^*, \phi \right\rangle + b \left\langle f^*, \psi \right\rangle \quad .
\end{aligned}$$

The functional continuity is also easily verified by those interested (see exercise 35.52 on page 626). Consequently, we can use equation (35.34) to define the complex conjugate of any generalized function f .

Let's make it official: For each generalized function f , the *(generalized) (analytic) complex conjugate* f^* is defined to be the generalized functions satisfying

$$\langle f^* , \phi \rangle = \langle f , \phi^* \rangle^* \qquad \text{for each } \phi \text{ in } \mathcal{G} \quad . \tag{35.35}$$

From the above discussion, it should be clear that

  1. f^* is a well-defined generalized function for each generalized function f ,

and

  2. if f is a classical, exponentially integrable function, then its generalized complex conjugate and its classical complex conjugate are the same.

When using equation (35.35), keep in mind that ϕ^* is the analytic complex conjugate discussed in chapter 32 (see the subsection starting on page 527). That is, as a function on $\mathbb{C}$, ϕ^* is the analytic function satisfying

$$\phi^*(z) = [\phi(z^*)]^* \qquad \text{for all } z \text{ in } \mathbb{C} \quad .$$

!▶Example 35.21:   *Let's find the complex conjugate of the delta function at i . For each Gaussian test function ϕ , we have*

$$\langle \delta_i{}^* , \phi \rangle = \langle \delta_i , \phi^* \rangle^* = [\phi^*(i)]^* = \left[[\phi(i^*)]^* \right]^* = \phi(-i) = \langle \delta_{-i} , \phi \rangle \quad .$$

Thus, $\delta_i{}^* = \delta_{-i}$.

?▶Exercise 35.30:   *Show that, in general, $\delta_a{}^* = \delta_{a^*}$.*

You should have little difficulty showing that the basic properties of classical conjugation hold for generalized conjugation.

?▶Exercise 35.31:   *Verify each of the following, assuming f and g are generalized functions, and α and β are constants:*

  a: $[\alpha f + \beta g]^* = \alpha^* f^* + \beta^* g^*$

  b: $(f^*)^* = f$

  c: *If g is a simple multiplier, so is g^* .*

  d: *If either f or g is a simple multiplier, then $(fg)^* = f^* g^*$.*

Verifying the identities in the next theorem requires a bit more work. For one thing, it is necessary to first show that a corresponding identity holds for each Gaussian test function using classical theory and the fact that $\phi^*(z) = [\phi(z^*)]^*$ when ϕ is in $\mathcal{G}$. We'll verify one, and leave verifying the rest as exercises (exercise 35.53 on page 627).

Theorem 35.17

Let f be any generalized function, σ any nonzero real number and a any complex number. Then

$$\mathcal{F}[f^*] = \left(\mathcal{F}^{-1}[f] \right)^* \quad , \qquad \mathcal{F}^{-1}[f^*] = \left(\mathcal{F}[f] \right)^* \quad ,$$

$$S_\sigma[f^*] = \left(S_\sigma[f] \right)^* \quad , \qquad T_a[f^*] = \left(T_{a^*}[f] \right)^* \quad \text{and} \quad D[f^*] = (Df)^* \quad .$$

PROOF (of $T_a[f^*] = (T_{a^*}[f])^*$):   Let ϕ be any Gaussian test function. By the definitions of generalized translation and conjugation,

$$\langle\, T_a[f^*]\,,\phi\,\rangle \;=\; \langle\, f^*\,,T_{-a}[\phi]\,\rangle \;=\; \langle\, f\,,\,(T_{-a}[\phi])^*\,\rangle^* \quad. \tag{35.36}$$

Now, for each z in $\mathbb{C}$,

$$(T_{-a}[\phi])^*\,(z) \;=\; \left(T_{-a}[\phi]|_{z^*}\right)^*$$
$$= \; \left[\phi(z^* + a)\right]^*$$
$$= \; \left[\phi\left([z + a^*]^*\right)\right]^*$$
$$= \; \phi^*(z + a^*) \;=\; T_{-a^*}\!\left[\phi^*\right]\big|_z \quad.$$

Thus, for the Gaussian test function ϕ,

$$(T_{-a}[\phi])^* \;=\; T_{-a^*}[\phi^*] \quad.$$

Plugging this into equation (35.36) and applying the definitions once more, we obtain

$$\langle\, T_a[f^*]\,,\phi\,\rangle \;=\; \langle\, f\,,T_{-a^*}[\phi^*]\,\rangle^* \;=\; \langle\, T_{a^*}[f]\,,\phi^*\,\rangle^* \;=\; \langle\,(T_{a^*}[f])^*\,,\phi\,\rangle$$

for each Gaussian test function ϕ, verifying that $T_a[f^*] = (T_{a^*}[f])^*$. ∎

Additional Exercises

35.32. *Find the Fourier transform for each of the following:*

a. $\text{pulse}_1(x)$ b. 4 c. $4e^{i10\pi x}$

d. $4e^{-i10\pi x}$ e. δ_4 f. $4\sin^2(2\pi x)$

35.33. *Find the inverse Fourier transform for each of the following:*

a. $\text{pulse}_1(x)$ b. 4 c. $4e^{i10\pi x}$

d. $4e^{-i10\pi x}$ e. δ_4 f. $4\sin^2(2\pi x)$

35.34. *Using the cheap trick* "$t = \dfrac{1}{i2\pi}(\alpha + i2\pi t - \alpha)$", *compute the Fourier transforms of the following:*

a. $\dfrac{t}{8 + i2\pi t}$ b. $\dfrac{t}{8 - i2\pi t}$

35.35. *Let S_σ be the scaling operator where σ is any nonzero real number. Verify that:*

a. *S_1 is the identity mapping. That is, for every generalized function f,*

$$S_1[f] \;=\; f \quad.$$

b. *The inverse of S_σ is $S_{1/\sigma}$. That is, for every generalized function f,*

$$S_\sigma\!\left[S_{1/\sigma}[f]\right] \;=\; f \;=\; S_{1/\sigma}\!\left[S_\sigma[f]\right] \quad.$$

35.36. Verify that the modulation identities hold for the generalized Fourier transforms. That is, show that

$$\mathcal{F}[\sin(2\pi\alpha t) f(t)]|_{\omega} = \frac{i}{2}[F(\omega+\alpha) - F(\omega-\alpha)]$$

and

$$\mathcal{F}[\cos(2\pi\alpha t) f(t)]|_{\omega} = \frac{1}{2}[F(\omega+\alpha) + F(\omega-\alpha)]$$

for every positive value α and every pair of generalized functions f and F with $F = \mathcal{F}[f]$.

35.37. Find the Fourier transform for each of the following:

   a. $e^{2\pi x}$ **b.** δ_{3+2i} **c.** $\cosh(6\pi x)$ **d.** $\sinh(6\pi x)$

35.38. Find the Fourier transform for each of the following assuming α and β are two real numbers with $\alpha > 0$:

   a. $e^{2\pi\beta t}\sin(2\pi\alpha t)$

   b. $e^{2\pi\beta t}\cos(2\pi\alpha t)$

   c. $e^{2\pi\alpha t}\,\text{step}(t)$ (Hint: First verify that $e^{2\pi\alpha t}\,\text{step}(t) = e^{2\pi\alpha t} - e^{2\pi\alpha t}\,\text{step}(-t)$.)

   d. $e^{2\pi\alpha t}\sin(8\pi t)\,\text{step}(t)$

   e. $e^{-2\pi\alpha t}\,\text{step}(-t)$

   f. $e^{2\pi\alpha|t|}$

35.39. Sketch the graph of each of the following functions, and then find both the classical and the generalized derivatives:

   a. $\text{ramp}(x)$

   b. $g(x) = \begin{cases} x^3 & \text{if } -1 < x < 2 \\ 0 & \text{otherwise} \end{cases}$

   c. $h(x) = \begin{cases} x & \text{if } x < -1 \\ x^2 & \text{if } -1 < x < 2 \\ x^3 & \text{if } 2 < x \end{cases}$

35.40. Let n denote an arbitrary positive integer, and find the formula for each of the following transforms:

   a. $\mathcal{F}[D^n\delta_a]|_x$ **b.** $\mathcal{F}^{-1}[D^n\delta_a]|_x$ **c.** $\mathcal{F}[x^n]$ **d.** $\mathcal{F}^{-1}[x^n]$

35.41. Compute the Fourier transform of each of the following:

   a. $x^2 + 4x - 5$ **b.** $x\,e^{6\pi x}$ **c.** $x\sin(6\pi x)$ **d.** $x^2\sin(6\pi x)$

35.42 a. From example 35.12 and exercise 35.19 (see page 603), we know

$$\mathcal{F}[\text{step}(x)]|_y = \frac{1}{i2\pi}T_{-i}\left[\frac{1}{y-i}\right]$$

and

$$\mathcal{F}[\text{step}(-x)]|_y = \frac{-1}{i2\pi}T_i\left[\frac{1}{y+i}\right] \quad .$$

Now, for each positive integer n, find the Fourier transforms of the following using the above formulas and the Fourier differentiation identities:

i. $x^n \operatorname{step}(x)$ **ii.** $x^n \operatorname{step}(-x)$

b. Again, let n be any positive integer. Using the previous part, find a formula for the Fourier transform of each of the following:

i. $\operatorname{ramp}(x)$ **ii.** $|x|^n \operatorname{step}(-x)$ **iii.** $|x|^n$

35.43. Using the results from example 35.12 and exercise 35.19 (again, see page 603), show there is a nonzero value c such that

$$T_{-ia}\left[\frac{1}{x-ia}\right] = T_{-ib}\left[\frac{1}{x-ib}\right] + c\delta(x)$$

whenever $a > 0$ and $b < 0$. Also, find the value c.

35.44. For each generalized function f given below, show that $xf(x) = c$ where c is some constant, and find the value of that constant.

a. $f(x) = D \ln|x|$ **b.** $f(x) = \mathcal{F}\left[\operatorname{step}\right]\big|_x$

c. $f(x) = T_i\left[\dfrac{1}{x+i}\right]$ **d.** $f(x) = T_{-i}\left[\dfrac{1}{x-i}\right]$

35.45. The signum function sgn is the classical function on the real line given by

$$\operatorname{sgn}(x) = \begin{cases} -1 & \text{if } x < 0 \\ +1 & \text{if } 0 < x \end{cases}.$$

a. What is the relation between $\mathcal{F}\left[\operatorname{sgn}\right]$ and $\mathcal{F}\left[\operatorname{step}\right]$?

b. Evaluate $x\mathcal{F}\left[\operatorname{sgn}\right]\big|_x$.

35.46. Show that, for any complex value a,

$$\lim_{\gamma \to \infty} \sqrt{\gamma}\, e^{-\gamma\pi(x-a)^2} = \delta_a(x) \qquad \text{(in the generalized sense)} \qquad .$$

35.47. Prove theorem 35.10 on page 608. You should consider two cases:

a. The case where f has only a finite number of discontinuities.

b. The case where f has an infinite number of discontinuities.

You should verify that f is exponentially bounded in the first case (see lemma 30.10 on page 496). For the second case you may wish to use lemma 35.13 or 35.14.

35.48. In section 23.3, we derived the formula

$$\mathcal{F}\left[e^{-(\gamma+i\kappa)t^2}\right]\Big|_\omega = \sqrt{\frac{\pi}{|\gamma+i\kappa|}} \exp\left(-i\frac{\theta}{2}\right) \exp\left(\frac{-\pi^2}{\gamma+i\kappa}\omega^2\right)$$

where θ is the angle between $-\pi/2$ and $\pi/2$ given by $\theta = \arctan(\kappa/\gamma)$. This formula was derived assuming γ and κ are real numbers with $0 < \gamma$. In the following we will verify that it also holds when $\gamma = 0$ and $\kappa \neq 0$

a. Verify that

$$\lim_{\gamma \to 0^+} e^{-(\gamma+i\kappa)t^2} = e^{-i\kappa t^2} \qquad \text{(in the generalized sense)} \qquad .$$

b. Assume $\alpha > 0$. Using the above and lemma 35.13 on page 613, compute each of the following:

i. $\mathcal{F}\left[e^{i\pi\alpha t^2}\right]\Big|_\omega$

ii. $\mathcal{F}\left[e^{-i\pi\alpha t^2}\right]\Big|_\omega$

iii. $\mathcal{F}\left[\exp\left(\pm i\pi\left(t^2 - \frac{1}{8}\right)^2\right)\right]\Big|_\omega$

c. Again, let $\alpha > 0$. Using the formulas just derived, find each of the following:

i. $\mathcal{F}\left[\cos\left(\pi\alpha t^2\right)\right]\Big|_\omega$

ii. $\mathcal{F}\left[\sin\left(\pi\alpha t^2\right)\right]\Big|_\omega$

35.49. Prove lemma 35.8 on page 599. (Those who have had a course in complex variables should try using Cauchy's integral formula for analytic functions. Those who have not had a course in complex variables probably should not attempt this exercise.)

35.50. For the following, assume f is any given generalized function, ϕ is any given Gaussian test function, and h is the function defined by

$$h(t) = \langle\, f(x)\,,\, \phi(x - t)\,\rangle \qquad \text{for each } t \text{ in } \mathbb{R} \quad.$$

a. Letting $F = \mathcal{F}[f]$ and $\psi = \mathcal{F}^{-1}[\phi]$, verify that h is also given by

$$h(t) = \left\langle\, F(y)\,,\, e^{i2\pi ty}\,\psi(y)\,\right\rangle \quad.$$

b. Show that h is a smooth function on $\mathbb{R}$ with

$$h'(t) = -\langle\, f(x)\,,\, \phi'(x - t)\,\rangle \quad.$$

(Use the result from the previous part of this exercise along with lemma 33.3 on page 556.)

35.51. Let f be a generalized function, let (a, b) be any finite interval, and let η be any piecewise continuous function on (a, b). As shown in the previous exercise,

$$\langle\, f(x)\,,\, \phi(x - t)\,\rangle$$

is a smooth function of t for each ϕ in $\mathcal{G}$. Consequently, we can define a functional on $\mathcal{G}$ by

$$\langle\, \Gamma\,,\, \phi\,\rangle = \int_a^b \langle\, f(x)\,,\, \phi(x - t)\,\rangle\, \eta(t)\, dt \quad.$$

Show that Γ is a generalized function (i.e., a continuous, linear functional on $\mathcal{G}$) by verifying the following:

a. Γ is linear.

b. Γ is functionally continuous. (The result from exercise 32.10 on page 535 may help.)

(Note: The results from exercises 35.50 and 35.51 will be used in proving a major theorem on periodic functions in chapter 37.)

35.52. We showed that the complex conjugate of any generalized function (as defined by equation (35.35) on page 622) is a linear functional. Now verify that it is functionally continuous.

35.53. Let f be any generalized function, σ any nonzero real number and a any complex number. Verify the following:

a. $\mathcal{F}[f^*] = \left(\mathcal{F}^{-1}[f]\right)^*$ **b.** $\mathcal{F}^{-1}[f^*] = (\mathcal{F}[f])^*$

c. $S_\sigma[f^*] = (S_\sigma[f])^*$ **d.** $D[f^*] = (Df)^*$

35.54. In the following, we will generalize the operation of convolution with test functions via "adjoints". You will need either the results from exercise 32.16 on page 536 (in which case, let $B = A$), or the results from the exercise following it, exercise 32.17 (in which case, let B be the set of all piecewise continuous, exponentially bounded functions on the real line).

a. Let ψ be a fixed Gaussian test function, and, for convenience, let $\mathcal{K}_\psi$ be the operator on B given by
$$\mathcal{K}_\psi[f] = f * \psi \qquad \text{for each } f \text{ in } B \quad .$$

i. Verify that $\mathcal{K}_\psi[f]$ is an exponentially integrable function on $\mathbb{R}$ for each f in B.

ii. Verify that the adjoint of $\mathcal{K}_\psi$ exists and is given by
$$\mathcal{K}_\psi^A[\phi] = \psi^* \star \phi$$
where $\psi^* \star \phi$ is the classical correlation of the complex conjugate of ψ with ϕ.

b. Finish showing that the adjoint-defined generalization of convolution with a Gaussian test function ψ — which we will denote by $f * \psi$ for each generalized function f — is defined by the formula
$$\left\langle f * \psi , \phi \right\rangle = \left\langle f , \psi^* \star \phi \right\rangle \qquad \text{for each } \phi \text{ in } \mathcal{G} \quad .$$

c. Using the above definition for the convolution of a generalized function f with a Gaussian test function ψ, verify that
$$\mathcal{F}[f * \psi] = \mathcal{F}[f]\,\mathcal{F}[\psi] \qquad \text{and} \qquad \mathcal{F}[f\psi] = \mathcal{F}[f] * \mathcal{F}[\psi] \quad .$$

d. Using results from the previous parts of this exercise, do the following (assume ψ denotes an arbitrary Gaussian test function):

i. Show that
$$\delta_a * \psi = T_a[\psi]$$
for each real value a. (Thus, in particular, $\delta * \psi = \psi$.)

ii. Show that $f * \psi$ is in $\mathcal{G}$ if $\mathcal{F}[f]$ is a simple multiplier for $\mathcal{G}$.

iii. Let f be any generalized function, and let $\{\eta_\gamma\}_{\gamma=1}^\infty$ be the Gaussian identity sequence with
$$\eta_\gamma(x) = \sqrt{\gamma}\, e^{-\gamma \pi x^2} \quad .$$

Show that
$$\lim_{\gamma \to \infty} f * \eta_\gamma = f \qquad \text{(in the generalized sense)} \quad .$$

(Consider $\mathcal{F}\left[f * \eta_\gamma\right]$ and apply the result from example 34.2 on page 570.)

36

Generalized Products, Convolutions and Definite Integrals

Recall the classical definitions of multiplication and convolution for any two piecewise continuous functions f and g on the real line: The classical product fg is the piecewise continuous function given by

$$fg(x) = f(x)g(x)$$

for each x at which f and g are continuous, and the classical convolution is the function given by

$$f * g(x) = \int_{-\infty}^{\infty} f(x - s)g(s)\,ds$$

for all x in $\mathbb{R}$. Note that the product is always defined, while the existence of the convolution requires $f(x - s)g(s)$ to be a "sufficiently integrable" function of s for each real value x.

Our main goal in this chapter is to describe operations for generalized functions that can be viewed as the natural generalizations of classical multiplication and convolution. (The discussion of definite integrals will then follow naturally from our definition of convolution.) Unfortunately, there is a technical problem. Remember those comments about difficulties possibly arising because we are using the space of Gaussian test functions, $\mathcal{G}$, instead of the larger space of very rapidly decreasing test functions, $\mathcal{H}$ (see section 32.3 starting on page 521)? This is where those difficulties manifest themselves. While the final results are the same, it is much easier to rigorously develop reasonably complete generalizations of multiplication and convolution using $\mathcal{H}$ as the test function space instead of $\mathcal{G}$. This is because there are many more "simple multipliers" for $\mathcal{H}$ than for $\mathcal{G}$. However, even using $\mathcal{H}$ as our test function space, this development in quite nontrivial and would go, as they say, "beyond the scope of this text".

Still, saying nothing more about generalized multiplication and convolution can hardly be justified. These operations are just too useful. So, as a compromise, we will briefly develop the set of "simple multipliers" that we would have naturally obtained had we employed $\mathcal{H}$ as our test functions space, and then present a list of twelve or so rules for multiplying and convolving generalized functions. These rules will not completely describe multiplication and convolution, but they will characterize these operations well enough for many, if not most, practical applications.

36.1   Multiplication and Convolution
A Bigger Set of Simple Multipliers

We defined a simple multiplier for $\mathcal{G}$ to be any linear combination of functions of the form

$$x^n e^{cx} e^{-\gamma(x-\zeta)^2}$$

where n is any nonnegative integer, c and ζ are any pair of complex constants, and γ is any nonnegative real value (see page 523). The set of simple multipliers for $\mathcal{G}$ include

$$e^{iax} \quad , \quad \sin(ax) \quad , \quad e^{-ax^2} \quad , \quad x^3 \quad \text{and} \quad e^{ax}$$

where a is any positive constant. The important thing about a function h being a simple multiplier is that the product $h\phi$ is a Gaussian test function whenever ϕ is a Gaussian test function. Consequently, we were able to define the product fh for any generalized function f via the formula

$$\langle\, fh\,,\,\phi\,\rangle \;=\; \langle\, f\,,\,h\phi\,\rangle \qquad \text{for each } \phi \text{ in } \mathcal{G} \quad .$$

If we had instead developed our theory of generalized functions using the test function space $\mathcal{H}$, then we would have simply defined a function h to be a *simple multiplier* for $\mathcal{H}$ if and only if the product $h\phi$ is in $\mathcal{H}$ whenever ϕ is a function from $\mathcal{H}$. It is not hard to show that any analytic function on the complex plane that is also exponentially bounded on horizontal strips of $\mathbb{C}$ is a simple multiplier for $\mathcal{H}$ (see exercise 36.9 at the end of this chapter for details). From this, it is easily verified that the set of simple multipliers for $\mathcal{H}$ includes the following:

1.   all simple multipliers for $\mathcal{G}$

2.   all functions in $\mathcal{H}$

3. $\mathrm{sinc}(ax)$ where a is any constant (sinc functions will be important in dealing with delta function arrays)

4.   functions such as $e^{i\gamma x^2}$ and $\sin(\gamma x^2)$ where $\gamma \geq 0$

As you can see, the set of simple multipliers for $\mathcal{H}$ is a much larger than the set of simple multipliers for $\mathcal{G}$. As an exercise, you should also verify that all linear combinations and products of these simple multipliers are also simple multipliers.

?▶Exercise 36.1:   Let g and h be any two simple multipliers for $\mathcal{H}$, and let α and β be any two constants. Show that the linear combination $\alpha g + \beta h$ and the product gh are also simple multipliers for $\mathcal{H}$.

It is important to realize that, had we been willing to get even more deeply involved with complex variables, then we could have developed the theory described in the previous few chapters using $\mathcal{H}$ as our test function space instead of $\mathcal{G}$. In particular, if h is any simple multiplier for $\mathcal{H}$ and f is any generalized function, then the product fh is the generalized function satisfying

$$\langle\, fh\,,\,\phi\,\rangle \;=\; \langle\, f\,,\,h\phi\,\rangle \qquad \text{for each } \phi \text{ in } \mathcal{H} \quad .$$

It turns out that each result we derived involving "an arbitrary simple multiplier on $\mathcal{G}$" can also be derived with "an arbitrary simple multiplier on $\mathcal{G}$" replaced by "an arbitrary simple multiplier on $\mathcal{H}$". We will make this explicit in our list of rules characterizing multiplication and convolution. It also means that distinguishing between the simple multipliers for $\mathcal{G}$ and the simple multipliers for $\mathcal{H}$ is usually unnecessary. So let us agree that, henceforth, unless other indicated, whenever we refer to a function as a "simple multiplier", then it is a simple multiplier in the broadest sense, that is, a simple multiplier for $\mathcal{H}$.

The Rules

What follows are a dozen or so rules that, together, characterize the generalized product and generalized convolution well enough for most applications.

The first four can be viewed as definitions that apply when we already have a fairly natural idea as to what the product or convolution should be. Basically, these four rules assure us that the generalized definitions reduce to definitions we already know when those known definitions are valid. As a result, there will be little or no ambiguity if we use the classical notations fg (or, when clarity requires, $f \cdot g$) and $f * g$ to denote, respectively, the generalized product and the generalized convolution of two generalized functions. Accordingly, we will use this notation.

The rest of the rules give identities involving generalized multiplication and convolution. Through these identities we will also be able to find generalized products and convolutions for many pairs of generalized functions not directly covered by the first four rules.

Partial Definitions for Generalized Multiplication

Our first rule is simply that the generalized product reduces to the classical product when the classical product "makes sense".

Rule 1 (consistency with classical multiplication)
If f and g are two exponentially integrable functions on $\mathbb{R}$ whose classical product is also exponentially integrable, then the generalized product fg exists and is given by the classical product; that is, fg is the generalized function given by

$$\langle fg , \phi \rangle = \int_{-\infty}^{\infty} f(x)g(x)\phi(x)\,dx \qquad \textit{for each } \phi \textit{ in } \mathcal{G} \quad .$$

In the previous subsection, we briefly discussed "simple products" of generalized functions with simple multipliers. Naturally, any more generalized definition of multiplication should give the same result whenever one of the factors is a simple multiplier. That is our second rule.

Rule 2 (consistency with simple generalized multiplication)
If f and g are generalized functions and either is a simple multiplier, then the generalized product fg exists. Moreover, if either is a simple multiplier for $\mathcal{G}$, then the product is the simple generalized product defined in chapter 33.

Now recall that, if f is a simple multiplier for $\mathcal{G}$ and a is any real value, then

$$f\delta_a = f(a)\delta_a \quad .$$

Our next rule is that this formula holds for more general choices of f .

Rule 3 (products of classical functions with single delta functions)
If f is an exponentially integrable function on $\mathbb{R}$, and a is a point on $\mathbb{R}$ at which f is continuous, then the generalized product $f\delta_a$ exists and

$$f\delta_a = f(a)\delta_a \quad .$$

This last rule can be naively justified using an identity sequence. Recall that

$$\lim_{\gamma \to \infty} \sqrt{\gamma}e^{-\gamma\pi(x-a)^2} = \delta_a(x) \qquad \text{(in the generalized sense)}$$

(see exercise 35.46 on page 625). So it should seem reasonable[1] that

$$f(x)\delta_a(x) \;=\; \lim_{\gamma \to \infty} f(x)\sqrt{\gamma}\,e^{-\gamma\pi(x-a)^2} \qquad \text{(in the generalized sense)} \quad.$$

But we also know that, if g is any piecewise continuous and exponentially bounded function that is continuous at a, then

$$\lim_{\gamma \to \infty} \int_{-\infty}^{\infty} g(x)\sqrt{\gamma}\,e^{-\gamma\pi(x-a)^2}\,dx \;=\; g(a) \quad.$$

These last two equations, along with the fact that f, here, is a classical function, suggest that

$$\begin{aligned}
\left\langle\, f\delta_a \,,\, \phi \,\right\rangle &= \lim_{\gamma \to \infty} \left\langle\, f(x)\sqrt{\gamma}\,e^{-\gamma\pi(x-a)^2} \,,\, \phi(x) \,\right\rangle \\[2mm]
&= \lim_{\gamma \to \infty} \int_{-\infty}^{\infty} f(x)\phi(x)\sqrt{\gamma}\,e^{-\gamma\pi(x-a)^2}\,dx \\[2mm]
&= f(a)\phi(a) \\[2mm]
&= \left\langle\, f(a)\delta_a \,,\, \phi \,\right\rangle
\end{aligned}$$

for each Gaussian test function ϕ, just as the above rule asserts.

!▶ Example 36.1: *According to our last rule, if $a < 0$, then*

$$\text{step}\,\delta_a \;=\; \text{step}(a)\,\delta_a \;=\; 0 \cdot \delta_a \;=\; 0 \quad,$$

while if $0 < a$, then

$$\text{step}\,\delta_a \;=\; \text{step}(a)\,\delta_a \;=\; 1 \cdot \delta_a \;=\; \delta_a \quad.$$

We will not attempt to define $\text{step}\,\delta_0$, *since the step function is not continuous at 0.*

?▶ Exercise 36.2: *Assuming our last rule, verify that*

$$\text{ramp}\,\delta_a \;=\; \begin{cases} 0 & \text{if} \;\; a \le 0 \\[2mm] a\delta_a & \text{if} \;\; 0 < a \end{cases} \quad.$$

Partial Definition for Generalized Convolution

Naturally, to be a "generalization" of classical convolution, the generalized convolution must be equivalent to the classical convolution when the latter is well defined. That is rule number 4.

Rule 4 (consistency with classical convolution)
If f and g are two exponentially integrable (classical) functions whose classical convolution exists and is exponentially integrable, then the generalized convolution of f and g exists and is given by the classical convolution.

---

[1] Remember, though, that naive computations of generalized limits sometimes yield incorrect results (see the discussion starting on page 570). So don't take these computations as a proof that rule 3 holds, only that it does not contradict what we might expect.

Relation between Multiplication and Convolution

Convolution was originally derived (in chapter 24) as a way of computing the Fourier transform of the product of two classical functions. The next rule is that the relation between products and convolutions derived in that chapter holds in the more general sense.

Rule 5 *(Fourier transforms of products and convolutions)*
Let f, g, F and G be generalized functions with $F = \mathcal{F}[f]$ and $G = \mathcal{F}[g]$. Then:

1. *If either the generalized product fg or the generalized convolution $F * G$ exists, then both exist, and*
$$\mathcal{F}[fg] = F * G \quad .$$

2. *If either the generalized convolution $f * g$ or the generalized product FG exists, then both exist, and*
$$\mathcal{F}[f * g] = FG \quad .$$

Algebraic Properties of Products

From our discussion of simple products in chapter 33, we already know

$$0 \cdot f = 0 \qquad \text{and} \qquad 1 \cdot f = 1$$

for every generalized function f. That the generalized product satisfies two other standard laws of classical multiplication are the next two rules.

Rule 6 *(commutativity)*
Let f and g be two generalized functions. If either of the generalized products fg or gf exists, then both exist and are equal.

Rule 7 *(distributivity)*
Let f, g and h be generalized functions. Then

$$f(g + h) = (fg) + (fh)$$

whenever the generalized products in this equation exist. Moreover, if any two of the three products in this equation — $f(g + h)$, fg and fh — are known to exist, then all three of these generalized products exist.

This may come as a surprise, but we cannot insist that associativity always holds for the generalized product. That, as the next example shows, would be incompatible with rule 5 relating products and convolutions, and the fact that associativity does not always hold for classical convolution.

!▶**Example 36.2:** Let $F = \mathcal{F}[f]$, $G = \mathcal{F}[g]$ and $H = \mathcal{F}[h]$ where

$$f(x) = 1 \quad , \quad g(x) = x e^{-x^2} \quad \text{and} \quad h(x) = \text{step}(x) \quad .$$

In exercise 24.4 on page 381, we saw that

$$(f * g) * h \neq f * (g * h) \quad .$$

So, by the relation between products and convolutions (rule 5),

$$(FG)H = (\mathcal{F}[f * g])\mathcal{F}[h] = \mathcal{F}[(f * g) * h]$$
$$\neq \mathcal{F}[f * (g * h)] = \mathcal{F}[f](\mathcal{F}[g * h]) = F(GH) \quad .$$

There are cases where associativity does hold. Certainly, if f, g and h are piecewise continuous (classical) functions, then $f(gh) = (fg)h$ as generalized products simply because this equation is true using the corresponding classical definition of products. We also know, from exercise 33.13 on page 552, that this equality holds when any two of these functions are simple multipliers for $\mathcal{G}$. More generally, it turns out that this equality can be assumed when one of the generalized functions is a simple multiplier and the generalized product of the other two exists. That is something we can adopt as one of our rules.

Rule 8 (limited associativity)
Let f be a simple multiplier, and let g and h be any two generalized functions for which the generalized product gh exists. Then $f(gh)$ and $(fg)h$ exist, and

$$f(gh) = (fg)h \quad .$$

Relations with Other Operations

We certainly know that, if h, f and g are classical functions with $h = fg$, and if α is any real number, then, for each real value x,

$$h(\alpha x) = f(\alpha x)g(\alpha x) \qquad \text{and} \qquad h(x - \alpha) = f(x - \alpha)g(x - \alpha) \quad .$$

This last equation also holds if α is complex and both f and g are analytic functions on $\mathbb{C}$. Moreover, if f and g are suitably differentiable, then the product rule, $h' = f'g + fg'$, holds. These are all facts that, properly rephrased, can (and will) be taken as rules for the generalized product.

Rule 9 (products and scaling)
Let f and g be two generalized functions, and let σ be a nonzero real constant. Assume that either fg or $(S_\sigma[f])(S_\sigma[g])$ exists. Then both products exist, and

$$S_\sigma[fg] = (S_\sigma[f])(S_\sigma[g]) \quad .$$

Rule 10 (products and translation)
Let f and g be two generalized functions, and let a be any constant. Assume that either fg or $(T_a[f])(T_a[g])$ exists. Then both products exist, and

$$T_a[fg] = (T_a[f])(T_a[g]) \quad .$$

Rule 11 (product rule)
Let f and g be two generalized functions, and assume that any two of the following products exist:

$$fg \quad , \qquad [Df]g \qquad \text{and} \qquad f[Dg] \quad .$$

Then all three products exist, and

$$D[fg] = f(Dg) + (Df)g \quad .$$

Limits and Products

We know

$$h\left(\lim_{\gamma\to\alpha} f_\gamma\right) = \lim_{\gamma\to\alpha} hf_\gamma \quad \text{and} \quad h\sum_{k=-\infty}^{\infty} g_k = \sum_{k=-\infty}^{\infty} hg_k$$

provided the limit and infinite series are convergent (in the generalized sense) and h is a simple multiplier for $\mathcal{G}$ (see lemma 34.1 on page 568 and lemma 34.4 on page 576). Our next rule, which we will write in two parts, is that these equalities hold when h is any simple multiplier.

Rule 12.a (products involving limits)
Let h be a simple multiplier, and let $\{f_\gamma\}_{\gamma=\gamma_0}^{\alpha}$ be a convergent sequence of generalized functions. Then the corresponding the sequence of products $\{hf_\gamma\}_{\gamma=\gamma_0}^{\alpha}$ also converges. Moreover,

$$h\left(\lim_{\gamma\to\alpha} f_\gamma\right) = \lim_{\gamma\to\alpha} hf_\gamma \quad .$$

Rule 12.b (products involving summations)
Let h be a simple multiplier, and let $\sum_{k=-\infty}^{\infty} g_k$ be a convergent infinite series of generalized functions. Then the infinite series of products $\sum_{k=-\infty}^{\infty} hg_k$ also converges. Moreover,

$$h\sum_{k=-\infty}^{\infty} g_k = \sum_{k=-\infty}^{\infty} hg_k \quad .$$

Strictly speaking, we did not need to state rule 12.b as one of our basic rules. It follows directly from rule 12.a and the definition of infinite series of generalized functions. However, since the products of simple multipliers with infinite series will play rather important roles in many of our later computations, it seems appropriate to place this last rule in our basic list.

!▶**Example 36.3:** *Consider the product of $\mathrm{sinc}(\pi x)$ with $\mathrm{comb}_1(x)$. As previously noted, this sinc function is a simple multiplier, and so, using rule 12.b along with rule 3 on multiplying a classical function by a delta function,*

$$\mathrm{sinc}(\pi x)\,\mathrm{comb}_1(x) = \mathrm{sinc}(\pi x)\sum_{k=-\infty}^{\infty} \delta_k(x)$$

$$= \sum_{k=-\infty}^{\infty} \mathrm{sinc}(\pi x)\,\delta_k(x) = \sum_{k=-\infty}^{\infty} \mathrm{sinc}(\pi k)\,\delta_k(x) \quad . \tag{36.1}$$

Now, for $k = 0$,

$$\mathrm{sinc}(\pi k) = \mathrm{sinc}(0) = 1 \quad ,$$

while, for $k = \pm 1, \pm 2, \pm 3, \dots$,

$$\mathrm{sinc}(\pi k) = \frac{\sin(\pi k)}{\pi k} = 0 \quad .$$

So,

$$\mathrm{sinc}(\pi x)\,\mathrm{comb}_1(x) = \sum_{k=-\infty}^{\infty} \left\{\begin{array}{ll} 1 & \text{if } k = 0 \\ 0 & \text{if } k \neq 0 \end{array}\right\} \delta_k(x) = \delta_0 \quad .$$

It is worth noting that, whenever we have a simple multiplier h and a generalized function given by a regular array of delta functions,

$$f = \sum_{k=-\infty}^{\infty} f_k \, \delta_{k\Delta x} \quad ,$$

then, using both rules 12.b and 3 as we did in equation (36.1),

$$hf = h \sum_{k=-\infty}^{\infty} f_k \, \delta_{k\Delta x} = \sum_{k=-\infty}^{\infty} f_k \, h \, \delta_{k\Delta x} = \sum_{k=-\infty}^{\infty} f_k \, h(k\Delta x) \, \delta_{k\Delta x}$$

Since this sort of product arises every so often, let us restate this observation as a lemma.

Lemma 36.1
Suppose h is a simple multiplier and f is a generalized function given by a regular array of delta functions,

$$f = \sum_{k=-\infty}^{\infty} f_k \, \delta_{k\Delta x} \quad .$$

Then

$$hf = \sum_{k=-\infty}^{\infty} h(k\Delta x) f_k \, \delta_{k\Delta x} \quad .$$

Be careful, though, about assuming the last equation holds when h is not a simple multiplier (see exercise 36.16 at the end of this chapter).

Existence of the Generalized Product and Convolution

To make it official:

Theorem 36.2
The classical operations of multiplication and convolution can be generalized so that rules 1 through 12 are satisfied.

This theorem does *not* say that the generalized product and convolution exist for every pair of generalized functions. We will not, for example, pretend to take the product of the delta function with itself or the convolution of the constant 1 with itself.

?►Exercise 36.3: *Try to make sense of $\delta \cdot \delta$ and $1 * 1$. (Don't expect to succeed.)*

There are at least three ways to convince ourselves that theorem 36.2 is true:

1. Verify that there are no hidden inconsistencies in the given rules, and then simply let these rules define the generalized product and generalized convolution for the products and convolutions that can be computed from those rules.

2. Come up with a single definition for the generalized product and a single definition for the generalized convolution, and then verify that the rules are satisfied.

3. Quote an authority.

The practicality of the first is questionable. Nor could the author figure out a way to do the second without straying far beyond the spatial and mathematical bounds set for this text. That leaves the last as our only choice.

We will quote an authority — the author.[2] He says, "The theorem is true."

Those for whom this is not adequate should turn to section 36.3 starting on page 643. That contains both a brief outline of one procedure for obtaining "most general" definitions for generalized multiplication and convolution, along with references for more complete and rigorous treatments of the subject.

More Rules and Equalities for Convolution

Because of the relation between products and convolutions given in rule 5, the rules already given for generalized products have direct analogs for generalized convolutions.

For example, let f, g, F and G be generalized functions with $f = \mathcal{F}^{-1}[F]$ and $g = \mathcal{F}^{-1}[G]$. Assume further that either F or G is a simple multiplier. From rule 2 we know the product FG exists. Rule 5 then assures us that the convolution $f * g$ also exists and is related to the product FG by $f * g = \mathcal{F}^{-1}[FG]$. On the other hand, if we know the convolution $f * g$ exists, then rule 5 tells us that the product FG also exists and is related to the convolution $f * g$ by $FG = \mathcal{F}[f * g]$. Consequently, rule 2 is equivalent to:

Rule 2′ (consistency with simple convolution)
If f and g are generalized functions and either is a Fourier transform of a simple multiplier, then the generalized convolution $f * g$ exists.

Two particularly simple "Fourier transforms of simple multipliers" are 0 and δ_a where a can be any complex number. As you surely recall,

$$\mathcal{F}^{-1}[0] = 0 \qquad \text{and} \qquad \mathcal{F}^{-1}\left[e^{-i2\pi ax}\right] = \delta_a \quad .$$

So, letting $F = \mathcal{F}[f]$, we have

$$0 * f = \mathcal{F}^{-1}[0] * \mathcal{F}^{-1}[F] = \mathcal{F}^{-1}[0 \cdot F] = \mathcal{F}^{-1}[0] = 0 \quad ,$$

which probably comes as no surprise. Also, using the translation identities, we see that

$$\delta_a * f = \mathcal{F}^{-1}\left[e^{-i2\pi ax}\right] * \mathcal{F}^{-1}[F] = \mathcal{F}^{-1}\left[e^{-i2\pi ax} F(x)\right] = T_a[f] \quad .$$

In particular,

$$\delta * f = T_0[f] = f \quad .$$

These results are significant enough to be stated as a lemma.

Lemma 36.3
Let f be any generalized function and a any complex number. Then

$$0 * f = 0 \qquad \text{and} \qquad \delta_a * f = T_a[f] \quad .$$

In particular,

$$\delta * f = f \quad .$$

---

[2] No one said that a *respected* authority would be quoted.

For convenience, let us refer to a generalized function as a *simple convolver* if and only if it is a Fourier transform of a simple multiplier. By the above, we know the delta functions are simple convolvers. You can also show that rectangle functions on finite intervals are simple convolvers. In fact, since these rectangle functions will be playing significant roles in later computations, go ahead and do the next exercise.

?▶Exercise 36.4:   *Verify that the convolution* $\text{rect}_{(a,b)} * f$ *exists for each generalized function* f *by showing* $\text{rect}_{(a,b)}$ *is a simple convolver. (Suggestion: Rewrite the rectangle function as a shifted pulse function.)*

Other rules for convolution can be derived using arguments similar to those used to derive rule $2'$. Here are four particularly important ones:

Rule 6′ (commutativity)
Let f and g be two generalized functions. If either of the generalized convolutions $f * g$ or $g * f$ exist, then both exist and are equal.

Rule 7′ (distributivity)
Let f, g and h be generalized functions. Then

$$f * (g + h) = (f * g) + (f * h)$$

whenever the generalized convolutions in this equation exist. Moreover, if any two of the three convolutions in this equation — $f * (g + h)$, $f * g$ and $f * h$ — are known to exist, then all three of these generalized convolutions exist.

Rule 8′ (limited associativity)
Let f be a simple convolver, and let g and h be any two generalized functions for which the generalized convolution $g * h$ exists. Then $f * (g * h)$ and $(f * g) * h$ exist, and

$$f * (g * h) = (f * g) * h \quad .$$

Rule 12.b′ (convolutions involving summations)
Let f be a simple convolver, and let $\sum_{k=-\infty}^{\infty} h_k$ be a convergent infinite series of generalized functions. Then the infinite series of convolutions $\sum_{k=-\infty}^{\infty} f * h_k$ also converges. Moreover,

$$f * \sum_{k=-\infty}^{\infty} h_k = \sum_{k=-\infty}^{\infty} f * h_k \quad .$$

?▶Exercise 36.5:   *Using rules 1 through 12.b, verify each of the following statements:*

   a: *Rule 6′ is equivalent to rule 6.*

   b: *Rule 7′ is equivalent to rule 7.*

   c: *Rule 8′ is equivalent to rule 8.*

   d: *Rule 12.b′ is equivalent to rule 12.b.*

Another "rule" that is important for applications is given in the next lemma. It is not equivalent to any of the previously given rules but is easily derived from them and the differentiation identities.

Lemma 36.4 (derivatives of convolutions)
Let f and g be two generalized functions for which the convolution $f * g$ exists, and let n be any positive integer. Then $(D^n f) * g$ and $f * (D^n g)$ exist, and

$$D^n(f * g) = (D^n f) * g = f * (D^n g) \quad .$$

PROOF: Let $F = \mathcal{F}[f]$ and $G = \mathcal{F}[g]$. By the relation between products and convolutions, we know the generalized product FG exists and $f * g = \mathcal{F}^{-1}[FG]$. Using this, the differentiation identity and rule 8, we see that

$$\begin{aligned}
D^n(f * g) &= D^n(\mathcal{F}^{-1}[FG]) \\
&= \mathcal{F}^{-1}\big[(i2\pi x)^n\big(F(x)G(x)\big)\big] \\
&= \mathcal{F}^{-1}\big[\big((i2\pi x)^n F(x)\big)G(x)\big] \\
&= \mathcal{F}^{-1}\big[(i2\pi x)^n F(x)\big] * \mathcal{F}^{-1}[G] = (D^n f) * g \quad .
\end{aligned}$$

From this and the commutativity of convolution, we also have

$$D^n(f * g) = D^n(g * f) = (D^n g) * f = f * (D^n g) \quad . \qquad \blacksquare$$

As an immediate corollary to this and lemma 36.3, we have the following:

Corollary 36.5
Let f be any generalized function and n any nonnegative integer. Then $D^n f = f * (D^n \delta)$.

PROOF: Lemma 36.3 tells us that $f = f * \delta$, as claimed by the corollary when $n = 0$. Using this and lemma 36.4, we see that, for $n > 0$,

$$D^n f = D^n(f * \delta) = f * (D^n \delta) \quad . \qquad \blacksquare$$

36.2 Definite Integrals of Generalized Functions
Basic Definition

One advantage of having multiplication formally defined is that it allows us to formally identify certain expressions as being "integrals of generalized function". To see this, suppose we have a classical function f that is absolutely integrable over an interval (a, b). Using the classical definitions,

$$\int_a^b f(s)\, ds = \int_{-\infty}^\infty \text{rect}_{(a,b)}(s)\, f(s)\, ds$$

$$= \int_{-\infty}^\infty \text{rect}_{(-b,-a)}(0 - s)\, f(s) = \text{rect}_{(-b,-a)} * f(0) \quad .$$

Thanks to our more general notion of convolution, the expression on the right-hand side often makes sense when f is not as just supposed. To take advantage of this, we will say that any generalized function f is *integrable* (in the generalized sense) over the interval (a, b) whenever

1. the convolution of f with $\text{rect}_{(a,b)}$ exists,

2. this convolution is a classical function on the real line,

and

3. this classical function is continuous at 0.

If these three conditions hold, then the right-hand side of the above equation is a well-defined finite value, which, inspired by the above classical equation, we will call the (generalized) *definite integral* of f over (a, b), and denote by the more familiar expression on the left-hand side of the above equation. In other words, we are (re-)defining the definite integral notation by the formula

$$\int_a^b f(s) \, ds \;=\; \text{rect}_{(-b,-a)} * f(0) \quad . \tag{36.2}$$

In practice, of course, we will replace "s" with whatever symbol is convenient for the dummy variable in the integral.

By our conventions, (a, b) being some interval means $a < b$. Of course, if $a = b$, then $\text{rect}_{(a,b)}$ is $\text{rect}_{(a,a)}$, which (if you think about it for a moment) you will realize is just the zero function. Thus, by our definition, any generalized function f is integrable over any interval of length zero, and

$$\int_a^a f(s) \, ds \;=\; 0 * f(0) \;=\; 0 \qquad \text{for each } a \text{ in } \mathbb{R} \quad .$$

While we are at it, let's go ahead and agree that, as long as f is integrable in the generalized sense on (a, b), then the (generalized) definite integral of f from b to a is defined in the obvious manner; namely,

$$\int_b^a f(s) \, ds \;=\; -\int_a^b f(s) \, ds \quad .$$

Elementary Properties

Some useful and easily verified properties of the definite integral are given in the following lemmas. Consider the proofs as exercises.

Lemma 36.6

Let f and g both be integrable over the interval between a and b, and let α and β be any two constants. Then $\alpha f + \beta g$ is integrable over the interval between a and b. Furthermore,

$$\int_a^b [\alpha f(s) + \beta g(s)] \, ds \;=\; \alpha \int_a^b f(s) \, ds \;+\; \beta \int_a^b g(s) \, ds \quad .$$

Lemma 36.7

Let f be integrable over the intervals between a and b, and between b and c. Then f is integrable over the interval between a and c. Furthermore,

$$\int_a^b f(s) \, ds \;+\; \int_b^c f(s) \, ds \;=\; \int_a^c f(s) \, ds \quad .$$

Lemma 36.8
Suppose $f(x)$ is integrable over the interval between a and b, and let c be any nonzero real number. Then $f(cx)$ is integrable over the interval between ca and cb. Moreover,

$$\int_{ca}^{cb} f(cs)\,ds \; = \; \frac{1}{c} \int_{a}^{b} f(\sigma)\,d\sigma \quad .$$

Lemma 36.9
Assume $f(x)$ is integrable over (a, b), and let c be any real number. Then $f(x - c)$ is integrable over $(a - c, b - c)$, and

$$\int_{a-c}^{b-c} f(s - c)\,ds \; = \; \int_{a}^{b} f(\sigma)\,d\sigma \quad .$$

?►Exercise 36.6:   *Verify the claim made in*

   a: *lemma 36.6.*   **b:** *lemma 36.7.*   **c:** *lemma 36.8.*   **d:** *lemma 36.9.*

We'll prove the next result. It extends the classical fundamental theorem of calculus.

Theorem 36.10 (fundamental theorem of calculus, generalized)
Let (a, b) be a finite interval, and assume f is an exponentially integrable function on $\mathbb{R}$ that is continuous at both a and b. Then Df, the generalized derivative of f, is integrable on (a, b), and

$$\int_{a}^{b} Df(x)\,dx \; = \; f(b) - f(a) \quad .$$

PROOF: Applying lemmas 36.3 and 36.4 regarding convolutions involving derivatives and delta functions, we see that

$$\begin{aligned}
\text{rect}_{(-b,-a)} * Df(x) &= \left[D\,\text{rect}_{(-b,-a)} \right] * f(x) \\
&= \left[\delta_{-b} - \delta_{-a} \right] * f(x) \\
&= f(x - (-b)) - f(x - (-a)) = f(x + b) - f(x + a) \quad ,
\end{aligned}$$

which, by our choice of f and (a, b), is a classical function on $\mathbb{R}$ and is continuous at 0. So Df is integrable on (a, b), and

$$\int_{a}^{b} Df(x)\,dx \; = \; \text{rect}_{(-b,-a)} * Df(0) \; = \; f(b) - f(a) \quad . \qquad \blacksquare$$

Definite Integrals of Delta Functions and Arrays

In practice, we will usually find ourselves concerned with definite generalized integrals involving classical functions, delta functions and regular arrays of delta functions. There isn't much new to say about the generalized definite integral of a classical function; from the definition it is the same as the classical integral. So let's concentrate on definite integrals of delta functions and arrays.

Let (a, b) be any interval, and let c be any real number. Recalling that $f * \delta_c$ equals "f translated by a shift of c" (lemma 36.3), we see that

$$\text{rect}_{(-b,-a)} * \delta_c(x) = \text{rect}_{(-b,-a)}(x - c)$$

$$= \begin{cases} 1 & \text{if} \quad -b < x - c < -a \\ 0 & \text{if} \quad x - c < -b \text{ or } -a < x - c \end{cases}$$

$$= \begin{cases} 1 & \text{if} \quad c - b < x < c - a \\ 0 & \text{if} \quad x < c - b \text{ or } c - a < x \end{cases} = \text{rect}_{(c-b,c-a)}(x) \quad .$$

This is certainly a piecewise continuous function, and it is continuous at $x = 0$ whenever c is not one of the endpoints of (a, b). So, by our generalized definition of integrability, δ_c is integrable on (a, b) as long as c is a real number other than a or b. Moreover,

$$\int_a^b \delta_c(s) \, ds = \text{rect}_{(-b,-a)} * \delta_c(0) = \begin{cases} 1 & \text{if} \quad a < c < b \\ 0 & \text{if} \quad c < a \text{ or } b < c \end{cases} \quad .$$

For future reference, let us enshrine the results just derived in the following lemma.

Lemma 36.11 (integrals of delta functions)
Suppose (a, b) is an any interval, and c is a real number other than an endpoint of (a, b). Then δ_c is integrable on (a, b), and

$$\int_a^b \delta_c(s) \, ds = \begin{cases} 1 & \text{if} \quad a < c < b \\ 0 & \text{if} \quad c < a \text{ or } b < c \end{cases} \quad .$$

There are two things you should realize regarding the above calculations:

1. We are not requiring that (a, b) be a finite interval. The above is valid even if $a = -\infty$ or $b = \infty$.

2. We are only defining and computing $\int_a^b \delta_c(s) \, ds$ when c is a real number other than either a or b. We will not attempt to define this integral when $c = a$ or $c = b$. Nor will we attempt to define this integral when c is not a real number.

?▶ Exercise 36.7: *Verify that*

$$\text{step}(x - a) = \int_{-\infty}^x \delta_a(s) \, ds$$

whenever a is a fixed real number.

Extending the above to regular arrays of delta functions is easy provided the interval (a, b) is finite. Recalling that the rectangle function on a such an interval is a simple convolver, and then using the rule on convolving a simple convolver with infinite series (rule 12.b'), we get

$$\text{rect}_{(-b,-a)}(x) * \sum_{k=-\infty}^{\infty} f_k \, \delta_{k\Delta x}(x) = \sum_{k=-\infty}^{\infty} f_k \, \text{rect}_{(-b,-a)}(x) * \delta_{k\Delta x}(x)$$

$$= \sum_{k=-\infty}^{\infty} f_k \, \text{rect}_{(-b,-a)}(x - k\Delta x) \quad .$$

If a and b are finite, then only finitely many terms in the above series of shifted rectangle functions can be nonzero for each real value x. Because of this, you can readily confirm that this summation is a piecewise continuous function on $\mathbb{R}$ that can be computed term by term and which is continuous everywhere except at the points

$$a + k\Delta x \quad \text{and} \quad b + k\Delta x \quad \text{for} \quad k = 0, \pm 1, \pm 2, \pm 3, \ldots \quad .$$

So, by the definition of the generalized integral,

$$\int_a^b \sum_{k=-\infty}^{\infty} f_k \, \delta_{k\Delta x}(x) \, dx = \left[\text{rect}_{(-b,-a)} * \sum_{k=-\infty}^{\infty} f_k \, \delta_{k\Delta x} \right]\Bigg|_0$$

$$= \sum_{k=-\infty}^{\infty} f_k \, \text{rect}_{(-b,-a)}(0 - k\Delta x)$$

$$= \sum_{k=-\infty}^{\infty} f_k \, \text{rect}_{(-b,-a)} * \delta_{k\Delta x}(0) = \sum_{k=-\infty}^{\infty} f_k \int_a^b \delta_{k\Delta x}(x) \, dx \quad .$$

These calculations, of course, presuppose that the original array is a generalized function and that none of the $k\Delta x$'s equals a or b. Continuing these calculations through an invocation of the previous lemma on the integrals of delta functions then gives us our lemma on integrating arrays of delta functions.

Lemma 36.12 (integrals of delta function arrays)
Assume the regular array $\sum_{k=-\infty}^{\infty} f_k \, \delta_{k\Delta x}$ is a generalized function, and let (a, b) be any finite interval with neither a nor b being an integral multiple of the array's spacing, Δx. Then the array is integrable on (a, b), and

$$\int_a^b \sum_{k=-\infty}^{\infty} f_k \, \delta_{k\Delta x}(x) \, dx = \sum_{k=-\infty}^{\infty} f_k \int_a^b \delta_{k\Delta x}(x) \, dx$$

$$= \sum_{k=-\infty}^{\infty} f_k \begin{cases} 1 & \text{if} \quad a < k\Delta x < b \\ 0 & \text{otherwise} \end{cases} \quad .$$

36.3 Appendix: On Defining Generalized Products and Convolutions

Here is a brief outline of one way to define "most general" products and convolutions of generalized functions so that theorem 36.2 on page 636 holds. Strictly speaking it's an approach for generalizing products and convolutions using $\mathcal{H}$ as the test function space. This is because the richer set of simple multipliers for $\mathcal{H}$ allows more direct proofs of some of the following claims. However, as mentioned in section 32.3 (page 521) the set of generalized functions developed using $\mathcal{H}$ is the same as set of generalized functions developed using $\mathcal{G}$, so any product or convolution definition developed "using $\mathcal{H}$" automatically gives a valid product or convolution for generalized functions developed using $\mathcal{G}$.

1. First, define the product fh of any generalized function f with any simple multiplier h to be the generalized function given by

    $$\langle fh, \phi \rangle = \langle f, h\phi \rangle \qquad \text{for each } \phi \text{ in } \mathcal{H} \quad .$$

    (This is the simple product we were using before this chapter, and which we extended somewhat in this chapter by expanding our set of simple multipliers.)

2. Next, define the convolution of any generalized function f with any test function ψ via the "adjoint-identity" formula

    $$\langle f * \psi, \phi \rangle = \langle f, \psi^* \star \phi \rangle \qquad \text{for each } \phi \text{ in } \mathcal{H}$$

    (see exercise 35.54 on page 627).

3. It can then be shown that a generalized function h is a simple convolver if and only if $h * \phi$ and $h^* \star \phi$ are in $\mathcal{H}$ for each test function ϕ (see exercise 35.54 d ii on page 627 for a partial proof). Because of this, we can extend the last definition, and define $f * h$ for any generalized function f and any simple convolver h by the formula

    $$\langle f * h, \phi \rangle = \langle f, h^* \star \phi \rangle \qquad \text{for each } \phi \text{ in } \mathcal{H} \quad .$$

4. Remarkably, it can then be proven that, when g is any generalized function and ψ is any test function,

    (a) the convolution $f * \psi$ is a simple multiplier for $\mathcal{H}$, and

    (b) the product $f\psi$ is a simple convolver for $\mathcal{H}$.

5. The next steps are to attempt defining convolution and multiplication for arbitrary pairs of generalized functions f and g using sequences.

    (a) For each $\epsilon > 0$, let u_ϵ be the Gaussian

    $$u_\epsilon(x) = e^{-\epsilon \pi x^2} \quad .$$

    From step 4, above, we know each gu_ϵ is a simple convolver. Hence

    $$f * (gu_\epsilon)$$

    is a well-defined generalized function for each $\epsilon > 0$. Moreover, as noted in example 33.2 on page 554,

    $$\lim_{\epsilon \to 0^+} gu_\epsilon = ge^0 = g \qquad \text{(in the generalized sense)} \quad .$$

    This suggests using the generalized limit of $f * (gu_\epsilon)$ as $\epsilon \to 0^+$ for the convolution of f with g. However, because f and g play slightly different roles in this limit, it is not immediately clear that this "convolution" is commutative, so we refer to this limit as the "asymmetric convolution" of f with g, and denote it by $\operatorname{conv}(f, g)$. That is, the *asymmetric convolution* of f with g, $\operatorname{conv}(f, g)$, is defined by

    $$\operatorname{conv}(f, g) = \lim_{\epsilon \to 0^+} f * (gu_\epsilon) \quad ,$$

    provided the generalized limit exists. If the limit does not exist, then neither does the asymmetric convolution.

(b) For each $\gamma > 0$, let η_γ be the Gaussian

$$\eta_\gamma(x) = \sqrt{\gamma}e^{-\gamma\pi x^2} \quad .$$

From step 4, above, we know each $g * \eta_\gamma$ is a simple multiplier. Hence

$$f(g * \eta_\gamma)$$

is a well-defined generalized function for each $\gamma > 0$. Recall, also, that $\{\eta_\gamma\}_{\gamma=1}^\infty$ is a Gaussian identity sequence. Because of this, you can show that

$$\lim_{\gamma \to \infty} g * \eta_\gamma = g * \delta = g$$

(see exercise 35.54 d iii on page 627). This suggests using the limit of $f(g * \eta_\gamma)$ as $\gamma \to \infty$ for the product of f with g. Again however, because f and g play slightly different roles in this limit, it is best to refer to this limit as an asymmetric product and to denote it by, say, $\text{prod}(f, g)$. That is, we define the *asymmetric product* of f with g, $\text{prod}(f, g)$, by

$$\text{prod}(f, g) = \lim_{\gamma \to 0} f(g * \eta_\gamma) \quad ,$$

provided this limit exists. If the limit does not exist, then neither does this asymmetric product.

Analogs to all the rules for multiplication and convolution given in this chapter — excluding rules 6 and 6′ on commutativity — can be verified for the asymmetric product and convolution, $\text{prod}(f, g)$ and $\text{conv}(f, g)$. Commutativity cannot be verified because, in general, it is not true; generalized functions f and g can be found such that

$$\text{prod}(f, g) = \lim_{\gamma \to 0} f(g * \eta_\gamma) \neq \lim_{\gamma \to 0} g(f * \eta_\gamma) = \text{prod}(g, f) \quad .$$

Likewise, you can find an f and g such that $\text{conv}(f, g) \neq \text{conv}(g, f)$ (see exercise 36.22 at the end of this chapter).

6. Finally, let f and g be any two generalized functions. To obtain a commutative product fg and a commutative convolution $f * g$ from the asymmetric products and convolutions defined above:

(a) Define $f * g$ by

$$f * g = \frac{1}{2}\Big[\text{conv}(f, g) + \text{conv}(g, f)\Big]$$

whenever both $\text{conv}(f, g)$ and $\text{conv}(g, f)$ exist. If one or the other asymmetric convolutions does not exist, then, as far as this approach is concerned, neither does $f * g$. (Note that, if the asymmetric convolutions exist and $\text{conv}(f, g) = \text{conv}(g, f)$ — which is often the case — then $f * g = \text{conv}(f, g)$.)

(b) Define fg by

$$fg = \frac{1}{2}\Big[\text{prod}(f, g) + \text{prod}(g, f)\Big]$$

whenever both $\text{prod}(f, g)$ and $\text{prod}(g, f)$ exist. If one or the other asymmetric products does not exist, then, as far as this approach is concerned, neither does $f * g$. (Note that, if the asymmetric products exist and $\text{prod}(f, g) = \text{prod}(g, f)$ — which is often the case — then $fg = \text{prod}(f, g)$.)

For a rigorous development and analysis of the asymmetric product and convolution, you'll have to work your way through the original papers of the author (references [3] through [8]). Keep in mind the warnings previously made in footnote 6 on page 522. You should also be warned that the above definitions of prod(f, g) and conv(f, g) are simplifications of those in the papers. Only the most adventuresome and prepared readers should attempt these papers. Of course, those readers would then have no trouble confirming the validity of theorem 36.2 using the definitions for fg and $f * g$ given in the last step above. So maybe the payoff would justify the attempt.

By the way, the above is not the only way to generalize the notions of products and convolutions. Indeed, you may have wondered why we didn't just define $f * g$ and fg by

$$f * g = \lim_{\epsilon \to 0^+} (f u_\epsilon) * (g u_\epsilon) \quad \text{and} \quad fg = \lim_{\gamma \to \infty} (f * \eta_\gamma)(g * \eta_\gamma)$$

where u_ϵ and η_γ are as in step 5. They were not used above simply because the author of the aforementioned papers did not use them, and he didn't use them because he preferred using the simpler limits in step 5.

There are other ways of defining "most general" generalized products and convolutions. For somewhat different perspectives on this matter, look at the texts by Colombeau (reference [2]) and by Richards and Youn (reference [9]).

?▶ **Exercise 36.8:**   *Why is the author not particularly worried about how generalized products and convolutions are defined, as long as theorem 36.2 can be verified?*

Additional Exercises

36.9 a.   *Assume h is an analytic function on the entire complex plane and is exponentially bounded on horizontal strips of the complex plane. That is, for each horizontal strip of the complex plane*

$$S_{(a,b)} = \{x + iy : a < y < b\}$$

with $-\infty < a \le b < \infty$, there are corresponding finite positive constants M and c such that

$$|h(x + iy)| \le M e^{c|x|} \quad \text{for all } x + iy \text{ in } S_{(a,b)} \quad .$$

Show that h is a simple multiplier for $\mathcal{H}$. (This and the next part do require some knowledge of complex variables — such as can be found in section 6.4 starting on page 66 — and the definition of $\mathcal{H}$ given on page 522.)

b.   *Verify that each of the following is a simple multiplier for $\mathcal{H}$ (assume $\alpha > 0$):*

   i. *Any simple multiplier for $\mathcal{G}$*   **ii.** $\text{sinc}(\alpha x)$ **iii.** $\sin\left(\alpha x^2\right)$

36.10.   *Suppose f and F are classically transformable functions with $F = \mathcal{F}[f]$. Assume, further, that F is absolutely integrable. Using rules 4 and 5 — but not rule 3 — show that*

$$f \delta_a = f(a) \delta_a$$

for every a in $\mathbb{R}$. (Thus, rule 3 can be partially derived from rules 4 and 5.)

36.11. *Compute the following products (and simplify your answers):*

a. $\dfrac{1}{x+i}\,\delta(x)$

b. $T_i\left[\dfrac{1}{x+i}\right]\delta_i(x)$

c. $\sin(x)\,T_i\left[\dfrac{1}{x+i}\right]$

d. $\sin\left(\pi x^2\right)\mathrm{comb}_1(x)$

e. $\cos\left(\pi x^2\right)\mathrm{comb}_1(x)$

36.12 a. *Let a be a complex number. Using the rules for multiplication and convolution, show that $f\delta_a = f(a)\delta_a$ assuming f is an analytic function on $\mathbb{C}$ that is exponentially bounded on horizontal strips.*

b. *Simplify $\mathrm{sinc}(2\pi x)\,\delta_i$.*

36.13. *Let a be a real number, and assume f is smooth function on the real line.*

a. *Show that $f\,D\delta_a = f(a)D\delta_a - f'(a)\delta_a$.*

b. *What is the corresponding formula for $f\,D^2\delta_a$ assuming f' is also smooth?*

36.14. *For the following, let g be a generalized function given by a regular array*

$$g \;=\; \sum_{k=-\infty}^{\infty} g_k\,\delta_{k\Delta x} \quad .$$

a. *Verify that, for each integer N,*

$$\mathrm{sinc}\left(\frac{\pi}{\Delta x}(x - N\Delta x)\right) g(x) \;=\; g_N\,\delta_{N\Delta x} \quad .$$

b. *Show that $g = 0$ if and only if each g_k is 0.*

36.15. *Let f and g be two generalized functions given by regular arrays with the same spacing,*

$$f \;=\; \sum_{k=-\infty}^{\infty} f_k\,\delta_{k\Delta x} \qquad \text{and} \qquad g \;=\; \sum_{k=-\infty}^{\infty} g_k\,\delta_{k\Delta x} \quad .$$

Show that $f = g$ if and only if $f_k = g_k$ for each integer k. (Try using a result from the previous exercise.)

36.16. *Define the function f by*

$$f(x) \;=\; \sum_{k=-\infty}^{\infty} h_k(x)$$

where, for each integer k,

$$h_k(x) \;=\; \begin{cases} e^{k^2} & \text{if } -2^{-k}e^{-k^2} < x - k < 2^{-k}e^{-k^2} \\ 0 & \text{otherwise} \end{cases} \quad .$$

a. *Sketch the graph of f, and verify that f is not exponentially bounded, but is absolutely integrable (hence is exponentially integrable).*

b. *One might naively attempt to multiply f with comb_1 as follows:*

$$f(x)\sum_{k=-\infty}^{\infty} \delta_k(x) \;=\; \sum_{k=-\infty}^{\infty} f(x)\delta_k(x) \;=\; \sum_{k=-\infty}^{\infty} f(k)\delta_k(x) \quad .$$

Show that this is not valid by verifying that the last summation does not define a linear functional on $\mathcal{G}$ and, hence, cannot be treated as a generalized function.

36.17. Evaluate/simplify each of the following:

    a. $x^2 * \delta(x)$ **b.** $x^2 * \delta_3(x)$ **c.** $x^2 \delta_3(x)$

36.18. For the following, let

$$h(t) = -e^{6t} \operatorname{step}(-t)$$

    and treat all the derivatives as generalized derivatives.

    a. Verify that

$$\frac{dh}{dt} - 6h = \delta \quad .$$

    b. Let f be any generalized function such that $f * h$ exists, and show that $y = f * h$ satisfies

$$\frac{dy}{dt} - 6y = f \quad .$$

36.19. Assume h is a generalized function satisfying

$$a\frac{d^2h}{dt^2} + b\frac{dh}{dt} + ch = \delta$$

    where a, b and c are constants and the derivatives are treated as generalized derivatives. Verify that $y = f * h$ is then a solution to

$$a\frac{d^2y}{dt^2} + b\frac{dy}{dt} + cy = f$$

    for any generalized function f whose convolution with h exists.

36.20. Evaluate each of the following integrals:

    a. $\displaystyle\int_{-\infty}^{\infty} x^2 \delta(x)\, dx$ **b.** $\displaystyle\int_{-\infty}^{\infty} x^2 \delta_3(x)\, dx$ **c.** $\displaystyle\int_{-\infty}^{\infty} x^2 \delta_3(2x - 5)\, dx$

    d. $\displaystyle\int_{0}^{5} x^2 \delta_3(x)\, dx$ **e.** $\displaystyle\int_{5}^{\infty} x^2 \delta_3(x)\, dx$ **f.** $\displaystyle\int_{-1/2}^{7/2} \operatorname{comb}_1(x)\, dx$

    g. $\displaystyle\int_{-3/4}^{7/4} \sum_{k=-\infty}^{\infty} k^2\, \delta_{k/2}(x)\, dx$

36.21. Using the definition for $\operatorname{prod}(f, g)$ given in the last section, show that $\operatorname{prod}(\delta, \delta)$ does not exist.

36.22. The following involves the signum function,

$$\operatorname{sgn}(x) = \begin{cases} -1 & \text{if } x < 0 \\ +1 & \text{if } 0 < x \end{cases} \quad .$$

    a. Verify that the classical convolution of 1 with the signum function does not exist.

    b. Using the definition for $\operatorname{conv}(f, g)$ given above, show that the asymmetric convolutions $\operatorname{conv}(1, \operatorname{sgn})$ and $\operatorname{conv}(\operatorname{sgn}, 1)$ do exist, but are not equal.

37

Periodic Functions and Regular Arrays

In this chapter, we will tie together several different threads of Fourier analysis. We will extend our discussion from part II regarding periodic functions and their Fourier series, and, using our generalized theory, we will discover the intimate relation between the Fourier series and the Fourier transforms of periodic functions. In addition, we will develop material that will help form the framework for yet another "theory" of Fourier analysis — the "discrete theory" — which will be developed in later chapters.

37.1 Periodic Generalized Functions
Basics

When we say a generalized function f is *periodic* (in the generalized sense), we mean there is a positive value p such that, as generalized functions,

$$f(x - p) \;=\; f(x)$$

or, equivalently,

$$T_p[f] \;=\; f \quad .$$

The value p is called a (generalized) *period* of f. The corresponding *frequency* is $1/p$.

As you doubtlessly observed, the generalized definition of periodicity is almost identical to the classical definition of periodicity given in section 5.2 — the only difference being that the above equation is now interpreted as meaning

$$\big\langle\, f(x - p)\,,\,\phi(x)\,\big\rangle \;=\; \big\langle\, f(x)\,,\,\phi(x)\,\big\rangle \qquad \text{for each } \phi \text{ in } \mathcal{G}$$

instead of "the value of $f(x - p)$ is the same as the value of $f(x)$ at each point x of $\mathbb{R}$ where f is continuous." Still it should be clear that any exponentially integrable function that is periodic in the classical sense is also periodic in the generalized sense. After all, if f is such a classical function and p is its period (in the classical sense), then, for each Gaussian test function ϕ,

$$\big\langle\, f(x - p)\,,\,\phi(x)\,\big\rangle \;=\; \int_{-\infty}^{\infty} f(x - p)\phi(x)\,dx$$

$$= \int_{-\infty}^{\infty} f(x)\phi(x)\,dx \;=\; \big\langle\, f(x)\,,\,\phi(x)\,\big\rangle \quad .$$

So, classical periodicity should be viewed as a special case of generalized periodicity.

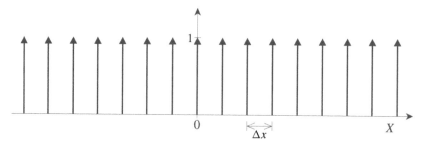

Figure 37.1: The comb function with spacing Δx.

On the other hand, there are certainly nonclassical generalized functions — such as the comb functions — whose graphical representations certainly look "periodic". Verifying that they are indeed periodic in the generalized sense is usually fairly easy.

!▶ **Example 37.1 (comb functions):** *Consider the comb function with spacing Δx,*

$$\text{comb}_{\Delta x} = \sum_{k=-\infty}^{\infty} \delta_{k\Delta x} \quad .$$

Its graphical representation (figure 37.1) certainly suggests that this comb function is periodic with a period equal to its spacing. For our verification of this, assume ϕ is any Gaussian test function. Now,

$$\left\langle \text{comb}_{\Delta x}(x) , \phi(x) \right\rangle = \left\langle \sum_{k=-\infty}^{\infty} \delta_{k\Delta x}(x) , \phi(x) \right\rangle$$

$$= \sum_{k=-\infty}^{\infty} \left\langle \delta_{k\Delta x}(x) , \phi(x) \right\rangle = \sum_{k=-\infty}^{\infty} \phi(k\Delta x) \quad .$$

Combining this with the generalized notion of translation and an obvious re-indexing, we get

$$\left\langle \text{comb}_{\Delta x}(x - \Delta x) , \phi(x) \right\rangle = \left\langle \text{comb}_{\Delta x}(x) , \phi(x + \Delta x) \right\rangle$$

$$= \sum_{k=-\infty}^{\infty} \phi(k\Delta x + \Delta x)$$

$$= \sum_{k=-\infty}^{\infty} \phi([k + 1]\Delta x)$$

$$= \sum_{n=-\infty}^{\infty} \phi(n\Delta x) = \left\langle \text{comb}_{\Delta x}(x) , \phi(x) \right\rangle \quad .$$

So, as generalized functions,

$$\text{comb}_{\Delta x}(x - \Delta x) = \text{comb}_{\Delta x}(x) \quad .$$

That is, $\text{comb}_{\Delta x}$ is periodic with period Δx in the generalized sense.

As noted a few paragraphs ago, classical periodicity is a special case of generalized periodicity. Moreover, it is easily verified that any exponentially integrable function that is periodic in the generalized sense must also be periodic in the classical sense (see exercise 37.9 at the end of this chapter). Since this is the case, and since we are mainly interested in the more general cases, let us agree that, henceforth, whenever a generalized function is referred to as "periodic" (i.e., without the quantifier "generalized") then it is periodic in the generalized sense. And, to save seemingly endless repetitions of the phrase "where p is some positive value", let us also agree that, for the rest of this chapter, p always denotes some positive value.

Before continuing, we should note the following analogs to a couple of well-known results regarding periodic classical functions. Their proofs are almost trivial and will be left as exercises.

Lemma 37.1
Assume f and g are two periodic generalized functions with a common period p. Then:

1. Each linear combination of f and g is periodic with period p.

2. If it exists, the product fg is periodic with period p.

Lemma 37.2
Let f and g be generalized functions with f being periodic with period p. If $g(x) = f(\sigma x)$ for some fixed nonzero real value σ, then g is also periodic and has period p/σ.

Lemma 37.3
If the generalized function f is periodic with period p, then, for any integer N,

$$f(x + Np) = f(x) \quad .$$

The last lemma points out that the period of a periodic generalized function is not unique; any positive integral multiple of one period is another period.

?▶Exercise 37.1: *Prove*

    a: *lemma 37.1.*   **b:** *lemma 37.2.*   **c:** *lemma 37.3.*

Educated Speculation on Transforms of Periodic Functions

Recall how we computed the transform of one simple periodic function, $\sin(2\pi t)$: First we expressed the function in terms of the complex exponential; then we took the transform of this expression using the fact that $\mathcal{F}\left[e^{\pm i 2\pi t}\right] = \delta_{\pm 1}$,

$$\mathcal{F}[\sin(2\pi t)] = \mathcal{F}\left[\frac{1}{2i}\left[e^{i2\pi t} - e^{-i2\pi t}\right]\right] = \frac{1}{2i}\left[\mathcal{F}\left[e^{i2\pi t}\right] - \mathcal{F}\left[e^{-i2\pi t}\right]\right] = \frac{1}{2i}[\delta_1 - \delta_{-1}] \quad .$$

Obviously, the same approach can be used to find the Fourier transform and inverse Fourier transform of any function that can similarly be expressed in terms of exponentials.

But now recall that most of part II of this text was spent discussing periodic classical functions on the real line, and a large part of that discussion was to verify that these functions can be expressed in terms of complex exponentials. Let's quickly review a few particulars from that discussion, and speculate (in an educated manner) on likely implications.

Let f be any periodic, classical function on $\mathbb{R}$ with period p. Assuming f is suitably integrable, its (complex exponential) Fourier series, $F.S.[f]$, is given by

$$F.S.[f]|_t = \sum_{k=-\infty}^{\infty} c_k\, e^{i2\pi\omega_k t}$$

where, for each k,

$$\omega_k = \frac{k}{p} \quad \text{and} \quad c_k = \frac{1}{p} \int_0^p f(t)\, e^{-i2\pi\omega_k t}\, dt \quad .$$

(The c_k's, you may recall, are called the Fourier coefficients for f.) Further assuming that f is "reasonable", we know from our work in part II that the Fourier series for f converges "in some sense" and "in that sense",

$$f(t) = F.S.[f]|_t = \sum_{k=-\infty}^{\infty} c_k\, e^{i2\pi\omega_k t} \quad . \tag{37.1}$$

If this last equation holds in the generalized sense, then finding the Fourier transform of f is almost trivial:

$$\mathcal{F}[f(t)] = \mathcal{F}\left[\sum_{k=-\infty}^{\infty} c_k\, e^{i2\pi\omega_k t}\right] = \sum_{k=-\infty}^{\infty} c_k \mathcal{F}\left[e^{i2\pi\omega_k t}\right] = \sum_{k=-\infty}^{\infty} c_k\, \delta_{\omega_k} \quad .$$

To make things a little more explicit, let

$$\Delta\omega = \frac{1}{p} \quad .$$

Then

$$\omega_k = \frac{k}{p} = k\Delta\omega \quad ,$$

and the formula derived above becomes

$$\mathcal{F}[f(t)] = \sum_{k=-\infty}^{\infty} c_k\, \delta_{k\Delta\omega} \quad .$$

Clearly then, if equation (37.1) holds treating everything as generalized functions, then the Fourier transform of f is a regular array of delta functions with spacing $\Delta\omega = 1/p$ and whose coefficients are the Fourier coefficients of the periodic function f.

In a similar manner, you can derive the regular array of delta functions that is likely to be the inverse Fourier transform of f.

?▶ Exercise 37.2:   *Assume equation (37.1) holds in the generalized sense, and show that*

$$\mathcal{F}^{-1}[f(t)] = \sum_{k=-\infty}^{\infty} c_{-k}\, \delta_{k\Delta\omega} \quad .$$

All of this is very fine, provided equation (37.1) holds within our generalized theory. Guessing that it does, we might go a bit further and even ask:

> *Can every periodic generalized function be represented as a summation of complex exponentials analogous to the classical Fourier series?*

If the answer to this question is *yes* then the computation of a Fourier transform of a periodic generalized function can be reduced to the computation of a Fourier transform of an infinite series of complex exponentials. This, as we saw above, results in a regular array of delta functions.

Let us try to gain a little more insight on the relation between periodic generalized functions and regular arrays by briefly examining Fourier transforms of such delta function arrays.

Transforms of Regular Arrays

Consider computing the inverse Fourier transform of a generalized function given by a regular array with spacing $\Delta\omega$

$$F = \sum_{k=-\infty}^{\infty} F_k \, \delta_{k\Delta\omega} \quad . \tag{37.2}$$

Letting $f = \mathcal{F}^{-1}[F]$, we have

$$f(t) = \mathcal{F}^{-1}\left[\sum_{k=-\infty}^{\infty} F_k \, \delta_{k\Delta\omega}\right]\Bigg|_t = \sum_{k=-\infty}^{\infty} F_k \mathcal{F}^{-1}[\delta_{k\Delta\omega}]\big|_t = \sum_{k=-\infty}^{\infty} F_k \, e^{i2\pi k\Delta\omega t} \quad .$$

Further letting

$$p = \frac{1}{\Delta\omega} \quad \text{and} \quad \omega_k = \frac{k}{p} \quad \text{(i.e., } \omega_k = k\Delta\omega\text{)} \quad,$$

this becomes

$$f(t) = \mathcal{F}^{-1}[F]\big|_t = \sum_{k=-\infty}^{\infty} F_k \, e^{i2\pi\omega_k t} \quad . \tag{37.3}$$

This last expression looks a lot like a Fourier series, prompting two questions:

Is f a periodic generalized function with period p ?

and, if so,

Is the series on the right side of equation (37.3) the Fourier series for f ?

To find the answer to the first question, observe that

$$T_p[f]\big|_t = T_p\left[\sum_{k=-\infty}^{\infty} F_k \, e^{i2\pi\omega_k t}\right] = \sum_{k=-\infty}^{\infty} F_k T_p\left[e^{i2\pi\omega_k t}\right] = \sum_{k=-\infty}^{\infty} F_k \, e^{i2\pi\omega_k(t-p)} \quad .$$

But, because $\omega_k = {}^{k}\!/_{p}$,

$$e^{i2\pi\omega_k(t-p)} = e^{i2\pi\omega_k t} e^{-i2\pi\omega_k p} = e^{i2\pi\omega_k t} e^{-i2\pi k} = e^{i2\pi\omega_k t} \cdot 1 = e^{i2\pi\omega_k} \quad .$$

So,

$$T_p[f]\big|_t = \sum_{k=-\infty}^{\infty} F_k \, e^{i2\pi\omega_k t} = f(t) \quad,$$

confirming that f is periodic with period p.

Since we will want to refer to these results later, let's summarize them in a lemma.

Lemma 37.4 (inverse transforms of regular arrays)
Suppose f and F are generalized functions with $F = \mathcal{F}[f]$. If F is a regular array with spacing $\Delta\omega$,

$$F = \sum_{k=-\infty}^{\infty} F_k \, \delta_{k\Delta\omega} \quad,$$

then f is periodic with period $p = {}^{1}\!/_{\Delta\omega}$ and is given by

$$f(t) = \sum_{k=-\infty}^{\infty} F_k \, e^{i2\pi k\Delta\omega t} \quad .$$

The second question — *Is the series on the right side of equation (37.3) the Fourier series for* f ? — cannot be adequately answered until we define just what is meant by a Fourier series for a nonclassical, periodic generalized function. We'll do this in the next section. Do note, however, that if

$$\sum_{k=-\infty}^{\infty} |F_k| < \infty \quad , \tag{37.4}$$

then you can easily verify that formula (37.3) defines f as a continuous, periodic function on the real line with period p and whose Fourier series is the series on the right side of equation (37.3) (the proof of this is similar to that of theorem 16.2 on page 239). There is even a remote possibility that you will recognize the resulting series as the Fourier series for some known function. On the other hand, it is quite possible that the F_k's do not shrink to zero fast enough as $k \to \pm\infty$ to ensure inequality (37.4). In fact, it is quite possible that the F_k's do not shrink to zero at all. In that case, while

$$\sum_{k=-\infty}^{\infty} F_k\, e^{i2\pi k \Delta\omega t}$$

defines a periodic generalized function, it cannot be the Fourier series for any of the classical functions we considered in part II of this text (see exercise 37.10 at the end of this chapter).

!▶ Example 37.2:   *Consider the regular array*

$$\sum_{k=-\infty}^{\infty} \frac{2}{\pi(1-4k^2)}\, \delta_{2k} \quad .$$

Taking the inverse Fourier transform of this, we get

$$\mathcal{F}^{-1}\left[\sum_{k=-\infty}^{\infty} \frac{2}{\pi(1-4k^2)}\, \delta_{2k} \right]\Bigg|_t = \sum_{k=-\infty}^{\infty} \frac{2}{\pi(1-4k^2)}\, \mathcal{F}^{-1}[\delta_{2k}]|_t$$

$$= \sum_{k=-\infty}^{\infty} \frac{2}{\pi(1-4k^2)}\, e^{i2\pi 2kt} \quad .$$

Checking back over the Fourier series computed in exercise 12.3 on page 152, we discover that

$$\mathcal{F}[|\sin(2\pi t)|] = \sum_{k=-\infty}^{\infty} \frac{2}{\pi(1-4k^2)}\, e^{i2\pi 2kt} \quad .$$

So,

$$\mathcal{F}^{-1}\left[\sum_{k=-\infty}^{\infty} \frac{2}{\pi(1-4k^2)}\, \delta_{2k} \right]\Bigg|_t = |\sin(2\pi t)| \quad .$$

!▶ Example 37.3:   *Consider the array*

$$F = \sum_{k=-\infty}^{\infty} \frac{k}{4}\, \delta_{k/2} \quad .$$

From our initial discussion of delta function arrays, we know this array is a generalized function. Taking its inverse Fourier transform, we have

$$\mathcal{F}^{-1}\left[\sum_{k=-\infty}^{\infty} \frac{k}{4}\, \delta_{k/2} \right]\Bigg|_t = \sum_{k=-\infty}^{\infty} \frac{k}{4}\, \mathcal{F}^{-1}[\delta_{k/2}]|_t = \sum_{k=-\infty}^{\infty} \frac{k}{4}\, e^{i2\pi(k/2)t} = \sum_{k=-\infty}^{\infty} \frac{k}{4}\, e^{i\pi kt} \quad .$$

As shown above, this last series defines a periodic generalized function with a period of $^1/_{(1/2)} = 2$. However, since the magnitude of the coefficients increases as $k \to \pm\infty$, this series cannot be the Fourier series for any piecewise continuous, periodic function on $\mathbb{R}$. Indeed, you can easily verify that, as an infinite series of numbers, this series is divergent for each real value t.

37.2 Fourier Series for Periodic Generalized Functions The Big Theorems

Here is what the discussion in the previous section has been leading to:

Theorem 37.5
Let p and $\Delta\omega$ be two positive real values with $p\Delta\omega = 1$, and let f and F be two generalized functions with $F = \mathcal{F}[f]$. Assume that either of the following conditions hold:

1. *f is periodic with period p.*

2. *F is a regular array with spacing $\Delta\omega$.*

Then both conditions hold, and there is a unique set of constants $\{\ldots, F_{-1}, F_0, F_1, F_2, \ldots\}$ such that

$$f(t) = \sum_{k=-\infty}^{\infty} F_k \, e^{i2\pi k \Delta\omega t} \tag{37.5}$$

and

$$F(\omega) = \sum_{k=-\infty}^{\infty} F_k \, \delta_{k\Delta\omega}(\omega) \quad . \tag{37.6}$$

Thanks to the near-equivalence of the Fourier transforms, we have the following statement as an immediate corollary.

Corollary 37.6 (relation between periodic functions and regular arrays)
Suppose f and F are two generalized functions with $F = \mathcal{F}[f]$. Then one is periodic if and only if the other is a regular array of delta functions. Moreover, the period of the periodic function and the spacing of the array are reciprocals of each other.

Part of the above theorem was proven when we proved lemma 37.4. The rest of the proof (which is somewhat lengthy and not for the mathematically faint-hearted) occupies section 37.3.

Following classical terminology, we will refer to the infinite series in equation (37.5) as the (generalized) (complex exponential) *Fourier series* for f, and we will denote this series by $F.S.[f]$. The coefficients in this series — the F_k's — will be called the *Fourier coefficients* for f, with F_k being the kth *Fourier coefficient*.

While the above theorem assures us of the existence of the Fourier coefficients, it does not tell us how to compute them. That issue is addressed in the next theorem.

Theorem 37.7 (computing Fourier coefficients)
Assume f is a periodic generalized function with period p, and let k be any integer and a be any real number. Then, letting $\omega_k = {}^k/_p$,

1. $f(t) e^{-i2\pi\omega_k t}$ is integrable (in the generalized sense) over the interval $(a, a + p)$, and

2. the kth Fourier coefficient for f is given by

$$F_k = \frac{1}{p} \int_a^{a+p} f(t) e^{-i2\pi\omega_k t} \, dt \quad .$$

PROOF: To verify the integrability of $f(t)e^{-i2\pi\omega_k t}$ over $(a, a + p)$, we must verify that

$$\text{rect}_{(-a-p,-a)}(t) * \left[f(t) e^{-i2\pi\omega_k t} \right]$$

defines a piecewise continuous, classical function continuous at 0. (This may be a good time to review "generalized integrals" as defined in section 36.2 starting on page 639.)

   First of all, because the above rectangle function is a simple convolver (see exercise 36.4 on page 638), we know the above convolution exists. Moreover, from theorem 37.5 we know

$$f(t) = \sum_{n=-\infty}^{\infty} F_n e^{i2\pi n \Delta\omega t}$$

where $\Delta\omega = {}^1/_p$ and the F_n's are the Fourier coefficients for f. Applying an obvious re-indexing, we also see that

$$f(t) e^{-i2\pi\omega_k t} = \left[\sum_{n=-\infty}^{\infty} F_n e^{i2\pi n \Delta\omega t} \right] e^{-i2\pi k \Delta\omega t}$$

$$= \sum_{n=-\infty}^{\infty} F_n e^{i2\pi(n-k)\Delta\omega t} = \sum_{m=-\infty}^{\infty} F_{m+k} e^{i2\pi m \Delta\omega t} \quad .$$

Using this and rule 12.b$'$ on convolution from chapter 36, we then obtain

$$\text{rect}_{(-a-p,-a)}(t) * \left[f(t) e^{-i2\pi\omega_k t} \right] = \text{rect}_{(-a-p,-a)}(t) * \sum_{m=-\infty}^{\infty} F_{m+k} e^{i2\pi m \Delta\omega t}$$

$$= \sum_{m=-\infty}^{\infty} F_{m+k} \, \text{rect}_{(-a-p,-a)}(t) * e^{i2\pi m \Delta\omega t} \quad . \qquad (37.7)$$

The convolution in each term of the last series is simply a classical convolution and is easily evaluated. For $m = 0$,

$$\text{rect}_{(-a-p,-a)}(t) * e^{i2\pi m \Delta\omega t} = \text{rect}_{(-a-p,-a)}(t) * 1 = \int_{-a-p}^{-a} 1 \, ds = p \quad .$$

On the other hand, if m is any other integer, then, because $p\Delta\omega = 1$ and $e^{i2\pi m} = 1$,

$$\text{rect}_{(-a-p,-a)}(t) * e^{i2\pi m \Delta\omega t} = \int_{-a-p}^{-a} e^{i2\pi m \Delta\omega (t-s)} \, ds$$

$$= \frac{i}{2\pi m \Delta\omega} \left[e^{i2\pi m \Delta\omega(t+a)} - e^{i2\pi m \Delta\omega(t+a+p)} \right]$$

$$= \frac{i}{2\pi m \Delta\omega} e^{i2\pi m \Delta\omega(t+a)} \left[1 - e^{i2\pi m \Delta\omega p} \right] = 0 \quad .$$

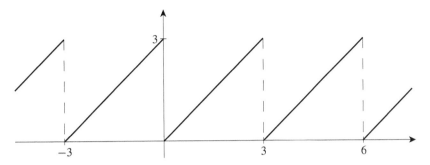

Figure 37.2: The graph of saw_3.

Thus, equation (37.7) reduces to

$$\text{rect}_{(-a-p,-a)}(t) * \left[f(t)\, e^{-i2\pi\omega_k t} \right] \;=\; \sum_{m=-\infty}^{\infty} F_{m+k} \begin{cases} p & \text{if} \quad m = 0 \\ 0 & \text{otherwise} \end{cases} \;=\; p F_k \quad .$$

Clearly then, the convolution on the left is a piecewise continuous function continuous at 0. In fact, it is the constant $p F_k$. So $f(t)\, e^{-i2\pi\omega_k t}$ is integrable on $(a, a+p)$, and

$$\frac{1}{p}\int_a^{a+p} f(t)\, e^{-i2\pi\omega_k t}\, dt \;=\; \frac{1}{p}\, \text{rect}_{(-a-p,-a)}(t) * \left[f(t)\, e^{-i2\pi\omega_k t} \right]\Big|_0 \;=\; F_k \quad . \qquad \blacksquare$$

Computing Fourier Series and Transforms of Periodic Functions

As long as f is any periodic generalized function with period p, the two theorems just discussed assure us that

$$f(t) \;=\; \sum_{k=-\infty}^{\infty} F_k\, e^{i2\pi\omega_k t}$$

where, using any convenient real value a,

$$\omega_k \;=\; \frac{k}{p} \qquad \text{and} \qquad F_k \;=\; \frac{1}{p}\int_a^{a+p} f(t)\, e^{-i2\pi\omega_k t}\, dt \quad .$$

The Fourier transforms of f can then be found by simply computing, term by term, the transforms of the function's Fourier series. We will illustrate this by finding the Fourier series and transforms of two periodic generalized functions — one a periodic, classical function and one a periodic, nonclassical generalized function. Then we will comment briefly on difficulties that may be encountered in computing the Fourier coefficients for certain generalized functions.

A Saw Function

Consider the classical saw function sketched in figure 37.2,

$$\text{saw}_3(t) \;=\; \begin{cases} t & \text{if} \quad 0 < t < 3 \\ \text{saw}_3(t-3) & \text{in general} \end{cases} \quad .$$

This is a piecewise continuous, periodic function with period $p = 3$. As just noted, our two big theorems on periodic generalized functions tell us that, in the generalized sense,

$$\text{saw}_3(t) = \sum_{k=-\infty}^{\infty} F_k \, e^{i2\pi\omega_k t} \tag{37.8}$$

where, using any convenient real value a,

$$\omega_k = \frac{k}{3} \quad \text{and} \quad F_k = \frac{1}{3}\int_a^{a+3} \text{saw}_3(t)\, e^{-i2\pi\omega_k t}\, dt \quad .$$

Choosing $a = 0$, the formula for the Fourier coefficients becomes

$$F_k = \frac{1}{3}\int_0^3 t\, e^{-i2\pi\omega_k t}\, dt \quad ,$$

which was evaluated when we computed the classical Fourier series of this saw function in example 12.1 on page 148. There we found that

$$c_0 = \frac{1}{3}\int_0^3 t\, e^{-i2\pi\cdot 0\cdot t}\, dt = \frac{3}{2}$$

and

$$F_k = \frac{1}{3}\int_0^3 t\, e^{-i2\pi\omega_k t}\, dt = \frac{3i}{2\pi k} \quad \text{for} \quad k = \pm 1,\ \pm 2,\ \pm 3,\ \dots \quad .$$

Plugging the above values for the ω_k's and F_k's into expression (37.8) yields

$$\text{saw}_3(t) = \frac{3}{2} + \sum_{\substack{k=-\infty \\ k\neq 0}}^{\infty} \frac{3i}{2\pi k} e^{i2\pi(k/3)t} \quad .$$

Thus,

$$\mathcal{F}[\text{saw}_3] = \mathcal{F}\left[\frac{3}{2} + \sum_{\substack{k=-\infty \\ k\neq 0}}^{\infty} \frac{3i}{2\pi k} e^{i2\pi(k/3)t} \right]$$

$$= \frac{3}{2}\mathcal{F}[1] + \sum_{\substack{k=-\infty \\ k\neq 0}}^{\infty} \frac{3i}{2\pi k}\mathcal{F}\left[e^{i2\pi(k/3)t} \right] = \frac{3}{2}\delta + \sum_{\substack{k=-\infty \\ k\neq 0}}^{\infty} \frac{3i}{2\pi k}\delta_{k/3} \quad .$$

Likewise,

$$\mathcal{F}^{-1}[\text{saw}_3] = \mathcal{F}^{-1}\left[\frac{3}{2} + \sum_{\substack{k=-\infty \\ k\neq 0}}^{\infty} \frac{3i}{2\pi k} e^{i2\pi(k/3)t} \right]$$

$$= \frac{3}{2}\mathcal{F}^{-1}[1] + \sum_{\substack{k=-\infty \\ k\neq 0}}^{\infty} \frac{3i}{2\pi k}\mathcal{F}^{-1}\left[e^{i2\pi(k/3)t} \right] = \frac{3}{2}\delta + \sum_{\substack{k=-\infty \\ k\neq 0}}^{\infty} \frac{3i}{2\pi k}\delta_{-k/3} \quad .$$

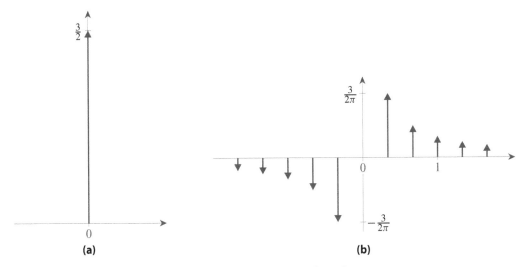

Figure 37.3: The (a) real part and (b) imaginary part of $\mathcal{F}[\text{saw}_3]$.

In summary:

$$F.S.\,[\text{saw}_3]|_t \;=\; \frac{3}{2} \;+\; \sum_{\substack{k=-\infty \\ k\neq 0}}^{\infty} \frac{3i}{2\pi k} e^{i2\pi(k/3)t} \quad,$$

$$\mathcal{F}[\text{saw}_3] \;=\; \frac{3}{2}\delta \;+\; \sum_{\substack{k=-\infty \\ k\neq 0}}^{\infty} \frac{3i}{2\pi k}\delta_{k/3}$$

and (after letting $n = -k$)

$$\mathcal{F}^{-1}[\text{saw}_3] \;=\; \frac{3}{2}\delta \;-\; \sum_{\substack{n=-\infty \\ n\neq 0}}^{\infty} \frac{3i}{2\pi n}\delta_{n/3} \quad.$$

The graphical representation for $\mathcal{F}[\text{saw}_3]$ has been sketched in figure 37.3.

The Comb Functions

In example 37.1 on page 650, we saw that any comb function is periodic with its period equaling its spacing. So let p and $\Delta\omega$ be any two positive numbers with $\Delta\omega = \frac{1}{p}$, and let us consider

$$\text{comb}_p \;=\; \sum_{n=-\infty}^{\infty} \delta_{np} \quad.$$

A few observations are in order before starting our computations: The first being that any comb function is both periodic and is a regular array of delta functions. But, according to the corollary on the relation between periodic functions and regular arrays (corollary 37.6),

comb_p *is periodic with period* $p \quad\Longrightarrow\quad \mathcal{F}[\text{comb}_p]$ *is a regular array with spacing* $\Delta\omega$

and

comb_p *is a regular array with spacing* $p \quad\Longrightarrow\quad \mathcal{F}[\text{comb}_p]$ *is periodic with period* $\Delta\omega \quad.$

So $\mathcal{F}\left[\text{comb}_p\right]$ is also both periodic and a regular array of delta functions, with the spacing and the period both being $\Delta\omega$. If you think about this a moment, you'll realize this must mean

$$\mathcal{F}\left[\text{comb}_p\right] = C \, \text{comb}_{\Delta\omega}$$

for some constant C. We'll verify this (and evaluate C) in our computations.

Since comb_p is periodic with period p, it can be written as

$$\text{comb}_p(t) = \sum_{k=-\infty}^{\infty} c_k \, e^{i2\pi\omega_k t}$$

where, using any real value a,

$$\omega_k = \frac{k}{p} \quad \text{and} \quad c_k = \frac{1}{p} \int_a^{a+p} \text{comb}_p(t) \, e^{-i2\pi\omega_k t} \, dt \quad .$$

Since integrals involving delta functions at the endpoints of the intervals of integration are somewhat problematic, it is not convenient to choose a to be 0 or any other integral multiple of p. Instead, let us choose a so that $(a, a+p)$ contains the origin, say, $a = -p/2$ (and hence, $a + p = p/2$). Let's also note that, for every pair of integers k and n,

$$\delta_{np}(t) \, e^{-i2\pi\omega_k t} = e^{-i2\pi\omega_k np} \, \delta_{np}(t) = e^{-i2\pi kn} \, \delta_{np}(t) = 1 \cdot \delta_{np}(t) \quad ,$$

and

$$\int_{-p/2}^{p/2} \delta_{np}(t) \, dt = \begin{cases} 1 & \text{if} \quad -p/2 < np < p/2 \\ 0 & \text{otherwise} \end{cases} = \begin{cases} 1 & \text{if} \quad n = 0 \\ 0 & \text{otherwise} \end{cases} \quad .$$

So, for each integer k,

$$\text{comb}_p(t) \, e^{-i2\pi\omega_k t} = \sum_{n=-\infty}^{\infty} \delta_{np}(t) \, e^{-i2\pi\omega_k t} = \sum_{n=-\infty}^{\infty} \delta_{np}(t) \quad ,$$

and

$$\int_a^{a+p} \text{comb}_p(t) \, e^{-i2\pi\omega_k t} \, dt = \int_{-p/2}^{p/2} \sum_{n=-\infty}^{\infty} \delta_{np}(t) \, dt$$

$$= \sum_{n=-\infty}^{\infty} \int_{-p/2}^{p/2} \delta_{np}(t) \, dt$$

$$= \sum_{n=-\infty}^{\infty} \begin{cases} 1 & \text{if} \quad n = 0 \\ 0 & \text{otherwise} \end{cases} = 1 \quad .$$

Consequently,

$$c_k = \frac{1}{p} \int_a^{a+p} \text{comb}_p(t) \, e^{-i2\pi\omega_k t} \, dt = \frac{1}{p} \quad \text{for} \quad k = 0, \pm 1, \pm 2, \pm 3, \dots \quad ,$$

and

$$\text{comb}_p(t) = \sum_{k=-\infty}^{\infty} c_k \, e^{i2\pi\omega_k t} = \sum_{k=-\infty}^{\infty} \frac{1}{p} e^{i2\pi(k/p)t} \quad .$$

Therefore,

$$
\mathcal{F}\big[\mathrm{comb}_p(t)\big] = \mathcal{F}\left[\sum_{k=-\infty}^{\infty} \frac{1}{p} e^{i2\pi(k/p)t} \right]
$$

$$
= \sum_{k=-\infty}^{\infty} \frac{1}{p} \mathcal{F}\left[e^{i2\pi(k/p)t} \right]
$$

$$
= \frac{1}{p} \sum_{k=-\infty}^{\infty} \delta_{k/p} = \frac{1}{p} \mathrm{comb}_{1/p} \quad .
$$

To summarize: Letting $\Delta\omega = {}^1\!/_p$,

$$
F.S.\big[\mathrm{comb}_p\big]\big|_t = \sum_{k=-\infty}^{\infty} \Delta\omega\, e^{i2\pi k\Delta\omega t} \qquad \text{and} \qquad \mathcal{F}\big[\mathrm{comb}_p\big] = \Delta\omega\, \mathrm{comb}_{\Delta\omega} \quad .
$$

In particular, with $p = 1$ we get the mildly interesting results that

$$
F.S.\,[\mathrm{comb}_1]\big|_t = \sum_{k=-\infty}^{\infty} e^{i2\pi kt} \qquad \text{and} \qquad \mathcal{F}[\mathrm{comb}_1] = \mathrm{comb}_1 \quad .
$$

?►Exercise 37.3: *Show that*

$$
\mathcal{F}^{-1}\big[\mathrm{comb}_p\big] = \frac{1}{p} \mathrm{comb}_{1/p} \quad .
$$

A Possible Difficulty in Computing Fourier Coefficients

Theorem 37.7 assures us that

$$
\int_{a}^{a+p} f(t)\, e^{-i2\pi\omega_k t}\, dt
$$

exists whenever f is a periodic generalized function with period p, a is a real value, and $\omega_k = {}^k\!/_p$ for some integer k. As we just saw in computing the Fourier coefficients for a comb function, however, there may be values of a for which the computation of this integral is "problematic". For the comb function, this was not much of a difficulty; there were many other values of a for which our integral was easily evaluated. However, as the next example shows, it is possible to have an f such that the computation of this integral is "problematic" no matter what real value we choose for a. The example also illustrates one alternative method for computing Fourier coefficients for some of these functions. Another method, based on computations used in proving theorem 37.5, is given in exercise 37.15 at the end of this chapter. Fortunately, such exotic generalized functions do not often arise in everyday applications.

!►Example 37.4: *Consider the comb function with spacing* 1 *shifted by* i,

$$
T_i[\mathrm{comb}_1] = \sum_{n=-\infty}^{\infty} \delta_{i+n} \quad .
$$

After recalling the properties of the translation operator and the fact that comb_1 *is periodic with period* 1, *we see that*

$$
T_1[T_i[\mathrm{comb}_1]] = T_i[T_1[\mathrm{comb}_1]] = T_i[\mathrm{comb}_1] \quad ,
$$

confirming that this generalized function is periodic with period $p = 1$. Thus,

$$F.S.\big[T_i[\text{comb}_1]\big]\big|_t \;=\; \sum_{k=-\infty}^{\infty} c_k \, e^{i2\pi kt}$$

where the c_k's are the Fourier coefficients for this shifted comb function. However, when we try to compute these coefficients, we get

$$c_k \;=\; \int_a^{a+1} \sum_{n=-\infty}^{\infty} \delta_{i+n}(t) \, e^{-i2\pi kt} \, dt$$

$$=\; \int_a^{a+1} \sum_{n=-\infty}^{\infty} \underbrace{e^{-i2\pi k(i+n)}}_{e^{2\pi k}e^{-i2\pi nk}} \delta_{i+n}(t)\,dt \;=\; e^{2\pi k}\int_a^{a+1} \sum_{n=-\infty}^{\infty} \delta_{i+n}(t)\,dt \quad.$$

This is an integral of an array of delta functions located off the real axis. Unfortunately for us, we've only discussed computing integrals of arrays of delta functions on the real axis.

On the other hand, we don't really need to compute these c_k's directly, because in the previous example, we discovered that

$$\text{comb}_1(t) \;=\; \sum_{k=-\infty}^{\infty} e^{i2\pi kt} \quad.$$

So,

$$T_i[\text{comb}_1(t)] \;=\; T_i\!\left[\sum_{k=-\infty}^{\infty} e^{i2\pi kt}\right]$$

$$=\; \sum_{k=-\infty}^{\infty} T_i\!\left[e^{i2\pi kt}\right]$$

$$=\; \sum_{k=-\infty}^{\infty} e^{i2\pi k(t-i)} \;=\; \sum_{k=-\infty}^{\infty} e^{2\pi k}\, e^{i2\pi kt} \quad.$$

According to theorem 37.5, this last series must be the Fourier series for $T_i[\text{comb}_1(t)]$. Thus, the kth Fourier coefficient for $T_i[\text{comb}_1(t)]$ is

$$c_k \;=\; e^{2\pi k} \quad,$$

and

$$\mathcal{F}\big[T_i[\text{comb}_1]\big] \;=\; \mathcal{F}\!\left[\sum_{k=-\infty}^{\infty} e^{2\pi k}\, e^{i2\pi kt}\right]$$

$$=\; \sum_{k=-\infty}^{\infty} e^{2\pi k}\, \mathcal{F}\!\left[e^{i2\pi kt}\right] \;=\; \sum_{k=-\infty}^{\infty} e^{2\pi k}\delta_k \quad.$$

Do note that this last expression is a regular array of delta functions, but is not periodic.

?▶Exercise 37.4:   *Sketch the graphical representation of $\mathcal{F}[T_i[\text{comb}_1]]$, computed above.*

?►Exercise 37.5:   *Using the results of the last example, show that, for each real value* a *,*

$$\int_a^{a+1} \sum_{n=-\infty}^{\infty} \delta_{i+n}(t) \, dt = 1 \quad .$$

?►Exercise 37.6:   *Find the inverse Fourier transform of* $T_i[\mathrm{comb}_1(t)]$ *.*

37.3 On Proving Theorem 37.5[*]
Preliminaries

The hard part of proving theorem 37.5 is proving that the transform of a periodic function is a regular array. Attempting to prove this all at once can be a bit overwhelming, so we will break it down to a series of lemmas.

The first, a corollary of lemmas 33.3 and 33.4 (see page 556), describes some computations that will be needed at various points in subsequent lemmas. Its confirmation is an exercise.

Lemma 37.8
Assume f *is some generalized function and* η *is any simple multiplier for* $\mathcal{G}$ *. Let* a *,* b *and* c *be any three constants. Then*

$$H(\lambda, \gamma, v) = \left\langle f(x) \,,\, e^{c\lambda} e^{-\gamma\pi(x-av-b)^2} \eta(x) \right\rangle$$

is an infinitely smooth function of the real variables λ *and* v *, and the positive variable* γ *. Moreover, for any positive integer* n *,*

$$\frac{\partial^n}{\partial\lambda^n} H(\lambda, \gamma, v) = \left\langle f(x) \,,\, \frac{\partial^n}{\partial\lambda^n} e^{c\lambda} e^{-\gamma\pi(x-av-b)^2} \eta(x) \right\rangle \quad ,$$

$$\frac{\partial^n}{\partial\gamma^n} H(\lambda, \gamma, v) = \left\langle f(x) \,,\, \frac{\partial^n}{\partial\gamma^n} e^{c\lambda} e^{-\gamma\pi(x-av-b)^2} \eta(x) \right\rangle$$

and

$$\frac{\partial^n}{\partial v^n} H(\lambda, \gamma, v) = \left\langle f(x) \,,\, \frac{\partial^n}{\partial v^n} e^{c\lambda} e^{-\gamma\pi(x-av-b)^2} \eta(x) \right\rangle \quad .$$

?►Exercise 37.7:   *Derive the above lemma from lemmas 33.3 and 33.4.*

The rest of the lemmas concern a classical function h constructed from a periodic generalized function f and an arbitrary Gaussian test function ϕ via

$$h(t) = \left\langle f(x) \,,\, \phi(x - t) \right\rangle \quad .$$

Lemma 37.9
Suppose f *is a periodic generalized function with period* p *, and* ϕ *is a Gaussian test function. Define the function* h *by*

$$h(t) = \left\langle f(x) \,,\, \phi(x - t) \right\rangle \quad .$$

Then h *is a smooth and classically periodic function on* $\mathbb{R}$ *with period* p *.*

---

[*] Warning: This material is only for the most dedicated of readers.

PROOF: That h is a smooth function on $\mathbb{R}$ was verified in exercise 35.50 on page 626. To confirm the claimed periodicity, simply use the periodicity of f with the definition of generalized translation:

$$
\begin{aligned}
h(t - p) &= \big\langle\, f(x)\,,\, \phi(x - (t - p))\,\big\rangle \\
&= \big\langle\, f(x)\,,\, \phi(x - t + p)\,\big\rangle \\
&= \big\langle\, f(x - p)\,,\, \phi(x - t)\,\big\rangle \\
&= \big\langle\, f(x)\,,\, \phi(x - t)\,\big\rangle = h(t) \quad .
\end{aligned}
$$

∎

Because h is a smooth and classically periodic function on the real line, we can express h as a classical Fourier series. By our assumptions, this means that, for each real t,

$$
h(t) = \sum_{k=-\infty}^{\infty} c_k\, e^{i2\pi\omega_k t}
$$

where, for each integer k,

$$
\omega_k = \frac{k}{p} \qquad \text{and} \qquad c_k = \frac{1}{p}\int_0^p h(t)\, e^{-i2\pi\omega_k t}\, dt \quad .
$$

With $h(t)$ replaced by its formula in terms of f, the formula for the c_k's becomes

$$
c_k = \frac{1}{p}\int_0^p \big\langle\, f(x)\,,\, \phi(x - t)\,\big\rangle\, e^{-i2\pi\omega_k t}\, dt \quad .
$$

The computation of these coefficients is addressed in the next two lemmas. Since the computations can be a little messy, and since we can always rescale our functions later, let us save ourselves a little labor here by assuming the period of f is 1. That way, the last equation simplifies to

$$
c_k = \int_0^1 \big\langle\, f(x)\,,\, \phi(x - t)\,\big\rangle\, e^{-i2\pi k t}\, dt \quad .
$$

Lemma 37.10

Assume f is a periodic generalized function with period 1. For each integer k, let Γ_k be defined by

$$
\big\langle\, \Gamma_k\,,\, \phi\,\big\rangle = \int_0^1 \big\langle\, f(x)\,,\, \phi(x - t)\,\big\rangle\, e^{-i2\pi k t}\, dt \qquad \text{for each ϕ in $\mathcal{G}$} \quad ,
$$

and let $\widehat{F}_k(\gamma)$ be defined by

$$
\widehat{F}_k(\gamma) = \sqrt{\gamma}\, \big\langle\, \Gamma_k(x)\,,\, e^{-\gamma\pi x^2}\,\big\rangle \exp\!\left(\frac{\pi}{\gamma}k^2\right) \qquad \text{for each $\gamma > 0$} \quad .
$$

Then:

1. *Each Γ_k is a well-defined generalized function.*

2. *Each $\widehat{F}_k$ is a smooth function on $(0, \infty)$.*

3. *For every complex value ξ and positive value γ,*

$$
\big\langle\, \Gamma_k(x)\,,\, e^{-\gamma\pi(x-\xi)^2}\,\big\rangle = \big\langle\, \widehat{F}_k(\gamma)\, e^{i2\pi k x}\,,\, e^{-\gamma\pi(x-\xi)^2}\,\big\rangle \quad .
$$

PROOF: The first claim — that each Γ_k is a generalized function — was verified more generally in exercise 35.51 on page 626. To verify the other claims, let k be any fixed integer, and, for each triple of real numbers (α, β, γ) with $\gamma > 0$, let $G_k(\alpha, \beta, \gamma)$ be given by

$$G_k(\alpha, \beta, \gamma) = \sqrt{\gamma}\left\langle \Gamma_k(x), e^{-\gamma\pi(x-\alpha-i\beta)^2} \right\rangle \exp\left(\frac{\pi}{\gamma}k^2 - i2\pi k(\alpha + i\beta)\right) \quad . \tag{37.9}$$

From lemma 37.8 and our knowledge of exponentials, we know $G_k(\alpha, \beta, \gamma)$ is a smooth function of each of these variables. Observe that, by this definition

$$\left\langle \Gamma_k(x), e^{-\gamma\pi(x-\alpha-i\beta)^2} \right\rangle = G_k(\alpha, \beta, \gamma)\frac{1}{\sqrt{\gamma}}\exp\left(-\frac{\pi}{\gamma}k^2 + i2\pi k(\alpha + i\beta)\right)$$

$$= G_k(\alpha, \beta, \gamma)\frac{1}{\sqrt{\gamma}}e^{i2\pi k(\alpha+i\beta)}\exp\left(-\frac{\pi}{\gamma}k^2\right) \quad .$$

But, letting $\xi = \alpha + i\beta$,

$$G_k(\alpha, \beta, \gamma)\frac{1}{\sqrt{\gamma}}e^{i2\pi k\xi}\exp\left(-\frac{\pi}{\gamma}k^2\right) = G_k(\alpha, \beta, \gamma)\mathcal{F}^{-1}\left[e^{-\gamma\pi(x-\xi)^2}\right]\Big|_k$$

$$= G_k(\alpha, \beta, \gamma)\int_{-\infty}^{\infty} e^{-\gamma\pi(x-\xi)^2}e^{i2\pi kx}\,dx$$

$$= \int_{-\infty}^{\infty} G_k(\alpha, \beta, \gamma)e^{i2\pi kx}e^{-\gamma\pi(x-\xi)^2}\,dx$$

$$= \left\langle G_k(\alpha, \beta, \gamma)e^{i2\pi kx}, e^{-\gamma\pi(x-\xi)^2} \right\rangle \quad .$$

So the preceding relation between Γ_k and G_k can be written as

$$\left\langle \Gamma_k(x), e^{-\gamma\pi(x-\alpha-i\beta)^2} \right\rangle = \left\langle G_k(\alpha, \beta, \gamma)e^{i2\pi kx}, e^{-\gamma\pi(x-\alpha-i\beta)^2} \right\rangle \quad . \tag{37.10}$$

Suppose we can now show

$$\frac{\partial}{\partial\alpha}G_k(\alpha, \beta, \gamma) = 0 \quad \text{and} \quad \frac{\partial}{\partial\beta}G_k(\alpha, \beta, \gamma) = 0 \quad . \tag{37.11}$$

Then $G_k(\alpha, \beta, \gamma)$ depends only on γ. Consequently, for all real values α and β,

$$G_k(\alpha, \beta, \gamma) = G_k(0, 0, \gamma) \quad .$$

After using formula (37.9) to compute $G_k(0, 0, \gamma)$ and recalling the definition of $\widehat{F}_k$, we find that

$$G_k(\alpha, \beta, \gamma) = \sqrt{\gamma}\left\langle \Gamma_k(x), e^{-\gamma\pi x^2} \right\rangle \exp\left(\frac{\pi}{\gamma}k^2\right) = \widehat{F}_k(\gamma) \quad .$$

From this and the known smoothness of G_k, the second claim of the lemma would immediately follow. Furthermore, since $\widehat{F}_k(\gamma) = G_k(\alpha, \beta, \gamma)$, equation (37.10) would immediately reduce to the equation in the third claim of this lemma.

Consequently, we can finish this proof by showing that the partial derivatives of G_k with respect to α and β vanish. Doing this is relatively simple. First of all, thanks to lemma 37.8, it is a simple exercise (exercise 37.8, to be precise) to show that

$$\frac{\partial G_k}{\partial\beta} = i\frac{\partial G_k}{\partial\alpha} \quad .$$

Thus, we only need to verify the vanishing of one of the partial derivatives.

Computing the partial with respect to α is especially easy after looking a little more closely at the factor in the formula for G_k involving Γ_k. By its definition and an obvious substitution,

$$\left\langle \Gamma_k(x) , e^{-\gamma(x-\alpha-i\beta)^2} \right\rangle = \int_0^1 \left\langle f(x) , e^{-\gamma\pi(x-t-\alpha-i\beta)^2} \right\rangle e^{-i2\pi kt} \, dt$$

$$= \int_\alpha^{\alpha+1} \left\langle f(x) , e^{-\gamma\pi(x-\tau-i\beta)^2} \right\rangle e^{-i2\pi k(\tau-\alpha)} \, d\tau$$

$$= e^{i2\pi k\alpha} \int_\alpha^{\alpha+1} \left\langle f(x) , e^{-\gamma\pi(x-\tau-i\beta)^2} \right\rangle e^{-i2\pi k\tau} \, d\tau \quad .$$

However, from lemma 37.9, we know the integrand in the last integral is a smooth, periodic function of τ with period 1. So this last integral can be evaluated using any interval of length 1, say, $(0, 1)$ (see lemma 5.1 on page 53). Using this interval, our last set of equalities reduces to

$$\left\langle \Gamma_k(x) , e^{-\gamma(x-\alpha-i\beta)^2} \right\rangle = e^{i2\pi k\alpha} H_k(\beta, \gamma)$$

where

$$H_k(\beta, \gamma) = \int_0^1 \left\langle f(x) , e^{-\gamma\pi(x-\tau-i\beta)^2} \right\rangle e^{i2\pi k\tau} \, d\tau \quad .$$

So,

$$G_k(\alpha, \beta, \gamma) = \sqrt{\gamma} \left\langle \Gamma_k(x) , e^{-\gamma\pi(x-\alpha-i\beta)^2} \right\rangle \exp\left(\frac{\pi}{\gamma}k^2 - i2\pi k(\alpha + i\beta)\right)$$

$$= \sqrt{\gamma} e^{i2\pi k\alpha} H_k(\beta, \gamma) \exp\left(\frac{\pi}{\gamma}k^2\right) e^{-i2\pi k(\alpha+i\beta)}$$

$$= \sqrt{\gamma} H_k(\beta, \gamma) \exp\left(\frac{\pi}{\gamma}k^2\right) e^{2\pi k\beta} \quad .$$

Since this last expression is completely independent of α, we clearly have

$$\frac{\partial G_k}{\partial \alpha} = \frac{\partial}{\partial \alpha}\left[\sqrt{\gamma} H_k(\beta, \gamma) \exp\left(\frac{\pi}{\gamma}k^2\right) e^{2\pi k\beta}\right] = 0 \quad . \qquad \blacksquare$$

?►Exercise 37.8:   *Using lemma 37.8, verify equation (37.10) is the above proof.*

The next lemma continues the previous lemma, and presents an even simpler formula for the generalized function Γ_k than described above.

Lemma 37.11
Assume f is a periodic generalized function with period 1. For each integer k, let Γ_k be the generalized function given by

$$\left\langle \Gamma_k , \phi \right\rangle = \int_0^1 \left\langle f(x), \phi(x - t) \right\rangle e^{i2\pi kt} \, dt \qquad \text{for each } \phi \text{ in } \mathcal{G} \quad ,$$

and let F_k be the constant

$$F_k = \left\langle \Gamma_k(x) , e^{-\pi x^2} \right\rangle e^{\pi k^2} \quad .$$

Then

$$\left\langle \Gamma_k(x) , \phi(x) \right\rangle = \left\langle F_k \, e^{i2\pi kx} , \phi(x) \right\rangle \qquad \text{for each } \phi \text{ in } \mathcal{G} \quad .$$

PROOF: Calling upon lemma 33.7 on page 557 again, we know it suffices to show that

$$\left\langle \Gamma_k(x) \, , \, e^{-\gamma\pi(x-\xi)^2} \right\rangle = \left\langle F_k \, e^{i2\pi kx} \, , \, e^{-\gamma\pi(x-\xi)^2} \right\rangle$$

for every complex value ξ, positive value γ and integer k.

So assume k is some integer, γ is some positive value and ξ is some complex number. Lemma 37.10 tells us that

$$\left\langle \Gamma_k(x) \, , \, e^{-\gamma\pi(x-\xi)^2} \right\rangle = \left\langle \widehat{F}_k(\gamma) \, e^{i2\pi kx} \, , \, e^{-\gamma\pi(x-\xi)^2} \right\rangle$$

where

$$\widehat{F}_k(\gamma) = \sqrt{\gamma} \left\langle \Gamma_k(x) \, , \, e^{-\gamma\pi x^2} \right\rangle \exp\left(\frac{\pi}{\gamma}k^2\right)$$

$$= \sqrt{\gamma} \int_0^1 \left\langle f(x) \, , \, e^{-\gamma\pi(x-t)^2} \right\rangle e^{-i2\pi kt} \, dt \quad .$$

Clearly then, all that needs to be shown is that $\widehat{F}_k(\gamma)$ is a constant, and, to show this, it suffices to verify that the derivative of $\widehat{F}_k$ vanishes on $(0, \infty)$.

For convenience, let

$$g(\gamma, t) = \sqrt{\gamma} \left\langle f(x) \, , \, e^{-\gamma\pi(x-t)^2} \right\rangle \quad \text{and} \quad \psi(\gamma) = \int_0^1 g(\gamma, t) \, e^{-i2\pi kt} \, dt \quad ,$$

so that

$$\widehat{F}_k(\gamma) = \exp\left(\frac{\pi}{\gamma}k^2\right) \psi(\gamma) \quad .$$

As lemma 37.9 pointed out, for each $\gamma > 0$, $g(\gamma, t)$ is a smooth, periodic function on $\mathbb{R}$ with period 1. This allows us to invoke well-known classical results, obtaining

$$\widehat{F}_k'(\gamma) = \frac{d}{d\gamma}\left[\exp\left(\frac{\pi}{\gamma}k^2\right) \psi(\gamma)\right] = \exp\left(\frac{\pi}{\gamma}k^2\right)\left[-\frac{\pi k^2}{\gamma^2}\psi(\gamma) + \psi'(\gamma)\right]$$

with

$$\psi'(\gamma) = \frac{d}{d\gamma}\int_0^1 g(\gamma, t) \, e^{-i2\pi kt} \, dt = \int_0^1 \frac{\partial}{\partial\gamma} g(\gamma, t) \, e^{-i2\pi kt} \, dt \quad . \tag{37.12}$$

Hence, to confirm that $\widehat{F}_k' = 0$ (thereby completing this proof), we need only show that

$$\psi'(\gamma) = \frac{\pi k^2}{\gamma^2}\psi(\gamma) \quad .$$

An alternate formula for g will simplify further computations. Letting $F = \mathcal{F}[f]$, and applying the generalized definition of $\mathcal{F}^{-1}[F]$ along with a translation identity, we see that

$$g(\gamma, t) = \sqrt{\gamma} \left\langle \mathcal{F}^{-1}[F]|_x \, , \, e^{-\gamma\pi(x-t)^2} \right\rangle$$

$$= \sqrt{\gamma} \left\langle F(y) \, , \, \mathcal{F}^{-1}\left[e^{-\gamma\pi(x-t)^2}\right]\Big|_y \right\rangle$$

$$= \sqrt{\gamma} \left\langle F(y) \, , \, \frac{1}{\sqrt{\gamma}} e^{i2\pi ty} \exp\left(-\frac{\pi}{\gamma}y^2\right) \right\rangle = \left\langle F(y) \, , \, e^{i2\pi ty} \exp\left(-\frac{\pi}{\gamma}y^2\right) \right\rangle \quad .$$

Using this, lemma 37.8 and the chain rule, we find that

$$\frac{\partial}{\partial \gamma} g(\gamma, t) = \left\langle F(y), \frac{\partial}{\partial \gamma} e^{i2\pi ty} \exp\left(-\frac{\pi}{\gamma} y^2\right) \right\rangle$$

$$= \frac{\pi}{\gamma^2} \left\langle F(y), y^2 e^{i2\pi ty} \exp\left(-\frac{\pi}{\gamma} y^2\right) \right\rangle \quad , \tag{37.13}$$

while

$$\frac{\partial^2}{\partial t^2} g(\gamma, t) = \left\langle F(y), \frac{\partial^2}{\partial t^2} e^{i2\pi ty} \exp\left(-\frac{\pi}{\gamma} y^2\right) \right\rangle$$

$$= -4\pi^2 \left\langle F(y), y^2 e^{2\pi ty} \exp\left(-\frac{\pi}{\gamma} y^2\right) \right\rangle \quad . \tag{37.14}$$

Comparing equations (37.13) and (37.14), above, we discover that

$$\frac{\partial}{\partial \gamma} g(\gamma, t) = \frac{\pi}{\gamma^2} \left[\frac{-1}{4\pi^2} \frac{\partial^2}{\partial t^2} g(\gamma, t) \right] = \frac{-1}{4\pi\gamma^2} \frac{\partial^2}{\partial t^2} g(\gamma, t) \quad .$$

Using this to replace the partial derivative in equation (37.12) yields

$$\psi'(\gamma) = \frac{-1}{4\pi\gamma^2} \int_0^1 \left[\frac{\partial^2}{\partial t^2} g(\gamma, t) \right] e^{-i2\pi kt} \, dt \quad .$$

Integrating this by parts twice, using the periodicity of $g(\gamma, t)$, and, finally, applying the formula for ψ gives

$$\psi'(\gamma) = \frac{-1}{4\pi\gamma^2} \left[e^{i2\pi kt} \frac{\partial}{\partial \gamma} g(\gamma, t) \Big|_{t=0}^1 + i2\pi k e^{-i2\pi kt} g(\gamma, t) \Big|_{t=0}^1 \right.$$

$$\left. - 4\pi^2 k^2 \int_0^1 g(\gamma, t) e^{-i2\pi kt} \, dt \right]$$

$$= \frac{-1}{4\pi\gamma^2} \left[0 + 0 - 4\pi^2 k^2 \int_0^1 g(\gamma, t) e^{-i2\pi kt} \, dt \right]$$

$$= \frac{\pi k^2}{\gamma^2} \int_0^1 g(\gamma, t) e^{-i2\pi kt} \, dt = \frac{\pi k^2}{\gamma^2} \psi(\gamma) \quad . \qquad \blacksquare$$

Lemma 37.12

Let f be a periodic generalized function with period 1. There are then constants $\ldots, F_{-1}, F_0, F_1, F_2, \ldots$ such that

$$f(x) = \sum_{k=-\infty}^{\infty} F_k e^{i2\pi kx} \qquad \text{and} \qquad \mathcal{F}[f] = \sum_{k=-\infty}^{\infty} F_k \delta_k \quad .$$

PROOF: Consider

$$h(t) = \left\langle f(x), \phi(x - t) \right\rangle$$

where ϕ is any given Gaussian test function. In lemma 37.9, we learned that h is a smooth, periodic function with period 1. From the classical theory of Fourier series (see theorem 13.1 on page 156 and theorem 13.7 on page 164), we then know that, for each real value t,

$$h(t) = \sum_{k=-\infty}^{\infty} c_k e^{i2\pi kt} \qquad \text{where} \qquad c_k = \int_0^1 h(t) e^{-i2\pi kt} \, dt \quad .$$

In particular then,

$$\langle f(x), \phi(x) \rangle = h(0) = \sum_{k=-\infty}^{\infty} c_k e^{i2\pi k \cdot 0} = \lim_{\substack{N \to \infty \\ M \to -\infty}} \sum_{k=M}^{N} c_k \quad . \tag{37.15}$$

Applying the previous lemma, we also see that, for each integer k,

$$c_k = \int_0^1 h(t) e^{-i2\pi kt} dt$$

$$= \int_0^1 \langle f(x), \phi(x-t) \rangle e^{-i2\pi kt} dt = \langle F_k e^{i2\pi kx}, \phi(x) \rangle$$

where

$$F_k = e^{\pi k^2} \int_0^1 \langle f(x), e^{-\pi(x-t)^2} \rangle e^{i2\pi kt} dt \quad .$$

With this,

$$\sum_{k=M}^{N} c_k = \sum_{k=M}^{N} \langle F_k e^{i2\pi kx}, \phi(x) \rangle = \left\langle \sum_{k=M}^{N} F_k e^{i2\pi kx}, \phi(x) \right\rangle \quad ,$$

and equation (37.15) becomes

$$\langle f(x), \phi(x) \rangle = \lim_{\substack{N \to \infty \\ M \to -\infty}} \left\langle \sum_{k=M}^{N} F_k e^{i2\pi kx}, \phi(x) \right\rangle \quad ,$$

showing us that, in the generalized sense,

$$f(x) = \sum_{k=-\infty}^{\infty} F_k e^{i2\pi kx} \quad .$$

Thus also,

$$\mathcal{F}[f] = \mathcal{F}\left[\sum_{k=-\infty}^{\infty} F_k e^{i2\pi kx} \right] = \sum_{k=-\infty}^{\infty} F_k \mathcal{F}\left[e^{i2\pi kx} \right] = \sum_{k=-\infty}^{\infty} F_k \delta_k \quad . \qquad \blacksquare$$

The Proof of Theorem 37.5

We are now assuming p and $\Delta\omega$ are two positive real values with $p\Delta\omega = 1$, and f and F are two generalized functions with $F = \mathcal{F}[f]$. Our goal is to confirm that f is periodic with period p if and only if F is a regular array with spacing $\Delta\omega$, and to verify the existence of uniquely determined constants $\ldots, F_{-1}, F_0, F_1, F_2, \ldots$ such that

$$f(t) = \sum_{k=-\infty}^{\infty} F_k e^{i2\pi k\Delta\omega t} \quad \text{and} \quad \mathcal{F}[f] = \sum_{k=-\infty}^{\infty} F_k \delta_{k\Delta\omega} \quad .$$

First of all, if F is a regular array, then the F_k's are simply the coefficients of the array, and, from lemma 37.4 on page 653, we know f is periodic with period $p = {}^1/_{\Delta\omega}$ and is given by

$$f(t) = \sum_{k=-\infty}^{\infty} F_k e^{i2\pi k\Delta\omega t} \quad .$$

Now, suppose f is periodic with period p. Let g be the generalized function obtained from f by variable scaling by p. In other words, g and f are related by

$$g(t) = f(pt) \quad \text{and} \quad f(t) = g\left(\frac{1}{p}t\right) = g(\Delta \omega \, t) \quad .$$

Equivalently, using the scaling operator,

$$g = S_p[f] \quad \text{and} \quad f = S_{\Delta \omega}[g] \quad .$$

By the periodicity of f,

$$g(t-1) = f(p(t-1)) = f(pt - p) = f(pt) = g(t) \quad .$$

So g is periodic with period 1, and lemma 37.12 tells us there are constants $\ldots, G_{-1}, G_0, G_1, G_2, \ldots$ such that

$$g(t) = \sum_{k=-\infty}^{\infty} G_k e^{i2\pi kt} \quad .$$

Thus, letting $F_k = G_k$,

$$f(t) = S_{\Delta \omega}[g(t)] = S_{\Delta \omega}\left[\sum_{k=-\infty}^{\infty} G_k e^{i2\pi kt}\right]$$

$$= \sum_{k=-\infty}^{\infty} G_k S_{\Delta \omega}\left[e^{i2\pi kt}\right] = \sum_{k=-\infty}^{\infty} F_k e^{i2\pi k \Delta \omega t} \quad ,$$

and

$$F = \mathcal{F}[f] = \mathcal{F}\left[\sum_{k=-\infty}^{\infty} F_k e^{i2\pi k \Delta \omega t}\right]$$

$$= \sum_{k=-\infty}^{\infty} F_k \mathcal{F}\left[e^{i2\pi k \Delta \omega t}\right] = \sum_{k=-\infty}^{\infty} F_k \delta_{k \Delta \omega} \quad ,$$

confirming the theorem's claim when f is periodic.

Finally, to show the claimed uniqueness, suppose $\{\ldots, \widehat{F}_{-1}, \widehat{F}_0, \widehat{F}_1, \widehat{F}_2, \ldots\}$ is any indexed set of constants such that

$$f(t) = \sum_{k=-\infty}^{\infty} \widehat{F}_k e^{i2\pi k \Delta \omega t} \quad \text{or} \quad \mathcal{F}[f] = \sum_{k=-\infty}^{\infty} \widehat{F}_k \delta_{k \Delta \omega} \quad .$$

Of course, either equation can be obtained from the other via the appropriate Fourier transform. So both of these equations must hold. Moreover, using the F_k's from above,

$$\sum_{k=-\infty}^{\infty} \widehat{F}_k \delta_{k \Delta \omega} = \mathcal{F}[f] = \sum_{k=-\infty}^{\infty} F_k \delta_{k \Delta \omega} \quad .$$

But, as you showed in exercise 36.15 on page 647, this last equation is possible if and only if

$$\widehat{F}_k = F_k \quad \text{for} \quad k = 0, \pm 1, \pm 2, \pm 3, \ldots \quad ,$$

completing the proof of theorem 37.5.   ∎

Additional Exercises

37.9. Let f be an exponentially integrable function on the real line, and suppose that, for some $p > 0$,

$$\big\langle\, f(x - p)\,,\, \phi(x)\,\big\rangle \;=\; \big\langle\, f(x)\,,\, \phi(x)\,\big\rangle \qquad \text{for each } \phi \text{ in } \mathcal{G} \;\; .$$

Show that, for any real value x_0 at which f is continuous,

$$f(x_0 - p) \;=\; f(x_0) \;\; .$$

(Gaussian identity sequences may be appropriate, here.)

37.10. Show that

$$\sum_{k=-\infty}^{\infty} F_k\, e^{i 2\pi k \Delta \omega t}$$

cannot be the Fourier series for some piecewise continuous and periodic classical function if the F_k's do not shrink to 0 as $k \to \pm\infty$. (Try using the Bessel's inequality from exercise 12.4 on page 152.)

37.11. For each of the following delta function arrays, determine the period p of the array's inverse Fourier transform and compute that inverse Fourier transform. Also, compute the Fourier transform of the given array.

a. $\displaystyle f = \sum_{k=-\infty}^{\infty} \frac{k}{1+k^2}\, \delta_{k/3}$ **b.** $\displaystyle g = \sum_{k=-\infty}^{\infty} \frac{k^2}{1+k^2}\, \delta_{4k}$ **c.** $\displaystyle h = \sum_{k=-\infty}^{\infty} (-k)^k\, \delta_k$

37.12. For each of the following periodic, classical functions:

    1. Find the Fourier series.[1]

    2. Find the Fourier transform.

    3. Sketch the real and imaginary parts of the Fourier transform.

    4. Finally, find the inverse Fourier transform,

a. $f(t) = \begin{cases} 0 & \text{if } -1 < t < 0 \\ 1 & \text{if } 0 < t < 1 \\ f(t-2) & \text{in general} \end{cases}$

b. $g(t) = \begin{cases} e^t & \text{if } 0 < t < 1 \\ g(t-1) & \text{in general} \end{cases}$

c. The even sawtooth function sketched in figure 37.4a

d. The odd sawtooth function sketched in figure 37.4b

e. $f(t) = \begin{cases} t^2 & \text{if } -1 < t < 1 \\ f(t-2) & \text{in general} \end{cases}$

---

[1] You may have already found the Fourier series for these functions in exercise 12.3 on page 152.

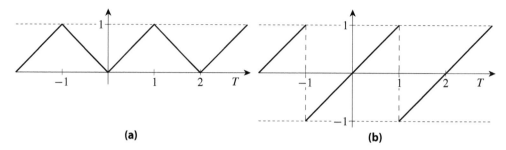

Figure 37.4: Two and one half periods of (a) evensaw(t), the even sawtooth function for exercise
37.12 c, and (b) oddsaw(t), the odd sawtooth function for exercise 37.12 d.

37.13. *For each of the following periodic, regular arrays of delta functions:*

    1. *Sketch the given array, and determine both the period and the spacing.*

    2. *Find the kth Fourier coefficient c_k for the array.*

    3. *Then find both the Fourier series and the Fourier transform for the given array, and
sketch the real and imaginary parts of the Fourier transform.*

    a. $f = \displaystyle\sum_{n=-\infty}^{\infty} (-1)^n \, \delta_n$

    b. $g = \displaystyle\sum_{n=-\infty}^{\infty} \begin{cases} 0 & \text{if } n = 0, \pm 4, \pm 8, \pm 12, \dots \\ 1 & \text{if } n = 1, 1 \pm 4, 1 \pm 8, 1 \pm 12, \dots \\ 2 & \text{if } n = 2, 2 \pm 4, 2 \pm 8, 2 \pm 12, \dots \\ 3 & \text{if } n = 3, 3 \pm 4, 3 \pm 8, 3 \pm 12, \dots \end{cases} \delta_n$

37.14. *Let f be a periodic generalized function with period p. Verify that, in the generalized
sense,*

$$f(x) = A_0 + \sum_{k=1}^{\infty} [a_k \cos(2\pi \omega_k x) + b_k \sin(2\pi \omega_k x)]$$

where $\omega_k = {}^{k}/_{p}$,

$$A_0 = \frac{1}{p} \int_0^p f(x)\, dx \quad , \quad a_k = \frac{2}{p} \int_0^p f(x) \cos(2\pi \omega_k x)\, dx \quad ,$$

and

$$b_k = \frac{2}{p} \int_0^p f(x) \sin(2\pi \omega_k x)\, dx \quad .$$

(The above series is the (generalized) trigonometric Fourier series for f.)

37.15. *Let f be a periodic, generalized function with period p, and let k be any integer. Show
that the kth Fourier coefficient of f can be computed from*

$$F_k = \frac{1}{\Theta} \int_0^p \langle f(s), \psi(s - \tau) \rangle e^{i2\pi \omega_k \tau}\, d\tau$$

*where $\omega_k = {}^{k}/_{p}$, ψ is any convenient Gaussian test function, and $\Theta = p\mathcal{F}^{-1}[\psi]\big|_{\omega_k}$.
(Try using scaling and the result from lemma 37.11.)*

38

Pole Functions and
General Solutions to Simple Equations

This chapter is about division. More precisely, it is about solving equations of the form $fu = g$ for u. To see why we might want to spend an entire chapter on this, let's look at some problems that arise when using Fourier transforms to solve the differential equations

$$\frac{dy}{dt} + 3y = \delta \quad \text{and} \quad \frac{d^2 y}{dt^2} = \delta \quad .$$

Taking the transform of both sides of the first equation and using a differentiation identity yields

$$i2\pi\omega Y(\omega) + 3Y(\omega) = 1$$

where $Y = \mathcal{F}[y]$. By elementary algebra, this can be written as

$$[3 + i2\pi\omega]Y(\omega) = 1 \quad .$$

Dividing through by $3 + i2\pi\omega$ then gives

$$Y(\omega) = \frac{1}{3 + i2\pi\omega} \quad ,$$

and from this we obtain

$$y(t) = \mathcal{F}^{-1}\left[\frac{1}{3 + i2\pi\omega}\right]\bigg|_t = e^{-3t} \, \text{step}(t) \quad .$$

You can easily verify that this is a solution to the differential equation. However, you can also easily verify that

$$e^{-3t} \, \text{step}(t) + 827e^{-3t} \quad \text{and} \quad e^{-3t} \, \text{step}(t) - 4e^{-3t}$$

are also solutions. This means that *our procedure only found one of many possible solutions* — a distinct problem if the solution needed is one of those other solutions.

Another problem arises when we attempt to solve the other differential equation,

$$\frac{d^2 y}{dt^2} = \delta \quad .$$

Taking the transform and applying the appropriate identities and algebraic manipulations leads to

$$-4\pi^2\omega^2 Y(\omega) = 1$$

where, again, $Y = \mathcal{F}[y]$. "Thus",

$$Y(\omega) = -\frac{1}{4\pi^2\omega^2} \quad .$$

The problem here is that this formula is not exponentially integrable and, so, does not give us a valid generalized function (see exercise 33.7 on page 545 and the discussion leading to that exercise). Thus, in fact, *we have not even found one solution to our differential equation.*

It turns out that our failure to obtain all (or any) solutions to our two differential equations can be traced to the way we attempted to solve the equations

$$[3 + i2\pi\omega]Y(\omega) = 1 \quad \text{and} \quad -4\pi^2\omega^2 Y(\omega) = 1 \quad .$$

So let us take a careful look at solving these types of equations.

38.1   Basics on Solving Simple Algebraic Equations

We are interested in the problem of determining every generalized function u that satisfies

$$fu = g$$

for some known pair of generalized functions f and g.

Sometimes, solving such an equation for u can be very easy.

!▶**Example 38.1:**   *Suppose we want to find the generalized function u satisfying the equation*

$$e^{3x}u(x) = \sin(x) + \delta_i(x) \quad . \tag{38.1}$$

The obvious approach is to multiply both sides by e^{-3x}. "Obviously",

$$e^{-3x}\left[e^{3x}u(x)\right] = \left[e^{-3x}e^{3x}\right]u(x) = 1 \cdot u(x) = u(x)$$

and

$$e^{-3x}[\sin(x) + \delta_i(x)] = e^{-3x}\sin(x) + e^{-3x}\delta_i(x)$$
$$= e^{-3x}\sin(x) + e^{-3i}\delta_i(x) \quad .$$

So,

$$u(x) = e^{-3x}\left[e^{3x}u(x)\right]$$
$$= e^{-3x}[\sin(x) + \delta_i(x)] = e^{-3x}\sin(x) + e^{-3i}\delta_i(x) \quad .$$

As a precaution, we should confirm that this function is a solution to our original equation. Plugging the above derived formula for $u(x)$ into the right side of equation (38.1), we see that

$$e^{3x}u(x) = e^{3x}\left[e^{-3x}\sin(x) + e^{-3i}\delta_i(x)\right]$$
$$= 1 \cdot \sin(x) + e^{-3i}\left[e^{3x}\delta_i(x)\right]$$
$$= \sin(x) + e^{-3i}\left[e^{3i}\delta_i(x)\right]$$
$$= \sin(x) + \delta_i(x) \quad .$$

The last line is the same as the right side of equation (38.1), and so, the above formula for $u(x)$ solves our original equation.

We probably should have been a little more careful in our last example. After all, we are dealing with generalized products, not the classical products we are so comfortable with. So we should verify that each computation in the example can be justified by some rule or theorem for generalized products.

Or we can verify our computations more generally by proving the next theorem.

Theorem 38.1

Let f and g be two generalized functions with f being a classical function whose reciprocal, f^{-1} (i.e., $^1/_f$), is a simple multiplier. If there is a generalized function u satisfying $fu = g$, then it is given by $u = f^{-1}g$.

PROOF: Assume a generalized function u does exist satisfying $fu = g$. Since f^{-1} is a simple multiplier, the generalized products $f^{-1}(fu)$ and $f^{-1}g$ exist, and

$$f^{-1}(fu) = f^{-1}g \quad .$$

In addition, because f^{-1} is a simple multiplier, we can apply the rule on associativity (rule 8 on page 634) to obtain

$$f^{-1}(fu) = (f^{-1}f)u \quad .$$

But f^{-1} and f are classical functions, and our first rule for the generalized product was that the generalized product is given by the classical product when the classical product defines a generalized product. That is certainly the case here, and that classical product, $f^{-1}f$, is the constant 1. Thus, using also rule 2 and the fact that the simple product of 1 with any generalized function equals that function,

$$(f^{-1}f)u = 1 \cdot u = u \quad .$$

Together, the above equations yield

$$u = (f^{-1}f)u = f^{-1}(fu) = f^{-1}g \quad . \qquad \blacksquare$$

Two things should be noted regarding this theorem. The first is its assurance, under the conditions assumed, that the equation $fu = g$ cannot have more than one solution. Every existing solution is given by the single formula $u = f^{-1}g$ (we'll soon discuss equations for which this is not true). The other thing is the qualifying statement: *If there is a generalized function u satisfying $fu = g$.* As the next example illustrates, this is an important proviso. An equation might not have a solution. Consequently, it is important to check that any formula obtained as a possible solution truly satisfies the original equation.

!▶ Example 38.2: *Consider the equation*

$$\frac{1}{x-i} u(x) = \delta_i \quad . \tag{38.2}$$

The reciprocal of $(x-i)^{-1}$ is $x-i$, a simple multiplier. According to theorem 38.1, if a solution exists, it is given by the product of $x-i$ with δ_i,

$$u(x) = (x-i)\delta_i = (i-i)\delta_i = 0 \cdot \delta_i = 0 \quad .$$

Could $u = 0$ be a solution to our equation? Checking,

$$\frac{1}{x-i} u(x) = \frac{1}{x-i} \cdot 0 = 0 \neq \delta_i \quad .$$

So, no, the formula we obtained, $u = 0$, does not solve our equation. And since we know this would be the solution if one existed, we must conclude that equation (38.2) has no solution.

As suggested at the start of this chapter, one class of algebraic equations of particular interest are those obtained by taking Fourier transforms of differential equations.

!▶ Example 38.3:   *Consider the following second order differential equation:*

$$\frac{d^2 y}{dt^2} + 4\frac{dy}{dt} + 3y = e^{i2\pi t} \quad .$$

Taking the Fourier transform of both sides and letting $Y = \mathcal{F}[y]$:

$$\mathcal{F}\left[\frac{d^2 y}{dt^2} + 4\frac{dy}{dt} + 3y\right]\bigg|_\omega = \mathcal{F}\left[e^{i2\pi t}\right]\bigg|_\omega$$

$$\hookrightarrow \quad (i2\pi\omega)^2 Y(\omega) + 4(i2\pi\omega)Y(\omega) + 3Y(\omega) = \delta_1(\omega)$$

$$\hookrightarrow \quad \left[(i2\pi\omega)^2 + 4(i2\pi\omega) + 3\right] Y(\omega) = \delta_1(\omega) \quad .$$

In the above example, we ended up with a polynomial multiplied by the unknown generalized function Y equaling an known generalized function. More generally, if we take the Fourier transform of any n^{th} order linear differential equation with constant coefficients,

$$a_n \frac{d^n y}{dt^n} + a_{n-1}\frac{d^{n-1} y}{dt^{n-1}} + \cdots + a_1 \frac{dy}{dt} + a_0 y = g \quad ,$$

then we clearly end up with an algebraic equation involving an nth degree polynomial,

$$\left[a_n(i2\pi\omega)^n + a_{n-1}(i2\pi\omega)^{n-1} + \cdots + a_1(i2\pi\omega) + a_0\right] Y(\omega) = G(\omega) \quad .$$

In practice, we are more likely to be interested in the differential equation. However, algebraic equations are usually considered to be easier to solve; moreover, if we can solve the algebraic equation for Y, then the solution to the differential equation can be obtained by taking the inverse transform of Y. Unfortunately, the reciprocal of a polynomial is not a simple multiplier (unless the polynomial is *very* simple, namely, a constant). So we cannot appeal to theorem 38.1. In fact, as illustrated at the beginning of this chapter, naively "dividing through by $f(x)$" can give us something that is *not* a generalized function. And even if the result is a valid generalized function, there still remains the question as to whether there might be other valid solutions.

Finding all the solutions to these equations will require more work than was needed to solve the equations considered in theorem 38.1. That work begins with the following little lemma.

Lemma 38.2
Let f, g, u and v be generalized functions with $fu = g$. Then $fv = g$ if and only if $v = u + w$ for some generalized function w satisfying $fw = 0$.

PROOF: Suppose $v = u + w$ where $fw = 0$. Then

$$fv = f \cdot (u + w) = fu + fw = g + 0 = g \quad ,$$

proving the "if" part of the lemma.

To prove the "only if" part, assume v is a generalized function satisfying $fv = g$, and let $w = v - u$. Then $v = u + w$, and

$$fw = f \cdot (v - u) = fv - fu = g - g = 0 \quad . \qquad \blacksquare$$

For convenience, we will adopt the following terminology when given any generalized function equation of the form

$$fu = g$$

in which u is the unknown:

This equation will be referred to as *homogeneous* if and only if g is the zero function.

The *corresponding homogeneous equation* is simply the equation $fu = 0$.

A *general solution* for $fu = g$ is a formula for u describing all possible solutions to the equation.

You should be familiar with this terminology from your studies of differential equations. You should also realize that the following is an immediate corollary of the above lemma and our newly adopted terminology.

Theorem 38.3
Let f and g be two (known) generalized functions. Then, if it exists, the general solution to $fu = g$ is given by

$$u = u_0 + w$$

where u_0 is any single solution to $fu = g$, and w is the general solution for the corresponding homogeneous equation, $fw = 0$.

For the rest of this chapter, we will confine ourselves to equations of the form $fu = g$ where f is a polynomial. In the next section we will discover how to find general solutions to homogeneous equations. Following that, we'll discuss methods for obtaining particular solutions to nonhomogeneous equations. Then, using the above theorem, we will be able to construct general solutions to the nonhomogeneous equations. Finally, we will discuss some solutions of particular interest with the intent of developing generalized analogs to classical functions of the form x^{-k} where k is a positive integer.

38.2 Homogeneous Equations with Polynomial Factors

Our goal is to derive a usable formula describing all possible solutions to the equation $fu = 0$ when f is a known polynomial and u is the unknown generalized function. We'll start with the simplest case (where $f(x) = x$) and work our way up to the most general case.

Monomial Factors

Observe that, if c is any constant and $u = c\delta$, then, because $f(x)\delta = f(0)\delta$ for any function continuous at 0,

$$xu(x) = xc\delta(x) = 0 \cdot c\delta(x) = 0 \quad .$$

Hence, the delta function multiplied by any constant is always a solution to $xu(x) = 0$. The next lemma assures us that there are no other possible solutions. It is this lemma from which all else in this section follows.

Lemma 38.4
A generalized function u satisfies $xu(x) = 0$ if and only if $u = c\delta$ for some constant c.

PROOF: Since we already know $xu(x) = 0$ if $u = c\delta$ for some constant c, let's assume $xu(x) = 0$ and try to show that $u = c\delta$.

According to lemma 33.6 on page 557 (with $\sigma = \alpha + i\beta$), $u = c\delta$ for some constant c if

$$\left\langle u(x)\,,\ e^{\alpha x}e^{i\beta x}e^{-\gamma x^2} \right\rangle = \left\langle c\delta(x)\,,\ e^{\alpha x}e^{i\beta x}e^{-\gamma x^2} \right\rangle$$

for every triple of real numbers (α, β, γ) with $\gamma > 0$. But

$$\left\langle c\delta(x)\,,\ e^{\alpha x}e^{i\beta x}e^{-\gamma x^2} \right\rangle = ce^{\alpha 0}e^{i\beta 0}e^{-\gamma 0^2} = c \quad .$$

Consequently, all we need to show is that there is a constant c such that

$$\left\langle u(x)\,,\ e^{\alpha x}e^{i\beta x}e^{-\gamma x^2} \right\rangle = c$$

for every triple of real numbers (α, β, γ) with $\gamma > 0$.

So consider the function

$$h(\alpha, \beta, \gamma) = \left\langle u(x)\,,\ e^{\alpha x}e^{i\beta x}e^{-\gamma x^2} \right\rangle$$

where α, β and γ are real variables with $\gamma > 0$. From lemma 33.3 on page 556, we know h is a smooth classical function of each of these variables. That lemma, along with the assumption that $xu(x) = 0$, also gives us

$$\frac{\partial}{\partial \alpha}h(\alpha, \beta, \gamma) = \left\langle u(x)\,,\ \frac{\partial}{\partial \alpha}e^{\alpha x}e^{i\beta x}e^{-\gamma x^2} \right\rangle$$

$$= \left\langle u(x)\,,\ xe^{\alpha x}e^{i\beta x}e^{-\gamma x^2} \right\rangle$$

$$= \left\langle xu(x)\,,\ e^{\alpha x}e^{i\beta x}e^{-\gamma x^2} \right\rangle$$

$$= \left\langle 0\,,\ e^{\alpha x}e^{i\beta x}e^{-\gamma x^2} \right\rangle = 0 \quad .$$

This tells us $h(\alpha, \beta, \gamma)$ does not vary as α varies. Thus

$$h(\alpha, \beta, \gamma) = h(0, \beta, \gamma) \qquad \text{for each } \alpha \text{ in } \mathbb{R} \quad .$$

Virtually identical computations confirm that

$$\frac{\partial}{\partial \beta}h(0, \beta, \gamma) = 0 \qquad \text{and} \qquad \frac{\partial}{\partial \gamma}h(0, 0, \gamma) = 0 \quad .$$

Hence, no matter what real numbers we choose for α and β, and what positive value we choose for γ,

$$h(\alpha, \beta, \gamma) = h(0, \beta, \gamma) = h(0, 0, \gamma) = h(0, 0, 1) \quad .$$

Cutting out the middle, replacing $h(\alpha, \beta, \gamma)$ with its formula, and letting $c = h(0, 0, 1)$ then yields

$$\left\langle u(x)\,,\ e^{\alpha x}e^{i\beta x}e^{-\gamma x^2} \right\rangle = c$$

for all real values α, β and γ (with $\gamma > 0$), just as we needed to show.   ∎

With this lemma and just a little trickery, we can now derive the general solution to any equation of the form $x^n u(x) = 0$ where n is a positive integer. The basic ideas can be illustrated by considering how we might solve

$$x^2 u(x) = 0 \quad . \tag{38.3}$$

Since x is a simple multiplier, associativity holds and one of the x's can be "factored out",

$$x[xu(x)] = 0 \quad .$$

Lemma 38.4 then tells us that

$$x\, u(x) = a\delta(x) \tag{38.4}$$

where a is an arbitrary constant.

Suppose, now, that we can find a generalized function u_1 satisfying

$$xu_1(x) = \delta(x) \quad . \tag{38.5}$$

Then $u_0 = au_1$ is clearly one solution to equation (38.4), and theorem 38.3 and lemma 38.4 then assure us that

$$u(x) = au_1(x) + c_0\,\delta(x) \tag{38.6}$$

is a general solution to equation (38.3) (where c_0 is an arbitrary constant).

Finding a formula for u_1 involves a small bit of cleverness, starting with the realization that, because $x\delta(x) = 0$,

$$D[x\delta(x)] = 0 \quad .$$

From this, the product rule and the fact that $D[x] = {}^{dx}\!/_{dx} = 1$, we obtain

$$0 = D[x\delta(x)] = (D[x])\,\delta(x) + x\,D\delta(x) = \delta(x) + x\,D\delta(x) \quad .$$

Thus,

$$x[-D\delta(x)] = \delta(x) \quad .$$

This tells us that one solution to equation (38.5) is $u_1 = -D\delta$. Plugging this solution back into formula (38.6) (and letting $c_1 = -a$) gives

$$u(x) = c_1\,D\delta(x) + c_0\,\delta(x)$$

as the general solution to

$$x^2 u(x) = 0$$

where c_1 and c_0 are arbitrary constants.

Continuing along these lines leads to the next lemma.

Lemma 38.5

For every positive integer n,

$$x^n D^{n-1}\delta(x) = 0 \qquad and \qquad x^n D^n \delta(x) = (-1)^n n!\,\delta(x) \quad .$$

Moreover, the general solution to $x^{n+1}u(x) = 0$ is

$$u = \sum_{k=0}^{n} c_k\, D^k \delta$$

where the c_k's are arbitrary constants.

If you check, you'll see that we have already confirmed the validity of this lemma when $n = 1$ and $n = 2$. In the next two exercises, you will finish its proof.

?►Exercise 38.1:   *Show that*

$$x^n D^{n-1} \delta(x) \; = \; 0 \qquad \text{and} \qquad x^n D^n \delta(x) \; = \; (-1)^n n! \, \delta(x)$$

   a: *when* $n = 3$ **b:** *when* $n = 4$ **c:** *when* $n = 5, 6, \dots$

?►Exercise 38.2:   *Letting* $c_0, c_1, \dots$ *and* c_n *denote arbitrary constants, show that*

$$u \; = \; \sum_{k=0}^{n} c_k \, D^k \delta$$

  is the general solution to $x^{n+1} u(x) = 0$

   a: *when* $n = 3$ **b:** *when* $n = 4$ **c:** *when* $n = 5, 6, \dots$

Linear Factors

The results given in lemma 38.5 are easily extended using the translation identity to cases where the factor is of the form $(x - \lambda)^M$ with λ being any complex number and M being any positive integer. Suppose, for example,

$$(x - \lambda)^M u(x) \; = \; 0 \quad .$$

Then,

$$T_{-\lambda}\left[(x - \lambda)^M u(x) \right] \; = \; T_{-\lambda}\left[(x - \lambda)^M \right] T_{-\lambda}\left[u(x) \right]$$

$$= \; \left[(x - (-\lambda) - \lambda)^M \right] T_{-\lambda}[u(x)] \; = \; x^M T_{-\lambda}[u(x)] \quad .$$

So,

$$x^M T_{-\lambda}[u(x)] \; = \; T_{-\lambda}\left[(x - \lambda)^M u(x) \right] \; = \; T_{-\lambda}[0] \; = \; 0 \quad ,$$

and lemma 38.5 (if $M > 1$) or lemma 38.4 (if $M = 1$) tells us that

$$T_{-\lambda}[u(x)] \; = \; \sum_{k=0}^{M-1} c_k \, D^k \delta$$

for some constants $c_0, c_1, \dots$ and c_{M-1}. "Translating back", we then see that

$$u(x) \; = \; T_\lambda\left[T_{-\lambda}[u(x)] \right] \; = \; T_\lambda\left[\sum_{k=0}^{M-1} c_k \, D^k \delta \right] \; = \; \sum_{k=0}^{M-1} c_k \, T_\lambda\left[D^k \delta \right] \; = \; \sum_{k=0}^{M-1} c_k \, D^k \delta_\lambda \quad .$$

This proves part of the following lemma. The other parts can be proven in a similar manner — apply $T_{-\lambda}$, invoke lemma 38.5, and then apply T_λ.

Lemma 38.6

Let λ be any complex number, and let M be any positive integer. Then

$$(x - \lambda)^M D^{M-1} \delta_\lambda(x) \; = \; 0 \qquad \text{and} \qquad (x - \lambda)^M D^M \delta_\lambda(x) \; = \; (-1)^M M! \, \delta_\lambda(x) \quad .$$

Moreover, the general solution to $(x - \lambda)^M u(x) = 0$ is

$$u \; = \; \sum_{k=0}^{M-1} c_k \, D^k \delta_\lambda$$

where the c_k's are arbitrary constants.

Arbitrary Polynomial Factors

Let us now extend our discussion to cover every case in which f is a polynomial, say,

$$f(x) = A_N x^N + A_{N-1} x^{N-1} + \cdots + A_2 x^2 + A_1 x + A_0$$

where N is some nonnegative integer (the *degree* of the polynomial) and the A_k's are constants with $A_N \neq 0$. These constants can be complex. Recall that a complex number λ is called a *root* for f if and only if $f(\lambda) = 0$.[1] Also recall that, being a polynomial, f can be written in a completely factored form,

$$f(x) = A_N (x - \lambda_1)^{M_1} (x - \lambda_2)^{M_2} \cdots (x - \lambda_K)^{M_K}$$

where λ_1, λ_2, ... and λ_K are all the distinct roots of the polynomial, and the M_k's are positive integers. Each M_k is usually referred to as the *multiplicity* of the corresponding root, λ_k.

?► Exercise 38.3: *Verify the following factorizations:*

$$3x + 5 = 3 \left(x - \left[-\frac{5}{3} \right] \right)$$

$$x^2 + 3x - 10 = (x - 2)(x - [-5])$$

$$x^2 + 1 = (x - i)(x - [-i])$$

$$x^3 - 3x^2 + x - 3 = (x - i)(x - [-i])(x - 3)$$

To gain some intuition as to the nature of the general solutions to these more general problems, we will derive, in the next example, the solution to a relatively simple example.

!► Example 38.4: *Consider the problem of finding all possible solutions to*

$$\left[x^3 - 14x^2 + 60x - 72 \right] u(x) = 0 \quad .$$

You can easily verify that

$$x^3 - 14x^2 + 60x - 72 = (x - 6)^2 (x - 2) \quad .$$

So we can rewrite the equation to be solved as

$$(x - 6)^2 \left[(x - 2)u(x) \right] = 0 \quad .$$

This, as was just derived (lemma 38.6), means that

$$(x - 2)u(x) = c_0 \, \delta_6(x) + c_1 D\delta_6(x) \tag{38.7}$$

for some arbitrary pair of constants c_0 and c_1.

The general solution to equation (38.7) is given by

$$u = v + w \tag{38.8}$$

where v is any one solution to equation (38.7) and w_g is the general solution to the corresponding homogeneous equation. But that homogeneous equation is

$$(x - 2)w(x) = 0 \quad ,$$

---

[1] Sometimes, a root of a polynomial is also referred to as a *zero* for that polynomial. This leads to the observation that, in practice, most zeros are nonzero.

whose general solution we already know to be

$$w = c_2 \delta_2 \tag{38.9}$$

where c_2 is an arbitrary constant.

Now, if we can find generalized functions v_0 and v_1 satisfying

$$(x - 2)v_0(x) = \delta_6(x) \quad\quad and \quad\quad (x - 2)v_1(x) = D\delta_6(x) \quad,$$

then

$$(x - 2)[c_0 v_0(x) + c_1 v_1(x)] = c_0(x - 2)v_0(x) + c_1(x - 2)v_1(x)$$
$$= c_0 \delta_6(x) + c_1 D\delta_6(x) \quad.$$

So $v = c_0 v_0 + c_1 v_1$ would be a particular solution to equation (38.7), and, combining equations (38.8) and (38.9), we would have

$$u(x) = c_0 v_0(x) + c_1 v_1(x) + c_2 \delta_2(x) \tag{38.10}$$

as a general solution to our equation.

We will use the "method of half-educated guess" for finding v_0 and v_1.[2]

The educated part of our method is our knowledge that a delta function at any point times any simple multiplier is just that delta function times the value of the function at that point. This suggests that a good guess for v_0 might be some (yet unknown) constant a times δ_6. Letting $v_0 = a\delta_6$, we see that

$$(x - 2)v_0(x) = (x - 2)a\delta_6(x) = a(x - 2)\delta_6(x) = a(6 - 2)\delta_6(x) = 4a\delta_6(x)$$

and the equation $(x - 2)v_0(x) = \delta_6(x)$ becomes

$$4a\delta_6(x) = \delta_6(x) \quad,$$

which is clearly satisfied when $a = \frac{1}{4}$. From this we see that

$$v_0(x) = \frac{1}{4}\delta_6(x) \quad \text{is a solution to} \quad (x - 2)v_0(x) = \delta_6(x) \quad.$$

Encouraged and inspired by our success with finding v_0, let's try assuming

$$v_1(x) = a\delta_6(x) + bD\delta_6(x)$$

where a and b are constants to be determined. With this assumption,

$$(x - 2)v_1(x) = (x - 2)[a\delta_6(x) + bD\delta_6(x)]$$
$$= a(x - 2)\delta_6(x) + b(x - 2)D\delta_6(x) \quad.$$

Again,

$$(x - 2)\delta_6(x) = (6 - 2)\delta_6(x) = 4\delta_6(x) \quad.$$

Also, using the results from lemma 38.6 and a little algebra, we find that

$$(x - 2)D\delta_6(x) = (x - 6 + 8)D\delta_6(x)$$
$$= (x - 6)D\delta_6(x) + 8D\delta_6(x) = -\delta_6(x) + 8D\delta_6(x) \quad.$$

---

[2] The "method of half-educated guess" is very similar to the "the method of undetermined coefficients" found in most introductory books on differential equations.

Thus,

$$(x - 2)v_1(x) = a(x - 2)\delta_6(x) + b(x - 2)\delta_6(x)$$
$$= a[4\delta_6(x)] + b[-\delta_6(x) + 8D\delta_6(x)]$$
$$= (4a - b)\delta_6(x) + 8bD\delta_6(x) \quad ,$$

and equation $(x - 2)v_1(x) = D\delta_6(x)$ becomes

$$(4a - b)\delta_6(x) + 8bD\delta_6(x) = D\delta_6(x) \quad ,$$

which is certainly satisfied if

$$4a - b = 0 \quad \text{and} \quad 8b = 1 \quad .$$

This last system is easy enough to solve:

$$b = \frac{1}{8} \quad \text{and} \quad a = \frac{1}{4}b = \frac{1}{32} \quad .$$

And so,

$$v_1(x) = \frac{1}{32}\delta_6(x) + \frac{1}{8}D\delta_6(x) \quad \text{is a solution to} \quad (x - 2)v_1(x) = D\delta_6(x) \quad .$$

Finally, combining our formulas for v_0 and v_1 with equation (38.10), we see that the general solution to our original equation is

$$u = c_0 v_0 + c_1 v_1 + c_2 \delta_2$$
$$= c_0\left[\frac{1}{4}\delta_6\right] + c_1\left[\frac{1}{32}\delta_6 + \frac{1}{8}D\delta_6\right] + c_2 \delta_2$$
$$= \left[\frac{1}{4}c_0 + \frac{1}{32}c_1\right]\delta_6 + \frac{1}{8}c_1 + c_2 \delta_2$$

where c_0, c_1 and c_2 are arbitrary constants. Clearly, though, if we let

$$a_0 = \frac{1}{4}c_0 + \frac{1}{32}c_1 \quad \text{and} \quad a_1 = \frac{1}{8}c_1 \quad ,$$

then a_0 and a_1 are also completely arbitrary constants. Hence, the general solution can be written more simply as

$$u = a_0 \delta_6 + a_1 D\delta_6 + c_2 \delta_2$$

where a_0, a_1 and c_2 are arbitrary constants.

Look at what we just derived in the last example. We found that, if

$$f(x) = x^3 - 14x^2 + 60x - 72 \quad ,$$

which, in factored form, is

$$f(x) = (x - 6)^2(x - 2) \quad ,$$

then a general solution to $fu = 0$ is

$$u = a_0 \delta_6 + a_1 D\delta_6 + c_2 \delta_2$$

where a_0, b_1 and c_2 are arbitrary constants. From this, you can probably guess the correct formula for a general solution to any equation of the form $fu = 0$ where f is a polynomial. Go ahead, make that guess, and compare it to the formula in the following theorem.

Theorem 38.7

Let f be a polynomial with distinct roots λ_1, λ_2, ... and λ_K (and no others). For each λ_k, let M_k be the corresponding multiplicity. Then the general solution to $fu = 0$ is

$$u = \sum_{k=1}^{K} \sum_{m=0}^{M_k-1} c_{k,m} D^m \delta_{\lambda_k}$$

where the $c_{k,m}$'s are arbitrary constants.

The complete proof of this theorem is somewhat lengthy and, admittedly, a little tedious. Because of this, we will set aside the next subsection for that proof. Still, if you followed our work in the example leading to this theorem (example 38.4), and can recall the method of undetermined coefficients for solving differential equations, then you will probably find the proof relatively straight-forward.

In the meantime, let's employ this theorem to solve a simple equation.

!▶**Example 38.5:** *Consider solving*

$$\left[x^2 - 4x + 13\right]u(x) = 0 \tag{38.11}$$

for $u(x)$. The roots to the polynomial, that is, the solutions to

$$\lambda^2 - 4\lambda + 13 = 0 \quad,$$

are easily found via the well-known quadratic formula,

$$\lambda = \frac{-(-4) \pm \sqrt{(-4)^2 - 4(1)(13)}}{2(1)} = 2 \pm 3i \quad.$$

The multiplicity of each root is 1, because

$$x^2 - 4x + 13 = \left(x - [2+3i]\right)^1 \left(x - [2-3i]\right)^1 \quad.$$

Theorem 38.7 then tells us that the general solution to equation (38.11) is

$$u = a\,\delta_{2+3i} + b\,\delta_{2-3i}$$

where a and b are arbitrary constants.

?▶**Exercise 38.4:** *Verify that the general solution to*

$$\left[x^2 + 1\right]u(x) = 0$$

is

$$u(x) = a\,\delta_i + b\,\delta_{-i}$$

where a and b are arbitrary constants.

Proof of Theorem 38.7[*]

A significant portion of theorem 38.7 is contained in the following lemma.

Lemma 38.8
Let λ and μ be two different complex numbers, let M and K be two nonnegative integers, and let b_0, b_1, $\ldots$ and b_K be constants. There are then constants a_0, a_1, $\ldots$ and a_K such that

$$v(x) = \sum_{k=0}^{K} a_k D^k \delta_\mu(x) \tag{38.12}$$

satisfies

$$(x - \lambda)^M v(x) = \sum_{k=0}^{K} b_k D^k \delta_\mu(x) \quad . \tag{38.13}$$

PROOF: To simplify matters, begin by using the translation operator $T_{-\lambda}$ on each side of equation (38.13). Letting $w = T_{-\lambda}[v]$ and $\gamma = \mu - \lambda$, we see that

$$T_{-\lambda}\left[(x - \lambda)^M v(x)\right] = T_{-\lambda}\left[(x - \lambda)^M\right] T_{-\lambda}\left[v(x)\right]$$

$$= (x + \lambda - \lambda)^M w(x) = x^M w(x)$$

and

$$T_{-\lambda}\left[\sum_{k=0}^{K} b_k D^k \delta_\mu(x)\right] = \sum_{k=0}^{K} b_k D^k T_{-\lambda}\left[\delta_\mu(x)\right]$$

$$= \sum_{k=0}^{K} b_k D^k \delta_{\mu-\lambda}(x) = \sum_{k=0}^{K} b_k D^k \delta_\gamma(x) \quad .$$

Thus, equation (38.12) is equivalent to

$$x^M w(x) = \sum_{k=0}^{K} b_k D^k \delta_\gamma(x) \quad . \tag{38.14}$$

Now take the Fourier transform of each side. On the left, letting $W = \mathcal{F}[w]$, we get

$$\mathcal{F}\left[x^M w(x)\right]\Big|_y = \left(\frac{i}{2\pi}\right)^M D^M W(y) \quad .$$

On the right, we get

$$\mathcal{F}\left[\sum_{k=0}^{K} b_k D^k \delta_\gamma(x)\right]\Bigg|_y = \sum_{k=0}^{K} b_k \mathcal{F}\left[D^k \delta_\gamma(x)\right]\Big|_y = \sum_{k=0}^{K} b_k (i2\pi y)^k e^{-i2\pi \mu y} \quad .$$

So, in terms of W, equation (38.12) is the generalized differential equation

$$\left(\frac{i}{2\pi}\right)^M D^M W(y) = \sum_{k=0}^{K} b_k (i2\pi y)^k e^{-i2\pi \gamma y} \quad .$$

---

[*] Those who view example 38.4 as adequate proof of theorem 38.7 might as well skip ahead to page 689 and start on the discussion of nonhomogeneous equations.

Equivalently,

$$D^M W(y) = \sum_{k=0}^{K} c_k \, y^k e^{-i2\pi\gamma y} \tag{38.15}$$

where

$$c_k = (-1)^M (i2\pi)^{k-M} \quad .$$

Consider, now, the corresponding classical differential equation

$$\frac{d^M W}{dy^M} = \sum_{k=0}^{K} c_k \, y^k e^{-i2\pi\gamma y} \quad . \tag{38.16}$$

This is a simple differential equation for which one solution W_0 can be found by simply anti-differentiating (integrating) the equation M times and setting the constants of integration equal to 0. Because γ is nonzero (since it is the difference of two different numbers), and because the integration by parts formula gives us

$$\int y^k e^{-i2\pi\gamma y} \, dy = \frac{i}{2\pi\gamma} y^k e^{-i2\pi\gamma y} - \frac{ik}{2\pi\gamma} \int y^{k-1} e^{-i2\pi\gamma y} \, dy$$

whenever $k \geq 1$, it is easily verified that the result of these M anti-differentiations will be

$$W_0(y) = \sum_{k=0}^{K} C_k \, y^k e^{-i2\pi\gamma y}$$

where the C_k's are constants we could compute if we were really interested in these values.

This last function, W_0, is infinitely smooth, and all of its derivatives are certainly exponentially bounded. Hence, its classical derivatives are the same as its generalized derivatives. Consequently, W_0 satisfies the generalized differential equation (38.15) as well as equation (38.16). From this it follows that $w_0 = \mathcal{F}^{-1}[W_0]$ satisfies equation (38.14), and $v = T_\lambda[w_0]$ satisfies equation (38.13), as desired. Inverting the transforms used to obtain equation (38.15), we find that

$$w_0 = \mathcal{F}^{-1}[W_0] = \mathcal{F}^{-1}\left[\sum_{k=0}^{K} C_k y^k e^{-i2\pi\gamma y} \right] = \sum_{k=0}^{K} C_k \left(\frac{1}{i2\pi} \right)^k D^k \delta_\gamma \quad .$$

Thus, letting

$$a_k = C_k \left(\frac{1}{i2\pi} \right)^k \quad ,$$

we have

$$v = T_\lambda[w_0] = T_\lambda\left[\sum_{k=0}^{K} a_k D^k \delta_\gamma \right] = \sum_{k=0}^{K} a_k D^k \delta_{\gamma+\lambda} = \sum_{k=0}^{K} a_k D^k \delta_\mu \quad . \qquad \blacksquare$$

Now for theorem 38.7, which, as you may recall, is:

Let f be a polynomial with distinct roots λ_1, λ_2, ... and λ_K (and no others). For each λ_k let M_k be the corresponding multiplicity. Then the general solution to $f u = 0$ is

$$u = \sum_{k=1}^{K} \sum_{m=0}^{M_k-1} c_{k,m} D^m \delta_{\lambda_k} \tag{38.17}$$

where the $c_{k,m}$'s are arbitrary constants.

PROOF (of theorem 38.7):   By the assumptions on f,

$$f(x) = A_N(x - \lambda_1)^{M_1}(x - \lambda_2)^{M_2} \cdots (x - \lambda_K)^{M_K}$$

for some nonzero constant A_N. The equation $fu = 0$ can then be written as

$$A_N(x - \lambda_1)^{M_1}(x - \lambda_2)^{M_2} \cdots (x - \lambda_K)^{M_K} u(x) = 0 \quad .$$

Dividing out A_N leaves us with

$$(x - \lambda_1)^{M_1}(x - \lambda_2)^{M_2} \cdots (x - \lambda_K)^{M_K} u(x) = 0 \tag{38.18}$$

as an equation equivalent to the equation $fu = 0$.

Now, assume u satisfies $fu = 0$. Then it satisfies equation (38.18), which, since the leading factors are all elementary multipliers, we can rewrite as

$$(x - \lambda_1)^{M_1} \left[(x - \lambda_2)^{M_2} \cdots (x - \lambda_K)^{M_K} u(x) \right] = 0 \quad .$$

As noted in lemma 38.6 on page 680, it follows that

$$(x - \lambda_2)^{M_2} \cdots (x - \lambda_K)^{M_K} u(x) = \sum_{m=0}^{M_1-1} c^1_{1,m} D^m \delta_{\lambda_1}(x) \tag{38.19}$$

where the $c^1_{1,m}$'s are constants (the superscript ^{1} just indicates that this is a "first set" of constants). This, according to theorem 38.3 on page 677, means that

$$(x - \lambda_3)^{M_3} \cdots (x - \lambda_K)^{M_K} u(x) = v(x) + w(x)$$

where v is any particular solution to

$$(x - \lambda_2)^{M_2} v(x) = \sum_{m=0}^{M_1-1} b_{1,m} D^m \delta_{\lambda_1}(x)$$

and w is a corresponding solution to the homogeneous equation

$$(x - \lambda_2)^{M_2} w(x) = 0 \quad .$$

From lemma 38.8 and our study of homogeneous equations, we know there are then two sets of constants

$$\left\{ c^2_{1,0}, c^2_{1,1}, \ldots, c^2_{1,M_1-1} \right\} \quad \text{and} \quad \left\{ c^2_{2,0}, c^2_{2,1}, \ldots, c^2_{2,M_2-1} \right\}$$

such that

$$v = \sum_{k=0}^{M_1-1} c^2_{1,m} D^k \delta_{\lambda_1} \quad \text{and} \quad w = \sum_{m=0}^{M_2-1} c^2_{2,m} D^m \delta_{\lambda_2} \quad .$$

So,

$$(x - \lambda_3)^{M_3} \cdots (x - \lambda_K)^{M_K} u(x) = v(x) + w(x)$$

$$= \sum_{m=0}^{M_1-1} c^2_{1,m} D^m \delta_{\lambda_1}(x) + \sum_{m=0}^{M_2-1} c^2_{2,m} D^m \delta_{\lambda_2}(x)$$

$$= \sum_{k=1}^{2} \sum_{m=0}^{M_k-1} c^2_{k,m} D^m \delta_{\lambda_k}(x) \quad .$$

Obviously, the process leading from equation (38.19) to the last line in the previous paragraph can be repeated, again and again, to obtain

$$(x - \lambda_4)^{M_4} \cdots (x - \lambda_K)^{M_K} u(x) = \sum_{k=1}^{3} \sum_{m=0}^{M_k-1} c_{k,m}^3 D^m \delta_{\lambda_k}(x) \quad ,$$

$$(x - \lambda_5)^{M_5} \cdots (x - \lambda_K)^{M_K} u(x) = \sum_{k=1}^{4} \sum_{m=0}^{M_k-1} c_{k,m}^4 D^m \delta_{\lambda_k}(x) \quad ,$$

$$\vdots$$

$$u(x) = \sum_{k=1}^{K} \sum_{m=0}^{M_k-1} c_{k,m}^K D^m \delta_{\lambda_k}(x)$$

where the $c_{k,m}^n$'s are all constants.

This confirms that every solution to $fu = 0$ is of the form given in formula (38.17). To complete the proof we need to show that, whenever u is of the form given in formula (38.17), then u satisfies $fu = 0$.

So assume

$$u(x) = \sum_{k=1}^{K} \sum_{m=0}^{M_k-1} c_{k,m} D^m \delta_{\lambda_k}(x)$$

where the $c_{k,m}$'s are constants. Then

$$fu = f \sum_{k=1}^{K} \sum_{m=0}^{M_k-1} c_{k,m} D^m \delta_{\lambda_k} = \sum_{k=1}^{K} \sum_{m=0}^{M_k-1} c_{k,m} f D^m \delta_{\lambda_k} \quad .$$

But, in each term with $k = 1$, we have

$$f(x) D^m \delta_{\lambda_1}(x) = \left[A_0 (x - \lambda_1)^{M_1} (x - \lambda_2)^{M_2} \cdots (x - \lambda_K)^{M_K} \right] D^m \delta_{\lambda_1}(x)$$

$$= \left[A_0 (x - \lambda_2)^{M_2} \cdots (x - \lambda_K)^{M_K} \right] \left[(x - \lambda_1)^{M_1} D^m \delta_{\lambda_1}(x) \right]$$

where $m < M_1$. Using an identity from lemma 38.6 on page 680, we discover that

$$(x - \lambda_1)^{M_1} D^m \delta_{\lambda_1}(x) = (x - \lambda_1)^{M_1 - m - 1} (x - \lambda_1)^{m+1} D^m \delta_{\lambda_1}(x)$$

$$= (x - \lambda_1)^{M_1 - m - 1} \cdot 0$$

$$= 0 \quad .$$

Plugging this into the previous string of equalities gives

$$f(x) D^m \delta_{\lambda_1}(x) = 0 \quad .$$

By similar arguments, we obtain

$$f(x) D^m \delta_{\lambda_k}(x) = 0 \qquad \text{for} \quad k = 2, 3, \ldots, K \quad .$$

Thus,

$$fu = \sum_{k=1}^{K} \sum_{m=0}^{M_k-1} c_{k,m} f D^m \delta_{\lambda_k} = \sum_{k=1}^{K} \sum_{m=0}^{M_k-1} c_{k,m} \cdot 0 = 0 \quad . \qquad \blacksquare$$

38.3 Nonhomogeneous Equations with Polynomial Factors

We now shift our attention to finding particular solutions to equations of the form $fu = g$ where f is a polynomial and g is something other than the zero function. For the most part, we will restrict g to being a classical function. This will simplify our discussions somewhat and still allow us to cover the sort of equations that tend to be of greatest practical interest. Besides, if you check back, you will see that we have already dealt with the equations in which g is a delta function or the derivative of a delta function in lemma 38.6 on page 680 and the proof of lemma 38.8 on page 685.

Three broad classes of equations will be discussed. They are distinguished by whether the roots of f are real or not, and whether or not g is a function that "shares roots" with f.

Before starting, let's observe that our main concern should be the case where g is the unit constant 1. After all, if u_0 is a generalized function satisfying $fu_0 = 1$ and g is some other generalized function, then, assuming the product $u_0 g$ exists,

$$f \cdot [u_0 g] = [fu_0] \cdot g = 1 \cdot g = g \quad .$$

Thus, given a solution u_0 to $fu = 1$, we can find a solution to $fu = g$ by simply setting $u = u_0 g$ (provided the product exists).

Also, in the discussions which follow, don't forget that the general solution to $fu = g$ is given by $u = u_0 + w$ where u_0 is any single solution to $fu = g$ and w is the general solution to the corresponding homogeneous equation, $fw = 0$. It will probably also be a good idea to remember how to find w from the roots of f (theorem 38.7 on page 684).

First Case (No Real Roots)

The easiest equations to solve are those in which the factor, f, has no real roots. This means that the value of $f(x)$ is never zero when x is real, and that a solution $fu = g$ can be found by simply "dividing through" by f.

!▶*Example 38.6:*   *Consider solving*
$$(x - i)\, u(x) = 1 \quad .$$

Classically, the solution to this equation is the classical reciprocal of $f(x) = x - i$,

$$u_0(x) = \frac{1}{x - i} \quad .$$

Fortunately, because $x - i$ has no real roots, its reciprocal, $u_0(x)$, is a continuous and bounded function on the real line. So u_0 defines a generalized function and, so, can be used as a generalized function solution to our equation.

To obtain a general solution for this equation, we add u_0 to the general solution w of the corresponding homogeneous equation. In this case, the corresponding homogeneous equation is $(x - i)w(x) = 0$, and, since $\lambda = i$ is the only root of $x - i$, we know that $w = c\delta_i$ where c is an arbitrary constant. Thus, the general solution to

$$(x - i)\, u(x) = 1$$

is

$$u(x) = \frac{1}{x - i} + c\,\delta_i \quad .$$

Whenever f is a polynomial with no real roots, its reciprocal

$$u_0(x) = \frac{1}{f(x)}$$

is a continuous and bounded function on the real line that satisfies $f u_0 = 1$. Thus, if g is any classical function, then the product $u_0 g$ is defined and, as we saw above, is a solution to $f u = g$.

?▶ Exercise 38.5:   *Verify that one solution to*

$$\left[x^2 + 1\right] u(x) = \sin(x)$$

is

$$\frac{\sin(x)}{x^2 + 1} \quad,$$

and that the general solution to this equation is

$$u(x) = \frac{\sin(x)}{x^2 + 1} + a\delta_i(x) + b\delta_{-i}(x)$$

where a and b are arbitrary constants.

Second Case (Shared Real Roots)

In general, dividing through by f when f has real roots will lead to an expression that does not define a generalized function. If we are lucky, though, the function on the right will have roots that just happen to "cancel out" the troublesome roots of f.

!▶ Example 38.7:   *Consider the equation*

$$xu(x) = \sin(x) \quad.$$

Here, $f(x) = x$, which vanishes at 0. Its classical reciprocal, $1/x$, is not integrable at $x = 0$ and, so, does not define a generalized function. Fortunately, however, $\sin(x)$ also vanishes at $x = 0$, and it, divided by f,

$$\frac{\sin(x)}{x} \quad,$$

is our well-known sinc function — a perfectly good generalized function (in fact, it's classically transformable). So, we can use $\mathrm{sinc}(x)$ as a solution to $xu(x) = \sin(x)$. And since $c\delta$ is the general solution to the corresponding homogeneous equation, $xu(x) = 0$, the general solution to $xu(x) = \sin(x)$ is given by

$$u(x) = \mathrm{sinc}(x) + c\,\delta(x)$$

where c is an arbitrary constant.

You can see that, whenever the classical quotient g/f gives an exponentially integrable function, then that quotient is a solution to $f u = g$. Clearly, for this to occur, any vanishing of f at a point on the real line must be suitably balanced by g vanishing at that same point. Otherwise, the quotient g/f will not be integrable around that point and will not define a generalized function.

?▶ Exercise 38.6:   *Why can we not use*

$$u_0(x) = \frac{\cos(x)}{x}$$

as a generalized function solution to $xu(x) = \cos(x)$?

Third Case (Uncanceled Real Roots)

Sometimes, when the classical quotient g/f does not give a legitimate generalized function, we can still find a legitimate generalized solution via complex translation.

!▶ **Example 38.8:** *Consider finding a generalized function solution u_0 to*

$$x u(x) = 1 \quad . \tag{38.20}$$

Here, the classical solution $1/x$ cannot be used as a generalized function solution because $1/x$ cannot be viewed as a classical function. But let's use the translation operator with a complex shift, say i, on both sides of this equation. On the left-hand side, we have

$$T_i[x u(x)] = T_i[x] \, T_i[u] = (x - i) T_i[u] = (x - i) \, \widehat{u}(x) \quad ,$$

where, for convenience, we are letting $\widehat{u} = T_i[u]$. On the right-hand side, we simply have $T_i[1] = 1$. So our equation becomes

$$(x - i) \, \widehat{u}(x) = 1 \quad .$$

Since $x - i$ has no real roots, we can divide through by $x - i$, obtaining

$$\widehat{u}_0(x) = \frac{1}{x - i}$$

as one solution to the previous equation. Translating back (i.e., translating by $-i$), then gives us

$$u_0(x) = T_{-i}[\widehat{u}_0]|_x = T_{-i}\left[\frac{1}{x - i}\right]$$

as a solution to $x u(x) = 1$.[3]

Adding the general solution for $x u(x) = 0$ to $u_0(x)$ then gives

$$u(x) = T_{-i}\left[\frac{1}{x - i}\right] + c\,\delta(x) \tag{38.21}$$

(with c being an arbitrary constant) as a general solution to equation (38.20).

In general, if f is any polynomial, then we can find a *complex* number a such that the variable translation of f by a yields a polynomial with no real roots. Applying that translation operator to the equation $f u = g$, and letting $\widehat{f} = T_a[f]$, $\widehat{u} = T_a[u]$ and $\widehat{g} = T_a[g]$, we get an equation $\widehat{f}\,\widehat{u} = \widehat{g}$ in which the classical reciprocal of $\widehat{f}$ is a legitimate generalized function. Thus, as we saw in the first case considered, a solution $\widehat{u}$ can be found by "dividing through by $\widehat{f}$": Well, to be more precise, a solution $\widehat{u}$ is given by the product of $\widehat{g}$ with the (classical) reciprocal of $\widehat{f}$, provided that product exists. We can then obtain a solution to our original equation by "translating back", that is, by setting $u_0 = T_{-a}[\widehat{u}]$.

The catch is that our translation of g, $\widehat{g} = T_a[g]$, might not be a classical function, making the existence of the product $\widehat{f}^{-1}\widehat{g}$ a significant issue. Fortunately, g is often a simple multiplier, and, as we already know (recall theorem 35.7 on page 599 and the exercise following it), any translation of such a function is another simple multiplier. In fact, we know that the generalized translation of such a function is just the classical translation. Thus, if g is a simple multiplier, then the product $\widehat{f}^{-1}\widehat{g}$ is given by the classical function

$$\widehat{u}_0(x) = \frac{g(x - a)}{f(x - a)} \quad .$$

---

[3] You may want to go back and re-read the discussion on the difference between classical and generalized translations starting on page 603.

Even so, "translating this back" to obtain u_0 will not generally give us a classical function unless some of the roots of g match all the real roots of f (in which case, there was no real need for any translation to begin with!).

!▶ Example 38.9:   *Consider solving*

$$xu(x) \;=\; \cos(x) \quad.$$

Here $f(x) = x$ and $g(x) = \cos(x)$. Translating the variable by i, we see that

$$\widehat{f}(x) \;=\; T_i[x] \;=\; x - i \qquad and \qquad \widehat{g}(x) \;=\; T_i[\cos(x)] \;=\; \cos(x - i) \quad.$$

So, applying T_i to both sides of our equation gives

$$(x - i)T_i[u(x)] \;=\; \cos(x - i) \quad.$$

Dividing through by $x - i$, we get as one solution

$$T_i[u_0(x)] \;=\; \frac{\cos(x - i)}{x - i} \quad,$$

which can be treated as a classical function since it is continuous and bounded on the real line. Translating back, we have

$$u_0(x) \;=\; T_{-i}\left[\frac{\cos(x - i)}{x - i}\right]$$

as a particular solution to $xu(x) = \cos(x)$.

The corresponding homogeneous equation is simply $xw(x) = 0$, whose general solution we already know to be $w = c\delta$ where c is an arbitrary constant. Adding this to u_0, we obtain

$$u(x) \;=\; T_{-i}\left[\frac{\cos(x - i)}{x - i}\right] + c\delta(x)$$

as a general solution to $xu(x) = \cos(x)$.

In the two examples above, we initially translated the variable by i. In practice, any shift a could have been used provided the shifted polynomial, $f(x - a) = x - a$, has no real roots. This means that any complex number with a nonzero imaginary part could have been used for the shift in the above. For example, if we had used $a = 2 + 3i$ in the last example, we would have obtained

$$u_0(x) \;=\; T_{-2-3i}\left[\frac{\cos(x - 2 - 3i)}{x - 2 - 3i}\right]$$

as a solution to $xu(x) = \cos(x)$.

More generally, let f be any polynomial. It will have, say, K distinct roots λ_1, λ_2, ... and λ_K with corresponding multiplicities M_1, M_2, ... and M_K. For some constant A, the polynomial can be written in factored form as

$$f(x) \;=\; A(x - \lambda_1)^{M_1}(x - \lambda_2)^{M_2} \cdots (x - \lambda_K)^{M_K} \quad.$$

Observe, then, that

$$f(x - a) \;=\; A(x - [a + \lambda_1])^{M_1}(x - [a + \lambda_2])^{M_2} \cdots (x - [a + \lambda_K])^{M_K} \quad.$$

So (and this should come as no surprise) every root of $\widehat{f}(x) = f(x - a)$ can be found by just adding a to a root of $f(x)$. Consequently, to ensure that $f(x - a)$ has no real roots, we must choose a so that the imaginary part of each $a + \lambda_k$ is nonzero. In other words, we should be sure to choose a so that

$$\text{Im}[a] \;\neq\; -\text{Im}[\lambda_k] \qquad for \quad k = 1, 2, \ldots, K \quad.$$

!►*Example 38.10:*   *Consider solving*

$$\left[x^3 + x\right] u(x) = 1 \quad . \tag{38.22}$$

Here, since

$$f(x) = x^3 + x = (x - 0)(x - i)(x - (-i)) \quad ,$$

we do not want to use variable translation by $a = i$ or by $a = -i$ or by any other value of a whose imaginary part is ± 1. (Nor do we want a to be a purely real value, since $f(x - a)$ would then have a real root, namely a !) On the other hand, we can use $a = 3i$. Doing so, we obtain

$$(x - 3i)(x - 4i)(x - 2i)T_{3i}[u(x)] = T_{3i}[1] = 1 \quad .$$

One solution to this last equation is

$$T_{3i}[u_0(x)] = \frac{1}{(x - 3i)(x - 4i)(x - 2i)} \quad .$$

Thus,

$$u_0(x) = T_{-3i}\left[\frac{1}{(x - 3i)(x - 4i)(x - 2i)}\right]$$

is a solution to our original equation.

By now, you may be able to just look at the factorization of $f(x)$ and tell that a general solution to the corresponding homogeneous equation is

$$a\delta(x) + b\delta_i(x) + c\delta_{-i}(x)$$

where a, b and c are arbitrary constants. Thus, a general solution to equation (38.22) is

$$u(x) = T_{-3i}\left[\frac{1}{(x - 3i)(x - 4i)(x - 2i)}\right] + a\delta(x) + b\delta_i(x) + c\delta_{-i}(x)$$

where a, b and c are arbitrary constants.

Aside from requiring that $f(x - a)$ has no real roots, there are no restrictions on the choice of the shift a. In practice, choose something simple, such as $a = i$ (assuming $f(x - i)$ has no real roots). This does not mean that the particular solution obtained does not depend on a. It does, to a certain extent. But by adding a general solution to the corresponding homogeneous problem, we take into account all the possible different particular solutions.

38.4   The Pole Functions

As has been noted previously, because the classical functions

$$\frac{1}{x} \quad , \quad \frac{1}{x^2} \quad , \quad \frac{1}{x^3} \quad , \quad \cdots$$

are not exponentially integrable, they do not define generalized functions (again, see exercise 33.7 on page 545). This has hampered us somewhat in our attempts to solve equations of the form $x^k u(x) = 1$. There are, however, generalized functions that can serve as natural analogs to these classical functions, and, in the following, we will develop these analogs.

The Basic Pole Function and Its Transform

Our first goal is to describe the generalized analog to the classical function x^{-1}. We will call this generalized analog the (basic) "pole function" and cleverly denote it by $\text{pole}(x)$.

To help define the pole function as the "natural" analog to x^{-1}, we will require $\text{pole}(x)$ to satisfy two properties: It must be an odd generalized function, and it must satisfy the equation

$$x\,\text{pole}(x) \;=\; 1 \quad . \tag{38.23}$$

These requirements come from the fact that the classical function x^{-1} itself is an odd classical function and

$$x \cdot \frac{1}{x} \;=\; 1 \quad .$$

The general solution to equation (38.23) was derived earlier, in example 38.8 on page 691. From that, we know

$$\text{pole}(x) \;=\; T_{-i}\!\left[\frac{1}{x-i}\right] + c\,\delta(x)$$

for some constant c. The trick now is to find the value c that makes the above a formula for an odd generalized function.

Taking the Fourier transform of the last expression (and applying well-known identities), we get

$$\mathcal{F}\big[\text{pole}(x)\big]\big|_y \;=\; \mathcal{F}\!\left[T_{-i}\!\left[\frac{1}{x-i}\right]\right]\bigg|_y + c\mathcal{F}[\delta(x)]|_y$$

$$=\; e^{-i2\pi(-i)y}\,\mathcal{F}\!\left[\frac{1}{x-i}\right]\bigg|_y + c$$

$$=\; e^{-2\pi y}\,\mathcal{F}\!\left[\frac{i2\pi}{2\pi+i2\pi x}\right]\bigg|_y + c$$

$$=\; e^{-2\pi y}\left[i2\pi\,e^{2\pi y}\,\text{step}(-y)\right] + c \quad .$$

The exponentials cancel, leaving us with

$$\mathcal{F}\big[\text{pole}(x)\big]\big|_y \;=\; i2\pi\,\text{step}(-y) + c \;=\; \begin{cases} i2\pi + c & \text{if } y < 0 \\[4pt] c & \text{if } 0 < y \end{cases} \quad .$$

Nicely enough, this is a rather simple, bounded and piecewise continuous function on the real line. Moreover, it must be an odd function since it is the Fourier transform of a generalized function we are requiring to be odd. This means that, for any $y > 0$, c must satisfy

$$i2\pi + c \;=\; \mathcal{F}\big[\text{pole}(x)\big]\big|_{-y} \;=\; -\mathcal{F}\big[\text{pole}(x)\big]\big|_y \;=\; -c \quad .$$

Solving for c gives

$$c \;=\; -\frac{i2\pi}{2} \;=\; -i\pi \quad .$$

Thus,

$$\mathcal{F}\big[\text{pole}(x)\big]\big|_y \;=\; \begin{cases} i\pi & \text{if } y < 0 \\[4pt] -i\pi & \text{if } 0 < y \end{cases} \quad .$$

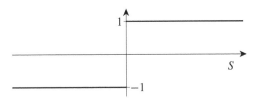

Figure 38.1: The signum function, sgn(s).

Our last formula can be expressed more concisely in terms of the *signum* function — denoted by sgn, sketched in figure 38.1, and defined by[4]

$$
\mathrm{sgn}(s) \;=\; \begin{cases} -1 & \text{if } s < 0 \\ +1 & \text{if } 0 < s \end{cases} \quad .
$$

In terms of this function,

$$
\mathcal{F}\big[\mathrm{pole}(x)\big]\big|_y \;=\; -i\pi\,\mathrm{sgn}(y) \quad .
$$

By near-equivalence and the fact that sgn is an odd function, we also have

$$
\mathcal{F}^{-1}\big[\mathrm{pole}(x)\big]\big|_y \;=\; -i\pi\,\mathrm{sgn}(-y) \;=\; i\pi\,\mathrm{sgn}(y) \quad .
$$

Thus,

$$
\mathrm{pole} \;=\; i\pi\,\mathcal{F}\big[\mathrm{sgn}\big] \quad . \tag{38.24}
$$

This formula describes the one generalized function that satisfies the two properties we were seeking to satisfy. Let us, therefore, make it official and proclaim equation (38.24) to be the definition of the (basic) *pole function*.

Higher Order Pole Functions

To obtain the generalized analog to x^{-2}, we might be tempted to multiply the pole function with itself. Unfortunately, attempting to compute the Fourier transform of this yields

$$
\mathcal{F}\big[\mathrm{pole}\cdot\mathrm{pole}\big]\big|_y \;=\; \mathcal{F}\big[\mathrm{pole}\big]\big|_y * \mathcal{F}\big[\mathrm{pole}\big]\big|_y
$$

$$
=\; [-i\pi\,\mathrm{sgn}(y)] * [-i\pi\,\mathrm{sgn}(y)] \;=\; -\pi^2 \int_{-\infty}^{\infty} \mathrm{sgn}(s)\,\mathrm{sgn}(y-s)\,ds \quad ,
$$

which, as you can easily confirm, is not a convergent integral. That tells us that our intended product, pole · pole, is somewhat problematic.

As an alternative, let us employ the observation that

$$
x^{-2} \;=\; -\frac{d}{dx}\big[x^{-1}\big] \quad .
$$

This suggests that the generalized analog to x^{-2} can be defined using the derivative of the analog to x^{-1}. Let us do so, and officially define the *second order pole function*, denoted pole^2, by

$$
\mathrm{pole}^2 \;=\; -D\,\mathrm{pole} \quad .
$$

---

[4] Because sgn(s) = "the sign of s", some call this the "sign" function (until, that is, they have to talk about expressions involving both sgn(x) and sin(x)).

Since this is the derivative of an odd generalized function, pole^2 is an even generalized function, just as x^{-2} is an even classical function. Moreover, using the product rule and fact that $x\,\text{pole}(x) = 1$, we have

$$x^2\,\text{pole}^2(x) \;=\; -x^2 D\,\text{pole}(x) \;=\; -\Big(D\big[x^2\,\text{pole}(x)\big] \;-\; D\big[x^2\big]\text{pole}(x)\Big)$$

$$= -D[x(x\,\text{pole}(x))] \;+\; 2x\,\text{pole}(x)$$

$$= -D[x \cdot 1] \;+\; 2 \cdot 1$$

$$= -1 + 2 \;=\; 1 \quad .$$

So, pole^2 satisfies

$$x^2\,\text{pole}^2(x) \;=\; 1 \quad ,$$

just as we should expect for our generalized analog of x^{-2}.

   More generally, we have the easily confirmed classical formula

$$x^{-k} \;=\; (-1)^{k-1}\frac{1}{(k-1)!}\frac{d^{k-1}}{dx^{k-1}}\big[x^{-1}\big] \qquad \text{for} \quad k = 2,\,3,\,4,\,\ldots \quad .$$

Accordingly, we define the kth *order pole function* pole^k by

$$\text{pole}^k \;=\; (-1)^{k-1}\frac{1}{(k-1)!}D^{k-1}\,\text{pole} \qquad \text{for} \quad k = 2,\,3,\,4,\,\ldots \quad .$$

For consistency with our notation, we'll go ahead and define the first order pole function to be the basic pole function,

$$\text{pole}^1 \;=\; \text{pole} \quad .$$

   A number of fundamental properties of these pole functions are listed in the next theorem. From these properties, we can see that $\text{pole}^k(x)$ is truly a "natural" generalized analog of the classical function x^{-k} for each positive integer k.

Theorem 38.9 (basic properties of pole functions)
For $k = 1, 2, 3, \ldots$,

1. $D\,\text{pole}^k \;=\; -k\,\text{pole}^{k+1}$,

2. $x^k\,\text{pole}^k(x) \;=\; 1$, *and*

3. $\text{pole}^k(-x) \;=\; (-1)^k\,\text{pole}^k(x)$.

   (It should be noted that the last claim tells us that pole^k is an even generalized function if k is even, and is an odd generalized function if k is odd.)
   We've already verified the claims of this theorem when $k = 1$. Verifying the claims in general is left as an exercise.

?►Exercise 38.7:   *Verify the claims of theorem 38.9 assuming*

   a: $k = 2$ **b:** $k = 3$ **c:** $k > 3$

   The Fourier transform of the kth order pole function is readily obtained from its definition, the already computed transform of the basic pole function, and the differentiation identities for the Fourier transform. For $k = 1$,

$$\mathcal{F}\Big[\text{pole}^1\Big]\Big|_y \;=\; \mathcal{F}[\text{pole}]\big|_y \;=\; -i\pi\,\text{sgn}(y) \quad ,$$

while for $k = 2, 3, 4, \ldots$,

$$\mathcal{F}\left[\text{pole}^k\right]\Big|_y = (-1)^{k-1}\frac{1}{(k-1)!}\mathcal{F}\left[D^{k-1}\text{pole}\right]\Big|_y$$

$$= (-1)^{k-1}\frac{1}{(k-1)!}(i2\pi y)^{k-1}\mathcal{F}\left[\text{pole}\right]\Big|_y$$

$$= (-1)^{k-1}\frac{1}{(k-1)!}(i2\pi y)^{k-1}(-i\pi\,\text{sgn}(y)) = \frac{(-i2\pi)^k}{2(k-1)!}y^{k-1}\,\text{sgn}(y) \quad .$$

Observe that, even though k was assumed to be greater than 1, this last formula reduces to the formula for the transform of pole^1 when $k = 1$. Thus,

$$\mathcal{F}\left[\text{pole}^k\right]\Big|_y = \frac{(-i2\pi)^k}{2(k-1)!}y^{k-1}\,\text{sgn}(y) \qquad \text{for} \quad k = 1,\ 2,\ 3,\ \ldots \quad . \tag{38.25}$$

By near-equivalence and part 3 of theorem 38.9, we have

$$\mathcal{F}^{-1}\left[\text{pole}^k(x)\right]\Big|_y = \mathcal{F}\left[\text{pole}^k(-x)\right]\Big|_y = (-1)^k\mathcal{F}\left[\text{pole}^k(x)\right]\Big|_y \quad .$$

Combined with equation (38.25), this gives

$$\mathcal{F}^{-1}\left[\text{pole}^k\right]\Big|_y = \frac{(i2\pi)^k}{2(k-1)!}y^{k-1}\,\text{sgn}(y) \quad . \tag{38.26}$$

Poles at Different Locations

The pole functions obtained thus far, being the generalized analogs of classical functions of the form $(x-0)^{-k}$, can be considered as the pole functions "at 0". To define corresponding pole functions at any other point $\zeta = a + ib$ in the complex plane, we will simply translate the already defined pole functions by ζ. That is, the *pole function at* ζ and the kth *order pole function at* ζ, denoted pole_ζ and pole_ζ^k, respectively, are defined by

$$\text{pole}_\zeta = T_\zeta[\text{pole}] \qquad \text{and} \qquad \text{pole}_\zeta^k = T_\zeta\left[\text{pole}^k\right] \quad .$$

We are assuming, of course, that k is some positive integer. In particular, then,

$$\text{pole}_0 = \text{pole} \qquad \text{and} \qquad \text{pole}_0^k = \text{pole}^k \quad .$$

The equations obtained earlier involving pole functions at 0 can be converted to corresponding equations involving pole functions at any point ζ in the complex plane through appropriate use of the translation operator. For example, since

$$T_\zeta\left[x^k\,\text{pole}(x)\right] = T_\zeta\left[x^k\right]\cdot T_\zeta\left[\text{pole}^k(x)\right] = (x-\zeta)^k\,\text{pole}_\zeta^k(x) \quad ,$$

the application of T_ζ to both sides of the equation in part 2 of theorem 38.9 yields

$$(x-\zeta)^k\,\text{pole}_\zeta^k(x) = T_\zeta\left[x^k\,\text{pole}(x)\right] = T_\zeta[1] = 1 \quad ,$$

confirming that, in some sense at least, $\text{pole}_\zeta^k(x)$ is a generalized analog of the classical function $(x-\zeta)^{-k}$.

?►Exercise 38.8:   *Verify that, for any complex number ζ and any positive integer k,*

$$\text{pole}_\zeta^k = (-1)^{k-1} \frac{1}{(k-1)!} D^{k-1} \text{pole}_\zeta \quad .$$

The Fourier transforms of the pole functions at ζ can be obtained from equations (38.25) and (38.26), and the Fourier translation identities:

$$\mathcal{F}\left[\text{pole}_\zeta^k\right]\Big|_y = \mathcal{F}\left[T_\zeta\left[\text{pole}^k\right]\right]\Big|_y$$

$$= e^{-i2\pi\zeta y} \mathcal{F}\left[\text{pole}^k\right]\Big|_y = \frac{(-i2\pi)^k}{2(k-1)!} y^{k-1} \text{sgn}(y) e^{-i2\pi\zeta y} \tag{38.27}$$

and

$$\mathcal{F}^{-1}\left[\text{pole}_\zeta^k\right]\Big|_y = \mathcal{F}^{-1}\left[T_\zeta\left[\text{pole}^k\right]\right]\Big|_y$$

$$= e^{i2\pi\zeta y} \mathcal{F}^{-1}\left[\text{pole}^k\right]\Big|_y = \frac{(i2\pi)^k}{2(k-1)!} y^{k-1} \text{sgn}(y) e^{i2\pi\zeta y} \quad . \tag{38.28}$$

Let us now consider the basic pole function at ζ, pole_ζ, when the imaginary part of $\zeta = a+ib$ is not 0. Remember, we are viewing $\text{pole}_\zeta(x)$ as a generalized analog of the classical function

$$f(x) = \frac{1}{x-\zeta} = \frac{1}{x-[a+ib]} \quad .$$

However, when $b \neq 0$, the function f is a well-known classically transformable function. The question naturally arises as to whether these two functions, f and pole_ζ, are the same generalized function. The answer, as illustrated in the next example and exercise, is *no*.

!►Example 38.11:   *We will compare*

$$\text{pole}_{3+i4}(x) \qquad with \qquad f(x) = \frac{1}{x-[3+i4]}$$

by first comparing the Fourier transforms of these functions (because these transforms are easily compared classical functions).

Using equation (38.27) and recalling that $\text{pole}_\zeta = \text{pole}_\zeta^1$, we obtain

$$\mathcal{F}\left[\text{pole}_{3+i4}(x)\right]\Big|_y = \frac{-i2\pi}{2} \text{sgn}(y) e^{-i2\pi[3+i4]y}$$

$$= \begin{cases} i\pi\, e^{-i2\pi[3+i4]y} & if \quad y < 0 \\ -i\pi\, e^{-i2\pi[3+i4]y} & if \quad 0 < y \end{cases} \quad .$$

On the other hand, using a little algebra and formulas derived for the classical transforms in part III, we find that

$$\mathcal{F}\left[\frac{1}{x-[3+i4]}\right]\Big|_y = i2\pi \mathcal{F}\left[\frac{1}{2\pi[4-i3]+i2\pi x}\right]\Big|_y$$

$$= i2\pi\, e^{2\pi[4-i3]y} \text{step}(-y)$$

$$= \begin{cases} i2\pi\, e^{-i2\pi[3+4i]y} & if \quad y < 0 \\ 0 & if \quad 0 < y \end{cases} \quad .$$

These two transforms are clearly not the same. Taking their difference, we see that

$$\mathcal{F}\left[\frac{1}{x - [3 + i4]} - \text{pole}_{3+i4}(x)\right]\bigg|_y$$

$$= \mathcal{F}\left[\frac{1}{x - [3 + i4]}\right]\bigg|_y - \mathcal{F}\left[\text{pole}_{3+i4}(x)\right]\big|_y$$

$$= \begin{cases} i2\pi\, e^{-i2\pi[3+4i]y} - i\pi\, e^{-i2\pi[3+i4]y} & \text{if } y < 0 \\ 0 + i\pi\, e^{-i2\pi[3+i4]y} & \text{if } 0 < y \end{cases}$$

$$= \begin{cases} i\pi\, e^{-i2\pi[3+i4]y} & \text{if } y < 0 \\ i\pi\, e^{-i2\pi[3+i4]y} & \text{if } 0 < y \end{cases} = i\pi\, e^{-i2\pi[3+i4]y} \quad.$$

Thus,

$$\frac{1}{x - [3 + i4]} - \text{pole}_{3+i4}(x) = \mathcal{F}^{-1}\left[i\pi\, e^{-i2\pi[3+i4]y}\right]\bigg|_x = i\pi\, \delta_{3+i4}$$

and hence,

$$\text{pole}_{3+i4}(x) = \frac{1}{x - [3 + i4]} - i\pi\, \delta_{3+i4} \quad.$$

?▶Exercise 38.9: *Show that*

$$\frac{1}{x - [3 - i4]} - \text{pole}_{3-i4}(x) = -i\pi\, \delta_{3-i4} \quad.$$

Of course, the calculations in the above example and exercise can be repeated with the numbers 3 and ± 4 replaced by any pair of real numbers a and b. Doing so gives us

$$\frac{1}{x - [a + ib]} - \text{pole}_{a+ib}(x) = \begin{cases} i\pi\, \delta_{a+ib} & \text{if } 0 < b \\ -i\pi\, \delta_{a+ib} & \text{if } b < 0 \end{cases} \quad. \tag{38.29}$$

To obtain the more general relation involving the pole functions at $\zeta = a + ib$ of positive integral order k, we use the facts that

$$\frac{1}{(x - \zeta)^k} = (-1)^{k-1} \frac{1}{(k-1)!} D^{k-1}\left[\frac{1}{x - \zeta}\right]$$

and (see exercise 38.8)

$$\text{pole}_{\zeta}^k = (-1)^{k-1} \frac{1}{(k-1)!} D^{k-1} \text{pole}_{\zeta} \quad.$$

Combined with equation (38.29), these formulas yield

$$\frac{1}{(x - [a + ib])^k} - \text{pole}_{a+ib}^k(x) = \begin{cases} (-1)^k \dfrac{-i\pi}{(k-1)!} D^{k-1}\delta_{a+ib} & \text{if } 0 < b \\ (-1)^k \dfrac{i\pi}{(k-1)!} D^{k-1}\delta_{a+ib} & \text{if } b < 0 \end{cases} \quad.$$

So,

$$\text{pole}_{a+ib}^k(x) = \frac{1}{(x - [a + ib])^k} - \begin{cases} (-1)^k \dfrac{-i\pi}{(k-1)!} D^{k-1}\delta_{a+ib} & \text{if } 0 < b \\ (-1)^k \dfrac{i\pi}{(k-1)!} D^{k-1}\delta_{a+ib} & \text{if } b < 0 \end{cases} \quad.$$

38.5 Pole Functions in Transforms, Products and Solutions
Pole and Step Functions

Remember, the basic pole function is defined by

$$\text{pole} = i\pi \mathcal{F}[\text{sgn}] \tag{38.30}$$

where sgn, the signum function, is an odd function defined by

$$\text{sgn}(x) = \begin{cases} -1 & \text{if } x < 0 \\ +1 & \text{if } 0 < x \end{cases} \quad .$$

By applying a little algebra, near-equivalence, and the fact that $\text{sgn}(-x) = -\text{sgn}(x)$, we can convert formula (38.30) to formulas for the Fourier transforms for the signum function,

$$\mathcal{F}[\text{sgn}] = \frac{1}{i\pi} \text{pole} \quad \text{and} \quad \mathcal{F}^{-1}[\text{sgn}] = -\frac{1}{i\pi} \text{pole} \quad .$$

Additional Fourier transform identities involving the signum function and the step functions can then be derived using any of a number of easily derived relations between the signum function and the step functions.

!▶**Example 38.12:**   *Observe that*

$$\text{sgn}(x) = \begin{cases} -1 & \text{if } x < 0 \\ +1 & \text{if } 0 < x \end{cases} = -1 + \begin{cases} 0 & \text{if } x < 0 \\ +2 & \text{if } 0 < x \end{cases} = -1 + 2\,\text{step}(x) \quad .$$

Solving this for the step function yields

$$\text{step}(x) = \frac{1}{2}\text{sgn}(x) + \frac{1}{2} \quad .$$

Hence,

$$\mathcal{F}[\text{step}(x)]\big|_y = \mathcal{F}\left[\frac{1}{2}\text{sgn}(x) + \frac{1}{2}\right]\bigg|_y$$

$$= \frac{1}{2}\mathcal{F}[\text{sgn}]\big|_y + \frac{1}{2}\mathcal{F}[1]\big|_y = \frac{1}{i2\pi}\text{pole}(y) + \frac{1}{2}\delta(y) \quad , \tag{38.31}$$

giving us a formula for the Fourier transform of the step function in terms of the pole function. (Compare this to the formula previously obtained for $\mathcal{F}[\text{step}]$ in example 35.12 on page 603.)

?▶**Exercise 38.10:**   *Verify that*

$$\mathcal{F}[\text{step}(-x)]\big|_y = \frac{-1}{i2\pi}\text{pole}(y) + \frac{1}{2}\delta(y) \quad .$$

These calculations can be expanded upon, and the Fourier transforms of such functions as $x^2\,\text{step}(x)$ and $|x|$ can be obtained in terms of pole functions by using either the above results with the Fourier differential identities, or by using equations (38.25) and (38.26) on page 697. You may have the pleasure of doing these computations yourself (exercise 38.18 at the end of this chapter).

Products Involving Pole Functions

As noted earlier, the product pole · pole is problematical. Attempting to compute its Fourier transform leads to a convolution integral that is not convergent. Similarly, as you can verify yourself, we obtain a divergent convolution integral whenever we attempt to compute a Fourier transform of the supposed product of any two pole functions. Consequently, we will not attempt such products.

On the other hand, if $f(x)$ is a simple multiplier, k is a positive integer, and ζ is any complex value, then we know the product $f(x) \cdot \text{pole}_\zeta^k(x)$ is well defined, simply because of the nature of simple multipliers.

One class of simple multipliers of particular interest to us are the polynomials, especially those polynomials of the form $(x - \zeta)^n$ for some positive integer n. Let us first examine a very simple case.

Lemma 38.10
Let k be a positive integer greater than 1. Then

$$x \, \text{pole}^k(x) = \text{pole}^{k-1}(x) \quad .$$

PROOF: It helps to consider the Fourier transform of $x \, \text{pole}^k(x)$. Using a differential identity, formula (38.25) for the transform of pole^k, and the product rule yields

$$\mathcal{F}\left[x \, \text{pole}^k(x)\right]\Big|_y = \frac{-1}{i2\pi} D\mathcal{F}\left[\text{pole}^k(x)\right]\Big|_y$$

$$= \frac{-1}{i2\pi} D\left[\frac{(-i2\pi)^k}{2(k-1)!} y^{k-1} \, \text{sgn}(y)\right]$$

$$= \frac{(-i2\pi)^{k-1}}{2(k-1)!} \left(\left[Dy^{k-1}\right]\text{sgn}(y) + y^{k-1} D \, \text{sgn}(y)\right) \quad . \tag{38.32}$$

Keeping in mind that $k \geq 2$, we have

$$Dy^{k-1} = \frac{d}{dy}y^{k-1} = (k-1)y^{k-2} \quad ,$$

$$\frac{k-1}{(k-1)!} = \frac{k-1}{(k-1)(k-2)\cdots 3 \cdot 2 \cdot 1} = \frac{1}{(k-2)\cdots 3 \cdot 2 \cdot 1} = \frac{1}{(k-2)!}$$

and

$$D \, \text{sgn}(y) = D[2 \, \text{step}(y) - 1] = 2D \, \text{step}(y) = 2\delta(y) \quad .$$

Thus,

$$y^{k-1} D \, \text{sgn}(y) = 2y^{k-1}\delta(y) = 2 \cdot 0^{k-1}\delta(y) = 0 \quad ,$$

and equation (38.32) reduces to

$$\mathcal{F}\left[x \, \text{pole}^k(x)\right]\Big|_y = \frac{(-i2\pi)^{k-1}}{2(k-1)!}\left((k-1)y^{k-2}\,\text{sgn}(y) + 0\right) = \frac{(-i2\pi)^{k-1}}{2(k-2)!}y^{k-2}\,\text{sgn}(y) \quad .$$

On the other hand, again applying formula (38.25), we see that

$$\mathcal{F}\left[\text{pole}^{k-1}(x)\right]\Big|_y = \frac{(-i2\pi)^{k-1}}{2([k-1]-1)!}y^{[k-1]-1}\,\text{sgn}(y) = \frac{(-i2\pi)^{k-1}}{2(k-2)!}y^{k-2}\,\text{sgn}(y) \quad .$$

So,

$$\mathcal{F}\left[x\,\text{pole}^k(x)\right]\Big|_y = \frac{(-i2\pi)^{k-1}}{2(k-2)!}y^{k-2}\,\text{sgn}(y) = \mathcal{F}\left[\text{pole}^{k-1}(x)\right]\Big|_y \quad,$$

telling us that

$$x\,\text{pole}^k(x) = \text{pole}^{k-1}(x) \quad.\qquad\blacksquare$$

More general products follow easily from this lemma. Consider, for example, $x^n\,\text{pole}^k(x)$ where n is some positive integer. If $n < k$, then repeated applications of the lemma gives

$$\begin{aligned}
x^n\,\text{pole}^k(x) &= x^{n-1}\left[x\,\text{pole}^k(x)\right] \\
&= x^{n-1}\,\text{pole}^{k-1}(x) \\
&= x^{n-2}\left[x\,\text{pole}^{k-1}(x)\right] \\
&= x^{n-2}\,\text{pole}^{k-2}(x) \\
&\quad\vdots \\
&= x^{n-n}\,\text{pole}^{k-n}(x) = \text{pole}^{k-n}(x) \quad.
\end{aligned}$$

If $n = k$, then, as noted in theorem 38.9 on page 696,

$$x^n\,\text{pole}^k(x) = x^k\,\text{pole}^k(x) = 1 \quad.$$

And if $n > k$, then

$$x^n\,\text{pole}^k(x) = x^{n-k}\left[x^k\,\text{pole}^k(x)\right] = x^{n-k}\cdot 1 = x^{n-k} \quad.$$

The corresponding equations involving $(x - \zeta)^n\,\text{pole}_\zeta^k(x)$ for nonzero values of ζ can then be obtained by applying T_ζ to the above. The result is summarized below.

Theorem 38.11
Let n and k be positive integers, and let ζ be any point in the complex plane. Then

$$(x - \zeta)^n\,\text{pole}_\zeta^k(x) = \begin{cases} \text{pole}_\zeta^{k-n}(x) & \text{if } n < k \\ 1 & \text{if } n = k \\ (x - \zeta)^{n-k} & \text{if } n > k \end{cases} \quad.$$

Using the above theorem and a little algebra, it is a simple matter to reduce the product of any polynomial with any pole function to the sum of another polynomial with a linear combination of pole functions.

!▶Example 38.13: *Consider $\left[x^3 + 4\right]\text{pole}_1^2(x)$. The little algebra involved is*

$$\begin{aligned}
x^3 + 4 &= (x - 1 + 1)^3 + 4 \\
&= (x - 1)^3 + 3(x - 1)^2 + 3(x - 1) + 5 \quad.
\end{aligned}$$

Employing this and theorem 38.11,

$$\left[x^3 + 4\right] \text{pole}_1^2(x) = \left[(x-1)^3 + 3(x-1)^2 + 3(x-1) + 5\right] \text{pole}_1^2(x)$$

$$= (x-1)^3 \, \text{pole}_1^2(x) + 3(x-1)^2 \, \text{pole}_1^2(x)$$

$$+ \; 3(x-1) \, \text{pole}_1^2(x) + 5 \, \text{pole}_1^2(x)$$

$$= (x-1) + 3 + 3 \, \text{pole}_1^1(x) + 5 \, \text{pole}_1^2(x) \quad .$$

?►Exercise 38.11:   *Show that*

$$x \, \text{pole}_i(x) = 1 + i \, \text{pole}_i(x) \quad .$$

This example and exercise illustrate the following corollary of theorem 38.11. Its proof will be left as an exercise.

Corollary 38.12
Let k and ζ be, respectively, a positive integer and a complex number, and assume $f(x)$ is a polynomial of degree n. If $n < k$, then there are constants A_1, A_2, ... and A_{k-n} such that

$$f(x) \, \text{pole}^k(x) = \sum_{j=1}^{k-n} A_j \, \text{pole}_\zeta^j(x) \quad .$$

If, on the other hand, $n \geq k$, then there is a polynomial $h(x)$ of degree $n - k$ and constants A_1, A_2, ... and A_k such that

$$f(x) \, \text{pole}^k(x) = h(x) + \sum_{j=1}^{k} A_j \, \text{pole}_\zeta^j(x) \quad .$$

?►Exercise 38.12:   *Prove the above corollary.*

Finally, we should at least mention the following corollary of theorem 38.11.

Corollary 38.13
Suppose g is a simple multiplier, n is a positive integer, and ζ is a complex number. Then

$$\text{pole}_\zeta^k(x) \left[(x-\zeta)^n g(x)\right] = (x-\zeta)^{n-k} g(x) \qquad \text{for} \quad k = 1, 2, 3, \ldots, n \quad .$$

?►Exercise 38.13:   *Prove the last corollary above.*

!►Example 38.14:   *Because*

$$\sin(x) = x \, \text{sinc}(x) \quad ,$$

corollary 38.13 assures us that

$$\text{pole}(x) \sin(x) = \text{sinc}(x) \quad .$$

Pole Function Solutions to Equations

Earlier in this chapter (section 38.3), we developed one way to find a particular solution u_0 to

$$f(x)\,u(x) \;=\; 1$$

when $f(x)$ is a polynomial. Since that particular solution involved functions and translations of functions of the form $(x - \zeta)^{-k}$, it should come as no surprise to learn that an alternative particular solution can be expressed in terms of the associated pole functions. That solution (which corresponds to a classical partial fraction expansion) is described in the next theorem. The details of that theorem's proof are left to you.

Theorem 38.14
Suppose f is a polynomial with distinct roots λ_1, λ_2, ... and λ_K (and no others). For each λ_k, let M_k denote the corresponding multiplicity, and let $A_{k,1}$, $A_{k,2}$, ... and A_{k,M_k} be the coefficients in the classical partial fraction expansion

$$\frac{1}{f(x)} \;=\; \sum_{k=1}^{K}\sum_{m=1}^{M_k} \frac{A_{k,m}}{(x - \lambda_k)^m} \quad .$$

Then the generalized function

$$u_0(x) \;=\; \sum_{k=1}^{K}\sum_{m=1}^{M_k} A_{k,m}\,\mathrm{pole}^m_{\lambda_k}(x)$$

satisfies

$$f(x)\,u_0(x) \;=\; 1 \quad .$$

?►Exercise 38.14:   *Verify the above theorem.*

!►Example 38.15:   *Consider the equation*

$$\left[(x-1)^2(x-3)\right]u(x) \;=\; 1 \quad . \tag{38.33}$$

Theorem 38.14 tells us that one solution to this equation is

$$u_0(x) \;=\; A\,\mathrm{pole}^1_1(x) \;+\; B\,\mathrm{pole}^2_1(x) \;+\; C\,\mathrm{pole}^1_3(x)$$

where A, B and C are the constants in the corresponding partial fraction expansion

$$\frac{1}{(x-1)^2(x-3)} \;=\; \frac{A}{x-1} \;+\; \frac{B}{(x-1)^2} \;+\; \frac{C}{x-3} \quad .$$

These constants are easily determined by, say, first multiplying the above equation through by $(x-1)^2(x-3)$, which gives us

$$1 \;=\; A(x-3)(x-1) \;+\; B(x-3) \;+\; C(x-1)^2 \quad ,$$

and then replacing x in this equation with 3, 1 and 0, respectively, giving us the easily solved system

$$4C \;=\; 1 \quad , \qquad -2B \;=\; 1 \qquad and \qquad 3A \;-\; 3B + C \;=\; 1 \quad .$$

Thus,

$$C \;=\; \frac{1}{4} \quad , \qquad B \;=\; -\frac{1}{2} \qquad and \qquad A \;=\; \frac{1}{3}[1 + 3B - C] \;=\; -\frac{1}{4} \quad ,$$

which, in turn, means that

$$u_0(x) = -\frac{1}{4}\operatorname{pole}_i^1(x) - \frac{1}{2}\operatorname{pole}_i^2(x) + \frac{1}{4}\operatorname{pole}_3^1(x)$$

is one solution to equation (38.33).

Additional Exercises

38.15. Find the solution $u(x)$ to each of the following if it exists:

a. $\dfrac{e^{2x}}{x - 3i} u(x) = e^{6x}$
b. $\dfrac{e^{2x}}{x - 3i} u(x) = 0$

c. $\dfrac{e^{2x}}{x - 3i} u(x) = \delta(x)$
d. $\dfrac{e^{2x}}{x - 3i} u(x) = \delta_{3i}(x)$

38.16. Find the general solution $u(x)$ to each of the following homogeneous equations:

a. $(x - i2\pi) u(x) = 0$
b. $(x - 4) u(x) = 0$

c. $(x - 4)^3 u(x) = 0$
d. $\left[x^3 - x^2\right] u(x) = 0$

e. $\left[x^2 + 5x + 6\right] u(x) = 0$
f. $\left[x^4 - 1\right] u(x) = 0$

g. $(x - 1)^4 u(x) = 0$
h. $\left[x^2 - 6x + 25\right] u(x) = 0$

38.17. Find a particular solution $u_0(x)$ and a general solution $u(x)$ to each of the following nonhomogeneous equations:

a. $(x - i2\pi) u(x) = 1$
b. $(x - i2\pi) u(x) = \delta_{i2\pi}(x)$

c. $\left[x^2 + 1\right] u(x) = 1$
d. $\left[x^2 + 5x + 6\right] u(x) = \delta$

e. $\left[x^2 + 5x + 6\right] u(x) = 1$
f. $\left[x^2 + 5x + 6\right] u(x) = (x + 2)^2$

38.18 a. Let k denote an arbitrary positive integer, and compute the following transforms in terms of pole functions:

i. $\mathcal{F}\left[y^{k-1}\operatorname{sgn}(y)\right]$
ii. $\mathcal{F}^{-1}\left[y^{k-1}\operatorname{sgn}(y)\right]$

iii. $\mathcal{F}\left[y^k\operatorname{step}(y)\right]$
iv. $\mathcal{F}\left[y^k\operatorname{step}(-y)\right]$

b. Using the above, find the following transforms:

i. $\mathcal{F}\left[\operatorname{ramp}(y)\right]$
ii. $\mathcal{F}[|y|]$

iii. $\mathcal{F}\left[|y|^k\right]$ for $k = 1, 2, 3, \ldots$

38.19. Find a particular solution to each of the following in terms of pole functions:

    a. $\left[x^2 + 5x + 6\right]u(x) = 1$ **b.** $\left[x^3 - x^2\right]u(x) = 1$

    c. $\left[x^2 + 9\right]u(x) = 1$ **d.** $\left[x^2 - 6x + 25\right]u(x) = 1$

38.20. Verify that each of the following generalized functions equals $\text{pole}(x)$:

    a. $D \ln|x|$ **b.** $T_{-i}\left[\dfrac{1}{x-i}\right] - i\pi\,\delta(x)$

    c. $\dfrac{1}{2}T_i\left[\dfrac{1}{i+x}\right] - \dfrac{1}{2}T_{-i}\left[\dfrac{1}{i-x}\right]$ **d.** $\displaystyle\lim_{\alpha\to\infty} \mathcal{F}[2i\,\text{Arctan}(\alpha y)]|_x$

38.21 a. Assume f is a generalized function whose Fourier transform F is an exponentially integrable function on the real line. Show that the convolution $f * \text{pole}_\zeta^k$ is well defined for each positive integer k and each complex number ζ .

    b. Compute the following convolutions:

        i. $\text{pole} * \text{pole}$ **ii.** $\text{pole} * \text{pole}^2$

        iii. $\text{pole}(x) * \dfrac{1}{1+x^2}$

38.22. Using Fourier transforms and the material developed in this chapter, find general solutions to each of the following differential equations:

    a. $\dfrac{dy}{dt} + 3y = \delta$ **b.** $\dfrac{d^2y}{dt^2} = \delta$

    (These are the equations we tried to solve at the beginning of this chapter.)

Part V

The Discrete Theory

39

Periodic, Regular Arrays

In earlier chapters, we computed Fourier transforms of periodic functions (obtaining regular arrays of delta functions) and Fourier transforms of regular arrays (obtaining periodic functions). You may have even computed a Fourier transform or Fourier series for one or two regular arrays that were also periodic (such as a comb function). In this chapter we will look more closely at the computation of transforms of those delta function arrays that are both periodic and regular. In the process, we will obtain results and formulas that will be particularly useful in developing the "discrete theory" of Fourier analysis in the next chapter.

39.1 The Index Period and Other Basic Notions

Let us consider an arbitrary periodic, regular array of delta functions, which, knowing no better, we will call f. Remember what these terms mean:

1. "f is a regular array of delta functions" means that f can be expressed as

$$f(t) \; = \; \sum_{k=-\infty}^{\infty} f_k \, \delta_{k\Delta t}(t)$$

    where Δt (the *spacing* of the array) is some positive value, and the f_k's (the *coefficients* of the array) are constants.

2. "f is periodic" means that, for some positive value p (the *period* of f),

$$f(t - p) \; = \; f(t) \quad .$$

One such f has been sketched in figure 39.1. Looking at this figure, you can see that the period must clearly be an integral multiple of the spacing; that is,

$$p \; = \; N\Delta t \qquad \text{for some positive integer } N \quad . \tag{39.1}$$

It should also be clear that the f_k's must form a repeating sequence with

$$\ldots \quad , \quad f_N = f_0 \quad , \quad f_{N+1} = f_1 \quad , \quad f_{N+2} = f_2 \quad , \quad f_{N+3} = f_3 \quad , \quad \ldots \quad .$$

In general,

$$f_{k+N} \; = \; f_k \qquad \text{for} \quad k = 0, \pm 1, \pm 2, \pm 3, \ldots \quad .$$

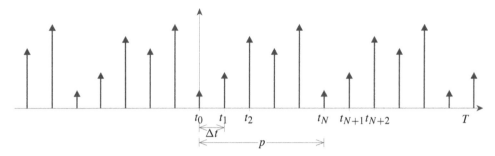

Figure 39.1: A periodic, regular array with spacing Δt, period p and index period N. In this figure $t_k = k\Delta t$ for $k = 0, 1, 2, \ldots$.

This equation will occasionally be referred as the (N^{th} order) *recursion relation* for the array coefficients of f, and the integer N will be called the *index period* for f.

The index period will turn out to be a rather important parameter. For future reference, let us rewrite equation (39.1) as

$$\text{index period of } f \;=\; \frac{\text{period of } f}{\text{spacing of } f} \quad . \tag{39.2}$$

!▶ Example 39.1:   *Consider the periodic, regular array f sketched in figure 39.2 and given by*

$$f \;=\; \cdots + 1\delta_{-4/3} + 2\delta_{-1} + 3\delta_{-2/3} + 3\delta_{-1/3}$$
$$+ \; 1\delta_0 + 2\delta_{1/3} + 3\delta_{2/3} + 3\delta_1$$
$$+ \; 1\delta_{4/3} + 2\delta_{5/3} + 3\delta_2 + 3\delta_{7/3} + \cdots \quad .$$

Equivalently,

$$f \;=\; \sum_{k=-\infty}^{\infty} f_k \, \delta_{k/3}$$

with

$$f_0 \;=\; 1 \quad , \quad f_1 \;=\; 2 \quad , \quad f_2 = 3 \quad , \quad f_3 \;=\; 3 \quad ,$$

and, for every integer k,

$$f_{k+4} \;=\; f_k \quad .$$

By inspection, we see that

the spacing of f is $\Delta t \;=\; \dfrac{1}{3}$,

the index period of f is $N \;=\; 4$,

and

the period of f is $p \;=\; 4 \cdot \dfrac{1}{3} \;=\; \dfrac{4}{3}$.

?▶ Exercise 39.1:   *What is the index period for the array sketched in figure 39.1?*

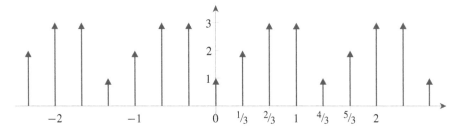

Figure 39.2: The periodic, regular array for example 39.1.

39.2 Fourier Series and Transforms of Periodic, Regular Arrays

Still assuming f is a periodic, regular array with spacing Δt, period p, index period N and formula

$$f(t) = \sum_{k=-\infty}^{\infty} f_k \, \delta_{k\Delta t}(t) \quad , \tag{39.3}$$

let's consider the corresponding Fourier transform $F = \mathcal{F}[f]$. However, for reasons that will soon be clear, let us not simply compute F by taking the transform of the above series.

Qualitative Results

From what we already know about transforms of regular arrays and periodic functions:

$$f \text{ is a regular array with spacing } \Delta t \quad \Longrightarrow \quad F \text{ is periodic with period } \frac{1}{\Delta t} \quad,$$

and

$$f \text{ is periodic with period } p \quad \Longrightarrow \quad F \text{ is a regular array with spacing } \frac{1}{p} \quad.$$

So F is also a periodic, regular array of delta functions, but with

$$\text{spacing of } F = \frac{1}{\text{period of } f} \quad \text{and} \quad \text{period of } F = \frac{1}{\text{spacing of } f} \quad.$$

Computing the index period of F yields

$$\text{index period of } F = \frac{\text{period of } F}{\text{spacing of } F}$$

$$= \frac{\dfrac{1}{\text{spacing of } f}}{\dfrac{1}{\text{period of } f}}$$

$$= \frac{\text{period of } f}{\text{spacing of } f} = \text{index period of } f \quad.$$

Thus,

$$F(\omega) = \sum_{n=-\infty}^{\infty} F_n \, \delta_{n\Delta\omega}(\omega) \quad \text{where} \quad \Delta\omega = \frac{1}{p} \tag{39.4}$$

and the F_n's are constants which, because the index period of F is N (just as for f), must satisfy the recursion formula

$$F_{k+N} = F_k \qquad \text{for} \quad k = 0, \pm1, \pm2, \pm3, \ldots \quad .$$

Relatively simple formulas for computing these coefficients will be derived in the next subsection.

One more thing: Since $F = \mathcal{F}[f]$, we can recover f from F by taking the inverse transform. But if we compute this inverse transform using formula (39.4), we get

$$
\begin{aligned}
f(t) &= \mathcal{F}^{-1}[F]|_t \\
&= \mathcal{F}^{-1}\left[\sum_{n=-\infty}^{\infty} F_n \, \delta_{n\Delta\omega} \right]\Bigg|_t \\
&= \sum_{n=-\infty}^{\infty} F_n \, \mathcal{F}^{-1}[\delta_{n\Delta\omega}]|_t = \sum_{n=-\infty}^{\infty} F_n \, e^{i2\pi n\Delta\omega t} \quad .
\end{aligned}
$$

Thus, since $\Delta\omega = {}^1/_p$,

$$f(t) = \sum_{n=-\infty}^{\infty} F_n \, e^{i2\pi \omega_n t} \qquad \text{where} \quad \omega_n = \frac{n}{p} \quad .$$

While this is not the formula for f we started with (formula (39.3)), you should recognize it as a Fourier series representation for f. Indeed, by the uniqueness of the Fourier series representations (see theorem 37.5 on page 655), this must be *the* Fourier series representation for the periodic, regular array f. Hence, the F_n's must be the Fourier coefficients for f.

What would we have gotten if we had simply computed F by taking the transform of the series in the original formula for f, equation (39.3)? This:

$$F(\omega) = \mathcal{F}\left[\sum_{k=-\infty}^{\infty} f_k \, \delta_{k\Delta t} \right]\Bigg|_{\omega} = \sum_{k=-\infty}^{\infty} f_k \, \mathcal{F}[\delta_{k\Delta t}]|_{\omega} = \sum_{k=-\infty}^{\infty} f_k \, e^{-i2\pi k\Delta t\, \omega} \quad ,$$

which, after letting $n = -k$, must clearly give the Fourier series expansion for F.

This is a good point for us to stop and summarize what we've just derived in a theorem. While we are at it, let us also note that completely analogous results would have been obtained if we had been seeking the inverse Fourier transform of a periodic, regular array F.

Theorem 39.1

The Fourier transform and inverse Fourier transform of a periodic, regular array of delta functions are also periodic, regular arrays, and all three have the same index period. Moreover, if

$$f = \sum_{k=-\infty}^{\infty} f_k \, \delta_{k\Delta t} \qquad \text{and} \qquad F = \sum_{n=-\infty}^{\infty} F_n \, \delta_{n\Delta\omega}$$

are periodic, regular arrays with $F = \mathcal{F}[f]$, then, with p denoting the period of f and P denoting the period of F,

1. $\Delta\omega = {}^1/_p$ *and* $\Delta t = {}^1/_P$,

and

2. F_n *is the nth Fourier coefficient for f, while f_{-k} is the kth Fourier coefficient for F.*

In other words, the Fourier series representations for f and F are

$$f(t) = \sum_{n=-\infty}^{\infty} F_n e^{i2\pi\omega_n t} \qquad \text{where} \quad \omega_n = \frac{n}{p} = n\Delta\omega$$

and

$$F(\omega) = \sum_{k=-\infty}^{\infty} f_{-k} e^{i2\pi t_k \omega} \qquad \text{where} \quad t_k = \frac{k}{P} = k\Delta t \quad .$$

!▶**Example 39.2:** Let $F = \mathcal{F}[f]$ where f is the array of delta functions discussed in example 39.1 and sketched in figure 39.2. Since f is a periodic, regular array, so is F. Moreover, since we already know the spacing Δt, index period N and period p of f are

$$\Delta t = \frac{1}{3} \quad , \qquad N = 4 \qquad \text{and} \qquad p = \frac{4}{3} \quad ,$$

we also know (by the above discussion) that

the spacing of F is $\Delta\omega = \dfrac{1}{\text{period of } f} = \dfrac{1}{p} = \dfrac{1}{4/3} = \dfrac{3}{4}$,

the index period of F is $N = 4$,

and

the period of F is $P = \dfrac{1}{\text{spacing of } f} = \dfrac{1}{\Delta t} = \dfrac{1}{1/3} = 3$.

Computing the Coefficients
Reducing Integration to Summation

Computing the F_n's from the f_k's is fairly straightforward (still assuming these are the coefficients of the arrays from the previous subsection). Since the F_n's are the Fourier coefficients of f, they are given by the integral formula

$$F_n = \frac{1}{p} \int_a^{a+p} f(t) e^{-i2\pi\omega_n t} \, dt$$

where a is any conveniently chosen real number and $\omega_n = n\Delta\omega$.

From figure 39.3, we see that any point between 0 and $-\Delta t$ would be a convenient choice for a. Let us choose $a = -\Delta t/2$. Then, by the relation between N, p and Δt,

$$a + p = -\frac{1}{2}\Delta t + N\Delta t$$

$$= \left(N - \frac{1}{2}\right)\Delta t \quad .$$

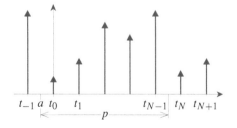

Figure 39.3: Choosing a ($t_k = k\Delta t$).

With this choice for a, the array formula for f and our knowledge of how to compute integrals of arrays and delta functions, the computation of the integral formula for F_n becomes straightforward:

$$
F_n = \frac{1}{p} \int_{-\frac{1}{2}\Delta t}^{\left(N-\frac{1}{2}\right)\Delta t} \left[\sum_{k=-\infty}^{\infty} f_k \, \delta_{k\Delta t}(t) \right] e^{-i2\pi\omega_n t} \, dt
$$

$$
= \frac{1}{p} \sum_{k=-\infty}^{\infty} \int_{-\frac{1}{2}\Delta t}^{\left(N-\frac{1}{2}\right)\Delta t} f_k \, e^{-i2\pi\omega_n t} \delta_{k\Delta t}(t) \, dt
$$

$$
= \frac{1}{p} \sum_{k=-\infty}^{\infty} \begin{cases} f_k \, e^{-i2\pi\omega_n k\Delta t} & \text{if} \quad -\Delta t/2 < k\Delta t < \left(N - \tfrac{1}{2}\right)\Delta t \\ 0 & \text{otherwise} \end{cases}
$$

$$
= \frac{1}{p} \sum_{k=-\infty}^{\infty} \begin{cases} f_k \, e^{-i2\pi\omega_n k\Delta t} & \text{if} \quad k = 0,\, 1,\, 2,\, \ldots,\, N-1 \\ 0 & \text{otherwise} \end{cases} \quad.
$$

Thus,

$$
F_n = \frac{1}{p} \sum_{k=0}^{N-1} f_k \, e^{-i2\pi\omega_n k\Delta t} \qquad \text{for} \quad n = 0,\, \pm 1,\, \pm 2,\, \pm 3,\, \ldots \quad.
$$

A similar formula allows us to compute the f_k's from the F_n's. Before deriving that formula, though, let us express this last formula in a form that will turn out to be more convenient (and will help prevent the confusion that may arise due to our foolishly denoting the periods of f and F by the similar-looking symbols p and P). Recalling the relations between N, p, Δt and $\Delta\omega$, we see that

$$
\Delta\omega = \frac{1}{p} = \frac{1}{N\Delta t} \qquad \text{and} \qquad \Delta\omega\,\Delta t = \frac{1}{N\Delta t}\Delta t = \frac{1}{N} \quad. \tag{39.5}
$$

With this,

$$
2\pi\omega_n k\Delta t = 2\pi n \Delta\omega\, k\Delta t = 2\pi(\Delta\omega\,\Delta t)nk = \frac{2\pi}{N}nk \quad,
$$

and the last formula for F_n becomes

$$
F_n = \frac{1}{N\Delta t} \sum_{k=0}^{N-1} f_k \, e^{-i\frac{2\pi}{N}nk} \qquad \text{for} \quad n = 0,\, \pm 1,\, \pm 2,\, \pm 3,\, \ldots \quad. \tag{39.6}
$$

!▶ *Example 39.3:* *Assuming*

$$
F = \sum_{n=-\infty}^{\infty} F_n \, \delta_{n\Delta\omega}
$$

is the Fourier transform of the periodic, regular array f from example 39.1 (and sketched in figure 39.2), let us compute F_1.

Here, $N = 4$, $\Delta t = {}^1\!/_3$ and the "first four" coefficients of the array f are

$$
f_0 = 1 \quad,\quad f_1 = 2 \quad,\quad f_2 = 3 \quad \text{and} \quad f_3 = 3 \quad.
$$

Plugging this into formula (39.6), with $n = 1$, we get

$$
F_1 = \frac{1}{4\left({}^1\!/_3\right)} \sum_{k=0}^{4-1} f_k \, e^{-i\frac{2\pi}{4}1\cdot k}
$$

$$
= \frac{3}{4}\left[f_0 \, e^{-i\frac{2\pi}{4}1\cdot 0} + f_1 \, e^{-i\frac{2\pi}{4}1\cdot 1} + f_2 \, e^{-i\frac{2\pi}{4}1\cdot 2} + f_3 \, e^{-i\frac{2\pi}{4}1\cdot 3} \right]
$$

$$= \frac{3}{4}\left[1\,e^0 + 2\,e^{-i\frac{\pi}{2}} + 3\,e^{-i\pi} + 3\,e^{-i\frac{3\pi}{2}}\right]$$

$$= \frac{3}{4}[1 + 2(-i) + 3(-1) + 3(i)] = -\frac{3}{2} + \frac{3}{4}i \quad .$$

?►Exercise 39.2: Let f and F be the arrays from which equation (39.6) was derived. Show that

$$f_k = \frac{1}{N\,\Delta\omega}\sum_{n=0}^{N-1} F_n\,e^{i\frac{2\pi}{N}kn} \qquad \text{for}\quad k = 0, \pm 1, \pm 2, \pm 3, \ldots \quad . \tag{39.7}$$

The Matrices of Exponentials for the Transforms

Here are a few observations regarding formula (39.6):

First of all, because we are computing the coefficients of the periodic, regular array

$$F = \sum_{n=-\infty}^{\infty} F_n\,\delta_{n\Delta\omega}$$

having index period N, we need only use formula (39.6) to compute N of these coefficients, say, F_0, F_1, F_2, $\ldots$ and F_{N-1}. The rest can be computed from the recursion relation $F_{N+k} = F_k$.

The second observation is that formula (39.6) can be written as a matrix product:

$$F_n = \frac{1}{N\,\Delta t}\sum_{k=0}^{N-1} f_k\,e^{-i\frac{2\pi}{N}nk}$$

$$= \frac{1}{N\,\Delta t}\left[e^{-i\frac{2\pi}{N}n\cdot 0}f_0 + e^{-i\frac{2\pi}{N}n\cdot 1}f_1 + e^{-i\frac{2\pi}{N}n\cdot 2}f_2 + \cdots + e^{-i\frac{2\pi}{N}n(N-1)}f_{N-1}\right]$$

$$= \frac{1}{N\,\Delta t}\left[e^{-i\frac{2\pi}{N}n\cdot 0}\quad e^{-i\frac{2\pi}{N}n\cdot 1}\quad e^{-i\frac{2\pi}{N}n\cdot 2}\quad \cdots \quad e^{-i\frac{2\pi}{N}n(N-1)}\right]\begin{bmatrix} f_0 \\ f_1 \\ f_2 \\ \vdots \\ f_{N-1}\end{bmatrix}\quad.$$

Since $e^0 = 1$, this gives

$$F_0 = \frac{1}{N\,\Delta t}\begin{bmatrix} 1 & 1 & 1 & \cdots & 1\end{bmatrix}\begin{bmatrix} f_0 \\ f_1 \\ f_2 \\ \vdots \\ f_{N-1}\end{bmatrix}\quad,$$

$$F_1 = \frac{1}{N\,\Delta t}\begin{bmatrix} 1 & e^{-i\frac{2\pi}{N}1\cdot 1} & e^{-i\frac{2\pi}{N}1\cdot 2} & \cdots & e^{-i\frac{2\pi}{N}1(N-1)}\end{bmatrix}\begin{bmatrix} f_0 \\ f_1 \\ f_2 \\ \vdots \\ f_{N-1}\end{bmatrix}\quad,$$

$$F_2 = \frac{1}{N\Delta t} \begin{bmatrix} 1 & e^{-i\frac{2\pi}{N}2\cdot1} & e^{-i\frac{2\pi}{N}2\cdot2} & \cdots & e^{-i\frac{2\pi}{N}2(N-1)} \end{bmatrix} \begin{bmatrix} f_0 \\ f_1 \\ f_2 \\ \vdots \\ f_{N-1} \end{bmatrix} \quad ,$$

and so on.

All of this can be written somewhat more concisely as

$$\begin{bmatrix} F_0 \\ F_1 \\ F_2 \\ \vdots \\ F_{N-1} \end{bmatrix} = \frac{1}{N\Delta t} \begin{bmatrix} 1 & 1 & 1 & \cdots & 1 \\ 1 & e^{-i\frac{2\pi}{N}1\cdot1} & e^{-i\frac{2\pi}{N}1\cdot2} & \cdots & e^{-i\frac{2\pi}{N}1(N-1)} \\ 1 & e^{-i\frac{2\pi}{N}2\cdot1} & e^{-i\frac{2\pi}{N}2\cdot2} & \cdots & e^{-i\frac{2\pi}{N}2(N-1)} \\ \vdots & \vdots & \vdots & \ddots & \vdots \\ 1 & e^{-i\frac{2\pi}{N}(N-1)1} & e^{-i\frac{2\pi}{N}(N-1)2} & \cdots & e^{-i\frac{2\pi}{N}(N-1)(N-1)} \end{bmatrix} \begin{bmatrix} f_0 \\ f_1 \\ f_2 \\ \vdots \\ f_{N-1} \end{bmatrix}$$

and even more concisely as

$$\mathbf{F} = \frac{1}{N\Delta t} \mathbf{M}_N \mathbf{f} \tag{39.8a}$$

provided it is understood that

$$\mathbf{F} = \begin{bmatrix} F_0 \\ F_1 \\ F_2 \\ \vdots \\ F_{N-1} \end{bmatrix} \quad , \quad \mathbf{f} = \begin{bmatrix} f_0 \\ f_1 \\ f_2 \\ \vdots \\ f_{N-1} \end{bmatrix} \tag{39.8b}$$

and

$$\mathbf{M}_N = \begin{bmatrix} 1 & 1 & 1 & \cdots & 1 \\ 1 & e^{-i\frac{2\pi}{N}1\cdot1} & e^{-i\frac{2\pi}{N}1\cdot2} & \cdots & e^{-i\frac{2\pi}{N}1(N-1)} \\ 1 & e^{-i\frac{2\pi}{N}2\cdot1} & e^{-i\frac{2\pi}{N}2\cdot2} & \cdots & e^{-i\frac{2\pi}{N}2(N-1)} \\ \vdots & \vdots & \vdots & \ddots & \vdots \\ 1 & e^{-i\frac{2\pi}{N}(N-1)1} & e^{-i\frac{2\pi}{N}(N-1)2} & \cdots & e^{-i\frac{2\pi}{N}(N-1)(N-1)} \end{bmatrix} \quad . \tag{39.8c}$$

Because $\mathbf{M}_N$ plays a significant role in the next chapter, we will give it a name — the Nth *matrix of exponentials for* $\mathcal{F}$. Do note the following:

1. $\mathbf{M}_N$ is an $N \times N$ matrix that depends only on the integer N.

2. We will index the rows and columns of this matrix using $0, 1, 2, \ldots, N - 1$ (instead of $1, 2, 3, \ldots, N$). Consequently, for us, the upper leftmost component is the $(0, 0)^{th}$ component, the lower rightmost component is the $(N - 1, N - 1)^{th}$ component, and the $(n, k)^{th}$ component is

$$\left[\mathbf{M}_N\right]_{n,k} = e^{-i\frac{2\pi}{N}nk} \quad .$$

3. Since

$$\left[\mathbf{M}_N\right]_{n,k} = e^{-i\frac{2\pi}{N}nk} = e^{-i\frac{2\pi}{N}kn} = \left[\mathbf{M}_N\right]_{k,n} \quad ,$$

$\mathbf{M}_N$ is a symmetric matrix.

Computing each component of $\mathbf{M}_N$ is fairly simple provided you recall some basic facts about complex exponentials (and N is not too large). For example, to compute each

$$\left[\mathbf{M}_N\right]_{n,k} = e^{-i\frac{2\pi}{N}nk} \quad ,$$

you can use Euler's formula ($e^{-i\theta} = \cos(\theta) - i\sin(\theta)$) or a sketch similar to figure 6.3 on page 63. Or you can compute just $e^{-i\theta}$ with $\theta = {}^{2\pi}/_N$, followed by the elementary calculation

$$e^{-i\theta nk} = \left(e^{-i\theta}\right)^{nk} \quad .$$

Naturally, when N is large, it is a good idea to combine these basic facts with a computer.

!▶ **Example 39.4:** *The 4th matrix of exponentials for $\mathcal{F}$, $\mathbf{M}_4$, is given by*

$$\mathbf{M}_4 = \begin{bmatrix} 1 & 1 & 1 & 1 \\ 1 & e^{-i\frac{2\pi}{4}1\cdot 1} & e^{-i\frac{2\pi}{4}1\cdot 2} & e^{-i\frac{2\pi}{4}1\cdot 3} \\ 1 & e^{-i\frac{2\pi}{4}2\cdot 1} & e^{-i\frac{2\pi}{4}2\cdot 2} & e^{-i\frac{2\pi}{4}2\cdot 3} \\ 1 & e^{-i\frac{2\pi}{4}3\cdot 1} & e^{-i\frac{2\pi}{4}3\cdot 2} & e^{-i\frac{2\pi}{4}3\cdot 3} \end{bmatrix} \quad .$$

In this case the computation of the components is especially simple because

$$e^{-i\frac{2\pi}{4}} = e^{-i\frac{\pi}{2}} = -i \quad .$$

So, for $n = 0$, 1, 2 and 3, and $k = 0$, 1, 2 and 3,

$$\left[\mathbf{M}_N\right]_{n,k} = e^{-i\frac{2\pi}{N}nk} = (-i)^{nk} \quad .$$

Thus,

$$\mathbf{M}_4 = \begin{bmatrix} 1 & 1 & 1 & 1 \\ 1 & (-i)^1 & (-i)^2 & (-i)^3 \\ 1 & (-i)^2 & (-i)^4 & (-i)^6 \\ 1 & (-i)^3 & (-i)^6 & (-i)^9 \end{bmatrix} = \begin{bmatrix} 1 & 1 & 1 & 1 \\ 1 & -i & -1 & i \\ 1 & -1 & 1 & -1 \\ 1 & i & -1 & -i \end{bmatrix} \quad . \tag{39.9}$$

!▶ **Example 39.5:** *Let's continue our earlier examples and finish computing the periodic, regular array*

$$F = \sum_{n=-\infty}^{\infty} F_n \, \delta_{n\Delta\omega}$$

obtained by taking the Fourier transform of the periodic, regular array f from example 39.1 and figure 39.2.

Remember, in the previous examples we saw or derived the following:

1. *The "first four" coefficients of the array f :*

$$f_0 = 1 \quad , \quad f_1 = 2 \quad , \quad f_2 = 3 \quad \text{and} \quad f_3 = 3 \quad .$$

2. *The index period N for both f and F , the spacing Δt of f and the spacing $\Delta\omega$ of F :*

$$N = 4 \quad , \quad \Delta t = \frac{1}{3} \quad \text{and} \quad \Delta\omega = \frac{3}{4} \quad .$$

3. *The formula for $\mathbf{M}_4$, the 4th matrix of exponentials for $\mathcal{F}$ (formula (39.9)).*

Plugging the appropriate quantities from above into equation (39.8a), we obtain

$$
\begin{bmatrix} F_0 \\ F_1 \\ F_2 \\ F_3 \end{bmatrix} = \frac{1}{4\left(\nicefrac{1}{3}\right)} \begin{bmatrix} 1 & 1 & 1 & 1 \\ 1 & -i & -1 & i \\ 1 & -1 & 1 & -1 \\ 1 & i & -1 & -i \end{bmatrix} \begin{bmatrix} 1 \\ 2 \\ 3 \\ 3 \end{bmatrix}
$$

$$
= \frac{3}{4} \begin{bmatrix} 1 \cdot 1 + 1 \cdot 2 + 1 \cdot 3 + 1 \cdot 3 \\ 1 \cdot 1 - i \cdot 2 - 1 \cdot 3 + i \cdot 3 \\ 1 \cdot 1 - 1 \cdot 2 + 1 \cdot 3 - 1 \cdot 3 \\ 1 \cdot 1 + i \cdot 2 - 1 \cdot 3 - i \cdot 3 \end{bmatrix} = \begin{bmatrix} \frac{27}{4} \\ -\frac{3}{2} + \frac{3}{4}i \\ -\frac{3}{4} \\ -\frac{3}{2} - \frac{3}{4}i \end{bmatrix} .
$$

So,

$$
F = \sum_{n=-\infty}^{\infty} F_n \, \delta_{3n/4}
$$

where

$$
F_0 = \frac{27}{4} \quad , \quad F_1 = -\frac{3}{2} + \frac{3}{4}i \quad , \quad F_2 = -\frac{3}{4} \quad \text{and} \quad F_3 = -\frac{3}{2} - \frac{3}{4}i \quad ,
$$

and the other F's can be computed from F_0, F_1, F_2 and F_3 via

$$
F_{k+4} = F_k \quad .
$$

In particular

$$
F_4 = F_{0+4} = F_0 = \frac{27}{4} \quad , \quad F_5 = F_{1+4} = F_1 = -\frac{3}{2} + \frac{3}{4}i
$$

and

$$
F_{-1} = F_{-1+4} = F_3 = -\frac{3}{2} - \frac{3}{4}i \quad .
$$

In a very similar fashion you can show that the f_k's can be computed from the F_n's using the recursion formula $f_{k+N} = f_k$ and the matrix/vector formula

$$
\mathbf{f} = \frac{1}{N \Delta \omega} \mathbf{M}_N^* \mathbf{F} \tag{39.10}
$$

where $\mathbf{f}$ and $\mathbf{F}$ are, again, as in line (39.8b), and

$$
\mathbf{M}_N^* = \begin{bmatrix} 1 & 1 & 1 & \cdots & 1 \\ 1 & e^{i\frac{2\pi}{N}1 \cdot 1} & e^{i\frac{2\pi}{N}1 \cdot 2} & \cdots & e^{i\frac{2\pi}{N}1(N-1)} \\ 1 & e^{i\frac{2\pi}{N}2 \cdot 1} & e^{i\frac{2\pi}{N}2 \cdot 2} & \cdots & e^{i\frac{2\pi}{N}2(N-1)} \\ \vdots & \vdots & \vdots & \ddots & \vdots \\ 1 & e^{i\frac{2\pi}{N}(N-1)1} & e^{i\frac{2\pi}{N}(N-1)2} & \cdots & e^{i\frac{2\pi}{N}(N-1)(N-1)} \end{bmatrix} .
$$

This matrix, which happens to be the complex conjugate of $\mathbf{M}_N$, will be called the Nth *matrix of exponentials for* $\mathcal{F}^{-1}$. Like $\mathbf{M}_N$, it is a symmetric $N \times N$ matrix depending only on the integer N. Its components are given by

$$
\left[\mathbf{M}_N^*\right]_{k,n} = \left([\mathbf{M}_N]_{k,n}\right)^* = e^{i\frac{2\pi}{N}kn}
$$

where, as with $\mathbf{M}_N$, we let the indices range from 0 to $N - 1$.

?▶Exercise 39.3: *Derive formula (39.10) from formula (39.7).*

Since we've spent so much effort deriving the formulas in this section, we should immortalize them in a theorem.

Theorem 39.2 (Fourier coefficients of periodic, regular arrays)
Assume

$$ f = \sum_{k=-\infty}^{\infty} f_k \, \delta_{k\Delta t} \quad \text{and} \quad F = \sum_{n=-\infty}^{\infty} F_n \, \delta_{n\Delta\omega} $$

are two periodic, regular arrays related by $F = \mathcal{F}[f]$. *Let* N *be their index period. Then*

$$ \mathbf{F} = \frac{1}{N\Delta t} \mathbf{M}_N \mathbf{f} \quad \text{and} \quad \mathbf{f} = \frac{1}{N\Delta\omega} \mathbf{M}_N^* \mathbf{F} \tag{39.11} $$

where $\mathbf{M}_N$ *and* $\mathbf{M}_N^*$ *are the* N^{th} *matrices of exponentials for* $\mathcal{F}$ *and* $\mathcal{F}^{-1}$, *respectively, and*

$$ \mathbf{F} = \begin{bmatrix} F_0 \\ F_1 \\ F_2 \\ \vdots \\ F_{N-1} \end{bmatrix} \quad \text{and} \quad \mathbf{f} = \begin{bmatrix} f_0 \\ f_1 \\ f_2 \\ \vdots \\ f_{N-1} \end{bmatrix} . $$

Finally, here are two corollaries that will be useful in the next chapter. The first describes transforms of periodic, regular arrays having spacing related to the index period in a certain way. It follows directly from the two theorems in this chapter and equation set (39.5) on page 714. The second, which can easily be obtained from either the first corollary or the above theorem, points out that the two matrices of exponentials, slightly rescaled, are inverses of each other.

Corollary 39.3
Let N, f, F, $\mathbf{f}$, $\mathbf{F}$, $\mathbf{M}_N$ *and* $\mathbf{M}_N^*$ *be as in theorem 39.2. Assume, further, that either*

$$ \text{spacing of } f = \frac{1}{\sqrt{N}} \quad \text{or} \quad \text{spacing of } F = \frac{1}{\sqrt{N}} . $$

Then

$$ \text{spacing of } f = \frac{1}{\sqrt{N}} = \text{spacing of } F \quad , $$

and

$$ \mathbf{F} = \frac{1}{\sqrt{N}} \mathbf{M}_N \mathbf{f} \quad \text{and} \quad \mathbf{f} = \frac{1}{\sqrt{N}} \mathbf{M}_N^* \mathbf{F} \quad . $$

Corollary 39.4 (inversion of exponential matrices)
Let $\mathbf{M}_N$ *and* $\mathbf{M}_N^*$ *be the* N^{th} *matrices of exponentials for* $\mathcal{F}$ *and* $\mathcal{F}^{-1}$, *respectively. Then, for any* N–*dimensional column vector* $\mathbf{x}$,

$$ \frac{1}{N} \mathbf{M}_N \mathbf{M}_N^* \mathbf{x} = \mathbf{x} = \frac{1}{N} \mathbf{M}_N^* \mathbf{M}_N \mathbf{x} \quad . $$

?▶Exercise 39.4 **a:** Prove corollary 39.3.

  b: Prove corollary 39.4.

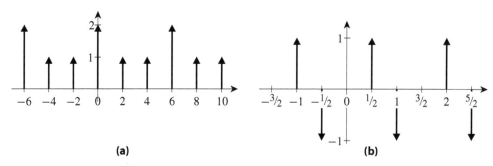

Figure 39.4: Two arrays of delta functions. (Two or more periods of each are sketched, and the coefficients are all integer valued.)

Additional Exercises

39.5. *Compute the following matrices of exponentials for $\mathcal{F}$ and $\mathcal{F}^{-1}$:*

    a. $\mathbf{M}_1$ **b.** $\mathbf{M}_2$ **c.** $\mathbf{M}_3$ **d.** $\mathbf{M}_2^*$ **e.** $\mathbf{M}_3^*$

39.6. *Let*

$$f = \sum_{k=-\infty}^{\infty} f_k\, \delta_{k\Delta t} \quad \text{and} \quad F = \sum_{n=-\infty}^{\infty} F_n\, \delta_{n\Delta\omega}$$

be two periodic, regular arrays with index period $N = 4$ and related by $F = \mathcal{F}[f]$. Also, assume

$$\Delta t = \frac{2}{3} \quad , \quad f_0 = 3 \quad , \quad f_1 = -1 \quad , \quad f_2 = 1 \quad \text{and} \quad f_3 = 5 \quad .$$

    a. *Determine the period of f and sketch the array.*

    b. *What is the spacing $\Delta\omega$ and period P of F ?*

    c. *Find the values F_0, F_1, F_2, F_3, F_4 and F_{-1}.*

39.7. *For each array f given below:*

    1. *Determine the spacing, index period and period of both the array f and the array's Fourier transform $F = \mathcal{F}[f]$.*

    2. *Sketch the array f (unless it is already sketched).*

    3. *Using the appropriate recursion formula and matrix of exponentials, compute the coefficients of $F = \mathcal{F}[f]$.*

    4. *Sketch the array F.*

    a. $\text{comb}_{1/3}$ **b.** $\displaystyle\sum_{k=-\infty}^{\infty} (-1)^k \delta_k$

    c. *the array in figure 39.4a*   **d.** *the array in figure 39.4b*

40

Sampling, Discrete Fourier Transforms and FFTs

There are many situations in which the functions we would like to analyze are mainly known through measurements. Consider, for example, the temperature at time t at some location. Whether this position is in a cup of coffee or a coupling in a rocket engine, it is overly idealistic to assume we can derive, from basic principles alone, a precise formula $f(t)$ describing how this temperature varies with time. Instead, we might measure that temperature at various times — t_0, t_1, t_2, ... — and then base further analysis on that sequence of measured values — $f(t_0)$, $f(t_1)$, $f(t_2)$, ... — along with, of course, our knowledge of thermodynamics.

Lack of knowledge is not the only reason to deal with a sequence of "sample" values for a function. Programming a computer to compute, say, the integral of an arbitrary function symbolically (as we normally do using pencil and paper) can be a challenging task. Programming a computer to do analogous computations with a list of sampled values (using, say, a Riemann sum to approximate an integral), however, can be a relatively straightforward job.

Here, we will develop the "discrete theory of Fourier analysis". This theory can be viewed as a collection of methods for analyzing and manipulating numerical sequences in ways that are analogous to well-known methods for analyzing and manipulating ordinary functions. Before that, though, we should discuss how to generate sequences from functions via "sampling", and how to construct "discrete approximations" from those sequences. It will be these discrete approximations that will link the theory we've already developed to the "discrete theory" being developed here.

40.1 Some General Conventions and Terminology

Throughout this chapter, N will denote some positive integer, and we will repeatedly find ourselves considering various ordered sets of N numbers. Unless it represents something special, such as a "sampling of some function" (defined in a few paragraphs), we will refer to any such set as either an ordered set, list or sequence, as the spirit moves us.

The *order* of a given set of numbers is simply the number of numbers in that set. Hence, any "N^{th} order sequence" is just a list of N numbers.

In the previous chapter, we found it convenient to index the rows and columns of certain $N \times N$ matrices using 0, 1, ..., $N-1$ instead of 1, 2, ..., N. We will extend that convention here, indexing all $N \times N$ matrices in that manner. To match this, the indexing of any N^{th} order sequence will also go from $index = 0$ to $index = N-1$. (This is a standard convention in this subject. It makes sense because, typically, the 0th element in each list corresponds to some variable — other than the index — being 0.)

40.2 Sampling and the Discrete Approximation
N^{th} Order Samplings

For the next few sections, we will be considering a function f defined over some interval of the real line. For now, let us assume f is at least piecewise continuous on that interval.

An (N^{th} order) *sampling* (or N^{th} order *sample*) of the function f consists of two sequences of N numbers. The first is an ordered set of points

$$\{x_0, x_1, x_2, \ldots, x_{N-1}\}$$

from the interval on which f is defined. These are the points "at which the sampling is taken" and are assumed to be points at which f is well defined. They are always chosen and indexed so that

$$x_0 < x_1 < x_2 < \cdots < x_{N-1} \quad .$$

The second sequence

$$\{f_0, f_1, f_2, \ldots, f_{N-1}\}$$

is the corresponding sequence of values for $f(x)$, that is,

$$f_k = f(x_k) \qquad \text{for} \quad k = 0, 1, 2, \ldots, N-1 \quad .$$

Any such sampling can be graphed by plotting the points

$$(x_0, f_0) \quad , \quad (x_1, f_1) \quad , \quad (x_2, f_2) \quad , \quad \cdots \quad \text{and} \quad (x_{N-1}, f_{N-1})$$

on an appropriate coordinate system, as illustrated in figure 40.1a (or on two coordinate systems, if both the real and imaginary parts of the f_k's are nonzero).

The *order* of the sampling is simply the number of samples taken, N. A *window* for the sampling is any interval containing the points at which the sampling is taken. In particular, the smallest possible window, $[x_0, x_{N-1}]$, will be referred to as the *minimal window* for the sampling, and, unless otherwise indicated, L will denote the length of this window,

$$L = \text{length of minimal window} = x_{N-1} - x_0 \quad .$$

A *uniform* sampling is one in which the spacing between the consecutive x_k's is uniform. To be more precise, our sampling is uniform if and only if there is a fixed positive distance Δx such that

$$x_k - x_{k-1} = \Delta x \qquad \text{for} \quad k = 1, 2, 3, \ldots, N-1 \quad .$$

Naturally enough, we will call this distance the *spacing* of the sampling. Clearly, assuming the above relation holds,

$$x_1 = x_0 + \Delta x \quad ,$$
$$x_2 = x_1 + \Delta x = x_0 + 2\Delta x \quad ,$$
$$x_3 = x_2 + \Delta x = x_0 + 3\Delta x \quad ,$$

$$and \; so \; on.$$

In general,

$$x_k = x_0 + k\Delta x \qquad \text{for} \quad k = 0, 1, 2, \ldots, N-1 \quad .$$

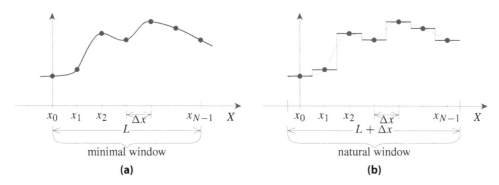

Figure 40.1: (a) The graph of an Nth order regular sampling with spacing Δx of a continuous, real-valued function f, with each dot representing a point (x_k, f_k). (b) A simple piecewise continuous approximation to the function on the natural window on the sampling. In these figures $x_k = k\Delta x$.

Consequently,

$$x_{N-1} \;=\; x_0 \;+\; (N-1)\Delta x$$

and

$$L \;=\; x_{N-1} \;-\; x_0 \;=\; (N-1)\Delta x \quad .$$

Thus, for a uniform sampling,

$$window\ length \;=\; (order \,-\, 1) \times spacing \quad . \tag{40.1}$$

A *regular* sampling, as illustrated in figure 40.1a, is a uniform sampling whose minimal window starts at 0. So, if we have a regular sampling of order N and spacing Δx, then the points at which the samples are taken are all given by

$$x_k \;=\; k\Delta x \qquad \text{for} \quad k = 0,\, 1,\, 2,\, \ldots,\, N-1 \quad , \tag{40.2}$$

and the corresponding samples of f are given by

$$f_k \;=\; f(k\Delta x) \qquad \text{for} \quad k = 0,\, 1,\, 2,\, \ldots,\, N-1 \quad .$$

Moreover, since $x_0 = 0$ and $x_{N-1} = (N-1)\Delta x = L$, the minimal window of the regular sampling is just $[0, L]$.

Most of the samplings we will consider are regular samplings. Because of this, the points in the first list (the x_k's) of most of our samplings can be computed from formula (40.2), and the minimal window length, spacing and order of the sampling can all be determined from equation (40.1), provided any two of these parameters are known. So we won't need to explicitly give all the parameters for our samplings, nor will we need to explicitly state the x_k's. This allows us (when our sampling is regular) to adopt the shorthand of referring to just the list of "sampled values", $\{f_0, f_1, \ldots, f_{N-1}\}$, as either the "$N^{th}$ order regular sampling of f over the minimal window $[0, L]$" or the "N^{th} order regular sampling of f with spacing Δx".

!► Example 40.1:   *Let's find the 4th order regular sampling for the function $f(x) = x^2$ over the window $[0, 2]$. Here, we know the order N is 4 and the window length L is 2. From equation (40.1), we find that the sample spacing Δx is*

$$\Delta x \;=\; \frac{L}{N-1} \;=\; \frac{2}{4-1} \;=\; \frac{2}{3} \quad .$$

So the four points at which f is sampled are

$$x_0 = 0 \quad , \quad x_1 = 1\Delta x = \frac{2}{3} \quad , \quad x_2 = 2\Delta x = \frac{4}{3} \quad \text{and} \quad x_3 = L = 2 \quad .$$

The corresponding sampled values of $f(x) = x^2$ *are*

$$f_0 = f(x_0) = (0)^2 = 0 \quad ,$$

$$f_1 = f(x_1) = \left(\frac{2}{3}\right)^2 = \frac{4}{9} \quad ,$$

$$f_2 = f(x_2) = \left(\frac{4}{3}\right)^2 = \frac{16}{9}$$

and

$$f_3 = f(x_3) = (2)^2 = 4 \quad ,$$

giving us

$$\left\{0, \frac{4}{9}, \frac{16}{9}, 4\right\}$$

as the 4th order sampling of $f(x) = x^2$ over the minimal window $[0, 2]$.

?▶ Exercise 40.1:   *What is the 3rd order regular sampling of $f(x) = x^2$ over the minimal window $[0, 1]$? (Be sure to also give the spacing for this sampling.)*

If our original function f is continuous, then it is natural to view each f_k in our sampling as being an approximation to the value of $f(x)$ for all x in a small interval about x_k. And if our sampling is uniform with spacing Δx, it is even more natural to take that small interval to be the interval of length Δx with x_k at its center. So, if f is continuous and $\{f_0, \ldots, f_{N-1}\}$ is its Nth order regular sampling with spacing Δx, then

$$f_0 \text{ approximates } f(x) \text{ over the interval } \left(-\frac{1}{2}\Delta x, \frac{1}{2}\Delta x\right) \quad ,$$

$$f_1 \text{ approximates } f(x) \text{ over the interval } \left(\frac{1}{2}\Delta x, \frac{3}{2}\Delta x\right) \quad ,$$

$$\vdots$$

and

$$f_{N-1} \text{ approximates } f(x) \text{ over the interval } \left(\left[N - 1 - \frac{1}{2}\right]\Delta x, \left[N - \frac{1}{2}\right]\Delta x\right)$$

(see figure 40.1b). All in all, this gives us a piecewise continuous approximation to $f(x)$ over the interval

$$\left(-\frac{1}{2}\Delta x, \left[N - \frac{1}{2}\right]\Delta x\right) \quad .$$

Let us call this interval the *natural window* for the sampling. Notice that

$$\text{length of natural window} = \left[N - \frac{1}{2}\right]\Delta x + \frac{1}{2}\Delta x = N\Delta x = L + \Delta x \quad .$$

So, in terms of the length of the minimal window, L, the natural window is

$$\left(-\frac{\Delta x}{2}, L + \frac{\Delta x}{2}\right) \quad .$$

For a graphical comparison of a minimal window with a corresponding natural window, see figure 40.1.

!▶Example 40.2:   *For the sampling in example 40.1, the minimal window is* $[0, 2]$ *and the spacing is* $\Delta x = \frac{2}{3}$. *So the natural window is*

$$\left(-\frac{1}{3}, 2 + \frac{1}{3}\right) = \left(-\frac{1}{3}, \frac{7}{3}\right) \quad .$$

?▶Exercise 40.2:   *What is the natural window for a* 3rd *order regular sampling over the minimal window* $[0, 1]$? *(Be sure to also give the spacing for this sampling.)*

40.3   The Discrete Approximation and Its Transforms
Derivation of the Discrete Approximation

Suppose
$$\{ f_0, f_1, f_2, \dots, f_{N-1} \}$$

is an Nth order regular sampling with spacing Δx of some piecewise continuous function f over the minimal window $[0, L]$. Suppose, further, that we want to do some analysis involving the Fourier transform of f. Of course, that is not possible if all we know about f is our sampling — Fourier transforms are taken of functions, not lists of numbers. So, as an alternative, we might try to construct a function (or generalized function) $\widehat{f}$ that can be used in place of the partially known f. This function would naturally have to satisfy the following three requirements:

1. $\widehat{f}$ can be constructed using only information from the sampling.

2. Over the natural window, $\widehat{f}$ approximates f in some way. In particular, since we are interested in doing Fourier analysis, we want some assurance that integrals of f can be reasonably approximated by the corresponding integrals of $\widehat{f}$.

3. The Fourier transforms of $\widehat{f}$ should be fairly easy to compute, as should the further transforms of $\mathcal{F}\left[\widehat{f}\right]$ and $\mathcal{F}^{-1}\left[\widehat{f}\right]$. If we can easily program a computer to do these computations, all the better.

We can break down the construction of $\widehat{f}$ to answering two basic questions: *What should* $\widehat{f}$ *be inside the natural window?* and *What should* $\widehat{f}$ *be outside the natural window?*

There are many ways of approximating f on the natural window using the given sampling. One obvious possibility, a piecewise continuous function whose graph consists of horizontal lines, is sketched in figure 40.1b. Unfortunately, computing the Fourier transform of such a function (and doing further computations with those transforms) can be downright tedious when N is large. So let's look a little more closely at our requirements to see if we can obtain a more convenient approximation $\widehat{f}$.

Converted to standard mathematical symbolism, the second requirement is that, as well as we can determine from our data,

$$\int_{-\frac{1}{2}\Delta x}^{\left(N-\frac{1}{2}\right)\Delta x} \widehat{f}(x)\phi(x)\,dx \approx \int_{-\frac{1}{2}\Delta x}^{\left(N-\frac{1}{2}\right)\Delta x} f(x)\phi(x)\,dx$$

for any continuous function ϕ on the natural window. True, computing the integral on the right is impossible if all we know about f is the sampling, but that sampling can be used to compute a corresponding Riemann sum approximation for the integral,

$$\int_{-\frac{1}{2}\Delta x}^{\left(N-\frac{1}{2}\right)\Delta x} f(x)\phi(x)\,dx \;\approx\; \sum_{k=0}^{N-1} f(k\Delta x)\phi(k\Delta x)\,\Delta x \;=\; \sum_{k=0}^{N-1} f_k\,\phi(k\Delta x)\Delta x \quad.$$

Notice, however, that

$$\sum_{k=0}^{N-1} f_k\,\phi(k\Delta x)\Delta x \;=\; \sum_{k=0}^{N-1} f_k \left[\int_{-\frac{1}{2}\Delta x}^{\left(N-\frac{1}{2}\right)\Delta x} \phi(x)\,\delta_{k\Delta x}(x)\,dx\right]\Delta x$$

$$=\; \int_{-\frac{1}{2}\Delta x}^{\left(N-\frac{1}{2}\right)\Delta x}\left[\sum_{k=0}^{N-1} f_k\,\Delta x\,\delta_{k\Delta x}(x)\right]\phi(x)\,dx \quad.$$

So,

$$\int_{-\frac{1}{2}\Delta x}^{\left(N-\frac{1}{2}\right)\Delta x} f(x)\phi(x)\,dx \;\approx\; \int_{-\frac{1}{2}\Delta x}^{\left(N-\frac{1}{2}\right)\Delta x}\left[\sum_{k=0}^{N-1} f_k\,\Delta x\,\delta_{k\Delta x}(x)\right]\phi(x)\,dx \quad,$$

suggesting that, within the natural window, we use

$$\widehat{f}(x) \;=\; \sum_{k=0}^{N-1} \widehat{f_k}\,\delta_{k\Delta x}(x)$$

where

$$\widehat{f_k} \;=\; f_k\,\Delta x \qquad \text{for} \quad k = 0,\,1,\,2,\,\ldots,\,N-1 \quad.$$

Since computing Fourier transforms of delta functions is a rather simple procedure, we'll take the suggestion.

Deciding how to extend our definition of $\widehat{f}$ beyond the natural window is a bit simpler. For one thing, we will not attempt to define $\widehat{f}$ as an approximation to f outside this interval because, in general, the sampling gives no data regarding $f(x)$ here. So we are free to extend our definition of $\widehat{f}$ however we wish, keeping in mind that we wish the computation of its Fourier transforms to be as simple as possible. There are two natural choices: Either define $\widehat{f}$ to be 0 outside the natural window, or define $\widehat{f}$ to be periodic with period p equaling the length of the natural window,

$$p \;=\; N\Delta x \quad.$$

Combined with our definition of $\widehat{f}$ on the natural window, the first choice would give

$$\widehat{f}(x) \;=\; \sum_{k=0}^{N-1} \widehat{f_k}\,\delta_{k\Delta x}(x) \quad,$$

while the second choice would give

$$\widehat{f}(x) \;=\; \sum_{k=-\infty}^{\infty} \widehat{f_k}\,\delta_{k\Delta x}(x) \qquad \text{with} \quad \widehat{f}_{k+N} \;=\; \widehat{f_k} \quad.$$

Clearly (given the material in the previous chapter), the latter is the more clever choice — it gives a periodic, regular array with index period N and with Fourier transforms that are also periodic, regular arrays with index period N which can be computed via simple matrix multiplications.

Let us make the more clever choice, and restate all that we have derived in a definition.

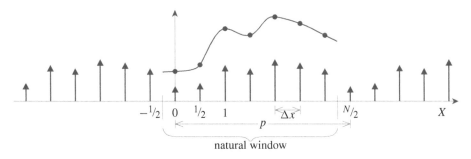

Figure 40.2: The discrete approximation for the function and sampling from figure 40.1 assuming that $\Delta x = \frac{1}{2}$. The sampling and the graph of the function over the natural window have also been sketched.

Definition of the Discrete Approximation

Assume $\{f_0, f_1, \ldots, f_{N-1}\}$ is the Nth order regular sampling with spacing Δx of some function f. The corresponding *discrete approximation* of f is the periodic, regular array

$$\widehat{f} = \sum_{k=-\infty}^{\infty} \widehat{f_k}\, \delta_{k\Delta x}$$

with spacing Δx, index period N and coefficients

$$\widehat{f_k} = \begin{cases} f_k\, \Delta x & \text{if } k = 0, 1, 2, \ldots, N-1 \\ \widehat{f_{k+N}} & \text{in general} \end{cases} \; .$$

An illustration of a discrete approximation has been given in figure 40.2.

The above array $\widehat{f}$ will also be referred to as the Nth *order discrete approximation to* f *with spacing* Δx. With this terminology, we need not explicitly state the sampling since that can be easily reconstructed from the array by dividing $\widehat{f_0}$, $\widehat{f_2}$, ... and $\widehat{f_{N-1}}$ by Δx.

?▶Exercise 40.3: *What would figure 40.2 look like if $\Delta x = 1$? if $\Delta x = 2$?*

!▶Example 40.3: *Consider the 4th order sampling of $f(x) = x^2$ over the minimal window $[0, 2]$. From example 40.1, we know the spacing for this sampling is $\Delta x = \frac{2}{3}$, and the sampling, itself, is*

$$\{f_0, f_1, f_2, f_3\} = \left\{ 0, \frac{4}{9}, \frac{16}{9}, 4 \right\} \quad .$$

The corresponding discrete approximation, then, is

$$\widehat{f} = \sum_{k=-\infty}^{\infty} \widehat{f_k}\, \delta_{2k/3}$$

where

$$\widehat{f_0} = 0 \cdot \frac{2}{3} = 0 \quad , \quad \widehat{f_1} = \frac{4}{9} \cdot \frac{2}{3} = \frac{8}{27} \quad ,$$

$$\widehat{f_2} = \frac{16}{9} \cdot \frac{2}{3} = \frac{32}{27} \quad , \quad \widehat{f_3} = 4 \cdot \frac{2}{3} = \frac{8}{3}$$

and

$$\widehat{f_{k+N}} = \widehat{f_k} \quad \text{for } k = 0, \pm 1, \pm 2, \pm 3, \ldots \quad .$$

?►Exercise 40.4: *What is the 3rd order discrete approximation of $f(x) = x^2$ over the minimal window $[0, 1]$? (This continues exercise 40.1.)*

Fourier Transforms of Discrete Approximations

Now consider computing the Fourier transform of an Nth order discrete approximation

$$\widehat{f} = \sum_{k=-\infty}^{\infty} \widehat{f}_k \, \delta_{k\Delta t}$$

of some function f. Behind this approximation is the Nth order regular sampling of f with spacing Δt,

$$\{ f_0, f_1, f_2, \ldots, f_{N-1} \} \quad .$$

The notation, unfortunately, can become awkward if we follow the conventions you are probably expecting. We would soon have far too many symbols based on the sixth letter of the alphabet, with some denoting two different entities. To avoid this, we will use $\widehat{G}$ to denote the Fourier transform of $\widehat{f}$,

$$\widehat{G} = \mathcal{F}[\widehat{f}] \quad .$$

By definition, $\widehat{f}$ is a periodic, regular array with index period N. As we saw in the previous chapter, so is its Fourier transform, $\widehat{G}$, but with spacing

$$\Delta\omega = \frac{1}{\text{period of } \widehat{f}} = \frac{1}{N\Delta t} \quad .$$

Thus,

$$\mathcal{F}[\widehat{f}] = \widehat{G} = \sum_{n=-\infty}^{\infty} \widehat{G}_n \, \delta_{n\Delta\omega} \quad \text{with} \quad \widehat{G}_{n+N} = \widehat{G}_n \quad .$$

Now let

$$\mathbf{f} = \begin{bmatrix} f_0 \\ f_1 \\ f_2 \\ \vdots \\ f_{N-1} \end{bmatrix} \quad , \quad \widehat{\mathbf{f}} = \begin{bmatrix} \widehat{f}_0 \\ \widehat{f}_1 \\ \widehat{f}_2 \\ \vdots \\ \widehat{f}_{N-1} \end{bmatrix} \quad \text{and} \quad \widehat{\mathbf{G}} = \begin{bmatrix} \widehat{G}_0 \\ \widehat{G}_1 \\ \widehat{G}_2 \\ \vdots \\ \widehat{G}_{N-1} \end{bmatrix} \quad .$$

From theorem 39.2 on page 719, we know the array coefficients of $\widehat{f}$ and $\widehat{G} = \mathcal{F}[\widehat{f}]$ are related by

$$\widehat{\mathbf{G}} = \frac{1}{N\Delta t}\mathbf{M}_N\widehat{\mathbf{f}} \quad \text{and} \quad \widehat{\mathbf{f}} = \frac{1}{N\Delta\omega}\mathbf{M}_N^*\widehat{\mathbf{G}} \tag{40.3}$$

where $\mathbf{M}_N$ and $\mathbf{M}_N^*$ are the Nth matrices of exponentials for $\mathcal{F}$ and $\mathcal{F}^{-1}$, respectively.

But, by the definition of the discrete approximation $\widehat{f}$,

$$\widehat{\mathbf{f}} = \begin{bmatrix} \widehat{f}_0 \\ \widehat{f}_1 \\ \widehat{f}_2 \\ \vdots \\ \widehat{f}_{N-1} \end{bmatrix} = \begin{bmatrix} \Delta t \, f_0 \\ \Delta t \, f_1 \\ \Delta t \, f_2 \\ \vdots \\ \Delta t \, f_{N-1} \end{bmatrix} = \Delta t \, \mathbf{f} \quad .$$

So

$$\widehat{\mathbf{G}} = \frac{1}{N\Delta t}\mathbf{M}_N\widehat{\mathbf{f}} = \frac{1}{N\Delta t}\mathbf{M}_N\Delta t\,\mathbf{f} = \frac{1}{N}\mathbf{M}_N\mathbf{f} \quad .$$

On the other hand, replacing $\widehat{\mathbf{f}}$ with $\Delta t\,\mathbf{f}$ in the second equation in equation set (40.3), and then dividing through by Δt yields

$$\mathbf{f} = \frac{1}{N\Delta t\Delta\omega}\mathbf{M}_N^*\widehat{\mathbf{G}} \quad ,$$

which, because

$$N\Delta t = \text{period of } f \qquad \text{and} \qquad \Delta\omega = \frac{1}{\text{period of } f} \quad ,$$

reduces to

$$\mathbf{f} = \mathbf{M}_N^*\widehat{\mathbf{G}} \quad .$$

All this gives us the following theorem.

Theorem 40.1
Let

$$\mathbf{f} = \begin{bmatrix} f_0 \\ f_1 \\ f_2 \\ \vdots \\ f_{N-1} \end{bmatrix} \qquad \text{and} \qquad \widehat{\mathbf{G}} = \begin{bmatrix} \widehat{G}_0 \\ \widehat{G}_1 \\ \widehat{G}_2 \\ \vdots \\ \widehat{G}_{N-1} \end{bmatrix}$$

where

$$\{ f_0, f_1, f_2, \dots, f_{N-1} \}$$

is the Nth order regular sampling with spacing Δt of some function f, and

$$\widehat{G} = \sum_{n=-\infty}^{\infty} \widehat{G}_n \delta_{n\Delta\omega} \qquad \text{with} \qquad \widehat{G}_{n+N} = \widehat{G}_n \quad \text{and} \quad \Delta\omega = \frac{1}{N\Delta t}$$

is the Fourier transform of the discrete approximation of f corresponding to the above sampling. Then

$$\widehat{\mathbf{G}} = \frac{1}{N}\mathbf{M}_N\mathbf{f} \qquad \text{and} \qquad \mathbf{f} = \mathbf{M}_N^*\widehat{\mathbf{G}} \tag{40.4}$$

where $\mathbf{M}_N$ and $\mathbf{M}_N^*$ are the Nth matrices of exponentials for $\mathcal{F}$ and $\mathcal{F}^{-1}$, respectively.

The corresponding theorem regarding the inverse transform of the discrete approximation can be obtained in a similar fashion or by using near-equivalence.

Theorem 40.2
Let

$$\mathbf{F} = \begin{bmatrix} F_0 \\ F_1 \\ F_2 \\ \vdots \\ F_{N-1} \end{bmatrix} \qquad \text{and} \qquad \widehat{\mathbf{g}} = \begin{bmatrix} \widehat{g}_0 \\ \widehat{g}_1 \\ \widehat{g}_2 \\ \vdots \\ \widehat{g}_{N-1} \end{bmatrix}$$

where

$$\{ F_0 , F_1 , F_2 , \dots , F_{N-1} \} \quad ,$$

is the Nth order regular sampling with spacing $\Delta\omega$ of some function F , and

$$\widehat{g} = \sum_{k=-\infty}^{\infty} \widehat{g}_k \, \delta_{k\Delta t} \quad \text{with} \quad \widehat{g}_{k+N} = \widehat{g}_k \quad \text{and} \quad \Delta t = \frac{1}{N\Delta\omega}$$

is the inverse Fourier transform of the discrete approximation of F corresponding to the above sampling. Then

$$\widehat{g} = \frac{1}{N}\mathbf{M}_N^* \mathbf{F} \quad \text{and} \quad \mathbf{F} = \mathbf{M}_N \widehat{g} \qquad (40.5)$$

where $\mathbf{M}_N$ and $\mathbf{M}_N^*$ are the Nth matrices of exponentials for $\mathcal{F}$ and $\mathcal{F}^{-1}$, respectively.

It is worth recalling that

$$\mathbf{M}_N = \begin{bmatrix} 1 & 1 & 1 & \cdots & 1 \\ 1 & e^{-i\frac{2\pi}{N}1\cdot 1} & e^{-i\frac{2\pi}{N}1\cdot 2} & \cdots & e^{-i\frac{2\pi}{N}1(N-1)} \\ 1 & e^{-i\frac{2\pi}{N}2\cdot 1} & e^{-i\frac{2\pi}{N}2\cdot 2} & \cdots & e^{-i\frac{2\pi}{N}2(N-1)} \\ \vdots & \vdots & \vdots & \ddots & \vdots \\ 1 & e^{-i\frac{2\pi}{N}(N-1)1} & e^{-i\frac{2\pi}{N}(N-1)2} & \cdots & e^{-i\frac{2\pi}{N}(N-1)(N-1)} \end{bmatrix}$$

and

$$\mathbf{M}_N^* = \begin{bmatrix} 1 & 1 & 1 & \cdots & 1 \\ 1 & e^{i\frac{2\pi}{N}1\cdot 1} & e^{i\frac{2\pi}{N}1\cdot 2} & \cdots & e^{i\frac{2\pi}{N}1(N-1)} \\ 1 & e^{i\frac{2\pi}{N}2\cdot 1} & e^{i\frac{2\pi}{N}2\cdot 2} & \cdots & e^{i\frac{2\pi}{N}2(N-1)} \\ \vdots & \vdots & \vdots & \ddots & \vdots \\ 1 & e^{i\frac{2\pi}{N}(N-1)1} & e^{i\frac{2\pi}{N}(N-1)2} & \cdots & e^{i\frac{2\pi}{N}(N-1)(N-1)} \end{bmatrix} .$$

Using these matrices, it is trivial to verify that the equations in set (40.4) can also be written as

$$\widehat{G}_n = \frac{1}{N} \sum_{k=0}^{N-1} f_k \, e^{-i\frac{2\pi}{N}kn} \quad \text{for} \quad n = 0, 1, 2, \dots, N-1$$

and

$$f_k = \sum_{n=0}^{N-1} \widehat{G}_n \, e^{i\frac{2\pi}{N}nk} \quad \text{for} \quad k = 0, 1, 2, \dots, N-1 \quad ,$$

while those in equation set (40.5) can be written as

$$\widehat{g}_k = \frac{1}{N} \sum_{n=0}^{N-1} F_n \, e^{i\frac{2\pi}{N}nk} \quad \text{for} \quad k = 0, 1, 2, \dots, N-1$$

and

$$F_n = \sum_{k=0}^{N-1} \widehat{g}_k \, e^{-i\frac{2\pi}{N}kn} \quad \text{for} \quad n = 0, 1, 2, \dots, N-1 \quad .$$

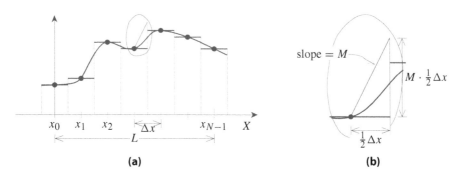

Figure 40.3: (a) A function h with the rectangles for the corresponding Riemann sum approximation and one of the "bounding triangles" (circled). (b) Enlargement of the circled bounding triangle. In these figures $x_k = k\Delta x$.

Are These Good Approximations?
Error Bounds on Integral Approximations

To get a better feeling for the extent to which we can trust our discrete approximation, let us briefly consider the possible error in using this approximation to compute those integrals for which we derived the discrete approximation. And since these computations involve Riemann sums, let us first try to estimate the error in using the Riemann sum

$$\sum_{k=0}^{N-1} h(k\Delta x)\Delta x \tag{40.6}$$

in place of

$$\int_{-\frac{\Delta x}{2}}^{L+\frac{\Delta x}{2}} h(x)\, dx \tag{40.7}$$

where N is some positive integer, Δx is some positive distance, $L = (N-1)\Delta x$ and h is some real-valued function such as that sketched in figure 40.3a. We will assume h is at least continuous and piecewise smooth over the interval of integration. This means there is some positive constant M such that

$$\left|h'(x)\right| < M \tag{40.8}$$

at every x in the interval of integration where f is differentiable.

Geometrically, integral (40.7) is the net area under the curve $y = h(x)$ over the interval of integration, while the Riemann sum is the net area in the collection of rectangles indicated in figure 40.3a. The difference between these two values is simply the total net area in the $2N$ regions of width $\Delta x/2$ between the tops of the rectangles and the graph of h. Each of these regions can certainly be contained in a right triangle with horizontal base of length $\Delta x/2$ and whose hypotenuse has a slope greater in magnitude than that of the derivative of h at any point of the interval of integration (see figure 40.3). So the magnitude of the error in using Riemann sum (40.6) for integral (40.7) must be, at worst, the total area in these $2N$ "bounding triangles".

For each of these bounding triangles, we can take the magnitude of the slope of the hypotenuse to be the value M from inequality (40.8). The area of each triangle can then be obtained using the well-known formula from elementary geometry,

$$\frac{1}{2}\, \text{base} \times \text{height} = \frac{1}{2}\left[\frac{1}{2}\Delta x\right]\left[M\frac{1}{2}\Delta x\right] = \frac{1}{8}M(\Delta x)^2 \quad .$$

Hence, the total area in these $2N$ triangles (and, hence also, a bound on the error in using Riemann sum (40.6) for integral (40.7)) is

$$2N \cdot \frac{1}{8}M(\Delta x)^2 = \frac{1}{4}M(N\Delta x)\Delta x \quad .$$

Letting $L = (N-1)\Delta x$ (as we have with our samplings), we can express this error bound equivalently as

$$\frac{1}{4}M(L + \Delta x)\Delta x \quad \text{or} \quad \frac{1}{4}ML^2\frac{N}{(N-1)^2} \quad .$$

If h is complex valued instead of real valued, then the error bound derived above applies to the real and the imaginary parts separately. Adding these bounds together yields the bound in the following theorem.

Theorem 40.3
Let N, Δx and L be a positive values with N being an integer and $L = (N-1)\Delta x$. Assume h is a continuous, piecewise smooth function on the interval $\left(-\frac{\Delta x}{2}, L + \frac{\Delta x}{2}\right)$, and let M be a constant such that, at each point x in this interval where h is differentiable,

$$\left|h'(x)\right| < M \quad .$$

Then

$$\left| \int_{-\frac{\Delta x}{2}}^{L+\frac{\Delta x}{2}} h(x)\,dx - \sum_{k=0}^{N-1} h(k\Delta x)\Delta x \right| \le E$$

where

$$E = \frac{1}{2}M(L + \Delta x)\Delta x = \frac{1}{2}ML^2\frac{N}{(N-1)^2} \quad .$$

Let us now suppose we have a discrete approximation $\widehat{f}$, generated from an Nth order regular sampling $\{f_0, f_1, \ldots, f_{N-1}\}$ of some function f. As usual, let $[0, L]$ be the minimal window for the sampling, and let Δx be the corresponding spacing. The integral of the product f with any other function ϕ over the natural window,

$$\int_{-\frac{\Delta x}{2}}^{L+\frac{\Delta x}{2}} f(x)\phi(x)\,dx \quad ,$$

is just the integral in theorem 40.3 with $h = f\phi$. Replacing f with its discrete approximation and reversing the computations previously done to derive $\widehat{f}$, we get

$$\int_{-\frac{\Delta x}{2}}^{L+\frac{\Delta x}{2}} \widehat{f}(x)\phi(x)\,dx = \int_{-\frac{1}{2}\Delta x}^{\left(N-\frac{1}{2}\right)\Delta x} \sum_{k=-\infty}^{\infty} f_k\,\Delta x\,\delta_{k\Delta x}(x)\,\phi(x)\,dx$$

$$= \sum_{k=0}^{N-1} f_k\,\Delta x\,\phi(k\Delta x) = \sum_{k=0}^{N-1} f(k\Delta x)\phi(k\Delta x)\,\Delta x \quad .$$

This last summation is just the Riemann sum in theorem 40.3 with $h = f\phi$. That, along with the product rule, $(f\phi)' = f'\phi + f\phi'$, then gives us the following corollary.

Corollary 40.4
Let $\widehat{f}$ be the Nth order discrete approximation with spacing Δx for some function f, and let ϕ be some other function. Assume, further, that f and ϕ are continuous and piecewise smooth on

the natural window of the corresponding sampling of f, and let A_0, A_1, B_0 and B_1 be constants such that

$$|f(x)| \leq A_0 \quad , \quad |f'(x)| \leq A_1 \quad , \quad |\phi(x)| \leq B_0 \quad \text{and} \quad |\phi'(x)| \leq B_1$$

on that window. Then, letting L be the length of the minimal window of the sampling,

$$\left| \int_{-\frac{\Delta x}{2}}^{L+\frac{\Delta x}{2}} f(x)\phi(x)\,dx - \int_{-\frac{\Delta x}{2}}^{L+\frac{\Delta x}{2}} \widehat{f}(x)\phi(x)\,dx \right| \leq E \tag{40.9a}$$

where

$$E = \frac{1}{2}(A_1 B_0 + A_0 B_1)(L + \Delta x)\Delta x = \frac{NL^2}{2(N-1)^2}(A_1 B_0 + A_0 B_1) \quad . \tag{40.9b}$$

Observe that, for fixed choices of f, ϕ and L, the error bound given by equation (40.9b) shrinks to 0 as N, the number of samples taken over the minimal window $[0, L]$, goes to ∞. This confirms the intuitive notion that we can improve our approximations by taking more samples of f over $[0, L]$. On the other hand, this error bound depends on the maximum values of the derivatives of f and ϕ (through the constants A_1 and B_1). Consequently, for a fixed choice of f, L and N, we can always find a ϕ for which the inequality

$$|\phi'(x)| \leq B_1$$

requires such a large value of B_1 that the error bound (40.9b) remains quite large.

?▶ **Exercise 40.5:**   Assume f is a continuous, piecewise smooth function on $(-1, 2)$ that satisfies

$$|f(x)| \leq 1 \quad \text{and} \quad |f'(x)| \leq 1 \quad \text{for} \quad -1 < x < 2 \quad .$$

Let $\widehat{f}$ be the Nth discrete approximation of f over the minimal window $[0, 1]$, and consider the error in the approximation

$$\int_{-\frac{\Delta x}{2}}^{L+\frac{\Delta x}{2}} f(x)\, e^{i\alpha x}\, dx \approx \int_{-\frac{\Delta x}{2}}^{L+\frac{\Delta x}{2}} \widehat{f}(x)\, e^{i\alpha x}\, dx$$

for some real value α. Using corollary 40.4:

a: Verify that this error shrinks to 0 as $N \to \infty$.

b: Determine how large N should be to ensure that this error is less than $1/10$ when $\alpha = 1$.

c: Determine a value of α such that the corresponding error bound from corollary 40.4 is greater than 1 even when $N = 100$.

Corollary 40.4 provides the information needed to choose the sampling parameters (i.e., the order and spacing) ensuring that your subsequent computations are within the accuracy desired. Of course, this presumes you have some idea of the largest values of the sampled function and its derivatives before sampling the function. If you have little *a priori* knowledge of f, then determining how small to make the sampling spacing can be a challenge, especially if the sampling is to be obtained by taking measurements of some phenomenon occurring in "real time". Then, not only must you guess at how rapidly f might vary, but you must also consider the resources available for doing the measurements along with the likely possibility that these measurements will be contaminated by some sort of random noise.

Approximating the Fourier Transform

One might naturally ask if the Fourier transform of f, $F = \mathcal{F}[f]$, is approximated by the Fourier transform of the discrete approximation, $\widehat{G} = \mathcal{F}\left[\widehat{f}\right]$.

In general, the answer is *no*.

Remember, $\mathcal{F}[f]$ depends on how f is defined "everywhere", while $\widehat{f}$ is only an approximation for f over the natural window of the sampling. Outside that window, $\widehat{f}$ and f may differ wildly. Indeed, f may not even exist outside the window. So, in general, we should not expect $\widehat{G}$ to accurately approximate $\mathcal{F}[f]$, even if this transform exists.

But let us not be so general. Let us assume f is a continuous and piecewise smooth function on $\mathbb{R}$ that vanishes outside the sampling's natural window. That is,

$$ f(t) = 0 \quad \text{if} \quad t < -\frac{1}{2}\Delta t \quad \text{or} \quad L + \frac{1}{2}\Delta t < t $$

where N and Δt are, respectively, the order and spacing of $\widehat{f}$, and, as usual, $L = (N-1)\Delta t$. From the classical theory, we then know $F = \mathcal{F}[f]$ is a continuous function on the real line, and is given by

$$ F(\omega) = \int_{-\infty}^{\infty} f(t)\, e^{-i2\pi\omega t}\, dt = \int_{-\frac{\Delta x}{2}}^{L+\frac{\Delta x}{2}} f(t)\, e^{-i2\pi\omega t}\, dt \quad . \tag{40.10} $$

This last integral is just the sort of integral $\widehat{f}$ was designed for. So, for this function,

$$ F(\omega) \approx \int_{-\frac{\Delta x}{2}}^{L+\frac{\Delta x}{2}} \widehat{f}(t)\, e^{-i2\pi\omega t}\, dt \quad . \tag{40.11} $$

Now,

$$ \int_{-\frac{\Delta x}{2}}^{L+\frac{\Delta x}{2}} \widehat{f}(t)\, e^{-i2\pi\omega t}\, dt = \int_{-\frac{1}{2}\Delta t}^{\left(N-\frac{1}{2}\right)\Delta t} \sum_{k=-\infty}^{\infty} f_k\, \Delta t\, \delta_{k\Delta t}(t)\, e^{-i2\pi\omega t}\, dt $$

$$ = \sum_{k=-\infty}^{\infty} f_k\, \Delta t \int_{-\frac{1}{2}\Delta t}^{\left(N-\frac{1}{2}\right)\Delta t} \delta_{k\Delta t}(t)\, e^{-i2\pi\omega t}\, dt $$

$$ = \sum_{k=-\infty}^{\infty} f_k\, \Delta t \left\{ \begin{array}{ll} e^{-i2\pi\omega k\Delta t} & \text{if} \quad k = 0,\ 1,\ 2,\ \ldots,\ N-1 \\ 0 & \text{otherwise} \end{array} \right\} $$

$$ = \sum_{k=0}^{N-1} f_k\, \Delta t\, e^{-i2\pi\omega k\Delta t} \quad . $$

So, approximation (40.11) reduces to

$$ F(\omega) \approx \sum_{k=0}^{N-1} f_k\, \Delta t\, e^{-i2\pi\omega k\Delta t} \quad . \tag{40.12} $$

Just how good an approximation this is can be estimated using our last corollary with $\phi(t) = e^{i2\pi\omega t}$. With this for ϕ,

$$ |\phi(t)| = \left| e^{i2\pi\omega t} \right| = 1 \quad \text{and} \quad |\phi'(t)| = \left| i2\pi\omega\, e^{i2\pi\omega t} \right| = 2\pi\,|\omega| \quad . $$

Corollary 40.4 then tells us that

$$\left| F(\omega) - \int_{-\frac{\Delta x}{2}}^{L+\frac{\Delta x}{2}} \widehat{f}(t)\, e^{-i2\pi\omega t}\, dt \right| \leq \frac{1}{2}(A_1 + 2\pi\, |\omega|\, A_0)L^2 \frac{N}{(N-1)^2}$$

where A_0 and A_1 are, respectively, upper bounds on the possible values of $f(t)$ and $f'(t)$ over the natural window of the sampling.

For a given function f and a given minimal window $[0, L]$, the above error bound depends on just the sample size, N, and the magnitude of ω. For any given value of $|\omega|$, this error bound shrinks steadily to 0 as $N \to \infty$. On the other hand, for any fixed choice for N, the error bound blows up as $|\omega| \to \infty$. Consequently, we can only be sure approximation (40.12) is reasonably accurate when $|\omega|$ is relatively small and N is relatively large (exactly how small and how large depends on the values of A_0, A_1 and L as well as our criteria for the approximation being "reasonably accurate").

?▶Exercise 40.6: *Assume f is a continuous, piecewise smooth function on $\mathbb{R}$ that vanishes outside $[0, 1]$ and satisfies*

$$|f(x)| \leq \frac{1}{2} \quad \text{and} \quad |f'(x)| \leq 1 \quad \text{for } 0 < x < 1 \quad .$$

Let $F = \mathcal{F}[f]$, and let $\widehat{f}$ be the Nth discrete approximation of f over the minimal window $[0, 1]$ (with spacing Δt), and, letting $L = (N-1)\Delta t$, consider the error in the approximation

$$F(\omega) \approx \int_{-\frac{\Delta x}{2}}^{L+\frac{\Delta x}{2}} \widehat{f}(x)\, e^{-i2\pi\omega t}\, dt \quad .$$

Based on the above discussion:

a: *What is an upper bound on the error when $N = 11$ and $\omega = \frac{1}{\pi}$?*

b: *What should N be to ensure that the error is less than $\frac{1}{10}$ when $\omega = \frac{1}{\pi}$?*

c: *For what values of ω can we be sure the error is less than $\frac{1}{100}$ when $N = 10$? when $N = 1000$?*

Is the Transform of a Discrete Approximation the Discrete Approximation of the Transform?

Let us continue our analysis (with you doing a bit more of the work) comparing $\widehat{G} = \mathcal{F}\left[\widehat{f}\right]$ with $F = \mathcal{F}[f]$, still assuming f is a continuous and piecewise smooth function on $\mathbb{R}$ that vanishes outside the sampling's natural window. Remember, $\widehat{G}$ is a regular, periodic array

$$\widehat{G} = \sum_{n=-\infty}^{\infty} \widehat{G}\, \delta_{n\Delta\omega} \quad .$$

The index period is N, the spacing $\Delta\omega$ is $(N\Delta t)^{-1}$, and the coefficients are given by the formulas in theorem 40.1 on page 729. This array, in turn, can be viewed as the Nth order discrete approximation of some function G with N^{th} order regular sampling

$$\{G_0, G_1, G_2, \ldots, G_{N-1}\}$$

over the minimal window $[0, (N-1)\Delta\omega]$.

?►Exercise 40.7:   *What is the relation between the G_n's and the $\widehat{G}_n$'s?*

On the other hand, since we know F is continuous, we can take its Nth order regular sampling

$$\{ F_0,\ F_1,\ F_2,\ \ldots,\ F_{N-1} \}$$

over the minimal window $[0, (N-1)\Delta\omega]$, and then construct $\widehat{F}$, the corresponding N^{th} order discrete approximation for F,

$$\widehat{F} = \sum_{n=-\infty}^{\infty} \widehat{F}\, \delta_{n\Delta\omega} \quad .$$

?►Exercise 40.8:   *How is each F_n computed, and how is it related to $\widehat{F}_n$?*

Since

$$\widehat{G} = \mathcal{F}\!\left[\widehat{f}\right] \qquad \text{and} \qquad \mathcal{F}[f] = F \quad,$$

and, in some sense,

$$\widehat{f} \approx f \qquad \text{and} \qquad F \approx \widehat{F} \quad,$$

it might seem reasonable that

$$\widehat{G} \approx \widehat{F} \quad .$$

As an exercise, why don't you investigate the validity of this last approximation?

?►Exercise 40.9   a: *Verify that*

$$\widehat{G} \approx \widehat{F}$$

by showing

$$\widehat{G}_n \approx \widehat{F}_n \qquad \text{for} \quad n = 0,\ 1,\ 2,\ \ldots,\ N-1 \quad .$$

(Use results from previous subsections.)

b: *From the above and your work in previous exercises, we have*

$$F(n\Delta\omega) \approx \frac{1}{\Delta\omega}\, \widehat{G}_n \qquad \text{for} \quad n = 0,\ \pm1,\ \pm2,\ \ldots, \quad .$$

Can we conclude that this will be a good approximation for every integer n?

c: *From the above and the periodicity of $\widehat{G}$,*

$$F(-\Delta\omega) \approx G_{-1} = G_{N-1} \approx F((N-1)\Delta\omega) \quad .$$

Based on the error bound from corollary 40.4, which is likely to be the "better" approximation:

$$F(-\Delta\omega) \approx G_{N-1} \qquad \text{or} \qquad F((N-1)\omega) \approx G_{N-1} \quad ?$$

Don't forget that the above analysis is for a continuous and piecewise smooth function f on $\mathbb{R}$ that vanishes outside the natural window for the original sampling $\{f_0, f_1, f_2, \ldots, f_{N-1}\}$. With a bit more work, you can derive similar results when f is merely a piecewise smooth function vanishing outside that natural window, though you should insist that N is significantly larger than the number of discontinuities in f. However, if f does not vanish outside that natural window, then equation

(40.10) on page 734 is not valid, and our arguments pretty well fall apart. Indeed, unless you can, somehow, show that

$$\int_{-\frac{1}{2}\Delta t}^{\left(N-\frac{1}{2}\right)\Delta t} f(t) \, e^{-i2\pi\omega t} \, dt \;\approx\; \int_{-\infty}^{\infty} f(t) \, e^{-i2\pi\omega t} \, dt \quad,$$

then you should not expect $\widehat{G}$ to be any sort of approximation to $\mathcal{F}[f]$.

Wait a moment. According to the next exercise, that last statement is not completely true.

?►Exercise 40.10: *Assume f is a continuous, piecewise smooth and periodic function on $\mathbb{R}$, and let $\widehat{f}$ be an Nth order discrete approximation of f. Suppose, further, that the length of the natural window for the corresponding sampling is a period for f. Derive approximate relationships between the coefficients of $\widehat{G} = \mathcal{F}[\widehat{f}]$ and the Fourier coefficients for f. Using this, compare $\widehat{G}$ with $\mathcal{F}[f]$.*

40.4   The Discrete Fourier Transforms

The "discrete theory" of Fourier analysis is mainly concerned with manipulations of finite sequences of numbers (such as commonly arise when we take samplings). Often, these manipulations can be nicely described by operations from elementary linear algebra. Because of this, it is convenient to formally adopt a notational convention relating sequences and vectors that we have been using informally for some time: Given any N^{th} order sequence $\{x_0, x_1, \ldots, x_{N-1}\}$, the corresponding ($N^{\text{th}}$ order) *sequence vector* $\mathbf{x}$ is the column vector

$$\mathbf{x} = \begin{bmatrix} x_0 \\ x_1 \\ x_2 \\ \vdots \\ x_{N-1} \end{bmatrix} \quad .$$

Conversely, if we refer to, say, $\mathbf{y}$ as an "N^{th} order sequence vector", then $\mathbf{y}$ should be assumed to be the sequence vector corresponding to an Nth order sequence $\{y_0, y_1, \ldots, y_{N-1}\}$.

The two "N^{th} order discrete Fourier transforms" are, essentially, the computational processes described by equation set (40.4) in theorem 40.1 on page 729 (slightly scaled). Each transforms an Nth order sequence into another N^{th} order sequence, and each can be described by a simple matrix. The matrix for the Nth order discrete Fourier transform and the matrix for the Nth order inverse discrete Fourier transform, denoted by $\mathcal{F}_N$ and $\mathcal{F}_N^{-1}$, respectively, are given by

$$\mathcal{F}_N = \frac{1}{\sqrt{N}}\mathbf{M}_N \quad \text{and} \quad \mathcal{F}_N^{-1} = \frac{1}{\sqrt{N}}\mathbf{M}_N^* \tag{40.13}$$

where $\mathbf{M}_N$ is the Nth matrix of exponentials for $\mathcal{F}$, and $\mathbf{M}_N^*$ is its conjugate, the Nth matrix of exponentials for $\mathcal{F}^{-1}$.

Now let $\mathbf{v}$ and $\mathbf{W}$ be two N^{th} order sequence vectors. The Nth order discrete transform of $\mathbf{v}$ is defined to be the Nth order sequence vector $\mathbf{V}$ given by

$$\mathbf{V} = \mathcal{F}_N \mathbf{v} = \frac{1}{\sqrt{N}}\mathbf{M}_N \mathbf{v} \quad , \tag{40.14}$$

while the Nth order inverse discrete transform of $\mathbf{W}$ is defined to be the Nth order sequence vector $\mathbf{w}$ given by

$$\mathbf{w} \;=\; \mathcal{F}_N^{-1}\mathbf{W} \;=\; \frac{1}{\sqrt{N}}\mathbf{M}_N^*\mathbf{W} \quad . \tag{40.15}$$

To actually compute discrete transforms, we'll need the full formulas for the above. Recalling the definitions of $\mathbf{M}$ and $\mathbf{M}_N^*$, we see that

$$\mathcal{F}_N \;=\; \frac{1}{\sqrt{N}}\begin{bmatrix} 1 & 1 & 1 & \cdots & 1 \\ 1 & e^{-i\frac{2\pi}{N}1\cdot1} & e^{-i\frac{2\pi}{N}1\cdot2} & \cdots & e^{-i\frac{2\pi}{N}1(N-1)} \\ 1 & e^{-i\frac{2\pi}{N}2\cdot1} & e^{-i\frac{2\pi}{N}2\cdot2} & \cdots & e^{-i\frac{2\pi}{N}2(N-1)} \\ \vdots & \vdots & \vdots & \ddots & \vdots \\ 1 & e^{-i\frac{2\pi}{N}(N-1)1} & e^{-i\frac{2\pi}{N}(N-1)2} & \cdots & e^{-i\frac{2\pi}{N}(N-1)(N-1)} \end{bmatrix} \tag{40.16}$$

and

$$\mathcal{F}_N^{-1} \;=\; \frac{1}{\sqrt{N}}\begin{bmatrix} 1 & 1 & 1 & \cdots & 1 \\ 1 & e^{i\frac{2\pi}{N}1\cdot1} & e^{i\frac{2\pi}{N}1\cdot2} & \cdots & e^{i\frac{2\pi}{N}1(N-1)} \\ 1 & e^{i\frac{2\pi}{N}2\cdot1} & e^{i\frac{2\pi}{N}2\cdot2} & \cdots & e^{i\frac{2\pi}{N}2(N-1)} \\ \vdots & \vdots & \vdots & \ddots & \vdots \\ 1 & e^{i\frac{2\pi}{N}(N-1)1} & e^{i\frac{2\pi}{N}(N-1)2} & \cdots & e^{i\frac{2\pi}{N}(N-1)(N-1)} \end{bmatrix} \quad . \tag{40.17}$$

So formulas (40.14) and (40.15) can be written as

$$\begin{bmatrix} V_0 \\ V_1 \\ V_2 \\ \vdots \\ V_{N-1} \end{bmatrix} = \frac{1}{\sqrt{N}}\begin{bmatrix} 1 & 1 & 1 & \cdots & 1 \\ 1 & e^{-i\frac{2\pi}{N}1\cdot1} & e^{-i\frac{2\pi}{N}1\cdot2} & \cdots & e^{-i\frac{2\pi}{N}1(N-1)} \\ 1 & e^{-i\frac{2\pi}{N}2\cdot1} & e^{-i\frac{2\pi}{N}2\cdot2} & \cdots & e^{-i\frac{2\pi}{N}2(N-1)} \\ \vdots & \vdots & \vdots & \ddots & \vdots \\ 1 & e^{-i\frac{2\pi}{N}(N-1)1} & e^{-i\frac{2\pi}{N}(N-1)2} & \cdots & e^{-i\frac{2\pi}{N}(N-1)(N-1)} \end{bmatrix}\begin{bmatrix} v_0 \\ v_1 \\ v_2 \\ \vdots \\ v_{N-1} \end{bmatrix}$$

and

$$\begin{bmatrix} W_0 \\ W_1 \\ W_2 \\ \vdots \\ W_{N-1} \end{bmatrix} = \frac{1}{\sqrt{N}}\begin{bmatrix} 1 & 1 & 1 & \cdots & 1 \\ 1 & e^{i\frac{2\pi}{N}1\cdot1} & e^{i\frac{2\pi}{N}1\cdot2} & \cdots & e^{i\frac{2\pi}{N}1(N-1)} \\ 1 & e^{i\frac{2\pi}{N}2\cdot1} & e^{i\frac{2\pi}{N}2\cdot2} & \cdots & e^{i\frac{2\pi}{N}2(N-1)} \\ \vdots & \vdots & \vdots & \ddots & \vdots \\ 1 & e^{i\frac{2\pi}{N}(N-1)1} & e^{i\frac{2\pi}{N}(N-1)2} & \cdots & e^{i\frac{2\pi}{N}(N-1)(N-1)} \end{bmatrix}\begin{bmatrix} w_0 \\ w_1 \\ w_2 \\ \vdots \\ w_{N-1} \end{bmatrix} \quad .$$

After examining the above matrix/vector formulas (or recalling how the matrix/vector formulas arose in the first place), it should be clear that formula (40.14), defining the discrete Fourier transform of $\mathbf{v}$, is completely equivalent to the component formulas

$$V_n \;=\; \frac{1}{\sqrt{N}}\sum_{k=0}^{N-1} e^{-i\frac{2\pi}{N}nk}v_k \qquad \text{for} \quad n = 0, 1, 2, \ldots N-1 \quad , \tag{40.18}$$

while formula (40.15), defining the inverse discrete Fourier transform of $\mathbf{W}$, is equivalent to the component formulas

$$w_k \;=\; \frac{1}{\sqrt{N}}\sum_{n=0}^{N-1} e^{i\frac{2\pi}{N}kn}W_n \qquad \text{for} \quad k = 0, 1, 2, \ldots N-1 \quad . \tag{40.19}$$

For our purposes, there is no reason to distinguish between "sequence vectors" and "sequences"; we will refer to both the vector $\mathbf{V}$ from equation (40.14) and the sequence $\{V_0, V_1, \ldots, V_{N-1}\}$ defined by equation (40.18) as the discrete Fourier transform of either the vector $\mathbf{v}$ or the sequence $\{v_0, v_1, \ldots, v_{N-1}\}$ as convenient. Likewise, only convenience will dictate whether $\mathbf{w}$ from equation (40.19) or $\{w_0, w_1, \ldots, w_{N-1}\}$ from equation (40.19) will be called the inverse discrete Fourier transform of $\mathbf{W}$ or $\{W_0, W_1, \ldots, W_{N-1}\}$.

!▶ **Example 40.4:** *Applying the results from example 39.4 on page 717, we obtain*

$$
\mathscr{F}_4 \;=\; \frac{1}{\sqrt{4}} \mathbf{M}_4 \;=\; \frac{1}{2}
\begin{bmatrix}
1 & 1 & 1 & 1 \\
1 & -i & -1 & i \\
1 & -1 & 1 & -1 \\
1 & i & -1 & -i
\end{bmatrix} .
$$

The discrete Fourier transform of the 4th order sequence $\{2, 4, 6, 8\}$ can be obtained using formula (40.14) with $N = 4$ and

$$
\mathbf{v} \;=\;
\begin{bmatrix}
v_0 \\ v_1 \\ v_2 \\ v_3
\end{bmatrix}
\;=\;
\begin{bmatrix}
2 \\ 4 \\ 6 \\ 8
\end{bmatrix} .
$$

Doing so:

$$
\begin{bmatrix}
V_0 \\ V_1 \\ V_2 \\ V_3
\end{bmatrix}
\;=\;
\frac{1}{2}
\begin{bmatrix}
1 & 1 & 1 & 1 \\
1 & -i & -1 & i \\
1 & -1 & 1 & -1 \\
1 & i & -1 & -i
\end{bmatrix}
\begin{bmatrix}
2 \\ 4 \\ 6 \\ 8
\end{bmatrix}
$$

$$
\;=\;
\frac{1}{2}
\begin{bmatrix}
1 \cdot 2 & 1 \cdot 4 & 1 \cdot 6 & 1 \cdot 8 \\
1 \cdot 2 & -i \cdot 4 & -1 \cdot 6 & i \cdot 8 \\
1 \cdot 2 & -1 \cdot 4 & 1 \cdot 6 & -1 \cdot 8 \\
1 \cdot 2 & i \cdot 4 & -1 \cdot 6 & -i \cdot 8
\end{bmatrix}
\;=\;
\begin{bmatrix}
10 \\ -2 + 2i \\ -2 \\ -2 - 2i
\end{bmatrix} .
$$

So the discrete Fourier transform of $\{2, 4, 6, 8\}$ is $\{10, -2 + 2i, -2, -2 - 2i\}$.

?▶ **Exercise 40.11:** *Find $\mathscr{F}_4^{-1}$, the matrix for the 4th order inverse discrete Fourier transform, and compute the inverse discrete Fourier transform of $\{2, 4, 6, 8\}$.*

As with the Fourier transforms, the term "discrete Fourier transform" (often abbreviated to "DFT") refers to both the results of the computations just described and the process of computing those sequences/vectors.

Alternate Definitions

Formulas (40.18) and (40.19) are close to the formulas you are likely to find in other texts and implemented in various software packages. Certain details of the definitions, however, can vary. Where we use formulas (40.18) and (40.19), another text or software package may use

$$
V_n \;=\; \frac{1}{N} \sum_{k=0}^{N-1} e^{-i\frac{2\pi}{N}nk} v_k
\qquad \text{and} \qquad
w_k \;=\; \sum_{n=0}^{N-1} e^{i\frac{2\pi}{N}kn} W_n .
$$

Yet another may use

$$V_n = \sum_{k=0}^{N-1} e^{-i\frac{2\pi}{N}nk} v_k \qquad \text{and} \qquad w_k = \frac{1}{N} \sum_{n=0}^{N-1} e^{i\frac{2\pi}{N}kn} W_n \quad .$$

In addition, you may find a few who run their indices from 1 to N (instead of from 0 to $N-1$) along with a few others who switch the signs in the exponentials.

Clearly, caution should be exercised when applying results or computations regarding discrete transforms based on any reference or software package. Always know how the discrete transforms are defined in the reference or software you are using, and make sure you can convert their results/computations to the analogous results/computations for the discrete transforms as you care to define them.

Relation with Samplings and Periodic, Regular Arrays

Our definitions for the discrete transforms were inspired by the formulas derived earlier for computing Fourier transforms of periodic, regular arrays and discrete approximations. Indeed, if you compare our discrete transform definitions with the formulas in corollary 39.3 on page 719, you will immediately see that those definitions exactly match the formulas for computing the array coefficients for the corresponding transforms of certain periodic, regular arrays. This fact, stated explicitly in the next lemma, provides a direct connection between the discrete transform theory and the theory already developed for Fourier transforms.

Lemma 40.5
Assume $\mathbf{f}$ *and* $\mathbf{F}$ *are two* N^{th} *order sequence vectors, and let*

$$f = \sum_{k=-\infty}^{\infty} f_k \, \delta_{k\Delta t} \qquad \text{and} \qquad F = \sum_{n=-\infty}^{\infty} F_n \, \delta_{n\Delta\omega}$$

be the two periodic, regular arrays with index period N *, spacing*

$$\Delta t = \frac{1}{\sqrt{N}} = \Delta\omega \quad ,$$

and related to $\mathbf{f}$ *and* $\mathbf{F}$ *by*

$$\mathbf{f} = \begin{bmatrix} f_0 \\ f_1 \\ f_2 \\ \vdots \\ f_{N-1} \end{bmatrix} \qquad \text{and} \qquad \mathbf{F} = \begin{bmatrix} F_0 \\ F_1 \\ F_2 \\ \vdots \\ F_{N-1} \end{bmatrix} \quad .$$

Then, if any one of the following statements is true, all are true:

1. $F = \mathcal{F}[f]$.

2. $f = \mathcal{F}^{-1}[F]$.

3. $\mathbf{F} = \mathcal{F}_N \mathbf{f}$.

4. $\mathbf{f} = \mathcal{F}_N^{-1} \mathbf{F}$.

The next theorem, which is nothing more than a combination of portions of theorems 40.1 and 40.2 with the formulas expressed in terms of the discrete transforms, describes the role of the discrete transforms in computing transforms based on samplings.

Theorem 40.6
Let

$$
\boldsymbol{\psi} = \begin{bmatrix} \psi_0 \\ \psi_1 \\ \psi_2 \\ \vdots \\ \psi_{N-1} \end{bmatrix} \quad , \quad \widehat{\mathbf{G}} = \begin{bmatrix} \widehat{G}_0 \\ \widehat{G}_1 \\ \widehat{G}_2 \\ \vdots \\ \widehat{G}_{N-1} \end{bmatrix} \quad \text{and} \quad \widehat{\mathbf{g}} = \begin{bmatrix} \widehat{g}_0 \\ \widehat{g}_1 \\ \widehat{g}_2 \\ \vdots \\ \widehat{g}_{N-1} \end{bmatrix}
$$

where

$$
\{\, \psi_0 \,, \, \psi_1 \,, \, \psi_2 \,, \, \ldots \,, \, \psi_{N-1} \,\}
$$

is the Nth order regular sampling with spacing Δx of some function ψ , and

$$
\widehat{G} = \sum_{n=-\infty}^{\infty} \widehat{G}_n \, \delta_{n\Delta\omega} \quad \text{and} \quad \widehat{g} = \sum_{n=-\infty}^{\infty} \widehat{g}_n \, \delta_{n\Delta t}
$$

are, respectively, the Fourier transform and the inverse Fourier transform of the discrete approximation of ψ corresponding to the above sampling. Then

$$
\widehat{\mathbf{G}} = \frac{1}{\sqrt{N}} \mathcal{F}_N \boldsymbol{\psi} \quad , \quad \boldsymbol{\psi} = \sqrt{N} \, \mathcal{F}_N^{-1} \widehat{\mathbf{G}} \quad ,
$$

$$
\widehat{\mathbf{g}} = \frac{1}{\sqrt{N}} \mathcal{F}_N^{-1} \boldsymbol{\psi} \quad \text{and} \quad \boldsymbol{\psi} = \sqrt{N} \, \mathcal{F}_N \widehat{\mathbf{g}} \quad .
$$

40.5 Discrete Transform Identities

Unsurprisingly, there are discrete analogs to many of the identities we derived for the Fourier transforms.

Linearity

Because the discrete transforms can be described as matrix multiplications, linearity is trivial to verify. By basic linear algebra,

$$
\mathcal{F}_N[a\boldsymbol{\phi} + b\boldsymbol{\psi}] = a\mathcal{F}_N\boldsymbol{\phi} + b\mathcal{F}_N\boldsymbol{\psi} \quad \text{and} \quad \mathcal{F}_N^{-1}[a\boldsymbol{\phi} + b\boldsymbol{\psi}] = a\mathcal{F}_N^{-1}\boldsymbol{\phi} + b\mathcal{F}_N^{-1}\boldsymbol{\psi}
$$

whenever $\boldsymbol{\phi}$ and $\boldsymbol{\psi}$ are two N^{th} order sequence vectors and a and b are two constants. Using the component formulas, these identities become the equally obvious equations

$$
\frac{1}{\sqrt{N}} \sum_{k=0}^{N-1} e^{-i\frac{2\pi}{N}nk} [a\phi_k + b\psi_k] = a \frac{1}{\sqrt{N}} \sum_{k=0}^{N-1} e^{-i\frac{2\pi}{N}nk}\phi_k + b \frac{1}{\sqrt{N}} \sum_{k=0}^{N-1} e^{-i\frac{2\pi}{N}nk}\psi_k
$$

$$
\text{for} \quad n = 0, 1, 2, \ldots, N-1
$$

and

$$\frac{1}{\sqrt{N}} \sum_{n=0}^{N-1} e^{i\frac{2\pi}{N}kn}[a\phi_n + b\psi_n] = a\frac{1}{\sqrt{N}} \sum_{n=0}^{N-1} e^{i\frac{2\pi}{N}kn}\phi_n + b\frac{1}{\sqrt{N}} \sum_{n=0}^{N-1} e^{i\frac{2\pi}{N}kn}\psi_n$$

$$\text{for} \quad k = 0, 1, 2, \ldots, N-1 \quad .$$

Invertibility

Verifying that $\mathscr{F}_N$ and $\mathscr{F}_N^{-1}$ are inverses of each other is most easily done by referring back to corollary 39.4 on page 719 on the inversion of exponential matrices. By definition of the discrete transforms, elementary linear algebra and the aforementioned corollary,

$$\mathscr{F}_N^{-1}\mathscr{F}_N \mathbf{v} = \left(\frac{1}{\sqrt{N}}\mathbf{M}_N\right)\left(\frac{1}{\sqrt{N}}\mathbf{M}_N^*\right)\mathbf{v} = \frac{1}{N}\mathbf{M}_N\mathbf{M}_N^*\mathbf{v} = \mathbf{v} \quad ,$$

for every N^{th} order sequence vector $\mathbf{v}$. Thus $\mathscr{F}_N$ and $\mathscr{F}_N^{-1}$ are inverses of each other. In terms of the component formulas, invertibility is:

Given any two N^{th} order sequences $\{v_0, v_1, \ldots, v_{N-1}\}$ and $\{V_0, V_1, \ldots, V_{N-1}\}$,

$$V_n = \frac{1}{\sqrt{N}} \sum_{k=0}^{N-1} e^{-i\frac{2\pi}{N}nk}v_k \qquad \text{for} \quad n = 0, 1, 2, \ldots, N-1$$

if and only if

$$v_k = \frac{1}{\sqrt{N}} \sum_{n=0}^{N-1} e^{i\frac{2\pi}{N}kn}V_n \qquad \text{for} \quad k = 0, 1, 2, \ldots, N-1 \quad .$$

Near-Equivalence

Compared to some of the other symmetries in the discrete transforms, the analog of near-equivalence is relatively unimportant. However, in deriving it, we will introduce two other notions that are relevant in other contexts.

First of all, given any N^{th} order sequence $\{x_0, x_1, \ldots, x_{N-1}\}$, we will automatically assume x_k is defined for every integer k by the recursion formula

$$x_{k+N} = x_k \quad .$$

This corresponds to viewing the x_k's as the coefficients of a periodic, regular array of delta functions with index period N. It is also consistent with the sequence formulas for the discrete transforms, as you can easily check.

?►Exercise 40.12: *Let $\{\phi_0, \phi_1, \ldots, \phi_{N-1}\}$ be an Nth order sequence, and, for each integer m, define u_m and w_m by the discrete transform formulas*

$$u_m = \frac{1}{\sqrt{N}} \sum_{k=0}^{N-1} e^{-i\frac{2\pi}{N}mk}\phi_k \qquad \text{and} \qquad w_m = \frac{1}{\sqrt{N}} \sum_{n=0}^{N-1} e^{i\frac{2\pi}{N}mn}\phi_n \quad .$$

Show that

$$u_{m+N} = u_m \qquad \text{and} \qquad w_{m+N} = w_m \qquad \text{for} \quad m = 0, \pm 1, \pm 2, \ldots \quad .$$

The other notion is that of "index negation". Given any N^{th} order sequence $\{x_0, x_1, \ldots, x_{N-1}\}$, the corresponding *index-negated* sequence is $\{\widetilde{x}_0, \widetilde{x}_1, \ldots, \widetilde{x}_{N-1}\}$ where

$$\widetilde{x}_k = x_{-k} \quad \text{for} \quad k = 0, \pm 1, \pm 2, \ldots \quad .$$

It's worth noting that, by the definition and the above recursion formula,

$$\{\widetilde{x}_0, \widetilde{x}_1, \widetilde{x}_2, \ldots, \widetilde{x}_{N-2}, \widetilde{x}_{N-1}\} = \{x_{-0}, x_{-1}, x_{-2}, \ldots, x_{-(N-2)}, x_{-(N-1)}\}$$

$$= \{x_0, x_{N-1}, x_{N-2}, \ldots, x_{N-(N-2)}, x_{N-(N-1)}\}$$

$$= \{x_0, x_{N-1}, x_{N-2}, \ldots, x_2, x_1\} \quad .$$

As just illustrated, index-negated sequences will be indicated by a " $\widetilde{}$ " above the sequence terms. Likewise, if $\mathbf{x}$ is a sequence vector corresponding to $\{x_0, x_1, \ldots, x_{N-1}\}$, then

$$\widetilde{\mathbf{x}} = \begin{bmatrix} \widetilde{x}_0 \\ \widetilde{x}_1 \\ \widetilde{x}_2 \\ \vdots \\ \widetilde{x}_{N-2} \\ \widetilde{x}_{N-1} \end{bmatrix} = \begin{bmatrix} x_{-0} \\ x_{-1} \\ x_{-2} \\ \vdots \\ x_{-(N-2)} \\ x_{-(N-1)} \end{bmatrix} = \begin{bmatrix} x_0 \\ x_{N-1} \\ x_{N-2} \\ \vdots \\ x_2 \\ x_1 \end{bmatrix} \quad .$$

!▶ Example 40.5:   *Let*

$$\{x_0, x_1, x_2, x_3\} = \{2, 4, 6, 8\} \quad .$$

Then

$$x_4 = x_{0+4} = x_0 = 2 \quad , \quad x_5 = x_{1+4} = x_1 = 4 \quad ,$$

and the sequence vector for the corresponding index-negated sequence is

$$\widetilde{\mathbf{x}} = \begin{bmatrix} \widetilde{x}_0 \\ \widetilde{x}_1 \\ \widetilde{x}_2 \\ \widetilde{x}_3 \end{bmatrix} = \begin{bmatrix} x_{-0} \\ x_{-1} \\ x_{-2} \\ x_{-3} \end{bmatrix} = \begin{bmatrix} x_0 \\ x_{4-1} \\ x_{4-2} \\ x_{4-3} \end{bmatrix} = \begin{bmatrix} x_0 \\ x_3 \\ x_2 \\ x_1 \end{bmatrix} = \begin{bmatrix} 2 \\ 8 \\ 6 \\ 4 \end{bmatrix} \quad .$$

Now we can state the near-equivalence identities for the discrete Fourier transforms. They are

$$\mathscr{F}_N \phi = \widetilde{\mathscr{F}_N^{-1} \phi} = \mathscr{F}_N^{-1} \widetilde{\phi} \quad \text{and} \quad \mathscr{F}_N^{-1} \phi = \widetilde{\mathscr{F}_N \phi} = \mathscr{F}_N \widetilde{\phi} \qquad (40.20)$$

where ϕ is any N^{th} order sequence vector. (Also, see exercise 40.27 at the end of this chapter.)

We'll prove one of these equations, leaving the verification of the rest to the interested reader.

PROOF (of the first near-equivalence identity):   Let

$$\mathbf{u} = \mathscr{F}_N \phi \quad \text{and} \quad \mathbf{w} = \mathscr{F}_N^{-1} \phi \quad ,$$

and define the periodic, regular arrays

$$\phi = \sum_{k=-\infty}^{\infty} \phi_k \, \delta_{k\Delta x} \quad , \quad u = \sum_{k=-\infty}^{\infty} u_k \, \delta_{k\Delta x} \quad \text{and} \quad w = \sum_{k=-\infty}^{\infty} w_k \, \delta_{k\Delta x}$$

where $\Delta x = N^{-1/2}$, and $\{\phi_0, \phi_1, \ldots, \phi_{N-1}\}$, $\{u_0, u_1, \ldots, u_{N-1}\}$ and $\{w_0, w_1, \ldots, w_{N-1}\}$ are the Nth order sequences corresponding to $\boldsymbol{\phi}$, $\mathbf{u}$ and $\mathbf{w}$, respectively. From lemma 40.5, we know

$$u = \mathcal{F}[\phi] \qquad \text{and} \qquad w = \mathcal{F}^{-1}[\phi] \quad .$$

Combined with one of the near-equivalence identities for the Fourier transforms, this gives

$$\sum_{n=-\infty}^{\infty} u_n \, \delta_{n\Delta x}(y) = \mathcal{F}\left[\sum_{k=-\infty}^{\infty} \phi_k \, \delta_{k\Delta x}\right]\Bigg|_y$$

$$= \mathcal{F}^{-1}\left[\sum_{k=-\infty}^{\infty} \phi_k \, \delta_{k\Delta x}\right]\Bigg|_{-y} = \sum_{m=-\infty}^{\infty} w_m \, \delta_{m\Delta x}(-y) \quad .$$

But, by known properties of the delta functions and a simple re-indexing,

$$\sum_{m=-\infty}^{\infty} w_m \, \delta_{m\Delta x}(-y) = \sum_{m=-\infty}^{\infty} w_m \, \delta_{-m\Delta x}(y)$$

$$= \sum_{n=-\infty}^{\infty} w_{-n} \, \delta_{n\Delta x}(y) = \sum_{n=-\infty}^{\infty} \widetilde{w}_n \, \delta_{n\Delta x}(y) \quad .$$

Together, these two sets of equations give us

$$\sum_{n=-\infty}^{\infty} u_n \, \delta_{n\Delta x}(y) = \sum_{n=-\infty}^{\infty} \widetilde{w}_n \, \delta_{n\Delta x}(y) \quad ,$$

which, in turn, means that $u_n = \widetilde{w}_n$ for each integer n. Therefore, $\mathbf{u} = \widetilde{\mathbf{w}}$, and

$$\mathcal{F}_N \boldsymbol{\phi} = \mathbf{u} = \widetilde{\mathbf{w}} = \widetilde{\mathcal{F}_N^{-1} \boldsymbol{\phi}} \quad . \qquad\qquad \blacksquare$$

?► Exercise 40.13:   Let $\boldsymbol{\phi}$ be any N^{th} order sequence, and verify that

$$\mathcal{F}_N^{-1} \boldsymbol{\phi} = \mathcal{F}_N \widetilde{\boldsymbol{\phi}} \quad .$$

?► Exercise 40.14:   *How are the near-equivalence identities expressed in terms of the sequence components?*

Translation

While the discrete translation identities can be derived from the generalized Fourier translation identities and lemma 40.5, it is easier to obtain them through the simple observation that

$$\sum_{k=0}^{N-1} e^{\pm i \frac{2\pi}{N}(n-M)k} v_k = \sum_{k=0}^{N-1} e^{\pm i \frac{2\pi}{N}nk} e^{\mp i \frac{2\pi}{N}Mk} v_k \quad .$$

So let's assume $\{V_0, V_1, \ldots, V_{N-1}\}$ is the discrete Fourier transform of $\{v_0, v_1, \ldots, v_{N-1}\}$. That is, let

$$V_n = \frac{1}{\sqrt{N}} \sum_{k=0}^{N-1} e^{-i \frac{2\pi}{N}nk} v_k \qquad \text{for} \quad n = 0, \pm 1, \pm 2, \ldots \quad . \qquad (40.21)$$

Equivalently, because of invertibility, we can assume $\{v_0, v_1, \ldots, v_{N-1}\}$ is the inverse discrete Fourier transform of $\{V_0, V_1, \ldots, V_{N-1}\}$. So we also have

$$v_k = \frac{1}{\sqrt{N}} \sum_{n=0}^{N-1} e^{i\frac{2\pi}{N}kn} V_n \qquad \text{for} \quad k = 0, \pm 1, \pm 2, \ldots \quad . \tag{40.22}$$

Now let M be some fixed integer. Replacing n with $n - M$ in equation set (40.21) gives us

$$V_{n-M} = \frac{1}{\sqrt{N}} \sum_{k=0}^{N-1} e^{-i\frac{2\pi}{N}(n-M)k} v_k = \frac{1}{\sqrt{N}} \sum_{k=0}^{N-1} e^{-i\frac{2\pi}{N}nk} \left[e^{i\frac{2\pi}{N}Mk} v_k \right] \quad .$$

Thus, the discrete Fourier transform of

$$\left\{ e^0 v_0 \, , \, e^{i\frac{2\pi}{N}M\cdot 1} v_1 \, , \, e^{i\frac{2\pi}{N}M\cdot 2} v_2 \, , \, \ldots \, , \, e^{i\frac{2\pi}{N}M(N-1)} v_{N-1} \right\}$$

is

$$\{ V_{0-M} \, , \, V_{1-M} \, , \, V_{2-M} \, , \, \ldots \, , \, V_{N-1-M} \} \quad .$$

By invertibility, it then follows that the inverse discrete Fourier transform of

$$\{ V_{0-M} \, , \, V_{1-M} \, , \, V_{2-M} \, , \, \ldots \, , \, V_{N-1-M} \}$$

is

$$\left\{ e^0 v_0 \, , \, e^{i\frac{2\pi}{N}M\cdot 1} v_1 \, , \, e^{i\frac{2\pi}{N}M\cdot 2} v_2 \, , \, \ldots \, , \, e^{i\frac{2\pi}{N}M(N-1)} v_{N-1} \right\} \quad .$$

Similar results come from replacing k with $k - N$ in formula set (40.22). You can derive those, yourself.

?▶ Exercise 40.15: *Let the v_k's and V_n's be as above.*

    a: *What is the discrete Fourier transform of*

$$\{ v_{0-M} \, , \, v_{1-M} \, , \, v_{2-M} \, , \, \ldots \, , \, v_{N-1-M} \} \quad ?$$

    b: *What is the inverse discrete Fourier transform of*

$$\left\{ e^0 V_0 \, , \, e^{i\frac{2\pi}{N}M\cdot 1} V_1 \, , \, e^{i\frac{2\pi}{N}M\cdot 2} V_2 \, , \, \ldots \, , \, e^{i\frac{2\pi}{N}M(N-1)} V_{N-1} \right\} \quad ?$$

For matrix equations describing the translation identities, see exercise 40.28 at the end of this chapter.

Multiplication, Convolution and Correlation

It turns out that products, convolutions and correlations of periodic, regular arrays are not well defined. Still, there are discrete analogs of these operations — we just cannot derive them from the theory for delta function arrays.

For our discussion here, assume

$$\{ v_0 \, , \, v_1 \, , \, v_2 \, , \, \ldots \, , \, v_{N-1} \} \qquad \text{and} \qquad \{ w_0 \, , \, w_1 \, , \, w_2 \, , \, \ldots \, , \, w_{N-1} \}$$

are two N^{th} order sequences with discrete Fourier transforms

$$\{ V_0 \, , \, V_1 \, , \, V_2 \, , \, \ldots \, , \, V_{N-1} \} \qquad \text{and} \qquad \{ W_0 \, , \, W_1 \, , \, W_2 \, , \, \ldots \, , \, W_{N-1} \} \quad ,$$

respectively. Thus, the v_k's and V_n's are related by

$$V_n = \frac{1}{\sqrt{N}} \sum_{k=0}^{N-1} e^{-i\frac{2\pi}{N}nk} v_k \qquad \text{and} \qquad v_k = \frac{1}{\sqrt{N}} \sum_{n=0}^{N-1} e^{i\frac{2\pi}{N}kn} V_n \quad , \tag{40.23}$$

while the w_k's and W_n's are related by

$$W_n = \frac{1}{\sqrt{N}} \sum_{k=0}^{N-1} e^{-i\frac{2\pi}{N}nk} w_k \qquad \text{and} \qquad w_k = \frac{1}{\sqrt{N}} \sum_{n=0}^{N-1} e^{i\frac{2\pi}{N}kn} W_n \quad . \tag{40.24}$$

The product of the two sequences $\{v_0, v_1, \ldots, v_{N-1}\}$ and $\{w_0, w_1, \ldots, w_{N-1}\}$ is defined as the sequence of products

$$\{ v_0 w_0 , \ v_1 w_1 , \ v_2 w_2 , \ \ldots , \ v_{N-1} w_{N-1} \} \quad .$$

That way, if $\{v_0, v_1, \ldots, v_{N-1}\}$ and $\{w_0, w_1, \ldots, w_{N-1}\}$ are regular samplings (with identical spacings) of two functions v and w, then the sequence product, $\{v_0 w_0, v_1 w_1 \ldots, v_{N-1} w_{N-1}\}$, is the corresponding regular sampling of the product vw. The discrete Fourier transform of this product is given by

$$\frac{1}{\sqrt{N}} \sum_{k=0}^{N-1} e^{-i\frac{2\pi}{N}nk} v_k w_k \qquad \text{for} \quad n = 0, 1, 2, \ldots, N-1 \quad .$$

Replacing w_k with its formula from equation set (40.24), rearranging the sums and then applying the formula for the V_n's from equation set (40.23) yields

$$\sum_{k=0}^{N-1} e^{-i\frac{2\pi}{N}nk} v_k w_k = \sum_{k=0}^{N-1} e^{-i\frac{2\pi}{N}nk} v_k \left[\frac{1}{\sqrt{N}} \sum_{j=0}^{N-1} e^{i\frac{2\pi}{N}kj} W_j \right]$$

$$= \frac{1}{\sqrt{N}} \sum_{k=0}^{N-1} \sum_{j=0}^{N-1} e^{-i\frac{2\pi}{N}(n-j)k} v_k W_j$$

$$= \sum_{j=0}^{N-1} \left[\frac{1}{\sqrt{N}} \sum_{k=0}^{N-1} e^{-i\frac{2\pi}{N}(n-j)k} v_k \right] W_j = \sum_{j=0}^{N-1} V_{n-j} W_j \quad .$$

Cutting out the middle and dividing by $\sqrt{N}$ leaves

$$\frac{1}{\sqrt{N}} \sum_{k=0}^{N-1} e^{-i\frac{2\pi}{N}nk} v_k w_k = \frac{1}{\sqrt{N}} \sum_{j=0}^{N-1} V_{n-j} W_j \qquad \text{for} \quad n = 0, 1, 2, \ldots, N-1 \quad . \tag{40.25}$$

Similarly, you can verify that

$$\frac{1}{\sqrt{N}} \sum_{n=0}^{N-1} e^{i\frac{2\pi}{N}kn} V_n W_n = \frac{1}{\sqrt{N}} \sum_{j=0}^{N-1} v_{k-j} w_j \qquad \text{for} \quad k = 0, 1, 2, \ldots, N-1 \quad . \tag{40.26}$$

?▶ **Exercise 40.16:**   *Verify equation (40.26).*

The (discrete) *convolution* of two N^{th} order sequences $\{x_0, x_1, \ldots, x_{N-1}\}$ and $\{y_0, y_1, \ldots, y_{N-1}\}$ is defined to be the sequence $\{z_0, z_1, \ldots, z_{N-1}\}$ given by

$$z_k = \frac{1}{\sqrt{N}} \sum_{j=0}^{N-1} x_{k-j} y_j \qquad \text{for} \quad k = 0, 1, 2, \ldots, N-1 \quad . \tag{40.27}$$

As equations (40.25) and (40.26) show, the discrete transform of the product of any two N^{th} order sequences is convolution of the corresponding discrete transforms of the individual sequences.

You may have noticed that the above formula for discrete convolution bears a superficial resemblance to

$$\int_{-\infty}^{\infty} x(t-s)y(s)\,ds \quad,$$

the integral formula for the convolution two functions x and y. This, and the recollection that the correlation of x and y is given by

$$\int_{-\infty}^{\infty} x^*(s-t)y(s)\,ds \quad,$$

may lead you to suspect that the discrete correlation of the sequences $\{x_0, x_1, \ldots, x_{N-1}\}$ and $\{y_0, y_1, \ldots, y_{N-1}\}$ is defined as the sequence $\{z_0, z_1, \ldots, z_{N-1}\}$ given by

$$z_k = \frac{1}{\sqrt{N}} \sum_{j=0}^{N-1} x^*_{j-k} y_j \qquad \text{for} \quad k = 0, 1, 2, \ldots, N-1 \quad.$$

You may then suspect that the discrete analogs to the correlation identities are

$$\frac{1}{\sqrt{N}} \sum_{k=0}^{N-1} e^{-i\frac{2\pi}{N}nk} v^*_k w_k = \frac{1}{\sqrt{N}} \sum_{j=0}^{N-1} V^*_{j-n} W_j \qquad \text{for} \quad n = 0, 1, 2, \ldots, N-1$$

and

$$\frac{1}{\sqrt{N}} \sum_{n=0}^{N-1} e^{i\frac{2\pi}{N}kn} V^*_n W_n = \frac{1}{\sqrt{N}} \sum_{j=0}^{N-1} v^*_{j-k} w_j \qquad \text{for} \quad k = 0, 1, 2, \ldots, N-1 \quad.$$

Well, your suspicions would be correct, and you can confirm these identities (along with a few others of mild interest) by doing exercise 40.29 at the end of this chapter.

40.6 Fast Fourier Transforms

A "fast Fourier transform" (often shortened to "FFT") is not a new type of transform, it is just a faster way of computing either of the discrete transforms we've been discussing. To be more precise, a fast Fourier transform is an implementation (usually on a computer!) of a particularly efficient algorithm for computing a discrete Fourier transform. With the development of these algorithms, it became practical to program computers to calculate and manipulate, in a reasonable amount of time, discrete Fourier transforms of extremely large samplings.

To achieve their efficiency, fast Fourier transforms take advantage of certain relations between various components of the discrete transforms. We will describe that which is exploited by some of the more commonly implemented fast Fourier transforms (the "radix 2" FFTs) and briefly discuss both how this relation is utilized and how this utilization reduces the number of operations in computing discrete transforms. This should give you the understanding needed to intelligently use and properly appreciate the FFTs already widely available in various software packages. We won't, however, carry out the development to the point where you can easily write a good FFT algorithm yourself. For that, you should consult a text specifically on fast Fourier transforms.[1]

---

[1] Such as the one by Chu and George (reference [1]) or the one by Walker (reference [12]).

Exploiting Some Simple Recursion Formulas

Take another look at the matrix for the discrete Fourier transform (formula 40.16 on page 738). You will find a number of symmetries and recursive structures that a clever person might use to shorten the process of computation of a discrete transform. One we will exploit is the relation between each horizontally adjacent pair of components,

$$[\mathscr{F}_N]_{n,k+1} = \frac{1}{\sqrt{N}} e^{-i\frac{2\pi}{N}n(k+1)} = \frac{1}{\sqrt{N}} e^{-i\frac{2\pi}{N}n\cdot k} e^{-i\frac{2\pi}{N}n\cdot 1} = e^{-i\frac{2\pi}{N}n} [\mathscr{F}_N]_{n,k} \quad .$$

To take advantage of this relation, we must assume N, the order of our sequences and discrete transforms, is even. Assume this, and let M be the integer $N/2$ (so $N = 2M$).

Now suppose

$$\{ V_0 , V_1 , V_2 , \ldots , V_{N-1} \}$$

is the Nth order discrete Fourier transform of

$$\{ v_0 , v_1 , v_2 , \ldots , v_{N-1} \} \quad .$$

Starting with the component formula for the discrete transform, rearranging the terms cleverly, and applying the relation noted above, we see that, for $n = 0, 1, 2, \ldots$ and $N - 1$,

$$V_n = \frac{1}{\sqrt{N}} \sum_{k=0}^{N-1} e^{-i\frac{2\pi}{N}nk} v_k$$

$$= \frac{1}{\sqrt{N}} \sum_{\substack{k=0 \\ k \text{ is even}}}^{N-1} e^{-i\frac{2\pi}{N}nk} v_k + \frac{1}{\sqrt{N}} \sum_{\substack{k=0 \\ k \text{ is odd}}}^{N-1} e^{-i\frac{2\pi}{N}nk} v_k$$

$$= \frac{1}{\sqrt{2M}} \sum_{j=0}^{M-1} e^{-i\frac{2\pi}{2M}n(2j)} v_{2j} + \frac{1}{\sqrt{2M}} \sum_{j=0}^{M-1} e^{-i\frac{2\pi}{2M}n(2j+1)} v_{2j+1}$$

$$= \frac{1}{\sqrt{2M}} \sum_{j=0}^{M-1} e^{-i\frac{2\pi}{M}nj} v_{2j} + e^{-i\frac{\pi}{M}n} \frac{1}{\sqrt{2M}} \sum_{j=0}^{M-1} e^{-i\frac{2\pi}{M}nj} v_{2j+1} \quad . \tag{40.28}$$

This shows that each component of our N^{th} order discrete transform is a relatively simple linear combination of corresponding components of two M^{th} order discrete transforms (don't forget, $M = N/2$). In more concise form, this last equation is

$$[\mathscr{F}_N \mathbf{v}]_n = \frac{1}{\sqrt{2}} [\mathscr{F}_M \mathbf{v}_E]_n + e^{-i\frac{\pi}{M}n} \frac{1}{\sqrt{2}} [\mathscr{F}_M \mathbf{v}_O]_n \tag{40.29}$$

where $\mathbf{v}$ is the sequence vector corresponding to $\{v_0, v_1, \ldots, v_{N-1}\}$, $\mathbf{v}_E$ is the sequence vector for the Mth order sequence of even-indexed v_k's,

$$\{ v_0 , v_2 , v_4 , \ldots , v_{N-2} \} \quad ,$$

and $\mathbf{v}_O$ is the sequence vector for the Mth order sequence of odd-indexed v_k's,

$$\{ v_1 , v_3 , v_5 , \ldots , v_{N-1} \} \quad .$$

Formula (40.28) can be used to calculate the V_n's for $n = 0$ through $n = N - 1$. However, we can reduce the number of computations explicitly performed after observing that, for integers j and ℓ,

$$e^{-i\frac{2\pi}{M}(M+\ell)j} = e^{-ij2\pi} e^{-i\frac{2\pi}{M}\ell j} = e^{-i\frac{2\pi}{M}\ell j}$$

and

$$e^{-i\frac{\pi}{M}(M+\ell)} = e^{-i\frac{\pi}{M}M}e^{-i\frac{\pi}{M}\ell} = e^{-i\pi}e^{-i\frac{\pi}{M}\ell} = -e^{-i\frac{\pi}{M}\ell} \quad .$$

Therefore, for $\ell = 0, 1, 2, \ldots$ and $M - 1$,

$$V_\ell = \frac{1}{\sqrt{2M}}\sum_{j=0}^{M-1} e^{-i\frac{2\pi}{M}\ell j}v_{2j} + e^{-i\frac{\pi}{M}\ell}\frac{1}{\sqrt{2M}}\sum_{j=0}^{M-1} e^{-i\frac{2\pi}{M}\ell j}v_{2j+1} \tag{40.30a}$$

$$= \frac{1}{\sqrt{2}}\left[\mathcal{F}_M \mathbf{v}_E\right]_\ell + e^{-i\frac{\pi}{M}\ell}\frac{1}{\sqrt{2}}\left[\mathcal{F}_M \mathbf{v}_O\right]_\ell \quad ,$$

and

$$V_{M+\ell} = \frac{1}{\sqrt{2M}}\sum_{j=0}^{M-1} e^{-i\frac{2\pi}{M}\ell j}v_{2j} - e^{-i\frac{\pi}{M}\ell}\frac{1}{\sqrt{2M}}\sum_{j=0}^{M-1} e^{-i\frac{2\pi}{M}\ell j}v_{2j+1} \tag{40.30b}$$

$$= \frac{1}{\sqrt{2}}\left[\mathcal{F}_M \mathbf{v}_E\right]_\ell - e^{-i\frac{\pi}{M}\ell}\frac{1}{\sqrt{2}}\left[\mathcal{F}_M \mathbf{v}_O\right]_\ell \quad .$$

At first glance, it may seem as if we have just complicated a relatively straightforward computation. However, formulas (40.30a) and (40.30b) contain common elements that can be computed once for each ℓ and then used to compute both V_ℓ and $V_{M+\ell}$. As we will see, this can significantly reduce the total number of computations to find all the V_n's.

The Algorithm

The following is an algorithm for computing an Nth order discrete transform using equation set (40.30). In the first few steps, the common elements in formulas (40.30a) and (40.30b) are identified and computed. The subsequent steps finish the computations using those common elements.

This algorithm is not a full fast Fourier transform algorithm (it's not even as efficient as possible — some efficiency was sacrificed to expedite later exposition). Consider it as a first approximation to a fast Fourier transform algorithm, and, accordingly, let's call it our *first level* fast Fourier transform algorithm (for computing an Nth order discrete Fourier transform).

First Level Algorithm for the FFT
Assume $N = 2M$ for some positive integer M. To compute the Nth order discrete Fourier transform $\{V_0, V_1, \ldots, V_{N-1}\}$ of $\{v_0, v_1, \ldots, v_{N-1}\}$:

1. Split the Nth order sequence $\{v_0, v_1, \ldots, v_{N-1}\}$ into the two M^{th} order sequences composed, respectively, of the even-indexed and the odd-indexed v_k's,

$$\{u_0, u_1, u_2, \ldots, u_{M-1}\} = \{v_0, v_2, v_4, \ldots, v_{2M-2}\}$$

   and

$$\{w_0, w_1, w_2, \ldots, w_{M-1}\} = \{v_1, v_3, v_5, \ldots, v_{2M-1}\} \quad .$$

2. For $m = 0, 1, 2, \ldots$ and $M - 1$, compute

$$U_m = \frac{1}{\sqrt{M}}\sum_{j=0}^{M-1} e^{-i\frac{2\pi}{M}mj}u_j \quad \text{and} \quad W_m = \frac{1}{\sqrt{M}}\sum_{j=0}^{M-1} e^{-i\frac{2\pi}{M}mj}w_j \quad .$$

3. For $m = 0, 1, 2, \ldots$ and $M - 1$, compute

$$X_m = \frac{1}{\sqrt{2}}U_m \quad \text{and} \quad Y_m = e^{-i\frac{\pi}{M}m}\frac{1}{\sqrt{2}}W_m \quad .$$

4.   For $n = 0, 1, 2, \ldots$ and $M - 1$, compute

$$V_n = X_n + Y_n \quad .$$

5.   For $n = M, M+1, M+2, \ldots$ and $N - 1$, let $m = n - M$ and compute

$$V_n = X_m - Y_m \quad .$$

?▶ Exercise 40.17:   *Convince yourself that the above is an algorithm for computing formulas (40.30a) and (40.30b).*

Keep in mind that the computations in step 2 are also described by

$$U_m = \left[\mathcal{F}_M \mathbf{u} \right]_m \qquad \text{and} \qquad W_m = \left[\mathcal{F}_M \mathbf{w} \right]_m$$

where $\mathbf{u}$ and $\mathbf{w}$ are, respectively, the sequence vectors from the sequences constructed in the first step and $M = {}^N\!/_2$. This means, we can view this procedure as one that reduces the computation of an Nth order discrete transform to the computation of two ${}^N\!/_2{}^{\text{th}}$ order sequences (plus a few other computations).

Consider, now, the computations of the Mth order discrete transforms — the U_m's and W_m's — in step 2. If M is also even, say, $M = 2L$ (so, $N = 2^2 L$), then those M^{th} order discrete transforms, themselves, can be computed using the above procedure. That is, we can replace step 2 with two more applications of the first order algorithm, one computing the U_m's and one computing the V_m's. This would mean that the discrete transforms directly computed using the basic definition are of order $L = 2^{-2}N$. Making this modification to the above algorithm turns it into our *second level* fast Fourier transform algorithm for computing the V_n's.

And if L is even, say, $L = 2P$ (so, $N = 2^3 P$), then the above can be further modified to use the second level algorithm to compute the U_m's and W_m's. This gives our *third level* fast Fourier transform algorithm for computing the V_n's. In this case the order of the discrete transforms computed using the basic definition is $P = 2^{-3}N$.

And if ${}^L\!/_2$ is even ...

Clearly, this process of "increasing levels" to our algorithm can be continued to the "K^{th} level" provided $N = 2^K N_0$ for some integer N_0. Moreover, that integer, $N_0 = 2^{-K} N$, will be the order of the discrete transforms computed directly using the basic definition.

In many implementations of the fast Fourier transform, the order N is required to equal 2^K for some integer K. This allows use of the Kth level algorithm, with the discrete transforms directly computed via the basic definition being of order 1 (which are awfully easy transforms to compute by the basic definition!). It is this final K^{th} level algorithm, fully implemented and refined, that is called a (radix 2) *fast Fourier transform.*

Naturally, the analogous development of the fast Fourier transform for the *inverse* discrete Fourier transform is left to you.

?▶ Exercise 40.18:   *By modifying the computations leading to our first level fast Fourier transform, derive the corresponding first level fast Fourier transform algorithm for computing the Nth order inverse discrete Fourier transform. (Start by deriving an equation similar to equation (40.29) relating an Nth order inverse discrete transform to two ${}^N\!/_2{}^{\text{th}}$ order inverse discrete transforms.)*

By the way, the derivation of the "radix 3" fast Fourier transform starts with the derivation of the formula

$$V_n = \frac{1}{\sqrt{3M}} \sum_{j=0}^{M-1} e^{-i\frac{2\pi}{M}nj} v_{3j} + e^{-i\frac{2\pi}{3M}n} \frac{1}{\sqrt{3M}} \sum_{j=0}^{M-1} e^{-i\frac{2\pi}{M}nj} v_{3j+1}$$

$$+ e^{-i\frac{4\pi}{3M}n} \frac{1}{\sqrt{3M}} \sum_{j=0}^{M-1} e^{-i\frac{2\pi}{M}nj} v_{3j+2}$$

under the assumptions that $N = 3M$ and that $\{V_0, V_1, \ldots, V_{N-1}\}$ is the Nth order discrete Fourier transform of $\{v_0, v_1, \ldots, v_{N-1}\}$.

How Is This an Improvement?

To get a feeling for the rate at which computation time can be speeded up by employing fast Fourier transform algorithms, let us count the number of arithmetic operations — complex-valued "adds", "subtracts", and "multiplies" — used to compute an Nth order discrete transform via (*1*) the basic definition, (*2*) our first level fast Fourier transform algorithm, and (*3*) the radix 2 fast Fourier transform algorithm just described.

Using the Basic Definition

By the basic definition, each component in the Nth order discrete transform $\{V_0, V_1, \ldots, V_{N-1}\}$ of $\{v_0, v_1, \ldots, v_{N-1}\}$ is given by

$$V_n = \frac{1}{\sqrt{N}} \sum_{k=0}^{N-1} e^{-i\frac{2\pi}{N}nk} v_k \quad .$$

When $n = 0$, things simplify. The exponential becomes $e^0 = 1$ and the above equation reduces to

$$V_0 = \frac{1}{\sqrt{N}} \sum_{k=0}^{N-1} 1 \cdot v_k = \frac{1}{\sqrt{N}} \left[v_0 + v_1 + v_2 + \cdots + v_{N-1} \right] \quad ,$$

which only requires $N - 1$ "adds" followed by 1 "multiply" (by $N^{-1/2}$), for a total of

$$N \quad \text{basic arithmetic operations (for } V_0 \text{)}.$$

When $n \neq 0$, the above formula for V_n can be written as

$$V_n = F_{n,0} \cdot v_0 + F_{n,1} \cdot v_1 + F_{n,2} \cdot v_2 + \cdots + F_{n,N-1} \cdot v_{N-1}$$

where

$$F_{n,k} = \frac{1}{\sqrt{N}} e^{-i\frac{2\pi}{N}nk} \quad .$$

(Since the same formula is used to compute every such transform, we can assume the $F_{n,k}$'s have already been computed and are readily available.) With $n \neq 0$, the $F_{n,k}$'s do not reduce to a single common factor. So the computation of V_n using the above formula requires that we first multiply every v_k by the corresponding $F_{n,k}$ (using N "multiplies"), and then add up those products (using $N - 1$ "adds"), for a total of

$$N + N - 1 = 2N - 1 \quad \text{basic arithmetic operations (for each } V_n \text{ with } n \neq 0 \text{)}.$$

Altogether then, the above formulas for computing V_0 and the $N - 1$ other V_n's require

$$N + (N - 1)(2N - 1) \quad \text{basic arithmetic operations}.$$

Thus, after a little algebra, we have:

> *The number of basic arithmetic operations to compute an Nth order discrete Fourier transform using just the defining formulas is*
>
> $$2N^2 - 2N + 1 \quad.$$

Using Our First Level Algorithm

(Note: As we count the number of arithmetic operations done in carrying out the first level algorithm, we will explicitly note the consequences of reducing an Nth order discrete transform to two $N/2^{th}$ order discrete transforms. This will give us results we can later employ in counting the operations done in the fast Fourier transform.)

In the first level algorithm, the first arithmetic operations occur in step 2, when the U_m's and W_m's are calculated. Since these are the components of two $N/2^{th}$ order discrete Fourier transforms, the total number of arithmetic operations to compute these U_m's and W_m's is

$$2 \times \text{ number of operations to compute an } N/2^{th} \text{ order discrete transform} \quad.$$

After computing the U_m's and W_m's, each X_n and Y_m is computed by multiplying each corresponding U_m and W_m by an appropriate constant. Assuming these constants have been pre-computed, this uses

$$2 \times M \text{ "multiplies"} \quad.$$

In step 4,

$$M \text{ "adds"}$$

are performed, one for each V_n computed. In step 5, however, two "adds" are done for each V_n computed, $m = n - M$ and $V_n = X_m - Y_m$. So step 5 uses[2]

$$2M \text{ "adds"} \quad.$$

Thus, the total number of arithmetic operations done after computing the U_m's and W_m's is

$$2M + M + 2M = 5M = \frac{5}{2}N \quad.$$

Altogether, then, for our first level algorithm,

> *the number of operations to compute an Nth order discrete transform*
>
> $= 2 \times$ *the number of operations to compute an $N/2^{th}$ order discrete transform*
>
> $+ \dfrac{5}{2}N \quad.$

(40.31)

This equation holds no matter how we compute the U_m's and W_m's. If we compute them as step 2 explicitly states (i.e., using the basic definition for the $N/2^{th}$ order discrete transform), then the results of the previous subsection apply (with $N/2$ replacing N), giving us

$2 \times$ *the number of operations to compute an $N/2^{th}$ order discrete transform* $+ \dfrac{5}{2}N$

$$= 2 \times \left[1 + 2\left(\frac{N}{2} - 1\right)\frac{N}{2}\right] + \frac{5}{2}N \quad.$$

Thus, after performing the elementary algebra, we have:

---

[2] Like us, most computers can do the integer calculation $m = n - M$ more easily than the other calculations being considered here. Because we are counting these integer calculations as if they were as difficult as the others, the numbers we derive past this point will give slightly inflated measures of the computational work involved.

The number of basic arithmetic operations required to compute an Nth order discrete Fourier transform using our first level algorithm is

$$N^2 + \frac{1}{2}N + 2 \quad .$$

Using the Fast Fourier Transform Algorithm

To see how many arithmetic operations are used to compute an Nth order discrete transform using our radix 2 fast Fourier transform, we need to assume $N = 2^K$ for some positive integer K. It will also help if we let $\mathcal{O}(N)$ denote the number of arithmetic operations this "highest level" algorithm uses to compute a discrete Fourier transform of order N.

Start with the lowest K possible, $K = 1$. Then our "highest level" algorithm is simply the first level algorithm for computing a discrete transform of order $N = 2^1 = 2$. By the analysis in the previous subsection, we know

$$\mathcal{O}(2^1) = 2 + \left(2 + \frac{1}{2}\right)2 = 7 \quad .$$

For $K > 1$, equation (40.31) applies. Rewritten in terms of our "$\mathcal{O}$" notation with $N = 2^K$, this equation is

$$\mathcal{O}(2^K) = 2 \times \mathcal{O}\left(\frac{1}{2} \cdot 2^K\right) + \frac{5}{2} \cdot 2^K \quad .$$

Thus,

$$\mathcal{O}(2^K) = 2\mathcal{O}(2^{K-1}) + 5 \cdot 2^{K-1} \quad . \tag{40.32}$$

Consequently,

$$\mathcal{O}(2^2) = 2\mathcal{O}(2^1) + 5 \cdot 2^1$$
$$= 2[7] + 5 \cdot 2 = (7 + 5)2 \quad .$$

Since we want to find a pattern for the values of the $\mathcal{O}(2^K)$'s, and since each application of equation (40.32) adds a term of the form $5 \cdot 2^{K-1}$, we won't carry out the arithmetic in the last equation any further. Continuing, we see that

$$\mathcal{O}(2^3) = 2\mathcal{O}(2^2) + 5 \cdot 2^2$$
$$= 2[7 \cdot 2 + 5 \cdot 2] + 5 \cdot 2^2 = (7 + 2 \cdot 5)2^2 \quad ,$$

$$\mathcal{O}(2^4) = 2\mathcal{O}(2^3) + 5 \cdot 2^3$$
$$= 2\left[7 \cdot 2^2 + 2(5 \cdot 2^2)\right] + 5 \cdot 2^3 = (7 + 3 \cdot 5)2^3 \quad ,$$

and

$$\mathcal{O}(2^5) = 2\mathcal{O}(2^4) + 5 \cdot 2^4$$
$$= 2\left[7 \cdot 2^3 + 3(5 \cdot 2^3)\right] + 5 \cdot 2^4 = (7 + 4 \cdot 5)2^4 \quad .$$

Behold! A pattern seems to be emerging. These formulas are all described by

$$\mathcal{O}(2^K) = (7 + [K - 1]5)2^{K-1} \quad \text{for} \quad K = 1, 2, 3, 4 \text{ and } 5 \quad .$$

Equivalently,

$$\mathcal{O}(2^{K+1}) = (7 + 5K)2^K \quad \text{for} \quad K = 0, 1, 2, 3 \text{ and } 4 \quad .$$

Assuming this holds for any particular integer K and re-applying equation (40.32) yields

$$\mathcal{O}(2^{K+1}) = 2\,\mathcal{O}(2^K) + 5 \cdot 2^K$$
$$= 2\left[(7 + [K-1]5)2^{K-1}\right] + 5 \cdot 2^K$$
$$= (7 + 5K)2^K \quad,$$

showing (by induction) that this pattern continues to hold as the integer K increases. So, (applying a little arithmetic) we have

$$\mathcal{O}(2^K) = (7 + [K-1]5)2^{K-1} = (2 + 5K)2^{K-1} \qquad \text{for} \quad K = 1, 2, 3, 4, \ldots \quad.$$

There are a couple of points we should recall or observe at this stage:

1. The order N equals 2^K. Thus, $K = \log_2 N$ and, by the last equation above,

$$\mathcal{O}(N) = \mathcal{O}(2^K) = (2 + 5\log_2 N)\frac{1}{2} \cdot 2^K = \frac{5}{2}N\log_2 N + N \quad.$$

2. The numbers we just derived should be viewed as an upper limit on the number of basic arithmetic computations done in a "well-written and implemented" fast Fourier transform. Remember, "our" fast Fourier transform algorithm is simply our K^{th} level algorithm without any of the refinements that would be added by a competent numerical analyst and programmer. For example, if you actually construct our FFT according to the discussion on page 750, you will find the algorithm repeats a number of divisions by $\sqrt{2}$ (see step 3). That number can be reduced by, say, doing none of these divisions until the end, at which point the Nth order sequence computed so far is multiplied by $2^{-K/2}$.

    As it turns out, most of the efficiency in a fast Fourier transform comes from the process developed here in which an Nth order discrete transform computation is converted to a number of 1st order discrete transform computations. The main effect of those additional refinements on the number of computations is to reduce the first coefficient in the last equation from $5/2$ to some smaller constant.

Keeping in mind the above observations, we then have:

The number of basic arithmetic operations required to compute an Nth order discrete Fourier transform using a well-written radix 2 fast Fourier transform algorithm is no more than

$$\frac{5}{2}N\log_2 N + N \quad.$$

(*This assumes* $\log_2 N$ *is an integer.*)

Comparing the Computational Times

Summarizing what we just derived:

Computing an Nth order discrete Fourier transform using just the basic definition takes

$$2N^2 - 2N + 1 \quad \text{arithmetic operations} \quad.$$

Computing an Nth order discrete Fourier transform using the first level algorithm takes

$$N^2 + \frac{1}{2}N + 2 \quad \text{arithmetic operations} \quad.$$

Computing an Nth order discrete Fourier transform using a well-written (radix 2) fast Fourier transform takes at most

$$\frac{5}{2} N \log_2 N \; + \; N \;\; \text{arithmetic operations} \quad .$$

(Provisos: Use of the first level algorithm requires that N be even. Use of the radix 2 fast Fourier transform requires that $\log_2 N$ be an integer.)

These numbers, of course, do not tell the entire story. In deriving them, we did not distinguish between the computational effort required to do a complex-valued "add", a complex-valued "multiply" and an integer-valued "add". Nor did we take into account the extra overhead necessary for the bookkeeping in the fast Fourier transform algorithm. On the other hand, the above number of operations for the fast Fourier transform is based on "our" fast Fourier transform algorithm, which is relatively primitive and could be further improved had we the time and inclination.

Still, these numbers do give reasonable estimates of the number of time-consuming operations a computer would use to compute a discrete transform by each of the indicated methods, and rough comparisons of the computational times for these methods can be obtained by assuming the total computational time is directly proportional to the number of basic arithmetic operations involved.

!▶ **Example 40.6:** *Suppose $N = 2^{10} = 1024$. According to our calculations, the computation of a 1024th order discrete Fourier transform would take*

$$2 \cdot 1024^2 \; - \; 2 \cdot 1024 \; + \; 1 \; \approx \; 2.1 \times 10^6$$

arithmetic operations using the basic definition,

$$1024^2 \; + \; \frac{1}{2} \cdot 1024 \; + \; 2 \approx 1.05 \times 10^6$$

arithmetic operations using our first level algorithm, and no more than

$$\frac{5}{2} \cdot 1024 \cdot 10 \; + \; 1024 \; \approx \; 2.7 \times 10^4$$

arithmetic operations using a well-implemented fast Fourier transform algorithm.

Consider what this means in terms of the time it takes to compute a 1024th order discrete Fourier transform. Assuming a direct relation between the number of basic arithmetic operations and the computational time, the above figures tell us that using our first level algorithm requires half the time needed using the basic definition. This pales, however, alongside the improvement resulting from the fast Fourier transform — using that reduces the computational time to $^{27}/_{2100}$ (a little over one percent) of the time required using the basic definition.

As the example indicates, using a fast Fourier transform can greatly reduce the computational time for a discrete transform, especially when the order of the transform is large. The relation between the savings in time and the order is, perhaps, most clearly seen in the ratio of the times required to compute an Nth order discrete transform using a fast Fourier transform and using the basic definition. Assuming computational time is directly proportional to the number of basic arithmetic operations, our calculations yield

$$\frac{\text{time to compute an } N^{th} \text{ order transform using an FFT}}{\text{time to compute an } N^{th} \text{ order transform using the basic definition}} \; \leq \; \frac{\frac{5}{2} N \log_2 N \; + \; N}{2N^2 \; - \; 2N \; + \; 1} \quad .$$

A slightly simpler upper bound on this ratio can be obtained after observing that, when $N > 4$,

$$1 \; = \; \frac{1}{2} \log_2 4 \; < \; \frac{1}{2} \log_2 N \quad \text{and} \quad N \; < \; \frac{1}{4} N^2 \quad .$$

So, for $N > 4$,

$$\frac{5}{2}N \log_2 N + N < \frac{5}{2}N \log_2 N + \frac{1}{2}N \log_2 N = 3N \log_2 N \quad,$$

$$2N^2 - 2N + 1 > 2N^2 - 2N > 2N^2 - 2\left(\frac{1}{4}N^2\right) = \frac{3}{2}N^2$$

and

$$\frac{\frac{5}{2}N \log_2 N + N}{2N^2 - 2N + 1} < \frac{3N \log_2 N}{\frac{3}{2}N^2} = \frac{2 \log_2 N}{N} \quad.$$

Thus, for $N > 4$,

$$\frac{\text{time to compute an } N^{\text{th}} \text{ order transform using an FFT}}{\text{time to compute an } N^{\text{th}} \text{ order transform using the basic definition}} < \frac{2 \log_2 N}{N} \quad. \tag{40.33}$$

As you can easily verify, this upper bound becomes extremely small as N gets large.

?► Exercise 40.19: What, approximately, is the upper bound on the above ratio of computational times when $N = 2^{10}$? when $N = 2^{12}$? when $N = 2^{20}$?

Additional Exercises

40.20. Let $N = 4$, and, for each of the following choices of $f(t)$ and L,

 1. find the Nth order regular sampling of f over the minimal window,

and

 2. sketch the corresponding discrete approximation, $\widehat{f}$.

 a. $f(t) = 2t$ and $L = 1$ **b.** $f(t) = 2t$ and $L = 2$

 c. $f(t) = (5 - 3t)^2$ and $L = 1$ **d.** $f(t) = (5 - 3t)^2$ and $L = 2$

 e. $f(t) = 1$ and $L = 2$ **f.** $f(t) = \text{step}\left(t - {}^4/_5\right)$ and $L = 2$

40.21. Repeat exercise 40.20 with $N = 3$ replacing $N = 4$.

40.22. In the text, it was assumed that all the samplings were from piecewise smooth functions. In the following, you are to generate the samplings corresponding to delta functions by doing the following for each of the indicated choices of $f(t)$, N and Δt:

 1. Determine and sketch the regular, periodic array

$$\widehat{f} = \sum_{k=-\infty}^{\infty} \widehat{f}_k \, \delta_{k\Delta t}$$

 with index period N that "best approximates" f on the window $\left(-\frac{\Delta t}{2}, L + \frac{\Delta t}{2}\right)$ where $L = (N - 1)\Delta t$).

  2. *Determine the sampling* $\{f_0, f_1, \ldots, f_{N-1}\}$ *corresponding to* $\widehat{f}$.

a. $N = 4$, $\Delta t = \frac{1}{2}$ *and*

  i. $f(t) = \delta(t)$ **ii.** $f(t) = \delta_1(t)$

b. $N = 4$, $\Delta t = \frac{1}{3}$ *and*

  i. $f(t) = \delta(t)$ **ii.** $f(t) = \delta_1(t)$

c. $N = 32$, $\Delta t = \frac{1}{16}$ *and*

  i. $f(t) = \delta(t)$ **ii.** $f(t) = \delta_1(t)$

40.23. Using $\mathscr{F}_4$, the matrix for the 4th order discrete Fourier transform (see example 40.4 on page 739), find the discrete Fourier transform of each of the following 4th order sequences:

  a. $\{1, 0, 0, 0\}$ **b.** $\{0, 1, 0, 0\}$ **c.** $\{1, 1, 1, 1\}$

  d. $\{3, 2, 1, 0\}$ **e.** $\{1, -1, 1, -1\}$

40.24 a. Compute $\mathscr{F}_3$, the matrix for the 3th order discrete Fourier transform.

  b. Find the discrete Fourier transform of each of the following 3rd order sequences:

  i. $\{1, 0, 0\}$ **ii.** $\{0, 1, 0\}$ **iii.** $\{1, 1, 1\}$

  iv. $\{3, 2, 1\}$ **v.** $\{1, -1, 1\}$

40.25 a. Compute $\mathscr{F}_8$, the matrix for the 8th order discrete Fourier transform.

  b. Find the discrete Fourier transform of each of the following 8th order sequences:

  i. $\{1, 0, 0, 0, 0, 0, 0, 0\}$ **ii.** $\{0, 1, 0, 0, 0, 0, 0, 0\}$

  iii. $\{1, 1, 1, 1, 1, 1, 1, 1\}$ **iv.** $\{1, -1, 1, -1, 1, -1, 1, -1\}$

40.26 a. Find the discrete Fourier transform of each 4th order sampling you computed in exercise 40.20.

  b. Find the array coefficients $\widehat{G}_0$, $\widehat{G}_1$, $\widehat{G}_2$ and $\widehat{G}_3$ for the Fourier transform of each discrete approximation $\widehat{f}$ found in exercise 40.20. (Use theorem 40.6 on page 741 and your answer to the previous part.)

40.27. Let $\mathbf{N}_N$ be the $N \times N$ matrix whose $(j, k)^{th}$ component is

$$[\mathbf{N}_N]_{j,k} = \begin{cases} 1 & \text{if } j = k = 0 \\ 1 & \text{if } j \neq 0 \text{ and } k = N - j \\ 0 & \text{otherwise} \end{cases}.$$

  a. Write out the matrix $\mathbf{N}_4$ and verify that $\mathbf{N}_4\mathbf{x} = \widetilde{\mathbf{x}}$ where $\mathbf{x}$ is any 4th order sequence vector and $\widetilde{\mathbf{x}}$ is the corresponding vector obtained by index negation.

  b. Verify that, in general, $\mathbf{N}_N\mathbf{x} = \widetilde{\mathbf{x}}$ where $\mathbf{x}$ is any N^{th} order sequence vector and $\widetilde{\mathbf{x}}$ is the corresponding vector obtained by index negation.

  c. Show that the discrete near-equivalence identities (equations 40.20 on page 743) are equivalent to the matrix equations

$$\mathscr{F}_N = \mathbf{N}_N\mathscr{F}_N^{-1} = \mathscr{F}_N^{-1}\mathbf{N}_N \quad \text{and} \quad \mathscr{F}_N^{-1} = \mathbf{N}_N\mathscr{F}_N = \mathscr{F}_N\mathbf{N}_N.$$

40.28. For $M = 0, 1, 2, \ldots$ and $N - 1$, let $\mathbf{E}_{N,M}$ and $\mathbf{T}_{N,M}$ be the $N \times N$ matrices whose $(j, k)^{th}$ components are

$$\left[\mathbf{E}_{N,M}\right]_{j,k} = \begin{cases} e^{-i\frac{2\pi}{N}jM} & \text{if } j = k \\ 0 & \text{if } j \neq k \end{cases}$$

and

$$\left[\mathbf{T}_{N,M}\right]_{j,k} = \begin{cases} 1 & \text{if } j < M \text{ and } k = j + N - M \\ 1 & \text{if } M \leq j \text{ and } k = j - M \\ 0 & \text{otherwise} \end{cases} .$$

a. Assume $\{v_0, v_1, \ldots, v_{N-1}\}$ is an Nth order sequence with corresponding sequence vector $\mathbf{v}$, and let

$$\mathbf{u} = \mathbf{E}_{N,M}\mathbf{v} \qquad \text{and} \qquad \mathbf{w} = \mathbf{T}_{N,M}\mathbf{v} .$$

Verify that the components of $\mathbf{u}$ are given by

$$\{u_0, u_1, u_2, \ldots, u_{N-1}\}$$
$$= \{e^0 v_0, e^{-i\frac{2\pi}{N}M \cdot 1} v_1, e^{-i\frac{2\pi}{N}M \cdot 2} v_2, \ldots, e^{-i\frac{2\pi}{N}M(N-1)} v_{N-1}\} ,$$

while the components of $\mathbf{w}$ are given by

$$\{w_0, w_1, w_2, \ldots, w_{N-1}\} = \{v_{0-M}, v_{1-M}, v_{2-M}, \ldots, v_{N-1-M}\} .$$

b. Letting $\mathbf{E}_{N,M}^*$ be the complex conjugate of $\mathbf{E}_{N,M}$, show that the discrete translation identities are equivalent to the matrix equations

$$\mathbf{T}_{N,M}\mathscr{F}_N = \mathscr{F}_N\mathbf{E}_{N,M}^* \qquad , \qquad \mathbf{E}_{N,M}\mathscr{F}_N = \mathscr{F}_N\mathbf{T}_{N,M} \quad ,$$

$$\mathscr{F}_N^{-1}\mathbf{T}_{N,M} = \mathbf{E}_{N,M}^*\mathscr{F}_N^{-1} \qquad \text{and} \qquad \mathscr{F}_N^{-1}\mathbf{E}_{N,M} = \mathbf{T}_{N,M}\mathscr{F}_N^{-1} .$$

40.29. In the following, $*$ denotes complex conjugation, and $\tilde{\ }$ denotes index negation. In addition, assume

$$\{v_0, v_1, v_2, \ldots, v_{N-1}\} \qquad \text{and} \qquad \{w_0, w_1, w_2, \ldots, w_{N-1}\}$$

are two N^{th} order sequences with respective discrete Fourier transforms

$$\{V_0, V_1, V_2, \ldots, V_{N-1}\} \qquad \text{and} \qquad \{W_0, W_1, W_2, \ldots, W_{N-1}\} ,$$

and let $\mathbf{v}$, $\mathbf{w}$, $\mathbf{V}$ and $\mathbf{W}$ be the corresponding sequence vectors. (So $\mathbf{V} = \mathscr{F}_N\mathbf{v}$ and $\mathbf{W} = \mathscr{F}_N\mathbf{w}$.)

a. Confirm that $\mathscr{F}_N^* = \mathscr{F}_N^{-1}$ and $\mathscr{F}_N^{-1*} = \mathscr{F}_N$.

b. Show that

$$\mathscr{F}_N\mathbf{v}^* = \tilde{\mathbf{V}}^* \qquad \text{and} \qquad \mathscr{F}_N\tilde{\mathbf{v}}^* = \mathbf{V}^* .$$

c. Using the above and the appropriate discrete convolution identity, prove the correlation identity

$$\frac{1}{\sqrt{N}} \sum_{k=0}^{N-1} e^{-i\frac{2\pi}{N}nk} v_k^* w_k = \frac{1}{\sqrt{N}} \sum_{j=0}^{N-1} V_{j-n}^* W_j \qquad \text{for } n = 0, 1, 2, \ldots, N - 1 .$$

d. *Prove the other correlation identity,*

$$\frac{1}{\sqrt{N}} \sum_{n=0}^{N-1} e^{i\frac{2\pi}{N}kn} V_n^* W_n = \frac{1}{\sqrt{N}} \sum_{j=0}^{N-1} v_{j-k}^* w_j \qquad \text{for} \quad k = 0, 1, 2, \ldots, N-1 \quad .$$

e. *Using either correlation identity, show that*

$$\mathbf{V} \cdot \mathbf{W} = \mathbf{v} \cdot \mathbf{w} \qquad and \qquad \|\mathbf{V}\| = \|\mathbf{v}\| \quad .$$

(The first is the discrete version of Parseval's equality, and the second is the discrete version of Bessel's equality.)

40.30 a. *Using a computer math package such as Maple or Mathematica, write a "program" or "worksheet" for computing the Nth order discrete Fourier transform using the basic definitions. (Do not use any "Fourier routines" that are already part of the math package.)*

b. *Test your program/worksheet by using it to compute the discrete Fourier transform of each of the following 8th order sequences and then comparing the answer obtained to the corresponding answer obtained in doing exercise 40.25 b:*

   i. $\{1, 0, 0, 0, 0, 0, 0, 0\}$ **ii.** $\{0, 1, 0, 0, 0, 0, 0, 0\}$

 iii. $\{1, 1, 1, 1, 1, 1, 1, 1\}$ **iv.** $\{1, -1, 1, -1, 1, -1, 1, -1\}$

40.31 a. *Using a computer math package such as Maple or Mathematica, write a "program" or "worksheet" for computing the Nth order discrete Fourier transform using our first level fast Fourier transform algorithm. (Do not use any "Fourier routines" that are already part of the math package.)*

b. *Test your program/worksheet by using it to compute the discrete Fourier transform of each of the following 8th order sequences and then comparing the answer obtained to the corresponding answer obtained in doing exercise 40.25 b:*

   i. $\{1, 0, 0, 0, 0, 0, 0, 0\}$ **ii.** $\{0, 1, 0, 0, 0, 0, 0, 0\}$

 iii. $\{1, 1, 1, 1, 1, 1, 1, 1\}$ **iv.** $\{1, -1, 1, -1, 1, -1, 1, -1\}$

40.32. *Set $N = 2^K$ for, say, $K = 10$ (this may be changed in a moment), and let $\{v_0, v_1, \ldots, v_{N-1}\}$ be the Nth order sequence with*

$$v_k = \begin{cases} 3 & k \text{ is even} \\ -3 & k \text{ is odd} \end{cases}$$

(or you can generate your own sequence). Compute the Nth order discrete transform of this sequence using each of the following:

a. *Your program/worksheet from exercise 40.30.*

b. *Your "first level" program/worksheet from exercise 40.31.*

c. *The fast Fourier transform in the computer math package you are using.*

In each case, measure the length of time it takes to compute the discrete transform. Which is fastest? How much faster than the others is it? (If necessary, adjust the value of K to get a discrete transform that requires a measurable amount of time to compute.)

40.33. *How many basic arithmetic operations does it take to compute the convolution of two N^{th} order sequences using the basic definition, formula (40.27) on page 746?*

40.34. *The discrete convolution of two N^{th} order sequences $\{v_0, v_1, \ldots, v_{N-1}\}$ and $\{w_0, w_1, \ldots, w_{N-1}\}$ can also be computed by the following alternative procedure:*

    1. *Compute the discrete transforms $\{V_0, V_1, \ldots, V_{N-1}\}$ and $\{W_0, W_1, \ldots, W_{N-1}\}$ of $\{v_0, v_1, \ldots, v_{N-1}\}$ and $\{w_0, w_1, \ldots, w_{N-1}\}$.*

    2. *Compute the product of $\{V_0, V_1, \ldots, V_{N-1}\}$ and $\{W_0, W_1, \ldots, W_{N-1}\}$, letting $\{U_0, U_1, \ldots, U_{N-1}\}$ denote this product.*

    3. *Finally, compute the inverse discrete transform $\{u_0, u_1, \ldots, u_{N-1}\}$ of $\{U_0, U_1, \ldots, U_{N-1}\}$.*

  a. *Using equation (40.26), verify that $\{u_0, u_1, \ldots, u_{N-1}\}$ is the convolution of $\{v_0, v_1, \ldots, v_{N-1}\}$ and $\{w_0, w_1, \ldots, w_{N-1}\}$.*

  b. *At most, how many basic arithmetic operations does this procedure require? Assume N is large and equal to 2^K for some integer K, and that the discrete transforms are computed using fast Fourier transforms. You may also use the upper bound*

$$\text{number of basic arithmetic operations for an } N^{th} \text{ order FFT} \; < \; 3N \log_2 N \quad ,$$

which we obtained in deriving inequality (40.33).

40.35. *Compare your answer for exercise 40.33 to your answer to the final question in the last exercise above. Assuming N is large and equal to 2^K for some integer K, which way of computing the convolution of two N^{th} order sequences should be faster: (1) using just the basic definition discrete convolution or (2) the alternative procedure described in the previous exercise? Approximately how much faster is that method when $N = 2^{10}$?*

Tables, References
and Answers

Fourier Transforms of Some Common Functions

In the following:

α, β and γ all denote **real** numbers.

$f(t)$	$F(\omega) = \mathcal{F}[f(t)]\vert_{\omega}$	Restrictions
$\text{pulse}_{\alpha}(t)$	$2\alpha \,\text{sinc}(2\pi\alpha\omega)$	$0 < \alpha$
$\text{rect}_{(\alpha,\beta)}(t)$	$\dfrac{i}{2\pi\omega}\left[e^{-i2\pi\beta\omega} - e^{-i2\pi\alpha\omega}\right]$	$\alpha \leq \beta$
$\text{tri}(t)$	$\text{sinc}^2(\pi\omega)$	none
$\cos\left(\dfrac{\pi}{2\alpha}t\right)\text{pulse}_{\alpha}(t)$	$\alpha\left[\text{sinc}\left(2\pi\alpha\omega + \dfrac{\pi}{2}\right) + \text{sinc}\left(2\pi\alpha\omega - \dfrac{\pi}{2}\right)\right]$	$0 < \alpha$
$\text{sinc}(2\pi\alpha t)$	$\dfrac{1}{2\alpha}\text{pulse}_{\alpha}(\omega)$	$0 < \alpha$
$\text{sinc}^2(2\pi\alpha t)$	$\dfrac{1}{2\alpha}\text{tri}\left(\dfrac{\omega}{2\alpha}\right)$	$0 < \alpha$
$e^{-(\alpha+i\beta)t}\,\text{step}(t)$	$\dfrac{1}{\alpha + i\beta + i2\pi\omega}$	$0 < \alpha$
$t^k e^{-(\alpha+i\beta)t}\,\text{step}(t)$	$\dfrac{k!}{(\alpha + i\beta + i2\pi\omega)^{k+1}}$	$0 < \alpha$ $k = 0, 1, 2, \ldots$
$e^{(\alpha+i\beta)t}\,\text{step}(-t)$	$\dfrac{1}{\alpha + i\beta - i2\pi\omega}$	$0 < \alpha$
$t^k e^{(\alpha+i\beta)t}\,\text{step}(-t)$	$(-1)^k \dfrac{k!}{(\alpha + i\beta - i2\pi\omega)^{k+1}}$	$0 < \alpha$ $k = 0, 1, 2, \ldots$
$\dfrac{1}{\alpha + i\beta + i\gamma t}$	$\dfrac{2\pi}{\gamma}e^{2\pi(\alpha+i\beta)\omega/\gamma}\,\text{step}(-\omega)$	$0 < \alpha, 0 < \gamma$
$\dfrac{1}{(\alpha + i\beta + it)^{k+1}}$	$\dfrac{2\pi}{k!}(2\pi\omega)^k e^{2\pi(\alpha+i\beta)\omega}\,\text{step}(-\omega)$	$k = 0, 1, 2, \ldots$

$f(t)$	$F(\omega) = \mathcal{F}[f(t)]\vert_\omega$	Restrictions
$\dfrac{1}{\alpha + i\beta - i\gamma t}$	$\dfrac{2\pi}{\gamma} e^{-2\pi(\alpha+i\beta)\omega/\gamma}\, \text{step}(\omega)$	$0 < \alpha,\, 0 < \gamma$
$\dfrac{1}{(\alpha + i\beta - it)^{k+1}}$	$\dfrac{2\pi}{k!}(-2\pi\omega)^k e^{-2\pi(\alpha+i\beta)\omega}\, \text{step}(\omega)$	$0 < \alpha$ $k = 0,\,1,\,2,\,\ldots$
$e^{-\alpha\vert t\vert}$	$\dfrac{2\alpha}{\alpha^2 + 4\pi^2\omega^2}$	$0 < \alpha$
$\dfrac{1}{\alpha^2 + \gamma^2 t^2}$	$\dfrac{\pi}{\alpha\gamma} e^{-2\pi\alpha\vert\omega\vert/\gamma}$	$0 < \alpha,\, 0 < \gamma$
$e^{-\alpha t^2 + \beta t}$	$\sqrt{\dfrac{\pi}{\alpha}}\, \exp\left(-\dfrac{1}{4\alpha}(2\pi\omega + i\beta)^2\right)$	$0 < \alpha$
$e^{-\lambda t^2}$	$\left[\sqrt{\vert\lambda\vert + \alpha}\; - \; i\,\text{sgn}(\beta)\sqrt{\vert\lambda\vert - \alpha}\right]$ $\times \dfrac{1}{\vert\lambda\vert}\sqrt{\dfrac{\pi}{2}}\, \exp\left(-\dfrac{\pi^2}{\lambda}\omega^2\right)$	$\lambda = \alpha + i\beta,$ $0 \le \alpha$
$e^{-\alpha t^2}\cos\left(\beta t^2\right)$	$\dfrac{1}{\vert\lambda\vert}\sqrt{\dfrac{\pi}{2}}\, \exp\left(-\dfrac{\pi^2\alpha}{\vert\lambda\vert^2}\omega^2\right) \times$ $\left[\sqrt{\vert\lambda\vert + \alpha}\,\cos\left(\dfrac{\pi^2\beta}{\vert\lambda\vert^2}\omega^2\right)\right.$ $\left. + \sqrt{\vert\lambda\vert - \alpha}\,\sin\left(\dfrac{\pi^2\beta}{\vert\lambda\vert^2}\omega^2\right)\right]$	$\lambda = \alpha + i\beta,$ $0 \le \alpha,\, 0 < \beta$
$e^{-\alpha t^2}\sin\left(\beta t^2\right)$	$\dfrac{1}{\vert\lambda\vert}\sqrt{\dfrac{\pi}{2}}\, \exp\left(-\dfrac{\pi^2\alpha}{\vert\lambda\vert^2}\omega^2\right) \times$ $\left[\sqrt{\vert\lambda\vert - \alpha}\,\cos\left(\dfrac{\pi^2\beta}{\vert\lambda\vert^2}\omega^2\right)\right.$ $\left. - \sqrt{\vert\lambda\vert + \alpha}\,\sin\left(\dfrac{\pi^2\beta}{\vert\lambda\vert^2}\omega^2\right)\right]$	$\lambda = \alpha + i\beta,$ $0 \le \alpha,\, 0 < \beta$
$e^{\pm i\pi\alpha t^2}$	$\dfrac{1}{\sqrt{\alpha}}\, \exp\left(\mp i\pi\left[\dfrac{\omega^2}{\alpha} - \dfrac{1}{4}\right]\right)$	$\alpha > 0$
$\cos\left(\pi\alpha t^2\right)$	$\dfrac{1}{\sqrt{\alpha}}\, \cos\left(\pi\left[\dfrac{\omega^2}{\alpha} - \dfrac{1}{4}\right]\right)$	$\alpha > 0$
$\sin\left(\pi\alpha t^2\right)$	$-\dfrac{1}{\sqrt{\alpha}}\, \sin\left(\pi\left[\dfrac{\omega^2}{\alpha} - \dfrac{1}{4}\right]\right)$	$\alpha > 0$

$f(t)$	$F(\omega) = \mathcal{F}[f(t)]\vert_\omega$	Restrictions
1	$\delta(\omega)$	none
t^k	$\left(\dfrac{i}{2\pi}\right)^k D^k \delta(\omega)$	$k = 0,\ 1,\ 2,\ \ldots$
$e^{i2\pi(\alpha+i\beta)t}$	$\delta_{\alpha+i\beta}(\omega)$	none
$\sin(2\pi\alpha t)$	$\dfrac{1}{2i}\left[\delta_\alpha(\omega) - \delta_{-\alpha}(\omega)\right]$	none
$\cos(2\pi\alpha t)$	$\dfrac{1}{2}\left[\delta_\alpha(\omega) + \delta_{-\alpha}(\omega)\right]$	none
$\text{step}(t)$	$\dfrac{1}{i2\pi} T_{-i}\left[\dfrac{1}{\omega - i}\right]$	none
$\text{step}(t)$	$\dfrac{1}{i2\pi}\,\text{pole}(\omega) + \dfrac{1}{2}\delta(\omega)$	none
$\text{step}(-t)$	$\dfrac{-1}{i2\pi} T_i\left[\dfrac{1}{\omega + i}\right]$	none
$\text{step}(-t)$	$\dfrac{-1}{i2\pi}\,\text{pole}(\omega) + \dfrac{1}{2}\delta(\omega)$	none
$\text{ramp}(t)$	$\dfrac{-1}{4\pi^2} T_{-i}\left[(\omega - i)^{-2}\right]$	none
$\text{ramp}(t)$	$\dfrac{-1}{4\pi^2}\left[\text{pole}^2(\omega) - i\pi D\delta(\omega)\right]$	none
$t^k\,\text{step}(t)$	$\left(\dfrac{1}{i2\pi}\right)^{k+1} k!\,T_{-i}\left[(\omega - i)^{-k-1}\right]$	$k = 0,\ 1,\ 2,\ \ldots$
$t^k\,\text{step}(t)$	$\left(\dfrac{-i}{2\pi}\right)^{k+1}\left[k!\,\text{pole}^{k+1}(\omega) + i(-1)^k\pi D^k\delta(\omega)\right]$	$k = 0,\ 1,\ 2,\ \ldots$
$t^k\,\text{step}(-t)$	$-\left(\dfrac{1}{i2\pi}\right)^{k+1} k!\,T_i\left[(\omega + i)^{-k-1}\right]$	$k = 0,\ 1,\ 2,\ \ldots$
$t^k\,\text{step}(-t)$	$\left(\dfrac{-i}{2\pi}\right)^{k+1}\left[-k!\,\text{pole}^{k+1}(\omega) + i(-1)^k\pi D^k\delta(\omega)\right]$	$k = 0,\ 1,\ 2,\ \ldots$

$f(t)$	$F(\omega) = \mathcal{F}[f(t)]\|_{\omega}$	Restrictions
$\operatorname{sgn}(t)$	$\dfrac{1}{i\pi} D \ln \|\omega\|$	none
$\operatorname{sgn}(t)$	$\dfrac{1}{i\pi} \operatorname{pole}(\omega)$	none
$\operatorname{pole}(t)$	$-i\pi \operatorname{sgn}(\omega)$	none
$\operatorname{pole}^k(t)$	$\dfrac{(-i2\pi)^k}{2(k-1)!} \omega^{k-1} \operatorname{sgn}(\omega)$	$k = 1,\ 2,\ \ldots$
$t^k \operatorname{sgn}(t)$	$\left(\dfrac{-i}{2\pi}\right)^{k+1} 2k!\,\operatorname{pole}^{k+1}(\omega)$	$k = 0,\ 1,\ 2,\ \ldots$
$\|t\|^n$	$\left(\dfrac{-i}{2\pi}\right)^{n+1} 2n!\,\operatorname{pole}^{n+1}(\omega)$	$n = 1,\ 3,\ 5,\ \ldots$
$\operatorname{comb}_\alpha(t)$	$\dfrac{1}{\alpha} \operatorname{comb}_{1/\alpha}(\omega)$	$0 < \alpha$
$\operatorname{saw}_\alpha(t)$	$\dfrac{\alpha}{2}\delta(\omega) + \displaystyle\sum_{\substack{k=-\infty \\ k\neq 0}}^{\infty} \dfrac{i\alpha}{2\pi k}\delta_{k/\alpha}(\omega)$	$0 < \alpha$

TABLE A.2
Identities for the Fourier Transforms

In the following:

$$\alpha = \text{any } \textbf{real} \text{ number, } F(\omega) = \mathcal{F}[f(t)]|_\omega \text{ , and } G(\omega) = \mathcal{F}[g(t)]|_\omega$$

$h(t)$	$H(\omega) = \mathcal{F}[h(t)]|_\omega$	Restrictions
$f(t)$	$\displaystyle\int_{-\infty}^{\infty} f(t)\, e^{-i2\pi\omega t}\, dt$	f in $\mathcal{A}$
$\displaystyle\int_{-\infty}^{\infty} F(\omega)\, e^{i2\pi\omega t}\, d\omega$	$F(\omega)$	F in $\mathcal{A}$
$f(\alpha t)$	$\dfrac{1}{|\alpha|} F\left(\dfrac{\omega}{\alpha}\right)$	$\alpha \neq 0$
$\dfrac{1}{|\alpha|} f\left(\dfrac{t}{\alpha}\right)$	$F(\alpha\omega)$	$\alpha \neq 0$
$f(t - \alpha)$	$e^{-i2\pi\alpha\omega} F(\omega)$	none
$T_{\alpha + i\beta}[f(t)]$	$e^{-i2\pi(\alpha + i\beta)\omega} F(\omega)$	none
$e^{i2\pi\alpha t} f(t)$	$F(\omega - \alpha)$	none
$e^{i2\pi(\alpha + i\beta)t} f(t)$	$T_{\alpha + i\beta}[F(\omega)]$	none
$\sin(2\pi\alpha t)\, f(t)$	$\dfrac{i}{2}[F(\omega + \alpha) - F(\omega - \alpha)]$	none
$\cos(2\pi\alpha t)\, f(t)$	$\dfrac{1}{2}[F(\omega + \alpha) + F(\omega - \alpha)]$	none
$\dfrac{1}{2}[f(t - \alpha) + f(t + \alpha)]$	$\cos(2\pi\alpha\omega)\, F(\omega)$	none
$\dfrac{i}{2}[f(t - \alpha) - f(t + \alpha)]$	$\sin(2\pi\alpha\omega)\, F(\omega)$	none
$\dfrac{df}{dt}$	$i2\pi\omega F(\omega)$	see chap. 22
$\dfrac{d^n f}{dt^n}$	$(i2\pi\omega)^n\, F(\omega)$	see chap. 22

$h(t)$	$H(\omega) = \mathcal{F}[h(t)]\vert_\omega$	Restrictions
$tf(t)$	$\dfrac{i}{2\pi}\dfrac{dF}{d\omega}$	see chap. 22
$t^n f(t)$	$\left(\dfrac{i}{2\pi}\right)^n \dfrac{d^n F}{d\omega^n}$	see chap. 22
Df	$i2\pi\omega F(\omega)$	none
$D^n f$	$(i2\pi\omega)^n F(\omega)$	none
$tf(t)$	$\dfrac{i}{2\pi}DF$	none
$t^n f(t)$	$\left(\dfrac{i}{2\pi}\right)^n D^n F$	none
fg	$F * G$	yes, see chap. 24
$f * g$	FG	see chap. 24
$f^* g$	$F \star G$	see chap. 25
$f \star g$	$F^* G$	see chap. 25

Other Useful Identities:

For all transformable functions:

Near-equivalence:

$$\mathcal{F}^{-1}[\phi(x)]\vert_y = \mathcal{F}[\phi(-x)]\vert_y = \mathcal{F}[\phi(x)]\vert_{-y}$$

and

$$\mathcal{F}[\phi(x)]\vert_y = \mathcal{F}^{-1}[\phi(-x)]\vert_y = \mathcal{F}^{-1}[\phi(x)]\vert_{-y}$$

Under suitable conditions (see chap. 25):

Fundamental Identity:

$$\int_{-\infty}^{\infty} F(x)g(x)\,dx = \int_{-\infty}^{\infty} f(y)G(y)\,dy$$

Parseval's Identity:

$$\int_{-\infty}^{\infty} F(x)G^*(x)\,dx = \int_{-\infty}^{\infty} f(y)g^*(y)\,dy$$

Bessel's Identity:

$$\int_{-\infty}^{\infty} |F(x)|^2\,dx = \int_{-\infty}^{\infty} |f(y)|^2\,dy$$

References

[1] E. Chu and A. George: *Inside the FFT Black Box*, CRC Press, Boca Raton, FL, 2000.

[2] J.F. Colombeau: *New Generalized Functions and Multiplication of Distributions*, North-Holland Mathematics Studies 90, Elsevier, Amsterdam, 1984.

[3] K.B. Howell: A New Theory for Fourier Analysis. Part I: The Space of Test Functions, *Journal of Mathematical Analysis and Applications*, vol. 168 (1992), 342–350.

[4] K.B. Howell: A New Theory for Fourier Analysis. Part II: Further Analysis on the Space of Test Functions, *Journal of Mathematical Analysis and Applications*, vol. 173 (1993), 419–429.

[5] K.B. Howell: A New Theory for Fourier Analysis. Part III: Basic Analysis on the Dual, *Journal of Mathematical Analysis and Applications*, vol. 175 (1993), 257–267.

[6] K.B. Howell: A New Theory for Fourier Analysis. Part IV: Basic Multiplication and Convolution on the Dual Space, *Journal of Mathematical Analysis and Applications*, vol. 180 (1993), 79–92.

[7] K.B. Howell: A New Theory for Fourier Analysis. Part V: Generalized Multiplication and Convolution on the Dual Space, *Journal of Mathematical Analysis and Applications*, vol. 187 (1994), 567–582.

[8] K.B. Howell: Exponential Bounds on Elementary Multipliers of Generalized Functions, *Journal of Mathematical Analysis and Applications*, vol. 193 (1995), 832–838.

[9] I. Richards and H.K. Youn: *Theory of Distributions: A Non-Technical Introduction*, Cambridge University Press, Cambridge, 1990.

[10] A.D. Poularikas, Ed.: *The Transforms and Applications Handbook, 2nd Edition*, CRC Press, Boca Raton, FL, 2000.

[11] R.S. Strichartz: *A Guide to Distribution Theory and Fourier Transforms*, World Scientific Publishing Company, 2003.

[12] J.S. Walker: *Fast Fourier Transforms (2nd ed.)*, CRC Press, Boca Raton, FL, 1996.

[13] G.N. Watson: *A Treatise on the Theory of Bessel Functions*, Cambridge Mathematical Library, Cambridge, 1922 (reprinted 1996).

Answers to Selected Exercises

Chapter 1

1. geometric definition: $\mathbf{v} \cdot \mathbf{w} = |\mathbf{v}|\,|\mathbf{w}| \cos(\text{angle between } \mathbf{v} \text{ and } \mathbf{w})$
component formula: $\mathbf{v} \cdot \mathbf{w} = v_1 w_1 + v_2 w_2 + v_3 w_3$

Chapter 6

7a. 2   **7b.** 3   **7c.** $\sqrt{13}$ **7d.** $\mathrm{Arctan}\left(\frac{3}{2}\right)$ **7e.** $-5 + 12i$
7f. $\frac{2}{13}$ **7g.** $-\frac{3}{13}$
10a. $\mathrm{Re}[f(t)] = \left(1 + t^2\right)^{-1}$, $\mathrm{Im}[f(t)] = t\left(1 + t^2\right)^{-1}$
13a. $\frac{1}{2}$ **13b.** $\left[-a + ae^{ax}\cos(bx) + be^{ax}\sin(bx)\right]\left(a^2 + b^2\right)^{-1}$ **13c.** 0   **13d.** $\frac{P}{2}$
16a. 1 , i , -1 and $-i$
16b. 1 , $\exp\left(i\frac{2\pi}{3}\right) = \frac{1}{2}\left[-1 + i\sqrt{3}\right]$ and $\exp\left(i\frac{4\pi}{3}\right) = \frac{1}{2}\left[-1 - i\sqrt{3}\right]$
16c. $\exp\left(i\frac{\pi}{3}\right) = \frac{1}{2}\left[1 + i\sqrt{3}\right]$, -1 and $\exp\left(i\frac{5\pi}{3}\right) = \frac{1}{2}\left[1 - i\sqrt{3}\right]$
16d. $2\exp\left(i\frac{\pi}{3}\right) = 1 + i\sqrt{3}$, -2 and $\exp\left(i\frac{5\pi}{3}\right) = 1 - i\sqrt{3}$
16e. $\exp\left(i\frac{\pi}{4}\right) = \frac{1}{\sqrt{2}}[1 + i]$ and $\exp\left(i\frac{5\pi}{4}\right) = -\frac{1}{\sqrt{2}}[1 + i]$
16f. $\exp\left(i\frac{3\pi}{4}\right) = \frac{1}{\sqrt{2}}[-1 + i]$ and $\exp\left(i\frac{7\pi}{4}\right) = \frac{1}{\sqrt{2}}[1 - i]$
16g. $\exp\left(i\frac{\pi}{6}\right) = \frac{1}{2}\left[\sqrt{3} + i\right]$, $\exp\left(i\frac{5\pi}{6}\right) = \frac{1}{2}\left[-\sqrt{3} + i\right]$ and $= -i$

Chapter 7

9b. 2   **9d.** $\frac{2}{3}$

Chapter 9

5a. fund. period $= 2$, fund. freq. $= \frac{1}{2}$, $F.S.[f]|_t = \frac{1}{2} + \sum_{k=1}^{\infty} \frac{1 - (-1)^k}{k\pi} \sin(k\pi t)$

5b. $F.S.[g]|_t = -\sum_{k=1}^{\infty} (-1)^k \frac{6}{k\pi} \sin\left(\frac{k\pi}{3}t\right)$

5c. $F.S.[h]|_t = e - 1 + \sum_{k=1}^{\infty} \left[\frac{2(e-1)}{1 + 4\pi^2 k^2} \cos(k2\pi t) - \frac{4\pi k(e-1)}{1 + 4\pi^2 k^2} \sin(k2\pi t)\right]$

5e. $F.S.\left[\cos^2(t)\right]\Big|_t = \frac{1}{2} + \frac{1}{2}\cos(2t)$ **5f.** $F.S.[|\sin(t)|]|_t = \frac{2}{\pi} + \sum_{k=1}^{\infty} \frac{4}{\pi\left(1 - 4k^2\right)} \cos(2kt)$

6a. odd, $F.S.[f]|_t = \sum_{k=1}^{\infty}\left[1-(-1)^k\right]\frac{2}{k\pi}\sin(k\pi t)$

6b. even, $F.S.[g]|_t = \frac{1}{2} + \sum_{k=1}^{\infty}\frac{2}{k\pi}\sin\left(\frac{k\pi}{2}\right)\cos\left(\frac{k\pi}{2}t\right)$

6c. $F.S.[\text{evensaw}]|_t = \frac{1}{2} + \sum_{k=1}^{\infty}\left[(-1)^k-1\right]\frac{2}{k^2\pi^2}\cos(k\pi t)$

6d. $F.S.[\text{oddsaw}]|_t = \sum_{k=1}^{\infty}(-1)^{k+1}\frac{2}{k\pi}\sin(k\pi t)$

6e. $F.S.[h]|_t = \frac{1}{3} + \sum_{k=1}^{\infty}(-1)^k\frac{4}{k^2\pi^2}\cos(k\pi t)$

6f. $F.S.[k]|_t = \sum_{k=1}^{\infty}\left(\left[(-1)^k-1\right]\frac{4}{k^3\pi^3} - (-1)^k\frac{2}{k\pi}\right)\sin(k\pi t)$

7a. $G(t) = 2g(t) - 1$, $F.S.[G]|_t = \sum_{k=1}^{\infty}\frac{4}{k\pi}\sin\left(\frac{k\pi}{2}\right)\cos\left(\frac{k\pi}{2}t\right)$

7b. $H(t) = 1 - h(t)$, $F.S.[H]|_t = \frac{2}{3} + \sum_{k=1}^{\infty}(-1)^{k+1}\frac{4}{k^2\pi^2}\cos(k\pi t)$

7c. $K(t) = h(t) - 2\,\text{oddsaw}(t) - 3$

7d. $\Phi(t) = \frac{1}{2}[\text{evensaw}(t) + \text{oddsaw}(t)]$,

$F.S.[\Phi]|_t = \frac{1}{4} + \sum_{k=1}^{\infty}\left\{\left[(-1)^k-1\right]\frac{1}{k^2\pi^2}\cos(k\pi t) + (-1)^{k+1}\frac{1}{k\pi}\sin(k\pi t)\right\}$

7e. $F.S.[\Psi_1]|_t = 1 + \sum_{k=1}^{\infty}\left[(-1)^k-1\right]\frac{4}{k^2\pi^2}\cos(k\pi t)$

7f. $F.S.[\Psi_2]|_t = 1 + \sum_{k=1}^{\infty}\left[(-1)^k-1\right]\frac{4}{k^2\pi^2}\cos(k\pi t)$

7g. $F.S.[\Psi_3]|_t = \frac{1}{2} + \sum_{k=1}^{\infty}\left[1-(-1)^k\right]\frac{2}{k^2\pi^2}\cos(k\pi t)$

Chapter 10

2. the sine series

3. $T.F.S.[f]|_t = 1$, $F.S.S.[f]|_t = \sum_{k=1}^{\infty}\frac{2}{k\pi}\left[1-(-1)^k\right]\sin(k\pi t)$, $F.C.S.[f]|_t = 1$

4. $T.F.S.[f]|_t = \frac{1}{2} + \sum_{k=1}^{\infty}\left[0\cos(2\pi kt) - \frac{1}{k\pi}\sin(2\pi kt)\right]$,

$F.S.S.[f]|_t = \sum_{k=1}^{\infty}(-1)^{k+1}\frac{2}{k\pi}\sin(k\pi t)$,

$F.C.S.[f]|_t = \frac{1}{2} + \sum_{k=1}^{\infty}\frac{2}{k^2\pi^2}\left[(-1)^k-1\right]\cos(k\pi t)$

5a. $\sum_{k=1}^{\infty}\left\{\frac{4}{3}\left(\frac{3}{k\pi}\right)^3\left[(-1)^k-1\right] - \frac{18}{k\pi}(-1)^k\right\}\sin\left(\frac{k\pi}{3}t\right)$ **5b.** $\sin(2t)$

6a. $3 + \sum_{k=1}^{\infty} (-1)^k \left(\frac{6}{k\pi}\right)^2 \cos\left(\frac{k\pi}{3} t\right)$

6b. $\sum_{k=1}^{\infty} a_k \cos(kt)$ where $a_k = \begin{cases} \frac{4}{\pi(4-k^2)}\left[1-(-1)^k\right] & \text{if } k \neq 2 \\ 0 & \text{if } k = 2 \end{cases}$

Chapter 11

5b. $\sqrt{\frac{L}{2}}$ **6b.** $\sqrt{L}$ **6c.** $\sqrt{\frac{L}{2}}$ **7b.** $\sqrt{p}$ **9a.** All real a and b with $b = -2a$

9b. All real a, b and c with $c = -b = 6a$ **10a.** $\|\phi_0\| = 1$ and $\|\phi_1\| = \sqrt{\frac{1}{3}}$

10b i. $c_0 = \frac{1}{2}$, $c_1 = -\frac{1}{2}$ **10b ii.** $c_0 = \frac{2}{3}$, $c_1 = -\frac{2}{5}$ **10b iii.** $c_0 = \frac{2}{\pi}$, $c_1 = 0$

11b. $\|\psi_0\| = 1$, $\|\psi_{1,0}\| = 1$, $\|\psi_{2,0}\| = \|\psi_{2,1}\| = \sqrt{\frac{1}{2}}$,

$\|\psi_{3,0}\| = \|\psi_{3,1}\| = \|\psi_{3,2}\| = \|\psi_{3,3}\| = \frac{1}{2}$ **12a.** $c_0 = 1$; every other $c_{j,k}$ is zero.

12b. $c_0 = c_{1,0} = \frac{1}{2}$; every other $c_{j,k}$ is zero.

12c. $c_0 = \frac{1}{2}$, $c_{1,0} = -\frac{1}{4}$, $c_{2,0} = c_{2,1} = -\frac{1}{8}$, $c_{3,0} = c_{3,1} = c_{3,2} = c_{3,3} = -\frac{1}{16}$

12d. $c_0 = \frac{1}{8}$, $c_{1,0} = \frac{1}{8}$, $c_{2,0} = -\frac{1}{8}$, $c_{3,0} = c_{3,1} = -\frac{1}{16}$, $c_{2,1} = c_{3,2} = c_{3,3} = 0$

Chapter 12

1a. $\sum_{k=-\infty}^{\infty} \frac{1+ik}{k^2+1} e^{i2\pi\omega_k t}$ **1b.** $\sum_{k=1}^{\infty} \frac{-2k}{k^2+4} \sin(2\pi\omega_k t)$

3a. $\frac{1}{2} + \sum_{\substack{k=-\infty \\ k\neq 0}}^{\infty} \frac{i}{2\pi k}\left[(-1)^k - 1\right] e^{i\pi k t}$ **3b.** $\sum_{k=-\infty}^{\infty} \frac{1+i2\pi k}{1+4\pi^2 k^2}[e-1] e^{i2\pi k t}$

3c. $\frac{1}{2} + \sum_{\substack{k=-\infty \\ k\neq 0}}^{\infty} \frac{1}{k^2\pi^2}\left[(-1)^k - 1\right] e^{ik\pi t}$ **3d.** $\sum_{\substack{k=-\infty \\ k\neq 0}}^{\infty} (-1)^k \frac{i}{k\pi} e^{ik\pi t}$ **3e.** $\frac{1}{2} - \frac{1}{4}e^{i2t} - \frac{1}{4}e^{-i2t}$

3f. $\sum_{k=-\infty}^{\infty} \frac{2}{\pi(1-4k^2)} e^{i4k\pi t}$ **3g.** $\sum_{\substack{k=-\infty \\ k\neq 0}}^{\infty} \frac{1}{k\pi} \sin\left(\frac{k\pi}{2}\right) e^{i2\pi\omega_k t}$ where $\omega_k = \frac{k}{4}$

3h. $\frac{1}{3} + \sum_{\substack{k=-\infty \\ k\neq 0}}^{\infty} (-1)^k \frac{2}{k^2\pi^2} e^{ik\pi t}$

Chapter 13

6a. $\frac{1}{2}, 1, \frac{1}{2}, \frac{1}{2}, 1$ **6b.** $(1+e)/2$, $\exp(\frac{1}{2})$, $(1+e)/2$, $(1+e)/2$, $\exp(\frac{1}{2})$
6c. $0, \frac{1}{4}, \frac{1}{2}, 1, \frac{3}{4}$ **6d.** $0, 0, 0, 0, 0$ **6e.** $1, 1, 0, -1, -1$
6f. $0, \frac{1}{4}, 1, 0, \frac{1}{4}$ **6g.** $0, \frac{1}{4}, \frac{1}{2}, 0, \frac{1}{4}$
8. The series converges uniformly for the even sawtooth, $|\sin(2\pi t)|$ and $k(t)$.
11a i. 0 and 0 **11a ii.** yes **11a iii.** no
11b i. 0 and 2 **11b ii.** yes **11b iii.** yes
14a. when $\lim_{t\to 0^+} f(t) \neq 0$ **14b.** never

Chapter 15

6a. $f' = 0$, $F.S.[f'] = 0$ **6b.** $g' = g$, $F.S.[g'] = F.S.[g]$

6c. $\displaystyle\sum_{\substack{k=-\infty \\ k \neq 0}}^{\infty} \frac{i}{k\pi}\left[(-1)^k - 1\right]e^{ik\pi t}$ **6d.** oddsaw$' = 1$, $F.S.[\text{oddsaw}'] = 1$

6e. $\displaystyle\sum_{k=-\infty}^{\infty} \frac{i8k}{1-4k^2}e^{i4k\pi t}$ **6f.** $\displaystyle\sum_{\substack{k=-\infty \\ k \neq 0}}^{\infty} (-1)^k \frac{2i}{k\pi}e^{ik\pi t}$ **8a.** $\dfrac{1}{2} - \dfrac{1}{2\pi^2}\displaystyle\sum_{\substack{k=-\infty \\ k \neq 0}}^{\infty} \frac{1}{k^2}$ **8b.** $1/6$

9a. $\dfrac{1}{2}t + \displaystyle\sum_{k=1}^{\infty} \frac{1}{2(\pi k)^2}\left[1 - (-1)^k\right] + \displaystyle\sum_{\substack{k=-\infty \\ k \neq 0}}^{\infty} \frac{1}{(2\pi k)^2}\left[(-1)^k - 1\right]e^{i\pi k t}$

9b. $\dfrac{1}{2}t + \displaystyle\sum_{\substack{k=-\infty \\ k \neq 0}}^{\infty} \frac{i}{(k\pi)^3}\left[1 - (-1)^k\right]e^{ik\pi t}$ **9c.** $\displaystyle\sum_{k=1}^{\infty}(-1)^{k+1}\frac{2}{(k\pi)^2} + \displaystyle\sum_{\substack{k=-\infty \\ k \neq 0}}^{\infty}(-1)^k\frac{1}{(k\pi)^2}e^{ik\pi t}$

9d. $\displaystyle\sum_{\substack{k=-\infty \\ k \neq 0}}^{\infty} \frac{-2i}{(k\pi)^2}\sin\left(\frac{k\pi}{2}\right)e^{i2\pi\omega_k t}$ where $\omega_k = \dfrac{k}{4}$

Chapter 16

1c. max. temp. < 128, max. temp. < 64, max. temp. < 0.25

10a. $5e^{-3\pi^2 t}\sin(\pi x)$ **10b.** $\displaystyle\sum_{k=1}^{\infty}(-1)^{k+1}\frac{4}{k\pi}e^{-3\lambda_k t}\sin\left(\frac{k\pi}{2}x\right)$ where $\lambda_k = \left(\dfrac{k\pi}{2}\right)^2$

11b. $A_0 + \displaystyle\sum_{k=1}^{\infty} a_k e^{-\lambda_k t}\cos\left(\frac{k\pi}{L}x\right)$ where $\lambda_k = \kappa\left(\dfrac{k\pi}{L}\right)^2$,

$A_0 = \dfrac{1}{L}\displaystyle\int_0^L f(x)\,dx$ and $a_k = \dfrac{2}{L}\displaystyle\int_0^L f(x)\cos\left(\frac{k\pi}{L}x\right)dx$

11c. $\dfrac{1}{2} + \displaystyle\sum_{k=1}^{\infty}\frac{2}{k\pi}\sin\left(\frac{k\pi}{2}\right)e^{-\lambda_k t}\cos\left(\frac{k\pi}{3}x\right)$ where $\lambda_k = 2\left(\dfrac{k\pi}{3}\right)^2$

12a. $\displaystyle\sum_{k=1}^{\infty}\frac{c_k}{\lambda_k}\left[1 - e^{-\lambda_k t}\right]\sin\left(\frac{k\pi}{L}x\right)$ where $\lambda_k = \kappa\left(\dfrac{k\pi}{L}\right)^2$

and $c_k = \dfrac{2}{L}\displaystyle\int_0^L f(x)\sin\left(\frac{k\pi}{L}x\right)dx$

13. $\displaystyle\sum_{k=1}^{\infty} a_k \sin\left(\frac{kc\pi}{L}t\right)\sin\frac{k\pi}{L}x$ where $a_k = \dfrac{2}{kc\pi}\displaystyle\int_0^L f(x)\sin\frac{k\pi}{L}x\,dx$

14a. $\displaystyle\sum_{k=1}^{\infty}[A_k\cos(v_k t) + B_k\sin(v_k t)]e^{-\beta t}\sin\left(\frac{k\pi}{L}x\right)$

where the A_k's and B_k's are arbitrary constants and $v_k = \dfrac{1}{L}\sqrt{(kc\pi)^2 - (\beta L)^2}$

Chapter 18

9a. abs. integrable   **9b.** not abs. integrable   **9c.** not abs. integrable
10b. $\gamma > 1$ **11b.** $\gamma < 1$ **12.** All are absolutely integrable if and only if $0 \leq \gamma$.
13a. abs. integrable   **13b.** not abs. integrable   **13c.** not abs. integrable
13d. abs. integrable   **18c.** $|x|^{-1/2}\text{rect}_{(-1,1)}(x)$ **18d.** It blows up.

Chapter 19

5. even, $\mathcal{F}_I^{-1}[\text{pulse}_a]\big|_x = \mathcal{F}_I[\text{pulse}_a]\big|_x = 2a\,\text{sinc}(2\pi ax)$ **10.** all except e^{-x^2}

14a. $(a - ib + i2\pi x)^{-1}$ **14b i.** $(2 - i3 + i2\pi x)^{-1}$ **14b ii.** $(2 + i(10\pi - 3))^{-1}$

14b iii. $(2 + 14\pi i)^{-1}$ **14b iv.** $(a + ib + i2\pi x)^{-1}$ **14c i.** $(a - ib - i2\pi x)^{-1}$

14c ii. $(a + ib - i2\pi x)^{-1}$ **14c iii.** $(a + ib + i2\pi x)^{-1}$ **14d i.** $2a\big(a^2 + (b - 2\pi x)^2\big)^{-1}$

14d ii. $\dfrac{1}{2}\Big[(a - ib + i2\pi x)^{-1} + (a + ib + i2\pi x)^{-1}\Big]$ **15a i.** $i\big(e^{-i2\pi x} - 1\big)(2\pi x)^{-1}$

15a ii. $\big(e^{-i2\pi x} + i2\pi x e^{-i2\pi x} - 1\big)(2\pi x)^{-2}$ **15b i.** $i\big(1 - e^{i2\pi x}\big)(2\pi x)^{-1}$

15b ii. $\big(1 - e^{i2\pi x} + i2\pi x e^{i2\pi x}\big)(2\pi x)^{-2}$ **15b iii.** $i\big(e^{-i2\pi x} - 1\big)(2\pi x)^{-1}$

15b iv. $\big(1 + i2\pi x - e^{i2\pi x}\big)(2\pi x)^{-2}$ **15b v.** $\big(1 - i2\pi x - e^{-i2\pi x}\big)(2\pi x)^{-2}$

16b i. $2a\,\text{sinc}(2\pi a[x + b])$ **16b ii.** $2a\,\text{sinc}(2\pi a[x + b])$

16b iii. $ia\,[\text{sinc}(2\pi a[x + b]) - \text{sinc}(2\pi a[x - b])]$

16b iv. $a\,[\text{sinc}(2\pi a[x + b]) + \text{sinc}(2\pi a[x - b])]$

16b v. $ia\,[\text{sinc}(2\pi a[x - b]) - \text{sinc}(2\pi a[x + b])]$

16b vi. $a\,[\text{sinc}(2\pi a[x + b]) + \text{sinc}(2\pi a[x - b])]$ **18a i.** $\dfrac{1}{2a}e^{-a|y|}$ **18a ii.** $\dfrac{1}{2a}e^{-a|y|}$

18a iii. $\dfrac{\pi}{a}e^{-2\pi a|y|}$ **18b i.** $\dfrac{1}{2}e^{-1}$ **18b ii.** $\dfrac{1}{6}e^{-6}$ **18b iii.** $\dfrac{1}{2}$ **18b iv.** 0

18b v. $\dfrac{1}{2}e^{-1}$

Chapter 20

9. any pulse function   **10.** the sinc function

12a. $\dfrac{1}{2a}\text{pulse}_a(t)$ **12b.** $e^{-at}\,\text{step}(t)$ **12c.** $e^{at}\,\text{step}(-t)$

12d. $e^{-(a+ib)t}\,\text{step}(t)$ **12e.** $e^{(a+ib)t}\,\text{step}(-t)$ **12f.** $e^{i2\pi bt}\,\text{pulse}_a(t)$

13a. $\dfrac{1}{2a}\text{pulse}_a(\omega)$ **13b.** $e^{a\omega}\,\text{step}(-\omega)$ **13c.** $e^{-a\omega}\,\text{step}(\omega)$ **13d.** $e^{(a+ib)\omega}\,\text{step}(-\omega)$

13e. $e^{-(a+ib)\omega}\,\text{step}(\omega)$ **13f.** $e^{-i2\pi bt}\,\text{pulse}_a(t)$ **14a.** $2\pi e^{2\pi a\omega}\,\text{step}(-\omega)$

14b. $\dfrac{2\pi}{c}e^{2\pi(a+ib)\omega/c}\,\text{step}(-\omega)$

15a. $2\pi e^{-2\pi a\omega}\,\text{step}(\omega)$ **15b.** $\dfrac{2\pi}{c}e^{-2\pi(a+ib)\omega/c}\,\text{step}(\omega)$ **15c.** $2\pi e^{-2\pi at}\,\text{step}(t)$

15d. $\dfrac{2\pi}{c}e^{-2\pi(a+ib)t/c}\,\text{step}(t)$ **15e.** $2\pi e^{2\pi at}\,\text{step}(-t)$ **15f.** $\dfrac{2\pi}{c}e^{2\pi(a+ib)t/c}\,\text{step}(-t)$

15g. $\dfrac{\pi}{ac}e^{-2\pi a|\omega|/c}$ **15h.** $\dfrac{\pi}{ac}e^{-2\pi a|t|/c}$

16a. $\dfrac{1}{10}\text{pulse}_5(\omega)$ **16b.** $e^{-3\omega}\,\text{step}(\omega)$ **16c.** $e^{2\omega}\,\text{step}(-\omega)$ **16d.** $e^{3t}\,\text{step}(-t)$

17a. $\dfrac{1}{5}\big[e^{-3\omega}\,\text{step}(\omega) + e^{2\omega}\,\text{step}(-\omega)\big]$ **17b.** $\dfrac{2}{13}\big[e^{-3\omega}\,\text{step}(\omega) + e^{4\omega/3}\,\text{step}(-\omega)\big]$

17c. $\dfrac{i}{4\pi}\big[e^{a\omega}\,\text{step}(-\omega) - e^{-a\omega}\,\text{step}(\omega)\big]$ **17d.** $i\pi c^{-2}\big[e^{2a\pi\omega/c}\,\text{step}(-\omega) - e^{-2a\pi\omega/c}\,\text{step}(\omega)\big]$

17e. $\dfrac{\pi(1 + i)}{a\sqrt{2}}\big[e^{a\pi\sqrt{2}(1+i)\omega}\,\text{step}(-\omega) + e^{-a\pi\sqrt{2}(1+i)\omega}\,\text{step}(\omega)\big]$

17f. $\dfrac{\pi(1 + i)}{ac\sqrt{2}}\big[e^{a\pi\sqrt{2}(1+i)\omega/c}\,\text{step}(-\omega) + e^{-a\pi\sqrt{2}(1+i)\omega/c}\,\text{step}(\omega)\big]$

17g. $\dfrac{\pi(1 - i)}{a\sqrt{2}}\big[e^{a\pi\sqrt{2}(1-i)\omega}\,\text{step}(-\omega) + e^{-a\pi\sqrt{2}(1-i)\omega}\,\text{step}(\omega)\big]$

17h. $\dfrac{\pi(1 - i)}{ac\sqrt{2}}\big[e^{a\pi\sqrt{2}(1-i)\omega/c}\,\text{step}(-\omega) + e^{-a\pi\sqrt{2}(1-i)\omega/c}\,\text{step}(\omega)\big]$

18. classically transformable: $x^{-2}\,\text{step}(x - 1)$, $e^{-ax}\,\text{step}(x)$, $(1 - ix)^{-1}$ and e^{-ax^2}
    not classically transformable: $\cos(ax)$, x^2 , x^{-2} , $\text{step}(x)$, $e^{ax}\,\text{step}(x)$ and $\dfrac{1}{1 - x}$

Chapter 21

11a. $6e^{-i8\pi\omega}\,\text{sinc}(6\pi\omega)$ **11b.** $6e^{i8\pi\omega}\,\text{sinc}(6\pi\omega)$ **11c.** $6\,\text{sinc}(6\pi(\omega - 4))$

11d. $6\,\mathrm{sinc}(6\pi(\omega+4))$ **11e.** $2\pi e^{-2\pi(3+4i)\omega}\,\mathrm{step}(\omega)$ **11f.** $e^{(3+i4\pi)t}\,\mathrm{step}(-t)$

11g. $10e^{-i4\pi\omega}\big(25+4\pi^2\omega^2\big)^{-1}$ **11h.** $6\big(9+4\pi^2(\omega-5)^2\big)^{-1}$ **11i.** $6\big(9+4\pi^2(t+5)^2\big)^{-1}$

11j. $e^{3(t+\frac{1}{2})}\,\mathrm{step}(-t-\nicefrac{1}{2})$ **11k.** $(3+i2\pi\omega)^{-1}e^{-12-i8\pi\omega}$ **11l.** $\dfrac{\pi}{4}e^{-i6\pi\omega}e^{-8\pi|\omega|}$

11m. $\dfrac{\pi}{3}e^{-i10\pi\omega}e^{-6\pi|\omega|}$ **11n.** $\dfrac{\pi}{5}e^{i4\pi\omega}e^{-10\pi|\omega|}$

11o. $\dfrac{1}{2}\big[(3+i2\pi(\omega+1))^{-1}+(3+i2\pi(\omega-1))^{-1}\big]$

11p. $5\left[\big(25+4\pi^2(\omega+3)^2\big)^{-1}+\big(25+4\pi^2(\omega-3)^2\big)^{-1}\right]$

11q. $5i\left[\big(25+4\pi^2(\omega+3)^2\big)^{-1}-\big(25+4\pi^2(\omega-3)^2\big)^{-1}\right]$

11r. $\dfrac{i}{4}\big[\mathrm{pulse}_1(\omega+1)-\mathrm{pulse}_1(\omega-1)\big]$ **11s.** $\alpha\left[\mathrm{sinc}\!\left(2\pi\alpha\omega+\dfrac{\pi}{2}\right)+\mathrm{sinc}\!\left(2\pi\alpha\omega-\dfrac{\pi}{2}\right)\right]$

12a. $\dfrac{2\pi}{3}e^{8\pi\omega}\,\mathrm{step}(-\omega)$ **12b.** $\dfrac{\pi}{\alpha\gamma}e^{-2\pi\alpha|\omega|/\gamma}$

13a. $\pi e^{-i6\pi(\omega+2)}\,\mathrm{pulse}_{1/2\pi}(\omega+2)$ **13b.** $-6e^{-i2\pi\omega}\,\mathrm{sinc}\!\left(6\pi\left[\omega-\dfrac{1}{2}\right]\right)$

13c. $i\pi e^{2\pi(6+3i)\omega}\big[e^{12\pi}\,\mathrm{step}(-\omega-1)-e^{-12\pi}\,\mathrm{step}(1-\omega)\big]$

13d. $\dfrac{2\pi}{3}e^{(1+3i)8\pi(\omega-3)/3}\,\mathrm{step}(3-\omega)$ **13e.** $\dfrac{2\pi}{\gamma}e^{2\pi(\alpha+i\beta)\omega/\gamma}\,\mathrm{step}(-\omega)$

13f. $\dfrac{2\pi}{\gamma}e^{-2\pi(\alpha+i\beta)\omega/\gamma}\,\mathrm{step}(\omega)$

14a. $\beta=\nicefrac{1}{\alpha}\,,\ b=-\nicefrac{1}{\alpha}$ **14b i.** $\dfrac{i}{2\pi\omega}\big[e^{-i2\pi\alpha\omega}-1\big]$ **14b ii.** $\dfrac{-i}{2\pi\omega}\big[e^{i2\pi\alpha\omega}-1\big]$

19a ii. $I_E=\mathcal{F}[v_E]$ and $I_O=-i\mathcal{F}[u_O]$

Chapter 22

4a. $(\alpha+i2\pi\omega)^{-2}$ **4b.** $2(\alpha+i2\pi\omega)^{-3}$ **4c.** $6(\alpha+i2\pi\omega)^{-4}$ **4d.** $(\alpha-i2\pi t)^{-2}$

4e. $-(\alpha-i2\pi\omega)^{-2}$ **4f.** $2(\alpha-i2\pi\omega)^{-3}$ **4g.** $-i8\alpha\pi\omega\big(\alpha^2+4\pi^2\omega^2\big)^{-2}$

4h. $i8\alpha\pi t\big(\alpha^2+4\pi^2t^2\big)^{-2}$ **4i.** $\dfrac{1}{2}\pi^{-3}\omega^{-3}\left[(2\pi^2\omega^2-1)\sin(2\pi\omega)+2\pi\omega\cos(2\pi\omega)\right]$

5a. $-(2\pi)^2\omega e^{2\pi\alpha\omega}\,\mathrm{step}(-\omega)$ **5b.** $4\pi^3\omega^2 e^{2\pi\alpha\omega}\,\mathrm{step}(-\omega)$ **5c.** $(2\pi)^2 t e^{-2\pi\alpha t}\,\mathrm{step}(t)$

5d. $4\pi^3 t^2 e^{-2\pi\alpha t}\,\mathrm{step}(t)$ **5e.** $(2\pi)^2\omega e^{-2\pi\alpha\omega}\,\mathrm{step}(\omega)$ **5f.** $4\pi^3\omega^2 e^{-2\pi\alpha\omega}\,\mathrm{step}(\omega)$

5g. $-i\dfrac{\pi^2\omega}{\alpha}e^{-2\pi\alpha|\omega|}$ **5h.** $i\dfrac{\pi^2 t}{\alpha}e^{-2\pi\alpha|t|}$

6c i. $(-1)^n n!(\alpha-i2\pi\omega)^{-n-1}$

6c ii. $-\dfrac{(-2\pi)^n}{(n-1)!}\omega^{n-1}e^{2\pi\alpha\omega}\,\mathrm{step}(-\omega)$ **6c iii.** $\dfrac{(2\pi)^n}{(n-1)!}\omega^{n-1}e^{-2\pi\alpha\omega}\,\mathrm{step}(\omega)$

7a. $2\left[(3+i2\pi\omega)\big(1+4\pi^2\omega^2\big)\right]^{-1}$ **9a.** $-2\,\mathrm{sinc}(2\pi\omega)\,\big(9+4\pi^2\omega^2\big)^{-1}$

9b. $\left[(1+i2\pi\omega)\big(3-i8\pi\omega-4\pi^2\omega^2\big)\right]^{-1}$

Chapter 23

6a. $\dfrac{\sqrt{\pi}}{2}\exp\!\left(-\dfrac{\pi^2}{4}\omega^2\right)$ **6b.** $\dfrac{\sqrt{\pi}}{2}\exp\!\left(-\dfrac{\pi^2}{4}t^2\right)$ **6c.** $\dfrac{1}{2}\exp\!\left(-\dfrac{\pi}{4}\omega^2\right)$

6d. $\sqrt{\dfrac{\pi}{\gamma}}\exp\!\left(-i6\pi\omega-\dfrac{\pi^2}{\gamma}\omega^2\right)$ **6e.** $\sqrt{\dfrac{\pi}{\gamma}}\exp\!\left(i6\pi t-\dfrac{\pi^2}{\gamma}t^2\right)$

6f. $\dfrac{\sqrt{\pi}}{3}\exp\!\left(-\dfrac{\pi^2}{9}(\omega-3)^2\right)$ **6g.** $i\dfrac{\sqrt{\pi}}{6}\left[\exp\!\left(-\dfrac{\pi^2}{9}(\omega+6)^2\right)-\exp\!\left(-\dfrac{\pi^2}{9}(\omega-6)^2\right)\right]$

6h. $-i\left(\dfrac{\pi}{\gamma}\right)^{3/2}\omega\exp\!\left(-\dfrac{\pi^2}{\gamma}\omega^2\right)$ **6i.** $\dfrac{1}{2\pi}\left(\dfrac{\pi}{\gamma}\right)^{3/2}\left[1-\dfrac{2\pi^2}{\gamma}\omega^2\right]\exp\!\left(-\dfrac{\pi^2}{\gamma}\omega^2\right)$

6j. $\left(4-i\dfrac{\omega}{8}\right)\exp\!\left(-i10\pi\omega-\dfrac{\pi}{4}\omega^2\right)$ **6k.** $\dfrac{1}{2}\exp\!\left(-i10\pi(\omega-3)-\dfrac{\pi}{4}(\omega-3)^2\right)$

7a. $e^{9\gamma}\sqrt{\dfrac{\pi}{\gamma}}\exp\left(-i6\pi\omega - \dfrac{\pi^2}{\gamma}\omega^2\right)$ **7b.** $\sqrt{\pi}\,\exp\left(25 - i\,10\pi\omega - \pi^2\omega^2\right)$

7c. $\sqrt{\dfrac{\pi}{3}}\exp\left(-i2\pi(2+6i)\omega - \dfrac{\pi^2}{3}\omega^2\right)$ **7d.** $\dfrac{1}{3}\exp\left(12\pi\omega - \dfrac{\pi}{9}\omega^2\right)$

10a. $\sqrt{\dfrac{\beta}{2}}e^{-\gamma\beta^2\omega^2}\left[\sqrt{1+\dfrac{\gamma}{|\lambda|}}\cos\left(\kappa\beta^2\omega^2\right) + \sqrt{1-\dfrac{\gamma}{|\lambda|}}\sin\left(\kappa\beta^2\omega^2\right)\right]$ with $\lambda = \gamma + i\kappa$ and $\beta = \dfrac{\pi}{|\lambda|}$

10b. $\sqrt{\dfrac{\beta}{2}}e^{-\gamma\beta^2\omega^2}\left[\sqrt{1-\dfrac{\gamma}{|\lambda|}}\cos\left(\kappa\beta^2\omega^2\right) - \sqrt{1+\dfrac{\gamma}{|\lambda|}}\sin\left(\kappa\beta^2\omega^2\right)\right]$ with $\lambda = \gamma + i\kappa$ and $\beta = \dfrac{\pi}{|\lambda|}$

Chapter 24

9a. e^{-2x} **9b.** does not exist   **9c.** $\dfrac{3}{10}\sin(x) - \dfrac{1}{10}\cos(x)$ **9d.** does not exist

9e. $\dfrac{1}{3}\left[(x+3)^3 - (x-3)^3\right] - 24$ **9f.** $\left[\dfrac{x}{20} - \dfrac{3}{5}\right](x+3)^4 - \left[\dfrac{x}{20} + \dfrac{3}{5}\right](x-3)^4$

9g. $\dfrac{\pi}{2} + \arctan(x)$ **9h.** $(3x+4)\sqrt{\pi}$ **9i.** $\sqrt{\pi}e^{4+4x}$ **9j.** $\dfrac{2\alpha}{\alpha^2 - \beta^2}e^{\beta x}$

10a. $\cos(x - \alpha) - \cos(x + \alpha)$ **10b.** No, because $\mathrm{pulse}_\pi * \sin(x) = 0$ **12b.** f is even.

12c i. $\sqrt{\pi}e^{-\pi^2}e^{i2\pi x}$ **12c ii.** $\sqrt{\pi}e^{-\pi^2}\cos(2\pi x)$

12c iii. $\sqrt{\pi}e^{-\pi^2}\sin(2\pi x)$ **12c iv.** $6(9 + \pi^2)^{-1}e^{i\pi x}$

12c v. $(1 + 36\pi^2)^{-1}[\cos(6\pi x) + 6\pi\sin(6\pi x)]$

15a. $\dfrac{1}{\beta - \alpha}\left[e^{-\alpha x} - e^{-\beta x}\right]\mathrm{step}(x)$ for $\beta \neq \alpha$; $xe^{-\beta x}\,\mathrm{step}(x)$ for $\beta = \alpha$

15b. $\dfrac{1}{\alpha + \beta}\left[e^{\beta x}\,\mathrm{step}(-x) + e^{-\alpha x}\,\mathrm{step}(x)\right]$ whenever $\beta > -\alpha$; does not exist if $\beta \le -\alpha$

16a. $(x+2)\,\mathrm{pulse}_2(x) + 4\,\mathrm{step}(x-2)$ **16b.** $\dfrac{1}{6}x^3\,\mathrm{step}(x)$

16c. $\dfrac{1}{3}\left[e^{5x} - e^{2x}\right]\mathrm{rect}_{(0,2)}(x) + \dfrac{1}{3}\left[e^6 - 1\right]e^{2x}\,\mathrm{step}(x-2)$

16d. $\dfrac{1}{6}e^{2x}\left[e^{6x} - e^{12}\right]\mathrm{rect}_{(2,3)}(x) + \dfrac{1}{6}e^{2x}\left[e^{18} - e^{12}\right]\mathrm{step}(x-3)$

16e. $\left[e^{5x} - e^{2x}\right]\mathrm{rect}_{(0,2)}(x) - e^{2x}\left[e^{6x} - e^{12} - e^6 + 1\right]\mathrm{rect}_{(2,3)}(x)$
$-e^{2x}\left[e^{18} - e^{12} - e^6 + 1\right]\mathrm{step}(x-3)$

16f. $\dfrac{1}{12}\left[4x + 3 + (x+3)(x-1)^3\right]\mathrm{rect}_{(0,2)}(x) + \dfrac{2}{3}x\,\mathrm{rect}_{(2,3)}(x)$
$+\dfrac{1}{12}\left[4x - 3 - (x+12)(x-4)^3\right]\mathrm{rect}_{(3,5)}(x)$

19a. $\pi\left[e^{4\pi\omega} - e^{8\pi\omega}\right]\mathrm{step}(-\omega)$ **19b.** $\dfrac{1}{8}\left[e^{5t}\,\mathrm{step}(-t) + e^{-3t}\,\mathrm{step}(t)\right]$

19c. $\dfrac{1}{36}\left[1 - e^{-6(t+3)}\right]\mathrm{pulse}_3(t) + \dfrac{1}{36}\left[e^{18} - e^{-18}\right]e^{-6t}\,\mathrm{step}(t-3)$

19d. $\dfrac{1}{4}(2 + \omega)\,\mathrm{rect}_{(-2,0)}(\omega) + \dfrac{1}{4}(2 - \omega)\,\mathrm{rect}_{(0,2)}(\omega)$

19e. $\dfrac{1}{42}e^{3\omega}\,\mathrm{step}(-\omega) + \left[\dfrac{1}{6}e^{-3\omega} - \dfrac{1}{7}e^{-4\omega}\right]\mathrm{step}(\omega)$

19f. $\omega e^{-\alpha\omega}\,\mathrm{step}(\omega)$ **19g.** $\left(\dfrac{1}{2\alpha}\right)^2\left[\dfrac{1}{\alpha} + |\omega|\right]e^{-\alpha|\omega|}$

19h. $\dfrac{1}{2\pi}[\arctan(t+1) - \arctan(t-1)]$

20a. $\dfrac{1}{4}\left[e^{4(t-1)} - e^{4(t+1)}\right]\mathrm{step}(-1-t) + \dfrac{1}{4}\left[e^{4(t+1)} - 1\right]\mathrm{pulse}_1(t)$

20b. $-\dfrac{1}{36}e^{3t}\,\mathrm{step}(-t) - \dfrac{1}{36}(1 + 6t)e^{-3t}\,\mathrm{step}(t)$

Chapter 25

5. $\alpha \ge \sqrt{10}$ **11a.** $(3 - i2\pi)^{-1}e^{i2\pi x}$ **11b.** $(3 + i2\pi)^{-1}e^{i2\pi x}$ **11c.** $\dfrac{1}{6}e^{-3|x|}$

13a. $(2\beta)^{-1}$ **13b.** $\dfrac{1}{2}\left[1 - e^{-2\pi}\right]$ **13c.** 0   **13d.** $(2\pi^2)^{-1}$ **13e.** $^1\!/_2$ **13f.** $^1\!/_8$

13g. 0 **15a.** $E = 2(2 + \epsilon)^{-1}$ and $(\Delta x)^2 = (2 + \epsilon)\epsilon^{-1}$ **16a.** $(\Delta t)^2 = 1/3$, $\Delta \omega = \infty$
16b. $(\Delta t)^2 = 1/16\pi$, $(\Delta \omega)^2 = 1/\pi$ **16c.** $\Delta t = \pi$, $\Delta \omega = (2\pi)^{-2}\sqrt{2}$

Chapter 26

1b. no **13a.** $1/3$ **13b.** $|1/b|$ **13c.** $c = |1/b|$, $\gamma = a/b$

14a. δ_4 **14b.** $\frac{1}{2}[\delta_3 + \delta_{-3}]$ **14c.** $\frac{i}{2}[\delta_{-3} - \delta_3]$ **14d.** $\frac{1}{4}[2\delta - \delta_6 - \delta_{-6}]$

14e. $\frac{1}{2}[\delta_6 + \delta]$ **14f.** $\delta(\omega) + 2 - 2(1 + 4\pi^2\omega^2)^{-1}$

15a. 0 **15b.** 9 **15c.** 8 **15d.** x^2 **15e.** $(x - 3)^2$ **15f.** 0
15g. $9\delta_3(x)$ **15h.** 9 **15i.** 0

16a. $\frac{1}{2}\delta(\omega) + \sum\limits_{\substack{k=-\infty \\ k \neq 0}}^{\infty} \frac{i}{2\pi k}[(-1)^k - 1]\delta\left(\omega - \frac{k}{2}\right)$ **16b.** $\sum\limits_{k=-\infty}^{\infty} \frac{1 + i2\pi k}{1 + 4\pi^2 k^2}[e - 1]\delta_k$

16c. $\frac{1}{2}\delta(\omega) + \sum\limits_{\substack{k=-\infty \\ k \neq 0}}^{\infty} \frac{1}{k^2\pi^2}[(-1)^k - 1]\delta\left(\omega - \frac{k}{2}\right)$ **16d.** $\sum\limits_{\substack{k=-\infty \\ k \neq 0}}^{\infty} (-1)^k \frac{i}{k\pi}\delta\left(\omega - \frac{k}{2}\right)$

16e. $\sum\limits_{k=-\infty}^{\infty} \frac{2}{\pi(1 - 4k^2)}\delta_{2k}$ **18a.** $-\sum\limits_{n=-\infty}^{\infty} n^3 e^{i6\pi n\omega}$ **18b.** $\sum\limits_{n=-\infty}^{\infty} \frac{4n^2}{1 + n^2}e^{i\pi n\omega}$

20. $F.S.[f]|_t = \sum\limits_{n=-\infty}^{\infty} \frac{1}{2}[1 - (-1)^n]e^{i\pi t}$, $\mathcal{F}[f] = \sum\limits_{n=-\infty}^{\infty} \frac{1}{2}[1 - (-1)^n]\delta_{n/2}$

23a. $f'(x) = Df(x) = \text{step}(x)$
23b. $g'(x) = 3x^2 \text{rect}_{(-1,2)}(x)$, $Dg(x) = 3x^2 \text{rect}_{(-1,2)}(x) - \delta_{-1}(x) - 8\delta_2(x)$
23c. $h'(x) = \text{step}(-x - 1) + 2x \text{rect}_{(-1,2)}(x) + 3x^2 \text{step}(x - 2)$,
$\quad Dh(x) = \text{step}(-x - 1) + 2x \text{rect}_{(-1,2)}(x) + 3x^2 \text{step}(x - 2) + 2\delta_{-1}(x) + 4\delta_2(x)$

23d. $f'(x) = 0$, $Df = \sum\limits_{k=-\infty}^{\infty}[\delta_{2k} - \delta_{2k+1}]$

24a. $-3e^{-3x} \text{step}(x) + \delta(x)$ **24b.** $3e^{3x} \text{step}(2 - x) - e^6\delta_2(x)$ **24c.** $\text{rect}_{(1,3)} + \delta_1 - 3\delta_3$
24d. $-i2\pi(3 + i2\pi x)^{-2} \text{pulse}_3(x) + (3 - i6\pi)^{-1}\delta_{-3}(x) - (3 + i6\pi)^{-1}\delta_3(x)$

24e. $\sum\limits_{k=-\infty}^{\infty}[\delta_{2k} - \delta_{2k+1}]$

25a. $6x \text{step}(2 - x) - 12\delta_2(x) - 8D\delta_2(x)$ **25b.** $2 \text{pulse}_2 - 4\delta_{-2} - 4\delta_2$
26a. $\frac{i}{2\pi}[\delta(\omega) - 3e^{-3\omega} \text{step}(\omega)]$ **26b.** $\frac{1}{4\pi}[D\delta_3 - D\delta_{-3}]$ **26c.** $\frac{-1}{4\pi^2}D^2\delta + i\frac{3}{2\pi}D\delta - 4\delta$
26d. $-4\pi^2\omega^2 e^{-i6\pi\omega}$ **27a.** $e^{-3t} \text{step}(t)$ **27b.** $-\frac{1}{6}e^{-3|t|}$ **27c.** $\frac{1}{2}\left[e^t - e^{3t}\right] \text{step}(-t)$

Chapter 27

6a. $H(\omega) = (i2\pi\omega - 6)^{-1}$, $h(t) = -e^{6t} \text{step}(-t)$ **6b i.** $-6e^{2t}$
6b ii. $-3\sin(2t) - \cos(2t)$ **6b iii.** $-e^{6t} \int_t^{\infty} e^{-6s} \text{sinc}(s)\, ds$
6b iv. $\text{step}(t) + e^{6t} \text{step}(-t)$
7a. $H(\omega) = \left[(i2\pi\omega)^2 + 7(i2\pi\omega) + 10\right]^{-1}$, $h(t) = \frac{1}{3}\left[e^{-2t} - e^{-5t}\right] \text{step}(t)$
7b. $H(\omega) = \left[(i2\pi\omega)^2 - 10(i2\pi\omega) + 29\right]^{-1}$, $h(t) = -\frac{1}{2}e^{5t} \sin(2t) \text{step}(-t)$
7c. $H(\omega) = (5 + i2\pi\omega)^{-2}$, $h(t) = te^{-5t} \text{step}(t)$ **8.** $h = \delta_\alpha$, $H(\omega) = e^{-i2\pi\alpha\omega}$

9c i. $6\sin(4\pi t)$ **9c ii.** 0 **9c iii.** $54\,\mathrm{sinc}(9\pi t)$ **9c iv.** $3 - \sum_{n=1}^{4} \frac{6}{n\pi}\sin(2\pi nt)$

10b. $h(t)=1$, $H=\delta$ **10c.** no

Chapter 28

1a. $2\,\mathrm{sinc}(2\pi\omega)\delta(\omega+\nu)$ **1b.** $\pi e^{-2\pi|\omega|}\delta(\nu-\omega)$ **1c.** $\delta(2\omega-\nu)$ **1d.** $e^{-i2\pi\omega\nu}$

1e. $e^{-i2\pi\omega\nu}e^{-3|\nu|}$ **1f.** $\dfrac{e^{i2\pi\omega\nu}}{3+i2\pi\nu}$ **1g.** $e^{-3\omega}\,\mathrm{step}(\omega)\delta_{-1}(\nu)$ **1h.** $\nu e^{(-3-i2\pi\omega)\nu}\,\mathrm{step}(\nu)$

1i. $\dfrac{i}{2\pi}\left[\delta(\nu)-(3+i2\pi\omega)e^{(-3-i2\pi\omega)\nu}\,\mathrm{step}(\nu)\right]$ **1j.** $\pi e^{-\pi^2(\omega^2+\nu^2)}$

1k. $e^{-3(\omega+1)}\,\mathrm{step}(\omega+1)\delta(\nu)$ **1l.** $6\delta(\omega)\,\mathrm{sinc}(6\pi\nu)$ **2a.** $6\,\mathrm{sinc}(6\pi\omega)\delta(\nu+\omega)\delta(\mu-\omega)$

2b. $e^{-i2\pi\omega\nu}\,\mathrm{pulse}_3(\nu)\delta(\mu+\nu)$ **2c.** $\frac{1}{8}\exp\!\left(\frac{\pi}{4}\left(\omega^2+\nu^2+\mu^2\right)\right)$

2d. $\dfrac{1}{\pi^2\left(1+\omega^2\right)\left(1+\nu^2\right)}\delta(\omega+\nu+\mu)$ **3.** $\frac{a}{\rho}J_1(2\pi a\rho)$

Chapter 29

2. all except e^{x^2} **10a.** $f(x)=3x$ **10b.** $g(x)=\cos(x)$ **10c.** $h(x)=\mathrm{step}(x)$
12a. $\alpha\geq 0$ **12b.** $\alpha>-1$ **14a.** $f(x)=4x$ **14b.** $g(x)=e^{-2\pi x}$ **18b.** π

Chapter 32

4b. $\psi(s)=\dfrac{1}{|a|}\phi\!\left(\frac{1}{a}s\right)$ **4c.** $\psi(s)=-\phi'(s)$
12a. in $\mathcal{G}$ **12b.** in $\mathcal{G}$ **12c.** in $\mathcal{G}$ **12d.** not in $\mathcal{G}$ **12e.** in $\mathcal{G}$ **12f.** not in $\mathcal{G}$
12g. in $\mathcal{G}$ **12h.** not in $\mathcal{G}$ **12i.** in $\mathcal{G}$ **12j.** not in $\mathcal{G}$ **12k.** in $\mathcal{G}$ **12l.** in $\mathcal{G}$
14a. $f(z^*)=u(x,-y)+iv(x,-y)$, $f^*(z)=u(x,y)-iv(x,y)$
  and $f^*(z^*)=u(x,-y)-iv(x,-y)$

Chapter 33

21a. yes **21b.** yes **21c.** no **21d.** yes **21e.** no **21f.** yes
23a. yes, it is exponentially integrable **23b.** yes **23c.** yes **23d.** yes **23e.** no
23f. yes **23g.** yes **23h.** yes **23i.** no **23j.** yes
26a. no **26b.** yes

Chapter 34

5a. 1 **5b.** 1 **5c.** 0 **5d.** $\mathrm{sgn}(x)$ (i.e., $\mathrm{step}(x)-\mathrm{step}(-x)$)
6a. δ **6b.** 0 **6c.** $\frac{1}{2}\delta$ **6d.** $2\pi\delta$ **6e.** $-2\pi\delta$
10b i. $h(x)=x^2$ **10b ii.** $h(x)=\cos(\pi x)$ **10b iii.** $g(x)=\cos(\pi x/\Delta x)$

Chapter 35

12. $\mathcal{F}^{-1}[\sin(2\pi at)]=\frac{i}{2}\left[\delta_a-\delta_{-a}\right]$ and $\mathcal{F}^{-1}[\cos(2\pi at)]=\frac{1}{2}\left[\delta_a+\delta_{-a}\right]$
32a. $2\,\mathrm{sinc}(2\pi y)$ **32b.** 4δ **32c.** $4\delta_5$ **32d.** $4\delta_{-5}$
32e. $e^{-i8\pi y}$ **32f.** $2\delta-\delta_2-\delta_{-2}$
33a. $2\,\mathrm{sinc}(2\pi y)$ **33b.** 4δ **33c.** $4\delta_{-5}$ **33d.** $4\delta_5$
33e. $e^{i8\pi y}$ **33f.** $2\delta-\delta_2-\delta_{-2}$
34a. $(i2\pi)^{-1}[\delta(\omega)-8e^{8\omega}\,\mathrm{step}(-\omega)]$ **34b.** $i(2\pi)^{-1}[\delta(\omega)-8e^{-8\omega}\,\mathrm{step}(\omega)]$ **37a.** δ_{-i}
37b. $e^{(2-3i)2\pi y}$ **37c.** $\frac{1}{2}[\delta_{-3i}+\delta_{3i}]$ **37d.** $\frac{1}{2}[\delta_{-3i}-\delta_{3i}]$ **38a.** $(2i)^{-1}[\delta_{\alpha-i\beta}-\delta_{-\alpha-i\beta}]$
38b. $2^{-1}[\delta_{\alpha-i\beta}+\delta_{-\alpha-i\beta}]$ **38c.** $\delta_{-i\alpha}(\omega)-(2\pi\alpha-i2\pi\omega)^{-1}$

38d. $(2i)^{-1}[\delta_{4-i\alpha}(\omega) - \delta_{-4-i\alpha}(\omega)] + (4\pi)^{-1}[(i\alpha + 4 + \omega)^{-1} - (i\alpha - 4 + \omega)^{-1}]$

38e. $\delta_{i\alpha}(\omega) - (2\pi\alpha + i2\pi\omega)^{-1}$ **38f.** $\delta_{i\alpha}(\omega) + \delta_{-i\alpha}(\omega) - \alpha(\pi\alpha^2 + \pi\omega^2)^{-1}$

39a. $f'(x) = Df(x) = \text{step}(x)$

39b. $g'(x) = 3x^2 \text{rect}_{(-1,2)}(x)$, $Dg(x) = 3x^2 \text{rect}_{(-1,2)}(x) - \delta_{-1}(x) - 8\delta_2(x)$

39c. $Dh(x) = \text{step}(-x - 1) + 2x \text{rect}_{(-1,2)}(x) + 3x^2 \text{step}(x - 2) + 2\delta_{-1}(x) + 4\delta_2(x)$

40a. $(i2\pi x)^n e^{-i2\pi ax}$ **40b.** $(-i2\pi x)^n e^{i2\pi ax}$ **40c.** $(-i2\pi)^{-n} D^n \delta$ **40d.** $(i2\pi)^{-n} D^n \delta$

41a. $(i2\pi)^{-2} D^2 \delta + i2(\pi)^{-1} D\delta - 5\delta$ **41b.** $i(2\pi)^{-1} D\delta_{-3i}$

41c. $(4\pi)^{-1}[D\delta_3 - D\delta_{-3}]$ **41d.** $i(8\pi^2)^{-1}[D^2\delta_3 - D^2\delta_{-3}]$

42a i. $(i2\pi)^{-n-1} n! T_{-i}\left[(y - i)^{-n-1}\right]$ **42a ii.** $-(i2\pi)^{-n-1} n! T_i\left[(y + i)^{-n-1}\right]$

42b i. $-(2\pi)^{-2} T_{-i}\left[(y - i)^{-2}\right]$ **42b ii.** $(-i2\pi)^{-n-1} n! T_i\left[(y + i)^{-n-1}\right]$

42b iii. $(i2\pi)^{-n-1} n!\left(T_{-i}\left[(y - i)^{-n-1}\right] - (-1)^n T_i\left[(y + i)^{-n-1}\right]\right)$

43. $c = i2\pi$ **44a.** 1   **44b.** $(i2\pi)^{-1}$ **44c.** 1   **44d.** 1

45a. $\mathcal{F}[\text{sgn}] = 2\mathcal{F}[\text{step}] - \delta$ **45b.** $(i\pi)^{-1}$ **48b i.** $\frac{1}{\sqrt{\alpha}} \exp\left(-i\pi\left[\frac{\omega^2}{\alpha} - \frac{1}{4}\right]\right)$

48b ii. $\frac{1}{\sqrt{\alpha}} \exp\left(i\pi\left[\frac{\omega^2}{\alpha} - \frac{1}{4}\right]\right)$ **48b iii.** $\exp\left(\mp i\pi\left(\omega^2 - \frac{1}{8}\right)^2\right)$

48c i. $\frac{1}{\sqrt{\alpha}} \cos\left(\pi\left[\frac{\omega^2}{\alpha} - \frac{1}{4}\right]\right)$ **48c ii.** $-\frac{1}{\sqrt{\alpha}} \sin\left(\pi\left[\frac{\omega^2}{\alpha} - \frac{1}{4}\right]\right)$

Chapter 36

11a. $-i\delta(x)$ **11b.** $-i\delta_i(x)$ **11c.** $\text{sinc}(x)$ **11d.** 0   **11e.** $\sum_{k=-\infty}^{\infty} (-1)^k \delta_k$

12b. $(2\pi)^{-1} \sinh(2\pi)\delta_i$ **17a.** x^2 **17b.** $(x - 3)^2$ **17c.** $3^2\delta_3(x)$

20a. 0   **20b.** 9   **20c.** 8   **20d.** 9   **20e.** 0   **20f.** 4   **20g.** 15

Chapter 37

11a. $p = 3$, $\mathcal{F}^{-1}[f]|_t = \sum_{k=-\infty}^{\infty} \frac{k}{1 + k^2} e^{i2\pi(k/3)t}$, $\mathcal{F}[f]|_\omega = \sum_{k=-\infty}^{\infty} \frac{-k}{1 + k^2} e^{i2\pi(k/3)\omega}$

11b. $p = 1/4$, $\mathcal{F}^{-1}[g]|_t = \sum_{k=-\infty}^{\infty} \frac{k^2}{1 + k^2} e^{i8k\pi t}$

11c. $p = 1$, $\mathcal{F}^{-1}[h]|_t = \sum_{k=-\infty}^{\infty} (-k)^k e^{i2\pi kt}$

12a. $F.S.[f]|_t = \frac{1}{2} + \sum_{\substack{k=-\infty \\ k\neq 0}}^{\infty} \frac{i}{2\pi k}\left[(-1)^k - 1\right] e^{i2\pi(k/2)t}$,

$$\mathcal{F}[f] = \frac{1}{2}\delta + \sum_{\substack{k=-\infty \\ k\neq 0}}^{\infty} \frac{i}{2\pi k}\left[(-1)^k - 1\right]\delta_{k/2} \qquad \textbf{12b.} \ \mathcal{F}[g] = \sum_{k=-\infty}^{\infty} \frac{1 + i2\pi k}{1 + 4\pi^2 k^2}[e - 1]\delta_k$$

12c. $\mathcal{F}[\text{evensaw}] = \frac{1}{2}\delta + \sum_{\substack{k=-\infty \\ k\neq 0}}^{\infty} \frac{1}{k^2\pi^2}\left[(-1)^k - 1\right]\delta_{k/2}$

12d. $\mathcal{F}[\text{oddsaw}] = \sum_{\substack{k=-\infty \\ k\neq 0}}^{\infty} (-1)^k \frac{i}{k\pi}\delta_{k/2}$ **12e.** $\mathcal{F}[f] = \frac{1}{3}\delta + \sum_{\substack{k=-\infty \\ k\neq 0}}^{\infty} (-1)^k \frac{2}{k^2\pi^2}\delta_{k/2}$

13a. $c_k = \frac{1}{2}\left[1 - (-1)^k\right]$, $F.S.[f]|_t = \sum_{k=-\infty}^{\infty} \frac{1}{2}\left[1 - (-1)^k\right] e^{i\pi kt}$,

$\mathcal{F}[f] = \sum_{k=-\infty}^{\infty} \frac{1}{2}\left[1 - (-1)^k\right]\delta_{k/2}$ **13b.** $c_k = \frac{1}{4}\left[(-i)^k + 2(-1)^k + 3(i)^k\right]$

Chapter 38

15a. $(x - 3i)e^{4x}$ **15b.** 0 **15c.** $-3i\delta(x)$ **15d.** does not exist

16a. $a\delta_{i2\pi}$ **16b.** $a\delta_4$ **16c.** $a\delta_4 + bD\delta_4 + cD^2\delta_4$ **16d.** $a\delta + bD\delta + c\delta_1$

16e. $a\delta_{-2} + b\delta_{-3}$ **16f.** $a\delta_1 + b\delta_{-1} + c\delta_i + d\delta_{-i}$ **16g.** $a\delta_1 + bD\delta_1 + cD^2\delta_1 + dD^3\delta_1$

16h. $a\delta_{3+4i} + b\delta_{3-4i}$ **17a.** $u_0(x) = (x - i2\pi)^{-1}$, $u(x) = (x - i2\pi)^{-1} + a\delta_{i2\pi}(x)$

17b. $u_0(x) = -D\delta_{i2\pi}(x)$ **17c.** $u_0(x) = (x^2 + 1)^{-1}$, $u(x) = (x^2 + 1)^{-1} + a\delta_i(x) + b\delta_{-i}(x)$

17d. $u_0(x) = \frac{1}{6}\delta$ **17e.** $u_0(x) = T_{-i}\left[\frac{1}{(x + 2 - i)(x + 3 - i)}\right]$

17f. $u_0(x) = (x + 2)T_{-i}\left[\frac{1}{x + 3 - i}\right]$

18a i. $\left(\frac{-i}{2\pi}\right)^k 2(k - 1)!\,\mathrm{pole}^k$ **18a ii.** $\left(\frac{i}{2\pi}\right)^k 2(k - 1)!\,\mathrm{pole}^k$

18a iii. $\left(\frac{-i}{2\pi}\right)^{k+1}\left[k!\,\mathrm{pole}^{k+1} + i(-1)^k\pi D^k\delta\right]$

18a iv. $\left(\frac{-i}{2\pi}\right)^{k+1}\left[-k!\,\mathrm{pole}^{k+1} + i(-1)^k\pi D^k\delta\right]$ **18b i.** $\frac{-1}{4\pi^2}\left[\mathrm{pole}^2 - i\pi D\delta\right]$

18b ii. $\frac{-1}{2\pi^2}\mathrm{pole}^2$ **18b iii.** $\left(\frac{-i}{2\pi}\right)^{k+1}\left[\left[1 - (-1)^k\right]k!\,\mathrm{pole}^{k+1} + i\left[1 + (-1)^k\right](-1)^k\pi D^k\delta\right]$

19a. $\mathrm{pole}_{-2}(x) - \mathrm{pole}_{-3}(x)$ **19b.** $\mathrm{pole}_1(x) - \mathrm{pole}(x) - \mathrm{pole}^2(x)$

19c. $\frac{i}{6}[\mathrm{pole}_{-3i} - \mathrm{pole}_{3i}]$ **19d.** $\frac{i}{8}[\mathrm{pole}_{3-4i} - \mathrm{pole}_{3+4i}]$

21b i. $-\pi^2\delta$ **21b ii.** $\pi^2 D\delta$ **21b iii.** $\frac{\pi x}{1 + x^2}$

22a. $e^{-3t}\,\mathrm{step}(t) + ce^{-3t}$ **22b.** $\mathrm{ramp}(t) + a + bt$

Chapter 39

1. 5

5a. $[1]$ **5b.** $\begin{bmatrix} 1 & 1 \\ 1 & -1 \end{bmatrix}$ **5c.** $\frac{1}{2}\begin{bmatrix} 2 & 2 & 2 \\ 2 & -1 - i\sqrt{3} & -1 + i\sqrt{3} \\ 2 & -1 + i\sqrt{3} & -1 - i\sqrt{3} \end{bmatrix}$

5d. $\begin{bmatrix} 1 & 1 \\ 1 & -1 \end{bmatrix}$ **5e.** $\frac{1}{2}\begin{bmatrix} 2 & 2 & 2 \\ 2 & -1 + i\sqrt{3} & -1 - i\sqrt{3} \\ 2 & -1 - i\sqrt{3} & -1 + i\sqrt{3} \end{bmatrix}$

6a. period $= {}^8\!/_3$ **6b.** $\Delta\omega = {}^3\!/_8$, $P = {}^3\!/_2$

6c. $F_0 = 3$, $F_1 = \frac{1}{4}(3 + 9i)$, $F_2 = 0$, $F_3 = \frac{1}{4}(3 - 9i)$, $F_4 = 3$, $F_{-1} = \frac{1}{4}(3 - 9i)$

7a. $3\,\mathrm{comb}_3$ **7b.** index period $= 2$, $F = \cdots + 0\delta_0 + \delta_{1/2} + \cdots$

7c. index period $= 3$, $F = \frac{1}{6}(\cdots + 4\delta_0 + \delta_{1/6} + \delta_{2/6} + \cdots)$

7d. index period $= 3$, $F = i\frac{2}{\sqrt{3}}(\cdots + 0\delta_0 - \delta_{2/3} + \delta_{4/3} + \cdots)$

Chapter 40

1. $\{0, {}^1\!/_4, 1\}$ **2.** $(-{}^1\!/_4, {}^5\!/_4)$ **4.** $\cdots + 0\delta_0 + \frac{1}{8}\delta_{1/2} + \frac{1}{2}\delta_1 + \cdots$

5b. $N \geq 12$ **5c.** 195.02 **6a.** 0.11 **6b.** $N \geq 12$ **6c.** none , $|\omega| \leq 6$ (approx.)

7. $\widehat{G}_n = \Delta\omega\, G_n$ **8.** $F_n = F(n\Delta\omega)$, $\widehat{F}_n = \Delta\omega\, F_n$ **9b.** No, just for "relatively small" n .

9c. $F(-\Delta\omega) \approx G_{N-1}$ **11.** $\mathscr{F}_4^{-1} = \dfrac{1}{2}\begin{bmatrix} 1 & 1 & 1 & 1 \\ 1 & i & -1 & -i \\ 1 & -1 & 1 & -1 \\ 1 & -i & -1 & i \end{bmatrix}$, $\{10, -2-2i, -2, -2+2i\}$

15a. $\left\{V_0,\ e^{-i\frac{2\pi}{N}M\cdot 1}V_1,\ e^{-i\frac{2\pi}{N}M\cdot 2}V_2,\ \ldots,\ e^{-i\frac{2\pi}{N}M(N-1)}V_{N-1}\right\}$

15b. $\{v_{0+M},\ v_{1+M},\ v_{2+M},\ \ldots,\ v_{N-1+M}\}$ **19.** $^2/_{100}$, $^6/_{1000}$, $^4/_{100,000}$ **20a.** $\{0,\ ^2/_3,\ ^4/_3,\ 2\}$

20b. $\{0,\ ^4/_3,\ ^8/_3,\ 4\}$ **20c.** $\{25, 16, 9, 4\}$ **20d.** $\{25, 9, 1, 1\}$ **20e.** $\{1, 1, 1, 1\}$

20f. $\{0, 0, 1, 1\}$

22a i. $\{2, 0, 0, 0\}$ **22a ii.** $\{0, 0, 2, 0\}$ **22b i.** $\{3, 0, 0, 0\}$ **22b ii.** $\{0, 0, 0, 3\}$

23a. $\left\{^1/_2,\ ^1/_2,\ ^1/_2,\ ^1/_2\right\}$ **23b.** $\left\{^1/_2,\ -^i/_2,\ -^1/_2,\ ^i/_2\right\}$ **23c.** $\{2, 0, 0, 0\}$

23d. $\{3,\ 1-i,\ 1,\ 1+i\}$ **23e.** $\{0, 0, 2, 0\}$

24b i. $\left\{\dfrac{1}{\sqrt{3}},\ \dfrac{1}{\sqrt{3}},\ \dfrac{1}{\sqrt{3}}\right\}$ **24b ii.** $\left\{\dfrac{1}{\sqrt{3}},\ \dfrac{-\sqrt{3}-3i}{6},\ \dfrac{-\sqrt{3}+3i}{6},\ \right\}$ **24b iii.** $\left\{\sqrt{3},\ 0,\ 0\right\}$

24b iv. $\left\{2\sqrt{3},\ \dfrac{\sqrt{3}-i}{2},\ \dfrac{\sqrt{3}+i}{2},\ \right\}$ **24b v.** $\left\{\dfrac{1}{\sqrt{3}},\ \dfrac{1+i\sqrt{3}}{\sqrt{3}},\ \dfrac{1-i\sqrt{3}}{\sqrt{3}},\ \right\}$

25b i. $\left\{\dfrac{1}{\sqrt{8}},\ \dfrac{1}{\sqrt{8}},\ \dfrac{1}{\sqrt{8}},\ \dfrac{1}{\sqrt{8}},\ \dfrac{1}{\sqrt{8}},\ \dfrac{1}{\sqrt{8}},\ \dfrac{1}{\sqrt{8}},\ \dfrac{1}{\sqrt{8}}\right\}$

25b ii. $\left\{\dfrac{1}{\sqrt{8}},\ \dfrac{1-i}{4},\ \dfrac{-i}{\sqrt{8}},\ \dfrac{-1-i}{4},\ \dfrac{-1}{\sqrt{8}},\ \dfrac{-1+i}{4},\ \dfrac{i}{\sqrt{8}},\ \dfrac{1+i}{4}\right\}$

25b iii. $\left\{\sqrt{8}, 0, 0, 0, 0, 0, 0, 0\right\}$ **25b iv.** $\left\{0, 0, 0, 0, \sqrt{8}, 0, 0, 0\right\}$

33. $3N^2$ (includes index arithmetic)   **34b.** $9N\log_2 N + N$

35. The alternative procedure should be over 33 times faster.

Index

$\mathcal{A}$, space of absolutely integrable functions, 261, 268, 289
addition, generalized, 548, 549, 551, 552
adjoint identity, 614
adjoints, 614, 615
algebraic equations (simple), 677
  general solutions, 677
  homogeneous, 677
    general solutions, 677, 679, 680, 684
  nonhomogeneous, 675, 677, 689–693
  pole function solutions, 704
analytic extensions, 71
analyticity, 67–71
  and Taylor series, 69
  test for, 67
arrays of delta functions, 429, 430, 579–582, 636
  coefficients, 429, 579, 709
    exponentially bounded, 582
  in integrals, 642, 643
  periodic, 431, 709
  regular, 429, 580, 709
    Fourier transforms of, 653–655
    relation with periodic functions, 428, 655
  spacing, 429, 580, 709
asymmetric convolution, 644
asymmetric product, 645
auto-correlation, 401, 402
  Fourier transform identities, 401

bandwidth, 311
  bounds, 311
    effective, 312, 316, 409
    theorem, 312, 316, 409, 411
Bessel functions, 469–470
Bessel's equality
  discrete, 759
  for Fourier series, 171, 173, 174, 246
  for Fourier transforms, 406
Bessel's inequality, 139, 142, 152
boundedness, 15
  exponential, 477
    on strips, 599

functional, 560
of Fourier integral transforms, 288
operational, 531, 532

$\mathbb{C}$, the complex plane, 57
Cauchy–Riemann equations, 67
$\mathrm{comb}_p(x)$, comb functions, 431, 581, 650, 659–661
  shifted, 661, 662
complex conjugation
  analytic, 527, 536
    generalized, 621, 622
  Fourier transform identities
    classical, 332
    discrete, 758
    generalized, 622
complex exponential function, 61–64
  relation with sines and cosines, 62–64
complex numbers, 57
  absolute value/magnitude/modulus, 58
  argument/phase/polar angle, 58
    principal, 58
  conjugate, 57
  parts, real and imaginary, 57
  polar form, 58
computer math packages, 115, 119, 120, 128
constants, 543
continuity, 60, 74
  at a point, 15, 60, 66
  functional, 540, 559–561, 620
  of convolution, 388, 389
  of Fourier integral transforms, 288, 290, 292
  of functions defined by integrals, 80–81, 84–88, 271, 272
  on intervals, 18, 60
  operational, 451, 452, 531, 620
  over a partitioning, 21
  piecewise, 20, 76–78
  sectional, 20
  uniform, 19, 74–76
    alternate definition, 20

convergence of Fourier series
  generalized, 584, 655
  in energy, 169
  in norm, 169–171, 173, 174, 246
  limited uniform, 167, 168, 186–189, 195
  mean-squared, 170
  nonuniform, 163–165
  pointwise, 156, 157, 174, 179
  strong, 239, 240
  uniform, 162–165, 206–209, 216
convergence of generalized sequences, 567, 568,
     613, 635
convergence of infinite series, 41
  absolute, 42
  conditional, 42
  generalized, 574–577, 585, 614, 635
  strong, 234–236
  symmetric, 46
  tests for, 43, 44
  two-sided, 46, 48
convolution, 377, 536
  alternate definition, 379
  and Fourier series, 395
  and LSI systems, 454, 455
  associativity, 381, 382, 398, 638
    myth of, 381
  asymmetric, 644
  discrete, 745–747, 759
    with the fast Fourier transform, 760
  for Laplace transform, 379
  Fourier transform identities
    classical, 393, 394, 507
    discrete, 746, 747
    generalized, 633
    multi-dimensional, 465
  generalized, 627, 636–639, 643–646
    compared with classical, 632
    rules for, 632–635
  integral, 377
  multi-dimensional, 465
correlation, 399
  discrete, 747
  Fourier transform identities
    classical, 400, 401
    discrete, 747
  relation with convolution, 400
$\cos(x)$, cosine function, 53, 54
  rectified, 109, 117
  tuncated, 335
curves of discontinuity, 76

de-modulation, 326
delta functions, 12, 541, 577–579, 756, 757
  and identity sequences, 423, 577–579, 625
  as a limit, 421–422

Fourier transforms of, 424, 590, 592, 602
  graphically visualized, 422–423
  in convolution, 424
  in integrals, 425, 426, 642
  in products, 425, 551, 631, 632, 677, 679, 680
  translated, 421, 597, 598
  visualizing, 421–423
  working definition, 420
derivative, 22, 60
  classical, 432
  complex, 67
  generalized, 433, 605
    elementary identities, 609
    higher order, 440, 605
    relation with classical, 608
differentiability, 22, 60, 67
  higher order / infinite, 24
  infinite, 25
  of functions defined by integrals, 81, 88, 89,
    273, 274, 554
    generalized, 556, 557
  of strongly convergent series, 236
differential equations, 343, 676
  as LSI systems, 452, 453, 457
differentiation, 432
  and convolutions, 390–392, 443, 639
  and Fourier series, 203–205, 215–216
  Fourier transform identities
    classical, 341, 342, 349, 350, 358, 495, 499,
    500
    generalized, 442, 611, 612
    misued, 344, 345
differention
  Fourier transform identities
    multi-dimensional, 465
Dirichlet kernel, 181, 182
discontinuity
  bad, 18
  jump, 17
  removable/trivial, 16, 17
discrete approximations, 727
  error in using, 732
  Fourier transforms of, 728–730
    and discrete transforms, 741
    error in, 734–737
  order and spacing, 727
  relation with samplings, 727
discrete Fourier transforms, 737–739
  alternate definitions, 739, 740
  and discrete approximations, 741
  and transforms of periodic arrays, 740
  component formulas, 738
  efficiency of, 751, 752, 754
    compared with FFTs, 755, 756

matrix formulas, 737, 738
   recursion formulas in, 748, 749
divergent Fourier series, 196–200
domain
  of a function, 8, 11
  of a transform, 13
duration, 310
  bounds, 311
   effective, 312, 316, 408, 409, 411
  interval of, 310

equality of functions, 11, 21
  tests for, 472, 491, 498, 511, 520, 521
equality of generalized functions, 547, 548
  tests for, 557, 559
errors, inexcusable, 329, 330
Euler's formula, 62
evaluation functionals, 538, 539, 541
exponential
  boundedness, 477
  integrability, 478
  order, 487

$f(x)$, meaning of, 9, 546, 547
fast Fourier transforms, 747, 750
  efficiency of, 753–755
   compared with DFTs, 755, 756
  first level algorithm, 749, 750
   efficiency of, 752, 754
  radix 2, 750
  radix 3, 751
Fejér, Leopold, 200
filter, low-pass, 462
Fourier coefficients
  complex exponential, 148, 149
  cosine, 125
  generalized, 135, 655, 656, 672
  sine, 124
  trigonometric, 101, 102, 108
Fourier integral transforms, 279–280
  alternative formulas, 286, 287
  as processes, 280, 281
Fourier integrals, 255, 279, 280
Fourier series
  and derivatives, 203–205, 215–216
  complex exponential, 147, 148
  cosine, 125, 215, 216
  divergent, 196–200
  generalized, 135, 651, 652, 654–656
  integrals of, 210–212
  partial sums, 115–117, 150
  sine, 124, 215, 216
  trigonometric, 101
   generalized, 672

Fourier transforms
  classical, 304, 305
  generalized, 587, 588
  integral formulas, 279, 280
  multi-dimensional, 463–464
   in polar coordinates, 468–470
  of functions in $\mathcal{A}$, 305
  of functions in $\mathcal{T}$, 299, 300
  real and imaginary parts, 333, 334
Fourier's bold conjecture, 4, 95, 155, 157, 160
frequency domain, 314
frequency response, 459
Fubini's theorem, 274
function values
  at interval endpoints, 21
  irrelevance at isolated points, 10
function, even and odd parts, 337, 338
functionals, 538, 540, 541
  and generalized functions, 540
  continuous, 540
  linear, 539
functions
  bandwidth limited, 312
  bounded, 15, 60, 406, 503
  classical, 541–543
  classically transformable, 302–304, 308–310,
   479, 503
   tests for non-transformability, 309, 310
  complex valued, 59–61
   absolute value/magnitude/modulus, 59
   conjugate, 59
   graphing, 59, 60
   parts, real and imaginary, 59, 66, 337, 338
  defined by an integral, 78–82
  differentiable, *see* differentiability
  duration limited, 312
  exponentially bounded, 477, 489, 496
   on strips, 646
  exponentially integrable, 478, 479, 491, 496,
   543, 544
   test for, 544
  finite energy/finite-normed, 403
  Gaussian-like, 368
  real valued, 8
  smooth, *see* smoothness
  spatially limited, 312
  square integrable, 403, 404, 406
  time limited, 312
  unbounded, 15
fundamental identity of Fourier analysis,
   412–415, 417, 494, 496, 503
fundamental theorem of calculus, 432
  generalized, 435, 641

$\mathcal{G}$, space of Gaussian test functions, 518, 553, 629, 630, 643

Gaussian functions, 264, 359, 493–495
  basic, 359, 360
    Fourier transforms of, 363
    integal of, 360, 361
  general, 364, 365
    Fourier transforms of, 366, 367

Gaussian test functions, 518, 519
  analyticity of, 524
  base, 515–517
  bounds on, 528, 530

generalized functions, 420, 540, 541, 545–547, 553
  as functionals, 540
  compared with classical functions, 541–543
  defined by functions, 538, 542–544
  formulas, 546, 547

geometric series, 42
  partial sum formula, 43, 180
  summation formula, 43

Gibbs phenomenon, 165–167, 190–196

$\mathcal{H}$, space of very rapidly decreasing test functions, 522, 553, 629, 630, 643

Haar wavelet set, 143

Hankel transforms, 470

harmonic series, 44, 47

heat equation, 219, 220

heat flow problem, 219, 220, 247, 248
  solution, 222, 243, 246

Heisenberg uncertainty principle, 312

hoppen, an example of what can happen, 329

identity sequences, 421, 473, 475
  and delta functions, 490, 577–579, 625
  Gaussian, 480, 491

impulse response functions, 455, 456

infinite series, 41
  divergent, 42
  generalized, 574–577, 635, 638
  two-sided, 45, 46

inner product, 129–131

input functions, 447, 455
  periodic, 459

integrability, 37, 501
  absolute, 257–259, 268, 502, 503
    tests for, 262–264
  conditional, 259
  exponential, 478
  generalized, 639, 640
  uniformly absolute, 270, 271

integrals, 37, 60
  double, 82–84, 268, 274, 275
  generalized, 640, 641

  well defined, 38, 257, 258

integration, 37
  and area, 37, 38, 258, 259
  Fourier transform identities, 352–354
  of Fourier series, 210–212

integration by parts, 40, 346, 347, 497
  generalized, 435

intervals, finite/bounded, 7

invertibility
  fundamental theorem on, 285, 490, 492
  of Fourier transforms
    classical, 300, 306
    discrete, 742
    generalized, 591, 592
    integral, 284–286, 492
    multi-dimensional, 465

jump in a function, 17, 163, 165, 433, 489, 490

Laplace transform, 12, 13, 379
  defined, 487
  inversion formula, 488, 490
  relation to Fourier transform, 488

limits, 27–28, 554
  generalized, 567–572, 635

linear combinations, 4, 552

linearity
  and Fourier series, 112–113, 149
  of Fourier transforms
    classical, 301, 305
    discrete, 741, 742
    generalized, 591
    integral, 283
  of systems, 449
  of transforms, 14, 617

matrices of exponentials, 716–719, 730
  for $\mathcal{F}$, 716
  for $\mathcal{F}^{-1}$, 718
  invertibility, 719

mean value theorem (for integrals), 40

modulation, 326
  Fourier transform identities, 325, 326

multiplication, generalized, 636, 643–646
  compared with classical, 631
  rules for, 631–635
  simple, 549–552, 630, 631

near-equivalence, principle of
  classical, 281, 287, 301, 306
  discrete, 742–744
  generalized, 596
  multi-dimensional, 464

noise, 402

norm functional, 538, 539

norms
  α-norms, 529, 530
  from inner product, 131, 132, 403, 406
Nyquist frequency, 312

orthogonal function expansions, 134, 135
orthogonality, 132
  for complex exponentials, 142, 147
  for sines and cosines, 54, 55, 133
  for sines or cosines, 141, 142
  use in deriving Fourier coefficients, 96–98,
        134, 147
orthonormality, 134
output functions, 447, 455

Parseval's equality
  discrete, 759
  for Fourier series, 171, 172, 485, 486
  for Fourier transforms, 407
partial fraction expansion, generalized, 704
partial sum approximations for Fourier series
  error in, 161, 163–165, 168, 176, 186, 189,
        195, 206–209
  limited uniform, 167, 168, 186–189, 195
  nonuniform, 163–165
  uniform, 162–165
partial sums, 41
  symmetric, 46
partitioning, 21
periodic extensions, 121, 123
  even, 125
  odd, 123, 124
periodic functions, 51, 260
  Fourier transforms of, 426–428, 651, 652, 655
  frequency, 51, 95, 649
    angular vs. circular, 51
    fundamental, 52
  generalized, 649–651
  period, 51, 52, 103, 106, 149, 649
    fundamental, 52
  relation with delta function arrays, 428, 430,
        655
periodic, regular arrays, 709, 710
  Fourier coefficients, 712, 719
  Fourier series of, 712, 713
  Fourier transforms of, 711–713, 715, 719
    relation with discrete transforms, 740
  index period, 709–712
    relation with period and spacing, 710
  period, 709–712
  recursion relation, 710
  spacing, 709–712
pole functions, 694–698
  as solutions to equations, 704
  compared with classical analogs, 698, 699

in products, 695, 701–703
  relations with step functions, 700
polynomials, 489, 681
  factoring / roots / multiplicity, 681
products
  asymmetric, 645
  of sequences, 746
products, generalized, 550
$\text{pulse}_a(x)$, pulse function, 280, 609

$\mathbb{R}$, the real line, 7
radar, 402
$\text{ramp}(x)$, ramp function, 8, 387, 625, 705
range of a function, 8
$\text{rect}_{(a,b)}(x)$, rectangle function, 259, 260, 336,
        609
region in the plane, 74
Riemann sum, 38, 39, 251–253
  error in using, 731, 732
Riemann–Lebesgue lemma, 141, 181, 289, 292
  and Fourier transforms, 289

$\mathcal{S}$, space of rapidly decreasing test functions, 521
sampling, 311
sampling theorem, 311
samplings, 722–724, 756, 757
  and discrete approximations, 727
  order and spacing, 722
  regular, 723
  uniform, 722
  windows, 722
    minimal, 722, 723
    natural, 723, 724
saw functions
  $\text{evensaw}_p(x)$, even saw function, 56, 118, 119,
        152, 175, 204, 217, 218, 444, 671
  $\text{oddsaw}_p(x)$, odd saw function, 56, 118, 152,
        217, 218, 444, 671
  $\text{saw}_p(x)$, basic saw function, 51, 52, 102, 103,
        117, 148, 158–159, 186, 195, 201, 213,
        427, 428, 657–659
scaling, 25
  and Fourier series, 114, 149
  Fourier transform identities
    classical, 327, 337
    generalized, 595
    multi-dimensional, 465
  generalized, 592, 593, 634
scaling operator, 593, 594
Schwartz, Laurent, 522
Schwarz inequality
  for finite sums, 30
  for infinite series, 45
  for inner products, 135, 144
  for square-integrable functions, 404

separable functions of several variables, 466
    Fourier transforms of, 467
sequence vectors, 737, 739
sequences, finite order, 721, 737, 739
    index negation, 743, 757
    order, 721
    recursion formula, 742
$\text{sgn}(x)$, signum function, 625, 648, 695, 700
shifting, *see* translation
signal, 314
simple convolvers, 638, 644
simple multipliers, 523, 630, 635, 644, 646, 675
$\sin(x)$, sine function, 53, 54
    rectified, 56, 118, 444
$\text{sinc}(x)$, sinc function, 16, 277, 280, 314, 315
singularities, 383
smoothness, 23, 60, 554, 556, 557
    of convolution, 390–392
    piecewise, 24
    under basic operations, 26
    uniform, 24
spatial domain, 314
stair function, 35
$\text{step}(x)$, step function, 10, 18, 260, 328, 387, 388, 589, 700
    derivative of, 23, 606
    Fourier transforms of, 603, 700
strips, 270, 599
sum, generalized, 548
support
    of a delta function array, 579
    of a function, 312
symmetry (even/odd functions), 49–51
    and Fourier series, 110, 149
    and Fourier transforms
        classical, 307, 333, 334, 338
        generalized, 596
        integral, 282, 283, 295
    generalized, 594
symmetry, meaning near-equivalence, 281
systems, 447
    linear / shift invariant, 449
    operationally continuous, 451, 452

$\mathcal{T}$, space of integral transforms, 298
test functions, 492, 515
    rapidly decreasing, 521
    very rapidly decreasing, 522
time domain, 314
transfer functions, 455, 456, 460, 461
transforms, 12–14
    adjoint defined, 616–620
    linear, 14
translation, 26
    and Fourier series, 114

complex, 367, 525, 535, 598
    in solving equations, 691–693
Fourier transform identities
    classical, 319, 322, 323
    discrete, 745, 758
    generalized, 601
    multi-dimensional, 465
    generalized, 597–599, 601, 634
        compared with classical, 599, 603–605
    imaginary, 336
    of sequences, 744, 745, 758
translation operator, 598, 599, 601
$\text{tri}(x)$, triangle function, 295, 351, 476
triangle inequality, 28, 29
    for infinite series, 42
    for complex numbers, 58

uncertainty principle, 312, 409, 411
unitary sequences, 501, 503

variables, 8
    dummy/internal, 9, 329
    in generalized functions, 547
vector
    components, 3, 4, 6
    dot product, 4, 137
    norm, 131, 137
vibrating string problem, 226, 227, 248
    harmonics, 233
    solution, 229, 230, 232

wave equation, 226
Wiener–Khintchine theorem, 401

Milton Keynes UK
Ingram Content Group UK Ltd.
UKHW051927141024
449569UK00027B/1388